SUSTAINABLE NANOSCALE ENGINEERING

SUSTAINABLE NANOSCALE ENGINEERING

FROM MATERIALS DESIGN TO CHEMICAL PROCESSING

Edited by

GYORGY SZEKELY
Advanced Membranes and Porous Materials Center,
Physical Science and Engineering Division (PSE),
King Abdullah University of Science and Technology (KAUST),
Thuwal, Saudi Arabia & School of Chemical Engineering & Analytical Science,
The University of Manchester, Manchester, United Kingdom

ANDREW LIVINGSTON
Barrer Centre, Department of Chemical Engineering,
Imperial College London, London, United Kingdom

ELSEVIER

Elsevier
Radarweg 29, PO Box 211, 1000 AE Amsterdam, Netherlands
The Boulevard, Langford Lane, Kidlington, Oxford OX5 1GB, United Kingdom
50 Hampshire Street, 5th Floor, Cambridge, MA 02139, United States

Library of Congress Cataloging-in-Publication Data
A catalog record for this book is available from the Library of Congress

British Library Cataloguing-in-Publication Data
A catalogue record for this book is available from the British Library

ISBN: 978-0-12-814681-1

For information on all Elsevier publications visit our website at https://www.elsevier.com/books-and-journals

Publisher: Joe Hayton
Acquisition Editor: Kostas KI Marinakis
Editorial Project Manager: Emma Hayes
Production Project Manager: Maria Bernard
Cover Designer: Mark Rogers

Typeset by TNQ Technologies

To my supporting parents, Gyongyi and Gyorgy, who were always there for me. To my wonderful wife, Mayamin, without whose never-failing patience this book would have not been completed. To my beloved baby girl, Alessia, who always cheers me up with her cuteness no matter what. To my friend, Péter, whose continuous criticism, strangely, motivates me to try turning the world a better place. To my friends and colleagues, Józsi and Jeong, for sharing their knowledge and wisdom beyond science. Without their love and support nothing would have been possible.

Gyorgy Szekely

One of the greatest pleasures in life is to imagine a world tomorrow, where things that are not possible today come to pass, and then to work to make those things real. To all my current and former research students and postdoc colleagues, without whom none of the pleasures that research has given me would have come to pass.

Andrew Livingston

Contents

Contributors

Zahra Abbasi Department of Chemical Engineering, Monash University, Clayton, VIC, Australia

Mohamed Hamada Abdel Kodous Center for Nanotechnology (CNT), School of Engineering and Applied Sciences, Nile University, Sheikh Zayed, Giza, Egypt

Callie W. Babbitt Golisano Institute for Sustainability, Rochester Institute of Technology, Rochester, NY, United States

Peter M. Budd School of Chemistry, University of Manchester, Manchester, United Kingdom

Zhen Lei Cheng Department of Chemical and Biomolecular Engineering, National University of Singapore, Singapore

Tai-Shung Chung Department of Chemical and Biomolecular Engineering, National University of Singapore, Singapore

Marc-Olivier Coppens Centre for Nature Inspired Engineering, Department of Chemical Engineering, University College London, London, United Kingdom

Joaquin Coronas Chemical and Environmental Engineering Department, Instituto de Nanociencia de Aragón (INA) and Instituto de Ciencia de Materiales de Aragón (ICMA), Universidad de Zaragoza-CSIC, Zaragoza, Spain

Levente Cseri School of Chemical Engineering & Analytical Science, University of Manchester, The Mill, Manchester, United Kingdom

Yue Cui Department of Chemical and Biomolecular Engineering, National University of Singapore, Singapore

Michael K. Danquah Chemical Engineering Department, University of Tennessee, Chattanooga, Tennessee, United States

Enrico Drioli Institute on Membrane Technology (ITM-CNR), Rende, Cosenza, Italy

Arzu Ersöz Faculty of Science, Chemistry Department, Eskisehir Technical University, Eskisehir, Turkey; Bionkit Ltd., Technology Park, Eskisehir, Turkey

Xiaolei Fan School of Chemical Engineering and Analytical Science, The University of Manchester, Manchester, Greater Manchester, United Kingdom

Anthony G. Fane UNESCO Centre for Membrane Science & Technology, University of New South Wales, Sydney, Australia; Singapore Membrane Technology Centre, Nanyang Technological University, Singapore

Maria-Chiara Ferrari School of Engineering, The University of Edinburgh, Edinburgh, United Kingdom

Wen Xiao Gai Department of Chemical and Biomolecular Engineering, National University of Singapore, Singapore

Ángel Galán-Martín Institute for Chemical and Bioengineering, Department of Chemistry and Applied Biosciences, ETH Zürich, Vladimir-Prelog-Weg 1, Zürich, Switzerland

Andrés González-Garay Department of Chemical Engineering, Centre for Process System Engineering, Imperial College London, South Kensington Campus, London, United Kingdom

Gonzalo Guillén-Gosálbez Institute for Chemical and Bioengineering, Department of Chemistry and Applied Biosciences, ETH Zürich, Vladimir-Prelog-Weg 1, Zürich, Switzerland

John D. Hayler Chemical Development, GlaxoSmithKline, Hertfordshire, United Kingdom

Istvan T. Horvath Department of Chemistry, City University of Hong Kong, Kowloon, Hong Kong

Jaison Jeevanandam Department of Chemical Engineering, Curtin University, CDT250 Miri, Sarawak, Malaysia

Yilai Jiao Shenyang National Laboratory for Materials Science, Institute of Metal Research, Chinese Academy of Sciences, Shenyang city, Liaoning province, P. R. China

Martin D. Johnson Small Molecule Design and Development, Eli Lilly and Company, Lilly Research Laboratory, Indianapolis, IN, United States

C. Oliver Kappe Research Center Pharmaceutical Engineering GmbH (RCPE), Center for Continuous Flow Synthesis and Processing (CC FLOW), Graz, Austria; Institute of Chemistry, University of Graz, Graz, Austria

Rüstem Keçili Yunus Emre Vocational School of Health Services, Department of Medical Services and Techniques, Anadolu University, Eskisehir, Turkey

Kenta Kokado Department of Chemistry, Faculty of Science, Hokkaido University, Sapporo, Hokkaido, Japan

Bradley P. Ladewig Department of Chemical Engineering, Imperial College London, London, United Kingdom; Institute for Micro Process Engineering (IMVT), Karlsruhe Institute for Technology, Eggenstein-Leopoldshafen, Germany

Alexei Lapkin Department of Chemical Engineering and Biotechnology, University of Cambridge, Cambridge, United Kingdom; Cambridge Centre for Advanced Research and Education in Singapore Ltd, Singapore

Elsa Lasseuguette School of Engineering, The University of Edinburgh, Edinburgh, United Kingdom

Phantisa Limleamthong Department of Chemical Engineering, Centre for Process System Engineering, Imperial College London, South Kensington Campus, London, United Kingdom

Ryan P. Lively School of Chemical & Biomolecular Engineering, Georgia Institute of Technology, Atlanta, GA, United States

Andrew Livingston Barrer Centre, Department of Chemical Engineering, Imperial College London, London, United Kingdom

Yao Ma School of Chemical & Biomolecular Engineering, Georgia Institute of Technology, Atlanta, GA, United States

Kiyoshi Matsuyama Fukuoka Institute of Technology, Fukuoka, Fukuoka-ken, Japan

Farhad Moghadam Department of Energy Engineering, Hanyang University, Seoul, Republic of Korea

Chandran Murugan SRM Research Institute, SRM Institute of Science and Technology, Kattankulathur, Kanchipuram, Tamil Nadu, India

Masami Naya Department of Chemistry, Faculty of Science, Hokkaido University, Sapporo, Hokkaido, Japan

Suzana P. Nunes King Abdullah University of Science and Technology (KAUST), Biological and Environmental Science and Engineering (BESE), Advanced Membranes and Porous Materials Center, Thuwal, Saudi Arabia

Kaushik Pal Department of Nanotechnology, Bharath University, BIHER Research Park, Selaiyur, Chennai, Tamil Nadu, India

Ho Bum Park Department of Energy Engineering, Hanyang University, Seoul, Republic of Korea

Camille Petit The Barrer Centre, Department of Chemical Engineering, Imperial College London, London, United Kingdom

Carlos Pozo Department of Chemical Engineering, Centre for Process System Engineering, Imperial College London, South Kensington Campus, London, United Kingdom

Thalappil Pradeep Department of Chemistry, Indian Institute of Technology Madras, Chennai, India

Ji Soo Roh Department of Energy Engineering, Hanyang University, Seoul, Republic of Korea

Kazuki Sada Department of Chemistry, Faculty of Science, Hokkaido University, Sapporo, Hokkaido, Japan

Rıdvan Say Bionkit Ltd., Technology Park, Eskisehir, Turkey

Giulia Schukraft The Barrer Centre, Department of Chemical Engineering, Imperial College London, London, United Kingdom

Varsha Sharma SRM Research Institute, SRM Institute of Science and Technology, Kattankulathur, Kanchipuram, Tamil Nadu, India

Jae Eun Shin Department of Energy Engineering, Hanyang University, Seoul, Republic of Korea

Richard L. Smith, Jr. Tohoku University, Sendai, Miyagi-ken, Japan

Anandhakumar Sundaramurthy SRM Research Institute, SRM Institute of Science and Technology, Kattankulathur, Kanchipuram, Tamil Nadu, India; Department of Physics and Nanotechnology, SRM Institute of Science and Technology, Kattankulathur, Kanchipuram, Tamil Nadu, India

Gyorgy Szekely Advanced Membranes and Porous Materials Center, Physical Science and Engineering Division (PSE), King Abdullah University of Science and Technology (KAUST), Thuwal, Saudi Arabia; School of Chemical Engineering & Analytical Science, The University of Manchester, Manchester, United Kingdom

Kam C. Tam Department of Chemical Engineering, Waterloo Institute for Nanotechnology, University of Waterloo, 200 University Avenue West, Waterloo, Canada

Panagiotis Trogadas Centre for Nature Inspired Engineering, Department of Chemical Engineering, University College London, London, United Kingdom

Bernhardt L. Trout Department of Chemical Engineering, Massachusetts Institute of Technology, Cambridge, MA, United States

Ikuo Ushiki Hiroshima University, Hiroshima, Hiroshima-ken, Japan

Luigi Vaccaro Laboratory of Green S.O.C., Dipartimento di Chimica, Biologia e Biotecnologie, Università degli Studi di Perugia Perugia, Italy

Chun Feng Wan Department of Chemical and Biomolecular Engineering, National University of Singapore, Singapore

Huanting Wang Department of Chemical Engineering, Monash University, Clayton, VIC, Australia

Ecevit Yılmaz YLS Consulting AB, Växtskyddsvägen, Alnarp, Sweden

Fengyi Zhang School of Chemical & Biomolecular Engineering, Georgia Institute of Technology, Atlanta, GA, United States

Xiwang Zhang Department of Chemical Engineering, Monash University, Clayton, VIC, Australia

Xiangliang Zhang Machine Intelligence & Knowledge Engineering; Computer, Electrical and Mathematical Science and Engineering Division (CEMSE), King Abdullah University of Science and Technology (KAUST), Thuwal, Saudi Arabia

Acknowledgment

Very special thanks to all the contributors to this book who tirelessly worked on the individual chapters, their revision, and proofreading. The quality of a book is entirely due to the authors and the referees who so kindly conducted thorough reviews of the component chapters. Their guidance and scrutiny throughout the process, and their wise counsel on how to improve this book, are much appreciated. The editors would like to gratefully acknowledge the contribution of the referees, who are listed in random order along with their institutional affiliations: Robert McNair, Hai Anh Le Phuong, Levente Cseri, Fan Fei, Christopher F Blanford (University of Manchester, United Kingdom); Jeong F Kim (Incheon National University, Korea); Jozsef Kupai, Peter Bagi (Budapest University of Technology & Economics, Hungary); Janet Lim Hong Ngee (Universiti Putra Malaysia, Malaysia); Fuat Topuz, Sang-Hee Park, Mahmoud Abdul Hamid (King Abdullah University of Science & Technology, Saudi Arabia); Sudhirkumar Shinde (Queen's University Belfast, United Kingdom); Tibor Holtzl (Furukawa Electric, Hungary); Peter Pogany (GlaxoSmithKline, United Kingdom); Jiaru Bai (University of Cambridge, United Kingdom); Mohamed H Abdellah (University of Melbourne, Australia); Ricardo Abejón (Universidad de Cantabria, Spain); Ahmed Elreedy (Tokyo Institute of Technology); Hatim Machrafi (Université de Liège, Belgium). Figure 1.3 was created by Ivan D Gromicho, scientific illustrator at KAUST. Thanks to everyone on the Elsevier team who helped us so much. Special thanks to Emma Hayes and Maria Bernadette Vidhya, our ever-patient Publishing Managers.

C H A P T E R

1

Challenges and Directions for Green Chemical Engineering—Role of Nanoscale Materials

Andrew Livingston[1], *Bernhardt L. Trout*[2], *Istvan T. Horvath*[3], *Martin D. Johnson*[4], *Luigi Vaccaro*[5], *Joaquin Coronas*[6], *Callie W. Babbitt*[7], *Xiangliang Zhang*[8], *Thalappil Pradeep*[9], *Enrico Drioli*[10], *John D. Hayler*[11], *Kam C. Tam*[12], *C. Oliver Kappe*[13,14], *Anthony G. Fane*[15,16], *Gyorgy Szekely*[17,18]

[1]Barrer Centre, Department of Chemical Engineering, Imperial College London, London, United Kingdom; [2]Department of Chemical Engineering, Massachusetts Institute of Technology, Cambridge, MA, United States; [3]Department of Chemistry, City University of Hong Kong, Kowloon, Hong Kong; [4]Small Molecule Design and Development, Eli Lilly and Company, Lilly Research Laboratory, Indianapolis, IN, United States; [5]Laboratory of Green S.O.C., Dipartimento di Chimica, Biologia e Biotecnologie, Università degli Studi di Perugia Perugia, Italy; [6]Chemical and Environmental Engineering Department, Instituto de Nanociencia de Aragón (INA) and Instituto de Ciencia de Materiales de Aragón (ICMA), Universidad de Zaragoza-CSIC, Zaragoza, Spain; [7]Golisano Institute for Sustainability, Rochester Institute of Technology, Rochester, NY, United States; [8]Machine Intelligence & kNowledge Engineering; Computer, Electrical and Mathematical Science and Engineering Division (CEMSE), King Abdullah University of Science and Technology (KAUST), Thuwal, Saudi Arabia; [9]Department of Chemistry, Indian Institute of Technology Madras, Chennai, India; [10]Institute on Membrane Technology (ITM-CNR), Rende, Cosenza, Italy; [11]Chemical Development, GlaxoSmithKline, Hertfordshire, United Kingdom; [12]Department of Chemical Engineering, Waterloo Institute for Nanotechnology, University of Waterloo, 200 University Avenue West, Waterloo, Canada; [13]Research Center Pharmaceutical Engineering GmbH (RCPE), Center for Continuous Flow Synthesis and Processing (CC FLOW), Graz, Austria; [14]Institute of Chemistry, University of Graz, Graz, Austria; [15]UNESCO Centre for Membrane Science & Technology, University of New South Wales, Sydney, Australia; [16]Singapore Membrane Technology Centre, Nanyang Technological University, Singapore; [17]Advanced Membranes and Porous Materials Center, Physical Science and Engineering Division (PSE), King Abdullah University of Science and Technology (KAUST), Thuwal, Saudi Arabia; [18]School of Chemical Engineering & Analytical Science, The University of Manchester, Manchester, United Kingdom

O U T L I N E

Sustainable Nanoscale Engineering
https://doi.org/10.1016/B978-0-12-814681-1.00001-1

1. Introduction

Nanotechnology and nanomaterials are among the most significant scientific and industrial research breakthroughs of the 21st century. With the rapid globalization of science, chemists, materials scientists, and chemical engineers are synergistically working together worldwide to understand how to manipulate matter for the benefit of humankind. The Sustainable Development Goals set by the United Nations provide a blueprint through which a thriving and more sustainable future can be achieved for all (Fig. 1.1) [1]. These goals address the global challenges we face, and most of them are directly affected by chemical manufacturing. Consequently, it is our responsibility to design, manufacture and recycle chemicals, and develop processes, considering sustainability. Although there is a lack of consensus on the detailed meaning of the concept [2], *sustainable manufacturing*, we take here a working definition by the United States' Environmental Protection Agency (EPA) as the creation of manufactured products through economically sound processes that minimize negative environmental impacts while conserving energy and natural resources [3].

There are several emerging areas of nanoscale engineering with great promise for sustainable chemical engineering. Enzymes, nature's biocatalysts, have outstanding selectivity and activity and facilitate a broad range of chemical transformations under mild reaction conditions. Besides their natural aqueous environment, there is a need for exploring and exploiting enzymes in organic solvents. The work on directed evolution to engineer enzymes earned the 2018 Nobel Prize in Chemistry for Frances Arnold. Enzymes engineered through directed evolution have great potential in the sustainable processing of a wide variety of chemical products, from pharmaceuticals to biomass. In parallel to the advancement of enzyme catalysis, the field of organocatalysis has emerged. Organocatalytic reactions exploit small-molecule enzyme mimics that are robust, safe, sustainable, metal-free, and scalable [4]. Further developments in the field of catalysis both at nanoscale and process scale are crucial to advancing sustainability because more than 90% of chemical engineering processes utilize catalysts globally [5].

FIGURE 1.1 The United Nations' Sustainable Development Goals covering 17 ambitions, most of which are impacted by chemical manufacturing. The chemicals, materials, and processes developed today will either make or break these ambitions. *Reprinted with permission from Sustainable Development Goals, United Nations, https://www.un.org/sustainabledevelopment/sustainable-development-goals/.*

There are a plethora of innovative methodologies, all with the potential to enable sustainable industrial development, on the rise. The World Economic Forum, the International Union of Pure and Applied Chemistry, and the MIT Technology Review have published their own selections of the top 10 emerging technologies improving sustainability [6–8]. The unique advantages offered by flow chemistry and flow reactors have already triggered companies to invest in research and development, pilot scale tests, and implementation in production lines. Solvent-free reactive extrusion for mechanochemical synthesis and 3D printing of advanced engineering materials are emerging fields, with implementation and scale-up challenges yet to be solved. For about half-a-century, there has been a race to develop artificial leaves to efficiently mimic photosynthesis and transforming carbon dioxide into liquid fuel [9,10]. The production of liquefiable hydrocarbons from excess carbon dioxide, water, and other sustainable resources such as sunlight will create new opportunities for energy storage. Despite the tremendous efforts, these ideas are in their infancy, and it is essential that the enabling process development keeps pace with the scientific breakthroughs. Speaking about his startup using hydrogen-producing artificial leaves, Nocera lamented that "I did a holy grail of science. Great! That doesn't mean I did a holy grail of technology," [9] highlighting the importance of scale and engineering.

Recent breakthroughs in artificial intelligence (AI), in particular deep learning and generative adversarial networks, have allowed machines to mimic imagination (Fig. 1.2), which is a big leap toward unsupervised learning [11]. These advanced methodologies have much to offer scientists and engineers working with large databases, albeit suffering from limitations such as heavy reliance on quality data. The power of these advancements in AI can only be exploited if research data are reported in a machine-readable manner and managed in online databases accessible to everyone. Most of the research data are mostly reported in image formats, which results in the loss of precise data points. Options for interactive plots are on the rise, and they ought to become mandatory in the near future.

There are numerous technical, engineering, and financial challenges associated with developing new, or repurposing conventional and existing, materials and processes to sustainable alternatives. Thanks to the increasing efforts of the industrial sector, in particular the pharmaceutical manufacturers, the "sustainability" buzz word has started to manifest in actions. Several companies have explicitly and publicly started using green chemistry and engineering as key drivers and to innovate around sustainable initiatives. The next subsections and chapters

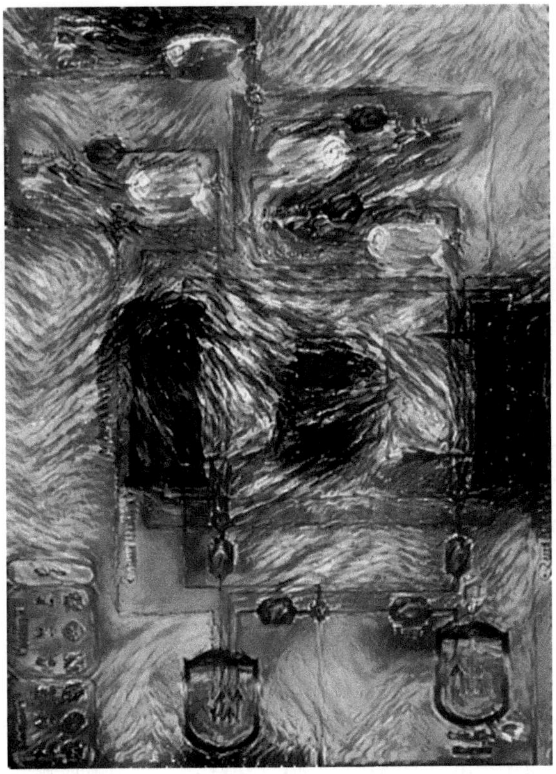

FIGURE 1.2 A schematic piping and instrumentation diagram (P&ID) consisting of stirred vessels, pumps and membrane modules, thermostats, and adsorption columns reimagined by an artificial intelligence (AI) using deep neural networks [12,13]. The weirdly fascinating result is due to the fact that the understanding of the ingested data does not precisely translate into the ability to generate similar data, which is one of the main limitations of AI.

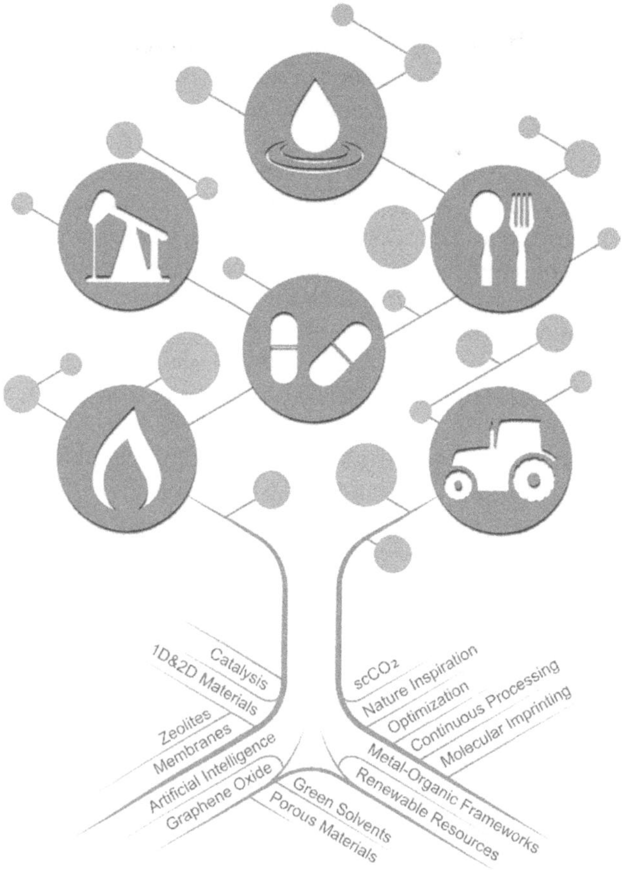

FIGURE 1.3 The relationship of industries and the state-of-the-art tools enabling them to be sustainable.

in Sustainable Nanoscale Engineering highlight the sustainability potential of state-of-the-art materials ranging from smart polymers, through 2D materials, to metal–organic frameworks (MOFs) considering both their fabrication and application (Fig. 1.3). Moreover, the potential of continuous flow processes, lifecycle optimization, and AI directed toward sustainable nanoscale engineering are also evaluated.

Besides the direct research and development initiatives aiming at sustainable solutions, it is equally important that we incorporate these initiatives into the chemistry, materials science, and chemical engineering curricula so as to equip our future generations to tackle challenges with a sustainable mindset. Surely, in the not-too-distant future, we will live in a sustainable world enabled by the fascinating materials, processes, and methodologies described in this book.

2. From Green Chemistry to Sustainable Chemical Engineering

In the green chemistry arena, scientists are contributing to creating chemical and technological innovations enabling more sustainable and competitive chemical production. To reach this goal, waste minimization is a priority and the adoption of catalytic protocols is preferred [14]. While much industrial manufacturing may currently still employ classic homogeneous methods [15], this tendency is mainly based on economic reasons and therefore on regulations and price/availability of materials. While not always economically sustainable at present, in the future, heterogeneous catalytic approaches will become the norm, enabling recovery of the catalytic system so that its durability is preserved over a sufficient number of runs. Accordingly, the design of novel materials for catalysis and suitable platform technologies for their use plays a pivotal role in the route toward a sustainable chemistry.

Owing to their easy separation from the reaction mixture, heterogeneous systems based on metals [16–18], acid [19], or base [20] catalysts have been developed using both organic and inorganic supports. Among the heterogeneous catalytic systems, nanomaterials and metal nanoparticles–based catalysts have been widely studied in the last 10 years, thanks to their high surface area. In fact, it is well-known now how particle size is related to surface energy, which in turn is directly connected with the catalytic activity. Thanks to this property, several non–noble metal catalyzed processes have been developed [21,22] to reduce the consumption of exhaustive metals. In these endeavors, the careful design of the support and the final catalyst has been found to be crucial for the catalytic process, indeed for example, the acid-base properties of inorganic supports can affect the reactivity, or the support can be compromised by the reaction conditions, including temperature or pH. We must also recognize that while heterogeneous catalysts should be totally recoverable and recyclable, it can happen that, due to the agitation and mixing used in batch conditions, the materials can be partially crushed, which makes their separation difficult.

Meeting this challenge to reduce attrition and provide thermal and mechanical stability of heterogeneous supported catalysts, flow technology is proving an effective tool [16,18,19,23,24]. In addition, the use of a flow technology has multiple advantages including safety, temperature control, mixing of reagents in two-phase systems, easy scale-up, and space economy. The use of flow reactors reduces exposure to toxic intermediates or dangerous reagents or additives, allows operation at high temperature and pressure in a safer manner, and guarantees a cleaner reaction mixture minimizing metal leaching of the catalysts with subsequent minimization of purification waste [16,18,23,24]. Moreover, by increasing the number of combined flow reactors (the "numbering-up" approach), it is possible to scale-up processes without any of the common problems found in batch scale-up scenarios. Heterogeneous catalytic systems, based on nanomaterials with high surface energy, are ideal partners for flow technologies, which can effectively be useful for the definition of greener processes.

3. The Promise of Continuous Processing and Monitoring

The International Union of Pure and Applied Chemistry (IUPAC) recently released their "Top 10" list of emerging technologies that will contribute to the sustainability of Planet Earth in the 21st century [25]. Flow chemistry is highlighted as a critical technology to achieve the United Nation's Sustainable Development Goals (SDGs) by 2030, in particular for tackling SDG12: responsible consumption and production [25,26].

Enhanced heat and mass transfer, precise residence time control, shorter process times, increased safety, reproducibility, better product quality, and easy scalability are the main advantages of continuous flow versus conventional batch processing and reason for the increasing implementation of continuous processes in the fine chemical manufacturing sector [27,28]. A green process needs to be scalable for the greatest environmental impact, and ideally a direct translation from lab to industrial scale is desirable [26]. Continuous flow processing specifically addresses these needs, and for a large number of processes, distinct advantages over batch processing have been demonstrated in terms of cost, equipment size, energy consumption, waste generation, safety, efficiency, and product quality [26–28].

In particular, for the highly regulated manufacturing of pharmaceuticals, there is currently a strong trend toward continuous manufacturing techniques [27–30]. This ideally entails the development of safe, robust, and cost-effective processes in the earliest design stages, by merging ideas stemming from both green chemistry and green engineering [26]. A key green engineering concept in this context is the telescoping of multistep syntheses accompanied by in-line separations to reduce solvent waste [31–34]. Therefore, virtually all pharma companies and contract manufacturing organizations are currently adopting this technology, as highlighted in a recent survey [30] and review [35].

Furthermore, continuous flow processing techniques are beginning to have a profound impact on the production of functional materials ranging from quantum dots, nanoparticles, and MOFs to polymers and dyes [31]. Flow chemistry provides robust procedures which not only enable accurate control of the product material's properties but also are ideally suited to conducting experiments on scale. Importantly, the modular nature of flow and continuous processing equipment facilitates rapid reaction optimization and variation in function of the products [31].

A more recent trend in continuous processing is the adoption of the "Industry 4.0" ethos, whereby simulation, system integration, and the generation of large datasets are key enablers for a greater focus on quality, safety, cost-effectiveness, and sustainability. Key to this initiative is the real-time acquisition of data for chemical process development and in-process monitoring by process analytical technology (PAT) [36]. Several recent reviews have highlighted the successful implementation of PAT within continuous flow environments, for example, for enabling feedback loops for process control, automated self-optimization, kinetic model discrimination, and parameter estimation [37]. The application of PAT principles to the monitoring of fine chemical synthesis is changing the way in which pharmaceutical processes are evaluated by regulatory authorities. These developments, although so far limited mainly to the pharmaceutical sector, will undoubtedly push the field of flow chemistry further in the years to come.

Continuous manufacturing of pharmaceuticals and the benefits it can provide are long overdue. In its true sense, it incorporates continuous flow with a systems approach to design and integrate model-based control. While there are now five approved continuous pharmaceutical processes in the United States, all of these focus on making the final product, as opposed to chemical synthesis, and they are all partial continuous processes. The benefits are immense, including higher quality, reduction of risk of stock-outs, flexibility, lower footprint, substantial savings of both CapEx and OpEx, and, of course, substantial green benefits. Solvent recycle seamlessly fits into continuous processes, and with more efficient energy utilization not to mention reduction of chemical waste and more efficient use of equipment, all interested in green manufacturing should be interested in continuous manufacturing of pharmaceuticals. Given all of the benefits, why has the industry been so slow to adopt? A major reason for this is mental inertia. The batch approach of the industry has led to procedures for process development and manufacturing that need to be modified to move to continuous, in addition to the need for new investments in continuous equipment. As managers embrace the benefits of continuous, more and more processes will get submitted and approved, and at one point a critical mass will be achieved leading to the transformation of the industry.

From an industrial perspective, 10 kg day^{-1} is a practical limit for fully continuous processes using mobile skids easily rearranged in laboratory fume hoods. Continuous formulations are more advanced than continuous synthesis in terms of implementations for marketed products. The US and EU regulatory authorities have approved five continuous processing of drug products: Vertex continuous OSD manufacturing [38], Vertex Symdeko [39], Janssen's Prezista [40,41], Lilly [42], Pfizer [43]. Fully continuous processes for production in laboratory fume hoods have been developed by researchers at MIT, Continuus Pharmaceuticals, Zaiput, Snapdragon, On-Demand Pharmaceuticals, among others [29,44,45]. Some of the key differences compared between the continuous processes for drug products and drug substances may have mean residence times and residence time distributions orders of magnitude longer, lower required analytical frequency, decoupling with surge vessels in the middle of the flow train, and a final batch drying step. From a manufacturing perspective, benefits of continuous flow synthesis have included chemical process safety, containment, elimination of isolations, quality assurance of online analytical, and chemistries or process separations not feasible batch. A continuous process developed at the Continuous Drug Substance Manufacturing facility of Eli Lilly and Company earned the 2019 FOYA Process Innovation Category Award [46].

4. Green and Sustainable Raw Materials for Chemical Manufacturing

Since the invention of the steam and internal combustion engines, the growth in the manufacturing of chemicals and materials from nonrenewable resources has exponentially increased. In a similar manner, the automotive (vehicles) and consumer electrical sectors (e.g., refrigerators) have also generated a huge demand for fossil fuels, resulting in the release and accumulation of greenhouse gases. These activities have impacted the environment and society in ways that we could not have imagined or anticipated. The industrial revolution has transformed the manner in which materials are derived and utilized to produce products that improve the lives of mankind. Unfortunately, many of these innovations focus on the transformation of fossil fuels into petrochemicals and engineering materials that have permeated all aspects of human activity. As a consequence, the modern world is now plagued with many new challenges, such as climate change, environmental pollution in oceans, rivers, and lakes. For instance, an increasing problem is the unsustainable practices in the generation and disposal of synthetic polymers, leading to alarming polymer pollution at global scale and significant materials value loss [47]. To mitigate these environmental and economic issues, the development of sustainable polymers with closed loop life cycles is emerging by shifting from a linear materials economy to circular materials economy (Fig. 1.4).

The concerns over climate change and global warming have propelled scientists and engineers to explore innovative solutions that minimize their impact on the environment. Hence, sustainable use of technology and materials has become a major consideration for governments, NGOs, and the private sectors. Consequently, a cradle-to-grave approach including the environment and health and safety aspects is necessary when developing and selecting nanomaterials. Owing to the ongoing international debate on how to regulate nanomaterials and the lack of risk assessment data, the market for nanomaterials is somewhat uncertain [48]. Although nanomaterials hold great promise for a wide variety of engineering fields, their long-term environmental and health impacts could

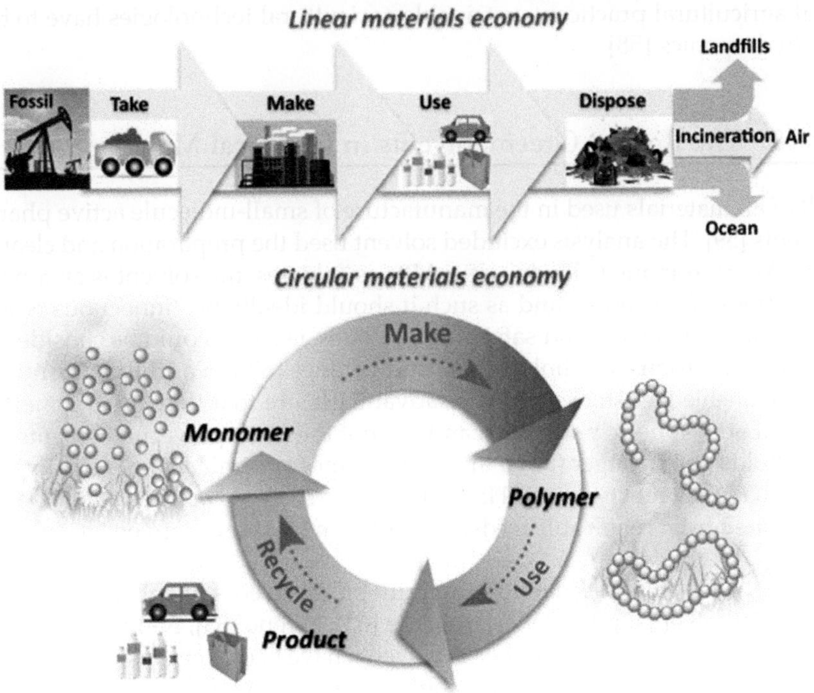

FIGURE 1.4 Toward circular materials economy frameworks by the development of sustainable polymers with full chemical recyclability. *Reprinted with permission from M. Hong, E. Y.-X. Chen, Future directions for sustainable polymers, Trends in Chemistry 1 (2019) 148–151.*

significantly undermine those advances. A framework for sustainable nanomaterials selection has been recently proposed with the aim of mitigating any unintended consequences [49,50].

Research on the utilization of renewable resources in the production of cellulose nanomaterials and lignin from the forest and agriculture biomass has intensified, and new innovations in these sectors have spawned many new opportunities which will contribute to the reduction of greenhouse gases. A major challenge that needs to be addressed is the role of government, academic institutions, industry, and nongovernmental organizations in finding a common platform that fosters the adoption of these innovations. Should the demands of sustainable nanomaterials increase because of these partnerships, investment in the manufacturing of sustainable nanomaterials will grow rapidly. It is heartening to know that industrial-scale applications of cellulose nanocrystals in sectors such as oil and gas, coatings, adhesives, and personal care have been reported. Additionally, the synthesis and production of sustainable nanomaterials has accelerated, and many demonstration plants for the production of cellulose nanocrystal, cellulose nanofibril, and lignin have been commissioned in countries including Canada, United States, Europe, Brazil, China, and Japan. Thus, the supply of these materials is now at a scale where it is economical to utilize them in product development.

The increasing rate of carbon dioxide emission to the atmosphere in the second half of the last century has resulted in ever faster expansion of the renewable energy portfolio [51]. Consequently, the amount of fossil resources used for energy applications will decrease, although the utilization of fossil-based energy for the production of renewable energy generating devices must be also replaced with renewable energy. Significantly, lower scale production and distribution of gasoline, diesel, and aviation fuel may result in increasing production costs to the point that fossil-based carbon fuels and chemicals will become uneconomical. Biomass-based carbon resources are available alternatives, but they can only replace fossil resources successfully if their production processes and facilities are based on green and sustainable chemistry and engineering concepts. Green chemicals, materials, and processes could be developed by applying the definitions and principles of green chemistry [52] and green engineering [53]. The sustainability of green chemicals, materials, and processes could be secured by using resources, including energy, at a rate at which they can be replaced naturally, and the generation of associated wastes should not be faster than the rate of their remediation [54]. Therefore, companies should use sustainable chemistry [55], ensure reliable biomass production by owning enough land to produce the required amount of biomass, and develop water management systems [56] for their production to avoid flooding or drought. If climate change threatens conventional agricultural practices, sustainable agricultural technologies have to be developed [57], such as intensification and hydroponics [58].

5. The Role of Green Solvents in Chemical Manufacturing

Approximately half of all materials used in the manufacture of small-molecule active pharmaceutical ingredients (APIs) are organic solvents [59]. The analysis excluded solvent used the preparation and cleaning of, typically, multipurpose batch reactors. Apart from inclusion in some APIs as solvates, the solvent is an auxiliary material that may be used in all unit operations of a process; and as such it should ideally be "innocuous when used" [52]. Classical solvents with few environmental, health, and safety (EHS) issues, i.e., that could be considered "greener" are generally oxygen-containing hydrocarbons: alcohols, esters, and ketones [60]. In addition, many of these solvents are, or can be derived from, sustainable feedstocks. Their disadvantages are that they are not inert for all chemistries; are volatile and thus potential atmospheric pollutants; and are not the most powerful solvents for many intermediates and APIs. The converse holds: the stronger (polar aprotic) and more chemically inert (chlorinated, hydrocarbon and ether) solvents generally have the most severe EHS issues. Some solvents are claimed to be green for a single issue, e.g., low volatility or obtained from renewable feedstock, but do not score as "green" in solvent guides when considered holistically [60,61].

Progress is being made to identify alternative, primarily bio-derived solvents for the most problematic solvent classes such as polar aprotic solvents: propylene carbonate [62], cyrene [63], *N*-butylpyrrolidinone [64], PolarClean [65], and hydrocarbons [66]. In particular, a number of solvents have been demonstrated as drop-in replacements for polar aprotic solvents in pharmaceutically relevant chemistries. Adoption of new solvents by the pharmaceutical industry is slow with several hurdles to overcome. The time taken to develop a complete understanding of human and/or environmental toxicology means they are considered with caution. The specified limit for the residual amount of a new solvent that is unclassified under ICH guidelines present in an API must be justified [67]. Many polar aprotic replacement solvents have high boiling points making removal by evaporation or recovery/purification by distillation energy intensive. Solvent recovery for reuse in API manufacture is permissible under ICH, provided the recovered solvent meets the appropriate specification for its intended use. Recovery for other industrial uses is also feasible. The cost of solvent recovery is a consideration for low-volume APIs [68]. Ultimately, the ability to recover solvent depends on process design, and while high water miscibility can aid separation by partition, it presents challenges for either solvent recovery or water treatment. In addition, security of supply is a concern. The use of water or alternative solvents, e.g., deep eutectic solvents [69] and natural deep eutectic solvents [70], prepared from natural products with fewer EHS issues is increasing, and these solvents are showing some surprising reactivity [71,72].

The application of chemo- and biocatalysis is aligned to green chemistry principles [52] and can be more compatible with the "greener" solvent classes. The use of nanoparticulate metal catalysts is emerging in the field of chemocatalysis and applied to reactions commonly used in API synthesis [73,74]. Such approaches offer the prospect of lower catalyst loadings, catalyst separation, and ligand-free catalysis which can be advantageous because, while precious metal recovery is achievable, ligands are often lost during workup. Like adoption of new solvents, the use of nanoparticulate catalysts presents some challenges to the industry, especially with concern about exposure to toxic metals during preparation and handling and decontamination of reaction vessels. Replacement of toxic metals with benign alternatives is a long-term green chemistry objective [75] where the use of nanoparticles and organocatalysts could be advantageous. The toxicity profile of any supporting media used in nanoparticle preparation or stabilization, if unknown, presents similar challenges to new solvents if present as a residue in API. Nanotechnology has been tested for materials sequestration, oxidative, and/or photodegradation in waste water treatment and extensively studied in areas of drug delivery and in therapeutic use for some cancers.

6. The Quest for Clean Water

Access to clean water is one of the most important indicators of development. This water has to be affordable to create a meaningful impact on the society. Chemistry of nanomaterials due to their enhanced surface area, confinement of the active nanostructures in nanoscale cages, exposure of specific surfaces with desired geometry for interaction with contaminants, optimum combination of properties such as surface area and electrical conductivity, control of phenomena such as droplet condensation, permeation, etc., have led to the creation of new methods for efficient water purification. Creation of affordable materials for constant release of silver ions [76] is one of the most promising ways to make water microbially safe. Combining the capacity of diverse nanocomposites to scavenge toxic species such as arsenic, lead, and other contaminants along with the above capability can result in affordable, all-inclusive drinking water purifiers that can function without electricity. A critical problem in achieving this is the complex species usually present in drinking water which deposit and cause scaling on high energy surfaces of nanomaterials. Such constant release/adsorbing materials can be synthesized in a simple and effective fashion in water itself, without the use of electrical power. The nanocomposite produced exhibits river sand-like properties, such as higher shear strength in loose and wet forms [75]. The ability to prepare nanostructured compositions at near-ambient temperature has wide relevance for adsorption-based water purification. Such properties are even more significant as active nanomaterials do not get into the flowing water, eliminating nanotoxicity concerns. Materials composed of metastable iron oxyhydroxides [77,78] are currently used to supply arsenic-free clean drinking water to about 1000,000 people in India daily, at a cost of US$0.3 per KL, delivered in the kitchen. Translation of such materials to affordable filters across the world could eliminate the arsenic menace, which affects close to 120 million people globally. Extension of such science to other contaminants such as fluoride, uranium, mercury, chromium, etc., can solve drinking water problems across the world in many resource-limited regions. Materials with high electro-adsorption properties have been used to develop capacitive deionization units [79,80]. New technologies using nanostructures have been used to create efficient water harvesters from air [81], while new nanocomposite structures with graphene are claimed to be able to reduce desalination costs substantially [82]. Nanopores made in MoS_2 nanosheets can produce H_2O_2 efficiently using visible light, causing disinfection of flowing water [83]. While new technologies for clean water are essential, contaminants have to be detected and measured quantitatively at ultratrace levels. Many nanostructures can effectively perform these functions [84,85], making affordable sensing and removal possible.

7. Membranes for a Greener Future

Innovations in process technology are strategically necessary for realizing sustainable industrial growth, minimizing environmental problems, energy consumption, costs, and increasing productivity and final high-quality production lines. The principle and goals of process intensification describe well the actions necessary for reaching these goals [86,87]. It is interesting to emphasize the important contribution that Membrane Engineering is offering today to this strategy [88,89]. In a recent report published in Chemical Engineering News, the example of the recent successes of "Made in China" in Chemical Process Technology describes the situation well. Most of the "case studies" reported are in fact related to the introduction of membrane operations in production lines [90].

Nanofiltration, zeolite membranes, membrane bioreactors (MBRs), microfiltration, pervaporation, and all pressure-driven membrane operations are typical examples of technologies becoming dominant in various fields, from desalination to wastewater treatment, from ethanol dewatering to gas purification. For instance, a membrane-assisted process for the conversion of carbon dioxide and methane into synthesis gas to reduce the carbon footprint of high-carbon fossil fuels was developed by Shanghai Advanced Research Institute (SARI) with the support of funds provided by Shell. Recently, Evonik Industries AG and the SINOPEC Beijing Research Institute of the Chemical Industry (BRICI) joined forces to build a process development laboratory for organic solvent nanofiltration membrane technology. Evonik stated that new technologies developed at the lab will be marketed to chemical, food, and drug producers in Asia. Recently, the Chinese Academy of Science finished the largest zeolite membrane separation system in the world for dewatering of ethanol with scale of 100,000 tons per annum. Approximately, 70% of the membrane plants are employed in petrochemical (38%), power generation (22%) and steel industries (9%), while textiles, paper, food, and mining grasp 17% of the remaining plants.

Progress in utilization of new green solvents in membrane preparation and in the use of advanced 2D materials including graphene, integrated with the development of emerging membrane operations including membrane distillation, membrane crystallizers, membrane condensers, membrane emulsifiers, offers important opportunities for overcoming existing limits to the traditional Chemical Process Engineering.

The most recent success of membrane engineering has not been in new membranes but in the development of new membrane systems and of integrated membrane operations. The most interesting and visible case is membrane systems in seawater desalination where membrane systems are today the dominant technology, including in the Gulf and MENA Countries. Further interesting progress may be realized in the next years based on the utilization of graphene and other 2D materials in membranes and mixed matrix membranes production.

The growth of the overall membranes and membrane systems market might be related to the realization of emerging membrane operations including membrane distillation, membrane crystallizers, membrane condensers, membrane emulsifiers, based mainly on the production of hydrophobic microporous or nanoporous membranes serving as membrane contactors. These systems could contribute to realization of a process intensification in chemical processes. In all these operations, hydrophobic microporous or nanoporous membranes with specific appropriate characteristic will have to be created because at present there is no large-scale production at industrial level for these hydrophobic porous membranes.

Membrane technology has a major role to play in sustainable access to clean water for humanity. Our demand for water continues to increase because of population growth and climate change. However, there are limited sources of clean surface waters and supplies need augmentation from alternative "infinite" sources, such as the oceans and wastewater. The dual-membrane processes of ultrafiltration (UF) pretreatment followed by reverse osmosis (RO) purification is now the established platform for seawater desalination and wastewater reclamation; in the latter case, the UF membrane could be the separation step in a wastewater MBR. In terms of sustainability criteria, membrane technologies are attractive with potential for modest cost and modest energy demand, resource recovery, low ecological impact, low human health impact, and high reliability and resilience. This particularly applies to decentralized systems [91] favored by the modular nature of membrane systems.

The modest energy demand required for sustainable processing is a potential feature of membranes in the water domain. For example, surface water treatment by UF can be achieved by the use of gravity-driven membranes (GDMs), which require very low energy use and generate no chemical residues [92]. Seawater desalination by UF+RO already operates relatively efficiently and there are prospects to almost halve energy demand by combining GDM pretreatment, use of next-generation "ultrapermeable" RO membranes in a staged cascade, and energy recovery from the brine using pressure retarded osmosis (PRO) [93]. For wastewater reclamation, the next step should be the anaerobic MBR (AnMBR) + RO process, with biogas production providing a net energy benefit for the MBR stage [94]. A feature of these examples is that there is a potential trade-off between capital and operating costs, where a more sustainable operation with lower energy and ecological impact typically requires more initial investment due to lower fluxes (GDM operation) and/or additional stages (PRO). For proper sustainability, accounting these additional investments needs to be balanced against the externalities [95]. Fig. 1.5, adapted from T. Asano (Stockholm Water Prize, 2001), illustrates how membrane technologies fit into the "engineered" water cycle. An important feature of this for the future is multiple reuse, which is made possible by sustainable membrane technologies.

8. Advanced Porous Materials for Energy-Efficient Processes

Porous materials consist of a solid matrix with interconnected pores. Porous structures are abundant in nature, with crucial functions across various aspects of our lives from biological tissues to rocks and soils to zeolites. The long and winding road exploring these natural materials has resulted in the exploitation of porosity in advanced

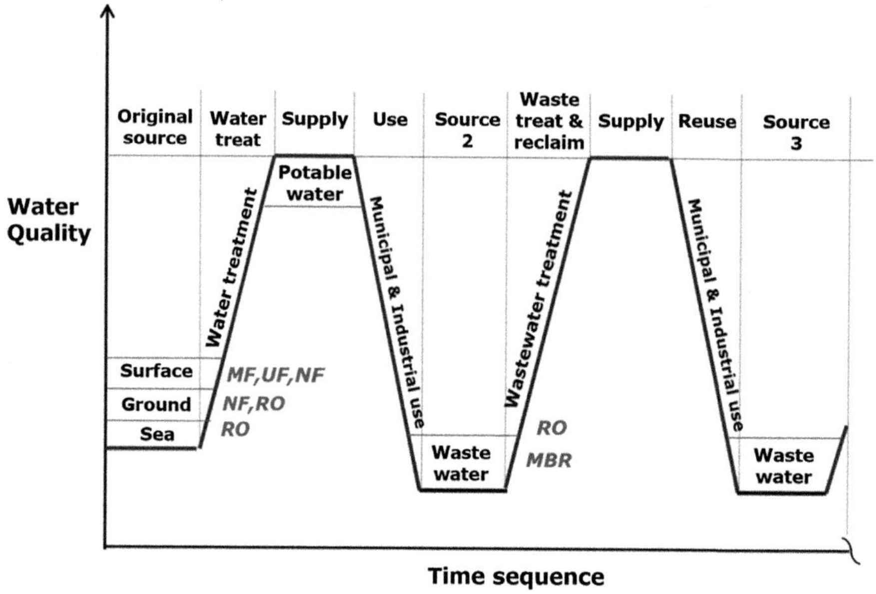

FIGURE 1.5 Water quality history for use and reuse—showing the role of membrane technologies.

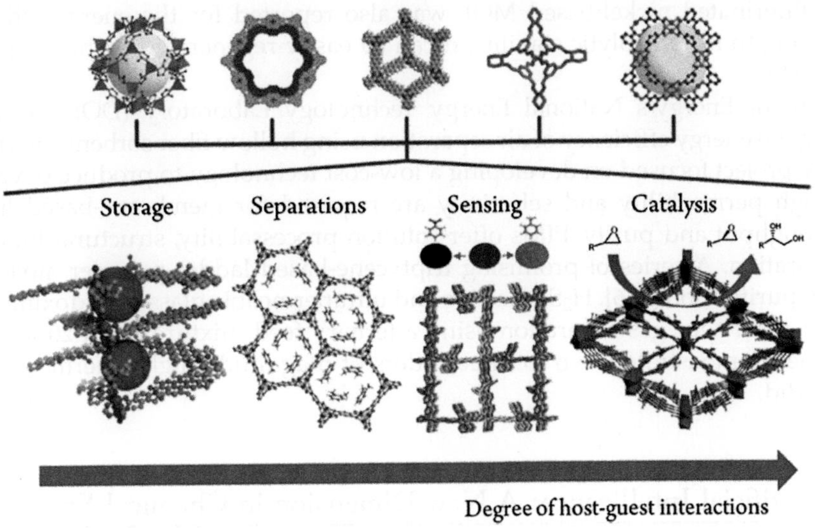

FIGURE 1.6 Rational design of porous materials has great potential to pioneer numerous engineering applications that are governed by host–guest interactions. *Reproduced with permission from D. Zhao, P.K. Thallapally, C. Petit, J. Gascon, Advanced porous materials: design, synthesis, and applications in sustainability, ACS Sustainable Chemistry & Engineering 7 (2019) 7997–7998.*

engineering materials covering separations, catalysis, sensing, petrochemicals, and beyond (Fig. 1.6). Precise control over the order and functionality of porous structures—via the utilization of structure-encoded molecular building blocks—enables the fine-tuning of materials for targeted applications [96]. These materials include zeolites, MOFs, polymers of intrinsic microporosity (PIMs), porous organic polymers, and covalent organic frameworks (COFs), all of which hold great promise for paradigm shifts in the energy, catalysis, separation, and environmental fields. However, the need and challenges for scalable and sustainable synthesis of these materials have been highlighted [97]. Moreover, understanding the synthetic mechanisms, catalyst-assisted synthesis, solvent-free fabrication, process modeling, and equipment design are all areas with room for improvement to achieve sustainable porous materials.

Zeolites are hydrated crystalline aluminosilicates with microporosity. There are more than 230 different zeolite-type structures and only a few of them are industrially applied. Most important types of zeolites (FAU, MOR, etc.) have their natural counterpart; even the most applied MFI-type structure (silicalite-1 and ZMS-5) has Mutinaite [98] as a counterpart, found in Antarctica years after its discovery in the middle 70s [99]. Thus, it can be said that current commercial zeolites are inspired by minerals. Nature-inspired synthetic zeolites and zeolites in turn have inspired the development of several organized porous materials with different chemical composition and properties, such us ordered mesoporous silica [100], octahedral–pentahedral–tetrahedral zeolitic materials [101], MOFs [102,103], and COFs [104]. Zeolites remain irreplaceable in fields requiring high performance (mainly high temperature and presence of moisture and acid gases) and optimum cost such as catalysis, and simultaneously zeolites have open new catalytic routes including those dealing with the transformation of renewable biofeedstocks into valuable platform compounds [105,106]. However, times are changing and no one knows what the destiny of petrochemistry is, as alternative fuels (H_2, biomethane, simple alcohols) coming from sustainable sources generating mechanical power and electricity are increasingly competing with oil derivative fuels. Of course, not only petrochemistry, related to the production of fuels (by FCC, hydrocracking, isomerization, etc.), but also other materials (e.g., polymers and fine chemicals) [107] and processes (e.g., separations, pollution abatement and emerging applications) [108] demand zeolites. Zeolite-supported bifunctional catalysts can be used for biomass valorization through the selective hydrodeoxygenation of platform molecules [109]. In this field, there is a need for more research on the rational design of the catalysts for reactions in biomass conversion and better understanding the synthesis–structure–performance relationships with regard to proximity and site balancing.

MOFs incorporating magnetic nanoparticles are emerging for the capture and energy-efficient triggered release of CO_2 via magnetic induction heating [110], which is called magnetic induction swing adsorption. The selectivity of MOF-based separation materials can be tuned via molecular building block approach and reticular chemistry, which was demonstrated for natural gas upgrading [111] and propane/propylene separation [112] performance. A promising class of emerging porous materials is zeolitic imidazolate frameworks with more than 150 synthesized structures. They are a thermally and chemically stable subclass of MOFs with a great potential for energy-efficient separations [113]. A fluorinated nickel-based MOF was also reported for the energy-efficient removal of water from gas streams. Owing to its hydrolytic stability, it can be easily regenerated by heating to only 105°C after the dehydration process [114].

The US Department of Energy's National Energy Technology Laboratory (DOE-NETL) recently reported a research on improving the energy efficiency of air separation using hollow fiber sorbents in a pressure swing adsorption process [115]. The project focused on developing a low-cost technology to produce oxygen for use in coal gasification processes. High permeability and selectivity are required for membrane-based hydrogen separation to achieve sufficient throughput and purity. PIMs offer solution processability, structural tunability, and mechanical flexibility for gas separation. A series of promising triptycene-based ladder-polymer molecular sieves have been reported for hydrogen purification [116]. H_2S-selective and ultrapermeable glassy amidoxime-functionalized porous polymers were reported for sour gas separation using a ternary feed mixture (with 20% H_2S:20% CO_2:60% CH_4) [117]. These membranes demonstrated two to three orders of magnitude higher permeability than commercially available competitors did.

9. Artificial Intelligence: A New Dimension in Chemical Engineering

In chemical engineering—from chemical, through materials, to process development and parallel maintenance and logistics—AI is emerging as a transformative power across the chemical manufacturing enterprise. AI has been attracting increasing attention in the chemical engineering community. Recent advances in AI (including the availability of large volume data, the development of sophisticated machine learning (ML) algorithms such as deep neural networks, and the support of powerful computing resources) have revolutionized many domains, one of which is chemical engineering [118]. Here we describe the chemical process as $f(x) = y$, where x are input variables (experimental parameters) and y is the target output variable to be optimized (e.g., yield, productivity, cost), which allows us to describe how AI has been introduced to chemical engineering, summarized in the following directions.

First, in most cases, f(x) is unknown. That is to say how y depends on x is unclear. ML (a subfield in AI) models are designed to have predictive power to map x to y by approximating the function f(.). Neural networks are popular ML models, which have been proved as general approximators of any nonlinear function with appropriate setting of network architecture and have been used successfully for prediction reaction results [119–121]. Data-driven chemical synthesis planning is another new trend to seek help of ML [122,123]. Reaction rules are extracted from large reaction corpora (such as the United States Patent and Trademark Office [USPTO], Reaxys, and SciFinder databases). ML models are trained on selected reactions and then employed to predict the transformation rule used to produce a molecule.

Second, a key problem in material design and laboratory exploration is to find the optimal set of parameters that lead to a satisfactory result, that is, finding x^* that can maximize $y = f(x)$. The traditional trial and error process optimization is cumbersome, takes a lot of time, and consumes a lot of chemicals. Even with chemists' expertise in screening a set of predetermined experimental conditions, it is still expensive due to the high number of required experiments. As f(x) is unknown, it is not feasible to use gradient-based optimization solutions. AI techniques open a new way to address the optimization problem in chemical laboratory studies, for example, using genetic algorithms, simulated annealing and direct search based on simplex to search the optimal parameter setting in recent work [124,125]. Also, the searching process can be integrated with the prediction model to provide a comprehensive materials discovery system using active learning [126].

Third, a direction of study explores published articles and extracts useful information for understanding chemical concepts and reaction process. The main motivation is the exponential increase in the number of published articles, which reading all relevant papers in a topic less and less feasible. With the help of the branch of AI known as natural language processing techniques, computers can be used to assimilate the information contained in these research papers. An example is their use to automatically extract synthesis information and trends from zeolite journal articles [127].

10. The Drawback of Nanomaterials

The rapid pace of innovation in the nanotechnology sector has led to a new sustainability conundrum. On one hand, engineered nanomaterials offer novel chemical and physical properties that enhance the performance of renewable energy generation and storage systems, environmental remediation technologies, and agricultural and health applications. These benefits will likely translate into reduced fossil fuel consumption, greater potential for carbon mitigation and sequestration, and improved technologies for treating air, water, and soil emissions. On the other hand, the same unique properties that make nanomaterials attractive for so many applications also introduce the potential for unanticipated environmental, economic, and social risks [128]. Ensuring that the potential benefits of nanotechnology use outweigh possible impacts requires a more proactive approach to nanoscale design and manufacturing and overall systems analyses.

Some sustainability risks stem from the potential for nanomaterials to be released during processes associated with their manufacturing, use, or decommissioning. Once released to the natural environment, the ultimate transport, transformation, fate, and resulting ecological impacts of these materials are still poorly understood [129]. These impacts may include aquatic ecotoxicity in freshwater or benthic organisms, which is widely studied, albeit through highly variable methodological approaches [130]. At a broader scale, long-term accumulation of nanomaterials in nature may result in impacts to ecosystem function and services. While these risks are still uncertain, early research suggests that they may include increased predation, decreased population growth and reproduction, and shifts in underlying microbial ecology underpinning ecosystem processes [131–133]. While direct risks from nanomaterial release have raised significant scientific and public attention, the broader perspective on environmental impacts of nanotechnology is more complex and challenging to understand.

Ongoing nanomaterial design and development must take into consideration the entire life cycle of these materials and the potential for environmental harm at this holistic scale. In particular, nanoscale material synthesis and nano-enabled manufacturing are critical points for environmental improvement. Carbon-based nanomaterials such as carbon nanotubes (CNTs), fullerenes (e.g., C60), and graphene have extremely energy-intense manufacturing processes [134]. As a point of comparison, 1 kilogram of functionalized fullerene C60-PCBM has an "embodied energy" of 65 GJ [135], which is over 300 times greater than that of aluminum produced from virgin resources ($\sim$200 MJ/kg^{-1}). Although these fullerene derivatives are only used in extremely small amounts in photovoltaic cells, their enormous energy footprint contributes almost 20% of the solar cell's own life cycle energy footprint [136].

Several life cycle assessment (LCA) studies have modeled the "cradle-to-grave" environmental impact of nanomaterial production, use, and disposal and have generally shown that the downstream impacts of nanomaterial release are often dwarfed by much greater energy, emission, and ecotoxicity impacts stemming from upstream processing. For nanomaterials characterized by low yields and high energy intensity, the key life cycle impact driver is typically energy use directly in the synthesis process or further upstream in precursor preparation. In the case of titania nanoparticle production, life cycle impact increased proportionally to the complexity of organic precursors as compared to simple inorganic precursors [137]. Energy intensity of single-walled CNT production has greater life cycle risks than the potential release of these materials into the environment, even under extreme scenarios [138]. Upstream processing steps often require significant electricity inputs, which come from fossil fuel combustion, a process that emits greenhouse gases, heavy metals, and other air toxics. Nanomaterial production also causes environmental impacts due to acid mine drainage and metals emissions during mining operations, particularly for metallic materials such as nanosilver [139]. Nanotechnology often has exacting technical specifications for materials used, resulting in significant organic solvent use during separation, cleaning, and purification processes [125]. In the limited literature yet to examine MOFs and zeolite membranes, solvent use was identified as the key environmental impact driver for several processing routes [140,141]. These findings collectively underscore the importance of greener nanoscale manufacturing routes, wherein increased yield, reduced energy demand, solvent recycling, and green solvents are critical factors to minimize environmental impacts of nanotechnology.

However, when considering the environmental trade-offs from nanomaterial manufacturing, it is essential to evaluate nano-enabled technology adoption through a holistic lens [50]. Modeling life cycle energy demand or ecotoxicity of manufacturing only begin to capture how material trade-offs may ripple out to broader techno-ecological scales. Consider the use of nanotechnology in renewable energy systems like solar photovoltaics. Nano-enabled improvements could increase efficiency, reduce costs, or extend system lifetimes, all of which may further catalyze more adoption of solar technology. Greater deployment of renewable energy may in turn displace fossil fuel use, thereby reducing heavy metal and greenhouse gas emissions associated with extracting and combusting coal and natural gas. Considering both benefits and costs of nanotechnology has the potential to point out opportunities where use of these materials can lead to the greatest overall system result [142–144]. Unfortunately, life cycle thinking is rarely mentioned, much less applied in conjunction with research devoted to creating, characterizing, and integrating novel nanomaterials, and many of these technologies have yet to be fully evaluated via LCA. To realize the full sustainability benefits of nanotechnology, future research must reflect greater integration of technological innovation with environmental assessment, ecodesign, and green manufacturing.

References

[1] Sustainable Development Goals, United Nations, https://www.un.org/sustainabledevelopment/sustainable-development-goals/.
[2] A. Moldavska, T. Welo, The concept of sustainable manufacturing and its definitions: a content-analysis based literature review, Journal of Cleaner Production 166 (2017) 744–755.
[3] United States' Environmental Protection Agency (EPA), https://www.epa.gov/sustainability/sustainable-manufacturing.
[4] C.P. Johnston, Safe, sustainable and scalable, Nature Chemistry 11 (2019) 400–401.
[5] R.A. Sheldon, H. van Bekkum (Eds.), Fine Chemicals Through Heterogeneous Catalysis, Wiley-VCH, Weinheim, Germany, 2001, https://doi.org/10.1002/9783527612963.
[6] Top 10 Emerging Technologies of 2017—Which 10 Disruptive Solutions Are Now Poised to Change the World? World Economic Forum Special Report, Scientific American, June 26 , 2017. https://www.scientificamerican.com/report/top-10-emerging-technologies-of-20171/.
[7] M. Droescher, Chemistry: the driving force for emerging technologies, Chemistry International 40 (2018) 14–17.
[8] 10 Breakthrough Technologies 2018, MIT Technology Review, 2018. https://www.technologyreview.com/lists/technologies/2018/.
[9] V. Sivaram, Taming the Sun: Innovations to Harness Solar Energy and Power the Planet, MIT Press, 2018. ISBN: 9780262037686.
[10] D.G. Nocera, The artificial leaf, Accounts of Chemical Research 45 (2012) 767–776.
[11] I.J. Goodfellow, J. Pouget-Abadie, M. Mirza, B. Xu, D. Warde-Farley, S. Ozair, A. Courville, Y. Bengio, Generative adversarial nets, NIPS Proceedings, 2014 arXiv:1406.2661v1.
[12] L.A. Gatys, A.S. Ecker, M. Bethge, A neural algorithm of artistic style, NIPS Proceedings (2015) arXiv:1508.06576.
[13] www.deepdreamgenerator.com.
[14] P. Anastas, N. Eghbali, Green chemistry: principles and practice, Chemical Society Review 39 (2010) 301–312.
[15] S. Hübner, J.G. de Vries, V. Farina, Why does industry not use immobilized transition metal complexes as catalysts? Advanced Synthesis & Catalysis 358 (2016) 3–25.
[16] H. Mahmoudi, F. Valentini, F. Ferlin, L.A. Bivona, I. Anastasiou, L. Fusaro, C. Aprile, A. Marrocchi, L. Vaccaro, A tailored polymeric cationic tag–anionic Pd(II) complex as a catalyst for the low-leaching Heck–Mizoroki coupling in flow and in biomass-derived GVL, Green Chemistry 21 (2019) 355–360.
[17] F. Valentini, H. Mahmoudi, L.A. Bivona, O. Piermatti, M. Bagherzadeh, L. Fusaro, C. Aprile, A. Marrocchi, L. Vaccaro, Polymer-supported bis-1,2,4-triazolium ionic tag framework for an efficient Pd(0) catalytic system in biomass derived γ-valerolactone, ACS Sustainable Chemistry & Engineering 7 (2019) 6939–6946.
[18] F. Ferlin, T. Giannoni, A. Zuliani, O. Piermatti, R. Luque, L. Vaccaro, Sustainable protocol for the reduction of nitroarenes by heterogeneous Au@SBA-15 with NaBH4 under flow conditions, ChemSusChem (2019), https://doi.org/10.1002/cssc.201802359.
[19] V. Trombettoni, D. Sciosci, M.P. Bracciale, F. Campana, M.L. Santarelli, A. Marrocchi, L. Vaccaro, Boosting biomass valorisation. Synergistic design of continuous flow reactors and water-tolerant polystyrene acid catalysts for a non-stop production of esters, Green Chemistry 20 (2019) 3222–3231.
[20] V. Kozell, F. Rahmani, O. Piermatti, D. Lanari, L. Vaccaro, A stereoselective organic base-catalyzed protocol for hydroamination of alkynes under solvent-free conditions, Molecular Catalysis 455 (2018) 188–191.
[21] G. Hahn, P. Kunnas, N. de Jonge, R. Kempe, General synthesis of primary amines via reductive amination employing a reusable nickel catalyst, Nature Catalysis 2 (2019) 71–77.
[22] W. Li, X. Cui, K. Junge, A.-E. Surkus, C. Kreyenschulte, S. Bartling, M. Beller, General and chemoselective copper oxide catalysts for hydrogenation reactions, ACS Catalysis 9 (2019) 4302–4307.
[23] D.L. Hughes, Applications of flow chemistry in drug development: highlights of recent patent literature, Organic Process Research & Development 22 (2018) 13–20.
[24] S. Santoro, F. Ferlin, L. Ackermann, L. Vaccaro, C–H functionalization reactions under flow conditions, Chemical Society Review 48 (2019) 2767–2782.
[25] F. Gomollón-Bel, Ten chemical innovations that will change our world, Chemistry International 41 (2019) 12–17.
[26] D. Dallinger, C.O. Kappe, Why flow means green – evaluating the merits of continuous processing in the context of sustainability, Current Opinion in Green and Sustainable Chemistry 7 (2017) 6–12.
[27] B. Gutmann, D. Cantillo, C.O. Kappe, Continuous-flow technology—a tool for the safe manufacturing of active pharmaceutical ingredients, Angewandte Chemie International Edition 54 (2015) 6688–6728.
[28] R. Mullin, Chemical Engineering News 97 (2019) 28.
[29] A. Adamo, R.L. Beingessner, M. Behnam, J. Chen, T.F. Jamison, K.F. Jensen, J.C. Monbaliu, A.S. Myerson, E.M. Revalor, D.R. Snead, T. Stelzer, N. Weeranoppanant, S.Y. Wong, P. Zhang, On-demand continuous-flow production of pharmaceuticals in a compact, reconfigurable system, Science 352 (2016) 61–67.
[30] J.C. McWilliams, D. Ayman, D. Allian, S.M. Opalka, S.M. May, M. Journet, T.M. Braden, The evolving state of continuous processing in pharmaceutical API manufacturing: a survey of pharmaceutical companies and contract manufacturing organizations, Organic Process Research & Development 22 (2018) 1143–1166.
[31] R.M. Myers, D.E. Fitzpatrick, R.M. Turner, S.V. Ley, Flow chemistry meets advanced functional materials, Chemical European Journal 20 (2014) 12348–12366.
[32] T. Fodi, C. Didaskalou, J. Kupai, G.T. Balogh, P. Huszthy, G. Szekely, Nanofiltration-enabled in situ solvent and reagent recycle for sustainable continuous-flow synthesis, ChemSusChem 10 (2017) 3435–3444.

[33] C. Didaskalou, J. Kupai, L. Cseri, J. Barabas, E. Vass, T. Holtzl, G. Szekely, Membrane-grafted asymmetric organocatalyst for an integrated synthesis–separation platform, ACS Catalysis 8 (2018) 7430–7438.
[34] P. Kisszekelyi, A. Alammar, J. Kupai, P. Huszthy, J. Barabas, T. Holtzl, L. Szente, C. Bawn, R. Adams, G. Szekely, Asymmetric synthesis with cinchona-decorated cyclodextrin in a continuous-flow membrane reactor, Journal of Catalysis 371 (2019) 255–261.
[35] L. Rogers, K.F. Jensen, Continuous manufacturing – the green chemistry promise? Green Chemistry 21 (2019) 3481–3498.
[36] L.L. Simon, H. Pataki, G. Marosi, F. Meemken, K. Hungerbühler, A. Baiker, S. Tummala, B. Glennon, M. Kuentz, G. Steele, H.J.M. Kramer, J.W. Rydzak, Z. Chen, J. Morris, F. Kjell, R. Singh, R. Gani, K.V. Gernaey, M. Louhi-Kultanen, J. O'Reilly, N. Sandler, O. Antikainen, J. Yliruusi, P. Frohberg, J. Ulrich, R.D. Braatz, T. Leyssens, M. von Stosch, R. Oliveira, R.B.H. Tan, H. Wu, M. Khan, D. O'Grady, A. Pandey, R. Westra, E. Delle-Case, D. Pape, D. Angelosante, Y. Maret, O. Steiger, M. Lenner, K. Abbou-Oucherif, Z.K. Nagy, J.D. Litster, V.K. Kamaraju, M.-S. Chiu, Assessment of recent process analytical technology (PAT) trends: a multiauthor review, Organic Process Research & Development 19 (2015) 3–62.
[37] S.V. Ley, D.E. Fitzpatrick, R.J. Ingham, R.M. Myers, Organic synthesis: march of the machines, Angewandte Chemie International Edition 54 (2015) 3449–3464.
[38] https://ispe.org/pharmaceutical-engineering/ispeak/continuous-osd-manufacturing-product-patient-perspective.
[39] http://www.pharmtech.com/vertex-receives-fda-approval-continuously-manufactured-drug-product.
[40] https://www.europeanpharmaceuticalreview.com/news/62587/ema-continuous-manufacturing.
[41] http://www.pharmtech.com/fda-approves-tablet-production-janssen-continuous-manufacturing-line.
[42] Pharm. Eng. 38 (January-February 2018). Number 1.
[43] http://ccpmj.org/downloads/1_Dr.Lee_rev(en).pdf.
[44] S. Mascia, P.L. Heider, H. Zhang, R. Lakerveld, B. Benyahia, P.I. Barton, R.D. Braatz, C.L. Cooney, J.M.B. Evans, T.F. Jamison, K.F. Jensen, A.S. Myerson, B.L. Trout, End-to-End continuous manufacturing of pharmaceuticals: integrated synthesis, purification, and final dosage formation, Angewandte Chemie International Edition 52 (2013) 12359–12363.
[45] R. Lakerveld, B. Benyahia, P.L. Heider, H. Zhang, A. Wolfe, C.J. Testa, S. Ogden, D.R. Hersey, S. Mascia, J.M.B. Evans, R.D. Braatz, P.I. Barton, The application of an automated control strategy for an integrated continuous pharmaceutical pilot plant, Organic Process Research & Development 19 (2015) 1088–1100.
[46] https://ispe.org/facility-year-awards/winners/2019/process-innovation.
[47] M. Hong, E.Y.-X. Chen, Future directions for sustainable polymers, Trends in Chemistry 1 (2019) 148–151.
[48] I. Malsch, M. Mullins, E. Semenzin, A. Zabeo, D. Hristozov, A. Marcomini, Decision support for international agreements regulating nanomaterials, NanoEthics 12 (2018) 39–54.
[49] M.M. Falinski, D.L. Plata, S.S. Chopra, T.L. Theis, L.M. Gilbertson, J.B. Zimmerman, A framework for sustainable nanomaterial selection and design based on performance, hazard, and economic considerations, Nature Nanotechnology 13 (2018) 708–714.
[50] C.W. Babbitt, E.A. Moore, Sustainable nanomaterials by design, Nature Nanotechnology 13 (2018) 621–629.
[51] U. Aswathanarayana, R.S. Divi (Eds.), Energy Portfolios, CRC Press, Boca Raton, 2009.
[52] P.T. Anastas, J.C. Warner, Green Chemistry: Theory and Practice, Oxford University Press, Oxford, UK, 1998.
[53] P.T. Anastas, J.B. Zimmerman, Peer reviewed: design through the 12 principles of green engineering, Environmental Science & Technology 37 (2003) 94A–101A.
[54] E. Csefalvay, G.R. Akien, L. Qi, I.T. Horvath, Definition and application of ethanol equivalent: sustainability performance metrics for biomass conversion to carbon-based fuels and chemicals, Catalysis Today 239 (2015) 50–55.
[55] I.T. Horvath, Introduction: sustainable chemistry, Chemical Reviews 118 (2018) 369–371.
[56] E.B. Barbier, The Water Paradox: Overcoming the Global Crisis in Water Management, Yale University Press, New Haven, CT, 2019.
[57] C. Campanhola, S. Pandey, Sustainable Food and Agriculture: An Integrated Approach, Academic Press, London, UK, 2019.
[58] M. Asaduzzaman, Controlled Environment Agriculture: Production of Specialty Crops Providing Human Health Benefits through Hydroponics, Nova Science Publishers, Inc., Hauppauge, NY, 2016.
[59] C. Jimenez-Gonzalez, C.S. Ponder, Q.B. Broxterman, J.B. Manley, Using the right green yardstick: why process mass intensity is used in the pharmaceutical industry to drive more sustainable processes, Organic Process Research & Development 15 (2011) 912–917.
[60] D. Prat, J. Hayler, A. Wells, A survey of solvent selection guides showed broad consensus in the solvents having most and least EHS issues, Green Chemistry 16 (2014) 4546–4551.
[61] D. Prat, A. Wells, J. Hayler, H. Sneddon, C.R. McElroy, S. Abou-Shehada, P.J. Dunn, CHEM21 selection guide of classical- and less classical-solvents, Green Chemistry 18 (2016) 288–296.
[62] J.W. Comerford, I.D.V. Ingram, M. North, X. Wu, Sustainable metal-based catalysts for the synthesis of cyclic carbonates containing five-membered rings, Green Chemistry 17 (2015) 1966–1987.
[63] J.E. Camp, Bio-available solvent cyrene: synthesis, derivatization, and applications, ChemSusChem 11 (2018) 3048–3055.
[64] J. Sherwood, H.L. Parker, K. Moonen, T.J. Farmer, A.J. Hunt, N-Butylpyrrolidinone as a dipolar aprotic solvent for organic synthesis, Green Chemistry 18 (2016) 3990–3996.
[65] L. Cseri, G. Szekely, Towards cleaner PolarClean: efficient synthesis and extended applications of the polar aprotic solvent methyl 5-(Dimethylamino)-2-methyl-5-oxopentanoate, Green Chemistry 21 (2019) 4178–4188.
[66] F. Byrne, B. Forier, G. Bossaert, C. Hoebers, T.J. Farmer, J.H. Clark, A.J. Hunt, 2,2,5,5-Tetramethyltetrahydrofuran (TMTHF): a non-polar, non-peroxide forming ether replacement for hazardous hydrocarbon solvents, Green Chemistry 19 (2017) 3671–3678.
[67] ICH Harmonised Guideline, Impurities: Guideline for Residual Solvents, Q3C(R6): https://www.ich.org/fileadmin/Public_Web_Site/ICH_Products/Guidelines/Quality/Q3C/Q3C__R6___Step_4.pdf.
[68] P. Thomas, F. Urbanski, Pharma Manufacturing: https://www.pharmamanufacturing.com/articles/2011/150.
[69] Q. Zhang, K. De O. Vigier, S. Royer, F. Jerome, Deep eutectic solvents: syntheses, properties and applications, Chemical Society Reviews 41 (2012) 7108–7146.
[70] Y. Liu, J.B. Friesen, J.B. McAlpine, D.C. Lankin, S.-N. Chen, G.F. Pauli, Natural deep eutectic solvents: properties, applications, and perspectives, Journal of Natural Products 81 (2018) 679–690.
[71] M. Obst, B. König, Organic synthesis without conventional solvents, European Journal of Organic Chemistry 2018 (2018) 4213–4232.

[72] F. Gallou, N.A. Isley, A. Ganic, U. Onken, M. Parmentier, Surfactant technology applied toward an active pharmaceutical ingredient: more than a simple green chemistry advance, Green Chemistry 18 (2016) 14–19.
[73] P.J. Ellis, I.J.S. Fairlamb, S.F.J. Hackett, K. Wilson, A.F. Lee, Evidence for the surface-catalyzed suzuki–miyaura reaction over palladium nanoparticles: an operando XAS study, Angewandte Chemie International Edition 49 (2010) 1820–1824.
[74] R.S. Shelkar, S.H. Gund, J.M. Nagarkar, Nano Pd–Fe3O4@Alg beads: as an efficient and magnetically separable catalyst for Suzuki, Heck and Buchwald–Hartwig coupling reactions, RSC Advances 4 (2014) 53387–53396.
[75] M.C. Bryan, P.J. Dunn, D. Entwistle, F. Gallou, S.G. Koenig, J.D. Hayler, M.R. Hickey, S. Hughes, M.E. Kopach, G. Moine, P. Richardson, F. Roschangar, A. Steven, F.J. Weiberth, Key Green Chemistry research areas from a pharmaceutical manufacturers' perspective revisited, Green Chemitry 20 (2018) 5082–5103.
[76] M.U. Sankar, S. Aigal, A. Chaudhary, A. Shihabudheen, M. Maliyekkal, A.A. Kumar, K. Chaudhari, T. Pradeep, Biopolymer reinforced synthetic granular nanocomposites for affordable point-of-use water purification, Proceedings of the National Academy of Sciences of the United States of America 110 (2013) 8459–8464.
[77] A.K. Avula, A. Som, P. Longo, C. Sudhakar, R.G. Bhuin, S.S. Gupta, M.U. Sankar, A. Chaudhary, R. Kumar, T. Pradeep, Confined metastable 2-line ferrihydrite for affordable point-of-use arsenic free drinking water, Advanced Materials 29 (2017) 1604260.
[78] C. Sudhakar, A.A. Kumar, R.G. Bhuin, S.S. Gupta, G. Natarajan, T. Pradeep, Species-specific uptake of arsenic on confined metastable 2-line ferrihydrite: a combined Raman-XPS investigation of the adsorption mechanism, ACS Sustainable Chemistry & Engineering (2018), https://doi.org/10.1021/acssuschemeng.8b01217.
[79] N. Pugazhenthiran, S.S. Gupta, A. Prabhath, M. Manikandan, J.R. Swathy, V.K. Raman, T. Pradeep, Cellulose derived graphenic fibers for capacitive desalination of brackish water, ACS Applied Materials & Interfaces 7 (2015) 20156–20163.
[80] S. Gupta, M. Islam, T. Pradeep, Capacitive deionization (CDI): an alternative cost-efficient desalination technique, in: S. Ahuja (Ed.), Advances in Water Purification Techniques, Elsevier, Amsterdam, 2019, pp. 165–202.
[81] D. Sarkar, A. Mahapatra, A. Som, R. Kumar, A. Nagar, A. Baidya, T. Pradeep, Patterned nanobrush nature mimics with unprecedented water harvesting efficiency, Advanced Materials Interfaces 5 (2018) 1800667.
[82] J. Abraham, K.S. Vasu, C.D. Williams, K. Gopinadhan, Y. Su, C.T. Cherian, J. Dix, E. Prestat, S.J. Haigh, I.V. Grigorieva, P. Carbone, A.K. Geim, R.R. Nair, Tunable sieving of ions using graphene oxide membranes, Nature Nanotechnology 12 (2017), 546–512.
[83] D. Sarkar, B. Mondal, A. Som, S.J. Ravindran, S.K. Jana, C.K. Manju, T. Pradeep, Holey MoS_2 nanosheets with photocatalytic metal rich edges by ambient electrospray deposition for solar water disinfection, Global Challenges 2 (2018) 1800052.
[84] A. Ghosh, V. Jeseentharani, M.A. Ganayee, R. Hemalatha, K. Chaudhari, C. Vijayan, T. Pradeep, Approaching sensitivity of tens of ions using atomically precise cluster-nanofiber composites, Analytical Chemistry 86 (2014) 10996–11001.
[85] A. Mathew, P.R. Sajanlal, T. Pradeep, Selective visual detection of TNT at the sub-zeptomole level, Angewandte Chemie International Edition 51 (2012) 9596–9600.
[86] E. Drioli, A. Stankiewicz, F. Macedonio, Membrane Engineering in process intensification-an overview, Journal of Membrane Science 380 (2011) 1–8.
[87] E. Drioli, L. Giorno, E. Fontananova (Eds.), Comprehensive Membrane Science and Engineering, second ed., Elsevier, 2017. ISBN: 9780444637758.
[88] F. Macedonio, E. Drioli, Membrane engineering for green process engineering, Engineering 3 (2017) 290–298.
[89] A. Brunetti, F. Macedonio, G. Barbieri, E. Drioli, Membrane engineering for environmental protection and sustainable industrial growth: options for water and gas treatment, Environmental Engineering Research 20 (2015) 307–328.
[90] J.F. Tremblay, "Made in China" now extends to chemical process technology, Chemical & Engineering News 95 (2017) 16–17.
[91] A.G. Fane, The role of membrane technology in sustainable decentralized wastewater systems, Water Science and Technology 51 (2005) 317–325.
[92] W. Pronk, A. Ding, E. Morgenroth, N. Derlon, P. Desmond, M. Burkhardt, B. Wu, A.G. Fane, Gravity-driven membrane filtration for water and wastewater treatment: a review, Water Research 149 (2019) 553–565.
[93] A. Fane, A grand challenge for membrane desal: more water, less carbon, Desalination 401 (2017) 99–111.
[94] P.L. McCarty, J. Bae, J. Kim, Domestic wastewater treatment as a net energy producer – can this be achieved? Environmental Science & Technology 45 (2011) 7100–7106.
[95] N. Lior, Sustainability as quantitative norm for desalination impacts, Desalination 401 (2017) 99–111.
[96] D. Zhao, P.K. Thallapally, C. Petit, J. Gascon, Advanced porous materials: design, synthesis, and applications in sustainability, ACS Sustainable Chemistry & Engineering 7 (2019) 7997–7998.
[97] S.B. Peh, Y. Wang, D. Zhao, Scalable and sustainable synthesis of advanced porous materials, ACS Sustainable Chemistry & Engineering 7 (2019) 3647–3670.
[98] E. Galli, G. Vezzalini, S. Quartieri, A. Alberti, M. Franzini, Mutinaite, a new zeolite from Antarctica: the natural counterpart of ZSM-5, Zeolites 19 (1997) 318–322.
[99] E.M. Flanigen, J.M. Bennett, R.W. Grose, J.P. Cohen, R.L. Patton, R.M. Kirchner, J.V. Smith, Silicalite, a new hydrophobic crystalline silica molecular sieve, Nature 271 (1978) 512–516.
[100] C.T. Kresge, M.E. Leonowicz, W.J. Roth, J.C. Vartuli, J.S. Beck, Ordered mesoporous molecular sieves synthesized by a liquid-crystal template mechanism, Nature 359 (1992) 710–712.
[101] J. Rocha, M.W. Anderson, European Journal of Inorganic Chemistry 2000 (2000) 801–818.
[102] H. Li, M. Eddaoudi, M. O'Keeffe, O.M. Yaghi, Design and synthesis of an exceptionally stable and highly porous metal-organic framework, Nature 402 (1999) 276–279.
[103] G. Liu, V. Chernikova, Y. Liu, K. Zhang, Y. Belmabkhout, O. Shekhah, C. Zhang, S. Yi, M. Eddaoudi, W.J. Koros, Mixed matrix formulations with MOF molecular sieving for key energy-intensive separations, Nature Materials 17 (2018) 283–289.
[104] A.P. Cote, A.I. Benin, N.W. Ockwig, M. O'Keeffe, A.J. Matzger, O.M. Yaghi, Porous, crystalline, covalent organic frameworks, Science 310 (2005) 1166–1170.

[105] A. Corma, S. Iborra, A. Velty, Chemical routes for the transformation of biomass into chemicals, Chemical Reviews 107 (2007) 2411–2502.

[106] M.S. Holm, S. Saravanamurugan, E. Taarning, Conversion of sugars to lactic acid derivatives using heterogeneous zeotype catalysts, Science 328 (2010) 602–605.

[107] M. Moliner, State of the art of Lewis acid-containing zeolites: lessons from fine chemistry to new biomass transformation processes, Dalton Trans. 43 (2014) 4197–4208.

[108] L. Bacakova, M. Vandrovcova, I. Kopova, I. Jirka, Applications of zeolites in biotechnology and medicine – a review, Biomaterials Science 6 (2018) 974–989.

[109] W. Luo, W. Cao, P.C.A. Bruijnincx, L. Lin, A. Wang, T. Zhang, Zeolite-supported metal catalysts for selective hydrodeoxygenation of biomass-derived platform molecules, Green Chemistry (2019), https://doi.org/10.1039/C9GC01216H.

[110] M.m. Sadiq, K. Suzuki, M.R. Hill, Towards energy efficient separations with metal organic frameworks, Chemical Communications 54 (2018) 2825–2837.

[111] Y. Belmabkhout, P.M. Bhatt, K. Adil, R.S. Pillai, A. Cadiau, A. Shkurenko, G. Maurin, G. Liu, W.J. Koros, M. Eddaoudi, Natural gas upgrading using a fluorinated MOF with tuned H_2S and CO_2 adsorption selectivity, Nature Energy 3 (2018) 1059–1066.

[112] D.-X. Xue, A. Cadiau, Ł.J. Weseliński, H. Jiang, P.M. Bhatt, A. Shkurenko, L. Wojtas, C. Zhijie, Y. Belmabkhout, K. Adila, M. Eddaoudi, Topology meets MOF chemistry for pore-aperture fine tuning: ftw-MOF platform for energy-efficient separations via adsorption kinetics or molecular sieving, Chemical Communications 54 (2018) 6404–6407.

[113] B.R. Pimentel, A. Parulkar, E. Zhou, N.A. Brunelli, R.P. Lively, Zeolitic imidazolate frameworks: next-generation materials for energy-efficient gas separations, ChemSusChem 7 (2014) 3202–3240.

[114] A. Cadiau, Y. Belmabkhout, K. Adil, P.M. Bhatt, R.S. Pillai, A. Shkurenko, C. Martineau-Corcos, G. Maurin, M. Eddaoudi, Hydrolytically stable fluorinated metal-organic frameworks for energy-efficient dehydration, Science 356 (2017) 731–735.

[115] A. Rosinski, G. Swindlehurst, H. Du, R. Lively, Improving Energy Efficiency of Air Separation via Hollow Fiber Sorbents, United States, 2019. https://doi.org/10.2172/1503603.

[116] B.S. Ghanem, R. Swaidan, X. Ma, E. Litwiller, I. Pinnau, Energy-effi cient hydrogen separation by AB-type ladder-polymer molecular sieves, Advanced Materials 26 (2014) 6696–6700.

[117] S. Yi, B. Ghanem, Y. Liu, I. Pinnau, W.J. Koros, Ultraselective glassy polymer membranes with unprecedented performance for energy-efficient sour gas separation, Science Advances 5 (2019) eaaw5459.

[118] V. Venkatasubramanian, The promise of artificial intelligence in chemical engineering: is it here, finally? AIChE Journal 65 (2019) 466–478.

[119] C.W. Coley, R. Barzilay, T.S. Jaakkola, W.H. Green, K.F. Jensen, Prediction of organic reaction outcomes using machine learning, ACS Central Science 3 (2017) 434–443.

[120] Y. Yu, X. Tan, S. Ning, Y. Wu, Machine learning for understanding compatibility of organic–inorganic hybrid perovskites with post-treatment amines, ACS Energy Letters 4 (2019) 397–404.

[121] H. Gao, T.J. Struble, C.W. Coley, Y. Wang, W.H. Green, K.F. Jensen, Using machine learning to predict suitable conditions for organic reactions, ACS Central Science 4 (2018) 1465–1476.

[122] J.L. Baylon, N.A. Cilfone, J.R. Gulcher, T.W. Chittenden, Enhancing retrosynthetic reaction prediction with deep learning using multiscale reaction classification, Journal of Chemical Information and Model 59 (2019) 673–688.

[123] C.W. Coley, W.H. Green, K.F. Jensen, Machine learning in computer-aided synthesis planning, Accounts of Chemical Research 51 (2018) 1281–1289.

[124] J.S. Moore, K.F. Jensen, Automation in microreactor systems, in: T. Wirth (Ed.), Microreactors in Organic Chemistry and Catalysis, Wiley-VCH, 2013, pp. 81–100.

[125] D. Cortés-Borda, E. Wimmer, B. Gouilleux, E. Barré, N. Oger, L. Goulamaly, L. Peault, B. Charrier, C. Truchet, P. Giraudeau, M. Rodriguez-Zubiri, An autonomous self-optimizing flow reactor for the synthesis of natural product carpanone, The Journal of Organic Chemistry 83 (2018) 14286–14299.

[126] J.M. Caruthers, J.A. Lauterbach, K.T. Thomson, V. Venkatasubramanian, C.M. Snively, A. Bhan, S. Katare, G. Oskarsdottir, Catalyst design: knowledge extraction from high-throughput experimentation, Journal of Catalysis 216 (2003) 98–109.

[127] Z. Jensen, E. Kim, S. Kwon, T.Z. Gani, Y. Roman-Leshkov, M. Moliner, A. Corma, E. Olivetti, A machine learning approach to zeolite synthesis enabled by automatic literature data extraction, ACS Central Science 5 (2019) 892–899.

[128] V.L. Colvin, The potential environmental impact of engineered nanomaterials, Nature Biotechnology 21 (2003) 1166–1170.

[129] M. Baalousha, G. Cornelis, T. Kuhlbusch, I. Lynch, C. Nickel, W. Peijnenburg, N. Van Den Brink, Modeling nanomaterial fate and uptake in the environment: current knowledge and future trends, Environmental Science: Nano 3 (2016) 323–345.

[130] S. Jahan, I.B. Yusoff, Y.B. Alias, A.F.B.A. Bakar, Reviews of the toxicity behavior of five potential engineered nanomaterials (ENMs) into the aquatic ecosystem, Toxicology Reports 4 (2017) 211–220.

[131] D.A. Arndt, J. Chen, M. Moua, R.D. Klaper, Multigeneration impacts on Daphnia magna of carbon nanomaterials with differing core structures and functionalizations, Environmental Toxicology and Chemistry 33 (2014) 541–547.

[132] E.A. Moore, C.W. Babbitt, S.J. Connelly, A.C. Tyler, G. Rogalskyj, Cascading ecological impacts of fullerenes in freshwater ecosystems, Environmental Toxicology and Chemistry (2019) (in press): https://doi.org/10.1002/etc.4465.

[133] S.P. Cabral, E.A. Moore, C.T. Border, C.W. Babbitt, A.C. Tyler, Fullerene toxicity in the benthos with implications for freshwater ecosystem services, Science of The Total Environment (2019) (in press): https://doi.org/10.1002/etc.4465.

[134] M.L. Healy, L.J. Dahlben, J.A. Isaacs, Environmental assessment of single-walled carbon nanotube processes, J. Ind. Ecol. 12 (2008) 376–393.

[135] A. Anctil, C.W. Babbitt, R.P. Raffaelle, B.J. Landi, Material and energy intensity of fullerene production, Environmental Science & Technology 45 (2011) 2353–2359.

[136] A. Anctil, C.W. Babbitt, R.P. Raffaelle, B.J. Landi, Cumulative energy demand for small molecule and polymer photovoltaics, Progress in Photovoltaics: Research and Applications 21 (2013) 1541–1554.

[137] P. Caramazana-González, P.W. Dunne, M. Gimeno-Fabra, M. Zilka, M. Ticha, B. Stieberova, F. Freiberg, J. McKechnie, E.H. Lester, Assessing the life cycle environmental impacts of titania nanoparticle production by continuous flow solvo/hydrothermal syntheses, Green Chemistry 19 (2017) 1536–1547.
[138] M.J. Eckelman, M.S. Mauter, J.A. Isaacs, M. Elimelech, New perspectives on nanomaterial aquatic ecotoxicity: production impacts exceed direct exposure impacts for carbon nanotoubes, Environmental Science & Technology 46 (2012) 2902–2910.
[139] T. Garvey, E.A. Moore, C.W. Babbitt, G. Gaustad, Comparing ecotoxicity risks for nanomaterial production and release under uncertainty, Clean Technologies and Environmental Policy 21 (2019) 229–242.
[140] C.A. Grand, R. Blom, A. Spjelkavik, V. Moreau, J. Payet, Life-cycle assessment as a tool for eco-design of metal-organic frameworks (MOFs), Sustainable Materials and Technologies 14 (2017) 11–18.
[141] A. Navajas, N. Mittal, N. Rangnekar, H. Zhang, A. Cornejo, L.M. Gandía, M. Tsapatsis, ACS Sustainable Chemistry & Engineering 6 (2019) 15773–15780.
[142] L.M. Gilbertson, A.A. Busnaina, J.A. Isaacs, J.B. Zimmerman, M.J. Eckelman, Life cycle impacts and benefits of a carbon nanotube-enabled chemical gas sensor, Environmental Science & Technology 48 (2014) 11360–11368.
[143] P. Zhai, J.A. Isaacs, M.J. Eckelman, Net energy benefits of carbon nanotube applications, Applied Energy 173 (2016) 624–634.
[144] E.A. Moore, C.W. Babbitt, G. Gaustad, S.T. Moore, Portfolio optimization of nanomaterial use in clean energy technologies, Environmental Science & Technology 52 (2018) 4440–4448.

C H A P T E R

2

Nature-Inspired Chemical Engineering: A New Design Methodology for Sustainability

Panagiotis Trogadas, Marc-Olivier Coppens

Centre for Nature Inspired Engineering, Department of Chemical Engineering, University College London, London, United Kingdom

O U T L I N E

1. SUSTAINABILITY IN CHEMICAL ENGINEERING

The concept of sustainability was first mentioned in the scientific literature by H. C. von Carlowitz referring to sustainable forestry in *Sylvicultura oeconomica* in 1713 [1]. The meaning of the term sustainability in this book was to cut only as much timber as was regrowing, applying forestry practices to ensure that soil fertility was maintained or increased. In the following decades, the emerging idea of sustainable development was engendered among visionaries such as R. Carson in her book *Silent Spring* during the 1960s [2]. During the 1980s, the idea of sustainability was embraced by the international scientific community with the publication of the *Our Common Future* report by the Brundtland Commission in 1987 [3,4], which defined sustainable development as "development which meets the needs of the present without compromising the ability of future generations to meet their own needs" [4].

Purposely, the definition of sustainability provided by the commission left space for various interpretations. At the Rio Conference in 1992 and over the following years, the term "sustainability" was defined more precisely, and it has been confirmed since through a wide range of agreements, national programs, and scientific studies [1]. Sustainable development is a continuing process, during which definitions and resulting activities have been constantly evolving. The main aim of this process is to ensure a bright future for our descendants, although there is the underlying risk that theory is put into practice in different ways, creating a highly complex situation. It is for this reason that there are many different interpretations of what constitutes sustainability and sustainable development, a term that is often misused to serve particular interests.

Sustainable Nanoscale Engineering
https://doi.org/10.1016/B978-0-12-814681-1.00002-3

One of the most widely used definitions for sustainability was proposed by Liverman [5]: "Sustainability is the indefinite survival of the human species through the maintenance of basic life support systems (air, water, land, biota) and the existence of infrastructure and institutions which distribute and protect the components of these systems."

Currently, the goal of sustainability would not be perceived as attractive to many stakeholders if it were not exceedingly advantageous [1,6,7]. Sustainability implies a threefold added value—economic, social, and environmental profit—all in one. Only few companies would have promoted sustainable production system and process development, without economic incentives.

Moving toward sustainability in the design of products, processes, and production systems entails to reinforce the foundations of chemical engineering, adding novel disciplines, design guidelines, and creating dynamic interconnectivity among them. During the last decade, new disciplines and design philosophies for materials and reactor systems have been developed to serve as inspiration for sustainable design in chemical engineering; examples include biomimicry, green chemistry, resilience engineering, and ecological design [1]. In the following sections, we introduce the concept of nature-inspired chemical engineering (NICE), a new design philosophy leading to the redesign of catalytic materials and processes, displacing conventional blueprints and promoting sustainable solutions to engineering challenges.

2. SUSTAINABILITY: DESIGN PHILOSOPHY

2.1 Why Should We Use Nature as a Source for Inspiration?

Scientific research becomes increasingly demanding as we attempt to solve, with a sense of urgency, many more technical problems than in previous decades and aim to answer more challenging scientific questions, regardless of the advances made in synthesis and manufacturing [8]. Usually, the number of variables in these complex problems of the problem under examination, with limited time and resources. Developing methodologies to optimize the management of time, cost of energy and materials, synthesis of materials, man power, and maintenance cost is crucial in research. Computational approaches have their limitations (such as number of processors and storage available, number of variables used in a specific problem, etc.) and do not necessarily provide fundamental understanding. If there were to be a guideline for strategies and methodologies that could increase the efficiency of research and development, and lead to more sustainable solutions, its impact on scientific advancement would be immeasurable. One powerful way is to study nature more closely. Natural systems contain many characteristics and mechanisms that can be studied and used to give us hints to the solution of critical problems [9–11].

Careful examination of the structure and dynamics of natural systems reveals certain patterns that are key to underlying desirable properties. For example, the branching of trees, riverbeds, the vascular network, and the human lung contains repeated divisions that are similar over many length scales. In many cases, such features result from optimization processes of energy, matter, and space in natural systems, via self-organization or evolution. Processes in the aforementioned examples are scalable as a result of the self-similar branching. Just like the periodic repetition of unit cells in crystals helps to predict their physical and chemical properties, also the observation of fractal and other patterns in nature helps us to describe objects and dynamic behavior mathematically, and start investigating the underlying reasons for the observations. This, in turn, is the basis for developing a systematic approach to understand natural materials and processes, in relation to function, which, for the engineer, carries additional interest when these functions are superior to what is observed in current artificial materials and processes. Hence, deeper insights in nature can form the basis for nature-inspired solutions for engineering.

2.2 Ways to Connect Nature to Design: Inspiration Versus Imitation

Nature is an excellent guide to redesign processes and catalytic materials, as it exemplifies hierarchical structures that are intrinsically scaling, efficient, and robust. This, however, should not reduce to mindless imitation: the biological example needs to be properly chosen, and the different context of technological applications should be accounted for. Hence, there are two distinct categories of research, based on how the natural component is used: (i) nature-inspired and (ii) nature-imitating design. The term "nature" infers a broader definition than "bio," as features of living and nonliving natural systems are included for materials and process design. We avoid the term "biomimetics," as we feel it is important to distinguish nature-inspired and nature-imitating designs, and thus avoid misrepresentations or misinterpretations.

FIGURE 2.1 The design of the front end of the Shinkansen 500 Series bullet train in Japan was inspired by the action of a kingfisher diving into the water [1]. *Copyright © 2007 Elsevier Ltd.*

Research based on the imitation of nature tends to mimic isolated features of biological or nonbiological structures for product or process design, which often leads to suboptimal results. In most cases, the actual physical processes or mechanistic features that govern the system are neglected. Despite this, there are successful examples of implementation of bio-imitating design for the solution of engineering problems. Indeed, it does not *mean* that imitating nature is not leading to a better design, it just *limits* the design space and often ignores the difference in application context. In Japan, severe noise is caused when a high-speed (~320 km/h) train exits a tunnel. At such high speed, sudden pressure changes occur, leading to the creation of low frequency waves and eventually to a sonic boom when the train exits the tunnel. The kingfisher, a skillful bird that hunts its prey by diving into the water, provided the solution to this noise challenge. Due to the specific shape of its beak and head, a kingfisher slides through the water with extraordinary ease without creating any turbulence. It is a remarkably efficient animal in terms of transition from a low pressure (air) to a high pressure medium (water); its long beak increases gradually in diameter from the tip to the back as it plunges to catch its next meal. Hence, by imitating the shape of the kingfisher's beak, engineers equipped high-speed trains with a 50-foot-long tapering nose, resulting in the elimination of the sonic boom issue, faster operating speed, and lower consumption of electricity (Fig. 2.1). Another successful example of bio-imitating design is based on the unique characteristics of the external structure of shark skin, which led to the attachment of 3D printed, bio-mimicking shark skin surfaces on swimsuits and on the hull of boats and surfaces of hospitals and airplanes to prevent bacterial growth [12–15]. Shark skin comprises numerous overlapping scales, called dermal denticles, which have grooves across their length that align with the water flow. These riblets can prevent the formation of vortices, allowing the water to flow through the shark skin quickly. The rough shape of dermal denticles discourages parasitic growth, such as algae and barnacles.

The difference in context between nature and technology, for example, in size, medium, and velocity, is not explicitly accounted for in the above examples. On the contrary, nature-inspired engineering is based on taking a scientific approach to uncover fundamental mechanisms underlying desirable traits. These mechanisms are subsequently applied to design and synthesize artificial systems that borrow the traits of the natural model [11]. Thus, nature-inspired designs may not even resemble their natural counterparts, but rather function as such. Such designs adopt some of the features of the systematic model, after suitable adaptation to fulfill the different contexts of nature and technology.

A great example for inspiration from nature when designing chemical reactors is a typical tree (Fig. 2.2). At the macroscale, its roots branch and spread out widely below the ground to anchor the tree and extract nutrients from the soil. These nutrients are transferred via its fractal root network throughout its volume. Above ground, the crown of the tree divides into increasingly smaller branches and twigs following fractal, self-similar scaling. The twigs bear leaves with a veinal architecture at the mesoscale for further chemical transport, and they contain, at the microscale, molecular complexes that are used to capture sunlight and convert CO_2 into sugars via photosynthesis at the nanoscale, providing the food for the growth of the tree.

The Centre for Nature Inspired Engineering (CNIE) at University College London (UCL) draws inspiration from such natural processes to create innovative solutions to engineering challenges in energy, water purification, functional materials, health, and living space [8,16–19]. Research in the CNIE is currently based on three themes, corresponding to three fundamental mechanisms, of wide applicability: (i) hierarchical transport networks, (ii) force balancing, and (iii) dynamic self-organization (Fig. 2.3). Employing these mechanisms can lead to the development of new materials and technologies with minimal resource consumption and environmental impact. As a result, the reach of nature-inspired design in science and engineering is vast and expanding. A few examples of our research at CNIE, ranging from nano- to macroscale, are presented in the following sections.

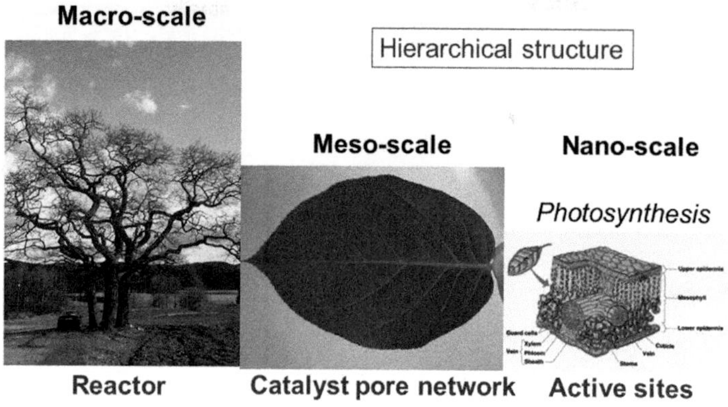

FIGURE 2.2 A tree as an example of nature inspiration for the design of chemical reactors.

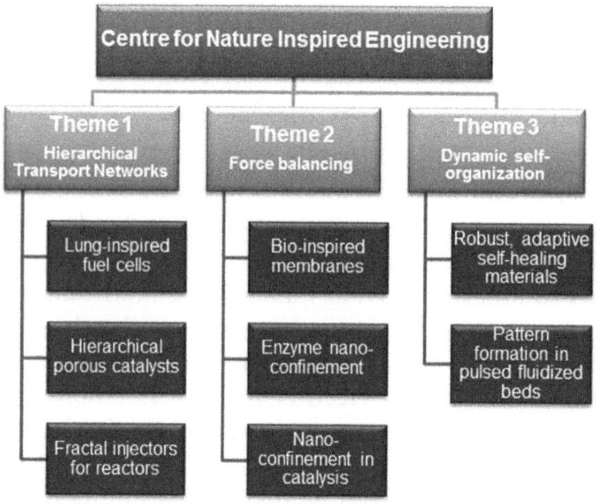

FIGURE 2.3 The broad scope of application of the nature-inspired chemical engineering (NICE) approach, as demonstrated by the research being conducted at the UCL Centre for Nature Inspired Engineering.

3. NATURE-INSPIRED STRUCTURING AT THE NANOSCALE: CONFINEMENT EFFECTS

Nature's catalysts, enzymes, are known to be extremely efficient catalysts under mild conditions. For many reactions under equivalent conditions, enzymes outperform synthetic catalysts by a wide margin, both in activity and selectivity. Enzymes have evolved over millennia to operate optimally within their host organisms in a typically close to neutral, aqueous solution at moderate temperature. However, these reaction conditions are incompatible with most industrial processes, and exposure to extreme conditions (high temperature, acidic pH, detergents, etc.) will result in the denaturation of the enzyme.

Enzyme immobilization (such as covalent binding, cross-linking, and physical adsorption) [20–23] is the most widely used solution to enhance its stability, because it allows the enzyme to be easily removed and recycled. Even though immobilization usually results in a decrease in enzymatic activity and selectivity, this attenuation can be significantly decreased or even inverted [24,25].

Molecular chaperonins can inspire the design of an effective enzyme immobilization material. Chaperonins are proteins that bind unfolded polypeptide chains to help them fold correctly [26]. Their effectiveness is based on their following unique characteristics: (i) a narrow, cylindrical pore structure large enough to fit a single protein, (ii) an electrostatic environment promoting facile protein adsorption, and (iii) a hydrophilic core allowing the correct folding of newly synthesized proteins [27,28].

Inspired by this natural solution, mesoporous silica SBA-15 with different pore diameters was used as a support material for immobilizing and protecting enzymes [26]. Myoglobin and lysozyme have been used as model enzymes, physically adsorbed to SBA-15 and exposed to a range of buffered pH conditions. Their activity was

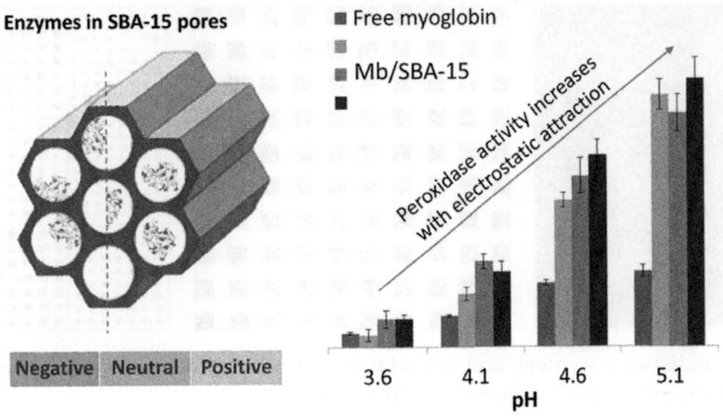

FIGURE 2.4 Peroxidase activities of free myoglobin and myoglobin immobilized to mesoporous silica SBA-15, with pore diameters of 6.1 (green [light gray in print version] color), 6.6 (light blue [gray in print version] color), and 8.1 nm (dark blue [dark gray in print version] color), in solutions of varying pH (3.6-5.1) at ionic strength 0.1 [26]. *Copyright © 2016 American Chemical Society.*

compared against nonimmobilized enzymes in the same solution conditions. It has been observed that myoglobin immobilized on SBA-15 is protected from acidic denaturation for pH values between 3.6 and 5.1 (Fig. 2.4), exhibiting a relative activity of up to 350%. Immobilized lysozyme is protected from unfavorable conditions from pH 6.6 to 7.6, with a relative activity of up to 200% [26]. These results indicate that the electrostatic attraction of the enzymes to the surface of SBA-15 is responsible for their stability in acidic conditions. The pore diameter of SBA-15 also affects the stability of these immobilized enzymes, but its contribution remains unclear at different pH values [26].

Rhodium (Rh) complexes were also immobilized by anchoring them to the pore surface of SBA-15 [17]. Conservation or enhancement of hydroformylation activity was observed when compared with the activity of the homogeneous Rh complex in solution [29]. The Rh complex immobilized on amorphous silica exhibited the lowest activity, larger than one order of magnitude lower than that of the homogeneous complex. On the contrary, the activity of immobilized Rh on SBA-15 was similar to the homogeneous complex, while exhibiting high stability and selectivity, showcasing the positive impact of controlled confinement of these Rh catalysts [29].

Silica imprinting is another method of creating confinement with the additional feature of pore functionalization [30,31]. This technique was used to generate microporous platforms with isolated amine sites with varying acidity and dielectric constants [30,31].

The above examples demonstrate the positive effect of confinement on the activity and stability of a catalyst; such confinement effects are also generated within an enzyme itself, induced by its three-dimensional structure. This raises the question whether structural aspects of an enzyme around its active site can be constructed using organic molecules bound to a metal cluster surface? To realize this, calixarene bound via functional groups on gold clusters has been synthesized [32–34]. Calixarene creates points of accessibility on the surface of gold clusters for substrate binding and tunes their electronic state through phosphine groups binding [33]. Such preliminary results demonstrate the great potential of this method toward the synthesis of new catalysts with improved activity, selectivity, and stability.

4. BRIDGING NANO- AND MACROSCALE: HIERARCHICAL TRANSPORT NETWORKS

Porous materials are used in several processes in chemical engineering, such as in separations and heterogeneous catalysis. The transport properties of these materials depend on the structure of the pore network, which has to be optimized to minimize transport limitations and increase their catalytic performance. Mass transport in porous media occurs predominantly via convective flow in wide pore channels and diffusion in narrower channels [35]. Continuum and discrete theoretical models have been developed to describe these transport processes: continuum models consider the porous space as an effective continuum of reduced permeability, and discrete models specifically account for the pores [35].

Can the design of these pore networks be improved to facilitate transport? Nature can serve as a guide, as it is full of hierarchical structures, bridging nano- to macroscale, with, in several cases in biology, optimized networks of wide and narrow pores, essential to the functioning of the organism [35]. An example (discussed in detail in the following Section 5) is the unique architectural design of the lung, which achieves scalable, uniform distribution of oxygen to the alveoli, in thermodynamically optimal way.

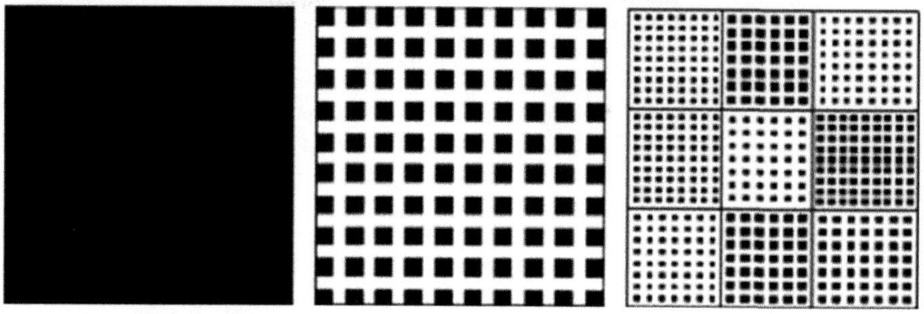

FIGURE 2.5 Monodisperse (left), bidisperse (center), and bimodal (right) pore networks in a nanostructured catalyst (nanoporous catalytic material: black; large diffusion channels: white) [36]. *Copyright © 2007 Elsevier Ltd.*

Inspired by the transport networks in lungs and trees, diffusion limitations in nanostructured catalysts can be minimized via an optimized network of broad and narrow pores. Oftentimes, the size of porous catalyst pellets is determined by hydrodynamic constraints in chemical reactor applications, e.g., pressure drop in a fixed bed reactor. A high catalytic activity per unit volume also requires a high specific surface area; this is realized by nanopores, covered by active sites. Molecular transport through nanopores occurs by diffusion, however, and is very slow. Thus, broad pore channels are required to reduce diffusion limitations in a catalyst pellet.

A 2D theoretical model was developed to investigate the effect of a broad pore size distribution in a nanostructured catalyst pellet for a first-order, isothermal chemical reaction [37]. A fractal-like network of large pores led to a higher conversion than a uniform pore network: ~40% and 5% increase for molecular and Knudsen diffusion, respectively [37]. This fractal-like structure also achieved the highest effectiveness factor, which was indistinguishable from the factor corresponding to a globally optimized structure [37]. However, the number of pores that could be considered in the optimization of the pore networks was limited, due to the high computational cost of the overall geometric optimization. Therefore, the performance of a nanostructured catalyst with varying large pore network structures was also investigated for a first-order, isothermal chemical reaction via a continuum modeling approach [36], and later generalized to arbitrary kinetics [38]. Monodisperse, bidisperse, and bimodal pore size distributions were considered. The monodisperse network contained only nanopores of known size and specific catalytic activity; the bidisperse network also contained large diffusion channels of the same size, within the nanoporous catalytic medium, whereas the bimodal network contained large diffusion channels of variable size distribution (Fig. 2.5) [36]. Global optimization showed that the yield of catalysts with a bidisperse pore distribution was very similar to that of catalysts with a bimodal distribution (<5% difference in yield), but that it could be one order of magnitude higher than that of catalysts with a monodisperse distribution of nanopores, due to the optimized network of large pore channels [36,38]. The optimization of porosity and large average pore size are crucial, while the distribution around the optimum average is shallow, hence less important.

This is similar to the distribution of veins in leaves, capillaries in blood vessel networks, and alveoli in the lung, where the channel size matters, but channels and cells tend to be uniformly distributed when diffusion is the limiting transport mechanism—different from the fractal networks observed when convection dominates transport [35]. The image of a tree, shown in Fig. 2.2, epitomizes the parallel between optimal hierarchical transport networks in nature and in catalysis and reaction engineering. At large length scales, discussed in the next section, fractal networks appear, but at intermediate scales, the channel distribution is much more uniform. These results show that the utilization of a nature-inspired design can decrease diffusion limitations in nanostructured catalysts via an optimized network of broad and narrow pores.

Continuum models are unsuitable to model multiphase reactions in hierarchically structured catalysts. Recently, a discrete pore network modeling approach has been developed to examine multiphase reactions in porous catalysts. Pore blocking was taken into account using a Hoshen–Kopelman algorithm [39]. Simulation results demonstrated that pore blocking plays an important role in multiphase reactions, because it contributed up to 50% to the hysteresis in effectiveness factor [39]. This cannot be observed using continuum models. Such pore-blocking effects are less pronounced when the connectivity of the pore network is high and for narrow pore size distributions [39,40]. However, the hysteresis loop in catalysts with monomodal, narrow pore size distributions and large volume-averaged pore radius is also steeper, which could pose difficulties for the operation of multiphase reactors. For catalysts with a bimodal pore size distribution, changes in hysteresis loop could be more significant, as hysteresis is affected by pore blocking and capillary condensation phenomena. The spatial distribution of the large pores plays a much more important role than for single-phase reactions [40].

Such simulation-based design criteria for optimal porosity and pore size distributions of hierarchically structured catalysts are a helpful guide to catalyst synthesis, as they can lead to significantly superior performance over catalysts with monodisperse pore distribution or nonoptimized macroporosity. Further details are provided in the cited articles and a recent book chapter [35]. From the perspective of ease of synthesis, it is attractive that catalysts with a bidisperse pore network can be close to optimal, with no need to further optimize the large pore channel size distribution around the optimal average large pore channel size. However, as discussed, this only strictly holds for single-phase reactions, as the situation for multiphase reactions is more complicated. A wide range of synthesis techniques (such as templating, scaffolding, etc.) is available to synthesize these catalysts [41–45]. Recently, the pore network representation of hierarchical materials has been greatly improved, thanks to easier access to spatially resolved imaging techniques, such as X-ray tomography, which allow for 3D reconstruction of porous materials, showcasing pore connectivity, pore volume, throat lengths, and radii [46–48]. Imaging and spectroscopic techniques help the structural investigation of porous materials, and inform the design of hierarchical structures with improved transport properties, higher catalytic yields, better selectivities, and stability. For example, hierarchically structured, sponge-like silicoaluminophosphates could be synthesized with very high isomer yield (79%) in the hydroisomerization of *n*-heptane. This was achieved by redesigning the hierarchical structure in such a way that the intrinsic reaction controlled regime is widened, and overall diffusion is enhanced, which reduces the cracking probability [49].

5. NATURE-INSPIRED STRUCTURING AT THE MACROSCALE

A representative example of the NICE approach at the macroscale is illustrated through the design of flow fields for proton exchange membrane fuel cells (PEMFCs) [50]. Prior to our work on electrochemical devices, the scalable, fractal architecture of tree crowns was used as the basis to design a fractal fluid injector that facilitates the scale-up of fluidized bed reactors and increases their efficiency, due to reduced bubble size and improved, more uniform mixing [51,52]. In the case of fuel cells, we use a similar concept, based on the main characteristics of the human lung to address transport limitations and scalability issues [50]. Lessons from nature to design fluidized beds with better-controlled hydrodynamics hence serve a different application in energy technology.

Flow field design is vital to fuel cell performance, as it is responsible for the transport of reactants and products to and from the membrane electrode assembly (MEA). The commercial flow fields employed currently (e.g., serpentine, parallel, and interdigitated) result in nonuniform distribution of reactants across the MEA and large pressure drops between their inlets and outlets, leading to fuel cell performance losses.

To circumvent these issues, we employed a nature-inspired design based on key characteristics of the human lung, including the fractal structure of the upper airway tree and its thermodynamic optimality (minimal

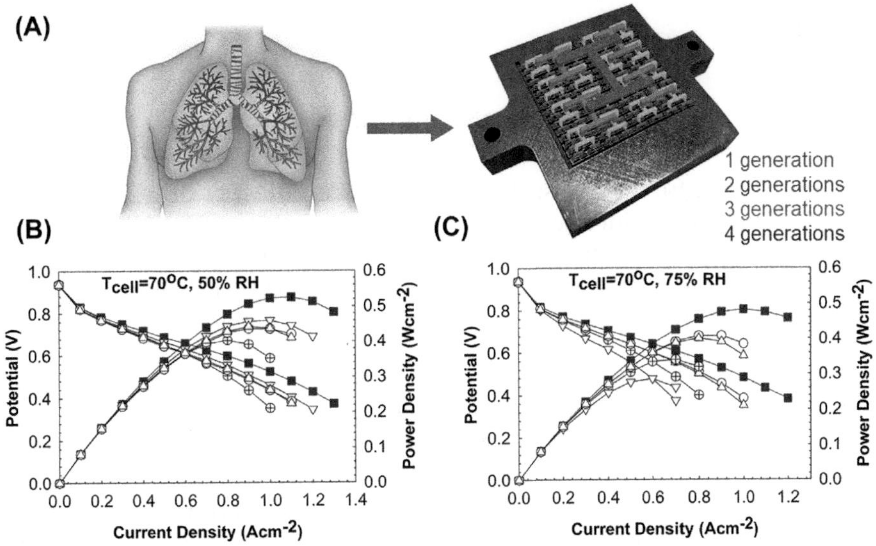

FIGURE 2.6 (A) The unique characteristics of the lung (fractal structure proportioned to lead to minimum entropy production) are implemented into the design of lung-inspired flow fields for proton exchange membrane fuel cells (PEMFCs). Experimental validation of the lung-inspired flow field–based PEMFCs (10 cm^2 flow field area) demonstrate improved performance at 50% and 75% RH ($N = 4$) compared with conventional, serpentine flow field–based PEMFCs (B and C) [50]. Symbols ○△⊕■▽ represent single serpentine, double serpentine, and lung inspired with $N = 3$, 4, and 5 generations, respectively. *Copyright © 2018 The Royal Society of Chemistry.*

entropy production) during air transport, properties that are desired for fuel cell flow fields as well (Fig. 2.6A). In the lung, air is transported uniformly through a scalable, fractal architecture, from the trachea via the bronchioles to the alveoli, and into the bloodstream to oxygenate the blood cells [53–55]. In addition, the upper airway's fractal branching structure obeys Murray's law, according to which the cube of the diameter (d_i, $i = 1 \ldots n$) of a parent vessel is equal to the sum of the cubes of the diameters of the two daughter vessels at each level of bifurcation, i, leading to minimal mechanical energy losses: $d_i^3 = 2d_{i+1}^3$. As a result, air gradually slows down when flowing from the trachea to the bronchioles. At some generation, n (14–16 in an adult human lung), the convection driven air flow in the bronchioles is equal to the diffusion driven air flow entering the acini, so that the Péclet number becomes unity [55]. Further reduction in channel diameter would lead to additional flow resistance, which is undesirable; interestingly, just at this level where Pé ~ 1, further branching within the acini does not lead to appreciable reduction in channel size, as diffusion is now the dominating transport mechanism. Furthermore, the length of each channel in the upper airway tree is such that the pressure drop is equal over each channel, while in the lower airway tree concentration drop is constant. This leads to equipartition of entropy production, making the lung thermodynamically optimally proportioned [52,55].

The abovementioned, remarkable characteristics of the lung served as the design principle to manufacture 3D printed, lung-inspired flow fields (10 cm^2) for PEMFCs with $N = 3$, 4, and 5 generations. The proportioning of channels in the fractal flow field results in constant entropy production at each branching level, and in minimal overall entropy production over the lung-inspired structure [44,50,56]. Flow fields with $N = 4$ generations demonstrated a ~30% increase in fuel cell performance and power density over commercial serpentine flow fields at 50% and 75% RH (Fig. 2.6 B and C) [50]. In terms of pressure drop, these lung-inspired flow fields exhibited ~75% lower values (<2 kPa) than commercial serpentine flow fields (5 kPa) for all RH tested, resulting in the minimization of parasitic losses, and thus to increased fuel cell performance. These improvements in pressure drop and fuel cell performance were maintained at large scale (25 cm^2), whereas the large pressure drop (25 kPa) in serpentine flow field–based PEMFCs is detrimental to their fuel cell performance.

Lung-inspired flow field–based PEMFCs with $N = 3$ generations exhibited the worst performance under all experimental conditions, due to the large spacing between adjacent outlets leading to insufficient oxygen concentration across the MEA. The flow field design with $N = 5$ generations demonstrated lower fuel cell performance than commercial serpentine flow fields, because its narrow channels in the final generation were prone to flooding [50].

At very high RH (100%) conditions, all lung-inspired flow field–based PEMFCs exhibited flooding, which lowered corresponding fuel cell performance. Hence, utilization of new effective water removal mechanisms is needed to overcome this issue, because current antiflooding measures cannot be easily incorporated into these lung-inspired flow fields [57]. However, this extreme humidity condition is rarely employed in practice, favoring the use of lung-inspired flow fields. Table 2.1 provides a detailed comparison of the most common flow field designs currently used in PEMFCs against lung-inspired design. Finally, this proposed nature-inspired approach is not limited to PEMFCs but extends to all electrochemical devices in which flooding is alleviated, such as redox flow batteries and high temperature fuel cells.

6. CONCLUSIONS

At the beginning of the 21st century, when sustainability has become a priority, business as usual is untenable, and we must develop novel design guidelines, methods, and procedures. The NICE approach discussed in the previous sections offers opportunities to innovate by redesigning materials and processes, replacing conventional designs, and leading to sustainable solutions to engineering challenges. NICE provides an avenue to sustainable design that combines creativity with engineering fundamentals, allowing for environmental, economic, and social benefits.

Currently, biomimetic research is too often focused on imitating isolated features of biological structures to synthesize products, like heterogeneous catalysts or flow fields for electrochemical devices, with suboptimal results. In most cases, the actual physical processes that govern the system are neglected, indicating that the full potential of learning lessons from nature has yet to be realized. In following the NICE methodology, solutions to the above applications are devised by not directly mimicking nature, but rather learning first from it and working within the constraints of each system. The nature-inspired engineering methodology provides a step-by-step approach of deriving mechanisms from the natural system as a model to create a nature-inspired concept, leading to a proposed nature-inspired design, and its implementation, conscious of the differences between the natural context and that of engineering applications. Introducing these nature-inspired design guidelines can lead to sustainable

TABLE 2.1 Comparison of Different Flow Field Designs for Proton Exchange Membrane Fuel Cells.

Flow Field Design	Advantages	Disadvantages
Serpentine	• Efficient water removal • Designed for small active area	• High pressure drop • Water build-up in bends • Uneven reactant distribution
Interdigitated	• Efficient water removal • High fuel cell performance at high current density	• Highest pressure drop of all flow field designs • Uneven reactant distribution • Difficult to scale
Parallel	• Increased power density at high current density	• Uneven reactant distribution • Poor water management • High pressure drop in stack configuration
Pin	• Low pressure drop	• Uneven reactant distribution • Inadequate water removal • Poor fuel cell performance
Lung inspired	• Low pressure drop • Uniform reactant distribution for low and high temperature fuel cells • Higher electrocatalyst stability than serpentine • Easily scalable	• More prone to flooding than serpentine at high relative humidity • High cost of the flow fields made by additive manufacturing, requiring cheaper alternatives

processes and products and, hereby, to a sustainable future. We especially suggest focusing research on the fundamental understanding of hierarchical structure/function relationships in biological systems, as well as properties emerging from synergistic systems behavior. As illustrated here, this understanding can then, for example, guide the development of novel, nature-inspired catalysts and chemical and electrochemical reactors.

Furthermore, the engineering principles of NICE can greatly benefit additional research areas, such as biomaterials, tissue engineering [58–62], and healthcare technologies [63], as well as architectural design for the built environment. A review on the prospects of nature-inspired design of biomaterials and devices for healthcare applications is beyond the scope of this chapter, but has been recently published, with further references therein [8]. Some extensively studied examples include materials that mimic shark skin (antifouling and drag reduction) [15,64–67], nacre (remarkable mechanical properties) [67–80], or gecko feet (reversible adhesion) [67,81–85]. Such properties are induced by hierarchical, composite structures, and the surface geometry. Current applications of biomimetic and bio-inspired materials include antimicrobial coatings and biocompatible adhesives. By adopting the systematic approach of NICE, we see opportunities that go beyond case-by-case mimicry for innovation in tissue engineering, drug delivery, osseointegration, and other biomedical applications [8]. As to applications in the built environment, the NICE methodology could aid in the development of building components, which self-regulate the urban environment, absorbing pollutants and providing environmental control in each room (humidity, temperature, light) [86–98].

These additional examples showcase the diversity of applications of the NICE methodology and how embracing this methodology could provide innovative solutions to engineering challenges, complemented by parallel advances in additive manufacturing, computation, and robotics. In the case of fluid injection devices, hierarchical catalyst architectures, and certain biomedical materials, NICE is already leading to products that are ready for industrial adoption, or are in an advanced stage of development. The timeframe of large-scale incorporation of NICE as a design framework for industrial applications will depend on the maturity of the sector, and how risk averse or progressive that sector is. In bulk process engineering, several years might be required until NICE designs are implemented, due to the fear of adopting unconventional technology, unless thoroughly tested. Nevertheless, even there, NICE is a toolbox that may result in great benefits with little modification, as the examples in catalysis and fluidization demonstrate. Cost reduction, ease of scale-up, and stable, reliable operations are drivers to meet environmental, economical, and safety challenges. In the distributed energy, specialties, fine chemicals, and biomedical sectors, NICE is an enabler for new technology, as well as offering significant improvement over existing processes and products. The fact that nature contains examples so much ahead of current technology (materials properties, systems behavior, scalability, resilience, or efficiency) is enough of a reason to probe the underlying mechanisms to inspire innovation, and NICE offers a uniquely systematic methodology to do so.

Acknowledgments

The authors gratefully acknowledge the financial support from an EPSRC "Frontier Engineering" Award (EP/K038656/1).

References

[1] J. García-Serna, L. Pérez-Barrigón, M.J. Cocero, New trends for design towards sustainability in chemical engineering: green engineering, Chemical Engineering Journal 133 (1) (2007) 7–30.
[2] R. Carson, Silent Spring, Hamish Hamilton, London, 1963 xix + 304 pp.
[3] R.J. Batterham, Sustainability—the next chapter, Chemical Engineering Science 61 (13) (2006) 4188–4193.
[4] G.H. Brundtland, UN World Commission on Environment and Development: Our Common Future, Oxford University Press, Oxford; New York, 1987.
[5] D.M. Liverman, M.E. Hanson, B.J. Brown, R.W. Merideth, Global sustainability: toward measurement, Environmental Management 12 (2) (1988) 133–143.
[6] A.S. McKim, Overcoming sustainability barriers within the chemical industry, Current Opinion in Green and Sustainable Chemistry 14 (2018) 10–13.
[7] M. Narodoslawsky, Chemical engineering in a sustainable economy, Chemical Engineering Research and Design 91 (10) (2013) 2021–2028.
[8] A.S. Perera, M.-O. Coppens, Re-designing materials for biomedical applications: from biomimicry to nature-inspired chemical engineering, Philosophical Transactions of the Royal Society A 377 (2018) 20180268.
[9] J. Aizenberg, P. Fratzl, Biological and biomimetic materials, Advanced Materials 21 (4) (2009) 387–388.
[10] E. Dujardin, S. Mann, Bio-inspired materials chemistry, Advanced Materials 14 (11) (2002) 775–788.
[11] P. Fratzl, Biomimetic materials research: what can we really learn from nature's structural materials? Journal of The Royal Society Interface 4 (15) (2007) 637–642.
[12] Y.F. Fu, C.Q. Yuan, X.Q. Bai, Marine drag reduction of shark skin inspired riblet surfaces, Biosurface and Biotribology 3 (1) (2017) 11–24.
[13] J. Oeffner, G.V. Lauder, The hydrodynamic function of shark skin and two biomimetic applications, The Journal of Experimental Biology 215 (5) (2012) 785–795.
[14] X. Pu, G. Li, Y. Liu, Progress and perspective of studies on biomimetic shark skin drag reduction, ChemBioEng Reviews 3 (1) (2016) 26–40.
[15] L. Wen, J.C. Weaver, G.V. Lauder, Biomimetic shark skin: design, fabrication and hydrodynamic function, The Journal of Experimental Biology 217 (10) (2014) 1656–1666.
[16] M.-O. Coppens, A nature-inspired approach to reactor and catalysis engineering, Current Opinion in Chemical Engineering 1 (3) (2012) 281–289.
[17] P. Trogadas, M.M. Nigra, M.-O. Coppens, Nature-inspired optimization of hierarchical porous media for catalytic and separation processes, New Journal of Chemistry 40 (5) (2016) 4016–4026.
[18] E. Arzt, Biological and artificial attachment devices: lessons for materials scientists from flies and geckos, Materials Science and Engineering: C 26 (8) (2006) 1245–1250.
[19] M.A. Meyers, P.-Y. Chen, M.I. Lopez, Y. Seki, A.Y.M. Lin, Biological materials: a materials science approach, Journal of the Mechanical Behavior of Biomedical Materials 4 (5) (2011) 626–657.
[20] U. Hanefeld, L. Cao, E. Magner, Enzyme immobilisation: fundamentals and application, Chemical Society Reviews 42 (15) (2013) 6211–6212.
[21] U. Hanefeld, L. Gardossi, E. Magner, Understanding enzyme immobilisation, Chemical Society Reviews 38 (2) (2009) 453–468.
[22] R.A. Sheldon, S. van Pelt, Enzyme immobilisation in biocatalysis: why, what and how, Chemical Society Reviews 42 (15) (2013) 6223–6235.
[23] D.N. Tran, K.J. Balkus, Perspective of recent progress in immobilization of enzymes, ACS Catalysis 1 (8) (2011) 956–968.
[24] L. Cao, Immobilised enzymes: science or art? Current Opinion in Chemical Biology 9 (2) (2005) 217–226.
[25] R.C. Rodrigues, C. Ortiz, Á. Berenguer-Murcia, R. Torres, R. Fernández-Lafuente, Modifying enzyme activity and selectivity by immobilization, Chemical Society Reviews 42 (15) (2013) 6290–6307.

[26] M.M. Lynch, J. Liu, M. Nigra, M.-O. Coppens, Chaperonin-Inspired pH protection by mesoporous silica SBA-15 on myoglobin and lysozyme, Langmuir 32 (37) (2016) 9604–9610.
[27] K. Braig, Z. Otwinowski, R. Hegde, D.C. Boisvert, A. Joachimiak, A.L. Horwich, P.B. Sigler, The crystal structure of the bacterial chaperonin GroEL at 2.8 Å, Nature 371 (1994) 578.
[28] G.G. Tartaglia, C.M. Dobson, F.U. Hartl, M. Vendruscolo, Physicochemical determinants of chaperone requirements, Journal of Molecular Biology 400 (3) (2010) 579–588.
[29] F. Marras, J. Wang, M.-O. Coppens, J.N.H. Reek, Ordered mesoporous materials as solid supports for rhodium–diphosphine catalysts with remarkable hydroformylation activity, Chemical Communications 46 (35) (2010) 6587–6589.
[30] J.D. Bass, A. Solovyov, A.J. Pascall, A. Katz, Acid–Base bifunctional and dielectric outer-sphere effects in heterogeneous catalysis: a comparative investigation of model primary amine catalysts, Journal of the American Chemical Society 128 (11) (2006) 3737–3747.
[31] A. Katz, M.E. Davis, Molecular imprinting of bulk, microporous silica, Nature 403 (2000) 286.
[32] J.-M. Ha, A. Solovyov, A. Katz, Synthesis and characterization of accessible metal surfaces in calixarene-bound gold nanoparticles, Langmuir 25 (18) (2009) 10548–10553.
[33] N. de Silva, J.-M. Ha, A. Solovyov, M.M. Nigra, I. Ogino, S.W. Yeh, K.A. Durkin, A. Katz, A bioinspired approach for controlling accessibility in calix[4]arene-bound metal cluster catalysts, Nature Chemistry 2 (2010) 1062.
[34] M.M. Nigra, A.J. Yeh, A. Okrut, A.G. DiPasquale, S.W. Yeh, A. Solovyov, A. Katz, Accessible gold clusters using calix[4]arene N-heterocyclic carbene and phosphine ligands, Dalton Transactions 42 (35) (2013) 12762–12771.
[35] M.-O. Coppens, G. Ye, Nature-Inspired optimization of transport in porous media, in: A. Bunde, et al. (Eds.), Diffusive Spreading in Nature, Technology and Society, Springer International Publishing, Cham, 2018, pp. 203–232.
[36] G. Wang, E. Johannessen, C.R. Kleijn, S.W. de Leeuw, M.-O. Coppens, Optimizing transport in nanostructured catalysts: a computational study, Chemical Engineering Science 62 (18) (2007) 5110–5116.
[37] S. Gheorghiu, M.-O. Coppens, Optimal bimodal pore networks for heterogeneous catalysis, AIChE Journal 50 (4) (2004) 812–820.
[38] G. Wang, M.-O. Coppens, Calculation of the optimal macropore size in nanoporous catalysts and its application to DeNOx catalysis, Industrial & Engineering Chemistry Research 47 (11) (2008) 3847–3855.
[39] G. Ye, X. Zhou, W. Yuan, M.-O. Coppens, Probing pore blocking effects on multiphase reactions within porous catalyst particles using a discrete model, AIChE Journal 62 (2) (2016) 451–460.
[40] G. Ye, X. Zhou, J. Zhou, W. Yuan, M.-O. Coppens, Influence of catalyst pore network structure on the hysteresis of multiphase reactions, AIChE Journal 63 (1) (2017) 78–86.
[41] Y. Liu, J. Goebl, Y. Yin, Templated synthesis of nanostructured materials, Chemical Society Reviews 42 (7) (2013) 2610–2653.
[42] C. Perego, R. Millini, Porous materials in catalysis: challenges for mesoporous materials, Chemical Society Reviews 42 (9) (2013) 3956–3976.
[43] J. Pérez-Ramírez, C.H. Christensen, K. Egeblad, C.H. Christensen, J.C. Groen, Hierarchical zeolites: enhanced utilisation of microporous crystals in catalysis by advances in materials design, Chemical Society Reviews 37 (11) (2008) 2530–2542.
[44] P. Trogadas, V. Ramani, P. Strasser, T.F. Fuller, M.-O. Coppens, Hierarchically structured nanomaterials for electrochemical energy conversion, Angewandte Chemie International Edition 55 (1) (2015) 122–148.
[45] A. Walcarius, Mesoporous materials and electrochemistry, Chemical Society Reviews 42 (9) (2013) 4098–4140.
[46] F. Larachi, R. Hannaoui, P. Horgue, F. Augier, Y. Haroun, S. Youssef, E. Rosenberg, M. Prat, M. Quintard, X-ray micro-tomography and pore network modeling of single-phase fixed-bed reactors, Chemical Engineering Journal 240 (2014) 290–306.
[47] V. Novák, F. Štěpánek, P. Kočí, M. Marek, M. Kubíček, Evaluation of local pore sizes and transport properties in porous catalysts, Chemical Engineering Science 65 (7) (2010) 2352–2360.
[48] P. Trogadas, O.O. Taiwo, B. Tjaden, T.P. Neville, S. Yun, J. Parrondo, V. Ramani, M.-O. Coppens, D.J.L. Brett, P.R. Shearing, X-ray micro-tomography as a diagnostic tool for the electrode degradation in vanadium redox flow batteries, Electrochemistry Communications 48 (2014) 155–159.
[49] D. Jin, G. Ye, J. Zheng, W. Yang, K. Zhu, M.-O. Coppens, X. Zhou, Hierarchical silicoaluminophosphate catalysts with enhanced hydroisomerization selectivity by directing the orientated assembly of premanufactured building blocks, ACS Catalysis 7 (9) (2017) 5887–5902.
[50] P. Trogadas, J.I.S. Cho, T.P. Neville, J. Marquis, B. Wu, D.J.L. Brett, M.O. Coppens, A lung-inspired approach to scalable and robust fuel cell design, Energy & Environmental Science 11 (1) (2018) 136–143.
[51] K. Wu, L. de Martín, M.-O. Coppens, Pattern formation in pulsed gas-solid fluidized beds – the role of granular solid mechanics, Chemical Engineering Journal 329 (2017) 4–14.
[52] K. Wu, L. de Martín, L. Mazzei, M.-O. Coppens, Pattern formation in fluidized beds as a tool for model validation: a two-fluid model based study, Powder Technology 295 (2016) 35–42.
[53] S. Gheorghiu, S. Kjelstrup, P. Pfeifer, M.O. Coppens, Is the lung an optimal gas exchanger? in: G.A. Losa, et al. (Eds.), Fractals in Biology and Medicine Birkhäuser Basel, Basel, 2005, pp. 31–42.
[54] B. Mauroy, M. Filoche, E.R. Weibel, B. Sapoval, An optimal bronchial tree may be dangerous, Nature 427 (2004) 633.
[55] B. Sapoval, M. Filoche, E.R. Weibel, Smaller is better—but not too small: a physical scale for the design of the mammalian pulmonary acinus, Proceedings of the National Academy of Sciences 99 (16) (2002) 10411–10416.
[56] S. Kjelstrup, M.-O. Coppens, J.G. Pharoah, P. Pfeifer, Nature-Inspired energy- and material-efficient design of a polymer electrolyte membrane fuel cell, Energy & Fuels 24 (9) (2010) 5097–5108.
[57] T. Yoshida, K. Kojima, Toyota MIRAI fuel cell vehicle and progress toward a future hydrogen society, The Electrochemical Society Interface 24 (2) (2015) 45–49.
[58] A.K. Capulli, M.Y. Emmert, F.S. Pasqualini, D. Kehl, E. Caliskan, J.U. Lind, S.P. Sheehy, S.J. Park, S. Ahn, B. Weber, J.A. Goss, S.P. Hoerstrup, K.K. Parker, JetValve: rapid manufacturing of biohybrid scaffolds for biomimetic heart valve replacement, Biomaterials 133 (2017) 229–241.
[59] D.E. Ingber, Developmentally inspired human 'organs on chips', Development 145 (16) (2018) p. dev156125.
[60] B.M. Maoz, A. Herland, E.A. FitzGerald, T. Grevesse, C. Vidoudez, A.R. Pacheco, S.P. Sheehy, T.-E. Park, S. Dauth, R. Mannix, N. Budnik, K. Shores, A. Cho, J.C. Nawroth, D. Segrè, B. Budnik, D.E. Ingber, K.K. Parker, A linked organ-on-chip model of the human neurovascular unit reveals the metabolic coupling of endothelial and neuronal cells, Nature Biotechnology 36 (2018) 865.

[61] S. Musah, N. Dimitrakakis, D.M. Camacho, G.M. Church, D.E. Ingber, Directed differentiation of human induced pluripotent stem cells into mature kidney podocytes and establishment of a Glomerulus Chip, Nature Protocols 13 (7) (2018) 1662–1685.
[62] A. Sydney Gladman, E.A. Matsumoto, R.G. Nuzzo, L. Mahadevan, J.A. Lewis, Biomimetic 4D printing, Nature Materials 15 (2016) 413.
[63] B.S. Gomes, B. Simões, P.M. Mendes, The increasing dynamic, functional complexity of bio-interface materials, Nature Reviews Chemistry 2 (2018) 0120.
[64] G. Domel August, M. Saadat, C. Weaver James, H. Haj-Hariri, K. Bertoldi, V. Lauder George, Shark skin-inspired designs that improve aerodynamic performance, Journal of The Royal Society Interface 15 (139) (2018) 20170828.
[65] O.Z. Fisher, A. Khademhosseini, R. Langer, N.A. Peppas, Bioinspired materials for controlling stem cell fate, Accounts of Chemical Research 43 (3) (2010) 419–428.
[66] L. Wen, J.C. Weaver, P.J.M. Thornycroft, G.V. Lauder, Hydrodynamic function of biomimetic shark skin: effect of denticle pattern and spacing, Bioinspiration & Biomimetics 10 (6) (2015) 066010.
[67] B. Bhushan, Bioinspired structured surfaces, Langmuir 28 (3) (2012) 1698–1714.
[68] R. Chen, C.-a. Wang, Y. Huang, H. Le, An efficient biomimetic process for fabrication of artificial nacre with ordered-nanostructure, Materials Science and Engineering: C 28 (2) (2008) 218–222.
[69] A. Finnemore, P. Cunha, T. Shean, S. Vignolini, S. Guldin, M. Oyen, U. Steiner, Biomimetic layer-by-layer assembly of artificial nacre, Nature Communications 3 (2012) 966.
[70] G.X. Gu, F. Libonati, S.D. Wettermark, M.J. Buehler, Printing nature: unraveling the role of nacre's mineral bridges, Journal of the Mechanical Behavior of Biomedical Materials 76 (2017) 135–144.
[71] T. Kokubo, H.-M. Kim, M. Kawashita, Novel bioactive materials with different mechanical properties, Biomaterials 24 (13) (2003) 2161–2175.
[72] G.M. Luz, J.F. Mano, Biomimetic design of materials and biomaterials inspired by the structure of nacre, Philosophical Transactions of the Royal Society A: Mathematical, Physical and Engineering Sciences 367 (1893) (2009) 1587–1605.
[73] E. Munch, M.E. Launey, D.H. Alsem, E. Saiz, A.P. Tomsia, R.O. Ritchie, Tough, bio-inspired hybrid materials, Science 322 (5907) (2008) 1516–1520.
[74] J. Wang, Q. Cheng, Z. Tang, Layered nanocomposites inspired by the structure and mechanical properties of nacre, Chemical Society Reviews 41 (3) (2012) 1111–1129.
[75] S. Xia, Z. Wang, H. Chen, W. Fu, J. Wang, Z. Li, L. Jiang, Nanoasperity: structure origin of nacre-inspired nanocomposites, ACS Nano 9 (2) (2015) 2167–2172.
[76] H.-B. Yao, H.-Y. Fang, X.-H. Wang, S.-H. Yu, Hierarchical assembly of micro-/nano-building blocks: bio-inspired rigid structural functional materials, Chemical Society Reviews 40 (7) (2011) 3764–3785.
[77] J. Zhang, W. Feng, H. Zhang, Z. Wang, H.A. Calcaterra, B. Yeom, P.A. Hu, N.A. Kotov, Multiscale deformations lead to high toughness and circularly polarized emission in helical nacre-like fibres, Nature Communications 7 (2016) 10701.
[78] D. Zhao, V. Gimenez-Pinto, A.M. Jimenez, L. Zhao, J. Jestin, S.K. Kumar, B. Kuei, E.D. Gomez, A.S. Prasad, L.S. Schadler, M.M. Khani, B.C. Benicewicz, Tunable multiscale nanoparticle ordering by polymer crystallization, ACS Central Science 3 (7) (2017) 751–758.
[79] H. Zhao, L. Guo, Nacre-Inspired structural composites: performance-enhancement strategy and perspective, Advanced Materials 29 (45) (2017) 1702903.
[80] H. Zhao, Z. Yang, L. Guo, Nacre-inspired composites with different macroscopic dimensions: strategies for improved mechanical performance and applications, NPG Asia Materials 10 (4) (2018) 1–22.
[81] A. del Campo, C. Greiner, I. Álvarez, E. Arzt, Patterned surfaces with pillars with controlled 3d tip geometry mimicking bioattachment devices, Advanced Materials 19 (15) (2007) 1973–1977.
[82] C. Greiner, E. Arzt, A. del Campo, Hierarchical gecko-like adhesives, Advanced Materials 21 (4) (2009) 479–482.
[83] A. Mahdavi, L. Ferreira, C. Sundback, J.W. Nichol, E.P. Chan, D.J.D. Carter, C.J. Bettinger, S. Patanavanich, L. Chignozha, E. Ben-Joseph, A. Galakatos, H. Pryor, I. Pomerantseva, P.T. Masiakos, W. Faquin, A. Zumbuehl, S. Hong, J. Borenstein, J. Vacanti, R. Langer, J.M. Karp, A biodegradable and biocompatible gecko-inspired tissue adhesive, Proceedings of the National Academy of Sciences 105 (7) (2008) 2307–2312.
[84] S. Sethi, L. Ge, L. Ci, P.M. Ajayan, A. Dhinojwala, Gecko-Inspired carbon nanotube-based self-cleaning adhesives, Nano Letters 8 (3) (2008) 822–825.
[85] R. Cutkosky Mark, Climbing with adhesion: from bioinspiration to biounderstanding, Interface Focus 5 (4) (2015) 20150015.
[86] L. Ripley Renee, B. Bhushan, Bioarchitecture: bioinspired art and architecture—a perspective, Philosophical Transactions of the Royal Society A: Mathematical, Physical and Engineering Sciences 374 (2073) (2016) 20160192.
[87] C. Audibert, J. Chaves-Jacob, J.-M. Linares, Q.-A. Lopez, Bio-inspired method based on bone architecture to optimize the structure of mechanical workspieces, Materials & Design 160 (2018) 708–717.
[88] A. Cannavale, F. Fiorito, M. Manca, G. Tortorici, R. Cingolani, G. Gigli, Multifunctional bioinspired sol-gel coatings for architectural glasses, Building and Environment 45 (5) (2010) 1233–1243.
[89] A. Gamage, R. Hyde, A model based on Biomimicry to enhance ecologically sustainable design, Architectural Science Review 55 (3) (2012) 224–235.
[90] B. Han, L. Zhang, J. Ou, Future developments and challenges of smart and multifunctional concrete, in: Smart and Multifunctional Concrete toward Sustainable Infrastructures, Springer Singapore, Singapore, 2017, pp. 391–400.
[91] G. John, D. Clements-Croome, G. Jeronimidis, Sustainable building solutions: a review of lessons from the natural world, Building and Environment 40 (3) (2005) 319–328.
[92] K. Klang, G. Bauer, N. Toader, C. Lauer, K. Termin, S. Schmier, D. Kovaleva, W. Haase, C. Berthold, K.G. Nickel, T. Speck, W. Sobek, Plants and animals as source of inspiration for energy dissipation in load bearing systems and facades, in: J. Knippers, K.G. Nickel, T. Speck (Eds.), Biomimetic Research for Architecture and Building Construction: Biological Design and Integrative Structures, Springer International Publishing, Cham, 2016, pp. 109–133.
[93] M. Mirkhalaf, A.K. Dastjerdi, F. Barthelat, Overcoming the brittleness of glass through bio-inspiration and micro-architecture, Nature Communications 5 (2014) 3166.

[94] F. Natalio, T.P. Corrales, M. Panthöfer, D. Schollmeyer, I. Lieberwirth, W.E.G. Müller, M. Kappl, H.-J. Butt, W. Tremel, Flexible minerals: self-assembled calcite spicules with extreme bending strength, Science 339 (6125) (2013) 1298–1302.
[95] M. Seifan, A.K. Samani, A. Berenjian, Bioconcrete: next generation of self-healing concrete, Applied Microbiology and Biotechnology 100 (6) (2016) 2591–2602.
[96] P. Vukusic, J.R. Sambles, Photonic structures in biology, Nature 424 (2003) 852.
[97] Y. Xing, P. Jones, M. Bosch, I. Donnison, M. Spear, G. Ormondroyd, Exploring design principles of biological and living building envelopes: what can we learn from plant cell walls? Intelligent Buildings International 10 (2) (2018) 78–102.
[98] M.P. Zari, Biomimetic design for climate change adaptation and mitigation, Architectural Science Review 53 (2) (2010) 172–183.

CHAPTER

3

Manufacturing Nanoporous Materials for Energy-Efficient Separations: Application and Challenges

Yao Ma, Fengyi Zhang, Ryan P. Lively

School of Chemical & Biomolecular Engineering, Georgia Institute of Technology, Atlanta, GA, United States

OUTLINE

Sustainable Nanoscale Engineering
https://doi.org/10.1016/B978-0-12-814681-1.00003-5

List of Abbreviations

APTES Aminopropyltriethoxysilane
BPDA Biphenyl-tetracarboxylic acid dianhydride
CMP Conjugated microporous polymer
CMS Carbon molecular sieve
COF Covalent organic framework
CTF Covalent triazine–based framework
EOF Elemental organic framework
FVV Fractional free volume
GPU Gas permeance unit
HCP Hypercrosslinked polymer
HDS Hydroxyl double salt
IUPAC International Union of Pure and Applied Chemistry
IZA International Zeolite Association
MMM Mixed matrix membrane
MOF Metal-organic framework
MP Methanopentacene
MPD M-phenylenediamine
NMP N-methyl-2-pyrrolidone
OSN Organic solvent nanofiltration
OSRO Organic solvent reverse osmosis
PAE Poly(aryleneethynylene)
PAF Porous aromatic framework
PAN Polyacrylonitrile
PAR Polyarylate
PDA Polydopamine
PEI Polyethylenimine
PFA Polyfurfuryl alcohol
PIM Polymers of intrinsic microporosity
POC Porous organic cage
PPB Poly(phenylene butadiynylene)
PPV Poly(phenylenevinylene)
PSA Pressure swing adsorption
PTMSP Polytrimethylsilylpropyne
PVDC-AC Polyvinylidene chloride–acrylate terpolymer
PVDF Poly(vinylidene fluoride)
RSA Rigid star amphiphile
SMB Simulated moving bed
SMR Steam methane reforming
TFN Thin film nanocomposite membrane
THF Tetrahydrofuran
TMC Trimesoyl chloride
WGS Water gas shift
ZIF Zeolitic imidazolate framework

List of Symbols

Roman Letters

D_A Diffusion coefficient
D_{0A} Preexponential factors of diffusion
$E_{D,A}$ Activation energy for diffusion
$E_{\mathbb{P},A}$ Activation energy for permeation
Δf_A Transmembrane fugacity difference
$\Delta H_{\mathbb{S},A}$ Apparent heat of sorption
ℓ Membrane thickness
N_A Penetrant flux
$\mathbb{P}_A$ Permeability of species A
P_A Permeance of species A
$\mathbb{P}_e$ Effective permeability in the mixed matrix material
$\mathbb{P}_f$ Permeability in fillers (molecular particles)
$\mathbb{P}_{fe}$ Effective permeability of inorganic particles and gap
$\mathbb{P}_m$ Permeability in the polymer matrix

$\mathbb{P}_{0A}$ Preexponential factors of permeation
R Universal gas constant
$\mathbb{S}_A$ Solubility or sorption coefficient of species A
$\mathbb{S}_{0A}$ Preexponential factors of sorption
T Absolute temperature

Greek Letters

α_{AB} Ideal permselectivity for guest molecule A versus B
$\varnothing_f$ Volume fraction of molecular sieve particles in the mixed matrix material
$\varnothing_{fe}$ Effective volume fraction of molecular sieve particles and gaps in the mixed matrix material

1. INTRODUCTION

The last two decades have witnessed a rapid growth in the study and development of nanoscale technology, especially in the area of nanoporous materials. This important class of nanostructured materials has attracted tremendous interest, investment, and research effort because of their unique characteristics, such as a large specific surface area, high pore volume, uniform pore size, and rich surface chemistry.

Nanoporous materials refer to materials with permanent nanosized pores. The permanent pores can be provided in a variety of ways, for example, by periodically connecting organic molecular "struts" to metal atom "nodes" to assemble repeating structures. Nanoporous materials include, but are not limited to, zeolites, metal-organic frameworks (MOFs) [1], covalent organic frameworks (COFs) [2], certain polymers [3], and carbon molecular sieves (CMSs) [4].

Separations are one of the oldest process technologies mastered by human beings and refer to techniques that turn mixtures into more pure components. In the modern world, separations are widely applied in industrial applications such as refining natural resources, postproduction purification of new chemicals, water treatment, etc. The solutions to the wide array of different separation challenges are largely dominated by energy-intensive, thermally driven techniques such as distillation, sublimation, and evaporation. Although these techniques are reliable and mature, there is a thermodynamic, economic, and environmental driving force motivating the continued exploration of alternative separation methods. For example, thermally driven separation techniques differentiate molecules via phase change and result in 10%−15% worldwide energy consumption as shown in Fig. 3.1 [5]. Moreover, separation techniques based on phase change are often not suitable for applications involving temperature-sensitive products, and in these cases, new separation paradigms are sorely needed.

Pressure swing adsorption (PSA) is one alternative to traditional thermally driven separation processes involving gases; concentration swing adsorption (like that found in a simulated moving bed (SMB) cycle) can be used in the case of liquids, but here another separation step is needed to remove the desorbate from the adsorbate. Typical adsorption processes separate different components based on their different equilibrium interactions with the adsorbent, which is mainly determined by the functional sites possessed by the adsorbent, nature of the adsorbate, and the surface area of adsorbent. By contrast, kinetically based adsorption separations can be achieved by utilizing differences in adsorbate diffusivity within the adsorbent materials [6,7].

Membrane-based separations are another attractive alternative to traditional separation techniques due to the potential to increase flexibility, lower operating costs, and reduce environmental emissions. Membranes can separate molecules using a variety of different equilibrium and kinetic mechanisms, but in the area of small molecule separations, the vast majority of membrane materials separate guest molecules based on the combination of molecule's sorptivity (or solubility) and diffusivity in the membrane material [8].

Adsorption and membrane technology contribute to sustainability via reducing the consumption of energy (and associated release of CO_2) in separation and purification processes. Thermally driven separation processes like distillation require energy (supplied via fossil fuel combustion) to induce a phase change of the components. This energy is significantly higher than the demixing energy of mixtures. For instance, the ideal energy cost of producing pure water from seawater via thermodynamically perfect demixing is approximately 1 kW-h m^{-3}. Highly optimized and heat-integrated multistage flash processes consume around 50 KW-h m^{-3}. Conversely, membrane-based desalination systems usually require approximately 4.5 kW-h m^{-3}, which is roughly the energy required to pressurize the intake fluid [9]. This order-of-magnitude reduction in energy and intensity that has been observed in the area of seawater desalination is expected to apply in other separation challenges where the membrane or adsorption unit can avoid phase change.

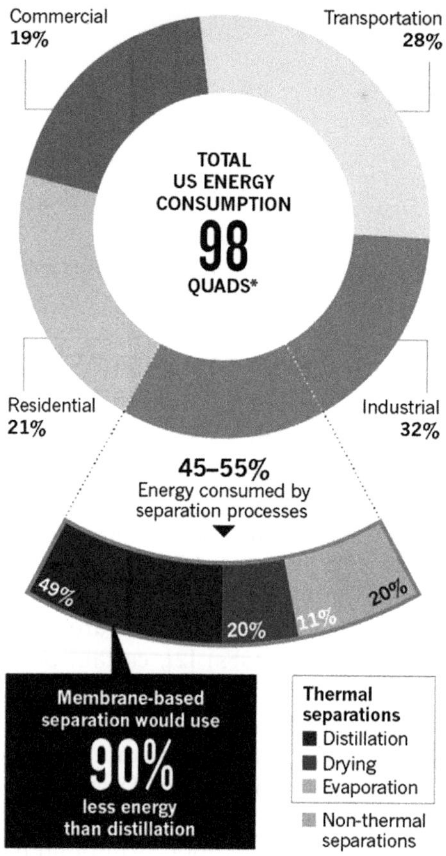

FIGURE 3.1 Chemical separation account for about half of US industrial energy use and 10%–15% of the nation's total energy consumption. *Reprinted by permission from D.S. Sholl, R.P. Lively, Seven chemical separations: to change the world: purifying mixtures without using heat would lower global energy use, emissions and pollution-and open up new routes to resources, Nature 532 (7600) (2016) 435–438. Copyright 2016 Springer Nature Publishing Group.*

Besides reductions in carbon and energy intensity relative to thermally driven processes, membrane and adsorption units also contribute to green chemistry by reducing the E (environmental) factor. The E factor, defined by the ratio of the mass of waste per mass of product during the process, provides the impetus for developing cleaner, more sustainable processes [10]. Industrial chemical synthesis produces a large amount of wastes (e.g., organic solvents, unreacted reactants, by-products, etc.). The more complex the reaction is, the higher the E factor could be (e.g., E factor for the bulk chemical industry is usually smaller than 5, while E factor for pharmaceutical industry ranges from 25 to over 100). Conventional thermally driven processes differentiate molecules based on the difference between their boiling points, which are often not effective for the separation of complex organic solvent waste streams, as azeotropes and isomers can significantly complicate the distillation process. Membranes and adsorbents made of nanoporous materials provide versatile energy-efficient molecular differentiation methods based on size differences and different interactions between guest and host molecules [11–13]. Membrane and adsorption technology can effectively reduce the E factor by enabling the separation of complex waste mixtures and transferring these wastes to useful products.

This chapter begins with a brief introduction of the current state-of-the-art in nanoporous materials research for energy-efficient separation. Following this introduction, the fabrication methods of membranes and adsorbents based on nanoporous materials will be presented before major applications in energy-efficient separations are discussed. Finally, in this chapter, key scientific and engineering issues and future directions are identified and presented as challenges and opportunities for researchers in this field.

2. NANOPOROUS MATERIALS

2.1 Zeolites

Zeolites are crystalline hydrated aluminosilicates with a periodic arrangement of nanopores (4–13 Å in diameter). Zeolite frameworks are formed by *T* atoms (primarily Si and Al) in tetrahedral coordination linked through oxygen atoms that produce a three-dimensional (3D) network defined by channels and cavities of molecular dimensions [14].

Based on their Si/Al ratio, zeolites can be classified into three categories: high-silica zeolites (Si/Al > 5), intermediate-silica zeolites (2 < Si/Al < 5), and low-silica zeolites (1 ≤ Si/Al < 2). The pore architectures of the zeolites can be tuned by the introduction of organic molecules, which act as pore-filling agents.

The Swedish chemist Axel Cronstedt was the first researcher to note that the mineral stilbite appeared to boil when heated and thus named this kind of materials zeolites or "boiling stones" in 1756 [15]. Currently, there are more than 200 different zeolite framework types that have been synthesized and accepted by the International Zeolite Association (IZA), which are classified according to the channel system (monodirectional, bidirectional, or three-directional) or to the pore size (small-8-MR, medium-10-MR, large-12-MR, or extra large-12-MR). It should be noted that theoretical estimates suggest that millions of other zeolite structures can be constructed from the primary building blocks, but only a small fraction of these have proven to be accessible using modern synthetic techniques.

The uniform distribution of well-defined nanopores imbues zeolites with the ability to separate molecules with only minor differences in molecule size (i.e., ~0.1 Å differences) and shape (i.e., exquisite ethane/ethylene selectivities have been observed in zeolites). Moreover, zeolites possess excellent thermal and pressure stability (e.g., zeolite 13X can operate at 400°C under 7 bar) [16], as well as the presence of exchangeable ions ("counterbalancing ions") within the framework, which enables the guest transport and sorption properties to be tuned. All of these features position zeolites as promising materials for the fabrication of molecular sieving membranes and adsorbents [17,18].

Zeolite structures and topologies are named using three-letter codes (see the IZA website: http://www.iza-structure.org/databases/ for further details). As shown in Fig. 3.2, various pore architectures, which are suitable for different separation targets, can be achieved by versatile zeolite materials.

Most zeolites contain nanopores with diameters below 2 nm. Pores less than 2 nm are defined by IUPAC as "micropores." Such micropores are capable of realizing shape selectivity in separation and catalytic modalities involved in various industries (e.g., petrochemistry) [19,20]. Beyond simply microporous zeolites, there is a drive to develop zeolites with larger pores to increase mass transfer rates into and out of the zeolite structure [21]; indeed, a significant effort has been made to develop zeolites containing mesopores (by IUPAC definition, pores with diameter ranging from 2 to 50 nm). Mesopores can be introduced within zeolite crystals during synthesis or postsynthesis by introducing gaps between intergrown zeolite nanocrystals. Intracrystalline mesopores are usually created via demetallization, i.e., leaching out silica or aluminum from zeolite networks [22]. Common dealumination methods are high-temperature steaming and acid leaching, which breaks the Si–O–Al bonds and free aluminum from the zeolite networks. Such dealumination process results in vacancies or amorphous materials, releasing mobile silicon species. Mobile silicon species tend to migrate and condense at other sites resulting in mesopores. Similarly, desilication can be achieved through base leaching [23]. Intercrystalline mesopores can be created via various zeolite nanoparticle assembly methods. Another straightforward method is synthesizing zeolites with mesoporous templates, which can be removed afterword. Such templates can either be hard material (e.g., carbon nanotubes, carbon particles, aerogels, etc.) or soft materials (e.g., surfactants) [23]. Besides, assembly of nanoparticles can also be achieved without expensive mesoporous templates via steam-assisted crystallization [24–26], solid-phase crystallization [27], nanofusion [28], and repetitive branching [29].

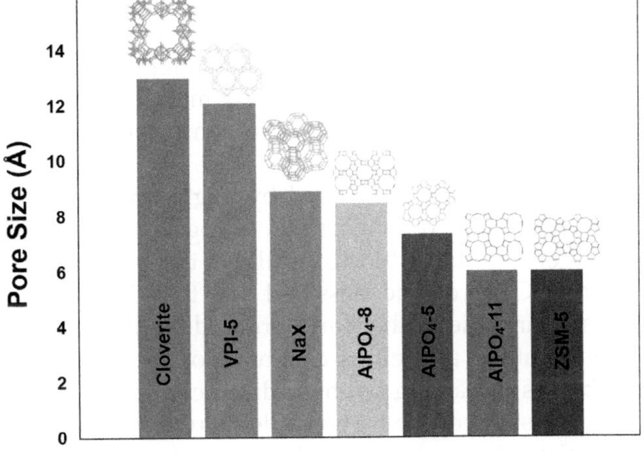

FIGURE 3.2 Structure and pore size of some common zeolite materials.

2.2 Metal-Organic Frameworks

MOFs refer to crystalline porous materials formed by metal ions and organic ligands that are connected via coordinative bonds [30]. To date, over 20,000 MOFs have been experimentally reported for various applications that are founded on molecular capture and differentiation (i.e., adsorption, membrane, sensing, catalysis, etc.) [31,32].

MOFs have a variety of tailorable structural characteristics that translate into tunable performance properties. For instance, MOF apertures can be varied from a few angstroms to several nanometers while maintaining uniform pore size distribution. The resulting molecular sieving effect has been applied for size-selective sensing in luminescent Zn_3btc_2 and Cd_2^{+}-based MOFs [33,34]. This feature suggests that MOFs may be useful for separating molecules with similar size [1,35–37]. Beyond changes in topology, MOF apertures can be further tuned by mixing linkers of different lengths; it is possible to do this while retaining the topology of the MOF comprised of the starting linker [38,39]. Indeed, the immense number of possible metal–ligand combination possibilities provides a rich design space with access to a variety of topologies and customized functionality. MOF functionality can derive from the organic linkers, the metal nodes/clusters, or both [40]. For example, by changing the metal centers of MOF-74, the CO_2 capacity can be increased by 50% owing to the enhanced interaction with CO_2 [41]. Via changing linkers with different functional groups, BET surface areas of UIO-66 derivates can be tuned from 540 to 1580 $m^2\ g^{-1}$ [42]. Mixed-linker hybrid zeolitic imidazolate frameworks were also developed with tunable gate-opening effect for target separation processes [43]. Besides engineering metal nodes of MOFs, organic linkers can also be tuned via either presynthesis or postsynthesis modification [1].

Mesoporous MOFs are a recently emerging area of nanoporous materials research. The vast majority of existing MOFs exhibit cavities and channels with diameters smaller than 2 nm. Although micropores (<2 nm) provide large surface areas, mesopores (according to IUPAC definition, pores with a diameter ranging from 2 to 50 nm) are attractive for energy-efficient separations, and advanced catalysis because the diffusivities of guest molecules in mesopores are orders of magnitudes higher than those in micropores [44]. This increase in the diffusivity improves the overall mass transfer rate of reactants/products into/out of the catalyst material, thus further improving the efficiency of the catalytic process. Mesoporous MOFs can be categorized into three types: (1) cage-type mesoporous MOFs, in which mesopores are connected via microporous channels, (2) channel-type mesoporous MOFs, in which mesopores are accessible as channels, and (3) mesoporous particles of MOFs, in which mesopores are formed via surfactant templating [44]. For instance, Férey and coworkers noted that MIL-100 exhibits mesopores with a diameter ranging from 2.5 to 3 nm and high surface area of 3100 $m^2\ g^{-1}$, and these fall within the Type 1 categorization of mesoporous MOFs [45]. Subsequent efforts resulted in the creation of MIL-101 with mesopores (2.9–3.4 nm), which resulted in even higher surface areas (5900 $m^2\ g^{-1}$) [46]. One typical Type 2 channel-type mesoporous MOF is NU-1000 featuring pores with diameters ranging from 1 to 3 nm and versatile functionalization [47]; moreover, a variety of other Type 2 mesoporous MOFs (e.g., mesoMOF-1, JUC-48, UMCM-1, etc.) were also developed [48–50]. The Type 3 mesoporous MOFs are created via a surfactant-templated strategy, which results in the formation of mesopores via surfactant micelles within microporous MOFs. After template removal, MOFs with hierarchical meso-/micropores are synthesized. Such strategies have been successfully applied to MOFs such as HKUST-1 [51–53].

2.3 Organic Nanoporous Materials

Organic nanoporous materials are formed from nonmetal elements via covalent bonding. Compared with MOFs, organic nanoporous materials often exhibit superior thermal stability and chemical stability due to the replacement of coordination bonds with covalent bonds. Crystalline organic nanoporous materials include COFs and porous aromatic frameworks (PAFs) [54–56]. Amorphous nanoporous materials consist of hypercrosslinked polymers (HCPs), conjugated microporous polymers (CMPs), elemental organic frameworks, and polymers of intrinsic microporosity (PIMs) [3,57,58].

COFs are crystalline porous materials firstly developed by Yaghi and coworkers in 2005 [59]. COFs are constructed by light elements, such as hydrogen, boron, carbon, nitrogen, and oxygen, connected by strong covalent bonds. Combination of different organic building blocks results in the formation of COFs with various topologies, functionalities, and pore dimensions. COFs can be classified into two-dimensional (2D) COFs and 3D COFs. 2D COFs, such as COF-5, consist of stacking sheets that form ordered columns. These columns provide surfaces for adsorption and chemical functionalization, as well as channels for rapid molecular transfer. 3D COFs, such as COF-1, is formed by 3D building blocks containing sp^3-hybridized carbon or silane [60]. The resulting crystal structures contain ordered pores and large surface areas. The earliest published boron-based COFs, COF-1 and COF-5, suffer from low water stability, which triggered the development of other connectors, such as imine, hydrazine, and triazine [61]. COFs linked by triazine are also called covalent triazine–based frameworks, which possess higher thermal, chemical and mechanical stability than the boron-based materials [2]. To date, COF-103 exhibits the highest surface areas, 4210 $m^2\ g^{-1}$, among all porous polymers [60].

CMPs refer to polymers consisting of rigid conjugated π-systems that result in the formation of micropores [3]. Conjugated polymers have been intensively studied due to their electrical conductivity properties [3]. In 2007, Cooper and coworkers developed poly(aryleneethynylene) (PAE) networks that exhibit BET surface areas as high as 834 $m^2\ g^{-1}$, which inspired additional research in the area of conjugated polymers with permanent microporosity [62]. A typical CMP network is constructed by a series of benzene rings connected via phenyleneethynylene struts at the 1,3,5 positions on the phenyl groups. By varying the length of the phenyleneethynylene groups, the micropore size, micropore volume, and surface area can be tuned. The 1,4- and 1,3,5-substituted PAE networks possess 3D structures instead of planar structures due to the rotation of the alkyne bonds within the network. Besides PAE networks, other CMPs such as poly(phenylene butadiynylene) and poly(phenylenevinylene) networks were also developed following a similar strategy [63,64].

Semipermanent nanopores within linear polymers can also be induced via insufficient packing of the polymer backbones. This hypothesis was the foundation for the development of amorphous nanoporous linear polymers such as the so-called "polymers of intrinsic microporosity" (PIM), which were developed by Budd and McKeown in 2004. The rigid backbones of PIMs hinder efficient packing of adjacent polymer chains thus providing semipermanent nanoscopic voids within the polymer matrix. These materials are somewhat different from traditional glassy polymers that contain small amounts of isolated voids: the contortion centers of PIMs effectively support micropores that are interconnected over several nanometer length scales. It is important to note that there is still debate in the literature regarding the nature of the voids within PIMs. These voids provide useful molecular transport channels for selective adsorption and membrane separation. The prototypical PIM, PIM-1, has attracted wide interest due to its relatively easy synthesis procedure, inexpensive monomers, and solution processability. Besides the conventional backbone design based on spirobiindane contortion center and benzodioxin connection, more rigid backbones, such as bridged bicyclic units, have also been applied to novel PIMs. For example, PIM-MP-TB composed of methanopentacene (MP) units connected by Tröger's base demonstrates exceptional gas selectivity and permeability, surpassing the 2008 Robeson upper bound for H_2/CH_4 and O_2/N_2 (e.g., $\mathbb{P}_{O_2} = 999$ Barrer; $\alpha_{O_2/N_2} = 5.0$) [65].

Porous organic cages (POCs) are another class of organic nanoporous materials and were discovered in 2009 [66,67]. A unique feature of POCs is that each molecule contains a nanopore, but the POCs can be packed in the solid state to induce a continuous nanoporous network (both ordered and disordered). These materials are one of the few classes of materials that can be solution processed while still providing uniform pore size distributions at the subnanometer level. The BET surface areas of existing POCs are also comparable with state-of-the-art nanoporous materials. For instance, CC-3 exhibits BET surface areas up to 624 $m^2\ g^{-1}$. By modifying the building blocks of POCs, Cooper and coworkers also achieved pore size control. By introducing methyl groups, CC-15R, a derivate of CC-3, effectively prevents large molecules such as nitrogen from entering the intrinsic pores while remaining accessible to small molecules such as hydrogen [68].

Besides the aforementioned organic materials with intrinsic pores, block copolymers can also be fabricated with nanoporous structures. Nanochannels can be created via the self-assembly of amphiphilic macromolecules on mesoscopic length scales [69]. Depending on the block nature and polymer chain topology, various morphological patterns (e.g., spheres, cylinders, interpenetrating networks, etc.) can be formed. The self-assembled block copolymers can then be translated into nanoporous materials via etching one of the blocks, thus removing additives or physically reconstructing the morphology [56]. These types of materials create either very large micropores (e.g., 1.5–2 nm) or small mesopores (e.g., 2–6 nm). These nanopores are useful in filtration applications where large macromolecules are to be removed from solvent molecules, but these pores are much too large to enable molecular separations such as those found in gas separations.

Bio-inspired or bio-derived nanoporous materials are also attractive for sustainable separation processes. Polydopamine (PDA) is a versatile bio-inspired material produced by oxidized self-polymerization of dopamine [70]. Dopamine can polymerize and aggregate into PDA nanoparticles in the presence of oxidizing agents. A higher oxidation degree leads to larger nanoparticle size. PDA nanoparticle suspensions can be coated onto substrates (e.g., catalysis particles, support membranes, etc.). The random packing of PDA nanoparticles generates interconnected molecular sieving channels, sizes of which can be tuned by the oxidizing conditions [71]. Monomers (e.g., morin, tannin, etc.) collected or generated from biomass have also been used for nanoporous polymer synthesis [72,73].

2.4 Carbon-Based Nanoporous Materials

CMSs, a type of nanoporous materials, are produced via the pyrolysis or carbonization of a polymeric, well-defined, and carbonaceous precursor under controlled temperature and atmosphere [74,75]. Often thought of as an inorganic material, CMS structures have excellent thermal and chemical stability (e.g., resistance to organic solvents, acids, bases, etc.) in comparison with the starting polymer precursors. Most CMS materials are believed to

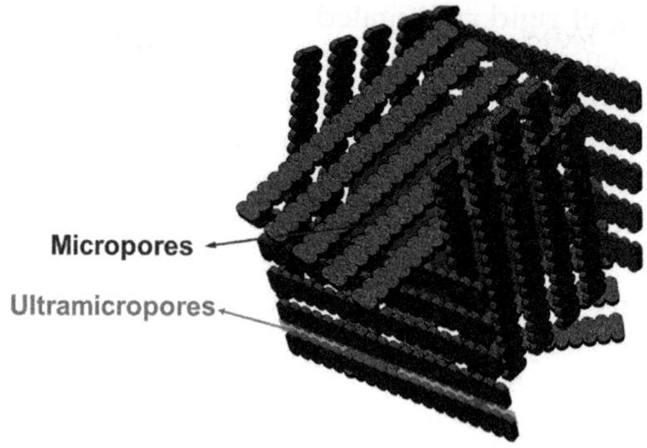

FIGURE 3.3 Hypothetical short-range turbostratic carbon structure of the carbon molecular sieve membranes.

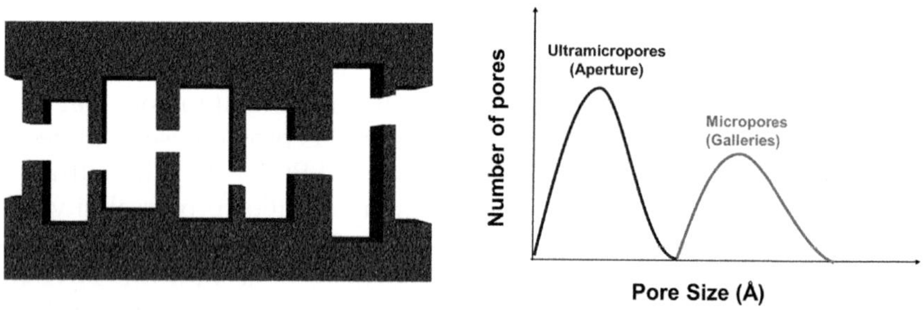

FIGURE 3.4 Hypothetical bimodal distribution of pores in carbon molecular sieve membranes [77].

possess a turbostratic lamellar structure resulting in an amorphous, isotropic, and microporous material. The carbon lamellae contain sp^2 hybridized condensed hexagonal carbon sheets, which exhibit short-range order via the alignment of the sheets parallel to each other but with a random rotational angle. The basic structural units of the CMS are believed to comprise a kinetically trapped array of plates formed from carbonaceous strands created during aromatization and fragmentation of the precursor backbone as shown in Fig. 3.3 [76]. Over the long range, these lamellae arrange randomly, bend, and twist to form an amorphous structure.

A simplified, idealized pore structure of CMS materials can be described as a "slit-like" bimodal pore model as shown in Fig. 3.4 [78–80]. CMS is distinct from activated carbon; the former has a well-defined narrow pore size distribution, whereas the latter as a broad pore size distribution ranging from microporous to macroporous [4,81]. This bimodal pore size distribution with larger micropores connected by smaller ultramicropores derives from packing imperfections of the carbon sheets. Ultramicropores enable molecular sieving, while micropores provide abundant sorption sites. This combination allows CMS membranes to sustain high permeability and high selectivity at the same time, which is attractive for separations [82].

Because of their narrow ultramicropores and high micropore volume, CMS materials enable the kinetic-based separation of various gas mixtures via PSA [81]. For instance, CMS produced by Takeda Chemical Industries Ltd. (Shirasagi MSC 3A) has been widely applied in PSA separation of nitrogen from the air [83]. Besides these successful commercial CMS adsorbents, researchers are also developing new CMS materials for other adsorptive applications such as CO_2 and H_2S separations [84].

Nanoporous graphene materials usually refer to laminated graphene planes with nanosized defects [85]. Synthesis of nanoporous graphene materials can be categorized as top-down methods and bottom-up methods. Top-down methods create pores on existing graphene sheets via electron beams or ultraviolet etching. Ordered nanopore patterns can be achieved using templates formed by block copolymers (block copolymer lithography) [86] or nanospheres (nanospheres lithography) [87]. Subnanopores can be formed by oxidation in the presence of concentrated sulfuric acid, sodium nitrate, and potassium permanganate, although it is difficult to simultaneously control pore size and a number of pores in these materials [88]. Conversely, bottom-up methods directly synthesize graphene sheets containing nanopores formed by barrier components [85,89] or the gaps between graphenic ribbons [89]. The nanopores on individual graphene sheets selectively permeate molecules with kinetic diameters smaller than the size of nanopores but reject larger molecules [90]. Graphene oxide (GO) is one of the most popular emerging nanoporous graphenic materials in the field of membrane science due to the ease of large-scale preparation and the versatility of assembly strategies [91–93]. Graphene-based membranes can be classified into two categories: porous graphene layers and assembled graphene laminates [94]. Nanoporous graphene sheets of one-atom thickness can be prepared by the aforementioned top-down methods (e.g., creating pores on graphite sheets) or bottom-up methods (e.g., chemical vapor deposition), and then transferred onto supporting substrates before being used as membranes. For instance, graphene monolayers with subnanometer (0.5–1 nm) pores were employed in desalination, exhibiting nearly 100% salt rejection and water permeance of 250 L m^{-2} h^{-1} bar^{-1} [95]. Although monolayer graphene membranes show great potential, assembled graphene laminates are more practical considering the technical difficulty in large-scale preparation and modularization of individual layers of graphene [91]. In one recent report, GO laminates (lateral length of 10–20 μm) have been assembled into thin membranes (~10 nm) and exhibited high water and solvent permeance while rejecting small dye molecules by >99.9% [96]. A common issue with these types of membranes is their stability during service, as the interplate distance between the sheets reduces slowly over time. Nanoporous graphene and GO can also be used as versatile adsorbents due to the high surface areas and tunable functional groups. For instance, polyurethane sponges have been coated with nanoporous graphene and GO for selective oil removal from water [97]. Owing to the superhydrophobic surface created by graphene coating, the composite sponge exhibits oil capacity as high as 70 g g^{-1} with little additional cost. The primary challenge in this area is to achieve sufficiently high performance to justify the additional cost of the graphenic materials beyond traditional carbon adsorbents such as activated carbon and CMSs.

3. FABRICATION OF MEMBRANES CONTAINING NANOPOROUS MATERIALS

A variety of membrane geometries and architectures exist, but nanoporous membranes can be broadly generalized into three categories: (1) integrally skinned asymmetric membranes, (2) supported crystalline membranes, and (3) thin film composite membranes (Fig. 3.5).

The majority of nanoporous materials are crystalline and are relatively brittle. As such, they are rarely fabricated as free-standing membranes or adsorbent beds. One way to address this in the case of membrane separations is to form a selective layer of nanoporous materials on a highly porous support structure, which will form so-called composite membranes. As pores of the substrates in the composite membranes are a few orders of magnitude larger than those of the nanoporous materials, the apparent separation performance is dominated by nanoporous materials. Another approach is fabricating mixed matrix membrane (MMM), where nanoporous materials are distributed in a polymeric matrix membrane (note that these types of membranes can be fabricated as integrally skinned asymmetric or thin film composite structures). The apparent mass transfer property of MMMs is governed by the combined effect of nanoporous materials and polymers. MMMs have the advantage of being easier to fabricate than supported crystalline membranes. PIMs and CMS materials are unique in the fact that they can be fabricated without the use of substrates. The homogenous membranes have streamlined fabrication processes and corresponding cost-effectiveness over nonhomogeneous membranes.

3.1 Molecular Transport Through Membranes

3.1.1 Sorption Diffusion Mechanism

The intrinsic transport properties of sorption-diffusion type membranes are described by two main parameters: "permeability," a measurement of intrinsic productivity, and "selectivity," α a measurement of separation efficiency. Permeability ($\mathbb{P}_A$) is equal to the transmembrane fugacity and thickness normalized flux:

$$\mathbb{P}_A = \frac{N_A \ell}{\Delta f_A} \tag{3.1}$$

where N_A is the penetrant flux through the membrane of thickness (ℓ) under a transmembrane fugacity difference (Δf_A). For homogeneous dense membranes, the membrane thickness (ℓ) can be measured directly. However, in the case of the asymmetric membrane, the actual membrane thickness is not readily known. Thus, the term of permeance (P_A), which is simply the fugacity-normalized flux, is commonly used to describe the productivity as defined below:

$$P_A = \frac{\mathbb{P}_A}{\ell} = \frac{N_A}{\Delta f_A} \tag{3.2}$$

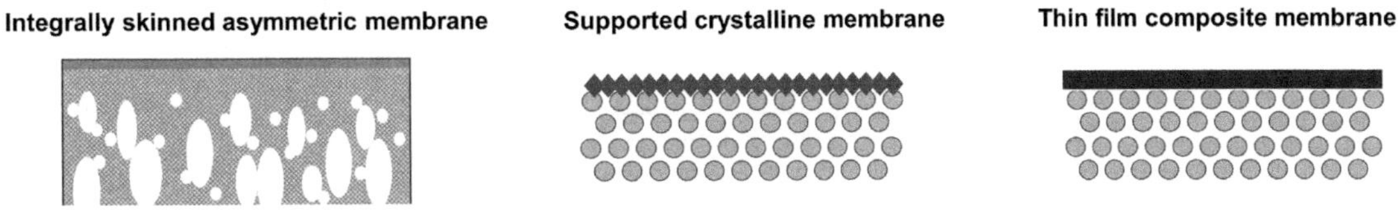

FIGURE 3.5 A schematic diagram of an integrally skinned asymmetric membrane, supported crystalline membrane, and thin film composite membrane.

In this sorption-diffusion transport mechanism, guest molecules sorb into the upstream side of the membrane and diffuse through it due to the presence of a chemical potential gradient and desorb at the downstream side. The permeability can be expressed as the product of D_A, named the transport diffusion coefficient, and $\mathbb{S}_A$, named the solubility or sorption coefficient.

$$\mathbb{P}_A = D_A \times \mathbb{S}_A \tag{3.3}$$

Sorption coefficient, $\mathbb{S}_A$, is a thermodynamic factor governed primarily by the condensability of a gas penetrant and the membrane–penetrant interaction. Diffusion coefficient, D_A, is a kinetic property, related to the ability of a guest molecule to jump within the membrane.

The ideal permselectivity for guest molecule A versus B, α_{AB}, reflects the separation efficiency of the membrane and is defined as the ratio of the permeability of the fast component to the slow component. Moreover, the selectivity can be further defined using the sorption-diffusion model as the product of the diffusive selectivity D_A/D_B and sorptive selectivity $\mathbb{S}_A/\mathbb{S}_B$.

$$\alpha_{AB} = \frac{\mathbb{P}_A}{\mathbb{P}_B} = \left(\frac{D_A}{D_B}\right) \times \left(\frac{\mathbb{S}_A}{\mathbb{S}_B}\right) \tag{3.4}$$

Permeation and diffusion of guest molecules across microporous membranes are accurately described as activated processes. The activation energies of permeation and diffusion follow Arrhenius relationships, while sorption can be described by a Van't Hoff expression [98,99].

$$\mathbb{P}_A = \mathbb{P}_{0A} \exp\left(\frac{-E_{\mathbb{P},A}}{RT}\right) \tag{3.5}$$

$$D_A = D_{0A} \exp\left(\frac{-E_{D,A}}{RT}\right) \tag{3.6}$$

$$\mathbb{S}_A = \mathbb{S}_{0A} \exp\left(\frac{-\Delta H_{\mathbb{S},A}}{RT}\right) \tag{3.7}$$

R is the universal gas constant, T is the absolute temperature, and $\mathbb{P}_{0A}$, D_{0A} and $\mathbb{S}_{0A}$ are preexponential factors of permeation, diffusion, and sorption, respectively. $E_{\mathbb{P},A}$ is the activation energy for permeation, $E_{D,A}$ is the activation energy for diffusion, and $\Delta H_{\mathbb{S},A}$ is the apparent heat of sorption, which is always negative.

Rearrangement of Eqs. (3.3), (3.5)–(3.7) reveals that activation energy for permeation is the combination of activation energy for diffusion and apparent heat of sorption. Moreover, the preexponential factors of permeation are the product of that for diffusion and sorption.

$$\mathbb{P}_{0A} = D_{0A} \times \mathbb{S}_{0A} \tag{3.8}$$

$$E_{\mathbb{P},A} = E_{D,A} + \Delta H_{\mathbb{S},A} \tag{3.9}$$

3.1.2 Molecular Transport in Mixed Matrix Membrane

MMMs are obtained by distributing nanoporous materials into a continuous polymeric matrix [100]. The advantage of the MMM concept is that it combines the ease of polymer film processing with the high selectivity and permeability of inorganic materials. Many nanoporous materials have been explored in this approach such as zeolites [101–103], MOFs [104–106], COFs [107–109], PAFs [110], POCs [111], PIMs [112], CMS [113], etc.

The gas transport in the MMM can be mathematically described by the Maxwell model. This model was initially derived by James C. Maxwell to describe dielectric properties in a conducting dilution suspension of identical particles and is used to describe gas transport in mixed matrix materials based on the close analogy between conductivity (heat) and permeability (mass). It should be noted that the model is based on the assumption that spheres of material are dispersed in an infinite medium (Fig. 3.6). The Maxwell model is given by:

$$\mathbb{P}_e = \mathbb{P}_m \frac{\mathbb{P}_f + 2\mathbb{P}_m + 2(\mathbb{P}_f - \mathbb{P}_m)\varnothing_f}{\mathbb{P}_f + 2\mathbb{P}_m - (\mathbb{P}_f - \mathbb{P}_m)\varnothing_f} \tag{3.10}$$

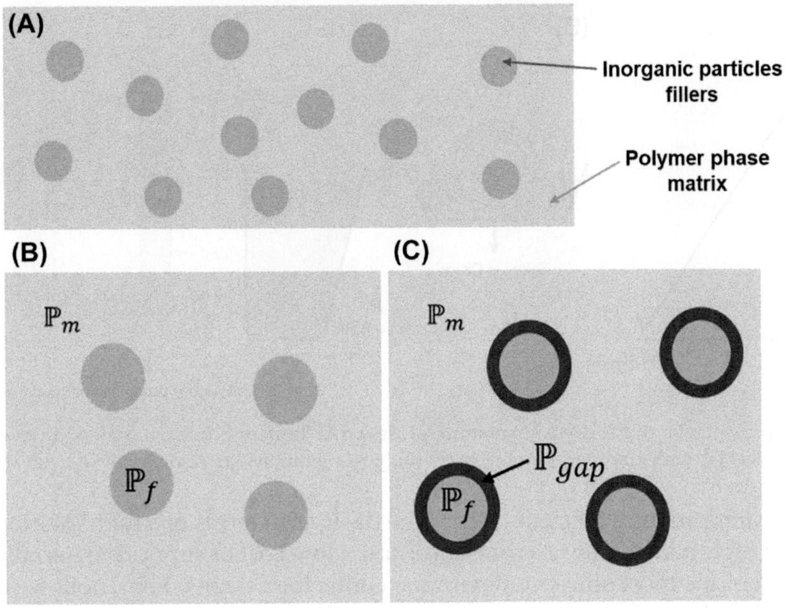

FIGURE 3.6 Schematic diagram of (A) mixed matrix membrane; (B) Maxwell model for defect-free mixed matrix membrane, and (C) Maxwell model for defective mixed matrix membrane.

where $\mathbb{P}_e$ is the effective permeability in the mixed matrix material, $\mathbb{P}_f$ is the permeability in fillers (molecular particles), $\mathbb{P}_m$ is the permeability in the polymer matrix, and $\varnothing_f$ is the volume fraction of molecular sieve particles in the mixed matrix material.

Moreover, results from the Maxwell model can also suggest the presence of defects inside the mixed matrix material. These defects take many forms: (1) gaps (voids or dilated interfaces) generated from the poor adhesion between the filler particles and the polymer matrix; (2) densely packed interfaces caused by more densely packed polymer chains around the sieve relative to the bulk; and (3) plugged sieves which result in a barrier-like filler phase [114].

In the case of nanoscale "gaps," the Maxwell model is given by:

$$\mathbb{P}_e = \mathbb{P}_m \frac{\mathbb{P}_{fe} + 2\mathbb{P}_m + 2(\mathbb{P}_{fe} - \mathbb{P}_m)\varnothing_{fe}}{\mathbb{P}_{fe} + 2\mathbb{P}_m - (\mathbb{P}_{fe} - \mathbb{P}_m)\varnothing_{fe}} \tag{3.11}$$

where $\mathbb{P}_{fe}$ is the effective permeability of inorganic particles and gap, and $\varnothing_{fe}$ is effective volume fraction of molecular sieve particles and gaps in the mixed matrix material.

$\varnothing_{fe}$ can be calculated by:

$$\varnothing_{fe} = \frac{4}{3}\pi(r+h)^3 n = \frac{4}{3}\pi r^3 n\left(1+\frac{h}{r}\right)^3 = \varnothing_f(1+\beta)^3 \tag{3.12}$$

where r is the radius of a bare molecular sieve particle, h is the thickness of gas between particles and the polymer matrix, and β is the ratio between h and r.

$\mathbb{P}_{fe}$ can be calculated by:

$$\mathbb{P}_{fe} = \mathbb{P}_f \frac{\left[2(1-\gamma) + (1+\beta)^3(1+2\gamma)\right]\gamma}{(1+\beta)^3(1+2\gamma) - (1-\gamma)} \tag{3.13}$$

where $\gamma = \mathbb{P}_{gap}/\mathbb{P}_f$ and $\mathbb{P}_{gap}$, the permeability of the gaps can be estimated by using the Knudsen diffusivity in a gap of thickness h.

3.2 Integrally Skinned Asymmetric Membranes

3.2.1 *Asymmetric Polymers of Intrinsic Microporosity Membrane*

Some linear PIMs (e.g., PIM-1) are soluble in common membrane casting solvents [58]. Free-standing PIM-1 membrane coupons are routinely tested to understand the intrinsic properties of PIM-1 [58,115,116]. Modern industrial

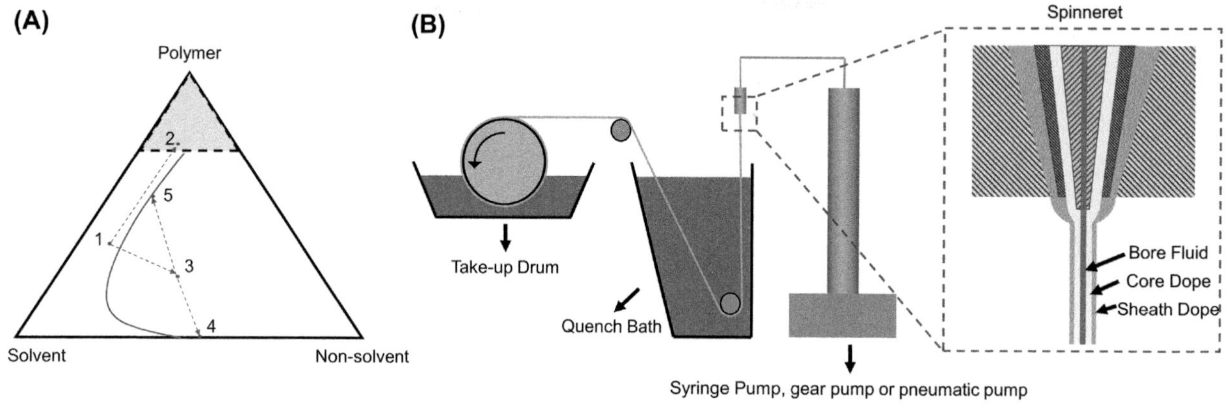

FIGURE 3.7 Schematic representation of the dry-wet spinning process. (A) Ternary phase diagram of a typical polymer, solvent, and nonsolvent system and (B) one typical lab-scale dry-wet spinning setup (polymer solutions can also be driven by gear pumps or compressed gases).

membrane separation systems require efficient membrane designs in terms of a high interfacial area to volume ratio, a capability to withstand high transmembrane pressure, and a low cost of support materials, etc. Hollow fiber membranes are viewed as one of the most efficient membrane contactor designs. First, hollow fiber membranes provide the highest ratio of interfacial area to volume among all practical membrane designs: common hollow fiber membranes with a 250 μm outer diameter and 125 μm inner diameter provide up to 8000 m^2 surface area per 1 m^3 of module volume [117], while spiral-wound membranes with the same wall thickness (62.5 μm) using a spacer of common thickness (700 μm) [118] only provide up to 1300 m^2 surface area per 1 m^3 of module volume. Second, the cylindrical design relates the transmembrane pressure to the compression of the membrane materials. There are many reported examples of hollow fiber membranes operating at extreme transmembrane pressures; for instance, CMS hollow fiber membrane can withstand hydraulic pressure >100 bar without failure [80]. Third, self-supported hollow fiber membranes do not require supporting and spacing materials, which are necessary for the widely applied spiral-wound membranes.

Hollow fiber membranes are typically fabricated via dry-wet spinning (Fig. 3.7). This technique processes ternary solution (referred as "dope") containing polymers, solvent, and nonsolvent with a composition close to the boundary curve (Fig. 3.7A). Syringe pumps deliver the homogenous dopes and bore liquid through a spinneret (Fig. 3.7B). Liquid filament then travels through an air gap, immersed into a nonsolvent (usually water) quench bath, and eventually collected onto a rotating drum. Although different spinning parameter combinations result in the different complex mass transfer process and different final membrane structures, an ideal dope composition is described in Fig. 3.7A. Skin of the filament undergoes evaporation induced vitrification and form dense (e.g., without macrovoids) selective layer, while the dope beneath the skin undergoes phase inversion and results in porous support layers. Point 1 represents the composition of initial dope. Trajectory 1–2 represents the hypothetical composition evolution of skin layer during evaporation in the air gap, in which point 2 represents the composition of the vitrified skin layer. Trajectory 1–3 represents the hypothetical composition evolution of the dope beneath the skin during phase inversion in the quench bath, in which point 3 represents the unstable composition. The unstable point 3 then undergoes phase inversion, separating into a polymer-rich phase (point 4) and a polymer-lean phase (point 5). After the sequential solvent exchange, polymer-lean phases are eliminated, and the porous structure is formed by polymer-rich phases. The spinning of solution-processable nanoporous polymers, PIMs, is not easy as they can only be dissolved by volatile solvents (e.g., tetrahydrofuran [THF], chloroform, and dimethylchloride), which are typically not used as the sole solvent in dry-wet spinning. The relatively rapid evaporation of volatile solvents tends to induce the formation of thick and defective skin layers; the lack of high boiling point solvents confounds the spinning process of PIM-1 and other polymers with similar solubility issues. Yong et al. firstly fabricated hollow fiber membranes made of a PIM-1/Matrimid blend [119]. After blending with a high fraction of Matrimid (85–95 wt%), the two polymers were successfully dissolved by a mixture of THF and N-methyl-2-pyrrolidone (NMP). Although the technique of polymer blend spinning managed to use NMP—a traditional solvent for dry-wet spinning—the narrow miscibility between Matrimid and PIM-1 limited the performance of the obtained membranes. Jue et al. adapted a dual-bath method to fabricate a pure PIM-1 hollow fiber membrane (Fig. 3.7) [120]. In their approach, an immiscible butanol sheath layer was applied to isolate the PIM-1 solution from the air and thus slowing THF evaporation. The resulting defect-free PIM-1 hollow fiber membrane exhibited high permeances (e.g., CO_2 permeance up to 360 GPUs). Due to shear-induced polymer chain alignment, the PIM-1 hollow fiber membranes were observed to have improved selectivity compared with PIM-1 flat membrane fabricated via traditional blade-casting. As shown in this report, owing to the solution processability of PIM-1, well-developed solution-based membrane fabrication techniques can be utilized to manufacture PIM-based membranes. However, the current high cost of PIMs restricts the wide application of membranes solely made of PIMs. Multilayer spinning techniques (i.e., a selective sheath of PIM-1 cospun with a low-cost support material) can ultimately drive these costs down, but more research is needed in this area.

3.2.2 *Asymmetric Carbon Molecular Sieve Membranes*

CMS membranes are fabricated by the pyrolysis of polymer precursor materials in a controlled environment. The carbonization process is quite complex and intricate, and several reactions may take place at the same time such as cleavage, dehydrogenation, condensation, isomerization, etc. [121,122]. Even though the pyrolysis process by which a polymer precursor is transformed into a CMS material is complex, it results in the reproducible production of carbon materials when pyrolysis conditions are carefully controlled. One proposed mechanism of translation of a polymer coil into rigid CMS membranes under inert atmosphere (nonvacuum) was put forth by Koros and coworkers and is discussed in detail in Fig. 3.8 using 6FDA:BPDA-DAM polymer precursor as an example [77,123,124]. In this theory, CMS materials are believed to be comprised of aromatized strands arranged to form plates, which are further organized into amorphous cellular structures. The narrow slits (typically <7 Å) between aromatized strands are the ultramicropores, which enable precise angstrom-level discrimination between molecules and is related with the sieving effect of the membrane. The voids (typically between 7–20 Å) generated from imperfect packing of the carbon plates are believed to be the micropores that provide abundant sorption sites for penetrant molecules. The transformation from polymer coil to CMS structure is achieved over the course of three periods: thermal ramp,

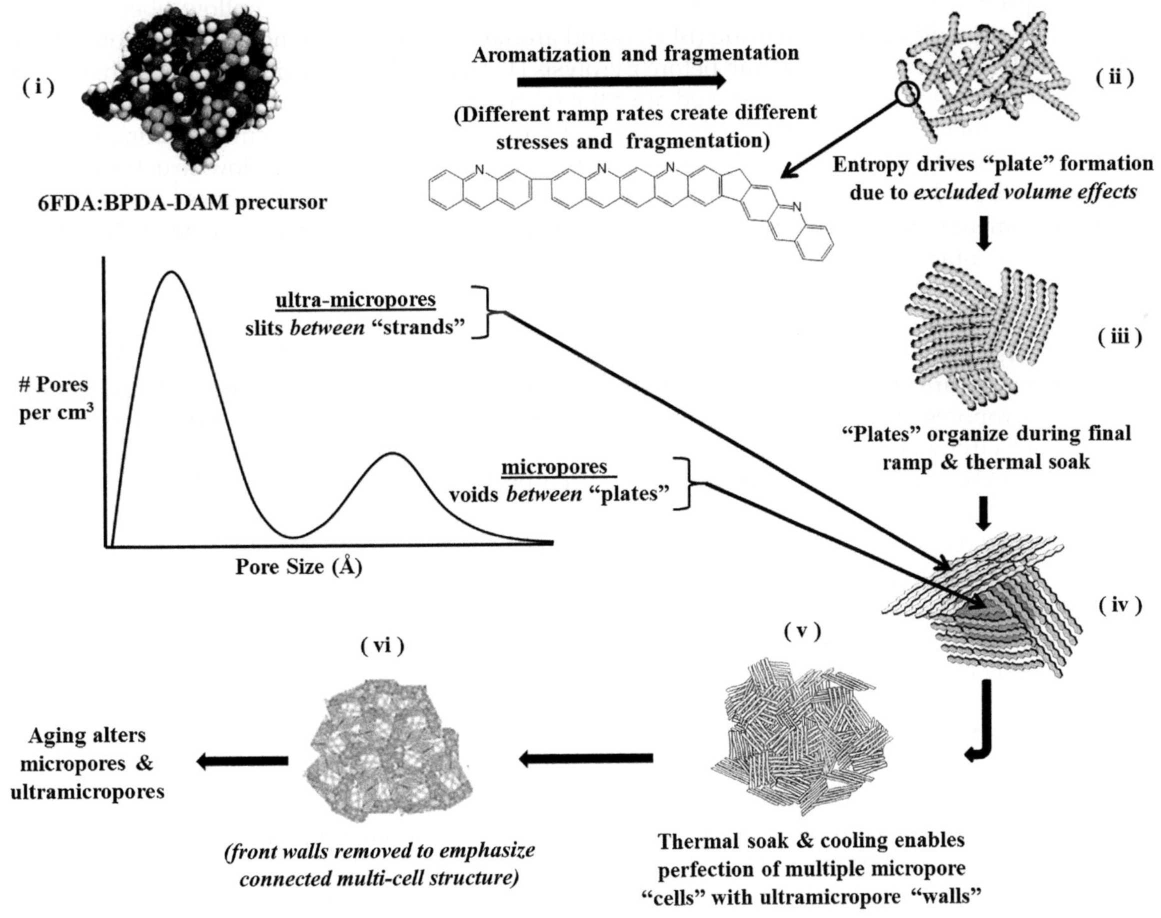

FIGURE 3.8 Envisioned steps in the transformation from random coil precursor polyimide to organized amorphous CMS material. *Reprinted by permission from M. Rungta, G.B. Wenz, C. Zhang, L. Xu, W. Qiu, J.S. Adams, W.J. Koros, Carbon molecular sieve structure development and membrane performance relationships, Carbon 115 (2017) 237–248. Copyright 2017 Elsevier Ltd.*

thermal soak, and cooling. During the thermal ramp process, a highly aromatic strand type structure forms because of the aromatization of the polymer and the evolution of volatiles as illustrated in Fig. 3.8 (i) and (ii). Aromatized strands will then be further aligned into plates driven by entropy to avoid the presence of excluded volume, as shown in Fig. 3.8 (iii) and (iv). The thermal soak and cooling process collate multiple neighboring cells, resulting in the sharing of ultramicropore "walls" between cells as illustrated in Fig. 3.8 (v) and (vi).

The performance of CMS membranes is mainly determined by the limiting pore size distribution (i.e., the pore size in the skin layer of the asymmetric membrane). Several parameters play crucial roles in affecting the structural properties and pore sizes of carbon membrane and further enhancing the separation performance of the CMS membrane such as the starting polymer precursor, pretreatment conditions, pyrolysis conditions, and posttreatment conditions [77,125,126].

The choice of the polymeric precursor is the first important factor because pyrolysis of different precursors may bring about different kinds of carbon membranes. Suitable polymer precursors for CMS membranes should firstly not melt or flow before or during pyrolysis conditions [127]. Numerous thermosetting precursors have been used to form carbon membranes, such as polyimide and derivatives [128,129], polyacrylonitrile (PAN) [130,131], phenolic resin [132,133], polyfurfuryl alcohol (PFA) [134,135], polyvinylidene chloride–acrylate terpolymer, phenol formaldehyde, cellulose derivatives [127], polymer of intrinsic micropores [131,136], and others. The intrinsic properties of the starting polymer precursor such as fractional free volume (FVV), chain mobility, chain structure, and glass transition temperature have significant effects on the performance of the resulting CMS membranes [137,138]. The studies by Williams [139] and Park [140] showed that the increase of FVV in the polyimides precursor would lead to a higher permeability in the resulting CMS membranes. The choice of the starting polymer is critical for the fabrication of CMS membranes that will be used to solve specific separation problems.

To ensure the stability of the precursor and preserve its structure to some degree during the pyrolysis process, the polymer precursor is often pretreated before pyrolysis. Thermostabilization and preoxidation are one of the most commonly used pretreatment methods. For example, Kusuki et al. found that the hollow fiber precursors based on biphenyl-tetracarboxylic acid dianhydride (BPDA) and aromatic diamines will not soften or collapse during pyrolysis after heated in air at 400°C for 30 min before pyrolysis [141]. David and Ismail also found that the stability of PAN hollow fiber precursor would be improved after the pretreatment in air or oxygen at 250°C for 30 min before pyrolysis. A nonsolvent chemical agent is another method for pretreatment of the polymer precursors [130]. Tin et al. found that the nonsolvent pretreatment will weaken the intermolecular interactions, allow structural rearrangement of carbon chains, and lead to smaller pore size and higher selectivity of the CMS membranes [142]. It is worth noting that all the pretreatment methods need to be optimized for different polymer precursor, which will be further transformed into CMS membranes for specific separation jobs.

Besides using different polymeric precursor, the specific pyrolysis conditions also offer additional tools to tailor the structure of CMS membranes such as final pyrolysis temperature, heating rate, soak time at final pyrolysis temperature, and pyrolysis environment. Final pyrolysis temperature is always chosen based on the decomposition and graphitization temperature of the polymeric precursor. Higher final pyrolysis temperature will lead to smaller ultramicropore and micropores inside the CMS membranes, and lower diffusion coefficient and permeability; however, larger diffusion selectivity and permselectivity are also often observed [143,144]. Koros and coworkers found that the 6FDA/BPDA-DAM-derived CMS membranes produced under vacuum pyrolysis environment showed higher selectivity than those pyrolyzed under inert atmosphere [145,146]. For example, polyimide precursor–derived CMS membranes produced using a vacuum pyrolysis at 550°C showed higher gas separation selectivities (7.4–9.0 for O_2/N_2 and 64–110 H_2/N_2) than those produced in argon (2.8–6.1 for O_2/N_2 and 6.8–31.2 H_2/N_2) or helium (4.7–6.1 for O_2/N_2 and 15.2–35.7 H_2/N_2) atmospheres at the same temperature [145]. This change of selectivity might be due to the different mechanism of the carbonization reaction by varying the type of pyrolysis atmosphere [145]. It has been shown that the activation energy of degradation decreased as the pressure of the inert pyrolysis atmosphere increased [147,148]. That is to say, the polyimide degradation process was "enhanced" when an inert gas was used. By accelerating the carbonization reaction, the inert gas pyrolysis environment appeared to produce a more "open" porous matrix in the CMS membranes resulting in a higher permeability and less selective pore structure [145,146]. A second explanation is that vacuum pyrolysis can "pull in" ppm levels of O_2 from the surrounding atmosphere, and these trace quantities of O_2 are known to dope the ultramicropores within CMS materials, thus improving the membrane performance [146].

Studies by Williams [128] and Kiyono et al. [146] showed that even trace amounts of oxygen in the pyrolysis environment could significantly affect the performance of the CMS membrane. The oxygen in the inert pyrolysis gas will selectively chemisorb in the ultramicroporous regions within the CMS. Intentional O_2 doping of the edges of the ultramicroporous slits has been successfully utilized to fine-tune the diffusion selectivity. The optimum O_2 doping level varies with separation target and CMS membrane. For example, over doping of oxygen (30 ppm) results in loss of productivity and selectivity [146]. Suda and Haraya found that the smaller heating rate during pyrolysis would lead to smaller pores and lower permeability, but higher selectivity [74]. Finally, soak time at the final pyrolysis temperature promotes pore sintering, thus increasing selectivity and decreasing the permeability of the resulting CMS membranes.

Several posttreatment methods such as postoxidation, chemical vapor deposition, and coating could be used to further fine-tune the pore size distribution or repair cracks and defects inside the CMS membrane after pyrolysis. Soffer et al. found that the permeability of cellulose-derived CMS membranes increases significantly after posttreated in the air at 400°C for 15 min [149]. Hayashi posttreated the BPDA-ODA–derived CMS membranes by chemical vapor deposition using a propylene carbon source at 650°C, which narrowed the pore size and increased the selectivity of the CMS membranes.

3.2.3 *Asymmetric Mixed Matrix Membrane*

As shown in Fig. 3.9, fabrication of MMMs starts by mixing polymer solution and particle dispersions. A primary dope containing polymer and part of the solvent is prepared in a sealed container. Nanoporous particles are then dispersed in remaining solvents under sonication. The primary dope and nanoporous particles are then mixed. Such suspension mixing rationale is believed to prevent agglomeration and enhance interaction between particles and polymers; however, one can also prepare the dopes in one step (e.g., adding polymers into a particle suspension). After further sonication, the final dope can be either cast into a liquid film or spun into hollow fiber membranes. The liquid of specific thickness undergoes a short exposure to air to allow for solvent evaporation to form a concentrated solid-like surface. After that, the liquid is transferred into a quench bath, which is usually water and undergoes a phase transition. The concentrated liquid surface is transferred into a defect-free (e.g., without holes or cracks larger than intrinsic pores of nanoporous materials) skin layer. The bulk of the liquid is transferred into a porous support layer with interconnected macropores. Although spiral wound flat membranes are widely applied in modern industry, hollow fiber membranes are preferred for their high surface area to volume ratio and better resistance to high transmembrane pressure [150]. Husain and Koros reported pioneering works outside of the patent literature demonstrating the fabrication of mixed matrix hollow fiber membranes containing HSSZ-13 zeolite and its application in gas separation [150]. As shown in this report, the solution-processability of mixed matrix fluids can be scalably translated into MMMs.

MOFs are currently the most popular porous filler for the creation of MMMs. Compared with zeolites, which have also been widely studied in MMMs, MOFs are often (but not always) found to be compatible with the polymer matrix, although the proper choice of organic ligand is needed to significantly eliminate the formation of voids between crystals and polymer matrix [103]. For example, amine-functionalized ligands have been found to form hydrogen bonding with the polymer matrix and improve the compatibility between MOFs and polymer. A large number of potential combinations of ligands and metal nodes also provide high tunability to enable MMM fabrication capable of separating target molecular pairs [151]. For example, polar functional groups (e.g., hydroxyl groups, amine

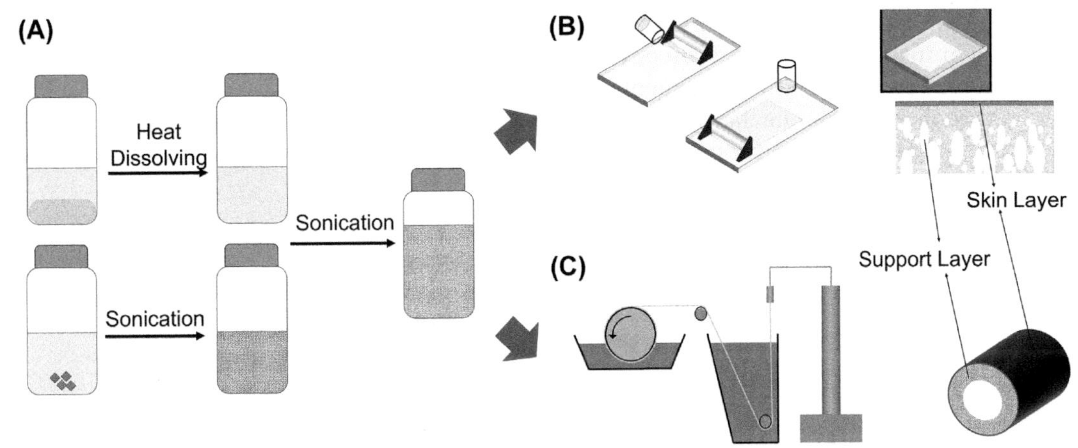

FIGURE 3.9 A schematic of the fabrication of asymmetric mixed matrix membranes: (A) preparation of polymer-particle solution; (B) asymmetric mixed matrix flat membrane cast by a blade followed by phase inversion, and (C) mixed matrix hollow fiber membrane fabricated via dry-wet spinning.

groups, etc.) can enhance the interaction with CO_2 [151–154]. Besides zeolites and MOFs, novel nanoporous materials, such as COFs, CMS, and HCPs, have also been studied as porous fillers in MMMs [107,113].

3.3 Supported Crystalline Membranes

3.3.1 *Zeolite Composite Membrane*

Initially, zeolite composite membranes are fabricated via an in situ synthesis method, which refers to the direct nucleation, growth, and intergrowth of zeolite crystals on a substrate [155]. Typically, a bare porous substrate (e.g., porous alumina) is firstly placed in a homogenous synthesis mixture containing a silica source, an aluminum source, structure-directing agent, base, and water [17]. A hydrothermal reaction is then conducted to produce a selective zeolite layer. Depending on the synthesis conditions, two membrane morphologies, thin zeolite membrane or infiltrated composite membrane (also referred to as pore plugging), can be formed [155–166]. The former structure theoretically exhibits better permeance due to the thin selective layer, while the latter one exhibits better intergrowth within the substrate and, as a result, better mechanical stability. In situ synthesis is a straightforward method and requires fewer posttreatments; however, this approach has a low reproducibility due to the difficulty in controlling the nucleation and crystal growth process [167,168].

Secondary growth is another promising strategy to fabricate defect-free zeolite membranes (Fig. 3.10) [168]. Typically, secondary growth of zeolite membranes consists of three critical steps: first, zeolite nanocrystals or sheets are prepared; second, presynthesized zeolite seeds are deposited onto the substrate; and third, the seeded substrate is immersed in synthesis solution where a thin zeolite layer is formed as a result of seed growth and intergrowth.

Zeolite seeds can either be prepared via direct synthesis with shape control or by disassembling, templating, or exfoliating parent zeolites [169–176]. By varying the structure directing agents and synthesis conditions, zeolite seeds with specific orientation and shape can be synthesized via traditional hydrothermal reactions [18,177–180]. For example, with carbon templates, agglomerated zeolites can be synthesized and then dissembled into nanocrystals with a diameter of 10–50 nm [169,173,174]. Zeolite nanosheets can also be exfoliated from parent zeolites and used for secondary growth or oriented zeolite membranes [181].

The resulting seeds can be deposited onto the substrate via manual rubbing, Langmuir trough assembly, or dip coating. A seed monolayer can be self-assembled by simple manual rubbing of the seeds on the support due to the interaction between seeds and substrates [182,183]. Several strategies have also been proposed to anchor seeds onto the substrate via groups like bidentate di-isocyanate and aminopropyl-triethoxy-silane [183–186]. A Langmuir trough assembly of particle monolayers on the fluid interface was firstly developed in the last century and adapted to zeolite membrane fabrication in recent years [187–197]. Despite these successful lab-scale demonstrations, further research is needed to scale this fabrication process up. Dip coating of zeolite seeds is more reproducible than the former two seeding methods; however, it usually results in thick membranes [198,199].

The seeded substrate is usually exposed to a synthesis solution containing a silica source, an aluminum source, structure-directing agent, base, and water. The hydrothermal reaction results in the secondary growth of the seeded zeolite. Ideally, the intergrowth of seeded zeolites will seal the gaps and result in a defect-free membrane. The synthesis conditions (e.g., structure-directing agents and pH) also play important roles in the finally membrane morphology [200–208]. There are also reports showing that secondary growth can be achieved without a synthesis solution [209,210]. For instance, Pham and coworkers prepared b-oriented MFI membrane by converting silica substrate in the presence of a structure-directing agent and steam [209]. Such solvent-free secondary growth method is a greener route for zeolite membrane fabrication compared with solvothermal synthesis, which requires large amounts of organic solvents. The existing vapor-phase synthesis is mostly conducted in lab-scale ovens and has

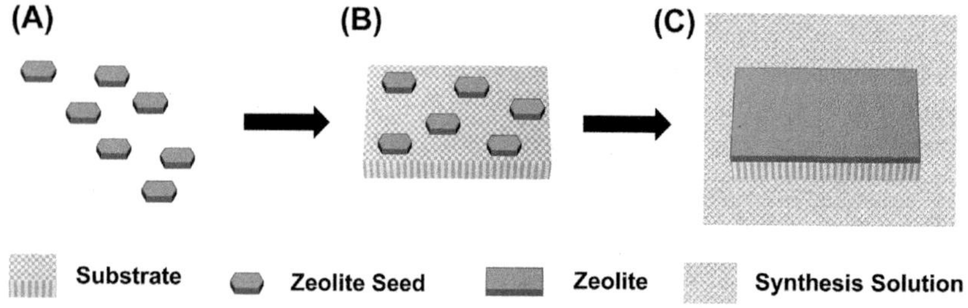

FIGURE 3.10 Secondary growth of zeolite membranes. (A) Synthesis of zeolite nanocrystals or nanosheets; (B) zeolite seeds are deposited onto a substrate; and (C) the seeded membrane was immersed in a synthesis solution containing precursors, structure-directing agents, and solvents.

not yet resulted in the creation of large membrane areas. Scale-up syntheses in large-scale vessels should be further investigated to better understand the challenges and hurdles associated with translating these materials from the lab and out into the field.

3.3.2 MOF Composite Membrane

Similar to zeolite membrane fabrication, MOF membrane fabrication techniques can be generally categorized into two strategies: in situ growth or secondary growth [37]. In situ growth means directly immersing a bare substrate (without MOF crystal attached) into the synthesis solution containing metal sources, ligands, additives, and solvents. During the reaction, MOFs will nucleate and grow on the substrate spontaneously. By contrast, secondary growth utilizes a substrate with nanoscopic presynthesized MOF or metal oxide seeds. During the reaction, growth and intergrowth of the seed crystals ideally result in the formation of a defect-free membrane. It should be pointed out that fabricating a membrane is more difficult than simply synthesizing a continuous MOF film. Although there were several reports about synthesizing of thin MOF films since the first development of MOFs in 1999, none of them reported any separation performance of the obtained MOF films [211–213]. To obtain a defect-free MOF membrane, researchers should not only consider the intergrowth of MOFs but also the thermal expansion of the substrate, capillary stresses during drying processes, and compatibility between substrate and MOFs, etc. [214].

Similar to the zeolite case, the original in situ growth method puts unmodified support into a synthesis solution and proceeds with a hydrothermal reaction. The first continuous MOF membrane was prepared by growing MOF-5 on pristine alumina support [215]. Unfortunately, such unmodified supports usually suffer from unfavored/uncontrolled heterogeneous growth and results in dense defective MOF "sponges" [214]. A CuBTC membrane was then prepared on a heat-treated copper net [216]. The oxide surface provides a metal source and the resulting CuBTC membrane exhibits promising hydrogen separation performance. These aforementioned in situ growth methods have no spatial control of nucleation sites and usually result in thick defective membranes. Jeong et al. developed a counterdiffusion method to fabricate a 1.5-μm ZIF-8 membrane on the surface of alumina support [217]. The alumina support was first soaked with a zinc source and subsequently immersed into a linker solution. By restricting the reaction region to the edge of the substrate, ZIF-7 and SIM-1 thin membranes were synthesized. Such methods can also be utilized to heal defective membranes. This feature has significant potential to reduce the costs of large-scale MOF membranes, as it reduces the need to create initially defect-free membranes during the membrane synthesis (i.e., slightly defective membranes are often easier to produce, and this posttreatment method allows for those defects to be healed in a facile manner). This counterdiffusion method was further modified by Wang and coworkers in 2011 (Fig. 3.11) [218]. Here, a nylon membrane was placed in a side-by-side diffusion cell separating metal ion solutions and linker solutions. As metal ions and linkers diffuse in opposite directions, an MOF layer was observed to form on the membrane surface where metal ions and linkers meet and react. Such modified counterdiffusion methods can form MOF membranes on both sides as the two solutions are miscible and there is no clear boundary

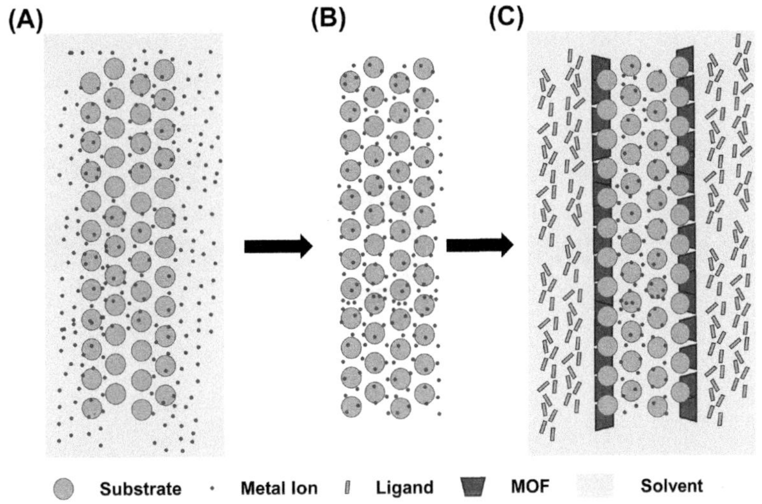

FIGURE 3.11 The process of in situ growth of a metal-organic framework (MOF) membrane via counterdiffusion method: (A) the substrate is firstly immersed into a metal ion solution; (B) the substrate soaked with metal ions is transferred into a ligand solution; and (C) MOF membranes constructed by metal ions and ligands.

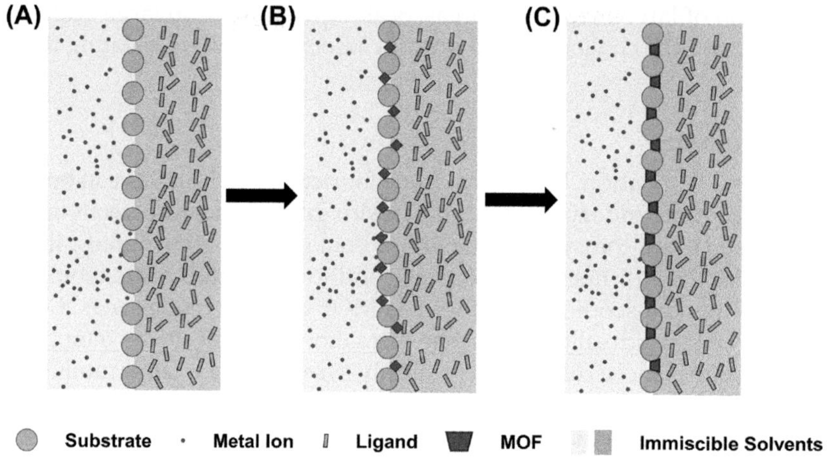

FIGURE 3.12 The process of in situ growth of a metal-organic framework (MOF) membrane via interfacial microfluidic processing: (A) the substrate is inserted at the interface of the immiscible metal solution and ligand solution; (B) MOF crystals grow at the interface; and (C) the resulting MOF membrane prohibits further assembly of MOFs.

between the two reagents. The next generation of the counterdiffusion method was the interfacial microfluidic processing developed by Nair and coworkers [219]. By dissolving metal salts and ligands in immiscible solvents, the reaction region of the MOF can be precisely limited to the interface of the two liquids (Fig. 3.12). In their work, an 8.8-μm defect-free ZIF-8 layer was generated on the inner surface of hollow fiber support. The resulting ZIF-8 membrane exhibited excellent H_2/C_3H_8 and C_3H_6/C_3H_8 separation performance with separation factors as high as 370 and 12, respectively. The synthesis of MOF selective layers on the inner surface of hollow fiber membranes can be directly conducted within a membrane module, which avoids additional equipment cost associated with crystal synthesis. As the synthesis is carried out in individual hollow fibers, this interfacial microfluidic processing technique can be theoretically performed at any scale. However, the existing approach requires repeated and long exposures to the organic solutions containing the MOF ligands and has not yet been scaled beyond a few fibers within a module. Overall, this approach has the potential to enable scalable MOF membranes, but additional research is needed to improve the yields, reduce the synthesis times, and drive down the costs. Besides counterdiffusion methods and related derivatives, a layer-by-layer assembly (also called step-by-step deposition) is another promising approach to produce thin MOF membranes [220,221]. By exposing the substrate sequentially to a metal salt solution and a linker solution, ultrathin MOF membrane can be generated (Fig. 3.13). Eddaoudi et al. applied layer-by-layer assembly to fabricate a ZIF-8 membrane with a thickness of approximately 0.5–1 μm. Kharul and coworkers also utilized a similar procedure to produce CuBTC and ZIF-8 membranes on polysulfone-based

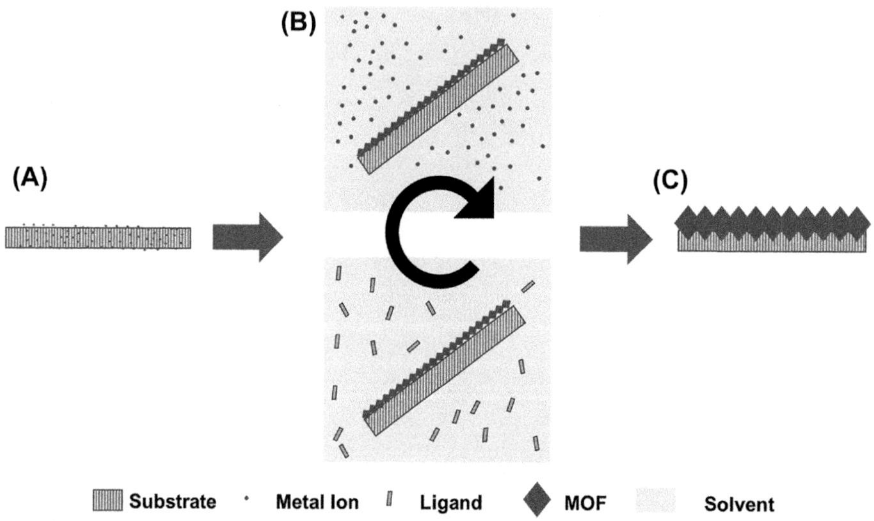

FIGURE 3.13 The process of in situ growth of a metal-organic framework (MOF) membrane via layer-by-layer assembly: (A) the substrate is soaked with metal ions; (B) the substrate is exposed to ligands and metals reciprocally; and (C) an MOF membrane is assembled layer by layer.

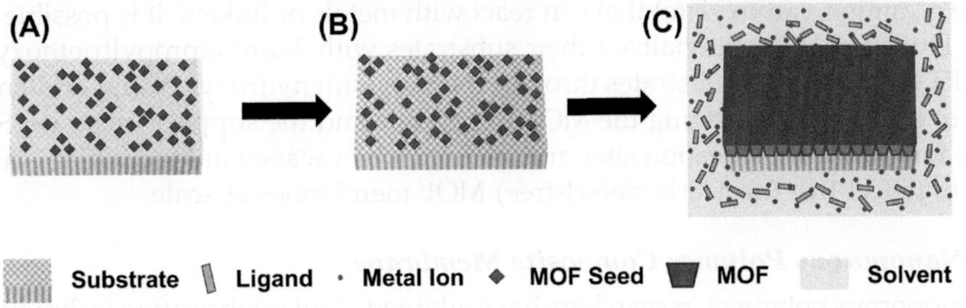

FIGURE 3.14 The process of secondary growth of a metal-organic framework (MOF) membrane via dip coating: (A) MOF nanocrystals are deposited onto the substrate via dip coating; (B) the seeded substrate is ready for secondary growth after drying; and (C) the MOF crystals grow into an MOF membrane.

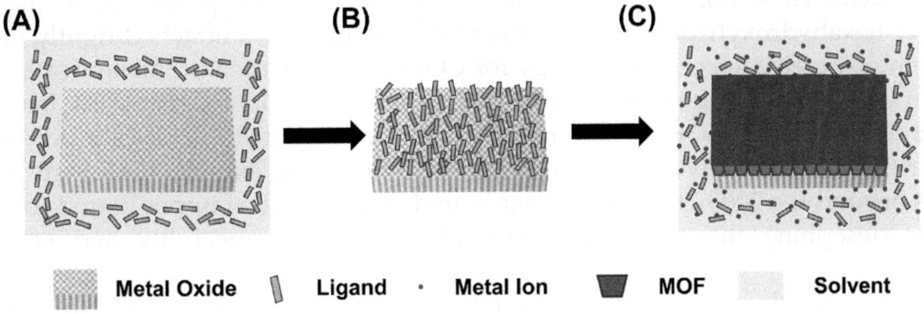

FIGURE 3.15 The process of secondary growth of a metal-organic framework (MOF) membrane via reactive seeding: (A) the metal oxide substrate functionalized with ligands; (B) the substrate modified by ligands facilitates MOF nucleation; and (C) an MOF membrane synthesized on the substrate in the synthesis solution.

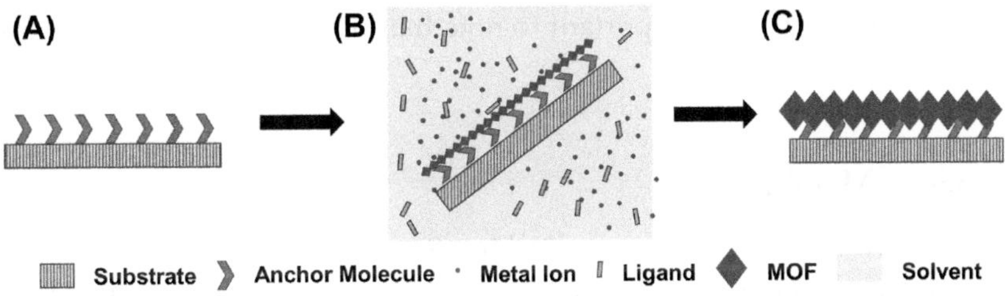

FIGURE 3.16 The process of secondary growth of a metal-organic framework (MOF) membrane on a substrate functionalized with anchor molecules: (A) the substrate functionalized with anchor molecules (e.g., APTES); (B) secondary growth of MOF membrane on the functionalized substrate; and (C) an MOF membrane synthesized on the substrate.

supports [222]. Layer-by-layer assembly is a promising strategy to control the thickness of MOF membranes; however, the cyclic exposure to different solutions is labor-intensive and difficult to scale up.

By seeding supports with different nucleation sites, secondary growth can easily prompt growth of MOF on substrates. Typically, nanosized MOF crystals are coated onto substrates via various approaches, such as dip coating (Fig. 3.14). Formation of a uniform seeding layer is critical to the secondary growth. In 2009, Tsapatsis et al. prepared copper-based MOF membranes on an alumina substrate [223]. By manually assembling monolayers of MOFs onto the polyethylenimine (PEI)-coated alumina substrate, a preferentially oriented MOF crystal layer was prepared. Carreon et al. seeded dry ZIF-8 crystals onto tubular alumina support by rubbing [224]. Zhang et al. seeded colloidal NH_2-MIL-53(Al) onto a glass frit disc [225]. Although these seeding approaches successfully produce high-performance membranes in the lab, such operator-dependent techniques are not easy to scale up for industrial application, and thus deserve additional attention. One alternative secondary growth method is reactive seeding (Fig. 3.15) [226]. Jin et al. modified alumina disk supports with 1,4-benzenedicarboxylic acid for nucleation of MIL-53. The resulting MIL-53 membranes were utilized for hydrogen separation. Jeong et al. modified alumina support with imidazolate ligands at 200°C. These anchored ligands bond with zinc ions and facilitated nucleation of ZIF-8. Zhang et al. grew vertically aligned ZnO nanorods on the alumina support and then activated these ZnO rods with imidazolate for nucleation of ZIF-8 [227]. Besides modifying substrates with ligands, it is also feasible to use another molecule to bridge or attach MOFs and substrates (Fig. 3.16). By grafting substrates with specific functional groups (e.g., amine, carboxylate) that can react with metals or linkers, it is possible to facilitate the nucleation process [228]. Huang et al. functionalized their substrates with 3-aminopropyltriethoxysilane (APTES) [229]. Ethoxy groups of APTES bond to the substrates through reaction with hydroxyl groups and amino groups of APTES connect the linkers or metals, thus bridging the MOF membrane and the support substrate. Such substrate modification facilitates the formation of nucleation sites and prevents the vacancy in the MOF selective layer, which helps improve the yields of defect-free (or nearly defect-free) MOF membranes at scale.

3.3.3 *Crystalline Nanoporous Polymer Composite Membrane*

For crystalline nanoporous polymers, researchers have adapted similar fabrication techniques utilized for zeolites and MOFs.

COF-5 membranes supported by alumina ceramic supports were firstly fabricated via in situ growth by Hao et al. [230]. In their approach, α-Al_2O_3 supports were functionalized sequentially by 3-aminopropytriethoxysilane and formylphenylboronic acid. These boronic acid groups anchored on the supports react with 1,4-benzenediboronic acid and 2,3,6,7,10,11-hexahydroxytriphenylene, forming a layer of COF-5 attached tightly onto the alumina support. Their work demonstrated a successful strategy for COF composite membrane fabrication. However, the large 2.7 nm mesopores of COF-5 have limited its potential in the area of molecular separations. Instead of using COF as a selective layer, Fu et al. fabricated a COF-MOF composite membrane where the COF-300 (pore size is 0.72 nm) serves as a gutter layer for the thin MOF layer [231]. The resulting membrane exhibits both high permeability and high selectivity and goes beyond the 2008 "Robeson upper bound" (an empirical upper bound for gas separation membrane performance, describing the state-of-art trade-off between permeability and selectivity) for H_2/CO_2 separations.

Besides in situ growth of COFs, Li et al. also adapted the strategy of the assembly of exfoliated COF 2D nanosheets [232]. In their approach, COF-1 particles were firstly synthesized through typical solvothermal condensation. The resulting particles were then exfoliated into flat-sheet morphology via sonication. The exfoliated COF-1 nanosheets were then coated onto alumina support with a SiO_2—ZrO_2 intermediate layer. Although the demonstrative membrane in their preliminary work did not show a combination of high permeability and high selectivity, it exhibits excellent thermal stability until 400°C. It is important to note that many MOFs cannot operate under temperatures higher than 200°C [36].

3.4 Thin Film Composite Membrane

3.4.1 *Nanoporous Polymeric Thin Film Composite Membrane*

Thin film composite polymeric membranes consist of an ultrathin (~100 nm) selective layer, a highly porous support layer, and, in some cases, an intermediate layer for better adhesion or less infiltration into support layer. Compared with the aforementioned integrally skinned asymmetric membrane, thin film composite significantly reduced the amount of expensive nanoporous materials used per unit area of membranes [233].

Polymeric thin film composite membranes can be formed via interfacial polymerization [233]. A porous substrate (e.g., sintered stainless steel, polymer support, etc.) is sequentially soaked in different monomer solutions. By selecting immiscible solvents, two monomers can only react and connect with each other at the interfaces of the two solutions and thus form ultrathin polymer skin layers. For instance, Livingston and coworkers fabricated polyarylate (PAR) (aromatic polyester) nanofilms via interfacial polymerization of trimesoyl chloride (TMC) and contorted phenols (e.g., 5,5′,6,6′-tetrahydroxy-3,3,3′,3′-tetramethylspirobisindane and cardo-structured 9,9-bis(4-hydroxyphenyl) fluorene) [234]. By introducing contortion sites, the resulting nanofilm possessed intrinsic microporosity and exhibited good organic solvent nanofiltration (OSN) performance. In particular, the permeance of THF in PAR/PI membranes made from contorted monomers is as high as 4.0 $L\,m^{-2}\,h^{-1}\,bar^{-1}$, which is 100 times higher than the membranes from noncontorted monomers with the similar dye rejection. McCutcheon and coworkers developed a printing strategy to produce thin polyamide film on PAN substrates in a scalable manner [235]. In their report, the best polyamide composite membranes exhibit salt rejection of 94% and water permeance of 14.7 $L\,m^{-2}\,h^{-1}\,bar^{-1}$. Owing to the fast reaction kinetics, interfacial polymerization of nanoporous polyamide can be easily adapted by large-scale industry application. The combination of the techniques described by Livingston and McCutcheon could potentially result in the formation of low-cost, high-performance thin film composite membranes. Additional work is needed in this area to tweak the molecular weight cutoff for different organic solvent separation situations.

Beside interfacial polymerization, thin film composite membrane can also be fabricated via coating of solution processable nanoporous polymers, such as PIM-1. Typical roll-to-roll dip coating processes for flat membranes

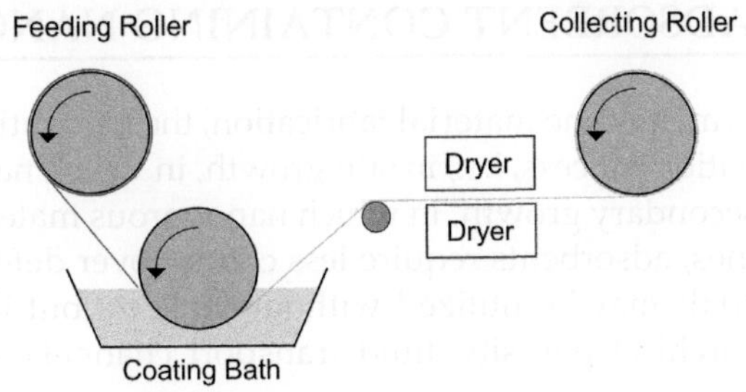

FIGURE 3.17 Typical roll-to-roll dip coating process for a flat membrane.

are illustrated in Fig. 3.17. Preprepared support membranes are continuously drawn from the feed roller to the collection roller at a certain drawing speed. During this process, the support membrane is immersed in a polymer coating solution bath and sequentially passes through several coagulations baths and dryers. Fritsch and coworkers have coated PAN supports with PIM-1 and infused the membrane with PEI [236]. The resulting thin composite membranes possess selective layers of around 100 nm thickness and exhibit steeper polystyrene retention curves compared with state-of-art OSN membranes. Livingston and coworkers roll-to-roll dip-coated PAN and cross-linked Ultem 1000 support membranes with PIM-1, PIM-7, and PIM-8 to fabricate scalable, thin film composite membranes for OSN [237]. Owing to the solution processability, PIMs can be successfully fabricated into large-scale membranes via traditional membrane processing techniques.

3.4.2 Mixed Matrix Thin Film Composite Membrane

One strategy to fabricate mixed matrix thin film composite membrane is interfacial polymerization on a support membrane predecorated with nanoporous particles. Such thin film nanocomposite membranes (TFN) were initially reported by Hoek and coworkers as illustrated in Fig. 3.18 [238]. In their approach, polysulfone support membranes were soaked with aqueous m-phenylenediamine (MPD) solution. The wet supports were then transferred into a dispersion of NaY zeolite nanoparticles in TMC/hexane solution. MPD and TMC rapidly react at the interface of hexane and water, forming a zeolite/polyamide mixed matrix skin layer. The hydrophilic NaY zeolite nanoparticles embedded within the polyamide provide rapid mass transfer channels for water, and the resulting TFN membrane exhibits excellent reverse osmosis (RO) performance (i.e., NaCl rejection up to 93.9% with a water permeability of $3.8 \times 10^{12} \frac{\text{m}}{\text{Pa S}}$). Such membrane fabrication strategy can also be applied to other nanoporous crystals. For example, research groups of Coronas and Livingston have demonstrated the fabrication of TFN membranes containing ZIF-8 and MIL-53 derivatives, which exhibits superior performance for OSN [239].

By dispersing nanoporous crystals into polymer solutions, mixed matrix thin film composite membranes can also be fabricated via classical solution-processing techniques (e.g., blade casting and dip coating). For instance, Lai and coworkers dispersed ZIF-7 nanoparticles in Pebax 1657. The suspension is then coated onto the intermediate layer of polytrimethylsilylpropyne on a PAN porous substrate. The resulting membrane exhibits enhanced CO_2 selectivity and permeance while reducing the material cost significantly [240].

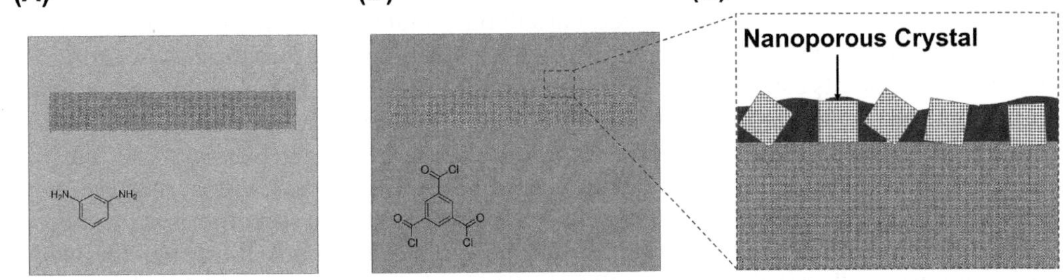

FIGURE 3.18 Fabrication of thin film nanocomposite membranes (TFN) via interfacial polymerization. (A) A porous substrate is firstly immersed into one polymer precursor solution; (B) the wet substrate is then transferred into the other polymer precursor solution, in which nanoparticles of nanoporous crystal are suspended; and (C) the polymer film synthesized at the interface cages nanoporous particles, forming thin mixed matrix membrane.

4. FABRICATION OF ADSORBENT CONTAINING NANOPOROUS MATERIALS

Similar to the case of nanoporous membrane material fabrication, the fabrication of nanoporous adsorbents can be categorized according to their formation process, i.e., in situ growth, in which nanoporous materials are synthesized directly in reactant solutions, and secondary growth, in which nanoporous materials are grown on seeds (e.g., nanocrystals). Compared with membranes, adsorbents require less control over defects (e.g., cracks or holes larger than the pores of the nanoporous materials may be utilized without concern) but instead focus more on the structure of the adsorbent device (e.g., hierarchical porosity, fluid transport channels, etc.) to enhance mass transfer and minimize pressure drop [241–249].

4.1 Scale-Up of Nanoporous Powders

Scale-up of nanoporous materials is an essential step in their utilization in industrial adsorption applications. Mainstream adsorption units are comprised of packed beds of nanoporous pellets owing to the simplicity and well-understood transfer of heat and mass in such systems. Large-scale synthesis of zeolites and MOFs has been intensively studied in both industry and academia.

Synthetic zeolites are commonly prepared via hydrothermal routes. Amorphous reactants (e.g., SiO_2 and Al_2O_3), structure-directing agents, cation sources, and alkaline medium are sealed in a high-pressure vessel, which is usually heated to 100–200°C. Synthesis parameters, such as temperature, reactants, and structure-directing agents, could significantly influence the product properties [250]. Zeolites fabricated using neutral media (e.g., fluoride-mediated routes) have been more difficult to scale-up due to the complexities of reactor corrosion at large scales.

Various new techniques have also been applied to modern zeolite industries. For example, microwave zeolite synthesis was firstly reported in 1988 [251]. Compared with traditional heating methods, microwave provides much higher heating rate and rapid synthesis of zeolites with high crystallinity [250]. There are also reports that the high heating rate generated by microwaves could influence the formation of different phases high selectivity [252,253].

Hydrothermal synthesis utilizes a large quantity of water, which results in not only environmental pollution but also an undesired loss of reactant in wastewater [254]. Therefore, synthesis routes of zeolite using less or no solvent have been developed. For example, dry gel conversion and dry synthesis have been successfully developed to produce zeolites without the addition of water [255,256]. Xiao and coworkers report the generalized mechanochemical synthesis of zeolites in the presence of NH_4F using anhydrous reactants [254]. Various zeolites (e.g., MFI, BEA, EUO, and TON) have been successfully synthesized with enhanced yields.

Coal power plants exhaust a huge amount of fly ash, resulting in severe environmental pollution. Coal fly ash has a high content of silica and alumina, which makes it a suitable reactant source for zeolite synthesis [257–259]. Many zeolites (e.g., FAU, GIS, MFI, PHI, etc.) have been synthesized via hydrothermal alkaline conversion of fly ash [259–263]. Classical hydrothermal alkaline conversion combines NaOH or KOH solutions with fly ash under varying temperatures and pressures to synthesize zeolites. Such a synthesis strategy has also been applied at pilot scale [264]. Although the classic method usually yields low zeolite content (40%–75%), Hollman and coworkers also developed a two-step process to produce high purity zeolites [259,265].

Since the first report of MOFs in 1995, academia has witnessed tremendous progress in the development of MOFs [266,267]. Although a large variety of them exhibited superior performance, only a few have been produced in quantities necessary for the large-scale application. Issues such as instability in water and heat and expensive precursors have hindered the widespread scale-up of these materials.

In laboratory settings, most MOFs are synthesized via solvothermal routes, which dissolve all reagents in solvents and conduct the reaction under controlled temperatures. With the help of a pressurized reactor, elevated temperatures above solvent boiling points can be applied to promote solubilization of reagents or growth of crystal samples. The solvothermal method is the dominating synthesis technique for lab-scale MOF production owing to its simplicity and versatility; companies (e.g., BASF) have also commercialized some MOFs via the solvothermal method in the early days. While solvothermal synthesis is a well-understood traditional reaction technique in industry, large-scale solvothermal synthesis of MOFs is not easy. During traditional solvothermal synthesis, heat is usually transferred to the bulk solution from the hot reactor wall, which provides nucleation sites for MOF particles. With the same reactor geometry, wall area decreases as the reaction scales up. Besides, efficient heat transfer and ease of posttreatments (e.g., purification, activation, etc.) should also be considered [268].

To solve the challenges of large-scale MOF synthesis, novel synthesis routes such as microwave synthesis, spray-drying (SD) synthesis, and flow chemistry have been developed [266].

Microwave synthesis provides reaction energy via electromagnetic waves [266,269]. The interaction between electromagnetic waves with molecules containing mobile electric charges (e.g., polar molecules) generates heats for MOF synthesis. Different from solvothermal synthesis with hot reactor walls, highly efficient microwave interacts directly with reactants and creates molecular hot spots, in other words, nucleation sites, throughout the reactant solution. Instead of nucleates on the reactor walls, MOF particles nucleate in the bulk solution during microwave synthesis. As a result, microwave synthesis achieves a higher reaction rate and smaller particle sizes. However, the development of microwaves capable of producing industrial scales of MOF powders is a remaining barrier for this technique.

A SD process produces dispersed powders from a liquid or slurry ejected from nozzles. These liquid droplets undergo rapid evaporation during travel through hot gases. This process was first designed for the drying process (i.e., milk powder); however, the formation of submicrometer droplets and rapid heat transfer during SD are ideal for MOF synthesis [270,271]. Maspoch and coworkers did the pioneering work to synthesize MOFs via the SD process [271]. In their approach, atomized microdroplets of reactant solutions were formed by specialized nozzles. Simultaneously, hot compressed gas (e.g., nitrogen) was ejected into and heated the spray of microdroplets, thus promoting the formation of MOF nanoparticles. During solvent evaporation, these nanoparticles further agglomerate into a compact or hollow MOF beads.

Flow chemistry is a continuous synthesis technology, where chemical reactions are conducted in streams flowing through tubes [266]. Compared with the classical solvothermal batch reactor, flow reactors made of tubes have a much higher surface area to volume ratios. Benefiting from this high surface area ratio, flow reactors exhibit high efficiency in heat transfer and nucleation operations, thus resulting in rapid synthesis. By installing multiple flow reactors, the production rate of the synthesis units can be scaled up arbitrarily with the same surface area per volume and nucleation rates in each reaction column. Moreover, flow reactors enable versatile engineering of fluid delivery, precise online control and monitoring over the reaction, and integrated automatic postsynthesis treatment. Continuous reaction effectively eliminated product quality variation between different batches. The first flow chemistry MOF synthesis was reported in 2011 [272]. Ameloot and coworkers synthesized HKUST-1 particles utilizing in a microscale microfluidics reactor [272]. Other MOFs, such as ZIF-8, UIO-66, MIL-53, etc., have also been synthesized via flow chemistry [272–281].

4.2 Monolithic Adsorbents

A monolithic structure is one of the most popular mass-transfer contactor designs. It has been widely adapted in various applications, such as three-way catalytic converters and adsorption columns [244,245,282]. Monolithic structures possess macropores and thin honeycomb-like channels, which significantly reduce pressure drop across the adsorber while providing high surface areas for mass transfer with processing fluids (e.g., gas mixtures). State-of-art monoliths with high mechanical strength minimize the influence of fluid-induced abrasion, which can be a severe issue in pellet-based packed beds [241,245]. By tuning porosity, channel sizes, and material types, monolithic adsorbents can be optimized for different operation conditions [249]. Generally, monolithic adsorbents can be fabricated via two approaches: (i) direct fabrication, in which pastes containing in situ synthesized adsorbents are extruded through a nozzle with specific opening, and (ii) postfabrication synthesis, in which secondary growth of adsorbent materials is conducted on seeded monolithic substrates [283].

Direct fabrication of monolithic adsorbents involves the extrusion of a paste-like mixture of nanoporous particles, binders, additives, and solvents into a specific monolithic structure. Posttreatments like drying are also required for the enhanced mechanical strength of the adsorbents. For example, Kaskel et al. fabricated HKUST-1 monolith by extruding HKUST-1 particles with silicone resin as binder and methyl hydroxyl propyl cellulose as the plasticizer [245]. In their approach, the partly cross-linked liquid silicone was chosen on purpose to create a viscoelastic suspension. The following microwave-assisted heating fully cross-linked the silicone resin and generated a robust monolith. Besides HKUST-1, a variety of MOFs monoliths, such as MIL-101(Cr), have also been fabricated through direct extrusion method [245,284–289].

Monolith fabrication by direct extrusion typically results in higher adsorbent loading than those fabricated via posttreatment methods but requires suitable binders, additives, and posttreatments [283]. By contrast, secondary growth utilizes prefabricated or commercially available ceramic monoliths and requires less modification to existing synthesis procedures. Ceramic monolith with high mechanical strength and high stability can be fabricated via extrusion of cordierite, water, and agglomerating agents [290]. For instance, Gascon et al. immobilized MIL-101(Cr) onto a commercially available cordierite monolith [291]. Their approach consists of three steps: first, the monolith is activated by NaOH and α-alumina particles; second, the resulting alumina-coated monolith is dip coated with MIL-101(Cr) seeds; and third, secondary growth of uniform MOF layer is achieved by subjecting seeded monolith to synthesis solution under rotation.

4.3 Fiber Adsorbents

Beside the monolith structure, the fiber adsorbent is another promising mass transfer contactor design featuring easy fabrication, low pressure drop, high surface area to volume ratio, and heat integration [292]. A typical fiber adsorbent consists of macroporous polymeric matrix and adsorbent particles distributed throughout the wall of the fiber. Fabrication of fiber adsorbents is mainly accomplished by dry-wet fiber spinning, which is widely applied in the fabrication of hollow fiber membranes.

Polymer, adsorbent particles (or seeding materials), solvent, nonsolvent, and other additives are first mixed. The resulting mixture, which is also referred to as "dope" in literature, is then extruded through a spinneret as shown in Fig. 3.7. Drawn by a rolling drum, the extruded viscoelastic filaments go through an air gap into a nonsolvent bath (typically water). Complicated mass transfer during this process triggers the phase inversion of the polymer, which turns into a hierarchical porous matrix that supports adsorbent particles [293,294]. After complete solvent exchange and drying, the composite fibers are ready for use (in situ synthesis) or further treatments (secondary growth).

Although fiber adsorbent spinning is inspired by hollow fiber membrane spinning, they have different fabrication strategies due to their distinct requirements for target morphologies. First, fiber adsorbent spinning utilizes inhomogeneous suspensions of solid particles instead of a homogenous polymer solution for membrane spinning. A primary dope containing solid particles and a small amount of polymer is first produced to ensure uniform distribution of particles within the mixture, and then the remaining materials are subsequently added. Repeated sonication, stirring, and shear mixing are also critical for the distribution of the particles. The resulting dopes for fiber adsorbent spinning should be handled as soon as possible to avoid sedimentation, while dopes for membrane spinning undergo additional treatments such as degassing. Second, fiber adsorbents should be free of skin layers that significantly retard mass transfer. By contrast, the skin layer is the most critical feature for separation membranes. Therefore, dopes for adsorbent fibers seldom contain volatile solvents and the air gap for adsorbent fiber spinning is minimized to avoid vitrification at the fiber surface.

Depending on the formation methods of adsorbent particles, fabrication of fiber adsorbents can be categorized into in situ synthesis and secondary growth. Straightforward in situ synthesis is applicable for adsorbents that remain active after immersion in solvents (e.g., water), sonication, and posttreatment (e.g., drying, oxidation in ambient condition). Until now, fiber adsorbents containing zeolite LiX, NaX, NaY, and UIO-66 have been reported with adsorbent loadings as high as 75 wt% [294–297]. Secondary growth after fabrication of porous polymer fibers is utilized to process nanoporous materials incompatible with the dry-wet spinning process. For instance, Pimentel et al. converted cellulose acetate/ZnO fibers into HKUST-1 fibers with 85 wt% loading of the MOF [298]. This type of secondary growth strategy was firstly described by Zhao et al. [299]. In both cases, the ZnO was transferred into a hydroxyl double salt (HDS) intermediate containing Zn and Cu, which was then rapidly converted into HKUST-1. Such HDS intermediate approach can also be used to synthesize other MOFs, such as ZIF-8 and IRMOF-3. The secondary growth process can be conducted inside the adsorber shell and isolated from moisture, which is a promising fabrication strategy for water-sensitive nanoporous materials.

4.4 Additive Manufactured Adsorbents

Additive manufacturing, also called 3D printing, is capable of fabricating complex structures for adsorbents. Such a technique enables precise control of the mass transfer throughout the adsorber. Existing preliminary approaches mainly process pastes containing adsorbent particles, solvent, binders, and additives. The viscoelastic mixture is then extruded through a micronozzle onto a substrate. The monolith structure is achieved by the relative movement between the substrate and the nozzle. The 3D printing ink preparation strategy is the same as that for direct fabrication of monolithic adsorbents. One of the first 3D-printed monoliths fabricated by Thakkar et al. is made of aminosilica [300]. In their approach, bentonite clay was used as a binder and methyl cellulose was used as an additive for proper rheology property. Besides this material, zeolite 13X, SAPO-34, ZSM-5, MOF-74 (Ni), and UTSA-16 (Co) have also been 3D printed into highly efficient adsorbents [301–304].

Compared with the 3D printing of pasty ink consisting nanoporous crystals, 3D printing of nanoporous polymers is more challenging from a rheological point of view. Typical binary polymer solutions (e.g., polymer and solvent) tend to undergo gravity-driven deformation before evaporation-induced solidification. Lively and coworkers demonstrated solution-based direct-ink-write 3D printing of PIM-1 [305]. In their approach, ternary inks consisting of PIM-1, THF (solvent), and dimethylacetamide (nonsolvent) were used. Evaporation of THF triggers rapid phase inversion resulting in a hierarchically porous structure without deformation. The printed prototypical adsorber exhibits superior performance in toluene vapor breakthrough compared with packed beds and fiber adsorbents.

Additive manufacturing, still a burgeoning technology, has shown its potential in the production of customized configuration for different adsorptive applications. It has been shown in the aforementioned reports that 3D printing enables precise design of individual fluid channels thus eliminating fluid bypasses and improving breakthrough behavior. For instance, PIM-1 adsorber fabricated by solution-based 3D printing possesses a well-defined fluid distribution system, which successfully eliminates the bypasses presented in traditional packed beds [305]. The 3D-printed miniature contactors (1.27 cm) can fully purify (<1 ppm) toluene vapor (10,000 ppm) in N_2 gas for 1.7 h, which is six times longer than PIM-1 in traditional structures, and more than 4000 times the residence time of gas (1.5 s) in the contactor. Existing research has shown the potential of 3D printing in mass transfer applications; however, more efforts should be made to investigate the relationship between breakthrough behavior and fluid distribution design. As additive manufacturing can produce complex structure without significantly increased cost, one should think beyond traditional monolith-type structures.

5. APPLICATION OF NANOPOROUS MATERIALS IN ENERGY-EFFICIENT SEPARATIONS

Although the majority of this chapter focuses on fabrication methods for nanoporous materials, it is instructive to provide some illustrative examples of these nanoporous materials in action in the area of molecular separations. Two broad classes will be briefly discussed: membrane separations and adsorptive separations. Beyond these illustrative examples, the reader is referred to the following detailed reviews in the areas of membrane separations [306–311] and adsorptive separations [30,35,36].

5.1 Membrane Separations

The transport properties of guest molecules through the nanoporous membranes can be tuned via the selection of suitable nanoporous materials and fabrication conditions. As a result, nanoporous membranes have become more prominent in the areas of gas separations, organic solvents separations, and desalination.

5.1.1 Gas Separation

Membranes separate gas molecules based on their molecular size, shape, and affinity to the membrane material [309,312]. Exemplary gas separations include H_2 purification, CO_2 capture, hydrocarbon separation, among others.

5.1.1.1 H_2 Purification

Hydrogen is an important precursor and fuel in industry and is often associated with various chemical processes [36]. From the perspective of sustainable development, hydrogen is also an efficient and clean energy source. Large-scale hydrogen production (e.g., water-gas shift, steam-methane reforming, etc.) and recovery (e.g., ammonia reactor purge gases, refinery streams) require efficient hydrogen separation techniques. Currently, H_2 can be purified through one (or a combination) of three major processes: (i) PSA, (ii) cryogenic distillation, and (iii) membrane separation. Although the previous two are commercially available, they are expensive and energy-intensive [307], primarily due to the need to cool the synthesis gas down to the separation conditions. Owing to the small kinetic diameter, hydrogen molecules can be effectively separated from other gases via molecular sieving effect. Nanoporous membranes used for hydrogen separation usually possess preferential hydrogen permeation.

Due to their thermal stability, various zeolites (e.g., MFI, FAU) have been studied for hydrogen separation in the past few decades, but real applications of these materials have been limited due to economic feasibility and difficulty of defect control [313]. For membranes made of as-prepared zeolites, NaA zeolite membranes with small nanopores (0.41 nm) have been reported to have the highest H_2/N_2 separation factor (~24) [314]. This somewhat low selectivity has been attributed to the intercrystalline spaces within the zeolite layers. One alternative strategy to improve the selectivity is postsynthesis silylation, which can reduce the pore size of the as-prepared zeolites [315,316]. For instance, Masuda and coworkers reduced the pore size of MFI via postsynthesis catalytic cracking of silane [315]. In their approach, silane compounds are infiltrated into the nanopores of zeolites, catalytically cracked into coke on active sites, and finally calcined into SiO_2 units. After catalytic cracking of silane, the H_2/N_2 separation factor of the MFI membranes increased from 1.4–4.5 to 90–140.

Hydrogen separation is also one of the most popular separations for research on MOF membranes [36]. The great potential of various MOFs (e.g., HKUST-1, MOF-5, MMOF, ZIF, etc.) has been demonstrated for hydrogen separations due to their tunable aperture size and sorption selectivities [215,223,317–319]. To efficiently separate H_2 from other industrial gases, MOFs with small aperture sizes are desired. Huang et al. created an imine-functionalized ZIF-90 membrane and removed defects using an ethanolamine posttreatment, which effectively enhanced the separation ability of H_2 from CO_2 and other gases (stable H_2/CO_2 mixture selectivity of 15 with H_2 permeance of 2 mol $m^{-2}\ s^{-1}\ Pa^{-1}$) [320].

Some of the outstanding challenges in the area of membrane-based hydrogen purification are (i) long-term stability under operating pressure and temperature, (ii) mixed gas selectivity, and (iii) resistance to contaminants (e.g., H_2S).

Industrial-scale hydrogen separation processes usually operate at high pressure (e.g., 120 bar for ammonia purge gases) and high temperature (e.g., 350°C for high-temperature water-gas shift reaction) to provide permeation driving force, accelerate mass transfer, and avoid organic vapor condensation [307,309]. Therefore, development of a robust membrane module and nanoporous materials with high thermal stability is required. High operation temperature also requires the development of thermally stable sealing materials or module design free of sealing materials. For instance, in the case of thermally stable inorganic nanoporous materials that can be regenerated under oxidizing high-temperature treatment, the development of a fully ceramic module would provide a path forward for long-term usage of the membrane [313].

Ideally, the separation performance can be predicted based on single-gas permeance. However, the real mixture gas selectivity can be significantly influenced by the presence of a plasticizer, organic contaminants, competitive adsorption, etc. To better benchmark the novel nanoporous membranes, mixture gas selectivity should be evaluated at operating pressure and temperature.

Contaminants in the hydrogen mixture could also influence the membranes. When exposed to gases containing organic vapors or strongly adsorbing vapors, nanoporous membranes can potentially exhibit performance loss [321]. For some nanoporous materials like HKUST-1, irreversible reactive adsorption of acidic gases (e.g., H_2S) could lead to permanent loss of porosity [322]. At the process level, these materials can be avoided via additional pretreatment, but the ideal membrane or sorbent material will be able to operate in these aggressive conditions without extensive pretreatment processes.

5.1.1.2 Light Hydrocarbon Separation

As crucial building blocks for petrochemicals and polymers, light olefins, such as C_2H_4 and C_3H_6, are in high demand. For instance, the global annual production of ethene and propene exceeds 200 million tonnes [323]. The separations of olefins from the corresponding paraffin pairs like C_3H_6/C_3H_8 or C_2H_4/C_2H_6 are required for subsequent use [35,36]. The industrial separation of light hydrocarbons typically relies on energy-intensive high-pressure cryogenic distillation. For example, 0.3% global energy is consumed in the purification of C_3H_6 and C_2H_4.

CMS membranes exhibit the most promising light hydrocarbon separation performance. CMS membranes with angstrom-level molecular discrimination properties have been extensively studied and have shown great potential for hydrocarbon separations [129]. Kim et al. prepared Matrimid 5218-derived CMS membranes on alumina hollow fibers and obtained a 69.2 GPU (GPU, gas permeance unit, is a measurement of gas permeance, $1\ \text{GPU} = 10^{-6}\frac{cm^3_{STP}}{cm^2\ s\ cm\ Hg} = 7.501 \times 10^{-12}\frac{m^3_{STP}}{m^2\ s\ Pa}$) C_3H_6 performance and 18.0 C_3H_6/C_3H_8 selectivity with a feed gas composition of 50:50 mol% C_3H_6/C_3H_8 at 2 bar and 25°C [324]. The Dow Chemical Company synthesized CMS membranes with a C_3H_6/C_3H_8 separation factor of approximately 27 by a facile pyrolysis process from a gel-type strong acid cation exchange resin [325]. Swaidan et al. utilized the PIM-6FDA-OH polyimide to fabricate CMS membranes at 600°C, which showed a pure-gas C_3H_6/C_3H_8 selectivity of 33 and a C_3H_6 permeability of 45 Barrer (Barrer is a non-SI unit of gas permeability named after Richard Barrer,1 Barrer $= 10^{-10}\frac{cm^3_{STP}\ cm}{cm^2\ s\ cm\ Hg}$) [326]. Chu et al. prepared Fe-containing CMS membranes by incorporating 1.1–3.2 wt% Fe ions in the 6FDA-DAM:DABA (3:2) polymer precursor before pyrolysis, achieving a C_2H_4/C_2H_6 permselectivity of 8.53 with ~100 Barrer C_2H_4 permeability, which is above the upper bound [327]. Rungta et al. made CMS membranes from commercial polyimide Matrimid with a C_2H_4 permeability of 14–15 Barrer and a high C_2H_4/C_2H_6 selectivity of ~12 [328]. Salinas et al. fabricated the CMS membranes from the precursor of PIM or intrinsically microporous PIM-6FDA-OH and obtained a pure-gas C_2H_4/C_2H_6 permselectivity of ~13 to 17.5 [329,330].

Similar to hydrogen separation membranes, membranes used for light hydrocarbon separation should be able to operate at high temperature and pressure to achieve comparable throughput with cryogenic distillation. The high operation temperature and transmembrane pressure require thermally stable nanoporous materials and robust membrane design. CMS membranes possess excellent thermal stability and can be fabricated into hollow fiber membrane with high-pressure resistance. However, large-scale CMS membrane production still requires the development of cost-effective and high-throughput furnace design.

5.1.1.3 CO_2 Capture

Carbon dioxide produced as a result of fossil fuel combustion is a primary cause of human-induced climate change. One method for reducing the rate of global climate change is to capture CO_2 from dilute sources such as power station flue gas (precombustion or postcombustion) or even from the air. In this regard, nanoporous materials have the potential to play an enormous role in addressing global carbon emissions, as many of the low-energy carbon capture techniques rely on this class of materials.

FAU zeolite membranes exhibit excellent performance in the CO_2/N_2 separation, exceeding the Robeson upper bound [331]. For instance, Morooka et al. produced polycrystalline Y-type zeolite membranes supported on alumina substrates, which achieved CO_2/N_2 selectivities ranging from 20 to 100. As for CO_2/CH_4 separation, several zeolite membranes (e.g., DDR, SAPO-34, etc.) exhibit a good balance between selectivity and CO_2 permeability [310]. For instance, Noble and coworkers prepared SAPO-34 membranes on alumina supports, which exhibit CO_2/CH_4 selectivity higher than 170 and CO_2 permeance of 2.0×10^{-6} mol m^{-2} S^{-1} Pa^{-1} [332].

Some MOF (e.g., SIM-1, ZIF-8, etc.) composite membranes have been successfully fabricated for CO_2 separation from industrial gases, typically methane [224,333,334]. For example, a ZIF-8/alumina membrane prepared via in situ crystallization exhibits a CO_2/CH_4 selectivity of 7 but a high CO_2 permeability of 2.4×10^{-5} mol m^{-2} S^{-1} Pa^{-1}, exceeding the Robeson upper bound [224]. A variety of MMMs using MOFs as fillers for CO_2 separation (e.g., 2D CuBDC/polyimide, ZIF-90/polyimide, etc.) were also reported [335–337]. However, most of them are limited by the trade-off between CO_2/CH_4 selectivity and CO_2 permeability, thus approaching the Robeson upper bound.

Moreover, there is extensive work focusing on the application of CMS membranes in the separation of CO_2 from CH_4 as it is industrially important especially for the purification and economical use of natural gas. Unlike the precursor polymeric membranes, carbon membrane will not suffer from plasticization induced by CO_2 molecules. Researchers in the Koros group have utilized different kinds of polymer precursors such as Matrimid 5218, 6FDA/BPDADAM, 6FDA/DETDA, and DABE polyimide to fabricate the CMS membrane, achieving a CO_2/CH_4 separation factors around 60 [338–342]. Tin et al. fabricated P84 polyimide–derived CMS membrane with an ideal selectivity of 89 for CO_2/CH_4 [343]. Fu et al. utilized the polyetherimide/polyimide blends as precursor to fabricate the CMS membranes and obtained a CO_2 permeance of 40 GPU and a CO_2/N_2 selectivity of 39 [344]. Owning to the coexistence of ultramicropores (excellent molecular sieving effect to gas molecules) and micropores (fast mass transfer pathways), CMS membranes can successfully overcome the Robeson upper bound.

Some of the remaining challenges in the area of membrane-based CO_2 capture are (i) stability in the operation condition (e.g., high temperature, contaminants, etc.), (ii) energy consumption, and (iii) equipment availability (e.g., compressors, pumps, etc.).

Membranes used for CO_2 capture from flue gases should have good stability against containments. For instance, flue gases from fossil fuel combustion contribute to 87% SO_2 and 67% NO_x emission in the United States [345]. These acidic containments may have deleterious effects on nanoporous materials or the seals in the membrane module. For instance, ZIF-8 undergoes degradation when exposed to humid SO_2 due to the acid attack on the coordination bonds. Nanoporous material with sufficiently strong bonds (e.g., COFs) or inorganic nanoporous materials (e.g., CMS) can potentially address these issues.

Energy consumption and equipment requirement should be carefully evaluated for CO_2 separation membranes. Most postcombustion CO_2 capture processes envisage the use of vacuum pumps on the permeate side of the membrane to supply the driving force for separation. Large-scale vacuum systems are difficult to maintain and often not capable of practically pulling below 0.2–0.3 bar. The most likely path for deploying these vacuum systems is to have smaller pumps supplying the vacuum for "banks" of modules rather than two to three massive pumps supplying vacuum to the entire CO_2 capture system. This modular vacuum approach increases capital and complexity, but addresses issues with downtimes and pump availability (e.g., massive vacuum pumps are likely not commercially available and would need to be custom-built). The final issue with postcombustion CO_2 capture via membranes is that most process schemes are "pressure ratio limited," which suggests that maximizing membrane permeance while attaining some nominal CO_2/N_2 selectivity (30 is often given as a target) is the route forward to drive down the energy costs.

5.1.2 Organic Solvents Separation

Organic solvent separation techniques are required in a variety of applications (e.g., biofuel refinery, pharmaceutical industry, petrochemical industry, etc.) and as a class are the most prevalent type of separations in industry [346–348]. Examples include alkane separations (e.g., n-heptane/isooctane), aromatic separations (e.g., p-xylene/o-xylene), and biofuel separations (e.g., glycerol/fatty acid) [311], among many others.

Within the broad context of nanoporous materials, zeolites possess the most rigid and well-defined pore structures, which is ideal for differentiating molecules of similar kinetic diameters. One successful commercial application of zeolite membranes is the dehydration of organic solvents via pervaporation. Due to the hydrophilic nature of the zeolite, LTA-type zeolites are suitable for dehydration processes. Tubular NaA membranes were first used in Japan to dehydrate solvent mixtures (EtOH, IPA, MtOH, etc.) via pervaporation at 120°C [349]. Some other dehydration plants were installed in China, Brazil, India, etc. [313,350,351]. Owing to the microporosity and oleophilic nature, various nanoporous polymers (e.g., PIMs, microporous PAR, etc.) have been used for organic solvent separation [234,352]. For example, thin film composites made of PIMs exhibits both high solvent permeance and solute rejection (e.g., dyes and polyester oligomers) [236,353]. Despite the good OSN performance, most polymeric membranes (e.g., PIMs, cross-linked polymers) are inherently unsuitable for challenging organic solvent separations (e.g., xylene isomers) due to their relative flexible backbones and bad pore size control [234,354–358]. Instead, separation of organic solvents of similar size (e.g., xylene isomers) can be achieved with nanoporous polymers with well-defined, rigid pores such as those in molecular organic cages [359].

Beyond organic cages, CMS membranes will not swell or plasticize significantly under organic solvent environment, which is a common problem with polymer membranes. This outstanding chemical and thermal stability of CMS membranes inspires researchers to utilize CMS membranes in organic solvent reverse osmosis (OSRO) separations. The Lively group reported the successful formation of free-standing CMS membranes with a diffusion selectivity up to 30 for para/ortho-xylene isomers by pyrolysis of cross-linked porous poly(vinylidene fluoride) hollow fibers [80].

Solvent-stable nanoporous membranes can be integrated with reactors to lower the E factor by effectively removing specific products. For instance, Szekely grafted polybenzimidazole membranes with the cinchona-squaramide bifunctional catalysts and designed a catalytic membrane cascade reactor with the solvent recovery capability [360]. Owing to the conformational changes induced by covalent grafting, enantioselective aza-Michael reaction of pyrazoles and triazoles was achieved with 99% enantiomeric excess. The integrated membrane separation unit exhibited at least 98% product recovery and 99% unreacted reactant recovery. Such membrane-based in situ solvent and reagent recovery is a promising route to reduce the E factor and carbon footprint of continuous-flow synthesis processes [361,362].

Membrane-based organic solvent separations possess the potential to debottleneck or even replace thermally driven processes. Several critical issues must be solved to fully realize the industrial application of membrane-based organic solvent separation: (i) pore size control of nanoporous materials, (ii) scalability of nanoporous membranes, and (iii) development of testing protocols and mass transfer models [311].

To differentiate organic solvent molecules, nanoporous materials should possess "midrange" micropores (6–15 Å). However, current research mostly focuses on nanoporous material with smaller micropores (smaller than 6 Å) for gas separation or nanoporous materials with larger pores (larger than 15 Å) for OSN. Development of templated nanoporous material with tunable porosity (e.g., mixed-linker MOFs, CMS, etc.) is of great interest. Moreover, it is important that nanoporous materials with "midrange" micropores maintain their structure under operation condition (e.g., organic solvent swelling, high pressure, etc.). For instance, some large-pore MOFs (e.g., ZIF-8) undergo structural transitions under a certain pressure, which should be considered during membrane design and operation.

Cost-effective large-scale fabrication of membrane devices plays an equally important role as the creation of new high-performance materials. Moreover, the rate of formation of defects must be studied and minimized for materials to be successfully scaled into membrane modules. The translation of materials into manufacturing processes should be considered during ideation and design.

Consistent and universal testing protocols for nanoporous membranes should be developed to benchmark the performance. Current testing protocols, such as molecular weight cutoff, quantify the rejection of chemically similar solutes of differing sizes. However, molecular weight cutoff highly depends on solute–solute, solute–solvent, solute–polymer, and solvent–polymer interactions, and these can all vary significantly [363]. Moreover, in the case of OSRO separations of isomers, it is not clear that the concept of a molecular weight cutoff is useful at all. It is essential to test the real selectivity of organic solvent mixtures to properly benchmark the organic solvent separation performance of novel nanoporous materials.

5.1.3 Desalination

Water desalination can be achieved by RO, electrodialysis, multieffect distillation, and multistage flash [364–367]. Owing to the high energy efficiency, RO is the most prevalent desalination technique today. Current RO membranes made of polymers (e.g., polyamides) are widely applied in commercial desalination projects but nanoporous materials can potentially exhibit better performance [365,368]. Theoretically, nanoporous membranes with pore sizes smaller than the sizes of the hydrated ions (e.g., hydrated sodium ion is around 0.76 nm) can exclusively transport water ions and achieve even higher efficiency desalination processes [365,368].

RO membranes made of zeolites (e.g., MFI, LTA) have been investigated for several decades and tremendous improvement has been made [369–372]. However, they have not outperformed the best commercial RO membranes. Recently, water-stable MOFs (e.g., ZIF-8, UiO-66, etc.) have also been applied to the fabrication of RO membranes [373–376]. For example. Liu et al. fabricated UIO-66 membranes supported on alumina hollow fibers which accomplished good ion rejections (e.g., 86.3% for Ca^{2+}, 98.0% for Mg^{2+}, and 99.3% for Al^{3+}) with a good water permeability (0.28 $Lm^{-2}\,h^{-1}\,bar^{-1}\,\mu m$) [376].

Nanoporous polymers (e.g., hypercrosslinked polymers, COFs, etc.) have also been developed and evaluated for desalination [377–379]. For instance, Moore et al. prepared membranes based on rigid star amphiphiles. Owing to the nanopores induced by insufficient packing of polymer chains, the resulting membranes exhibited a comparable ion rejection with commercial membranes and a doubled water permeance [378].

CMS membranes are another candidate for desalination. Kim et al. reported the preparation of free-standing carbonaceous membranes by the controlled carbonization of thin films of PIM; this membrane exhibits high water flux (13.30 LMH/Bar) and good $MgSO_3$ rejection (77.38%) as well as antifouling properties [380].

Although state-of-art nanoporous membranes are still less attractive than existing polymeric membranes in terms of performance and economics, nanoporous RO membranes with both exceptional performance and less investment should be made in the future. On the one hand, nanoporous materials should be translated into hollow fiber membranes or spiral wound membranes to withstand high transmembrane pressure. On the other hand, nanoporous materials should be fabricated into thin film composites to lower the overall price.

5.2 Adsorption Separations

5.2.1 Gas Separation

5.2.1.1 H_2 Separation

As mentioned in previous sections, hydrogen is an important energy source in the pursuit of a sustainable energy system. With advanced nanoporous materials, adsorption, especially PSA, a widely commercialized technique used in hydrogen separation, can be more cost-effective [381,382]. Owing to their tunable pores, various MOFs have exhibited potential for preferential adsorption of H_2. As H_2 is smaller than other gases (CO_2, CO, N_2, and O_2) involved in common H_2 generating processes (e.g., water gas shift [WGS], steam-methane reforming, ammonia synthesis, etc.), H_2 can be separated from most of them via molecular sieving effect [36,383,384]. For instance, H_2 can be separated from CH_4 using MOFs (e.g., $Mn(HCO_2)_2$, $[Ni(cyclam)]_2(mtb)$, etc.) with aperture size close to the kinetic diameter of H_2 (2.8 Å) [36,383,384]. Selective adsorption of H_2 over N_2 was also reported in several MOFs (e.g., PCN-13, $Mg_3(2,6\text{-ndc})_3$, MAF-26, etc.) [383–399]. Generally, preferential adsorption of H_2 achieved by MOFs over O_2, N_2, and CO is based on their kinetic diameter difference. However, owing to the high polarity and quadruple moment, CO_2 has stronger interactions with open metal sites in MOFs, which significantly enhances adsorption of CO_2 [36,400–402]. To achieve precise separation of H_2 out of a mixture containing CO_2, gate-type flexible MOFs (e.g., Zn(3,5-pydc) (DMA), $CoNa_2(1,3\text{-bdc})_2$, Mg-MOF-74, etc.) can be applied. Interactions between MOFs and CO_2 are strong enough to open the gates and allow further adsorption of CO_2 while H_2 is excluded in this process [400–402].

Compared with membrane-based hydrogen separation, nanoporous adsorbent materials are more tolerant to defects (although recent results suggest that defects may be initiation points for humid acid gas degradation [403]), which makes the large-scale fabrication of nanoporous adsorbent cost-effective. However, adsorption-based hydrogen separation still faces challenges like long-term stability under operation temperature, mechanical abrasion, poisoning by containments, etc.

5.2.1.2 Light Hydrocarbon Separation

Currently, the separation of light hydrocarbons (e.g., olefin and paraffin) is mainly achieved by energy-intensive cryogenic distillation [404]. However, alternative energy-efficient separation techniques using advanced nanoporous materials have also been investigated for several decades [35].

Owing to the versatile structure, aperture size, and functionality, MOFs can serve as attractive materials to separate light hydrocarbons via the method of adsorption. Separation of paraffin and olefin (e.g., C_3H_6/C_3H_8, C_2H_4/C_2H_6, etc.) has been tested with various MOFs (e.g., MIL-100, CuBTC, MOF-74, HKUST-7, etc.) [405–411]. Adsorptive separation of paraffin and olefin is usually achieved by interaction between hydrocarbons and open metal sites. Active metal sites in MOFs can selectively attract paraffin due to the π electron back donation [35]. For instance, Yoon et al. achieved C_3H_6/C_3H_8 separation (selectivity as high as 28.9) using partially reduced MIL-100(Fe) [412]. It was found that, with the presence of Fe(II), the adsorption heat of C_3H_6 increased from -30 kJ/mol to -70 kJ/mol while no influence to C_3H_8 was observed. Unsaturated Fe(III) is a Lewis acid site with moderate interactions with electron-donating guest molecules; for example, electrons from the filled π orbital or an electron lone pair of the unsaturated alkene donate into an empty *d* orbital of Fe(III). In contrast, unsaturated Fe(II) exhibits stronger interactions with electron donors through π back donation: delocalization of the additional *d* electron of the Fe(II) to the π^* antibonding orbitals of the unsaturated guest molecule. By exposing more open metal sites per surface area, adsorption preference of MOFs to alkene can be significantly enhanced [410]. Despite the common preference to alkene due to the interaction between open metal sites and alkene induced by π back donation, some MOFs (e.g., ZIF-7, MAF-49, etc.) also exhibit selective adsorption of alkane [413]. For instance, ZIF-7 prefers olefins due to the interaction between alkanes and benzimidazole linker [414,415].

Some of the remaining challenges for adsorptive separations of light hydrocarbons using nanoporous materials are the adsorbent stability and system design. Besides the long-term stability at operation temperature and resistance to mechanical abrasion, nanoporous materials with open metal sites, which are helpful in the separation of alkane and alkene, are susceptible to poisoning by sulfur species, acetylene, etc. Designing enormous adsorption systems with industry-scale throughput is not easy and requires intensive research in material production, contactor processing, process optimization, etc.

5.2.2 *Organic Solvent Separation*

Separation of organic solvents, for example, C8 alkyl aromatic isomers (o-xylene, m-xylene, p-xylene, and ethylbenzene), is challenging for separation techniques based on phase change due to the similarity of the chemical properties among them. By contrast, nanoporous materials with ordered pore structures and versatile functionalities are good candidates to differentiate molecules based on the difference in diffusivity, steric effects, interaction with adsorbents, etc. [35].

Zeolites (e.g., FAU, MFI, etc.) have been applied and tested for separation of C8 alkyl aromatic isomers [416,417]. Several MOFs, such as MIL-47, MIL-53, BIF-20, UIO-66, etc., have also been investigated in laboratories for organic molecules' separation [418–426]. Owing to the versatile structures of MOFs, different MOFs are believed to have different hypothesized separation mechanisms. For instance, the separation of xylene isomers achieved by MIL-47 is believed to be contributed by the differences in the packing modes of xylene isomers in nanopores, while Zn(bdc) $(dabco)_{0.5}$ accomplishes the separation based on the difference in interactions between xylene isomers and the adsorbents [420,426]. Owing to the relatively efficient packing of p-xylene, the selectivity of p-xylene over m-xylene in MIL-47 at 70°C is up to 2.07.

Besides the preferential adsorption, the difference in diffusivities of different molecules inside the same nanoporous adsorbents can also be used to separate hydrocarbon species. Well-defined nanopores can effectively differentiate isomers with sub-ångstrom diameter difference. For instance, BIF-20 displays high xylene uptake and p-xylene/o-xylene diffusion selectivity of 3 [419]. SMB chromatography is developed to separate molecules based on the difference in both the interaction with adsorbent and the diffusivity in nanopores. In traditional chromatography, mixtures can be eluted at different times but at the same position (outlets of the column). SMB systems have multiple inlets and outlets, the function of which alter at different operation times, so that different eluents can be collected continuously. Compared with fixed-bed systems, SMB is a steady countercurrent continuous process with higher separation performance. For instance, SMB systems utilizing MOFs (e.g., MIL-125-NH_2, MIL-140b, ZIF-8, and MOF-48) for aromatic C8 mixture separation have been developed [427].

Similar to gas separation, the remaining challenges of nanoporous materials for adsorptive separation of organic solvent are long-term stability at operation temperature, mechanical abrasion, poisoning by containments, etc.

5.2.3 *Pollutant Removal*

Low-concentration pollutant removal can be achieved by nanoporous materials with high surface areas and tunable pollutant affinity. For instance, Fryxell and coworkers grafted mesoporous silica with cationic metal via ethylenediamine-terminated silane. The resulting nanoporous adsorbents exhibit both high sorption rates (99% phosphate removal within 1min) and large phosphate capacities (up to 43.3 mg g^{-1}) [428]. Besides, biomass-derived nanoporous adsorbents have also shown great potential in pollutant removal. Hakkarainen and coworkers converted coffee-ground waste into quantum dots, which assist the aqueous fabrication of nanoporous graphite microfiltration membranes. Stabilized by the quantum dots, graphene nanosheets can be well-exfoliated and processed into membranes featuring fast water permeance ($\sim 1 \text{ L m}^{-2} \text{h}^{-1}$) and rapid sorption removal of metal ions and aromatic compounds [429].

Different from the separation of high-value chemicals where nanoporous materials with well-defined pores, engineered active sites, and control over defect concentrations are required, the requirements for nanoporous materials in the area of pollutant removal are environment-friendly syntheses, high surface areas, and suitable adsorption sites. The main challenge for nanoporous materials in pollutant removal is the cost. Especially for large-scale pollutant removal (e.g., oil spill removal from the ocean), a huge amount of adsorbent material is required and is often difficult to reuse. Therefore, cost-effective synthesis route of nanoporous materials should be developed, such as biomass conversion or waste material reuse.

6. FUTURE CHALLENGES

As discussed in previous sections, nanoporous materials exhibit improved performance (e.g., higher permeability and selectivity, thermal and chemical stability, etc.) compared with traditional separation materials in various applications (Fig. 3.19) [306,308,433,434]. However, the current membrane separation industry is still dominated by typical polymeric membranes (e.g., without interconnected nanopores), due to their relatively low cost, ease of fabrication, and high reproducibility [435–437]. Although some nanoporous materials such as zeolites have been investigated for several decades, there are few industrial membrane applications [167,438–446]. Zeolites and some MOFs have been applied in a series of adsorption processes; however, nanoporous materials like COFs are still limited to lab-scale experiments. Further research is required to deal with the following challenges associated with translating lab-scale nanoporous material performance to real industrial applications: cost reduction, mechanical strength, stability, and reproducibility [167].

6.1 Cost Reduction of Nanoporous Membrane

The most serious drawback of nanoporous materials is the relatively high cost, especially in the area of membrane-based separations [309,311]. Due to the fragile nature of crystalline materials, most nanoporous materials (except solution-processable microporous polymer and polymer-derived CMSs) cannot be handled as freestanding membranes [167]. Instead, the majority of the aforementioned nanoporous membranes supported on smooth alumina or stainless steel ultrafiltration membranes, which are expensive. For instance, in a common zeolite membrane, supports may contribute 70% of fabrication cost [167]. Four strategies have been proposed to reduce the cost incurred by the substrates. The first solution is to utilize inorganic substrates more efficiently. By developing

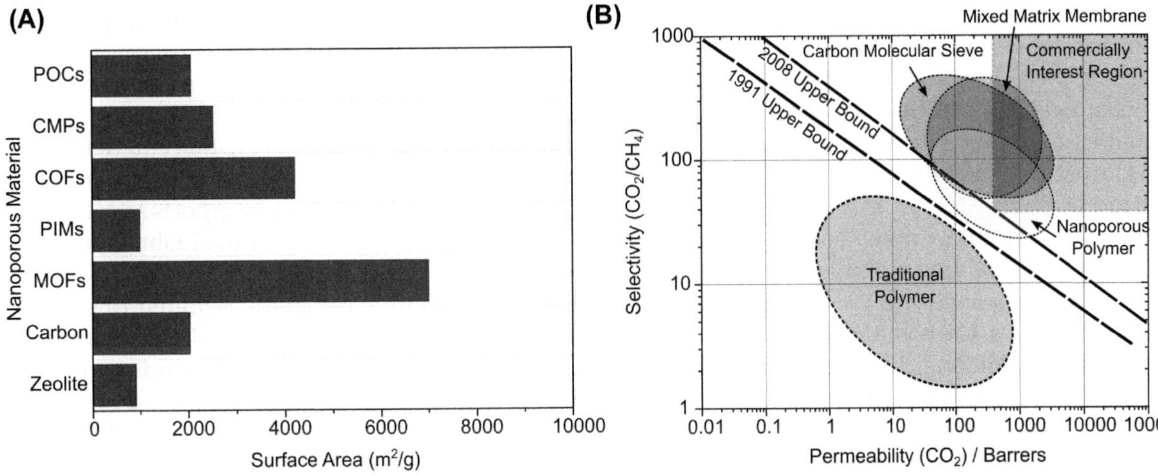

FIGURE 3.19 Nanoporous materials used as (A) the adsorbents [58,60,430–432] and (B) the membranes [306,308,433,434] with significantly improved performance.

advanced membrane modules (e.g., hollow fibers, multichannel modules, etc.), the ratio between membrane surface area and volume can be further improved [307,309]. Second, with optimized secondary growth techniques, defect-free nanoporous membranes may be fabricated on rough substrates that can be fabricated via inexpensive methods like extrusion [167]. Third, many researchers focus on replacing alumina and stainless-steel substrates with inexpensive substrates. For instance, ZIF membrane has been synthesized on polymeric and macroporous carbon membranes via interfacial microfluidic processing [217,219,447]. Suitable synthesis conditions (e.g., compatible solvents, mild temperature, less mechanical impact, etc.) must be designed to utilize these alternative substrates. Last but not least, CMS membranes and solution processable PIMs can be fabricated into robust freestanding hollow fiber membrane [80,120]. By engineering the pore sizes and functionalities, it is possible to use such kind of membranes for various separations without expensive substrates.

Beside substrates, production of nanoporous materials is also more expensive than conventional materials (e.g., cellulose acetate, active carbon, etc.), which inhibits the application of nanoporous materials in both membrane and adsorption separation areas [167]. The high synthesis cost of nanoporous materials is mainly contributed by components of nanoporous materials (e.g., ligands) and assisting materials involved in reaction (e.g., structure-directing structures, solvents, etc.). For example, many nanoporous materials are constructed by rare chemicals (e.g., some MOFs, COF, etc.), which are expensive due to the synthesis difficulty or limited supply. The latter reason can be easily eliminated via large-scale production, while the former one requires a revolution in synthesis route or development of inexpensive analogues. Besides, synthesis without solvents or organic SDAs can further decrease the price of nanoporous materials [207,208,448–451].

For materials prepared by special routes, equipment investment should also be considered. For instance, CMS hollow fiber membranes derived from polymers can be handled without expensive substrate; however, pyrolysis furnaces with inert gas purging or vacuum are required [80]. Compared with polymeric membranes, the cost of CMS membranes per unit area is always 1–3 times higher due to the difficulty of scaling up to rapid production rates for modules [128,307]. This high investment cost makes CMS membranes more attractive only when they can achieve a much better performance than polymeric membranes.

6.2 Mechanical Strength

Mechanical strength is another important factor influencing the application of nanoporous materials. For adsorption, adsorbent should be able to survive abrasion with fluid and thermal expansion. For membrane separation, the material should be robust enough to resist the transmembrane pressure, bending, shaking, and any other variance during operation. As adsorption process is tolerant to the defects (e.g., cracks and holes), adsorbents with high mechanical strength can be fabricated in the shape of monoliths, fibers, or customized 3D structures as discussed in former sections. However, more researches are needed to improve the mechanical strength of nanoporous membranes for energy-efficient separations.

To achieve high permeance, industrial membranes should withstand high transmembrane pressure (e.g., 120 bar for hydrogen recovery from ammonia reactor purge gas). Crystalline nanoporous materials are susceptible to failure at high transmembrane pressures. Robust metallic and ceramic membranes are usually utilized to support fragile selective layers made of nanoporous materials [313]. One of the prominent commercial ceramic support is designed by Inocermic GmbH (Germany): each ceramic cylinder possesses multiple bores, on the surface of which zeolite membranes grow. Alternative inexpensive supports have also been developed, such as alumina hollow fibers, SiC multichannel elements, stainless-steel reinforced ceramic foil, ceramic capillary, etc. [313].

Crystalline nanoporous membranes fabricated on inorganic substrates have good resistance to deformation like bending. However, scratching induced by improper handling and adjacent membranes can destroy the selectivity of the entire module. One strategy is to grow selective layer on the inner surface of tubular supports sacrificing the ratio of surface area to module volume [447]. Alternatively, there is also a development of novel fabrication techniques within the module shielding avoiding further handling before operation [217,219,447].

For freestanding nanoporous membranes, like CMS membranes, careful handling is required due to their poor mechanical properties. Carbon Membranes Ltd. was the first company to produce high-quality hollow fiber carbon membranes with 0.3 m long, 170 μm outer diameter, and a wall thickness of about 9 μm in modules of 4 m^2 membrane size on an industrial scale [452,453]. The brittleness of the carbon membranes was a major problem, and the company stopped production in 2001. In the future, this problem of brittleness can be prevented to a certain degree by optimizing precursors and fabrication methods. For example, supported carbon layers can be prepared thinner than self-supporting carbon membranes, which provides higher mechanical stability. Porous stainless steel and α-alumina supports can be used to stabilize the carbon layers.

6.3 Stability

A robust membrane or absorber should be able to operate under industrial conditions, which could be much harsher than the preliminary lab-scale test. For instance, many industrial processes (e.g., WGS, steam methane reforming, methanol synthesis, flue gas) involve steam at high temperature. Therefore, hydrothermal stability is an important criterion for novel nanoporous materials [168,454].

Zeolites are not the most thermodynamically stable materials and may collapse into dense or amorphous phases due to recrystallization [455]. However, such a reaction has quite a high energy barrier [168]. There are various detailed reports about hydrothermal stability regarding structure, morphology, interaction with the substrate, grain boundaries, etc. [455]. In some cases, steam exposure can also reduce the defects in frameworks (e.g., MCM-22) [456–459]. Although the loss of permeability or selectivity is reported in some membrane studies involving high temperatures steams [460], zeolites (e.g., MFI, SOD, FAU, etc.) are generally more stable than the other nanoporous materials when exposed to high temperature steam environment [315,461–466].

For MOFs, hydrothermal stability is critical due to the relatively weak coordination bonding [467]. For instance, MOF-5 decomposes at room temperature in atmospheric air by hydrolysis. A variety of promising MOFs (e.g., HKUST-1, MIL-53, etc.) degrade during exposure to ambient moisture [468–470]. Low et al. examined the steam stability of several MOFs by comparing XRD patterns after exposure to steam for several hours [467]. This pioneering work reveals that metal–ligand bond strength plays a determining role in the hydrothermal stability of MOFs. It is worth noting that contradicting results about the hydrothermal stability of MOFs exist in literature and imply that there must be some other factors contributing to the hydrothermal stability, which requires further investigation [467,471–473].

Nanoporous polymers are expected to exhibit outstanding stabilities due to their strong covalent bonds [2,54,56,59,60]. However, some prototypical nanoporous polymers do surfer from low hydrothermal stability. For instance, the earliest published boron-based COFs, COF-1 and COF-5, undergo nearly complete hydrolysis when exposed to aqueous media [61]. By functionalization and development of stable linkage, nanoporous polymers with better hydrothermal stability can be invented [2,61].

Although CMSs are inert to most chemicals except oxygen, the packing structure of carbon sheets can still be altered in different conditions. Like most of the disordered materials, CMS membranes can undergo the problem of physical aging which may be caused by either chemisorption of oxygen or physical adsorption of water and organics in the pore structures [474]. Imperfectly stacked carbon plates may be densified to achieve thermodynamically more stable states, which result in the loss of micropores and the decrease of the permeability through CMS membranes. Fortunately, the physical aging problem of the CMS membrane can be effectively "quenched." For example, the performance of the CMS membrane for CO_2/CH_4 separation remains stable if the membranes can be stored under 100 psig CO_2 environment [475].

6.4 Reproducibility

It is fair to admit that poor reproducibility is still one of the major problems inhibiting large-scale production of nanoporous membranes to meet industrial and commercial requirements.

Taking zeolites, which have been intensively studied over several decades, as an example, MFI zeolite membranes of different qualities have been reported with distinct separation performances [476–484]. There are some fabricated MFI zeolite membranes that enable the separation of n/i-butane mixture at a high temperature [483], some can only accomplish at low temperature [481] and others cannot do this separation at all [476,479]. On the one hand, such diverse performances result from the distinct growth conditions (e.g., orientation) of zeolite membranes, which can be difficult (but not impossible) to control at scale. For instance, the quality of substrate seeding via manual rubbing heavily depends on the operator. On the other hand, the presence of even few intercrystalline defects, like cracks and pinholes, can significantly impair the performance of the entire membrane. Moreover, this poor reproducibility may be due to the influence of the porosity and chemical composition of the support. Even at lab-scale, reproduction of defect-free zeolite membrane is still challenging [482]. In this sense, fabrication techniques for both zeolites and supports with less personal influences should be developed in parallel.

A poor reproducibility problem has also occurred in the case of CMS membranes. A few groups found that CMS membranes fabricated under what appear to be the same pyrolysis conditions tend to have a range of separation performance, especially selectivity [128]. Singh-Ghosal and Koros found that the O_2/N_2 selectivity of CMS membranes fabricated even at the same pyrolysis temperature varied by as much as a factor of 5 [485]. Foley et al. studied the variability in separation performance of supported PFA-derived CMS membranes, which are fabricated by several different techniques such as spray coating, brush coating, and ultrasonic deposition in detail [134,135,486,487]. Alhough fairly similar pyrolysis conditions, the permeance and selectivity for these obtained membranes differed a lot. Strano and Foley have pointed out the poor reproducibility for CMS membranes may be due to the defective nanopores formed during the pyrolysis process [488]. However, continued research has further improved the reproducibility of this class of membranes [4].

The difficulty of reproducible fabrication of supported crystalline membranes (e.g., zeolites, MOFs, etc.) drives researchers to find alternative materials compatible with existing reproducible membrane fabrication techniques, such as hollow fiber spinning. One alternative material is a solution processable nanoporous polymer (e.g., PIMs, POCs, etc.) [58,489]. For instance, defect-free PIM-1 hollow fiber membranes can be produced with state-of-art dry-wet spinning equipment with ease [120].

6.5 Sustainability of Nanoporous Material Fabrication

Nanoporous materials can be fabricated into highly efficient adsorbents and membranes, reducing the energy consumption and waste generated during various separation and chemical production processes. Before these nanoporous materials are used in operation, they must go through a fabrication process (as discussed in this chapter). This fabrication process can also be improved via the utilization of various green chemistry and green technology approaches.

As introduced in previous sections, most of the nanoporous materials are synthesized via solvothermal reactions, which usually involve a large amount of organic solvents, usage of organic templates, high pressure, etc. [490]. For instance, state-of-art synthesis routes of zeolites widely utilize organic molecules to direct the assembly, which are then removed by combustion [491]. Even though these additives play significant roles in the production of nanoporous materials, they are wastes in the aspect of green chemistry. Green synthesis route for nanoporous materials has become an important research topic with many examples in the literature. For instance, nanoporous materials have been made from converted biomass or wastes [492], unreacted reagent recycling [493], solvent-free synthesis [209,210], synthesis in ionic liquids [494], template-free zeolites, zeolite synthesis with less toxic, recyclable, or no organic templates, etc. [490,493].

Fabrication of adsorbents and membranes generates additional organic solvent wastes [11,495]. For instance, the dry-wet spinning of fiber sorbents or hollow fiber membranes requires toxic organic solvents for polymer dissolution, generates wastewater contaminated by these organic solvents, and consumes a huge amount of organic solvents during postspinning solvent exchange [496]. Assuming that contaminant concentration in quench baths and exchanging solvents is less than 1 wt% for consistency in membrane quality, lab-scale fabrication of 1 kg of PIM-1 hollow fiber membranes generates 1600 kg of wastewater, and consumes 8.4 kg THF, 4 kg dimethylacetamide, 0.7 kg ethanol, 3.5 kg 1-butanol, 3 kg methanol, and 3 kg n-hexane [120]. These solvent intensities are likely to decrease during scale-up and commercialization. For instance, solvent recovery and recycle from quench baths is a promising method for reducing the waste intensity of fiber spinning processes [11,12]. Beyond solvent recovery, the use of environment-friendly alternative solvents (e.g., supercritical CO_2, ionic liquids, degradable solvents, etc.) should also be investigated to replace the toxic organic solvents currently utilized in commercial spinning processes (e.g., N,N-dimethylformamide, N,N-dimethylacetamide, NMP, THF, etc.) [496–498].

7. CONCLUSIONS

Separations are critical processes that are applied to various areas, such as the petrochemical industry, the pharmaceutical industry, etc. Currently, industrial separation processes are dominated by thermally driven processes such as distillation, which purify mixtures via phase change. Thermally driven separation techniques inevitably result in a huge consumption of energy and emission of pollutants and greenhouse gases [5]. Such energy-intensive and environmental-unfriendly processes significantly inhibit sustainable development. In the past several decades, engineers all over the world have made tremendous progress in the development of energy-efficient alternative separation techniques, among which membrane separation and adsorption are the most promising two. Different from traditional thermal-driven techniques, these alternatives separate molecules based on the difference in diffusivities (e.g., membrane separation) or difference in interaction with solid materials (e.g., adsorption). Nanoporous materials with well-defined pore structure, large surface area, and versatile active sites have been developed to facilitate precise molecular differentiation.

Currently, the following state-of-art nanoporous materials are intensively studied: (i) zeolites, (ii) MOFs, (iii) nanoporous polymers, and (iv) CMSs. These nanoporous materials have exhibited promising performance in various applications, like gas separation, organic solvent separation, desalination, etc. Owing to the tunable pore sizes and versatile functionalities of nanoporous materials, engineers can customize the membranes and adsorbents to achieve target separation. Although the industrial application of nanoporous materials is still challenged by the problem of high cost, low mechanical strength, low stability, low reproducibility, etc., there is no doubt that nanoporous materials play an indispensable role in sustainable development by achieving energy-efficient separation.

Acknowledgments

The authors thank ExxonMobil Research & Engineering for funding this work. The authors also thank Dr. Dhaval Bhandari and Dr. JR Johnson for their helpful suggestions.

References

[1] S. Kitagawa, Metal-organic frameworks (MOFs), Chemical Society Reviews 43 (16) (2014) 5415–5418.
[2] X. Feng, X. Ding, D. Jiang, Covalent organic frameworks, Chemical Society Reviews 41 (18) (2012) 6010–6022.
[3] A.I. Cooper, Conjugated microporous polymers, Advanced Materials 21 (12) (2009) 1291–1295.
[4] W.J. Koros, C. Zhang, Materials for next-generation molecularly selective synthetic membranes, Nature Materials 16 (3) (2017) 289.
[5] D.S. Sholl, R.P. Lively, Seven chemical separations: to change the world: purifying mixtures without using heat would lower global energy use, emissions and pollution-and open up new routes to resources, Nature 532 (7600) (2016) 435–438.
[6] M.W. Ackley, R.T. Yang, Kinetic separation by pressure swing adsorption: method of characteristics model, AIChE Journal 36 (8) (1990) 1229–1238.
[7] K. Li, D.H. Olson, J. Seidel, T.J. Emge, H. Gong, H. Zeng, J. Li, Zeolitic imidazolate frameworks for kinetic separation of propane and propene, Journal of the American Chemical Society 131 (30) (2009) 10368–10369.
[8] W. Koros, G. Fleming, Membrane-based gas separation, Journal of Membrane Science 83 (1) (1993) 1–80.
[9] W.J. Koros, R.P. Lively, Water and beyond: expanding the spectrum of large-scale energy efficient separation processes, AIChE Journal 58 (9) (2012) 2624–2633.
[10] R.A. Sheldon, The E factor 25 years on: the rise of green chemistry and sustainability, Green Chemistry 19 (1) (2017) 18–43.
[11] M. Razali, J.F. Kim, M. Attfield, P.M. Budd, E. Drioli, Y.M. Lee, G. Szekely, Sustainable wastewater treatment and recycling in membrane manufacturing, Green Chemistry 17 (12) (2015) 5196–5205.
[12] G. Szekely, M.F. Jimenez-Solomon, P. Marchetti, J.F. Kim, A.G. Livingston, Sustainability assessment of organic solvent nanofiltration: from fabrication to application, Green Chemistry 16 (10) (2014) 4440–4473.
[13] J.F. Kim, G. Szekely, M. Schaepertoens, I.B. Valtcheva, M.F. Jimenez-Solomon, A.G. Livingston, In situ solvent recovery by organic solvent nanofiltration, ACS Sustainable Chemistry and Engineering 2 (10) (2014) 2371–2379.
[14] D. Breck, Zeolite Molecular Sieves, John Wiley & Sons, New York, 1974.
[15] A. Cronstedt, Observation and description of an unknown kind of rock to be named zeolites, Kongl Vetenskaps Acad Handl Stockh 17 (1756) 120–123 (in Swedish).
[16] A. Lee, G. Xiao, P. Xiao, K. Joshi, R. Singh, P.A. Webley, High temperature adsorption materials and their performance for pre-combustion capture of carbon dioxide, Energy Procedia 4 (2011) 1199–1206.
[17] N. Kosinov, J. Gascon, F. Kapteijn, E.J. Hensen, Recent developments in zeolite membranes for gas separation, Journal of Membrane Science 499 (2016) 65–79.
[18] M.A. Snyder, M. Tsapatsis, Hierarchical nanomanufacturing: from shaped zeolite nanoparticles to high-performance separation membranes, Angewandte Chemie International Edition 46 (40) (2007) 7560–7573.
[19] A. Corma, Inorganic solid acids and their use in acid-catalyzed hydrocarbon reactions, Chemical Reviews 95 (3) (1995) 559–614.
[20] J. Weitkamp, Zeolites and catalysis, Solid State Ionics 131 (1–2) (2000) 175–188.
[21] D. Zhao, J. Feng, Q. Huo, N. Melosh, G.H. Fredrickson, B.F. Chmelka, G.D. Stucky, Triblock copolymer syntheses of mesoporous silica with periodic 50 to 300 angstrom pores, Science 279 (5350) (1998) 548–552.
[22] S. Van Donk, A.H. Janssen, J.H. Bitter, K.P. de Jong, Generation, characterization, and impact of mesopores in zeolite catalysts, Catalysis Reviews 45 (2) (2003) 297–319.
[23] Y. Wei, T.E. Parmentier, K.P. de Jong, J. Zečević, Tailoring and visualizing the pore architecture of hierarchical zeolites, Chemical Society Reviews 44 (20) (2015) 7234–7261.
[24] K. Möller, B. Yilmaz, R.M. Jacubinas, U. Müller, T. Bein, One-step synthesis of hierarchical zeolite beta via network formation of uniform nanocrystals, Journal of the American Chemical Society 133 (14) (2011) 5284–5295.
[25] C. Li, Y. Wang, B. Shi, J. Ren, X. Liu, Y. Wang, Y. Guo, Y. Guo, G. Lu, Synthesis of hierarchical MFI zeolite microspheres with stacking nanocrystals, Microporous and Mesoporous Materials 117 (1–2) (2009) 104–110.
[26] G. Majano, S. Mintova, O. Ovsitser, B. Mihailova, T. Bein, Zeolite beta nanosized assemblies, Microporous and Mesoporous Materials 80 (1–3) (2005) 227–235.
[27] L.H. Chen, X.Y. Li, G. Tian, Y. Li, J.C. Rooke, G.S. Zhu, S.L. Qiu, X.Y. Yang, B.L. Su, Highly stable and reusable multimodal zeolite TS-1 based catalysts with hierarchically interconnected three-level micro-meso-macroporous structure, Angewandte Chemie International Edition 50 (47) (2011) 11156–11161.
[28] K. Möller, B. Yilmaz, U. Müller, T. Bein, Nanofusion: mesoporous zeolites made easy, Chemistry-A European Journal 18 (25) (2012) 7671–7674.

[29] X. Zhang, D. Liu, D. Xu, S. Asahina, K.A. Cychosz, K.V. Agrawal, Y. Al Wahedi, A. Bhan, S. Al Hashimi, O. Terasaki, Synthesis of self-pillared zeolite nanosheets by repetitive branching, Science 336 (6089) (2012) 1684–1687.
[30] A.U. Czaja, N. Trukhan, U. Müller, Industrial applications of metal-organic frameworks, Chemical Society Reviews 38 (5) (2009) 1284–1293.
[31] H. Furukawa, K.E. Cordova, M. O'Keeffe, O.M. Yaghi, The chemistry and applications of metal-organic frameworks, Science 341 (6149) (2013) 1230444.
[32] L.E. Kreno, K. Leong, O.K. Farha, M. Allendorf, R.P. Van Duyne, J.T. Hupp, Metal-organic framework materials as chemical sensors, Chemical Reviews 112 (2) (2011) 1105–1125.
[33] L.-G. Qiu, Z.-Q. Li, Y. Wu, W. Wang, T. Xu, X. Jiang, Facile synthesis of nanocrystals of a microporous metal-organic framework by an ultrasonic method and selective sensing of organoamines, Chemical Communications (31) (2008) 3642–3644.
[34] X. Zou, G. Zhu, I.J. Hewitt, F. Sun, S. Qiu, Synthesis of a metal-organic framework film by direct conversion technique for VOCs sensing, Dalton Transactions 16 (2009) 3009–3013.
[35] X. Zhao, Y. Wang, D. Li, X. Bu, P. Feng, Metal-organic frameworks for separation, Advanced Materials (30) (2018), 1705189.
[36] J.-R. Li, J. Sculley, H.-C. Zhou, Metal-organic frameworks for separations, Chemical Reviews 112 (2) (2011) 869–932.
[37] S. Qiu, M. Xue, G. Zhu, Metal-organic framework membranes: from synthesis to separation application, Chemical Society Reviews 43 (16) (2014) 6116–6140.
[38] W. Kleist, F. Jutz, M. Maciejewski, A. Baiker, Mixed-linker metal-organic frameworks as catalysts for the synthesis of propylene carbonate from propylene oxide and CO_2, European Journal of Inorganic Chemistry 2009 (24) (2009) 3552–3561.
[39] H. Deng, C.J. Doonan, H. Furukawa, R.B. Ferreira, J. Towne, C.B. Knobler, B. Wang, O.M. Yaghi, Multiple functional groups of varying ratios in metal-organic frameworks, Science 327 (5967) (2010) 846–850.
[40] P. Deria, J.E. Mondloch, O. Karagiaridi, W. Bury, J.T. Hupp, O.K. Farha, Beyond post-synthesis modification: evolution of metal-organic frameworks via building block replacement, Chemical Society Reviews 43 (16) (2014) 5896–5912.
[41] S.R. Caskey, A.G. Wong-Foy, A.J. Matzger, Dramatic tuning of carbon dioxide uptake via metal substitution in a coordination polymer with cylindrical pores, Journal of the American Chemical Society 130 (33) (2008) 10870–10871.
[42] M.J. Katz, Z.J. Brown, Y.J. Colón, P.W. Siu, K.A. Scheidt, R.Q. Snurr, J.T. Hupp, O.K. Farha, A facile synthesis of UiO-66, UiO-67 and their derivatives, Chemical Communications 49 (82) (2013) 9449–9451.
[43] J.A. Thompson, C.R. Blad, N.A. Brunelli, M.E. Lydon, R.P. Lively, C.W. Jones, S. Nair, Hybrid zeolitic imidazolate frameworks: controlling framework porosity and functionality by mixed-linker synthesis, Chemistry of Materials 24 (10) (2012) 1930–1936.
[44] W. Xuan, C. Zhu, Y. Liu, Y. Cui, Mesoporous metal-organic framework materials, Chemical Society Reviews 41 (5) (2012) 1677–1695.
[45] G. Férey, C. Serre, C. Mellot-Draznieks, F. Millange, S. Surblé, J. Dutour, I. Margiolaki, A hybrid solid with giant pores prepared by a combination of targeted chemistry, simulation, and powder diffraction, Angewandte Chemie 116 (46) (2004) 6456–6461.
[46] G. Férey, C. Mellot-Draznieks, C. Serre, F. Millange, J. Dutour, S. Surblé, I. Margiolaki, A chromium terephthalate-based solid with unusually large pore volumes and surface area, Science 309 (5743) (2005) 2040–2042.
[47] J.E. Mondloch, W. Bury, D. Fairen-Jimenez, S. Kwon, E.J. DeMarco, M.H. Weston, A.A. Sarjeant, S.T. Nguyen, P.C. Stair, R.Q. Snurr, Vapor-phase metalation by atomic layer deposition in a metal-organic framework, Journal of the American Chemical Society 135 (28) (2013) 10294–10297.
[48] X.-S. Wang, S. Ma, D. Sun, S. Parkin, H.-C. Zhou, A mesoporous metal-organic framework with permanent porosity, Journal of the American Chemical Society 128 (51) (2006) 16474–16475.
[49] Q.R. Fang, G.S. Zhu, Z. Jin, Y.Y. Ji, J.W. Ye, M. Xue, H. Yang, Y. Wang, S.L. Qiu, Mesoporous metal-organic framework with rare etb topology for hydrogen storage and dye assembly, Angewandte Chemie 119 (35) (2007) 6758–6762.
[50] K. Koh, A.G. Wong-Foy, A.J. Matzger, A crystalline mesoporous coordination copolymer with high microporosity, Angewandte Chemie International Edition 47 (4) (2008) 677–680.
[51] L.G. Qiu, T. Xu, Z.Q. Li, W. Wang, Y. Wu, X. Jiang, X.Y. Tian, L.D. Zhang, Hierarchically micro-and mesoporous metal-organic frameworks with tunable porosity, Angewandte Chemie International Edition 47 (49) (2008) 9487–9491.
[52] X. Roy, L.K. Thompson, N. Coombs, M.J. MacLachlan, Mesostructured Prussian blue analogues, Angewandte Chemie International Edition 47 (3) (2008) 511–514.
[53] Y. Zhao, J. Zhang, B. Han, J. Song, J. Li, Q. Wang, Metal-organic framework nanospheres with well-ordered mesopores synthesized in an ionic liquid/CO_2/surfactant system, Angewandte Chemie International Edition 50 (3) (2011) 636–639.
[54] A.P. Cote, A.I. Benin, N.W. Ockwig, M. O'keeffe, A.J. Matzger, O.M. Yaghi, Porous, crystalline, covalent organic frameworks, Science 310 (5751) (2005) 1166–1170.
[55] T. Ben, H. Ren, S. Ma, D. Cao, J. Lan, X. Jing, W. Wang, J. Xu, F. Deng, J.M. Simmons, Targeted synthesis of a porous aromatic framework with high stability and exceptionally high surface area, Angewandte Chemie 121 (50) (2009) 9621–9624.
[56] D. Wu, F. Xu, B. Sun, R. Fu, H. He, K. Matyjaszewski, Design and preparation of porous polymers, Chemical Reviews 112 (7) (2012) 3959–4015.
[57] M. Tsyurupa, V. Davankov, Hypercrosslinked polymers: basic principle of preparing the new class of polymeric materials, Reactive and Functional Polymers 53 (2–3) (2002) 193–203.
[58] N.B. McKeown, P.M. Budd, Polymers of intrinsic microporosity (PIMs): organic materials for membrane separations, heterogeneous catalysis and hydrogen storage, Chemical Society Reviews 35 (8) (2006) 675–683.
[59] A.P. Côté, A.I. Benin, N.W. Ockwig, M. O'Keeffe, A.J. Matzger, O.M. Yaghi, Porous, crystalline, covalent organic frameworks, Science 310 (5751) (2005) 1166–1170.
[60] H.M. El-Kaderi, J.R. Hunt, J.L. Mendoza-Cortés, A.P. Côté, R.E. Taylor, M. O'keeffe, O.M. Yaghi, Designed synthesis of 3D covalent organic frameworks, Science 316 (5822) (2007) 268–272.
[61] L.M. Lanni, R.W. Tilford, M. Bharathy, J.J. Lavigne, Enhanced hydrolytic stability of self-assembling alkylated two-dimensional covalent organic frameworks, Journal of the American Chemical Society 133 (35) (2011) 13975–13983.
[62] J.X. Jiang, F. Su, A. Trewin, C.D. Wood, N.L. Campbell, H. Niu, C. Dickinson, A.Y. Ganin, M.J. Rosseinsky, Y.Z. Khimyak, Conjugated microporous poly (aryleneethynylene) networks, Angewandte Chemie International Edition 46 (45) (2007) 8574–8578.

[63] R. Dawson, F. Su, H. Niu, C.D. Wood, J.T. Jones, Y.Z. Khimyak, A.I. Cooper, Mesoporous poly (phenylenevinylene) networks, Macromolecules 41 (5) (2008) 1591–1593.

[64] J.-X. Jiang, F. Su, H. Niu, C.D. Wood, N.L. Campbell, Y.Z. Khimyak, A.I. Cooper, Conjugated microporous poly (phenylene butadiynylene) s, Chemical Communications (4) (2008) 486–488.

[65] R. Williams, L.A. Burt, E. Esposito, J.C. Jansen, E. Tocci, C. Rizzuto, M. Lanč, M. Carta, N.B. McKeown, A highly rigid and gas selective methanopentacene-based polymer of intrinsic microporosity derived from Tröger's base polymerization, Journal of Materials Chemistry A 6 (14) (2018) 5661–5667.

[66] T. Tozawa, J.T. Jones, S.I. Swamy, S. Jiang, D.J. Adams, S. Shakespeare, R. Clowes, D. Bradshaw, T. Hasell, S.Y. Chong, Porous organic cages, Nature materials 8 (12) (2009) 973.

[67] J.R. Holst, A. Trewin, A.I. Cooper, Porous organic molecules, Nature chemistry 2 (11) (2010) 915.

[68] A.G. Slater, P.S. Reiss, A. Pulido, M.A. Little, D.L. Holden, L. Chen, S.Y. Chong, B.M. Alston, R. Clowes, M. Haranczyk, Computationally-guided synthetic control over pore size in isostructural porous organic cages, ACS Central Science 3 (7) (2017) 734–742.

[69] K.-V. Peinemann, V. Abetz, P.F. Simon, Asymmetric superstructure formed in a block copolymer via phase separation, Nature Materials 6 (12) (2007) 992.

[70] H. Lee, S.M. Dellatore, W.M. Miller, P.B. Messersmith, Mussel-inspired surface chemistry for multifunctional coatings, Science 318 (5849) (2007) 426–430.

[71] C. Zhang, Y. Lv, W.-Z. Qiu, A. He, Z.-K. Xu, Polydopamine coatings with nanopores for versatile molecular separation, ACS Applied Materials and Interfaces 9 (16) (2017) 14437–14444.

[72] L. Pérez-Manríquez, P. Neelakanda, K.-V. Peinemann, Morin-based nanofiltration membranes for organic solvent separation processes, Journal of Membrane Science 554 (2018) 1–5.

[73] L. Pérez-Manríquez, P. Neelakanda, K.-V. Peinemann, Tannin-based thin-film composite membranes for solvent nanofiltration, Journal of Membrane Science 541 (2017) 137–142.

[74] H. Suda, K. Haraya, Gas permeation through micropores of carbon molecular sieve membranes derived from Kapton polyimide, The Journal of Physical Chemistry B 101 (20) (1997) 3988–3994.

[75] B. Rand, A. Hosty, S. West, H. Marsh, Introduction to Carbon Science, Butterworths, London, 1989, pp. 75–92.

[76] H.O. Pierson, Handbook of Carbon, Graphite, Diamond, and Fullerenes, Processing, Properties and Applications, Noyes Publications, New York, 1993.

[77] M. Rungta, G.B. Wenz, C. Zhang, L. Xu, W. Qiu, J.S. Adams, W.J. Koros, Carbon molecular sieve structure development and membrane performance relationships, Carbon 115 (2017) 237–248.

[78] K.M. Steel, W.J. Koros, Investigation of porosity of carbon materials and related effects on gas separation properties, Carbon 41 (2) (2003) 253–266.

[79] L. Shao, T.-S. Chung, G. Wensley, S.H. Goh, K.P. Pramoda, Casting solvent effects on morphologies, gas transport properties of a novel 6FDA/PMDA-TMMDA copolyimide membrane and its derived carbon membranes, Journal of Membrane Science 244 (1) (2004) 77–87.

[80] D.-Y. Koh, B.A. McCool, H.W. Deckman, R.P. Lively, Reverse osmosis molecular differentiation of organic liquids using carbon molecular sieve membranes, Science 353 (6301) (2016) 804–807.

[81] A. Samanta, A. Zhao, G.K. Shimizu, P. Sarkar, R. Gupta, Post-combustion CO_2 capture using solid sorbents: a review, Industrial and Engineering Chemistry Research 51 (4) (2011) 1438–1463.

[82] M. Das, J.D. Perry, W.J. Koros, Gas-transport-property performance of hybrid carbon molecular sieve– polymer materials, Industrial and Engineering Chemistry Research 49 (19) (2010) 9310–9321.

[83] H. Qinglin, S. Sundaram, S. Farooq, Revisiting transport of gases in the micropores of carbon molecular sieves, Langmuir 19 (2) (2003) 393–405.

[84] T. Burchell, R. Judkins, M. Rogers, A. Williams, A novel process and material for the separation of carbon dioxide and hydrogen sulfide gas mixtures, Carbon 35 (9) (1997) 1279–1294.

[85] W. Yuan, J. Chen, G. Shi, Nanoporous graphene materials, Materials Today 17 (2) (2014) 77–85.

[86] B.H. Kim, J.Y. Kim, S.-J. Jeong, J.O. Hwang, D.H. Lee, D.O. Shin, S.-Y. Choi, S.O. Kim, Surface energy modification by spin-cast, large-area graphene film for block copolymer lithography, ACS Nano 4 (9) (2010) 5464–5470.

[87] N.S. Safron, A.S. Brewer, M.S. Arnold, Semiconducting two-dimensional graphene nanoconstriction arrays, Small 7 (4) (2011) 492–498.

[88] W.S. Hummers Jr., R.E. Offeman, Preparation of graphitic oxide, Journal of the American Chemical Society 80 (6) (1958), 1339-1339.

[89] C. Moreno, M. Vilas-Varela, B. Kretz, A. Garcia-Lekue, M.V. Costache, M. Paradinas, M. Panighel, G. Ceballos, S.O. Valenzuela, D. Peña, Bottom-up synthesis of multifunctional nanoporous graphene, Science 360 (6385) (2018) 199–203.

[90] R. Nair, H. Wu, P. Jayaram, I. Grigorieva, A. Geim, Unimpeded permeation of water through helium-leak-tight graphene-based membranes, Science 335 (6067) (2012) 442–444.

[91] G. Liu, W. Jin, N. Xu, Graphene-based membranes, Chemical Society Reviews 44 (15) (2015) 5016–5030.

[92] T. Gao, H. Wu, L. Tao, L. Qu, C. Li, Enhanced stability and separation efficiency of graphene oxide membranes in organic solvent nanofiltration, Journal of Materials Chemistry A 6 (40) (2018) 19563–19569.

[93] D. Cohen-Tanugi, L.-C. Lin, J.C. Grossman, Multilayer nanoporous graphene membranes for water desalination, Nano Letters 16 (2) (2016) 1027–1033.

[94] Y.-H. Xi, J.-Q. Hu, Z. Liu, R. Xie, X.-J. Ju, W. Wang, L.-Y. Chu, Graphene oxide membranes with strong stability in aqueous solutions and controllable lamellar spacing, ACS Applied Materials and Interfaces 8 (24) (2016) 15557–15566.

[95] S.P. Surwade, S.N. Smirnov, I.V. Vlassiouk, R.R. Unocic, G.M. Veith, S. Dai, S.M. Mahurin, Water desalination using nanoporous single-layer graphene, Nature Nanotechnology 10 (5) (2015) 459–464.

[96] Q. Yang, Y. Su, C. Chi, C. Cherian, K. Huang, V. Kravets, F. Wang, J. Zhang, A. Pratt, A. Grigorenko, Ultrathin graphene-based membrane with precise molecular sieving and ultrafast solvent permeation, Nature Materials 16 (12) (2017) 1198.

[97] Z. Rahmani, M.T. Samadi, A. Kazemi, A.M. Rashidi, A.R. Rahmani, Nanoporous graphene and graphene oxide-coated polyurethane sponge as a highly efficient, superhydrophobic, and reusable oil spill absorbent, Journal of Environmental Chemical Engineering 5 (5) (2017) 5025–5032.

[98] G.J. Van Amerongen, The permeability of different rubbers to gases and its relation to diffusivity and solubility, Journal of Applied Physics 17 (11) (1946) 972–985.
[99] W.J. Koros (Ed.), Barrier polymers and structures, American Chemical Society, Washington, DC, 1990.
[100] P. Goh, A. Ismail, S. Sanip, B. Ng, M. Aziz, Recent advances of inorganic fillers in mixed matrix membrane for gas separation, Separation and Purification Technology 81 (3) (2011) 243–264.
[101] M.G. Süer, N. BaC, L. Yilmaz, Gas permeation characteristics of polymer-zeolite mixed matrix membranes, Journal of Membrane Science 91 (1–2) (1994) 77–86.
[102] T.W. Pechar, M. Tsapatsis, E. Marand, R. Davis, Preparation and characterization of a glassy fluorinated polyimide zeolite-mixed matrix membrane, Desalination 146 (1–3) (2002) 3–9.
[103] D. Bastani, N. Esmaeili, M. Asadollahi, Polymeric mixed matrix membranes containing zeolites as a filler for gas separation applications: a review, Journal of Industrial and Engineering Chemistry 19 (2) (2013) 375–393.
[104] E.V. Perez, K.J. Balkus Jr., J.P. Ferraris, I.H. Musselman, Mixed-matrix membranes containing MOF-5 for gas separations, Journal of Membrane Science 328 (1–2) (2009) 165–173.
[105] J. Hu, H. Cai, H. Ren, Y. Wei, Z. Xu, H. Liu, Y. Hu, Mixed-matrix membrane hollow fibers of Cu_3 $(BTC)_2$ MOF and polyimide for gas separation and adsorption, Industrial and Engineering Chemistry Research 49 (24) (2010) 12605–12612.
[106] H.B.T. Jeazet, C. Staudt, C. Janiak, Metal-organic frameworks in mixed-matrix membranes for gas separation, Dalton Transactions 41 (46) (2012) 14003–14027.
[107] Z. Kang, Y. Peng, Y. Qian, D. Yuan, M.A. Addicoat, T. Heine, Z. Hu, L. Tee, Z. Guo, D. Zhao, Mixed matrix membranes (MMMs) comprising exfoliated 2D covalent organic frameworks (COFs) for efficient CO_2 separation, Chemistry of Materials 28 (5) (2016) 1277–1285.
[108] M. Shan, B. Seoane, E. Rozhko, A. Dikhtiarenko, G. Clet, F. Kapteijn, J. Gascon, Azine-linked covalent organic framework (COF)-Based mixed-matrix membranes for CO_2/CH_4 separation, Chemistry-A European Journal 22 (41) (2016) 14467–14470.
[109] B.P. Biswal, H.D. Chaudhari, R. Banerjee, U.K. Kharul, Chemically stable covalent organic framework (COF)-polybenzimidazole hybrid membranes: enhanced gas separation through pore modulation, Chemistry-A European Journal 22 (14) (2016) 4695–4699.
[110] L. Meng, X. Zou, S. Guo, H. Ma, Y. Zhao, G. Zhu, Self-supported fibrous porous aromatic membranes for efficient CO_2/N_2 separations, ACS Applied Materials and Interfaces 7 (28) (2015) 15561–15569.
[111] J.D. Evans, D.M. Huang, M.R. Hill, C.J. Sumby, A.W. Thornton, C.J. Doonan, Feasibility of mixed matrix membrane gas separations employing porous organic cages, The Journal of Physical Chemistry C 118 (3) (2014) 1523–1529.
[112] M.M. Khan, V. Filiz, G. Bengtson, S. Shishatskiy, M. Rahman, V. Abetz, Functionalized carbon nanotubes mixed matrix membranes of polymers of intrinsic microporosity for gas separation, Nanoscale Research Letters 7 (1) (2012) 504.
[113] D.Q. Vu, W.J. Koros, S.J. Miller, Mixed matrix membranes using carbon molecular sieves: I. Preparation and experimental results, Journal of Membrane Science 211 (2) (2003) 311–334.
[114] R. Adams, J.R. Johnson, C. Zhang, R. Lively, Y. Dai, O. Esekhile, J. Liu, W.J. Koros, Mixed-matrix membranes, Encyclopedia of Membrane Science and Technology (2013) 1–34.
[115] P.M. Budd, E.S. Elabas, B.S. Ghanem, S. Makhseed, N.B. McKeown, K.J. Msayib, C.E. Tattershall, D. Wang, Solution-processed, organophilic membrane derived from a polymer of intrinsic microporosity, Advanced Materials 16 (5) (2004) 456–459.
[116] P.M. Budd, N.B. McKeown, B.S. Ghanem, K.J. Msayib, D. Fritsch, L. Starannikova, N. Belov, O. Sanfirova, Y. Yampolskii, V. Shantarovich, Gas permeation parameters and other physicochemical properties of a polymer of intrinsic microporosity: polybenzodioxane PIM-1, Journal of Membrane Science 325 (2) (2008) 851–860.
[117] D.T. Clausi, W.J. Koros, Formation of defect-free polyimide hollow fiber membranes for gas separations, Journal of Membrane Science 167 (1) (2000) 79–89.
[118] S.S. Sablani, M.F. Goosen, R. Al-Belushi, V. Gerardos, Influence of spacer thickness on permeate flux in spiral-wound seawater reverse osmosis systems, Desalination 146 (1–3) (2002) 225–230.
[119] W.F. Yong, F.Y. Li, Y.C. Xiao, T.S. Chung, Y.W. Tong, High performance PIM-1/Matrimid hollow fiber membranes for CO_2/CH_4, O_2/N_2 and CO_2/N_2 separation, Journal of Membrane Science 443 (2013) 156–169.
[120] M.L. Jue, V. Breedveld, R.P. Lively, Defect-free PIM-1 hollow fiber membranes, Journal of Membrane Science 530 (2017) 33–41.
[121] H. Hatori, Y. Yamada, M. Shiraishi, M. Yoshihara, T. Kimura, The mechanism of polyimide pyrolysis in the early stage, Carbon 34 (2) (1996) 201–208.
[122] G. Ehlers, K. Fisch, W. Powell, Thermal degradation of polymers with phenylene units in the chain. IV. Aromatic polyamides and polyimides, Journal of Polymer Science Part A: Polymer Chemistry 8 (12) (1970) 3511–3527.
[123] O. Sanyal, C. Zhang, G.B. Wenz, S. Fu, N. Bhuwania, L. Xu, M. Rungta, W.J. Koros, Next generation membranes-using tailored carbon, Carbon 127 (2018) 688–698.
[124] H.O. Pierson, Handbook of Carbon, Graphite, Diamonds and Fullerenes: Processing, Properties and Applications, William Andrew, 2012.
[125] C. Zhang, W.J. Koros, Ultraselective carbon molecular sieve membranes with tailored synergistic sorption selective properties, Advanced Materials 29 (33) (2017).
[126] S. Fu, G.B. Wenz, E.S. Sanders, S.S. Kulkarni, W. Qiu, C. Ma, W.J. Koros, Effects of pyrolysis conditions on gas separation properties of 6FDA/DETDA: DABA (3: 2) derived carbon molecular sieve membranes, Journal of Membrane Science 520 (2016) 699–711.
[127] J.E. Koresh, A. Sofer, Molecular sieve carbon permselective membrane. Part I. Presentation of a new device for gas mixture separation, Separation Science and Technology 18 (8) (1983) 723–734.
[128] P.J. Williams, W.J. Koros, Gas Separation by Carbon Membranes. Advanced Membrane Technology and Applications, 2008, pp. 599–631.
[129] C.W. Jones, W.J. Koros, Carbon molecular sieve gas separation membranes-I. Preparation and characterization based on polyimide precursors, Carbon 32 (8) (1994) 1419–1425.
[130] L. David, A. Ismail, Influence of the thermastabilization process and soak time during pyrolysis process on the polyacrylonitrile carbon membranes for O_2/N_2 separation, Journal of Membrane Science 213 (1–2) (2003) 285–291.
[131] Y. Ma, F. Zhang, S. Yang, R.P. Lively, Evidence for entropic diffusion selection of xylene isomers in carbon molecular sieve membranes, Journal of Membrane Science (564) (2018) 404–414.

[132] A.B. Fuertes, I. Menendez, Separation of hydrocarbon gas mixtures using phenolic resin-based carbon membranes, Separation and Purification Technology 28 (1) (2002) 29–41.
[133] W. Wei, H. Hu, L. You, G. Chen, Preparation of carbon molecular sieve membrane from phenol-formaldehyde Novolac resin, Carbon 40 (3) (2002) 465–467.
[134] M. Acharya, H.C. Foley, Spray-coating of nanoporous carbon membranes for air separation, Journal of Membrane Science 161 (1–2) (1999) 1–5.
[135] M. Acharya, B.A. Raich, H.C. Foley, M.P. Harold, J.J. Lerou, Metal-supported carbogenic molecular sieve membranes: synthesis and applications, Industrial and Engineering Chemistry Research 36 (8) (1997) 2924–2930.
[136] H.J. Kim, D.-G. Kim, K. Lee, Y. Baek, Y. Yoo, Y.S. Kim, B.G. Kim, J.-C. Lee, A carbonaceous membrane based on a polymer of intrinsic microporosity (PIM-1) for water treatment, Scientific Reports 6 (2016) 36078.
[137] Y.K. Kim, H.B. Park, Y.M. Lee, Gas separation properties of carbon molecular sieve membranes derived from polyimide/polyvinylpyrrolidone blends: effect of the molecular weight of polyvinylpyrrolidone, Journal of Membrane Science 251 (1–2) (2005) 159–167.
[138] Y.K. Kim, H.B. Park, Y.M. Lee, Preparation and characterization of carbon molecular sieve membranes derived from BTDA-ODA polyimide and their gas separation properties, Journal of Membrane Science 255 (1–2) (2005) 265–273.
[139] P.J. Williams, Analysis of Factors Influencing the Performance of CMS Membranes for Gas Separation, Georgia Institute of Technology, 2006.
[140] H.B. Park, Y.K. Kim, J.M. Lee, S.Y. Lee, Y.M. Lee, Relationship between chemical structure of aromatic polyimides and gas permeation properties of their carbon molecular sieve membranes, Journal of Membrane Science 229 (1–2) (2004) 117–127.
[141] Y. Kusuki, H. Shimazaki, N. Tanihara, S. Nakanishi, T. Yoshinaga, Gas permeation properties and characterization of asymmetric carbon membranes prepared by pyrolyzing asymmetric polyimide hollow fiber membrane, Journal of Membrane Science 134 (2) (1997) 245–253.
[142] P.S. Tin, T.-S. Chung, A.J. Hill, Advanced fabrication of carbon molecular sieve membranes by nonsolvent pretreatment of precursor polymers, Industrial and Engineering Chemistry Research 43 (20) (2004) 6476–6483.
[143] A. Singh, W. Koros, Significance of entropic selectivity for advanced gas separation membranes, Industrial and Engineering Chemistry Research 35 (4) (1996) 1231–1234.
[144] A. Singh-Ghosal, W. Koros, Energetic and entropic contributions to mobility selectivity in glassy polymers for gas separation membranes, Industrial and Engineering Chemistry Research 38 (10) (1999) 3647–3654.
[145] V.C. Geiszler, W.J. Koros, Effects of polyimide pyrolysis conditions on carbon molecular sieve membrane properties, Industrial and Engineering Chemistry Research 35 (9) (1996) 2999–3003.
[146] M. Kiyono, P.J. Williams, W.J. Koros, Effect of pyrolysis atmosphere on separation performance of carbon molecular sieve membranes, Journal of Membrane Science 359 (1–2) (2010) 2–10.
[147] B. Dickens, Thermally degrading polyethylene studied by means of factor-jump thermogravimetry, Journal of Polymer Science: Polymer Chemistry Edition 20 (4) (1982) 1065–1087.
[148] B. Dickens, Thermal degradation study of isotactic polypropylene using factor-jump thermogravimetry, Journal of Polymer Science: Polymer Chemistry Edition 20 (5) (1982) 1169–1183.
[149] J.E. Koresh, A. Soffer, The carbon molecular sieve membranes. General properties and the permeability of CH_4/H_2 mixture, Separation Science and Technology 22 (2–3) (1987) 973–982.
[150] S. Husain, W.J. Koros, Mixed matrix hollow fiber membranes made with modified HSSZ-13 zeolite in polyetherimide polymer matrix for gas separation, Journal of Membrane Science 288 (1–2) (2007) 195–207.
[151] J. Dechnik, J. Gascon, C.J. Doonan, C. Janiak, C.J. Sumby, Mixed-matrix membranes, Angewandte Chemie International Edition 56 (32) (2017) 9292–9310.
[152] C. Zlotea, D. Phanon, M. Mazaj, D. Heurtaux, V. Guillerm, C. Serre, P. Horcajada, T. Devic, E. Magnier, F. Cuevas, Effect of NH_2 and CF_3 functionalization on the hydrogen sorption properties of MOFs, Dalton Transactions 40 (18) (2011) 4879–4881.
[153] M.G. Goesten, J. Juan-Alcañiz, E.V. Ramos-Fernandez, K.S.S. Gupta, E. Stavitski, H. van Bekkum, J. Gascon, F. Kapteijn, Sulfation of metal-organic frameworks: opportunities for acid catalysis and proton conductivity, Journal of Catalysis 281 (1) (2011) 177–187.
[154] A. Torrisi, C. Mellot-Draznieks, R.G. Bell, Impact of ligands on CO_2 adsorption in metal-organic frameworks: first principles study of the interaction of CO_2 with functionalized benzenes. II. Effect of polar and acidic substituents, The Journal of Chemical Physics 132 (4) (2010) 044705.
[155] E.R. Geus, M.J. Den Exter, H. van Bekkum, Synthesis and characterization of zeolite (MFI) membranes on porous ceramic supports, Journal of the Chemical Society, Faraday Transactions 88 (20) (1992) 3101–3109.
[156] B.S. Kang, G.R. Gavalas, Intrapore synthesis of silicalite membranes at temperatures below 100 C, Industrial and Engineering Chemistry Research 41 (13) (2002) 3145–3150.
[157] A. Julbe, Zeolite membranes-synthesis, characterization and application, in: Introduction to Zeolite Science and Practice, 168, 2007, pp. 181–219.
[158] L.A. Truter, V.V. Ordomsky, T.A. Nijhuis, J.C. Schouten, Preparation of ZSM-5 zeolite coatings within capillary microchannels, Journal of Materials Chemistry 22 (31) (2012) 15976–15980.
[159] Y. Li, M. Pera-Titus, G. Xiong, W. Yang, E. Landrivon, S. Miachon, J.-A. Dalmon, Nanocomposite MFI-alumina membranes via pore-plugging synthesis: genesis of the zeolite material, Journal of Membrane Science 325 (2) (2008) 973–981.
[160] S. Miachon, P. Ciavarella, L. Van Dyk, I. Kumakiri, K. Fiaty, Y. Schuurman, J.-A. Dalmon, Nanocomposite MFI-alumina membranes via pore-plugging synthesis: specific transport and separation properties, Journal of Membrane Science 298 (1–2) (2007) 71–79.
[161] S. Miachon, E. Landrivon, M. Aouine, Y. Sun, I. Kumakiri, Y. Li, O.P. Prokopová, N. Guilhaume, A. Giroir-Fendler, H. Mozzanega, Nanocomposite MFI-alumina membranes via pore-plugging synthesis: preparation and morphological characterisation, Journal of Membrane Science 281 (1–2) (2006) 228–238.
[162] M. Tawalbeh, F. Tezel, B. Kruczek, S. Letaief, C. Detellier, Synthesis and characterization of silicalite-1 membrane prepared on a novel support by the pore plugging method, Journal of Porous Materials 20 (6) (2013) 1407–1421.
[163] S. Aguado, E. McLeary, A. Nijmeijer, M. Luiten, J.C. Jansen, F. Kapteijn, b-Oriented MFI membranes prepared from porous silica coatings, Microporous and Mesoporous Materials 120 (1–2) (2009) 165–169.

[164] A. Alshebani, M. Pera-Titus, E. Landrivon, T. Schiestel, S. Miachon, J.-A. Dalmon, Nanocomposite MFI-ceramic hollow fibres: prospects for CO_2 separation, Microporous and Mesoporous Materials 115 (1–2) (2008) 197–205.

[165] J. Hedlund, J. Sterte, M. Anthonis, A.-J. Bons, B. Carstensen, N. Corcoran, D. Cox, H. Deckman, W. De Gijnst, P.-P. de Moor, High-flux MFI membranes, Microporous and Mesoporous Materials 52 (3) (2002) 179–189.

[166] M. Matsukata, N. Nishiyama, K. Ueyama, Preparation of a thin zeolitic membrane, in: Studies in Surface Science and Catalysis, 84, Elsevier, 1994, pp. 1183–1190.

[167] J. Gascon, F. Kapteijn, B. Zornoza, V. Sebastian, C. Casado, J. Coronas, Practical approach to zeolitic membranes and coatings: state of the art, opportunities, barriers, and future perspectives, Chemistry of Materials 24 (15) (2012) 2829–2844.

[168] N. Rangnekar, N. Mittal, B. Elyassi, J. Caro, M. Tsapatsis, Zeolite membranes-a review and comparison with MOFs, Chemical Society Reviews 44 (20) (2015) 7128–7154.

[169] W. Fan, M.A. Snyder, S. Kumar, P.-S. Lee, W.C. Yoo, A.V. McCormick, R.L. Penn, A. Stein, M. Tsapatsis, Hierarchical nanofabrication of microporous crystals with ordered mesoporosity, Nature Materials 7 (12) (2008) 984.

[170] M. Tsapatsis, 2-dimensional zeolites, AIChE Journal 60 (7) (2014) 2374–2381.

[171] W.J. Roth, P. Nachtigall, R.E. Morris, J. Cejka, Two-dimensional zeolites: current status and perspectives, Chemical Reviews 114 (9) (2014) 4807–4837.

[172] U. Díaz, A. Corma, Layered zeolitic materials: an approach to designing versatile functional solids, Dalton Transactions 43 (27) (2014) 10292–10316.

[173] H. Chen, J. Wydra, X. Zhang, P.-S. Lee, Z. Wang, W. Fan, M. Tsapatsis, Hydrothermal synthesis of zeolites with three-dimensionally ordered mesoporous-imprinted structure, Journal of the American Chemical Society 133 (32) (2011) 12390–12393.

[174] P.-S. Lee, X. Zhang, J.A. Stoeger, A. Malek, W. Fan, S. Kumar, W.C. Yoo, S. Al Hashimi, R.L. Penn, A. Stein, Sub-40 nm zeolite suspensions via disassembly of three-dimensionally ordered mesoporous-imprinted silicalite-1, Journal of the American Chemical Society 133 (3) (2010) 493–502.

[175] A. Corma, V. Fornes, S. Pergher, T.L. Maesen, J. Buglass, Delaminated zeolite precursors as selective acidic catalysts, Nature 396 (6709) (1998) 353.

[176] K. Varoon, X. Zhang, B. Elyassi, D.D. Brewer, M. Gettel, S. Kumar, J.A. Lee, S. Maheshwari, A. Mittal, C.-Y. Sung, Dispersible exfoliated zeolite nanosheets and their application as a selective membrane, Science 334 (6052) (2011) 72–75.

[177] S. Mintova, J.-P. Gilson, V. Valtchev, Advances in nanosized zeolites, Nanoscale 5 (15) (2013) 6693–6703.

[178] G. Bonilla, I. Díaz, M. Tsapatsis, H.-K. Jeong, Y. Lee, D.G. Vlachos, Zeolite (MFI) crystal morphology control using organic structure-directing agents, Chemistry of Materials 16 (26) (2004) 5697–5705.

[179] G.N. Karanikolos, J.W. Wydra, J.A. Stoeger, H. García, A. Corma, M. Tsapatsis, Continuous c-oriented AlPO4-5 films by tertiary growth, Chemistry of Materials 19 (4) (2007) 792–797.

[180] H. Awala, J.-P. Gilson, R. Retoux, P. Boullay, J.-M. Goupil, V. Valtchev, S. Mintova, Template-free nanosized faujasite-type zeolites, Nature Materials 14 (4) (2015) 447.

[181] K.V. Agrawal, B. Topuz, T.C.T. Pham, T.H. Nguyen, N. Sauer, N. Rangnekar, H. Zhang, K. Narasimharao, S.N. Basahel, L.F. Francis, Oriented MFI membranes by gel-less secondary growth of sub-100 nm MFI-nanosheet seed layers, Advanced Materials 27 (21) (2015) 3243–3249.

[182] K.B. Yoon, Organization of zeolite microcrystals for production of functional materials, Accounts of Chemical Research 40 (1) (2007) 29–40.

[183] Y.S. Chun, K. Ha, Y.-J. Lee, J.S. Lee, H.S. Kim, Y.S. Park, K.B. Yoon, Diisocyanates as novel molecular binders for monolayer assembly of zeolite crystals on glass, Chemical Communications 17 (2002) 1846–1847.

[184] A. Huang, J. Caro, Facile synthesis of LTA molecular sieve membranes on covalently functionalized supports by using diisocyanates as molecular linkers, Journal of Materials Chemistry 21 (30) (2011) 11424–11429.

[185] A. Huang, F. Liang, F. Steinbach, J. Caro, Preparation and separation properties of LTA membranes by using 3-aminopropyltriethoxysilane as covalent linker, Journal of Membrane Science 350 (1–2) (2010) 5–9.

[186] A. Huang, J.R. Caro, Cationic polymer used to capture zeolite precursor particles for the facile synthesis of oriented zeolite LTA molecular sieve membrane, Chemistry of Materials 22 (15) (2010) 4353–4355.

[187] M.C. Petty, Langmuir-Blodgett Films: An Introduction, Cambridge University Press, 1996.

[188] K. Ariga, Y. Yamauchi, T. Mori, J.P. Hill, 25th anniversary article: what can Be done with the Langmuir-Blodgett method? Recent developments and its critical role in materials science, Advanced Materials 25 (45) (2013) 6477–6512.

[189] Z. Wang, L.H. Wee, B. Mihailova, K.J. Edler, A.M. Doyle, Langmuir-Blodgett deposited monolayers of silicalite-1 seeds for secondary growth of continuous zeolite films, Chemistry of Materials 19 (24) (2007) 5806–5808.

[190] K. Morawetz, J. Reiche, H. Kamusewitz, H. Kosmella, R. Ries, M. Noack, L. Brehmer, Zeolite films prepared via the Langmuir-Blodgett technique, Colloids and Surfaces A: Physicochemical and Engineering Aspects 198 (2002) 409–414.

[191] L. Tosheva, V.P. Valtchev, B. Mihailova, A.M. Doyle, Zeolite beta films prepared via the Langmuir-Blodgett technique, The Journal of Physical Chemistry C 111 (32) (2007) 12052–12057.

[192] K.B. Blodgett, Monomolecular films of fatty acids on glass, Journal of the American Chemical Society 56 (2) (1934) 495.

[193] I. Langmuir, V.J. Schaefer, Activities of urease and pepsin monolayers, Journal of the American Chemical Society 60 (6) (1938) 1351–1360.

[194] L. Imperiali, K.-H. Liao, C. Clasen, J. Fransaer, C.W. Macosko, J. Vermant, Interfacial rheology and structure of tiled graphene oxide sheets, Langmuir 28 (21) (2012) 7990–8000.

[195] Z. Wang, T. Yu, P. Nian, Q. Zhang, J. Yao, S. Li, Z. Gao, X. Yue, Fabrication of a highly b-oriented MFI-type zeolite film by the Langmuir-Blodgett method, Langmuir 30 (16) (2014) 4531–4534.

[196] J. Li, X. Liu, X. Lv, B. Zhang, Reliable fabrication of preferentially b-oriented silicalite-1 monolayers on various substrates using the Langmuir-Blodgett technique, Materials Letters 124 (2014) 299–301.

[197] L.H. Wee, Z. Wang, L. Tosheva, L. Itani, V. Valtchev, A.M. Doyle, Silicalite-1 films with preferred orientation, Microporous and Mesoporous Materials 116 (1–3) (2008) 22–27.

[198] X. Chen, J. Wang, D. Yin, J. Yang, J. Lu, Y. Zhang, Z. Chen, High-performance zeolite T membrane for dehydration of organics by a new varying temperature hot-dip coating method, AIChE Journal 59 (3) (2013) 936–947.

[199] H. Li, J. Wang, J. Xu, X. Meng, B. Xu, J. Yang, S. Li, J. Lu, Y. Zhang, X. He, Synthesis of zeolite NaA membranes with high performance and high reproducibility on coarse macroporous supports, Journal of Membrane Science 444 (2013) 513–522.
[200] X. Li, Y. Peng, Z. Wang, Y. Yan, Synthesis of highly b-oriented zeolite MFI films by suppressing twin crystal growth during the secondary growth, CrystEngComm 13 (11) (2011) 3657–3660.
[201] X. Li, Y. Yan, Z. Wang, Continuity control of b-oriented MFI zeolite films by microwave synthesis, Industrial and Engineering Chemistry Research 49 (12) (2010) 5933–5938.
[202] Y. Liu, Y. Li, R. Cai, W. Yang, Suppression of twins in b-oriented MFI molecular sieve films under microwave irradiation, Chemical Communications 48 (54) (2012) 6782–6784.
[203] A.I. Lupulescu, J.D. Rimer, In situ imaging of silicalite-1 surface growth reveals the mechanism of crystallization, Science 344 (6185) (2014) 729–732.
[204] H. Greer, Non-classical crystal growth of inorganic and organic materials, Materials Science and Technology 30 (6) (2014) 611–626.
[205] M.L. Gualtieri, Synthesis of MFI films on α-alumina at neutral pH, Microporous and Mesoporous Materials 117 (1–2) (2009) 508–510.
[206] M.E. Davis, Zeolites from a materials chemistry perspective, Chemistry of Materials 26 (1) (2013) 239–245.
[207] K. Iyoki, K. Itabashi, T. Okubo, Progress in seed-assisted synthesis of zeolites without using organic structure-directing agents, Microporous and Mesoporous Materials 189 (2014) 22–30.
[208] K. Itabashi, Y. Kamimura, K. Iyoki, A. Shimojima, T. Okubo, A working hypothesis for broadening framework types of zeolites in seed-assisted synthesis without organic structure-directing agent, Journal of the American Chemical Society 134 (28) (2012) 11542–11549.
[209] T.C.T. Pham, T.H. Nguyen, K.B. Yoon, Gel-free secondary growth of uniformly oriented silica MFI zeolite films and application for xylene separation, Angewandte Chemie International Edition 52 (33) (2013) 8693–8698.
[210] W. Chaikittisilp, M.E. Davis, T. Okubo, TPA+-mediated conversion of silicon wafer into preferentially-oriented MFI zeolite film under steaming, Chemistry of Materials 19 (17) (2007) 4120–4122.
[211] J. Gascon, S. Aguado, F. Kapteijn, Manufacture of dense coatings of Cu_3 $(BTC)_2$ (HKUST-1) on α-alumina, Microporous and Mesoporous Materials 113 (1–3) (2008) 132–138.
[212] S. Hermes, F. Schröder, R. Chelmowski, C. Wöll, R.A. Fischer, Selective nucleation and growth of metal-organic open framework thin films on patterned COOH/CF_3-terminated self-assembled monolayers on Au (111), Journal of the American Chemical Society 127 (40) (2005) 13744–13745.
[213] M. Arnold, P. Kortunov, D.J. Jones, Y. Nedellec, J. Kärger, J. Caro, Oriented crystallisation on supports and anisotropic mass transport of the metal-organic framework manganese formate, European Journal of Inorganic Chemistry 2007 (1) (2007) 60–64.
[214] Y. Lin, Metal organic framework membranes for separation applications, Current Opinion in Chemical Engineering 8 (2015) 21–28.
[215] Y. Liu, Z. Ng, E.A. Khan, H.-K. Jeong, C.-b. Ching, Z. Lai, Synthesis of continuous MOF-5 membranes on porous α-alumina substrates, Microporous and Mesoporous Materials 118 (1) (2009) 296–301.
[216] H. Guo, G. Zhu, I.J. Hewitt, S. Qiu, "Twin copper source" growth of metal-organic framework membrane: $Cu_3(BTC)_2$ with high permeability and selectivity for recycling H_2, Journal of the American Chemical Society 131 (5) (2009) 1646–1647.
[217] H.T. Kwon, H.-K. Jeong, In situ synthesis of thin zeolitic-imidazolate framework ZIF-8 membranes exhibiting exceptionally high propylene/propane separation, Journal of the American Chemical Society 135 (29) (2013) 10763–10768.
[218] J. Yao, D. Dong, D. Li, L. He, G. Xu, H. Wang, Contra-diffusion synthesis of ZIF-8 films on a polymer substrate, Chemical Communications 47 (9) (2011) 2559–2561.
[219] A.J. Brown, N.A. Brunelli, K. Eum, F. Rashidi, J. Johnson, W.J. Koros, C.W. Jones, S. Nair, Interfacial microfluidic processing of metal-organic framework hollow fiber membranes, Science 345 (6192) (2014) 72–75.
[220] R. Zhang, S. Ji, N. Wang, L. Wang, G. Zhang, J.R. Li, Coordination-driven in situ self-assembly strategy for the preparation of metal-organic framework hybrid membranes, Angewandte Chemie International Edition 53 (37) (2014) 9775–9779.
[221] O. Shekhah, R. Swaidan, Y. Belmabkhout, M. Du Plessis, T. Jacobs, L.J. Barbour, I. Pinnau, M. Eddaoudi, The liquid phase epitaxy approach for the successful construction of ultra-thin and defect-free ZIF-8 membranes: pure and mixed gas transport study, Chemical Communications 50 (17) (2014) 2089–2092.
[222] D. Nagaraju, D.G. Bhagat, R. Banerjee, U.K. Kharul, In situ growth of metal-organic frameworks on a porous ultrafiltration membrane for gas separation, Journal of Materials Chemistry A 1 (31) (2013) 8828–8835.
[223] R. Ranjan, M. Tsapatsis, Microporous metal organic framework membrane on porous support using the seeded growth method, Chemistry of Materials 21 (20) (2009) 4920–4924.
[224] S.R. Venna, M.A. Carreon, Highly permeable zeolite imidazolate framework-8 membranes for CO_2/CH_4 separation, Journal of the American Chemical Society 132 (1) (2009) 76–78.
[225] F. Zhang, X. Zou, X. Gao, S. Fan, F. Sun, H. Ren, G. Zhu, Hydrogen selective NH_2-MIL-53 (Al) MOF membranes with high permeability, Advanced Functional Materials 22 (17) (2012) 3583–3590.
[226] Y. Hu, X. Dong, J. Nan, W. Jin, X. Ren, N. Xu, Y.M. Lee, Metal-organic framework membranes fabricated via reactive seeding, Chemical Communications 47 (2) (2011) 737–739.
[227] X. Zhang, Y. Liu, S. Li, L. Kong, H. Liu, Y. Li, W. Han, K.L. Yeung, W. Zhu, W. Yang, New membrane architecture with high performance: ZIF-8 membrane supported on vertically aligned ZnO nanorods for gas permeation and separation, Chemistry of Materials 26 (5) (2014) 1975–1981.
[228] D. Bradshaw, A. Garai, J. Huo, Metal-organic framework growth at functional interfaces: thin films and composites for diverse applications, Chemical Society Reviews 41 (6) (2012) 2344–2381.
[229] A. Huang, H. Bux, F. Steinbach, J. Caro, Molecular-sieve membrane with hydrogen permselectivity: ZIF-22 in LTA topology prepared with 3-aminopropyltriethoxysilane as covalent linker, Angewandte Chemie 122 (29) (2010) 5078–5081.
[230] D. Hao, J. Zhang, H. Lu, W. Leng, R. Ge, X. Dai, Y. Gao, Fabrication of a COF-5 membrane on a functionalized α-Al_2O_3 ceramic support using a microwave irradiation method, Chemical Communications 50 (12) (2014) 1462–1464.
[231] J. Fu, S. Das, G. Xing, T. Ben, V. Valtchev, S. Qiu, Fabrication of COF-MOF composite membranes and their highly selective separation of H_2/CO_2, Journal of the American Chemical Society 138 (24) (2016) 7673–7680.

[232] G. Li, K. Zhang, T. Tsuru, Two-dimensional covalent organic framework (COF) membranes fabricated via the assembly of exfoliated COF nanosheets, ACS Applied Materials and Interfaces 9 (10) (2017) 8433–8436.
[233] W. Lau, A. Ismail, N. Misdan, M. Kassim, A recent progress in thin film composite membrane: a review, Desalination 287 (2012) 190–199.
[234] M.F. Jimenez-Solomon, Q. Song, K.E. Jelfs, M. Munoz-Ibanez, A.G. Livingston, Polymer nanofilms with enhanced microporosity by interfacial polymerization, Nature Materials 15 (7) (2016) 760.
[235] M.R. Chowdhury, J. Steffes, B.D. Huey, J.R. McCutcheon, 3D printed polyamide membranes for desalination, Science 361 (6403) (2018) 682–686.
[236] D. Fritsch, P. Merten, K. Heinrich, M. Lazar, M. Priske, High performance organic solvent nanofiltration membranes: development and thorough testing of thin film composite membranes made of polymers of intrinsic microporosity (PIMs), Journal of Membrane Science 401 (2012) 222–231.
[237] M. Cook, P.R. Gaffney, L.G. Peeva, A.G. Livingston, Roll-to-roll dip coating of three different PIMs for organic solvent nanofiltration, Journal of Membrane Science 558 (2018) 52–63.
[238] B.-H. Jeong, E.M. Hoek, Y. Yan, A. Subramani, X. Huang, G. Hurwitz, A.K. Ghosh, A. Jawor, Interfacial polymerization of thin film nanocomposites: a new concept for reverse osmosis membranes, Journal of Membrane Science 294 (1–2) (2007) 1–7.
[239] S. Sorribas, P. Gorgojo, C. Tellez, J. Coronas, A.G. Livingston, High flux thin film nanocomposite membranes based on metal-organic frameworks for organic solvent nanofiltration, Journal of the American Chemical Society 135 (40) (2013) 15201–15208.
[240] T. Li, Y. Pan, K.-V. Peinemann, Z. Lai, Carbon dioxide selective mixed matrix composite membrane containing ZIF-7 nano-fillers, Journal of Membrane Science 425 (2013) 235–242.
[241] S.J. DeWitt, A. Sinha, J. Kalyanaraman, F. Zhang, M.J. Realff, R.P. Lively, Critical comparison of structured contactors for adsorption-based gas separations, Annual Review of Chemical and Biomolecular Engineering (9) (2018) 129–152.
[242] A. Ahmed, M. Forster, R. Clowes, P. Myers, H. Zhang, Hierarchical porous metal-organic framework monoliths, Chemical Communications 50 (92) (2014) 14314–14316.
[243] M.I. Nandasiri, S.R. Jambovane, B.P. McGrail, H.T. Schaef, S.K. Nune, Adsorption, separation, and catalytic properties of densified metal-organic frameworks, Coordination Chemistry Reviews 311 (2016) 38–52.
[244] T. Tian, J. Velazquez-Garcia, T.D. Bennett, D. Fairen-Jimenez, Mechanically and chemically robust ZIF-8 monoliths with high volumetric adsorption capacity, Journal of Materials Chemistry A 3 (6) (2015) 2999–3005.
[245] P. Küsgens, A. Zgaverdea, H.G. Fritz, S. Siegle, S. Kaskel, Metal-organic frameworks in monolithic structures, Journal of the American Ceramic Society 93 (9) (2010) 2476–2479.
[246] O.M. Yaghi, H. Li, C. Davis, D. Richardson, T.L. Groy, Synthetic strategies, structure patterns, and emerging properties in the chemistry of modular porous solids, Accounts of Chemical Research 31 (8) (1998) 474–484.
[247] K. Schlichte, T. Kratzke, S. Kaskel, Improved synthesis, thermal stability and catalytic properties of the metal-organic framework compound $Cu_3(BTC)_2$, Microporous and Mesoporous Materials 73 (1–2) (2004) 81–88.
[248] T. Tian, Z. Zeng, D. Vulpe, M.E. Casco, G. Divitini, P.A. Midgley, J. Silvestre-Albero, J.-C. Tan, P.Z. Moghadam, D. Fairen-Jimenez, A sol-gel monolithic metal-organic framework with enhanced methane uptake, Nature Materials 17 (2) (2018) 174.
[249] F. Rezaei, P. Webley, Optimum structured adsorbents for gas separation processes, Chemical Engineering Science 64 (24) (2009) 5182–5191.
[250] C.S. Cundy, P.A. Cox, The hydrothermal synthesis of zeolites: history and development from the earliest days to the present time, Chemical Reviews 103 (3) (2003) 663–702.
[251] P. Chu, F. Dwyer, J. Vartuli, Patent 4,778,666, 1988. Eur. Pat. Appl. 0,358,827, 1990.
[252] A. Arafat, J. Jansen, A. Ebaid, H. Van Bekkum, Microwave preparation of zeolite Y and ZSM-5, Zeolites 13 (3) (1993) 162–165.
[253] C.S. Cundy, Microwave techniques in the synthesis and modification of zeolite catalysts. A review, Collection of Czechoslovak Chemical Communications 63 (11) (1998) 1699–1723.
[254] Q. Wu, X. Liu, L. Zhu, L. Ding, P. Gao, X. Wang, S. Pan, C. Bian, X. Meng, J. Xu, Solvent-free synthesis of zeolites from anhydrous starting raw solids, Journal of the American Chemical Society 137 (3) (2015) 1052–1055.
[255] W. Xu, J. Dong, J. Li, J. Li, F. Wu, A novel method for the preparation of zeolite ZSM-5, Journal of the Chemical Society, Chemical Communications 10 (1990) 755–756.
[256] U. Deforth, K. Unger, F. Schüth, Dry synthesis of B-MFI, MTN-and MTW-type materials, Microporous Materials 9 (5–6) (1997) 287–290.
[257] Y.-R. Lee, J.T. Soe, S. Zhang, J.-W. Ahn, M.B. Park, W.-S. Ahn, Synthesis of nanoporous materials via recycling coal fly ash and other solid wastes: a mini review, Chemical Engineering Journal 317 (2017) 821–843.
[258] K. Ojha, N.C. Pradhan, A.N. Samanta, Zeolite from fly ash: synthesis and characterization, Bulletin of Materials Science 27 (6) (2004) 555–564.
[259] X. Querol, N. Moreno, J. t. Umaña, A. Alastuey, E. Hernández, A. Lopez-Soler, F. Plana, Synthesis of zeolites from coal fly ash: an overview, International Journal of Coal Geology 50 (1–4) (2002) 413–423.
[260] H.-L. Chang, W.-H. Shih, Synthesis of zeolites A and X from fly ashes and their ion-exchange behavior with cobalt ions, Industrial and Engineering Chemistry Research 39 (11) (2000) 4185–4191.
[261] S. Rayalu, S. Meshram, M. Hasan, Highly crystalline faujasitic zeolites from flyash, Journal of Hazardous Materials 77 (1–3) (2000) 123–131.
[262] Y. Yaping, Z. Xiaoqiang, Q. Weilan, W. Mingwen, Synthesis of pure zeolites from supersaturated silicon and aluminum alkali extracts from fused coal fly ash, Fuel 87 (10–11) (2008) 1880–1886.
[263] S. Vichaphund, D. Aht-Ong, V. Sricharoenchaikul, D. Atong, Characteristic of fly ash derived-zeolite and its catalytic performance for fast pyrolysis of Jatropha waste, Environmental Technology 35 (17) (2014) 2254–2261.
[264] X. Querol, J.C. Umaña, F. Plana Llevat, A. Alastuey, A. López Soler, A. Medinaceli, A. Valero, M.J. Domingo, E. Garcia-Rojo, Synthesis of Zeolites from Fly Ash in a Pilot Plant Scale. Examples of Potential Environmental Applications, 1999.
[265] G. Hollman, G. Steenbruggen, M. Janssen-Jurkovičová, A two-step process for the synthesis of zeolites from coal fly ash, Fuel 78 (10) (1999) 1225–1230.
[266] M. Rubio-Martinez, C. Avci-Camur, A.W. Thornton, I. Imaz, D. Maspoch, M.R. Hill, New synthetic routes towards MOF production at scale, Chemical Society Reviews 46 (11) (2017) 3453–3480.
[267] O.M. Yaghi, Crystalline Metal-Organic Microporous Materials, Google Patents, 1997.

[268] N. Stock, S. Biswas, Synthesis of metal-organic frameworks (MOFs): routes to various MOF topologies, morphologies, and composites, Chemical Reviews 112 (2) (2011) 933–969.
[269] I. Bilecka, M. Niederberger, Microwave chemistry for inorganic nanomaterials synthesis, Nanoscale 2 (8) (2010) 1358–1374.
[270] K. Okuyama, M. Abdullah, I.W. Lenggoro, F. Iskandar, Preparation of functional nanostructured particles by spray drying, Advanced Powder Technology 17 (6) (2006) 587–611.
[271] A. Carné-Sánchez, I. Imaz, M. Cano-Sarabia, D. Maspoch, A spray-drying strategy for synthesis of nanoscale metal-organic frameworks and their assembly into hollow superstructures, Nature Chemistry 5 (3) (2013) 203.
[272] R. Ameloot, F. Vermoortele, W. Vanhove, M.B. Roeffaers, B.F. Sels, D.E. De Vos, Interfacial synthesis of hollow metal-organic framework capsules demonstrating selective permeability, Nature Chemistry 3 (5) (2011) 382.
[273] M. Faustini, J. Kim, G.-Y. Jeong, J.Y. Kim, H.R. Moon, W.-S. Ahn, D.-P. Kim, Microfluidic approach toward continuous and ultrafast synthesis of metal-organic framework crystals and hetero structures in confined microdroplets, Journal of the American Chemical Society 135 (39) (2013) 14619–14626.
[274] L. D'Arras, C. Sassoye, L. Rozes, C. Sanchez, J. Marrot, S. Marre, C. Aymonier, Fast and continuous processing of a new sub-micronic lanthanide-based metal-organic framework, New Journal of Chemistry 38 (4) (2014) 1477–1483.
[275] A. Polyzoidis, T. Altenburg, M. Schwarzer, S. Loebbecke, S. Kaskel, Continuous microreactor synthesis of ZIF-8 with high space-time-yield and tunable particle size, Chemical Engineering Journal 283 (2016) 971–977.
[276] S. Tai, W. Zhang, J. Zhang, G. Luo, Y. Jia, M. Deng, Y. Ling, Facile preparation of UiO-66 nanoparticles with tunable sizes in a continuous flow microreactor and its application in drug delivery, Microporous and Mesoporous Materials 220 (2016) 148–154.
[277] M. Gimeno-Fabra, A.S. Munn, L.A. Stevens, T.C. Drage, D.M. Grant, R.J. Kashtiban, J. Sloan, E. Lester, R.I. Walton, Instant MOFs: continuous synthesis of metal-organic frameworks by rapid solvent mixing, Chemical Communications 48 (86) (2012) 10642–10644.
[278] A. Munn, P.W. Dunne, S. Tang, E. Lester, Large-scale continuous hydrothermal production and activation of ZIF-8, Chemical Communications 51 (64) (2015) 12811–12814.
[279] P.A. Bayliss, I.A. Ibarra, E. Pérez, S. Yang, C.C. Tang, M. Poliakoff, M. Schröder, Synthesis of metal-organic frameworks by continuous flow, Green Chemistry 16 (8) (2014) 3796–3802.
[280] K.-J. Kim, Y.J. Li, P.B. Kreider, C.-H. Chang, N. Wannenmacher, P.K. Thallapally, H.-G. Ahn, High-rate synthesis of Cu-BTC metal-organic frameworks, Chemical Communications 49 (98) (2013) 11518–11520.
[281] G.H. Albuquerque, R.C. Fitzmorris, M. Ahmadi, N. Wannenmacher, P.K. Thallapally, B.P. McGrail, G.S. Herman, Gas-liquid segmented flow microwave-assisted synthesis of MOF-74 (Ni) under moderate pressures, CrystEngComm 17 (29) (2015) 5502–5510.
[282] A. Cybulski, J.A. Moulijn, Structured Catalysts and Reactors, CRC press, 2005.
[283] F. Akhtar, L. Andersson, S. Ogunwumi, N. Hedin, L. Bergström, Structuring adsorbents and catalysts by processing of porous powders, Journal of the European Ceramic Society 34 (7) (2014) 1643–1666.
[284] W.Y. Hong, S.P. Perera, A.D. Burrows, Manufacturing of metal-organic framework monoliths and their application in CO_2 adsorption, Microporous and Mesoporous Materials 214 (2015) 149–155.
[285] Y. Li, S. Perera, B. Crittenden, J. Bridgwater, The effect of the binder on the manufacture of a 5A zeolite monolith, Powder Technology 116 (1) (2001) 85–96.
[286] Z. Chen, S. Deng, H. Wei, B. Wang, J. Huang, G. Yu, Activated carbons and amine-modified materials for carbon dioxide capture-a review, Frontiers of Environmental Science and Engineering 7 (3) (2013) 326–340.
[287] G.-P. Hao, W.-C. Li, D. Qian, G.-H. Wang, W.-P. Zhang, T. Zhang, A.-Q. Wang, F. Schüth, H.-J. Bongard, A.-H. Lu, Structurally designed synthesis of mechanically stable poly (benzoxazine-co-resol)-based porous carbon monoliths and their application as high-performance CO_2 capture sorbents, Journal of the American Chemical Society 133 (29) (2011) 11378–11388.
[288] T. Jordan, Z.-L. Yu, S.-H. Yu, M. Antonietti, N. Fechler, Porous nitrogen-doped carbon monoliths derived from biopolymer-structured liquid precursors, Microporous and Mesoporous Materials 255 (2018) 53–60.
[289] L. Liu, Z. Liu, J. Yang, Z. Huang, Z. Liu, Effect of preparation conditions on the properties of a coal-derived activated carbon honeycomb monolith, Carbon 45 (14) (2007) 2836–2842.
[290] P. Avila, M. Montes, E.E. Miró, Monolithic reactors for environmental applications: a review on preparation technologies, Chemical Engineering Journal 109 (1–3) (2005) 11–36.
[291] E.V. Ramos-Fernandez, M. Garcia-Domingos, J. Juan-Alcañiz, J. Gascon, F. Kapteijn, MOFs meet monoliths: hierarchical structuring metal organic framework catalysts, Applied Catalysis A: General 391 (1–2) (2011) 261–267.
[292] S.J. DeWitt, A. Sinha, J. Kalyanaraman, F. Zhang, M.J. Realff, R.P. Lively, Critical comparison of structured contactors for adsorption-based gas separations, Annual Review of Chemical and Biomolecular Engineering 9 (2018) 129–152.
[293] R. Lively, R.R. Chance, W.J. Koros, H.W. Deckman, B.T. Kelley, Sorbent Fiber Compositions and Methods of Temperature Swing Adsorption, 2012.
[294] R.P. Lively, R.R. Chance, B. Kelley, H.W. Deckman, J.H. Drese, C.W. Jones, W.J. Koros, Hollow fiber adsorbents for CO_2 removal from flue gas, Industrial and Engineering Chemistry Research 48 (15) (2009) 7314–7324.
[295] D.A. Bhandari, N. Bessho, W.J. Koros, Hollow fiber sorbents for desulfurization of natural gas, Industrial and Engineering Chemistry Research 49 (23) (2010) 12038–12050.
[296] G. Chen, W.J. Koros, C.W. Jones, Hybrid polymer/UiO-66 (Zr) and polymer/NaY fiber sorbents for mercaptan removal from natural gas, ACS Applied Materials and Interfaces 8 (15) (2016) 9700–9709.
[297] A.R. Sujan, D.-Y. Koh, G. Zhu, V.P. Babu, N. Stephenson, A. Rosinski, H. Du, Y. Luo, W.J. Koros, R.P. Lively, High temperature activation of zeolite-loaded fiber sorbents, Industrial and Engineering Chemistry Research 57 (34) (2018) 11757–11766.
[298] B.R. Pimentel, A.W. Fultz, K.V. Presnell, R.P. Lively, Synthesis of water-sensitive metal-organic frameworks within fiber sorbent modules, Industrial and Engineering Chemistry Research 56 (17) (2017) 5070–5077.
[299] J. Zhao, W.T. Nunn, P.C. Lemaire, Y. Lin, M.D. Dickey, C.J. Oldham, H.J. Walls, G.W. Peterson, M.D. Losego, G.N. Parsons, Facile conversion of hydroxy double salts to metal-organic frameworks using metal oxide particles and atomic layer deposition thin-film templates, Journal of the American Chemical Society 137 (43) (2015) 13756–13759.

[300] H. Thakkar, S. Eastman, A. Al-Mamoori, A. Hajari, A.A. Rownaghi, F. Rezaei, Formulation of aminosilica adsorbents into 3d-printed monoliths and evaluation of their CO_2 capture performance, ACS Applied Materials and Interfaces 9 (8) (2017) 7489–7498.
[301] S. Couck, J. Cousin-Saint-Remi, S. Van der Perre, G.V. Baron, C. Minas, P. Ruch, J.F. Denayer, 3D-printed SAPO-34 monoliths for gas separation, Microporous and Mesoporous Materials 255 (2018) 185–191.
[302] S. Couck, J. Lefevere, S. Mullens, L. Protasova, V. Meynen, G. Desmet, G.V. Baron, J.F. Denayer, CO_2, CH_4 and N_2 separation with a 3DFD-printed ZSM-5 monolith, Chemical Engineering Journal 308 (2017) 719–726.
[303] H. Thakkar, S. Eastman, A. Hajari, A.A. Rownaghi, J.C. Knox, F. Rezaei, 3D-printed zeolite monoliths for CO_2 removal from enclosed environments, ACS Applied Materials and Interfaces 8 (41) (2016) 27753–27761.
[304] H. Thakkar, S. Eastman, Q. Al-Naddaf, A.A. Rownaghi, F. Rezaei, 3D-Printed metal-organic framework monoliths for gas adsorption processes, ACS Applied Materials and Interfaces 9 (41) (2017) 35908–35916.
[305] F. Zhang, Y. Ma, J. Liao, V. Breedveld, R.P. Lively, Solution-based 3D printing of polymers of intrinsic microporosity, Macromolecular Rapid Communications 39 (2018) 1800274.
[306] B.D. Freeman, Basis of permeability/selectivity tradeoff relations in polymeric gas separation membranes, Macromolecules 32 (2) (1999) 375–380.
[307] N.W. Ockwig, T.M. Nenoff, Membranes for hydrogen separation, Chemical Reviews 107 (10) (2007) 4078–4110.
[308] L.M. Robeson, The upper bound revisited, Journal of Membrane Science 320 (1–2) (2008) 390–400.
[309] P. Bernardo, E. Drioli, G. Golemme, Membrane gas separation: a review/state of the art, Industrial and Engineering Chemistry Research 48 (10) (2009) 4638–4663.
[310] M. Pera-Titus, Porous inorganic membranes for CO_2 capture: present and prospects, Chemical Reviews 114 (2) (2013) 1413–1492.
[311] M.L. Jue, D.-Y. Koh, B.A. McCool, R.P. Lively, Enabling widespread use of microporous materials for challenging organic solvent separations, Chemistry of Materials 29 (23) (2017) 9863–9876.
[312] J. Caro, M. Noack, P. Kölsch, R. Schäfer, Zeolite membranes-state of their development and perspective, Microporous and Mesoporous Materials 38 (1) (2000) 3–24.
[313] J. Caro, M. Noack, P. Kölsch, Zeolite membranes: from the laboratory scale to technical applications, Adsorption 11 (3–4) (2005) 215–227.
[314] X. Xu, W. Yang, J. Liu, X. Chen, L. Lin, N. Stroh, H. Brunner, Synthesis and gas permeation properties of an NaA zeolite membrane, Chemical Communications (7) (2000) 603–604.
[315] T. Masuda, N. Fukumoto, M. Kitamura, S.R. Mukai, K. Hashimoto, T. Tanaka, T. Funabiki, Modification of pore size of MFI-type zeolite by catalytic cracking of silane and application to preparation of H2-separating zeolite membrane, Microporous and Mesoporous Materials 48 (1–3) (2001) 239–245.
[316] M. Hong, J.L. Falconer, R.D. Noble, Modification of zeolite membranes for H_2 separation by catalytic cracking of methyldiethoxysilane, Industrial and Engineering Chemistry Research 44 (11) (2005) 4035–4041.
[317] V.V. Guerrero, Y. Yoo, M.C. McCarthy, H.-K. Jeong, HKUST-1 membranes on porous supports using secondary growth, Journal of Materials Chemistry 20 (19) (2010) 3938–3943.
[318] Y. Li, F. Liang, H. Bux, W. Yang, J. Caro, Zeolitic imidazolate framework ZIF-7 based molecular sieve membrane for hydrogen separation, Journal of Membrane Science 354 (1–2) (2010) 48–54.
[319] M.C. McCarthy, V. Varela-Guerrero, G.V. Barnett, H.-K. Jeong, Synthesis of zeolitic imidazolate framework films and membranes with controlled microstructures, Langmuir 26 (18) (2010) 14636–14641.
[320] A. Huang, J. Caro, Covalent post-functionalization of zeolitic imidazolate framework ZIF-90 membrane for enhanced hydrogen selectivity, Angewandte Chemie International Edition 50 (21) (2011) 4979–4982.
[321] S. Kluiters, Status Review on Membrane Systems for Hydrogen Separation, Energy Center of the Netherlands, Petten, The Netherlands, 2004.
[322] C. Petit, B. Mendoza, T.J. Bandosz, Hydrogen sulfide adsorption on MOFs and MOF/graphite oxide composites, ChemPhysChem 11 (17) (2010) 3678–3684.
[323] D.S. Sholl, R.P. Lively, Seven chemical separations to change the world, Nature 532 (7600) (2016) 435.
[324] S.-J. Kim, P.S. Lee, J.-S. Chang, S.-E. Nam, Y.-I. Park, Preparation of carbon molecular sieve membranes on low-cost alumina hollow fibers for use in C_3H_6/C_3H_8 separation, Separation and Purification Technology 194 (2018) 443–450.
[325] J. Liu, Y. Liu, D.K. Talay, E. Calverley, M. Brayden, M. Martinez, A new carbon molecular sieve for propylene/propane separations, Carbon 85 (2015) 201–211.
[326] R.J. Swaidan, X. Ma, I. Pinnau, Spirobisindane-based polyimide as efficient precursor of thermally-rearranged and carbon molecular sieve membranes for enhanced propylene/propane separation, Journal of Membrane Science 520 (2016) 983–989.
[327] Y.-H. Chu, D. Yancey, L. Xu, M. Martinez, M. Brayden, W. Koros, Iron-containing carbon molecular sieve membranes for advanced olefin/paraffin separations, Journal of Membrane Science 548 (2018) 609–620.
[328] M. Rungta, L. Xu, W.J. Koros, Carbon molecular sieve dense film membranes derived from Matrimid® for ethylene/ethane separation, Carbon 50 (4) (2012) 1488–1502.
[329] O. Salinas, X. Ma, E. Litwiller, I. Pinnau, High-performance carbon molecular sieve membranes for ethylene/ethane separation derived from an intrinsically microporous polyimide, Journal of Membrane Science 500 (2016) 115–123.
[330] O. Salinas, X. Ma, E. Litwiller, I. Pinnau, Ethylene/ethane permeation, diffusion and gas sorption properties of carbon molecular sieve membranes derived from the prototype ladder polymer of intrinsic microporosity (PIM-1), Journal of Membrane Science 504 (2016) 133–140.
[331] K. Kusakabe, T. Kuroda, A. Murata, S. Morooka, Formation of a Y-type zeolite membrane on a porous α-alumina tube for gas separation, Industrial and Engineering Chemistry Research 36 (3) (1997) 649–655.
[332] M.A. Carreon, S. Li, J.L. Falconer, R.D. Noble, Alumina-supported SAPO-34 membranes for CO_2/CH_4 separation, Journal of the American Chemical Society 130 (16) (2008) 5412–5413.
[333] S. Aguado, C.-H. Nicolas, V. Moizan-Baslé, C. Nieto, H. Amrouche, N. Bats, N. Audebrand, D. Farrusseng, Facile synthesis of an ultramicroporous MOF tubular membrane with selectivity towards CO_2, New Journal of Chemistry 35 (1) (2011) 41–44.
[334] A. Demessence, C. Boissiere, D. Grosso, P. Horcajada, C. Serre, G. Férey, G.J. Soler-Illia, C. Sanchez, Adsorption properties in high optical quality nanoZIF-8 thin films with tunable thickness, Journal of Materials Chemistry 20 (36) (2010) 7676–7681.

[335] T. Rodenas, I. Luz, G. Prieto, B. Seoane, H. Miro, A. Corma, F. Kapteijn, F.X.L. i Xamena, J. Gascon, Metal-organic framework nanosheets in polymer composite materials for gas separation, Nature Materials 14 (1) (2015) 48.
[336] T.H. Bae, J.S. Lee, W. Qiu, W.J. Koros, C.W. Jones, S. Nair, A high-performance gas-separation membrane containing submicrometer-sized metal-organic framework crystals, Angewandte Chemie International Edition 49 (51) (2010) 9863–9866.
[337] Y. Ban, Z. Li, Y. Li, Y. Peng, H. Jin, W. Jiao, A. Guo, P. Wang, Q. Yang, C. Zhong, Confinement of ionic liquids in nanocages: tailoring the molecular sieving properties of ZIF-8 for membrane-based CO_2 capture, Angewandte Chemie 127 (51) (2015) 15703–15707.
[338] D.Q. Vu, W.J. Koros, S.J. Miller, High pressure CO_2/CH_4 separation using carbon molecular sieve hollow fiber membranes, Industrial and Engineering Chemistry Research 41 (3) (2002) 367–380.
[339] M.G. Kamath, S. Fu, A.K. Itta, W. Qiu, G. Liu, R. Swaidan, W.J. Koros, 6FDA-DETDA: DABE polyimide-derived carbon molecular sieve hollow fiber membranes: circumventing unusual aging phenomena, Journal of Membrane Science 546 (2018) 197–205.
[340] C. Zhang, G.B. Wenz, P.J. Williams, J.M. Mayne, G. Liu, W.J. Koros, Purification of aggressive supercritical natural gas using carbon molecular sieve hollow fiber membranes, Industrial and Engineering Chemistry Research 56 (37) (2017) 10482–10490.
[341] S. Fu, E.S. Sanders, S.S. Kulkarni, G.B. Wenz, W.J. Koros, Temperature dependence of gas transport and sorption in carbon molecular sieve membranes derived from four 6FDA based polyimides: entropic selectivity evaluation, Carbon 95 (2015) 995–1006.
[342] X. Ning, W.J. Koros, Carbon molecular sieve membranes derived from Matrimid® polyimide for nitrogen/methane separation, Carbon 66 (2014) 511–522.
[343] P.S. Tin, T.-S. Chung, Y. Liu, R. Wang, Separation of CO_2/CH_4 through carbon molecular sieve membranes derived from P84 polyimide, Carbon 42 (15) (2004) 3123–3131.
[344] Y.-J. Fu, C.-C. Hu, D.-W. Lin, H.-A. Tsai, S.-H. Huang, W.-S. Hung, K.-R. Lee, J.-Y. Lai, Adjustable microstructure carbon molecular sieve membranes derived from thermally stable polyetherimide/polyimide blends for gas separation, Carbon 113 (2017) 10–17.
[345] S. Bhattacharyya, R. Han, W.-G. Kim, Y. Chiang, K.C. Jayachandrababu, J.T. Hungerford, M.R. Dutzer, C. Ma, K.S. Walton, D.S. Sholl, Acid gas stability of zeolitic imidazolate frameworks: generalized kinetic and thermodynamic characteristics, Chemistry of Materials 30 (12) (2018) 4089–4101.
[346] P. Marchetti, M.F. Jimenez Solomon, G. Szekely, A.G. Livingston, Molecular separation with organic solvent nanofiltration: a critical review, Chemical Reviews 114 (21) (2014) 10735–10806.
[347] I. Atadashi, M. Aroua, A.A. Aziz, Biodiesel separation and purification: a review, Renewable Energy 36 (2) (2011) 437–443.
[348] R.M. Gould, L.S. White, C.R. Wildemuth, Membrane separation in solvent lube dewaxing, Environmental Progress and Sustainable Energy 20 (1) (2001) 12–16.
[349] Y. Morigami, M. Kondo, J. Abe, H. Kita, K. Okamoto, The first large-scale pervaporation plant using tubular-type module with zeolite NaA membrane, Separation and Purification Technology 25 (1–3) (2001) 251–260.
[350] E. McLeary, J. Jansen, F. Kapteijn, Zeolite based films, membranes and membrane reactors: progress and prospects, Microporous and Mesoporous Materials 90 (1–3) (2006) 198–220.
[351] X. Wang, Y. Chen, C. Zhang, X. Gu, N. Xu, Preparation and characterization of high-flux T-type zeolite membranes supported on YSZ hollow fibers, Journal of Membrane Science 455 (2014) 294–304.
[352] S. Tsarkov, V. Khotimskiy, P.M. Budd, V. Volkov, J. Kukushkina, A. Volkov, Solvent nanofiltration through high permeability glassy polymers: effect of polymer and solute nature, Journal of Membrane Science 423 (2012) 65–72.
[353] P. Gorgojo, S. Karan, H.C. Wong, M.F. Jimenez-Solomon, J.T. Cabral, A.G. Livingston, Ultrathin polymer films with intrinsic microporosity: anomalous solvent permeation and high flux membranes, Advanced Functional Materials 24 (30) (2014) 4729–4737.
[354] M.L. Jue, C.S. McKay, B.A. McCool, M. Finn, R.P. Lively, Effect of nonsolvent treatments on the microstructure of PIM-1, Macromolecules 48 (16) (2015) 5780–5790.
[355] W. Ogieglo, B. Ghanem, X. Ma, I. Pinnau, M. Wessling, How much do ultrathin polymers with intrinsic microporosity swell in liquids? The Journal of Physical Chemistry B 120 (39) (2016) 10403–10410.
[356] N. Du, M.M.D. Cin, I. Pinnau, A. Nicalek, G.P. Robertson, M.D. Guiver, Azide-based cross-linking of polymers of intrinsic microporosity (PIMs) for condensable gas separation, Macromolecular Rapid Communications 32 (8) (2011) 631–636.
[357] N. Du, M.M. Dal-Cin, G.P. Robertson, M.D. Guiver, Decarboxylation-induced cross-linking of polymers of intrinsic microporosity (PIMs) for membrane gas separation, Macromolecules 45 (12) (2012) 5134–5139.
[358] Q. Song, S. Cao, R.H. Pritchard, B. Ghalei, S.A. Al-Muhtaseb, E.M. Terentjev, A.K. Cheetham, E. Sivaniah, Controlled thermal oxidative crosslinking of polymers of intrinsic microporosity towards tunable molecular sieve membranes, Nature Communications 5 (2014) 4813.
[359] T. Mitra, K.E. Jelfs, M. Schmidtmann, A. Ahmed, S.Y. Chong, D.J. Adams, A.I. Cooper, Molecular shape sorting using molecular organic cages, Nature Chemistry 5 (4) (2013) 276–281.
[360] C. Didaskalou, J. Kupai, L. Cseri, J. Barabas, E. Vass, T. Holtzl, G. Szekely, Membrane-grafted asymmetric organocatalyst for an integrated synthesis-separation platform, ACS Catalysis 8 (8) (2018) 7430–7438.
[361] T. Fodi, C. Didaskalou, J. Kupai, G.T. Balogh, P. Huszthy, G. Szekely, Nanofiltration-enabled in situ solvent and reagent recycle for sustainable continuous-flow synthesis, ChemSusChem 10 (17) (2017) 3435–3444.
[362] C. Didaskalou, S. Buyuktiryaki, R. Kecili, C.P. Fonte, G. Szekely, Valorisation of agricultural waste with an adsorption/nanofiltration hybrid process: from materials to sustainable process design, Green Chemistry 19 (13) (2017) 3116–3125.
[363] D.R. Paul, Reformulation of the solution-diffusion theory of reverse osmosis, Journal of Membrane Science 241 (2) (2004) 371–386.
[364] C. Fritzmann, J. Löwenberg, T. Wintgens, T. Melin, State-of-the-art of reverse osmosis desalination, Desalination 216 (1–3) (2007) 1–76.
[365] T. Humplik, J. Lee, S. O'hern, B. Fellman, M. Baig, S. Hassan, M. Atieh, F. Rahman, T. Laoui, R. Karnik, Nanostructured materials for water desalination, Nanotechnology 22 (29) (2011) 292001.
[366] Z. Yang, Z.-W. Zhou, H. Guo, Z. Yao, X.-H. Ma, X. Song, S.-P. Feng, C.Y. Tang, Tannic acid/Fe^{3+} nanoscaffold for interfacial polymerization: toward enhanced nanofiltration performance, Environmental Science and Technology 52 (16) (2018) 9341–9349.
[367] C. Boo, Y. Wang, I. Zucker, Y. Choo, C.O. Osuji, M. Elimelech, High performance nanofiltration membrane for effective removal of perfluoroalkyl substances at high water recovery, Environmental Science and Technology 52 (13) (2018) 7279–7288.
[368] K.P. Lee, T.C. Arnot, D. Mattia, A review of reverse osmosis membrane materials for desalination-development to date and future potential, Journal of Membrane Science 370 (1–2) (2011) 1–22.

[369] L. Li, J. Dong, T.M. Nenoff, R. Lee, Desalination by reverse osmosis using MFI zeolite membranes, Journal of Membrane Science 243 (1–2) (2004) 401–404.

[370] F. Jareman, J. Hedlund, J. Sterte, Effects of aluminum content on the separation properties of MFI membranes, Separation and Purification Technology 32 (1–3) (2003) 159–163.

[371] M.C. Duke, J. O'Brien-Abraham, N. Milne, B. Zhu, J.Y. Lin, J.C.D. da Costa, Seawater desalination performance of MFI type membranes made by secondary growth, Separation and Purification Technology 68 (3) (2009) 343–350.

[372] H. Dong, L. Zhao, L. Zhang, H. Chen, C. Gao, W.W. Ho, High-flux reverse osmosis membranes incorporated with NaY zeolite nanoparticles for brackish water desalination, Journal of Membrane Science 476 (2015) 373–383.

[373] X. Liu, C. Wang, B. Wang, K. Li, Novel organic-dehydration membranes prepared from zirconium metal-organic frameworks, Advanced Functional Materials 27 (3) (2017).

[374] Z. Hu, Y. Chen, J. Jiang, Zeolitic imidazolate framework-8 as a reverse osmosis membrane for water desalination: insight from molecular simulation, The Journal of Chemical Physics 134 (13) (2011) 134705.

[375] J. Duan, Y. Pan, F. Pacheco, E. Litwiller, Z. Lai, I. Pinnau, High-performance polyamide thin-film-nanocomposite reverse osmosis membranes containing hydrophobic zeolitic imidazolate framework-8, Journal of Membrane Science 476 (2015) 303–310.

[376] X. Liu, N.K. Demir, Z. Wu, K. Li, Highly water-stable zirconium metal-organic framework UiO-66 membranes supported on alumina hollow fibers for desalination, Journal of the American Chemical Society 137 (22) (2015) 6999–7002.

[377] Y. Lu, T. Suzuki, W. Zhang, J.S. Moore, B.J. Mariñas, Nanofiltration membranes based on rigid star amphiphiles, Chemistry of Materials 19 (13) (2007) 3194–3204.

[378] T. Suzuki, Y. Lu, W. Zhang, J.S. Moore, B.J. Mariñas, Performance characterization of nanofiltration membranes based on rigid star amphiphiles, Environmental Science and Technology 41 (17) (2007) 6246–6252.

[379] H. Fan, J. Gu, H. Meng, A. Knebel, J. Caro, High-flux membranes based on the covalent organic framework COF-LZU1 for selective dye separation by nanofiltration, Angewandte Chemie International Edition 57 (2018) 4083–4087.

[380] H.J. Kim, D.-G. Kim, K. Lee, Y. Baek, Y. Yoo, Y.S. Kim, B.G. Kim, J.-C. Lee, A carbonaceous membrane based on a polymer of intrinsic microporosity (PIM-1) for water treatment, Scientific Reports 6 (2016), 36078.

[381] S. Sircar, T. Golden, Purification of hydrogen by pressure swing adsorption, Separation Science and Technology 35 (5) (2000) 667–687.

[382] J. Stocker, M. Whysall, G. Miller, Years of PSA Technology for Hydrogen Purification, UOP LLC, 1998.

[383] Y.E. Cheon, M.P. Suh, Multifunctional fourfold interpenetrating diamondoid network: gas separation and fabrication of palladium nanoparticles, Chemistry-A European Journal 14 (13) (2008) 3961–3967.

[384] D.N. Dybtsev, H. Chun, S.H. Yoon, D. Kim, K. Kim, Microporous manganese formate: a simple metal-organic porous material with high framework stability and highly selective gas sorption properties, Journal of the American Chemical Society 126 (1) (2004) 32–33.

[385] C.-J. Li, Z.-j. Lin, M.-X. Peng, J.-D. Leng, M.-M. Yang, M.-L. Tong, Novel three-dimensional 3d-4f microporous magnets exhibiting selective gas adsorption behavior, Chemical Communications 47 (2008) 6348–6350.

[386] M. Xue, Z. Zhang, S. Xiang, Z. Jin, C. Liang, G.-S. Zhu, S.-L. Qiu, B. Chen, Selective gas adsorption within a five-connected porous metal-organic framework, Journal of Materials Chemistry 20 (19) (2010) 3984–3988.

[387] M. Xue, S. Ma, Z. Jin, R.M. Schaffino, G.-S. Zhu, E.B. Lobkovsky, S.-L. Qiu, B. Chen, Robust metal-organic framework enforced by triple-framework interpenetration exhibiting high H_2 storage density, Inorganic Chemistry 47 (15) (2008) 6825–6828.

[388] A. Mallick, S. Saha, P. Pachfule, S. Roy, R. Banerjee, Selective CO_2 and H_2 adsorption in a chiral magnesium-based metal organic framework (Mg-MOF) with open metal sites, Journal of Materials Chemistry 20 (41) (2010) 9073–9080.

[389] Y.-J. Zhang, T. Liu, S. Kanegawa, O. Sato, Interconversion between a nonporous nanocluster and a microporous coordination polymer showing selective gas adsorption, Journal of the American Chemical Society 132 (3) (2009) 912–913.

[390] S. Biswas, M. Grzywa, H.P. Nayek, S. Dehnen, I. Senkovska, S. Kaskel, D. Volkmer, A cubic coordination framework constructed from benzobistriazolate ligands and zinc ions having selective gas sorption properties, Dalton Transactions (33) (2009) 6487–6495.

[391] B. Chen, S. Ma, E.J. Hurtado, E.B. Lobkovsky, C. Liang, H. Zhu, S. Dai, Selective gas sorption within a dynamic metal-organic framework, Inorganic Chemistry 46 (21) (2007) 8705–8709.

[392] S. Ma, X.S. Wang, D. Yuan, H.C. Zhou, A coordinatively linked Yb metal-organic framework demonstrates high thermal stability and uncommon gas-adsorption selectivity, Angewandte Chemie 120 (22) (2008) 4198–4201.

[393] S. Ma, D. Yuan, X.-S. Wang, H.-C. Zhou, Microporous lanthanide metal-organic frameworks containing coordinatively linked interpenetration: syntheses, gas adsorption studies, thermal stability analysis, and photoluminescence investigation, Inorganic Chemistry 48 (5) (2009) 2072–2077.

[394] S. Ma, X.-S. Wang, C.D. Collier, E.S. Manis, H.-C. Zhou, Ultramicroporous metal-organic framework based on 9, 10-anthracenedicarboxylate for selective gas adsorption, Inorganic Chemistry 46 (21) (2007) 8499–8501.

[395] S. Ma, D. Sun, X.S. Wang, H.C. Zhou, A mesh-adjustable molecular sieve for general use in gas separation, Angewandte Chemie International Edition 46 (14) (2007) 2458–2462.

[396] S. Ma, D. Sun, D. Yuan, X.-S. Wang, H.-C. Zhou, Preparation and gas adsorption studies of three mesh-adjustable molecular sieves with a common structure, Journal of the American Chemical Society 131 (18) (2009) 6445–6451.

[397] Y.E. Cheon, M.P. Suh, Selective gas adsorption in a microporous metal-organic framework constructed of Co II 4 clusters, Chemical Communications (17) (2009) 2296–2298.

[398] M. Dinca, J.R. Long, Strong H_2 binding and selective gas adsorption within the microporous coordination solid $Mg_3(O_2C\text{-}C_{10}H_6\text{-}CO_2)_3$, Journal of the American Chemical Society 127 (26) (2005) 9376–9377.

[399] D.-C. Zhong, W.-G. Lu, L. Jiang, X.-L. Feng, T.-B. Lu, Three coordination polymers based on 1 H-Tetrazole (HTz) generated via in situ decarboxylation: synthesis, structures, and selective gas adsorption properties, Crystal Growth and Design 10 (2) (2009) 739–746.

[400] H. Chun, J. Seo, Discrimination of small gas molecules through adsorption: reverse selectivity for hydrogen in a flexible metal-organic framework, Inorganic Chemistry 48 (21) (2009) 9980–9982.

[401] J.-P. Zhang, S.K. Ghosh, J.-B. Lin, S. Kitagawa, New heterometallic carboxylate frameworks: synthesis, structure, robustness, flexibility, and porosity, Inorganic Chemistry 48 (16) (2009) 7970–7976.

[402] Z.R. Herm, J.A. Swisher, B. Smit, R. Krishna, J.R. Long, Metal-organic frameworks as adsorbents for hydrogen purification and precombustion carbon dioxide capture, Journal of the American Chemical Society 133 (15) (2011) 5664–5667.

[403] A. Dutta, N. Tyminska, G. Zhu, J. Collins, R.P. Lively, J.R. Schmidt, S. Vasenkov, Influence of hydrogen sulfide exposure on the transport and structural properties of the metal-organic framework ZIF-8, The Journal of Physical Chemistry C 122 (13) (2018) 7278–7287.

[404] R.B. Eldridge, Olefin/paraffin separation technology: a review, Industrial and Engineering Chemistry Research 32 (10) (1993) 2208–2212.

[405] N. Lamia, M. Jorge, M.A. Granato, F.A.A. Paz, H. Chevreau, A.E. Rodrigues, Adsorption of propane, propylene and isobutane on a metal–organic framework: molecular simulation and experiment, Chemical Engineering Science 64 (14) (2009) 3246–3259.

[406] J.-W. Yoon, I.-T. Jang, K.-Y. Lee, Y.-K. Hwang, J.-S. Chang, Adsorptive separation of propylene and propane on a porous metal-organic framework, copper trimesate, Bulletin of the Korean Chemical Society 31 (1) (2010) 220–223.

[407] Z. Bao, S. Alnemrat, L. Yu, I. Vasiliev, Q. Ren, X. Lu, S. Deng, Adsorption of ethane, ethylene, propane, and propylene on a magnesium-based metal-organic framework, Langmuir 27 (22) (2011) 13554–13562.

[408] Y.S. Bae, C.Y. Lee, K.C. Kim, O.K. Farha, P. Nickias, J.T. Hupp, S.T. Nguyen, R.Q. Snurr, High propene/propane selectivity in isostructural metal-organic frameworks with high densities of open metal sites, Angewandte Chemie International Edition 51 (8) (2012) 1857–1860.

[409] E.D. Bloch, W.L. Queen, R. Krishna, J.M. Zadrozny, C.M. Brown, J.R. Long, Hydrocarbon separations in a metal-organic framework with open iron (II) coordination sites, Science 335 (6076) (2012) 1606–1610.

[410] S. Yang, A.J. Ramirez-Cuesta, R. Newby, V. Garcia-Sakai, P. Manuel, S.K. Callear, S.I. Campbell, C.C. Tang, M. Schröder, Supramolecular binding and separation of hydrocarbons within a functionalized porous metal-organic framework, Nature Chemistry 7 (2) (2015) 121.

[411] J.Y. Lin, Molecular sieves for gas separation, Science 353 (6295) (2016) 121–122.

[412] J.W. Yoon, Y.K. Seo, Y.K. Hwang, J.S. Chang, H. Leclerc, S. Wuttke, P. Bazin, A. Vimont, M. Daturi, E. Bloch, Controlled reducibility of a metal-organic framework with coordinatively unsaturated sites for preferential gas sorption, Angewandte Chemie 122 (34) (2010) 6085–6088.

[413] P.-Q. Liao, W.-X. Zhang, J.-P. Zhang, X.-M. Chen, Efficient purification of ethene by an ethane-trapping metal-organic framework, Nature Communications 6 (2015) 8697.

[414] C. Gücüyener, J. van den Bergh, J. Gascon, F. Kapteijn, Ethane/ethene separation turned on its head: selective ethane adsorption on the metal-organic Framework ZIF-7 through a gate-opening mechanism, Journal of the American Chemical Society 132 (50) (2010) 17704–17706.

[415] J. van den Bergh, C. Gücüyener, E.A. Pidko, E.J. Hensen, J. Gascon, F. Kapteijn, Understanding the anomalous alkane selectivity of ZIF-7 in the separation of light alkane/alkene mixtures, Chemistry-A European Journal 17 (32) (2011) 8832–8840.

[416] M. Guisnet, J.-P. Gilson, Zeolites for Cleaner Technologies, vol. 3, Imperial College Press, London, 2002.

[417] G.-Q. Guo, H. Chen, Y.-C. Long, Separation of p-xylene from C8 aromatics on binder-free hydrophobic adsorbent of MFI zeolite. I. Studies on static equilibrium, Microporous and Mesoporous Materials 39 (1–2) (2000) 149–161.

[418] V. Finsy, C.E. Kirschhock, G. Vedts, M. Maes, L. Alaerts, D.E. De Vos, G.V. Baron, J.F. Denayer, Framework breathing in the vapour-phase adsorption and separation of xylene isomers with the metal-organic framework MIL-53, Chemistry-A European Journal 15 (31) (2009) 7724–7731.

[419] R. Lyndon, J. Bacsa, M.L. Jue, R.P. Lively, Direct Structural Evidence of Molecular Packing Effects of Xylene Isomers Adsorbed in BIF-20, Crystal Growth & Design, 2018.

[420] V. Finsy, H. Verelst, L. Alaerts, D. De Vos, P.A. Jacobs, G.V. Baron, J.F. Denayer, Pore-filling-dependent selectivity effects in the vapor-phase separation of xylene isomers on the metal-organic framework MIL-47, Journal of the American Chemical Society 130 (22) (2008) 7110–7118.

[421] C. Serre, F. Millange, C. Thouvenot, M. Nogues, G. Marsolier, D. Louër, G. Férey, Very large breathing effect in the first nanoporous chromium (III)-Based solids: MIL-53 or $Cr^{III}(OH)\cdot\{O_2C\text{-}C_6H_4\text{-}CO_2\}\cdot\{HO_2C\text{-}C_6H_4\text{-}CO_2H\}_x\cdot H_2O_y$, Journal of the American Chemical Society 124 (45) (2002) 13519–13526.

[422] M.A. Moreira, J.O.C. Santos, A.F. Ferreira, J.M. Loureiro, F. Ragon, P. Horcajada, K.-E. Shim, Y.-K. Hwang, U.-H. Lee, J.-S. Chang, Reverse shape selectivity in the liquid-phase adsorption of xylene isomers in zirconium terephthalate MOF UiO-66, Langmuir 28 (13) (2012) 5715–5723.

[423] H. Li, M. Eddaoudi, M. O'Keeffe, O.M. Yaghi, Design and synthesis of an exceptionally stable and highly porous metal-organic framework, Nature 402 (6759) (1999) 276.

[424] Z.Y. Gu, X.P. Yan, Metal-organic framework MIL-101 for high-resolution gas-chromatographic separation of xylene isomers and ethylbenzene, Angewandte Chemie International Edition 49 (8) (2010) 1477–1480.

[425] M. Edgar, R. Mitchell, A.M. Slawin, P. Lightfoot, P.A. Wright, Solid-state transformations of zinc 1, 4-benzenedicarboxylates mediated by hydrogen-bond-forming molecules, Chemistry-A European Journal 7 (23) (2001) 5168–5175.

[426] Z.-Y. Gu, D.-Q. Jiang, H.-F. Wang, X.-Y. Cui, X.-P. Yan, Adsorption and separation of xylene isomers and ethylbenzene on two Zn-terephthalate metal-organic frameworks, The Journal of Physical Chemistry C 114 (1) (2009) 311–316.

[427] M. Rungta, J.S. Abichandani, D.L. Pilliod, R.G. Tinger, A. Go, K. Zhang, S. Nair, J. Gee, D. Sholl, Process for Recovering Para-Xylene Using a Metal Organic Framework Adsorbent in a Simulated Moving-Bed Process, US Patents, 2018.

[428] W. Chouyyok, R.J. Wiacek, K. Pattamakomsan, T. Sangvanich, R.M. Grudzien, G.E. Fryxell, W. Yantasee, Phosphate removal by anion binding on functionalized nanoporous sorbents, Environmental Science and Technology 44 (8) (2010) 3073–3078.

[429] H. Xu, L. Xie, M. Hakkarainen, Coffee-ground-derived quantum dots for aqueous processable nanoporous graphene membranes, ACS Sustainable Chemistry and Engineering 5 (6) (2017) 5360–5367.

[430] H.K. Chae, D.Y. Siberio-Perez, J. Kim, Y. Go, M. Eddaoudi, A.J. Matzger, M. O'keeffe, O.M. Yaghi, A route to high surface area, porosity and inclusion of large molecules in crystals, Nature 427 (6974) (2004) 523.

[431] Y. Xu, S. Jin, H. Xu, A. Nagai, D. Jiang, Conjugated microporous polymers: design, synthesis and application, Chemical Society Reviews 42 (20) (2013) 8012–8031.

[432] T. Hasell, A.I. Cooper, Porous organic cages: soluble, modular and molecular pores, Nature Reviews Materials 1 (9) (2016) 16053.

[433] R.K. Kemmerlin, Carbon Molecular Sieve Membranes for Aggressive Sour Gas Separations, Georgia Institute of Technology, 2012.

[434] A.M. Hillock, S.J. Miller, W.J. Koros, Crosslinked mixed matrix membranes for the purification of natural gas: effects of sieve surface modification, Journal of Membrane Science 314 (1–2) (2008) 193–199.

[435] D.F. Sanders, Z.P. Smith, R. Guo, L.M. Robeson, J.E. McGrath, D.R. Paul, B.D. Freeman, Energy-efficient polymeric gas separation membranes for a sustainable future: a review, Polymer 54 (18) (2013) 4729–4761.

[436] S.E. Kentish, C.A. Scholes, G.W. Stevens, Carbon dioxide separation through polymeric membrane systems for flue gas applications, Recent Patents on Chemical Engineering 1 (1) (2008) 52–66.

[437] H.A. Mannan, H. Mukhtar, T. Murugesan, R. Nasir, D.F. Mohshim, A. Mushtaq, Recent applications of polymer blends in gas separation membranes, Chemical Engineering and Technology 36 (11) (2013) 1838–1846.

[438] G. Meindersma, A. De Haan, Economical feasibility of zeolite membranes for industrial scale separations of aromatic hydrocarbons, Desalination 149 (1–3) (2002) 29–34.

[439] A. Urtiaga, E. Gorri, C. Casado, I. Ortiz, Pervaporative dehydration of industrial solvents using a zeolite NaA commercial membrane, Separation and Purification Technology 32 (1–3) (2003) 207–213.

[440] D. Gorri, A. Urtiaga, I. Ortiz, Pervaporative recovery of acetic acid from an acetylation industrial effluent using commercial membranes, Industrial and Engineering Chemistry Research 44 (4) (2005) 977–985.

[441] S. Sommer, T. Melin, Performance evaluation of microporous inorganic membranes in the dehydration of industrial solvents, Chemical Engineering and Processing: Process Intensification 44 (10) (2005) 1138–1156.

[442] A.L. Deshayes, E.E. Miró, G.I. Horowitz, Xylene isomerization in a membrane reactor: part II. Simulation of an industrial reactor, Chemical Engineering Journal 122 (3) (2006) 149–157.

[443] A. Urtiaga, E. Gorri, P. Gómez, C. Casado, R. Ibáñez, I. Ortiz, Pervaporation technology for the dehydration of solvents and raw materials in the process industry, Drying Technology 25 (11) (2007) 1819–1828.

[444] K. Sato, K. Sugimoto, T. Nakane, Separation of ethanol/ethyl acetate mixture by pervaporation at 100-130°C through NaY zeolite membrane for industrial purpose, Microporous and Mesoporous Materials 115 (1–2) (2008) 170–175.

[445] Y. Cui, H. Kita, K.-I. Okamoto, Zeolite T membrane: preparation, characterization, pervaporation of water/organic liquid mixtures and acid stability, Journal of Membrane Science 236 (1–2) (2004) 17–27.

[446] K. Sato, K. Sugimoto, T. Nakane, Synthesis of industrial scale NaY zeolite membranes and ethanol permeating performance in pervaporation and vapor permeation up to 130°C and 570 kPa, Journal of Membrane Science 310 (1–2) (2008) 161–173.

[447] K. Eum, C. Ma, D.Y. Koh, F. Rashidi, Z. Li, C.W. Jones, R.P. Lively, S. Nair, Zeolitic imidazolate framework membranes supported on macroporous carbon hollow fibers by fluidic processing techniques, Advanced Materials Interfaces 4 (12) (2017).

[448] T. Yokoi, M. Yoshioka, H. Imai, T. Tatsumi, Diversification of RTH-type zeolite and its catalytic application, Angewandte Chemie International Edition 48 (52) (2009) 9884–9887.

[449] B. Xie, H. Zhang, C. Yang, S. Liu, L. Ren, L. Zhang, X. Meng, B. Yilmaz, U. Müller, F.-S. Xiao, Seed-directed synthesis of zeolites with enhanced performance in the absence of organic templates, Chemical Communications 47 (13) (2011) 3945–3947.

[450] H. Shi, Organic template-free synthesis of SAPO-34 molecular sieve membranes for CO_2-CH_4 separation, RSC Advances 5 (48) (2015) 38330–38333.

[451] I. Stassen, N. Campagnol, J. Fransaer, P. Vereecken, D. De Vos, R. Ameloot, Solvent-free synthesis of supported ZIF-8 films and patterns through transformation of deposited zinc oxide precursors, CrystEngComm 15 (45) (2013) 9308–9311.

[452] S. Lagorsse, F. Magalhaes, A. Mendes, Carbon molecular sieve membranes: sorption, kinetic and structural characterization, Journal of Membrane Science 241 (2) (2004) 275–287.

[453] S. Lagorsse, A. Leite, F. Magalhaes, N. Bischofberger, J. Rathenow, A. Mendes, Novel carbon molecular sieve honeycomb membrane module: configuration and membrane characterization, Carbon 43 (4) (2005) 809–819.

[454] S.M. Maier, A. Jentys, J.A. Lercher, Steaming of zeolite BEA and its effect on acidity: a comparative NMR and IR spectroscopic study, The Journal of Physical Chemistry C 115 (16) (2011) 8005–8013.

[455] H.G. Karge, H.K. Beyer, Solid-state ion exchange in microporous and mesoporous materials, in: Post-Synthesis Modification I, Springer, 2002, pp. 43–201.

[456] G. Kennedy, S. Lawton, A. Fung, M. Rubin, S. Steuernagel, Multinuclear MAS NMR studies of zeolites MCM-22 and MCM-49, Catalysis Today 49 (4) (1999) 385–399.

[457] G.J. Kennedy, S.L. Lawton, M.K. Rubin, 29Si MAS NMR studies of a high silica form of the novel molecular sieve: MCM-22, Journal of the American Chemical Society 116 (24) (1994) 11000–11003.

[458] D. Ma, F. Deng, R. Fu, X. Han, X. Bao, MAS NMR studies on the dealumination of zeolite MCM-22, The Journal of Physical Chemistry B 105 (9) (2001) 1770–1779.

[459] B. Elyassi, X. Zhang, M. Tsapatsis, Long-term steam stability of MWW structure zeolites (MCM-22 and ITQ-1), Microporous and Mesoporous Materials 193 (2014) 134–144.

[460] H. Wang, Y. Lin, Effects of water vapor on gas permeation and separation properties of MFI zeolite membranes at high temperatures, AIChE Journal 58 (1) (2012) 153–162.

[461] K. Sato, K. Sugimoto, Y. Sekine, M. Takada, M. Matsukata, T. Nakane, Application of FAU-type zeolite membranes to vapor/gas separation under high pressure and high temperature up to 5 MPa and 180 C, Microporous and Mesoporous Materials 101 (1–2) (2007) 312–318.

[462] K.-I. Sawamura, T. Shirai, M. Takada, Y. Sekine, E. Kikuchi, M. Matsukata, Selective permeation and separation of steam from water-methanol-hydrogen gas mixtures through mordenite membrane, Catalysis Today 132 (1–4) (2008) 182–187.

[463] S.A.S. Rezai, J. Lindmark, C. Andersson, F. Jareman, K. Möller, J. Hedlund, Water/hydrogen/hexane multicomponent selectivity of thin MFI membranes with different Si/Al ratios, Microporous and Mesoporous Materials 108 (1–3) (2008) 136–142.

[464] S.-J. Kim, Z. Xu, G.K. Reddy, P. Smirniotis, J. Dong, Effect of pressure on high-temperature water gas shift reaction in microporous zeolite membrane reactor, Industrial and Engineering Chemistry Research 51 (3) (2012) 1364–1375.

[465] M. Bernal, E. Piera, J. Coronas, M. Menendez, J. Santamarıa, Mordenite and ZSM-5 hydrophilic tubular membranes for the separation of gas phase mixtures, Catalysis Today 56 (1–3) (2000) 221–227.

[466] C. Günther, H. Richter, I. Voigt, Zeolite membranes for hydrogen and water separation under harsh conditions, Chemical Engineering Transactions 32 (2013) 1963–1968.

[467] J.J. Low, A.I. Benin, P. Jakubczak, J.F. Abrahamian, S.A. Faheem, R.R. Willis, Virtual high throughput screening confirmed experimentally: porous coordination polymer hydration, Journal of the American Chemical Society 131 (43) (2009) 15834–15842.

[468] J. Canivet, A. Fateeva, Y. Guo, B. Coasne, D. Farrusseng, Water adsorption in MOFs: fundamentals and applications, Chemical Society Reviews 43 (16) (2014) 5594–5617.

[469] N.C. Burtch, H. Jasuja, K.S. Walton, Water stability and adsorption in metal-organic frameworks, Chemical Reviews 114 (20) (2014) 10575–10612.

[470] K.A. Cychosz, A.J. Matzger, Water stability of microporous coordination polymers and the adsorption of pharmaceuticals from water, Langmuir 26 (22) (2010) 17198–17202.

[471] P. Küsgens, M. Rose, I. Senkovska, H. Fröde, A. Henschel, S. Siegle, S. Kaskel, Characterization of metal-organic frameworks by water adsorption, Microporous and Mesoporous Materials 120 (3) (2009) 325–330.

[472] J.L. Rowsell, O.M. Yaghi, Effects of functionalization, catenation, and variation of the metal oxide and organic linking units on the low-pressure hydrogen adsorption properties of metal-organic frameworks, Journal of the American Chemical Society 128 (4) (2006) 1304–1315.

[473] C. Janiak, S.K. Henninger, Porous coordination polymers as novel sorption materials for heat transformation processes, CHIMIA International Journal for Chemistry 67 (6) (2013) 419–424.

[474] L. Xu, M. Rungta, J. Hessler, W. Qiu, M. Brayden, M. Martinez, G. Barbay, W.J. Koros, Physical aging in carbon molecular sieve membranes, Carbon 80 (2014) 155–166.

[475] G.B. Wenz, Tuning Carbon Molecular Sieve Membrane Performance for Challenging Gas Separations, Georgia Institute of Technology, 2017.

[476] Z. Lai, M. Tsapatsis, J.P. Nicolich, Siliceous ZSM-5 membranes by secondary growth of b-oriented seed layers, Advanced Functional Materials 14 (7) (2004) 716–729.

[477] P. Ciavarella, H. Moueddeb, S. Miachon, K. Fiaty, J.-A. Dalmon, Experimental study and numerical simulation of hydrogen/isobutane permeation and separation using MFI-zeolite membrane reactor, Catalysis Today 56 (1–3) (2000) 253–264.

[478] M. Lassinantti, F. Jareman, J. Hedlund, D. Creaser, J. Sterte, Preparation and evaluation of thin ZSM-5 membranes synthesized in the absence of organic template molecules, Catalysis Today 67 (1–3) (2001) 109–119.

[479] C.J. Gump, V.A. Tuan, R.D. Noble, J.L. Falconer, Aromatic permeation through crystalline molecular sieve membranes, Industrial and Engineering Chemistry Research 40 (2) (2001) 565–577.

[480] T.Q. Gardner, J.G. Martinek, J.L. Falconer, R.D. Noble, Enhanced flux through double-sided zeolite membranes, Journal of Membrane Science 304 (1–2) (2007) 112–117.

[481] J. Hedlund, M. Noack, P. Kölsch, D. Creaser, J. Caro, J. Sterte, ZSM-5 membranes synthesized without organic templates using a seeding technique, Journal of Membrane Science 159 (1–2) (1999) 263–273.

[482] A. Navajas, R. Mallada, C. Téllez, J. Coronas, M. Menéndez, J. Santamaría, Study on the reproducibility of mordenite tubular membranes used in the dehydration of ethanol, Journal of Membrane Science 299 (1–2) (2007) 166–173.

[483] M. Bernal, J. Coronas, M. Menendez, J. Santamarıa, On the effect of morphological features on the properties of MFI zeolite membranes, Microporous and Mesoporous Materials 60 (1–3) (2003) 99–110.

[484] Z. Lai, G. Bonilla, I. Diaz, J.G. Nery, K. Sujaoti, M.A. Amat, E. Kokkoli, O. Terasaki, R.W. Thompson, M. Tsapatsis, Microstructural optimization of a zeolite membrane for organic vapor separation, Science 300 (5618) (2003) 456–460.

[485] A. Singh-Ghosal, W. Koros, Air separation properties of flat sheet homogeneous pyrolytic carbon membranes, Journal of Membrane Science 174 (2) (2000) 177–188.

[486] M.S. Strano, H.C. Foley, Temperature-and pressure-dependent transient analysis of single component permeation through nanoporous carbon membranes, Carbon 40 (7) (2002) 1029–1041.

[487] M.B. Shiflett, H.C. Foley, On the preparation of supported nanoporous carbon membranes, Journal of Membrane Science 179 (1–2) (2000) 275–282.

[488] M.S. Strano, H.C. Foley, Modeling ideal selectivity variation in nanoporous membranes, Chemical Engineering Science 58 (12) (2003) 2745–2758.

[489] Q. Song, S. Jiang, T. Hasell, M. Liu, S. Sun, A.K. Cheetham, E. Sivaniah, A.I. Cooper, Porous organic cage thin films and molecular-sieving membranes, Advanced Materials 28 (13) (2016) 2629–2637.

[490] X. Meng, F.-S. Xiao, Green routes for synthesis of zeolites, Chemical Reviews 114 (2) (2013) 1521–1543.

[491] R. Xu, W. Pang, J. Yu, Q. Huo, J. Chen, Structural Chemistry of Microporous Materials, Chemistry of Zeolites and Related Porous Materials, John Wiley & Sons (Asia) Pte Ltd., 2007.

[492] J. Park, N.F. Attia, M. Jung, M.E. Lee, K. Lee, J. Chung, H. Oh, Sustainable nanoporous carbon for CO_2, CH_4, N_2, H_2 adsorption and CO_2/CH_4 and CO_2/N_2 separation, Energy 158 (1) (2018) 9–16.

[493] E.-P. Ng, J.-Y. Goh, T.C. Ling, R.R. Mukti, Eco-friendly synthesis for MCM-41 nanoporous materials using the non-reacted reagents in mother liquor, Nanoscale Research Letters 8 (1) (2013) 120.

[494] I. Azcune, I. García, P.M. Carrasco, A. Genua, M. Tanczyk, M. Jaschik, K. Warmuzinski, G. Cabañero, I. Odriozola, Facile and scalable synthesis of nanoporous materials based on poly (ionic liquid) s, ChemSusChem 7 (12) (2014) 3407–3412.

[495] L. Cseri, M. Razali, P. Pogany, G. Szekely, Organic solvents in sustainable synthesis and engineering, in: Green Chemistry, Elsevier, 2017, pp. 513–553.

[496] A. Figoli, T. Marino, S. Simone, E. Di Nicolo, X.-M. Li, T. He, S. Tornaghi, E. Drioli, Towards non-toxic solvents for membrane preparation: a review, Green Chemistry 16 (9) (2014) 4034–4059.

[497] R.K. Henderson, C. Jiménez-González, D.J. Constable, S.R. Alston, G.G. Inglis, G. Fisher, J. Sherwood, S.P. Binks, A.D. Curzons, Expanding GSK's solvent selection guide-embedding sustainability into solvent selection starting at medicinal chemistry, Green Chemistry 13 (4) (2011) 854–862.

[498] B. Torok, T. Dransfield, Green Chemistry: An Inclusive Approach, Elsevier, 2017.

CHAPTER

4

Sustainability of One-Dimensional Nanostructures: Fabrication and Industrial Applications

Jaison Jeevanandam[1], *Anandhakumar Sundaramurthy*[2,3], *Varsha Sharma*[2], *Chandran Murugan*[2], *Kaushik Pal*[4], *Mohamed Hamada Abdel Kodous*[5], *Michael K. Danquah*[6]

[1]Department of Chemical Engineering, Curtin University, CDT250 Miri, Sarawak, Malaysia; [2]SRM Research Institute, SRM Institute of Science and Technology, Kattankulathur, Kanchipuram, Tamil Nadu, India; [3]Department of Physics and Nanotechnology, SRM Institute of Science and Technology, Kattankulathur, Kanchipuram, Tamil Nadu, India; [4]Department of Nanotechnology, Bharath University, BIHER Research Park, Selaiyur, Chennai, Tamil Nadu, India; [5]Center for Nanotechnology (CNT), School of Engineering and Applied Sciences, Nile University, Sheikh Zayed, Giza, Egypt; [6]Chemical Engineering Department, University of Tennessee, Chattanooga, Tennessee, United States

OUTLINE

Sustainable Nanoscale Engineering
https://doi.org/10.1016/B978-0-12-814681-1.00004-7

1. INTRODUCTION

Since the industrial revolution, artificial acceleration of swift growth in the global economy has resulted in continuous and ever-increasing energy demand. Fuel consumption, especially conventional fossil fuels, has correspondingly increased to meet the escalating energy demand and this has resulted in an increase in carbon dioxide (CO_2) emissions. Carbon is the basic element that is present in all the living organisms. Atomic carbon is present as the basic building blocks in deoxyribonucleic acid (DNA) and proteins of all the living cells [1]. Even in nonliving structures, carbon atom plays a significant role in binding with other atoms and in the fabrication of strong and stable materials [2]. Volcanoes and oceanic beds are the natural means of releasing carbon as CO_2 and their derivatives into the atmosphere [3]. Likewise, burning of several nonrenewable resources and fossil fuels such as petrol, diesel, and gasoline has also released CO_2 and their derivatives into the atmosphere [4]. CO_2 released through natural activities is usually balanced by other natural events such as afforestation. However, in recent times, there is an imbalance in the atmospheric CO_2 due to the increase of CO_2-releasing vehicles, pollution, and deforestation activities, especially after the industrial revolution. Recent reports showed that there is a 2% increase in the global CO_2 emissions in 2017, which is rapidly increasing, since the year 2000 [5]. These carbon gases released via anthropogenic activities are the causative factors in increasing the atmospheric greenhouse gases. Among various anthropogenic greenhouse gas emissions, thermal power plants have solely contributed to about 80% and the remaining 20% is emitted from other industrial sources such as cement production plants (~7%), oil refineries (~6%), and iron and steel industry (~5%). Greenhouse gases possess the ability to absorb and emit radiant energy within the thermal infrared region [6]. The presence of large volumes of greenhouse gases in the atmosphere traps heat from the sun and increases ground temperature. Furthermore, it is projected that global emission of greenhouse gases will rise by 25%–90% in 2030 to bring the total CO_2 level to >2000 ppm [7]. Thus, climate change and global warming are the consequents of the greenhouse gas effect [8]. Hence, it is highly necessary to capture the carbon that is released into the atmosphere or avoid its release into the atmosphere.

There are several regulations and laws drafted and passed by international bodies to regulate the emission of greenhouse gases, especially carbon-based gases. Carbon credit is a term or certificate that is issued as a right to a company or a country to emit one ton of CO_2 or other greenhouse gas mass that is equivalent to one ton of CO_2 [9]. Similarly, carbon emission trading is a protocol for emissions during trading that targets only CO_2 and helps to reduce future climate change [10]. Likewise, low-carbon economy, carbon neutral fuel, carbon pricing, carbon offset, personal carbon trading, carbon tax, carbon finance, carbon diet, and carbon footprint are the distinct regulatory rules that are used to reduce anthropogenic carbon emission into the atmosphere [11]. Out of which carbon neutrality or net zero carbon footprint is a widely practiced and researched content which denotes achieving of net zero emissions of carbon by matching a calculated carbon quantity that is released with an equal sequestered carbon quantity to make up the difference [12]. Industries and several countries are trying to achieve carbon neutrality by adopting and practicing various regulatory measures. Carbon capturing is one such measure which aims at capturing carbon-based gases, even before releasing into the atmosphere or from the atmosphere. Conventional carbon capturing methods are based on capturing carbon from fossil fuel power plants and depositing them in a place (underground) where it will not enter the atmosphere or use chemicals to bind these gases to form beneficial by-products [13]. However, these methods lead to other side effects such as toxicity in the environment and pollution [14]. Thus, search for alternative methods to capture and store carbon has been continued by researchers to find a technique or material with improved carbon capturing efficiency.

Nanostructures have received much interest during the past decade due to their enhanced volume to surface area ratio, unique structures, and superior properties compared with their bulk counterparts. The best example is the nanosized gold, which has a lower melting point than bulk gold [15]. In addition, nanomaterials possess enhanced mechanical properties, when compared with bulk materials [16]. Zinc oxide nanowires are the best examples because they have five times more sophisticated fracture strains than the conventional zinc oxides [17]. Other nanocomposites such as carbon-doped carbon dots, titanium dioxide [18], and nanometals, such as gold, platinum, and ruthenium, have displayed greater behavior as a catalyst for numerous applications and reactions [19]. Nanomaterials are the foundation for any application that is based on nanotechnology and are usually explained as the materials with a minimum of one nanoscale dimension from 1 to 100 nm. It is noteworthy that attention has been given to research that bothers on the characterization and preparation of diverse nanomaterials types. This cumulative interest toward nanomaterials is ascribed to the necessity of their optical, chemical, electrical, and mechanical properties by altering their morphology and size. Various experiment-based studies have proved that the nanomaterials' morphology has substantial impact on their properties [20]. The sheer control of these properties is responsible for using nanomaterials in novel large-scale industrial applications and in the development of new technologies. Among different nanomaterial morphology, one-dimensional (1D) nanostructures showed potential to be highly useful in carbon capturing applications, which determines the efficiency of an industry in maintaining environment sustainability and reducing greenhouse gases emissions. Due to improvements in the field of nanotechnology, these nanostructures are highly explored to effectively capture carbon. There are several 1D nanostructures, namely nanorods, nanofibers, nanowires, nanotubes, nanocylinders, and nanowhiskers [21]. Each morphology exhibits a unique carbon capturing property which makes them superior to other nanoparticles in reducing carbon emission. This chapter is an overview of various 1D nanostructures that are synthesized through several methods. In addition, this chapter also gives information about different 1D nanostructures that are used for the large-scale industrial application and in maintaining environment sustainability.

2. ONE-DIMENSIONAL NANOSTRUCTURES AND THEIR PROPERTIES

1D nanostructures have received significant attention in the past decade. These unique nanostructures are a good platform to investigate the impact of the reduction in dimension and size toward thermal, electrical transport, or mechanical properties. They play a significant role in the fabrication of functional units and deliver nanoscale properties to electronic, electromechanical, electrochemical, and optoelectronic devices compared with other nanoscale dimensions [22]. Novel 1D nanostructures with unique properties such as nanorods, nanowires, nanofibers, and nanotubes have one dimension (x-axis) in the nanodimension. The difference between these 1D nanostructures is their aspect ratio (length to diameter or width ratio). A nanorod possesses a length to width aspect ratio of 3–5, whereas a nanowire possesses an aspect ratio above 1000. Nanotubes have a similar aspect ratio as nanowires but are hollow. Nanofibers are similar to nanorods in terms of aspect ratio, but they are different in terms of their fibrous structure and method of fabrication [23]. These nanomaterials are most promising in optical excitations and transport of electrons which are crucial factors in nanoscale devices due to their smaller size as compared with zero and two-dimensional (2D) nanostructures [24]. 1D nanostructures are further classified into carbon allotrope, metal, and novel materials depending on the precursor used for their fabrication method as shown in Fig. 4.1.

2.1 One-Dimensional Carbon Allotrope Nanostructures

Carbon is an element in the periodic table which is the base for all the life that exists on earth. Furthermore, this element possesses a high potential in several technical and technological applications, ranging from pharmaceutical, biomedical to electronic applications [25]. Diamond with sp^3 hybridization and graphite with sp^2 hybridized carbon atoms are the two natural allotropes of elemental carbon that are well known [26]. Only after 1985, fullerene was

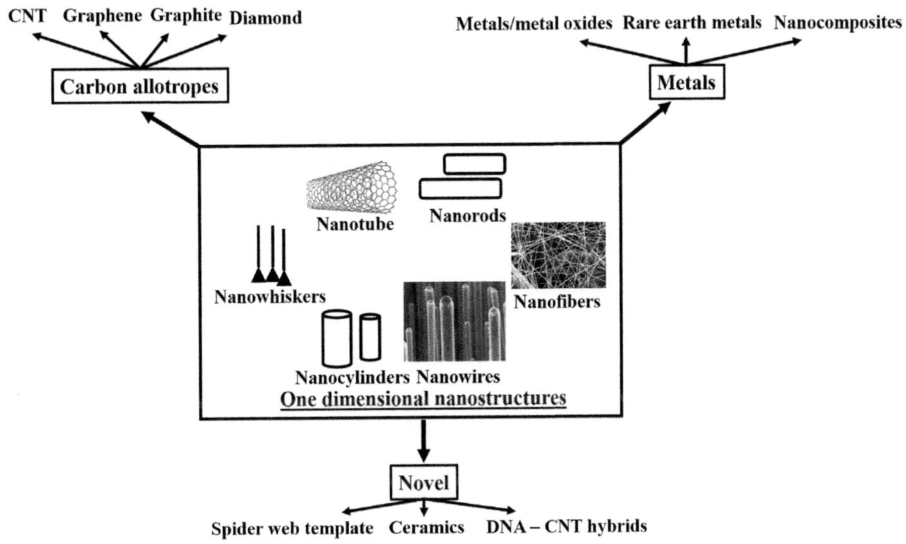

FIGURE 4.1 Schematics of several one-dimensional nanostructure types.

synthesized, which marks a new era of artificial carbon allotrope materials [27]. Fullerene is a zero-dimensional carbon nanostructure that is electronically confined in all the directions [28]. Later, carbon nanotubes (CNTs), graphene, and graphyne 20 were developed to match the advancements in the technologies to fabricate nanostructures [29]. Among these nanostructures, CNT is a 1D nanostructure, which is electronically confined in two directions and one dimension is in nanosize [30]. Compared with other allotropes, CNTs are generally utilized in various purposes due to their enhanced surface area [31] to accommodate several molecules in them which facilitate electron transfer and drug encapsulation [32].

2.1.1 Nanotubes

CNTs are 1D tubular nanostructures that have an interior hollow cavity that are composed of carbon and were discovered in 1991 [33]. It received substantial importance in applications due to their exclusive optical, electronic, and mechanical properties [34]. CNTs are fabricated with 0.4–40 nm in diameter with lengths of micron to centimeter range and are 100 times stronger compared with steel with four times advanced electrical conductivity compared with copper [15]. CNTs have several structures with unique electrical characteristics depending on their length, the number of layers, type of helicity, and thickness [35]. CNTs can perform as either semiconductors or conductors (metals), based on the twist or chirality of the nanotube [36]. The arm-chair type of CNT is metallic in nature [37], whereas chiral and zigzag types of CNTs are semiconductors [38] as shown in Fig. 4.2. These nanostructures are further divided into two main classes according to the number of walls, namely single-walled carbon nanotubes (SWCNTs) [39] and multiwalled carbon nanotubes (MWCNTs) [40]. SWCNTs possess a diameter of 1 nanometer and its structure can be achieved by rolling graphene of one-atom-thick layer into a morphology of cylinder [41]. Meanwhile, MWCNTs are fabricated by rolling multiple graphene layers with an inner layer distance of 3.4 Å [42]. A detailed classification of SWCNTs and MWCNTs displaying their properties [43] is presented in Table 4.1.

Apart from CNTs, other allotropes of carbon have also been fabricated into nanotubes. Currently, the simple wet chemical method was used to fabricate hierarchical manganese dioxide nanoparticles—multiwalled fullerene nanotubes—graphene hybrid films. The hybrid film was subjected to electrochemical studies and the results showed that the film possesses enhanced electrocatalytic activity compared with manganese dioxide—graphene films. In addition, the hybrid film also showed a swift response to hydrogen peroxide with high sensitivity and excellent selectivity [45]. It can be noted that fullerene nanotubes possess potential attention among researchers for several applications [46]. Similarly, helix-shaped graphite nanotubes have been synthesized using spatial-velocity hodograph concept for quantitative extrusion of carbon tubule from catalytic particle [47]. Several other synthesis

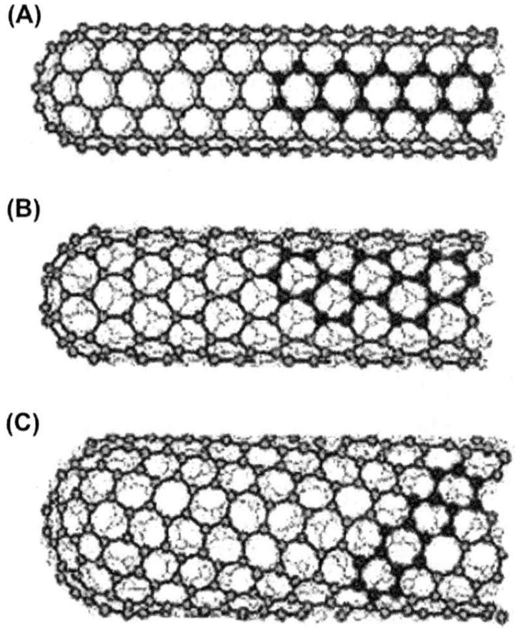

FIGURE 4.2 Types of carbon nanotubes: (A) arm-chair, (B) zig-zag, and (C) chiral [44].

TABLE 4.1 Properties of Single- and Multiwalled Carbon Nanotubes.

Characteristics	Single-walled Carbon Nanotube	Multiwalled Carbon Nanotube
Structure	Graphene rolled into a single layer as a tube	Graphene rolled into multiple layers as a tube
Requirement of catalyst	During synthesis, catalyst is needed	During synthesis, catalyst is not needed
Synthesis	Large-scale synthesis is tedious as it involves suitable control over atmospheric and growth conditions	Large synthesis is easy
Purity	Poorly pure	Highly pure
Formation of defects	High chances of defect formation while functionalization process	Low chance of defect formation; however, once formed, it is tedious to recover
Accumulation in living cells	Accumulation is less in living cells	Accumulation is more in living cells
Characterization	Easy to characterize and evaluate	Difficult to characterize due to its complex structure
Twist	Easy to twist and are highly pliable	Difficult to twist
Type	Chirality decides their conductivity	Always metallic

methods have been reported for the fabrication of graphite nanotubes that possess unique properties to be used in various applications [48]. Likewise, large-scale template free synthesis was used to produce graphene nanotubes doped with nitrogen and nitrogen-doped silicon dioxide graphene nanotubes with enhanced field emission property [49]. The literature available also reveals that graphene nanotubes are widely synthesized for numerous applications [50]. Thus, it is clear that all the allotropes of carbon can be fabricated into nanotubes with unique properties that are highly beneficial in industrial applications.

2.1.2 Nanowires

Nanowires are another set of 1D carbon nanostructures as shown in Fig. 4.3A that have potential in enormous applications. Amorphous carbon nanowires are prepared by heating graphite powder and nickel at 1200°C in a quartz tube [51]. Later, nitrogen-enriched carbon nanowires with superior-specific capacitance and current density were prepared by polyaniline nanowire–mediated direct carbonization method [52]. Apart from normal carbon nanowires, hollow carbon nanowires are synthesized using polyaniline nanowire as precursor using the pyrolyzation method [53]. Several different types of carbon nanowires with unique properties have been synthesized in recent times and found to be useful in industrial, environmental, and biomedical applications [54].

Other than carbon, allotropes of carbon such as graphene, graphite, fullerene, and diamond were also fabricated into nanowires. Recently, reduced graphene oxide nanowires were fabricated using the three-dimensional printing technique at room temperature by chemical or thermal treatment with the ability to grow at selected sites and in any direction [55]. Meanwhile, hierarchical porous iron oxide–graphene nanowires were developed into electrode

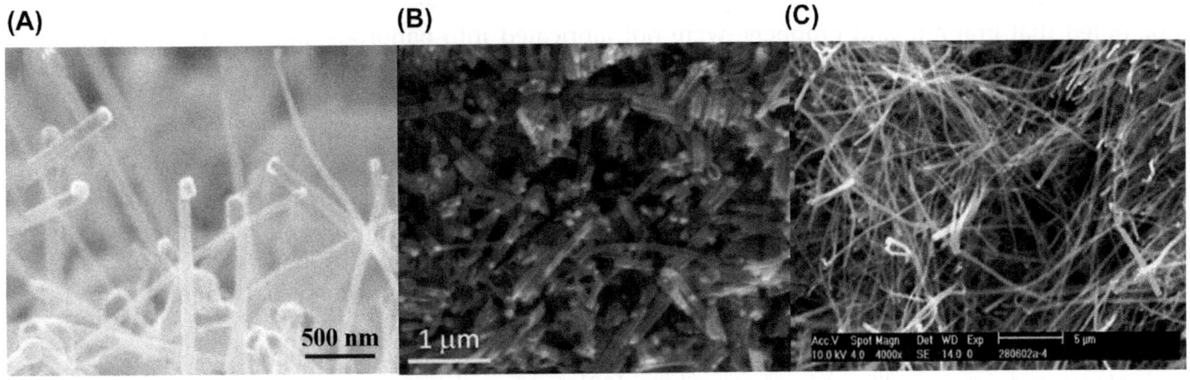

FIGURE 4.3 SEM images of (A) graphene nanowires [57a], (B) carbon nanorods [88], and (C) carbon nanofibers [89].

materials with the support of amorphous vanadium oxide matrixes which possess high power and energy densities. It is worthy of note that even at a high density of current, high reversible capacity was maintained by the porous nanowires, which are highly beneficial as anodes in energy storage applications [56]. There are several studies that have been conducted to fabricate different sizes and properties of graphene nanowires [57]. Similarly, graphite nanowires were on the radar of researchers for their unique adsorption properties. Presently, graphite-silicon carbide hybrid nanowires were synthesized via electrospinning and annealing processes with tunable electromagnetic absorption ability [58]. Most of the several literatures available prove that graphite acts as a growth pattern for other nanowires [59]. Diamond which is a popular allotrope of carbon is also fabricated into nanowires for desired applications [60]. Atomistic simulation, reactive ion etching and biofunctionalization are the recent methods that are used for fabricating diamond nanowires with unique properties. Hybrid nanowires of diamonds along with graphite [61] and single diamond nanowire crystals that are used for nanoelectromechanical switches [62] were also available due to advancements in nanoscopes and nanotechnology. In the recent past, diamond nanowires are synthesized for applications ranging from gas sensing [63] to catalysis [64]. Fullerenes are also fabricated into nanowires to be used in organic electronics and other similar applications [65]. Because carbon nanowires have shown potentially unique properties compared with fullerene and fullerene is preferred to be useful as zero-dimensional nanostructures, very few reports are available on fullerene nanowires after the year 2014. All these studies showed that nanowires also possess benefits in several applications, other than nanotubes.

2.1.3 *Nanorods*

Nanosized rods are another significant 1D morphology that can be fabricated using carbon atoms. Similar to nanotubes and nanowires, carbon nanorods as displayed in Fig. 4.3B are extensively synthesized and used for a wide range of applications [66]. Carbon nanorods were synthesized at a comparatively low temperature of 200°C using different metallic catalysts and tetrachloroethylene as feedstock. The result revealed that the catalyst and the reaction time show a substantial impact in the nucleation and nanorods growth [67]. Likewise, simple nanocasting through chemical vapor deposition method was utilized for the fabrication of mesostructured carbon nanorods with mesoporous silica SBA-15 rods as rigid templates. The study demonstrated that the pore size of silica template was vital in carbon nanorod formation [68]. Recently, self-templated carbon nanorods using catalyst-free strategy was fabricated via morphology-maintained thermal conversion of metal-organic framework (MOF). The obtained CNTs were further utilized to synthesize two- to six-layered graphene ribbons through chemical-activated graphene nanoribbons. These nanoribbons and nanorods are proposed to be apt for scalable industrial purposes and exhibit excellent supercapacitor electrode properties [69]. Additionally, there are several literatures that show a variety of techniques to fabricate CNTs [70]. However, doping and composites enhance the efficiency of carbon nanorods in supercapacitor and lithium ion battery applications [71].

Apart from carbon, fullerene has also been fabricated into nanorods and has been investigated to bring out their unique properties. Heat treatment–mediated solvent-induced self-assembly [72], ultrarapid interfacial formed at room temperature [73], controlled solvent vapor for homogenous distribution in polymer matrix [74], and solution-based self-assembly using graphene substrates [75] were some of the synthesis procedures used to produce fullerene nanorods. Currently, fullerene nanorods are fabricated for several intrinsic applications. In a recent study, fullerene nanorods were fabricated using the surfactant-assisted self-assembly method, and they exhibited unusual thermal transport behavior [76]. Similarly, fullerene nanorods were synthesized through liquid–liquid interfacial precipitation (LLIP) method and the study proved that LLIP possesses enhanced optoelectronic properties [77]. It can be noted that graphite and graphene were not fabricated into nanorods, and future research can focus in this regime to increase the number of carbon allotropes with nanorod morphology that can be utilized for several applications.

2.1.4 *Nanofibers*

Nanofibers, similar to other 1D morphologies, have been extensively fabricated using carbon as shown in Fig. 4.3C. Currently, the carbonization method via a mixture of bacterial cellulose and potassium citrate was used to bridge porous carbon nanosheets and carbon nanofibers. This nanostructure possesses a high specific surface area, precise capacitance, exceptional rate capacity, and supreme cycling constancy as compared with pure porous carbon nanosheets [78]. In addition, several doped and composite materials were synthesized using carbon nanofibers for supercapacitor applications such as carbon nanofibers, manganese dioxide [79], bubble nanorod–structured iron oxide, carbon nanofibers [80], and mesoporous iron encapsulated in iron carbide nanoparticles and nitrogen-doped carbon nanofibers [81]. These composites and doped nanostructures showed enhanced properties compared with conventional carbon nanofibers.

Fullerenes are the other set of carbon allotrope that are widely fabricated in nanofiber morphology. The LLIP precipitation method was initially used for the synthesis of 1D fullerene nanofibers. The study proved that the solvent and precipitation agent plays a crucial role in the nanofibers development [82]. In addition, fullerene nanofiber electrodes with platinum support were also synthesized for enhanced electrochemical applications [83]. Apart from nanofibers, fullerenes were fabricated extensively into rodlike nanowhiskers [84] and nanocylinder morphology [85]. Nanopipes [86] and nanocylinders [87] are other unique 1D nanostructures that are fabricated using allotropes of carbons. All these studies showed that the change in morphology of 1D nanostructures leads to a variation in property which is highly helpful in using them for desired applications.

2.2 One-Dimensional Metal Nanostructures

There are several 1D metal nanostructures that are synthesized for application ranging from electronics to biomedical, as an alternative to carbon allotropes. Additionally, 1D metal nanostructures are classified into metals, metal oxides, and composites with each category consisting of nanotubes, nanofibers, nanorods, and nanowires.

2.2.1 Metals

Gold and silver are the most prominent metals that are fabricated into 1D nanostructures like nanosized fibers, wires, rods, and tubes. Gold nanotubes are generally synthesized using the template method [90], electroless synthesis through powerful auxiliary reagents such as 4-(dimethylamino) pyridine, cobalt nanoparticles as templates, and electron beam thinning of gold thin foil [91]. Recently, methods such as galvanic displacement of silver wires [92], anodic aluminum oxide template [93], and enhanced methanol oxidation catalysis [94] were used for gold nanotube fabrication. Gold was also synthesized into nanorods using techniques such as wet chemical, electrochemical, photochemical, and biosynthesis via bacterial enzymes [95]. In addition, gold nanowires were synthesized using the wormlike micelles as soft templates, liquid phase, self-assembly by nanoimprinting, atomic force microscopy–based nanoxerography, and green synthesis methods using pomegranate juice [96]. In addition, nanofibers [97], nanocylinders [98], nanowhiskers [99], and nanotapers [100] are other 1D morphologies that are fabricated using gold for desired large-scale, sustainable applications.

Silver is an important metal that shows a wide variety of properties in nanodimensions and is highly helpful in terms of applications, especially as an antimicrobial agent. Electroless deposition in porous aluminum oxide templates [101], bolaamphiphile–based organogel template [102], and functionalized silica rod as a template [103] are the general methods that are used for silver nanotube fabrication. Lately, synthesis of silver nanotubes has been less in number; however, silver is added or combined with CNT to enhance the property of both materials [104]. Meanwhile, silver nanorods are widely synthesized using the wet chemical method [105], natural tubule clay template [106], and sonochemical [107] methods. In recent times, several green synthesis methods using *Piper nigrum* [108], *Citrus medica* fruit extract [109], and germinated fenugreek seed extract [110] have been introduced for less toxic silver nanorod fabrication for several sustainable industrial applications. Other metals such as magnesium, copper, aluminum, and platinum were also fabricated into 1D morphologies due to advancements in nanomaterial fabrication methods for unique industrial applications with sustainability [111].

2.2.2 Metal Oxides

1D oxide materials are highly stable nanostructures with enormous unique properties and application importance. There are several 1D metal oxide nanostructures available and are used in various industries. Thus, only few important widely synthesized 1D metal oxide nanostructures that are utilized in diverse applications are mentioned in this section. Zinc oxide is a common oxide that is synthesized in various morphologies, especially 1D nanostructures. At the time of this research, zinc oxide nanotubes were synthesized using anodized zinc foils in aqueous washing and baking soda solution at room temperature. The resultant nanotube was used in the detection of breast cancer–related volatile organics [112]. Similarly, methanol-sensing zinc oxide nanotubes were extensively studied through quantum chemical methods [113]. Additionally, nanorods [114], nanofibers [115], and nanowhiskers [116] of zinc oxides were also fabricated to enhance their property to use them in unique applications.

Other than zinc, iron and titanium are the second widely synthesized 1D nanostructures in the group of metal oxides. Iron oxide nanowires as magnetic photocatalysts [117], for lithium-promoted conversion reaction [118] and filter for airborne bacterial inactivation [119], were extensively researched. In addition, titanium oxide nanotubes were also largely produced for photocatalytic [120], hydrogen extraction [121], and water desalination applications [122]. Likewise, nanofibers of titanium oxide were synthesized for effectual solar cells of the dye-sensitized model [123] and hydrogen extraction [124]. Copper, silicon, and magnesium are other sets of oxides that are fabricated in one dimension for desired applications ranging from biomedical to electronics. Copper oxide nanowires were synthesized for photothermochemical reduction [125], supercapacitor electrodes [126], and water disinfection [127]. Similarly, copper oxide nanorods were fabricated for supercapacitor [128], photoelectrode [129], and lithium-ion batteries [130]. Silicon dioxide or silica such as silica-based nanofibers [131], nanowires [132], and nanocylinders [133] were also fabricated with promising ability to be useful for sustainable applications in biomedical and electrical material manufacturing industries.

2.2.3 *Rare Earth Metals and Nanocomposites*

Various rare earth metals are also synthesized into 1D nanostructures for electronic and several unique applications. Gallium has been fabricated into nanowires in the form of gallium phosphide [134] and gallium nitride [135], whereas nanorods are synthesized with gallium nitride [136] and gallium oxide [137]. Likewise, thallium, lanthanum, europium, gadolinium, thorium, and uranium were also fabricated into 1D nanostructures via exclusive synthesis approaches for their enhanced sustainability properties. In addition, rubidium, beryllium, strontium, vanadium, yttrium, niobium, molybdenum, ruthenium, rhodium, palladium, hafnium, tantalum, rhenium, and iridium were some of the rare earth metals that were extensively fabricated into 1D nanostructures intended for peculiar large-scale, industrial applications [138].

Nanocomposites are nanosized composites that are fabricated to compensate the properties of two materials in a single novel structure [139]. There are various types of material combinations that are used for fabricating nanocomposites. These nanocomposites are widely classified into magnetic and nonmagnetic materials, based on the absence or presence of iron oxide in the composite combination. There are various studies that show the synthesis routes, characteristics, properties, and applications of iron oxide and carbon allotrope–based 1D nanocomposites such as iron oxide–CNT [140] and iron oxide–graphene [56]. Likewise, metals, metal oxides, and polymers were also incorporated into 1D iron oxides to form nanocomposites [141]. Nonmagnetic nanocomposites are formed using the metal–metal [142], metal–metal oxide [143], metal–polymer [144], and sometimes metal–carbon allotropes [145]. These nanocomposites are significant in improving the properties of 1D nanostructures to explore their ability to be used in a variety of applications.

2.3 Novel One-Dimensional Nanostructures

Apart from the group of 1D nanostructures, there are certain novel nanostructures that possess the ability to be used in various applications in the future. Spiders, especially cribellate spiders, produce nanosized fibers to form their web to trap insects for their food. Thus, various studies have been performed using spiders for producing nanofibers without any artificial viscous chemicals or electrospinning [146]. Cerium oxide [147], carbon nanofibers [148] as shown in Fig. 4.4A, and also several nanocomposite nanofibers [149] were also synthesized using spider-based nanofiber fabrication technique. Likewise, DNA–CNT hybrid nanocomposite combination was fabricated for enzyme detection applications [150]. Fluorescence quenching [151] and atomistic simulations using quantum mechanics along with the molecular mechanistic approach [152] were used to bind DNA with SWCNT and DNA nucleotide with CNT hybrids, respectively. Protein–porphyrin hybrids wrapped around SWCNT as displayed in Fig. 4.4B are the new research trends in bionanomaterials for large-scale, sustainable bioelectronic applications

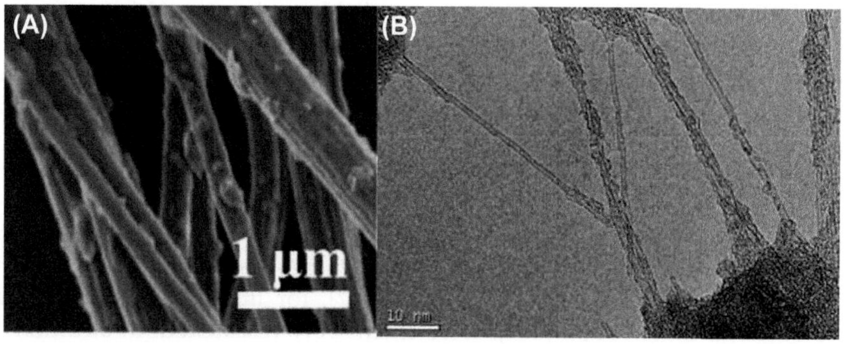

FIGURE 4.4 (A) SEM image of carbon nanofibers produced by spider silk [148] and (B) TEM image of protein–porphyrin hybrids wrapped around single-walled carbon nanotubes [160].

[153]. Boron nitride is fabricated into exclusive nanotube structures via several chemical synthesis approaches as they have gained attention of research for sustainable, less toxic, large-scale applications in pharmaceutical and hydrogen storage [154]. Currently, numerous 1D ceramic nanostructures are being studied extensively in various applications because of their exceptional properties.

This section is the overview of numerous 1D nanoparticles that are fabricated, since the emergence of nanotechnological advancements. These nanostructures are superior than other nanomaterials due to their large surface to volume ratio in one dimension and enhanced surface area on the other dimensions. This unique surface property of 1D nanostructures can adsorb or hold numerous molecules per particle which makes them to be beneficial in sustainable industrial applications, waste management, and carbon capture applications, compared with other nanomaterials [155]. This section also reveals that there are wide variety of 1D nanostructures with several elements which can be picked depending on the necessity of the desired purposes. The availability of this diverse 1D nanostructures is possible only due to the progressions in their fabrication processes. The selection of synthesis approach is highly significant to obtain nanostructures with desired morphology, size, and property [156]. It is noteworthy from the section that the 1D nanostructures are mostly fabricated via chemical and physical structures. In synthesis approaches, the selection of precursor, solvents, and their quantities play a major role in deciding the structural properties of 1D materials, toxicity toward environment, and the ability to utilize those structures in applications with sustainability. However, most of the conventional organic and alcoholic solvents used as reducing and stabilizing agent for 1D nanomaterial fabrication are reported to be toxic to both environment and other living organisms [157]. Thus, green solvents namely imidazolium-based ionic liquids and recycling of solvents are introduced as an alternate to conventional toxic solvents to reduce the toxicity and maintain sustainability [158]. Moreover, among 90% solvent used for the 1D nanostructure fabrication, only few percentages involve in the reaction and the rest are considered as waste. This leads to the enhanced recycling of solvents via nanofiltration process for filtering toxic organic solvents which contributes to the sustainable continuous-flow synthesis and lowering of carbon footprints [159].

3. APPLICATION OF 1D NANOSTRUCTURES IN INDUSTRIES AND FOR ENVIRONMENTAL SUSTAINABILITY MAINTENANCE

1D nanostructures are used in several applications due to their unique dimensional properties as mentioned in the previous section. Two methods are being adopted and considered by different countries to lessen their CO_2 emission from industries. First is the passive approach that involves (a) improving proficiency of energy and encouraging conservation of energy, (b) usage of fuels emitting low carbon (e.g., hydrogen, nuclear energy, and natural gas), (c) improving the consumption of renewable energy (wind, solar, hydropower, and bioenergy), and (d) adopting a greener approach to increase the forest area. Secondly, the active approach that involves capturing carbon and storing it in other forms. Capturing of carbon and converting them into beneficial products is a widely investigated application of 1D nanostructures with significance in large-scale industrial operations. A schematic of various carbon capturing methods and the involvement of 1D nanostructures in this application was displayed in Fig. 4.5. In addition, the sustainable carbon capturing ability of nanostructures makes them beneficial in unique applications such as carbon storage, sustainable energy production, and pollution control as displayed in Fig. 4.6.

3.1 CO_2 Capturing Methods in Industries

There are three ways to capture CO_2 from the industrial processes: (a) pre-, (b) post-, and (c) rich oxy-combustion capture.

3.1.1 *Precombustion Route*

It helps in converting emission of CO_2 into hydrogen-rich gas, which can be further used either as fuel directly [161] or used for production of synthetic chemicals. In this route, the fuel is oxidized partially at low levels of oxygen using a gasifier for syngas formation (a mixture of hydrogen and carbon monoxide), which undergoes water gas shift reactions and results in CO_2 and hydrogen-rich gas stream. The high concentration (>20%) of CO_2 and absence of other impurities in CO_2/hydrogen-rich gas forms hydrogen-rich gas after capture of CO_2.

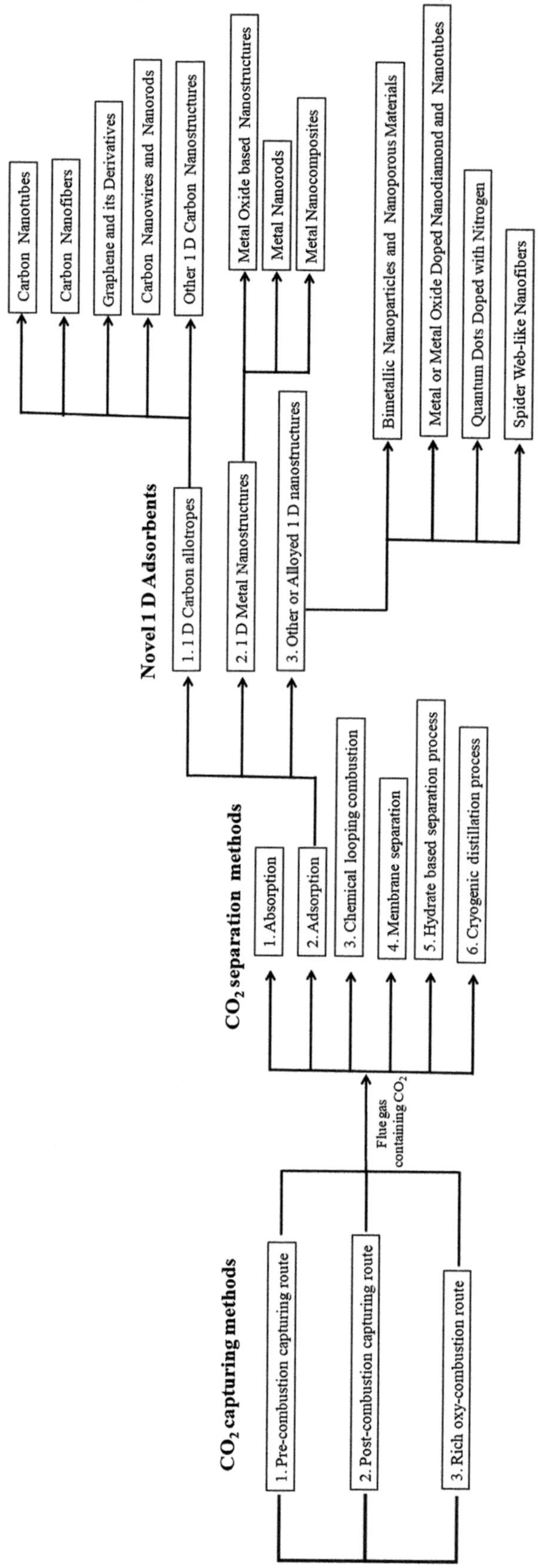

FIGURE 4.5 Schematics showing carbon capturing, carbon separation methods, and the involvement of 1D nanostructures.

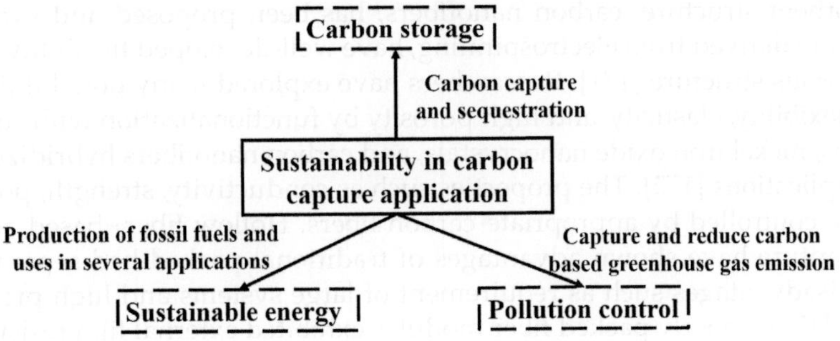

FIGURE 4.6 Schematics of various carbon capture applications in maintaining sustainability.

3.1.2 *Postcombustion Capturing Route*

It is a direct method and provides low technical risk, wherein flue gases coming out of the combustion chamber are subjected to processes of CO_2 capturing to absorb flue gas CO_2. As CO_2 concentration in the flue gas is generally low, the energy penalty and associated cost of capturing as well as transportation is quite high [162].

3.1.3 *Rich Oxy-Combustion Route*

In oxy-fuel-based combustion, oxygen rather than air is utilized for fuel combustion, which avoids nitrogen contamination and simplifies the CO_2 capture process. The important aspect for oxy-fuel combustion method is the cost involved in cryogenic separation of oxygen and retrofit for the existing boilers to withstand the oxy-fuel combustion. It is noteworthy that carbon capture is different from carbon storage applications. Capturing carbon is an activity of binding (capture) carbon molecules from the atmosphere via molecular linkages of 1D nanostructures. Contrarily, storing carbon in the hollow spaces of nanostructures is called carbon storage. The aspect ratio of 1D nanostructures plays a key role in determining their carbon capture and storage performances.

3.2 1D Nanostructures for Industrial Carbon Capture Applications

Carbon-based and metal-based 1D nanomaterials have been thoroughly investigated for captivating flue gas CO_2. Several distinct 1D materials that specifically contain carbon and metal atoms such as nanotubes, nanorods, nanofibers, and nanowires are highlighted in this section. Microsized porosity particularly acts as a predominant characteristic for the uptake of carbon in 1D materials by physisorption [163]. In addition, the synthesis and modification process might result in interactions based on chemisorption. Furthermore, due to their hydrophobic nature, such materials are highly affected by moisture, which decreases its performance as compared with dry conditions [164]. The presence of carbon matrix roots on the surface of oxygen and nitrogen functional groups leads to an upsurge in their elementary performance, which alters the distribution of charge in the carbon graphene allotrope layers, thereby increasing its adsorption capacity [165]. The adsorption interactions can be improved by functionalizing the 1D nanostructure with nitrogen groups. These are considered as virtuous materials for carbon capture applications due to their stable chemistry, tailorable porous structure, large surface area, and surface functionalization.

3.2.1 *1D Carbon Nanostructures*

CO_2 has been found to have stronger interactions with CNT walls, creating a concentrated environment of CO_2 in the core [166]. According to reports, SWCNTs displayed 150% increase in CO_2 adsorption capacity (29.97 g of CO_2 was adsorbed per kg) due to its good geometric structure when compared to MWCNTs (12.09 g of CO_2 was adsorbed per kg) which had poor porosity [167]. Currently, a new form of C50 fullerene–based peanut-shaped CNT has been discovered to have better CO_2 adsorption ability than SWCNT [168]. In recent times, MWCNT impregnated with polyethylenimine (PEI), a maximum CO_2 uptake of 93 mg/g of fiber at 20% by weight of PEI, was observed [169]. Due to chemical and thermal stability of CNT matrix and PEI sorbent, the applicability of these microtubes for CO_2 capture is anticipated. With all the physicochemical properties and adsorption behavior studies undertaken, modified CNTs have proved to be a better adsorption option than that of modified zeolites and modified granular–activated carbons, and thus, they possess latent application for capturing CO_2 from gas streams [170].

Another novel adsorbent structure, carbon nanofibers, has been proposed and evaluated for CO_2 capture. Carbon nanofibers, mostly derived from electrospinning, have well-developed flexibility, high stability, hydrophobicity, and tailorable porous structure [171]. Researchers have explored many possibilities of using carbon nanofibers with improved flexibility, elasticity, and high porosity by functionalization with cobalt oxide nanoparticles, nickel/cobalt nanotubes, nickel iron oxide nanocrystals, and carbon nanofibers hybridized by tin oxide nanoclusters for CO_2 capture applications [172]. The properties such as conductivity, strength, pore size, electrical properties, and density can be controlled by appropriate carbon fibers. Hollow fiber–based adsorption systems in the postcombustion CO_2 capture have shown advantages of traditional packed bed separation systems on the one hand and reduce the disadvantages such as requirement of large systems and high pressure drops at high flow rates on the other hand [173]. Loosely packed fiber modules exhibited external film resistances over a range of velocities of 0–50 cm/s, whereas tightly packed fiber modules exhibited reduced film resistances only at higher gas velocities.

Graphene-based nanomaterials and its derivatives are the unique set of 1D nanomaterials, as they have large surface area compared with other carbon materials as discussed above. Major interactions such as Van der Waals interaction or pole–ion and pole–pole interaction are present in graphene between CO_2 and solid surface which makes them highly stable and strong [174]. Reduced graphene oxide had shown remarkable application in CO_2 capture capacity of 248% of the samples reduced at 300°C, which was 7-fold higher than zeolites and 3.5-fold higher than other activated carbons [175]. In a separate study, novel permeable composite glassy polymer membranes, few graphene layers, and poly (2,6-dimethyl-1,4-phenylene oxide) was fabricated [176]. Spheres of carbon that are hollow and doped with nitrogen possess a high surface area of 762 m^2g^{-1} and was demonstrated as an effective material for capturing CO_2 with 2.67 mmol g^{-1} capacity with improved selectivity in mixed gases [177].

3.2.2 *One-Dimensional Metal Nanostructures*

During combustion, metal oxides are abridged to metal and the fuel gets converted into water and CO_2. Pure CO_2 is then obtained by further reduction of metal and removal of water by condensation. Thus, oxides of metals such as copper, iron, manganese, and nickel are used as oxygen carriers in the combustion process. They are promising carbon capture materials as they possess high adsorption capacities at higher temperatures (>300°C). A sorbent with the ability to regenerate, select, and be stable in several temperatures was formed by mesoporous magnesium oxide, which demonstrated 10 times greater CO_2 removal than nonporous conventional magnesium [178]. Magnesium oxide [111] nanosheets as faceted sorbent for carbon capture has displayed 65% increment in the capacity of CO_2 despite 30% decrement in the surface area after the sintering process. Apart from that, calcium oxide has also been widely used as an adsorbent material. However, its major drawback lies in sorbent throughput in multiple adsorption–desorption cycles [179]. Novel pod-like nanohollow calcium oxide particles were fabricated by Yang et al. through calcination of calcium carbonate nanopods during precipitation process [180]. Its adsorption capacity (17.5 mol CO_2/kg) was found to be greater than common calcium oxide. It was able to retain 50% of CO_2 through adsorption even after 50 adsorption–desorption cycles, which was also much better than commercial calcium oxide (25%).

Titanium dioxide nanorods were synthesized hydrothermally using PEI as surfactant and copper nanoparticles were immobilized onto its surface to influence optical properties. This resulting nanocomposite with LSPR (localized surface plasmon resonance) effect was used for photoreduction of CO_2 to methane with the yield of up to 2.91 ppm/g per hour [181]. Synthesis of a nanorod with γ-alumina-graphene oxide composite along with a high surface area was achieved as a solid adsorbent amine for capturing CO_2 (200.6 mg g PEI^{-1}) at 75°C and release at 100°C with recyclability. The reduced graphene oxide helped to improve the thermal conductivity and increase the surface area [182]. More recently, microwave-assisted sol–gel synthesis approach was demonstrated for lithium silicate oxide particle with a nanorod morphology, which showed excellent sorption kinetics due to simple reaction in the surface with CO_2 by virtue of smaller lithium diffusion from bulk to nanorod surface [183]. By mixing sodium and potassium to the particles, an extraordinary CO_2 adsorption rate of 0.72 wt% s^{-1} at 100% of CO_2/923K was observed with remarkable performance even at lower temperatures and at lower CO_2 pressure. In addition, 1D calcium oxide–titanium dioxide nanotube composite was formulated using different concentrations of calcium nitrate tetrahydrate by a technique named swift oxidation of anode using the electrochemical anodization [184]. The maximum capacity of CO_2, which is 2.45 mmol g^{-1}, was seen in nanocomposite having 0.01 M calcium nitrate tetrahydrate due to precise calcium ion concentration with increased surface area. In another study, a composite material formulated by functionalization of PEI and tetraethylenepentamine (TEPA) through capsules of mesoporous silica showed rapid capture kinetics of 7.93 mmol g^{-1} and reached total capacity of 90% in a short time period with good stability and reversibility [185].

3.2.3 *Novel One-Dimensional Metal Nanostructure in CO_2 Capture*

In a different approach to effectively capture CO_2 by silver/polyoxometalate (Ag/PMo) nanocomposite, Guo et al. developed the silver nanostructures for CO_2 reduction by electrodeposition of silver ions in the presence of a Keggin type polyoxometalate (PMo) for stabilizing the final nanostructure [186]. The formed Ag/PMo nanocomposite efficiently speeds up the process of CO_2 reduction into carbon monoxide with a potential of −1.70 V with saturated dimethylformamide medium containing 0.1 M [n-Bu_4N] PF_6. Another approach developed by Hoang et al. a bimetallic nanoporous copper–silver (<10% Ag content) hybrid material as catalyst significantly improves the conversion of CO_2 into carbon products such as ethanol and ethene and selectivity of CO_2 [187]. When the CO_2 is electroreduced at a low potential (−0.7 V) and high current density (−300 mA/cm^2), the copper–silver bimetal catalyst exhibits high selectivity toward ethene (~60%) and ethanol production (~25%) in an alkaline flow electrolyzer.

The gold–iron alloy as catalyst transforms rapidly into a constant gold–iron core-shell nanoparticles (AuFe-CSNP) after the process of leaching iron particles on the surface. This core-shell nanoparticle displays inimitable selectivity of CO_2, stability for a long term, a 100-fold increment in the mass activity toward reduction of CO_2 than gold nanoparticles, and 0.2 V lesser in overpotential. Zhang et al. developed a novel electroactive polymeric film with highly dispersed copper–palladium alloy nanoparticles on its surface [188]. Liu et al. demonstrated a method to fabricate nanodiamond/silicon rod array (NDD/Si RA) that are doped with nitrogen for reducing CO_2 [189]. It is selectively converted CO_2 into acetate at 0.8 ~ −1.0 V and Faradaic efficiency of ~91.2–91.8% has been attained. Its greater performance in reducing CO_2 is ascribed to its elevated over the potential for evolution of hydrogen and doping of nitrogen, wherein nitrogen–sp^3 carbon species were impressively active in reducing CO_2. Alternatively, Iqbal et al. fabricated a controllable synthesis of carbon nanofibers that are doped into nickel cobaltite/carbon nanotube ($NiCo_2O_4$/CNTs CNFs) mesoporous membrane that showed well-established flexibility, hydrophobic character, personalized pore structure, and elevated stability [190]. These membranes, due to these unique features, showed elevated capturing ability of CO_2 (1.54 mmol/g) at 25°C and 1.0 bar.

Wu et al. fabricated graphene quantum dots that are doped with nitrogen (NGQDs). It speeds up the CO_2 reduction into oxygenates and multicarbon hydrocarbons using the electrochemical methods at elevated Faradaic efficiencies (up to 90%), low overpotentials, and high current densities with discernment for conversion of ethanol and ethylene, reaching up to 45% [191]. The carbon production and distribution rate for NGQD-catalyzed reduction of CO_2 is similar to those attained via electrocatalysts of copper nanoparticles. While Zainab et al. developed an effective composite spider web–like carbon nanotube/polyamide-6 (CNT/PA) nanofiber/net membrane including mutual nanofibers that are electrospun and nanonets that are 2D and spider web–like were also utilized as a substrate with pores for the impregnation of PEI. It displays certain vital characteristics, such as good interconnectivity and open porosity [172a]. The ideal loading of PEI on composite of CNT/PA nanofiber/nets is found to be 75 wt% with a capacity to adsorb CO_2 at 25°C of 51 mg/g sorbent. Furthermore, it exhibits excellent regeneration and stability. Other 1D nanostructures that are used for industrial carbon capture applications are listed in Table 4.2.

3.3 Carbon Storage

Carbon capture followed by storage of carbon is a technology to capture and store CO_2 up to 90% that are emitted due to anthropogenic processes such as fossil fuel utilization in industries and electricity generation [211]. The captured CO_2 is stored underground in deep saline aquifer formations [212] and also in depleted oil and gas fields [213]. The process of storing CO_2 artificially in geological rock formation [214], coastal wetlands [215], or soil [216], deep below surface of the earth, is called carbon sequestration [217]. Several nanostructured materials are recently employed in capturing of carbon as stated in earlier sections and are also utilized in artificial carbon sequestration application to store carbon [218]. 1D nanostructures possess hollow spaces [219] and active terminal ends [220] which makes them a better candidate for long-term carbon storage applications.

1D carbon nanostructures such as CNT are initially used for the carbon sequestration process. Currently, CNTs were used as a selective agent for CO_2 sequestration. The objective of the work is to examine the capability of SWCNT, doped with silicon, to store CO_2. Measurements based on density functional theory are utilized where [5] SWCNTs are mixed with 2%, 33%, and 50% of silicon to enhance their carbon storage ability. It can be observed from the experiment that the pristine carbon tubes and CO_2 showed an adsorption route via endergonic reaction, whereas exergonic reactions are detected in silicon-doped materials. The results showed that silicon-doped CNTs are beneficial in efficient room temperature, CO_2 capture, and also proved that one CNT can sequester and store up

TABLE 4.2 List of 1D Nanostructures Used for Carbon Capturing Applications.

1D Nanostructures	Carbon Capture Application	References
Single-walled CNT	~30 g of CO_2 adsorption per kg	[167]
C50 fullerene–based peanut-shaped CNT	At 298 K and 10,000 kPa, the amount of CO_2 adsorption is 8.30 mmol/g which is greater compared with single-walled CNT	[168]
Multiwalled CNT functionalized with 3-aminopropyl-triethoxysilane	Prolonged cyclic operation for cost-effective CO_2 sorbent with effective 15% adsorption (55.1–67.2 mg/g) of CO_2 at 60°C with water content 0%–7%	[192]
Multiwalled CNT impregnated with polyethylenimine	A maximum CO_2 uptake of 93 mg/g	[169]
Carbon nanofiber functionalization with cobalt oxide nanoparticles, nickel/cobalt nanotubes, nickel iron oxide nanocrystals, and carbon nanofibers hybridized by tin oxide nanoclusters	Functionalized carbon nanofibers show highest CO_2 adsorption capacity of 51 mg/g at 25°C and 1 bar, nickel/cobalt nanotubes possess high CO_2 adsorption capacity of 5.4 mmol/g at 1 bar and room temperature, nickel iron oxide nanocrystal complex possess high electrical capacitance of 343 F/g, good reversibility, and cycling efficiency of 97.4% after 10,000 cycles for CO_2 capture and storage	[172a,193]
Nitrogen-doped porous carbon nanofibers film with nickel oxide nanoparticles	Favorable catalytic activity of CO_2 and 10.6 wt% higher CO_2 uptake	[194]
Carbon fiber nanocomposite with phenolic resin and polyacrylonitrile fiber	10.2–11.5 wt% of postcombustion CO_2 capture at 760 mm Hg	[195]
Reduced graphene oxide	CO_2 capture capacity of 248 wt% at 25°C and 30 bar pressure condition	[175]
1D graphene nanoribbons	Strong CO_2 interactions of 11.5 mg/g (0.26 mmol/g) at 0.01 Pa and 303 K	[196]
Graphitic carbon nitride nanotubes doped with phosphorus	3.14 times increase in CO_2 adsorption capacity with 63.6 mg of CO_2/g of ionic liquid and 22.4 mg of CO_2/g absorption peak rate of ionic liquid/min	[176]
Permeable glassy polymer membranes, few graphene layers, and poly (2,6-dimethyl-1,4-phenylene oxide) composite	High permeability and selectivity of 0.3–15 wt% for helium and CO_2 at 35 and 65°C	[197]
Graphite oxide onto metal oxide framework with urea	Enhanced atmospheric CO_2 uptake of 13.41 mmol/g in surface of nanoparticles at room temperature and 1.5 MPa with complete reversibility	[198]
B80 fullerenes	Carbon capture with adsorption energies in the range of 17.19 –19.13 kcal/mol using Lewis acid–base reaction	[199]
Nitrogen-doped carbon spheres	Capture of 2.67 mmol g^{-1} capacity of CO_2 with improved selectivity in mixed gases	[177]
Carbon nanofiber–titanium dioxide core-shell nanocomposite	Catalytic capturing of 17 wt% carbon at 468–600°C using light energy	[200]
MCM-41 with PEI	Carbon capture with 24 times greater sorbent capacity along with productivity of 1.9 mol kg^{-1} h^{-1} and maximum of 97% CO_2 recovery	[201]
Magnesium oxide nanosheets	$CaO–MgAl_2O_4$ possess 34 wt% of CO_2 holding capacity after 65 carbonation—decarbonation cycles, MgO nanoparticles show faceted sorbent for 65% increment (0.77 mmol/g) in carbon capture at 227 m_2/g	[179,202]
Novel hollow pod–like calcium oxide nanoparticles	>50% CO_2 absorption capacity after 50 CO_2 capture and release cycles which is greater than common calcium oxide	[180]
Lithium silicate nanowires	Ultrafast adsorption of CO_2 at a rate of 0.22 g/g per min at 650–700°C with enhanced desorption reversibility	[203]

TABLE 4.2 List of 1D Nanostructures Used for Carbon Capturing Applications.—cont'd

1D Nanostructures	Carbon Capture Application	References
Titanium dioxide nanorod and copper nanoparticles composite	Photoreduction of CO_2 to methane at a rate of ~2.91 ppm/$g_{catal.}$ h	[181]
Mesoporous titanium dioxide–graphene oxide nanocomposite	Inclined CO_2 adsorption with adsorption capacity of up to 1.88 mmol/g at 25°C	[182]
γ-Alumina-graphene oxide nanorod composite	Solid adsorbent amine for capturing CO_2 with adsorption capacity of 200.6 mg/g_{PEI}, 0.22 amine efficiency, and enhanced recyclability	[204]
Lithium silicate oxide nanorod	Excellent CO_2 sorption kinetics with absorption rates of 0.72 wt%/s at 100% CO_2/923 K	[183]
Calcium oxide–titanium dioxide nanotube	Enhanced CO_2 adsorption of 2.45 mmol/g at 400°C	[184]
1D lithium orthosilicate–based ceramic nanocomposite	CO_2 absorption capacity of 270 mg CO_2/g sorbent at higher temperatures with faster kinetics and reasonable durability	[205]
Magnesium nanocrystals with diamines and mixed-matrix membranes	Increased permeability of CO_2 at 1 mbar adsorption threshold and high selectivity of CO_2/nitrogen at 35–75°C	[206]
Mesoporous silica capsules with functionalization of PEI and TEPA	Rapid CO_2 capture kinetics with capture capacity of 7.9 mmol/g under simulated flue gas conditions, good stability, and reversibility	[185]
Silver/polyoxometalate nanocomposite	Effective CO_2 capture with 90% of faradic efficiency	[186]
Bimetallic nanoporous copper–silver hybrid material	Conversion of CO_2 into ethanol and ethene, selectivity of CO_2	[187]
Monodispersed copper and silver bimetallic nanoparticles	High CO_2 reduction capability with 21.2% of faradic efficiency	[207]
Gold–iron alloy nanoparticles	Improved CO_2 reduction with CO selectivity, long-term stability, and nearly a 100-fold increase in mass activity toward CO_2 reduction compared with gold nanoparticles and 0.2 V low overpotential	[208]
Copper–palladium alloy nanoparticles	Reduction of CO_2 into methane with a faradic efficiency of 33% which is a doubling yield compared with nanocopper	[188]
Nanodiamond/silicon rod array	Nitrogen–sp^3 carbon active species for reducing CO_2 with faradic efficiency of 91.2%–91.8% at −0.8–1.0 V	[189]
Cobalt phthalocyanine–CNTs	Carbon monoxide production from CO_2 with selectivity, improved durability, and >95% faradic efficiency at 15.0 mA/cm^2, 4.1/s turnover frequency, and 0.52 V overpotential in a near-neutral aqueous solution	[209]
Hybrid carbon nanofibers—cobalt oxide nanoparticles	Brilliant CO_2 capturing performance of 5.4 mmol/g at 1 bar at room temperature	[210]
Carbon nanofibers doped with nickel cobaltite/carbon nanotube	Elevated CO_2 capturing ability of 1.54 mmol/g at 25°C and 1.0 bar	[190]
Nitrogen-doped graphene quantum dots	CO_2 reduction into oxygenates and multicarbon hydrocarbons with faradic efficiency of 90% and ethylene, ethanol conversion selectivity up to 45%	[191]
Spider web–like carbon nanotube/polyamide-6 nanofiber/net membrane	Highest CO_2 adsorption capacity of 51 mg/g at 25°C and 1 bar	[172a]

to seven CO_2 molecule [221]. A typical schematic representing the capture of CO_2 by CNT was represented in Fig. 4.7. Another study compared the efficient CO_2 capture ability of CNTs, activated carbons, and zeolites. The results revealed that modified CNT with 3-aminopropyl-triethoxysilane shows effective CO_2 adsorption, compared with other modified carbon and silica adsorbents, and are potential candidates for capturing CO_2 from gas streams [170]. Similarly, a mixed matrix membrane that is fabricated with MWCNTs was also used to separate CO_2 and nitrogen. Cellulose acetate butyrate (CAB) polymer with outstanding CO_2 sorption ability

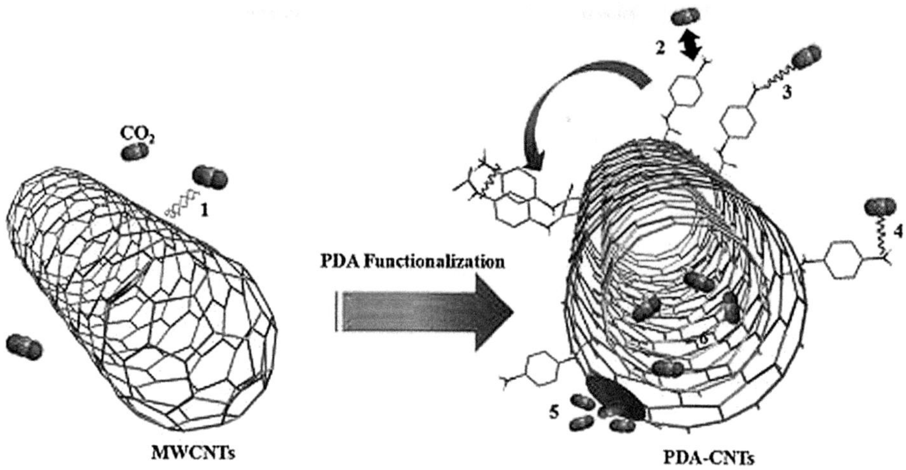

FIGURE 4.7 Schematics showing defect sites in the carbon nanotube that helps to trap and capture CO_2 [224].

was incorporated into functionalized MWCNT to form a mixed matrix membrane. The results revealed that the separation process of nitrogen and CO_2 has been increased drastically due to the incorporation of CAB polymer [222]. Thus, these types of CNTs can potentially be useful in unique separation and storing of CO_2 in the future. However, it can be noted that SWCNTs possess high surface to volume ratio than MWCNTs which is crucial in carbon storage applications [223]. Hence, various researchers are trying to improve the surface area of MWCNTs, similar to MWCNTs, to use them for carbon separation, capture, and storage applications.

MOFs are the latest natural and synthetic porous organic materials that are fabricated using metal oxide clusters and organic linkers [225]. In 2005, MOFs, including 1D cylinders, were reported to show high potential for room temperature CO_2 storage applications [226]. In 2008, isoreticular MOFs were fabricated for enhanced CO_2 storage application which showed better CO_2 storage capacity, compared with cation-exchanged and aluminum-free Zeolite Socony Mobil-5 (ZSM-5) and CNT bundle [227]. Meanwhile, 1D nanofibers were also used for capturing, sequestering, and storing carbon. In 2006, titanium oxide and zinc oxide nanofibers were fabricated using electrospinning method for CO_2 sequestration on pyrolysis. The results showed that CO_2 can be stored in electrospinned nanofibers, irrespective of the type of metal oxide used for the fabrication [228]. In addition, several nanocomposites were also reported for CO_2 capture and storage application with high efficacy. In 2016, mesoporous silica nanotube impregnated with amine was identified to be an efficient nanocomposite for enhanced capturing of CO_2. The results showed that the nanocomposite is highly stable with CO_2 adsorption capacity up to 70% within 2 min which can be utilized possibly in CO_2 capture and storage application [229]. Likewise, novel hierarchical nanocomposite was fabricated by the combination of amorphous carbon, graphene oxide, and magnesium oxide nanocrystallites for enhanced capturing of CO_2. The results exposed that the sandwich-like structure helps in high CO_2 capture which can further be enhanced through the treatment of promoting metal salt [230]. In future, the fabrication of similar type of hierarchical nanostructure in 1D cylindrical nanostructures will help in producing highly efficient CO_2 storage and separation systems.

There is an increasing trend in capturing CO_2 and converting them into 1D carbon nanostructures that are useful for various purposes. In 2013, a novel integrated chemical vapor deposition process was introduced to catalytic capture of CO_2 as an alternate to carbon sequestration. In this work, CO_2 is used as carbon source which is later converted into a solid powder of CNT via chemical vapor deposition. The results served as a novel alternate to conventional sequestration of carbon by converting CO_2 into useful nanoparticles, instead of storing them [231]. Likewise, a molten electrolyte system of lithium carbonate (Li_2CO_3) in a power plant that runs using natural gas in combined cycles are thermodynamically proved to be able to yield carbon nanofibers and oxygen. It was reported that 219 KJ is required by a single molten electrolyte system to convert 1M of CO_2. This approach is highly efficient in generating electricity, splitting CO_2, and converting all the CO_2 into valuable CNTs [232]. Similarly, electrosynthesis of the CNT was achieved by effective capturing of CO_2 directly from the air using the dissolution and electrosplitting molten carbonate [233]. In 2017, amino acids were also investigated to react with CO_2 to form amino acid nanofibers and bicarbonate nanoflowers that are further used for capturing CO_2 [234]. In the recent past, a novel electrochemical route was introduced for sustainable capture, sequestration, and conversion of CO_2 from the atmosphere and low-value metal scraps into valuable MWCNTs. The results showed that this approach is highly promising in capturing, sequestering, and transforming CO_2 into pure metals and MWCNT, which eventually helps in efficient carbon capturing applications [235]. Thus, carbon will be sequestrated and stored in the form of carbon nanostructures in the future, which will be beneficial in other applications and helps to reduce atmospheric CO_2 content which is a major greenhouse gas.

3.4 Sustainable Energy

CO_2 is the source of several activities in earth, ranging from photosynthesis to fossil fuel production. Carbon that is captured and stored has higher benefits in contributing as a sustainable energy source. The capturing of carbon using 1D nanostructures will eventually lead to their storage for future applications. These stored carbons are useful in applications such as biological conversions, extractants, mineralization, chemicals, refrigerant, inerting agent, fire suppression, plastics, and enhanced fuel recovery [236]. Stored CO_2 in 1D nanostructures is helpful as a source for the growth of anaerobic algae [237] and in photosynthesis of plants [238]. In turn, these organisms help to utilize CO_2 as a fossil fuel source. The plants and algae utilize CO_2 for their growth and development which decomposes after their death and remains in the soil for several years. The CO_2 present in these organisms will be converted into carbon source due to geochemical reactions and form fossil fuel [239]. Conversion or storage of these carbon sources into 1D nanostructures will increase their efficiency as fuel and increase their carbon storage ability for long term.

Stored CO_2 was also used as extractants of flavors or fragrances and as decaffeination agent. Supercritical and liquid CO_2 has been used to extract perfumes and flavors from plant materials [240]. Furthermore, this unique CO_2 has been used to extract other phytochemicals from plants and metabolites from microorganisms [241]. The industries that depend on CO_2 as extractants have to spend more cost on processing and converting these CO_2 into supercritical and liquid phases [242]. In these cases, 1D nanostructures with more surface area can help to incorporate and store these CO_2, which can be highly beneficial as extractants. The CO_2 stored in confined structures can also help in the decaffeination application [243]. The process of decaffeination is defined as the removal of caffeine from cocoa, tea leaves, coffee beans, and other materials that are caffeinated [244]. Recently, supercritical CO_2 was utilized for extracting caffeine from coffee beans [245], tea stalk [246], green tea [247], and black tea [248]. Supercritical and liquid CO_2 that is stored in 1D nanostructures can be utilized in the decaffeination process to extract caffeine, which will contribute to the cost reduction involved in using supercritical and liquid CO_2.

Captured gaseous CO_2 also involved in the mineralization process through carbonate formation. Mineralization via CO_2 is explained as the CO_2 reaction to yield carbonates with alkaline-earth oxides such as calcium oxide [249]. Toxic chemicals in wastewater can be converted into their carbonate substituents by using CO_2. The process of carbonate formation will help in the agglomeration of toxic ions in the water to settle on the surface of water beds. Such carbonates will turn into minerals, which helps in improving the fertility of the soil [250]. CO_2 captured and stored in 1D nanostructures can be useful in sustained release of CO_2, which can maintain CO_2 and oxygen level in the water as well as carbonate toxic ions into useful minerals and improve soil quality. In addition, captured CO_2 is highly beneficial in the formation of valuable chemicals such as methanol, urea, carbon monoxide, and methane [251]. Hydrogenation of CO_2 leads to the production of methanol that can be effectively used as fuel and a source of energy in several industries. Presently, 1D nanostructures such as CNTs [252], titanium oxide nanotubes [253], nitrogen doped—carbon nanofibers [254], catalysts of copper/zirconia in carbon nanofibers [255], copper decorated zinc oxide nanorod [256], and silica nanowires encapsulated ruthenium nanoparticles were used for hydrogenating CO_2 for methanol production. Similarly, urea was produced by combining CO_2 with ammonia in which captured CO_2 plays a significant role [257]. 1D hematite nanowires [258], nickel nanowires [259], hematite-modified titanium dioxide nanotubes [260], and carbon nanofibers [261] were used for the production of urea by facilitating the CO_2 that is captured along with ammonia. Furthermore, urea is also used as a fertilizer to improve the fertility and quality of the agricultural soil to facilitate plant growth [262].

Likewise, carbon monoxide and methane were also produced by capturing CO_2 from the atmosphere. 1D CNTs [263], nitrogen-doped carbon nanofibers [264], zinc oxide nanorods [265], and ultrathin gold nanowires [266] are recently being extensively used for producing carbon monoxide through CO_2 reaction. Meanwhile, gallium nitride nanowires are used to reduce CO_2 into methane and carbon monoxide using the water-level artificial photosynthesis method as shown in Fig. 4.8 [267]. In addition, copper nanowires [268], cohesive nanowire of gallium nitride and photocathode of silicon-based solar cell [269], copper oxide—modified titanium dioxide nanotube arrays [270], titanium dioxide arrays grafted with gold, ruthenium, and zinc palladium [253], nitrogen-doped titania nanotube arrays [271], titanium dioxide nanofibers [272], cerium oxide nanorods [273], and silver- or platinum nanoparticle—deposited titanium dioxide nanorods [274] are used for the reduction of CO_2 into methane. It can be noted that all the

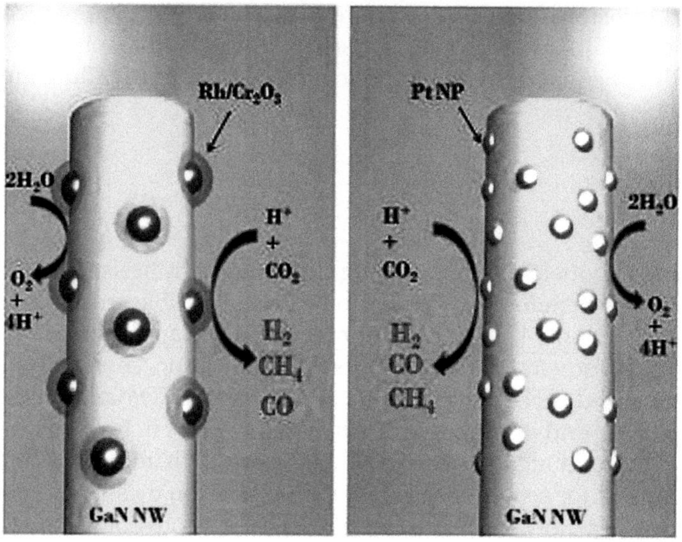

FIGURE 4.8 Schematic showing photoreduction process of CO_2 and its conversion into methane and carbon monoxide [267].

four by-products formed by reduction of CO_2 are used as fuels in various industries [275]. Thus, 1D nanostructures with carbon capture and storage ability with CO_2 reduction capacity are highly efficient in sustainable fuel production in future.

The CO_2 that is captured using 1D nanostructures is also used as refrigerants for refrigeration purposes and for dry ice production. Such refrigerants are widely used for household refrigeration as well as industrial purposes [276], and dry ice is highly used in ice cream production and storage [277]. The CO_2 that is stored in 1D nanostructures can also be helpful in reduction of the agglomeration between dry ice molecules and enhance their refrigerant property. In addition, CO_2 captured from the atmosphere is extensively used as an inert gas in the production of blankets [278], protection of carbon powder [279], and in welding as a shield gas [280]. The captured and stored carbon in 1D nanostructures will increase the efficiency of these CO_2 as inert agent which adversely inclines their utilization in a variety of welding and mechanical applications. CO_2 is widely used in fire extinguishers for suppressing fire. The liquid CO_2 in fire extinguishers is sprinkled to suppress the fire as the CO_2 will neutralize the oxygen on which the fire feeds on and disable the spread ability of fire [281]. Fire extinguishers with a liquid CO_2 encapsulated or formulated in 1D nanostructures will be the future of fire extinguishers as they can store numerous CO_2 molecule and control the release to regulate long-lasting and forest fires.

In addition to all these unique applications, CO_2 that is captured and stored in 1D nanostructures are helpful in applications such as fabrication of polycarbonate and polymers for the production of plastics [282] and in the oil and gas industry for recovery. The captured CO_2 is used in enhanced oil recovery [283], enhanced gas recovery [284], and enhanced coal bed methane recovery [285], which potentially helps in recovering fossil fuels. Furthermore, green routes to fabricate 1D nanostructures were also utilized to yield sustainable products. Batteries produced by ambient CO_2 synthesized CNTs [286], waste carbon fiber for silicon nitride nanowires [287], thin cellulose nanoribbons for catalytic carbonization [288], and environmentally friendly fabrication of ultrathin metal telluride nanowires [289] are examples of sustainable green methods for 1D nanostructure fabrication for energy production. Thus, it is evident that capturing and storing CO_2 is useful in various applications to maintain sustainability in energy production. Because 1D nanostructures have possessed high surface areas to withhold several carbon molecules, it is highly beneficial in the production of sustainable energy. The future of maintaining sustainability in energy production depends on the improvements in developing 1D nanostructures to capture and store more carbon molecules.

3.5 Pollution Control

Carbon, especially CO_2 and carbon monoxide, is the major effluent gas that is termed as greenhouse gas [290]. Due to the industrial revolution and carbon emissions in large quantities, the level of atmospheric greenhouse gases has increased in the past decades and it has been an issue to address in recent times [291]. Carbon emissions in the form of CO_2, carbon monoxide, methane, chlorofluorocarbon (CFC) 11, CFC-12, CFC-113, hydrochlorofluorocarbon

(HCFC) 141b, HCFC-142b, hydrofluorocarbon (HFC) 134a, halon 1211, carbon tetrachloride, and halon 1301 are released into the atmosphere through industrial and human activities [292]. The presence of excess greenhouse gases in the atmosphere will lead to climate change [293], global warming [294], and will affect food production [295]. Additionally, increasing anthropogenic activities and deforestation also leads to the upsurge emission of greenhouse gases [296]. In this case, carbon capture is crucial in reducing the atmospheric greenhouse gases, which are harmful to the environment. The carbon emitted from the industries and other source is captured in three substantial approaches such as pre- and postcombustion capture and capture of carbon through fossil fuel combustion in a pure oxygen environment [297].

1D nanostructures are utilized in capturing carbon from both pre- and postcombustion process. In precombustion, the emission of carbon is converted into useful nanotubes as shown in Fig. 4.9 [298], nanofibers [299], and nanorods [300], whereas in postcombustion, 1D nanostructures are used to capture emitted carbon [168]. CNTs are the widely fabricated 1D nanostructure from the process of fossil fuel combustion and from industrial carbon emissions. In the recent past, CNTs were synthesized and extracted from a diesel engine after the combustion process, after adding certain catalytic sources into the fuel such as a mixture of dodecane/ethanol, 1-decanol/ethanol, and methyl laurate/ethanol. The results showed that carbon monoxide is released from the combustion process of hydrocarbons, which is later served as a precursor for the CNT formation [301]. Another experiment was patented recently with a system for producing CNTs from combustion engine exhausts [302]. Meanwhile, the CNT was synthesized by using fatty acid methyl ester as biodiesel and ethanol as fuel in diesel engines. The results revealed that the carbon monoxide was formed due to combustion and is an important precursor for the CNT formation along with ferrocene as catalyst source and molybdenum/sulfur as promoter in the fuel [303]. In addition, CNTs are extensively used to capture several carbon molecules from the atmosphere and from industrial effluents [304]. In addition, CNTs are used for CO_2 conversion into renewable and value-added fuels [305] which further helps to reduce their excess emission into the atmosphere and in developing sustainable energy.

Apart from CNTs, carbon nanofibers, nanorods, and nanowires were also highly beneficial in capturing carbon, reducing their level in the atmosphere, and avoiding the greenhouse gas pollution and environment depletion. It is proven now that carbon nanofibers can be fabricated using captured atmospheric CO_2 that are dissolved in molten carbonates. The conversion is also proved to be efficient by using solar or conventional energy along with inexpensive nickel or steel electrodes. Thus, this study reported the importance of converting CO_2 and other greenhouse gas into useful 1D nanofibers to mitigate climate change and global warming [306]. Likewise, carbon nanofibers that are doped with amine-functionalized CNTs [307] and flexible nickel-cobalt oxide/CNTs–doped carbon nanofibers [172b] are used for efficient CO_2 capture and adsorption, respectively. Similarly, amino-rich phosphorus–doped graphitic carbon nitride nanotubes for efficient carbon capture [308], copper nanorods/glassy carbon electrode for electroenzymatic CO_2 reduction as displayed in Fig. 4.10 [309], and hydrogen titanate nanotubes and nanorods for CO_2 uptake [310] reveal that other 1D carbon nanostructures also possess high carbon capture property to reduce pollution in the environment. Additionally, bismuth vanadate/carbon-coated copper oxide nanowire arrays also possess the ability for CO_2 conversion under visible light via photocatalysts [311]. All these studies give an insight that 1D carbon nanostructures play a significant role in the carbon capture application which eventually reduces carbon footprints by converting them into useful fuels and structures to reduce environmental pollution caused by carbon-based greenhouse gas emission.

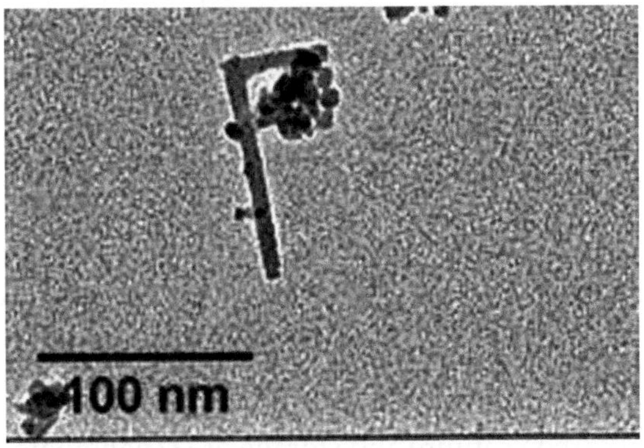

FIGURE 4.9 TEM image of carbon nanotube–like structures produced in laboratory diesel and gasoline exhaust [298].

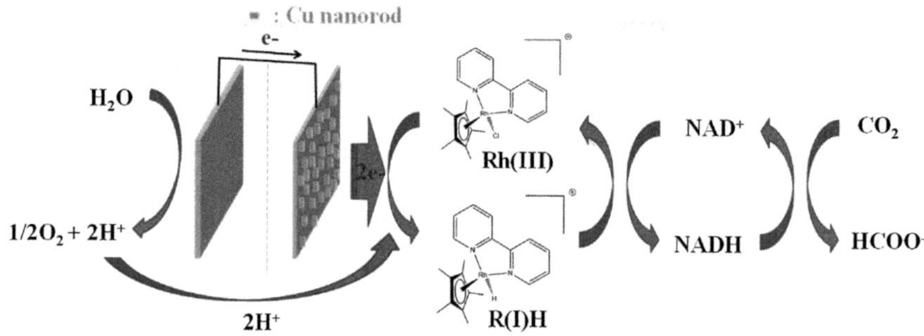

FIGURE 4.10 Schematic showing electroenzymatic CO_2 reduction on copper nanorods/glassy carbon electrode by cyclic deposition [309].

Other than 1D carbon nanostructures, diamine-appended magnesium (dobpdc) nanorods [206], lithium silicate nanorods [183], copper decorated zinc oxide nanorod [256], fibrous polyvinylidene fluoride/siloxane membranes [312], and TEPA-modified protonated titanate nanotubes [313] also showed enhanced carbon capture properties. In addition, novel bacteria and nanowire hybrid were used for unassisted solar, CO_2 fixation for converting them into value-added chemicals [314]. Likewise, copper nanowires possess the ability for low overpotential CO_2 reduction [315], and ultrathin gold nanowires are proved to possess the property for active and selective conversion of CO_2 to carbon monoxide [266]. Ultrafast CO_2 sorption via lithium silicate nanowires [203], copper oxide nanowires with zinc oxide islands for CO_2 photoreduction [316], copper nanowires for electroreduction of CO_2 to carbon monoxide [317], Pd@H_yWO_{3-x} nanowires for heterogenous reduction of CO_2 [318], carbon nanofiber webs that are porous and doped with nitrogen for efficient CO_2 capture and conversion [319], and polymeric gC_3N_4-coupled $NaNbO_3$ nanowires for upgraded CO_2 reduction into renewable energy using photocatalysts [320] are some of the recent works that portray the ability of 1D noncarbon nanostructures with carbon capture application to reduce greenhouse gas level in environment.

The previous sections clearly show that 1D nanostructures possess the ability to capture as well as store carbon and even convert them into beneficial gases. Thus, it is highly recommended to use these nanostructures in industries to reduce carbon emission, carbon footprints, and carbon neutrality. Utilization of 1D nanostructure in the process of industrial combustion and power plants will enhance the capture of carbon and help to reduce greenhouse gas emission and carbon footprints [321]. Coal-based energy-intensive industrial plants consist of gasification plant, pulverized coal boiler, cement production plant, integrated steel mill, air separation unit, condenser, and heat exchangers are the possible components to release carbon. Incorporation of 1D nanostructure in these components will potentially increase in situ carbon capture to reduce carbon-based greenhouse gas emission [322]. Hence, environmental pollution caused by the release of carbon gases into the atmosphere can be reduced by 1D nanostructures which eventually reduces the carbon footprint of an industry and bring carbon neutrality in a country.

4. CURRENT CHALLENGES AND FUTURE DIRECTIONS

Increase in the concentration of CO_2 emitted by process industries has led to adverse effects on the environment. 1D nanostructures are proposed as highly effective for carbon capture applications. This chapter discusses several 1D nanostructures for capturing carbon applications with viable scale-up potentials for commercial scale applications. As most of the reported carbon capture applications are focused on lab-scale analysis using artificial flue gas mixtures, hence, future work should focus on the use of industrial flue gases and explore the potential of emerging 1D nanostructures for efficient and direct environmental applications [323]. The captured carbon can be converted into beneficial products such as methane and ethanol. The efficiency of carbon conversion into useful products by 1D nanostructure is based on their enhanced functionality. Presently, 1D nanostructures that have demonstrated enhanced carbon capture ability are mostly prepared chemically using precursors that are toxic to the environment. Green chemistry plays a significant role in reducing environmental toxicity. It is proposed that methods to synthesize 1D nanostructures should incorporate green chemistry approaches and regents to minimize impacts on the environment. Promoting sustainability in large-scale industrial applications via the use of 1D nanostructures in future will require the combined efforts of policy makers in the development of new carbon technologies, commercialization of carbon capture and storage of 1D nanomaterials, and researchers in developing new assessment approaches for low carbon technologies [324].

5. CONCLUSION

This is a critical overview of using 1D nanostructures for carbon capturing applications. It can be noted that a wide variety of nanostructures were used in capturing and storing carbon to reduce and balance the atmospheric carbon. The utilization of nanostructure helps to maintain environmental sustainability through carbon storage, reducing pollution and by producing sustainable energy. The major drawback relates to the use of toxic chemical solvents and precursors for the synthesis of these nanostructures. These conventional toxic chemical solvents can be replaced via green chemistry to promote environmental sustainability. In future, there will be an escalating increase in the number of vehicles, industries, and equipment that release carbon. Thus, novel 1D carbon capturing nanostructures will be in high demand in near future. Novel nanostructures such as DNA and protein-bound 1D nanostructures with the potential to capture carbon before it is released into the atmosphere will capture the market of future machineries. These novel nanostructures will ultimately reduce carbon emission which will eventually reduce greenhouse gas emission and mitigate climate change and global warming.

References

[1] E. Rosenberg, Origin of nucleic acids and the first cells. It's in your DNA: from discovery to structure, function and role in evolution, cancer and aging, Elsevier, 2017, pp. 129–138.
[2] W.D. Callister, Fundamentals of Materials Science and Engineering, vol. 471660817, Wiley London, 2000.
[3] M.M. Halmann, M. Steinberg, Greenhouse Gas Carbon Dioxide Mitigation: Science and Technology, CRC press, 1998.
[4] A. Demirbas, Biomass resource facilities and biomass conversion processing for fuels and chemicals, Energy Conversion and Management 42 (11) (2001) 1357–1378.
[5] R. Jackson, C. Le Quéré, R. Andrew, J. Canadell, G. Peters, J. Roy, L. Wu, Warning signs for stabilizing global CO_2 emissions, Environmental Research Letters 12 (11) (2017) 110202.
[6] A. Baede, P. Van Der Linden, A. Verbruggen, IPCC AR4 SYR Appendix Glossary, 2008. Retrieved.
[7] N. Nakicenovic, J. Alcamo, A. Grubler, K. Riahi, R. Roehrl, H.-H. Rogner, N. Victor, Special Report on Emissions Scenarios (SRES), a Special Report of Working Group III of the Intergovernmental Panel on Climate Change, Cambridge University Press, 2000.
[8] P.J. Crutzen, A.R. Mosier, K.A. Smith, W. Winiwarter, N_2O release from agro-biofuel production negates global warming reduction by replacing fossil fuels, in: P.J. Crutzen (Ed.), A Pioneer on Atmospheric Chemistry and Climate Change in the Anthropocene, Springer, 2016, pp. 227–238.
[9] G. Vacchiano, R. Berretti, R. Romano, R. Motta, Voluntary carbon credits from improved forest management: policy guidelines and case study, iForest-Biogeosciences and Forestry 11 (1) (2018) 1–10.
[10] X. Yang, H. Liao, X. Feng, X. Yao, Analysis and tests on weak-form efficiency of the EU carbon emission trading market, Low Carbon Economy 9 (01) (2018) 1–17.
[11] R. Duarte, J. Sánchez-Chóliz, C. Sarasa, Consumer-side actions in a low-carbon economy: a dynamic CGE analysis for Spain, Energy Policy 118 (2018) 199–210.
[12] X. Hao, D. Batstone, J.S. Guest, Carbon neutrality: an ultimate goal towards sustainable wastewater treatment plants, Water Research 87 (2015) 413–415.
[13] H. Singh, P. Gupta, A. Soni, R. Joshi, R.J. Yadav, A. Singh, Capturing Carbon Dioxide from Air by Using Sodium Hydroxide (CO_2 Trapper), International Research Journal of Engineering and Technology 5 (4) (2018) 870–876.
[14] S. Omwoma, C.T. Gore, Y. Ji, C. Hu, Y.-F. Song, Environmentally benign polyoxometalate materials, Coordination Chemistry Reviews 286 (2015) 17–29.
[15] G. Schmid, B. Corain, Nanoparticulated gold: syntheses, structures, electronics, and reactivities, European Journal of Inorganic Chemistry 2003 (17) (2003) 3081–3098.
[16] H.D. Espinosa, R.A. Bernal, M. Minary-Jolandan, A review of mechanical and electromechanical properties of piezoelectric nanowires, Advanced Materials 24 (34) (2012) 4656–4675.
[17] R. Agrawal, B. Peng, H.D. Espinosa, Experimental-computational investigation of ZnO nanowires strength and fracture, Nano Letters 9 (12) (2009) 4177–4183.
[18] M.H.A. Elkodous, A. Hassaan, A. Ghoneim, Z. Abdeen, C-dots dispersed macro-mesoporous TiO2 phtocatalyst for effective waste water treatment, Characterization and Application of Nanomaterials 1 (2) (2018).
[19] V.C. Corberán, V. Rives, V. Stathopoulos, Recent applications of nanometal oxide catalysts in oxidation reactions, in: Advanced Nanomaterials for Catalysis and Energy, Elsevier, 2019, pp. 227–293.
[20] J. Jeevanandam, A. Barhoum, Y.S. Chan, A. Dufresne, M.K. Danquah, Review on nanoparticles and nanostructured materials: history, sources, toxicity and regulations, Beilstein Journal of Nanotechnology 9 (2018) 1050–1074.
[21] Z.M. Wang, One-dimensional Nanostructures, vol. 3, Springer Science & Business Media, 2008.
[22] Y. Xia, P. Yang, Y. Sun, Y. Wu, B. Mayers, B. Gates, Y. Yin, F. Kim, H. Yan, One-dimensional nanostructures: synthesis, characterization, and applications, Advanced Materials 15 (5) (2003) 353–389.
[23] C.J. Murphy, N.R. Jana, Controlling the aspect ratio of inorganic nanorods and nanowires, Advanced Materials 14 (1) (2002) 80–82.
[24] J. Hu, T.W. Odom, C.M. Lieber, Chemistry and physics in one dimension: synthesis and properties of nanowires and nanotubes, Accounts of Chemical Research 32 (5) (1999) 435–445.
[25] M.F.L. De Volder, S.H. Tawfick, R.H. Baughman, A.J. Hart, Carbon nanotubes: present and future commercial applications, Science 339 (6119) (2013) 535–539.

[26] H.O. Pierson, Handbook of Carbon, Graphite, Diamonds and Fullerenes: Processing, Properties and Applications, William Andrew, 2012.

[27] H. Kroto, The stability of the fullerenes Cn, with n= 24, 28, 32, 36, 50, 60 and 70, Nature 329 (6139) (1987) 529.

[28] R. Ruoff, D.S. Tse, R. Malhotra, D.C. Lorents, Solubility of fullerene (C60) in a variety of solvents, The Journal of Physical Chemistry 97 (13) (1993) 3379–3383.

[29] V. Georgakilas, J.A. Perman, J. Tucek, R. Zboril, Broad family of carbon nanoallotropes: classification, chemistry, and applications of fullerenes, carbon dots, nanotubes, graphene, nanodiamonds, and combined superstructures, Chemical Reviews 115 (11) (2015) 4744–4822.

[30] E.A. Laird, F. Kuemmeth, G.A. Steele, K. Grove-Rasmussen, J. Nygård, K. Flensberg, L.P. Kouwenhoven, Quantum transport in carbon nanotubes, Reviews of Modern Physics 87 (3) (2015) 703.

[31] R. Bacsa, C. Laurent, A. Peigney, W. Bacsa, T. Vaugien, A. Rousset, High specific surface area carbon nanotubes from catalytic chemical vapor deposition process, Chemical Physics Letters 323 (5–6) (2000) 566–571.

[32] A. Hirsch, The era of carbon allotropes, Nature Materials 9 (11) (2010) 868.

[33] S. Iijima, Helical microtubules of graphitic carbon, Nature 354 (6348) (1991) 56.

[34] [34a] M. Bar-Sadan, I. Kaplan-Ashiri, R. Tenne, Inorganic fullerenes and nanotubes: wealth of materials and morphologies, The European Physical Journal Special Topics 149 (1) (2007) 71–101;
[34b] R. Kreizman, A.N. Enyashin, F.L. Deepak, A. Albu-Yaron, R. Popovitz-Biro, G. Seifert, R. Tenne, Synthesis of core–shell inorganic nanotubes, Advanced Functional Materials 20 (15) (2010) 2459–2468.

[35] M. Nihei, A. Kawabata, D. Kondo, M. Horibe, S. Sato, Y. Awano, Electrical properties of carbon nanotube bundles for future via interconnects, Japanese Journal of Applied Physics 44 (4R) (2005) 1626.

[36] F. Ding, A.R. Harutyunyan, B.I. Yakobson, Dislocation theory of chirality-controlled nanotube growth, Proceedings of the National Academy of Sciences 106 (8) (2009) 2506–2509.

[37] E.H. Hároz, J.G. Duque, E.B. Barros, H. Telg, J.R. Simpson, A.R.H. Walker, C.Y. Khripin, J.A. Fagan, X. Tu, M. Zheng, Asymmetric excitation profiles in the resonance Raman response of armchair carbon nanotubes, Physical Review B 91 (20) (2015) 205446.

[38] H. Wang, Y. Yuan, L. Wei, K. Goh, D. Yu, Y. Chen, Catalysts for chirality selective synthesis of single-walled carbon nanotubes, Carbon 81 (2015) 1–19.

[39] T.W. Odom, J.L. Huang, C.M. Lieber, Single-walled carbon nanotubes, Annals of the New York Academy of Sciences 960 (1) (2002) 203–215.

[40] C. Schönenberger, L. Forro, Multiwall carbon nanotubes, Physics World 13 (6) (2000) 37.

[41] M. Baratpour, A. Karimipour, M. Afrand, S. Wongwises, Effects of temperature and concentration on the viscosity of nanofluids made of single-wall carbon nanotubes in ethylene glycol, International Communications in Heat and Mass Transfer 74 (2016) 108–113.

[42] X. Lu, W.-L. Yim, B.H. Suryanto, C. Zhao, Electrocatalytic oxygen evolution at surface-oxidized multiwall carbon nanotubes, Journal of the American Chemical Society 137 (8) (2015) 2901–2907.

[43] K. Varshney, Carbon nanotubes: a review on synthesis, properties and applications, International Journal of Engineering Research 2 (4) (2014) 660–677.

[44] A.L. Kalamkarov, A.V. Georgiades, S.K. Rokkam, V.P. Veedu, M.N. Ghasemi-Nejhad, Analytical and numerical techniques to predict carbon nanotubes properties, International Journal of Solids and Structures 43 (22) (2006) 6832–6854.

[45] Y. Pan, Z. Hou, W. Yi, W. Zhu, F. Zeng, Y.-N. Liu, Hierarchical hybrid film of MnO2 nanoparticles/multi-walled fullerene nanotubes–graphene for highly selective sensing of hydrogen peroxide, Talanta 141 (2015) 86–91.

[46] [46a] R.E. Smalley, D.T. Colbert, H. Dai, J. Liu, A.G. Rinzler, J.H. Hafner, K. Smith, T. Guo, P. Nikolaev, A. Thess, Method for Forming a Patterned Array of Fullerene Nanotubes, 2009. Google Patents;
[46b] K. i. Miyazawa, J.-i. Minato, T. Yoshii, M. Fujino, T. Suga, Structural characterization of the fullerene nanotubes prepared by the liquid–liquid interfacial precipitation method, Journal of materials research 20 (3) (2005) 688–695.

[47] S. Amelinckx, X.B. Zhang, D. Bernaerts, X.F. Zhang, V. Ivanov, J.B. Nagy, A formation mechanism for catalytically grown helix-shaped graphite nanotubes, Science 265 (5172) (1994) 635–639.

[48] [48a] O.P. Krivoruchko, N.I. Maksimova, V.I. Zaikovskii, A.N. Salanov, Study of multiwalled graphite nanotubes and filaments formation from carbonized products of polyvinyl alcohol via catalytic graphitization at 600–800 C in nitrogen atmosphere, Carbon 38 (7) (2000) 1075–1082;
[48b] T. Wang, Y. Guo, Z. Zhou, X. Chang, J. Zheng, X. Li, Ni–Mo nanocatalysts on N-doped graphite nanotubes for highly efficient electrochemical hydrogen evolution in acid, ACS Nano 10 (11) (2016) 10397–10403.

[49] G. Song, Z. Li, A. Meng, M. Zhang, K. Li, K. Zhu, Large-scale template-free synthesis of N-doped graphene nanotubes and N-doped SiO2-coated graphene nanotubes: growth mechanism and field-emission property, Journal of Alloys and Compounds 706 (2017) 147–155.

[50] [50a] S. Datta, C.B. Chaudhuri, S. Chakraborty, A. Mookerjee, Effects of Extended Stone-Wales Defects on the Electronic Structure of Graphene Nanotubes: An Ab-Initio Real Space Recursion Approach, 2015 arXiv preprint arXiv:1604.07768;
[50b] S.M. Unni, R. Illathvalappil, S.N. Bhange, H. Puthenpediakkal, S. Kurungot, Carbon nanohorn-derived graphene nanotubes as a platinum-free fuel cell cathode, ACS Applied Materials & Interfaces 7 (43) (2015) 24256–24264.

[51] Y. Tang, N. Wang, Y. Zhang, C. Lee, I. Bello, S. Lee, Synthesis and characterization of amorphous carbon nanowires, Applied Physics Letters 75 (19) (1999) 2921–2923.

[52] D.-s. Yuan, T.-x. Zhou, S.-l. Zhou, W.-j. Zou, S.-s. Mo, N.-n. Xia, Nitrogen-enriched carbon nanowires from the direct carbonization of polyaniline nanowires and its electrochemical properties, Electrochemistry Communications 13 (3) (2011) 242–246.

[53] Y. Cao, L. Xiao, M.L. Sushko, W. Wang, B. Schwenzer, J. Xiao, Z. Nie, L.V. Saraf, Z. Yang, J. Liu, Sodium ion insertion in hollow carbon nanowires for battery applications, Nano Letters 12 (7) (2012) 3783–3787.

[54] [54a] K. Zhou, J. He, X. Wang, J. Lin, Y. Jing, W. Zhang, Y. Chen, Self-assembled CoSe2 nanocrystals embedded into carbon nanowires as highly efficient catalyst for hydrogen evolution reaction, Electrochimica Acta 231 (2017) 626–631;
[54b] H. Pan, X. Yin, J. Xue, L. Cheng, L. Zhang, In-situ synthesis of hierarchically porous and polycrystalline carbon nanowires with excellent microwave absorption performance, Carbon 107 (2016) 36–45.

[55] J.H. Kim, W.S. Chang, D. Kim, J.R. Yang, J.T. Han, G.W. Lee, J.T. Kim, S.K. Seol, 3D printing of reduced graphene oxide nanowires, Advanced Materials 27 (1) (2015) 157–161.

[56] Q. An, F. Lv, Q. Liu, C. Han, K. Zhao, J. Sheng, Q. Wei, M. Yan, L. Mai, Amorphous vanadium oxide matrixes supporting hierarchical porous Fe_3O_4/graphene nanowires as a high-rate lithium storage anode, Nano Letters 14 (11) (2014) 6250–6256.

[57] [57a] X. Liu, D. Chao, D. Su, S. Liu, L. Chen, C. Chi, J. Lin, Z.X. Shen, J. Zhao, L. Mai, Graphene nanowires anchored to 3D graphene foam via self-assembly for high performance Li and Na ion storage, Nano Energy 37 (2017) 108–117;
[57b] J. Xu, Nanoporous Graphene Nanowires and Producing Methods and Applications of Same, 2018. Google Patents.

[58] P. Wang, L. Cheng, Y. Zhang, W. Yuan, H. Pan, H. Wu, Electrospinning of graphite/SiC hybrid nanowires with tunable dielectric and microwave absorption characteristics, Composites Part A: Applied Science and Manufacturing 104 (2018) 68–80.

[59] [59a] Y. Chen, Z. Zhang, Z. Sui, Z. Liu, J. Zhou, X. Zhou, Ni (OH) 2 nanowires/graphite foam composite as an advanced supercapacitor electrode with improved cycle performance, International Journal of Hydrogen Energy 41 (28) (2016) 12136–12145;
[59b] E. Anyebe, A. Sanchez, S. Hindmarsh, X. Chen, J. Shao, M.K. Rajpalke, T.D. Veal, B. Robinson, O. Kolosov, F. Anderson, Realization of vertically aligned, ultrahigh aspect ratio InAsSb nanowires on graphite, Nano Letters 15 (7) (2015) 4348–4355.

[60] N. Yang, H. Uetsuka, C.E. Nebel, Biofunctionalization of vertically aligned diamond nanowires, Advanced Functional Materials 19 (6) (2009) 887–893.

[61] I. Vlasov, O.I. Lebedev, V.G. Ralchenko, E. Goovaerts, G. Bertoni, G. Van Tendeloo, V.I. Konov, Hybrid diamond-graphite nanowires produced by microwave plasma chemical vapor deposition, Advanced Materials 19 (22) (2007) 4058–4062.

[62] M. Liao, S. Hishita, E. Watanabe, S. Koizumi, Y. Koide, Suspended single-crystal diamond nanowires for high-performance nanoelectromechanical switches, Advanced Materials 22 (47) (2010) 5393–5397.

[63] X. Peng, J. Chu, L. Wang, S. Duan, P. Feng, Boron-doped diamond nanowires for CO gas sensing application, Sensors and Actuators B: Chemical 241 (2017) 383–389.

[64] [64a] F. Gao, R. Thomann, C.E. Nebel, Aligned Pt-diamond core-shell nanowires for electrochemical catalysis, Electrochemistry Communications 50 (2015) 32–35;
[64b] Y. Szunerits, R. Coffinier, Boukherroub, Diamond Nanowires: A Recent Success Story for Biosensing, in: C. Kranz (Ed.), Carbon-Based Nanosensor Technology. Springer Series on Chemical Sensors and Biosensors (Methods and Applications), 17, Springer, 2017, pp. 1–18.

[65] [65a] Y. Maeyoshi, A. Saeki, S. Suwa, M. Omichi, H. Marui, A. Asano, S. Tsukuda, M. Sugimoto, A. Kishimura, K. Kataoka, Fullerene nanowires as a versatile platform for organic electronics, Scientific Reports 2 (2012) 600;
[65b] T. Nakanishi, H. Takahashi, T. Michinobu, M. Takeuchi, T. Teranishi, K. Ariga, Fullerene nanowires on graphite: epitaxial self-organizations of a fullerene bearing double long-aliphatic chains, Colloids and Surfaces A: Physicochemical and Engineering Aspects 321 (1–3) (2008) 99–105.

[66] B. Tüzün, S. Erkoç, Structural and electronic properties of unusual carbon Nanorods, Quantum Matter 1 (2) (2012) 136–148.

[67] X. Wang, J. Lu, Y. Xie, G. Du, Q. Guo, S. Zhang, A novel route to multiwalled carbon nanotubes and carbon nanorods at low temperature, The Journal of Physical Chemistry B 106 (5) (2002) 933–937.

[68] Y. Xia, Z. Yang, R. Mokaya, Simultaneous control of morphology and porosity in nanoporous Carbon: graphitic mesoporous carbon nanorods and nanotubules with tunable pore size, Chemistry of Materials 18 (1) (2006) 140–148.

[69] P. Pachfule, D. Shinde, M. Majumder, Q. Xu, Fabrication of carbon nanorods and graphene nanoribbons from a metal–organic framework, Nature Chemistry 8 (2016) 718.

[70] [70a] X. Gu, J. Yue, L. Chen, S. Liu, H. Xu, J. Yang, Y. Qian, X. Zhao, Coaxial MnO/N-doped carbon nanorods for advanced lithium-ion battery anodes, Journal of Materials Chemistry A 3 (3) (2015) 1037–1041;
[70b] M. Prabu, P. Ramakrishnan, P. Ganesan, A. Manthiram, S. Shanmugam, LaTi0.65Fe0. 35O3– δ nanoparticle-decorated nitrogen-doped carbon nanorods as an advanced hierarchical air electrode for rechargeable metal-air batteries, Nano Energy 15 (2015) 92–103.

[71] [71a] L. Wang, J. Yu, X. Dong, X. Li, Y. Xie, S. Chen, P. Li, H. Hou, Y. Song, Three-dimensional macroporous carbon/Fe_3O_4-doped porous carbon nanorods for high-performance supercapacitor, ACS Sustainable Chemistry & Engineering 4 (3) (2016) 1531–1537;
[71b] X. Gao, J. Wang, D. Zhang, K. Nie, Y. Ma, J. Zhong, X. Sun, Hollow $NiFe_2O_4$ nanospheres on carbon nanorods as a highly efficient anode material for lithium ion batteries, Journal of Materials Chemistry A 5 (10) (2017) 5007–5012.

[72] H.-X. Ji, J.-S. Hu, L.-J. Wan, Q.-X. Tang, W.-P. Hu, Controllable crystalline structure of fullerene nanorods and transport properties of an individual nanorod, Journal of Materials Chemistry 18 (3) (2008) 328–332.

[73] R.G. Shrestha, L.K. Shrestha, A.H. Khan, G.S. Kumar, S. Acharya, K. Ariga, Demonstration of ultrarapid interfacial formation of 1D fullerene nanorods with photovoltaic properties, ACS Applied Materials & Interfaces 6 (17) (2014) 15597–15603.

[74] G. Lu, L. Li, X. Yang, Creating a uniform distribution of fullerene C60 nanorods in a polymer matrix and its photovoltaic applications, Small 4 (5) (2008) 601–606.

[75] T.J. Gnanaprakasa, D. Sridhar, W.J. Beck, K. Runge, B.G. Potter, T.J. Zega, P.A. Deymier, S. Raghavan, K. Muralidharan, Graphene mediated self-assembly of fullerene nanorods, Chemical Communications 51 (10) (2015) 1858–1861.

[76] H. Tang, K. Dou, Y. Xiong, F. Wang, Y. Zhao, X. Wang, Q. Fu, J. Yang, N. Zhao, D. Xu, Unusual thermal transport behavior in self-assembled fullerene nanorods, RSC Advances 6 (72) (2016) 67509–67513.

[77] X. Jiang, M. Yuan, H. Liu, L. Li, Z. Du, Optoelectronic properties of one-dimensional fullerene Nanorods, Materials Letters 176 (2016) 52–55.

[78] Y. Jiang, J. Yan, X. Wu, D. Shan, Q. Zhou, L. Jiang, D. Yang, Z. Fan, Facile synthesis of carbon nanofibers-bridged porous carbon nanosheets for high-performance supercapacitors, Journal of Power Sources 307 (2016) 190–198.

[79] P. Ning, X. Duan, X. Ju, X. Lin, X. Tong, X. Pan, T. Wang, Q. Li, Facile synthesis of carbon nanofibers/MnO_2 nanosheets as high-performance electrodes for asymmetric supercapacitors, Electrochimica Acta 210 (2016) 754–761.

[80] J.S. Cho, Y.J. Hong, Y.C. Kang, Design and synthesis of bubble-nanorod-structured Fe_2O_3–carbon nanofibers as advanced anode material for Li-ion batteries, ACS Nano 9 (4) (2015) 4026–4035.

[81] Z.Y. Wu, X.X. Xu, B.C. Hu, H.W. Liang, Y. Lin, L.F. Chen, S.H. Yu, Iron carbide nanoparticles encapsulated in mesoporous Fe-N-doped carbon nanofibers for efficient electrocatalysis, Angewandte Chemie 127 (28) (2015) 8297–8301.

[82] Y. Qu, S. Liang, K. Zou, S. Li, L. Liu, J. Zhao, G. Piao, Effect of solvent type on the formation of tubular fullerene nanofibers, Materials Letters 65 (3) (2011) 562–564.

[83] Q. Wang, Y. Zhang, K. Miyazawa, R. Kato, K. Hotta, T. Wakahara, In Improved fullerene nanofiber electrodes used in direct methanol fuel cells, in: Journal of Physics: Conference Series, IOP Publishing, 2009, p. 012023.

[84] K. i. Miyazawa, M. Tachibana, T. Mashino, T. Kizuka, Y. Ochiai, Fullerene Nanowhiskers, CRC Press, 2016.

[85] T. Yamazaki, K. Kuramochi, D. Takagi, Y. Homma, F. Nishimura, N. Hori, K. Watanabe, S. Suzuki, Y. Kobayashi, Ordered fullerene nanocylinders in large-diameter carbon nanotubes, Nanotechnology 19 (4) (2008) 045702.

[86] L. Li, J. Zhu, Y. Niu, Z. Chen, Y. Liu, S. Liu, M. Xu, C.M. Li, J. Jiang, Efficient production of coaxial core–shell MnO@ carbon nanopipes for sustainable electrochemical energy storage applications, ACS Sustainable Chemistry & Engineering 5 (7) (2017) 6288–6296.

[87] H. Masuda, T. Yanagishita, K. Yasui, K. Nishio, I. Yagi, T.N. Rao, A. Fujishima, Synthesis of well-aligned diamond nanocylinders, Advanced Materials 13 (4) (2001) 247–249.

[88] M. Zhang, C. Yu, C. Zhao, X. Song, X. Han, S. Liu, C. Hao, J. Qiu, Cobalt-embedded nitrogen-doped hollow carbon nanorods for synergistically immobilizing the discharge products in lithium–sulfur battery, Energy Storage Materials 5 (2016) 223–229.

[89] E. Hammel, X. Tang, M. Trampert, T. Schmitt, K. Mauthner, A. Eder, P. Pötschke, Carbon nanofibers for composite applications, Carbon 42 (5–6) (2004) 1153–1158.

[90] P. Kohli, J.E. Wharton, O. Braide, C.R. Martin, Template synthesis of gold nanotubes in an anodic alumina membrane, Journal of Nanoscience and Nanotechnology 4 (6) (2004) 605–610.

[91] F. Muench, S. Schaefer, L. Hagelüken, L. Molina-Luna, M. Duerrschnabel, H.-J. Kleebe, J. Brötz, A. Vaskevich, I. Rubinstein, W. Ensinger, Template-free electroless plating of gold nanowires: direct surface functionalization with shape-selective nanostructures for electrochemical applications, ACS Applied Materials & Interfaces 9 (36) (2017) 31142–31152.

[92] S.M. Alia, D. Kathlynne, L. Toby, J. Kurt, Y. Yushan, Palladium and gold nanotubes as oxygen reduction reaction and alcohol oxidation reaction catalysts in base, ChemSusChem 7 (6) (2014) 1739–1744.

[93] J. Wang, C. Zhang, J. Zhang, H. Song, P. Wang, Y. Lu, G. Fei, W. Xu, W. Xu, L. Zhang, Hybrid plasmonic cavity modes in arrays of gold nanotubes, Advanced Optical Materials 5 (4) (2017) 1600731.

[94] X. Yin, N.L. Teradal, R. Jelinek, Porous gold nanotubes for enhanced methanol oxidation catalysis, ChemistrySelect 2 (34) (2017) 10961–10964.

[95] V.S. Marangoni, J. Cancino-Bernardi, V. Zucolotto, Synthesis, physico-chemical properties, and biomedical applications of gold nanorods—a review, Journal of Biomedical Nanotechnology 12 (6) (2016) 1136–1158.

[96] M. Kang, H. Lee, T. Kang, B. Kim, Synthesis, properties, and biological application of perfect crystal gold nanowires: a review, Journal of Materials Science & Technology 31 (6) (2015) 573–580.

[97] D. Serbezeanu, A.M. Popa, I. Sava, I.-D. Carja, M. Amberg, R.M. Rossi, G. Fortunato, Design and synthesis of polyimide–gold nanofibers with tunable optical properties, European Polymer Journal 64 (2015) 10–20.

[98] G. Bi, L. Wang, C. Cai, K. Ueno, H. Misawa, J. Qiu, Surface-enhanced Raman scattering of crystal violets from periodic array of gold nanocylinders, Journal of Modern Optics 61 (15) (2014) 1231–1235.

[99] E. Bitzek, Atomistic study of twinning in gold nanowhiskers, Journal of Solid Mechanics and Materials Engineering 6 (1) (2012) 99–105.

[100] B. Piglosiewicz, J. Vogelsang, S. Schmidt, D.J. Park, P. Groß, C. Lienau, Electron photoemission and acceleration from sharp gold nanotapers in the strong-field, few-cycle regime, Quantum Matter 3 (4) (2014) 297–306.

[101] S.-H. Zhang, Z.-X. Xie, Z.-Y. Jiang, X. Xu, J. Xiang, R.-B. Huang, L.-S. Zheng, Synthesis of silver nanotubes by electroless deposition in porous anodic aluminium oxide templates, Chemical Communications (9) (2004) 1106–1107.

[102] P. Gao, C. Zhan, M. Liu, Controlled synthesis of double-and multiwall silver nanotubes with template organogel from a bolaamphiphile, Langmuir 22 (2) (2006) 775–779.

[103] J.-H. Park, S.-G. Oh, B.-W. Jo, Fabrication of silver nanotubes using functionalized silica rod as templates, Materials Chemistry and Physics 87 (2–3) (2004) 301–310.

[104] F. Lorestani, Z. Shahnavaz, P. Mn, Y. Alias, N.S. Manan, One-step hydrothermal green synthesis of silver nanoparticle-carbon nanotube reduced-graphene oxide composite and its application as hydrogen peroxide sensor, Sensors and Actuators B: Chemical 208 (2015) 389–398.

[105] N.R. Jana, L. Gearheart, C.J. Murphy, Wet chemical synthesis of silver nanorods and nanowires of controllable aspect ratioElectronic supplementary information (ESI) available: UV–VIS spectra of silver nanorods, Chemical Communications (7) (2001) 617–618. See, http://www.rsc.org/suppdata/cc/b1/b100521i.

[106] E. Abdullayev, K. Sakakibara, K. Okamoto, W. Wei, K. Ariga, Y. Lvov, Natural tubule clay template synthesis of silver nanorods for antibacterial composite coating, ACS Applied Materials & Interfaces 3 (10) (2011) 4040–4046.

[107] Y.-p. Zhu, X.-k. Wang, W.-l. Guo, J.-g. Wang, C. Wang, Sonochemical synthesis of silver nanorods by reduction of sliver nitrate in aqueous solution, Ultrasonics Sonochemistry 17 (4) (2010) 675–679.

[108] B. Mohapatra, S. Kuriakose, S. Mohapatra, Rapid green synthesis of silver nanoparticles and nanorods using Piper nigrum extract, Journal of Alloys and Compounds 637 (2015) 119–126.

[109] T. Jeyapragasam, R.S. Kannan, Microwave assisted green synthesis of silver nanorods as catalysts for rhodamine B degradation, Russian Journal of Physical Chemistry A 90 (7) (2016) 1334–1337.

[110] T. Suganya, T. Devasena, Green synthesis of silver nanorods and optimization of its therapeutic cum toxic dose, Journal of Nanoscience and Nanotechnology 15 (12) (2015) 9565–9570.

[111] [111a] H. Wang, X. Song, L. You, B. Zhang, Oriented growth of magnesium nanowires by physical vapor deposition in high vacuum, Scripta Materialia 108 (2015) 68–71;
[111b] A.V. Singh, S. Baylan, B.-W. Park, G. Richter, M. Sitti, Hydrophobic pinning with copper nanowhiskers leads to bactericidal properties, PloS One 12 (4) (2017) e0175428;
[111c] E.R. Encina, N. Passarelli, E.A. Coronado, Plasmon enhanced light absorption in aluminium@ Hematite core shell hybrid nanocylinders: the critical role of length, RSC Advances 7 (5) (2017) 2857–2868;
[111d] L. Pu, H. Zhang, T. Yuan, Z. Zou, L. Zou, X.-M. Li, H. Yang, High performance platinum nanorod assemblies based double-layered cathode for passive direct methanol fuel cells, Journal of Power Sources 276 (2015) 95–101.

[112] G. Katwal, M. Paulose, I.A. Rusakova, J.E. Martinez, O.K. Varghese, Rapid growth of zinc oxide nanotube–nanowire hybrid architectures and their use in breast cancer-related volatile organics detection, Nano Letters 16 (5) (2016) 3014–3021.

[113] A.A. Peyghan, S.A. Aslanzadeh, H. Soleymanabadi, Methanol-sensing characteristics of zinc oxide nanotubes: quantum chemical study, Monatshefte für Chemie – Chemical Monthly 145 (8) (2014) 1253–1257.

[114] E.A. Dil, M. Ghaedi, A. Ghaedi, A. Asfaram, M. Jamshidi, M.K. Purkait, Application of artificial neural network and response surface methodology for the removal of crystal violet by zinc oxide nanorods loaded on activate carbon: kinetics and equilibrium study, Journal of the Taiwan Institute of Chemical Engineers 59 (2016) 210–220.

[115] M.A. Ali, K. Mondal, C. Singh, B.D. Malhotra, A. Sharma, Anti-epidermal growth factor receptor conjugated mesoporous zinc oxide nanofibers for breast cancer diagnostics, Nanoscale 7 (16) (2015) 7234–7245.

[116] C. Xu, X. Sun, Multipod zinc oxide nanowhiskers, Journal of Crystal Growth 277 (1–4) (2005) 330–334.

[117] L. Wang, T. Kumeria, A. Santos, P. Forward, M.F. Lambert, D. Losic, Iron oxide nanowires from bacteria biofilm as an efficient visible-light magnetic photocatalyst, ACS Applied Materials & Interfaces 8 (31) (2016) 20110–20119.

[118] I. Hong, M. Angelucci, R. Verrelli, M.G. Betti, S. Panero, F. Croce, C. Mariani, B. Scrosati, J. Hassoun, Electrochemical characteristics of iron oxide nanowires during lithium-promoted conversion reaction, Journal of Power Sources 256 (2014) 133–136.

[119] D. Wang, B. Zhu, X. He, Z. Zhu, G. Hutchins, P. Xu, W.-N. Wang, Iron oxide nanowire-based filter for inactivation of airborne bacteria, Environmental Science: Nano 5 (5) (2018) 1096–1106.

[120] M.M. Momeni, Fabrication of copper decorated tungsten oxide–titanium oxide nanotubes by photochemical deposition technique and their photocatalytic application under visible light, Applied Surface Science 357 (2015) 160–166.

[121] N.A. Barakat, A. Zaki, E. Ahmed, A. Farghali, F.S. Al-Mubaddel, FexCo1− x-doped titanium oxide nanotubes as effective photocatalysts for hydrogen extraction from ammonium phosphate, International Journal of Hydrogen Energy 43 (16) (2018) 7990–7997.

[122] H. Abdallah, A. Moustafa, A.A. AlAnezi, H. El-Sayed, Performance of a newly developed titanium oxide nanotubes/polyethersulfone blend membrane for water desalination using vacuum membrane distillation, Desalination 346 (2014) 30–36.

[123] S.S. Mali, C.S. Shim, H. Kim, J.V. Patil, D.H. Ahn, P.S. Patil, C.K. Hong, Evaluation of various diameters of titanium oxide nanofibers for efficient dye sensitized solar cells synthesized by electrospinning technique: a systematic study and their application, Electrochimica Acta 166 (2015) 356–366.

[124] N.A. Barakat, E. Ahmed, M.A. Abdelkareem, A. Farghali, M. Nassar, M.H. El-Newehy, S.S. Al-Deyab, Ag, Zn and Cd-doped titanium oxide nanofibers as effective photocatalysts for hydrogen extraction from ammonium phosphates, Journal of Molecular Catalysis A: Chemical 409 (2015) 117–126.

[125] S. Han, S. Hong, J. Yeo, D. Kim, B. Kang, M.Y. Yang, S.H. Ko, Nanorecycling: monolithic integration of copper and copper oxide nanowire network electrode through selective reversible photothermochemical reduction, Advanced Materials 27 (41) (2015) 6397–6403.

[126] B. Vidhyadharan, I.I. Misnon, R.A. Aziz, K. Padmasree, M.M. Yusoff, R. Jose, Superior supercapacitive performance in electrospun copper oxide nanowire electrodes, Journal of Materials Chemistry A 2 (18) (2014) 6578–6588.

[127] C. Liu, X. Xie, W. Zhao, J. Yao, D. Kong, A.B. Boehm, Y. Cui, Static electricity powered copper oxide nanowire microbicidal electroporation for water disinfection, Nano Letters 14 (10) (2014) 5603–5608.

[128] B. Ameri, S.S.H. Davarani, R. Roshani, H.R. Moazami, A. Tadjarodi, A flexible mechanochemical route for the synthesis of copper oxide nanorods/nanoparticles/nanowires for supercapacitor applications: the effect of morphology on the charge storage ability, Journal of Alloys and Compounds 695 (2017) 114–123.

[129] H.-b. Oh, H. Ryu, W.-J. Lee, Effects of copper precursor concentration on the growth of cupric oxide nanorods for photoelectrode using a modified chemical bath deposition method, Journal of Alloys and Compounds 620 (2015) 55–59.

[130] Z. Yang, S. Bai, H. Yue, X. Li, D. Liu, S. Lin, F. Li, D. He, Germanium anode with lithiated-copper-oxide nanorods as an electronic-conductor for high-performance lithium-ion batteries, Materials Letters 136 (2014) 107–110.

[131] L. Ren, K. Pashayi, H.R. Fard, S.P. Kotha, T. Borca-Tasciuc, R. Ozisik, Engineering the coefficient of thermal expansion and thermal conductivity of polymers filled with high aspect ratio silica nanofibers, Composites Part B: Engineering 58 (2014) 228–234.

[132] Z.S. Al-Garawi, J.R. Thorpe, L.C. Serpell, Silica nanowires templated by amyloid-like fibrils, Angewandte Chemie International Edition 54 (45) (2015) 13327–13331.

[133] A.H. Thilsted, J.Y. Pan, K. Wu, K. Zór, T. Rindzevicius, M.S. Schmidt, A. Boisen, Lithography-free fabrication of silica nanocylinders with suspended gold nanorings for LSPR-based sensing, Small 12 (48) (2016) 6745–6752.

[134] A. Standing, S. Assali, L. Gao, M.A. Verheijen, D. van Dam, Y. Cui, P.H.L. Notten, J.E.M. Haverkort, E.P.A.M. Bakkers, Efficient water reduction with gallium phosphide nanowires, Nature Communications 6 (2015) 7824.

[135] M.G. Kibria, S. Zhao, F.A. Chowdhury, Q. Wang, H.P.T. Nguyen, M.L. Trudeau, H. Guo, Z. Mi, Tuning the surface Fermi level on p-type gallium nitride nanowires for efficient overall water splitting, Nature Communications 5 (2014) 3825.

[136] F. Richard Webster, C. David, V. Novikov Sergei, F.C. Thomas, Transmission electron microscopy of indium gallium nitride nanorods grown by molecular beam epitaxy, Physica Status Solidi C 11 (3-4) (2014) 417–420.

[137] G. Meligrana, W. Lueangchaichaweng, F. Colò, M. Destro, S. Fiorilli, P.P. Pescarmona, C. Gerbaldi, Gallium oxide nanorods as novel, safe and durable anode material for Li- and Na-ion batteries, Electrochimica Acta 235 (2017) 143–149.

[138] [138a] X. Wang, Y. Li, Rare-earth-compound nanowires, nanotubes, and fullerene-like nanoparticles: synthesis, characterization, and properties, Chemistry–A European Journal 9 (22) (2003) 5627–5635;
[138b] Z.-G. Yan, C.-H. Yan, Controlled synthesis of rare earth nanostructures, Journal of Materials Chemistry 18 (42) (2008) 5046–5059;
[138c] T.T.T. Yang, Rare Earth Nanotechnology, CRC Press, 2012.

[139] M. Safdari, M.S. Al-Haik, 5 - a review on polymeric nanocomposites: effect of hybridization and synergy on electrical properties A2 - Ismail, ahmad fauzi, in: P.S. Goh (Ed.), Carbon-Based Polymer Nanocomposites for Environmental and Energy Applications, Elsevier, 2018, pp. 113–146.

[140] Q. Lin, J. Wang, Y. Zhong, J. Sunarso, M.O. Tadé, L. Li, Z. Shao, High performance porous iron oxide-carbon nanotube nanocomposite as an anode material for lithium-ion batteries, Electrochimica Acta 212 (2016) 179–186.

[141] L. Zhao, Q. Cao, A. Wang, J. Duan, W. Zhou, Y. Sang, H. Liu, Iron oxide embedded titania nanowires–An active and stable electrocatalyst for oxygen evolution in acidic media, Nano Energy 45 (2018) 118–126.

[142] A. Kumar, R. Varshney, M. Kumar, Calculation of propagation characteristics of surface plasmons in gold/silver nanowires, Applied Optics 54 (12) (2015) 3715–3719.

[143] D. Sahu, N. Panda, B. Acharya, A. Panda, Enhanced UV absorbance and photoluminescence properties of ultrasound assisted synthesized gold doped ZnO nanorods, Optical Materials 36 (8) (2014) 1402–1407.

[144] E. Yan, M. Cao, Y. Wang, X. Hao, S. Pei, J. Gao, Y. Wang, Z. Zhang, D. Zhang, Gold nanorods contained polyvinyl alcohol/chitosan nanofiber matrix for cell imaging and drug delivery, Materials Science and Engineering: C 58 (2016) 1090–1097.

[145] S. Ni, J. Zhu, M.A. Mezour, R.B. Lennox, Single wall carbon nanotube (SWCNT)–gold nanorod (AuNR) conjugates via thermally-mild reaction conditions, New Journal of Chemistry 41 (21) (2017) 12392–12396.

[146] A.-C. Joel, W. Baumgartner, Nanofibre production in spiders without electric charge, The Journal of Experimental Biology 220 (12) (2017) 2243.

[147] I. Kandas, N. Shehata, I. Hassounah, P. Sobolčiak, I. Krupa, R. Lewis, Optical fluorescent spider silk electrospun nanofibers with embedded cerium oxide nanoparticles, Journal of Nanophotonics 12 (2) (2018) 026016.

[148] L. Zhou, P. Fu, X. Cai, S. Zhou, Y. Yuan, Naturally derived carbon nanofibers as sustainable electrocatalysts for microbial energy harvesting: a new application of spider silk, Applied Catalysis B: Environmental 188 (2016) 31–38.

[149] T. Amna, M.S. Hassan, F.A. Sheikh, H.K. Lee, K.-S. Seo, D. Yoon, I. Hwang, Zinc oxide-doped poly (urethane) spider web nanofibrous scaffold via one-step electrospinning: a novel matrix for tissue engineering, Applied Microbiology and Biotechnology 97 (4) (2013) 1725–1734.

[150] X. Qiu, F. Ke, R. Timsina, C.Y. Khripin, M. Zheng, Attractive interactions between DNA–carbon nanotube hybrids in monovalent salts, The Journal of Physical Chemistry C 120 (25) (2016) 13831–13835.

[151] K. Umemura, S. Sato, G. Bustamante, J.Y. Ye, Using a fluorescence quenching method to detect DNA adsorption onto single-walled carbon nanotube surfaces, Colloids and Surfaces B: Biointerfaces 160 (2017) 201–206.

[152] M.C. Amirani, T. Tang, Electrostatics of DNA nucleotide–carbon nanotube hybrids evaluated from QM: MM simulations, Nanoscale 7 (46) (2015) 19586–19595.

[153] J. López-Andarias, S.H. Mejías, T. Sakurai, W. Matsuda, S. Seki, F. Feixas, S. Osuna, C. Atienza, N. Martín, A.L. Cortajarena, Toward bioelectronic nanomaterials: photoconductivity in protein–porphyrin hybrids wrapped around SWCNT, Advanced Functional Materials 28 (24) (2017) 1704031.

[154] S. Kalay, Z. Yilmaz, O. Sen, M. Emanet, E. Kazanc, M. Çulha, Synthesis of boron nitride nanotubes and their applications, Beilstein Journal of Nanotechnology 6 (1) (2015) 84–102.

[155] S. Liu, Z.-R. Tang, Y. Sun, J.C. Colmenares, Y.-J. Xu, One-dimension-based spatially ordered architectures for solar energy conversion, Chemical Society Reviews 44 (15) (2015) 5053–5075.

[156] J. Jeevanandam, Y. San Chan, M.K. Danquah, Nano-formulations of drugs: recent developments, impact and challenges, Biochimie 128 (2016) 99–112.

[157] [157a] N. Kobayashi, H. Izumi, Y. Morimoto, Review of toxicity studies of carbon nanotubes, Journal of occupational health 59 (5) (2017) 394–407;
[157b] L. Cseri, M. Razali, P. Pogany, G. Szekely, Chapter 3.15 - organic solvents in sustainable synthesis and engineering, in: B. Török, T. Dransfield (Eds.), Green Chemistry, Elsevier, 2018, pp. 513–553.

[158] [158a] Z. Tshemese, S.C. Masikane, S. Mlowe, N. Revaprasadu, Progress in green solvents for the stabilisation of nanomaterials: imidazolium based ionic liquids, in: Recent Advances in Ionic Liquids, IntechOpen, 2018;
[158b] M. Schaepertoens, C. Didaskalou, J.F. Kim, A.G. Livingston, G. Szekely, Solvent recycle with imperfect membranes: a semi-continuous workaround for diafiltration, Journal of Membrane Science 514 (2016) 646–658.

[159] [159a] T. Fodi, C. Didaskalou, J. Kupai, G.T. Balogh, P. Huszthy, G. Szekely, Nanofiltration-enabled In Situ solvent and reagent recycle for sustainable continuous-flow synthesis, ChemSusChem 10 (17) (2017) 3435–3444;
[159b] J.F. Kim, G. Szekely, M. Schaepertoens, I.B. Valtcheva, M.F. Jimenez-Solomon, A.G. Livingston, In situ solvent recovery by organic solvent nanofiltration, ACS Sustainable Chemistry & Engineering 2 (10) (2014) 2371–2379.

[160] J. López-Andarias, S.H. Mejías, T. Sakurai, W. Matsuda, S. Seki, F. Feixas, S. Osuna, C. Atienza, N. Martín, A.L. Cortajarena, Toward bioelectronic nanomaterials: photoconductivity in protein–porphyrin hybrids wrapped around SWCNT, Advanced Functional Materials 28 (24) (2018) 1704031.

[161] D.Y. Leung, G. Caramanna, M.M. Maroto-Valer, An overview of current status of carbon dioxide capture and storage technologies, Renewable and Sustainable Energy Reviews 39 (2014) 426–443.

[162] D. Bhattacharyya, D.C. Miller, Post-combustion CO_2 capture technologies — a review of processes for solvent-based and sorbent-based CO_2 capture, Current Opinion in Chemical Engineering 17 (2017) 78–92.

[163] M. Bui, C.S. Adjiman, A. Bardow, E.J. Anthony, A. Boston, S. Brown, P.S. Fennell, S. Fuss, A. Galindo, L.A. Hackett, Carbon capture and storage (CCS): the way forward, Energy & Environmental Science 11 (5) (2018) 1062–1176.

[164] S. Choi, J.H. Drese, C.W. Jones, Adsorbent materials for carbon dioxide capture from large anthropogenic point sources, ChemSusChem 2 (9) (2009) 796–854.

[165] A.A. Olajire, CO_2 capture and separation technologies for end-of-pipe applications–a review, Energy 35 (6) (2010) 2610–2628.

[166] S. Legoas, V. Coluci, S. Braga, P. Coura, S. Dantas, D.S. Galvao, Molecular-dynamics simulations of carbon nanotubes as gigahertz oscillators, Physical Review Letters 90 (5) (2003) 055504.

[167] K. Osler, D. Dheda, J. Ngoy, N. Wagner, M.O. Daramola, Synthesis and evaluation of carbon nanotubes composite adsorbent for CO_2 capture: a comparative study of CO_2 adsorption capacity of single-walled and multi-walled carbon nanotubes, International Journal of Coal Science & Technology 4 (1) (2017) 41–49.

[168] T. Zhao, Q. Wang, Y. Kawazoe, P. Jena, A metallic peanut-shaped carbon nanotube and its potential for CO_2 capture, Carbon 132 (2018) 249–256.

[169] L. Keller, B. Ohs, J. Lenhart, L. Abduly, P. Blanke, M. Wessling, High capacity polyethylenimine impregnated microtubes made of carbon nanotubes for CO_2 capture, Carbon 126 (2018) 338–345.

[170] C. Lu, H. Bai, B. Wu, F. Su, J.F. Hwang, Comparative study of CO_2 capture by carbon nanotubes, activated carbons, and zeolites, Energy & Fuels 22 (5) (2008) 3050–3056.

[171] I.M. Szilágyi, D. Nagy, In Review on one-dimensional nanostructures prepared by electrospinning and atomic layer deposition, in: Journal of Physics: Conference Series, IOP Publishing, 2014, p. 012010.

[172] [172a] G. Zainab, N. Iqbal, A.A. Babar, C. Huang, X. Wang, J. Yu, B. Ding, Free-standing, spider-web-like polyamide/carbon nanotube composite nanofibrous membrane impregnated with polyethyleneimine for CO_2 capture, Composites Communications 6 (2017) 41–47;

[172b] N. Iqbal, X. Wang, A.A. Babar, J. Yu, B. Ding, Highly flexible $NiCo_2O_4$/CNTs doped carbon nanofibers for CO_2 adsorption and supercapacitor electrodes, Journal of colloid and interface science 476 (2016) 87–93.

[173] R.P. Lively, D.P. Leta, B.A. DeRites, R.R. Chance, W.J. Koros, Hollow fiber adsorbents for CO_2 capture: kinetic sorption performance, Chemical engineering journal 171 (3) (2011) 801–810.

[174] [174a] A. Yaumi, M.A. Bakar, B. Hameed, Recent advances in functionalized composite solid materials for carbon dioxide capture, Energy 124 (2017) 461–480;
[174b] Q. Lu, F. Gao, S. Komarneni, A green chemical approach to the synthesis of tellurium nanowires, Langmuir 21 (13) (2005) 6002–6005.

[175] L.-Y. Meng, S.-J. Park, Effect of exfoliation temperature on carbon dioxide capture of graphene nanoplates, Journal of Colloid and Interface Science 386 (1) (2012) 285–290.

[176] Y. Li, J. Cheng, L. Hu, J. Liu, J. Zhou, K. Cen, Graphene nanoplatelet and reduced graphene oxide functionalized by ionic liquid for CO_2 capture, Energy & Fuels 32 (6) (2018) 6918–6925.

[177] S. Feng, W. Li, Q. Shi, Y. Li, J. Chen, Y. Ling, A.M. Asiri, D. Zhao, Synthesis of nitrogen-doped hollow carbon nanospheres for CO_2 capture, Chemical Communications 50 (3) (2014) 329–331.

[178] M. Bhagiyalakshmi, J.Y. Lee, H.T. Jang, Synthesis of mesoporous magnesium oxide: its application to CO_2 chemisorption, International Journal of Greenhouse Gas Control 4 (1) (2010) 51–56.

[179] L. Li, D.L. King, Z. Nie, X.S. Li, C. Howard, $MgAl_2O_4$ spinel-stabilized calcium oxide absorbents with improved durability for high-temperature CO_2 capture, Energy & Fuels 24 (6) (2010) 3698–3703.

[180] Z. Yang, M. Zhao, N.H. Florin, A.T. Harris, Synthesis and characterization of CaO nanopods for high temperature CO_2 capture, Industrial & Engineering Chemistry Research 48 (24) (2009) 10765–10770.

[181] J.Z. Tan, Y. Fernández, D. Liu, M. Maroto-Valer, J. Bian, X. Zhang, Photoreduction of CO_2 using copper-decorated TiO_2 nanorod films with localized surface plasmon behavior, Chemical Physics Letters 531 (2012) 149–154.

[182] S. Chowdhury, G.K. Parshetti, R. Balasubramanian, Post-combustion CO_2 capture using mesoporous TiO_2/graphene oxide nanocomposites, Chemical Engineering Journal 263 (2015) 374–384.

[183] P. Subha, B.N. Nair, A.P. Mohamed, G. Anilkumar, K. Warrier, T. Yamaguchi, U. Hareesh, Morphologically and compositionally tuned lithium silicate nanorods as high-performance carbon dioxide sorbents, Journal of Materials Chemistry A 4 (43) (2016) 16928–16935.

[184] C.W. Lai, Anodic CaO-TiO_2 nanotubes composite film for low temperature CO_2 adsorption, Journal of Nanomaterials 2014 (2014), 5-5.

[185] G. Qi, Y. Wang, L. Estevez, X. Duan, N. Anako, A.-H.A. Park, W. Li, C.W. Jones, E.P. Giannelis, High efficiency nanocomposite sorbents for CO_2 capture based on amine-functionalized mesoporous capsules, Energy & Environmental Science 4 (2) (2011) 444–452.

[186] S.-X. Guo, F. Li, L. Chen, D.R. MacFarlane, J. Zhang, Polyoxometalate-promoted electrocatalytic CO_2 reduction at nanostructured silver in dimethylformamide, ACS Applied Materials & Interfaces 10 (15) (2018) 12690–12697.

[187] T.T.H. Hoang, S. Verma, S. Ma, T.T. Fister, J. Timoshenko, A.I. Frenkel, P.J.A. Kenis, A.A. Gewirth, Nanoporous copper–silver alloys by additive-controlled electrodeposition for the selective electroreduction of CO_2 to ethylene and ethanol, Journal of the American Chemical Society 140 (17) (2018) 5791–5797.

[188] S. Zhang, P. Kang, M. Bakir, A.M. Lapides, C.J. Dares, T.J. Meyer, Polymer-supported CuPd nanoalloy as a synergistic catalyst for electrocatalytic reduction of carbon dioxide to methane, Proceedings of the National Academy of Sciences 112 (52) (2015) 15809.

[189] Y. Liu, S. Chen, X. Quan, H. Yu, Efficient electrochemical reduction of carbon dioxide to acetate on nitrogen-doped nanodiamond, Journal of the American Chemical Society 137 (36) (2015) 11631–11636.

[190] N. Iqbal, X. Wang, A. Ahmed Babar, J. Yu, B. Ding, Highly flexible $NiCo_2O_4$/CNTs doped carbon nanofibers for CO_2 adsorption and supercapacitor electrodes, Journal of Colloid and Interface Science 476 (2016) 87–93.

[191] J. Wu, S. Ma, J. Sun, J.I. Gold, C. Tiwary, B. Kim, L. Zhu, N. Chopra, I.N. Odeh, R. Vajtai, A.Z. Yu, R. Luo, J. Lou, G. Ding, P.J.A. Kenis, P.M. Ajayan, A metal-free electrocatalyst for carbon dioxide reduction to multi-carbon hydrocarbons and oxygenates, Nature Communications 7 (2016) 13869.

[192] F. Su, C. Lu, W. Cnen, H. Bai, J.F. Hwang, Capture of CO_2 from flue gas via multiwalled carbon nanotubes, Science of the total environment 407 (8) (2009) 3017–3023.

[193] [193a] J. Ge, G. Fan, Y. Si, J. He, H.-Y. Kim, B. Ding, S.S. Al-Deyab, M. El-Newehy, J. Yu, Elastic and hierarchical porous carbon nanofibrous membranes incorporated with NiFe2O4 nanocrystals for highly efficient capacitive energy storage, Nanoscale 8 (4) (2016) 2195–2204;
[193b] N. Iqbal, X. Wang, J. Ge, J. Yu, H.-Y. Kim, S.S. Al-Deyab, M. El-Newehy, B. Ding, Cobalt oxide nanoparticles embedded in flexible carbon nanofibers: attractive material for supercapacitor electrodes and CO_2 adsorption, RSC Advances 6 (57) (2016) 52171–52179.

[194] Q. Li, J. Guo, D. Xu, J. Guo, X. Ou, Y. Hu, H. Qi, F. Yan, Electrospun N-doped porous carbon nanofibers incorporated with NiO nanoparticles as free-standing film electrodes for high-performance supercapacitors and CO_2 capture, Small 14 (15) (2018) 1704203.

[195] R. Thiruvenkatachari, S. Su, H. An, X.X. Yu, Post combustion CO_2 capture by carbon fibre monolithic adsorbents, Progress in Energy and Combustion Science 35 (5) (2009) 438–455.

[196] M. Asai, T. Ohba, T. Iwanaga, H. Kanoh, M. Endo, J. Campos-Delgado, M. Terrones, K. Nakai, K. Kaneko, Marked adsorption irreversibility of graphitic nanoribbons for CO_2 and H_2O, Journal of the American Chemical Society 133 (38) (2011) 14880–14883.

[197] R. Rea, S. Ligi, M. Christian, V. Morandi, M. Giacinti Baschetti, G.M. De Angelis, Permeability and selectivity of PPO/graphene composites as mixed matrix membranes for CO_2 capture and gas separation, Polymers 10 (2) (2018) 129.

[198] A. Policicchio, Y. Zhao, Q. Zhong, R.G. Agostino, T.J. Bandosz, Cu-BTC/aminated graphite oxide composites as high-efficiency CO_2 capture media, ACS Applied Materials & Interfaces 6 (1) (2013) 101–108.

[199] Q. Sun, M. Wang, Z. Li, A. Du, D.J. Searles, Carbon dioxide capture and gas separation on B80 fullerene, The Journal of Physical Chemistry C 118 (4) (2014) 2170–2177.

[200] J. Zhang, J. Fu, S. Chen, J. Lv, K. Dai, 1D carbon nanofibers@ TiO_2 core-shell nanocomposites with enhanced photocatalytic activity toward CO_2 reduction, Journal of Alloys and Compounds 746 (2018) 168–176.

[201] M. Songolzadeh, M.T. Ravanchi, M. Soleimani, Carbon dioxide capture and storage: a general review on adsorbents, World Academy of Science, Engineering and Technology 70 (2012) 225–232.

[202] G.A. Mutch, S. Shulda, A.J. McCue, M.J. Menart, C.V. Ciobanu, C. Ngo, J.A. Anderson, R.M. Richards, D. Vega-Maza, Carbon capture by metal oxides: unleashing the potential of the (111) facet, Journal of the American Chemical Society 140 (13) (2018) 4736–4742.

[203] A. Nambo, J. He, T.Q. Nguyen, V. Atla, T. Druffel, M. Sunkara, Ultrafast carbon dioxide sorption kinetics using lithium silicate nanowires, Nano letters 17 (6) (2017) 3327–3333.

[204] K. Bhowmik, A. Chakravarty, S. Bysakh, G. De, γ-Alumina nanorod/reduced graphene oxide as support for poly (ethylenimine) to capture carbon dioxide from flue gas, Energy Technology 4 (11) (2016) 1409–1419.

[205] M. Seggiani, M. Puccini, S. Vitolo, High-temperature and low concentration CO_2 sorption on Li_4SiO_4 based sorbents: study of the used silica and doping method effects, International Journal of Greenhouse Gas Control 5 (4) (2011) 741–748.

[206] L. Maserati, S.M. Meckler, J.E. Bachman, J.R. Long, B.A. Helms, Diamine-appended Mg2(dobpdc) nanorods as phase-change fillers in mixed-matrix membranes for efficient CO2/N2 separations, Nano Letters 17 (11) (2017) 6828–6832.

[207] Y. Wang, D. Wang, C.J. Dares, S.L. Marquard, M.V. Sheridan, T.J. Meyer, CO_2 reduction to acetate in mixtures of ultrasmall (Cu)n,(Ag)m bimetallic nanoparticles, Proceedings of the National Academy of Sciences 115 (2) (2018) 278–283.

[208] K. Sun, T. Cheng, L. Wu, Y. Hu, J. Zhou, A. Maclennan, Z. Jiang, Y. Gao, W.A. Goddard, Z. Wang, Ultrahigh mass activity for carbon dioxide reduction enabled by gold–iron core–shell nanoparticles, Journal of the American Chemical Society 139 (44) (2017) 15608–15611.

[209] X. Zhang, Z. Wu, X. Zhang, L. Li, Y. Li, H. Xu, X. Li, X. Yu, Z. Zhang, Y. Liang, H. Wang, Highly selective and active CO_2 reduction electrocatalysts based on cobalt phthalocyanine/carbon nanotube hybrid structures, Nature Communications 8 (2017) 14675.

[210] N. Iqbal, X. Wang, J. Ge, J. Yu, H.-Y. Kim, S.S. Al-Deyab, M. El-Newehy, B. Ding, Cobalt oxide nanoparticles embedded in flexible carbon nanofibers: attractive material for supercapacitor electrodes and CO_2 adsorption, RSC Advances 6 (57) (2016) 52171–52179.

[211] M.E. Boot-Handford, J.C. Abanades, E.J. Anthony, M.J. Blunt, S. Brandani, N. Mac Dowell, J.R. Fernández, M.-C. Ferrari, R. Gross, J.P. Hallett, Carbon capture and storage update, Energy & Environmental Science 7 (1) (2014) 130–189.

[212] S. Bachu, Review of CO_2 storage efficiency in deep saline aquifers, International Journal of Greenhouse Gas Control 40 (2015) 188–202.

[213] M. Godec, V. Kuuskraa, T. Van Leeuwen, L.S. Melzer, N. Wildgust, CO_2 storage in depleted oil fields: the worldwide potential for carbon dioxide enhanced oil recovery, Energy Procedia 4 (2011) 2162–2169.

[214] K.G. Knauss, J.W. Johnson, C.I. Steefel, Evaluation of the impact of CO_2, co-contaminant gas, aqueous fluid and reservoir rock interactions on the geologic sequestration of CO_2, Chemical Geology 217 (3–4) (2005) 339–350.

[215] W. Mitsch, K. Schafer, A. Cabezas, B. Bernal, In balancing methane emissions and carbon sequestration in tropical/subtropical coastal wetlands: a review, American Geophysical Union, Ocean Sciences Meeting (2016) abstract# EC54B-1325, 2016.

[216] R. Lal, W. Negassa, K. Lorenz, Carbon sequestration in soil, Current Opinion in Environmental Sustainability 15 (2015) 79–86.

[217] L.M. Schile, J.B. Kauffman, S. Crooks, J.W. Fourqurean, J. Glavan, J.P. Megonigal, Limits on carbon sequestration in arid blue carbon ecosystems, Ecological Applications 27 (3) (2017) 859–874.

[218] [218a] A.E. Creamer, B. Gao, Carbon-based adsorbents for postcombustion CO2 capture: a critical review, Environmental science & technology 50 (14) (2016) 7276–7289;
[218b] S. Patwardhan, In Bioinspired green nanomaterials (GN) design and manufacture for post-combustion carbon capture and sequestration, in: Abstracts of Papers of the American Chemical Society, Amer Chemical Soc 1155 16th St, Nw, Washington, Dc 20036 Usa, 2015.

[219] Y. Son, S. Sim, H. Ma, M. Choi, Y. Son, N. Park, J. Cho, M. Park, Exploring critical factors affecting strain distribution in 1D silicon-based nanostructures for lithium-ion battery anodes, Advanced Materials 30 (15) (2018) 1705430.

[220] M.G. Hahm, H. Wang, H.Y. Jung, S. Hong, S.-G. Lee, S.-R. Kim, M. Upmanyu, Y.J. Jung, Bundling dynamics regulates the active mechanics and transport in carbon nanotube networks and their nanocomposites, Nanoscale 4 (11) (2012) 3584–3590.

[221] M. Francisco-Marquez, A. Galano, Silicon-doped carbon nanotubes: promising CO_2/N_2 selective agents for Sequestering carbon dioxide, The Journal of Physical Chemistry C 120 (42) (2016) 24476–24481.

[222] R.J. Lee, Z.A. Jawad, A.L. Ahmad, H.B. Chua, Incorporation of functionalized multi-walled carbon nanotubes (MWCNTs) into cellulose acetate butyrate (CAB) polymeric matrix to improve the CO_2/N_2 separation, Process Safety and Environmental Protection 117 (2018) 159–167.

[223] M.C. Ncibi, M. Sillanpää, Optimized removal of antibiotic drugs from aqueous solutions using single, double and multi-walled carbon nanotubes, Journal of hazardous materials 298 (2015) 102–110.

[224] H. Hu, T. Zhang, S. Yuan, S. Tang, Functionalization of multi-walled carbon nanotubes with phenylenediamine for enhanced CO_2 adsorption, Adsorption 23 (1) (2017) 73–85.

[225] O.M. Yaghi, M. O'keeffe, N.W. Ockwig, H.K. Chae, M. Eddaoudi, J. Kim, Reticular synthesis and the design of new materials, Nature 423 (6941) (2003) 705.

[226] A.R. Millward, O.M. Yaghi, Metal– organic frameworks with exceptionally high capacity for storage of carbon dioxide at room temperature, Journal of the American Chemical Society 127 (51) (2005) 17998–17999.

[227] R. Babarao, J. Jiang, Molecular screening of metal– organic frameworks for CO_2 storage, Langmuir 24 (12) (2008) 6270–6278.

[228] E.T. Bender, P. Katta, A. Lotus, S.J. Park, G.G. Chase, R.D. Ramsier, Identification of CO_2 sequestered in electrospun metal oxide nanofibers, Chemical Physics Letters 423 (4) (2006) 302–305.

[229] M. Niu, H. Yang, X. Zhang, Y. Wang, A. Tang, Amine-impregnated mesoporous silica nanotube as an emerging nanocomposite for CO_2 capture, ACS Applied Materials & Interfaces 8 (27) (2016) 17312–17320.

[230] P. Li, H.C. Zeng, Hierarchical nanocomposite by the integration of reduced graphene oxide and amorphous carbon with ultrafine MgO nanocrystallites for enhanced CO_2 capture, Environmental Science & Technology 51 (21) (2017) 12998–13007.

[231] W. Chu, M. Ran, X. Zhang, N. Wang, Y. Wang, H. Xie, X. Zhao, Remarkable carbon dioxide catalytic capture (CDCC) leading to solid-form carbon material via a new CVD integrated process (CVD-IP): an alternative route for CO_2 sequestration, Journal of Energy Chemistry 22 (1) (2013) 136–144.

[232] J. Lau, G. Dey, S. Licht, Thermodynamic assessment of CO_2 to carbon nanofiber transformation for carbon sequestration in a combined cycle gas or a coal power plant, Energy Conversion and Management 122 (2016) 400–410.

[233] S. Licht, J. Ren, M.J. Lefler, M. Johnson, J. Vicini, In carbon nanotubes electrosynthesized from CO_2 for efficient greenhouse gas removal, In *Meeting Abstracts* (No. 9), The Electrochemical Society (2017), 650-650.

[234] B. Li, X. Wang, D. Hopkinson, J. Hoffman, A. Egbebi, K.P. Resnik, Amino acids react with carbon dioxide (CO_2) and form nanofibers and nanoflowers, US Patent App 15/609 (2017) 373.

[235] A. Douglas, N. Muralidharan, R. Carter, C.L. Pint, Sustainable capture and conversion of carbon dioxide into valuable multiwalled carbon nanotubes using metal scrap materials, ACS Sustainable Chemistry & Engineering 5 (8) (2017) 7104–7110.

[236] F.A. Rahman, M.M.A. Aziz, R. Saidur, W.A.W.A. Bakar, M. Hainin, R. Putrajaya, N.A. Hassan, Pollution to solution: capture and sequestration of carbon dioxide (CO_2) and its utilization as a renewable energy source for a sustainable future, Renewable and Sustainable Energy Reviews 71 (2017) 112–126.

[237] Z. Chi, Y. Xie, F. Elloy, Y. Zheng, Y. Hu, S. Chen, Bicarbonate-based integrated carbon capture and algae production system with alkalihalophilic cyanobacterium, Bioresource Technology 133 (2013) 513–521.

[238] S.A. Rackley, Carbon Capture and Storage, Butterworth-Heinemann, 2017.

[239] R.A. Berner, The long-term carbon cycle, fossil fuels and atmospheric composition, Nature 426 (6964) (2003) 323.

[240] S.N. Naik, H. Lentz, R.C. Maheshwari, Extraction of perfumes and flavours from plant materials with liquid carbon dioxide under liquid—vapor equilibrium conditions, Fluid Phase Equilibria 49 (1989) 115–126.

[241] M. Goto, H. Kanda, S. Machmudah, Extraction of carotenoids and lipids from algae by supercritical CO_2 and subcritical dimethyl ether, The Journal of Supercritical Fluids 96 (2015) 245–251.

[242] W.-w. Liu, B. Zhang, Y.-z. Li, Y.-m. He, H.-c. Zhang, An environmentally friendly approach for contaminants removal using supercritical CO_2 for remanufacturing industry, Applied Surface Science 292 (2014) 142–148.

[243] A. Pietsch, Decaffeination—process and quality, in: The Craft and Science of Coffee, Elsevier, 2017, pp. 225–243.

[244] M.R. Molina, G. De la Fuente, M.A. Batten, R. Bressani, Decaffeination. Process to detoxify coffe pulp, Journal of Agricultural and Food Chemistry 22 (6) (1974) 1055–1059.

[245] I. De Marco, S. Riemma, R. Iannone, Life cycle assessment of supercritical CO_2 extraction of caffeine from coffee beans, The Journal of Supercritical Fluids 133 (2018) 393–400.

[246] H. İçen, M. Gürü, Determination of mass transfer coefficients on the obtaining of caffeine from tea stalk by supercritical carbon dioxide with and without ethanol, Arabian Journal for Science and Engineering 43 (5) (2018) 2257–2262.

[247] D.V. Bermejo, E. Ibáñez, G. Reglero, T. Fornari, Effect of cosolvents (ethyl lactate, ethyl acetate and ethanol) on the supercritical CO_2 extraction of caffeine from green tea, The Journal of Supercritical Fluids 107 (2016) 507–512.

[248] S. Ilgaz, I.G. Sat, A. Polat, Effects of processing parameters on the caffeine extraction yield during decaffeination of black tea using pilot-scale supercritical carbon dioxide extraction technique, Journal of food science and technology 55 (4) (2018) 1407–1415.

[249] A.A. Abd-Rabou, H.H. Ahmed, CS-PEG decorated PLGA nano-prototype for delivery of bioactive compounds: a novel approach for induction of apoptosis in HepG2 cell line, Advances in Medical Sciences 62 (2) (2017) 357–367.

[250] S. Partey, R. Preziosi, G. Robson, Improving maize residue use in soil fertility restoration by mixing with residues of low C-to-N ratio: effects on C and N mineralization and soil microbial biomass, Journal of Soil Science and Plant Nutrition 14 (3) (2014) 518–531.

[251] N.H.M. Yasin, T. Maeda, A. Hu, C.-P. Yu, T.K. Wood, CO_2 sequestration by methanogens in activated sludge for methane production, Applied Energy 142 (2015) 426–434.

[252] L.M. Chew, P. Kangvansura, H. Ruland, H.J. Schulte, C. Somsen, W. Xia, G. Eggeler, A. Worayingyong, M. Muhler, Effect of nitrogen doping on the reducibility, activity and selectivity of carbon nanotube-supported iron catalysts applied in CO_2 hydrogenation, Applied Catalysis A: General 482 (2014) 163–170.

[253] P. Kar, S. Farsinezhad, N. Mahdi, Y. Zhang, U. Obuekwe, H. Sharma, J. Shen, N. Semagina, K. Shankar, Enhanced CH_4 yield by photocatalytic CO_2 reduction using TiO_2 nanotube arrays grafted with Au, Ru, and ZnPd nanoparticles, Nano Research 9 (11) (2016) 3478–3493.

[254] L. Roldán, Y. Marco, E. García-Bordejé, Origin of the excellent performance of Ru on nitrogen-doped carbon nanofibers for CO_2 hydrogenation to CH_4, ChemSusChem 10 (6) (2017) 1139–1144.

[255] I.U. Din, M.S. Shaharun, A. Naeem, S. Tasleem, M.R. Johan, Carbon nanofibers based copper/zirconia catalysts for carbon dioxide hydrogenation to methanol: effect of copper concentration, Chemical Engineering Journal 334 (2018) 619–629.

[256] S. Du, W. Tang, X. Lu, S. Wang, Y. Guo, P.X. Gao, Cu-Decorated ZnO nanorod array integrated structured catalysts for low-pressure CO_2 hydrogenation to methanol, Advanced Materials Interfaces 5 (3) (2018) 1700730.

[257] A. Bose, K. Jana, D. Mitra, S. De, Co-production of power and urea from coal with CO_2 capture: performance assessment, Clean Technologies and Environmental Policy 17 (5) (2015) 1271–1280.

[258] N. Yahya, S. Qureshi, Z. ur Rehman, B. Alqasem, C.F. Kait, Green urea synthesis catalyzed by hematite nanowires in magnetic field, Journal of Magnetism and Magnetic Materials 428 (2017) 469–480.

[259] W. Yan, D. Wang, L.A. Diaz, G.G. Botte, Nickel nanowires as effective catalysts for urea electro-oxidation, Electrochimica Acta 134 (2014) 266–271.

[260] W.M. Omymen, J.R. Rogan, B.Z. Jugović, M.M. Gvozdenović, B.N. Grgur, Photo-assisted electrochemical oxidation of the urea onto TiO_2-nanotubes modified by hematite, Journal of Saudi Chemical Society 21 (8) (2017) 990–997.

[261] N.A. Barakat, M.H. El-Newehy, A.S. Yasin, Z.K. Ghouri, S.S. Al-Deyab, Ni&Mn nanoparticles-decorated carbon nanofibers as effective electrocatalyst for urea oxidation, Applied Catalysis A: General 510 (2016) 180–188.

[262] P. Rychter, M. Kot, K. Bajer, D. Rogacz, A. Šišková, J. Kapuśniak, Utilization of starch films plasticized with urea as fertilizer for improvement of plant growth, Carbohydrate Polymers 137 (2016) 127–138.

[263] S. Aoi, K. Mase, K. Ohkubo, T. Suenobu, S. Fukuzumi, Selective CO production in photoelectrochemical reduction of CO_2 with a cobalt chlorin complex adsorbed on multiwalled carbon nanotubes in water, ACS Energy Letters 2 (3) (2017) 532–536.

[264] A.S. Varela, N. Ranjbar Sahraie, J. Steinberg, W. Ju, H.S. Oh, P. Strasser, Metal-doped nitrogenated carbon as an efficient catalyst for direct CO_2 electroreduction to CO and hydrocarbons, Angewandte Chemie International Edition 54 (37) (2015) 10758–10762.

[265] Y.J. Jang, I. Jeong, J. Lee, J. Lee, M.J. Ko, J.S. Lee, Unbiased sunlight-driven artificial photosynthesis of carbon monoxide from CO_2 using a ZnTe-based photocathode and a perovskite solar cell in tandem, ACS Nano 10 (7) (2016) 6980–6987.

[266] W. Zhu, Y.-J. Zhang, H. Zhang, H. Lv, Q. Li, R. Michalsky, A.A. Peterson, S. Sun, Active and selective conversion of CO_2 to CO on ultrathin Au nanowires, Journal of the American Chemical Society 136 (46) (2014) 16132–16135.

[267] B. AlOtaibi, S. Fan, D. Wang, J. Ye, Z. Mi, Wafer-level artificial photosynthesis for CO_2 reduction into CH_4 and CO using GaN nanowires, ACS Catalysis 5 (9) (2015) 5342–5348.

[268] Y. Li, F. Cui, M.B. Ross, D. Kim, Y. Sun, P. Yang, Structure-sensitive CO_2 electroreduction to hydrocarbons on ultrathin 5-fold twinned copper nanowires, Nano Letters 17 (2) (2017) 1312–1317.

[269] Y. Wang, S. Fan, B. AlOtaibi, Y. Wang, L. Li, Z. Mi, A monolithically integrated gallium nitride nanowire/silicon solar cell photocathode for selective carbon dioxide reduction to methane, Chemistry-A European Journal 22 (26) (2016) 8809–8813.
[270] Y. Li, W. Zhang, X. Shen, P. Peng, L. Xiong, Y. Yu, Octahedral Cu_2O-modified TiO_2 nanotube arrays for efficient photocatalytic reduction of CO_2, Chinese Journal of Catalysis 36 (12) (2015) 2229–2236.
[271] S. Delavari, N.A.S. Amin, M. Ghaedi, Photocatalytic conversion and kinetic study of CO_2 and CH_4 over nitrogen-doped titania nanotube arrays, Journal of Cleaner Production 111 (2016) 143–154.
[272] P. Reñones, A. Moya, F. Fresno, L. Collado, J.J. Vilatela, A. Víctor, Hierarchical TiO_2 nanofibres as photocatalyst for CO_2 reduction: influence of morphology and phase composition on catalytic activity, Journal of CO_2 Utilization 15 (2016) 24–31.
[273] Y. Li, M. Mao, Q. Zhang, Y. Yang, H. Huang, Z. Jiang, Q. Hu, X. Zhao, Solar-light-driven CO_2 reduction by methane on Pt nanocrystals partially embedded in mesoporous CeO_2 nanorods with high light-to-fuel efficiency, Green Chemistry 20 (12) (2018) 2857–2869.
[274] Q. Wang, P. Dong, Z. Huang, X. Zhang, Synthesis of Ag or Pt nanoparticle-deposited TiO_2 nanorods for the highly efficient photoreduction of CO_2 to CH_4, Chemical Physics Letters 639 (2015) 11–16.
[275] [275a] Y. Okuyama, S. Kawano, G. Sakai, N. Matsunaga, Y. Mizutani, A direct methane fuel cell with double-layered electrolyte using proton conducting oxide, ECS Transactions 78 (1) (2017) 1953–1961;
[275b] C. Thongchanthep, S. Thountom, The synthesis of Ba0. 7Sr0. $3TiO_3$ ceramics prepared by sol–gel combustion method with urea as fuel, Ceramics International 41 (2015) S95–S99.
[276] S. Jahar, Performance optimization of transcritical CO_2 refrigeration cycle with thermoelectric subcooler, International Journal of Energy Research 37 (2) (2013) 121–128.
[277] Y.-H. Liu, G. Calvert, C. Hare, M. Ghadiri, S. Matsusaka, Size measurement of dry ice particles produced from liquid carbon dioxide, Journal of Aerosol Science 48 (2012) 1–9.
[278] J.I. Linares, A. Cantizano, E. Arenas, B.Y. Moratilla, V. Martín-Palacios, L. Batet, Recuperated versus single-recuperator re-compressed supercritical CO_2 Brayton power cycles for DEMO fusion reactor based on dual coolant lithium lead blanket, Energy 140 (2017) 307–317.
[279] T. Mousavinejad, M.R. Bagherzadeh, E. Akbarinezhad, M. Ahmadi, M.J.F. Guinel, A novel water-based epoxy coating using self-doped polyaniline–clay synthesized under supercritical CO_2 condition for the protection of carbon steel against corrosion, Progress in Organic Coatings 79 (2015) 90–97.
[280] M. Gao, X. Zeng, Q. Hu, Effects of gas shielding parameters on weld penetration of CO_2 laser-TIG hybrid welding, Journal of Materials Processing Technology 184 (1) (2007) 177–183.
[281] P.N. Reddy, P. Basarkod, S. Manvi, Wireless sensor network based fire monitoring and extinguishing system in real time environment, International Journal of Advanced Networking and Applications 3 (2) (2011) 1070.
[282] V. Mittal, Manufacturing of Nanocomposites with Engineering Plastics, Woodhead Publishing, 2015.
[283] M.M. Yousufi, S. Hashmi, Carbon nanotubes laced superabsorbent polymer for enhanced oil recovery technology, Professor Dr. Muhammad Tufail Convener 268 (2016) 268.
[284] P. Esmaeilzadeh, M.T. Sadeghi, Z. Fakhroueian, A. Bahramian, R. Norouzbeigi, Wettability alteration of carbonate rocks from liquid-wetting to ultra gas-wetting using TiO_2, SiO_2 and CNT nanofluids containing fluorochemicals, for enhanced gas recovery, Journal of Natural Gas Science and Engineering 26 (2015) 1294–1305.
[285] L. Huang, Z. Ning, Q. Wang, W. Zhang, Z. Cheng, X. Wu, H. Qin, Effect of organic type and moisture on CO_2/CH_4 competitive adsorption in kerogen with implications for CO_2 sequestration and enhanced CH_4 recovery, Applied Energy 210 (2018) 28–43.
[286] S. Licht, A. Douglas, J. Ren, R. Carter, M. Lefler, C.L. Pint, Carbon nanotubes produced from ambient carbon dioxide for environmentally sustainable lithium-ion and sodium-ion battery anodes, ACS Central Science 2 (3) (2016) 162–168.
[287] R. Farzana, R. Rajarao, I. Mansuri, V. Sahajwalla, Sustainable synthesis of silicon nitride nanowires using waste carbon fibre reinforced polymer (CFRP), Journal of Cleaner Production 188 (2018) 371–377.
[288] Y. Lu, G. Ye, X. She, S. Wang, D. Yang, Y. Yin, Sustainable route for molecularly thin cellulose nanoribbons and derived nitrogen-doped carbon electrocatalysts, ACS Sustainable Chemistry & Engineering 5 (10) (2017) 8729–8737.
[289] H. Yang, S.W. Finefrock, J.D. Albarracin Caballero, Y. Wu, Environmentally benign synthesis of ultrathin metal telluride nanowires, Journal of the American Chemical Society 136 (29) (2014) 10242–10245.
[290] D. Ehhalt, M. Prather, F. Dentener, R. Derwent, E.J. Dlugokencky, E. Holland, I. Isaksen, J. Katima, V. Kirchhoff, P. Matson, Atmospheric Chemistry and Greenhouse Gases, Pacific Northwest National Laboratory (PNNL), Richland, WA (US), 2001.
[291] K. Chatterjee, Causes of greenhouse gas emissions, in: Climate Change: An Integrated Perspective, Springer, 1999, pp. 143–200.
[292] T. Blasing, Recent Greenhouse Gas Concentrations; ESS-DIVE (Environmental System Science Data Infrastructure for a Virtual Ecosystem), Oak Ridge National Laboratory (ORNL), Oak Ridge, TN (United States), 2016.
[293] J. Kiehl, B. Briegleb, The relative roles of sulfate aerosols and greenhouse gases in climate forcing, Science 260 (5106) (1993) 311–314.
[294] S. Weller, B. Janz, L. Jörg, D. Kraus, H.S. Racela, R. Wassmann, K. Butterbach-Bahl, R. Kiese, Greenhouse gas emissions and global warming potential of traditional and diversified tropical rice rotation systems, Global Change Biology 22 (1) (2016) 432–448.
[295] U. Sonesson, J. Davis, F. Ziegler, Food production and emissions of greenhouse gases: an overview of the climate impact of different product groups, SIK Institutet för livsmedel och bioteknik 802 (2010) 1–25.
[296] J.K. Casper, Greenhouse Gases: Worldwide Impacts, Infobase Publishing, 2010.
[297] S.M. Benson, F.M. Orr, Carbon dioxide capture and storage, Mrs Bulletin 33 (4) (2008) 303–305.
[298] J.J. Swanson, R. Febo, A.M. Boies, D.B. Kittelson, Fuel sulfur and iron additives contribute to the formation of carbon nanotube-like structures in an internal combustion engine, Environmental Science & Technology Letters 3 (10) (2016) 364–368.
[299] S. Licht, Carbon Nanofibers, Precious Commodities from Sunlight & CO2 to Ameliorate Global Warming, and Supplement: Carbon Nanofibers (From Fossil Fuel) Electric Power Plants, 2015 arXiv preprint arXiv:1503.06727.
[300] P. Sudarsanam, B. Hillary, M.H. Amin, N. Rockstroh, U. Bentrup, A. Brückner, S.K. Bhargava, Heterostructured copper–ceria and iron–ceria nanorods: role of morphology, redox, and acid properties in catalytic diesel soot combustion, Langmuir 34 (8) (2018) 2663–2673.
[301] S. Suzuki, S. Mori, Considerations on the key precursor for the growth of carbon nanotubes using a diesel engine as a reactor, Chemical Engineering Science 186 (2018) 62–73.
[302] Y. Haik, S. Aldajah, System for Producing Carbon Nanotubes from Combustion Engine Exhausts, 2018. Google Patents.

[303] S. Suzuki, S. Mori, Synthesis of carbon nanotubes from biofuel as a carbon source through a diesel engine, Diamond and Related Materials 82 (2018) 79–86.
[304] Z. Zhou, C. He, J. Xiu, L. Yang, C. Duan, Metal–organic polymers containing discrete single-walled nanotube as a heterogeneous catalyst for the cycloaddition of carbon dioxide to epoxides, Journal of the American Chemical Society 137 (48) (2015) 15066–15069.
[305] H.R. Kim, A. Razzaq, C.A. Grimes, S.-I. In, Heterojunction pnp $Cu_2O/S\text{-}TiO_2/CuO$: synthesis and application to photocatalytic conversion of CO_2 to methane, Journal of CO_2 Utilization 20 (2017) 91–96.
[306] J. Ren, F.-F. Li, J. Lau, L. González-Urbina, S. Licht, One-pot synthesis of carbon nanofibers from CO_2, Nano Letters 15 (9) (2015) 6142–6148.
[307] N. Iqbal, X. Wang, J. Yu, B. Ding, Robust and flexible carbon nanofibers doped with amine functionalized carbon nanotubes for efficient CO_2 capture, Advanced Sustainable Systems 1 (3–4) (2017).
[308] B. Liu, L. Ye, R. Wang, J. Yang, Y. Zhang, R. Guan, L. Tian, X. Chen, Phosphorus-doped graphitic carbon nitride nanotubes with amino-rich surface for efficient CO_2 capture, enhanced photocatalytic activity, and product selectivity, ACS Applied Materials & Interfaces 10 (4) (2018) 4001–4009.
[309] S.-H. Kim, G.-Y. Chung, S.-H. Kim, G. Vinothkumar, S.-H. Yoon, K.-D. Jung, Electrochemical NADH regeneration and electroenzymatic CO_2 reduction on Cu nanorods/glassy carbon electrode prepared by cyclic deposition, Electrochimica Acta 210 (2016) 837–845.
[310] S. Sim, E.-B. Cho, S. Chatterjee, H_2 and CO_2 uptake for hydrogen titanate ($H_2Ti_3O_7$) nanotubes and nanorods at ambient temperature and pressure, Chemical Engineering Journal 303 (2016) 64–72.
[311] C. Kim, K.M. Cho, A. Al-Saggaf, I. Gereige, H.-T. Jung, Z-scheme photocatalytic CO_2 conversion on three-dimensional $BiVO_4$/carbon-coated Cu_2O nanowire arrays under visible light, ACS Catalysis 8 (5) (2018) 4170–4177.
[312] Y.F. Lin, C.S. Wang, C.C. Ko, C.H. Chen, K.S. Chang, K.L. Tung, K.R. Lee, Polyvinylidene fluoride/siloxane nanofibrous membranes for long-term continuous CO_2-capture with large absorption-flux enhancement, ChemSusChem 7 (2) (2014) 604–609.
[313] L. Guo, X. Hu, G. Hu, J. Chen, Z. Li, W. Dai, H.F. Dacosta, M. Fan, Tetraethylenepentamine modified protonated titanate nanotubes for CO_2 capture, Fuel Processing Technology 138 (2015) 663–669.
[314] C. Liu, J.J. Gallagher, K.K. Sakimoto, E.M. Nichols, C.J. Chang, M.C. Chang, P. Yang, Nanowire–bacteria hybrids for unassisted solar carbon dioxide fixation to value-added chemicals, Nano Letters 15 (5) (2015) 3634–3639.
[315] D. Raciti, K.J. Livi, C. Wang, Highly dense Cu nanowires for low-overpotential CO_2 reduction, Nano Letters 15 (10) (2015) 6829–6835.
[316] W.-N. Wang, F. Wu, Y. Myung, D.M. Niedzwiedzki, H.S. Im, J. Park, P. Banerjee, P. Biswas, Surface engineered CuO nanowires with ZnO islands for CO_2 photoreduction, ACS Applied Materials & Interfaces 7 (10) (2015) 5685–5692.
[317] L. Cao, D. Raciti, C. Li, K.J. Livi, P.F. Rottmann, K.J. Hemker, T. Mueller, C. Wang, Mechanistic insights for low-overpotential electroreduction of CO_2 to CO on copper nanowires, ACS Catalysis 7 (12) (2017) 8578–8587.
[318] Y.F. Li, N. Soheilnia, M. Greiner, U. Ulmer, T. Wood, A.A. Jelle, Y. Dong, A.P. Yin Wong, J. Jia, G.A. Ozin, Pd@ H_yWO_{3-x} nanowires efficiently catalyze the CO_2 heterogeneous reduction reaction with a pronounced light effect, ACS Applied Materials & Interfaces 11 (6) (2018) 5610–5615.
[319] Y. Li, B. Zou, C. Hu, M. Cao, Nitrogen-doped porous carbon nanofiber webs for efficient CO_2 capture and conversion, Carbon 99 (2016) 79–89.
[320] H. Shi, G. Chen, C. Zhang, Z. Zou, Polymeric $g\text{-}C_3N_4$ coupled with $NaNbO_3$ nanowires toward enhanced photocatalytic reduction of CO_2 into renewable fuel, Acs Catalysis 4 (10) (2014) 3637–3643.
[321] A.B. Rao, E.S. Rubin, A technical, economic, and environmental assessment of amine-based CO_2 capture technology for power plant greenhouse gas control, Environmental Science & Technology 36 (20) (2002) 4467–4475.
[322] A.-M. Cormos, C. Dinca, L. Petrescu, D.A. Chisalita, S. Szima, C.-C. Cormos, Carbon capture and utilisation technologies applied to energy conversion systems and other energy-intensive industrial applications, Fuel 211 (2018) 883–890.
[323] D. Fransaer, Nanomaterials and direct air capture of CO_2, Nanotechnology for Energy Sustainability (2017) 1161–1178.
[324] J. Watson, F. Kern, M. Gross, R. Gross, P. Heptonstall, F. Jones, S. Haszeldine, F. Ascui, H. Chalmers, N. Ghaleigh, J. Gibbins, N. Markusson, W. Marsden, D. Rossati, S. Russell, M. Winskel, P. Pearson, S. Arapostathis, Carbon Capture and Storage: Realising the Potential? Energy Research Centre (UKERC), London: UK, 2012.

C H A P T E R

5

Porous Materials for Catalysis: Toward Sustainable Synthesis and Applications of Zeolites

Xiaolei Fan[1], *Yilai Jiao*[2]

[1]School of Chemical Engineering and Analytical Science, The University of Manchester, Manchester, Greater Manchester, United Kingdom; [2]Shenyang National Laboratory for Materials Science, Institute of Metal Research, Chinese Academy of Sciences, Shenyang city, Liaoning province, P. R. China

O U T L I N E

1. INTRODUCTION

To address the challenge of sustainability, due to an increasing global population and energy demand, as well as the increased living standards, green chemistry principles need to be implemented in any scientific innovations [1,2]. Catalysis, especially heterogeneous catalysis, plays a vital role in the sustainable synthesis of various chemicals (including commodity, fine and specialty chemicals, as well as pharmaceuticals) and sustainable production of energy and fuels, representing one of the most important principles of green chemistry (i.e., the ninth principle catalytic reagents (as selective as possible) are superior to stoichiometric reagents [2]). Currently, about 90% of chemical manufacturing processes and more than 20% of all industrial products involve catalytic steps [3]. Porous materials lie at the heart of the heterogeneous catalysis. The use of porous materials, either as catalysts or catalyst supports, has contributed significantly to the modern development of chemical processes, leading to not only the increased economic benefits but also the reduced environmental impact, thus improving the sustainability of the chemical industry. For example, in petrochemistry, the use of crystalline microporous zeolites as catalysts (to replace the conventional catalysts such as low-selective harmful mineral) has significantly improved the selectivities and yields of the reactions (such as cracking, alkylation, and isomerization), the quality of the products, and the overall life of the catalytic systems, reducing, at the same time, the energy consumption [4].

Sustainable Nanoscale Engineering
https://doi.org/10.1016/B978-0-12-814681-1.00005-9

Porous materials are normally solids, but not limited to solids nowadays [5], containing permanent interconnected networks of pores at different length scales. In addition to their presences in industries, porous materials (herein porous materials refer to solid porous materials) are everywhere as functional materials in our daily life, such as natural substances of rocks, clays, and biological tissues (e.g., bones) and synthetic materials of ceramics, metal oxides, and carbonaceous materials [6]. Classification of solid porous materials for catalysis can be based on their pore sizes (i.e., micropores of <2 nm, mesopores of 2−50 nm, or macropores of >50 nm by IUPAC based on the nitrogen physisorption analysis at −196°C) [7] or the nature of their solid phases (i.e., crystalline or amorphous). In this chapter, the sustainable aspects of various porous materials (classified according to the nature of their solid phases of crystalline and amorphous) are discussed briefly at first. After the general discussion of the strengths and disadvantages regarding their overall sustainability toward practical applications, a critical overview on the recent progress of postsynthesis modification of zeolites and hierarchically structured zeolite catalysts is presented, outlining the potential of these methods for preparing new zeolitic materials and catalysts with improved physicochemical and textural properties for sustainable catalytic applications.

2. CRYSTALLINE POROUS MATERIALS

Zeolites, especially synthetic zeolites (Fig. 5.1), represent the most important microporous crystalline materials for industrial catalytic applications, as well as fascinating inorganic materials that stimulate a wide range of research in their synthesis, modification, structures, and properties [8]. Zeolites are inorganic aluminosilicates with an ordered framework structure built by corner-sharing SiO_4 and AlO_4 units in different fractions but with a silicon-to-aluminum ratio (Si/Al) greater than unity [9]. The SiO_4 and AlO_4 units in zeolites are interconnected in various ways, resulting in a large portfolio of zeolites with different pore sizes (normally <1 nm) and pore topologies, i.e., 230 fully ordered framework types and 9 partially disordered type materials [10]. These microporous crystalline materials possess varied physical and chemical properties, which have been allocated unique three-letter codes by the International Zeolite Association, such as FAU (faujasite) and MFI (mordenite framework inverted−type). Among these zeolites, nearly 20 are implemented at the industrial scale, but only five constitute the so-called "Big Five" zeolites, i.e., MFI, BEA, MOR, FAU, and FER [11], due to their significances in the modern chemical industry, especially the petrochemical conversions such as the production of gasoline, propylene, and aromatics productions (e.g., cumene and ethylbenzene). Synthetic zeolites emerged from c.1948−55 [12], then followed by their major breakthroughs in industrial catalysis exemplified by the synthetic FAU zeolite Y in 1962 (for fluid catalytic cracking, FCC) [13] and pentasil MFI ZSM-5 zeolite in 1988 (for selective toluene disproportionation) [14]. Because of the bond strength of oxygen bridge linking Al and Si and the high framework density, zeolites demonstrate very high thermal and hydrothermal stability (as high as c.800°C), which is one of the most important technical parameters required by industries for gas/vapor phase reactions.

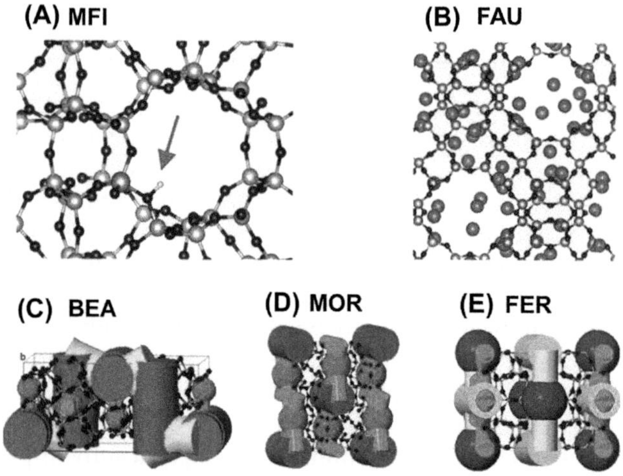

FIGURE 5.1 Structures of the "Big Five" synthetic zeolites. Si is shown in yellow (light gray in print version), Al in gray, O in red (black gray in print version), and Na in magenta (gray in print version). (A) H-ZSM-5, with the MFI topology and one Brønsted acid site indicated with the green (gray arrow in print version) arrow; (B) zeolite Y, with the FAU topology and Na^+ compensating the negative charges of Al in the framework; (C) zeolite beta (all-silica form), with the BEA topology; (D) mordenite (MOR, all-silica form); and (E) ferrierite (FER, all-silica form) [8]. *Reproduced with permission from The Royal Society of Chemistry.*

The development of zeolites was a revolution toward the rationalization of the surface and solid-state chemistry [9], and zeolite-based catalysts are standard catalysts in many industrial applications nowadays. Beyond the traditional applications in chemical and petrochemical sectors, zeolites are playing an increasingly important role in many sustainable conversions and applications, especially their promising uses in the fields of renewable energy [15], biomass conversion [16], fuel cells [15], emission control [17], and water purification/treatment [18,19]. From a sustainability point of view, the catalytic applications of zeolites have demonstrated their potentials for several catalytic processes that will potentially impact the energy, environmental, and chemical sectors of our society. Zeolites are playing an important role in the world's transition to renewable fuels and chemicals. The need to reduce the dependence of the world on fossil fuels and their contributions to man-made climate change are among the most important grand challenges of this century. Biofuels derived from nonfood source lignocellulosic biomass such as wood and straw will help to diversify renewable energy resources, thus improving energy security and boosting agriculture, forestry, waste management, and transport. Zeolites, as the cracking catalysts, have been proposed as one of the enabling technologies to upgrade the bio-oils (from fast pyrolysis processes) or vegetable oils (e.g., palm oil), via catalytic deoxygenation, to produce biofuels [20]. HZSM-5 zeolite and mesoporous MCM-41 composites were formulated for the catalytic cracking of palm oil, yielding organic liquid products of hydrocarbons corresponding to gasoline, kerosene, and diesel boiling point range. The maximum conversion of palm oil at 99 wt.% was achieved with gasoline yield of 48 wt.%, which is very promising [21]. Catalytic upgrading of fast pyrolysis bio-oils (mainly model compounds) and catalytic fast pyrolysis of biomasses using zeolites (mainly ZSM-5) were also heavily researched, showing promising results as well [20]. However, the catalyst deactivation due to coke formation is still a challenging issue for zeolite catalysts for biofuel applications. Zeolites also see many applications in the production and synthesis of key platform commodity, fine and specialty chemicals from renewable sources. For example, the catalytic dehydration of bioethanol into ethylene [22] and propylene [23] has been performed successfully over HZSM-5 zeolites, especially the nanosized HZSM-5 for bioethanol dehydration to ethylene, which showed excellent conversion (of 95% bioethanol by volume in water) and selectivity to ethylene at >98.40% and >98.43%, respectively, for a time on stream of 630 h [22]. Other examples also include the production of acrylic acid via the catalytic hydration of biolactic acid over the modified NaY zeolites (using alkali phosphates) [24] and the catalytic esterification reactions over hierarchical FAU zeolites (with different Si/Al ratios of 2.6–30) for the pretreatment of pyrolysis bio-oil components (e.g., acetic acid) and the real bio-oil (with methanol) [16]. For environmental protection applications, zeolites are also influential catalyst for a wide range of applications including (i) the removal and decomposition of indoor volatile organic compound (VOC) pollutants (such as toluene and formaldehyde) over TiO_2-ZSM-5 composites [25,26], (ii) the catalytic advanced oxidation processes for gas- (e.g., ozonation of chlorinated VOCs on ZSM-5) [27] and liquid-phase organics (e.g., catalytic wet peroxide oxidation of phenolic compounds over FER zeolites) [19] abatement, and (iii) the selective catalytic reduction of NOx using small-pore zeolites such as SAPO-34 and SSZ-13 for emission control applications [17,28]. Also, zeolites also find their potential applications in the mission of tackling the plastic pollution problem (via cracking reactions for depolymerization) [29,30]. The comprehensive application of zeolites and zeolite-based materials in areas of sustainable chemistry (Fig. 5.2) has been recently reviewed by Yu et al. [15].

Therefore, one can see that zeolites are a class of the most important inorganic microporous materials for heterogeneous catalysis, being increasingly important in the sustainable use of resources and environmental remediation and protection. Hence, from the chemical, physical, and practical points of view, any newly proposed or developed solid porous materials cannot be discussed independently from the zeolites. Hence, in the following discussion of other types of porous materials, zeolites are used as the benchmark for evaluating the physicochemical and textural properties, as well as the sustainable features of other porous materials. Zeolites are intrinsically microporous with pore openings smaller than 1 nm, making them only effective to deal with molecules with kinetic diameters of <1 nm. Therefore, the transport/diffusion limitations can restrict molecules, especially bulky ones, to go into and get out the crystal domain of zeolites, hence limiting yields, process efficiency, profits, and ultimately sustainability. To answer this challenge, various solutions are proposed including the development of hierarchical zeolites and the exploration of new classes of porous materials with large pores (>1 nm).

Metal–organic frameworks (MOFs) are one of the state-of-the-art "star" materials since the successful synthesis of a stable MOF of $[Zn_4O(TPA)_3]_n$ (i.e., MOF-5, TPA = terephthalic acid) with a permanent porosity after the full desolvation [31]. MOFs are a class of crystalline hybrid (organic–inorganic) structures constructed by binding metal or metal clusters using organic ligands (such as carboxylates, pyridines, and amines), hence forming one-(1D), two-(2D), or three-dimensional (3D) structures with pore sizes ranging from microporous to mesoporous scales. In comparison with the microporous inorganic zeolites, MOFs possess not only the exceptionally high surface areas (>10,000 $m^2\ g^{-1}$) [32] but also the flexibility and versatility of their pore chemistry and architecture

FIGURE 5.2 Examples of applications of zeolites in sustainable chemistry. *Reprinted with permission from Y. Li, L. Li, J. Yu, Chem 3 (6) (2017) 928–949.*

(>20,000 MOFs are reported to date), which enable the rational design of novel catalysts for synthesis [9,33]. Experimental exploration of MOFs for heterogeneous catalysis has been carried out since 1990s and then expanded to a wide catalog of model reactions (such as cyanosilylation, condensation, epoxidation, etc.) and strategies (based on metal nodes, metal nodes/organic linkers engineering, encapsulation of other catalyst, etc.), which has been extensively reviewed previously [9,33–35]. However, because of their organic nature and the weak coordination bonding between metal nodes and organic ligands, MOFs suffer issues concerning stability (for example, MOF-5 is only stable up to 300°C [31]) and moisture sensitivity (for example, HKUST-1 decomposes readily and rapidly in contact with moisture [36,37]). Although MOFs' stability can be improved using metals with high oxidation states such as Zr^{4+} and Al^{3+} (to reinforce the coordinative bonds, e.g., UiO-66 is stable above 500°C [38,39]) and MOFs-catalysis can be activated and sustained by low-temperature plasma activation [40], MOFs are not likely to be alternative catalysts for zeolite catalysis. More importantly, the nature of MOFs synthesis (mainly solvent-based wet methods) and cost of starting organic materials make MOFs far less sustainable than the inorganic zeolites. Therefore, unless the catalytic niche of MOFs can be clearly identified and robustly established (such as the synthesis highly value-added specialty chemicals, delicate molecules, individual enantiomers, etc.) [33], the application of MOF-based catalysts for sustainable synthesis seems challenging.

Based on covalent bonding, another class of crystalline porous organic polymers was developed in 2005 (e.g., 2D COF-1 and COF-5) [41] and 2007 (e.g., 3D COF-102 and COF-103) [42], which is named as covalent organic frameworks (COFs). Because of the versatile covalent bonding of organic building units consisting of light elements (C, Si, O, B, and N) only, COFs demonstrate the advantages similar to that by MOFs, large surface area (up to 4,200 $m^2\ g^{-1}$), tunable pore size and structure, and facilely tailored functionality [43]. Additionally, COFs also show the properties cannot be matched by MOFs such as low density (e.g., 0.17 g cm^{-3} for COF-108 [42] vs. 0.42 g cm^{-3} for MOF-177 [44]) and relatively high thermal stabilities (stable up to 400–500°C) [42,43]. However, the efficient synthesis of COFs at increased scales is still very challenging, together with the nature of COFs synthesis and raw materials used, and COFs cannot be considered as sustainable candidates for their envisioned applications in practical settings (at least for the next 20 years), including catalysis.

3. AMORPHOUS POROUS MATERIALS

The discovery of amorphous ordered mesoporous materials (OMMs) in 1971 (being described as low-bulk density silica) [4,45], followed by the synthesis of Mobil Composition of Matter (MCM) materials in 1992 [46,47], offers new opportunities for developing novel mesoporous catalysts to mitigate the diffusion limitations occurred in microporous zeolites in heterogeneous catalysis, as well as for other applications such as sensors and separation [45,48]. The synthesis of OMMs normally requires the use of organic surfactant micelles, supramolecules, or polymers as organizing agents (or templates) [45,48,49] and depends on the self-assembly of organic templates, which defines OMMs' porous structures. The formation of highly ordered mesostructures is achieved by assembling inorganic silica around organic templates, followed by the removal of organic templates mainly by calcination (but not exclusively) [48]. For example, cationic cetyltrimethylammonium (CTA+) surfactants used in the synthesis of MCM-41 [47] and poly(alkylene oxide) triblock copolymers used in the synthesis of Santa Barbara Amorphous type material (i.e., SBA-15) [50]. The synthesis procedure of OMMs is generally similar, and thanks to the availability of various templates, pore sizes of OMMs can be tuned to give a wide range of mesopores (up to 42 nm for siliceous mesostructured cellular foams [51]). However, the amorphous nature of OMMs results in the relatively poor thermal and hydrothermal stability in reference to microporous crystalline zeolites, which limits their industrial values in practice. The main reason of OMMs' poor thermal and hydrothermal stability is believed to lie in the nature of the structural units in their amorphous walls (rather than the presence of mesopores), in which the connectivity of the SiO_4 tetrahedra is often incomplete [4,45]. Although postsynthesis modifications of OMMs such as crystallization of amorphous silica walls of OMMs have been proposed to improve the quality of OMMs, the relevant procedures increase the complexity of synthesis (e.g., the use of additional components and synthesis steps) [4], making the resulting materials not sustainable and suitable for further exploitation toward the industrial use. Additionally and more importantly, the use of large amount of organic templates in structuring OMMs, although effective and versatile, imposes the significant cost and environmental impact on the applications employing OMMs. For example, considering a large-scale synthesis of OMMs, the emission of dangerous gaseous products will be associated with the production of large volumes of flue gases due to template removal, being less attractive from the suitability point of view. However, it is worth mentioning that the pilot-scale synthesis of MCM-41 (at 600 kg scale) and pilot-scale FCC demonstration using MCM-41–based catalysts (5 wt.%) have been reported in 2008, showing the improved yields of diesel (by 1.85%) and lighter oils (by 3.47%) and reduced coke formation (by 0.29%) in comparison with a conventional FCC catalyst [52].

Amorphous porous carbons such as activated carbons, carbon black, and templated carbons are another class of porous materials with significant importance in heterogeneous catalysis. Their applications in catalysis have been focusing primarily on being used as the catalyst support [53–55] with recent developments of doped porous carbons to be used directly as catalysts, e.g., the nitrogen-doped porous carbons for oxygen reduction reaction [56]. Being different from MOFs and OMMs, which are compared directly with zeolites, the area of application of porous carbons is quite dissimilar to zeolites, i.e., as the catalyst supports to disperse catalytically active metals or metal compounds using their pore systems. In addition to their porous structures and high specific surface areas, porous carbons (mainly activated carbons) also possess good stability (under reaction and regeneration conditions) and adequate mechanical strength, making them suitable catalyst supports (along with alumina and silica) for preparing industrial supported catalysts (e.g., Pt/C for hydrogenation) [53]. However, the application of metal catalysts supported on porous material is limited by the reaction conditions, especially the temperature, i.e., porous carbons cannot be used in hydrogenation reactions at about >400°C or in the presence of oxygen at about >200°C [53]. Furthermore, the manufacturing process or synthesis methods of porous carbons are not relatively sustainable, e.g., the requirement of high temperatures and/or templates (for preparing ordered microporous carbons) [54]. For instance, modern manufacturing processes for making porous activated carbons generally involve steps of carbonization and activation of raw materials, either physical or chemical [55], which require high-temperature treatments (at >600°C) or the use of acids, strong bases, or salts.

Another recent development of porous carbons is based on the derivation of MOFs, mainly via thermal treatments such as pyrolysis [57], generally known as MOFs-derived porous carbons. By using MOFs as the carbon precursors and pore templates, the resulting MOFs-derived porous carbons demonstrate various expected advantages [57] such as large surface area (up to 5,500 $m^2 g^{-1}$) [58], high porosity, adjustable morphology, good stability, and conductivity. Especially, considering the metals/metal oxides remained in the carbonized framework, as well as the possibility of incorporating other substrates such as graphene oxide [59], MOFs-derived porous carbons are proposed exclusively for heterogeneous catalysis, as summarized in a recent review [57]. Thanks to the highly porous framework, together with the highly dispersed catalytic sites (even at the atomic level), good performances of transition metal catalysts (such as Fe, Co, and Ni) supported on MOFs-derived porous carbons are achieved for electrocatalytic reactions of oxygen evolution reaction [60,61] and hydrogen evolution reduction [62,63], demonstrating their potentials as the alternative catalytic systems to the conventional systems using noble metals (e.g., Pt and Pd). Other catalytic applications such as oxidation, hydrogenation, and photocatalysis are also reported using such materials [57]. However, without the unprecedented catalytic enhancement (such as at least two order of magnitude increase in activity and extraordinary recyclability), the use of MOFs-derived porous carbons for practical catalysis seems not feasible and sustainable. Specifically, when considering the thermal treatment (normally at harsh conditions) of expensive and organic-containing precursors (MOFs), the sustainability of MOFs-derived porous carbons cannot be well justified, especially at larger scales.

Built on foundations of polymer sciences, amorphous polymeric microporous materials, particularly polymers of intrinsic microporosity (PIMs) [64,65], are successfully synthesized. In comparison with amorphous activated carbons, which normally possess a wide distribution of pore sizes, PIMs demonstrate well-defined micropores and surface chemistries [65]. Consider the pure organic nature of PIMs, their applications in heterogeneous catalysis are very limited in the liquid phase [66,67], as well as being not possible in the gas-phase catalysis.

Based on the discussion above, in comparison with zeolites, other classes of crystalline and amorphous materials are undoubtedly advanced materials from the scientific point of view. However, it is also necessary to consider the relatively high cost of raw materials, the complex synthesis methods, and the health, safety, and environment (HSE) issues (associated with the synthesis and postsynthesis treatments) in the proposals of using them for heterogeneous catalysis. Therefore, although the exciting properties and promising catalytic results of some of these advanced porous materials are reported in literature, they have not yet demonstrated significant advantages over the current conventional catalysts such as zeolites, especially from the sustainable point of view. However, one has to admit that some of these novel porous materials are still in their infancies today, and with the significant technological progress in the future, they should be promising to establish robust applications in certain areas (may not necessarily be in catalysis, especially the strong solid acid catalyzed and shape selective reactions [4]). By comparing crystalline to amorphous porous materials at an angle of catalyst design, crystalline porous materials are more advantageous than amorphous porous materials for catalysis because the accurate determination of the fine structure of a porous material is essential to understand the catalysis at elementary levels using spectroscopic characterization and theoretical modeling, hence allowing the smart design of new materials, catalysts, and catalytic systems [8].

4. HIERARCHICAL ZEOLITES

Current challenges faced by chemical industries, specifically petrochemistry, require sustainable and highly efficient processes to maximize the use of finite fossil resources and to mitigate the associated environmental impact to the largest extent. In this regard, zeolites have brought many step changes to these fields (e.g., FCC). As the possibility of using the newly proposed porous materials for practical heterogeneous catalysis is at least remote, zeolites are still in the central position as the key and mature class of porous materials to provide technological solutions toward relatively efficient yet sustainable catalytic processes. On this subject, hierarchical zeolites (i.e., zeolites with the hierarchical meso–microporous structures) are probably the best examples. For the chemical transformation in hierarchical zeolites, diffusion and reaction can be improved significantly, i.e., (i) feed molecules can readily access the zeolite crystal for reactions over the active sites and (ii) product molecules can quickly exist the zeolite crystal before their overreaction to undesirable products and even coke formation. Diffusion limitation in zeolites with intrinsic micropores is one of the most important driving forces in the zeolite community to develop novel zeolites and modify the current zeolites. This is particularly true in the case, among others, of refining industry, which is facing the significant problem of treating heavy feedstocks and residues, with the aim to reach the total conversion of the barrel into valuable products, as stated by Perego and Millini [4]. There are ever-increasing reports on solutions to address this issue fall mainly into the following categories of nanosizing [68], synthesizing multipore zeolite composites [69], creating hierarchical mesopores (via the templating methods [or bottom-up] or postsynthesis methods [or top-down], Fig. 5.3) [69,70], and making novel bulk crystal structures such as core-shell and hollow spheres [71], etc., which has been critically reviewed in previous literature [72,73], even from the industrial perspective [74]. Considering the main short-/middle-term requirement in industry, practical and suitable methods that can be economically realized on a multiton scale, as well as being able to improve the performance of commercial off-the-shelf zeolites [75], are probably sustainable solutions for solving the problems associated with the diffusion limitation. Therefore, the postsynthesis modifications of zeolites (for preparing hierarchical zeolites), although they are already the current standard methods used by the industry (e.g., in preparation of ultrastable zeolite Y, USY), are still undoubtedly the most practical approach to tackle the diffusion limitation.

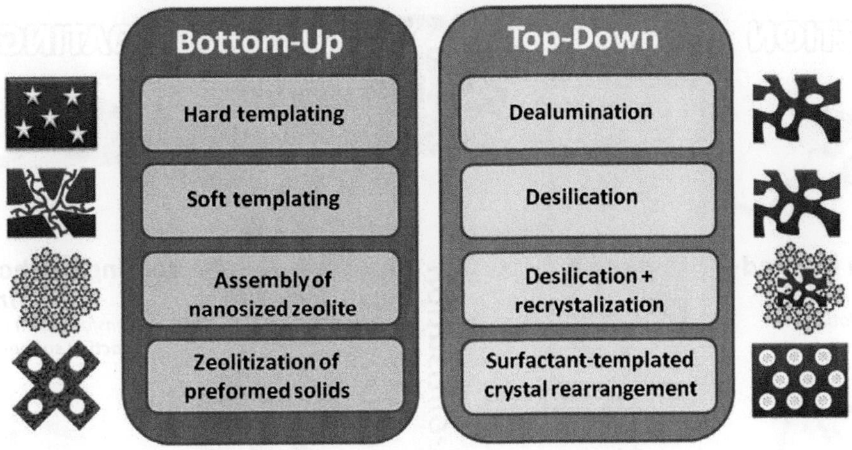

FIGURE 5.3 Templating methods (or bottom-up) or postsynthesis methods (or top-down) for producing hierarchical zeolites [69]. *Reproduced with permission from Wiley-VCH Verlag.*

5. A MULTISCALE SYSTEM IN ZEOLITE CATALYSIS

The application of zeolites in industrial catalysis is a representative multiscale process, which involves different length scales from the active atomic sites, porous networks, crystals, shaped catalyst bodies, packing of catalysts, to reactors. Thus, it is a challenging task to bridge such various entities with different chemical (e.g., the nature of catalytic sites) and physical properties (e.g., shape [pellets or monoliths] and strength of the packing) over more than 10 orders of magnitude (Fig. 5.4). In addition to these chemical and physical aspects, the fluid phase (carrying the reactants and products) also has to find their flow and diffusion path across the whole catalytic system, making the process even more complex [76]. Zeolite catalysts developed in laboratories are normally assessed in forms of bulk powders and pressed/extruded particles, whereas technical catalysts used by industry should be shaped into macroscopic entities with other components, meaning that the preservation of the intrinsic property of zeolites can be difficult. Therefore, in scaling up the developed zeolites from the laboratory to the practical application in industry, one of the key challenges is to mitigate the multiscale effects and reserve the intrinsic property of the zeolite in industrial applications [77].

Shaped zeolitic bodies are conventionally prepared by aggregating a pasty mixture of desired bulk zeolites and binders via compacting and shaping techniques to obtain pellets, beads, cylinders, etc., as depicted in Fig. 5.5A. The resulting catalyst particles can be considered as granular systems and are widely used as unconsolidated

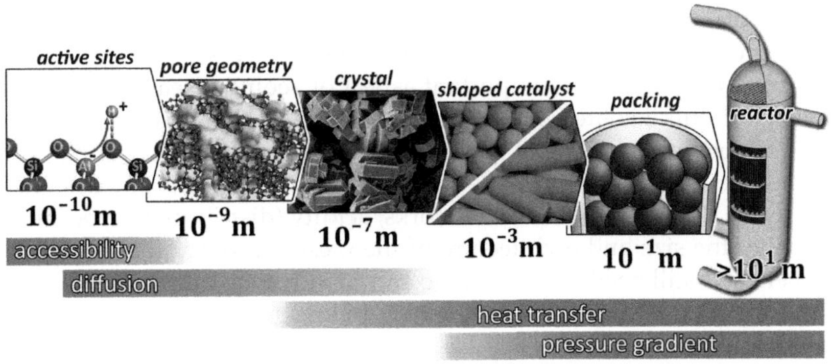

FIGURE 5.4 Multiscale situation in translating zeolites into practical catalysis in a catalytic reactor [76]. *Reproduced with permission from The Royal Society of Chemistry.*

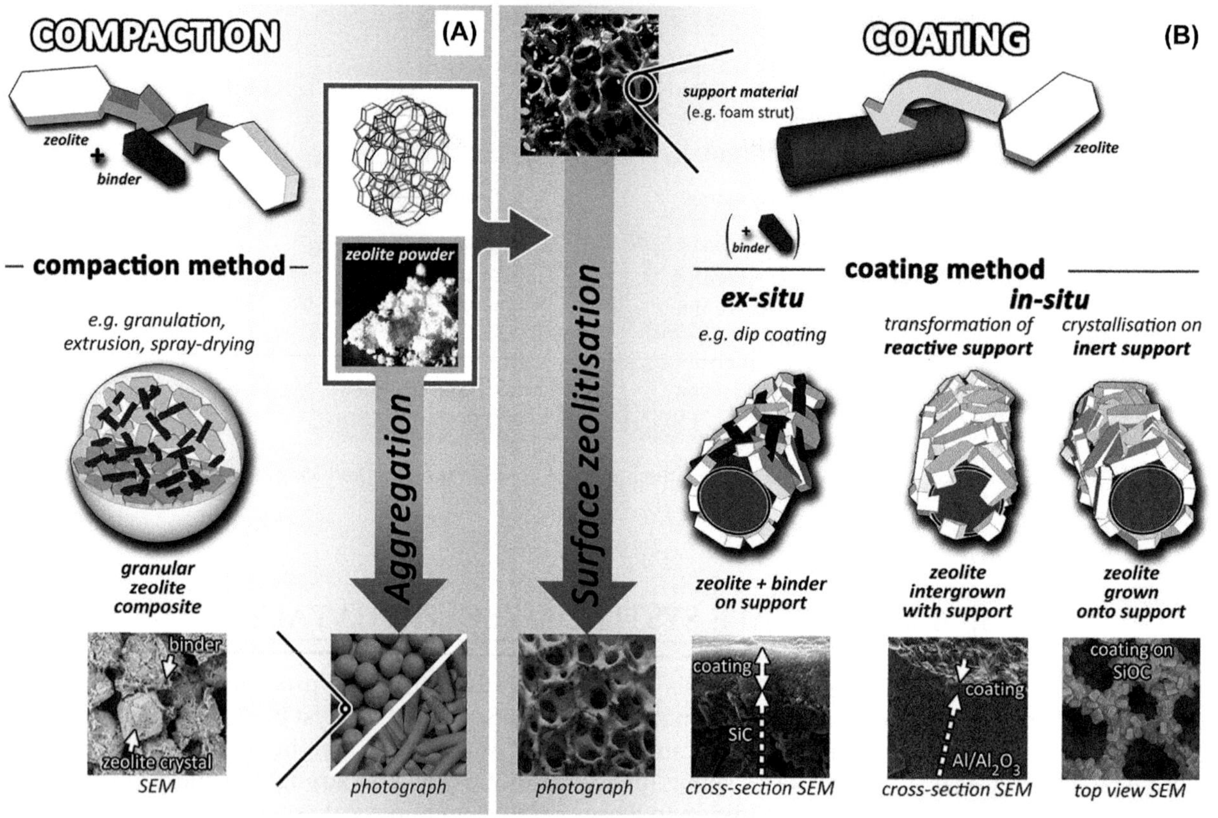

FIGURE 5.5 Schematic comparison of compaction and coating approaches: preparation routes for hierarchical zeolitic composites either based on zeolite shaping methods (left) or on ex situ and in situ zeolite coating on consolidated supports (right, exemplarily shown for open-cell solid foam supports). The characteristics of the resulting products are shown schematically (middle, black: binder, gray: support, zeolite: white) and by photographs and scanning electron microscopy (SEM) images (bottom) [76]. *Reproduced with permission from The Royal Society of Chemistry.*

packings in reactors/columns for different industrial applications. In a packed bed configuration, catalysis across the shaped zeolitic bodies contains multiscale mass transfer steps due to the presence of interparticle voids at the packed bed level, intercrystal porosity at the catalyst particle level, and the intracrystalline porosity at the zeolite crystal level. Conversely, in systems containing structured zeolite catalysts, i.e., zeolites supported on macroscopic supports such as honeycombs (or monoliths) and solid open-cell foams, mostly referred to as the structured monoliths, as shown in Fig. 5.5B [76], the mass transfer limitation can be improved to some extent, especially at the "global" level (i.e., across the bed and reactor) because of the elimination of interparticle mass transfer. Therefore, in addition to the engineering of individual zeolite crystals, strategies of improving structured zeolite catalysts can also be beneficial to enhance the efficiency and hence the sustainability of the catalytic process employing these catalysts. In summary, the process intensification of zeolites-related processes (to improve the process efficiency and reduce the associated energy uses) is another important aspect of zeolite research.

Postsynthesis treatments of bulk zeolites include, generally, steaming, acid leaching (dealumination), alkaline treatment (desilication), and/or combinations of the three procedures. In general, all these treatments are effective, depending on the type and nature of zeolites, but the conditions should be carefully designed and controlled to produce hierarchical zeolites with the desired porous networks, acidity (distribution, type, and strength), and stability. However, accurate control of the size and architecture of the mesoporous system is difficult if not impossible [4]. Therefore, the dedicated effect is still needed to optimize and/or advance the postsynthesis methods for improving their capability of controlling the hierarchical mesoporosity to be generated and the sustainability. As introduced previously, conventional postsynthesis treatments of zeolite for creating hierarchical zeolites have been extensively reviewed [68–74,78,79]. Apart from the engineering of intracrystalline porosity in zeolites at microscopic levels, the design and optimization of shaped zeolite catalysts at macroscopic levels are other important aspects of applying zeolites in practical catalysis. Additional mass transfer steps may be present within a catalyst particle containing zeolites, between catalyst particles, and across the reactor bed; therefore, a careful design of the shaped catalyst and catalytic system according to the application's requirements may significantly improve the process efficiency and hence sustainability. In the following context, the discussion will be focused on the selected recent development of postsynthesis/preparation methods, i.e., the microwave (MW)-assisted postsynthesis treatment of bulk zeolites (i.e., crystals) and postshaping dry gel conversion (DGC) modification of structured zeolite catalysts, which are proposed to address the limitations (such as mass transfer limitations) of zeolites and zeolite-based catalysts at different length scales.

6. MICROWAVE-ASSISTED POSTSYNTHESIS TREATMENT OF ZEOLITES

The life cycle of a chemical molecule via catalytic synthesis includes the preparation of the catalyst, in this context the preparation of porous materials, which may be the "hot spot" contributing significantly to the environmental impact and sustainability of the product molecule. Therefore, the sustainable aspects of porous materials, concerning aspects of raw materials, synthesis, and/or modification, should be considered for the sustainability of their catalytic applications thereafter. The manufacturing and/or modification of porous materials are important aspects as well, as discussed above, accounting for the sustainable feature of porous material–catalyzed processes. Taking zeolite catalysis as the example, both synthesis and postsynthetic modification of zeolites are relatively energy-intensive and time-consuming. Therefore, new developments to intensify these processes are continuously reported over the years. As heating is one of the key elements in both synthesizing and modifying zeolites, MW irradiation as one of the effective heating methods is proposed to intensify the zeolite chemistry [80–82].

The MW-assisted postsynthetic treatment of zeolite was first explored by Schmidt et al. [83,84] for a zeotype, i.e., the titanosilicate ETS-10 with ca. 0.8 nm intrinsic micropore opening. ETS-10 in hydrogen peroxide (H_2O_2) solutions (10–30 wt.%) was heated at 150°C (for 15 min) in a multimode MW oven (CEM MARS-5). The process yielded the modified ETS-10 with a trimodal pore structure including the intrinsic micropores, the supermicropores at about 0.76×1.98 nm, and the mesopores at 5–20 nm (Fig. 5.6). In comparison with the conventional method using H_2O_2 (5–30 wt.% solutions, at the room temperature for 5–30 min) [85], MW process enabled the generation of mesopores in addition to the supermicropores. The supermicropores in ETS-10 were created due to the extraction of Ti atoms together with the adjacent SiO_4 tetrahedra, forming elliptic channels with 24 oxygen-bridged Si atoms (theoretical dimensions of ca. 0.76×1.98). MW treatment was likely to intensify the detitanation process, removing Ti–O–Ti effectively and excessively and enlarging the supermicropores to mesopores [83]. The resulting mesoporous ETS-10 showed high external surface area (at ca. 70 $m^2\ g^{-1}$) and high activity in the Beckmann rearrangement of cyclohexanone oxime to ε-caprolactam under vapor-phase conditions (about $1.9 \times 10^{-1}\ g_{\varepsilon\text{-caprolactam}}\ g_{\text{ETS-10}}^{-1}\ h^{-1}$), whereas for the parent ETS-10, it was only about $0.6 \times 10^{-1}\ g_{\varepsilon\text{-caprolactam}}\ g_{\text{ETS-10}}^{-1}\ h^{-1}$. Although the effective volume heating by MW was accounted for the rapid creation of mesopores in ETS-10, there were no control experiments (e.g., under comparable conventional heating conditions) to support the claim.

Later the concept of the MW-assisted dealumination of zeolites using mineral and organic acids was investigated, exemplified by the MW-hydrochloride acid (HCl) treatment of mordenite, beta, and ZSM-5 [86–88] and MW-*p*-toluenesulfonic acid (*p*-TSA) treatment of beta polymorph A (BEA) [89]. MW-induced thermal effect was the main rationale behind these investigations. Treating mordenite and beta zeolites (Na forms) using HCl (6 M) under MW (in a multimode MW oven) resulted in an increase of the bulk Si/Al ratio (by XRF, e.g., from 10.00 for the parent beta to 121.9 for the treated sample). However, the HCL leaching was not effective in treating ZSM-5 zeolite (Na forms, Si/Al = 20.0, Table 5.1). By comparing the MW irradiation with the conventional

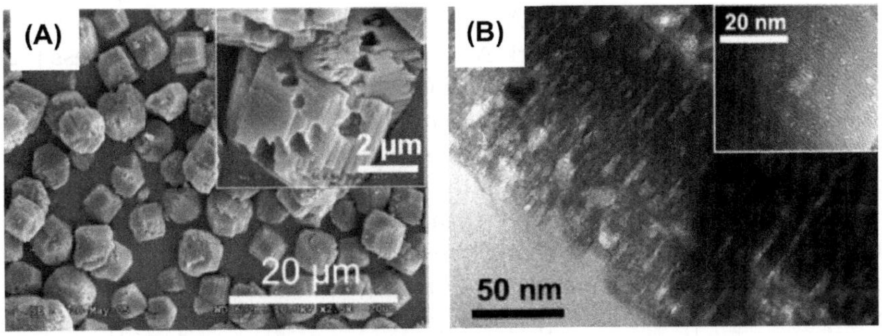

FIGURE 5.6 (A) Scanning electron microscopy (SEM) and (B) transmission electron microscopy (TEM) analysis of microwave–H_2O_2–treated ETS-10 zeotypes [83]. *Reproduced with permission from The Royal Society of Chemistry.*

TABLE 5.1 Dealumination Using Mineral/Organic Acids by Microwave (MW)-Induced and Conventional Heating.

	Dealumination Conditions					Properties by N_2 Physisorption			
Zeolite	**Heating Method**	**T [°C]**	**t [min]**	**Acid**	**Si/Al [−]**	**S_{BET} [$m^2 g^{-1}$]**	**V_{pore} [$cm^3 g^{-1}$]**	**S_{micro}/$S_{nonmicro}$ [−]**	**References**
Parent Na-mordenite (Si/Al = 6.5)	n/a	n/a	n/a	n/a	6.5	303	0.059	8.9	[86–88]
Dealuminated mordenite	Heating under reflux	100	15	6 M HCl	14.9	384	0.109	5.7	
			120		18.2	412	0.110	5.1	
	Autoclaving		15		11.2	376	0.096	6.1	
			120		17.1	397	0.106	5.3	
	MW irradiation under reflux		15		15.5	388	0.102	5.1	
			120		20.2	409	0.121	5.2	
	MW autoclaving		15		15.8	419	0.110	5.9	
			120		23.6	430	0.120	4.9	
Parent Na-beta (Si/Al = 10)	n/a	n/a	n/a	n/a	10.0	573	0.228	1.8	[87,88]
Dealuminated beta	Autoclaving	100	15	6 M HCl	110.8	554	0.244	1.8	
	MW autoclaving				121.9	451	0.242	1.5	
	Autoclaving		5		84.7	527	0.32	1.6	
	MW autoclaving				98.9	507	0.359	1.5	
Parent Na-ZSM-5 (Si/Al = 20)	n/a	n/a	n/a	n/a	20.0	300	0.063	2.4	
Dealuminated ZSM-5	Autoclaving	100	15	6 M HCl	21.3	306	0.072	2.0	
	MW autoclaving				22.4	300	0.073	1.7	
Parent H-beta BEA (Si/Al = 30)	n/a	n/a	n/a	n/a	30	555	0.43	–	[89]
Dealuminated beta	MW irradiation	110	2	0.1 M *p*-TSA	38	564	0.45	–	
				0.2 M *p*-TSA	48	579	0.46	–	
				0.5 M *p*-TSA	58	568	0.46	–	
				1.0 M *p*-TSA	58	566	0.46	–	

hydrothermal treatment methods such as the hydrothermal reflux and autoclaving methods, the former showed insignificant improvements on enhancing the Si/Al ratio and modifying the textural property (Table 5.1) [86–88]. Acid leaching is effective to remove nonframework Al species after the steaming process [72], which produces the nonframework Al species. Therefore, the significant variation of zeolite framework, and hence the creation of mesopores, is accomplished primarily during steam treatment. Acid treatment is subsequently performed to dissolve and wash out the debris, i.e., extraframework Al species deposited onto the internal and external surfaces [69]. Direct acid leaching can also be effective to regulate the Si/Al ratio and create mesopores, depending on (i) the nature of the acid used, (ii) the initial Si/Al ratio of the zeolite, and (iii) the pretreatment such as calcination and ion exchange. However, in these investigations [86–88], these aspects in the direct acid leaching of zeolites under MW heating were not thoroughly considered. Comparative catalytic tests were performed using isomerization of styrene oxide, and styrene oxide ring-opening reactions (in the liquid phase at atmospheric pressure and room temperature) on the parent, conventionally treated and MW-treated zeolites, showing no strong evidence of the benefit by the MW treatment.

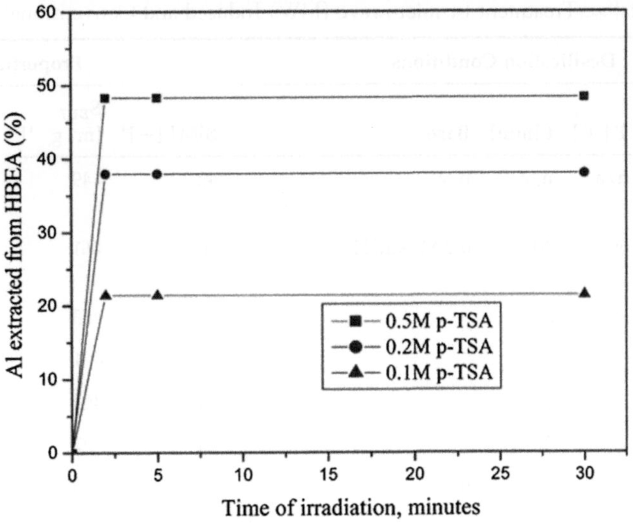

FIGURE 5.7 Microwave (MW)-assisted dealumination of BEA zeolite using *p*-toluenesulfonic acid (*p*-TSA). Effect of the MW treatment time on Al extraction from BEA zeolite [89]. *Reproduced with permission from American Chemical Society.*

Bhat et al. modified BEA zeolite (Si/Al = 30, H form) using *p*-toluenesulfonic acid (*p*-TSA) under MW irradiation (in a multimode MW system) systematically by varying the concentration of *p*-TSA (0.1–1 M), type of sulfonic acid (toluene-4-sulfonic acid and methanesulfonic acid), treatment temperature (30–110°C), and time (2–30 min) [89]. In the MW treatment with *p*-TSA (2 min treatment time), the extent of dealumination increased (from about 21% to about 47%) with an increase in the *p*-TSA concentration of 0.1–0.5 M, then plateaued with 1 M *p*-TSA. A similar trend was also measured for the MW treatment with methanesulfonic acid, which was explained ambiguously by the special interaction between the sulfonic group and framework aluminum. Interestingly, with MW heating, the extent of dealumination did not depend on the treatment time, showing that 2 min was sufficient to achieve the goal, as seen in Fig. 5.7. Conversely, the conventional hydrothermal method required at least 120 min to achieve the same extent of dealumination. The acidic property of the MW-*p*-TSA–treated BEA zeolites correlates well to the extent of dealumination. However, the process did not modify too much of the textural property of materials, as shown in Table 5.1. The subsequent catalytic evaluation (i.e., esterification of *p*-cresol with phenylacetic acid to give *p*-cresyl phenyl acetate) of the developed zeolite catalysts in reference to the parent BEA showed that 0.1 M is the optimum concentration of *p*-TSA in the MW-*p*-TSA treatment, producing optimum acid site distribution in the resulting zeolite. In summary, the work by Bhat et al. demonstrated that the use of MW irradiation in the postsynthetic treatment of BEA zeolite for dealumination was more efficient and controllable than the conventional method. However, the lack of comprehensive and fair comparison of the materials by the two methods regarding the detailed analysis of textural, structural, and acidic properties of the resulting materials, as well as the relevant comparative catalytic tests regarding the relevant use of BEA zeolites such as alkylation reactions for ethylbenzene and cumene production, makes the claims not fully justified.

Desilication (i.e., the extraction of silicon from zeolite frameworks) represents another "top-down" method for creating mesoporosity in zeolites, especially zeolites with relatively high Si/Al ratios such as MFI zeolites. Conventional desilication requires the alkaline media (typically sodium hydroxide, NaOH) and hydrothermal conditions, which is effective and simple. Inspired by the rapid and uniform heating induced by MW radiation, Abelló and Javier Pérez-Ramírez, for the first time, explored the MW-assisted desilication of ZSM-5 zeolites using a monomode MW reactor (CEM Explorer) [75]. Comparative experiments (i.e., conventional and MW-assisted hydrothermal treatments at 65°C with 0.2 M NaOH) were performed using commercial NH_4-form ZSM-5 zeolites with Si/Al ratios of 42 and 37, respectively (CBV 8014 and 8020 from Zeolyst International). It was found that, in comparison with the conventional method (30 min alkaline treatment), MW heating can significantly reduce the treatment time to 3 minutes (for CBV 8020), creating considerable mesoporosity (with about 10 nm pore sizes) in ZSM-5 zeolites comparable with that by the conventional treatment (Table 5.2). Both the conventional and MW-assisted desilication (of an NH_4-form ZSM-5, Si/Al = 37) depended on the treatment time (Fig. 5.8). By extending the treatment time over the optimum point, the excessive silicon leaching could create large mesopores of >20 nm, compromising the mesopore surface area (Table 5.2). The intensification effect due to MW irradiation was briefly discussed

TABLE 5.2 Desilication via Alkaline Treatment by microwave (MW)-Induced and Conventional Heating.

	Desilication Conditions					Properties by N_2 Physisorption			
Zeolite	**Heating Method**	**T [°C]**	**t [min]**	**Base**	**Si/Al [−][a]**	**S_{BET} [$m^2 g^{-1}$]**	**$S_{ext.}$ [$m^2 g^{-1}$]**	**V_{meso} [$cm^3 g^{-1}$]**	**References**
Parent NH_4-ZSM-5 (Si/Al = 42)	n/a	n/a	n/a	n/a	42	449	58	0.11	[75]
Desilicated ZSM-5	Conventional heating	65	30	0.2 M NaOH	26	461	208	0.40	
	MW heating	65	3		28	463	156	0.39	
			5		26	471	212	0.43	
			15		26	452	224	0.53	
			30		26	427	141	0.43	
Parent NH_4-ZSM-5 (Si/Al = 37)	n/a	n/a	n/a	n/a	37	430	40	0.09	
Desilicated ZSM-5	Conventional heating	65	30	0.2 M NaOH	24	450	225	0.40	
	MW heating	65	3		24	492	230	0.39	
			5		24	500	204	0.37	
			15		23	502	196	0.43	
			30		23	471	166	0.44	
Parent NH_4-MOR (Si/Al = 9.5)	n/a	n/a	n/a	n/a	9.5	–	24	0.05	[90]
Desilicated MOR	Conventional heating	85	30	0.2 M NaOH	9.1	–	92	0.11	
			60		9.1	–	122	0.15	
			120		8.6	–	133	0.19	
			240		-	–	115	0.17	
			600		9.1	–	107	0.14	
	MW heating	85	5		9.2	–	99	0.12	
			15		8.8	–	96	0.12	
			30		7.3	–	90	0.15	
Parent NH_4-ZSM-5 (Si/Al = 46)	n/a	n/a	n/a	n/a	46	391	84	0.071	[91]
Desilicated ZSM-5	Conventional heating	100	240	0.05 M NaOH	36	392	155	0.181	
		170	1440	9 M NH_4OH	32	396	132	0.165	
		150	720	3.6 M $NH_2CH_2CH_2NH_2$	–	395	124	0.159	
	MW heating	100	20	0.05 M NaOH	–	394	151	0.179	
		170	60	9 M NH_4OH	–	397	129	0.158	
		150	60	3.6 M $NH_2CH_2CH_2NH_2$	–	397	127	0.168	

[a]*By ICP-OES or ICP-AES.*

and attributed to the selective heating of zeolites (as the MW absorber in the system) and ionic conduction (NaOH as the ionic species in the system). Abelló and Javier Pérez-Ramírez suggested the possible generic feature of the MW-assisted desilication for applications to other zeolite topologies such as MOR and BEA, which was later proved true for ferrierite (Zeolyst International, CP 914, Si/Al = 27.5) [75] and mordenite (Zeolyst International, CBV21A, Si/Al = 9.5) [90].

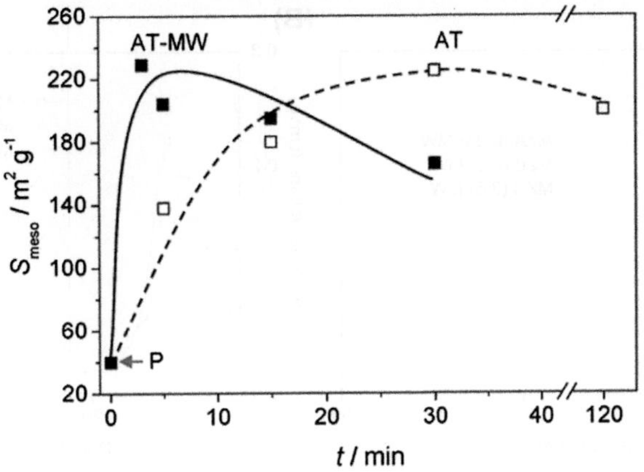

FIGURE 5.8 Mesopore surface area (S_{meso}) as a functions of the treatment time in desilication of ZSM-5 by MW-assisted (solid symbols) and conventional heating (open symbols) [75]. *Reproduced with permission from The Royal Society of Chemistry.*

To speed up the desilication of mordenite (MOR) for developing intracrystalline mesoporosity, MW-assisted desilication was carried out by Paixão et al. (at 85°C with 0.2 M NaOH in a CEM Discover monomode MW reactor) [90]. The treatment time was the main parameter studied, which was varied from 30 to 600 min and from 5 to 30 min, respectively, in experiments by conventional and MW-induced heating. Again, regarding values of specific external surface area ($S_{ext.}$) and mesopores volume (V_{meso}), the MW-assisted alkaline treatment at 5 min was sufficient to produce mesoporous mordenite zeolites comparable with that by the conventional heating at 30 min ($S_{ext.}$: 99 vs. 92 m^2 g^{-1} and V_{meso}: 0.12 vs. 0.11 cm^3 g^{-1} [92], Table 5.2). However, by decomposing the micropores in the treated mordenite into ultramicropores (pore width: <0.7 nm) and supermicropores (pore width: 0.7–2.0 nm), the authors pointed out the difference between the conventional and MW-induced heating in the alkaline treatment. Desilication by conventional heating is primarily due to the alkali corrosion of the outer crystal surface toward the interior of zeolites [92,93], resulting in the relatively high external surface areas ($S_{ext.}$, Table 5.2). Conversely, by MW radiation, an intracrystalline transformation also results in desilication, transforming a fraction of the intrinsic micropores of mordenite into supermicropores [90]. In summary, according to findings by Paixão et al., MW-assisted desilication promoted the formation of supermicropores along with a less significant mesoporosity than the conventional hydrothermal alkaline treatment. Subsequently, catalytic *m*-xylene isomerization over the parent-desilicated MOR zeolites was performed to understand the effect of different desilication methods on the catalytic property of the resulting zeolites. Both isomerization and transalkylation reactions occurred in the system. By analyzing the ratio between the sum of transalkylation products (T) to the sum of isomerization products (I), at similar conversion (i.e., T/I ratio), it was found that the catalytic performance depends strongly on the mesoporous textural properties of MOR zeolites, as well as acidic properties (which was not characterized for the zeolites studied). In general, desilicatied MOR zeolites with high external surface areas and mesopore volumes presented relatively insignificant stereochemical constraints, thus promoting transalkylation reactions involving the production of bulky intermediates. All MOR zeolites studied deactivated over the course of the reaction. The coke analysis by thermogravimetry also showed distinct heat flow profiles for the parent and desilicatied MOR zeolites (by conventional and MW heating, respectively), suggesting different deactivation mechanisms due to the properties introduced by the postsynthesis desilication.

In the work conducted by Hasan et al., the necessity of additional steps (such as ion exchange and calcination for converting the desilicated zeolites into their H-forms) after the desilication using NaOH, KOH, and LiOH was emphasized [91]. Therefore, they performed the comparative MW-assisted and conventional desilication using various bases including NaOH, ammonium hydroxide (NH_4OH), and ethylenediamine ($NH_2CH_2CH_2NH_2$). Under both conventional hydrothermal and MW irradiation conditions, NaOH was found much more effective to enable the desilication of ZSM-5 (CBV 8014 from Zeolyst International) than NH_4OH and ethylenediamine, using less concentrated solutions and shorter treatment time. In comparison to the conventional heating, MW irradiation shortened the treatment time of desilication significantly (Table 5.2), as well as promoting the formation of well-defined mesopores with narrower pore size distributions (Fig. 5.9). Because of the difference in pore sizes of desilicatied MOR zeolites prepared using conventional and MW-assisted methods, their application in liquid-phase butylation (of *t*-butanol with phenol) resulted in the variation in corresponding selectivities. Desilicatied ZSM-5 zeolites by MW irradiation were more selective to less bulky products of 2-*t*-butylphenol and 4-*t*-butylphenol than the zeolites

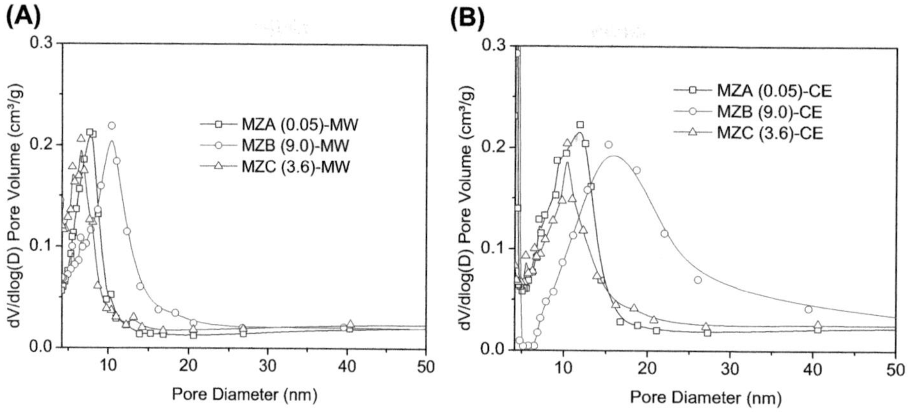

FIGURE 5.9 Comparison of pore size distributions of ZSM-5 zeolites treated by different bases (A for NaOH; B for NH_4OH; C for $NH_2CH_2CH_2NH_2$; numbers in brackets represent the concentration in M) (A) from MW and (B) conventional heating. *Reprinted with permission from Z. Hasan, J.W. Jun, C.-U. Kim, K.-E. Jeong, S.-Y. Jeong, S.H. Jhung, Materials Research Bulletin 61 (Suppl. C) (2015) 469–474.*

obtained using the conventional hydrothermal method, showing the consequence of shape selectivity. The outcomes summarized above all suggest that the possible rapid heating promoted by MW irradiation is accountable for the effective mesopore creation in the postsynthesis treatment of zeolites via demetalation and desilication.

7. STRUCTURED BINDERLESS ZEOLITE CATALYSTS BY DRY GEL CONVERSION (DGC)

Refractory materials such as inorganic oxides of silica, alumina, and clay are used as essential additives, i.e., binders, in shaping bulk zeolite crystals into robust macroscopic bodies. Compared to zeolites, binders are catalytically less active components with proportions of usually 20%–40% in shaped catalysts, diluting the availability of catalytic active components in practical catalysts. Therefore, the additional porous and textural properties, as well as the associated diffusion phenomena, in composite catalyst particles can make the catalytic performance of shaped catalytic bodies significantly different from that of bulk zeolites. Recrystallizing zeolites/binder mixtures into the full-zeolitic phase by conventional hydrothermal methods were originally proposed as the solution to address the inert binder issue, however, resulting in compromised mechanical strength of catalyst grains and/or foreign crystal phase in the crystallized bodies [94]. DGC is another method to transform amorphous precursors into crystalline zeolitic phase without directly contacting with a liquid phase [95,96]. In DGC, either steam or steam/volatile amine (e.g., ethylenediamine and triethylamine) mixtures are used to convert amorphous precursor dry gels (with or without structure-directing agents, SDAs [95]) into crystalline phases at elevated temperatures (normally >100°C) and autogenous pressures. DGC is also referred as (i) the steam-assisted conversion method if pure steam is used or (ii) the vapor-phase transport crystallization method if volatile SDAs are used in the vapor [97,98]. The benefits of DGC include the high degree of conversion with high yield of crystalline phases, reduced waste production, and reduced consumption of expensive organic templates (in some cases) [99], which are sustainable. However, the treated zeolite catalysts via DGC also suffer the loss of mechanical properties such as the mechanical strength [99].

To overcome the loss of mechanical strength associated with the binder conversion via DGC, the postshaping treatment of zeolite-based monoliths (which have a good mechanical strength) was proposed by Zhou et al. [100]. Zeolite-based monoliths (denoted as Con-Cat as in Fig. 5.10, containing MFI ZSM-5 zeolites and 40% silica sol as the binder) were prepared by extrusion molding, then treated by a DGC method (at 200°C for 120 h) employing steam/*n*-butyl amine (20% aqueous solution, as SDA). After the postshaping DGC treatment, the monolith was intact; but the binder silica sol in the technical body was transformed into zeolitic phase, giving fully crystalline zeolite monoliths (denoted as Fch-Cat). Moreover, in addition to extrazeolitic phases (via the recrystallization of binders) with catalytic sites, intracrystalline mesoporosity was also etched in zeolite crystals by the DGC treatment (Fig. 5.11), improving the internal diffusion in zeolite crystals. Similar phenomena were also reported by others in their effort to make binderless zeolite catalysts via DGC [101,102]. Con-Cat and Fch-Cat monoliths were assessed comparatively using catalytic methanol-to-hydrocarbons (MTH) and C4 olefin catalytic cracking for propylene production (OCC) reactions, in which Fch-Cat monoliths demonstrated superior performance regarding activity,

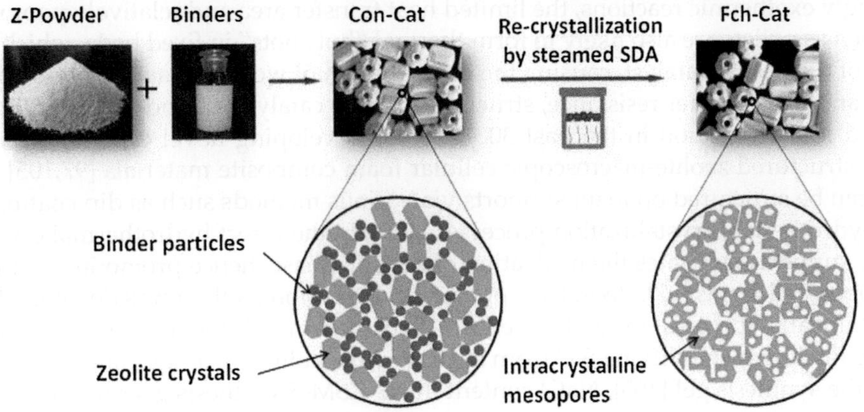

FIGURE 5.10 Schematic illustration of a procedure for preparing hierarchical full-zeolitic monolith. *Reprinted with permission from J. Zhou, J. Teng, L. Ren, Y. Wang, Z. Liu, W. Liu, W. Yang, Z. Xie, Journal of Catalysis (340) (2016) 166–176.*

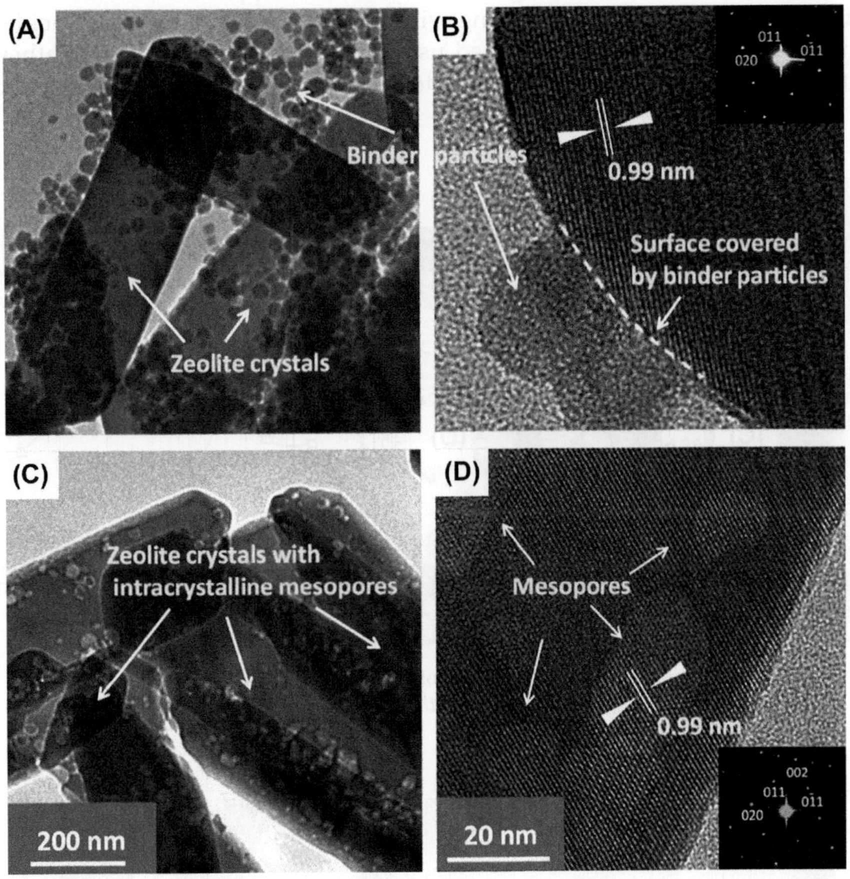

FIGURE 5.11 Transmission electron microscopy micrographs of Con-Cat monolith (A and B, inset in b: the corresponding ED pattern) and recrystallized Fch-Cat monolith (C and D, inset in d: the corresponding ED pattern). *Reprinted with permission from J. Zhou, J. Teng, L. Ren, Y. Wang, Z. Liu, W. Liu, W. Yang, Z. Xie, Journal of Catalysis (340) (2016) 166–176.*

propylene selectivity, and stability over Con-Cat monoliths. For example, in MTH at 480°C, Con-Cat gradually deactivated after 780 h on stream, whereas, under the same condition, Fch-Cat only showed the sign of deactivation (i.e., methanol conversion <95%) after 2030 h on stream.

Considering the scenario of using shaped zeolite catalysts in fixed bed configurations, such as ZSM-5 zeolites in methanol-to-propylene (MTP) reaction [103,104], the catalyst bed is very likely to increase the diffusion resistance for the products departed from active sites, subsequently resulting in their further reaction to form carbonous coke.

Moreover, for strongly exothermic reactions, the limited heat transfer area and relatively poor radial heat transfer of catalyst packing such as pellets are also likely to form thermal "hot spots" in fixed beds, which can contribute to the rapid deactivation of the zeolite catalyst, causing the deterioration of work rate and process efficiency. Therefore, to diminish the mass and heat transfer resistance, structured zeolite catalysts supported on macroscopic cellular supports have attracted much attention in the past 30 years for developing novel catalytic processes using zeolites, exemplified by the structured zeolite-macroscopic cellular foam composite materials [97,105].

Zeolitic phases can be structured on foam supports via various methods such as dip coating, slurry coating, and in situ (or direct) hydrothermal crystallization processes [105]. In the direct hydrothermal crystallization synthesis, the surface of foam supports facilitates the nucleation of zeolitic phase, hence promoting a strong anchoring of the zeolite coating on open-cell foams [97]. Recently, a precursor dispersion method was developed (based on the in situ hydrothermal crystallization process) to synthesize ZSM-5 coatings on SiC foams with controllable morphology, Si/Al ratio, and zeolite loading by modifying SiC foam surface with zeolite precursors and control the amount sodium chloride (NaCl) in the synthesis gel [103]. NaCl content in the ZSM-5 synthesis gel can affect the ZSM-5 nucleation significantly, thus the amount of zeolite coating formed on the foam support (Fig. 5.12).

The open-cell structure of foam supports is proved to improve the transport phenomena across the reactor (or at the "global" level) [106]. To improve the "local" mass transfer (i.e., within the supported zeolite coating) of the structured zeolite/foam catalysts, hierarchical coatings of ZSM-5 zeolite supported on nano SiC whiskers on SiC foam catalyst (as shown in Fig. 5.13) [107] and MFI coating with the intrazeolitic aluminum (acidic) gradient supported on SiC foams (as shown in Fig. 5.14) were developed using the direct hydrothermal synthesis [108], showing the improved antideactivation properties and selectivity to propylene in MTP reactions in comparison with the

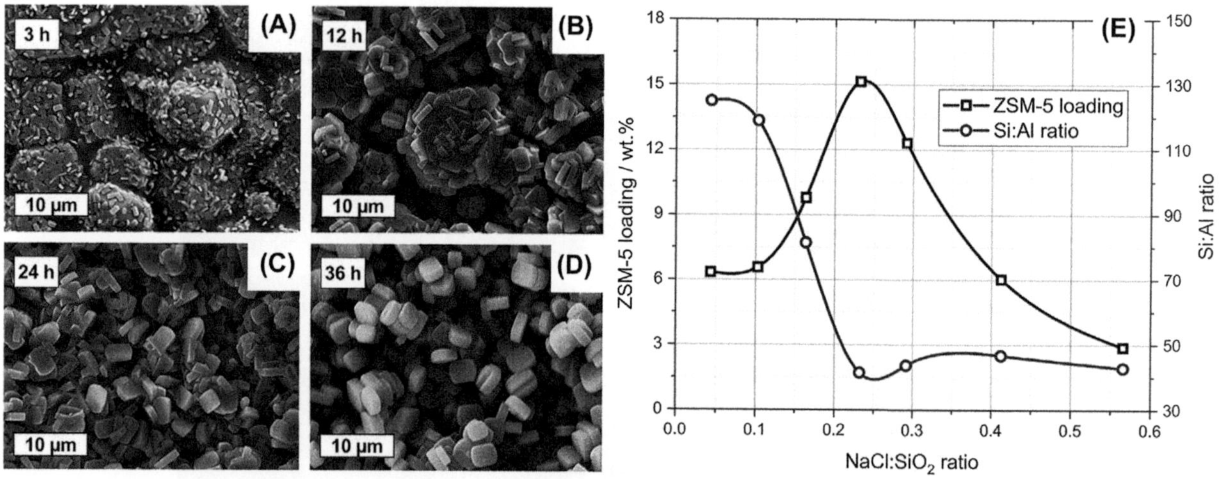

FIGURE 5.12 Effect of the crystallization time on the direct growth of ZSM-5 on SiC ceramic foams (A)–(D); ZSM-5 loading and Si:Al ratio as a function of the NaCl concentration in the synthesis gel (E). *Reprinted with permission from Y. Jiao, C. Jiang, Z. Yang, J. Zhang, Microporous Mesoporous Materials 162 (2012) 152–158.*

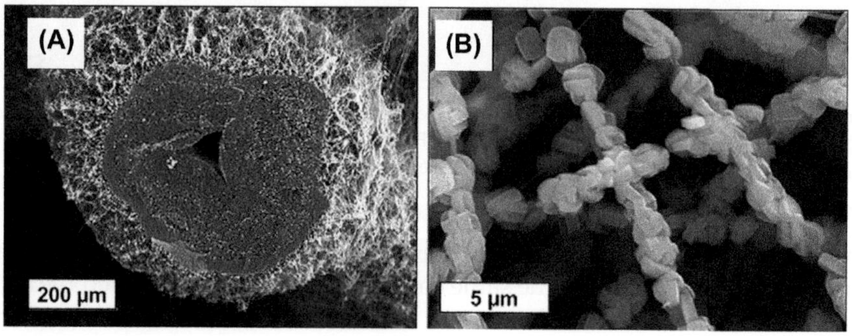

FIGURE 5.13 (A) Scanning electron microscopy (SEM) micrograph of SiC nanowhiskers/SiC foam composites; (B) SEM micrograph of ZSM-5 coatings on SiC nanowhiskers/SiC foam composites. *Reprinted with permission from Y. Jiao, X. Yang, C. Jiang, C. Tian, Z. Yang, J. Zhang, Journal of Catalysis 332 (2015) 70–76.*

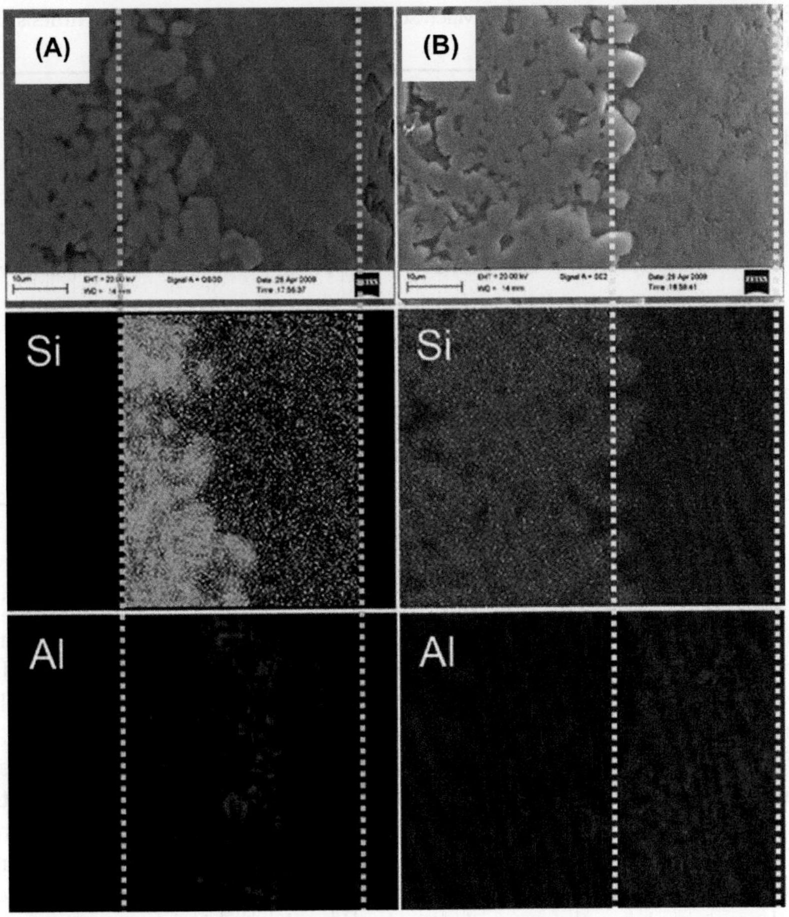

FIGURE 5.14 Scanning electron microscopy micrographs and Si and Al EDX mapping of the cross-sectional surface of (A) MFI coating with the intrazeolitic aluminum (acidic) gradient supported on SiC foams and (B) conventional ZSM-5/SiC foam catalyst. *Reprinted with permission from Y. Jiao, S. Xu, C. Jiang, M. Perdjon, X. Fan, J. Zhang, Applied Catalysis A: General 559 (2018) 1–9.*

conventional ZSM-5/SiC foam catalyst. For example, compared with the conventional ZSM-5/SiC foam catalyst, the gradient MFI/SiC foam catalyst showed excellent performance in MTP with good catalytic longevity (8 vs. 76 h for >95% methanol conversion) and low coke deposition (6.7×10^{-3} wt.% h^{-1} vs. 0.26 wt.% h^{-1}), as well as high propylene selectivity (c.36% vs. 46%). To enable the sustainable synthesis of zeolite coatings on foam supports, MW irradiation was employed to enable the energy-efficient hydrothermal process [109,110]. Specifically, SiC foams have the high MW-adsorbing ability (with a high loss tangent of 1.71 at 2.45 GHz at 25 °C [111]), which leads to the differential heating on the support and the fast yet selective assembly of ZSM-5 zeolites on SiC foams (i.e., 4 vs. 48 h for the conventional hydrothermal synthesis).

Although direct zeolite crystallization methods provide direct routes of coating foam supports with catalytically active zeolite coatings, dip-coating methods are probably the practical method for preparing such structured catalysts for applications at scales. For the structured zeolites/foam catalyst prepared by dip-coating methods, in which inert binders are used to provide adhesion between the zeolite coating and foam supports, the same scenario related to the drawbacks expected by conventionally shaped zeolite catalysts may present as well, as discussed above. Recently, postsynthesis steam-assisted crystallization methods were proposed to improve zeolite coatings prepared by the dip-coating methods, in which a zeolite precursor gel was first deposited on the support and then crystallized to zeolite coating under a mild steam-assisted hydrothermal condition in an autoclave. As the crystallization only takes place on the support and the hydrothermal synthesis conditions are quite mild, an insignificant degradation of the mechanical strength of the support and a high adherence between the zeolite coating and support can be obtained [112]. However, the zeolite coating prepared by the postsynthesis method is thin and dense; and the intercrystal porosity and zeolite loading cannot be optimized.

To mitigate the adverse effects due to the use of inert binders in the dip-coating methods of preparing structured zeolite supported on foams, as well as to achieve the coating with high porosity, controlled loading amount and good

TABLE 5.3 Structured Binderless Zeolites Supported on Macroscopic Supports Prepared by Postsynthesis Dry gel conversion (DGC) Methods.

Zeolite/Dispersing Agent	Support	DGC Conditions	Vapor Phase	Application	References
ZSM-5/zeolite precursor sol	SiC foams	170°C for 24 h	Steam	MTP	[104]
ZSM-5/zeolite precursor sol	SiC foams	160°C for 48 h	Steam/TPAOH	MTP	[97]
ZSM-5/silica sol	SS tubes	175°C for 120 h	Steam/EDA/TEA	*n*-Dodecane cracking	[98]
SAPO-34/synthesis gel	α-Al_2O_3 tubes	200°C for 24 h	Steam	H_2 separation	[114]
Silicalite-1 gel	SS fibers	180°C for 120 h	Steam/EDA/TEA	MTP	[115]
ZSM-5 precursor gel	SS fibers	180°C for 12 h	Steam	MTP	[116]
ZSM-5 or silicalite-1 seeding gel	SS fibers	180°C for 72 h	Steam/EDA	MTP	[117]
ZSM-5 precursor sol	SS fibers	180°C for 120 h	Steam/EDA	MTP/*n*-hexane cracking	[118]
ZSM-11 synthesis sol	SS fibers	180°C for 48 h	Steam	MTP	[119]

interfacial bonding strength with support, a combined wash-coating and postsynthesis DGC method was developed [104]. The method makes use of a sol containing ZSM-5 crystals and precursor in preparing the zeolite coating, which has several functions (i) as dispersing agent to disperse bulk ZSM-5 crystals (obtained by a hydrothermal synthesis from the same precursor sol) on foam supports and (ii) as the precoating on the cellular supports for the subsequent phase transformation (i.e., from the precursor binder to zeolitic phases via DGC). With an aim to further improve the MTP performance of such structured zeolite/SiC foam catalysts, the DGC method was optimized by adding tetrapropylammonium hydroxide (TPAOH, as the SDA) in the steam to increase the relative crystallinity, mesopore volume, and Si/Al ratio of the zeolite coating on SiC foams [97]. In comparison with other SDAs such as ethylenediamine, EDA, and trimethylamine (TEA), TPAOH is a less toxic SDA and more effective template than EDA and TEA to promote the formation of MFI zeolites [113]. By using the proposed postsynthesis method employing TPAOH in the vapor phase, the amorphous aluminosilicate binder was converted to the zeolitic phase, hence improving the catalytic activity and reducing the coke selectivity in MTP reactions [97].

DGC methods employing either pure steam or volatile SDAs are generic to prepare high-quality zeolitic coatings on surfaces of structured supports such as tubes [98,114], foams [97,104], and fibers [115–119] with different materials and geometric structures (Table 5.3). For example, Liu et al. [98] used amorphous SiO_2 as the binder to precoat micro-HZSM-5 on stainless steel (SS) tubes and then treated by DGC with EDA/TEA vapor to prepare mechanically stable zeolite coatings. DGC promoted the crystallization of amorphous SiO_2 binders (<20 wt.%) into MFI zeolitic phases, which interlocked micro-HZSM-5 zeolite particles and enhanced the adherence strength of the zeolite coatings considerably (more than tenfold according to the sonication test). The resulting structured zeolite supported on tubes catalysts also showed improved catalytic performance concerning activity and antideactivation in catalytic cracking of supercritical *n*-dodecane (at 500°C and 4 MPa). Zhou et al. prepared SAPO-34/synthesis solution paste (after the thermal treatment at 60°C) for coating macroporous α-Al_2O_3 tubes (average pore size of 2 mm and porosity of 30%–40%). The precoated α-Al_2O_3 tubes were then treated by DGC using pure steam (at 200°C for 24 h) to give a high-quality SAPO-34 membrane (thickness about 4 mm) with good performances in hydrogen separation applications (H_2 permeance $= 6.96 \times 10^{-6}$ mol m^{-2} s^{-1} Pa^{-1} at room temperature) from various binary systems such as H_2/CO_2, H_2/N_2, etc. [114] Lu et al. also expended this method to prepare various zeolite coatings (e.g., ZSM-5 and ZSM-11) supported on SS fibers (316-L, 20 μm diameter and 85 vol% voidage) and used the developed structured catalysts in catalytic MTP and cracking (of *n*-hexane) reactions [115–119].

In the latest review by Betke and Lieb [120], which summarizes the preparation methods and properties of microporous–macroporous composite materials based on cellular foam supports such as ceramic, metal, and polymer foams with the supported zeolite coatings, a fair comparison of dispersion-based and in situ hydrothermal crystallization (i.e., DGC) techniques in the past decades. They highlighted a generic coating strategy could be derived from the methods discussed above [97,104], which combines the advantages of both in situ and postsynthesis methods, being able to improve the coating layer adherence and the overall amount of deposited zeolite coatings and to control of the crystal morphology and properties of coatings.

8. CONCLUSION

Zeolites are industrially important porous materials for catalysis, ion exchange, and adsorption, especially the emerging and developing applications in sustainable chemistry, contributing significantly to the sustainable development of our industry and society. They are also generally robust inorganic materials with the relatively long life span in industrial applications (e.g., >200 h for gas-phase reactions at elevated temperatures and about 10 years for gas adsorption applications), as well as being recyclable and regeneratable [121,122]. For example, FAU Y zeolites (as most import active component in the FCC catalysts) can be regenerated several times under harsh conditions (at ca. 650–760°C) before being discharged for recycling [123]. The recently developed other classes of porous solids such as MOFs, COFs, and OMMs have demonstrated some interesting chemical and physical properties and seem to be more advantageous than zeolites in some aspects of properties and applications. However, the synthesis (especially at scales for the potential industrial adoption) and practical application of some of these porous materials need additional developments. Therefore, the further development of zeolites and zeolite-based catalysts/processes are undoubtedly the present industrial focus to improve the sustainability of current chemical conversion processes, especially heterogeneous catalysis.

It should be noted that the sustainability of the current syntheses and applications of zeolites is not optimum, and it requires further effort to be improved to meet the sustainable requests raised by the society and industry. For example, the conventional hydrothermal synthesis processes for making zeolites are not sustainable because of the use of organic structure–directing agents (or templates), high energy consumption and the related cost, and HSE issues (such as the generation of toxic and greenhouse gases from the postsynthesis calcination for template removal and the high temperature [during the calcination] and high autogenous pressure [during the synthesis] required by the synthesis) [124,125]. Therefore, enormous efforts have been being made by the relevant communities to improve the sustainability of zeolites synthesis, including but not limited to (i) the use of sustainable energy sources such as MW [125,126] and ultrasonic irradiation [127]; (ii) the synthesis of zeolites using the low-cost templates (such as amines [128] and copper complexes [129]), natural templates (such as yeast) [130], and recyclable templates [131]; (iii) the synthesis of zeolites using the low-cost and sustainable precursors (such as rice husk ash [132] and kaolinite [133]); (iv) template-free (such as the seeding methods) [134,135] and solvent-free methods for zeolite synthesis [132,136], etc.

Hierarchical zeolites with mesoporosity have been regarded as an important development in zeolite science and technology, thanks to their combinational advantages of both zeolites with strong acidity and good hydrothermal stability and OMMs with good mass transfer and accessible bulky molecules. Therefore, sustainable synthesis of hierarchical zeolites via novel strategies with less waste and energy consumption should be the imminent focus of research and development, which has direct relevance to the chemical industry. MW-assisted postsynthesis treatment of zeolites for creating mesoporosity can be a promising method for achieving this goal. However, without revisiting the fundamental mechanisms of demetalation and desilication processes, the state-of-the-art developments do not show a significant improvement in this aspect. Structured zeolite/macroscopic catalysts can be other alternatives to replace the conventional practice using pelletized zeolite catalysts in a wide range of industrial processes, specifically in cases of using the packed-bed configurations. Such structured zeolite catalysts offer a good opportunity to intensify the catalytic processes due to the enhanced mass and heat transfer promoted by the macroscopic supports, particularly at the 'global' level. Current research in academia also shows the possibility of optimizing the zeolite coatings on the structured supports and hence the catalytic performance, exemplified by MTP, MTH, and cracking reactions. Therefore, such structured zeolite catalysts deserve further investigation for feasibility studies at scales.

References

[1] M.M. Kirchhoff, Promoting sustainability through green chemistry, Resources, Conservation and Recycling 44 (3) (2005) 237–243.

[2] P. Anastas, N. Eghbali, Green chemistry: principles and practice, Chemical Society Reviews 39 (1) (2010) 301–312.

[3] J. Zhao, Y. Yin, Y. Li, W. Chen, B. Liu, Synthesis and characterization of mesoporous zeolite Y by using block copolymers as templates, Chemical Engineering Journal 284 (Suppl. C) (2016) 405–411.

[4] C. Perego, R. Millini, Porous materials in catalysis: challenges for mesoporous materials, Chemical Society Reviews 42 (9) (2013) 3956–3976.

[5] N. Giri, M.G. Del Pópolo, G. Melaugh, R.L. Greenaway, K. Rätzke, T. Koschine, L. Pison, M.F.C. Gomes, A.I. Cooper, S.L. James, Liquids with permanent porosity, Nature 527 (2015) 216.

[6] R.J. White, R. Luque, V.L. Budarin, J.H. Clark, D.J. Macquarrie, Supported metal nanoparticles on porous materials. Methods and applications, Chemical Society Reviews 38 (2) (2009) 481–494.
[7] A.D. McNaught, A. Wilkinson, Compendium of Chemical Terminology, IUPAC. Compendium of Chemical Terminology, second ed., Blackwell Scientific Publications, Oxford, 1997.
[8] V. Van Speybroeck, K. Hemelsoet, L. Joos, M. Waroquier, R.G. Bell, C.R.A. Catlow, Advances in theory and their application within the field of zeolite chemistry, Chemical Society Reviews 44 (20) (2015) 7044–7111.
[9] M. Ranocchiari, J.A.V. Bokhoven, Catalysis by metal–organic frameworks: fundamentals and opportunities, Physical Chemistry Chemical Physics 13 (14) (2011) 6388–6396.
[10] Database of Zeolite Structures. Database of Zeolite Structures, http://www.iza-structure.org/databases/.
[11] R. Bingre, B. Louis, P. Nguyen, An overview on zeolite shaping technology and solutions to overcome diffusion limitations, Catalysts 8 (4) (2018) 163.
[12] J. Weitkamp, Zeolites and catalysis, Solid State Ionics 131 (1) (2000) 175–188.
[13] A.D. Reichle, Cat cracking. A retrospective, Chemical Engineering Progress (1990) 70–74.
[14] Y.Y. Huang, M.P. Nicoletti, R.A. Sailor, The Mobil Selective Toluene Disproportionation Process, Dewitt 1990 Petrochemical Review, Houston, TX, 1990. March 27–29.
[15] Y. Li, L. Li, J. Yu, Applications of zeolites in sustainable chemistry, Chem 3 (6) (2017) 928–949.
[16] A. Osatiashtiani, B. Puértolas, C.C.S. Oliveira, J.C. Manayil, B. Barbero, M. Isaacs, C. Michailof, E. Heracleous, J. Pérez-Ramírez, A.F. Lee, K. Wilson, On the influence of Si:Al ratio and hierarchical porosity of FAU zeolites in solid acid catalysed esterification pretreatment of bio-oil, Biomass Conversion and Biorefinery 7 (3) (2017) 331–342.
[17] D.W. Fickel, E. D'Addio, J.A. Lauterbach, R.F. Lobo, The ammonia selective catalytic reduction activity of copper-exchanged small-pore zeolites, Applied Catalysis B: Environmental 102 (3) (2011) 441–448.
[18] Y. Yan, S. Jiang, H. Zhang, X. Zhang, Preparation of novel Fe-ZSM-5 zeolite membrane catalysts for catalytic wet peroxide oxidation of phenol in a membrane reactor, Chemical Engineering Journal 259 (2015) 243–251.
[19] X. Ou, F. Pilitsis, Y. Jiao, Y. Zhang, S. Xu, M. Jennings, Y. Yang, S.F.R. Taylor, A. Garforth, H. Zhang, C. Hardacre, Y. Yan, X. Fan, Hierarchical Fe-ZSM-5/SiC foam catalyst as the foam bed catalytic reactor (FBCR) for catalytic wet peroxide oxidation (CWPO), Chemical Engineering Journal 362 (2019) 53–62.
[20] R. French, S. Czernik, Catalytic pyrolysis of biomass for biofuels production, Fuel Processing Technology 91 (1) (2010) 25–32.
[21] O.Y. Sang, Biofuel production from catalytic cracking of palm oil, Energy Sources 25 (9) (2003) 859–869.
[22] J. Bi, X. Guo, M. Liu, X. Wang, High effective dehydration of bio-ethanol into ethylene over nanoscale HZSM-5 zeolite catalysts, Catalysis Today 149 (1) (2010) 143–147.
[23] J. Huangfu, D. Mao, X. Zhai, Q. Guo, Remarkably enhanced stability of HZSM-5 zeolite co-modified with alkaline and phosphorous for the selective conversion of bio-ethanol to propylene, Applied Catalysis A: General 520 (2016) 99–104.
[24] J. Zhang, Y. Zhao, M. Pan, X. Feng, W. Ji, C.-T. Au, Efficient acrylic acid production through bio lactic acid dehydration over NaY zeolite modified by alkali phosphates, ACS Catalysis 1 (1) (2010) 32–41.
[25] H. Ichiura, T. Kitaoka, H. Tanaka, Removal of indoor pollutants under UV irradiation by a composite TiO_2–zeolite sheet prepared using a papermaking technique, Chemosphere 50 (1) (2003) 79–83.
[26] E.P. Reddy, L. Davydov, P. Smirniotis, TiO_2-loaded zeolites and mesoporous materials in the sonophotocatalytic decomposition of aqueous organic pollutants: the role of the support, Applied Catalysis B: Environmental 42 (1) (2003) 1–11.
[27] A. Ikhlaq, B. Kasprzyk-Hordern, Catalytic ozonation of chlorinated VOCs on ZSM-5 zeolites and alumina: formation of chlorides, Applied Catalysis B: Environmental 200 (2017) 274–282.
[28] J.H. Kwak, R.G. Tonkyn, D.H. Kim, J. Szanyi, C.H.F. Peden, Excellent activity and selectivity of Cu-SSZ-13 in the selective catalytic reduction of NOx with NH_3, Journal of Catalysis 275 (2) (2010) 187–190.
[29] K. Gobin, G. Manos, Polymer degradation to fuels over microporous catalysts as a novel tertiary plastic recycling method, Polymer Degradation and Stability 83 (2) (2004) 267–279.
[30] A.A. Garforth, S. Ali, J. Hernández-Martínez, A. Akah, Feedstock recycling of polymer wastes, Current Opinion in Solid State & Materials Science 8 (6) (2004) 419–425.
[31] H. Li, M. Eddaoudi, M. O'Keeffe, O.M. Yaghi, Design and synthesis of an exceptionally stable and highly porous metal-organic framework, Nature 402 (1999) 276.
[32] H. Furukawa, N. Ko, Y.B. Go, N. Aratani, S.B. Choi, E. Choi, A.Ö. Yazaydin, R.Q. Snurr, M. O'Keeffe, J. Kim, O.M. Yaghi, Ultrahigh porosity in metal-organic frameworks, Science 329 (5990) (2010) 424–428.
[33] J. Lee, O.K. Farha, J. Roberts, K.A. Scheidt, S.T. Nguyen, J.T. Hupp, Metal-organic framework materials as catalysts, Chemical Society Reviews 38 (5) (2009) 1450–1459.
[34] L. Zhu, X.-Q. Liu, H.-L. Jiang, L.-B. Sun, Metal–organic frameworks for heterogeneous basic catalysis, Chemical Reviews 117 (12) (2017) 8129–8176.
[35] L. Jiao, Y. Wang, H.-L. Jiang, Q. Xu, Metal–organic frameworks as platforms for catalytic applications, Advanced Materials 30 (37) (2018) 1703663.
[36] N. Al-Janabi, P. Hill, L. Torrente-Murciano, A. Garforth, P. Gorgojo, F. Siperstein, X. Fan, Mapping the Cu-BTC metal–organic framework (HKUST-1) stability envelope in the presence of water vapour for CO_2 adsorption from flue gases, Chemical Engineering Journal 281 (2015) 669–677.
[37] N. Al-Janabi, V. Martis, N. Servi, F.R. Siperstein, X. Fan, Cyclic adsorption of water vapour on CuBTC MOF: sustaining the hydrothermal stability under non-equilibrium conditions, Chemical Engineering Journal 333 (2018) 594–602.
[38] J.H. Cavka, S. Jakobsen, U. Olsbye, N. Guillou, C. Lamberti, S. Bordiga, K.P. Lillerud, A new zirconium inorganic building brick forming metal organic frameworks with exceptional stability, Journal of the American Chemical Society 130 (42) (2008) 13850–13851.
[39] R. Vakili, S. Xu, N. Al-Janabi, P. Gorgojo, S.M. Holmes, X. Fan, Microwave-assisted synthesis of zirconium-based metal organic frameworks (MOFs): optimization and gas adsorption, Microporous and Mesoporous Materials 260 (Suppl. C) (2018) 45–53.

[40] S. Xu, S. Chansai, C. Stere, B. Inceesungvorn, A. Goguet, K. Wangkawong, S.F.R. Taylor, N. Al-Janabi, C. Hardacre, P.A. Martin, X. Fan, Sustaining metal−organic frameworks for water−gas shift catalysis by non-thermal plasma, Nature Catalysis 2 (2019) 142−148.

[41] A.P. Côté, A.I. Benin, N.W. Ockwig, M. O'Keeffe, A.J. Matzger, O.M. Yaghi, Porous, crystalline, covalent organic frameworks, Science 310 (5751) (2005) 1166−1170.

[42] H.M. El-Kaderi, J.R. Hunt, J.L. Mendoza-Cortés, A.P. Côté, R.E. Taylor, M. O'Keeffe, O.M. Yaghi, Designed synthesis of 3D covalent organic frameworks, Science 316 (5822) (2007) 268−272.

[43] S.-Y. Ding, W. Wang, Covalent organic frameworks (COFs): from design to applications, Chemical Society Reviews 42 (2) (2013) 548−568.

[44] H.K. Chae, D.Y. Siberio-Pérez, J. Kim, Y. Go, M. Eddaoudi, A.J. Matzger, M. O'Keeffe, O.M. Yaghi, A route to high surface area, porosity and inclusion of large molecules in crystals, Nature 427 (2004) 523.

[45] M.E. Davis, Ordered porous materials for emerging applications, Nature 417 (2002) 813.

[46] C.T. Kresge, M.E. Leonowicz, W.J. Roth, J.C. Vartuli, J.S. Beck, Ordered mesoporous molecular sieves synthesized by a liquid-crystal template mechanism, Nature 359 (1992) 710.

[47] J.S. Beck, J.C. Vartuli, W.J. Roth, M.E. Leonowicz, C.T. Kresge, K.D. Schmitt, C.T.W. Chu, D.H. Olson, E.W. Sheppard, S.B. McCullen, J.B. Higgins, J.L. Schlenker, A new family of mesoporous molecular sieves prepared with liquid crystal templates, Journal of the American Chemical Society 114 (27) (1992) 10834−10843.

[48] Y. Wan, Zhao, On the controllable soft-templating approach to mesoporous silicates, Chemical Reviews 107 (7) (2007) 2821−2860.

[49] X.S. Zhao, X.Y. Bao, W. Guo, F.Y. Lee, Immobilizing catalysts on porous materials, Materials Today 9 (3) (2006) 32−39.

[50] D. Zhao, J. Feng, Q. Huo, N. Melosh, G.H. Fredrickson, B.F. Chmelka, G.D. Stucky, Triblock copolymer syntheses of mesoporous silica with periodic 50 to 300 angstrom pores, Science 279 (5350) (1998) 548−552.

[51] P. Schmidt-Winkel, W.W. Lukens, P. Yang, D.I. Margolese, J.S. Lettow, J.Y. Ying, G.D. Stucky, Microemulsion templating of siliceous mesostructured cellular foams with well-defined ultralarge mesopores, Chemistry of Materials 12 (3) (2000) 686−696.

[52] D. Jian-Wen, S. Chun-Min, Q. Ling, S. Jian-Ming, X. Xing-Zhong, Pilot synthesis and commercial application in FCC catalyst of MCM-41 zeolite, Journal of Porous Materials 15 (2) (2008) 189−197.

[53] F. Rodríguez-reinoso, The role of carbon materials in heterogeneous catalysis, Carbon 36 (3) (1998) 159−175.

[54] J. Lee, J. Kim, T. Hyeon, Recent progress in the synthesis of porous carbon materials, Advanced Materials 18 (16) (2006) 2073−2094.

[55] Y. Yang, K. Chiang, N. Burke, Porous carbon-supported catalysts for energy and environmental applications: a short review, Catalysis Today 178 (1) (2011) 197−205.

[56] H.-W. Liang, X. Zhuang, S. Brüller, X. Feng, K. Müllen, Hierarchically porous carbons with optimized nitrogen doping as highly active electrocatalysts for oxygen reduction, Nature Communications 5 (2014) 4973.

[57] Y.-Z. Chen, R. Zhang, L. Jiao, H.-L. Jiang, Metal−organic framework-derived porous materials for catalysis, Coordination Chemistry Reviews 362 (2018) 1−23.

[58] M. Hu, J. Reboul, S. Furukawa, N.L. Torad, Q. Ji, P. Srinivasu, K. Ariga, S. Kitagawa, Y. Yamauchi, Direct carbonization of Al-based porous coordination polymer for synthesis of nanoporous carbon, Journal of the American Chemical Society 134 (6) (2012) 2864−2867.

[59] H.-X. Zhong, J. Wang, Y.-W. Zhang, W.-L. Xu, W. Xing, D. Xu, Y.-F. Zhang, X.-B. Zhang, ZIF-8 derived graphene-based nitrogen-doped porous carbon sheets as highly efficient and durable oxygen reduction electrocatalysts, Angewandte Chemie International Edition 53 (51) (2014) 14235−14239.

[60] D. Zhao, J.-L. Shui, C. Chen, X. Chen, B.M. Reprogle, D. Wang, D.-J. Liu, Iron imidazolate framework as precursor for electrocatalysts in polymer electrolyte membrane fuel cells, Chemical Science 3 (11) (2012) 3200−3205.

[61] W. Xia, R. Zou, L. An, D. Xia, S. Guo, A metal−organic framework route to in situ encapsulation of Co@Co_3O_4@C core@bishell nanoparticles into a highly ordered porous carbon matrix for oxygen reduction, Energy & Environmental Science 8 (2) (2015) 568−576.

[62] X.-Y. Yu, L. Yu, H.B. Wu, X.W. Lou, Formation of nickel sulfide nanoframes from metal−organic frameworks with enhanced pseudocapacitive and electrocatalytic properties, Angewandte Chemie International Edition 54 (18) (2015) 5331−5335.

[63] L. Fan, P.F. Liu, X. Yan, L. Gu, Z.Z. Yang, H.G. Yang, S. Qiu, X. Yao, Atomically isolated nickel species anchored on graphitized carbon for efficient hydrogen evolution electrocatalysis, Nature Communications 7 (2016) 10667.

[64] N.B. McKeown, S. Makhseed, P.M. Budd, Polymers of intrinsic microporosity (PIMs): organic materials for membrane separations, heterogeneous catalysis and hydrogen storage, Chemical Communications (23) (2002) 2780−2781.

[65] N.B. McKeown, P.M. Budd, Polymers of intrinsic microporosity (PIMs): organic materials for membrane separations, heterogeneous catalysis and hydrogen storage, Chemical Society Reviews 35 (8) (2006) 675−683.

[66] P.M. Budd, B. Ghanem, K. Msayib, N.B. McKeown, C. Tattershall, A nanoporous network polymer derived from hexaazatrinaphthylene with potential as an adsorbent and catalyst support, Journal of Materials Chemistry 13 (11) (2003) 2721−2726.

[67] N.B. McKeown, B. Gahnem, K.J. Msayib, P.M. Budd, C.E. Tattershall, K. Mahmood, S. Tan, D. Book, H.W. Langmi, A. Walton, Towards polymer-based hydrogen storage materials: engineering ultramicroporous cavities within polymers of intrinsic microporosity, Angewandte Chemie International Edition 45 (11) (2006) 1804−1807.

[68] E. Koohsaryan, M. Anbia, Nanosized and hierarchical zeolites: a short review, Chinese Journal of Catalysis 37 (4) (2016) 447−467.

[69] K. Li, J. Valla, J. Garcia-Martinez, Realizing the commercial potential of hierarchical zeolites: new opportunities in catalytic cracking, ChemCatChem 6 (1) (2014) 46e66.

[70] D. Verboekend, N. Nuttens, R. Locus, J. Van Aelst, P. Verolme, J.C. Groen, J. Perez-Ramirez, B.F. Sels, Synthesis, characterisation, and catalytic evaluation of hierarchical faujasite zeolites: milestones, challenges, and future directions, Chemical Society Reviews 45 (12) (2016) 3331−3352.

[71] N. Masoumifard, R. Guillet-Nicolas, F. Kleitz, Synthesis of engineered zeolitic materials: from classical zeolites to hierarchical core−shell materials, Advanced Materials 30 (16) (2018) 1704439.

[72] S. van Donk, A.H. Janssen, J.H. Bitter, K.P. de Jong, Generation, characterization, and impact of mesopores in zeolite catalysts, Catalysis Reviews 45 (2) (2003) 297−319.

[73] J. Cejka, S. Mintova, Perspectives of micro/mesoporous composites in catalysis, Catalysis Reviews 49 (4) (2007) 457−509.

[74] R. Chal, C. Gérardin, M. Bulut, S. van Donk, Overview and industrial assessment of synthesis strategies towards zeolites with mesopores, ChemCatChem 3 (1) (2011) 67–81.
[75] S. Abello, J. Perez-Ramirez, Accelerated generation of intracrystalline mesoporosity in zeolites by microwave-mediated desilication, Physical Chemistry Chemical Physics 11 (16) (2009) 2959–2963.
[76] W. Schwieger, A.G. Machoke, T. Weissenberger, A. Inayat, T. Selvam, M. Klumpp, A. Inayat, Hierarchy concepts: classification and preparation strategies for zeolite containing materials with hierarchical porosity, Chemical Society Reviews 45 (12) (2016) 3353–3376.
[77] S. Mitchell, N.-L. Michels, J. Pérez-Ramírez, From powder to technical body: the undervalued science of catalyst scale up, Chemical Society Reviews 42 (14) (2013) 6094–6112.
[78] M.S. Holm, E. Taarning, K. Egeblad, C.H. Christensen, Catalysis with hierarchical zeolites, Catalysis Today 168 (1) (2011) 3–16.
[79] K. Möller, T. Bein, Mesoporosity – a new dimension for zeolites, Chemical Society Reviews 42 (9) (2013) 3689–3707.
[80] C.S. Cundy, Microwave techniques in the synthesis and modification of zeolite catalysts: a review, Collection of Czechoslovak Chemical Communications 63 (11) (1998) 1699–1723.
[81] Y. Li, W. Yang, Microwave synthesis of zeolite membranes: a review, Journal of Membrane Science 316 (1) (2008) 3–17.
[82] X. Meng, F.-S. Xiao, Green routes for synthesis of zeolites, Chemical Reviews 114 (2) (2014) 1521–1543.
[83] C.C. Pavel, W. Schmidt, Generation of hierarchical pore systems in the titanosilicate ETS-10 by hydrogen peroxide treatment under microwave irradiation, Chemical Communications (8) (2006) 882–884.
[84] C.C. Pavel, R. Palkovits, F. Schüth, W. Schmidt, The benefit of mesopores in ETS-10 on the vapor-phase Beckmann rearrangement of cyclohexanone oxime, Journal of Catalysis 254 (1) (2008) 84–90.
[85] C.C. Pavel, S.-H. Park, A. Dreier, B. Tesche, W. Schmidt, Structural defects induced in ETS-10 by postsynthesis treatment with H_2O_2 solution, Chemistry of Materials 18 (16) (2006) 3813–3820.
[86] M.D. González, Y. Cesteros, P. Salagre, F. Medina, J.E. Sueiras, Effect of microwaves in the dealumination of mordenite on its surface and acidic properties, Microporous and Mesoporous Materials 118 (1) (2009) 341–347.
[87] M.D. González, Y. Cesteros, P. Salagre, Effect of microwaves on the surface and acidic properties of dealuminated zeolites, Physics Procedia 8 (Suppl. C) (2010) 104–108.
[88] M.D. González, Y. Cesteros, P. Salagre, Comparison of dealumination of zeolites beta, mordenite and ZSM-5 by treatment with acid under microwave irradiation, Microporous and Mesoporous Materials 144 (1) (2011) 162–170.
[89] B.M. Chandra Shekara, B.S. Jai Prakash, Y.S. Bhat, Dealumination of zeolite BEA under microwave irradiation, ACS Catalysis 1 (3) (2011) 193–199.
[90] V. Paixão, R. Monteiro, M. Andrade, A. Fernandes, J. Rocha, A.P. Carvalho, A. Martins, Desilication of MOR zeolite: conventional versus microwave assisted heating, Applied Catalysis A: General 402 (1) (2011) 59–68.
[91] Z. Hasan, J.W. Jun, C.-U. Kim, K.-E. Jeong, S.-Y. Jeong, S.H. Jhung, Desilication of ZSM-5 zeolites for mesoporosity development using microwave irradiation, Materials Research Bulletin 61 (Suppl. C) (2015) 469–474.
[92] A.N.C. van laak, R.W. Gosselink, S.L. Sagala, J.D. Meeldijk, P.E. de Jongh, K.P. de Jong, Alkaline treatment on commercially available aluminum rich mordenite, Applied Catalysis A: General 382 (1) (2010) 65–72.
[93] J.C. Groen, T. Sano, J.A. Moulijn, J. Pérez-Ramírez, Alkaline-mediated mesoporous mordenite zeolites for acid-catalyzed conversions, Journal of Catalysis 251 (1) (2007) 21–27.
[94] T. Fakin, A. Ristić, V. Mavrodinova, N. Zabukovec Logar, Highly crystalline binder-free ZSM-5 granules preparation, Microporous and Mesoporous Materials 213 (2015) 108–117.
[95] A. Corma, M.T. Navarro, in: R. Aiello, G. Giordano, F. Testa (Eds.), From Micro to Mesoporous Molecular Sieves: Adapting Composition and Structure for Catalysis, Studies in Surface Science and Catalysis, vol. 142, Elsevier, 2002, pp. 487–501.
[96] W. Zhang, S. Gao, S. Xie, H. Liu, X. Zhu, Y. Shang, S. Liu, L. Xu, Y. Zhang, A shaped binderless ZSM-11 zeolite catalyst for direct amination of isobutene to tert-butylamine, Chinese Journal of Catalysis 38 (1) (2017) 168–175.
[97] Y. Jiao, X. Fan, M. Perdjon, Z. Yang, J. Zhang, Vapor-phase transport (VPT) modification of ZSM-5/SiC foam catalyst using TPAOH vapor to improve the methanol-to-propylene (MTP) reaction, Applied Catalysis A: General 545 (2017) 104–112.
[98] G. Liu, L. Zhang, Z. You, L. Wang, X. Zhang, Vapor-phase transport crystallization of micro-HZSM-5 zeolite coatings, Journal of the American Ceramic Society 97 (10) (2014) 3320–3327.
[99] G. Majano, S. Mintova, O. Ovsitser, B. Mihailova, T. Bein, Zeolite beta nanosized assemblies, Microporous and Mesoporous Materials 80 (1) (2005) 227–235.
[100] J. Zhou, J. Teng, L. Ren, Y. Wang, Z. Liu, W. Liu, W. Yang, Z. Xie, Full-crystalline hierarchical monolithic ZSM-5 zeolites as superiorly active and long-lived practical catalysts in methanol-to-hydrocarbons reaction, Journal of Catalysis 340 (2016) 166–176.
[101] M.B. Yue, N. Yang, W.Q. Jiao, Y.M. Wang, M.-Y. He, Dry-gel synthesis of shaped binderless zeolites composed of nanosized ZSM-5, Solid State Science 20 (2013) 1–7.
[102] J. Wang, X. Cheng, J. Guo, X. Chen, H. He, Y. Long, Catalytic performances of binder-free ZSM-5 catalysts for dehydration of crude methanol to dimethyl ether, Chinese Journal of Chemical Physics 28 (2) (2010) 183–188.
[103] Y. Jiao, C. Jiang, Z. Yang, J. Zhang, Controllable synthesis of ZSM-5 coatings on SiC foam support for MTP application, Microporous and Mesoporous Materials 162 (2012) 152–158.
[104] Y. Jiao, C. Jiang, Z. Yang, J. Liu, J. Zhang, Synthesis of highly accessible ZSM-5 coatings on SiC foam support for MTP reaction, Microporous and Mesoporous Materials 181 (2013) 201–207.
[105] C. Duong-Viet, H. Ba, Z. El-Berrichi, J.-M. Nhut, M.J. Ledoux, Y. Liu, C. Pham-Huu, Pham-Huu, Silicon carbide foam as a porous support platform for catalytic applications, New Journal of Chemistry 40 (5) (2016) 4285–4299.
[106] Y. Jiao, X. Ou, J. Zhang, X. Fan, Structured ZSM-5 coated SiC foam catalysts for process intensification in catalytic cracking of n-hexane, Reaction Chemistry & Engineering 4 (2) (2019) 427–435.
[107] Y. Jiao, X. Yang, C. Jiang, C. Tian, Z. Yang, J. Zhang, Hierarchical ZSM-5/SiC nano-whisker/SiC foam composites: preparation and application in MTP reactions, Journal of Catalysis 332 (2015) 70–76.
[108] Y. Jiao, S. Xu, C. Jiang, M. Perdjon, X. Fan, J. Zhang, MFI zeolite coating with intrazeolitic aluminum (acidic) gradient supported on SiC foams to improve the methanol-to-propylene (MTP) reaction, Applied Catalysis A: General 559 (2018) 1–9.

[109] X. Ou, S. Xu, J.M. Warnett, S.M. Holmes, A. Zaheer, A.A. Garforth, M.A. Williams, Y. Jiao, X. Fan, Creating hierarchies promptly: microwave-accelerated synthesis of ZSM-5 zeolites on macrocellular silicon carbide (SiC) foams, Chemical Engineering Journal 312 (2017) 1–9.
[110] M.M. Elamin, O. Muraza, Z. Malaibari, H. Ba, J.-M. Nhut, C. Pham-Huu, Microwave assisted growth of SAPO-34 on β-SiC foams for methanol dehydration to dimethyl ether, Chemical Engineering Journal 274 (2015) 113–122.
[111] G. Zou, M. Cao, H. Lin, H. Jin, Y. Kang, Y. Chen, Nickel layer deposition on SiC nanoparticles by simple electroless plating and its dielectric behaviors, Powder Technology 168 (2) (2006) 84–88.
[112] S. Alfaro, M. Arruebo, J.N. Coronas, M. Menéndez, J. Santamaria, Preparation of MFI type tubular membranes by steam-assisted crystallization, Microporous and Mesoporous Materials 50 (2) (2001) 195–200.
[113] Y. He, M. Liu, C. Dai, S. Xu, Y. Wei, Z. Liu, X. Guo, Modification of nanocrystalline HZSM-5 zeolite with tetrapropylammonium hydroxide and its catalytic performance in methanol to gasoline reaction, Chinese Journal of Catalysis 34 (6) (2013) 1148–1158.
[114] L. Zhou, J. Yang, G. Li, J. Wang, Y. Zhang, J. Lu, D. Yin, Highly H_2 permeable SAPO-34 membranes by steam-assisted conversion seeding, International Journal of Hydrogen Energy 39 (27) (2014) 14949–14954.
[115] J. Ding, S. Fan, P. Chen, T. Deng, Y. Liu, Y. Lu, Vapor-phase transport synthesis of microfibrous-structured SS-fiber@ZSM-5 catalyst with improved selectivity and stability for methanol-to-propylene, Catalysis Science & Technology 7 (10) (2017) 2087–2100.
[116] J. Ding, P. Chen, S. Fan, Z. Zhang, L. Han, G. Zhao, Y. Liu, Y. Lu, Microfibrous-Structured SS-fiber@meso-HZSM-5 catalyst for methanol-to-propylene: steam-assisted crystallization synthesis and insight into the stability enhancement, ACS Sustainable Chemistry & Engineering 5 (2) (2017) 1840–1853.
[117] J. Ding, P. Chen, G. Zhao, Y. Liu, Y. Lu, High-performance thin-felt SS-fiber@HZSM-5 catalysts synthesized via seed-assisted vapor phase transport for methanol-to-propylene reaction: effects of crystal size, mesoporosity and aluminum uniformity, Journal of Catalysis 360 (2018) 40–50.
[118] J. Ding, Y. Jia, P. Chen, G. Zhao, Y. Liu, Y. Lu, Microfibrous-structured hollow-ZSM-5/SS-fiber catalyst with mesoporosity development dependent lifetime improvement for MTP reaction, Microporous and Mesoporous Materials 261 (2018) 1–8.
[119] J. Ding, Z. Zhang, C. Meng, G. Zhao, Y. Liu, Y. Lu, From nano- to macro-engineering of ZSM-11 onto thin-felt stainless-steel-fiber: steam-assisted crystallization synthesis and methanol-to-propylene performance, Catalysis Today (2018). https://doi.org/10.1016/j.cattod.2018.02.058.
[120] U. Betke, A. Lieb, Micro-macroporous composite Materials – Preparation techniques and selected applications: a review, Advanced Engineering Materials 20 (2018) 1800252.
[121] A. Cabrera-Codony, A. Georgi, R. Gonzalez-Olmos, H. Valdés, M.J. Martín, Zeolites as recyclable adsorbents/catalysts for biogas upgrading: removal of octamethylcyclotetrasiloxane, Chemical Engineering Journal 307 (2017) 820–827.
[122] J. Meshram, P. Ali, V. Tiwari, Zeolite as an efficient and recyclable activation surface for the synthesis of bis-thiazolidinones: theoretical screening owing to experimental biology, Green Chemistry Letters and Reviews 3 (3) (2010) 195–200.
[123] E.T.C. Vogt, B.M. Weckhuysen, Fluid catalytic cracking: recent developments on the grand old lady of zeolite catalysis, Chemical Society Reviews 44 (20) (2015) 7342–7370.
[124] X. Meng, F. Xiao, Green routes for synthesis of zeolites, Chemical Reviews 114 (2013) 1521–1543.
[125] T. Abdullahi, Z. Harun, M.H.D. Othman, A review on sustainable synthesis of zeolite from kaolinite resources via hydrothermal process, Advanced Powder Technology 28 (8) (2017) 1827–1840.
[126] J. Zhao, W. Yan, R. Xu, W. Pang, Q. Huo, Modern Inorganic Synthetic Chemistry, Modern Inorganic Synthetic Chemistry, Elsevier, Amsterdam, 2011.
[127] D. Reinoso, M. Adrover, M. Pedernera, Green synthesis of nanocrystalline faujasite zeolite, Ultrasonics Sonochemistry 42 (2018) 303–309.
[128] X. Meng, L. Wang, F.-S. Xiao, in: F.-S. Xiao, X. Meng (Eds.), Zeolites in Sustainable Chemistry: Synthesis, Characterization and Catalytic Applications, Zeolites in Sustainable Chemistry: Synthesis, Characterization and Catalytic Applications, Springer Berlin Heidelberg, Berlin, Heidelberg, 2016, pp. 3–35.
[129] R. Limin, Y. Zhang, Z. Shangjing, Z. Longfeng, S. Qi, H. Zhang, Y. Chengguang, M. Xiangju, Y. Xiangguang, X. Feng-Shou, Design and synthesis of a catalytically active Cu-SSZ-13 zeolite from a copper-amine complex template, Chinese Journal of Catalysis 33 (1) (2012) 92–105.
[130] P. Rani, B. Satpati, R. Srivastava, Natural template mediated sustainable synthesis of nanocrystalline zeolite with significantly improved catalytic activity, Chemistry 2 (9) (2017) 2870–2879.
[131] H. Lee, S.I. Zones, M.E. Davis, A combustion-free methodology for synthesizing zeolites and zeolite-like materials, Nature 425 (2003) 385.
[132] C. Zhang, S. Li, S. Bao, Sustainable synthesis of ZSM-5 zeolite from rice husk ash without addition of solvents, Waste and Biomass Valorization (2018).
[133] E.B.G. Johnson, S.E. Arshad, Hydrothermally synthesized zeolites based on kaolinite: a review, Applied Clay Science 97–98 (2014) 215–221.
[134] B. Xie, J. Song, L. Ren, Y. Ji, J. Li, F.-S. Xiao, Organotemplate-free and fast route for synthesizing beta zeolite, Chemistry of Materials 20 (14) (2008) 4533–4535.
[135] H. Zhang, B. Xie, X. Meng, U. Müller, B. Yilmaz, M. Feyen, S. Maurer, H. Gies, T. Tatsumi, X. Bao, W. Zhang, D. De Vos, F.-S. Xiao, Rational synthesis of beta zeolite with improved quality by decreasing crystallization temperature in organotemplate-free route, Microporous and Mesoporous Materials 180 (2013) 123e129.
[136] Q. Wu, X. Meng, X. Gao, F.-S. Xiao, Solvent-Free synthesis of zeolites: mechanism and utility, Accounts of Chemical Research 51 (6) (2018) 1396–1403.

CHAPTER

6

Green Synthesis and Engineering Applications of Metal–Organic Frameworks

Giulia Schukraft, Camille Petit

The Barrer Centre, Department of Chemical Engineering, Imperial College London, London, United Kingdom

OUTLINE

1. INTRODUCTION

Like only few materials before, metal–organic frameworks (MOFs) have enthused both chemists and engineers since their discovery in the 1990s [1,2]. Their synthesis is relatively straightforward and operates through a self-assembly process of organic and inorganic moieties, leading to a wide library of well-defined structures. These frameworks exhibit a remarkable versatility in terms of chemistry, structure, porosity, morphology, and reactivity [3]. For this reason, MOFs have unlocked unprecedented opportunities in a number of areas, e.g., gas storage [4], gas separation [5], heterogeneous catalysis [6], solar energy conversion [7], drug delivery [8], or energy storage [9]. A schematic of an MOF structure and its formation is shown in Fig. 6.1.

For many years, research in the field focused on understanding and optimizing the unique properties of MOFs. More recently though, the interest toward the commercialization of these materials has created the need to develop innovative, scalable, and environmentally friendly MOF synthesis routes [10–13]. Although the need to develop greener and safer materials and synthesis routes has been implemented in organic chemistry, such development has been considerably slower in the field of MOF production [10].

The green aspect of MOFs covers not only their synthesis but also their applications. Increasingly, their use has moved from end-of-the-pipe technologies (e.g., CO_2 capture, water and air pollutants removal) to pollution prevention approaches such as solar energy conversion and energy storage [4,7].

Sustainable Nanoscale Engineering
https://doi.org/10.1016/B978-0-12-814681-1.00006-0

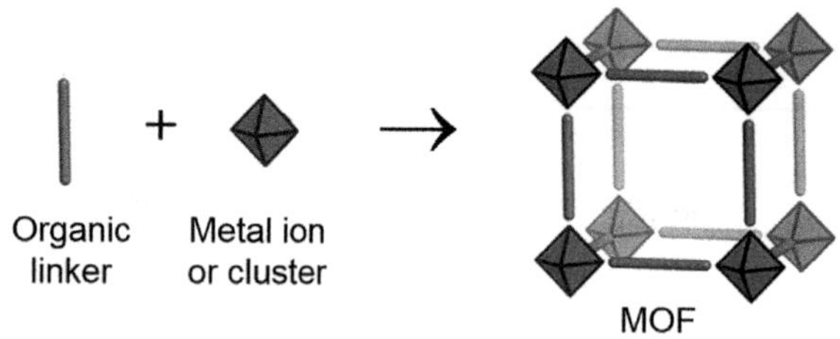

FIGURE 6.1 Schematic of a metal–organic framework (MOF) structure and its formation.

This chapter covers both the green synthesis and green engineering applications of MOFs. The first part discusses aspects of MOF synthesis through the lenses of the 12 Principles of Green Chemistry, pointing to recent reports in this area and highlighting needs for future research. The second part of the chapter focuses on describing the potential role of MOFs in solar energy conversion and energy storage through selected examples.

Note on MOF nomenclature for the rest of the chapter: We note that each MOF structure is given a particular name. This name is generally of the type MOF-X with X being a number. One may also encounter names like ZIF-X, with again X being a number. ZIF stands for zeolitic imidazolate framework and refers to a specific type of MOF, composed of tetrahedrally coordinated transition metal ions connected by imidazolate linkers. In addition, some MOFs are named after the institute where they were first synthesized. Examples include HKUST-1 (HKSUT = Hong Kong University of Science and Technology), UiO-66 (UiO = Universitetet i Oslo), MIL-53 (MIL = Materiaux de l'Institut Lavoisier). Finally, one can find MOFs referred to by their chemical formula. One given MOF structure can, therefore, be designated in various ways. For instance, MOF-199 and HKUST-1 both correspond to the MOF structure whose chemical formula is $Cu_3(BTC)_2(H2O)_3$.

2. GREEN SYNTHESIS OF METAL–ORGANIC FRAMEWORKS

As indicated above, MOF synthesis is evaluated here taking into account the 12 Principles of Green Chemistry. These principles have been developed to promote the design of products and chemical processes that minimize the use of hazardous chemicals and favor resource-efficient and safer processes [14]. These 12 principles are listed in Fig. 6.2 [15,16], and their descriptions have been adapted to the context of MOFs synthesis. They can be used as guidelines toward the green production of these materials [12]. Applying such guidelines is particularly complex and challenging for MOFs as one needs to account for both the environmental impacts of organic chemicals and those associated with metal ions and metal salts. This section focuses on the main areas of growth for MOF synthesis research that would yield safe and environmental-friendly MOFs synthesis. Examples of recent advances in the field are included.

2.1 Selection of Metals and Organic Linkers

The use of sustainable building blocks—i.e., metals nodes and organic ligands—is central to MOF synthesis scale-up. Here, we discuss their selection and the related impact.

2.1.1 Metals

The use of nitrate salts, and occasionally perchlorate, is very common at the laboratory scale owing to their high solubility and the weak counter-ion coordination that facilitate framework formation [10,17]. Such precursors should be avoided for scale-up because of their strong oxidative properties and high toxicity (Principle 3, Fig. 6.2). Similarly, halides salts such as chlorides are undesirable for large-scale green synthesis because of their strong tendency to metal coordination and corrosion [17].

In contrast, acetate and sulfate precursors represent safer and greener alternatives. In 2011, BASF SE developed the large-scale synthesis of an Al-based MOF (Basolite 520, space-time yield > 3600 kg m^{-3} day^{-1}) using an Al-sulfate salt as a precursor [18]. Despite the advantages of acetate and sulfates compared with nitrates and chloride salts, the poor solubility of sulfate and the possible hydrolysis of acetate are limiting their use [10].

Nowadays, oxides and hydroxides are seen as the greener precursors as they are nontoxic and only form water as a by-product. Yet, we note that these precursors call for special attention in terms of synthesis conditions due to

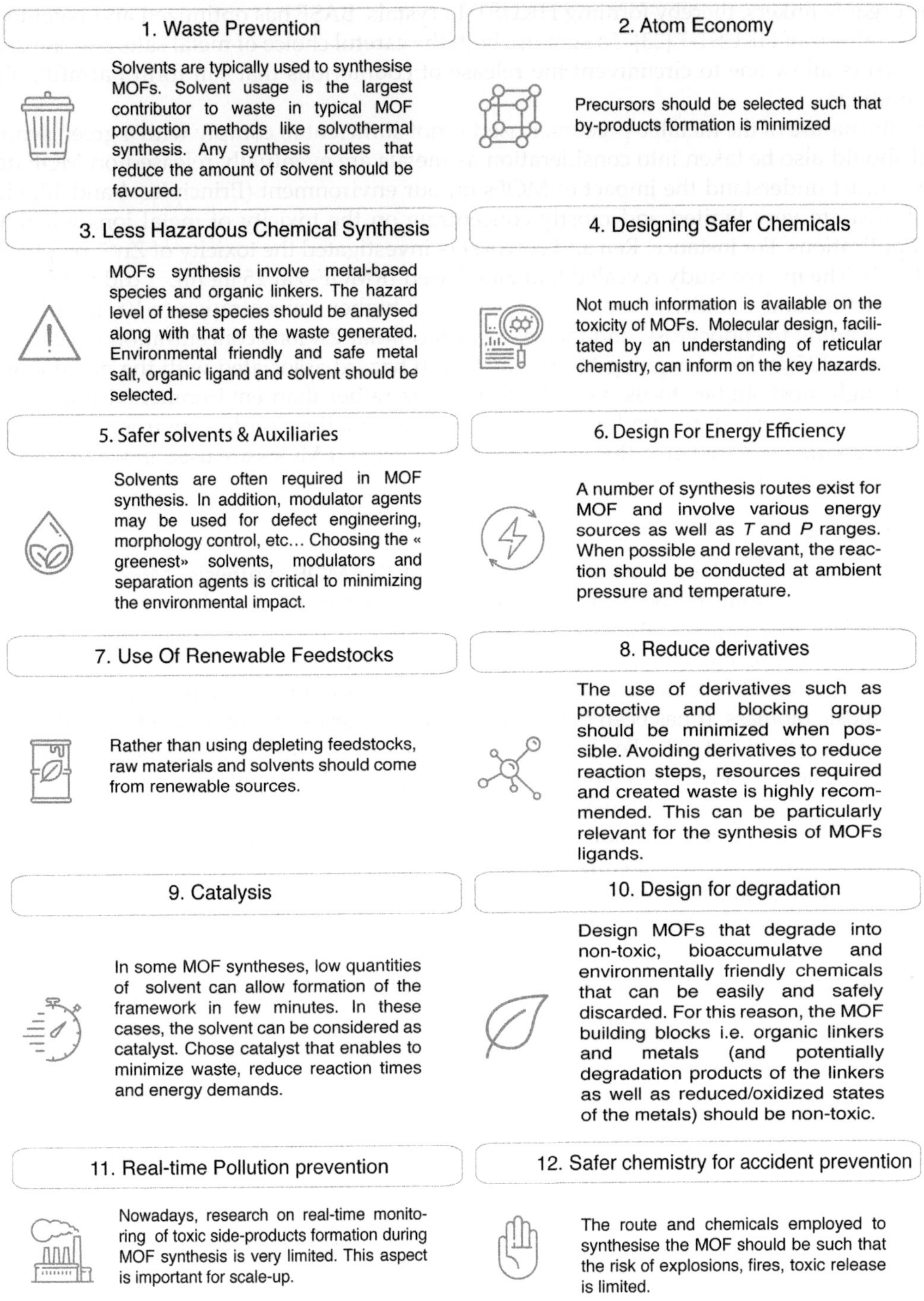

FIGURE 6.2 The 12 Principles of Green Chemistry applied to metal–organic framework (MOF) synthesis.

their low solubility under commonly used solvothermal conditions. A typical example of the use of these precursors is for the synthesis of ZIF-8 (ZIF = zeolite imidazole framework), using as precursors $Zn(OH)_2$ and imidazole linkers [19]. Although the synthesis involves ammonia, the only by-product generated is water.

Using elementary metal is also promising. An example is the synthesis of the iconic MOF, HKUST-1, via an electrochemical route. The first reported synthesis utilized copper nitrate [20], which led to the formation of copper ions and cuprous oxides as by-products, hindering any scope for scale-up. Scientists and engineers from BASF have patented an electrochemical synthesis approach to produce HKUST-1 [21,22], which relies on the use of a copper plate electrode as metal precursor. On application of a current, the electrode releases Cu^{2+} into the solution containing the organic linkers, thereby forming HKUST-1 crystals. BASF has optimized and patented this approach for large-scale synthesis of HKUST-1 [23]. To sum up, both the careful choice of metal salts and the use of elementary metals as precursors allow one to circumvent the release of counterions that can form harmful, highly toxic, and corrosive by-products.

Considering the nature of the metallic precursor used is not sufficient to comply with a green synthesis approach. MOF disposal should also be taken into consideration as metals are eventually released on MOF degradation. For this reason, one must understand the impact of MOFs on our environment (Principles 4 and 10, Fig. 6.2). To date, studies in that area are very limited and mostly concentrate on the toxicity of metal ions when MOFs are used for biological applications. For instance, Ren and coworkers investigated the toxicity of Zn^{2+} on pheochromocytoma (PC12) cells [10,24]. The in vivo study revealed that zinc-based IRMOF-3 at 25 $\mu g\ mL^{-1}$ did not have any significant impact on the structure and viability of the cells. However, at higher concentration (>100 $\mu g\ mL^{-1}$), Zn ions show high cytotoxicity, leading to endocytosis and increased intracellular Zn ions concentration [10,24]. As illustrated by this study and others [25], when designing MOFs, the health hazard associated with the metal ions must be taken into account. Though most studies focus on biological effects rather than environmental impacts (e.g., effects on aquatic life, plant growth, etc…), the findings can be used as guidelines on the environmental impact associated to framework degradation. For instance, this is particularly relevant if MOFs are used in water purification as metal leaching may harm aquatic life.

2.1.2 *Organic Linkers*

The choice of the organic linkers is another key aspect of the MOF synthesis. Most linkers used to date are carboxylic linkers derived from the petrochemical sector [26], hence not compliant with the principles of green chemistry listed before (see Fig. 6.2, Principle 7). For this reason, researchers now tend to favor and develop inexpensive and nontoxic organic ligands obtained from *renewable resources*.

For instance, using biomass represents a sustainable route. 2,5-furandicarboxylic acid is an example of a linker derived from cellulosic biomass. It has been demonstrated to be a cost-efficient, environmentally friendly building block for the green synthesis of Al-, Fe- and Zr-based MOFs [27]. In 2017, Dreischarf and coworkers used 2,5-furanedicarboxylic acid (H_2FDC) as a biorenewable building block to synthesize Zr-CAU-28 [28]. H_2FDC is an environmentally friendly alternative to terephthalic and isophthalic acid owing to its similar functionalities as it is produced from glucose and fructose. Another interesting example is γ-cyclodextrin (Fig. 6.3), which is generated via the degradation of potato starch and corn [29].

In addition to the use of biomass as renewable source, scientists have exploited *waste recycling* to produce greener organic linkers. An increasing number of studies report the production of biomass-derived p-xylene, a precursor in the synthesis of terephthalic acid (also called benzene dicarboxylic acid, BDC) [30]. Deleu and coworkers reported the use of waste polyethylene terephthalate (PET) bottle for the synthesis of BDC linkers [31]. By combining PET

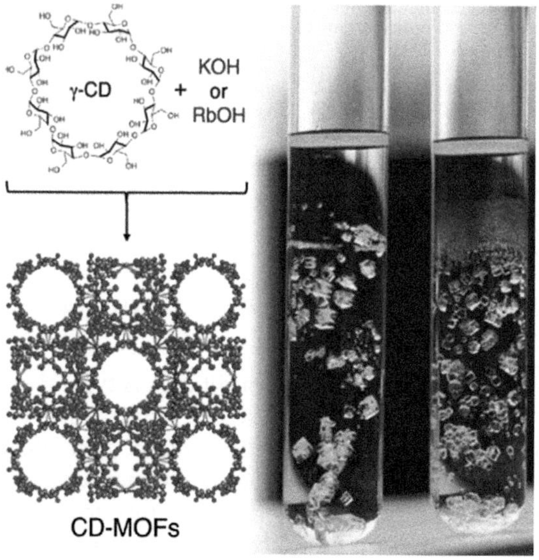

FIGURE 6.3 Cyclization of γ-cyclodextrin (CD) in the presence of KOH and RbOH to yield a porous metal–organic framework (MOF) called CD-MOF. *Reprinted with permission from M. K. Smith, S. R. Angle, B. H. Northrop, Preparation and Analysis of Cyclodextrin-Based Metal-Organic Frameworks: Laboratory Experiments Adaptable for High School through Advanced Undergraduate Students, Journal of Chemical Education 92 (2015) 368–372.*

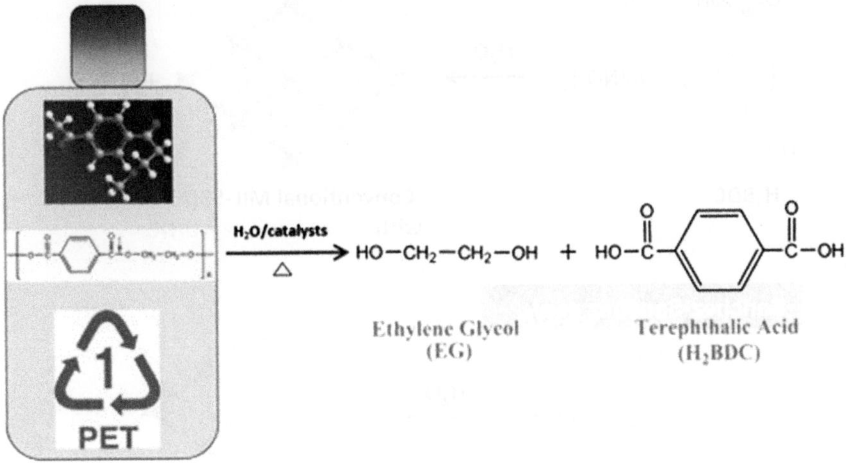

FIGURE 6.4 Catalyzed depolymerization of PET to benzene dicarboxylic acid (BDC), which can be used as a linker to produce metal–organic framework (MOF). *Reprinted with permission from J. W. Ren, X. Dyosiba, N. M. Musyoka, H. W. Langmi, B. C. North, M. Mathe, M. S. Onyango, Green synthesis of chromium-based metalorganic framework (Cr-MOF) from waste polyethylene terephthalate (PET) bottles for hydrogen storage applications, International Journal of Hydrogen Energy 41 (2016) 18141–18146.*

flakes with the corresponding metal salts under hydrothermal condition in a microwave, PET hydrolysis and MOF synthesis are achieved simultaneously. Following this strategy, the authors used PET waste bottle as BDC precursor for the synthesis of highly stable Al- and Fe-based MOFs [31]. Lo and coworkers extended this concept to Cr-based MOFs [32]. Notably though, the two aforementioned studies used HCl or HF as modulator, two highly toxic and harmful chemicals. Taking this into account, Ren and coworkers investigated a greener approach to replace HF and HCl with formic acid (Fig. 6.4) [33]. As illustrated by these examples, the successful development of such process favored recycling of PET bottle while allowing the synthesis of stable MOFs.

The selection of the ligand cannot be done in complete isolation as it affects that of the solvents and synthesis conditions. Ligands that favor milder synthetic conditions and produce by-products as safe as possible are desirable. To date most linkers are used in their protonated state, which typically requires harsh solvothermal conditions and high boiling point solvents. The by-products derived from protonated linkers are often hazardous mineral acids such as HCl, HNO_3, and H_2SO_4. For this reason, researchers favor the use of *alkali-based organic salts.* Using salts increases the ligands' water compatibility, avoids the formation of mineral acids, lowers the reaction temperature, and in some cases, decreases the reaction time (Fig. 6.5) [34]. Despite the advantages of organic salts over their protonated counterparts, the synthesis of the salts requires an additional step. A more holistic approach (cf. life cycle assessment type of study) is therefore needed to determine the dominant factors impacting the greenness of the synthesis approach.

Ligand's toxicity and environmental impact are equally important to those of metal precursors. As most commonly used ligands are commercially available, such information can easily be found and should be taken into consideration while designing MOFs.

2.2 Solvent Selection and Management

Like for many chemical processes, solvents play a major role on the overall greenness of the synthesis process (Fig. 6.2, Principle 5). In fact, solvents—when they are used—are the main source of waste in MOF synthesis. The use of sustainable and nontoxic solvents is central to any manufacturing process. In MOF chemistry, owing to the poor solubility of ligands, conventional MOF syntheses commonly require the use of high boiling point solvents (i.e., dimethylformamide [DMF] and diethylformamide), which are toxic, harmful, and environmentally detrimental. Finding sustainable yet workable solvent alternatives requires an understanding of their physicochemical properties. Hence, it is useful to consult the selection guidelines developed by pharmaceutical companies that consider environmental, safety, and health characteristics (Fig. 6.6)[35,36].

Unsurprisingly, water is the ideal solvent, and several studies have focused on developing new water-based MOF synthesis [18]. A typical example of such strategy is the optimized synthesis of Basolite A520 by BASF SE [18]. To scale up Basolite A520 synthesis from laboratory to industrial production, BASF SE minimized the use of organic solvents and replaced them by inexpensive and greener counterparts, switching from a solvothermal synthesis

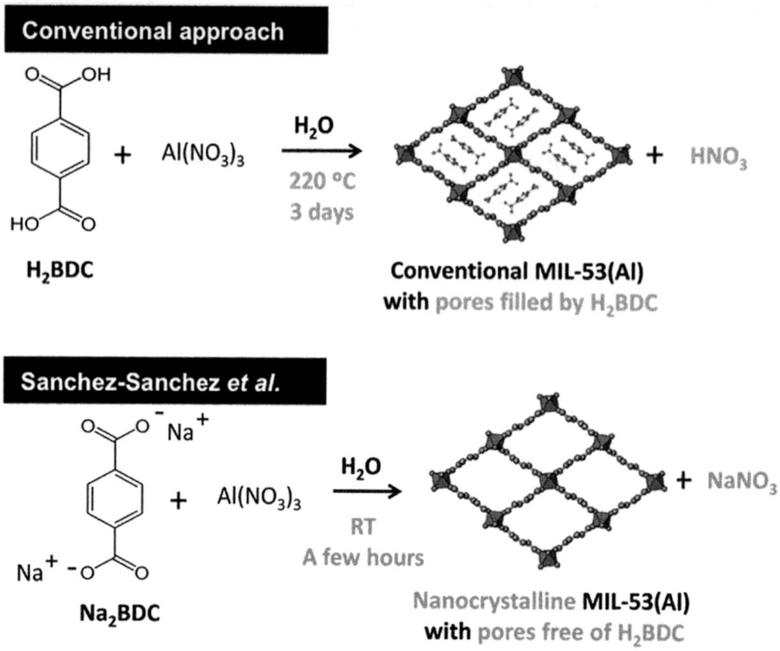

FIGURE 6.5 Schematic illustration of MIL-55(Al) synthesis using a conventional (top) and a new ligand salt (bottom) approach. *Reproduced from M. Sanchez-Sanchez, N. Getachew, K. Diaz, M. Diaz-Garcia, Y. Chebude, I. Diaz, Synthesis of metal-organic frameworks in water at room temperature: salts as linker sources, Green Chemistry 17 (2015) 1500–1509 with permission from The Royal Society of Chemistry.*

Classification	Solvent	CAS number	Melting point °C	Boiling point °C	Waste recycling, incineration, VOC, and biotreatment issues	Environmental Impact fate and effects on the environment	Health acute and chronic effects on human health and exposure potential	Flammability & Explosion storage and handing	Reactivity/ Stability factors affecting the stability of the solvent	Life Cycle Score Environmental impacts to produce the solvent	Legislation Flag alerts regulatory restrictions
Greenest	Water	7732-18-5	0	100	4	10	10	10	10	10	
Alcohols	1-Butanol	71-36-3	-89	118	5	7	5	8	9	5	
	2-Butanol	78-92-2	-115	100	4	6	8	7	9	6	
	Ethanol/IMS	64-17-5	-114	78	3	8	8	6	9	9	
	t-Butanol	75-65-0	25	82	3	9	6	6	10	8	
	Methanol	67-56-1	-98	65	4	9	5	5	10	9	
	2-Propanol	67-63-0	-88	82	3	9	8	6	8	4	
	1-Propanol	71-23-8	-127	97	4	7	5	7	10	7	
	2-Methoxyethanol	109-86-4	-85	124	3	8	2	7	6	7	
Ester	t-Butyl acetate	540-88-5	-78	95	6	9	8	6	10	8	
	Isopropyl acetate	108-21-4	-73	89	5	7	7	6	9	7	
	Propyl acetate	109-60-4	-92	102	5	7	8	6	10	4	
	Dimethyl carbonate	616-38-6	-1	91	4	8	7	6	10	8	
	Ethyl acetate	141-78-6	-84	77	4	8	8	4	8	6	
	Methyl acetate	79-20-9	-98	57	3	9	7	4	9	7	
Ketone	Methylisobutyl ketone	108-10-1	-84	117	6	6	6	7	8	2	
	Acetone	67-64-1	-95	56	3	9	8	4	9	7	
	Methylethyl ketone	78-93-3	-87	80	3	7	8	4	8	3	
Organic Acids	Propionic acid	79-09-4	-21	141	4	8	6	8	8	7	
	Acetic acid (glacial)	64-19-7	17	118	4	8	6	8	7	8	
Aromatics	p-Xylene	106-42-3	-13	138	7	2	6	5	10	7	
	Toluene	108-88-3	-95	111	6	3	4	4	10	7	
	Benzene	71-43-2	6	80	5	6	1	3	10	7	
Hydrocarbons	Isooctane	540-84-1	-107	99	6	4	8	3	10	7	
	Cyclohexane	110-82-7	7	81	5	5	7	2	10	7	
	Heptane	142-82-5	-91	98	6	3	8	3	10	7	
	2-Methylpentane	107-83-5	-153	60	5	4	7	2	10	7	
	Hexane	110-54-3	-95	69	5	3	4	2	10	7	
	Petroleum spirit	8032-32-4	-73	55	6	2	2	3	10	7	
Ethers	t-Butyl methyl ether	1634-04-4	-109	55	4	5	5	3	9	8	
	Cyclopentyl methyl ether	5614-37-9	-140	106	6	4	4	5	8	4	
	2-Methyl THF	96-47-9	-137	78	4	5	4	3	6	4	
	Diethyl ether	60-29-7	-116	35	4	4	5	2	4	6	
	Bis(2-methoxyethyl) ether	111-96-6	-68	162	4	5	2	8	4	6	
	1,4-Dioxane	123-91-1	12	102	3	4	4	4	5	6	
	Tetrahydrofuran	109-99-9	-108	65	3	5	6	3	4	4	
	1,2-Dimethoxyethane	110-71-4	-58	85	4	5	2	4	4	7	
	Diisopropyl ether	108-20-3	-86	68	4	3	8	1	1	9	
Dipolar aprotics	Dimethyl sulfoxide	67-68-5	19	189	5	5	7	9	2	6	
	Dimethyl formamide	68-12-2	-61	153	4	6	2	9	9	7	
	N-Methylformamide	123-39-7	-4	200	4	6	2	10	10	7	
	N-Methyl pyrrolidone	872-50-4	-24	202	5	6	3	9	8	4	
	Dimethyl acetamide	127-19-5	-20	165	5	6	2	10	8	2	
	Acetonitrile	75-05-8	-45	82	2	6	6	6	10	3	
Chlorinated	Carbon tetrachloride	56-23-5	-23	77	4	5	3	4	10	7	
	Dichloromethane	75-09-2	-95	40	3	6	4	6	9	7	
	Chloroform	67-66-3	-64	61	3	6	3	6	9	6	
	1,2-Dichloroethane	107-06-2	-36	84	4	4	2	6	10	7	

Legislation Flag	
	Substitution recommended - There are no current restrictions but future regulatory restrictions may apply
	Substitution recommended - existing regulatory restrictions apply
	Must be substituted - A regulatory ban applies

FIGURE 6.6 Solvent selection guide published by GSK in 2010. *Reproduced from M. Sanchez-Sanchez, N. Getachew, K. Diaz, M. Diaz-Garcia, Y. Chebude, I. Diaz, Synthesis of metal-organic frameworks in water at room temperature: salts as linker sources, Green Chemistry 17 (2015) 1500–1509 with permission from The Royal Society of Chemistry.*

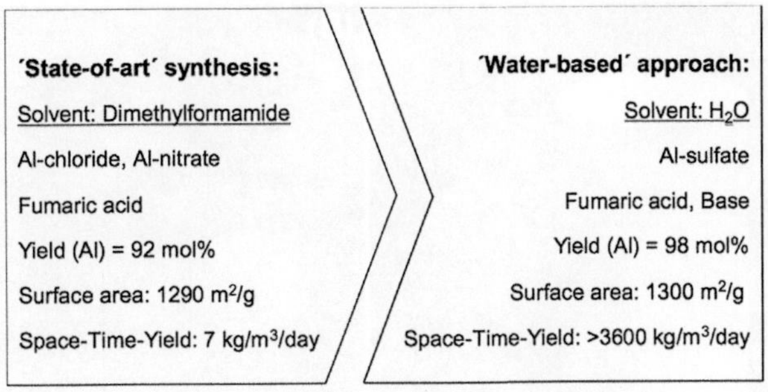

FIGURE 6.7 Optimization of Basolite A520 proposed by BASF SE. *Reprinted with permission from M. Gaab, N. Trukhan, S. Maurer, R. Gummaraju, U. Muller, The progression of Al-based metal-organic frameworks - From academic research to industrial production and applications, Microporous and Mesoporous Materials 157 (2012) 131–136.*

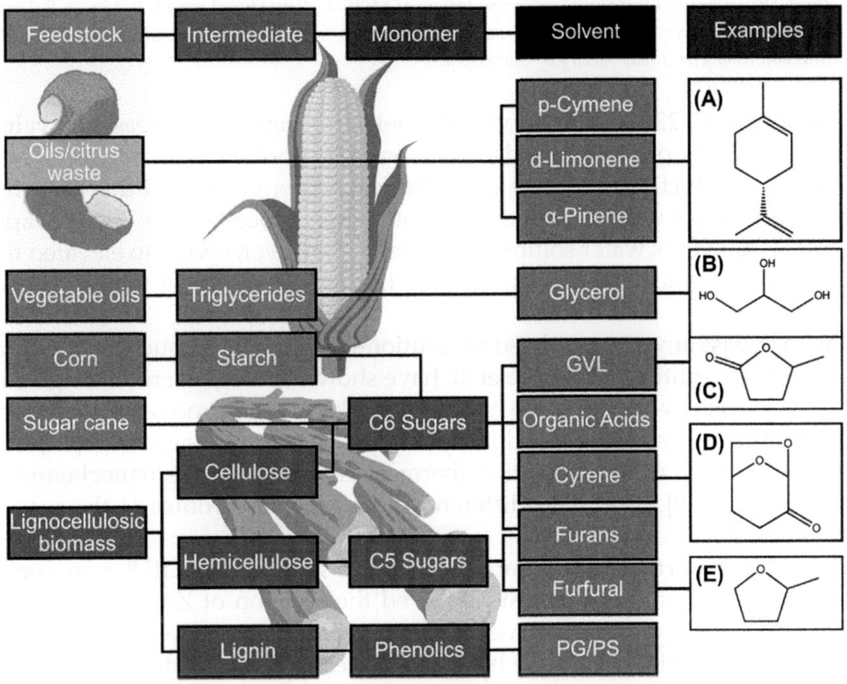

FIGURE 6.8 Examples of solvents that can be produced from biomass. *Reprinted with permission from C. J. Clarke, W. C. Tu, O. Levers, A. Brohl, J. P. Hallett, Green and Sustainable Solvents in Chemical Processes Chemical Reviews 118 (2018) 747–800.*

that used DMF as main solvent to a water-based approach (Fig. 6.7) [18]. Although attractive and relevant in many instances, a hydrothermal approach is not suitable for all MOF structures. This is related to the limited water solubility of certain linkers and the water instability of some frameworks.

For this reason, the use of renewable solvents has been investigated (Fig. 6.2, Principles 5 and 7). As shown in Fig. 6.8, a number of bio-based solvents have been developed. Yet, their implementation in MOF synthesis remains limited. One of the few examples is the synthesis of HKUST-1 using Cyrene [38]. Cyrene is a bio-derived solvent obtained from waste cellulose [38,39]. More of this class of solvents must be studied to synthesize MOFs and progressively move away from the use of petrochemical solvents.

Using ionic liquids (ILs) is another way to enhance the greenness of MOF synthesis [40–42]. ILs are polar and can easily dissolve metal salts and ligands. Despite their relative ease of recyclability and their low volatility, ILs recycling methods require large amounts of energy and produce large quantities of waste [43]. Such shortcoming represents a limiting factor to their use at industrial scale. In addition, some ILs exhibit stability and toxicity issues [44,45].

Aside from ionothermal synthesis, the use of supercritical or near-critical solvents is attractive. As shown by Ibarra and coworkers, who pioneered research in that field [46], at the near-critical point, the properties of solvents dramatically change. In their first study, they demonstrated that near-critical high-temperature water can be used as reaction media for the synthesis of MOFs whose precursors are not soluble in room temperature water [46]. Indeed,

FIGURE 6.9 Thermochemical synthesis of a metal–organic framework (MOF) proceeds via partial or complete melt phase; (A) photography of an MOF-based shaped object, (B) scanning electron microscopy image of MOF-4. *Reproduced from J. B. Lin, R. B. Lin, X. N. Cheng, J. P. Zhang, X. M. Chen, Solvent/additive-free synthesis of porous/zeolitic metal azolate frameworks from metal oxide/hydroxide, Chemical Communications 47 (2011) 9185–9187Ref. [48] with permission from The Royal Society of Chemistry.*

at the water critical point (374°C, 220 bar), the solvent's dielectric constant decreases to values similar to those of nonpolar solvents, hence favoring organic ligands solvation [17]. Another key change is the 1000-fold increase in H^+ and OH^- concentration. Both changes create a unique reaction environment that dissolves the organic ligand while favoring faster reaction rates via protonation/deprotonation processes [10,17]. This approach eliminates the use of toxic solvents and circumvents water solubility problems; however, owing to elevated temperatures and pressures, cost and safety remain significant concerns. Again, a life cycle assessment would be a way to simultaneously evaluate all these aspects.

An obvious solution to bypass any issues related to solution-based reaction, such as toxicity, safety, solubility, and costs concerns, is a solvent-free synthesis. Grande et al. have shown through a cradle-to-gate life cycle analysis that not using organic solvents lowers by 50% the CO_2 emissions of the overall process while reducing the toxicity of the water produced and minimizing resource depletion—both by an order of magnitude [47]. To date, there are three different solvent-free routes to synthesize MOFs: thermochemical [48], mechanochemical [49], and diffused-controlled "accelerated aging" [50]. The main difference between these routes is the way energy is introduced into the reaction [10].

In 2011, Lin and coworkers reported a multigram solvent-free synthesis of ZIF-8, now commercially available as Basolite Z1200 [48]. The thermochemical synthesis involved the reaction of ZnO with molten 2-methylimidazole. The main strengths of such strategy are the solvent-free nature of the reaction and the possibility to synthesize structured MOF objects (Fig. 6.9). This technique has been successfully expanded to Co-based MOFs [51]. Despite these advantages, the process requires organic linkers to be stable at or above the melting temperature of the linker. Additionally, as the reaction proceeds at high temperature, energy efficiency should to be assessed.

Another solvent-free method to synthesize MOFs is via chemical transformation induced by grinding and milling. Such syntheses are called mechanochemical syntheses. To date, three main mechanochemical paths have been explored: solvent-free grinding (SFG) [52], liquid-assisted grinding (LAG) [49], and ion- and liquid- assisted grinding (ILAG) [53]. The SFG approach is considered as the most environmentally friendly approach as no solvent is used during the MOF synthesis. It also allows using metal salts exhibiting low solubility such as metal oxides, hydroxides, and carbonates, which have been identified as the greenest metal precursors [54,55]. Moreover, the use of such precursors, when combined with protons, allows the formation of only water as a by-product, hence eliminating the purification process. Yet, the scope of this technique is limited as it can lead to incomplete conversion and amorphization (caused by structural collapse) [56]. In contrast, the LAG approach, which uses little solvent compared with typical solvothermal approaches (between 0 and 1 mL mg^{-1} of the sample), favors the synthesis of highly crystalline MOF frameworks [57]. In the LAG method, solvent molecules act as space-filling agents and are incorporated as guests into the final MOF structure, enabling to template the formation of a porous structure [49]. Adding small amounts of solvent also favor faster reaction rates [58]. LAG has been widely used to synthesize distinct MOFs from a wide range of starting materials [59–61]. In the case of metal oxides, the ILAG approach is particularly interesting [17,62,63]. The use of a catalytic quantity of solvents (*ca.* 5 mol%) and salt additives favors the fast and quantitative incorporation of the metal salts into the MOF structure [53]. The additives also help control on reaction outcome. Namely, in the synthesis of zinc-pillared MOFs, the use of nitrate additives such as KNO_3 and NH_4NO_3 favors the synthesis of square-grid zinc terephthalate sheets [53], whereas sulfate additives favor

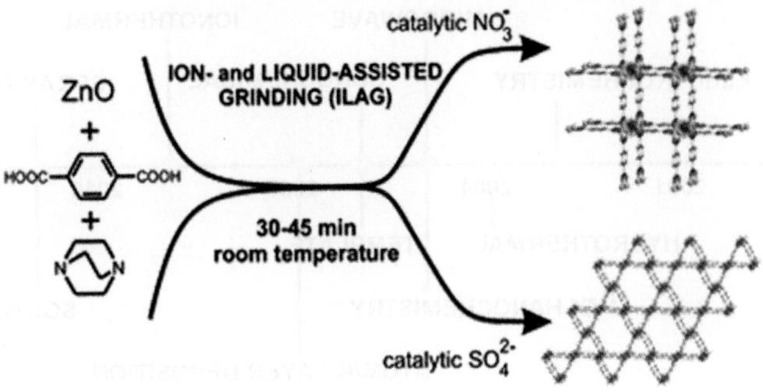

FIGURE 6.10 (A) Rapid room temperature synthesis of a porous metal–organic framework (MOF) using the liquid-assisted grinding (LAG) approach. *Reproduced from T. Friscic, New opportunities for materials synthesis using mechanochemistry, Journal of Materials Chemistry 20 (2010) 7599–7605 with permission from The Royal Society of Chemistry.*

Kagome-topology sheets (Fig. 6.10) [64]. Depending on the precursors, the role of additives can vary. In some cases, the salt additives can also act as protic catalyst [62]. As shown by the SFG, LAG, and ILAG approaches, mechanochemistry can be a safe, efficient, and versatile method that can yield the most common MOF structures. However, because of the milling process, the scalability of such synthesis remains a challenge, and most of the mechanochemical MOF synthesis studies do not report synthesis higher than the multigram scale. In addition, even though the LAG approach uses little solvent, it still requires about 1000 more solvent than the sample it produces.

Another solvent-free route to MOF synthesis is "*accelerated aging*" [50,65]. The main concept of such approach is to convert a static physical mixture of the metal oxide and organic ligands into an MOF structure under suitable/controlled high humidity atmosphere and mild temperature (around 45°C, > 95% relative humidity) [10]. This approach enables the conversion of inert and poorly soluble reagents that are typically activated at high temperature. To date, accelerated aging has been developed for transition metals (MnO, CoO, NiO, CdO, CuO) and selected main-group elements such as PdO and MgO [17,65,66]. C. Montillo and coworkers showed how accelerating aging can be used and optimized for multigram MOF synthesis. In this study, the authors demonstrated that 5 mg of ZIF-8 could be synthesized under mild conditions in 4 days [65]. Overall, this approach has similar environmental, safety, and health benefits to mechanochemical studies with the added benefit that it does not require energy intensive ball milling. However, the long reaction times could hinder future large-scale production.

2.3 Energy Input

When designing MOFs for large-scale production, the energy consumed during synthesis is a key aspect to consider. Depending on the approach used, the carbon footprint of the synthesis can vary significantly and negatively impact the environment. Nowadays, one can synthesize MOFs following any of these five approaches: conventional solvothermal, microwave-assisted, electrochemical, sonochemical, and mechanochemical synthesis [10,67].

As mentioned in the previous section, the solvothermal approach is the most popular MOF synthesis route. It easily yields highly crystalline MOFs, yet requires harsh synthesis conditions [68]. Owing to their long reaction time and their high temperature, solvothermal syntheses are considered as highly energy-intensive techniques (Principal 6, Fig. 6.2). Efforts should be directed to optimizing the conditions toward near ambient temperature/pressure and decrease the reaction time.

An efficient solution to reduce energy demand is the use of microwave heating. It allows reduced reaction time, and all the power input is directed specifically on the reaction mixture [10,69]. For instance, microwave-assisted synthesis of UiO-66 can be achieved in 2 h at 100°C [70], whereas conventional synthesis requires 120°C for 24 h [71]. Scale-up of microwave-assisted reactions are, however, not straightforward and would require either running batches in parallel or running in continuous flow. Both approaches are emerging rather than mature technologies. Electrochemical syntheses also display advantages over solvothermal reactions as they are performed at ambient temperature/pressure, and, as mentioned earlier, they circumvent the release of counterions that can form harmful, highly toxic, and corrosive by-products. Hence, this approach favors energy-efficient and safer MOF production.

Sonochemical synthesis operates under mild/room temperature and represents another potential sustainable alternative to solvothermal MOF synthesis [10]. The principle behind this technique is to use ultrasonic irradiations that create alternating high and low pressure environments in the solvents. The cavitation generates localized high

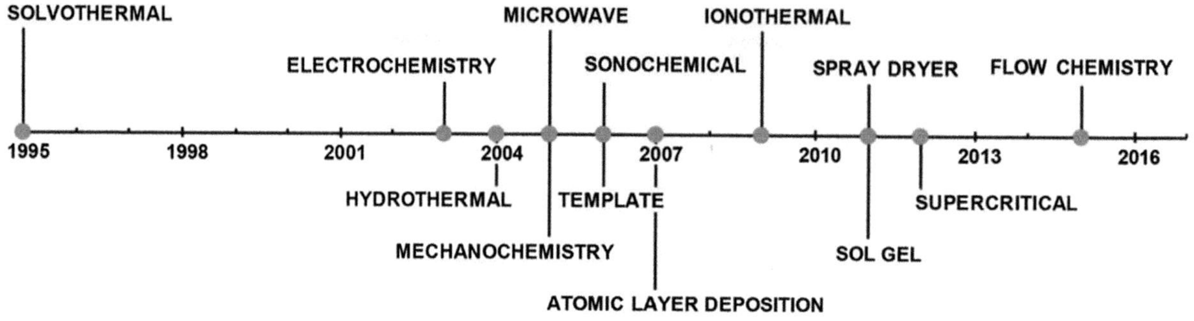

FIGURE 6.11 Timeline of the most popular synthetic approaches patented for metal–organic frameworks (MOFs) synthesis. *Reproduced from M. Rubio-Martinez, C. Avci-Camur, A. W. Thornton, I. Imaz, D. Maspoch, M. R. Hill, New synthetic routes towards MOF production at scale, Chemical Society Reviews 46 (2017) 3453–3480 with permission from The Royal Society of Chemistry.*

temperature and pressure called "hot spots," that favor MOFs synthesis [10,67]. Sonochemical routes offer fast reaction rates and require low energy input.

To date, the most promising approach is mechanochemical synthesis [59]. As explained in Section 2.2, mechanochemical synthesis require no or minimal amount of solvents. In addition, the ambient temperature and pressure conditions and the fast reaction time (few minutes or hours) lower the energy requirement compared with conventional solvothermal synthesis. Today, mechanochemical synthesis is considered the most cost-efficient and sustainable MOF synthesis approach.

Of course, the synthesis approach, while minimizing energy demand, selected should maintain the desired physicochemical properties of the material. For large-scale production, energy consumption of the overall synthesis is crucial as it influences the cost and the environmental impact. Researchers should continue to identify synthetic approaches that optimize energy demand and reaction time while minimizing the use of toxic compounds and solvents and favoring the desired MOF features (Fig. 6.11) [55].

2.4 Flow Chemistry and Downstream Processes

2.4.1 *Flow Chemistry*

The previous sections explained the main strategies employed to yield sustainable and safe MOFs. These strategies included (i) using renewable and environmentally friendly precursors and solvents and (ii) minimizing energy consumption. Recently, the concept of flow chemistry has been applied to MOF synthesis allowing further progress toward scalable and sustainable MOF production. Flow chemistry refers to a continuous process and is routinely used in industry for producing chemicals (Fig. 6.12) [13,55]. In contrast to batch reactions, the MOF synthesis occurs in a continuously flowing stream. Switching to flow chemistry eliminates time-consuming cyclic processes such as heating, cooling, and opening the reactor. Additionally, this method allows higher surface-to volume ratio compared with batch-type, faster synthesis rates, minimization of solvents, lower energy consumption by enhancing mass/heat transfer, and enables a constant and consistent MOF production [13,55]. Flow chemistry is also generally safer

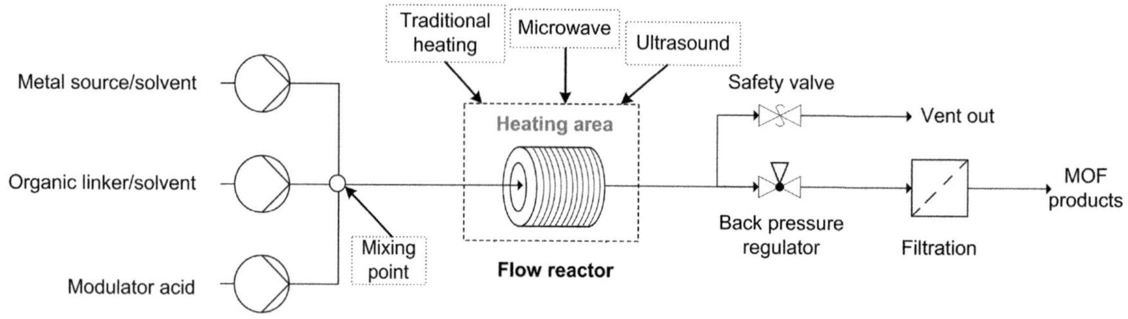

FIGURE 6.12 Schematic illustration of a typical flow chemistry setup for metal–organic framework (MOF) production. *Reprinted with permission from J. W. Ren, X. Dyosiba, N. M. Musyoka, H. W. Langmi, M. Mathe, S. J. Liao, Review on the current practices and efforts towards pilot-scale production of metal-organic frameworks (MOFs), Coordination Chemistry Reviews 352 (2017) 187–219.*

TABLE 6.1 Examples of Metal–Organic Frameworks (MOFs) Synthesis Using a Plug Flow Reactor.

Method	MOF	Residence Time	Temperature (°C)	Space-Time Yield (kg m^{-3} day$-^{1}$)
Plug Flow Reactor : reagents are pumped through a pipe and are consumed as they flow down the length of the reactor (Fig. 6.12) [55]	ZIF-8 [76]	<5 s	100	11,625
	HKUST-1 [77]	<5 s	300	4399
	UiO-66 [78]	10 min	130	1186
	MIL-53(Al) [79]	20 min	250	1021

Reproduced from M. Rubio-Martinez, C. Avci-Camur, A. W. Thornton, I. Imaz, D. Maspoch, M. R. Hill, New synthetic routes towards MOF production at scale, Chemical Society Reviews 46 (2017) 3453–3480 with permission from The Royal Society of Chemistry.

compared with batch processes as it handles smaller amounts of reactants and allows a greater control of the reactor temperature. Owing to these advantages, flow chemistry has been identified as a cost-efficient and scalable method that complies with the principles of green chemistry (Fig. 6.2) and satisfies the requirements for industrial/scale-up production of MOFs. Examples of MOFs that have been produced using this approach include UiO-66, HKUST-1, and ZIF-8 (Table 6.1).

As shown in Fig. 6.12, a typical flow setup comprises two or three pumps that enable to transfer and mix the fluid reagents before being transferred into the flow reactor [13]. The efficient mixing of reagents is crucial as it can affect the reaction kinetics. After mixing, a heating and cooling process is undertaken, followed by a filtration process to recover the synthesized MOF crystals. Depending on the temperature and the residence time, a fine control over the reaction rate and the crystal size can be achieved. In such a setup, a backpressure regulator is inserted at the tubing outlet [13]. As it is well known, most common MOF syntheses are conducted at very high temperatures; however, in flow chemistry, solvent vapors should be avoided as they can interfere with the flow steam. For this reason, back-pressure regulators are usually used to increase pressure inside the tubing. To date, several flow chemistry methods have been developed. Depending on the MOF, it is possible to use traditional [72] or microwave [73] heating processes. New flow processes based on ultrasounds have also been developed [74]. Avci-Camur et al. demonstrated the production of MOFs using a continuous-flow spray-drying method [75].

As illustrated by Table 6.1, [55], flow chemistry has proven to be a highly attractive technique to continuously produce MOFs in a cost-efficient, green, and safe manner. Nowadays, this technique is considered the most promising MOF synthesis approach as it complies with all the requirements needed to produce green MOFs at the ton scale. However, to apply such technology to real-world applications, it is necessary to couple it with efficient downstream processes (see next section).

2.4.2 Downstream Processes

Most of the cost, waste, and energy consumption related to MOF manufacturing are associated to downstream processes [55]. Here, we refer to downstream processing as the steps following MOF crystallization, i.e., washing, filtration, activation, and casting. In laboratories, such processes are well established and can be readily completed; however, these methods are not well suited for ton scale quantities and do not necessarily comply with environmental and industrial legislations [55].

Again, downstream processes must be evaluated following the guidance of the 12 Principles of Green Chemistry. Similar to flow synthesis, such approach can easily be scaled and allows a constant and consistent MOF processing (Fig. 6.13). In addition, optimization of the system—i.e., optimization of the quantity of liquid necessary for the

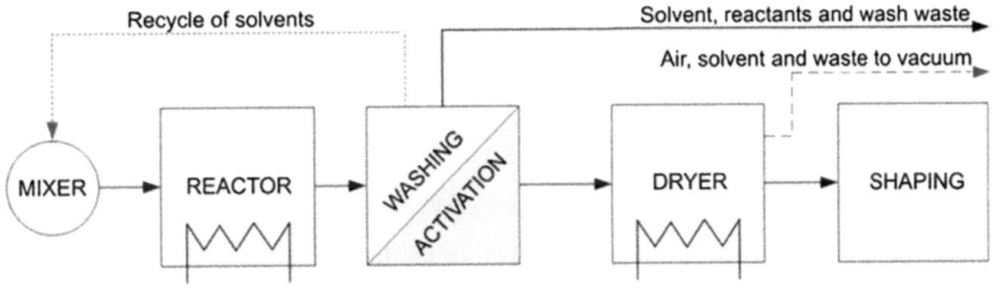

FIGURE 6.13 Schematic representation of a continuous downstream metal–organic framework (MOF) process. *Reproduced from M. Rubio-Martinez, C. Avci-Camur, A. W. Thornton, I. Imaz, D. Maspoch, M. R. Hill, New synthetic routes towards MOF production at scale, Chemical Society Reviews 46 (2017) 3453–3480 with permission from The Royal Society of Chemistry.*

washing step, conscious selection of the different activation [80,81], filtration [55], and shaping [82,83] techniques—enables the development of an efficient setup that reduces the usage of toxic and harmful solvents and/or compounds, hence assisting sustainable MOF synthesis.

Optimization of all the abovementioned section has been thoroughly investigated by different laboratories and companies. The first MOF-based product (i.e., a structured MOF "object" rather than a powder) commercially available was released in 2016 by MOF Technologies and NuMat Technoloigies [84]. The ION-x gas storage tank enables to store specific gases used in the electronics manufacturing industry. In the same year, Decco Post-Harvest and MOF technologies developed NT-7815 (another MOF-based product) that extends the storage life of fruits and vegetables [84].

2.5 Commercial Development and Future Perspectives

Nowadays many commonly researched MOFs are commercially available. For instance, UiO-66, ZIF-8, and HKUST-1 can be easily obtained (Table 6.2)[55].

TABLE 6.2 Sales Catalogs of Some Metal–Organic Frameworks (MOFs) Available by Different Manufacturers.

MOF	Manufacturer
AI(OH) fumarate	MOF Apps
	MOF Technologies
CAU-10	ProfMOF
Cu-BTC	BASF
	MOF Apps
	MOF Technologies
Fe-BTC	BASF
Magnesium formate	BASF
	MOF Technologies
Mg-MOF-74	MOF Technologies
MIL-100	KRICT
	MOF Apps
MIL-101-NH2	MOF Apps
MIL-53	BASF
MIL-68	MOF Apps
MOF-177	BASF
PCN-250(Fe)	framergy
UiO-66 series	Inven2
	MOF Apps
	ProfMOF
ZIF-67	MOF Apps
	MOF Technologies
ZIF-8	BASF
	MOF Apps
	MOF Technologies
	STREM Chemicals Inc.
Zn-SIFSIX-pyrazine	MOF Technologies

Reproduced from M. Rubio-Martinez, C. Avci-Camur, A. W. Thornton, I. Imaz, D. Maspoch, M. R. Hill, New synthetic routes towards MOF production at scale, Chemical Society Reviews 46 (2017) 3453–3480 with permission from The Royal Society of Chemistry.

The market availability of these MOFs accelerates research in the field and favors the development of MOF-based products. The number of such MOF structures remains limited compared with the number of existing ones. This imbalance is likely due to (i) the absence of a well-defined market/application for the materials and (ii) remaining concerns in terms of safety, toxicity, and environmental impact associated with the synthesis of the materials or the materials themselves [10].

3. GREEN APPLICATIONS OF METAL–ORGANIC FRAMEWORKS

The term "green" is used here for applications whose aim is to foster a reduced and minimal harm on ecosystems or the environment either in an absolute sense or in comparison to existing technologies. Taking this into account, such applications include heterogeneous catalysis [11], molecular separations [85], photocatalysis [86,87], electrocatalysis [88], and energy storage [85,89]. The former two areas are covered in Chapters 9 and 15 and are therefore excluded from the discussion here. The latter three areas on the other hand are included here. We note that applying the principles highlighted in the former subsection for the synthesis of MOFs is necessary for the whole process to be green. For each application discussed, we concentrate essentially on describing the materials requirements and highlighting how MOFs meet all or some of these criteria. A number of studies reporting the use of the MOFs are included to illustrate the main concepts but do not provide an exhaustive list. For this, review articles should be consulted [11,85,87,90–92].

3.1 Solar Energy Conversion

Building a more sustainable future relies in part on the diversification of our energy portfolio. The share of renewables and particularly solar energy is projected to increase in the years to come. Solar energy can be used either to produce electricity via the deployment of solar cells (SCs) or it can be used to produce fuels (chemical energy). The former route has significantly matured over the past decades, while the latter is still at a relatively low technology readiness level. Both are discussed below, highlighting the role of MOFs in progressing these technologies.

3.1.1 Solar Fuel Photocatalytic Production

Solar energy can be harvested to form chemical fuels. Specifically, water can be split via action of sunlight to produce H_2 (and O_2). Similarly, CO_2 can be reduced to produce C_1 and C_{1+} energized molecules, e.g., CO, CH_4, CH_3OH, etc.... These two classes of reaction are referred to as solar fuel production. To proceed, they require a catalyst—typically an electrocatalyst or a photocatalyst. We note that both H_2 production from water and CO_2 conversion into fuels relate to reduction reactions. The following sections therefore focus on catalysis of reduction reactions.

3.1.1.1 Photocatalysis

Absorption of light by a photocatalyst leads to the creation of an electron–hole pair. These charges are the ones controlling the reduction and oxidation reactions. The photocatalytic process occurs in three main steps: (i) electron–hole pair is formed on light absorption as an electron moves from the valence to the conduction band of the catalyst, (ii) charges migrate to the surface of the photocatalyst, and (iii) charge transfer occurs from the photocatalyst to the preadsorbed reactants. In light of this, we describe below the requirements of a photocatalyst and how these apply to MOFs.

- The photocatalyst should be visible light active, i.e., it should absorb efficiently the solar light, which comprises 47% of visible light. This dictates the band gap of the photocatalyst. Owing to their chemical modularity, MOFs can be made active under visible light, via functionalization of the ligand, attachment or encapsulation of a sensitizer, and by appropriate selection of the metal [87,93].
- The valence and conduction bands of the photocatalyst should be of appropriate energy levels. Indeed, the band energies should be such that they are more negative or more positive than the redox potentials of the species they are supposed to reduce or oxidize, respectively. Hence, the band structure of the photocatalyst must be tuneable. In MOFs, this aspect is possible, and particularly the position of the conduction band is typically dictated by the type of metal.
- The photocatalyst should exhibit controlled adsorption/desorption properties and favor easy access to catalytic sites. This is because only reactants adsorbed on the surface of the catalyst can be effectively reduced/oxidized.

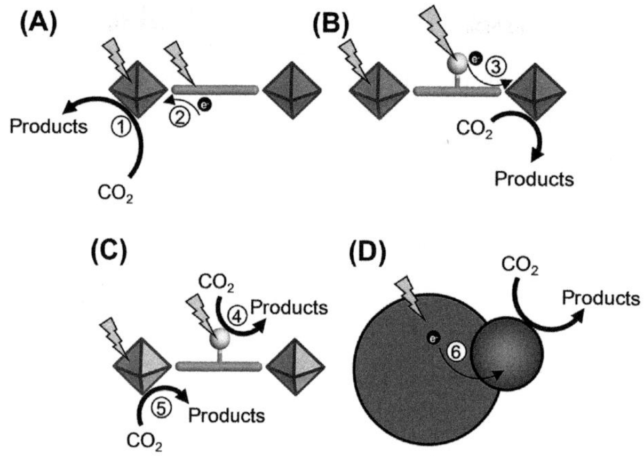

FIGURE 6.14 Metal–organic framework (MOF) photocatalytic reduction of CO_2 reaction pathways (numbered 1–6) for the various types of MOF-based materials: (A) unmodified MOF, (B) MOF with ligand functionalization, (C) MOF subjected to metal ion exchange, and (D) MOF-based composites. *Reproduced with permission from A. Crake, Materials Science and Technology London 33 (2017) 1737–1749.*

In addition, the products formed should easily desorb from the surface to be collected. The high porosity and surface area of MOFs is particularly relevant here. In addition, the chemical and structural versatility of MOFs allow one to tailor the energy of adsorption and desorption of specific chemicals.

- Charge recombination in a photocatalyst should be minimized. As much as possible, the materials/system should be such that it prevents recombination of the electrons and holes formed on light irradiation. Reduced charge recombination can be achieved in the case of MOFs by building an MOF-based composite, adding a cocatalyst, or via metal substitution. All these routes allow positive charge transfer between the particles (composites and cocatalyst) or between metals (metal substitution).
- A photocatalyst should be robust and easily processed. This relates to the long-term usage of the photocatalyst and the ease of manufacturing. It is important to note that while H_2 production is a liquid-phase reaction with the MOF acting as a heterogeneous photocatalyst, CO_2 photoreduction can be carried out either in a liquid phase or gas phase. Hence, in addition to exhibiting good light stability, an MOF must be water-stable. Recent work has provided directions toward improving the water stability of MOFs, and in fact, a number of structures are water stable [94,95].
- Finally, a photocatalyst must be cost effective and sustainable. The sustainability aspect is directly connected to the principles introduced in the first part of this chapter. The cost aspect relates to the cost of the precursors, solvents, and production route selected for the synthesis and manufacturing. These considerations are not included here and are only starting to be analyzed in the literature [26].

Fig. 6.14 summarizes the pathways in which MOF can act as a photocatalyst. MOFs can be directly catalytically active as the metal cluster is excited leading to the generation of an exciton pair (Fig. 6.14, pathway 1). Alternatively, it can be indirectly active if another moiety first absorbs light and then the generated electron is transferred to the metal cluster. Light absorption can be improved by ligand functionalization (Fig. 6.14, pathway 3). The organic ligand itself can also be intrinsically active (Fig. 6.14, pathway 4). Postsynthesis metal substitution of the metal cluster (partial substitution) favors charge separation, as electrons transfer between each metal in the metal cluster (Fig. 6.14, pathway 4). Following the same mechanism of charge transfer, charge separation is also improved by the addition of a cocatalyst or the formation of a composite material (Fig. 6.14, pathways 6).

Examples of MOF-based systems for H_2 evolution include among others Ti- and Zr-based MOFs [97,98] and related composites as well as porphyrin-based MOFs [99]. For CO_2 photoreduction, we can name, for instance, Ti-MIL-125 and its amine-functionalized derivative [100]. MOF-based composites are also increasingly reported [101].

3.1.1.2 Solar Fuel Electrocatalytic Production

In an electrocatalytic process, an electrocatalyst is deposited on the surface of an electrode. The role of the electrocatalyst is to lower the overpotential of a reaction. The key steps of an electrocatalytic process are first the electron transfer from the electrode to the electrocatalyst and then the transfer of the electron from the electrocatalyst to the reactant, thereby leading to a reduction reaction. Similar to a photocatalyst, an electrocatalyst must display a number of features—some of which are in fact similar to the ones highlighted above for a photocatalyst:

- The electrocatalyst should be highly catalytically active and display low electrochemical overpotential. The catalytic performance is heavily linked to the activity of the catalytic sites and the density of such sites. The chemical versatility of MOF can be used to address the former point by selecting an MOF with the desired metallic sites. The high porosity of MOFs is beneficial in that a large number of catalytic sites are easily accessible.
- The electrocatalyst should be selective toward a given reaction/product. This aspect is in part directed by the sorption behavior of the electrocatalyst. As for any catalyst, reactants and reaction intermediates should adsorb with strength commensurate with the expected reaction. Here again, the modularity—chemical and structural—of MOFs is a key enabler for controlling the adsorption and desorption of species on the surface of MOFs.
- An electrocatalyst should be robust and withstand long cyclic usage. If the MOF is to be used directly as the electrocatalyst, this calls for structural and chemical stability of the framework. More and more of such stable MOFs structures are being reported [94,95]. MOFs can also be used to immobilize catalysts [102]. Acting as a support, they prevent the agglomeration of the catalyst and hence deactivation. Allowing dispersion, they also enhance the density of accessible catalytic sites.
- Similar to photocatalysts, electrocatalysts should be scalable and sustainable (see section 2).
- If the electrocatalyst is to act as the electrode as well, the material should be conductive. An increasing number of conductive MOFs are being reported [103,104].

Some MOFs have been used directly as the electrocatalyst [105]. This is the case for instance of a porphyrin-based MOF, which has been tested for CO_2 electroreduction (Fig. 6.15) [105]. MOF can also serve as the electrocatalyst indirectly as they are used as template to produce a porous material. MOFs have also been employed as a host for an electrocatalyst, either directly (i.e., as synthesized) or indirectly by first forming a porous template that was used to support the electrocatalyst (Fig. 6.16)[106].

3.1.2 Dye-Sensitized Solar Cells

Solar energy can also be used to produce electricity. This is what SCs, also called photovoltaic cells (PV), allow. In a typical SC, a semiconductor is sandwiched between two electrodes. The semiconductor layer is in fact made of two layers: an n-type and a p-type version of the semiconductor. When light of sufficient energy strikes the surface of the semiconductor, electron–hole pairs are formed: electrons sit in the n-type layer and the holes in the p-type layer. When the unit is connected to an external circuit, electrons can move back to the p-type through the external circuit, thereby generating electricity. In the particular case of dye-sensitized solar cells (DSSCs), the light absorption and charge transport phenomena are decoupled. The anode is a photoanode, which absorbs light via use of a dye, leading to the creation of an electron-hole pair. The electron is then transferred from the dye to the semiconductor.

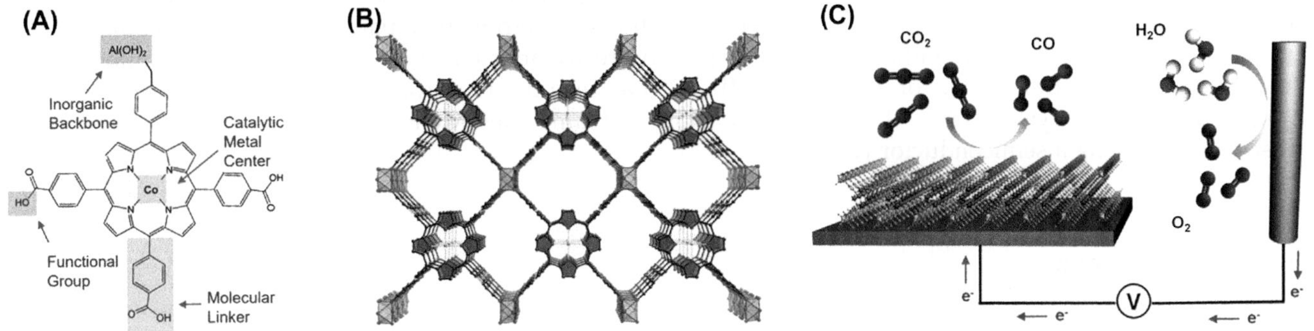

FIGURE 6.15 Our metal–organic framework (MOF) catalyst allows for modulation of metal centers, molecular linkers, and functional groups at the molecular level (A). The organic building units, in the form of cobalt-metalated TCPP, are assembled into a 3D MOF, $Al_2(OH)_2$TCPP-Co with variable inorganic building blocks (B). Co, *orange (light gray in print version) spheres*; O, *red (dark gray in print version) spheres*; C, *black spheres*; N, *blue (gray in print version) spheres*; Al, *light blue (gray in print version) octahedra*; and *pyrrole ring*, blue (gray in print version). In this structure, each carboxylate from (A) is bound to the aluminum inorganic backbone. The MOF is integrated with a conductive substrate to achieve a functional CO_2 electrochemical reduction system (C). (For interpretation of the references to colour in this figure legend, the reader is referred to the web version of this article.) *Reprinted with permission from N. Kornienko, Y. B. Zhao, C. S. Kiley, C. H. Zhu, D. Kim, S. Lin, C. J. Chang, O. M. Yaghi, P. D. Yang, MetaleOrganic Frameworks for Electrocatalytic Reduction of Carbon Dioxide, Journal of the American Chemical Society 137 (2015) 14129–14135. Copyright (2015) American Chemical Society.*

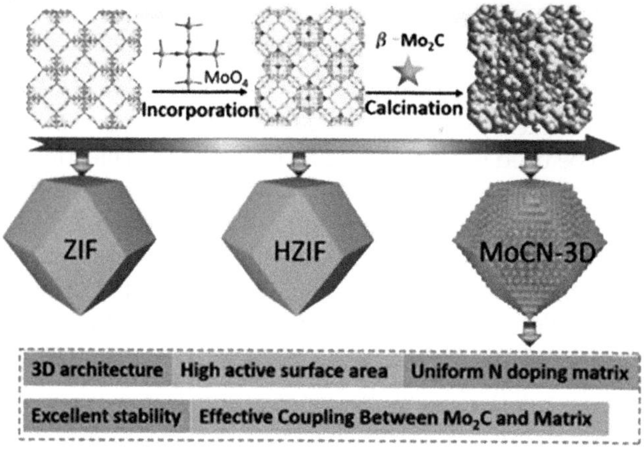

FIGURE 6.16 Schematic illustration for the construction of MoCN-3D for a highly efficient hydrogen evolution reaction. *Reprinted with permission from H. Zhang, Z. Ma, G. Liu, L. Shi, J. Tang, H. Pang, K. Wu, T. Takei, J. Zhang, Y. Yamauchi, J. Ye, Highly active nonprecious metal hydrogen evolution electrocatalyst: ultrafine molybdenum carbide nanoparticles embedded into a 3D nitrogen-implanted carbon matrix, NPG Asia Materials 8 (2016) e293.*

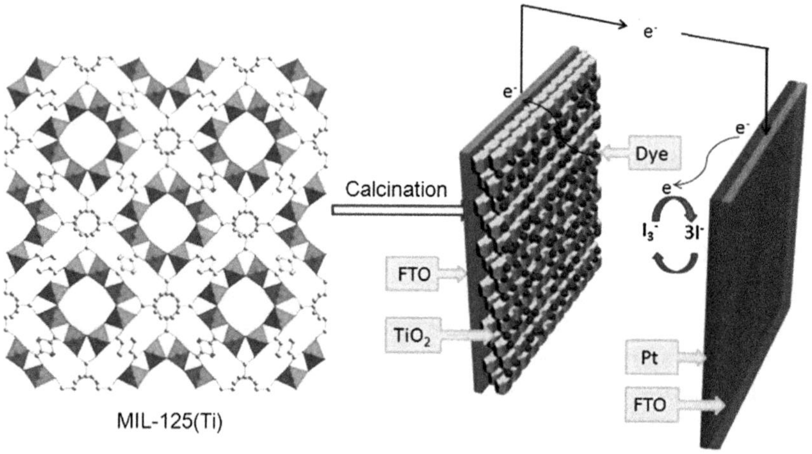

FIGURE 6.17 Schematic of the dye-sensitized solar cell (DSSC) based on a porous TiO_2 photoanode derived from Ti-MIL-125. *Reprinted with permission from J. Dou, Y. Li, F. Xie, X. Ding, M. Wei, MetaleOrganic Framework Derived Hierarchical Porous Anatase TiO_2 as a Photoanode for Dye-Sensitized Solar Cell, Crystal Growth and Design 16 (2016) 121–125. Copyright (2016) American Chemical Society.*

A photoanode should combine some of the properties of a photocatalyst and an electrocatalyst. Of particular importance is the high light absorption and efficient creation of electron–hole pairs. This can be accomplished by using MOFs as hosts for dyes either directly [107] or indirectly after conversion of the structure into a porous semiconducting substrate [108]. For instance, Ti-MIL-125 has been used to produce porous TiO_2, which was subsequently used to host dye in a DSSC (Fig. 6.17) [108]. Along the same lines, MOFs have also been used as host for dye in addition to using a semiconductor [109].

3.2 Energy Storage

3.2.1 Gas Storage

CH_4 and H_2 are attractive fuels from a specific energy standpoint compared with hydrocarbons such as gasoline, diesel, or ethanol. This feature can only be exploited if the gases are to be stored efficiently to address their low energy density. This is particularly important for portable applications. Although an elevated pressure, i.e., using compressed gas cylinders, is a straightforward way to go, it does not allow maximal energy density to be reached under the safest and greenest conditions. By comparison, under similar T and P conditions, gas storage in a porous media leads to a more compact packing of the gases and therefore higher energy density at lower pressures. This is due to the thermodynamically favored adsorption of the gases in the pores of the so-called adsorbents, which leads to the adsorbed gases to reach a quasi-liquid state. A central feature of a good adsorbent for H_2 and CH_4 storage is therefore a high porosity/surface area. For this reason, MOFs are ideal candidates for H_2 and CH_4 storage.

H_2 storage in MOFs has been studied at both cryogenic (77K) and room temperature conditions. Many structures have been reported for this application [110,111], though none of them meet the Department of Energy (DOE) target of 9wt% and 81 g/L at 313–343K and 100 bar. Studies show that as expected pore size and pore shape play a dominant role at low temperature, 6 Å has been identified as an optimal pore size to maximize hydrogen uptake at 77K and 100 bar on carbon nanotubes [112]. This is due to the overlap of the potential fields from multiple walls leading to a high affinity for H_2. At room temperature, unsaturated metal sites govern the adsorption mechanism. For this reason, one should (i) use low coordination metals that favorably interact with H_2 [113] and (ii) increase the number of metal sites via addition of extraframework cations [114]. Today, Mn-BTT's (H_3BTT = 1,3,5-tris(2H-tetrazol-5-yl)benzene) hydrogen capacity of 12.1 g/L at 298K and 90 bar exceeds that of a high-pressure hydrogen tank at the same conditions. Other promising compounds for H_2 storage at cryogenic conditions include MOF-210, PCN-68, NOTT-400, and ZIF-8.

One of the first applications envisioned for MOFs was CH_4 storage, with BASF actively showcasing the potential of these materials early on. CH_4 storage has been studied at room temperature for a number of MOFs [110,115,116] with one structure breaking the DOE target of 0.5 g g^{-1} (700 cm^3 g^{-1} or 350 cm^3 cm^{-3}) at 298K and 250 bar [117], a densified form of HKUST-1. The three controlling factors of CH_4 adsorption in MOFs are electrostatic interactions, van der Waals interactions, and network flexibility. The former mechanism is enhanced by providing open metal sites, as seen with MOF-74 [118,119]. The second mechanism is favored through optimization of the pore size and shape, as reported for Al-soc-1, HKUST-1, PCN-250(Fe_2Co), MAF-38, and UTSA-76 [110,120,121]. Finally, flexible MOFs with a steep decrease in gas uptake below 5 bar are ideal as they allow high working capacity without the need for a vacuum swing. In fact, one such structure is Co(bdp) (H_2bdp = 1,4-benzenedi(4′-pyrazole)), which comes second to the densified HKUST-1 in terms of CH_4 working capacity[122].

It is worth noting that modeling studies have an important role to play in screening a large number of MOF structures for gas storage and identifying promising structures [123]. Their importance is probably greater than for other applications described here as the adsorption process is relatively straightforward to model and does not involve complex or even any reactions. Hence, as long as the adsorbent–adsorbate interactions are understood and appropriately modeled, effective screening can be performed.

3.2.2 Supercapacitors and Batteries

Although gas storage belongs to chemical energy storage, supercapacitors and batteries both represent electrochemical routes of energy storage. Supercapacitors store potential energy in an electric field. They are made of conducting plates/electrodes separated by an "unconventional" dielectric material that uses electrostatic double-layer capacitance (EDLC, Helmotz double layer) and electrochemical pseudocapacitance. In the former, the charge storage occurs via adsorption of electrolyte ions on the surface of the electrode. In the latter, the charge storage occurs via electron transfer through redox reactions (Faradaic electron transfer).

EDLC is favored by the use of a porous support for the electrolyte ions. In addition, the morphology of the pore is important as it can influence the penetration of the electrolyte [124]. Having a porous support with tailorable porosity is therefore a key. This combined to the fact that carbon-based materials make good supercapacitor (cf. low weight, fast charge, and discharge rates) has led researchers to use MOFs as template to produce porous carbon materials with controlled nanoarchitecture [125]. An example of a synthesis route is shown in Fig. 6.18. Depending on the linker selected, N-functional groups can be incorporated into the carbon matrix leading to enhanced performance [127].

To promote Faradaic charge transfer, a battery-like electrode is needed, calling for a conductive material to be used. In addition to this, fast ion transport is a key to accelerate charge and discharge processes. This can be facilitated using nanostructures as transport distances are shortened. For this reason, MOFs have also been used as template to produce nanostructured metal oxide/sulfide electrodes [128]. An example of a synthesis route is shown in Fig. 6.19.

Taking into account, both the EDLC and electrochemical pseudocapacitance call for a combination of high electrical conductivity and high surface area. Although many MOFs are not conductive, conductive MOFs still do exist, and in fact Sheberla et al. have reported the use of a pure MOF as supercapacitor taking advantage of its conductivity and porosity [129].

Batteries are made of two electrodes, an electrolyte, a separator, and redox species. They generate energy through electrochemical reactions. Here, we focus our attention on Li^+ and Na^+ batteries, as well as Li-S batteries. The former typically calls for the use of a graphite electrode in which the Li^+ ions are intercalated. Higher surface area allows greater Li^+ storage. For this reason, MOFs have been employed as the anode and porous support for Li^+. Similar to

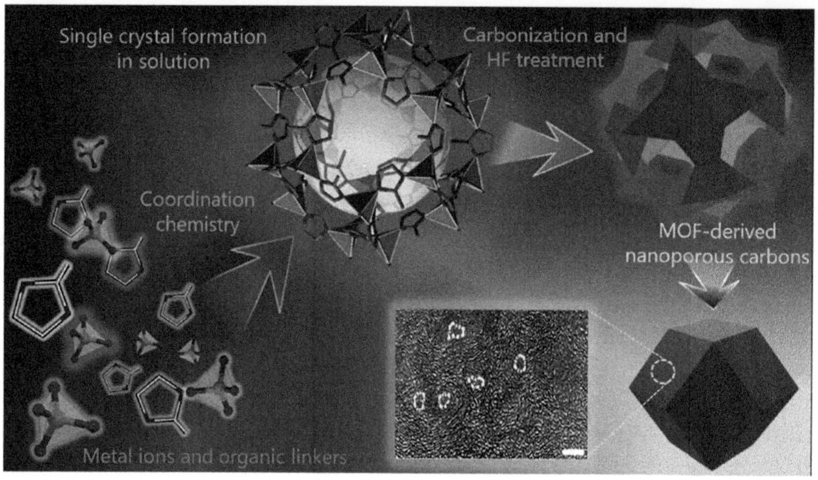

FIGURE 6.18 The metal–organic frameworks (MOFs) are formed in the solution through a coordination reaction between their organic and inorganic constituents. Calcination in a nitrogen atmosphere followed by HF treatment to remove metal particles forms the MOF-derived NPCs. An actual high-resolution TEM image of ZIF-8–derived NPC is shown as well. The scale bar is 5 nm. The dotted lines in the TEM image show the presence of pores. *Reprinted with permission from R. R. Salunkhe, Y. V. Kaneti, J. Kim, J. H. Kim, Y. Yamauchi, Nanoarchitectures for MetaleOrganic Framework-Derived Nanoporous Carbons toward Supercapacitor Applications, Accounts of Chemical Research 49 (2016) 2796–2806.*

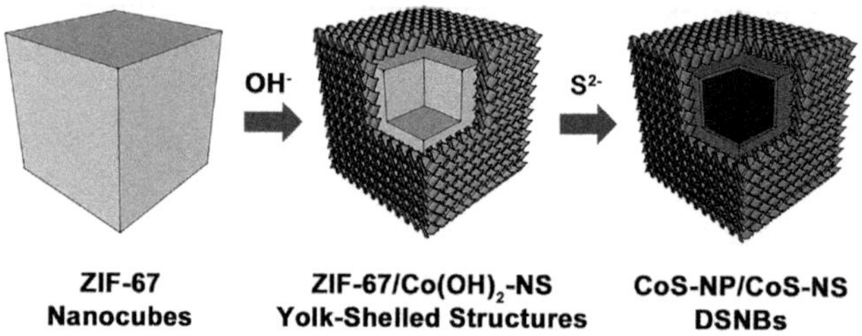

FIGURE 6.19 Fabrication of hierarchical double-shelled hollow structures consisting of CoS-NP nanoboxes surrounded by CoS nanosheets (CoS-NP/CoS-NS DSNBs). *Reprinted with permission from H. Hu, B.Y. Guan, X.W. Lou, Construction of Complex CoS Hollow Structures with Enhanced Electrochemical Properties for Hybrid Supercapacitors, Chemistry 1 (2016) 102–113.*

other applications described above, MOFs have also been used as templates to create porous structures that can host Li^+ ions. In fact, the trend is toward achieving complex carbon/metal oxide nanostructures that combine robustness, conductivity, and dispersion of the metallic species. This is for instance the case of a Co-based oxide/carbon nanotube composite derived from ZIF-67 [130]. Li-S batteries represent the next generation of batteries. The anode contains lithium, whereas the cathode is made of sulfur. As described before for electrocatalyst, the surface area of the electrode is important. Again here, MOFs have been used as template to form porous carbon and serve as support for sulfur. This is for instance the case of an MOF-5 derived porous carbon [131]. MOF can also be used directly as the separator for the electrochemical system. This is to address one of the key challenges of the Li-S batteries, i.e., the polysulfide shuttling effect. Polysulfides, formed on reaction between Li^+ and the sulfur-based electrodes, can dissolve in some electrolytes leading to them not being able to react further after formation and causing a capacity loss. MOFs can be used directly to form the porous separator. They act as molecular sieve and prevent the polysulfides from reaching the anode compartment based on size effect [132]. We note that using MOFs or MOF-derived materials as hosts for sulfur can also limit the shuttling effect by immobilizing the polysulfides in their structure.

3.2.3 *Thermal Storage*

Similar to gas storage, thermal storage in MOFs (or other materials) takes advantage of the sorption potential of these porous systems. Specifically, adsorbents can be used to build heat pumps or refrigerators/chillers. This is owing to the fact that adsorption is an exothermic process; hence, heat is released on sorption of vapors. Conversely, desorption is an endothermic process, leading to a cooling effect. Broadly speaking, an adsorbent-based heat pump/chiller is made of two compartments (Fig. 6.20): one that contains the adsorbent and is called the generator, the other is where the adsorbate or refrigerant (e.g., water, ammonia, ethanol) is evaporated or condensed and is called the receiver. At low temperature, the adsorbent is loaded. It is then heated, which causes desorption of the

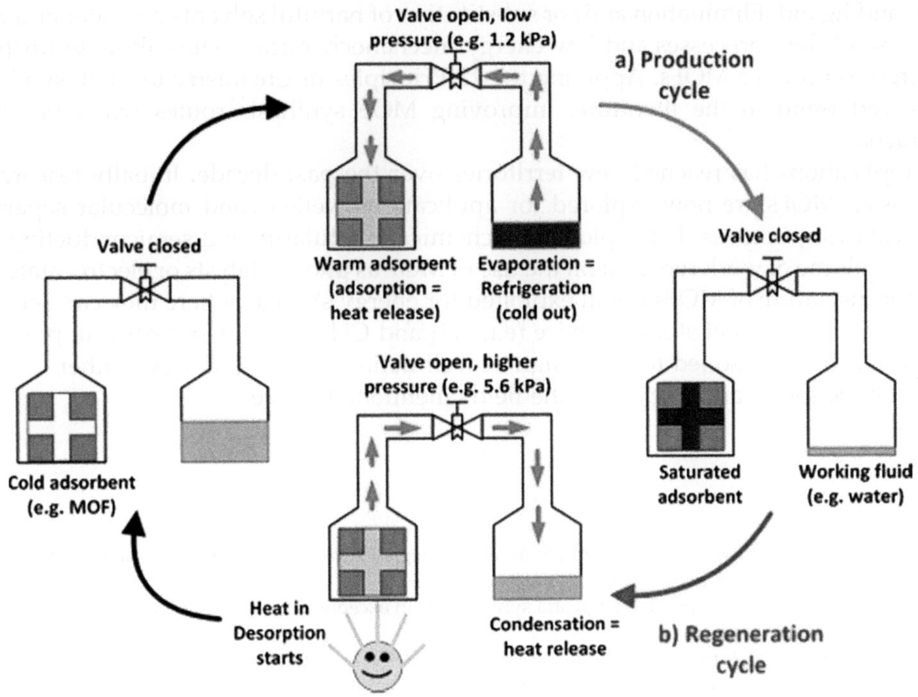

FIGURE 6.20 Illustration of the basic principle for adsorption chillers or heat pumps. *Reprinted with permission from S. K. Henninger, F. Jeremias, H. Kummer, C. Janiak, MOFs for Use in Adsorption Heat Pump Processes, European Journal of Inorganic Chemistry 2012 (2012) 2625–2634. Copyright (2011), John Wiley and Sons.*

vapor. The vapor condenses in the receiver thereby releasing heat. Conversely, when the generator is cooled down, adsorption occurs, and therefore the refrigerant vaporizes, which leads to cooling. Overall, adsorption-based thermal storage calls for porous materials that

- exhibit large working capacity for the refrigerant. This aspect can be satisfied owing to the porosity and chemistry of MOFs. In fact, a number of studies have been reported on NH_3 adsorption in MOFs and more recently on water adsorption in MOFs [85].
- can adsorb the refrigerant at relatively low vapor pressure. This can be facilitated in MOFs, which allow chemical interactions with the refrigerant (as well as strong physisorption interactions). It should be noted that such interactions should not be too strong, not to increase too much the energy needed to desorb the molecules. Similar to gas storage, modeling studies have a key role to play here in predicting the adsorption and desorption performance of MOFs for a given refrigerant under set conditions and identifying the best structures.
- display high thermal conductivity and be easily processed. The goal here is to allow efficient heat transfer. This is in part dictated by the nature of the material and the possibility to shape it so that it efficiently coats the surface of the heat exchanger. The thermal conductivity of MOFs is expected to be lower than for metals. Yet, the processability of MOFs is increasingly promising with studies reporting the ability to form films, monoliths, and nanostructures of MOFs [85].
- exhibit good recyclability. The material should be robust and withstand many sorption/desorption cycles. This is linked to the chemical and structural stability of the materials and also the conditions required to run the process. Lower desorption temperatures are expected to be less damaging. For this reason, the design of MOF structures exhibiting moderate adsorption enthalpies is important.

MOFs typically perform better in terms of working capacity compared with the benchmarks in the field, i.e., silica gel and zeolites [89,133]. Examples of structures that have been tested so far for thermal storage include, among others, MIL-100/101 series, MIL-53, HKUST-1, and ZIFs.

4. CONCLUSIONS

As illustrated through the first part of this chapter, recent efforts have been directed toward developing scalable sustainable MOF synthesis routes. In particular, scientists and engineers have investigated the use of nontoxic metal ions, counterions, and ligand. Elimination and/or substitution of harmful solvents by water and renewable solvents are possible. The use of flow processes and low energy mechanochemical routes allow to further reduce the cost and energy required to produce MOFs. Applying the 12 Principles of Chemistry to MOF synthesis is a recent yet increasingly observed trend in the literature. Improving MOF synthesis routes will favor implementation of MOF-based products.

The range of applications has reached new territories over the past decade. Initially researched for their high and tuneable porosity, MOFs are now explored for applications well-beyond molecular separation, gas storage, or heterogeneous catalysis, in areas that exploit their chemical modularity and semiconducting property. This has led to an increasing volume of work reported on the use of MOF as photocatalysts or electrocatalysts for solar energy conversion. The porous nature of MOFs is still exploited for energy storage, where they can act as support (e.g., for batteries and supercapacitors), fuel storage device (e.g., H_2 and CH_4 storage), or heat pump/chillers. The ordered nature of MOF structures has also led to the design of well-defined nanostructures—either carbon-based or metal oxide–based—that have found applications in the fields mentioned above.

References

[1] O.M. Yaghi, H.L. Li, C. Davis, D. Richardson, T.L. Groy, Synthetic strategies, structure patterns, and emerging properties in the chemistry of modular porous solids, Accounts of Chemical Research 31 (1998) 474–484.

[2] H. Li, M. Eddaoudi, M. O'Keeffe, O.M. Yaghi, Design and synthesis of an exceptionally stable and highly porous metal-organic framework, Nature 402 (1999) 276–279.

[3] H.C. Zhou, J.R. Long, O.M. Yaghi, Introduction to Metal-Organic Frameworks, Chemical Reviews 112 (2012) 673–674.

[4] S.Q. Ma, H.C. Zhou, Gas storage in porous metal-organic frameworks for clean energy applications, Chemical Communications 46 (2010) 44–53.

[5] M.S. Denny, J.C. Moreton, L. Benz, S.M. Cohen, Metal-organic frameworks for membrane-based separations, Nature Reviews Materials 1 (2016) 16078.

[6] J. Lee, O.K. Farha, J. Roberts, K.A. Scheidt, S.T. Nguyen, J.T. Hupp, Metal-organic framework materials as catalysts, Chemical Society Reviews 38 (2009) 1450–1459.

[7] T. Zhang, W.B. Lin, Metal-organic frameworks for artificial photosynthesis and photocatalysis, Chemical Society Reviews 43 (2014) 5982–5993.

[8] P. Horcajada, T. Chalati, C. Serre, B. Gillet, C. Sebrie, T. Baati, J.F. Eubank, D. Heurtaux, P. Clayette, C. Kreuz, J.S. Chang, Y.K. Hwang, V. Marsaud, P.N. Bories, L. Cynober, S. Gil, G. Ferey, P. Couvreur, R. Gref, Porous metal-organic-framework nanoscale carriers as a potential platform for drug delivery and imaging, Nature Materials 9 (2010) 172–178.

[9] L. Wang, Y.Z. Han, X. Feng, J.W. Zhou, P.F. Qi, B. Wang, Metal-organic frameworks for energy storage: Batteries and supercapacitors, Coordination Chemistry Reviews 307 (2016) 361–381.

[10] P.A. Julien, C. Mottillo, T. Friscic, Metal-organic frameworks meet scalable and sustainable synthesis, Green Chemistry 19 (2017) 2729–2747.

[11] J.Y. Chen, K. Shen, Y.W. Li, Greening the Processes of Metal-Organic Framework Synthesis and their Use in Sustainable Catalysis, ChemSusChem 10 (2017) 3165–3187.

[12] H. Reinsch, "Green" Synthesis of Metal-Organic Frameworks, European Journal of Inorganic Chemistry (2016) 4290–4299.

[13] J.W. Ren, X. Dyosiba, N.M. Musyoka, H.W. Langmi, M. Mathe, S.J. Liao, Review on the current practices and efforts towards pilot-scale production of metal-organic frameworks (MOFs), Coordination Chemistry Reviews 352 (2017) 187–219.

[14] IUPAC, Green Chemistry Directory, 2016. http://www.incaweb.org/transit/iupacgcdir/overview.htm.

[15] W.J.C. Anastas P. T, Oxford University Press, 2000, p. 132.

[16] W. Leitner, Green chemistry - Theory and practice, Science 284 (1999) 1780–1781.

[17] P.A.J. Tomislav Friščić, C. Mottillo, Environmentally-Friendly Designs and Syntheses of Metal-Organic Frameworks (MOFs), American Chemical Society 1186 (2014) 161–183. Chap 9.

[18] M. Gaab, N. Trukhan, S. Maurer, R. Gummaraju, U. Muller, The progression of Al-based metal-organic frameworks - From academic research to industrial production and applications, Microporous and Mesoporous Materials 157 (2012) 131–136.

[19] C.T. He, L. Jiang, Z.M. Ye, R. Krishna, Z.S. Zhong, P.Q. Liao, J.Q. Xu, G.F. Ouyang, J.P. Zhang, X.M. Chen, Exceptional Hydrophobicity of a Large-Pore Metal-Organic Zeolite, Journal of the American Chemical Society 137 (2015) 7217–7223.

[20] S.S.Y. Chui, S.M.F. Lo, J.P.H. Charmant, A.G. Orpen, I.D. Williams, A chemically functionalizable nanoporous material [Cu-3(TMA)(2)(H2O)(3)](n), Science 283 (1999) 1148–1150.

[21] A.U. Czaja, N. Trukhan, U. Muller, Industrial applications of metal-organic frameworks, Chemical Society Reviews 38 (2009) 1284–1293.

[22] U. Mueller, M. Schubert, F. Teich, H. Puetter, K. Schierle-Arndt, J. Pastre, Metal-organic frameworks - prospective industrial applications, Journal of Materials Chemistry 16 (2006) 626–636.

[23] U. Mueller, H. Puetter, M. Hesse, M. Schubert, Method for electrochemical production of a crystalline porous metal organic skeleton material, American Patent No. US7968739B2, 2011.

[24] F. Ren, B.C. Yang, J. Cai, Y.D. Jiang, J. Xu, S. Wang, Toxic effect of zinc nanoscale metal-organic frameworks on rat pheochromocytoma (PC12) cells in vitro, Journal of Hazardous Materials 271 (2014) 283–291.

[25] M. Filippousi, S. Turner, K. Leus, P.I. Siafaka, E.D. Tseligka, M. Vandichel, S.G. Nanaki, I.S. Vizirianakis, D.N. Bikiaris, P. Van der Voort, G. Van Tendeloo, Biocompatible Zr-based nanoscale MOFs coated with modified poly (epsilon-caprolactone) as anticancer drug carriers, International Journal of Pharmaceutics 509 (2016) 208–218.

[26] D. DeSantis, J.A. Mason, B.D. James, C. Houchins, J.R. Long, M. Veenstra, Techno-economic Analysis of Metal-Organic Frameworks for Hydrogen and Natural Gas Storage, Energy and Fuel 31 (2017) 2024–2032.
[27] M. Rose, D. Weber, B.V. Lotsch, R.K. Kremer, R. Goddard, R. Palkovits, Biogenic metal-organic frameworks: 2,5-Furandicarboxylic acid as versatile building block, Microporous and Mesoporous Materials 181 (2013) 217–221.
[28] A.C. Dreischarf, M. Lammert, N. Stock, H. Reinsch, Green Synthesis of Zr-CAU-28: Structure and Properties of the First Zr-MOF Based on 2,5-Furandicarboxylic Acid, Inorganic Chemistry 56 (2017) 2270–2277.
[29] M.K. Smith, S.R. Angle, B.H. Northrop, Preparation and Analysis of Cyclodextrin-Based Metal-Organic Frameworks: Laboratory Experiments Adaptable for High School through Advanced Undergraduate Students, Journal of Chemical Education 92 (2015) 368–372.
[30] A. Maneffa, P. Priecel, J.A. Lopez-Sanchez, Biomass-Derived Renewable Aromatics: Selective Routes and Outlook for p-Xylene Commercialisation, ChemSusChem 9 (2016) 2736–2748.
[31] W.P.R. Deleu, I. Stassen, D. Jonckheere, R. Ameloot, D.E. De Vos, Waste PET (bottles) as a resource or substrate for MOF synthesis, Journal of Materials Chemistry A 4 (2016) 9519–9525.
[32] S.H. Lo, D.S. Raja, C.W. Chen, Y.H. Kang, J.J. Chen, C.H. Lin, Waste polyethylene terephthalate (PET) materials as sustainable precursors for the synthesis of nanoporous MOFs, MIL-47, MIL-53(Cr, Al, Ga) and MIL-101(Cr), Dalton Transactions 45 (2016) 9565–9573.
[33] J.W. Ren, X. Dyosiba, N.M. Musyoka, H.W. Langmi, B.C. North, M. Mathe, M.S. Onyango, Green synthesis of chromium-based metal-organic framework (Cr-MOF) from waste polyethylene terephthalate (PET) bottles for hydrogen storage applications, International Journal of Hydrogen Energy 41 (2016) 18141–18146.
[34] M. Sanchez-Sanchez, N. Getachew, K. Diaz, M. Diaz-Garcia, Y. Chebude, I. Diaz, Synthesis of metal-organic frameworks in water at room temperature: salts as linker sources, Green Chemistry 17 (2015) 1500–1509.
[35] R.K. Henderson, C. Jimenez-Gonzalez, D.J.C. Constable, S.R. Alston, G.G.A. Inglis, G. Fisher, J. Sherwood, S.P. Binks, A.D. Curzons, Expanding GSK's solvent selection guide - embedding sustainability into solvent selection starting at medicinal chemistry, Green Chemistry 13 (2011) 854–862.
[36] GKS, Solvent Selection Guide, 2009. http://solventguide.gsk.com/.
[37] C.J. Clarke, W.C. Tu, O. Levers, A. Brohl, J.P. Hallett, Green and Sustainable Solvents in Chemical Processes, Chemical Reviews 118 (2018) 747–800.
[38] J.F. Zhang, G.B. White, M.D. Ryan, A.J. Hunt, M.J. Katz, Dihydrolevoglucosenone (Cyrene) As a Green Alternative to N,N-Dimethylformamide (DMF) in MOF Synthesis, ACS Sustainable Chemistry and Engineering 4 (2016) 7186–7192.
[39] J. Sherwood, M. De Bruyn, A. Constantinou, L. Moity, C.R. McElroy, T.J. Farmer, T. Duncan, W. Raverty, A.J. Hunt, J.H. Clark, Dihydrolevoglucosenone (Cyrene) as a bio-based alternative for dipolar aprotic solvents, Chemical Communications 50 (2014) 9650–9652.
[40] E.R. Parnham, R.E. Morris, Ionothermal synthesis of zeolites, metal-organic frameworks, and inorganic-organic hybrids, Accounts in Chemical Research 40 (2007) 1005–1013.
[41] D.N. Dybtsev, H. Chun, K. Kim, Three-dimensional metal-organic framework with (3,4)-connected net, synthesized from an ionic liquid medium, Chemical Communications (2004) 1594–1595.
[42] G.A.V. Martins, P.J. Byrne, P. Allan, S.J. Teat, A.M.Z. Slawin, Y. Li, R.E. Morris, The use of ionic liquids in the synthesis of zinc imidazolate frameworks, Dalton Transactions 39 (2010) 1758–1762.
[43] N.R. Baral, A. Shah, Techno-economic analysis of cellulose dissolving ionic liquid pretreatment of lignocellulosic biomass for fermentable sugars production (vol 10, pg 70, 2016), Biofuels, Bioproducts and Biorefining 10 (2016), 664-664.
[44] B.S. Wang, L. Qin, T.C. Mu, Z.M. Xue, G.H. Gao, Are Ionic Liquids Chemically Stable? Chemical Reviews 117 (2017) 7113–7131.
[45] P.T.P. Thi, C.W. Cho, Y.S. Yun, Environmental fate and toxicity of ionic liquids: A review Water Research 44 (2010) 352–372.
[46] I.A. Ibarra, P.A. Bayliss, E. Perez, S.H. Yang, A.J. Blake, H. Nowell, D.R. Allan, M. Poliakoff, M. Schroder, Near-critical water, a cleaner solvent for the synthesis of a metal-organic framework, Green Chemistry 14 (2012) 117–122.
[47] C.A. Grande, R. Blom, A. Spjelkavik, V. Moreau, J. Payet, Life-cycle assessment as a tool for eco-design of metal-organic frameworks (MOFs), Sustainable Materials Technology 14 (2017) 11–18.
[48] J.B. Lin, R.B. Lin, X.N. Cheng, J.P. Zhang, X.M. Chen, Solvent/additive-free synthesis of porous/zeolitic metal azolate frameworks from metal oxide/hydroxide, Chemical Communications 47 (2011) 9185–9187.
[49] T. Friscic, L. Fabian, Mechanochemical conversion of a metal oxide into coordination polymers and porous frameworks using liquid-assisted grinding (LAG), CrystEngComm 11 (2009) 743–745.
[50] M.J. Cliffe, C. Mottillo, R.S. Stein, D.K. Bucar, T. Friscic, Accelerated aging: a low energy, solvent-free alternative to solvothermal and mechanochemical synthesis of metal-organic materials, Chemical Science 3 (2012) 2495–2500.
[51] M. Lanchas, D. Vallejo-Sanchez, G. Beobide, O. Castillo, A.T. Aguayo, A. Luque, P. Roman, A direct reaction approach for the synthesis of zeolitic imidazolate frameworks: template and temperature mediated control on network topology and crystal size, Chemical Communications 48 (2012) 9930–9932.
[52] A. Pichon, A. Lazuen-Garay, S.L. James, Solvent-free synthesis of a microporous metal-organic framework, CrystEngComm 8 (2006) 211–214.
[53] T. Friscic, D.G. Reid, I. Halasz, R.S. Stein, R.E. Dinnebier, M.J. Duer, Ion- and Liquid-Assisted Grinding: Improved Mechanochemical Synthesis of Metal-Organic Frameworks Reveals Salt Inclusion and Anion Templating, Angewandte Chemie International Edition 49 (2010) 712–715.
[54] C.J. Adams, M.A. Kurawa, M. Lusi, A.G. Orpen, Solid state synthesis of coordination compounds from basic metal salts, CrystEngComm 10 (2008) 1790–1795.
[55] M. Rubio-Martinez, C. Avci-Camur, A.W. Thornton, I. Imaz, D. Maspoch, M.R. Hill, New synthetic routes towards MOF production at scale, Chemical Society Reviews 46 (2017) 3453–3480.
[56] T.D. Bennett, S. Cao, J.C. Tan, D.A. Keen, E.G. Bithell, P.J. Beldon, T. Friscic, A.K. Cheetham, Facile Mechanosynthesis of Amorphous Zeolitic Imidazolate Frameworks, Journal of the American Chemical Society 133 (2011) 14546–14549.
[57] T. Friscic, S.L. Childs, S.A.A. Rizvi, W. Jones, The role of solvent in mechanochemical and sonochemical cocrystal formation: a solubility-based approach for predicting cocrystallisation outcome, CrystEngComm 11 (2009) 418–426.

[58] T. Friscic, I. Halasz, P.J. Beldon, A.M. Belenguer, F. Adams, S.A.J. Kimber, V. Honkimaki, R.E. Dinnebier, Real-time and in situ monitoring of mechanochemical milling reactions (vol 5, pg 66, 2013), Nature Chemistry 5 (2013), 145-145.

[59] I. Halasz, T. Friscic, S.A.J. Kimber, K. Uzarevic, A. Puskaric, C. Mottillo, P. Julien, V. Strukil, V. Honkimaki, R.E. Dinnebier, Quantitative in situ and real-time monitoring of mechanochemical reactions, Faraday Discussions 170 (2014) 203–221.

[60] P.A. Julien, K. Uzarevic, A.D. Katsenis, S.A.J. Kimber, T. Wang, O.K. Farha, Y.C. Zhang, J. Casaban, L.S. Germann, M. Etter, R.E. Dinnebier, S.L. James, I. Halasz, T. Friscic, In Situ Monitoring and Mechanism of the Mechanochemical Formation of a Microporous MOF-74 Framework, Journal of the American Chemical Society 138 (2016) 2929–2932.

[61] K. Uzarevic, T.C. Wang, S.Y. Moon, A.M. Fidelli, J.T. Hupp, O.K. Farha, T. Friscic, Mechanochemical and solvent-free assembly of zirconium-based metal-organic frameworks, Chemical Communications 52 (2016) 2133–2136.

[62] P.J. Beldon, L. Fabian, R.S. Stein, A. Thirumurugan, A.K. Cheetham, T. Friscic, Rapid Room-Temperature Synthesis of Zeolitic Imidazolate Frameworks by Using Mechanochemistry, Angewandte Chemie International Edition 49 (2010) 9640–9643.

[63] A. Delori, T. Friscic, W. Jones, The role of mechanochemistry and supramolecular design in the development of pharmaceutical materials, CrystEngComm 14 (2012) 2350–2362.

[64] T. Friscic, New opportunities for materials synthesis using mechanochemistry, Journal of Materials Chemistry 20 (2010) 7599–7605.

[65] C. Mottillo, Y.N. Lu, M.H. Pham, M.J. Cliffe, T.O. Do, T. Friscic, Mineral neogenesis as an inspiration for mild, solvent-free synthesis of bulk microporous metal-organic frameworks from metal (Zn, Co) oxides, Green Chemistry 15 (2013) 2121–2131.

[66] F. Qi, R.S. Stein, T. Friscic, Mimicking mineral neogenesis for the clean synthesis of metal-organic materials from mineral feedstocks: coordination polymers, MOFs and metal oxide separation, Green Chemistry 16 (2014) 121–132.

[67] N. Stock, S. Biswas, Synthesis of Metal-Organic Frameworks (MOFs): Routes to Various MOF Topologies, Morphologies, and Composites, Chemical Reviews 112 (2012) 933–969.

[68] R. Banerjee, A. Phan, B. Wang, C. Knobler, H. Furukawa, M. O'Keeffe, O.M. Yaghi, High-throughput synthesis of zeolitic imidazolate frameworks and application to CO2 capture, Science 319 (2008) 939–943.

[69] J. Klinowski, F.A.A. Paz, P. Silva, J. Rocha, Microwave-Assisted Synthesis of Metal-Organic Frameworks, Dalton Transactions 40 (2011) 321–330.

[70] Y.F. Li, Y. Liu, W.Y. Gao, L.M. Zhang, W. Liu, J.J. Lu, Z. Wang, Y.J. Deng, Microwave-assisted synthesis of UIO-66 and its adsorption performance towards dyes, CrystEngComm 16 (2014) 7037–7042.

[71] Y.T. Han, M. Liu, K.Y. Li, Y. Zuo, Y.X. Wei, S.T. Xu, G.L. Zhang, C.S. Song, Z.C. Zhang, X.W. Guo, Facile synthesis of morphology and size-controlled zirconium metal-organic framework UiO-66: the role of hydrofluoric acid in crystallization, CrystEngComm 17 (2015) 6434–6440.

[72] C. McKinstry, R.J. Cathcart, E.J. Cussen, A.J. Fletcher, S.V. Patwardhan, J. Sefcik, Scalable continuous solvothermal synthesis of metal organic framework (MOF-5) crystals, Chemical Engineering Journal 285 (2016) 718–725.

[73] M. Taddei, D.A. Steitz, J.A. van Bokhoven, M. Ranocchiari, Continuous-Flow Microwave Synthesis of Metal-Organic Frameworks: A Highly Efficient Method for Large-Scale Production, Chemistry: A European Journal 22 (2016) 3245–3249.

[74] J. Gordon, H. Kazemian, S. Rohani, Rapid and efficient crystallization of MIL-53(Fe) by ultrasound and microwave irradiation, Microporous and Mesoporous Materials 162 (2012) 36–43.

[75] C. Avci-Camur, J. Troyano, J. Perez-Carvajal, A. Legrand, D. Farrusseng, I. Imaz, D. Maspoch, Aqueous production of spherical Zr-MOF beads via continuous-flow spray-drying, Green Chemistry 20 (2018) 873–878.

[76] A.S. Munn, P.W. Dunne, S.V.Y. Tang, E.H. Lester, Large-scale continuous hydrothermal production and activation of ZIF-8, Chemical Communications 51 (2015) 12811–12814.

[77] M. Gimeno-Fabra, A.S. Munn, L.A. Stevens, T.C. Drage, D.M. Grant, R.J. Kashtiban, J. Sloan, E. Lester, R.I. Walton, Instant MOFs: continuous synthesis of metal-organic frameworks by rapid solvent mixing, Chemical Communications 48 (2012) 10642–10644.

[78] M. Rubio-Martinez, M.P. Batten, A. Polyzos, K.C. Carey, J.I. Mardel, K.S. Lim, M.R. Hill, Versatile, High Quality and Scalable Continuous Flow Production of Metal-Organic Frameworks, Scientific Reports 4 (2014) 1–5.

[79] P.A. Bayliss, I.A. Ibarra, E. Perez, S.H. Yang, C.C. Tang, M. Poliakoff, M. Schroder, Synthesis of metal-organic frameworks by continuous flow, Green Chemistry 16 (2014) 3796–3802.

[80] H. Furukawa, N. Ko, Y.B. Go, N. Aratani, S.B. Choi, E. Choi, A.O. Yazaydin, R.Q. Snurr, M. O'Keeffe, J. Kim, O.M. Yaghi, Ultrahigh Porosity in Metal-Organic Frameworks, Science 329 (2010) 424–428.

[81] K. Koh, J.D. Van Oosterhout, S. Roy, A.G. Wong-Foy, A.J. Matzger, Exceptional surface area from coordination copolymers derived from two linear linkers of differing lengths, Chemical Science 3 (2012) 2429–2432.

[82] B. Garai, A. Mallick, R. Banerjee, Photochromic metal-organic frameworks for inkless and erasable printing, Chemical Science 7 (2016) 2195–2200.

[83] Y.F. Chen, X.Q. Huang, S.H. Zhang, S.Q. Li, S.J. Cao, X.K. Pei, J.W. Zhou, X. Feng, B. Wang, Shaping of Metal-Organic Frameworks: From Fluid to Shaped Bodies and Robust Foams, Journal of the American Chemical Society 138 (2016) 10810–10813.

[84] Frameworks for commercial success, Nature Chemistry 8 (2016), 987-987.

[85] C. Petit, Present and future of MOF research in the field of adsorption and molecular separation, Current Opinion in Chemical Engineering 20 (2018) 132–142.

[86] E.M. Dias, C. Petit, Towards the use of metal–organic frameworks for water reuse: a review of the recent advances in the field of organic pollutants removal and degradation and the next steps in the field, Journal of Materials Chemistry A 3 (2015) 22484–22506.

[87] A. Dhakshinamoorthy, A.M. Asiri, H. Garcia, Metal-Organic Framework (MOF) Compounds: Photocatalysts for Redox Reactions and Solar Fuel Production, Angewandte Chemie International Edition 55 (2016) 5414–5445.

[88] A. Mahmood, W. Guo, H. Tabassum, R. Zou, Metal-Organic Framework-Based Nanomaterials for Electrocatalysis, Advanced Energy Materials 6 (2016) 1600423.

[89] S. Chaemchuen, X. Xiao, N. Klomkliang, M. Yusubov, F. Verpoort, Tunable Metal–Organic Frameworks for Heat Transformation Applications, Nanomaterials 8 (2018) 661–686.

[90] A. Mahmood, W.H. Guo, H. Tabassum, R.Q. Zou, Metal-Organic Framework-Based Nanomaterials for Electrocatalysis, Advanced Energy Materials 6 (2016) 1600423–1600449.

[91] H.B. Zhang, J.W. Nai, L. Yu, X.W. Lou, Metal-Organic-Framework-Based Materials as Platforms for Renewable Energy and Environmental Applications, Joule 1 (2017) 77–107.
[92] S. Chaemchuen, X. Xiao, N. Klomkliang, M.S. Yusubov, F. Verpoort, Tunable Metal-Organic Frameworks for Heat Transformation Applications, Nanomaterials 8 (2018) 661.
[93] C.H. Hendon, D. Tiana, M. Fontecave, C. Sanchez, L. D'arras, C. Sassoye, L. Rozes, C. Mellot-Draznieks, A. Walsh, Engineering the Optical Response of the Titanium-MIL-125 Metal–Organic Framework through Ligand Functionalization, Journal of the American Chemical Society 135 (2013) 10942–10945.
[94] N.C. Burtch, H. Jasuja, K.S. Walton, Water Stability and Adsorption in Metal-Organic Frameworks, Chemical Reviews 114 (2014) 10575–10612.
[95] C. Wang, X. Liu, N. Keser Demir, J.P. Chen, K. Li, Applications of water stable metal–organic frameworks, Chemical Society Reviews 45 (2016) 5107–5134.
[96] A. Crake, Metal-organic frameworks based materials for photocatalytic CO2 reduction, Materials Science and Technology London 33 (2017) 1737–1749.
[97] C. Gomes Silva, I. Luz, F.X. LlabrésiXamena, A. Corma, H. García, Water Stable Zr–Benzenedicarboxylate Metal–Organic Frameworks as Photocatalysts for Hydrogen Generation, Chemistry – A European Journal 16 (2010) 11133–11138.
[98] Y. Horiuchi, T. Toyao, M. Saito, K. Mochizuki, M. Iwata, H. Higashimura, M. Anpo, M. Matsuoka, Visible-Light-Promoted Photocatalytic Hydrogen Production by Using an Amino-Functionalized Ti(IV) Metal–Organic Framework, The Journal of Physical Chemistry C 116 (2012) 20848–20853.
[99] A. Fateeva, P.A. Chater, C.P. Ireland, A.A. Tahir, Y.Z. Khimyak, P.V. Wiper, J.R. Darwent, M.J. Rosseinsky, A Water-Stable Porphyrin-Based Metal–Organic Framework Active for Visible-Light Photocatalysis, Angewandte Chemie International Edition 51 (2012) 7440–7444.
[100] Y. Fu, D. Sun, Y. Chen, R. Huang, Z. Ding, X. Fu, Z. Li, An Amine-Functionalized Titanium Metal–Organic Framework Photocatalyst with Visible-Light-Induced Activity for CO2 Reduction, Angewandte Chemie International Edition 51 (2012) 3364–3367.
[101] C.P. Angus Crake, K.C. Christoforidis, C. Petit, CO2 capture and photocatalytic reduction using bifunctional TiO2/MOF nanocomposites under UV–vis irradiation, Applied Catalysis B-Environmental 210 (2017) 131–140.
[102] I. Hod, P. Deria, W. Bury, J.E. Mondloch, C.-W. Kung, M. So, M.D. Sampson, A.W. Peters, C.P. Kubiak, O.K. Farha, J.T. Hupp, A porous proton-relaying metal-organic framework material that accelerates electrochemical hydrogen evolution, Nature Communications 6 (2015) 8304.
[103] C.H. Hendon, D. Tiana, A. Walsh, Conductive metal–organic frameworks and networks: fact or fantasy? Physical Chemistry Chemical Physics 14 (2012) 13120–13132.
[104] M.A. Nasalevich, M. van der Veen, F. Kapteijn, J. Gascon, Metal–organic frameworks as heterogeneous photocatalysts: advantages and challenges, CrystEngComm 16 (2014) 4919–4926.
[105] N. Kornienko, Y.B. Zhao, C.S. Kiley, C.H. Zhu, D. Kim, S. Lin, C.J. Chang, O.M. Yaghi, P.D. Yang, Metal–Organic Frameworks for Electrocatalytic Reduction of Carbon Dioxide, Journal of the American Chemical Society 137 (2015) 14129–14135.
[106] H. Zhang, Z. Ma, G. Liu, L. Shi, J. Tang, H. Pang, K. Wu, T. Takei, J. Zhang, Y. Yamauchi, J. Ye, Highly active nonprecious metal hydrogen evolution electrocatalyst: ultrafine molybdenum carbide nanoparticles embedded into a 3D nitrogen-implanted carbon matrix, NPG Asia Materials 8 (2016) e293.
[107] H.A. Lopez, A. Dhakshinamoorthy, B. Ferrer, P. Atienzar, M. Alvaro, H. Garcia, Photochemical Response of Commercial MOFs: Al2(BDC)3 and Its Use As Active Material in Photovoltaic Devices, The Journal of Physical Chemistry C 115 (2011) 22200–22206.
[108] J. Dou, Y. Li, F. Xie, X. Ding, M. Wei, Metal–Organic Framework Derived Hierarchical Porous Anatase TiO2 as a Photoanode for Dye-Sensitized Solar Cell, Crystal Growth and Design 16 (2016) 121–125.
[109] G. Férey, F. Millange, M. Morcrette, C. Serre, M.-L. Doublet, J.-M. Grenèche, J.-M. Tarascon, Mixed-Valence Li/Fe-Based Metal–Organic Frameworks with Both Reversible Redox and Sorption Properties, Angewandte Chemie International Edition 46 (2007) 3259–3263.
[110] H. Li, K. Wang, Y. Sun, C.T. Lollar, J. Li, H.-C. Zhou, Recent advances in gas storage and separation using metal–organic frameworks, Materials Today 21 (2018) 108–121.
[111] H.W. Langmi, J. Ren, B. North, M. Mathe, D. Bessarabov, Hydrogen Storage in Metal-Organic Frameworks: A Review, Electrochimica Acta 128 (2014) 368–392.
[112] V.V. Simonyan, P. Diep, J.K. Johnson, Molecular simulation of hydrogen adsorption in charged single-walled carbon nanotubes, The Journal of Chemical Physics 111 (1999) 9778–9783.
[113] T. Runčevski, M.T. Kapelewski, R.M. Torres-Gavosto, J.D. Tarver, C.M. Brown, J.R. Long, Adsorption of two gas molecules at a single metal site in a metal–organic framework, Chemical Communications 52 (2016) 8251–8254.
[114] M. Dincă, A. Dailly, Y. Liu, C.M. Brown, D.A. Neumann, J.R. Long, Hydrogen Storage in a Microporous Metal–Organic Framework with Exposed Mn2+ Coordination Sites, Journal of the American Chemical Society 128 (2006) 16876–16883.
[115] Y. He, F. Chen, B. Li, G. Qian, W. Zhou, B. Chen, Porous metal–organic frameworks for fuel storage, Coordination Chemistry Reviews 373 (2018) 167–198.
[116] Y. Peng, V. Krungleviciute, I. Eryazici, J.T. Hupp, O.K. Farha, T. Yildirim, Methane Storage in Metal–Organic Frameworks: Current Records, Surprise Findings, and Challenges, Journal of the American Chemical Society 135 (2013) 11887–11894.
[117] T. Tian, Z. Zeng, D. Vulpe, M.E. Casco, G. Divitini, P.A. Midgley, J. Silvestre-Albero, J.-C. Tan, P.Z. Moghadam, D. Fairen-Jimenez, A sol–gel monolithic metal–organic framework with enhanced methane uptake, Nature Materials 17 (2017) 174.
[118] P.D.C. Dietzel, V. Besikiotis, R. Blom, Application of metal–organic frameworks with coordinatively unsaturated metal sites in storage and separation of methane and carbon dioxide, Journal of Materials Chemistry 19 (2009) 7362–7370.
[119] J.A. Mason, M. Veenstra, J.R. Long, Evaluating metal–organic frameworks for natural gas storage, Journal of Chemical Science 5 (2014) 32–51.
[120] D. Alezi, Y. Belmabkhout, M. Suyetin, P.M. Bhatt, Ł.J. Weseliński, V. Solovyeva, K. Adil, I. Spanopoulos, P.N. Trikalitis, A.-H. Emwas, M. Eddaoudi, MOF Crystal Chemistry Paving the Way to Gas Storage Needs: Aluminum-Based soc-MOF for CH4, O2, and CO2 Storage, Journal of the American Chemical Society 137 (2015) 13308–13318.
[121] J.-M. Lin, C.-T. He, Y. Liu, P.-Q. Liao, D.-D. Zhou, J.-P. Zhang, X.-M. Chen, A Metal–Organic Framework with a Pore Size/Shape Suitable for Strong Binding and Close Packing of Methane, Angewandte Chemical International Edition 55 (2016) 4674–4678.

[122] J.A. Mason, J. Oktawiec, M.K. Taylor, M.R. Hudson, J. Rodriguez, J.E. Bachman, M.I. Gonzalez, A. Cervellino, A. Guagliardi, C.M. Brown, P.L. Llewellyn, N. Masciocchi, J.R. Long, Methane storage in flexible metal–organic frameworks with intrinsic thermal management, Nature 527 (2015) 357.

[123] T. Tian, Z.X. Zeng, D. Vulpe, M.E. Casco, G. Divitini, P.A. Midgley, J. Silvestre-Albero, J.C. Tan, P.Z. Moghadam, D. Fairen-Jimenez, A sol-gel monolithic metal-organic framework with enhanced methane uptake, Nature Materials 17 (2018) 174–179.

[124] J.W. Long, B. Dunn, D.R. Rolison, H.S. White, Three-Dimensional Battery Architectures, Chemical Reviews 104 (2004) 4463–4492.

[125] B. Liu, H. Shioyama, T. Akita, Q. Xu, Metal-Organic Framework as a Template for Porous Carbon Synthesis, Journal of the American Chemical Society 130 (2008) 5390–5391.

[126] R.R. Salunkhe, Y.V. Kaneti, J. Kim, J.H. Kim, Y. Yamauchi, Nanoarchitectures for Metal–Organic Framework-Derived Nanoporous Carbons toward Supercapacitor Applications, Accounts of Chemical Research 49 (2016) 2796–2806.

[127] L.-F. Chen, Y. Lu, L. Yu, X.W. Lou, Designed formation of hollow particle-based nitrogen-doped carbon nanofibers for high-performance supercapacitors, Energy and Environmental Science 10 (2017) 1777–1783.

[128] H. Hu, B.Y. Guan, X.W. Lou, Construction of Complex CoS Hollow Structures with Enhanced Electrochemical Properties for Hybrid Supercapacitors, Chemistry 1 (2016) 102–113.

[129] D. Sheberla, J.C. Bachman, J.S. Elias, C.-J. Sun, Y. Shao-Horn, M. Dincă, Conductive MOF electrodes for stable supercapacitors with high areal capacitance, Nature Materials 16 (2016) 220.

[130] Y.M. Chen, L. Yu, X.W. Lou, Hierarchical Tubular Structures Composed of Co3O4 Hollow Nanoparticles and Carbon Nanotubes for Lithium Storage, Angewandte Chemie International Edition 55 (2016) 5990–5993.

[131] G. Xu, B. Ding, L. Shen, P. Nie, J. Han, X. Zhang, Sulfur embedded in metal organic framework-derived hierarchically porous carbon nanoplates for high performance lithium–sulfur battery, Journal of Materials Chemistry A 1 (2013) 4490–4496.

[132] S. Bai, X. Liu, K. Zhu, S. Wu, H. Zhou, Metal–organic framework-based separator for lithium–sulfur batteries, Nature Energy 1 (2016) 16094.

[133] S.K. Henninger, F. Jeremias, H. Kummer, C. Janiak, MOFs for Use in Adsorption Heat Pump Processes, European Journal of Inorganic Chemistry 2012 (2012) 2625–2634.

CHAPTER

7

Metal–Organic Frameworks (MOFs) and MOF-Derived Porous Carbon Materials for Sustainable Adsorptive Wastewater Treatment

Zahra Abbasi[1], *Levente Cseri*[2], *Xiwang Zhang*[1], *Bradley P. Ladewig*[3,4], *Huanting Wang*[1]

[1]Department of Chemical Engineering, Monash University, Clayton, VIC, Australia; [2]School of Chemical Engineering & Analytical Science, University of Manchester, The Mill, Manchester, United Kingdom; [3]Department of Chemical Engineering, Imperial College London, London, United Kingdom; [4]Institute for Micro Process Engineering (IMVT), Karlsruhe Institute for Technology, Eggenstein-Leopoldshafen, Germany

OUTLINE

Sustainable Nanoscale Engineering
https://doi.org/10.1016/B978-0-12-814681-1.00007-2

1. INTRODUCTION TO WASTEWATER TREATMENT

1.1 Industrial Waste and Water Pollution

Clean water demand is increasing because of rapid population and industrial growth worldwide. The limited availability of new water sources makes essential the efficient and sustainable treatment and reuse of wastewater. A variety of contaminants such as organics (solvents, hydrocarbons, dyes, etc.) and heavy metals (cadmium, chromium, copper, etc.) are discharged from industries, which need to be treated before entering the water bodies. Efficient and low-cost technologies are required to decontaminate the polluted waters [1,2].

Most of the industries generate certain amount of wastewater containing various pollutants. One of the biggest contaminated waste streams is in the textile industry where aqueous solutions of organic dyes are generated in the dying process. If inefficiently treated, the dyes can cause serious pollution of natural waters. They cause coloration of water bodies even in low concentration, decreasing the sunlight reaching the aqueous plants. These dyes that are often toxic for microbial organisms also decrease the dissolved oxygen levels in water. Dyes can be also present in the wastewater of industrial laundries alongside with various surfactants and softeners. Solvents and other organic compounds can be found in the wastewater from the pharmaceutical and fine chemical manufacturing. Heavy metals have nondegradable nature, and they exist in the wastewaters from mining activities, tannery, steel, and battery industries, as well as thermal power plants. They impose serious environmental problems and threats on human life and ecosystem [3]. There are countless other examples of pollutants in wastewater streams, which have to be efficiently removed before release to the environment to meet sustainability requirements.

1.2 Current State of Wastewater Treatment

Owing to the variety of contaminants present in different wastewaters, there is no single generally applicable superior strategy for water treatment and reuse. The main broad approaches include physical, chemical, and biological methods. Regarding the removal of organic pollutants from wastewaters, biological treatment is considered the most cost-effective approach [4]. However, the presence of complicated and toxic chemicals, dyes, and heavy metals may impose limitation for utilizing this treatment process as these hazardous contaminants are not biodegradable [5]. Therefore, a wide range of physicochemical techniques such as filtration, coagulation, and flocculation, advanced oxidation methods such as Fenton's reaction, ozone, and UV, and chemical oxidation using chlorine, ozone, and hydrogen peroxide are employed for the treatment of industrial wastewaters. Nevertheless, the usage of these techniques is restricted in practical applications as they are mostly expensive or technically complicated. For instance, using filtration method is rarely enough to target the discharge limits within the environmental standards. In addition, coagulation and flocculation produce a large amount of sludge, which causes handling and disposal problems. In addition, in chemical oxidation, a large amount of chemical and hazardous substance is required. Finally, advanced oxidation imposes high investment costs [5].

The major disadvantage of physicochemical processes is related to the high cost, low efficiency, interference with other wastewater constituents, as well as handling of the generated waste. Biological treatment involves the biological activity for the removal of organic substances [6]. However, the biological treatment process does not always provide satisfactory results for the treatment of industrial wastewater because many organic pollutants are toxic or resistant to biological treatment [7–10]. Therefore, it is essential to consider innovative water treatment techniques with high performance, low cost, and that can operate sustainably. The main techniques that are used for the removal of heavy metals form effluents are ion exchange, membrane process, precipitation, and electrocatalytic technology [11]. However, most of these technologies are expensive and require complicated approaches and equipment. Therefore, there is a need for an effective, simple, and inexpensive treatment method that can be developed, installed, and utilized easily everywhere.

The efficiency of physical adsorption is determined by adsorption capacity of the adsorbents, selectivity for specific compounds, durability, and regenerability of the adsorbents [12]. Adsorption has been considered to be superior to other decontamination techniques because of its low cost, wide range of applications, simplicity of design, easy operation, insensitivity to toxic pollutants, low harmful secondary products, and easy regeneration of the adsorbents [12,13]. This technique is specifically efficient if adsorbents with suitable characteristics are employed. For this reason, the investigation of adsorbents' textural, physicochemical, and morphological properties is highly important to provide useful information for selecting appropriate adsorbents for future applications. Adsorption is an economically feasible technology for the removal of hazardous toxic metals from wastewater [14]. The adsorption process has the advantages of flexible design and easy operation. Moreover, adsorption is in most cases reversible, and the adsorbents can be recycled by appropriate desorption technique [11].

This chapter focuses on two specific classes of materials that have been developed extensively in recent years for sustainable wastewater adsorptive treatment, namely metal–organic frameworks (MOFs) and porous carbons derived from them. The selection of MOFs is primarily due to the significant research interest in this field, especially relating to the chemical and structural tailorability of these materials. The authors consider these materials are promising candidates for highly specialized water treatment where a highly customized adsorbent (for example targeting the removal or persistent organic chemicals or endocrine disruptors at low concentration) is required. MOFs may be particularly suited to this task, as the pore opening and pore size can be tailored toward the target size of the adsorbate, which the pore chemistry can be likewise compatibilized. From the perspective of nanoporous carbons derived from MOFs, these materials have special interest because they usually retain the attractive porosity and pore structure of MOFs while also possessing enhanced hydrophobicity and aqueous stability. In the perspective of the scope of this book with a focus on sustainability, using readily available, low cost, biological precursors greatly enhance the sustainability of water treatment using porous carbons.

2. ADSORPTION FOR WASTEWATER TREATMENT

2.1 Theory of Adsorption

Adsorption refers to the accumulation of a compound at the interface between two phases: liquid–solid interface or gas–solid interface. The substance that builds up at the interface is called adsorbate, and the solid on which adsorption process takes place is adsorbent [15]. Adsorption in water treatment always concerns the aqueous liquid phase and a solid adsorbent; therefore, the discussion here will be limited to liquid–solid interface. Adsorption can be considered as a chemical or physical process according to the nature of interactions between adsorbates and adsorbents. Chemical adsorption or chemisorption occurs through chemical reaction that results in primary bond (including covalent, dative, and ionic bonds) formation between molecules of adsorbate to adsorbent surface. Chemical adsorption is usually irreversible or the chemical equilibrium strongly favors the adsorption. In this case, regeneration is carried out with chemical treatment or calcination [12,13]. In case of physical adsorption, the adsorbates are attached to the pores of the solid adsorbents by secondary bonds including hydrogen bonding, dipole–dipole, and van der Waals interaction. Therefore, physical adsorption is reversible in most cases, and the adsorbent can be recovered by simple methods such as solvent exchange or physical treatments such as temperature swing [16].

The efficiency of adsorption is typically described by adsorption equilibrium isotherms as well as kinetic models. Adsorption isotherm demonstrates the amount of material that is adsorbed per unit mass of adsorbent as a function of the equilibrium concentration of the adsorbate. In adsorption experiments, a specified mass of adsorbent reaches equilibrium by certain volume of solution at specific concentration of an adsorbate, and the mass balance is introduced in Eq. (7.1) [17].

$$Q_e = \frac{(C_0 - C_e) \times V}{m} \tag{7.1}$$

where C_0 and C_e are the initial and equilibrium concentration of the adsorbate in solution, V is the solution volume, and m stands for the mass of adsorbent while Q_e is the equilibrium adsorption amount. The maximum of Q_e measured in saturated solution is the adsorption capacity of the adsorbent for the solute.

From the equilibrium studies, adsorption isotherms and their constants are determined. These constant values express the surface properties and affinity of the adsorbents. To understand, interpret, and predict the extent of the adsorption for particular applications, it is required to realize the relationship between equilibrium data and theoretical or practical equations [18]. Among several adsorption isotherms that have been used for the adsorption process, Langmuir and Freundlich models are considered the most common because of their simplicity and capability to describe experimental results in a broad range of concentrations [19]. Kinetic studies are performed to have a better understanding of adsorption mechanisms. The kinetic studies determine the rate of adsorption, and kinetic models are beneficial to determine the feasible application of the adsorbent in adsorption process. A slow adsorption rate shows significant limitations and issues for the practical applications.

The adsorption mechanism includes three stages, namely film diffusion, pore diffusion, and adhesion of the solute molecules to the adsorbent surface. Film diffusion includes the diffusion of the adsorbate molecule through the films surrounding the particle's surface. Pore diffusion involves the penetration of the molecules into the pores to the adsorption site. When the adsorbate molecule reaches the surface of the pore, adhesion occurs and the molecule attaches to the surface [20].

2.2 Influencial Factors on Adsorption

The adsorption capacity of different adsorbents depends on a number of parameters. These parameters include adsorbate features such as molecular size, functional groups, polarity, and solubility, as well as adsorbent properties such as surface area, pore structure, particle size, and surface chemistry [21,22]. It is worth noting that the process conditions such as solution pH, ionic strength, temperature, and contact time also play a key role in the adsorption process [23,24].

The molecular size of the adsorbates is important in terms of surface interactions and accessibility of the adsorbent pores [25]. Bigger molecules tend to get adsorbed more efficiently as the bigger size also means better polarizability which in turn gives rise to stronger dispersion forces with the surface. On the other hand, the access of big molecules to the adsorbent pores is restricted due to steric effects that can lead to slow, diffusion limited adsorption, pore blockage, and significant reduction of the effective surface area. For example, Lachheb et al. have reported that the weak adsorption of Congo red dye in comparison with other dyes such as methylene blue (MB) or methyl red is attributed to the large steric hindrance because of large aromatic ensembles such as central biphenyl groups and symmetric naphthyl groups [26]. Moreover, Valix et al. reported that there is steric hindrance related to the size of the acid dye molecules that restrict the diffusion of the molecules into the porous structure of the carbon material [27].

The functional groups of the adsorbate determine many of its properties including polarity, solubility, ionic charge, acidic–basic character, and hydrogen-bonding properties. The acidity and basicity of an adsorbate is described by the logarithmic equilibrium constant (p*K*a). As the deprotonated or protonated form of a molecule show distinct adsorption properties from the parent molecule, the p*K*a of adsorbates has to be considered in process design. Generally, molecules get adsorbed strongly on adsorbents with complementary functional groups: positive and negative ionic charge, acidic and basic behavior, or hydrogen bond donor and acceptor functional groups. Pereira et al. [28] studied the effect of surface chemical groups on activated carbons for dye adsorption. They reported that surface chemistry of activated carbon remarkably influences the adsorption of dyes from solution. It was found that for anionic dyes such as reactive, direct, and acid dyes, the surface basicity plays an important role in dye adsorption. The main adsorption mechanism was the interaction between the oxygen-free Lewis basic sites and the free electrons of dye molecule. For the anionic dyes, it was shown that oxygen containing groups that are acidic had a negative impact on the adsorption uptake of these dyes. While, for the adsorption of cationic (basic) dyes, the acid oxygen-containing surface groups (mainly carboxylic) exhibited a positive effect on adsorption.

Radovic et al. reported that π-π dispersion interactions seems to be dominant in the adsorption of aromatic solutes over the carbon compounds [29]. They concluded that both electrostatic and dispersive interactions can affect and control the equilibrium uptakes of weak aromatic systems. Another study by Ghaffar and Younis [30] shows that for the adsorption of MB over carbon nanotubes (CNTs), π-π electron donor–acceptor, hydrophobic interaction, hydrogen bonding, electrostatic and covalent interactions, and van der Waals forces might affect the adsorption process simultaneously or individually.

An adsorbent material is required to have high internal volume accessible to the compounds that are going to be removed from the solvent. Surface area, pore size distribution, and the nature of the pores highly affect the adsorption process. The suitable pore size distribution is required for providing the adsorption sites and channels for transporting the adsorbates [31,32]. The small molecules such as phenol have the ability to access micropores (<2 nm), natural polymeric compounds such as proteins can access mesopores (2–50 nm), and bacteria are able to access macropores (>50 nm) only [31].

The pH conditions of a solution affect the chemical properties on the surface of the adsorbents as well as the forms of the adsorbate molecules. The surface chemistry of the adsorbent can change depending on the interaction of acidic or basic species with the surface groups.

Zeta potential (ζ) is a pH-related parameter that determines the surface charge of solids in aqueous solutions. Lower pH results in more positive or less negative surface, whereas higher pH leads to more negative or less positive surface charge. This changes the zeta potential of the adsorbent and impacts the electrostatic interaction of adsorption process. For the adsorption of negatively charged molecules such as blue-106 dye, increase of the pH reduces the adsorption of molecule because of the electrostatic repulsion [33]. The point of zero charge (pzc) is the pH at which the surface charge is zero and is generally used to quantify or define the electrokinetic properties of the adsorbent surface.

The temperature of a solution affects the adsorption of water as well as the hydration degree of the molecules [34]. Gas-phase adsorption to unoccupied sites is always exothermic as determined by the van't Hoff equation of adsorption. In liquid phase processes involving wetted surfaces, such as water treatment, there are no truly empty adsorption sites. The adsorption of a solute from liquid phase is always associated with the desorption of solvent molecules occupying the site. Therefore, the thermodynamic effects are more complex in these processes. Endothermic behavior is commonly reported for the adsorption of dyes in solution [35]. The solution temperature also determines kinetic factors such as viscosity and diffusion rate. As a consequence, adsorption equilibria are faster achieved at higher temperatures. In addition, certain adsorbents exhibit specific thermoresponsive behavior that makes them attractive for temperature swing adsorption [16].

3. ADSORBENT MATERIALS

3.1 Current State of Adsorbent Materials

Porosity, pore geometry, and specific adsorption sites of adsorbents play a key role in efficient adsorptive removal as discussed in the previous sections. The pores of solids can be classified based on their sizes. Pore sizes below 2 nm are called micropores. The pores between 2 and 50 nm are mesopores, and those above 50 nm are considered as macropores. Fig. 7.1 illustrates micro-, meso-, and macropores in porous materials and shows how pore size can be used to exlude larger particles as well as the difference between connected and dead-end pores.

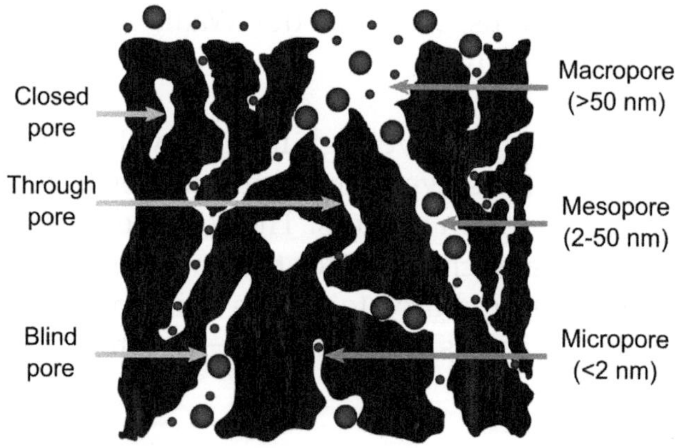

FIGURE 7.1 Schematic illustration of the pores and pore connectivity in a porous material, ranging from macropores to mesopores and micropores [36].

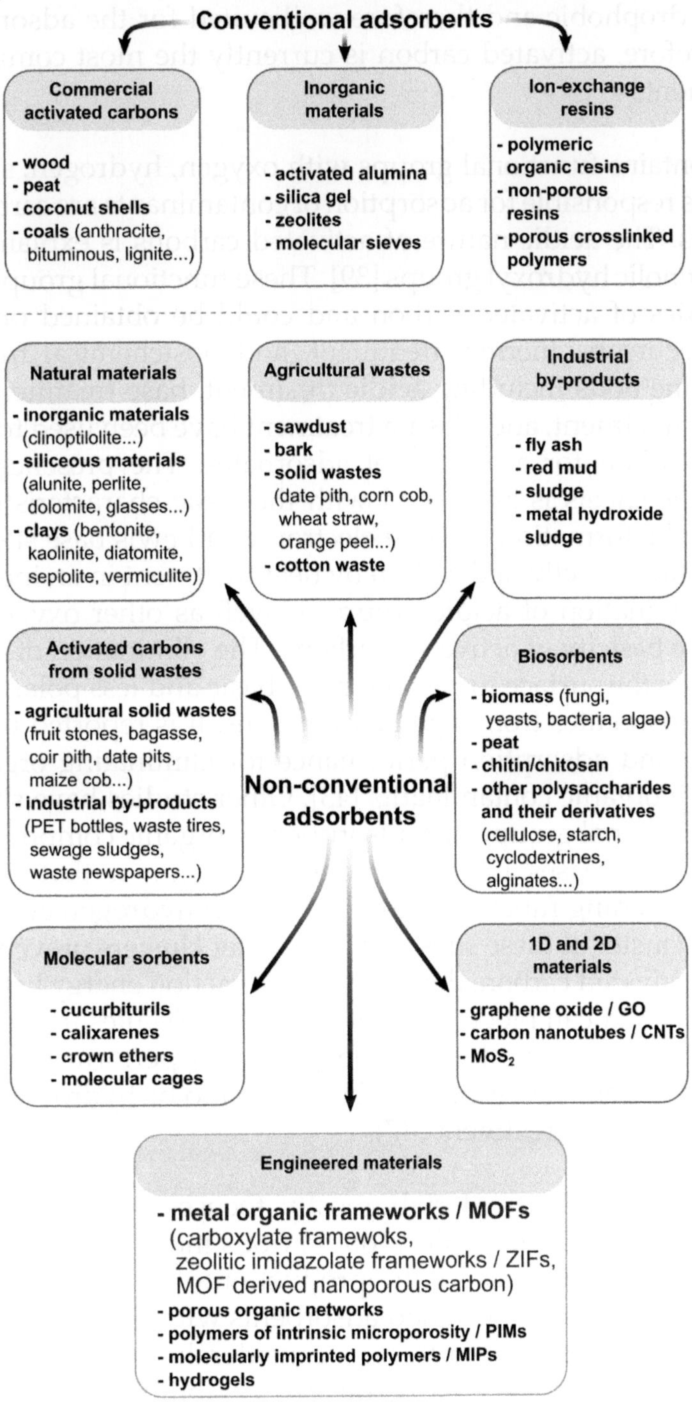

FIGURE 7.2 Overview of conventional and nonconventional adsorbent materials for water treatment based on Crini et al. expanded with further categories [38].

A variety of porous materials have been studied for adsorptive removal of hazardous substances as summarized in Fig. 7.2. Conventionally, the landscape of adsorbents has been dominated by a few materials such as activated carbon, alumina, silica, zeolites, and resins. These materials offer a reasonable compromise between cost, capacity, and selectivity, which has been the key in their success. Most untreated alumina, silica, and aluminosilicate zeolite materials are hydrophilic and have the ability to adsorb water from organic solvents [37]. On the other hand, activated carbon is hydrophobic and therefore well suited for the adsorption of organic pollutants from water. Therefore, activated carbon is currently the most common adsorbent material for wastewater treatment.

Activated carbon contains functional groups with oxygen, hydrogen, sulfur, and nitrogen. The main functional groups responsible for adsorption of contaminants are carboxyl, carbonyl, phenols, lactones, and quinones. The acidic nature of activated carbons is explained by the formation of carboxylic acid and phenolic hydroxyl groups [39]. These functional groups have significant effects on adsorption properties of activated carbon and could be obtained or modified by means of activation process, precursors, thermal treatment, and postchemical treatments [40]. Different types of modification methods including acidic treatment, base treatment, impregnation, microwave treatment, ozone treatment, and plasma treatment have been used to increase the adsorption capacity of activated carbon for a variety of adsorbates. The presence of π electrons on the condensed polyaromatic sheets is associated with the basic characteristics of activated carbons [41,42]. The reason for the formation of these electron-rich Lewis base sites is due to the removal of oxygen from the surface of activated carbon by heat treatment in an inert atmosphere. The heat treatment results in elimination of acidic groups as well as other oxygen-containing functional groups that increase the basicity of activated carbons. The elimination of oxygen from the surface of activated carbon helps the surface to become more basic and less polar. This is favorable for the adsorption of organic adsorbates from aqueous solutions. It is reported that activated carbon has much higher tendency and adsorption performance for eliminating organic compounds rather than metals and other inorganic contaminants [43]. Other studies have shown that by increasing the oxygen or acidic functional groups, the adsorption of organic compounds on activated carbon from aqueous solution decreases [35,44]. Studies have shown that water molecules can be adsorbed on oxygen-containing functional groups through hydrogen bonding and then the additional water molecules cluster at these sites [45]. The water clusters prevent the pollutant access to hydrophobic areas on activated carbon, decrease the interaction energy between contaminants and the surface of adsorbent, and can block the contaminant molecules to access the micropores [46]. Therefore, adsorbents having stable hydrophobic surfaces perform much better in removal of organic pollutants from water. However, adsorbents need to be hydrophilic enough to get wetted by water to be useful for water treatment [47].

Current research in adsorption materials mostly focuses on the production of specialized adsorbents to optimize them for the application. Various industrial, agricultural, and municipal waste streams have been considered for the production of low-cost adsorbents. On the other hand, there is also an emergence of designed adsorbents with exceptional capacity or selectivity. Recently, 1D and 2D materials have attracted considerable scientific attention as adsorbent materials due to their high surface area, good processability, and unique surface chemistry. The organic nature and surface functionality of CNTs and graphene oxide enables further functionalization for more effective application. However, these adsorbents are still in their infancy mostly because of the high costs and batch-to-batch inconsistency of their production. The example of MoS_2 nanosheets may also feed the skepticism toward some new adsorbents. Although it was proposed for dye adsorption, it was later revealed that its partial solubility makes it impractical for aqueous applications [48]. Molecular sorbents are macrocyclic molecules that are capable of adsorbing compounds in their internal cavity. The matching size of the cavity and adsorbate and the multipoint interactions make these molecules excellent candidates for the selective adsorption of pollutants [49–51]. These compounds can be applied for water treatment after immobilization on a suitable support.

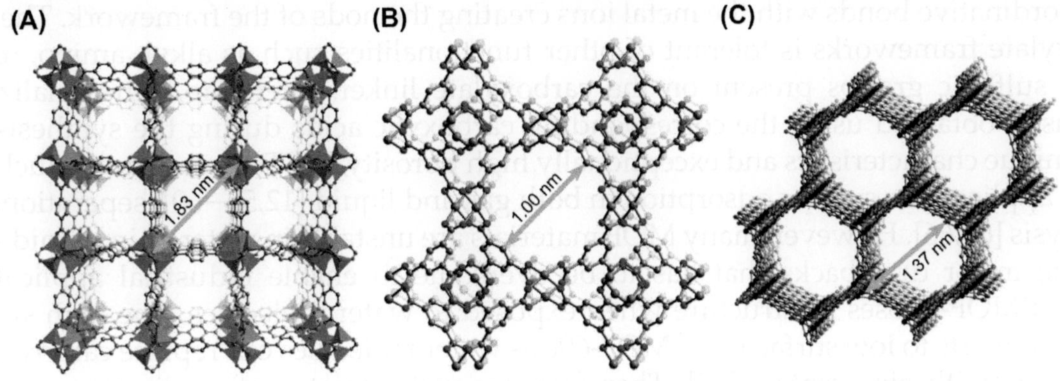

FIGURE 7.3 Molecular structures of some typical metal–organic frameworks (MOFs). (A) MOF-5, (B) Cu-BTC, and (C) CPO-27 [54].

Other nonconventional adsorbents can be referred to and grouped as engineered materials. These adsorbents can be characterized by the extensive and precise control over their porosity, pore structure, and surface chemistry. The manipulation of the properties is realized by the selection of building blocks or postsynthetic modification. Among these materials, different classes have unique properties that make them superior in an aspect over other adsorbents. Molecularly imprinted polymers (MIPs) contain molecular cavities that are complementary in shape and size to the imprinted compound. Therefore, MIPs show unparalleled selectivity in the adsorption of the imprinted compound even from complex matrices. Polymers of intrinsic microporosity are solvent-processable materials that have clear advantages in adsorbent formulation. Nevertheless, among these nonconventional materials, MOFs have arguably attracted the most research interest in the past years because of their unique properties.

3.2 Metal–Organic Frameworks

3.2.1 *Carboxylate Frameworks*

MOFs are emerging as new functional porous materials with interesting features such as high porosity, pore functionality, and open metal sites. In comparison with inorganic microporous materials such as zeolites, MOF structures have the potential for more flexible design by controlling the functionalization of the pores [52]. Therefore, MOFs have attracted much attention as nanoporous materials because of designable framework structures [53]. Fig. 7.3 shows the molecular structure of some well-known MOFs.

An MOF molecular structure includes metal atoms or a cluster of metal ions in combination with an organic linker. One of the key defining characteristics of MOFs is their porosity. MOFs are similar to zeolites in terms of crystal structures although zeolites are purely inorganic materials. In spite of the fact that MOFs and zeolites have some features in common like large internal surface areas as well as uniform pore and cavity sizes, they are different in some important aspects. For example, MOFs can be synthesized in greater variety compared with zeolites [55]. A large number of MOFs can be synthesized by varying the metal centers and organic linkers to adjust surface area as well as pore size and surface functionality. Recently reported numbers for MOF structures deposited to the Cambridge Crystallographic Data Center are around 70–75,000 [56,57]. Among these, most structures can be classified as carboxylate frameworks that contain linkers with two or more carboxylate functionalities. The carboxylate groups form strong coordinative bonds with the metal ions creating the nods of the framework. The synthesis of carboxylate frameworks is tolerant of other functionalities such as alkyl, amino, nitro, halogenic, or sulfonic groups present on the carboxylate linker. Therefore, functionalized MOFs can be easily obtained using the corresponding carboxylic acids during the synthesis. Because of their unique characteristics and exceptionally high porosity, MOFs have drawn much attention for many applications such as adsorption in both gas and liquid [12,58–60], separations [61–64], and catalysis [65,66]. However, many MOF materials are unstable in water or in humid air. This is one of the major drawbacks that has to be overcome to enable industrial applications. For example, IRMOF-1 loses its structure when exposed to water molecules. The high surface area IRMOF-1 converts to low surface area MOF-69c as water molecules can replace carboxylic groups to coordinate with zinc centers [67]. Therefore, for water treatment applications, water-stable MOFs should be selected, such as MIL-100 or UiO-66. Alternatively, functionalization of MOFs has proved to be an effective strategy to control wettability and improve water stability [68].

3.2.2 *Zeolitic Imidazolate Frameworks*

Zeolitic imidazolate frameworks (ZIFs) are a subcategory of MOFs that have permanent porosity and because of comparatively high thermal and chemical stability have the potential for development and modification [69,70]. ZIFs are composed of metal atoms including Co, Cu, Zinc, etc., which are linked by nitrogen atoms of imidazolate linkers. ZIFs and zeolites have similar frameworks in which the T-O-T bridges with T being Si, P, or Al are substituted by M-Im-M bridges by M being Zn, Cu, or Co with similar bond angle around 145 degree [71]. Different types of ZIF structures can be made by changing the metal ions as well as imidazolate organic molecules [72]. Among the ZIF structures, ZIF-8 and ZIF-11 have shown high exceptional thermal and chemical stability even in aqueous alkaline solutions and water [73–75]. Experiment showed that both ZIF-8 and ZIF-11 maintained their structures in water at 50°C for 7 days. Nevertheless, ZIF-8 was the only material that was stable in boiling water and retained the original structure after 7 days, while ZIF-11 structure was converted to another material after 3 days. Moreover, ZIF-8 was also stable in 0.1 and 8 M aqueous sodium hydroxide at 100°C for up to 24 h [73]. This proves the remarkable resistance of ZIF-8 among the MOF solids. This is attributed to the fact that imidazolate linkers are more basic compared with the carboxylate linkers, and this leads to stronger bonds between metal atoms and ligands. The other reason for superior stability of ZIF-8 in comparison with other ZIFs is its hydrophobicity [76]. ZIF-8 has the formula of $Zn(2\text{-methylimidazolate})_2$ with a sodalite-related zeolite structure that consists of six-membered ring pore window of 0.34 nm and much larger pore size of 1.14 nm. ZIF-8 has interesting characteristics that make it important for adsorption applications, including high porosity and high surface area of 1300–1600 $m^2\ g^{-1}$ [77].

3.2.3 *Nanoporous Carbons Derived From Metal–Organic Frameworks*

Nanostructured porous carbon materials are very interesting because of their unique physical and chemical properties that make them suitable to be used as sorbents in gas or liquid adsorption, catalyst support, electrodes, and fuel cells [78–80]. Because these materials have affinity to organic contaminations, they have the potential to be used in environmental applications such as adsorptive removal of hazardous contaminants [81,82]. These porous materials have high surface area and narrow pore size distribution as well as good thermal and chemical stability. The nanoporous carbons are produced by carbonization or chemical vapor deposition of carbon sources using the hard templates. They can also be made by carbonization of polymeric carbon gels using an organic–organic soft templating method [53]. Some materials are resistant to modifications such

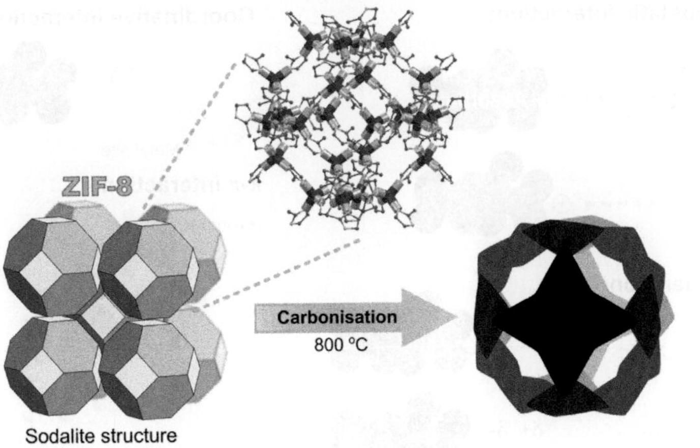

FIGURE 7.4 Direct carbonization of zeolitic imidazolate framework (ZIF)-8 at 800°C, resulting in a highly nanoporous carbon templated by the structure of the original ZIF-8 [86].

as hydrothermal carbonization; therefore, the produced carbonaceous network is nonporous exhibiting a low surface area. In particular, these materials lack microporosity that is necessary for high adsorption capacities. These materials need to be treated by traditional activation methods to increase the specific surface area and improve the pore volume. An example of these materials is crystalline cellulose and lignin of lignocellulosic biomass that are resistant to hydrothermal carbonization conditions [83]. MOFs have been considered as alternative resources for the production of nanoporous carbons as they have diverse structures, as well as large pore volumes and high surface areas. Different kinds of MOFs including MOF-5, Al-PCP, and ZIF-8 have been used to produce nanoporous carbons and demonstrated suitable properties in applications such as gas adsorption, sensing, catalysis, and electrochemical capacitance [52,53,73,84]. Because MOFs have substantial carbon content, they can be converted to nanoporous carbons through direct carbonization without the need for any additional carbon source. Direct carbonization method is an easy and one-step procedure and has been utilized to convert Al-PCP to nanoporous carbon at 800°C. The study showed the obtained nanoporous carbon had exceptional surface area of more than 5000 $m^2\ g^{-1}$ and the large pore volume of 4.3 $cm^3\ g^{-1}$. It was concluded that the carbonization temperature was critical to reach the desired high surface area as well as large pore volume [85]. Fig. 7.4 illustrates the synthesis of nanoporous carbon derived from ZIF-8 via direct carbonization.

Recently, Lim et al. [87] reported the direct carbonization of MOFs containing zinc atoms. They highlighted that even the nonporous MOF materials could give highly nanoporous carbons because of the reorganization of carbon atoms. This is an interesting finding that shows the superior characteristics and performance of MOF materials in comparison with other compounds.

3.2.4 Adsorption Mechanisms in Metal–Organic Frameworks

The adsorption mechanism on MOFs is similar to the general adsorption mechanisms over the common traditional adsorbents except that MOFs have open metal sites that might act as an extra adsorption site. In this section, the common adsorption mechanisms are explained with specific focus on MOFs. The overview of different adsorption mechanism for the removal of hazardous pollutants over MOF is shown in Fig. 7.5.

Electrostatic interactions are one of the most common pathways for adsorptive removal of contaminants from water. Haque et al. [89] studied the adsorption of methyl orange (anionic dye) over chromium-based MOFs, MIL-101-Cr, and MIL-53-Cr. It was observed that MIL-101-Cr

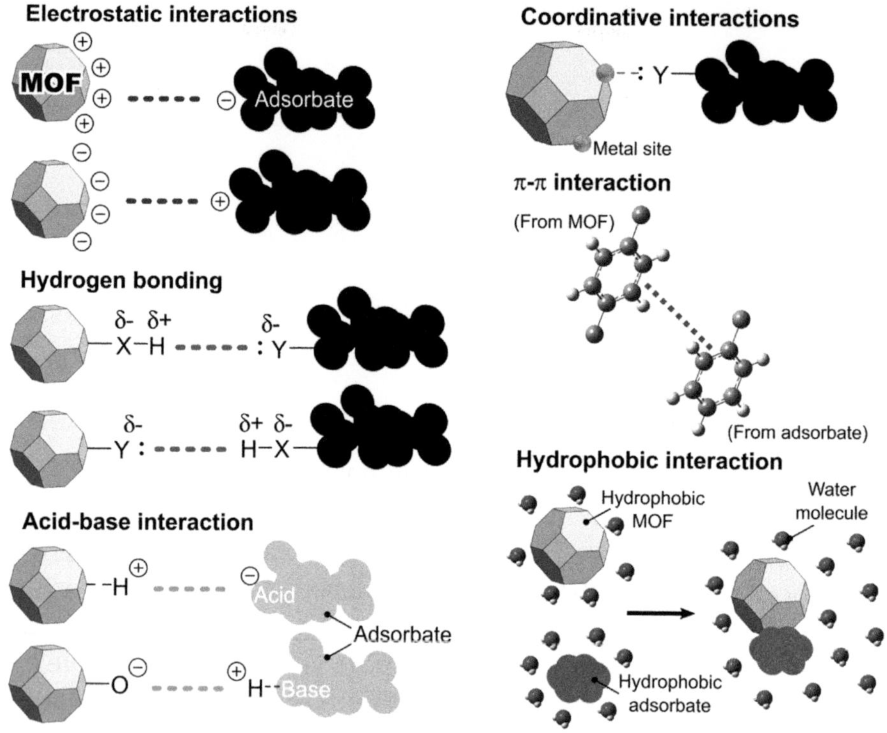

FIGURE 7.5 Schematic outline of the adsorption mechanisms for adsorption of contaminants using metal–organic frameworks [88].

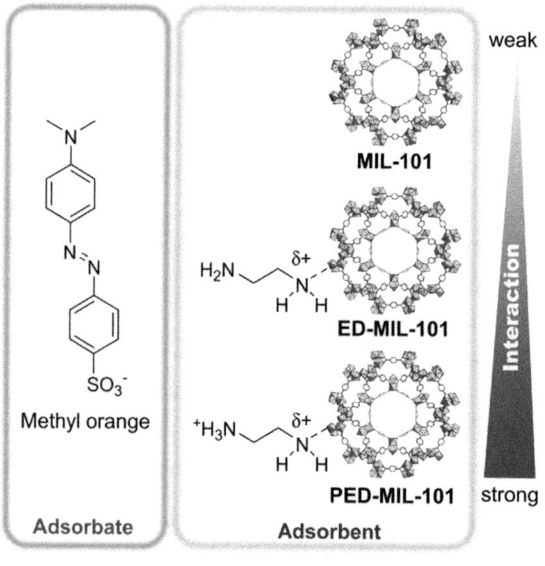

FIGURE 7.6 Electrostatic interaction between methyl orange and adsorbent (MIL-101) [89].

had better adsorption performance compared with MIL-53-Cr, which demonstrates the effect of porosity and pore size distribution. Nevertheless, by functionalizing MIL-101-Cr with ethylene diamine (ED) to induce positive surface charge, higher adsorption capacity was observed despite the slight reduction of porosity and pore size after grafting. The schematic of electrostatic interaction is shown in Fig. 7.6, which shows the adsorption performance sequence as MIL-101-Cr < ED-MIL-101-Cr < PED-MIL-101-Cr [90].

In another study by Lin et al. [92], the adsorption of MB on copper (II)-benzene-1,3,5-tricarboxylate (Cu-BTC) was investigated. It was reported that pH of the solution had a significant effect on the electrostatic interactions between the MB molecules and adsorbent. It was shown that increasing the pH above the point of zero charge induced negative surface charge, thus creating favorable electrostatic interactions between negatively charged Cu-BTC and positively charged MB molecules leading to higher adsorption capacity.

Hydrogen bonding between adsorbate and the adsorbent also commonly occurs, and it can often be the governing interaction of the adsorption. Nevertheless, in aqueous media, water molecules compete for the hydrogen bonding sites. Therefore, in these systems, multipoint interaction is necessary for the efficient adsorption. The adsorption of phenol and *p*-nitrophenol (PNP) on different MOF materials (MIL-100-Fe, Cr and NH_2-MIL-101-Al) from water has been investigated, and the results showed low adsorption capacity of these MOFs for adsorption of phenol. However, NH_2-MIL-101-Al exhibited high adsorption performance of PNP with 4.3 and 1.9 times higher than capacities of MIL-100-Fe and MIL-100-Cr, respectively [93]. This high affinity of NH_2-MIL-101-Al toward PNP was attributed to the hydrogen bonding between PNP and the amino groups in NH_2-MIL-101-Al as illustrated in Fig. 7.7.

Although acid—base interactions are not observed very often in adsorption of contaminants form water, there are some studies reporting this mechanism. For example, Khan et el [94]. studied the adsorption of phthalic acid (H_2-PA) from aqueous solution on different MOFs. The results of the study revealed that ZIF-8 and NH_2-UiO-66 showed higher adsorption for H_2-PA especially at neutral pH compared with MOFs without imidazole or amino groups.

There are a few reports regarding the influence of metal ions in MOFs for dye adsorption from aqueous media. For instance, Tong and coworkers [95] reported the adsorption of MO and MB dyes on MIL-100-Fe and MIL-100-Cr. They observed that despite having similar surface areas and pore volumes, these two adsorbents show different behavior for adsorption of dyes. MIL-100-Fe exhibited much higher adsorption capacity in comparison with MIL-100-Cr, which was attributed to the greater bonding energy of water molecules with metal centers of MIL-100-Cr. This results in competitive adsorption of MO and water, and the adsorption of water restricts the cages of MIL-100-Cr for adsorption of MO. However, for the adsorption of MB, both MOF materials exhibited high adsorption due to electrostatic interactions between positively charged MB and negatively charged MOF materials.

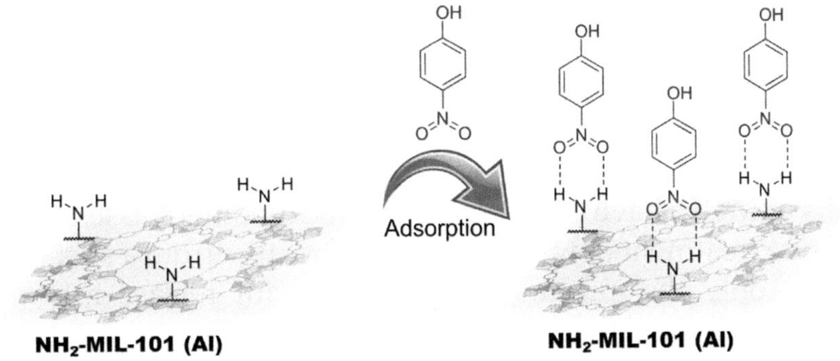

FIGURE 7.7 Adsorption mechanism of *p*-nitrophenol (PNP) on NH_2-MIL-101-Al through hydrogen bonding [93].

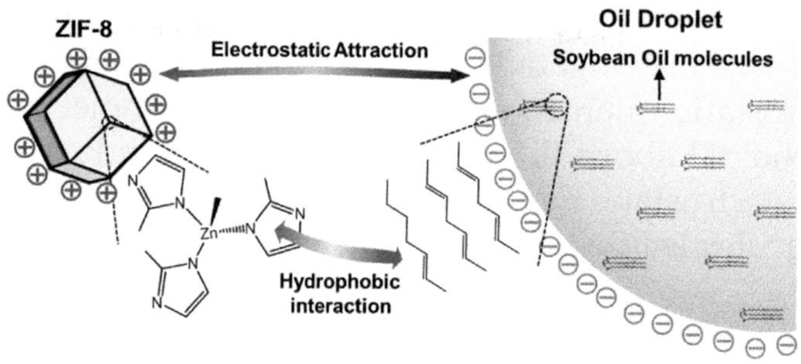

FIGURE 7.8 Mechanisms of oil removal by zeolitic imidazolate framework (ZIF)-8 adsorbent [101]. *Adapted from K.-Y.A. Lin, Y.-C. Chen, S. Phattarapattamawong, Efficient demulsification of oil-in-water emulsions using a zeolitic imidazolate framework: adsorptive removal of oil droplets from water, Journal of Colloid and Interface Science. 478 (2016) 97–106. https://doi:10.1016/j.jcis.2016.05.057 with permission from Elsevier.*

π-π interactions may be observed for the removal of aromatic compounds from water. Qin et al. [96] reported adsorption of bisphenol-A (BPA) on MIL-101-Cr and MIL-100-Fe. It was shown MIL-101-Cr exhibited the highest adsorption capacity for BPA, and it was suggested that the π-π interactions between the benzene rings of BPA and MIL-101 played a role in addition to hydrogen bonding. Other studies have suggested π-π interactions during the adsorptive removal of malachite green (MG) using MIL-100-Fe [97] and for the adsorption of BPA on MIL-53 [98]. It is important to note that all of these reports hypothesize the presence of π-π interactions without experimental evidence. While it is plausible that this interaction occurs between simple aromatic moieties such as benzene ring, it would be too weak to be the governing factor of adsorption. In a more thorough attempt, Akpinar et al. demonstrated the adsorptive removal of atrazine from aqueous samples by MOFs containing pyrene-derived linkers [99]. The systematic investigation of structure–property correlations supported the role of π-π interactions as primary source of adsorbent efficiency.

Hydrophobic interactions are usually more prevalent in the adsorption of organic molecules from aqueous solution. MOFs have been utilized for oil spill cleanup as well as demulsification of oil droplets. Yang et al. [100] investigated the fluorous MOF (FMOF-1) for oil spill cleanup. This MOF material is highly hydrophobic, which is due to the perfluorinated inner surface. This hydrophobicity leads to interactions between FMOF-1 and the nonpolar hydrocarbon adsorbates. Lin et al. [101] also reported adsorption of soybean oil and demulsification of oil droplets from water, and the mechanism was attributed to the hydrophobic interaction as well as electrostatic interactions as shown in Fig. 7.8.

4. SUSTAINABILITY ASPECTS OF METAL–ORGANIC FRAMEWORK–BASED ADSORBENT MATERIALS

MOFs and MOF-based materials have demonstrated excellent and often unparalleled performance in many field of science over the past decades. The research into these materials opened up immense possibilities for commercial applications. In addition to their performance, the sustainability aspects of MOFs have also become the focal point of assessment and further development. The adsorptive treatment of wastewaters is attractive from a sustainability point of view. It makes possible not only the minimization of pollutant release into the environment but also the recovery and recycling of the often valuable chemicals from the waste stream. However, the sustainability of an adsorbent material does not only depend on its performance during the application but also its synthesis, formulation, and regeneration.

Several review articles aimed to summarize and evaluate the progress in sustainable MOF technologies. Reinsch reviewed MOF syntheses from a green chemistry point of view [102] while reflecting on the 12 principles established by Anastas et al. [103]. Julien et al. took a broader approach and summarized the sustainability aspects of MOF design, synthesis, formulation, and application [104]. In their review, evaluative criteria for green, sustainable, and scalable metal–organic chemistry was highlighted, along with six key areas of experimental and theoretical progress that are relevant for developing cleaner MOF syntheses: (1) using safe and biocompatible building blocks, (2) reducing energy consumption, (3) using water as reaction media, (4) minimal amount or no solvent at all, (5) continuous production, and (6) theoretical prediction and design of MOF performance for given applications. Connolly et al. covered some key development in synthetic procedures but mostly focused on the sustainability aspects of various MOF applications in their minireview [105]. Mon et al. reviewed and evaluated MOF-based materials for sustainable water remediation [90]. Although there are many examples when MOFs outperformed the current technologies in the field, further research into low-cost and long-lifetime materials was encouraged. Li et al. focused their review on water contaminant elimination with MOF-based materials via adsorption and oxidation of membrane separation [106]. In this review, MOF stability was identified as key area of development for more sustainable materials and processes. In addition, the importance of real life applications was emphasized to validate research in the field.

4.1 Production

MOFs are traditionally synthesized under solvothermal conditions using toxic polar aprotic solvents such as N,N-dimethylformamide (DMF) or *N,N*-dimethylacetamide (DMAc) between 100 and 200°C under autogenous pressure over 24–72 h. The transition metal component is usually added as a nitrate or chloride salt, which in turn results in the production of nitric acid or hydrochloric acid. Therefore, after the filtration of the synthesized MOFs, the remaining mother liquor is a complex waste stream containing high-boiling organic solvent, transition metal residues, and strong acids. In light of this, it is clear that the development of benign MOF synthesis is a key area toward truly sustainable adsorption processes.

One direction of research in greener synthesis is the selection of the metal source. Sulfate salts have been proposed as less corrosive alternatives but solubility of these salts is lower. Acetate salts are also milder alternatives that cause fast nucleation rate and thus smaller particles than in case of the corresponding nitrate salt [107]. The application of metal oxides, hydroxydes, or carbonates for MOF production is desirable to generate only harmless water and/or carbon dioxide as by-product. However, these salts generally have extremely poor solubility and therefore require different methodologies from the conventional solvothermal route. An interesting approach is the direct use of metal in electrochemical conditions because no by-product is generated. BASF reported the production of highly porous HKUST-1 ($Cu_3(btc)_2$) from copper cathode in this manner [108].

Besides the salt form, the selection of the metal itself should be considered. Especially, in water treatment, the release of transition metals from the adsorbent through leaching or erosion can have serious ramifications. Among these metals, Mn, Ti, Fe, and Zr may be favored due to their lower toxicity [104]. The development of alkaline earth metal–based MOFs is also an interesting direction toward less toxic MOF production.

The organic linkers used in MOFs are not of particular concern due to their toxicity. However, their current production heavily relies on fossil resources. Although there are examples of their replacement in the literature, this effort is limited by the strong dependence of the MOF crystal structure on the linker compounds. Terephthalic acid is a common linker for MOF materials, which is usually produced from *p*-xylene. Ren et al. utilized the similar reaction conditions of poly(ethylene terephthalate) (PET) hydrolysis and MOF synthesis [109]. In their approach, waste PET bottles were used as direct source of terephthalic acid linker during the synthesis of Cr-MOF. There is also an interest of bio-based linker compounds for carboxylate networks. A bio-based terephthalic acid analogue, 2,5-furandicarboxylic acid, and the natural amino acid, asparagine acid, were considered as building blocks for Zr-based MOFs [110,111]. The inexpensive, renewable, and nontoxic fumaric acid has been widely used as a linker material for example for the synthesis of the aluminum fumarate network that has been commercialized by BASF (Basolite A520) [112].

Solvothermal process was the first approach for the synthesis of MOFs, and it still remained the most common technique for the laboratory synthesis on gram scale. Its popularity can be attributed to two main factors. It does not require expensive instrumentation but reliably produces MOFs with high crystallinity. However, the solvothermal process suffers from high energy demand and waste generation, health and safety concerns, and low space time yield. Some of these issues arise from the use of toxic polar aprotic solvents; therefore, attempts have been made to replace them with greener alternatives. Solvent selection guides have been published to highlight problematic solvents and provide greener alternatives [113,114]. One of the emerging solvents to replace polar aprotic solvents is the bio-based dihydrolevoglucosenone (Cyrene) that has been also proposed for MOF synthesis [115]. The results were satisfactory; however, the instability of the solvent among the applied synthetic conditions can undermine its more widespread application for MOF production. The use of ionic liquids has been also studied as their low volatility can be advantageous at the applied high temperature [116]. Nevertheless, the generally low yields, the incorporation of the ionic liquid in the crystal structure, and the recycling problems are likely to prevent the general use of ionothermic processes in MOF synthesis.

Water, which is arguably the greenest solvent, has also attracted interest for MOF synthesis. Hydrothermal synthesis has been developed for the commercial production of Basolite A520 [112]. The room temperature synthesis of ZIF-8 in aqueous solution has also been demonstrated [117]. Although there are a few other examples of aqueous MOF synthesis, the approach is limited by the solubility of some organic linkers in water as well as by the aqueous instability of many MOFs. Synthesis in near-critical water is a promising approach to overcome these problems, and it has been demonstrated for the production of a Zn-based MOF [118]. However, the harsh synthetic conditions result in high energy demand and safety concerns.

The rocketing interest in MOF materials inspired research in more scalable syntheses with high productivity and low waste generation. There are several review articles dedicated to summarize these efforts toward scalable synthesis and in turn more widespread industrial applications [119,120]. Flow chemistry has been frequently used as an effective tool for the continuous production of MOFs. Polyzoidis et al. demonstrated the production of ZIF-8 in a continuous microreactor with theoretical space–time yield of 210,000 kg m^{-3} day^{-1} while maintaining precise control over the particle size. Mechanochemical processes aim to minimize the solvent consumption of MOF syntheses. These syntheses involve milling, grinding, or extrusion under neat, liquid-assisted or ion- and liquid-assisted conditions. The mechanochemistry also allows the use of

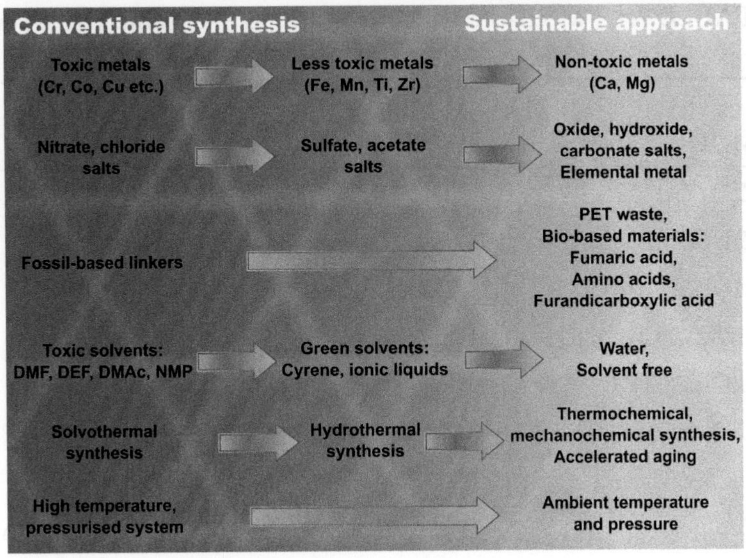

FIGURE 7.9 Directions of research for more sustainable metal–organic framework (MOF) syntheses. (*DEF*, *N,N*-diethylformamide; *DMAc*, *N,N*-dimethylacetamide; *DMF*, *N,N*-dimethylformamide; *NMP*, *N*-methylpyrrolidone; *PET*, poly(ethylene terephthalate)).

metal oxides and hydroxides, thus further improving the sustainability of these procedures. Although the completely solvent-free production is most desirable, often inadequate conversion and low crystallinity is observed. Tanaka et al. overcame these problems by the addition of zinc acetate as auxiliary Zn source and catalyst in the solvent-free mechanosynthesis of hierarchically porous ZIF-8 from ZnO [121]. Liquid and ionic additives in mechanochemical synthesis have proved to improve reaction rates and crystallinity [122]. The selection of these additives can provide control over the crystal form of the produced MOFs. Other interesting approaches to sustainable and scalable syntheses include accelerated aging [123,124] and aerosol synthesis [125].

Using MOFs as starting material for nanoporous carbons has certain advantages from a sustainability point of view. Many materials that undergo heat treatment and carbonization do not possess microporosity and high surface area after carbonization. These materials need to be treated by traditional activation methods to increase the specific surface area and improve the pore volume. However, MOFs have been used for producing nanoporous carbons by one-step and direct carbonization method without the need for further physical or chemical activation [85]. The resulting nanoporous carbons had high porosity and high surface area as well as excellent performance in many applications, particularly in water treatment (Fig. 7.9).

4.2 Formulation and Shaping

Shaping MOF powders into macroscopic bodies is of great importance especially in regard to practical applications such as adsorption, separation, and catalysis [126–128]. The resulting shaped MOFs need to have good properties such as high porosity and surface area, crystallinity, and suitable performance similar to the original MOF powder [108]. Therefore, it is crucial to select an appropriate shaping method that is simple and effective and at the same time preserves the MOF's unique characteristics. The formulation and shaping of the adsorbents must be taken into account during the sustainability assessment of these materials. Some common techniques for preparation of the shaped catalysts and adsorbents are pressing, extrusion, granulation, and spray drying, as well as casting and coating can also be employed for preparing the shaped

TABLE 7.1 Mechanical Strength of Metal–Organic Frameworks (MOFs), Zeolites, and Ceramics [132].

MOF	Hardness/Shear Stability (GPa)	Mechanical Strength
ZIFs(Zn)	0.2–1.1	Weak
ZIF-8(Zn)	0.94–1.4	Weak
MOF-5(Zn)	1.16	Weak
HKUST(Cu)	1.04	Weak
UiO-67(Zr)	5.69	Medium
UiO-68(Zr)	4.18	Medium
UiO-66(Zr)	13.75	High
Zeolites	1–10	Medium to high
Ceramics	>10	High

MOFs in the forms of monoliths, granules, pellets, and tablets [129–131]. But the challenge is that the prepared shaped MOFs need to have enough mechanical and chemical stability with the original powder's characteristics and properties. For shaping the porous materials, a general methodology is to mix the porous powder with inorganic and organic additives, shape the powder to the required form, and finally eliminate the additives by thermal treatment to obtain a mechanically strong framework [132]. Nevertheless, most MOFs suffer from poor mechanical robustness under high pressure conditions as reported in Table 7.1.

Exposing MOF materials to external pressure can often result in a collapse of their pore structures and amorphization. Moreover, many MOF materials are not thermally stable at high temperatures; therefore, they cannot be exposed to high temperatures for the removal of the additives and binders in the same way as zeolites and ceramics [133].

Cu-BTC in the form of spherical granules has been prepared by granulation shaping method. The Cu-BTC spheres were obtained with the size range between 1 and 3 mm [134]. In addition, Zr-MOF powders have been shaped into spheres with diameters of 0.5–15 mm with the usage of sucrose as a binder [135]. However, the Zr-MOF spheres showed 50% decrease in the BET surface area and pore volume, which was attributed to pore blocking with the binder molecules. It was therefore shown that it is difficult to obtain spheres with equal sizes or appropriate properties with the granulation method. In shaping MOF powders, it is important to get spheres with uniform size in terms of practical applications to maximize the utilization of adsorptive sites. Moreover, for preparing MOF shapes with granulation process, many parameters must be controlled regarding the formulation as well as the equipment. The reason for this can be addressed to the limitation of granulation method, which is due to the balance of collision stress applied to granules relative to the fracture stress of MOF materials. Under collision stress, MOF granules often agglomerate or erode [132].

Spray drying is another method for producing MOF bodies for industrial products such as catalysts and adsorbents. This method can be utilized for the production of small particles in micrometer range to large particles in millimeter range. However, the millimeter-sized droplets might not be free-flowing because of the existence of large particles or powders [132]. Additives and binders are added to the feed liquid in this process, which may cause problems during the process as they increase the viscosity of suspensions. In addition, the interaction between the additives and MOF materials is very complicated; therefore, the general design rules do not apply. The use of spray drying technology for shaping MOFs is more complex compared with other common materials [132]. Avci-Camur et al. reported the aqueous production of spherical Zr-MOF beads via continuous-flow spray drying [136]. Their approach is especially attractive from a sustainability point of view because of the green solvent selection and continuous one-step fabrication of shaped MOF microbeads.

Extrusion is another technique that has been used for shaping MOF powders. Crawford et al. [137] have utilized extrusion method to shape Cu-BTC into monoliths with millimeter sizes. The problem with this method is that the formed bodies are not uniform and not resistant to abrasion in comparison with pressed shapes. The mixture paste (mixture of MOF, additive, and solvent) must have special characteristics including plasticity to enable the paste to be extruded and, at the same time, exhibit cohesion to avoid creation of surface and bulk defects. Küsgens et al. extruded Cu-BTC monoliths on a centimeter scale with satisfactory stability [138]. The MOF monolith outperformed its zeolite counterpart in terms of kinetic performance and total adsorption capacity.

Pressing strategy includes compressing and pelleting the powders. The resulting pellets or tablets have high mechanical strength. For pressing MOF materials, the quality of pellets depends on the mechanical property of MOF powders. The deformation characteristics of MOF powders have a significant effect on the pressing process. It is noteworthy that the main factor that controls the properties of the resulting shaped bodies (pellets) is the response of MOF material to the pressure [139–141]. The problem with this method is that the crystallinity of MOF degrades during pressing, and it gets close to an amorphous structure. Tagliabue et al. [142] investigated the adsorption performance of CPO-27(Ni) pellets without the usage of binder. Their study revealed that after pressing the sample at 1 GPa the crystalline structure was lost because of breakdown of the MOF structure.

Bazer-Bachi et al. [143] prepared tablets of several MOFs and ZIF-8 using direct compression. The results of the study showed these materials undergo amorphization depending on the amount of pressure applied and the nature of material. Therefore, the properties and performances of MOFs may be affected; hence, this is not a suitable approach. In particular, as only very few MOFs such as NH_2-MIL-53(In) show amorphization resistance [144,145], the direct compression cannot be considered a general and universal strategy for shaping MOFs. Other strategies include in situ growth of MOFs on porous substrates such as silica and aerogel using methods such as seeding growth and layer-by-layer deposition [146], which requires several steps and often gives unsatisfactory coverage [138].

Phase inversion method can also be employed for producing spheres and beads for many applications in particular adsorption. In a recent study, Abbasi et al. [147] utilized phase inversion method to produce ZIF-8/polymer composite beads for adsorption of different oils from water. The composite beads exhibited high surface areas and low bulk densities, and they were capable of adsorbing serval oils and organic compounds from water. In other studies, Zeng et al. [148] and Gu et al. [149] prepared composite beads and converted them to carbon and utilized the produced spheres and beads for oil and dye sorption from water. These investigations revealed that phase inversion could be an effective method for shaping MOF powders into spheres and beads without affecting their crystalline structures.

3D printing can provide an interesting platform for the versatile formulation of MOF materials. Monolith adsorbents with complex shapes were obtained by this technique without compromising the physical properties or mechanical integrity of the MOFs [150]. The potential geometries available to 3D-printed materials coupled to the variety of MOF structures can lead to a number of possibilities for optimizing any printed adsorbent [151].

4.3 Application for the Adsorptive Removal of Organic Pollutants From Water

4.3.1 *Adsorption of Organic Contaminants from Water Using Metal–Organic Frameworks*

A variety of modified and unmodified MOFs have been investigated for the removal of different kinds of organic compounds from water. The important parameters affecting the adsorption over MOF adsorbents are including the pore structure, open metal sites, and charge interaction between the adsorbate and adsorbent [12]. MIL-100 (Fe) and MIL-101 (Cr) were studied for the removal of naproxen and clofibric acid. The study showed these adsorbents can effectively take up the organic contaminants from water and exhibited superior capacity compared with activated carbon [152].

In a recent work, Jiang and coworkers [77] studied the adsorption of benzotriazoles on ZIF-8 and found out the adsorption is very fast and high adsorption of capacity of the pollutant compounds was achieved. Khan et al. [94] also reported the removal of phthalic acid and diethyl phthalate from water using ZIF-8. It was revealed that ZIF-8 adsorbs phthalic acid significantly. This high adsorption uptake was attributed to electrostatic interaction between ZIF-8 (having a positively charged surface) and negatively charged phthalic acid molecules at high pH.

MIL-100 (Fe) was reported for the adsorption of MG, and it was found that the open metal sites were the active species for interaction with MG because of the interaction between the Lewis base-$N(CH_3)_2$ in MG and Lewis acid of MIL-100(Fe). It was also proposed that the π-π interaction between the benzene rings in MG and MIL-100(Fe) is effective in the adsorption process [97]. A hierarchically mesostructured MIL-101(Cr) was synthesized by cetyltrimethylammonium bromide as a surfactant, and it was shown that the prepared material adsorbed MB dye quickly [153]. Haque et al. studied the adsorption of MB and methyl orange by MOF-235. The adsorption capacities were reported as 187 and 477 mg g^{-1} for MB and MO uptake, respectively [91]. Recently, Lin and Chang [154] investigated the adsorption properties of ZIF-67 for the adsorption of MG from water. Their study showed ZIF-67 had the ultrahigh adsorption capacity of 2430 mg g^{-1} for MG removal. They suggested the high sorption capacity was probably due to π-π interactions between MG and ZIF-67.

Hydrophobic MOFs also have been used for oil spill cleanup. As mentioned previously, Yang et al. [100] investigated the oil sorption properties of the FMOF-1 (constructed from silver (1) 3,5-bis(trifluoromethyl)-1,2,4-triazolate) to remove oil spill from the surface of water. This MOF is exceptionally hydrophobic because of perfluorinated inner surface. Another study shows that soybean oil droplets can be removed from the water surface using Cu-BTC MOF. The high adsorption was addressed to the interactions the soybean oil and benzene rings in Cu-BTC [155]. Another study by Lin et al. [101] reveals the oil sorption capability of ZIF-8 for demulsification of oil-in-water emulsions. The results of their investigations show that ZIF-8 is a promising adsorbent for removal of emulsified oil droplets from water.

4.3.2 *Adsorption of Organic Pollutants From Water by Nanoporous Carbons*

Nanoporous carbon from different sources has been investigated for the removal of organic compounds including dyes from water. Han et al. [156] developed a nanoporous carbon material using silica sol particles as templates, which exhibited faster and greater adsorption of humic acid molecules compared with commercial activated carbons. In addition, Sandeman et al. [157] investigated the adsorption properties and porous characteristics of activated carbons and activated carbon/polyvinyl alcohol composites for the removal of cationic and anionic dyes. They reported that the interaction of dyes with porous activated carbons depends on two factors. First, the interaction is associated with the properties of dyes such as molecular size, charge sign, and value, as well as the symmetry of molecule. Second, it is related to the nature of the carbon surface as well as structural and textural characteristics. Nitrogen, oxygen, and hydrogen atoms provide dispersion, electrostatic, and hydrogen bonding of adsorption complexes to heterogeneous surface of carbon materials [158]. The bonding depends on the properties of the solvent molecules [159,160].

Ghaedi et al. [161] reported the synthesis of activated carbon loaded with silver nanoparticles, which showed the adsorption capacity of 34.5 mg g^{-1} for the removal of MB from water. Faria et al. [162] investigated the adsorption of anionic and cationic dyes on activated carbons and reported that the adsorption of reactive and acid dyes is associated with the surface basicity of adsorbents. They revealed that the interaction between oxygen-free Lewis basic sites and the free electron of dye molecules in addition to the electrostatic interactions between the protonated sites and anions play the main role in adsorption mechanism.

Activated carbon has been widely used for the removal of organic pollutants especially dyes from water [28,33,163–165]. For example, Hameed and Daud [166] studied the adsorption of a basic dye on activated carbon derived from agricultural waste. The activated carbon showed maximum adsorption capacity of 227 mg g^{-1}. In spite of the usefulness of activated carbon as an effective adsorbent for heavy metals, the high cost of activated carbon and the difficulty of its regeneration have limited its more predominant use.

However, there are only few reports in literature addressing the adsorption of organic contaminants including dyes over nanoporous carbon-derived from MOF materials. Torad et al. [81] synthesized a magnetic nanoporous carbon MOF without the usage of any extra precursor by one-step carbonization for the removal of MB from water. In their study, the MOF ZIF-67 was carbonized at 600 and 800°C and showed high adsorption capacity of 302.1 and 502.5 mg g^{-1} for MB removal, respectively. However, higher temperatures were not investigated in this work. Xiao et al. [167] synthesized magnetic porous carbons from MIL-100(Fe) using a second precursor of furfuryl alcohol via microwave-enhanced high temperature ionothermal method. The resulting magnetic carbon exhibited adsorption capacity of 303.95 mg g^{-1} for the removal of MB from aqueous solution. This synthesis method required an expensive microwave irradiation along with the usage of an extra carbon precursor. In another study by Jiao et al. [168], a nanoporous carbon was produced from MOF-5, and the magnetic property was introduced using chemical coprecipitation of iron ions on the resulting nanoporous carbon. The resulting magnetic porous carbon showed the maximum adsorption capacity of 292.4 mg g^{-1}. Recent works have been done for carbonization of ZIF-8 at 800°C [169], and the porous carbon showed low adsorption capacities of 59 mg g^{-1} for the adsorption of MB from water. Furthermore, in this work, higher carbonization temperatures were not tested.

In a recent study, Abbasi and coworkers [147] investigated the effect of temperature on adsorption properties of nanoporous carbons derived from ZIF-8. They carbonized ZIF-8 at temperatures of 600, 1000, and 1200°C and investigated the adsorption of MB dye on nanoporous carbons derived at abovementioned temperatures. They concluded that carbonization temperature had significant effect on adsorption properties of nanoporous carbons as the carbon derived at 1000°C had the highest adsorption capacity of 186.3^{-1}, while the nanoporous carbon ZIF-8 at 600 and 1200°C showed much lower adsorption capacities of 49.5 and 36.7 mg g^{-1}, respectively. This observation was attributed to the introduction of more negative surface charge and broader pore size distribution at carbonization temperature higher than boiling point of Zn (908°C).

5. CONCLUSIONS AND PERSPECTIVES

In summary, adsorption has advantages over many other water treatment methods as it provides an easy-to-operate, environmentally friendly, and energy efficient alternative approach for the removal of hazardous contaminants from wastewaters. Porous materials have interesting characteristics such as high porosity and high surface areas for many applications specifically in adsorption and water treatment. ZIFs are particularly interesting for adsorption applications because of high porosity and large carbon content. The review of the recent research shed some light on the progress achieved in the adsorptive removal of different contaminants from water by MOFs and ZIFs.

Nevertheless, there are still challenges regarding the application of these materials for adsorption. One challenge could be addressed to the fact that MOFs and ZIFs are generally prepared as powders that are not suitable for many practical applications including adsorption and catalysis. Therefore, shaping these materials is required for easy handling and recycling. There are many strategies for shaping MOF powders. However, there are many issues with regard to the mechanical stability of the shaped bodies obtained by these techniques including the lack of mechanical stability and loss of crystallinity. Therefore, a novel low cost and efficient method needs to be developed and investigated. The new method needs to be efficient enough to maintain the high surface area of the adsorbent while preserving its crystallinity, robustness, and stability. Therefore, different strategies for shaping MOF powders were critically discussed.

The second challenge was discussed based on the limitation of ZIF-8 in adsorption of large molecules such as dyes in the small pores. This issue could be resolved by some modifications such as heat treatment (carbonization). The presence of unique and interesting exceptionally porous framework and large carbon content in MOFs and ZIFs allows these materials to be used as efficient precursors for porous carbons. The pore size distribution and pore structure can be controlled by the carbonization temperature. Another challenge of MOFs is the stability in aqueous solutions. Many MOFs are not stable due to the leaching of either metal ions or organic ligands. This has to be solved before any possible real applications.

The third major challenge is regarding the sustainable production and use of MOFs and porous carbon materials. As highlighted throughout the text, MOFs are usually synthesized from complex organic ligand precursors that are not naturally derived but rather prepared synthetically. Most synthesis routes require substantial quantities of highly pure organic solvents, and as a result, there is substantial consumption of nonrenewable (i.e., fossil fuel–derived) feedstocks and generation of waste and by-products. While porous carbons can be produced from natural feedstocks, these often do not have the highly desirable pore structure and chemistry that can be obtained using MOFs as the carbon source and template, and in any case, porous carbon materials are often difficult to regenerate for repeated or cyclic use industrially.

To address the abovementioned challenges, the new approaches need to be fast, highly reproducible, and economically and environmentally viable. The study of the textural and performance change of the adsorbents under these new approaches will help us to understand and analyze the adsorptive behavior of materials for the successful selection of porous adsorbents and modifying/processing techniques for future adsorption applications specifically in water treatment.

References

[1] X. Liu, M. Wang, S. Zhang, B. Pan, Application potential of carbon nanotubes in water treatment: a review, Journal of Environmental Science (China) 25 (2013) 1263–1280, https://doi.org/10.1016/S1001-0742(12)60161-2.

[2] M.N. Chong, B. Jin, C.W.K. Chow, C. Saint, Recent developments in photocatalytic water treatment technology: a review, Water Research 44 (2010) 2997–3027, https://doi.org/10.1016/j.watres.2010.02.039.

[3] W.S. Wan Ngah, M.A.K.M. Hanafiah, Removal of heavy metal ions from wastewater by chemically modified plant wastes as adsorbents: a review, Bioresource Technology 99 (2008) 3935–3948, https://doi.org/10.1016/j.biortech.2007.06.011.

[4] C.H. Neoh, Z.Z. Noor, N.S.A. Mutamim, C.K. Lim, Green technology in wastewater treatment technologies: integration of membrane bioreactor with various wastewater treatment systems, Chemical Engineering Journal 283 (2016) 582–594, https://doi.org/10.1016/j.cej.2015.07.060.

[5] M. Panizza, G. Cerisola, Direct and mediated anodic oxidation of organic pollutants, Chemical Reviews 109 (2009) 6541–6569, https://doi.org/10.1021/cr9001319.

[6] G. Tchobanoglous, F.L. Burton, H.D. Stensel, Wastewater Engineering: Treatment and Reuse, McGraw-Hill, 2003.

[7] I. Oller, S. Malato, J.A. Sánchez-Pérez, Combination of advanced oxidation processes and biological treatments for wastewater decontamination-A review, Science of the Total Environment 409 (2011) 4141–4166, https://doi.org/10.1016/j.scitotenv.2010.08.061.

[8] M. Lapertot, C. Pulgarín, P. Fernández-Ibáñez, M.I. Maldonado, L. Pérez-Estrada, I. Oller, W. Gernjak, S. Malato, Enhancing biodegradability of priority substances (pesticides) by solar photo-Fenton, Water Research 40 (2006) 1086–1094, https://doi.org/10.1016/j.watres.2006.01.002.

[9] R. Muñoz, B. Guieysse, Algal-bacterial processes for the treatment of hazardous contaminants: a review, Water Research 40 (2006) 2799–2815, https://doi.org/10.1016/j.watres.2006.06.011.

[10] M.T. García, I. Ribosa, T. Guindulain, J. Sánchez-Leal, J. Vives-Rego, Fate and effect of monoalkyl quaternary ammonium surfactants in the aquatic environment, Environmental Pollution 111 (2001) 169–175, https://doi.org/10.1016/S0269-7491(99)00322-X.

[11] F. Fu, Q. Wang, Removal of heavy metal ions from wastewaters: a review, Journal of Environmental Management 92 (2011) 407–418, https://doi.org/10.1016/j.jenvman.2010.11.011.

[12] N.A. Khan, Z. Hasan, S.H. Jhung, Adsorptive removal of hazardous materials using metal-organic frameworks (MOFs): a review, Journal of Hazardous Materials 244–245 (2013) 444–456, https://doi.org/10.1016/j.jhazmat.2012.11.011.

[13] M.T. Yagub, T.K. Sen, S. Afroze, H.M. Ang, Dye and its removal from aqueous solution by adsorption: a review, Advances in Colloid and Interface Science 209 (2014) 172–184, https://doi.org/10.1016/j.cis.2014.04.002.

[14] S.J. Allen, P.A. Brown, Isotherm analyses for single component and multi-component metal sorption onto lignite, Journal of Chemical Technology and Biotechnology 62 (1995) 17–24, https://doi.org/10.1002/jctb.280620103.
[15] A. Dabrowski, Adsorption – from theory to practice, Advances in Colloid and Interface Science 93 (2001) 135–224, https://doi.org/10.1016/S0001-8686(00)00082-8.
[16] K. Yamagiwa, M. Katoh, M. Yoshida, A. Ohkawa, H. Ichijo, Potentiality of temperature-swing adsorption for removal of hydrophobic contaminants in water, Water Science and Technology 35 (1997) 213–218, https://doi.org/10.1016/S0273-1223(97)00133-9.
[17] T.L. Silva, A. Ronix, O. Pezoti, L.S. Souza, P.K.T. Leandro, K.C. Bedin, K.K. Beltrame, A.L. Cazetta, V.C. Almeida, Mesoporous activated carbon from industrial laundry sewage sludge: adsorption studies of reactive dye Remazol Brilliant Blue R, Chemical Engineering Journal 303 (2016) 467–476, https://doi.org/10.1016/j.cej.2016.06.009.
[18] E. Lorenc-Grabowska, G. Gryglewicz, Adsorption characteristics of Congo Red on coal-based mesoporous activated carbon, Dyes and Pigments 74 (2006) 34–40, https://doi.org/10.1016/j.dyepig.2006.01.027.
[19] I. Langmuir, The adsorption of gases on plane surfaces of glass, mica and platinum, Journal of the American Chemical Society 40 (1918) 1361–1403, https://doi.org/10.1021/ja02242a004.
[20] B. Okolo, C. Park, M.A. Keane, Interaction of phenol and chlorophenols with activated carbon and synthetic zeolites in aqueous media, Journal of Colloid and Interface Science 226 (2000) 308–317, https://doi.org/10.1006/jcis.2000.6796.
[21] A.P. Terzyk, Molecular properties and intermolecular forces – factors balancing the effect of carbon surface chemistry in adsorption of organics from dilute aqueous solutions, Journal of Colloid and Interface Science 275 (2004) 9–29, https://doi.org/10.1016/j.jcis.2004.02.011.
[22] X. Zhang, A. Li, Z. Jiang, Q. Zhang, Adsorption of dyes and phenol from water on resin adsorbents: effect of adsorbate size and pore size distribution, Journal of Hazardous Materials 137 (2006) 1115–1122, https://doi.org/10.1016/j.jhazmat.2006.03.061.
[23] P. Cañizares, M. Carmona, O. Baraza, A. Delgado, M.A. Rodrigo, Adsorption equilibrium of phenol onto chemically modified activated carbon F400, Journal of Hazardous Materials 131 (2006) 243–248, https://doi.org/10.1016/j.jhazmat.2005.09.037.
[24] Y.S. Al-Degs, M.I. El-Barghouthi, A.H. El-Sheikh, G.M. Walker, Effect of solution pH, ionic strength, and temperature on adsorption behavior of reactive dyes on activated carbon, Dyes and Pigments 77 (2008) 16–23, https://doi.org/10.1016/j.dyepig.2007.03.001.
[25] C. Moreno-Castilla, Adsorption of organic molecules from aqueous solutions on carbon materials, Carbon N. Y. 42 (2004) 83–94, https://doi.org/10.1016/j.carbon.2003.09.022.
[26] H. Lachheb, E. Puzenat, A. Houas, M. Ksibi, E. Elaloui, C. Guillard, J.-M. Herrmann, Photocatalytic degradation of various types of dyes (Alizarin S, Crocein orange G, methyl red, Congo red, MB) in water by UV-irradiated titania, Applied Catalysis B Environmental 39 (2002) 75–90, https://doi.org/10.1016/S0926-3373(02)00078-4.
[27] M. Valix, W.H. Cheung, G. McKay, Preparation of activated carbon using low temperature carbonisation and physical activation of high ash raw bagasse for acid dye adsorption, Chemosphere 56 (2004) 493–501, https://doi.org/10.1016/j.chemosphere.2004.04.004.
[28] M.F.R. Pereira, S.F. Soares, J.J.M. Órfão, J.L. Figueiredo, Adsorption of dyes on activated carbons: influence of surface chemical groups, Carbon N. Y. 41 (2003) 811–821, https://doi.org/10.1016/S0008-6223(02)00406-2.
[29] L.R. Radovic, I.F. Silva, J.I. Ume, J.A. Menéndez, C.A. Leon Y Leon, A.W. Scaroni, An experimental and theoretical study of the adsorption of aromatics possessing electron-withdrawing and electron-donating functional groups by chemically modified activated carbons, Carbon N. Y. 35 (1997) 1339–1348, https://doi.org/10.1016/S0008-6223(97)00072-9.
[30] A. Ghaffar, M.N. Younis, Interaction and thermodynamics of methylene blue adsorption on oxidized multi-walled carbon nanotubes, Green Processing and Synthesis 4 (2015) 209–217, https://doi.org/10.1515/gps-2015-0009.
[31] J.M. Dias, M.C.M. Alvim-Ferraz, M.F. Almeida, J. Rivera-Utrilla, M. Sánchez-Polo, Waste materials for activated carbon preparation and its use in aqueous-phase treatment: a review, Journal of Environmental Management 85 (2007) 833–846, https://doi.org/10.1016/j.jenvman.2007.07.031.
[32] E.M. Dias, C. Petit, Towards the use of metal–organic frameworks for water reuse: a review of the recent advances in the field of organic pollutants removal and degradation and the next steps in the field, Journal of Materials Chemistry A 3 (2015) 22484–22506, https://doi.org/10.1039/C5TA05440K.

[33] N.K. Amin, Removal of direct blue-106 dye from aqueous solution using new activated carbons developed from pomegranate peel: adsorption equilibrium and kinetics, Journal of Hazardous Materials 165 (2009) 52–62, https://doi.org/10.1016/j.jhazmat.2008.09.067.

[34] A. Dabrowski, P. Podkościelny, Z. Hubicki, M. Barczak, Adsorption of phenolic compounds by activated carbon – a critical review, Chemosphere 58 (2005) 1049–1070, https://doi.org/10.1016/j.chemosphere.2004.09.067.

[35] S. Hong, C. Wen, J. He, F. Gan, Y.-S. Ho, Adsorption thermodynamics of methylene blue onto bentonite, Journal of Hazardous Materials 167 (2009) 630–633, https://doi.org/10.1016/j.jhazmat.2009.01.014.

[36] R. Li, L. Zhang, P. Wang, Rational design of nanomaterials for water treatment, Nanoscale 7 (2015) 17167–17194, https://doi.org/10.1039/c5nr04870b.

[37] M.E. Davis, Ordered porous materials for emerging applications, Nature 417 (2002) 813–821, https://doi.org/10.1038/nature00785.

[38] G. Crini, E. Lichtfouse, L.D. Wilson, N. Morin-Crini, Conventional and non-conventional adsorbents for wastewater treatment, Environmental Chemistry Letters 17 (2019) 195–213, https://doi.org/10.1007/s10311-018-0786-8.

[39] C.A. Leon Y Leon, L. Radovic, Interfacial chemistry and electrochemistry of carbon surfaces, in: Chem. Phys. Carbon, vol. 24, CRC Press, 1993, pp. 213–310.

[40] A. Bhatnagar, W. Hogland, M. Marques, M. Sillanpää, An overview of the modification methods of activated carbon for its water treatment applications, Chemical Engineering Journal 219 (2013) 499–511, https://doi.org/10.1016/j.cej.2012.12.038.

[41] C.A. Leon y Leon, J.M. Solar, V. Calemma, L.R. Radovic, Evidence for the protonation of basal plane sites on carbon, Carbon N. Y. 30 (1992) 797–811, https://doi.org/10.1016/0008-6223(92)90164-R.

[42] S.S. Barton, M.J.B. Evans, E. Halliop, J.A.F. MacDonald, Acidic and basic sites on the surface of porous carbon, Carbon N. Y. 35 (1997) 1361–1366, https://doi.org/10.1016/S0008-6223(97)00080-8.

[43] L. Monser, N. Adhoum, Modified activated carbon for the removal of copper, zinc, chromium and cyanide from wastewater, Separation and Purification Technology 26 (2002) 137–146, https://doi.org/10.1016/S1383-5866(01)00155-1.

[44] Y. Yao, F. Xu, M. Chen, Z. Xu, Z. Zhu, Adsorption behavior of methylene blue on carbon nanotubes, Bioresource Technology 101 (2010) 3040–3046, https://doi.org/10.1016/j.biortech.2009.12.042.

[45] H.P. Boehm, Some aspects of the surface chemistry of carbon blacks and other carbons, Carbon N. Y. 32 (1994) 759–769, https://doi.org/10.1016/0008-6223(94)90031-0.

[46] R.W. Coughlin, F.S. Ezra, Role of surface acidity in the adsorption of organic pollutants on the surface of carbon, Environmental Science and Technology 2 (1968) 291–297, https://doi.org/10.1021/es60016a002.

[47] S.W. Davis, S.E. Powers, Alternative sorbents for removing MTBE from gasoline-contaminated ground water, Journal of Environmental Engineering 126 (2000) 354–360, https://doi.org/10.1061/(ASCE)0733-9372(2000)126:4(354).

[48] S. Roobakhsh, Z. Rostami, S. Azizian, Can MoS_2 nanosheets be used as adsorbent for water treatment? Separation and Purification Technology 200 (2018) 23–28, https://doi.org/10.1016/j.seppur.2018.02.015.

[49] D. Shetty, I. Jahovic, J. Raya, Z. Asfari, J.-C. Olsen, A. Trabolsi, Porous polycalix[4]arenes for fast and efficient removal of organic micropollutants from water, ACS Applied Materials and Interfaces 10 (2018) 2976–2981, https://doi.org/10.1021/acsami.7b16546.

[50] X. Sun, B. Li, D. Wan, N. Wang, Using a novel adsorbent macrocyclic compound cucurbit[8]uril for Pb^{2+} removal from aqueous solution, Journal of Environmental Science (China) 50 (2016) 3–12, https://doi.org/10.1016/j.jes.2016.04.029.

[51] R.-H. Chen, H.-T. Qiao, Y. Liu, Y.-H. Dong, P. Wang, Z. Zhang, T. Jin, Adsorption of methylene blue from an aqueous solution using a cucurbituril polymer, Environmental Progress and Sustainable Energy 34 (2015) 512–519, https://doi.org/10.1002/ep.12029.

[52] H. Li, M. Eddaoudi, M. O'Keeffe, O.M. Yaghi, Design and synthesis of an exceptionally stable and highly porous metal- organic framework, Nature 402 (1999) 276–279, https://doi.org/10.1038/46248.

[53] W. Chaikittisilp, K. Ariga, Y. Yamauchi, A new family of carbon materials: synthesis of MOF-derived nanoporous carbons and their promising applications, Journal of Materials Chemistry A 1 (2013) 14–19, https://doi.org/10.1039/c2ta00278g.

[54] K. Sumida, D.L. Rogow, J.A. Mason, T.M. McDonald, E.D. Bloch, Z.R. Herm, T.-H. Bae, J.R. Long, Carbon dioxide capture in metal-organic frameworks, Chemical Reviews 112 (2012) 724–781, https://doi.org/10.1021/cr2003272.

[55] J. Lee, O.K. Farha, J. Roberts, K.A. Scheidt, S.T. Nguyen, J.T. Hupp, Metal-organic framework materials as catalysts, Chemical Society Reviews 38 (2009) 1450–1459, https://doi.org/10.1039/b807080f.

[56] P.Z. Moghadam, A. Li, S.B. Wiggin, A. Tao, A.G.P. Maloney, P.A. Wood, S.C. Ward, D. Fairen-Jimenez, Development of a Cambridge structural database subset: a collection of metal–organic frameworks for past, present, and future, Chemistry of Materials 29 (2017) 2618–2625, https://doi.org/10.1021/acs.chemmater.7b00441.

[57] S. Øien-ØDegaard, G.C. Shearer, D.S. Wragg, K.P. Lillerud, Pitfalls in metal-organic framework crystallography: towards more accurate crystal structures, Chemical Society Reviews 46 (2017) 4867–4876, https://doi.org/10.1039/c6cs00533k.

[58] J. Ma, X. Guo, Y. Ying, D. Liu, C. Zhong, Composite ultrafiltration membrane tailored by MOF@GO with highly improved water purification performance, Chemical Engineering Journal 313 (2017) 890–898, https://doi.org/10.1016/j.cej.2016.10.127.

[59] W.P. Mounfield, U. Tumuluri, Y. Jiao, M. Li, S. Dai, Z. Wu, K.S. Walton, Role of defects and metal coordination on adsorption of acid gases in MOFs and metal oxides: an in situ IR spectroscopic study, Microporous and Mesoporous Materials 227 (2016) 65–75, https://doi.org/10.1016/j.micromeso.2016.02.023.

[60] S. Singha, S.K. Maity, S. Biswas, R. Saha, S. Kumar, A magnesium-based bifunctional MOF: studies on proton conductivity, gas and water adsorption, Inorganica Chimica Acta 453 (2016) 321–329, https://doi.org/10.1016/j.ica.2016.08.033.

[61] K. Huang, S. Liu, Q. Li, W. Jin, Preparation of novel metal-carboxylate system MOF membrane for gas separation, Separation and Purification Technology 119 (2013) 94–101, https://doi.org/10.1016/j.seppur.2013.09.008.

[62] S. Castarlenas, C. Téllez, J. Coronas, Gas separation with mixed matrix membranes obtained from MOF UiO-66-graphite oxide hybrids, Journal of Membrane Science 526 (2017) 205–211, https://doi.org/10.1016/j.memsci.2016.12.041.

[63] C. Huang, R. Sun, H. Lu, Q. Yang, J. Hu, H. Wang, H. Liu, A pilot trial for fast deep desulfurization on MOF-199 by simultaneous adsorption-separation via hydrocyclones, Separation and Purification Technology 182 (2017) 110–117, https://doi.org/10.1016/j.seppur.2017.03.046.

[64] Y. Chen, D. Lv, J. Wu, J. Xiao, H. Xi, Q. Xia, Z. Li, A new MOF-505@GO composite with high selectivity for CO_2/CH_4 and CO_2/N_2 separation, Chemical Engineering Journal 308 (2017) 1065–1072, https://doi.org/10.1016/j.cej.2016.09.138.

[65] J.L. Harding, J.M. Metz, M.M. Reynolds, A tunable, stable, and bioactive MOF catalyst for generating a localized therapeutic from endogenous sources, Advanced Functional Materials 24 (2014) 7503–7509, https://doi.org/10.1002/adfm.201402529.

[66] D. Wang, Z. Li, Coupling MOF-based photocatalysis with Pd catalysis over Pd@MIL-100(Fe) for efficient N-alkylation of amines with alcohols under visible light, Journal of Catalysis 342 (2016) 151–157, https://doi.org/10.1016/j.jcat.2016.07.021.

[67] T. Wu, L. Shen, M. Luebbers, C. Hu, Q. Chen, Z. Ni, R.I. Masel, Enhancing the stability of metal-organic frameworks in humid air by incorporating water repellent functional groups, Chemical Communications 46 (2010) 6120–6122, https://doi.org/10.1039/c0cc01170c.

[68] H.N. Rubin, M.M. Reynolds, Functionalization of metal-organic frameworks to achieve controllable wettability, Inorganic Chemistry 56 (2017) 5266–5274, https://doi.org/10.1021/acs.inorgchem.7b00373.

[69] J. Yao, H. Wang, Zeolitic imidazolate framework composite membranes and thin films: synthesis and applications, Chemical Society Reviews 43 (2014) 4470–4493, https://doi.org/10.1039/c3cs60480b.

[70] J.B. James, Y.S. Lin, Thermal stability of ZIF-8 membranes for gas separations, Journal of Membrane Science 532 (2017) 9–19, https://doi.org/10.1016/j.memsci.2017.02.017.

[71] H. Hayashi, A.P. Côté, H. Furukawa, M. O'Keeffe, O.M. Yaghi, Zeolite A imidazolate frameworks, Nature Materials 6 (2007) 501–506, https://doi.org/10.1038/nmat1927.

[72] R. Banerjee, A. Phan, B. Wang, C. Knobler, H. Furukawa, M. O'Keeffe, O.M. Yaghi, High-throughput synthesis of zeolitic imidazolate frameworks and application to CO_2 capture, Science 319 (2008) 939–943, https://doi.org/10.1126/science.1152516.

[73] K.S. Park, Z. Ni, A.P. Côté, J.Y. Choi, R. Huang, F.J. Uribe-Romo, H.K. Chae, M. O'Keeffe, O.M. Yaghi, Exceptional chemical and thermal stability of zeolitic imidazolate frameworks, Proceedings of the National Academy of Sciences of the United States of America 103 (2006) 10186–10191, https://doi.org/10.1073/pnas.0602439103.

[74] H. Zhang, D. Liu, Y. Yao, B. Zhang, Y.S. Lin, Stability of ZIF-8 membranes and crystalline powders in water at room temperature, Journal of Membrane Science 485 (2015) 103–111, https://doi.org/10.1016/j.memsci.2015.03.023.

[75] H. Jiang, Q. Yan, R. Chen, W. Xing, Synthesis of Pd@ZIF-8 via an assembly method: influence of the molar ratios of Pd/Zn^{2+} and 2-methylimidazole/Zn^{2+}, Microporous and Mesoporous Materials 225 (2016) 33–40, https://doi.org/10.1016/j.micromeso.2015.12.010.

[76] D. Ge, H.K. Lee, Water stability of zeolite imidazolate framework 8 and application to porous membrane-protected micro-solid-phase extraction of polycyclic aromatic hydrocarbons from environmental water samples, Journal of Chromatography A 1218 (2011) 8490–8495, https://doi.org/10.1016/j.chroma.2011.09.077.

[77] J.-Q. Jiang, C.-X. Yang, X.-P. Yan, Zeolitic imidazolate framework-8 for fast adsorption and removal of benzotriazoles from aqueous solution, ACS Applied Materials and Interfaces 5 (2013) 9837–9842, https://doi.org/10.1021/am403079n.

[78] H.-L. Jiang, B. Liu, Y.-Q. Lan, K. Kuratani, T. Akita, H. Shioyama, F. Zong, Q. Xu, From metal-organic framework to nanoporous carbon: toward a very high surface area and hydrogen uptake, Journal of the American Chemical Society 133 (2011) 11854–11857, https://doi.org/10.1021/ja203184k.

[79] J.M. Carlsson, M. Scheffler, Structural, electronic, and chemical properties of nanoporous carbon, Physical Review Letters 96 (2006), https://doi.org/10.1103/PhysRevLett.96.046806.

[80] M. Jiang, X. Cao, D. Zhu, Y. Duan, J. Zhang, Hierarchically porous N-doped carbon derived from ZIF-8 nanocomposites for electrochemical applications, Electrochimica Acta 196 (2016) 699–707, https://doi.org/10.1016/j.electacta.2016.02.094.

[81] N.L. Torad, M. Hu, S. Ishihara, H. Sukegawa, A.A. Belik, M. Imura, K. Ariga, Y. Sakka, Y. Yamauchi, Direct synthesis of MOF-derived nanoporous carbon with magnetic Co nanoparticles toward efficient water treatment, Small 10 (2014) 2096–2107, https://doi.org/10.1002/smll.201302910.

[82] M. Angamuthu, G. Satishkumar, M. V Landau, Precisely controlled encapsulation of Fe_3O_4 nanoparticles in mesoporous carbon nanodisk using iron based MOF precursor for effective dye removal, Microporous and Mesoporous Materials 251 (2017) 58–68, https://doi.org/10.1016/j.micromeso.2017.05.045.

[83] E. Unur, Functional nanoporous carbons from hydrothermally treated biomass for environmental purification, Microporous and Mesoporous Materials 168 (2013) 92–101, https://doi.org/10.1016/j.micromeso.2012.09.027.

[84] A. Comotti, S. Bracco, P. Sozzani, S. Horike, R. Matsuda, J. Chen, M. Takata, Y. Kubota, S. Kitagawa, Nanochannels of two distinct cross-sections in a porous Al-based coordination polymer, Journal of the American Chemical Society 130 (2008) 13664–13672, https://doi.org/10.1021/ja802589u.

[85] M. Hu, J. Reboul, S. Furukawa, N.L. Torad, Q. Ji, P. Srinivasu, K. Ariga, S. Kitagawa, Y. Yamauchi, Direct carbonization of Al-based porous coordination polymer for synthesis of nanoporous carbon, Journal of the American Chemical Society 134 (2012) 2864–2867, https://doi.org/10.1021/ja208940u.

[86] N.L. Torad, M. Hu, Y. Kamachi, K. Takai, M. Imura, M. Naito, Y. Yamauchi, Facile synthesis of nanoporous carbons with controlled particle sizes by direct carbonization of monodispersed ZIF-8 crystals, Chemical Communications 49 (2013) 2521–2523, https://doi.org/10.1039/c3cc38955c.

[87] S. Lim, K. Suh, Y. Kim, M. Yoon, H. Park, D.N. Dybtsev, K. Kim, Porous carbon materials with a controllable surface area synthesized from metal-organic frameworks, Chemical Communications 48 (2012) 7447–7449, https://doi.org/10.1039/c2cc33439a.

[88] Z. Hasan, S.H. Jhung, Removal of hazardous organics from water using metal-organic frameworks (MOFs): plausible mechanisms for selective adsorptions, Journal of Hazardous Materials 283 (2015) 329–339, https://doi.org/10.1016/j.jhazmat.2014.09.046.

[89] E. Haque, J.E. Lee, I.T. Jang, Y.K. Hwang, J.-S. Chang, J. Jegal, S.H. Jhung, Adsorptive removal of methyl orange from aqueous solution with metal-organic frameworks, porous chromium-benzenedicarboxylates, Journal of Hazardous Materials 181 (2010) 535–542, https://doi.org/10.1016/j.jhazmat.2010.05.047.

[90] M. Mon, R. Bruno, J. Ferrando-Soria, D. Armentano, E. Pardo, Metal-organic framework technologies for water remediation: towards a sustainable ecosystem, Journal of Materials Chemistry A 6 (2018) 4912–4947, https://doi.org/10.1039/c8ta00264a.

[91] E. Haque, J.W. Jun, S.H. Jhung, Adsorptive removal of methyl orange and methylene blue from aqueous solution with a metal-organic framework material, iron terephthalate (MOF-235), Journal of Hazardous Materials 185 (2011) 507–511, https://doi.org/10.1016/j.jhazmat.2010.09.035.

[92] S. Lin, Z. Song, G. Che, A. Ren, P. Li, C. Liu, J. Zhang, Adsorption behavior of metal-organic frameworks for methylene blue from aqueous solution, Microporous and Mesoporous Materials 193 (2014) 27–34, https://doi.org/10.1016/j.micromeso.2014.03.004.

[93] B. Liu, F. Yang, Y. Zou, Y. Peng, Adsorption of phenol and p -nitrophenol from aqueous solutions on metal-organic frameworks: effect of hydrogen bonding, Journal of Chemical Engineering Data 59 (2014) 1476–1482, https://doi.org/10.1021/je4010239.

[94] N.A. Khan, B.K. Jung, Z. Hasan, S.H. Jhung, Adsorption and removal of phthalic acid and diethyl phthalate from water with zeolitic imidazolate and metal-organic frameworks, Journal of Hazardous Materials 282 (2015) 194–200, https://doi.org/10.1016/j.jhazmat.2014.03.047.

[95] M. Tong, D. Liu, Q. Yang, S. Devautour-Vinot, G. Maurin, C. Zhong, Influence of framework metal ions on the dye capture behavior of MIL-100 (Fe, Cr) MOF type solids, Journal of Materials Chemistry A 1 (2013) 8534–8537, https://doi.org/10.1039/c3ta11807j.

[96] F.-X. Qin, S.-Y. Jia, Y. Liu, H.-Y. Li, S.-H. Wu, Adsorptive removal of bisphenol A from aqueous solution using metal-organic frameworks, Desalination Water Treatment 54 (2015) 93–102, https://doi.org/10.1080/19443994.2014.883331.

[97] S.-H. Huo, X.-P. Yan, Metal-organic framework MIL-100(Fe) for the adsorption of malachite green from aqueous solution, Journal of Materials Chemistry 22 (2012) 7449–7455, https://doi.org/10.1039/c2jm16513a.

[98] E.Y. Park, Z. Hasan, N.A. Khan, S.H. Jhung, Adsorptive removal of bisphenol-a from water with a metal-organic framework, a porous chromium-benzenedicarboxylate, Journal of Nanoscience Nanotechnology 13 (2013) 2789–2794, https://doi.org/10.1166/jnn.2013.7411.

[99] I. Akpinar, R.J. Drout, T. Islamoglu, S. Kato, J. Lyu, O.K. Farha, Exploiting π- π interactions to design an efficient sorbent for atrazine removal from water, ACS Applied Materials and Interfaces (2019), https://doi.org/10.1021/acsami.8b20355.

[100] C. Yang, U. Kaipa, Q.Z. Mather, X. Wang, V. Nesterov, A.F. Venero, M.A. Omary, Fluorous metal-organic frameworks with superior adsorption and hydrophobic properties toward oil spill cleanup and hydrocarbon storage, Journal of the American Chemical Society 133 (2011) 18094–18097, https://doi.org/10.1021/ja208408n.

[101] K.-Y.A. Lin, Y.-C. Chen, S. Phattarapattamawong, Efficient demulsification of oil-in-water emulsions using a zeolitic imidazolate framework: adsorptive removal of oil droplets from water, Journal of Colloid and Interface Science 478 (2016) 97–106, https://doi.org/10.1016/j.jcis.2016.05.057.

[102] H. Reinsch, "Green" synthesis of metal-organic frameworks, European Journal of Inorganic Chemistry 2016 (2016) 4290–4299, https://doi.org/10.1002/ejic.201600286.

[103] P. Anastas, N. Eghbali, Green chemistry: principles and practice, Chemical Society Reviews 39 (2010) 301–312, https://doi.org/10.1039/b918763b.

[104] P.A. Julien, C. Mottillo, T. Friščić, Metal-organic frameworks meet scalable and sustainable synthesis, Green Chemistry 19 (2017) 2729–2747, https://doi.org/10.1039/c7gc01078h.

[105] B.M. Connolly, J.P. Mehta, P.Z. Moghadam, A.E.H. Wheatley, D. Fairen-Jimenez, From synthesis to applications: metal–organic frameworks for an environmentally sustainable future, Current Opinion in Green and Sustainable Chemistry 12 (2018) 47–56, https://doi.org/10.1016/j.cogsc.2018.06.012.

[106] X. Li, B. Wang, Y. Cao, S. Zhao, H. Wang, X. Feng, J. Zhou, X. Ma, Water contaminant elimination based on metal-organic frameworks and perspective on their industrial applications, ACS Sustainable Chemistry and Engineering 7 (2019) 4548–4563, https://doi.org/10.1021/acssuschemeng.8b05751.

[107] E. Biemmi, S. Christian, N. Stock, T. Bein, High-throughput screening of synthesis parameters in the formation of the metal-organic frameworks MOF-5 and HKUST-1, Microporous and Mesoporous Materials 117 (2009) 111–117, https://doi.org/10.1016/j.micromeso.2008.06.040.

[108] U. Mueller, M. Schubert, F. Teich, H. Puetter, K. Schierle-Arndt, J. Pastré, Metal-organic frameworks – prospective industrial applications, Journal of Materials Chemistry 16 (2006) 626–636, https://doi.org/10.1039/b511962f.

[109] J. Ren, X. Dyosiba, N.M. Musyoka, H.W. Langmi, B.C. North, M. Mathe, M.S. Onyango, Green synthesis of chromium-based metal-organic framework (Cr-MOF) from waste polyethylene terephthalate (PET) bottles for hydrogen storage applications, International Journal of Hydrogen Energy 41 (2016) 18141–18146, https://doi.org/10.1016/j.ijhydene.2016.08.040.

[110] A.C. Dreischarf, M. Lammert, N. Stock, H. Reinsch, Green synthesis of Zr-CAU-28: structure and properties of the first Zr-MOF based on 2,5-furandicarboxylic acid, Inorganic Chemistry 56 (2017) 2270–2277, https://doi.org/10.1021/acs.inorgchem.6b02969.

[111] S. Wang, M. Wahiduzzaman, L. Davis, A. Tissot, W. Shepard, J. Marrot, C. Martineau-Corcos, D. Hamdane, G. Maurin, S. Devautour-Vinot, C. Serre, A robust zirconium amino acid metal-organic framework for proton conduction, Nature Communications 9 (2018), https://doi.org/10.1038/s41467-018-07414-4.

[112] M. Gaab, N. Trukhan, S. Maurer, R. Gummaraju, U. Müller, The progression of Al-based metal-organic frameworks - from academic research to industrial production and applications, Microporous and Mesoporous Materials 157 (2012) 131–136, https://doi.org/10.1016/j.micromeso.2011.08.016.

[113] R.K. Henderson, C. Jiménez-González, D.J.C. Constable, S.R. Alston, G.G.A. Inglis, G. Fisher, J. Sherwood, S.P. Binks, A.D. Curzons, Expanding GSK's solvent selection guide - embedding sustainability into solvent selection starting at medicinal chemistry, Green Chemistry 13 (2011) 854–862, https://doi.org/10.1039/c0gc00918k.

[114] L. Cseri, M. Razali, P. Pogany, G. Szekely, Organic solvents in sustainable synthesis and engineering, in: Green Chem. An Incl. Approach, 2017, pp. 513–553, https://doi.org/10.1016/B978-0-12-809270-5.00020-0.

[115] J. Zhang, G.B. White, M.D. Ryan, A.J. Hunt, M.J. Katz, Dihydrolevoglucosenone (Cyrene) as a green alternative to N,N-dimethylformamide (DMF) in MOF synthesis, ACS Sustainable Chemistry and Engineering 4 (2016) 7186–7192, https://doi.org/10.1021/acssuschemeng.6b02115.

[116] E.R. Parnham, R.E. Morris, Ionothermal synthesis of zeolites, metal-organic frameworks, and inorganic-organic hybrids, Accounts of Chemical Research 40 (2007) 1005–1013, https://doi.org/10.1021/ar700025k.

[117] J. Yao, L. Li, W.H. Benjamin Wong, C. Tan, D. Dong, H. Wang, Formation of ZIF-8 membranes and crystals in a diluted aqueous solution, Materials Chemistry and Physics 139 (2013) 1003–1008, https://doi.org/10.1016/j.matchemphys.2013.02.071.

[118] I.A. Ibarra, P.A. Bayliss, E. Pérez, S. Yang, A.J. Blake, H. Nowell, D.R. Allan, M. Poliakoff, M. Schröder, Near-critical water, a cleaner solvent for the synthesis of a metal–organic framework, Green Chemistry 14 (2012) 117–122, https://doi.org/10.1039/C1GC15726D.

[119] M. Rubio-Martinez, C. Avci-Camur, A.W. Thornton, I. Imaz, D. Maspoch, M.R. Hill, New synthetic routes towards MOF production at scale, Chemical Society Reviews 46 (2017) 3453–3480, https://doi.org/10.1039/c7cs00109f.

[120] J. Ren, X. Dyosiba, N.M. Musyoka, H.W. Langmi, M. Mathe, S. Liao, Review on the current practices and efforts towards pilot-scale production of metal-organic frameworks (MOFs), Coordination Chemistry Reviews 352 (2017) 187–219, https://doi.org/10.1016/j.ccr.2017.09.005.

[121] S. Tanaka, T. Nagaoka, A. Yasuyoshi, Y. Hasegawa, J.F.M. Denayer, Hierarchical pore development of ZIF-8 MOF by simple salt-assisted mechanosynthesis, Crystal Growth and Design 18 (2018) 274–279, https://doi.org/10.1021/acs.cgd.7b01211.

[122] B. Karadeniz, A.J. Howarth, T. Stolar, T. Islamoglu, I. Dejanović, M. Tireli, M.C. Wasson, S.-Y. Moon, O.K. Farha, T. Friščić, K. Užarević, Benign by design: green and scalable synthesis of zirconium UiO-metal-organic frameworks by water-assisted mechanochemistry, ACS Sustainable Chemistry and Engineering 6 (2018) 15841–15849, https://doi.org/10.1021/acssuschemeng.8b04458.

[123] M.J. Cliffe, C. Mottillo, R.S. Stein, D.-K. Bučar, T. Friščić, Accelerated aging: a low energy, solvent-free alternative to solvothermal and mechanochemical synthesis of metal-organic materials, Chemical Science 3 (2012) 2495–2500, https://doi.org/10.1039/c2sc20344h.

[124] C. Mottillo, Y. Lu, M.-H. Pham, M.J. Cliffe, T.-O. Do, T. Friščić, Mineral neogenesis as an inspiration for mild, solvent-free synthesis of bulk microporous metal-organic frameworks from metal (Zn, Co) oxides, Green Chemistry 15 (2013) 2121–2131, https://doi.org/10.1039/c3gc40520f.

[125] A. Garcia Marquez, P. Horcajada, D. Grosso, G. Ferey, C. Serre, C. Sanchez, C. Boissiere, Green scalable aerosol synthesis of porous metal-organic frameworks, Chemical Communications 49 (2013) 3848–3850, https://doi.org/10.1039/c3cc39191d.

[126] G. Férey, Hybrid porous solids: past, present, future, Chemical Society Reviews 37 (2008) 191–214, https://doi.org/10.1039/b618320b.

[127] D.J. Tranchemontagne, J.L. Mendoza-Cortés, M. O'Keeffe, O.M. Yaghi, Secondary building units, nets and bonding in the chemistry of metal-organic frameworks, Chemical Society Reviews 38 (2009) 1257–1283, https://doi.org/10.1039/b817735j.

[128] L. Alaerts, C.E.A. Kirschhock, M. Maes, M.A. Van Der Veen, V. Finsy, A. Depla, J.A. Martens, G. V Baron, P.A. Jacobs, J.F.M. Denayer, D.E. De Vos, Selective adsorption and separation of xylene isomers and ethylbenzene with the microporous vanadium(IV) terephthalate MIL-47, Angewandte Chemie International Edition 46 (2007) 4293–4297, https://doi.org/10.1002/anie.200700056.

[129] F. Akhtar, L. Andersson, S. Ogunwumi, N. Hedin, L. Bergström, Structuring adsorbents and catalysts by processing of porous powders, Journal of the European Ceramic Society 34 (2014) 1643–1666, https://doi.org/10.1016/j.jeurceramsoc.2014.01.008.

[130] D. Zacher, O. Shekhah, C. Wöll, R.A. Fischer, Thin films of metal-organic frameworks, Chemical Society Reviews 38 (2009) 1418–1429, https://doi.org/10.1039/b805038b.
[131] M.G. Schwab, I. Senkovska, M. Rose, M. Koch, J. Pahnke, G. Jonschker, S. Kaskel, MOF@PolyHIPEs, Advanced Engineering Materials 10 (2008) 1151–1155, https://doi.org/10.1002/adem.200800189.
[132] U.-H. Lee, A.H. Valekar, Y.K. Hwang, J.-S. Chang, Granulation and shaping of metal–organic frameworks, Chemistry of Metal-Organic Frameworks (2016), https://doi.org/10.1002/9783527693078.ch18.
[133] H. Yin, H. Kim, J. Choi, A.C.K. Yip, Thermal stability of ZIF-8 under oxidative and inert environments: a practical perspective on using ZIF-8 as a catalyst support, Chemical Engineering Journal 278 (2015) 293–300, https://doi.org/10.1016/j.cej.2014.08.075.
[134] I.T. Cameron, F.Y. Wang, C.D. Immanuel, F. Stepanek, Process systems modelling and applications in granulation: a review, Chemical Engineering Science 60 (2005) 3723–3750, https://doi.org/10.1016/j.ces.2005.02.004.
[135] J. Ren, N.M. Musyoka, H.W. Langmi, A. Swartbooi, B.C. North, M. Mathe, A more efficient way to shape metal-organic framework (MOF) powder materials for hydrogen storage applications, International Journal of Hydrogen Energy 40 (2015) 4617–4622, https://doi.org/10.1016/j.ijhydene.2015.02.011.
[136] C. Avci-Camur, J. Troyano, J. Pérez-Carvajal, A. Legrand, D. Farrusseng, I. Imaz, D. Maspoch, Aqueous production of spherical Zr-MOF beads: via continuous-flow spray-drying, Green Chemistry 20 (2018) 873–878, https://doi.org/10.1039/c7gc03132g.
[137] D. Crawford, J. Casaban, R. Haydon, N. Giri, T. McNally, S.L. James, Synthesis by extrusion: continuous, large-scale preparation of MOFs using little or no solvent, Chemical Science 6 (2015) 1645–1649, https://doi.org/10.1039/c4sc03217a.
[138] P. Küsgens, A. Zgaverdea, H.-G. Fritz, S. Siegle, S. Kaskel, Metal-organic frameworks in monolithic structures, Journal of the American Ceramic Society 93 (2010) 2476–2479, https://doi.org/10.1111/j.1551-2916.2010.03824.x.
[139] J. Alcañiz-Monge, G. Trautwein, M. Pérez-Cadenas, M.C. Román-Martínez, Effects of compression on the textural properties of porous solids, Microporous and Mesoporous Materials 126 (2009) 291–301, https://doi.org/10.1016/j.micromeso.2009.06.020.
[140] M.G. Plaza, A.F.P. Ferreira, J.C. Santos, A.M. Ribeiro, U. Müller, N. Trukhan, J.M. Loureiro, A.E. Rodrigues, Propane/propylene separation by adsorption using shaped copper trimesate MOF, Microporous and Mesoporous Materials 157 (2012) 101–111, https://doi.org/10.1016/j.micromeso.2011.06.024.
[141] U. Mueller, L. Lobree, M. Hesse, O. Yaghi, M. Eddaoudi, Shaped Bodies Containing Metal-Organic Frameworks 6 (2005) 893, 564.
[142] M. Tagliabue, C. Rizzo, R. Millini, P.D.C. Dietzel, R. Blom, S. Zanardi, Methane storage on CPO-27-Ni pellets, Journal of Porous Materials 18 (2011) 289–296, https://doi.org/10.1007/s10934-010-9378-0.
[143] D. Bazer-Bachi, L. Assié, V. Lecocq, B. Harbuzaru, V. Falk, Towards industrial use of metal-organic framework: impact of shaping on the MOF properties, Powder Technology 255 (2014) 52–59, https://doi.org/10.1016/j.powtec.2013.09.013.
[144] S. Couck, E. Gobechiya, C.E.A. Kirschhock, P. Serra-Crespo, J. Juan-Alcañiz, A. Martinez Joaristi, E. Stavitski, J. Gascon, F. Kapteijn, G. V Baron, J.F.M. Denayer, Adsorption and separation of light gases on an amino-functionalized metal-organic framework: an adsorption and in situ XRD study, ChemSusChem 5 (2012) 740–750, https://doi.org/10.1002/cssc.201100378.
[145] P. Serra-Crespo, E. Stavitski, F. Kapteijn, J. Gascon, High compressibility of a flexible metal-organic framework, RSC Advances 2 (2012) 5051–5053, https://doi.org/10.1039/c2ra20528a.
[146] Y. Hu, H. Lian, L. Zhou, G. Li, In situ solvothermal growth of metal-organic framework-5 supported on porous copper foam for noninvasive sampling of plant volatile sulfides, Analytical Chemistry 87 (2015) 406–412, https://doi.org/10.1021/ac502146c.
[147] Z. Abbasi, E. Shamsaei, X.-Y. Fang, B. Ladewig, H. Wang, Simple fabrication of zeolitic imidazolate framework ZIF-8/polymer composite beads by phase inversion method for efficient oil sorption, Journal of Colloid and Interface Science 493 (2017), https://doi.org/10.1016/j.jcis.2017.01.006.
[148] Y. Zeng, K. Wang, J. Yao, H. Wang, Hollow carbon beads fabricated by phase inversion method for efficient oil sorption, Carbon N. Y. 69 (2014) 25–31, https://doi.org/10.1016/j.carbon.2013.11.036.
[149] X. Gu, K. Zhou, Y. Li, J. Yao, Millimeter-sized carbon/TiO_2 beads fabricated by phase inversion method for oil and dye adsorption, RSC Advances 6 (2016) 16314–16318, https://doi.org/10.1039/c5ra28058c.

[150] H. Thakkar, S. Eastman, Q. Al-Naddaf, A.A. Rownaghi, F. Rezaei, 3D-Printed metal-organic framework monoliths for gas adsorption processes, ACS Applied Materials and Interfaces 9 (2017) 35908–35916, https://doi.org/10.1021/acsami.7b11626.

[151] M.C. Kreider, M. Sefa, J.A. Fedchak, J. Scherschligt, M. Bible, B. Natarajan, N.N. Klimov, A.E. Miller, Z. Ahmed, M.R. Hartings, Toward 3D printed hydrogen storage materials made with ABS-MOF composites, Polymers for Advanced Technologies 29 (2018) 867–873, https://doi.org/10.1002/pat.4197.

[152] Z. Hasan, J. Jeon, S.H. Jhung, Adsorptive removal of naproxen and clofibric acid from water using metal-organic frameworks, Journal of Hazardous Materials 209–210 (2012) 151–157, https://doi.org/10.1016/j.jhazmat.2012.01.005.

[153] X.-X. Huang, L.-G. Qiu, W. Zhang, Y.-P. Yuan, X. Jiang, A.-J. Xie, Y.-H. Shen, J.-F. Zhu, Hierarchically meso-structured MIL-101 metal-organic frameworks: supramolecular template-directed synthesis and accelerated adsorption kinetics for dye removal, CrystEngComm 14 (2012) 1613–1617, https://doi.org/10.1039/c1ce06138k.

[154] K.-Y.A. Lin, H.-A. Chang, Ultra-high adsorption capacity of zeolitic imidazole framework-67 (ZIF-67) for removal of malachite green from water, Chemosphere 139 (2015) 624–631, https://doi.org/10.1016/j.chemosphere.2015.01.041.

[155] H.-N. Wang, F.-H. Liu, X.-L. Wang, K.-Z. Shao, Z.-M. Su, Three neutral metal-organic frameworks with micro- and meso-pores for adsorption and separation of dyes, Journal of Materials Chemistry A 1 (2013) 13060–13063, https://doi.org/10.1039/c3ta13242k.

[156] S. Han, S. Kim, H. Lim, W. Choi, H. Park, J. Yoon, T. Hyeon, New nanoporous carbon materials with high adsorption capacity and rapid adsorption kinetics for removing humic acids, Microporous and Mesoporous Materials 58 (2003) 131–135, https://doi.org/10.1016/S1387-1811(02)00611-X.

[157] S.R. Sandeman, V.M. Gun'ko, O.M. Bakalinska, C.A. Howell, Y. Zheng, M.T. Kartel, G.J. Phillips, S. V Mikhalovsky, Adsorption of anionic and cationic dyes by activated carbons, PVA hydrogels, and PVA/AC composite, Journal of Colloid and Interface Science 358 (2011) 582–592, https://doi.org/10.1016/j.jcis.2011.02.031.

[158] P. Kowalczyk, A.P. Terzyk, P.A. Gauden, V.M. Gun'ko, L. Solarz, Evaluation of the structural and energetic heterogeneity of microporous carbons by means of novel numerical methods and genetic algorithms, Journal of Colloid and Interface Science 256 (2002) 378–395, https://doi.org/10.1006/jcis.2002.8680.

[159] R. Leboda, V.V. Turov, W. Tomaszewski, V.M. Gun'ko, J. Skubiszewska-Zieba, Effect of adsorption of nitroaromatic compounds on the characteristics of bound water layers in aqueous suspensions of activated carbons, Carbon N. Y. 40 (2002) 389–396, https://doi.org/10.1016/S0008-6223(01)00132-4.

[160] V.M. Gun'ko, V.V. Turov, J. Skubiszewska-Zieba, R. Leboda, M.D. Tsapko, D. Palijczuk, Structural characteristics of a carbon adsorbent and influence of organic solvents on interfacial water, Applied Surface Science 214 (2003) 178–189, https://doi.org/10.1016/S0169-4332(03)00345-3.

[161] M. Ghaedi, S. Heidarpour, S. Nasiri Kokhdan, R. Sahraie, A. Daneshfar, B. Brazesh, Comparison of silver and palladium nanoparticles loaded on activated carbon for efficient removal of Methylene blue: kinetic and isotherm study of removal process, Powder Technology 228 (2012) 18–25, https://doi.org/10.1016/j.powtec.2012.04.030.

[162] P.C.C. Faria, J.J.M. Órfão, M.F.R. Pereira, Adsorption of anionic and cationic dyes on activated carbons with different surface chemistries, Water Research 38 (2004) 2043–2052, https://doi.org/10.1016/j.watres.2004.01.034.

[163] Y. Gao, S. Xu, Q. Yue, Y. Wu, B. Gao, Chemical preparation of crab shell-based activated carbon with superior adsorption performance for dye removal from wastewater, Journal of the Taiwan Institute of Chemical Engineers 61 (2016) 327–335, https://doi.org/10.1016/j.jtice.2015.12.023.

[164] A. Khaled, A.E. Nemr, A. El-Sikaily, O. Abdelwahab, Removal of Direct N Blue-106 from artificial textile dye effluent using activated carbon from orange peel: adsorption isotherm and kinetic studies, Journal of Hazardous Materials 165 (2009) 100–110, https://doi.org/10.1016/j.jhazmat.2008.09.122.

[165] N. Hoda, E. Bayram, E. Ayranci, Kinetic and equilibrium studies on the removal of acid dyes from aqueous solutions by adsorption onto activated carbon cloth, Journal of Hazardous Materials 137 (2006) 344–351, https://doi.org/10.1016/j.jhazmat.2006.02.009.

[166] B.H. Hameed, F.B.M. Daud, Adsorption studies of basic dye on activated carbon derived from agricultural waste: *Hevea brasiliensis* seed coat, Chemical Engineering Journal 139 (2008) 48–55, https://doi.org/10.1016/j.cej.2007.07.089.

[167] J.-D. Xiao, L.-G. Qiu, X. Jiang, Y.-J. Zhu, S. Ye, X. Jiang, Magnetic porous carbons with high adsorption capacity synthesized by a microwave-enhanced high temperature ionothermal method from a Fe-based metal-organic framework, Carbon N. Y. 59 (2013) 372–382, https://doi.org/10.1016/j.carbon.2013.03.032.

[168] C. Jiao, Y. Wang, M. Li, Q. Wu, C. Wang, Z. Wang, Synthesis of magnetic nanoporous carbon from metal-organic framework for the fast removal of organic dye from aqueous solution, Journal of Magnetism and Magnetic Materials 407 (2016) 24–30, https://doi.org/10.1016/j.jmmm.2016.01.031.

[169] B. Chen, G. Ma, D. Kong, Y. Zhu, Y. Xia, Atomically homogeneous dispersed ZnO/N-doped nanoporous carbon composites with enhanced CO_2 uptake capacities and high efficient organic pollutants removal from water, Carbon N. Y. 95 (2015) 113–124, https://doi.org/10.1016/j.carbon.2015.08.015.

CHAPTER

8

Toward Sustainable Chemical Processing With Graphene-Based Materials

Farhad Moghadam, Ji Soo Roh, Jae Eun Shin, Ho Bum Park

Department of Energy Engineering, Hanyang University, Seoul, Republic of Korea

OUTLINE

1. INTRODUCTION

Separation processes generally account for large portion of costs in the chemical industry. Therefore, development of advanced separation technologies with viable economical costs is of critical needs. In this regard, membrane-based separation technology continues to attract intense interests in the broad area of the chemical industry because of its low energy and operating costs, ease of operation, and low environmental impacts, compared with conventional separation processes such as chemical absorption, physical adsorption, cryogenic systems, etc. [1]. Polymers are the most common materials that have been widely used for membrane fabrication in both gas- and water separation—based processes. However, polymers are mostly subjected to selectivity—permeability trade-off behavior, which means membranes with high selectivity suffer from low permeability and vice versa [2,3]. This intrinsic property has largely limited the potential application of polymers in membrane technology so that despite

Sustainable Nanoscale Engineering
https://doi.org/10.1016/B978-0-12-814681-1.00008-4

numerous polymer materials developed so far, the number of polymer materials exploited in industrial scale separation area is limited [4]. Hence, developing and designing new membrane materials with promising separation performance is vital for the chemical industry. Two-dimensional (2D) materials are an emerging class of nanomaterials that hold a great promise in membrane-based separation area due to their unique structural and physiochemical features [5–7]. As the permeance of a membrane is inversely proportional to its thickness, 2D materials with infinitesimal thickness offer a unique building block for realizing new membrane materials with outstanding gas and water separation performance. Recently, extensive studies have been particularly focused on fabrication of 2D material–based membranes such as graphene oxide (GO), hexagonal boron nitride, transition metal dichalcogenide, layered double hydroxide, and metal–organic framework (MOF) with promising separation performance in the lab-scale [5,7–9]. As a relatively mature and well-studied 2D material, graphene and its derivatives have been extensively explored in the membrane-based liquid and gas separation field. In this chapter, we aim to first describe the structural properties of graphene and then present the latest breakthrough advances in the synthesis of large area graphene. Then, we will particularly focus on fabrication of porous graphene (PG) and GO membranes. In addition, we highlight the performance of GO membranes, both pristine and mixed matrix GO membranes, in water purification, organic solvent nanofiltration (OSN), and gas separation area. The water and gas permeation mechanism will be also discussed in detail.

2. GRAPHENE

Graphene is a 2D nanomaterial with one atomic thickness and hexagonal carbon lattice, which is an individual layer of graphite. Since Novoselov et al. [10] discovered graphene by detaching it from graphite bulk, graphene has been studied intensively due to its promising potential applications. Carbon atoms in graphene have one free electron and also covalently bonded to three adjacent carbon atoms making a hexagonal lattice. The free electrons consist of dense sp^2 networks. The delocalized electrons along the graphene enable high carrier mobility, $2.0 \times 10^4\ cm^2\ V^{-1}\ s^{-1}$ in theoretical value and $1.5 \times 10^4\ cm^2\ V^{-1}\ s^{-1}$ in experimental [10–13]. Because phonons can pass through graphene lattice without hindrance, it shows high thermal conductivity, $\sim 5.30 \times 10^3\ WmK^{-1}$ at room temperature [14,15]. Graphene is widely known as the strongest material, and the Young's modulus of 1060 GPa has been reported for the suspended single layer graphene [16,17]. Along with the excellent physical properties, graphene has extremely high transparency, ~97.7% at 550 nm wavelength, making it a good candidate for transparent conductor [18]. In addition, graphene is highly thermally and chemically stable due to the strong carbon bonds in its hexagonal lattice. Of particular note, the electron clouds of sp^2 networks occupy the voids in the carbon ring, making all molecules and atoms impermeable through graphene at ambient condition [19,20]. However, graphene can be modified under some chemical treatment, so it is possible to tune its chemical properties or generate pores on the basal plane [21,22].

Form membrane application standpoint, graphene is an excellent platform due to its single atom thickness, high mechanical strength, and chemical stability. Using graphene and modified graphene as the building block, various types of graphene-based membranes have been fabricated [23]. Graphene layer with pores, stacked graphene laminates, and polymer matrix containing graphene flakes are the main categories of graphene-based membranes. Molecular separation can be achieved through nanopores in graphene sheet and/or available nanochannels between graphene layers. Depending on the type of graphene-based membranes, graphene with different structural and physiochemical features (different number of layers, flake size, defect density, and functional groups) is needed, which can be realized by various synthetic methods (Fig. 8.1).

2.1 Synthetic Methods for Graphene

2.1.1 Graphene Flakes: Graphite Exfoliation

Graphene flakes with the size of less than few hundreds of micrometer are usually prepared by exfoliating graphite based on solution processing (top-down) (Fig. 8.2A) [24,25]. Graphite is a cheap and largely available material, formed based on the metamorphosis of carbon-containing materials. Graphite has a layered structure and the individual layers called graphene. The graphene layers are stacked with the interlayer distance of 0.334 nm due to the weak van der Waals interactions between individual graphene layers [26]. These individual graphene layers can be separated easily by external energy. For example, the first graphene was obtained using adhesive tape by Geim and Novoselov [27]. The obtained graphene had high quality, but this method needs numerous exfoliation steps to

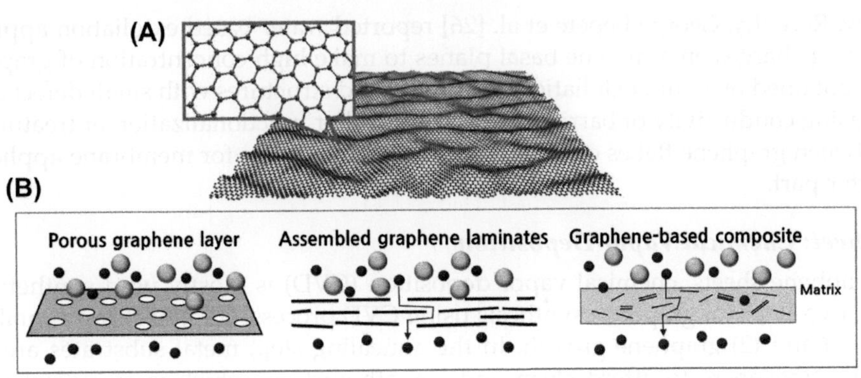

FIGURE 8.1 (A) Graphene structure with hexagonal carbon lattice and (B) types of graphene-based membranes: porous graphene layer, stacked graphene flakes, and graphene-containing polymer matrix.

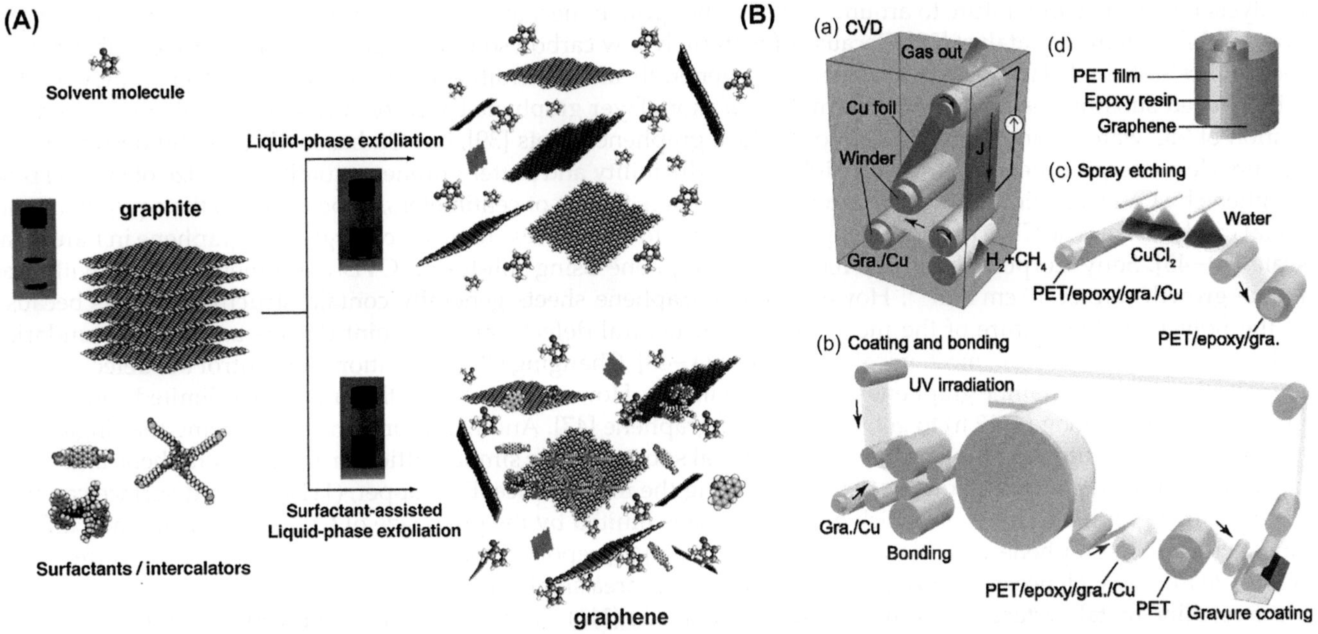

FIGURE 8.2 (A) Liquid exfoliation of graphite with organic solvent and water with surfactant. (B) Roll-to-roll process for graphene synthesis using chemical vapor deposition (CVD) method and also roll-to-roll wet transfer on PET films. *(A) Reprinted from A. Ciesielski, P. Samorì, Graphene via sonication assisted liquid-phase exfoliation, Chemical Society Reviews 43(1) (2014) 381–398, with permission from Royal Society of Chemistry. (B) Reprinted from T. Kobayashi, et al., Production of a 100-m-long high-quality graphene transparent conductive film by roll-to-roll chemical vapor deposition and transfer process, Applied Physics Letters 102(2) (2013) 023112, with permission from AIP publishing.*

get few pieces of monolayer graphene. For an efficient and scalable production of graphene flakes, liquid-phase exfoliation has been suggested. Graphite powder firstly is dispersed in a solvent, and external energy for the exfoliation is applied by means of exploiting ultrasound [24], shearing force [25], ball milling [28], or others [29,30]. Exfoliation approach strongly affects the layer uniformity, monolayer yield, and defect density of obtained graphene flakes. The liquid used for the exfoliation of bulk graphite is also of great importance [31]. Generally, the obtained graphene or few-layered graphene tends to be easily restacked, therefore the interfacial tension between the graphene and liquid molecules needs to be minimized. An organic solvent with surface energy in the range of 40–50 mJ m^{-2} is desirable for stable graphene/liquid interface, and the most widely used solvents are N-methyl-2-pyrrolidone (NMP) and ethanol [32,33]. However, these organic solvents need careful caution that limits their applicability. Known as the most environmentally benign solvent, water has high surface energy (72 mJ m^{-2}) to be used as exfoliation liquid. Therefore, some surfactant-based stabilizers such as sodium dodecyl sulfate and sodium deoxycholate can be utilized to tune the surface energy [34]. A purification step is usually required after the water-assisted exfoliation to remove the excessive surfactant. The liquid-phase exfoliation enables mass production of graphene flakes, but further improvements are still needed to improve the production yield, long-term stability of dispersed graphene, and graphene purity. Recently, George Bepete et al. [26] reported water-based exfoliation approach (without using surfactant) by inducing charges on graphene basal planes to make high concentration of graphene [35]. Generally, the graphene flakes obtained by liquid exfoliation has conjugated structures with small defect density, which can be used as fillers increasing conductivity or barrier properties. Further functionalization or treatment is needed to tune the channel size between graphene flakes or affinity to specific molecules for membrane application, which will be discussed in the latter part.

2.1.2 Graphene Sheet: Chemical Vapor Deposition

For large area graphene sheets, chemical vapor deposition (CVD) is mostly used synthetic method (Fig. 8.2B) [34,36,37]. There are two steps for graphene synthesis using CVD process: (1) annealing of catalytic metal substrates (e.g., Ni and Cu foils) and (2) graphene growth. In the annealing step, metal substrates are annealed at around 1000°C in hydrogen atmosphere to afford clean and smooth surface, which is of critical importance for high-quality graphene growth. As the common procedure, a carbon source such as methane and ethylene is introduced, then decomposes at the surface of the metal substrate, and affords activated carbon, forming graphene sheet along the metal surface. The C−H bonds in hydrocarbon need temperature over 2000°C to be broken, but transition metal catalysts lower the temperature to around 1000°C. The growth mechanisms for graphene sheets depend on the carbon solubility of metal catalyst [38]. Because of extremely low carbon solubility, Cu is the most proper substrate for the synthesis of single layer graphene, whereas Ni generally affords multilayer graphene sheets. Generally, Cu is the widely used substrate as it allows the formation of monolayer graphene based on a cost-affordable price. Carbonization of SiC wafer is another approach to produce graphene sheets [39], but Cu-based CVD is still desirable as it requires less energy, time, and cost with higher reproducibility and better graphene quality. The size of synthesized graphene by CVD technique varied from a few millimeters to tens of centimeters, depending on the equipment and metal catalyst size. The CVD method is applicable to roll-to-roll process to produce large-area graphene in industrial scale [40−43]. Sony Corporation made 100 m-long graphene using roll-to-roll CVD equipment in 2013, with graphene growth rate of 10 cm min^{-1}. However, CVD graphene sheets generally contain structural defects because of the polycrystalline nature of the metal catalyst. Structural defects such as point defects and grain boundaries (GBs) degrade the electrical, mechanical properties [44−46]. Changing CVD condition can control the defect density to make the single crystalline graphene with 1 cm diameter. However, the size of the graphene is limited and also the growth time is too long (~12 h) to get few pieces of graphene [47]. Another approach to producing the single crystalline graphene is utilizing the specific-oriented metal substrate. The similar lattice structure of graphene and metal substrate results in epitaxial growth of graphene along the metal lattice. For copper, (111)-oriented surface can produce GB-free graphene sheets, in which graphene size is limited by the grain size of Cu (111). This method affords single crystalline graphene with high efficiency. Xu et al. [40] reported the synthesis of a continuous GB-free sheet within 20 min [48]. Hence, to realize the high quality, large area single layer graphene, preparation of large area single crystalline metal surface (substrate) is crucial. So far, Cu (111) surface was obtained through sputtering Cu on a single crystalline substrate such as sapphire (0001) [49] or repeated heat cycle of Cu foil [50]. However, these approaches are not scalable or producible because of high cost and long process time. Recently, Xu et al. [40] and Jo et al. [43] reported a novel method to convert polycrystalline Cu to single crystalline one with (111) orientation by controlling the grain growth, although the mechanism of the single crystallization is not clear [48,51].

The graphene sheets grown on metal substrate need to be transferred to another substrate. Although various approaches have been suggested, it is challenging to transfer graphene on target substrate without causing defects or contaminants due to its one atomic thickness. To protect graphene, in the most of transfer techniques, supporting layer is used. The supporting layer is firstly introduced on the graphene surface to give mechanical strength and it will be removed when the transfer process is completed. The commonly used supporting layer is poly(methyl methacrylate) (PMMA) as it gives a uniform film on graphene and also has high mechanical strength. However, residues of PMMA layers after acetone treatment cause graphene wrinkles due to its low flexibility. Poly(ethylene-vinyl acetate) (EVA) was used to make up the disadvantages of PMMA [52]. Using EVA, the transferred graphene had fewer wrinkles due to its low glass transition temperature and fewer residues because of its higher solubility in solvents. To achieve perfectly clean graphene surface, polymer free−transfer methods have been reported [53,54]. In these cases, hexane and cyclohexane were used as sacrificial layers and they can be completely removed by sublimation owing to their solid phases in low temperature. Considering polymer-based supporting layer, after coating the polymer layer on the graphene, metal substrate is removed based on wet and/or dry methods [55,56]. In the wet transfer process, PMMA/graphene/metal layer is floated on the chemical etchant. The PMMA/graphene is rinsed in deionized water by scooping the layer after removing the metal and placed on the target substrate. Drying and annealing stepes to increase the adherence between graphene and target substrate, and finally, PMMA is removed using acetone. In this process, other materials can be also used as supporting layer for graphene. Wet transfer is an easy method due to its high reproducibility and relatively low defect density during the process. However, This process is not cost-affordable to be applied on the large area due to the large amount of wast liquid. As an alternative, dry transfer, in which a support layer with higher adhesion energy to graphene is used, can be a better approach for industrial scale. Deposited Ni layer or other polymer adhesives are widely used in this technique. For example, Ni has higher adhesion energy with graphene than Cu-graphene [57]. After the graphene delamination, Ni/graphene layer hot-pressed on the target substrate and 30–100 nm-thick Ni layer can be removed in 1 min with only a small amount of chemical etchant. The additional cost can also be reduced through recycling metal substrate. It is worth mentioning graphene damage during the physical delamination is the major issue in the dry transfer process. Electrochemical (EC) delamination process is another way to remove metal substrates from graphene [58,59]. In this process, H_2 bubbles formed at the metal surface detached the graphene layers. The defect density is lower than dry transfer method and Cu can be reused, but wrinkles and tearing are still induced by the H_2 bubbles. Regarding the aforementioned challenges on transfer process, recently some researchers have tried to directly synthesize graphene on insulator such as Ge, but the graphene quality was lower than that grown on Cu [60,61].

Graphene production using environmentally benign and energy-efficient processes is of particular importance to ensure a sustainable graphene supply. Life cycle assessment (LCA) is a reliable method to investigate the potential impacts of currently available graphene production routes [62–67]. Several factors such as energy use, resource scarcity, and environmental impacts should be taken into account to assess the sustainable supply of graphene-based materials. Graphene is highly desired as a material in transparent electrodes for liquid crystal displays and solar cell applications. Arvidsson et al. [63] compared the life cycle of resource and energy use of graphene layers and commercial indium tin oxide (ITO). They found out that production of graphene layers has lower life cycle energy use than ITO layers, 3–10 times reduction for the best-case scenario. Regarding the resource scarcity, indium in ITO production is more problematic than copper used for CVD-grown graphene [63]. The same group assessed the life cycle of two graphene synthesis routes: ultrasonic-based exfoliation route (USR) and chemical reduction route (CRR). Energy use and blue water footprint of USR were more than 50% lower than that of CRR. However, in terms of human- and ecotoxicity, CRR was the more efficient case [64]. Using process information and data in scientific papers and patents, LCA can provide valuable directions regarding the future application potential of graphene.

2.2 Porous Graphene Membranes

Despite being atomically thick, graphene is impermeable to liquids, ions, and even small molecular size gases such as helium [68]. On the other hand, graphene layer has very high mechanical strength. Wang et al. tested the mechanical strength of the free-standing graphene on porous polycarbonate membrane, by applying pressure [69]. As reported theoretically, graphene tends to withstand higher pressure with smaller pore size, and without defects, wrinkles and slacks. The highest pressure that graphene withstood was over 100 bar, proving its great potential for the preparation of single-layer graphene membrane. Numerous simulation studies underlined the great potential of perforated graphene, called PG, membranes for molecular separation.

2.2.1 *In-Plane Pore Generation Methods*

Graphene used for the fabrication of PG membranes is generally produced by exfoliation and CVD methods. The exfoliated graphene flakes usually have low structural defects in graphene plane, but the size is limited in hundreds of micrometer scale. In the case of CVD graphene, a graphene sheet can be produced, but the defect density is relatively high. For measuring gas or water permeance through graphene membrane, graphene layer should be transferred to porous substrate such as silicon nitride chips [70], track-etched polycarbonate membrane [71–73], and anodic aluminum oxide membrane [74]. In most cases, graphene was directly attached to the porous substrate and then the metal substrate was wet-etched. The process usually causes large defects on graphene layer such as tearing, resulting in poor separation performance of the membrane. The large extrinsic defects become unavoidable in the case of graphene sheets over millimeter size. Multilayer graphene layers, interfacial polymerization, and additional supporting layer were proposed to reduce the effect of large tearing [71,72], but the transfer processes without any defects are still huge challenges for ultimate PG membranes.

Pore generation in the pristine graphene is another way to prepare PG membrane. To separate molecules and ions based on kinetic size, the in-plane pore size should be carefully controlled with the size range of sub- to a few nanometers. Most of the pore generation methods are based on physical drilling. Various approaches such as electron

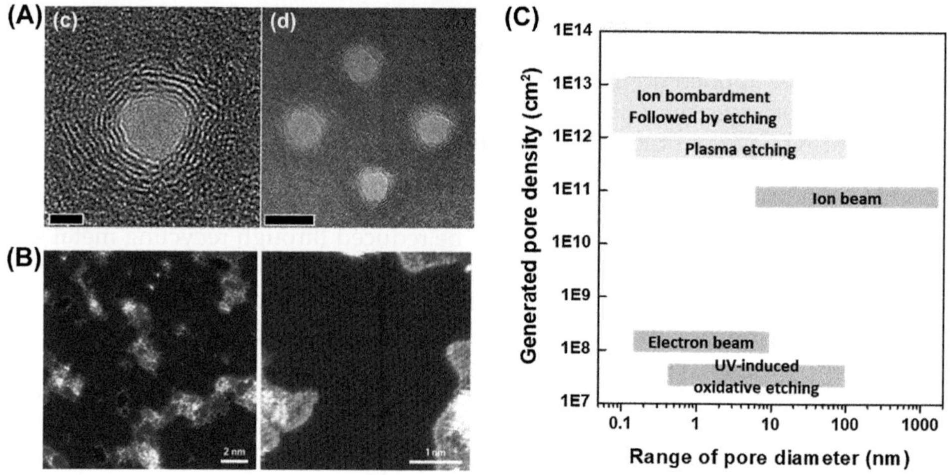

FIGURE 8.3 TEM images of single and multiple nanopores generated by (A) electron beam (scale bars are 2 and 10 nm, respectively, from left image) and (B) oxygen plasma. (C) Pore diameter versus approximately generated pore density that can be generated by various pore generation methods. Green bars are applicable to larger area over centimeter size and blue (gray in print version) colored bars are only for small area under millimeter membrane size. *(A) Reprinted from M.D. Fischbein, M. Drndić, Electron beam nanosculpting of suspended graphene sheets, Applied Physics Letters 93(11) (2008) 113107, with permission from American Institute of Physics (AIP) publication. (B) Reprinted from S.P. Surwade, et al., Water desalination using nanoporous single-layer graphene, Nature Nanotechnology 10(5) (2015) 459–464, with permission from Nature.*

beam irradiation [75,76], ion beam irradiation [77], UV-induced oxidative etching [70], oxygen plasma etching [78], and ion bombardment followed by chemical etching were suggested [71]. Characteristics of each method including possible pore size range and treatable membrane (graphene) area are summarized in Fig. 8.3. In the case of electron beam irradiation, the 3.5–5.9 nm of pore size was controllable with narrow size distribution, but this method is applicable to the only small area under few μm^2. Ion beam irradiation can control more wide range of the pore size, 5–100 nm, but also suffers from the size limitation. PG membrane fabricated by UV-induced oxidative etching has the pore size of 0.4–10 nm. This approach is applicable to the large area, but the pore size distribution is wide for molecular-sieving membrane. Ion bombardment with chemical etching and oxygen plasma can produce scalable, subnanometer pores, holding great promise for making the large-area PG membrane although further steps are required to increase the pore density. The main challenge of large area graphene treatment using UV-induced oxidative etching or chemical etching is the graphene uniformity induced by intrinsic defects. The intrinsic defects in CVD graphene also act as pores for molecular transport [71,79]. The structural defects can have larger pore than hexagonal carbon ring due to the graphene plane distortion and vacancies. The defect size is hard to exceed 1 nm, which is desirable for the molecular separation ,but generally, the density of intrinsic defects in graphene sheets is low, giving rise to low permeance.

Selecting a proper pore generation method largely depends on the processability and treatment area as well as the desired pore size and density. The effect of the pore generation method (i.e., pore size and density) on the membrane performance is hard to be considered because the fabricated PG membranes are different in terms of large defect density, membrane fabrication method, numbers of graphene layer, and type of support.

2.2.2 Gas Transport in Porous Graphene Membranes

The gas separation properties of PG membrane were first revealed by Jiang et al. in 2009 (Fig. 8.4A and B) [19]. They used density functional theory to calculate the interaction energy between PG and gas molecules in the designed pores. Based on the calculation, H-functionalized pore with diameter of 0.25 nm had H_2/CH_4 selectivity over 10^{23} at room temperature along with high H_2 permeance. In this study, only the relationship between the molecular size and pore size was taken into account to evaluate the gas transport properties. This work triggered extensive simulation studies on the gas transport properties of PG membranes.

Drahushuk et al. [80] demonstrated the role of adsorption of gas molecules on graphene surfaces. In the case of H_2 and N_2 molecules, H_2 has smaller kinetic diameter than N_2, 2.9 and 3.6 Å, respectively, but N_2 molecules adsorb more strongly onto graphene surface forming adsorbed layer. The small pores allowed H_2 to permeate while retaining N_2 molecules, resulting in high H_2/N_2 selectivity. In the case of large size pores, which can allow both gases to permeate, N_2 flux exceeds H_2 flux because of high N_2 concentration near the graphene pores. The adsorption of the gas molecules changes with the types of gas molecules, feed gas concentration, pore size, and pore distribution.

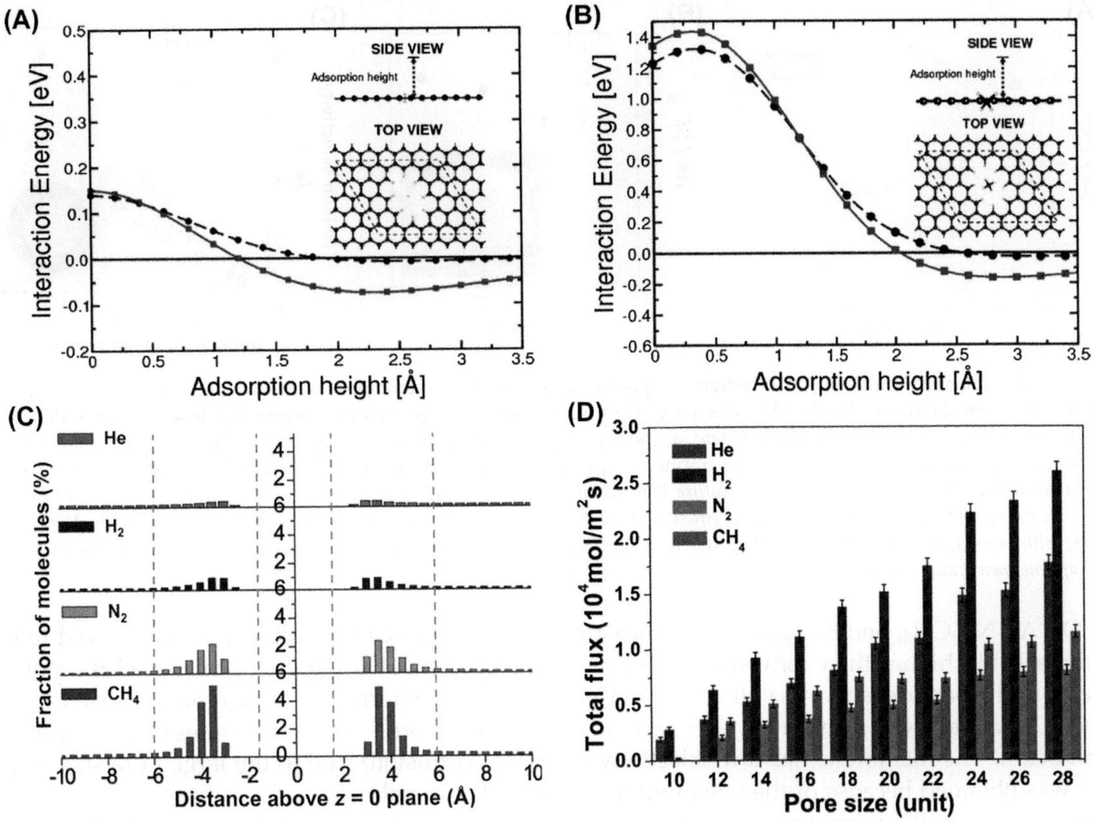

FIGURE 8.4 Interaction energy between (A) H_2, (B) CH_4, and H-passivated graphene (van der Waals density functional: *red (gary in print version) line with squares;* Perdew–Burke–Ernzerhof: *black line with circles*). (C) Probability density of gas molecules along the z-direction for He, H_2, N_2, and CH_4, respectively. The adsorption amount decreases in the order of CH_4, N_2, H_2, and He. (D) Permeation flux of the four different gases in (C), in units of mol m^{-2}·s. *(A–B) Reprinted from D.-E. Jiang, V.R. Cooper, S. Dai, Porous graphene as the ultimate membrane for gas separation, Nano Letters 9(12) (2009) 4019–4024, with permission from ACS Publications. (C–D) Reprinted from C. Sun, et al., Mechanisms of molecular permeation through nanoporous graphene membranes, Langmuir 30(2) (2014) 675–682, with permission from American Chemical Society (ACS).*

Considering the adsorption rate as the dominant permeation factor, larger molecules exhibited higher permeance over smaller molecules, for example, H_2 than He or CH_4 than N_2 (Fig. 8.4C and D) [81]. The surface adsorption (or surface diffusion) was enhanced using functionalized graphene basal plane [82] or ionic liquid coating [83] on the PG membrane.

Wang et al. [84] and other groups investigated the effect of functional groups at pore rim on the gas selectivity. They used density functional theory and molecular dynamics simulation to calculate the energy barrier of N- and F-terminated pores with various sizes. The electrostatic interaction between the gas and designed pores contributes to the different energy barrier. According to the result, N-terminated pores allow CO_2 and N_2 molecules to pass while posing high-energy barrier to CH_4 molecules. However, F-functionalized pores are more proper for CO_2 separation from N_2 or CH_4. The effect of the electrostatic interaction between pores and gas molecules was studied by Lei et al. [85]. The pore made by removing 12 carbon atoms and N atoms are on the pore rim. Charges of −0.241 e applied to the N atoms and system neutrality were kept by adding 0.241 e on neighbor C atoms. The pore size was 4.5 Å, which is larger than the used gas molecules, H_2S (3.6 Å) and CH_4 (3.8 Å). The selectivity of H_2S/CH_4 for charged pores increased to 8.01 from 2.44, compared with the pristine one. Because of the stronger electrostatic interaction between H_2S and the charged pore, the H_2S concentration in the vicinity of the pore was 9.53 times higher than that of CH_4.

All simulation studies emphasize the great potential of PG membranes for gas separation application. In addition, gas transport properties of PG membranes are tunable by functionality, size, temperature, and charge of pores. However, realizing PG membranes with well-designed pores with specific diameter, functional groups and pore distribution is challenging. First experimental work on PG membrane was done by Koenig et al. in 2012 (Fig. 8.5A and B) [70]. To make a PG membrane, exfoliated graphene was first transferred on 5 μm-diameter cavities. Then, subnanosized pores were generated using UV-oxidative etching. The gas permeation was measured by pressurized blister

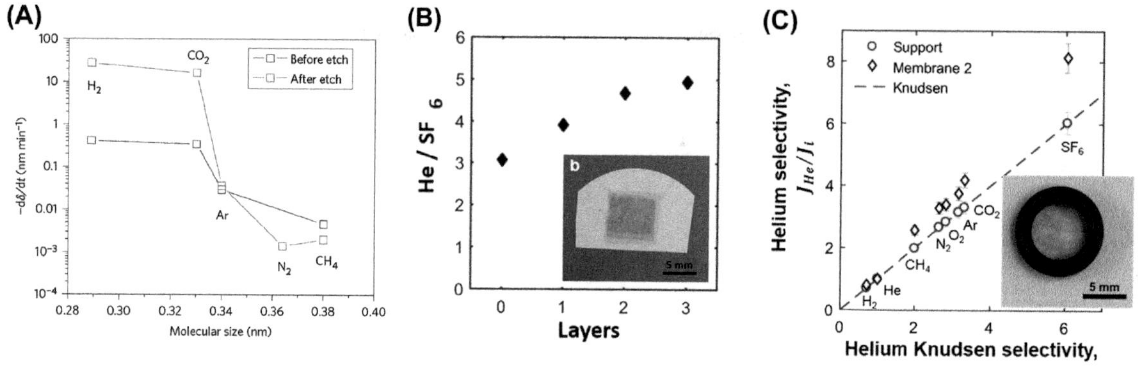

FIGURE 8.5 Experimental results on gas selectivity of porous graphene (PG) membrane. (A) Gas selectivity of single-layer graphene on microchamber with 5 μm diameter. (B) He/SF6 selectivity (Knudsen value) of large area PG membrane. Inset image is PG membrane with structure of PDMS/3-layered graphene/polycarbonate supporter. (C) Calculated gas selectivity (molecular sieving) from centimeter scale PG membrane. Inset image is pore-generated monolayer graphene supported by anodic aluminum oxide (AAO). *(A) Reprinted from S.P. Koenig, et al., Selective molecular sieving through porous graphene, Nature Nanotechnology 7(11) (2012) 728, with permission from Nature. (B) Reprinted from M.S. Boutilier, N.G. Hadjiconstantinou, R. Karnik, Knudsen effusion through polymer-coated three-layer porous graphene membranes, Nanotechnology 28(18) (2017) 184003, with permission from IOP Publishing. (C) Reprinted from M.S. Boutilier, et al., Molecular sieving across centimeter-scale single-layer nanoporous graphene membranes. ACS Nano 11(6) (2017) 5726–5736, with permission from American Chemical Society (ACS).*

test. H_2, CO_2, Ar, N_2, CH_4, and SF_6 gases were measured, and the H_2/CH_4 selectivity was observed to be over 10^4, which is remarkably higher than conventional polymer membranes. This work demonstrated the potential of PG membranes, but the pore density was still very low. Celebi et al. [77] reported a few micrometer-sized PG membrane. Bilayer of CVD graphene was transferred on SiNx chips and drilled 8 nm−1 μm-sized pores on graphene sheets using focused ion beam. The membrane has low selectivity (Knudsen diffusion) due to the large pore size, although high permeance was obtained because of the one-atom thickness of the membrane.

To realize large area PG membranes, CVD graphene containing intrinsic defects is a good alternative to liquid exfoliated graphene. The gas transport through intrinsic defects in CVD graphene was investigated by Kim et al. [79]. They transferred five layers of CVD graphene on dense poly[1-(trimethylsilyl)-1-propyne] (PTMSP) film. Because of the polycrystalline structure, graphene layers control the gas permeance and selectivity. Boutilier et al. [72] investigated the relationship between gas permeance and number of stacked graphene layers, as graphene layer has randomly distributed intrinsic and extrinsic defects. They transferred graphene on track-etched polycarbonate porous membrane by directly attaching graphene on the substrate followed by Cu etching. The gas permeance through the PG membranes exponentially decreased, while the gas selectivity followed an increasing trend with increasing the added graphene layer, implying that multistacking graphene layer can be one of the approaches to reducing the effects of intrinsic defects and large tearing.

Boutilier et al. [86] proposed a new approach, based on using additional graphene layers, to afford a centimeter-sized PG membrane with no large defects (Fig. 8.5C). The strategy was based on stacking three-layer graphene, controlling ion bombardment density and additional polymer coating. Large defects created during transfer process were covered by additional graphene layers, as confirmed via He permeance measurement and SEM images.

Another work from Boutilier et al. [74] investigated the gas permeance of monolayer PG membrane (Fig. 8.5D). In this work, CVD graphene was first transferred on polished porous alumina membrane (AAO) with 20 nm pores, and then pores were generated by ion bombardment followed by oxygen plasma. The small pore and polished surface of the AAO substrate improved the resistance to applied pressure, and high porosity of the substrate increased the gas permeance of the PG membrane. Regarding the parallel alignment of pores in the substrate, gas transport through nanopores and large defects could be isolated according to the resistance-based calculation. The calculated He/SF_6 selectivity was about 8, higher than the value predicted based on Knudsen mechanism implying that the gas transport through the PG membrane is based on molecular-sieving mechanism. Recently, Huang et al. [87] succeeded in fabricating crack-free PG membrane and reported the outstanding H_2/CH_4 selectivity of 25, demonstrating that the membrane mainly contains subnanometer pores. The strategy for the crack-free transfer of PG on track-etched polycarbonate is utilizing nanoporous carbon as a supporting layer. Removal of a supporting layer with high permeance is avoidable, minimizing the risk of graphene layer damaging. Similarly, Kidambi et al. [88] used porous PES as a nonremovable support layer, prepared by nonsolvent induced phase separation method, and evaluated its water permeability.

The size of the reported PG membrane for gas separation was around 1 cm^2, and the H_2/CH_4 selectivity was 25, which is not high enough compared with commercial polymeric membranes or theoretical value of PG membrane. To improve the performance of a PG membrane, generation of uniform, subnanometer size pore is of critical need.

However, as previously mentioned, preventing and healing the large size defects (over tens of nanometer) during membrane fabrication process is a great challenge that should be overcome.

PG membranes hold great promise for industrial application as they can considerably decrease the module size because of high permeance derived from their atomic thickness and also less susceptible to performance degradation under mixed gas or long-term operation.

3. GRAPHENE OXIDE

3.1 Synthesis and Chemistry

GO has a long history, and British chemist B. C. Brodie was the first to investigate the oxidative treatment of graphite by adding potassium chlorate ($KClO_3$) to a slurry of graphite and acid nitric (HNO_3). The resulting product was found to have the chemical formula of $C_{2.19}H_{0.80}O_{1.00}$ [89]. Afterward, different synthesis approaches, such as Staudenmaier [90], Hummers [91], modified Hummers [92,93], and Tour [94] were developed. In a modified route proposed by Staudenmaier, chlorate was added to graphite stepwise, and concentrated sulfuric acid was also utilized along with HNO_3 to make a more acidic environment for oxidation reaction [90]. The C/O ratio of resulting GO was about 2.1, slightly lower than that of Brodie method. In 1957, Hummers and Offeman [91] proposed a more versatile method using different oxidative agents including sodium nitrate, potassium permanganate ($KMNO_4$), and sulfuric acid (H_2SO_4). The oxidation reaction temperature was kept below 45°C for about 2 h. Currently, the modified Hummers' method has been the most common and widely used approach for GO synthesis. The reaction of potassium permanganate ($KMNO_4$) with sulfuric acid (H_2SO_4) results in a dark red oil, called diamanganese heptoxide (Mn_2O_7), which is more reactive at the temperature above 55°C. This oxidative agent can preferentially oxidize the unsaturated aliphatic double bonds over aromatic double bonds [92,93].

Structure of GO is still the subject of debate because of the complexity in oxidative reaction mechanisms. Hoffman and Holts [95] proposed a structure containing epoxy groups distributed over the basal planes of graphite with the C/O ratio of 2 (Fig. 8.6A). In 1946, Ruess [96] suggested a modified model through incorporating hydroxyl groups into the basal plane (Fig. 8.6B). This model contains ethers on cyclohexane ring with four positions hydroxylated and stoichiometrically ordered. In 1969, Scholz and Boehm [97] proposed a revised model without any epoxy and ether groups but including hydroxyl groups at the four-position of 1,2-oxidized cyclohexane rings, as shown in Fig. 8.6C. In 1988, Nakajima et al. [98] suggested a model based on a lattice framework similar to poly(dicarbon monofluoride), $(C_2F)_n$, as shown in (Fig. 8.6D), even though the most recent models rejected this lattice model. One of the well-known and widely adopted models is suggested by Lerf and Klinowski, in which the GO structure is considered nonstoichiometric and amorphous [99]. The main advantage of this model is that the GO structure was characterized using the precise solid-state C-nuclear magnetic resonance spectroscopy. The proposed structure for GO consists of epoxy and hydroxyl groups on the basal plane and ketone or aldehyde groups at the periphery (Fig. 8.6E). Szabo–DeKany suggested a model similar to the Ruess and Scholz–Boehm models, which considered a regular, corrugated nature of the carbon network (Fig. 8.6F) [100]. Despite extensive studies, understanding the exact chemical structure of GO is still a controversial topic due to the variability in sample-to-sample synthesis, various graphite sources, difficulty in controlling the oxidation degree, and metastability of GO.

GO and reduced GO (rGO) are highly desired for various applications such as electronics, catalysis, energy storage, membrane-based separation, etc. However, current synthesis methods of GO are based on the reaction of graphite with strong oxidants, mainly accompanied by environmental pollution and explosion risk. Therefore, developing scalable, safe, and environmentally benign synthesis methods is of urgent need. Pei et al. [101] proposed a green and scalable EC process to synthesize high quality GO. In this method, a flexible graphite paper is subjected to concentrated H_2SO_4 to tune the anodic electrocatalytic oxygen evolution reaction of water to enable the ultrafast oxidation of graphene lattice. Notably, GO synthesized by EC method has the benefits of no explosion risk and high oxidation rate (100 times faster), compared with the common Hummers' method. Reduction of GO by chemical, thermal, or ultraviolet-assisted methods is a common way to produce conductive graphene-based materials with the benefit of massive production [102–108]. However, using reductants (such as hydrazine, dimethylhydrazine, and sodium borohydride) is not only environmentally benign but also usually accompanied by some impurities in the final products, posing a big challenge on the commercial application of rGO. To address this issue, several environmentally friendly reducing agents such as alanine, ascorbic acid, and glucose have been proposed [109–112]. Wang et al. [109] suggested the green reduction of GO using alanine (ALA, 200 mL of ALA solution).

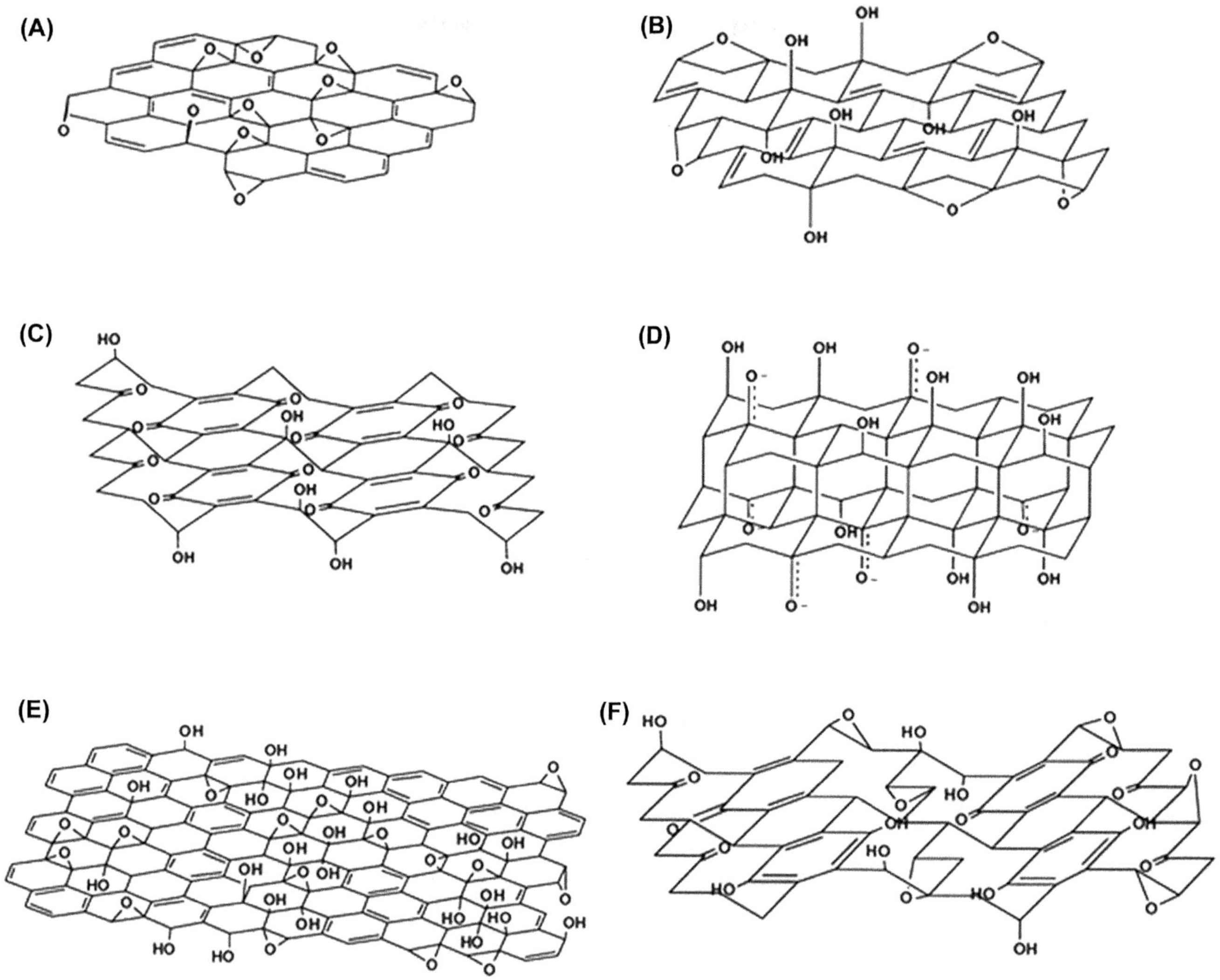

FIGURE 8.6 Proposed structural models of graphene oxide (GO): (A) Hoffman, (B) Ruess, (C) Scholz–Boehm, (D) Nakajima–Matsuo, (E) Lerf–Klinowski, and (F) Dekany models. *Reproduced from C. Bielawski, et al., The chemistry of graphene oxide, Chemical Society Reviews 39(1) (2010) 228–240, with permission from the Royal Society of Chemistry (RSC).*

The rGO was prepared using thermal treatment at 85°C with different treatment time (4–24 h). Zhang et al. [110] added L-ascorbic acid (50 mg) as reductant into GO dispersion (D.I. water, 0.1 mg mL^{-1}) followed by 48 h treatment at room temperature to make rGO, as confirmed by FTIR and XPS analyses. Glucose was also reported as a green reducing agent to make rGO used for composite membrane fabrication [111,112].

3.2 Fabrication of Graphene Oxide Membranes

Free-standing pristine GO membranes are generally fabricated through the filtration of GO suspension on the porous supports, which forms an assembled and tightly packed laminate structure [113]. GO flakes can be assembled into a laminate structure via highly ordered or randomly ordered patterns, using pressure-assisted or vacuum-assisted methods, respectively [114]. In the case of composite membranes, GO dispersion can be coated on the porous ultrafiltration (UF) supports using various approaches such as vacuum filtration [113], dip coating [115], spin coating [116], and layer-by-layer coating [117]. Kim et al. [79] demonstrated that the gas permeation mechanism of GO membranes strongly affected by fabrication route. They prepared GO membranes in two different manners: (1) contacting the surface of the porous support to the air–GO suspension interface, followed by spin coating and (2) direct spin coating of a GO dispersion on the porous support. Interestingly, the stacking manner of GO flakes was completely different because of the various interaction forces, where edge-to-edge repulsive forces were dominant in

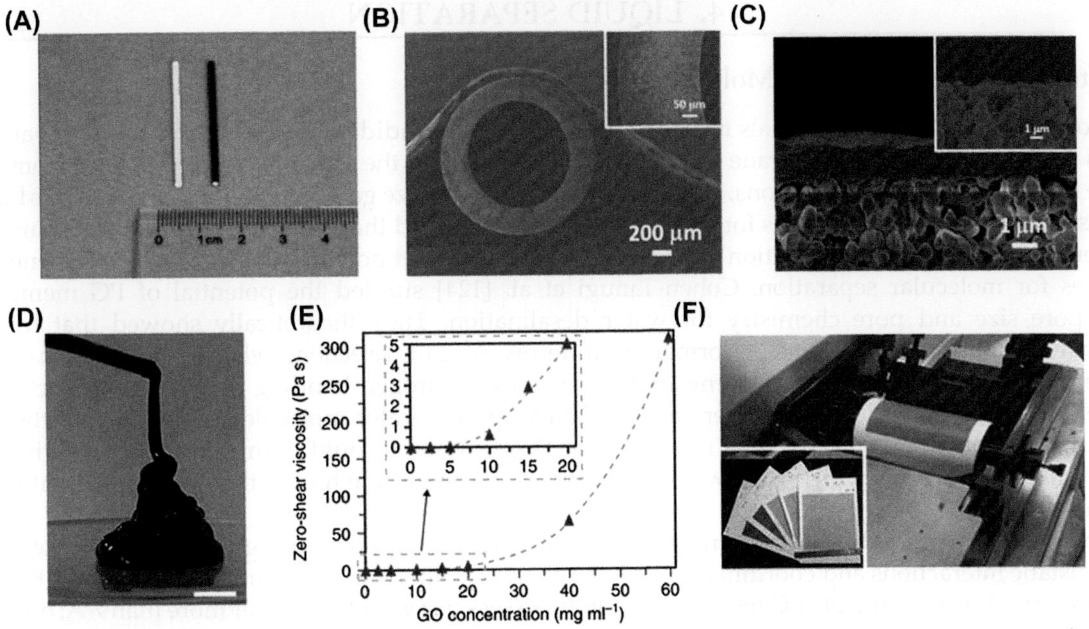

FIGURE 8.7 (A) Photographs of pristine hollow fiber support (white) and graphene oxide (GO)–coated hollow fiber membrane (black). SEM images of (B) GO-coated hollow fiber and (C) its cross-section (all inserts belong to pristine hollow fiber support). (D) GO with viscoelastic properties with concentration of 40 mg mL^{-1}. (E) Viscosity of GO suspension significantly increased on further concentrating with hydrogel beads. (F) Photograph of large area GO/nylon composite membranes fabricated by a gravure printing machine. *(A–C) reprinted from K. Huang, et al., A graphene oxide membrane with highly selective molecular separation of aqueous organic solution, Angewandte Chemie International Edition 53(27) (2014) 6929–6932, with permission from Wiley Online Library. (D–F) Reprinted from A. Akbari, et al., Large-area graphene-based nanofiltration membranes by shear alignment of discotic nematic liquid crystals of graphene oxide, Nature Communications 7 (2016) 10891, with permission from Nature.*

the former case, while attractive capillary forces between face-to-face GO nanosheets were determining for the second preparation method.

For practical application, GO-based membranes need to be packed in the large area, low-cost modules such as spiral wound, tubular, plate-and-frame, and hollow fiber. GO flakes can also be assembled into hollow fiber configuration using vacuum filtration [118–121]. The thickness of the thin GO layer coated on hollow fiber support can be adjusted preferentially by controlling the filtration time or concentration of flake suspension. Huang et al. [119] reported a GO membrane coated on ceramic hollow fiber support with highly selective water permeation in a water/dimethyl carbonate solution (Fig. 8.7A). For example, at 40°C and for a solution composed of 2.6 wt% of water, GO membrane had the water flux of 2200 g m^{-2} h^{-1}. Zhou et al. [121] demonstrated an amine-functionalized (piperazine) ultrathin GO membrane (thickness of 16 nm) coated into the interior surface of polyethersulfone (PES) fibers that exhibited high CO_2 permeance (1000 GPU) and CO_2/N_2 selectivity of 680 under humid condition. An ultrathin layer of amine-functionalized GO layer was readily fabricated by mixing GO suspension with amine agent followed by vacuum filtration through interior part of fibers.

All proposed approaches for fabrication of GO-based membranes so far rely on the coating of a diluted GO suspension on a porous support, which not only need a large volume of GO dispersion and significant time but also their separation performance significantly affected by the alignment of GO flakes. From the practical application standpoint, large area GO-based composite membranes should be fabricated based on a facile and high throughput manufacturing routes. Recently, Akbari et al. [122] demonstrated large-area and high-performance NF membranes fabricated by the shear alignment of discotic nematic liquid crystals of GO. In this method, GO suspension underwent a transition from isotropic to nematic liquid crystalline phases by concentrating the GO suspension (from 0.25 to 25 mg mL^{-1}) with the aid of superabsorbent polymer hydrogel beads. The resulting GO dispersion with discotic nematic phase and appropriate viscosity was readily cast on the porous UF support using a doctor blade (Fig. 8.7B). It is worth mentioning that the GO composite membrane showed rejection of over 90% for charged and uncharged organic probe molecules with a hydrated radius above 5 Å but a relatively low rejection of monovalent and divalent salts (<40%).

Despite substantial progress, to raise the potential of GO-based membranes for industrial application, there is no doubt that further developments in the manufacturing routes are required.

4. LIQUID SEPARATION

4.1 Water Purification and Ion Molecular Sieving

The atomic thickness of 2D materials makes them a promising candidate for membrane-based separation application, as the permeance of a membrane is inversely proportional to the thickness. Despite being atomically thick, graphene is impermeable to liquids, ions, and even small molecular size gases such as helium [68]. This barrier property arises from the fact that π orbitals form a dense, delocalized cloud that blocks the gap within the aromatic rings of graphene [123]. Numerous simulation studies underlined the great potential of perforated graphene, called PG, membranes for molecular separation. Cohen-Tanugi et al. [124] studied the potential of PG membranes with different pore size and pore chemistry for water desalination. They theoretically showed that hydrogenated (H-terminated) pores show better performance in terms of salt rejection, whereas the hydroxylated (OH-terminated) pores yield higher water permeance because the hydrogen bonding between OH functional groups and water molecules provides a smoother entropic landscape for water molecules to pass across the membrane (Fig. 8.8). Importantly, theoretical results revealed that PG membrane could permeate the water with the rate of 10–100 L $cm^{-2}day^{-1}MPa^{-1}$, which was two to three orders of magnitude higher than that of current polymer RO membranes [124].

Generally, transport of hydrated ions through PG membranes is affected by charges at pore edge, steric exclusion, and electrostatic interactions and coordination with functional groups [125]. Hydration shell of Na^+, K^+, and Cl^- is in the range of 6–7 Å; so a suitable PG membrane should possess pores with diameter more than 7 Å; otherwise, ions confronted a barrier, proportional to the energy required for the dehydration, to pass through the small size pores. Pore functionality (or chemistry) also plays an important role in permeation of hydrated ions, particularly for pores with the size smaller than that of ions. For example, Sun et al. [126] demonstrated that 5 Å-diameter graphene with N- or F-terminated pores offered more cation-selective characteristic. In contrast to simulation studies, few experimental studies on the applicability of PG membranes for gas and liquid separation have been reported so far. Koeing et al. [70] were one of the first groups to demonstrate a nanoporous PG membrane (pore size ~ 0.34 nm) using ultraviolet-induced oxidative etching with good performance for H_2 and CO_2 selective separation from N_2 and CH_4. O'Hern et al. [71] also demonstrated an ion selective PG membrane with the pore size of 0.4 ± 0.24 nm and pore density of 10^{12} cm^{-2} fabricated by ion bombardment of single graphene layer followed by oxidative etching. At a short time of oxidative etching, created pores showed cation selective behavior due to the electrostatic repulsion induced by the negative charges of functional groups decorated on pore edges. Further etching of PG was found to allow all salt ions to transport through pores, while larger organic molecules were retained as a result of steric size

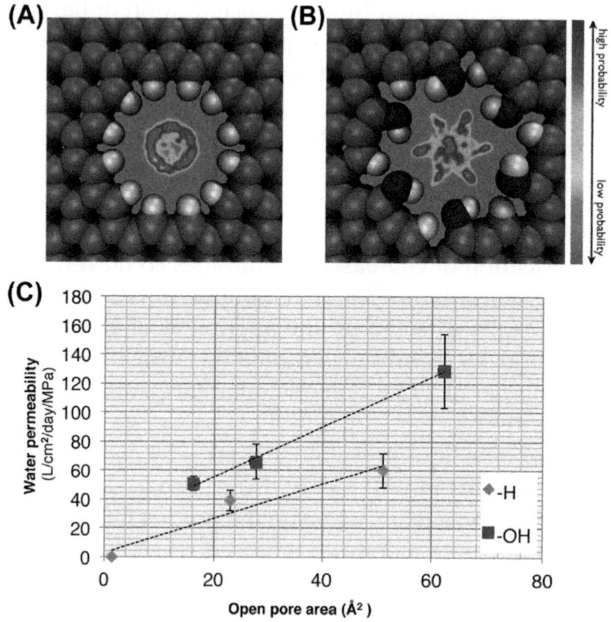

FIGURE 8.8 (A) Hydrogen-terminated and (B) hydroxyl-terminated graphene pores. (C) Computed water permeability values of porous graphene membranes with different pore chemistry. Water permeability showed a linear relationship with open area of pores. *Reprinted from D. Cohen-Tanugi, J.C. Grossman, Water desalination across nanoporous graphene, Nano Letters 12(7) (2012) 3602–3608, with permission from American Chemical Society (ACS).*

exclusion. PG membrane can also be fabricated by exposing to oxygen plasma [78]. The density of created pores was tailored by exposing time of graphene. In particular, PG membrane with the low defect density ($I_D/I_G \leq 0.5$) showed 100% of KCl rejection and an exceptional water permeation rate. Recently, there has been significant progress in the manufacturing of PG membranes based on scalable processes. Kidambi et al. [88] also demonstrated a continuous roll-to-roll process for fabricating a PG membrane with good selectivity toward large molecules such as protein. Despite exceptional water permeance and ion-sieving properties, engineering of the large scale of PG membranes with high-density, uniform pores using currently available perforation techniques remained a critical challenge.

GO membranes with laminar structure assembled from 2D nanosheets have been proposed as a feasible alternative to PG membranes. GO-based membranes can be readily fabricated from GO dispersion though vacuum filtration, layer-by-layer coating, spin coating, and spray-coating methods. Interlayer channels and structural defects on the basal plane of GO membranes provide ultrafast diffusion paths for water molecules. As a pioneer work, Nair et al. [68] discovered that micrometer-thick GO membranes are completely impermeable to organic solvents such as methanol, acetone, and hexane and even small gases such as helium, while showing ultrafast water vapor permeation rate, four to five orders of magnitude higher than of even small gases. They assumed that such an exceptional water vapor permeation rate is attributed to the frictionless water flow through graphitic regions, while oxygen functionalized (sp^3) regions play the role of a spacer to allow water molecules to intercalate into the interlayer spacing of GO sheets. In the next section, water permeation mechanism through GO-based membranes will be thoroughly discussed. Numerous subsequent studies have been focused on the fabrication of GO-based membranes for water purification application. Sun et al. [126] were first to report the Na^+ separation from Cu^{+2} due to the interaction between heavy metal ions and oxygen-based functional groups on GO nanosheets. More interestingly, organic solutes, like rhodamine B (RB), were completely rejected because of the repulsive forces with oxygen functional groups of GO sheets.

The interlayer spacing (d) and intrinsic defects strongly affect the water permeation and/or ion rejection of GO membranes. In general, a GO membrane with intrinsic hydrophilic characteristic due to the oxygen functional groups undergoes a significant swelling (hydration) in aqueous media, as reflected by increased interlayer spacing [127,128]. Joshi et al. [129] demonstrated a sharp molecular sieve cutoff of GO membranes at around 9 Å. Interlayer spacing (d) of GO laminates in the dry state is about 9 ± 1 Å, which is expanded to more than 13 Å thorough soaking in water. Taking into account the thickness of graphene (i.e., 3.4 Å), nanocapillaries in GO membranes sieve out ions with the hydrated radius larger than 4.5 Å. For example, regardless of charge, ions such as K^+, Mg^{2+}, and AsO_4^{3-} are allowed to pass through capillaries exceptionally faster than that is expected for simple diffusion, while glycerol, sucrose, and $[Fe(CN)_6]^{3-}$ were completely sieved out. Tailoring the interlayer spacing (d) by physical or chemical approaches is an efficient way to improve the water permeance and ion-sieving ability of GO membranes. Peng et al. [130] created 3–5 nm nanochannels in GO laminates using sacrificial templates. Positively charge $Cu(OH)_2$ nanostrands were decorated on GO sheets with the aid of electrostatic interactions with negative charge oxygen functional groups. The nanostrands were then sacrificed by treating with an acid solution that created nanochannels with the size distribution of 3–5 nm. The water permeance of the resulting membrane was about 700 $L\ m^{-2}\ h^{-1}\ bar^{-1}$ without any rejection loss, which was almost two orders of magnitude higher than that of commercial UF membranes. Chemically reduction was suggested as an efficient way to improve the water permeation rate of GO membranes. Han et al. [131] demonstrated the rGO NF membranes with ultrahigh water flux through creating nanostructure pores on the basal plan of GO nanosheets using sodium hydroxide (NaOH) treatment. As shown in Fig. 8.9, the generated holes provide more available pathways with less diffusion length for permeation of water molecules. The resulting membrane showed ultrahigh pure water flux (25 $L\ m^{-2}\ h^{-1}$) at 5 bar and more than 99% of organic dye rejection. Importantly, this type of rGO membranes exhibited good stability at aqueous media because of the reduction of oxygen functional groups.

Most of the reported research studies reveal the fact that interlayer spacing (d-spacing) of GO membranes in the aqueous media still is larger than the diameter of hydrated ions of salts, posing a big challenge on their application in water desalination. Physical confinement of GO laminates using epoxy is a proposed approach to control the interlayer spacing in the range of 6.4–9.8 Å, depending on relative humidity (Fig. 8.10A) [132]. In this method, 100 μm GO laminates were fabricated via vacuum filtration and stored in different relative humidities. The interlayer spacing was preferentially adjusted by storing the GO strips at different humidities. Then, individual GO strips were stacked together by a Stycast epoxy, which mechanically constrained the swelling of the laminates in aqueous solution (Fig. 8.10B). Most importantly, the permeation rate of water was weakly affected (less than two fold reduction) with decreasing the interlayer spacing, while the ions permeation rates exponentially reduced. Such a drastic reduction in the permeation rate of ions was explained based on the significant energy barrier required to overcome partial dehydration before entering the low interlayer spacing nanocapillaries. In the case of water, because of the weak interaction between water–water molecules, compared with hydration shells of ions, the energy barrier is considerably low; therefore, water permeation rate is weakly affected with respect to the interlayer spacing.

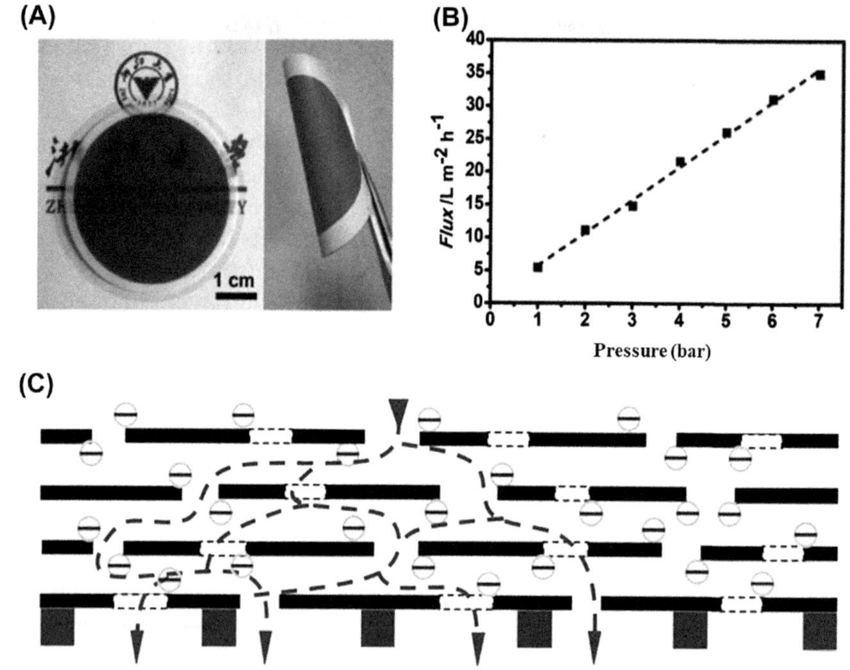

FIGURE 8.9 (A) Photograph of thin reduced graphene oxide (rGO) membrane coated on an anodic aluminum oxide (AAO). (B) Pure water flux of rGO membranes at different pressures. (C) Schematic view of water and ions permeation routes in rGO membrane; created extra pores provided more fast diffusion pathways for water molecules. *Reproduced from Y. Han, Z. Xu, C. Gao, Ultrathin graphene nanofiltration membrane for water purification, Advanced Functional Materials 23(29) (2013) 3693–3700, with permission from Wiley Online Library.*

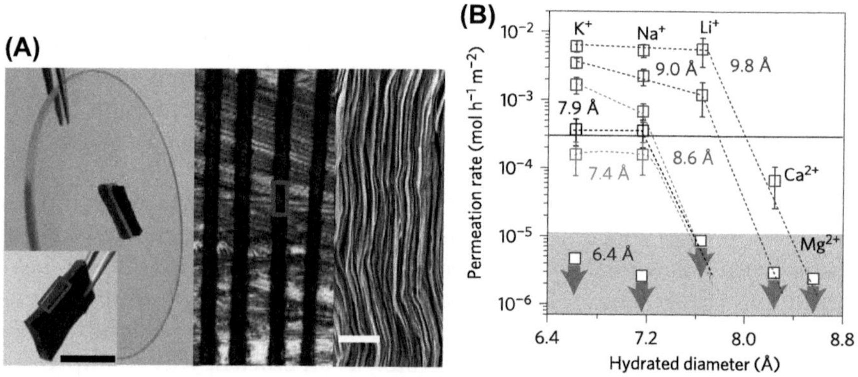

FIGURE 8.10 (A) Photograph of a strip of graphene oxide (GO) membrane glued in a slot (left), optical micrograph of the cross-sectional area (middle), which GO laminates (*black strip*) physically confined by epoxy (light yellow [dull white in print version]), and cross-section of GO strip marked by a red rectangular (right). (B) Ions permeation rates of physically confined GO membranes displayed a sharp cutoff. *Reprinted from J. Abraham, et al., Tunable sieving of ions using graphene oxide membranes, Nature Nanotechnology 12(6) (2017) 546, with the permission from Nature.*

Cation intercalation is another approach to control the interlayer spacing of GO membranes. It was demonstrated that as GO membranes were soaked in the KCL solution (0.25 M), cations intercalated into the interlayer of GO laminates and created strong cation$-\pi$ interaction with oxygen functional groups of GO sheets, leading to the precise control (in angstrom scale) of GO interlayer spacing [133]. The ion rejection of the K^+-intercalated GO membrane was over 99%, which was considerably higher than that of normal GO membrane. Molecular dynamic simulation revealed the physical interaction between cations and GO sheets. The lifetime of hydrogen bonding between the water molecules surrounded the hydrated cations and oxygen functional groups on GO sheets was considerably longer than that of bulk water molecules, implying that interaction between intercalated cations and GO nanosheets is stable.

Cross-linking with diamine monomers (ethylenediamine, butylenediamine, and p-phenylenediamine) was suggested as an efficient way to control the interlayer spacing of GO membranes [134]. Compared with hydrogen bonds and $\pi-\pi$ interaction between GO layers, C—N covalent bonds efficiently suppressed the swelling of GO films when soaked in an ethanol—water mixture. Despite pristine GO membranes, the diffraction pattern of amine-decorated GO membranes clearly revealed that their interlayer d-spacing remained unchanged. The resulting amine cross-linked GO membranes showed great performance for ethanol/water dehydration, water flux of 1400 g m^{-2} h^{-1}, and ethanol rejection of 100%. 1,3,5-benzenetricarbonyl trichloride (TMC) is another cross-linker that can be used to tune the interlayer spacing of GO membranes. Hu and Mi [117] demonstrated that cross-linking the GO nanosheets via TMC not only improved the structural integrity of GO membrane in the aqueous media but also led to the higher rejection of Methylene blue (MB) and Rhodamine WT.

As discussed above, the interlayer spacing plays a key role in determining the liquid permeance and ion-sieving ability of GO membranes. Hence, an accurate and reliable characterization method to quantify the d-spacing of GO membranes is of critical need. X-ray diffraction is a common characterization technique to measure the interlayer spacing of GO films under dry condition. However, d-spacing of GO membranes soaked in an organic solvent or aqueous media is significantly different from that in dry state. Zheng et al. [128] proposed an integrated quartz crystal microbalance with dissipation (QCM-D)/ellipsometry method to quantify the d-spacing of GO membranes when fully soaked in aqueous solution. QCM-D can precisely measure the mass of GO membrane through monitoring the changes in the frequency and dissipation of an acoustic resonator, while ellipsometry, an optical technique based on the light reflection, is responsible for the measurement of the thickness of GO films. By integrating the QCM-D and ellipsometry data, density and d-spacing of swelled GO membrane can be readily calculated. This technique showed a great promise for monitoring the structural changes of GO membranes in the organic solvent and aqueous media.

4.2 Water Transport Mechanism in Graphene Oxide Membranes

In spite of extensive studies on water permeation and ions-sieving properties of GO membranes, underlying permeation mechanism of water is still the subject of debate. This is mainly due to the complexity in layered, nanoporous structure and also uncertainty in the spatial distribution of oxygen-based functional groups on the basal plane of GO membranes [135–138]. In general, there are several possible pathways for water and/or ions permeation through GO membranes: (1) inside hydrophobic regions between interlayer galleries, (2) across defect pores, and (3) through interedges spaces (Fig. 8.11) [137]. Compared with other two cases, galleries between GO flakes are more favorable for water permeation as they are less energy intensive [139].

The unusual water permeation rate through GO membranes, reported by Nair et al. [68], was explained based on slip theory, in which the capillary-based driving force (i.e., capillary pressure) and frictionless flow of water confined between 2D pristine graphitic regions are mainly responsible for the high water permeation rate [68,140]. To study water transport in 2D nanochannels of GO membrane, by assuming the water as a classical liquid, Poiseuille's flow (viscous flow) was employed as the following equation:

$$J = \frac{d^4 \cdot \Delta P}{12L^2 \cdot \eta \cdot h} \tag{8.1}$$

where d is the channel thickness that is defined here as the vertical distance between adjacent GO sheets, L the average length of GO sheets, h the thickness of GO membrane, and η the viscosity of fluid (water, in this case).

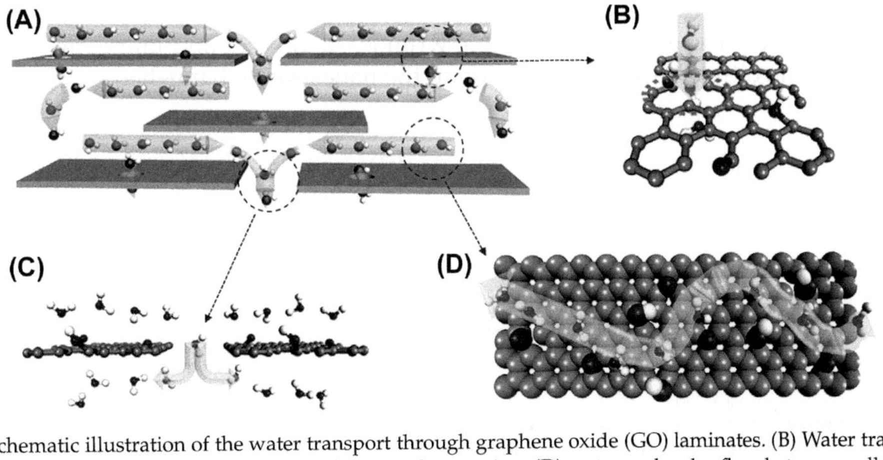

FIGURE 8.11 (A) Schematic illustration of the water transport through graphene oxide (GO) laminates. (B) Water transport through a defect pore. (C) Water transport between nanocapillaries through the interedge spacing; (D) water molecules flow between galleries. Gray color: carbon atoms; red (black in print version) color: oxygen atoms; white color: hydrogen atoms; *blue (light gray in print version) arrows* refer to the water flow direction. *Reproduced from J. Deng, et al., Mechanism of water transport in graphene oxide laminates, Chemical Science 8(3) (2017) 1701–1704, with the permission from Royal Society of Chemistry (RSC).*

Geim et al. [140] employed the Poiseuille flow to describe the water flow between GO layers. They found that water permeation rate in GO laminates is at least two orders of magnitude higher than that predicted based on classical flow. The reason behind this considerable deviation of experimental water flux from the theoretical value comes from the basic assumptions of Hagen–Poiseuille equation, which are the laminar flow in the interlayer galleries and no-slip flow (liquid flow with zero velocity) at the nanocapillaries wall. Slip length is a common parameter reflecting the level of water molecules interacting with the nanochannels of GO membranes [141,142]. The larger values of slip length indicated the nonzero velocity of water flow at the graphene wall. Theoretical studies revealed that the water has a slip length (67 ± 45 nm) on the graphene surface due to the hydrophobic nature of graphitic regions [143,144]. Therefore, such a frictionless water transport can be achieved in the nonoxidized regions of GO nanochannels. Similar frictionless water flow has been observed for carbon nanotube (CNT) membranes when the slip length is much larger than the pore radius of CNT [145]. Han et al. [131] also explained the exceptional water permeation rate in rGO nanofiltration (NF) membranes based on slip theory. Nanochannels in rGO membranes displayed more hydrophobic characteristics as they contained less oxygen-based functional groups, compared with GO membranes, leading to exceptionally high water flux (no-slip flow). In contrast, hydrophobic liquids, such as ethanol, hexane, and toluene, exhibited lower flux than that of water because of their strong hydrophobic interactions with graphitic nanochannels.

Water motion in the nanochannels is also of importance to gain more insight into the permeation mechanism of water in GO membranes. It was suggested that water transportation in the hydrophobic, pristine graphene regions of GO membranes may follow a single-file type pattern, in which water molecules confined into a chain-like configuration with minimum collisions between water molecules [136,137,146], similar to behavior observed in the CNTs when the slip length is largely higher than the pore radius [147]. As experimentally proved, confinement of water in graphene nanocapillaries form a square ice with symmetric phase (ice crystals) at room temperature, which is clearly different from the tetrahedral geometry of hydrogen bonding between water molecules [140]. These ice crystals can move fast through the nanochannels because of extremely weak interactions with hydrophobic walls of GO membranes. In addition, the simulation results showed that the formation of hexagonal ice bilayer crystals is more energetically favorable than monolayer ones. Even though interlayer channels of GO membranes are more energy-favorable paths for permeation, water molecules need to flow through interedge space to flow from one nanochannel to another one. The interedge spaces provide an energy barrier that should be overcome by water molecules. Simulation studies revealed that water molecules, in the form of ice crystals, would subject to the phase transition (i.e., melting) due to the interactions of hydroxyl groups on the edges and water molecules [136].

Although slip theory is recognized the main permeation mechanism of water in GO membranes, it is still controversial. A simulation investigation by Wei et al. [138] claimed that ultrafast water permeation through graphene regions was significantly prohibited by the side-pinning effect of oxidized regions in GO membranes. As shown in Fig. 8.12, compared with pristine graphene membranes, boundary slip in oxidized regions of GO membranes strongly reduce the water velocity in pristine graphene. The theoretical result indicated that the enhancement factor of water permeation, which was defined as the flux of water based on slip theory to water flux predicted by Poiseuille flow $\left(\frac{Q_{slip}}{Q_{nonslip}}\right)$, was in the range of 10–100, which is greatly lower than reported experimental values. They finally concluded that ultrafast water vapor permeation rate in GO membranes may be attributed to the other parameters such as expanded interlayer spacing, possible wrinkles, and holes and interedge spaces. These computational results along with the ultrafast water permeation in the nanostrand-channelled UF GO membranes and rGO

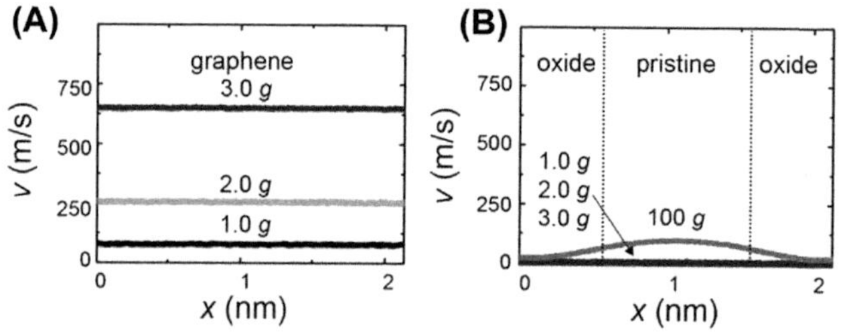

FIGURE 8.12 Velocity profiles for water flow between (A) pristine graphene sheets and (B) graphene oxide (GO) sheets 1.0, 2.0, and 3.0 g represent the applied pressure gradient. The velocity of water flow in pristine graphene regions of GO sheets is largely reduced because of the spinning effect of oxidized regions. *Reprinted from N. Wei, X. Peng, Z. Xu, Understanding water permeation in graphene oxide membranes, ACS Applied Materials and Interfaces 6(8) (2014) 5877–5883, with permission from American Chemical Society (ACS).*

NF membranes emphasize that porous structure of GO membranes may also play an important role in the water permeation through GO membranes.

Although significant progress has been achieved in understanding the main mechanism of water and ion transportation, further experimental studies are required to directly prove the proposed theoretical models.

4.3 Organic Solvent Nanofiltration Membranes

Besides water desalination and NF, separation and purification of products from organic solvents in chemical and pharmaceutical industries is another application area of GO-based membranes. In these systems, valuable products have to be separated from solvents and/or by-products [148]. Conventional separation processes, such as evaporation and distillation, account for more than 50% of capital and operating costs. In this regard, membrane technology continues to attract extreme interest as an environmentally friendly, cost- and energy-efficient alternative [148,149]. From the practical point of view, suitable membranes should not only be mechanically and chemically stable in the organic media but also show promising performance in terms of permeance and retention. Commercially available OSN membranes, including polymeric and ceramic membranes, still suffer from low permeance, and their applicability is limited to several solvents [148–150]. Recently, Karan et al. [151] used a controlled interfacial polymerization to fabricate a sub-10 nm polyamide membrane. By manipulating the morphology, they found that the crumpled polyamide nanofilms exhibited higher solvent permeances than that of smooth surface nanofilms. For example, methanol flux of crumpled polyamide nanofilm was 20 times higher than that of reported thin film composite (TFC) membranes. These results emphasize that ultrahigh solvent permeance can be realized through further decreasing the selective layer thickness. Taking the advantage of their infinitesimal thickness, GO-based membranes are promising candidate to achieve the ultrafast solvent permeances. In addition, good chemical stability of GO membranes makes them a good alternative to polymeric OSN membranes that generally suffer from poor stability in polar solvents.

Sun et al. [126] found that despite the entire permeation of sodium salts through free-standing GO membrane, organic contaminant such as RB can be completely rejected because of its strong interaction with oxygen-based functional groups of GO sheets. However, the layered structure of GO membranes is highly susceptible to disintegration in aqueous media because of swelling behavior resulted from their hydrophilic characteristic of GO membrane conferred by oxygen-based functional groups on its basal plane. To address this issue, chemical reduction of GO was suggested as an effective way to improve the stability of GO-based membranes. Huang et al. [152] demonstrated that rGO membranes were highly stable in organic solvents such as ethanol, acetone, toluene, and n-hexane and also completely rejected big molecules such as Lumogen Red 300.

Despite the aqueous media, the separation performance of OSN membranes is less predictable and strongly depends on solute–solvent–membrane interactions [148,149]. Van der Bruggen et al. [153] and Yang et al. [154] highlighted the role of solvent polarity, represented by the dielectric constant (ε), on the solvent flux through hydrophilic and hydrophobic polymer membranes. They found that the permeance of high polar solvents such as water in hydrophilic polymer is higher than that of ethanol [153]. In the case of GO membranes, permeance of organic solvents is affected by not only the solvent properties, such as molecular size, viscosity, and dielectric constant, but also the structural properties and surface chemistry of GO membranes including interlayer distance, thickness, carbon to oxygen (C/O) ratio of GO sheets, and density of pinholes. Huang et al. [152] showed that molecular size determined the permeance rate of organic solvents in GO membranes. In a different study, Yang et al. [155] claimed that permeance of organic solvents is inversely proportional to viscosity. For example, hexane with a kinetic diameter similar to 1-butanol but much lower viscosity showed higher permeance (18 vs. 2.5 $L\ m^{-2}\ h^{-1}\ bar^{-1}$) through GO membranes (Fig. 8.13A).

The interaction between solvent and membrane can also strongly affect the solvent permeance of GO-based membranes. In contrast to hydrophilic polymeric OSN membranes, Akbari et al. [156] demonstrated that GO membranes with inherent hydrophilic characteristics showed water permeance of 100 $L\ m^{-2}\ h^{-2}\ bar^{-1}$, which is lower than that of ethanol (200 $L\ m^{-2}\ h^{-2}\ bar^{-1}$) and IPA (150 $L\ m^{-2}\ h^{-2}\ bar^{-1}$), even though the interlayer spacing of GO membranes in aqueous media is larger. The main reason behind this behavior was that carboxylic groups, as the major source of negative charges in the GO nanosheets, could be dissociated easier in high polarity (or high dielectric constant) solvents such as water. The resulting strong ion-dipole forces between negatively charged GO sheets and water molecules create interfacial water layers in the vicinity of negatively charged GO sheets, leading to decrease of polar solvents permeance. A new scaling behavior, i.e., permeance $= f(1/\varepsilon\eta)$, was proposed to predict the permeance of polar solvents through GO membrane. Most importantly, GO-based membrane showed higher permeance for polar and nonpolar organic solvents and also more stability, compared with commercial polymeric NF membranes (NF270, Dow Chemical Company, USA) (Fig. 8.13B).

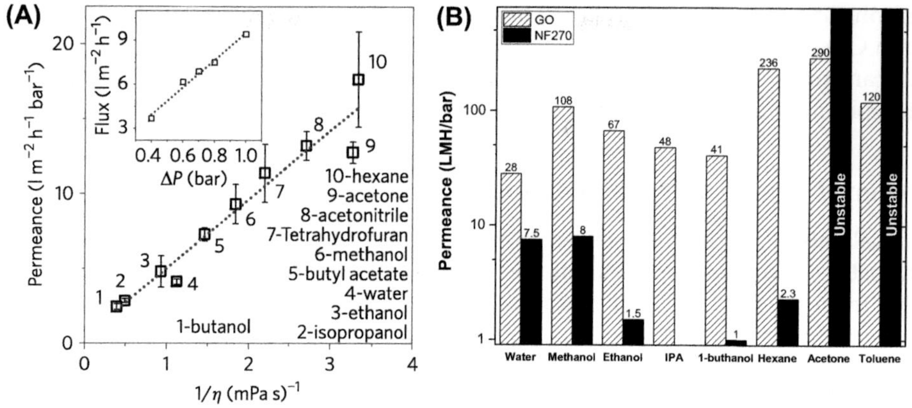

FIGURE 8.13 (A) Permeance of pure organic solvents through the highly laminated GO (HLGO) membrane is inversely proportional to their viscosity. (B) Comparison between the permeance of polar and nonpolar solvents in the graphene oxide (GO) membrane and commercial NF270 membrane. The NF270 membrane showed instability in acetone and toluene. *(A) Reprinted from Q. Yang, et al., Ultrathin graphene-based membrane with precise molecular sieving and ultrafast solvent permeation, Nature Materials 16(12) (2017) 1198, with permission from Nature. (B) Reproduced from A. Akbari, et al., Solvent transport behavior of shear aligned graphene oxide membranes and implications in organic solvent nanofiltration, ACS Applied Materials and Interfaces 10(2) (2018) 2067–2074, with permission from American Chemical Society (ACS).*

Interlayer spacing of GO membranes can significantly control the permeance of organic solvents. It was reported that acetone permeation rate of a GO membrane decreased from 1.5 to 0.03 mol m^{-2} h^{-1} when thermally annealed at 180°C because of the interlayer spacing reduction [152]. Thickness of GO membranes, which is mainly controlled by the loading of GO dispersion used for the membrane preparation, is another important factor determining the water and organic solvent permeances. A thinner GO membrane provides more diffusion pathways with less tortuosity for solvents permeation. Yang et al. [155] revealed that methanol and hexane permeance through highly laminated GO (HLGO) membranes exponentially decayed with increase in the thickness. This unusual behavior comes from that fact that ultrathin GO membranes include more randomly distributed pinholes in interedges of GO nanosheets, allowing fast and less tortuous permeation of solvent's molecules.

Chemical conversion of oxygen-based functional groups is a straightforward way to improve the stability and permeance of GO membranes. Han et al. [131] reported ultrathin rGO membranes (22–50 nm) with high water permeance (21.8 L m^{-2} h^{-1} bar^{-1}) and high retention (over 99%) for organic dyes due to its well-packed layered structure. In a different work, Huang et al. [157] demonstrated ultrafast solvent-solvated rGO-based OSN membranes with unprecedented acetone permeance of 215 L m^{-2} h^{-1} bar^{-1}, two order of magnitudes higher than that of commercially available OSN membranes [148], and methanol permeance of 75.3 L m^{-2} h^{-1} bar^{-1}, 1.5 times higher than that of polyamide nanofilms [151]. Notably, these rGO membranes are stable to organic solvents (such as methanol, acetone, and dimethylformamide), strong acid (sulfuric acid), alkaline (potassium hydroxide), and oxidative (nitric acid) media.

Effective diameter and charge of solute are two important parameters, which strongly affect a solute rejection of OSN membranes. Effective solute diameter in an organic solvent can be calculated based on the following empirical correlation [158]:

$$d_i^s = d_i^w \left(\frac{\varnothing_w MW_w}{\varnothing_s MW_s} \right)^{1/2} \tag{8.2}$$

where d_i^s and d_i^w are effective solute diameters in organic solvent and water, ϕ_s and ϕ_w are the association parameter of solvent and water, and MW_w and MW_s are the molecular weight of solvent and water, respectively. Van Der Bruggen et al. [159] proposed an empirical relationship for effect solute diameter in water as follow:

$$d_i^w = 0.065 MW_i^{0.438} \tag{8.3}$$

Solute rejection in GO-based OSN membranes, analogous to polymeric membranes, is largely affected by charges of solute and membrane surface. Generally, assembled GO nanosheets in the laminar structure are negatively charged due to the presence of carboxylic groups on the edge of the basal plane. This negative charge affects the solute rejection in GO-based membranes. For example, a rGO membrane with 18 nm thickness showed complete rejection (i.e., 100%) for Evans blue (EB, 3.4 nm) while 86.2% and 70.1% rejection for brilliant yellow (BY, 2.6 nm) and acid fuchsin (AF, 1.8 nm), respectively, while no retention was observed for positively charged molecules such as basic fuchsin and MB. These results clearly indicated that the rejection of organic molecules depends on both solute diameter and charges [131]. In addition, the interaction of solvent and GO layers plays a key role in retaining organic molecules. It was experimentally demonstrated that GO membranes show smaller hydrodynamic pore size when contacted with more polar solvents such as water because of the interaction of oxygen-based functional group with water molecules. Such interaction creates an interfacially immobilized layer of water that considerably decreases the available interlayer spacing for solute permeation [156].

In general, a GO flake is negatively charged nanosheet with abundant oxygen-based functional groups decorated on its basal plane and edges. Therefore, GO membranes can be easily reexfoliated in aqueous media and organic solvents, posing a big obstacle to their practical application. To address this issue, several strategies including intercalation of polyelectrolytes [160], self-assembled molecules [161,162], or cross-linked polymer [163] into the interlayer spacing of GO nanosheets have been proposed. Gao et al. [161] fabricated GO membranes cross-linked with boronic acid (BA) polymer exhibiting not only good stability in organic solvents such as methanol, ethanol, and DMF but also high separation performance for OSN. For example, addition of 0.5 wt% BA into a GO dispersion followed by 14 h thermal treatment (at 40°C) led to 27% increase in AF rejection (from 72% to 92%) without sacrificing methanol permeance [161]. The cross-linked GO membrane also exhibited high AF and EB rejection (>90% and 98%, respectively) in harsh DMF with good stability for more than 30 h. rGO membranes intercalated with cationic tetrakis(1-methyl-4-pyridinino)porphyrin showed the enhanced methanol and water permeance as well as high EB, By, and AF rejection while withstanding good stability for more than 25 h [163].

As a final note, organic solvent permeation and solute rejection in GO-based membranes depend on various parameters such as interlayer spacing, pore density, and chemistry of GO as well as the effective diameter and charge of solutes. Therefore, for a particular separation target, all abovementioned factors should be taken into consideration at the same time.

4.4 Polymer–Graphene Oxide Mixed Matrix Membranes

Rich surface chemistry conferred by abundant oxygen-based functional groups and the high aspect ratio of GO nanosheets make them a promising material as a filler for the fabrication of mixed matrix membranes (MMMs). Owing to its hydrophilic characteristic, GO nanosheet can be uniformly dispersed in the aqueous and organic phases. In addition, oxygen functional groups on high aspect ratio GO nanosheets can lead to strong interactions (covalent and noncovalent bonds) with the polymer matrix, which is of particular importance to realize high performance MMMs. Recently, extensive studies have been reported on the GO-based microfiltration (MF), UF, NF, and even RO MMMs [69–75]. Incorporating GO nanosheets into polymer matrices such as poly(vinylidene fluoride) (PVDF), PES, and polysulfone significantly improved their hydrophilicity, stability, antifouling properties, and water permeability.

Zinadini et al. [164,165] incorporated GO filler into PES polymer that significantly improved the water permeability and dye rejection. They fabricated asymmetric PES NF membranes via phase inversion method, in which PES polymer (20 wt%), polyvinylpyrrolidone (PVP), and GO were dissolved in dimethylacetamide as the solvent. The resulting PES/PVP NF membrane containing 0.5 wt% of GO showed 20.4 kg m^{-2} s^{-1} water flux and 96% dye rejection, about 150% and 7%, respectively, higher than that of pristine PES membrane. Such a significant increase in water permeability was mainly attributed to the enhanced membrane porosity (from 73% to 83%) and pore size (from 3.2 to 4.5 nm). This membrane exhibited not only 90% flux recovery but also significantly fouling resistance compared with pristine PES membrane due to the hydrophilic characteristic of GO filler [164]. Zhao et al. [166] reported the similar role of GO filler in improving the water flux, fouling resistance, and also mechanical strength of PVDF/PVP/GO UF and MF membranes. For instance, pure water flux of PVDF/PVP/GO UF membrane composed of 2 wt% of GO filler increased 80% compared with pristine PVDF membrane due to increased porosity and pore size. In addition, hydrophilic characteristic of GO-incorporated PVDF/PVP membrane, as reflected by contact angle measurement, resulted in 90% flux recovery ratio that was considerably higher than that of pristine PVDF/PVP membrane (i.e., 65%).

Taking the benefit of hydrogen bonding between oxygen functional groups of GO nanosheets and polar groups of chitosan polymer, Dharupaneedi et al. [167] fabricated the GO-based MMMs with high performance for water separation from alcohols. Incorporating only 2.5 wt% of functionalized graphene sheets into chitosan polymer matrix resulted in the outstanding selectivities of 7781 and 1093 for isopropanol–water and ethanol–water mixtures, respectively, because of the water-philic characteristic of GO-embedded MMMs, confirmed via contact angle measurement.

GO nanofillers can also be embedded into the TFC polyamide membranes to improve the water permeance. Through cross-linking with trimesoyl chloride (TMC) during interfacial polymerization, GO nanosheets can improve the hydrophilicity of membrane matrix. For example, Bano et al. [168] found that 0.3 wt% GO in polyamide matrix increased the water flux from 2.5 to 20 L $(m^2 h)^{-1}$ while maintaining high salt rejection about 90%. In addition, as confirmed by contact angle reduction from 88 to 60°C, more hydrophilic property of PA-based TFC membrane containing 0.3 wt% of GO gave rise to the improved antifouling property. In a different work, Chae et al. [169] reported the enhanced water permeability for PA-based TFC membranes containing GO nanosheets. Water permeance of polyamide membranes at elevated pressures usually decreases because of the compression of polymer layer. Incorporating GO and/or CNTs into the polyamide layer was found to be an effective way to mitigate the compression effect because of creating extra water diffusion channels (Fig. 8.14A). In addition, poor resistance to oxidizing agents such as chlorine is another critical issue of polyamide RO membranes that significantly deteriorate their salt rejection ability. Amide groups can be easily attacked by chlorine radical, hydroxyl radical, and hypochlorite, which degrade the structure of polyamide layer. Kim et al. [170] demonstrated that PA-based TFC membranes consisting of a mixture of GO and CNT showed increased water flux and also exceptionally improved chlorine resistance (Fig. 8.14B). Phenolic groups of GO were found to play an effective role in trapping oxidizing agents and therefore improving the resistance to chlorine. Therefore, it can be concluded that GO nanosheets as nanofiller play a major role to improve water flux, antifouling property, salt rejection, and chlorine resistance of polymeric membranes.

GO can play a key role in improving the stability, permeance, and solute rejection of polymer-based OSN membranes [171–174]. It was found that embedding GO into polypyrrole (PPy)/hydrolyzed-PAN (H-PAN) composite membrane significantly improved the solvent permeance. For example, incorporation of the small amount (0.05 wt%) of GO into PPy/H-PAN composite membrane increased the permeances of methanol, ethanol, and isopropanol by approximately 945%, 635%, and 302%, respectively, without compromising solute RB rejection. In addition, after initial reduction of the permeance due to the compaction, GO-PPy/H-PAN showed a stable performance for about 4 days.

In contrast, Ding et al. [172] reported that incorporation of GO nanosheets into polyethyleneimine (PEI) improved the polyethylene glycol (PEG) rejection while sacrificing the permeance of solvents even though still comparable to pristine PEI membrane. Generally, good interaction between dispersed phase (i.e., GO nanosheet) and continuous phase (i.e., polymer matrix) is a determining factor in the fabrication of high-performance OSN membranes. Fei et al. [174] reported that hydroxylation of polybenzimidazole (H-PBI) led to better interaction with GO nanosheets, giving rise to significant enhancement of acetone permeance (from 25 to 45 L $m^{-2} h^{-1} bar^{-1}$) without compromising solute rejection. ineresting1y, GO-embedded H-PBI membranes showed stable performance within 14 days, in contrast to pristine PBI membrane. Covalent cross-linking (via toluene 2,4-diisocyanate) of GO with H-PBI was also found to be a good strategy to further improve the solute rejection.

GO-embedded MMMs have also shown great potential for different application areas such as pervaporation [175,176] and membrane distillation (MD) [177] even though further studies are required.

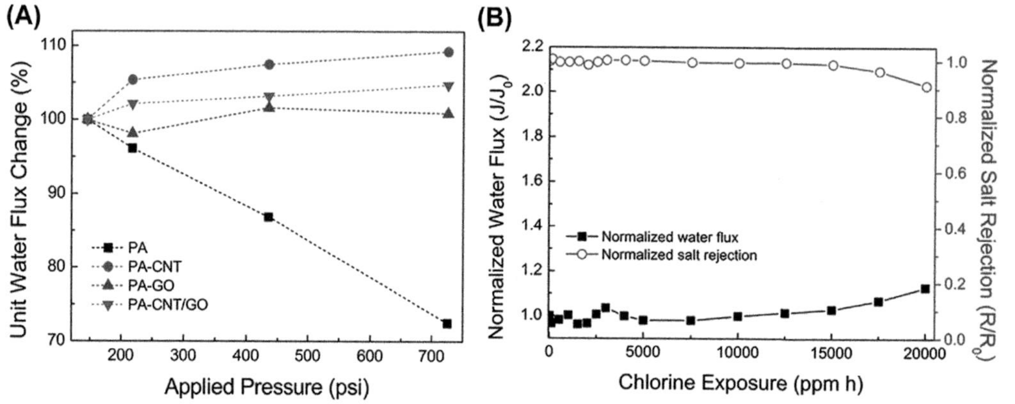

FIGURE 8.14 (A) Stability of graphene oxide (GO)– and carbon nanotube (CNT)–embedded PA membranes under pressurized condition. Water permeance of PA membrane dramatically decreased with increasing pressure, due to compression effect. (B) PA-CNT/GO membrane containing 0.02 wt% of CNT/GO was highly resistant to chlorine without loosing salt rejection, in contrast to the pristine PA membrane. Experimental conditions: cross-flow test, NaCl, and chlorine concentrations in feed solution were 2000, 500, and 2000 ppm, respectively, and feed pressure was 15.5 bar. *Reprinted from H.J. Kim, et al., High-performance reverse osmosis nanocomposite membranes containing the mixture of carbon nanotubes and graphene oxides, Journal of Materials Chemistry A 3(13) (2015) 6798–6809, with permission from Royal Society of Chemistry (RSC).*

5. GAS SEPARATION OF GRAPHENE OXIDE MEMBRANES

5.1 Gas Transport Mechanism in Graphene Oxide Membranes

In general, gas permeation through free-standing GO membranes is a controversial topic due to the complexity in GO structure. GO nanosheets contains abundant oxygen functional groups such as hydroxyl, epoxy, and carbonyl on the basal planes and carboxylic at the edges, which substantially affect the sorption property of gas molecules [178]. In addition, layered structure of GO laminates provides long and tortuous pathways for molecular diffusion. A gas permeation model for the 2D-layered structure was represented by Nielsen model, based on diffusion length (l') as the following equation [179–181]:

$$l' = lN\,L/2 \tag{8.4}$$

$$N = 1/D' + W \tag{8.5}$$

Where l and L are the membrane thickness and width of the graphene sheet, respectively, D' is the effective distance between graphene layers [110,182,183], and W is the thickness of the graphene sheet [184]. GO with a 2D structure and randomly stacked layers acts as a barrier for gas molecules [181]. However, the length of diffusion path can be preferentially controlled by flake size of GO sheet in membranes. As shown in Fig. 8.15, there are two pathways for diffusion of gas molecules through GO membranes: (1) between sheets edges and (2) between sheet edge and basal plane of adjacent GO sheet. The gas permeability through GO membranes can be improved via creating more interedge spaces in the laminar structure [185].

To realize this, controlling the average flake size of GO sheets was found a rational strategy to control the gas transport properties of GO membranes. Kim et al. [79] demonstrated that gas permeation through GO membranes strongly depends on the flake size of GO sheets. Basically, GO membrane made from the small size flakes shows higher gas permeability than that with larger flake size because it provides short diffusional pathways, as shown in Fig. 8.16A. It was observed that GO membranes with different GO flake size (300, 500, and 1000 nm) showed different values of gas permeabilities (Fig. 8.16B). GO membranes with larger flake size provide diffusion pathways with high tortuosity, which lead to lower gas permeabilities. Also, gas transport mechanism through GO membrane under the humid condition is completely different from that of dry condition. The order of gas permeability in a thick GO membrane under the dry state followed the kinetic diameter of the gases (He > H_2 > CO_2 > O_2 > N_2 > CH_4) (Fig. 8.16B), indicating that the separation mechanism can be explained by molecular-sieving mechanism, identical to the observed mechanism in carbon molecular sieve membranes [186,187]. The stacking manner of GO flakes also strongly affects the permeation mechanism through pristine GO membranes. Kim et al. [79] fabricated two types of thin GO membranes by coating the GO dispersion on the polymeric support (PES) with different coating methods, called method one and method two. The gas permeability of two types of GO membranes showed different gas transport behavior. The thin GO membrane prepared by using method one represented Knudsen transport behavior, indicating that nanopores formed between the edges of less interlocked GO sheet are hardly covered during the coating process (Fig. 8.16C).

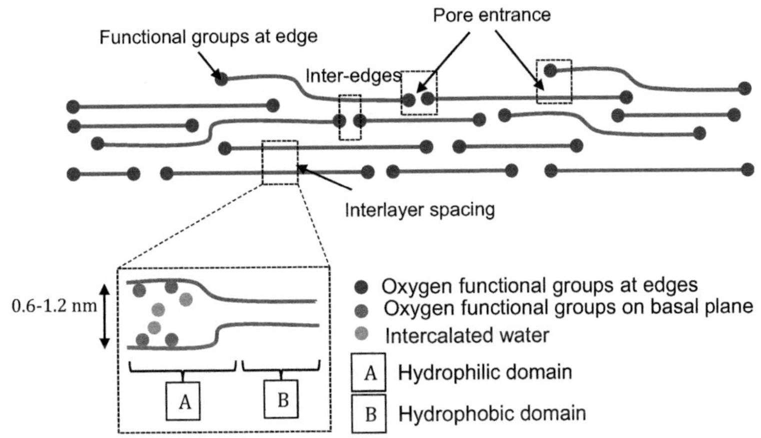

FIGURE 8.15 Simplified illustration for graphene oxide (GO) membrane structure to be considered for molecular separation. *Reprinted from H.B. Park, H.W. Yoon, Y.H. Cho, Graphene oxide membrane for molecular separation. Graphene Oxide: Fundamentals and Applications (2016) 296, with permission from Wiley Online Library.*

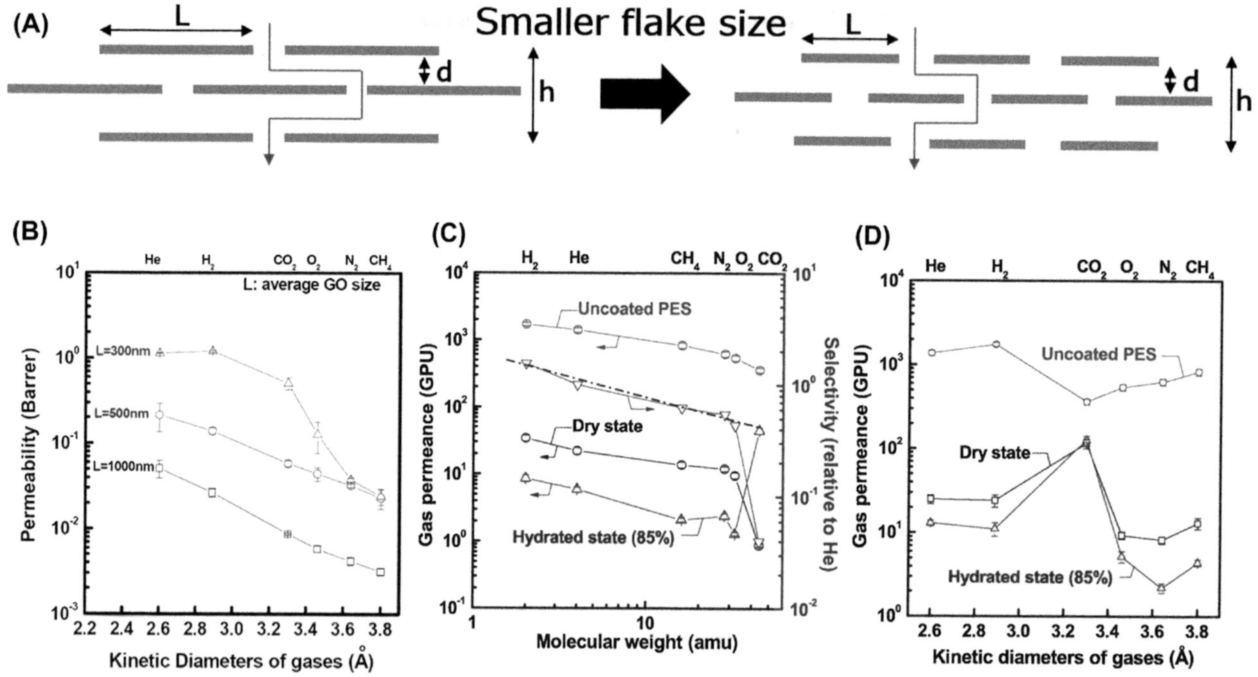

FIGURE 8.16 (A) Schematic illustration of graphene-layered structure. (B) Gas permeability of graphene oxide (GO) membrane as a function of flake size of GO sheets. Gas permeance of GO membranes prepared by method one (C) and method two (D) under dry and hydrated state. *(B–D) represented from H.W. Kim, et al., Selective gas transport through few-layered graphene and graphene oxide membranes, Science 342(6154) (2013) 91–95, with permission from AAAS.*

In contrast, GO membranes prepared by method two followed a different permeation mechanism based on the molecular sieving, except for CO_2 (Fig. 8.16D). In this case, during fabrication of GO membrane, water molecules trapped and remained between GO layers due to the interaction (hydrogen bonding) with oxygen functional groups on the basal plane of GO sheets [116].

In general, pristine GO membranes suffer from low gas permeance due to the long and tortuous diffusion pathways. To address this challenge, structural and chemical modifications of GO nanosheets are two possible strategies to improve the permeability and selectivity of target gases.

5.2 Gas Barrier Properties

Thick pristine GO membrane under dry condition is impermeable to gas molecules, even small gas molecules such as helium. Gas molecules should enter into GO layers and penetrate along the interlayer space with long diffusional pathway [188]. Therefore, GO membranes can be exploited as barrier materials. As it was mentioned before, GO flake size and thickness of membrane significantly affect the gas barrier properties of GO membranes. In addition, the layered stacking structure and amount of water intercalated between GO sheets also play an important role in improving the gas barrier properties of GO membranes. Fig. 8.17A shows the gas permeability of thick GO membranes with different flake sizes. In this work, the flake size was controlled by using different initial graphite sources [189]. Also, the average size of GO flakes is dependent on the extension of the oxidation or sonication time [190,191]. The gas permeation property of GO membranes composed of different flake sizes showed molecular-sieving behavior, and all gas permeabilities were decreased with increasing average platelet size of GO sheet because of long diffusional pathways in the laminar structure of GO membranes. He permeance of GO membranes as a function of feed pressure was shown in Fig. 8.17B. At low feed pressures, He molecules could not enter the GO interlayer because the potential energy barrier of the GO sheets is larger than the kinetic energy of the gas molecule. However, as the feed pressure is increased, the kinetic energy of gas molecules overcomes the potential gas barrier and therefore gas molecules can penetrate through the interlayer space of GO membranes. Furthermore, with increasing GO flake size, the total potential energy barrier of the GO sheets will be higher and thus higher pressure should be applied to provide the required kinetic energy for the penetration of gas molecules.

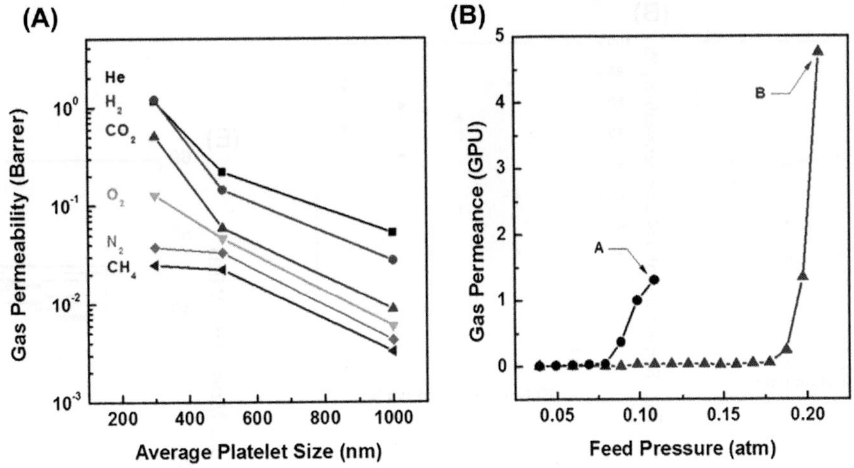

FIGURE 8.17 (A) Gas permeability of thick graphene oxide (GO) membranes with different flake sizes. (B) He permeance of thin GO membranes as increasing feed pressure. (Average flake size of A is 3 μm and B is 1 μm, respectively.) *(A–B) reprinted from C.-Y. Su, et al., Electrical and spectroscopic characterizations of ultra-large reduced graphene oxide monolayers, Chemistry of Materials 21(23) (2009) 5674–5680, with permission from ACS Publications.*

GO-reinforced polymers have been extensively studied to enhance the barrier, mechanical, and conductive properties of GO/polymer nanocomposites. Huang et al. [192,193] prepared GO/poly(vinyl alcohol) (GO/PVA) nanocomposites and GO/poly(lactic acid) (GO/PLA) used as gas barrier films. In case of GO/PLA nanocomposites, both O_2 and CO_2 permeability coefficients of PVA film declined by about 45% and 68% at 1.37 vol.% GO loading. GO has also been suggested to improve mechanical properties of several polymers such as poly(N-isopropylacrylamide), polyacrylamide, and PPy [194–196]. However, there are few studies about the effect of GO depending on its geometry in the polymer matrix. There is also an argument whether GO is oriented randomly or horizontally in GO/polymer nanocomposites through a simple solution mixing and casting method.

Shin et al. [197] studied the barrier property and also role of orientation of 2D GO sheets in cross-linked poly(-ethylene oxide) (XPEG) hydrogel. They measured gas transport properties of various gases, such as He, H_2, O_2, N_2, CH_4, and CO_2, and observed the 80%–90% reduction in gas permeability of XPEG hydrogel membrane with incorporation of only 4 wt.% GO sheets. Also, both Young's modulus and tensile strength of GO/XPEG composite membranes were measured as a function of GO content. It was found that adding impermeable GO sheets into XPEG matrix generates a brick-and-mortar structure that is composed of multiple platelets within the matrix. Based on the Nielsen's transport model, this structure could make the tortuous pathways that inhibit the mass transport through GO/XPEG nanocomposite. This barrier property of GO/XPEG nanocomposites was experimentally confirmed by fundamental studies of gas transport properties (i.e., permeability, diffusivity, and solubility) of various gases (Fig. 8.18A–D). For example, it was observed that gas permeabilities were dramatically decreased with increasing GO loading. CO_2 permeability reduction is mainly caused by diffusivity coefficient decline. In other words, GO played the role of barrier in XPEG matrix. The model proposed by Halpin-Tsai can predict mechanical properties of GO-embedded barrier membranes. This model can theoretically describe the role of GO sheets orientation in polymer matrix on the mechanical properties of barrier materials. As shown in Fig. 8.18E, experimental values of Young's modulus for GO/XPEG nanocomposites are in a good agreement with model's results predicted based on horizontally orientation of GO sheets. This clearly showed that GO sheets are aligned horizontally in XPEG matrix. This is mainly due to the van der Waals and hydrogen interactions between the GO and XPEG, leading to the ful dispersion of GO sheets in XPEG matrix. Beside outstanding barrier properties, incorporation of GO played a unique role to increase friction between the XPEG chains by sliding rigid GO sheet under external force, leading to improved Young's modulus. The rigid GO effectively prevents crack formation by suppressing the stress on nanocomposites.

5.3 Porous Graphene Oxide: Synthesis and Possibility for Membrane Application

Generally, pristine GO membranes suffer from low gas permeability due to the tortuous and layered structure of GO laminates. One of the strategies to compromise the diffusional penalty of layered GO structure is creating pores on individual GO sheets, which not only reduce the length of diffusion pathways but also offer more available pores for entering gas molecules. Recently, synthesis of porous GO has been proposed as a rational strategy to overcome the abovementioned issue [198–205]. Koinuma et al. [198] developed porous GO using photoreaction in O_2 under UV irradiation (Fig. 8.19A). Nanopores were created in areas of sp^3-oxidized region through reduction of oxygen

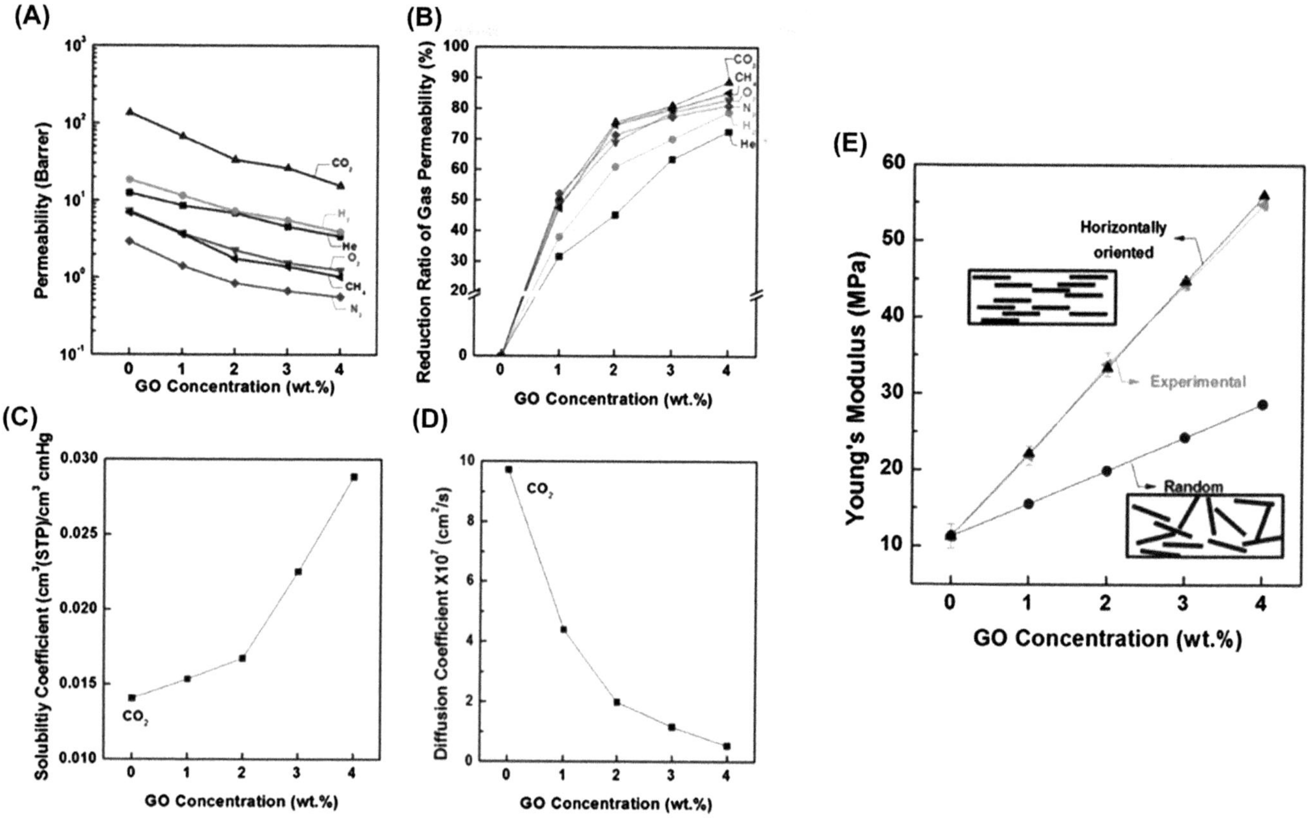

FIGURE 8.18 (A) Gas permeabilities, (B) reduction ratio, (C) CO_2 solubility coefficient, and (D) diffusivity coefficient of pristine XPEG hydrogel membrane and GO/XPEG nanocomposite membranes as a function of GO content. (E) The experimental and theoretical Young's modulus of GO/XPEG nanocomposite hydrogel membranes as a function of GO content. *(A–E) reprinted from J.E. Shin, et al., Graphene oxide nanosheet-embedded crosslinked poly (ethylene oxide) hydrogel, Journal of Applied Polymer Science 135(24) (2018) 4541, with permission from Wiley Online Library.*

functional groups on the GO basal plane. Interestingly, the ratio of epoxide and carbonyl groups were decreased, whereas hydroxyl and carboxyl groups were increased after photoinduced reduction in the presence of O_2 atmosphere. It is worth mentioning that even though the proposed approach for the synthesis of porous GO was facile and fast, the resulting porous GO underwent a reduction, so-called porous rGO.

Yu et al. [199] reported nanoporous GO using hydroxyl radicals. In this method, oxygen functional groups can be removed by hydroxyl radicals, leaving the pores on GO sheets [199]. Nanopores and "lacelike" edges were created via "etching" process. The pore size can be easily controlled by adjusting the irradiation time and rate. However, most of the pores were generated on the edge side of GO sheets that deteriorate the structure and size of GO sheets, as shown in Fig. 8.19B. This structural deformation during the reaction poses a big limitation on utilizing porous GO flakes for the preparation of multistacked porous GO membranes [199]. Han et al. [200] also suggested a steam etching method for creating defects on GO sheet by hydrothermal steaming at 200°C, which was accompanied by the reduction of oxygen functional groups on the basal plane of GO sheets. The degree of etching and defect porosity was controlled easily by adjusting the hydrothermal steaming time. This approach not only takes a long time but also generates cracks from edge to center of GO sheet that disintegrates large GO flakes into small pieces.

In 2017, Choo et al. [201] developed 3-dimensional (3D) porous GO/gold nanoparticle composites for EC detection of dopamine. Pores were formed via ultrasonication and heat energy in aqueous solution without using any chemicals during the reaction. This method requires long time (~12 h) for creating pores and is not feasible to control the pore size distribution and porosity of porous GO. In addition, applied high energy during sonication treatment decreased the size of GO flakes. Through controlling the flake size, pore size, and porosity, this porous GO can be applied to porous GO membrane fabrication.

Generating pores on GO basal plane can provide fast and less tortuous diffusional pathways for gas molecules. However, the currently available pore generation methods are generally accompanied by chemical and/or thermal reduction of GO. For membrane preparation purposes, a suitable synthesis approach should be able to create well-defined pores on GO sheets without any chemical or thermal reduction. Therefore, developing facile, scalable, and mild processes for generating pores on GO sheets is of critical need.

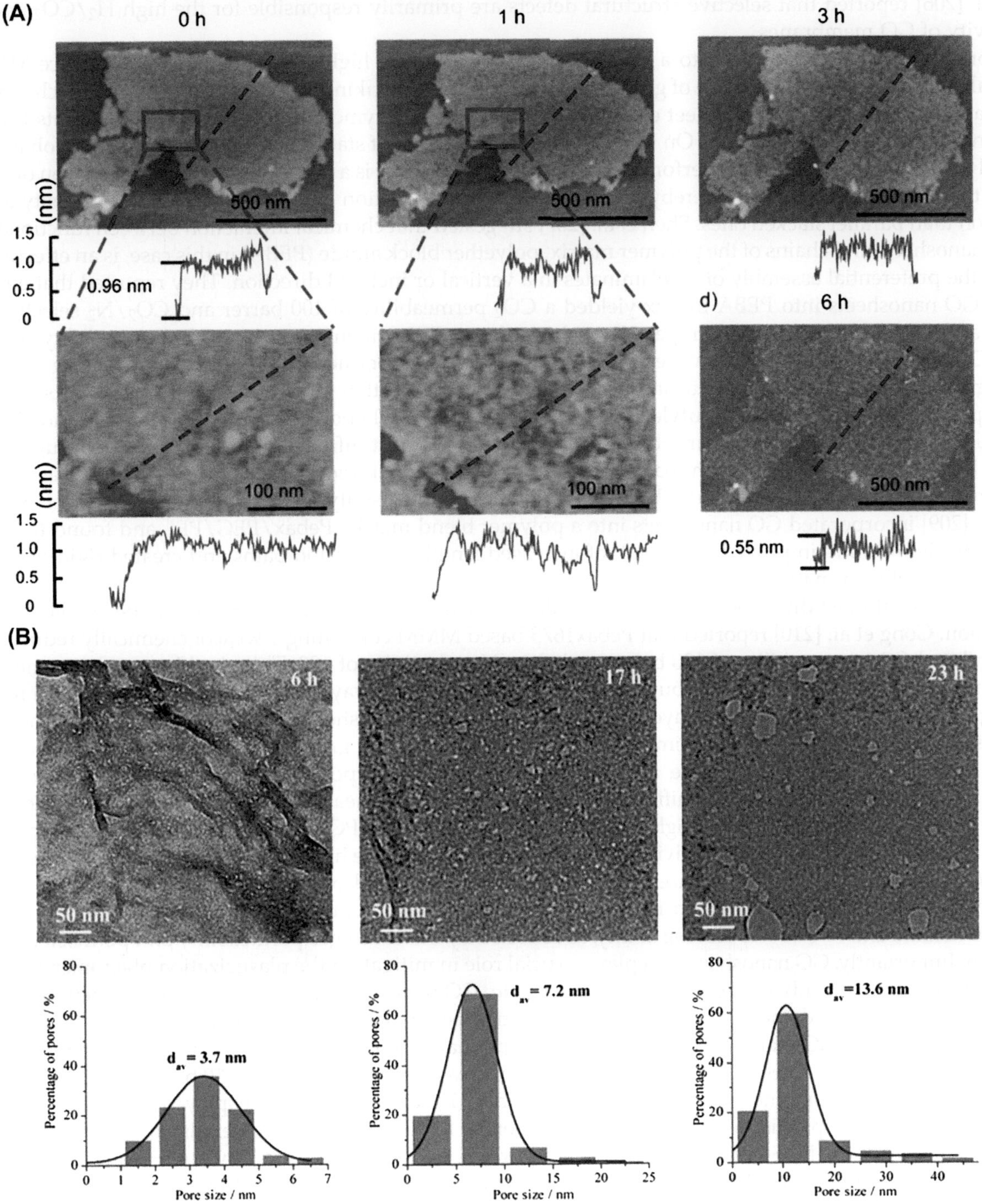

FIGURE 8.19 (A) Etched graphene oxide (GO) sheets via UV irradiation under O_2 atmosphere. (B) TEM images and pore size distribution of porous GO as a function of treatment time. *(A) Reprinted from M. Koinuma, et al., Photochemical engineering of graphene oxide nanosheets, The Journal of Physics Chemistry C 116(37) (2012) 19822–19827, with permission from ACS publications. (B) Reprinted from C. Yu, et al., Engineering nano-porous graphene oxide by hydroxyl radicals, Carbon 105 (2016) 291–296, with permission from Elsevier.*

5.4 Graphene Oxide–Based Mixed Matrix Membranes for Gas Separation

As it was discussed before, thin film processing of GO dispersion using versatile and scalable fabrication routes still remained a challenge. Additionally, the gas separation performance of GO-based membranes is extremely sensitive to the size, structural defects, and stacking manner of nanosheets. For instance, Kim et al. [79] contributed the high CO_2 and H_2 separation performance of GO membranes to interlayer galleries between stacked GO layers, while

Li et al. [206] reported that selective structural defects are primarily responsible for the high H_2/CO_2 and H_2/N_2 selectivity of GO membranes.

Incorporating GO nanosheets into a polymer matrix to produce high gas separation performance MMMs is a rational strategy with the advantage of good film processability. In striking contrast to 3D materials such as spherical nanoparticles, ultrathin GO nanosheet enable to create stronger polymer–filler interactions due to its high aspect ratio and distinct surface chemistry. On the other hand, orientation of stacked GO nanosheets in the polymer matrix strongly affects the gas separation performance. Of particular interest is a vertical or inclined orientation of GO nanosheets to the membrane surface, thereby providing straight permeation pathways with lower tortuosity and faster diffusion than parallel stacked ones. Shen et al. [207] suggested that chemical interaction between functional groups of GO nanosheets and chains of the polymer matrix, polyether block amide (PEBA) in this case, is an effective way to enable the preferential assembly of GO laminates in a vertical or inclined direction. They reported that embedding 1 wt% GO nanosheets into PEBA matrix yielded a CO_2 permeability of 100 barrer and CO_2/N_2 selectivity of 91, which are much higher than the comparative values of neat PEBA membrane (i.e., CO_2 permeability of 52 barrer and CO_2/N_2 selectivity of 50). However, at higher loadings, CO_2 permeability followed a declining trend due to the aggregation of GO nanosheet. Karunakaran et al. [208] revealed that incorporating GO nanosheets into the matrix of poly(ethylene oxide)–poly(butylene terephthalate) (PEO–PBT) polymer enhanced CO_2/N_2 and CO_2/H_2 selectivities, while decreasing the permeability. They assumed that fast diffusion of gas molecules through interlayer channels of vertical-oriented GO laminates was hindered by parallel oriented nanosheets. The high aspect ratio of GO nanosheets can play a significant role in improving the gas diffusivity selectivity in the polymer matrix. Xueqin Li et al. [209] incorporated GO nanosheets into a polymer blend matrix, Pebax/PEG/PEI, and found that CO_2/N_2 selectivity significantly improved as a result of increased length of diffusion paths and created rigidified interface between the polymer matrix and fillers.

Porous GO with fast diffusion pathways, provided by nanoscale pores, is a good candidate as filler for MMMs fabrication. Gong et al. [210] reported that Pebax1675-based MMM containing 5 wt% of chemically reduced porous GO exhibited CO_2 permeability of 120 barrer and CO_2/N_2 selectivity of 108. This significant enhancement in CO_2 transport properties was mainly attributed to the fast diffusion pathways (i.e., created pores) and also molecular-sieving characteristic (0.34 nm interlayer space) of porous GO nanosheets. Recently, we proposed novel nanocomposite filler containing zeolitic imidazole framework-8 (ZIF-8) grown on porous GO nanosheets, called ZPGO, exhibiting exceptional surface area (2170 m^2g^{-1}) [211]. Incorporating very low concentration of ZPGO (0.02 wt%) into Pebax1657 matrix significantly increased the CO_2 permeability from 80 to 180 barrer. Such a drastic improvement was mainly due to the high surface area and porosity of ZPGO filler creating short diffusion pathways.

Apart from the permeability/selectivity trade-off, most polymers are highly susceptible to the plasticization phenomenon induced by highly polar gases, such as CO_2 and heavy hydrocarbons, under high-pressure conditions. Condensable gases disrupt interchain interactions of the polymer matrix, resulting in increased chain mobility and free volume and loss of size-sieving ability, as reflected by increased gas permeability and permselectivity deterioration. Importantly, GO nanosheets can play a crucial role in mitigating the plasticization phenomena in polymer membranes. Taking the advantage of its high aspect ratio, GO nanosheets can create strong interactions with polymer chains, which considerably suppress the plasticization behavior. Our group already reported that embedding small amount of ZPGO nanosheets (0.02 wt%) exceptionally improved the CO_2 plasticization resistance of Pebax1657-based membranes [211]. We found that on increasing feed pressure to 10 bar, the CO_2/N_2 selectivity of pristine Pebax membrane severely decreased, while Pebax/GO MMM displayed a stable performance (Fig. 8.20). Similar antiplasticization function was reported for other 2D nanomaterials such as MOFs [212] and molybdenum disulfide (MoS_2) [213] nanosheets incorporated into polyimide and Pebax1657 polymers, respectively.

Generally, nonequilibrium glass polymers with high free volume, such as PTMSP and polymer of intrinsic microscopy (PIM-1), are highly prone to aging phenomenon over time, thus decreasing their gas permeability. Physical aging is mainly due to the rearrangement of polymer chains approaching an equilibrium state. Taking the advantage of high aspect ratio and surface area of 2D nanomaterials, GO can play a unique role in restricting the rearrangement of polymer chains due to the better interaction [214–217]. Adding less than 1 wt% GO into PTMSP matrix was found to significantly mitigate the aging process. For example, CO_2 permeability of pristine PTMSP underwent a 29% decrease after 9 months, while GO incorporated MMM showed a 17% reduction in permeability. Furthermore, embedded GO resulted in relative enhancement in both gas permeability and selectivity, compared with pristine PTMSP membrane [215,216]. GO also played a similar role in impeding physical aging in PIM-1 membranes for CO_2/CH_4 separation. Alberto et al. [217] reported that CO_2 permeability of pristine PIM-1 membrane decreased from 6400 to 2000 barrer (about 68% reduction) after 155 days. Interestingly, incorporation of low loading (0.05 wt %) of octadecylamine (ODA)-functionalized rGO into PIM-1 matrix substantially retarded the physical aging so that CO_2 permeability only declined about 39%. In fact, ODA-functionalized rGO nanosheets with high aspect ratio and surface area aligned with PIM-1 segments so that it largely constricted the motion of PIM-1 chains.

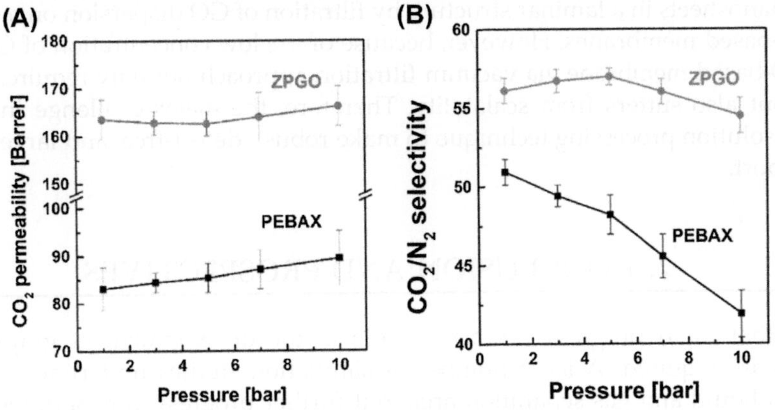

FIGURE 8.20 Gas permeation properties of pristine Pebax- and Pebax/ZPGO-mixed matrix membranes under mixed gas condition (50/50). (A) CO_2 permeability and (B) CO_2/N_2 selectivity as a function of transmembrane pressure. *Reprinted from G. Li, W.-C. Law, K.C. Chan, Floating, highly efficient, and scalable graphene membranes for seawater desalination using solar energy, Green Chemistry 20(16) (2018) 3689–3695, with the permission from Royal Society of Chemistry (RSC).*

6. LARGE AREA GRAPHENE-BASED MEMBRANES: CHALLENGES AND OPPORTUNITIES

It is more than one decade since the University of Manchester for the first time demonstrated fabrication of single-layer graphene, triggering numerous efforts on graphene-based materials for various applications such as electronics, energy storage, membranes, coating, etc. Nevertheless, there are challenges regarding the commercialization of graphene that should be addressed. Most importantly, synthesis of graphene by scalable, cost-affordable, and environmentally benign processes is of pivotal importance. Electronic companies such as Sony and Samsung developed roll-to-roll methods on the basis of a continuous CVD-based process for the synthesis of high quality and large area graphene films. However, these synthesis routes suffer from poor scalability in large scale due to the complexity in reactor designs and transfer methods [42,44,218]. Recently, Seo et al. [219] developed a novel technique for synthesizing graphene at the ambient air environment using a renewable natural precursor, soybean oil, in a single step. However, the graphene quality still needs to be further improved for industrial application.

From the practical application standpoint, a graphene-based selective layer should be coated on a commercial porous support to make a TFC membrane with good mechanical strength and high flux. Recently, a research group at MIT developed a novel roll-to-roll process to produce large-scale nanoporous atomically thin membranes (NATMs) [88]. In this method, graphene was grown on Cu foil, followed by PES coating, interfacial polymerization, and plasma etching. This membrane showed relatively good performance for molecular separation. Geim et al. [220] fabricated a graphene-based EC pump for efficient separation of hydrogen isotopes. In this method, graphene was grown on Cu foil and then hot-pressed against a nafion film/carbon cloth composite. After etching the copper, Pd nanoparticles (as a catalyst) were decorated on the graphene by electron-beam evaporation. Finally, a carbon cloth was coated on Pd-decorated graphene. The Pd-graphene membrane with 30 cm^2 surface area gave a separation factor of 1.8 for hydrogen isotope mixtures.

In a different work, a laser synthesis process was proposed to produce PG membranes [221]. Flexible and large size polyimide film was converted to carbon material by a laser scribing process. Although the fabricated PG membrane was used for seawater desalination using solar energy, this scalable method has a good potential to be applied for manufacturing of large area membranes for both gas separation and water purification purposes. As mentioned before, Nair et al. [132] demonstrated the great potential of GO-based membranes with tunable sieving properties for water desalination application. However, industrially adaptable membranes for water desalination should be fabricated in large area and packed in the form of plate-to-frame, hollow fiber, or spiral wound membrane modules to acquire high flux. Akbari et al. [122,222] recently proposed a scalable method to make GO-based membranes in a relatively large scale by casting a discotic nematic phase of GO on a porous nylon support. This type of membrane showed promising performance for OSN [156,222].

Despite considerable progress in fabricating large-area membranes for liquid purification and gas separation, there are still remained challenges for practical realization of graphene-based membranes that should be addressed. Regarding PG membranes, development of scalable, cost-effective, and advanced techniques to generate pores with uniform size and high density on large-area graphene is vital.

Assembling of GO nanosheets in a laminar structure by filtration of GO dispersion on a porous support is a common way to make GO-based membranes. However, because of the low concentration of GO in dispersion, fabrication of a large area GO-based membrane via vacuum filtration approach not only requires large volumes of liquid and significant time but also suffers from scalability. Therefore, the major challenge in this filed is to develop scalable, cost-effective solution processing technique to make robust, defect-free, and large area GO-based selective layer on a porous support.

7. CONCLUSION AND PROSPECTIVES

There is no doubt that rapid progress has been made in the development of graphene-based membranes, although further steps are required. A large number of simulation studies underlined the promising future for PG membranes in both liquid and gas separation area, but further progress in experimental level, particularly in developing large area PG membranes with controllable pore size and density based on facile and versatile approach, is inevitable. As an alternative to graphene, GO can be produced massively based on the cost-affordable approaches, which is of great importance for practical realization. GO-based membranes exhibited promising water permeation and ion-sieving properties in the laboratory scale. In terms of both separation efficiency and chemical stability, GO-based membranes also showed great performance for organic solvent NF, compared with commercial polymeric membranes. As a 2D nanofiller, GO can considerably improve the antifouling properties, water permeance, and even ion and organic molecules rejection of polymer membranes. Importantly, because of the rich surface chemistry and the high aspect ratio, GO nanofillers can act a unique role in mitigating the plasticization phenomena in polymeric membranes.

Despite such promising separation performance of graphene-based membranes, there are still some remaining challenges for both scientific community and industry need to be addressed. Water and gas permeation mechanism through graphene-based membranes is still a controversial topic. Although simulation studies have provided valuable information to gain more insight into the anomalous transport mechanism of molecules thorough complex subnanometer structure of graphene-based membranes, further conclusive experimental evidence supported by advanced characterization techniques are required.

As starting materials for the membrane preparation, both graphene and GO should be synthesized via facile and controllable routes. In the case of graphene, controlling the pore size and developing clean and scalable methods for transferring the large area of graphene are also of particular need. Controlling the degree of oxidation and understanding the real chemistry of GO are other facing challenges. Compared with the state-of-art polymeric membranes, fabrication of large area graphene or GO membranes suffers from scalability issue. Particular attention should be given to controlling the interlayer spacing as well as pore size and pore density of membranes.

Graphene and GO-based membranes exhibited promising performance on a lab-bench scale. However, most of the developed membranes in lab-scale have the size of several micrometers, several orders of magnitude smaller than target for practical applications. Always there is a question; can these membranes be formed into common industrial modules such as spiral wounds or hollow fiber? How is the stability of membranes under real operating conditions? Therefore, graphene- and GO-based membranes need to be fabricated on a porous flexible support to achieve enough mechanical strength as well as high flux.

Fabrication of large-area PG membranes based on techniques similar to those proposed by electronic companies (i.e., CVD-based continuous roll-to-roll processes) can be considered as a solution.

Large-scale GO membranes ($>10 \times 10\ cm^2$) fabricated via facile casting of a discotic nematic phase of GO on porous support holds a great promise for practical application. This type of membranes showed high separation performance for OSN application, compared with commercially polymer-based membranes. Engineering the interlayer spacing of GO-based selective layer raises the possibility of using this large-scale TFC membrane for NF and even water desalination application.

In the case of gas separation, GO-based MMMs are highly desired as GO can play a unique role in improving gas permeance and molecular-sieving properties as well as antiplasticization and antiphysical aging behavior. Compared with the pristine GO membranes, easy processing of GO-based MMMs into spiral wound and hollow fiber modules is beneficial. Of particular note, GO has a unique role to improve the stability and solute rejection ability of polymer-based membranes for OSN application. There is no doubt that further studies, in both lab- and bench-scales, are highly required to develop novel, scalable, and cost-effective techniques for manufacturing graphene- and GO-based membranes.

References

[1] R.W. Baker, Membrane Technology and Applications, Membrane Technology, 2004.
[2] B.D. Freeman, Basis of permeability/selectivity tradeoff relations in polymeric gas separation membranes, Macromolecules 32 (2) (1999) 375–380.
[3] L.M. Robeson, The upper bound revisited, Journal of Membrane Science 320 (1–2) (2008) 390–400.
[4] M. Galizia, W.S. Chi, Z.P. Smith, T.C. Merkel, R.W. Baker, B.D. Freeman, 50th anniversary perspective: polymers and mixed matrix membranes for gas and vapor separation: a review and prospective opportunities, Macromolecules 50 (20) (2017) 7809–7843.
[5] G. Liu, W. Jin, N. Xu, Two-dimensional-material membranes: a new family of high-performance separation membranes, Angewandte Chemie International Edition 55 (43) (2016) 13384–13397.
[6] C. Tan, X. Cao, X.-J. Wu, Q. He, J. Yang, X. Zhang, J. Chen, W. Zhao, S. Han, G.-H. Nam, Recent advances in ultrathin two-dimensional nanomaterials, Chemical Reviews 117 (9) (2017) 6225–6331.
[7] X. Zhu, C. Tian, C.L. Do-Thanh, S. Dai, Two-dimensional materials as prospective Scaffolds for mixed-matrix membrane-based CO_2 separation, ChemSusChem 10 (2017) 3304–3316.
[8] F. Moghadam, H.B. Park, Two-dimensional materials: an emerging platform for gas separation membranes, Current Opinion in Chemical Engineering 20 (2018) 28–38.
[9] Graphene and graphene oxide: advanced membranes for gas separation and water purification, Inorganic Chemistry Frontiers 2 (5) (2015) 417–424.
[10] K.S. Novoselov, A.K. Geim, S. Morozov, D. Jiang, M. Katsnelson, I. Grigorieva, S. Dubonos, A.A. Firsov, Two-dimensional gas of massless Dirac fermions in graphene, Nature 438 (7065) (2005) 197.
[11] S. Morozov, K. Novoselov, M. Katsnelson, F. Schedin, D. Elias, J.A. Jaszczak, A. Geim, Giant intrinsic carrier mobilities in graphene and its bilayer, Physical Review Letters 100 (1) (2008) 016602.
[12] J.-H. Chen, C. Jang, S. Xiao, M. Ishigami, M.S. Fuhrer, Intrinsic and extrinsic performance limits of graphene devices on SiO2, Nature Nanotechnology 3 (4) (2008) 206.
[13] A. Akturk, N. Goldsman, Electron transport and full-band electron-phonon interactions in graphene, Journal of Applied Physics 103 (5) (2008) 053702.
[14] Superior thermal conductivity of single-layer graphene, Nano Letters 8 (3) (2008) 902–907.
[15] J.H. Seol, I. Jo, A.L. Moore, L. Lindsay, Z.H. Aitken, M.T. Pettes, X. Li, Z. Yao, R. Huang, D. Broido, Two-dimensional phonon transport in supported graphene, Science 328 (5975) (2010) 213–216.
[16] C. Lee, X. Wei, J.W. Kysar, J. Hone, Measurement of the elastic properties and intrinsic strength of monolayer graphene, Science 321 (5887) (2008) 385–388.
[17] I. Frank, D.M. Tanenbaum, A.M. van der Zande, P.L. McEuen, Mechanical properties of suspended graphene sheets, Journal of Vacuum Science and Technology B: Microelectronics and Nanometer Structures Processing, Measurement, and Phenomena 25 (6) (2007) 2558–2561.
[18] R.R. Nair, P. Blake, A.N. Grigorenko, K.S. Novoselov, T.J. Booth, T. Stauber, N.M. Peres, A.K. Geim, Fine structure constant defines visual transparency of graphene, Science 320 (5881) (2008), p. 1308–1308.
[19] D.-E. Jiang, V.R. Cooper, S. Dai, Porous graphene as the ultimate membrane for gas separation, Nano Letters 9 (12) (2009) 4019–4024.
[20] J.S. Bunch, S.S. Verbridge, J.S. Alden, A.M. Van Der Zande, J.M. Parpia, H.G. Craighead, P.L. McEuen, Impermeable atomic membranes from graphene sheets, Nano Letters 8 (8) (2008) 2458–2462.
[21] V. Georgakilas, M. Otyepka, A.B. Bourlinos, V. Chandra, N. Kim, K.C. Kemp, P. Hobza, R. Zboril, K.S. Kim, Functionalization of graphene: covalent and non-covalent approaches, derivatives and applications, Chemical Reviews 112 (11) (2012) 6156–6214.
[22] H. Bai, C. Li, G. Shi, Functional composite materials based on chemically converted graphene, Advanced Materials 23 (9) (2011) 1089–1115.
[23] G. Liu, W. Jin, N. Xu, Graphene-based membranes, Chemical Society Reviews 44 (15) (2015) 5016–5030.
[24] A. Ciesielski, P. Samorì, Graphene via sonication assisted liquid-phase exfoliation, Chemical Society Reviews 43 (1) (2014) 381–398.
[25] K.R. Paton, E. Varrla, C. Backes, R.J. Smith, U. Khan, A. O'Neill, C. Boland, M. Lotya, O.M. Istrate, P. King, Scalable production of large quantities of defect-free few-layer graphene by shear exfoliation in liquids, Nature Materials 13 (6) (2014) 624.
[26] P. Delhaes, Graphite and Precursors, CRC Press, 2014.
[27] K.S. Novoselov, A.K. Geim, S.V. Morozov, D. Jiang, Y. Zhang, S.V. Dubonos, I.V. Grigorieva, A.A. Firsov, Electric field effect in atomically thin carbon films, Science 306 (5696) (2004) 666–669.
[28] W. Zhao, M. Fang, F. Wu, H. Wu, L. Wang, G. Chen, Preparation of graphene by exfoliation of graphite using wet ball milling, Journal of Materials Chemistry 20 (28) (2010) 5817–5819.
[29] Y. Zhu, S. Murali, M.D. Stoller, A. Velamakanni, R.D. Piner, R.S. Ruoff, Microwave assisted exfoliation and reduction of graphite oxide for ultracapacitors, Carbon 48 (7) (2010) 2118–2122.
[30] G.M. Morales, P. Schifani, G. Ellis, C. Ballesteros, G. Martínez, C. Barbero, H.J. Salavagione, High-quality few layer graphene produced by electrochemical intercalation and microwave-assisted expansion of graphite, Carbon 49 (8) (2011) 2809–2816.
[31] M. Monajjemi, Liquid-phase exfoliation (LPE) of graphite towards graphene: an ab initio study, Journal of Molecular Liquids 230 (2017) 461–472.
[32] Y. Hernandez, V. Nicolosi, M. Lotya, F.M. Blighe, Z. Sun, S. De, I. McGovern, B. Holland, M. Byrne, Y.K. Gun'Ko, High-yield production of graphene by liquid-phase exfoliation of graphite, Nature Nanotechnology 3 (9) (2008) 563.
[33] J.N. Coleman, Liquid exfoliation of defect-free graphene, Accounts of Chemical Research 46 (1) (2012) 14–22.
[34] M. Lotya, Y. Hernandez, P.J. King, R.J. Smith, V. Nicolosi, L.S. Karlsson, F.M. Blighe, S. De, Z. Wang, I. McGovern, Liquid phase production of graphene by exfoliation of graphite in surfactant/water solutions, Journal of the American Chemical Society 131 (10) (2009) 3611–3620.
[35] G. Bepete, E. Anglaret, L. Ortolani, V. Morandi, K. Huang, A. Pénicaud, C. Drummond, Surfactant-free single-layer graphene in water, Nature Chemistry 9 (4) (2017) 347.
[36] R. Muñoz, C. Gómez-Aleixandre, Review of CVD synthesis of graphene, Chemical Vapor Deposition 19 (10–11-12) (2013) 297–322.
[37] S. Bhaviripudi, X. Jia, M.S. Dresselhaus, J. Kong, Role of kinetic factors in chemical vapor deposition synthesis of uniform large area graphene using copper catalyst, Nano Letters 10 (10) (2010) 4128–4133.

[38] X. Li, W. Cai, L. Colombo, R.S. Ruoff, Evolution of graphene growth on Ni and Cu by carbon isotope labeling, Nano Letters 9 (12) (2009) 4268–4272.
[39] R. Tromp, J. Hannon, Thermodynamics and kinetics of graphene growth on SiC (0001), Physical Review Letters 102 (10) (2009) 106104.
[40] Y. Zhu, H. Ji, H.-M. Cheng, R.S. Ruoff, Mass production and industrial applications of graphene materials, National Science Review 5 (1) (2017) 90–101.
[41] S. Bae, H. Kim, Y. Lee, X. Xu, J.-S. Park, Y. Zheng, J. Balakrishnan, T. Lei, H.R. Kim, Y.I. Song, Roll-to-roll production of 30-inch graphene films for transparent electrodes, Nature Nanotechnology 5 (8) (2010) 574.
[42] T. Kobayashi, M. Bando, N. Kimura, K. Shimizu, K. Kadono, N. Umezu, K. Miyahara, S. Hayazaki, S. Nagai, Y. Mizuguchi, Production of a 100-m-long high-quality graphene transparent conductive film by roll-to-roll chemical vapor deposition and transfer process, Applied Physics Letters 102 (2) (2013) 023112.
[43] T. Yamada, M. Ishihara, J. Kim, M. Hasegawa, S. Iijima, A roll-to-roll microwave plasma chemical vapor deposition process for the production of 294 mm width graphene films at low temperature, Carbon 50 (7) (2012) 2615–2619.
[44] Q. Yu, L.A. Jauregui, W. Wu, R. Colby, J. Tian, Z. Su, H. Cao, Z. Liu, D. Pandey, D. Wei, Control and characterization of individual grains and grain boundaries in graphene grown by chemical vapour deposition, Nature Materials 10 (6) (2011) 443.
[45] J. Zhang, J. Zhao, J. Lu, Intrinsic strength and failure behaviors of graphene grain boundaries, ACS Nano 6 (3) (2012) 2704–2711.
[46] A.A. Balandin, Thermal properties of graphene and nanostructured carbon materials, Nature Materials 10 (8) (2011) 569.
[47] H. Zhou, W.J. Yu, L. Liu, R. Cheng, Y. Chen, X. Huang, Y. Liu, Y. Wang, Y. Huang, X. Duan, Chemical vapour deposition growth of large single crystals of monolayer and bilayer graphene, Nature Communications 4 (2013) 2096.
[48] X. Xu, Z. Zhang, J. Dong, D. Yi, J. Niu, M. Wu, L. Lin, R. Yin, M. Li, J. Zhou, Ultrafast epitaxial growth of metre-sized single-crystal graphene on industrial Cu foil, Science Bulletin 62 (15) (2017) 1074–1080.
[49] K.M. Reddy, A.D. Gledhill, C.-H. Chen, J.M. Drexler, N.P. Padture, High quality, transferrable graphene grown on single crystal Cu (111) thin films on basal-plane sapphire, Applied Physics Letters 98 (11) (2011) 113117.
[50] V.L. Nguyen, B.G. Shin, D.L. Duong, S.T. Kim, D. Perello, Y.J. Lim, Q.H. Yuan, F. Ding, H.Y. Jeong, H.S. Shin, Seamless stitching of graphene domains on polished copper (111) foil, Advanced Materials 27 (8) (2015) 1376–1382.
[51] I. Jo, S. Park, D.J. Kim, J. San Moon, W.B. Park, T.H. Kim, J.H. Kang, W. Lee, Y. Kim, D.N. Lee, Tension-controlled single-crystallization of copper foils for roll-to-roll synthesis of high-quality graphene films, 2D Materials 5 (2) (2018), 024002.
[52] J.Y. Hong, Y.C. Shin, A. Zubair, Y. Mao, T. Palacios, M.S. Dresselhaus, S.H. Kim, J. Kong, A rational strategy for graphene transfer on substrates with rough features, Advanced Materials 28 (12) (2016) 2382–2392.
[53] G. Zhang, A.G. Guell, P.M. Kirkman, R.A. Lazenby, T.S. Miller, P.R. Unwin, Versatile polymer-free graphene transfer method and applications, ACS Applied Materials and Interfaces 8 (12) (2016) 8008–8016.
[54] L.A. Belyaeva, W. Fu, H. Arjmandi-Tash, G.g.F. Schneider, Molecular caging of graphene with cyclohexane: transfer and electrical transport, ACS Central Science 2 (12) (2016) 904–909.
[55] J.W. Suk, A. Kitt, C.W. Magnuson, Y. Hao, S. Ahmed, J. An, A.K. Swan, B.B. Goldberg, R.S. Ruoff, Transfer of CVD-grown monolayer graphene onto arbitrary substrates, ACS Nano 5 (9) (2011) 6916–6924.
[56] M. Chen, R.C. Haddon, R. Yan, E. Bekyarova, Advances in transferring chemical vapour deposition graphene: a review, Materials Horizons 4 (6) (2017) 1054–1063.
[57] A.V. Zaretski, H. Moetazedi, C. Kong, E.J. Sawyer, S. Savagatrup, E. Valle, T.F. O'Connor, A.D. Printz, D.J. Lipomi, Metal-assisted exfoliation (MAE): green, roll-to-roll compatible method for transferring graphene to flexible substrates, Nanotechnology 26 (4) (2015) 045301.
[58] L. Gao, W. Ren, H. Xu, L. Jin, Z. Wang, T. Ma, L.-P. Ma, Z. Zhang, Q. Fu, L.-M. Peng, Repeated growth and bubbling transfer of graphene with millimetre-size single-crystal grains using platinum, Nature Communications 3 (2012) 699.
[59] Y. Wang, Y. Zheng, X. Xu, E. Dubuisson, Q. Bao, J. Lu, K.P. Loh, Electrochemical delamination of CVD-grown graphene film: toward the recyclable use of copper catalyst, ACS Nano 5 (12) (2011) 9927–9933.
[60] G. Lippert, J. Dąbrowski, T. Schroeder, M.A. Schubert, Y. Yamamoto, F. Herziger, J. Maultzsch, J. Baringhaus, C. Tegenkamp, M.C. Asensio, Graphene grown on Ge (0 0 1) from atomic source, Carbon 75 (2014) 104–112.
[61] G. Wang, M. Zhang, Y. Zhu, G. Ding, D. Jiang, Q. Guo, S. Liu, X. Xie, P.K. Chu, Z. Di, Direct growth of graphene film on germanium substrate, Scientific Reports 3 (2013) 2465.
[62] R. Arvidsson, Review of Environmental Life Cycle Assessment Studies of Graphene Production, 2017.
[63] R. Arvidsson, D. Kushnir, S. Molander, B.A. Sandén, Energy and resource use assessment of graphene as a substitute for indium tin oxide in transparent electrodes, Journal of Cleaner Production 132 (2016) 289–297.
[64] R. Arvidsson, D. Kushnir, B.r.A. Sanden, S. Molander, Prospective life cycle assessment of graphene production by ultrasonication and chemical reduction, Environmental Science and Technology 48 (8) (2014) 4529–4536.
[65] R. Arvidsson, S. Molander, Prospective life cycle assessment of epitaxial graphene production at different manufacturing scales and maturity, Journal of Industrial Ecology 21 (5) (2017) 1153–1164.
[66] A.L. Chun, Life-cycle assessment: should graphene replace ITO? Nature Nanotechnology (2015, June 3).
[67] M. Cossutta, J. McKechnie, S.J. Pickering, A comparative LCA of different graphene production routes, Green Chemistry 19 (24) (2017) 5874–5884.
[68] R. Nair, H. Wu, P. Jayaram, I. Grigorieva, A. Geim, Unimpeded permeation of water through helium-leak–tight graphene-based membranes, Science 335 (6067) (2012) 442–444.
[69] L. Wang, C.M. Williams, M.S. Boutilier, P.R. Kidambi, R. Karnik, Single-layer graphene membranes withstand ultrahigh applied pressure, Nano Letters 17 (5) (2017) 3081–3088.
[70] S.P. Koenig, L. Wang, J. Pellegrino, J.S. Bunch, Selective molecular sieving through porous graphene, Nature Nanotechnology 7 (11) (2012) 728.
[71] S.C. O'Hern, M.S. Boutilier, J.-C. Idrobo, Y. Song, J. Kong, T. Laoui, M. Atieh, R. Karnik, Selective ionic transport through tunable subnanometer pores in single-layer graphene membranes, Nano Letters 14 (3) (2014) 1234–1241.
[72] M.S. Boutilier, C. Sun, S.C. O'Hern, H. Au, N.G. Hadjiconstantinou, R. Karnik, Implications of permeation through intrinsic defects in graphene on the design of defect-tolerant membranes for gas separation, ACS Nano 8 (1) (2014) 841–849.

[73] S.C. O'Hern, C.A. Stewart, M.S. Boutilier, J.-C. Idrobo, S. Bhaviripudi, S.K. Das, J. Kong, T. Laoui, M. Atieh, R. Karnik, Selective molecular transport through intrinsic defects in a single layer of CVD graphene, ACS Nano 6 (11) (2012) 10130–10138.

[74] M.S. Boutilier, D. Jang, J.-C. Idrobo, P.R. Kidambi, N.G. Hadjiconstantinou, R. Karnik, Molecular sieving across centimeter-scale single-layer nanoporous graphene membranes, ACS Nano 11 (6) (2017) 5726–5736.

[75] M.D. Fischbein, M. Drndić, Electron beam nanosculpting of suspended graphene sheets, Applied Physics Letters 93 (11) (2008) 113107.

[76] D. Fox, A. O'Neill, D. Zhou, M. Boese, J. Coleman, H. Zhang, Nitrogen assisted etching of graphene layers in a scanning electron microscope, Applied Physics Letters 98 (24) (2011) 243117.

[77] K. Celebi, J. Buchheim, R.M. Wyss, A. Droudian, P. Gasser, I. Shorubalko, J.-I. Kye, C. Lee, H.G. Park, Ultimate permeation across atomically thin porous graphene, Science 344 (6181) (2014) 289–292.

[78] S.P. Surwade, S.N. Smirnov, I.V. Vlassiouk, R.R. Unocic, G.M. Veith, S. Dai, S.M. Mahurin, Water desalination using nanoporous single-layer graphene, Nature Nanotechnology 10 (5) (2015) 459–464.

[79] H.W. Kim, H.W. Yoon, S.-M. Yoon, B.M. Yoo, B.K. Ahn, Y.H. Cho, H.J. Shin, H. Yang, U. Paik, S. Kwon, Selective gas transport through few-layered graphene and graphene oxide membranes, Science 342 (6154) (2013) 91–95.

[80] L.W. Drahushuk, M.S. Strano, Mechanisms of gas permeation through single layer graphene membranes, Langmuir 28 (48) (2012) 16671–16678.

[81] C. Sun, M.S. Boutilier, H. Au, P. Poesio, B. Bai, R. Karnik, N.G. Hadjiconstantinou, Mechanisms of molecular permeation through nanoporous graphene membranes, Langmuir 30 (2) (2014) 675–682.

[82] M. Shan, Q. Xue, N. Jing, C. Ling, T. Zhang, Z. Yan, J. Zheng, Influence of chemical functionalization on the CO 2/N 2 separation performance of porous graphene membranes, Nanoscale 4 (17) (2012) 5477–5482.

[83] Z. Tian, S.M. Mahurin, S. Dai, D.-e. Jiang, Ion-gated gas separation through porous graphene, Nano Letters 17 (3) (2017) 1802–1807.

[84] Y. Wang, Q. Yang, C. Zhong, J. Li, Theoretical investigation of gas separation in functionalized nanoporous graphene membranes, Applied Surface Science 407 (2017) 532–539.

[85] G. Lei, C. Liu, H. Xie, F. Song, Separation of the hydrogen sulfide and methane mixture by the porous graphene membrane: effect of the charges, Chemical Physics Letters 599 (2014) 127–132.

[86] M.S. Boutilier, N.G. Hadjiconstantinou, R. Karnik, Knudsen effusion through polymer-coated three-layer porous graphene membranes, Nanotechnology 28 (18) (2017) 184003.

[87] S. Huang, M. Dakhchoune, W. Luo, E. Oveisi, G. He, M. Rezaei, J. Zhao, D.T. Alexander, A. Züttel, M.S. Strano, Single-layer graphene membranes by crack-free transfer for gas mixture separation, Nature Communications 9 (2018).

[88] P.R. Kidambi, D.D. Mariappan, N.T. Dee, A. Vyatskikh, S. Zhang, R. Karnik, A.J. Hart, A scalable route to nanoporous large-area atomically thin graphene membranes by roll-to-roll chemical vapor deposition and polymer support casting, ACS Applied Materials and Interfaces 10 (12) (2018) 10369–10378.

[89] B.C. Brodie, XIII. On the atomic weight of graphite, Philosophical Transactions of the Royal Society of London 149 (1859) 249–259.

[90] L. Staudenmaier, Verfahren zur darstellung der graphitsäure, European Journal of Inorganic Chemistry 32 (2) (1899) 1394–1399.

[91] W.S. Hummers Jr., R.E. Offeman, Preparation of graphitic oxide, Journal of the American Chemical Society 80 (6) (1958) 1339, 1339.

[92] N.I. Kovtyukhova, P.J. Ollivier, B.R. Martin, T.E. Mallouk, S.A. Chizhik, E.V. Buzaneva, A.D. Gorchinskiy, Layer-by-layer assembly of ultrathin composite films from micron-sized graphite oxide sheets and polycations, Chemistry of Materials 11 (3) (1999) 771–778.

[93] J. Chen, B. Yao, C. Li, G. Shi, An improved Hummers method for eco-friendly synthesis of graphene oxide, Carbon 64 (2013) 225–229.

[94] D.C. Marcano, D.V. Kosynkin, J.M. Berlin, A. Sinitskii, Z. Sun, A. Slesarev, L.B. Alemany, W. Lu, J.M. Tour, Improved synthesis of graphene oxide, ACS Nano 4 (8) (2010) 4806–4814.

[95] U. Hofmann, R. Holst, Über die Säurenatur und die Methylierung von Graphitoxyd, Berichte der deutschen chemischen Gesellschaft (A and B Series) 72 (4) (1939) 754–771.

[96] G. Ruess, Über das graphitoxyhydroxyd (graphitoxyd), Monatshefte für Chemie und verwandte Teile anderer Wissenschaften 76 (3–5) (1947) 381–417.

[97] W. Scholz, H. Boehm, Untersuchungen am graphitoxid. VI. Betrachtungen zur struktur des graphitoxids, Zeitschrift für anorganische und allgemeine Chemie 369 (3-6) (1969) 327–340.

[98] T. Nakajima, A. Mabuchi, R. Hagiwara, A new structure model of graphite oxide, Carbon 26 (3) (1988) 357–361.

[99] A. Lerf, H. He, M. Forster, J. Klinowski, Structure of graphite oxide revisited, The Journal of Physical Chemistry B 102 (23) (1998) 4477–4482.

[100] T. Szabó, O. Berkesi, P. Forgó, K. Josepovits, Y. Sanakis, D. Petridis, I. Dékány, Evolution of surface functional groups in a series of progressively oxidized graphite oxides, Chemistry of Materials 18 (11) (2006) 2740–2749.

[101] S. Pei, Q. Wei, K. Huang, H.-M. Cheng, W. Ren, Green synthesis of graphene oxide by seconds timescale water electrolytic oxidation, Nature Communications 9 (1) (2018) 145.

[102] S. Park, R.S. Ruoff, Chemical methods for the production of graphenes, Nature Nanotechnology 4 (4) (2009) 217.

[103] W. Gao, L.B. Alemany, L. Ci, P.M. Ajayan, New insights into the structure and reduction of graphite oxide, Nature Chemistry 1 (5) (2009) 403.

[104] K.H. Lee, B. Lee, S.-J. Hwang, J.-U. Lee, H. Cheong, O.-S. Kwon, K. Shin, N.H. Hur, Large scale production of highly conductive reduced graphene oxide sheets by a solvent-free low temperature reduction, Carbon 69 (2014) 327–335.

[105] S. Lee, S.H. Eom, J.S. Chung, S.H. Hur, Large-scale production of high-quality reduced graphene oxide, Chemical Engineering Journal 233 (2013) 297–304.

[106] D. Li, M.B. Müller, S. Gilje, R.B. Kaner, G.G. Wallace, Processable aqueous dispersions of graphene nanosheets, Nature Nanotechnology 3 (2) (2008) 101.

[107] Y. Zhu, S. Murali, W. Cai, X. Li, J.W. Suk, J.R. Potts, R.S. Ruoff, Graphene and graphene oxide: synthesis, properties, and applications, Advanced Materials 22 (35) (2010) 3906–3924.

[108] G. Williams, B. Seger, P.V. Kamat, TiO2-graphene nanocomposites. UV-assisted photocatalytic reduction of graphene oxide, ACS Nano 2 (7) (2008) 1487–1491.

[109] J. Wang, E.C. Salihi, L. Šiller, Green reduction of graphene oxide using alanine, Materials Science and Engineering: C 72 (2017) 1–6.

[110] J. Zhang, H. Yang, G. Shen, P. Cheng, J. Zhang, S. Guo, Reduction of graphene oxide via L-ascorbic acid, Chemical Communications 46 (7) (2010) 1112–1114.
[111] I. Akin, E. Zor, H. Bingol, M. Ersoz, Green synthesis of reduced graphene oxide/polyaniline composite and its application for salt rejection by polysulfone-based composite membranes, The Journal of Physical Chemistry B 118 (21) (2014) 5707–5716.
[112] P. Zhang, X. Zhang, S. Zhang, X. Lu, Q. Li, Z. Su, G. Wei, One-pot green synthesis, characterizations, and biosensor application of self-assembled reduced graphene oxide–gold nanoparticle hybrid membranes, Journal of Materials Chemistry B 1 (47) (2013) 6525–6531.
[113] D.A. Dikin, S. Stankovich, E.J. Zimney, R.D. Piner, G.H. Dommett, G. Evmenenko, S.T. Nguyen, R.S. Ruoff, Preparation and characterization of graphene oxide paper, Nature 448 (7152) (2007) 457.
[114] C.-H. Tsou, Q.-F. An, S.-C. Lo, M. De Guzman, W.-S. Hung, C.-C. Hu, K.-R. Lee, J.-Y. Lai, Effect of microstructure of graphene oxide fabricated through different self-assembly techniques on 1-butanol dehydration, Journal of Membrane Science 477 (2015) 93–100.
[115] Y. Lou, G. Liu, S. Liu, J. Shen, W. Jin, A facile way to prepare ceramic-supported graphene oxide composite membrane via silane-graft modification, Applied Surface Science 307 (2014) 631–637.
[116] H.W. Kim, H.W. Yoon, B.M. Yoo, J.S. Park, K.L. Gleason, B.D. Freeman, H.B. Park, High-performance CO_2-philic graphene oxide membranes under wet-conditions, Chemical Communications 50 (88) (2014) 13563–13566.
[117] M. Hu, B. Mi, Enabling graphene oxide nanosheets as water separation membranes, Environmental Science and Technology 47 (8) (2013) 3715–3723.
[118] N.F.D. Aba, J.Y. Chong, B. Wang, C. Mattevi, K. Li, Graphene oxide membranes on ceramic hollow fibers–Microstructural stability and nanofiltration performance, Journal of Membrane Science 484 (2015) 87–94.
[119] K. Huang, G. Liu, Y. Lou, Z. Dong, J. Shen, W. Jin, A graphene oxide membrane with highly selective molecular separation of aqueous organic solution, Angewandte Chemie International Edition 53 (27) (2014) 6929–6932.
[120] K. Huang, J. Yuan, G. Shen, G. Liu, W. Jin, Graphene oxide membranes supported on the ceramic hollow fibre for efficient H_2 recovery, Chinese Journal of Chemical Engineering 25 (6) (2017) 752–759.
[121] F. Zhou, H.N. Tien, W.L. Xu, J.-T. Chen, Q. Liu, E. Hicks, M. Fathizadeh, S. Li, M. Yu, Ultrathin graphene oxide-based hollow fiber membranes with brush-like CO_2-philic agent for highly efficient CO_2 capture, Nature Communications 8 (1) (2017) 2107.
[122] A. Akbari, P. Sheath, S.T. Martin, D.B. Shinde, M. Shaibani, P.C. Banerjee, R. Tkacz, D. Bhattacharyya, M. Majumder, Large-area graphene-based nanofiltration membranes by shear alignment of discotic nematic liquid crystals of graphene oxide, Nature Communications 7 (2016) 10891.
[123] V. Berry, Impermeability of graphene and its applications, Carbon 62 (2013) 1–10.
[124] D. Cohen-Tanugi, J.C. Grossman, Water desalination across nanoporous graphene, Nano Letters 12 (7) (2012) 3602–3608.
[125] L. Wang, M.S. Boutilier, P.R. Kidambi, D. Jang, N.G. Hadjiconstantinou, R. Karnik, Fundamental transport mechanisms, fabrication and potential applications of nanoporous atomically thin membranes, Nature Nanotechnology 12 (6) (2017) 509.
[126] P. Sun, M. Zhu, K. Wang, M. Zhong, J. Wei, D. Wu, Z. Xu, H. Zhu, Selective ion penetration of graphene oxide membranes, ACS Nano 7 (1) (2012) 428–437.
[127] A. Klechikov, J. Yu, D. Thomas, T. Sharifi, A.V. Talyzin, Structure of graphene oxide membranes in solvents and solutions, Nanoscale 7 (37) (2015) 15374–15384.
[128] S. Zheng, Q. Tu, J.J. Urban, S. Li, B. Mi, Swelling of graphene oxide membranes in aqueous solution: characterization of interlayer spacing and insight into water transport mechanisms, ACS Nano 11 (6) (2017) 6440–6450.
[129] R. Joshi, P. Carbone, F.-C. Wang, V.G. Kravets, Y. Su, I.V. Grigorieva, H. Wu, A.K. Geim, R.R. Nair, Precise and ultrafast molecular sieving through graphene oxide membranes, Science 343 (6172) (2014) 752–754.
[130] H. Huang, Z. Song, N. Wei, L. Shi, Y. Mao, Y. Ying, L. Sun, Z. Xu, X. Peng, Ultrafast viscous water flow through nanostrand-channelled graphene oxide membranes, Nature communications 4 (2013) 2979.
[131] Y. Han, Z. Xu, C. Gao, Ultrathin graphene nanofiltration membrane for water purification, Advanced Functional Materials 23 (29) (2013) 3693–3700.
[132] J. Abraham, K.S. Vasu, C.D. Williams, K. Gopinadhan, Y. Su, C.T. Cherian, J. Dix, E. Prestat, S.J. Haigh, I.V. Grigorieva, Tunable sieving of ions using graphene oxide membranes, Nature Nanotechnology 12 (6) (2017) 546.
[133] L. Chen, G. Shi, J. Shen, B. Peng, B. Zhang, Y. Wang, F. Bian, J. Wang, D. Li, Z. Qian, Ion sieving in graphene oxide membranes via cationic control of interlayer spacing, Nature 550 (7676) (2017) 380.
[134] W.-S. Hung, C.-H. Tsou, M. De Guzman, Q.-F. An, Y.-L. Liu, Y.-M. Zhang, C.-C. Hu, K.-R. Lee, J.-Y. Lai, Cross-linking with diamine monomers to prepare composite graphene oxide-framework membranes with varying d-spacing, Chemistry of Materials 26 (9) (2014) 2983–2990.
[135] R. Joshi, S. Alwarappan, M. Yoshimura, V. Sahajwalla, Y. Nishina, Graphene oxide: the new membrane material, Applied Materials Today 1 (1) (2015) 1–12.
[136] D.W. Boukhvalov, M.I. Katsnelson, Y.-W. Son, Origin of anomalous water permeation through graphene oxide membrane, Nano Letters 13 (8) (2013) 3930–3935.
[137] J. Deng, Y. You, H. Bustamante, V. Sahajwalla, R.K. Joshi, Mechanism of water transport in graphene oxide laminates, Chemical Science 8 (3) (2017) 1701–1704.
[138] N. Wei, X. Peng, Z. Xu, Understanding water permeation in graphene oxide membranes, ACS Applied Materials and Interfaces 6 (8) (2014) 5877–5883.
[139] B. Mi, Graphene oxide membranes for ionic and molecular sieving, Science 343 (6172) (2014) 740–742.
[140] B. Radha, A. Esfandiar, F. Wang, A. Rooney, K. Gopinadhan, A. Keerthi, A. Mishchenko, A. Janardanan, P. Blake, L. Fumagalli, Molecular transport through capillaries made with atomic-scale precision, Nature 538 (7624) (2016) 222.
[141] G. Algara-Siller, O. Lehtinen, F. Wang, R. Nair, U. Kaiser, H. Wu, A. Geim, I. Grigorieva, Square ice in graphene nanocapillaries, Nature 519 (7544) (2015) 443.
[142] X. Chen, G. Cao, A. Han, V.K. Punyamurtula, L. Liu, P.J. Culligan, T. Kim, Y. Qiao, Nanoscale fluid transport: size and rate effects, Nano Letters 8 (9) (2008) 2988–2992.
[143] S. Kumar Kannam, B. Todd, J.S. Hansen, P.J. Daivis, Slip length of water on graphene: limitations of non-equilibrium molecular dynamics simulations, The Journal of Chemical Physics 136 (2) (2012) 024705.

[144] G. Cicero, J.C. Grossman, E. Schwegler, F. Gygi, G. Galli, Water confined in nanotubes and between graphene sheets: a first principle study, Journal of the American Chemical Society 130 (6) (2008) 1871–1878.
[145] M. Majumder, N. Chopra, R. Andrews, B.J. Hinds, Nanoscale hydrodynamics: enhanced flow in carbon nanotubes, Nature 438 (7064) (2005) 44.
[146] A. Striolo, The mechanism of water diffusion in narrow carbon nanotubes, Nano Letters 6 (4) (2006) 633–639.
[147] X. Qin, Q. Yuan, Y. Zhao, S. Xie, Z. Liu, Measurement of the rate of water translocation through carbon nanotubes, Nano Letters 11 (5) (2011) 2173–2177.
[148] P. Marchetti, M.F. Jimenez Solomon, G. Szekely, A.G. Livingston, Molecular separation with organic solvent nanofiltration: a critical review, Chemical Reviews 114 (21) (2014) 10735–10806.
[149] P. Vandezande, L.E. Gevers, I.F. Vankelecom, Solvent resistant nanofiltration: separating on a molecular level, Chemical Society Reviews 37 (2) (2008) 365–405.
[150] J. Scarpello, D. Nair, L.F. Dos Santos, L. White, A. Livingston, The separation of homogeneous organometallic catalysts using solvent resistant nanofiltration, Journal of Membrane Science 203 (1–2) (2002) 71–85.
[151] S. Karan, Z. Jiang, A.G. Livingston, Sub–10 nm polyamide nanofilms with ultrafast solvent transport for molecular separation, Science 348 (6241) (2015) 1347–1351.
[152] L. Huang, Y. Li, Q. Zhou, W. Yuan, G. Shi, Graphene oxide membranes with tunable semipermeability in organic solvents, Advanced Materials 27 (25) (2015) 3797–3802.
[153] B. Van der Bruggen, J. Geens, C. Vandecasteele, Fluxes and rejections for nanofiltration with solvent stable polymeric membranes in water, ethanol and n-hexane, Chemical Engineering Science 57 (13) (2002) 2511–2518.
[154] X. Yang, A. Livingston, L.F. Dos Santos, Experimental observations of nanofiltration with organic solvents, Journal of Membrane Science 190 (1) (2001) 45–55.
[155] Q. Yang, Y. Su, C. Chi, C. Cherian, K. Huang, V. Kravets, F. Wang, J. Zhang, A. Pratt, A. Grigorenko, Ultrathin graphene-based membrane with precise molecular sieving and ultrafast solvent permeation, Nature Materials 16 (12) (2017) 1198.
[156] A. Akbari, S.E. Meragawi, S.T. Martin, B. Corry, E. Shamsaei, C.D. Easton, D. Bhattacharyya, M. Majumder, Solvent transport behavior of shear aligned graphene oxide membranes and implications in organic solvent nanofiltration, ACS Applied Materials and Interfaces 10 (2) (2018) 2067–2074.
[157] L. Huang, J. Chen, T. Gao, M. Zhang, Y. Li, L. Dai, L. Qu, G. Shi, Reduced graphene oxide membranes for ultrafast organic solvent nanofiltration, Advanced Materials 28 (39) (2016) 8669–8674.
[158] J. Geens, K. Boussu, C. Vandecasteele, B. Van der Bruggen, Modelling of solute transport in non-aqueous nanofiltration, Journal of Membrane Science 281 (1–2) (2006) 139–148.
[159] B. Van der Bruggen, C. Vandecasteele, Modelling of the retention of uncharged molecules with nanofiltration, Water Research 36 (5) (2002) 1360–1368.
[160] J.-T. Chen, Y.-J. Fu, Q.-F. An, S.-C. Lo, S.-H. Huang, W.-S. Hung, C.-C. Hu, K.-R. Lee, J.-Y. Lai, Tuning nanostructure of graphene oxide/polyelectrolyte LbL assemblies by controlling pH of GO suspension to fabricate transparent and super gas barrier films, Nanoscale 5 (19) (2013) 9081–9088.
[161] T. Gao, H. Wu, L. Tao, L. Qu, C. Li, Enhanced stability and separation efficiency of graphene oxide membranes in organic solvent nanofiltration, Journal of Materials Chemistry A 6 (2018) 19563–19569.
[162] X.-L. Xu, F.-W. Lin, Y. Du, X. Zhang, J. Wu, Z.-K. Xu, Graphene oxide nanofiltration membranes stabilized by cationic porphyrin for high salt rejection, ACS Applied Materials and Interfaces 8 (20) (2016) 12588–12593.
[163] T. Gao, L. Huang, C. Li, G. Xu, G. Shi, Graphene membranes with tuneable nanochannels by intercalating self-assembled porphyrin molecules for organic solvent nanofiltration, Carbon 124 (2017) 263–270.
[164] S. Zinadini, A.A. Zinatizadeh, M. Rahimi, V. Vatanpour, H. Zangeneh, Preparation of a novel antifouling mixed matrix PES membrane by embedding graphene oxide nanoplates, Journal of Membrane Science 453 (2014) 292–301.
[165] S. Zinadini, V. Vatanpour, A.A. Zinatizadeh, M. Rahimi, Z. Rahimi, M. Kian, Preparation and characterization of antifouling graphene oxide/polyethersulfone ultrafiltration membrane: application in MBR for dairy wastewater treatment, Journal of Water Process Engineering 7 (2015) 280–294.
[166] C. Zhao, X. Xu, J. Chen, F. Yang, Effect of graphene oxide concentration on the morphologies and antifouling properties of PVDF ultrafiltration membranes, Journal of Environmental Chemical Engineering 1 (3) (2013) 349–354.
[167] S.P. Dharupaneedi, R.V. Anjanapura, J.M. Han, T.M. Aminabhavi, Functionalized graphene sheets embedded in chitosan nanocomposite membranes for ethanol and isopropanol dehydration via pervaporation, Industrial and Engineering Chemistry Research 53 (37) (2014) 14474–14484.
[168] S. Bano, A. Mahmood, S.-J. Kim, K.-H. Lee, Graphene oxide modified polyamide nanofiltration membrane with improved flux and antifouling properties, Journal of Materials Chemistry A 3 (5) (2015) 2065–2071.
[169] H.-R. Chae, J. Lee, C.-H. Lee, I.-C. Kim, P.-K. Park, Graphene oxide-embedded thin-film composite reverse osmosis membrane with high flux, anti-biofouling, and chlorine resistance, Journal of Membrane Science 483 (2015) 128–135.
[170] H.J. Kim, M.-Y. Lim, K.H. Jung, D.-G. Kim, J.-C. Lee, High-performance reverse osmosis nanocomposite membranes containing the mixture of carbon nanotubes and graphene oxides, Journal of Materials Chemistry A 3 (13) (2015) 6798–6809.
[171] L. Shao, X. Cheng, Z. Wang, J. Ma, Z. Guo, Tuning the performance of polypyrrole-based solvent-resistant composite nanofiltration membranes by optimizing polymerization conditions and incorporating graphene oxide, Journal of Membrane Science 452 (2014) 82–89.
[172] R. Ding, H. Zhang, Y. Li, J. Wang, B. Shi, H. Mao, J. Dang, J. Liu, Graphene oxide-embedded nanocomposite membrane for solvent resistant nanofiltration with enhanced rejection ability, Chemical Engineering Science 138 (2015) 227–238.
[173] J. Aburabie, K.-V. Peinemann, Crosslinked poly (ether block amide) composite membranes for organic solvent nanofiltration applications, Journal of Membrane Science 523 (2017) 264–272.
[174] F. Fei, L. Cseri, G. Szekely, C.F. Blanford, Robust covalently cross-linked polybenzimidazole/graphene oxide membranes for high-flux organic solvent nanofiltration, ACS Applied Materials and Interfaces 10 (18) (2018) 16140–16147.

[175] M. Alberto, J.M. Luque-Alled, L. Gao, M. Iliut, E. Prestat, L. Newman, S.J. Haigh, A. Vijayaraghavan, P.M. Budd, P. Gorgojo, Enhanced organophilic separations with mixed matrix membranes of polymers of intrinsic microporosity and graphene-like fillers, Journal of Membrane Science 526 (2017) 437–449.
[176] M. Alberto, R. Bhavsar, J.M. Luque-Alled, E. Prestat, L. Gao, P.M. Budd, A. Vijayaraghavan, G. Szekely, S.M. Holmes, P. Gorgojo, Study on the formation of thin film nanocomposite (TFN) membranes of polymers of intrinsic microporosity and graphene-like fillers: effect of lateral flake size and chemical functionalization, Journal of Membrane Science 565 (2018) 390–401.
[177] S. Leaper, A. Abdel-Karim, B. Faki, J.M. Luque-Alled, M. Alberto, A. Vijayaraghavan, S.M. Holmes, G. Szekely, M.I. Badawy, N. Shokri, Flux-enhanced PVDF mixed matrix membranes incorporating APTS-functionalized graphene oxide for membrane distillation, Journal of Membrane Science 554 (2018) 309–323.
[178] A.M. Dimiev, L.B. Alemany, J.M. Tour, Graphene oxide. Origin of acidity, its instability in water, and a new dynamic structural model, ACS Nano 7 (1) (2012) 576–588.
[179] L.E. Nielsen, Models for the permeability of filled polymer systems, Journal of Macromolecular Science—Chemistry 1 (5) (1967) 929–942.
[180] G. Choudalakis, A. Gotsis, Permeability of polymer/clay nanocomposites: a review, European Polymer Journal 45 (4) (2009) 967–984.
[181] B.M. Yoo, H.J. Shin, H.W. Yoon, H.B. Park, Graphene and graphene oxide and their uses in barrier polymers, Journal of Applied Polymer Science 131 (1) (2014).
[182] Y. Xu, H. Bai, G. Lu, C. Li, G. Shi, Flexible graphene films via the filtration of water-soluble noncovalent functionalized graphene sheets, Journal of the American Chemical Society 130 (18) (2008) 5856–5857.
[183] F. Liu, S. Song, D. Xue, H. Zhang, Folded structured graphene paper for high performance electrode materials, Advanced Materials 24 (8) (2012) 1089–1094.
[184] Z. Luo, Y. Lu, L.A. Somers, A.C. Johnson, High yield preparation of macroscopic graphene oxide membranes, Journal of the American Chemical Society 131 (3) (2009) 898–899.
[185] A.M. Dimiev, S. Eigler, Graphene Oxide: Fundamentals and Applications, John Wiley & Sons, 2016.
[186] H.B. Park, I.Y. Suh, Y.M. Lee, Novel pyrolytic carbon membranes containing silica: preparation and characterization, Chemistry of Materials 14 (7) (2002) 3034–3046.
[187] H. Suda, K. Haraya, Gas permeation through micropores of carbon molecular sieve membranes derived from Kapton polyimide, The Journal of Physical Chemistry B 101 (20) (1997) 3988–3994.
[188] D.R. Paul, Creating new types of carbon-based membranes, Science 335 (6067) (2012) 413–414.
[189] C.-Y. Su, Y. Xu, W. Zhang, J. Zhao, X. Tang, C.-H. Tsai, L.-J. Li, Electrical and spectroscopic characterizations of ultra-large reduced graphene oxide monolayers, Chemistry of Materials 21 (23) (2009) 5674–5680.
[190] L. Zhang, J. Liang, Y. Huang, Y. Ma, Y. Wang, Y. Chen, Size-controlled synthesis of graphene oxide sheets on a large scale using chemical exfoliation, Carbon 47 (14) (2009) 3365–3368.
[191] G. Eda, M. Chhowalla, Graphene-based composite thin films for electronics, Nano Letters 9 (2) (2009) 814–818.
[192] H.-D. Huang, P.-G. Ren, J. Chen, W.-Q. Zhang, X. Ji, Z.-M. Li, High barrier graphene oxide nanosheet/poly (vinyl alcohol) nanocomposite films, Journal of Membrane Science 409 (2012) 156–163.
[193] H.-D. Huang, P.-G. Ren, J.-Z. Xu, L. Xu, G.-J. Zhong, B.S. Hsiao, Z.-M. Li, Improved barrier properties of poly (lactic acid) with randomly dispersed graphene oxide nanosheets, Journal of Membrane Science 464 (2014) 110–118.
[194] D. Nuvoli, L. Valentini, V. Alzari, S. Scognamillo, S.B. Bon, M. Piccinini, J. Illescas, A. Mariani, High concentration few-layer graphene sheets obtained by liquid phase exfoliation of graphite in ionic liquid, Journal of Materials Chemistry 21 (10) (2011) 3428–3431.
[195] H. Bai, K. Sheng, P. Zhang, C. Li, G. Shi, Graphene oxide/conducting polymer composite hydrogels, Journal of Materials Chemistry 21 (46) (2011) 18653–18658.
[196] N. Zhang, R. Li, L. Zhang, H. Chen, W. Wang, Y. Liu, T. Wu, X. Wang, W. Wang, Y. Li, Actuator materials based on graphene oxide/polyacrylamide composite hydrogels prepared by in situ polymerization, Soft Matter 7 (16) (2011) 7231–7239.
[197] J.E. Shin, H.W. Kim, B.M. Yoo, H.B. Park, Graphene oxide nanosheet-embedded crosslinked poly (ethylene oxide) hydrogel, Journal of Applied Polymer Science 135 (24) (2018) 45417.
[198] M. Koinuma, C. Ogata, Y. Kamei, K. Hatakeyama, H. Tateishi, Y. Watanabe, T. Taniguchi, K. Gezuhara, S. Hayami, A. Funatsu, Photochemical engineering of graphene oxide nanosheets, The Journal of Physical Chemistry C 116 (37) (2012) 19822–19827.
[199] Engineering nano-porous graphene oxide by hydroxyl radicals, Carbon 105 (2016) 291–296.
[200] T.H. Han, Y.-K. Huang, A.T. Tan, V.P. Dravid, J. Huang, Steam etched porous graphene oxide network for chemical sensing, Journal of the American Chemical Society 133 (39) (2011) 15264–15267.
[201] S.-S. Choo, E.-S. Kang, I. Song, D. Lee, J.-W. Choi, T.-H. Kim, Electrochemical detection of dopamine using 3D porous graphene oxide/gold nanoparticle composites, Sensors 17 (4) (2017) 861.
[202] D. Shan, S. Deng, J. Li, H. Wang, C. He, G. Cagnetta, B. Wang, Y. Wang, J. Huang, G. Yu, Preparation of porous graphene oxide by chemically intercalating a rigid molecule for enhanced removal of typical pharmaceuticals, Carbon 119 (2017) 101–109.
[203] G. Mercier, A. Klechikov, M. Hedenstrom, D. Johnels, I.A. Baburin, G. Seifert, R. Mysyk, A.V. Talyzin, Porous graphene oxide/diboronic acid materials: structure and hydrogen sorption, The Journal of Physical Chemistry C 119 (49) (2015) 27179–27191.
[204] A. Zakaria, E.S. Vasquez, K.B. Walters, D. Leszczynska, Functional holey graphene oxide: a new electrochemically transformed substrate material for dopamine sensing, RSC Advances 5 (129) (2015) 107123–107135.
[205] E. Drioli, L. Giorno, E. Fontananova, Comprehensive Membrane Science and Engineering, Elsevier, 2017.
[206] H. Li, Z. Song, X. Zhang, Y. Huang, S. Li, Y. Mao, H.J. Ploehn, Y. Bao, M. Yu, Ultrathin, molecular-sieving graphene oxide membranes for selective hydrogen separation, Science 342 (6154) (2013) 95–98.
[207] J. Shen, G. Liu, K. Huang, W. Jin, K.R. Lee, N. Xu, Membranes with fast and selective gas-transport channels of laminar graphene oxide for efficient CO2 capture, Angewandte Chemie 127 (2) (2015) 588–592.
[208] M. Karunakaran, R. Shevate, M. Kumar, K.-V. Peinemann, CO 2-selective PEO–PBT (PolyActive™)/graphene oxide composite membranes, Chemical Communications 51 (75) (2015) 14187–14190.
[209] A. Tena, S. Shishatskiy, V. Filiz, Poly (ether–amide) vs. poly (ether–imide) copolymers for post-combustion membrane separation processes, RSC Advances 5 (29) (2015) 22310–22318.

[210] G. Dong, J. Hou, J. Wang, Y. Zhang, V. Chen, J. Liu, Enhanced CO2/N2 separation by porous reduced graphene oxide/Pebax mixed matrix membranes, Journal of Membrane Science 520 (2016) 860–868.

[211] H. Lee, S.C. Park, J.S. Roh, G.H. Moon, J.E. Shin, Y.S. Kang, H.B. Park, Metal–organic frameworks grown on a porous planar template with an exceptionally high surface area: promising nanofiller platforms for CO 2 separation, Journal of Materials Chemistry A 5 (43) (2017) 22500–22505.

[212] T. Rodenas, I. Luz, G. Prieto, B. Seoane, H. Miro, A. Corma, F. Kapteijn, F.X.L. i Xamena, J. Gascon, Metal–organic framework nanosheets in polymer composite materials for gas separation, Nature Materials 14 (1) (2015) 48.

[213] Y. Shen, H. Wang, X. Zhang, Y. Zhang, MoS2 nanosheets functionalized composite mixed matrix membrane for enhanced CO2 capture via surface drop-coating method, ACS Applied Materials and Interfaces 8 (35) (2016) 23371–23378.

[214] Y. LAI, in: Annual Meeting October 28–November 2, 2012, 2014.

[215] L. Olivieri, S. Ligi, M.G. De Angelis, G. Cucca, A. Pettinau, Effect of graphene and graphene oxide nanoplatelets on the gas permselectivity and aging behavior of poly (trimethylsilyl propyne)(PTMSP), Industrial and Engineering Chemistry Research 54 (44) (2015) 11199–11211.

[216] L. Olivieri, S. Meneguzzo, S. Ligi, A. Saccani, L. Giorgini, A. Orsini, A. Pettinau, M.G. De Angelis, Reducing ageing of thin PTMSP films by incorporating graphene and graphene oxide: effect of thickness, gas type and temperature, Journal of Membrane Science 555 (2018) 258–267.

[217] M. Alberto, R. Bhavsar, J.M. Luque-Alled, A. Vijayaraghavan, P.M. Budd, P. Gorgojo, Impeded physical aging in PIM-1 membranes containing graphene-like fillers, Journal of Membrane Science 563 (2018) 513–520.

[218] B.H. Hong, Y.J. Kim, J. Choi, H.K. Kim, J. Kang, S.K. Bae, Graphene Roll-To-Roll Coating Apparatus and Graphene Roll-To-Roll Coating Method Using the Same, Google Patents, 2011.

[219] D.H. Seo, S. Pineda, J. Fang, Y. Gozukara, S. Yick, A. Bendavid, S.K.H. Lam, A.T. Murdock, A.B. Murphy, Z.J. Han, Single-step ambient-air synthesis of graphene from renewable precursors as electrochemical genosensor, Nature Communications 8 (2017) 14217.

[220] M. Lozada-Hidalgo, S. Zhang, S. Hu, A. Esfandiar, I. Grigorieva, A. Geim, Scalable and efficient separation of hydrogen isotopes using graphene-based electrochemical pumping, Nature Communications 8 (2017) 15215.

[221] G. Li, W.-C. Law, K.C. Chan, Floating, highly efficient, and scalable graphene membranes for seawater desalination using solar energy, Green Chemistry 20 (16) (2018) 3689–3695.

[222] M. Majumder, A. Akbarivakilabadi, Graphene Oxide Membranes and Methods Related Thereto, Google Patents, 2017.

[223] D.R. Dreyer, S. Park, C.W. Bielawski, R.S. Ruoff, The chemistry of graphene oxide, Chemical Society Reviews 39 (1) (2010) 228–240.

[224] H.B. Park, H.W. Yoon, Y.H. Cho, Graphene oxide membrane for molecular separation. Graphene Oxide: Fundamentals and Applications, 2016, p. 296.

CHAPTER

9

Polymers of Intrinsic Microporosity and Their Potential in Process Intensification

Peter M. Budd

School of Chemistry, University of Manchester, Manchester, United Kingdom

OUTLINE

Sustainable Nanoscale Engineering
https://doi.org/10.1016/B978-0-12-814681-1.00009-6

List of Abbreviations

6FDA 4,4′-(Hexafluoroisopropylidene)diphthalic anhydride
BET Brunauer—Emmett—Teller
FAV Fractional accessible volume
FFV Fractional free volume
GPU Gas permeation unit
MMM Mixed matrix membrane
MOF Metal-organic framework
NMP *N*-methyl-2-pyrrolidone
OLED Organic light-emitting diode
OSN Organic solvent nanofiltration
PAF Porous aromatic framework
PAN Polyacrylonitrile
PDMS Polydimethylsiloxane
PIM Polymer of intrinsic microporosity
PSI Pervaporation separation index
PTMSP Poly(1-trimethylsilyl-1-propyne)
PVDF Polyvinylidene fluoride
TB Tröger's base
TFC Thin film composite
TFN Thin film nanocomposite
TR Thermally rearranged

1. INTRODUCTION

The move toward cleaner, more compact, and more energy-efficient engineering processes—process intensification—is facilitated by the introduction of new, high-performance materials and by the development of innovative technologies that use those materials to the best advantage. Polymers of intrinsic microporosity (PIMs) represent a class of high free volume polymer that shows considerable promise in the development of membranes and adsorbents for highly efficient molecular separations. This chapter introduces the PIM concept, describes the variety of PIM structures that have been synthesized since the first membrane-forming PIMs were reported in 2004, and discusses applications that are being explored.

2. DEVELOPMENT OF THE POLYMER OF INTRINSIC MICROPOROSITY CONCEPT

2.1 Intrinsic Microporosity

Materials that exhibit porosity on the molecular scale, with pore sizes less than 2 nm, are referred to as "microporous" in the context of sorption studies [1]. Microporous materials exhibit enhanced physisorption of small molecules due to multiwall interactions. In polymers, "intrinsic microporosity" may be defined as "a continuous network of interconnected intermolecular voids, which forms as a direct consequence of the shape and rigidity of the component macromolecules" [2]. In 1999 the term "intrinsic microporosity" was applied by Ilinitch et al. [3] to poly(phenylene oxide) polymers that exhibited some evidence of microporous character in nitrogen sorption studies, although in that case the microporosity was not as well developed as in the types of polymer now described as PIMs.

2.2 Microporous Network Polymers

The PIM concept arose out of work at the University of Manchester that was aimed at incorporating metal-containing phthalocyanines into polymer networks in a way that prohibited the plate-like phthalocyanine units from aggregating together [4,5]. The idea was to ensure that the catalytic metal centers were accessible to reagents, so that the networks could be used as heterogeneous catalysts. McKeown et al. [6] used the phthalocyanine-forming reaction of a spirocyclic bisphthalonitrile to create a network in which phthalocyanine units were interconnected by rigid spirocyclic linkers (Fig. 9.1A). The spirocenter (two five-membered rings connected by a single carbon atom)

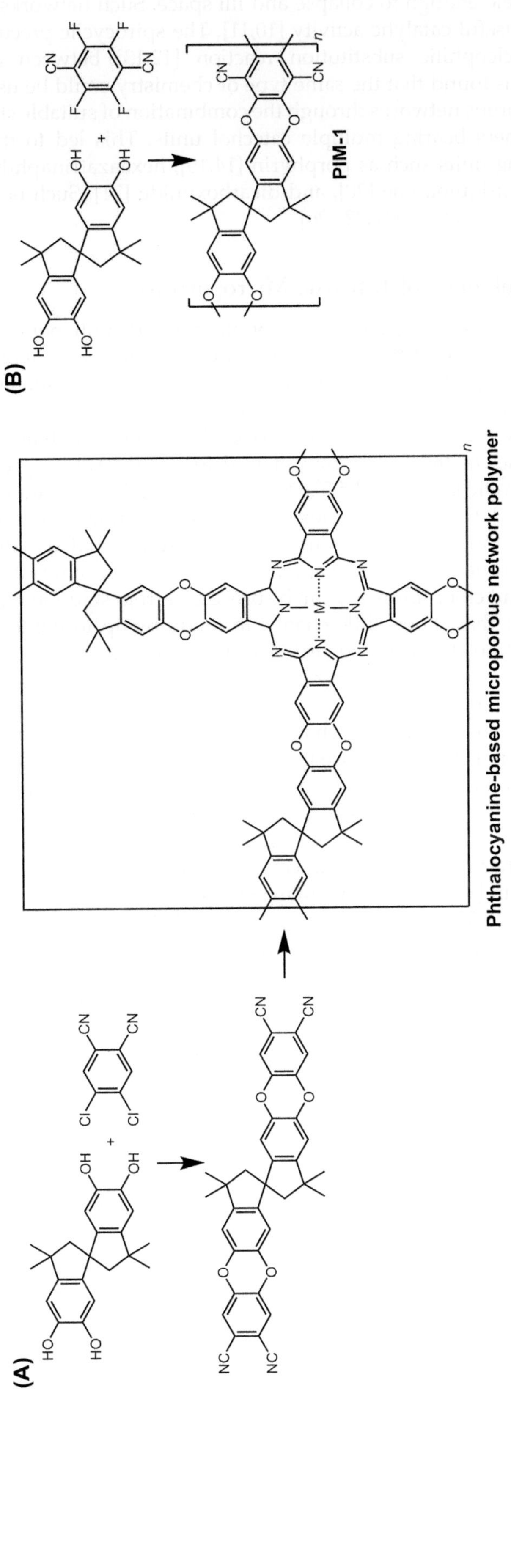

FIGURE 9.1 Preparation of (A) phthalocyanine-based microporous network polymer [6] and (B) solution-processable polybenzodioxane PIM-1 [7,8].

ensured that adjacent phthalocyanine units were orthogonal to each other, while the lack of any single bonds meant that the structure was not flexible enough to collapse and fill space. Such networks were shown to act as effective adsorbents [9] and to possess useful catalytic activity [10,11]. The spirocyclic precursor was prepared by a highly efficient double aromatic nucleophilic substitution reaction [12,13] between a spirocyclic biscatechol and 4,5-dichlorophthalonitrile. It was found that the same type of chemistry could be used in step-growth polymerization to create a range of microporous networks through the combination of suitable chlorinated or fluorinated monomers with appropriate monomers bearing multiple catechol units. This led to the preparation of microporous network polymers incorporating units such as porphyrin [14,15], hexaazatrinaphthylene [16], cyclotricatechylene [17], triptycene [18,19], tribenzotriquinacene [20], and dicarboximide [21]. Such networks have been investigated for applications such as hydrogen storage [17,22–26].

2.3 Solution-Processable Polymers of Intrinsic Microporosity

For many applications, the disadvantage of microporous network polymers, along with other microporous materials such as zeolites and activated carbons, is that they are insoluble and cannot easily be processed into useful forms such as films and fibers. A significant breakthrough came with the realization that the same type of chemistry used successfully in the creation of a range of network polymers—double nucleophilic aromatic substitution—can be utilized to synthesize nonnetwork, solution-processable polybenzodioxanes that exhibit microporous character in the solid state. The first polymer of this type, now referred to as PIM-1, was prepared from two commercially available monomers, the spirocyclic biscatechol 5,5′,6,6′,-tetrahydroxy-3,3,3′,3′-tetramethyl-1,1′-spirobisindane and tetrafluoroterephthalonitrile [7,8] (Fig. 9.1B). In low-temperature nitrogen adsorption experiments, powdered PIM-1 exhibits high uptake at low relative pressure, as typically associated with microporous materials, and Brunauer–Emmett–Teller (BET) analysis yields an apparent surface area around 800 $m^2 g^{-1}$. PIM-1 is soluble in solvents such as tetrahydrofuran and chloroform and can be processed from solution into membranes [8]. It is a yellow, fluorescent [27], glassy polymer that shows little change in modulus up to the temperature at which it starts to degrade, around 350°C. A glass transition can be detected at around 440°C by fast scanning calorimetry [28]. A lower temperature relaxation can be observed by dielectric spectroscopy [29]. A patent application filed in 2003 led to the granting of US and European patents [30].

In terms of chemical structure, the features that characterize a PIM are (1) a macromolecular backbone that cannot undergo large-scale conformational change because there are no single bonds in the backbone about which rotation can occur or because rotation is highly restricted about any single bonds that exist, and (2) a highly contorted backbone, arising from "sites of contortion" such as spirocenters (Fig. 9.2A). Such rigid, contorted macromolecules cannot easily pack together to fill space in the solid state, as demonstrated by computer simulation [31–34]. Consequently, they possess much higher free volume than conventional polymers and behave as molecular sieves with effective pore sizes less than 2 nm [35,36]. Experimental techniques used to characterize free volume and its distribution include sorption studies [9,37–41], hydrostatic weighing [42], positron annihilation lifetime spectroscopy [37,38,43–46], inverse gas chromatography [47], ^{129}Xe NMR spectroscopy [38,39,48], X-ray scattering [33,40], and inelastic neutron scattering [49]. Fractional free volume, FFV, is often calculated from the specific volume, V_{sp}, which is the reciprocal of density, and the van der Waals volume, V_w, which may be calculated by group contribution methods [50,51], using

$$\mathrm{FFV} = \frac{V_{sp} - 1.3V_w}{V_{sp}} \tag{9.1}$$

On this basis, the FFV for the prototypical polymer of intrinsic microporosity, PIM-1, is in the region of 0.26 [52]. When a vapor or liquid is taken up by a thin PIM-1 film, it may occupy free volume and/or lead to swelling of the polymer, as can be demonstrated by spectroscopic ellipsometry [53–57].

Following from the first report of PIM-1 and related polybenzodioxanes [7], a great variety of PIM structures have been produced by benzodioxane formation. Other forms of polymerization have also been employed. Polyimides have been extensively studied as membrane polymers, and a natural progression was the development of PIM-polyimides [58–70]. Although imide formation necessarily introduces a single bond into the backbone, PIM-like behavior can be achieved, provided that rotation about the imide linkage is sufficiently restricted. Aromatic polyimides with ortho-positioned –OH groups can be converted to polybenzoxazoles by thermal treatment [71], and this idea of a thermally rearranged polymer has been combined with the PIM concept [60,72,73]. Neil McKeown, who first recognized the potential of benzodioxane formation in polymerization, introduced another new type of polymerization, based on some chemistry first reported in 1887, Tröger's base (TB) formation [74]. This led to a family of TB and related PIMs [48,75–87].

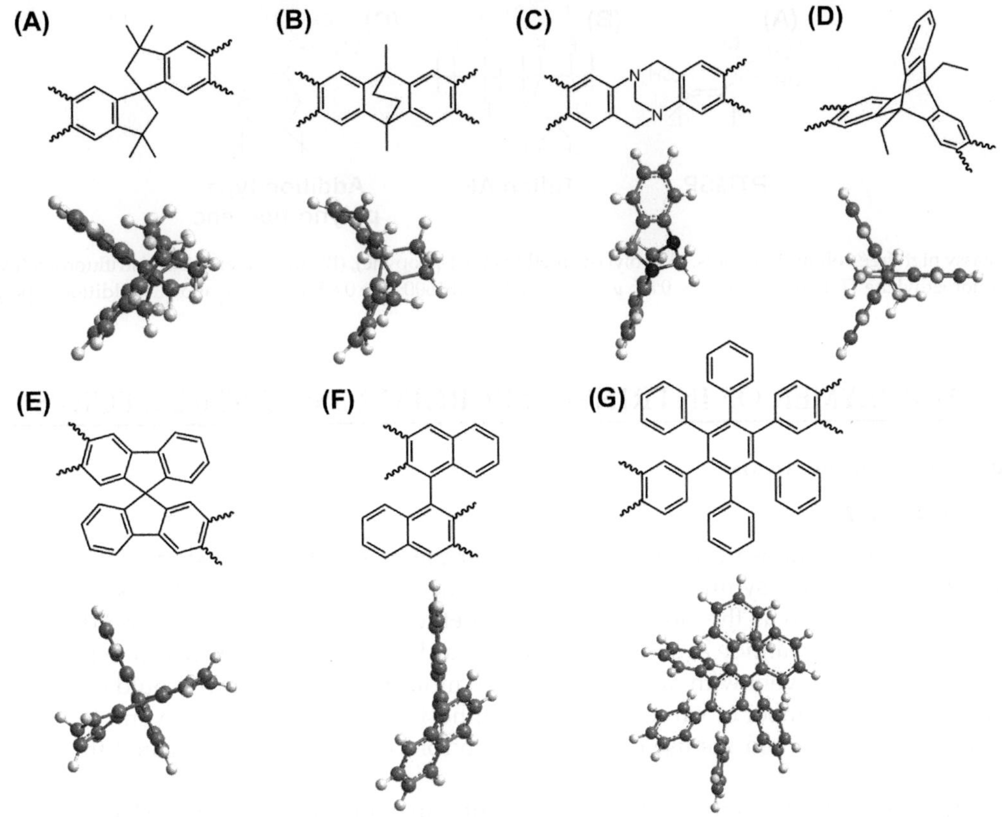

FIGURE 9.2 Chemical structures and molecular models of units utilized as sites of contortion in polymers of intrinsic microporosity (PIMs): (A) spirobisindane, (B) ethanoanthracene, (C) Tröger's base, (D) triptycene, (E) spirobifluorene, (F) binaphthyl, and (G) hexaphenylbenzene.

In addition to extending the PIM concept from polybenzodioxanes to other types of polymer, the range of "sites of contortion" has also been extended. The spirocenter in a spirobisindane unit exhibits significant flexibility, and the ethanoanthracene unit was explored as a more rigid site of contortion. Polymer properties can be fine-tuned by forming copolymers in which spirobisindane units are partially replaced with ethanoanthracene units [38,88] (Fig. 9.2B). TB PIMs incorporating ethanoanthracene units are very effective molecular sieves [48,74,89,90]. The TB unit (Fig. 9.2C) itself acts as a site of contortion, and TB units have been incorporated into polyimides [65,83,91]. Triptycene, with its fused-ring structure and threefold symmetry, was recognized by Swager as having "internal free volume" [92–97]. Triptycene-based units (Fig. 9.2D) have been utilized as sites of contortion in both microporous network polymers [18,19,76,98] and solution-processable PIMs [66,68,69,78,82,90,99]. Other units that have been used as sites of contortion in PIMs include spirobifluorene (Fig. 9.2E) [34,39,61,62,100], binaphthyl (Fig. 9.2F) [7,59,101,102], and hexaphenylbenzene (Fig. 9.2G) [103,104]. The various types of PIM structure are discussed further in the following section.

2.4 Other High Free Volume Polymers

A few other classes of glassy high free volume polymer are known, which may be regarded as having intrinsic microporosity but which do not possess specific sites of contortion in the backbone in the same way as the PIMs described above [105]. In 1983, Masuda et al. [106] reported that a substituted polyacetylene, poly(1-trimethylsilyl-1-propyne) (PTMSP) (Fig. 9.3A), exhibited extraordinarily high gas permeability. In PTMSP, the bulky side groups force the backbone into a twisted shape and reduce its ability to change conformation. This spawned a great deal of research into substituted polyacetylenes for membrane applications [107–120]. Certain perfluoropolymers, such as copolymers of 2,2-bistrifluoromethyl-4,5-difluoro-1,3-dioxole and tetrafluoroethylene (sold commercially as Teflon AF) (Fig. 9.3B), show high gas permeability [121–124], which may be attributed to weak interactions between the fluoropolymer chains coupled with a higher barrier to rotation between neighboring dioxolane rings. Addition-type polynorbornenes with bulky groups on the ring (Fig. 9.3C) form another class of highly permeable polymer [125–127].

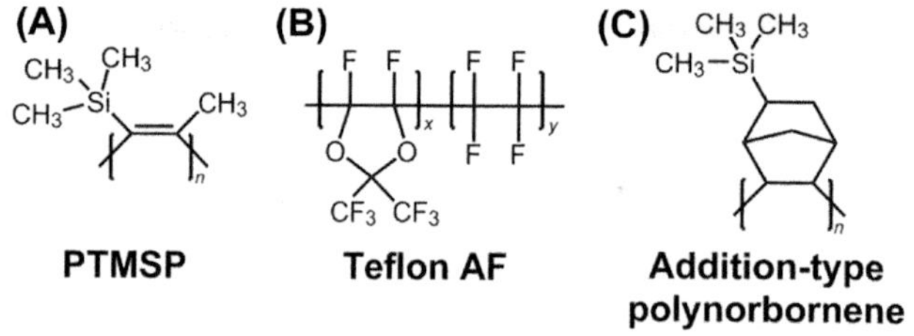

FIGURE 9.3 Glassy high free volume polymers: (A) poly(1-trimethylsilyl-1-propyne), (B) copolymer of 2,2-bistrifluoromethyl-4,5-difluoro-1,3-dioxole and tetrafluoroethylene (Teflon AF2400: $x = 0.87$, $y = 0.13$; Teflon AF1600: $x = 0.65$, $y = 0.35$), and (C) addition type poly(trimethysilyl norbornene).

3. POLYMER OF INTRINSIC MICROPOROSITY STRUCTURES

3.1 PIM-1 and Its Modifications

3.1.1 *Synthesis of PIM-1*

PIM-1, the prototypical solution-processable PIM, is by far the most extensively studied polymer of its type, not least because it is relatively easily synthesized from two commercially available monomers (Fig. 9.1B). The original synthesis, sometimes referred to as the "low-temperature method," was carried out over 3 days at a temperature around 65°C with dimethylformamide as solvent and K_2CO_3 as base [7,8]. Later, it was shown that at a higher temperature of 155°C in a mixed solvent with rapid stirring, high molecular weight PIM-1 could be produced in as short a time as 8 minutes [128]. A number of variations of the synthesis conditions have been employed [47,52,129–131]. Kricheldorf et al. [132,133] explored the use of a silylated spiromonomer and investigated the formation of cyclic PIM-1 structures.

From the perspective of sustainability, many of the solvents currently employed in PIM-1 synthesis are not ideal, and there is scope to explore greener chemistry in PIM preparation. Dimethyl sulfoxide has been suggested as a green solvent for PIM-1 synthesis [134]. Solvents can be avoided altogether and high molecular weight PIM-1 can be prepared by mechanochemistry, whereby the solid reactants are manually ground with a pestle and mortar or by ball milling [135,136].

The reaction scheme shown in Fig. 9.1B may be regarded as an AA + BB polycondensation, with A representing a pair of hydroxyls and B a pair of fluorines. This type of step-growth polymerization not only needs the reaction to be extremely efficient but also requires the stoichiometry to be carefully controlled if high molecular weight polymer is to be achieved, as indicated by Carothers theory. Even low levels of impurity in the monomers can disrupt the stoichiometry, leading to polymer that is too low in molecular weight to form a good membrane. There are potential advantages in employing an AB-type monomer that can undergo self-polycondensation. Jianyong Jin's group [137] has established a route to an AB-type monomer for PIM-1 synthesis. AB-type monomers have also been developed for the preparation of other types of PIM [99,138,139].

3.1.2 *Chemical Modification of PIM-1*

PIM-1 may be used as the starting point to generate a range of polymers with different characteristics through chemical modification. In particular, the nitrile groups in PIM-1 can be converted to a host of other functionalities, as indicated in Fig. 9.4.

Reactions involving polymers can prove more complicated than the equivalent small molecule reactions. This is illustrated by the base-catalyzed hydrolysis of PIM-1, which was originally assumed to yield a predominately carboxylated product [140–144]. However, careful characterization showed that, under the conditions commonly employed, a mixture of products is obtained, typically with >50% amide [145,146]. Conditions that enable mechanically strong and flexible hydrolyzed PIM-1 membranes to be formed have been reported [147]. Hydrolyzed PIM-1 samples can be prepared with improved solubility for electrospinning into ultrafine fibers [148], which may be further modified by cross-linking with hexamethylene diisocyanate [149]. Metal ions can also be incorporated into hydrolyzed PIM-1 [150].

A facile route to a nearly pure amide-PIM-1, using hydrogen peroxide as reagent, has been demonstrated [151]. Thioamide-PIM-1 can be generated using phosphorus pentasulfide as a thionating agent [152]. "Click chemistry" involving a [2 + 3] cycloaddition between the nitrile and an azide yields a tetrazole-PIM-1 that is favorable for CO_2 adsorption [153,154]. The tetrazole-PIM-1 can be converted to a methyl tetrazole polymer with improved

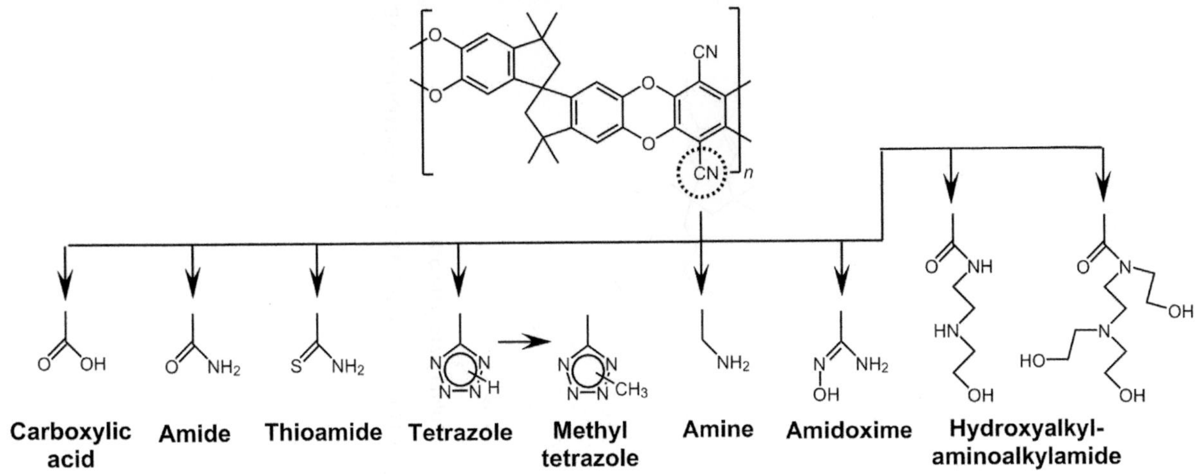

FIGURE 9.4 Chemical modifications of PIM-1.

solubility [155,156]. Amine-PIM-1 may be formed by reduction of the nitrile using borane complexes [157–159]. Reaction of the nitrile with hydroxylamine yields amidoxime-PIM-1 [160–163]. Hydroxyalkylaminoalkylamide structures are formed on reaction of PIM-1 with ethanolamine or diethanolamine [158,164]. Sulfonation of PIM-1 has been attempted but did not work well [165].

3.1.3 *Crosslinking of PIM-1*

The solubility that enables PIM-1 to be processed from solution may also be a disadvantage in applications requiring solvent-resistant materials. A number of methods have been proposed to cross-link PIM-1 postprocessing, rendering it insoluble and also modifying its properties. Chemical cross-linking can be achieved by a nitrene reaction with diazides [166,167]. Interactions with polycyclic aromatic hydrocarbons can effectively cross-link PIM-1 [168]. Stabilized membranes may be formed by thermal or chemical cross-linking of a blend of PIM-1 and polyethyleneimine [169]. Thermal treatment of PIM-1 at temperatures above 300°C under controlled conditions results in self–cross-linking [170–172]. Ultraviolet treatment has also been reported to cross-link PIM-1 [173,174], although the main effect in the presence of trace amounts of oxygen may be photooxidative chain scission near the polymer surface that collapses the polymer framework [175]. Cross-linked structures may be brought about by thermal decarboxylation of hydrolyzed PIM-1 [141] and by interactions of multivalent cations with hydrolyzed PIM-1 [176].

3.2 Polymers of Intrinsic Microporosity Incorporating Spirocenters

The spirocyclic biscatechol utilized in the synthesis of PIM-1, 5,5′,6,6′,-tetrahydroxy-3,3,3′,3′-tetramethyl-1,1′-spirobisindane, may be used as a starting point to create a wide range of PIMs. For example, it can be reacted with bespoke fluoromonomers to produce structures such as the phthalimide PIM-R1 [177] (Fig. 9.5A). A variety of chloromonomers may also be employed in PIM synthesis [181]. Bis(phenazyl) monomers can be prepared, which are then reacted with biscatechols to create polymers such as PIM-7 (Fig. 9.5B) and Cardo-PIM-1 [178,179] (Fig. 9.5C). An AB-type phenazine-containing monomer may be created, which self-polymerizes to give PSBI-AB [138] (Fig. 9.5D).

For practical application in membrane technology, very thin separation layers are generally needed to achieve high permeance. A highly porous substrate is required for mechanical support, which may comprise the same material as the separation layer (integrally skinned) or of a different material (thin film composite [TFC]). In the drive to achieve ever better performance, ultrathin PIM-1 films have been studied [182]. In the context of organic solvent nanofiltration (OSN), permeance increases as thickness decreases down to 140 nm, but in thinner films, there is a loss of permeance, which may be attributed to easier relaxation of the polymer chains in the vicinity of a surface, leading to better packing. One way to avoid such relaxation is to create a highly cross-linked structure in situ. An interfacial polymerization approach has been utilized to create ultrathin (down to 20 nm) highly cross-linked polyarylate nanofilms incorporating spirocenters (Fig. 9.5E) [180]. The use of a contorted monomer enhances microporosity and greatly increases permeance compared with conventional polyarylate structures.

Polymer properties may be tailored by copolymerizing the PIM-1 monomers with other monomers. For example, TFMPSPIM copolymers with phenylsulfone units [52] (Fig. 9.6A), DSPIM copolymers with disulfone units [183]

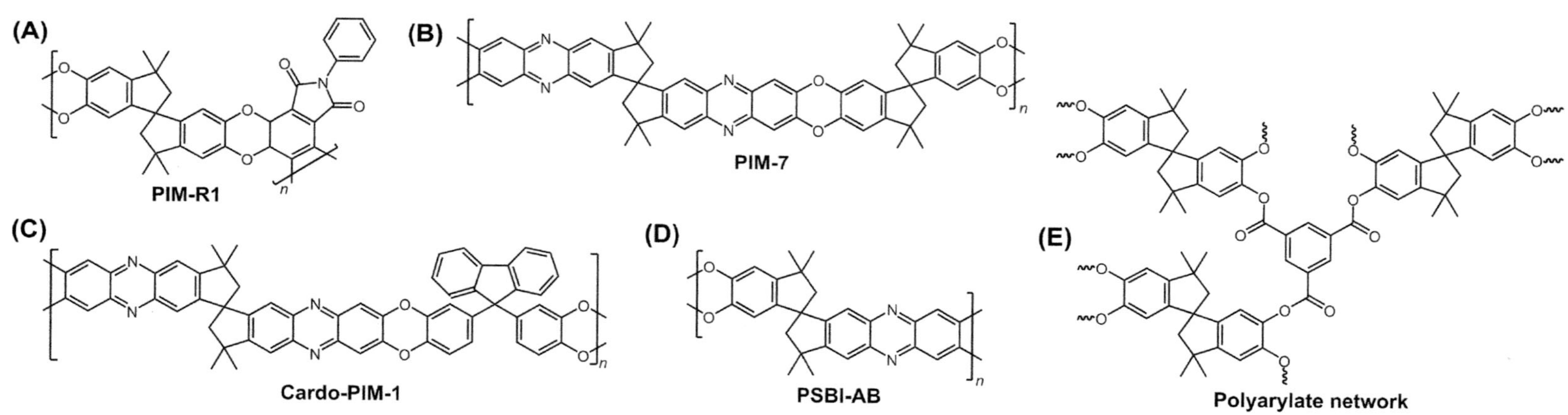

FIGURE 9.5 (A) Phthalimide-containing PIM-R1 [177]. Phenazine-containing polymers of intrinsic microporosity (PIMs): (B) PIM-7 [178,179], (C) Cardo-PIM-1 [179], and (D) PSBI-AB [138]. (E) Aromatic polyester (polyarylate) network incorporating spirobisindane [180].

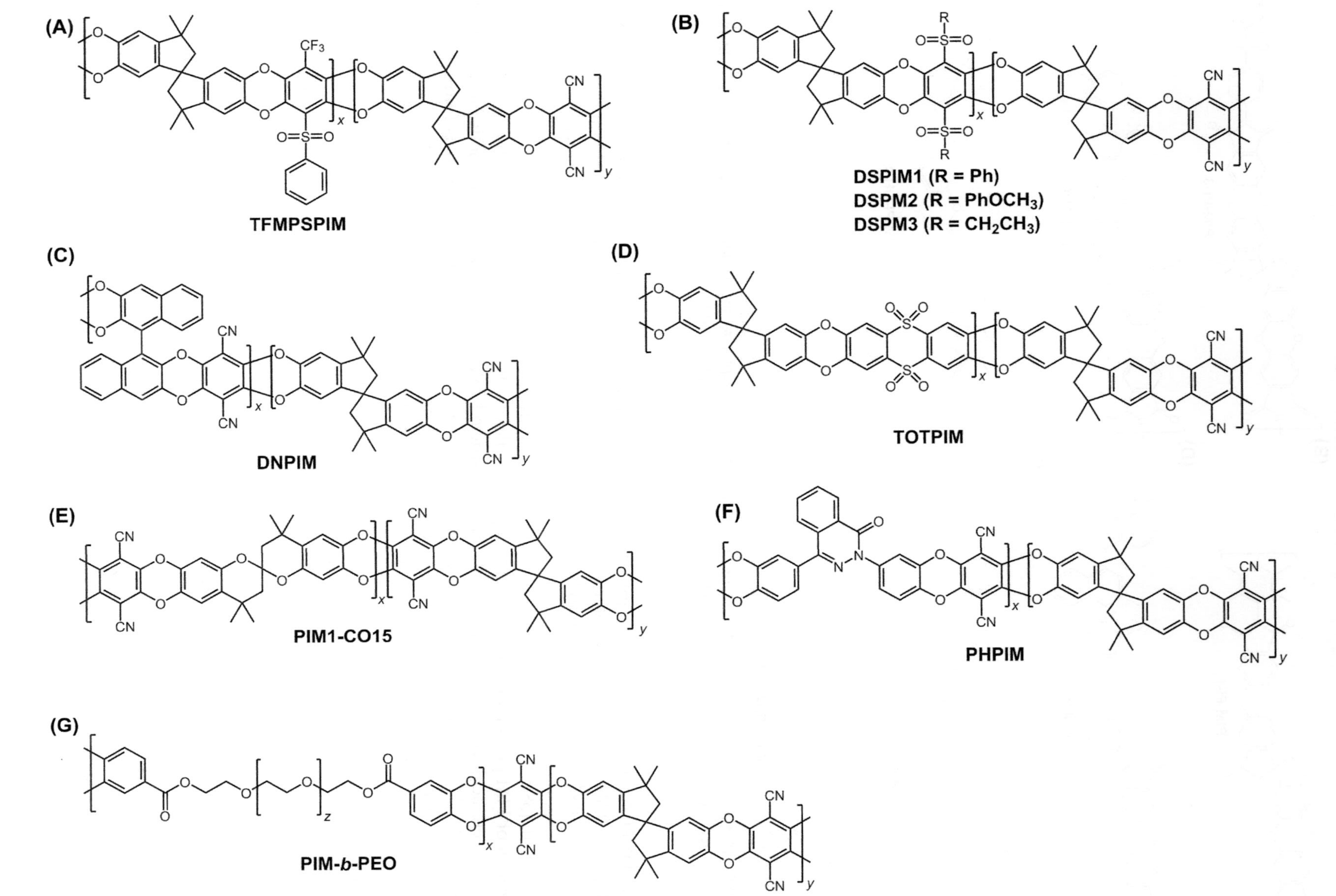

FIGURE 9.6 Copolymers based on PIM-1 with (A) phenylsulfone [52], (B) disulfone [183], (C) dinapthyl [101,102], (D) thianthrene [102], (E) spirobischromane [184], (F) phthalazinone [185], and (G) polyethylene oxide [186].

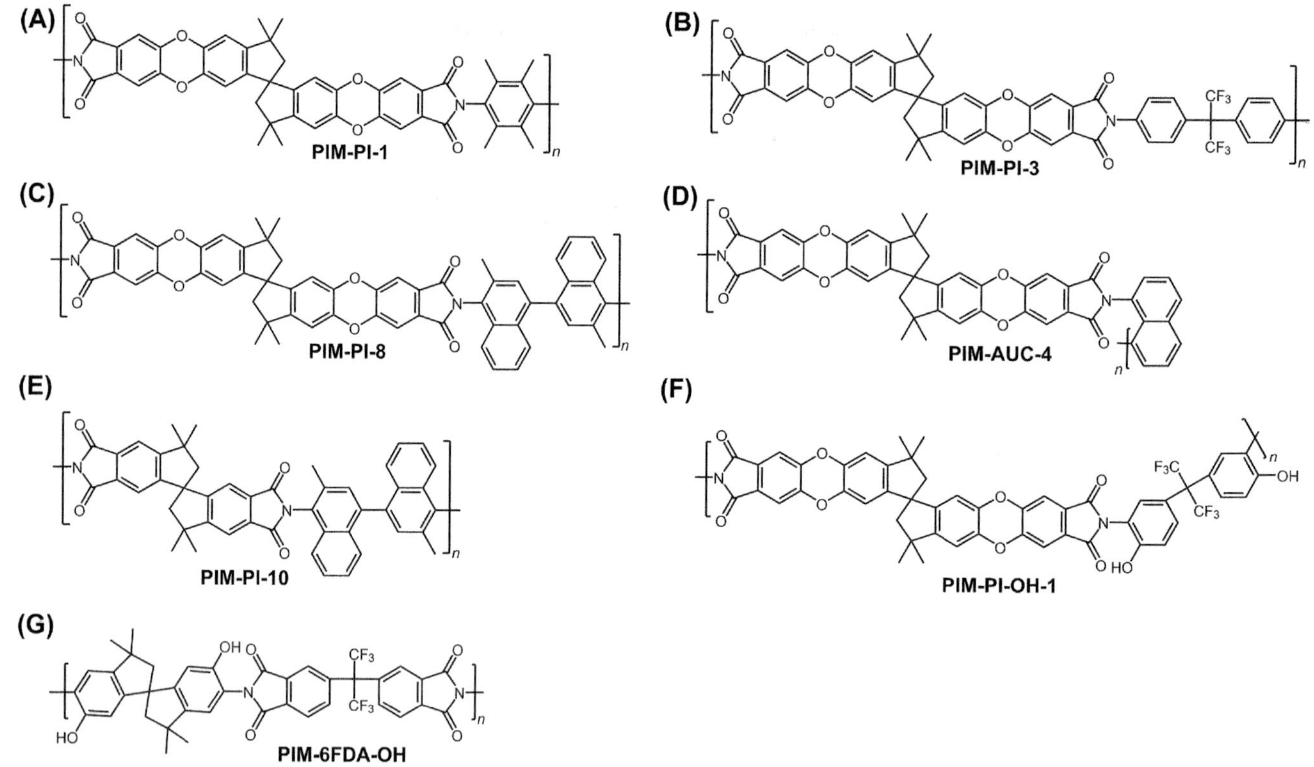

FIGURE 9.7 PIM-polyimides: (A) PIM-PI-1 [58,59], (B) PIM-PI-3 [58,59], (C) PIM-PI-8 [58,59], (D) PIM-AUC-4 [187], and (E) PIM-PI-10 [63]. Thermally rearrangeable hydroxyl-containing polyimides: (F) PIM-PI–OH–1 [73] and (G) PIM-6FDA-OH [60,72].

(Fig. 9.6B), DNPIM copolymers with dinapthyl units [101,102] (Fig. 9.6C), TOTPIM copolymers with thianthrene units [102] (Fig. 9.6D), PIM1-CO15 copolymers with spirobischromane units [184] (Fig. 9.6E), and PHPIMs with phthalazinone units [185] (Fig. 9.6F). Multiblock copolymers of PIM-1 and polyethylene oxide, PIM-*b*-PEO, have been reported [186] (Fig. 9.6G).

Ghanem et al. [58,59] demonstrated the preparation of a spirobisindane-containing dianhydride and used it to create a series of PIM-polyimides with varying degrees of flexibility of the backbone. Membranes with high gas permeability were achieved for structures with highly restricted rotation about backbone bonds, such as PIM-PI-1, PIM-PI-3, and PIM-PI-8 (Fig. 9.7A–C). The approach has been extended to structures such as PIM-AUC-4 [187] (Fig. 9.7D). A more compact spirobisindane-containing dianhydride has been developed, yielding polymers such as PIM-PI-10 [63] (Fig. 9.7E). Hydroxyl-functionalized PIM-polyimides may be prepared, such as PIM-PI-OH-1 [73] (Fig. 9.7F) and PIM-6FDA-OH [60,72] (Fig. 9.7G), which can be formed into membranes then "thermally rearranged" at temperatures of 400–500°C to give polybenzoxazole structures or thermally treated at higher temperatures to yield carbon molecular sieve membranes [64,188–191].

Spirobisindane (Fig. 9.2A) allows a significant degree of distortion about the spirocenter, and researchers have sought sites of contortion with less flexibility. Spirobifluorene (Fig. 9.2E), with its additional fused benzene rings, represents a more "rigid" unit than spirobisindane. A spirobifluorene equivalent of PIM-1, SBP-PIM-1 (Fig. 9.8A), has been prepared [100], along with a range of substituted variants [34]. Spirobifluorene has been incorporated into TB polymers [77] (Fig. 9.8B) and into polyimides such as P4 [61] (Fig. 9.8C) and 6FDA-SBF [62] (Fig. 9.8D) and the thermally rearrangeable polyimide 6FDA-HSBF [192] (Fig. 9.8E). A variety of spiro-fused fluorene-based monomers have been investigated for PIMs [193], as have a range of monomers based on 1,1-spiro-bis(1,2,3,4-tetrahydronaphthalene) [194]. It is possible to create a more rigid spirobisindane by forming an interlocked polycyclic structure that leads to "intramolecular locking" of the spirocarbon; a "locked" version of PIM-1, designated PIM-C1, exhibits enhanced gas permeability compared with PIM-1 [195]. PIMs based on spirobischromane have been investigated [184].

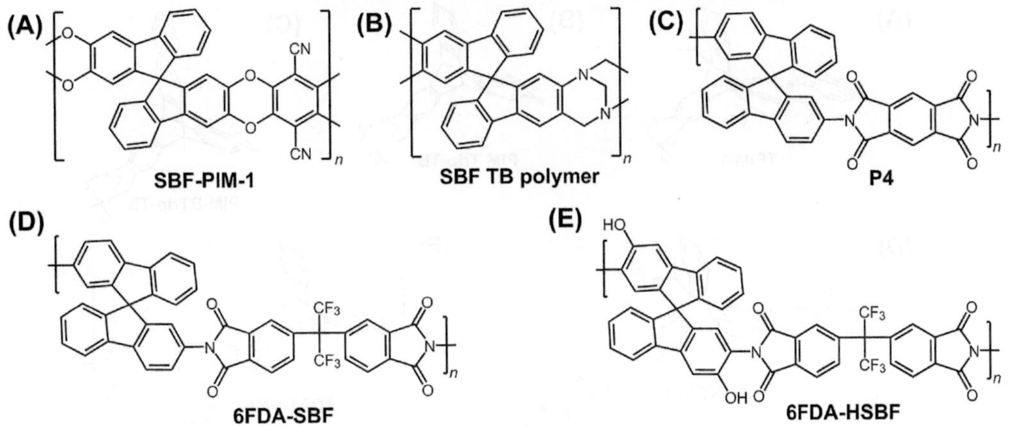

FIGURE 9.8 Spirobifluorene (SBF)-containing polymers of intrinsic microporosity (PIMs): (A) SBF-PIM-1 [34,100] and (B) SBF-containing Tröger's base polymer [77]. SBF containing PIM-polyimides (C) P4 [61], (D) 6FDA-SBF [62], and (E) 6FDA-HSBF [192].

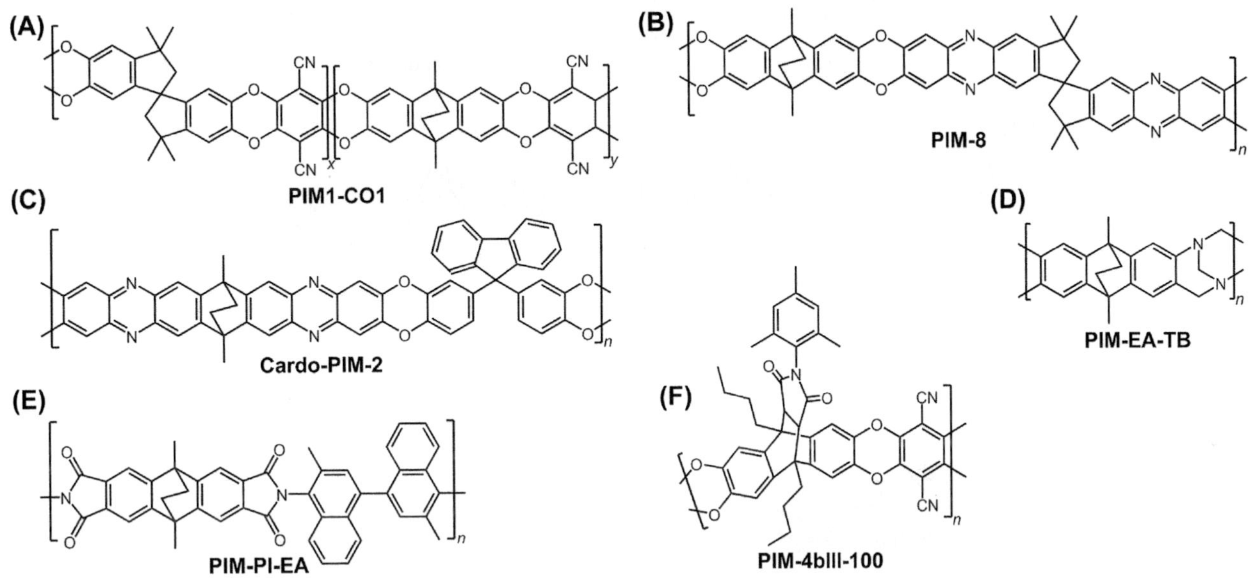

FIGURE 9.9 Polymers of intrinsic microporosity (PIMs) incorporating ethanoanthracene: (A) PIM1-CO1 copolymers in which the spirobisindane units of PIM-1 are partially replaced with ethanoanthracene [38,88], phenazine-containing (B) PIM-8 [179] and (C) Cardo-PIM-2 [179], (D) Tröger's base polymer PIM-EA-TB [48,74,89], and (E) polyimide PIM-PI-EA, also referred to as PIM-PI-12 [196] and EA-DMN [197]. (F) Anthracene maleimide-based PIM-4bIII-100 [198,199].

3.3 Polymers of Intrinsic Microporosity Incorporating Ethanoanthracene or Anthracene Maleimide Derivatives

Ethanoanthracene (Fig. 9.2B) has been used in a series of PIM1-CO1 copolymers to partially replace the more flexible spirobisindane [38] (Fig. 9.9A). A comparison has been made between alternating and random copolymers [88]. Ethanoanthracene has been built into phenazine-containing PIMs such as PIM-8 (Fig. 9.9B) and Cardo-PIM-2 (Fig. 9.9C) [179]. A TB polymer incorporating ethanoanthracene, PIM-EA-TB (Fig. 9.9D), acts as an effective molecular sieve [48,74,89]. A highly gas permeable ethanoanthracene-containing polyimide with good molecular sieving behavior has been formed, referred to as PIM-PI-EA, PIM-PI-12 [196], or EA-DMN [197] (Fig. 9.9E). Khan et al. [198,199] generated a range of anthracene-maleimide–based PIM homopolymers and copolymers, exemplified by PIM-4bIII-100 (Fig. 9.9F).

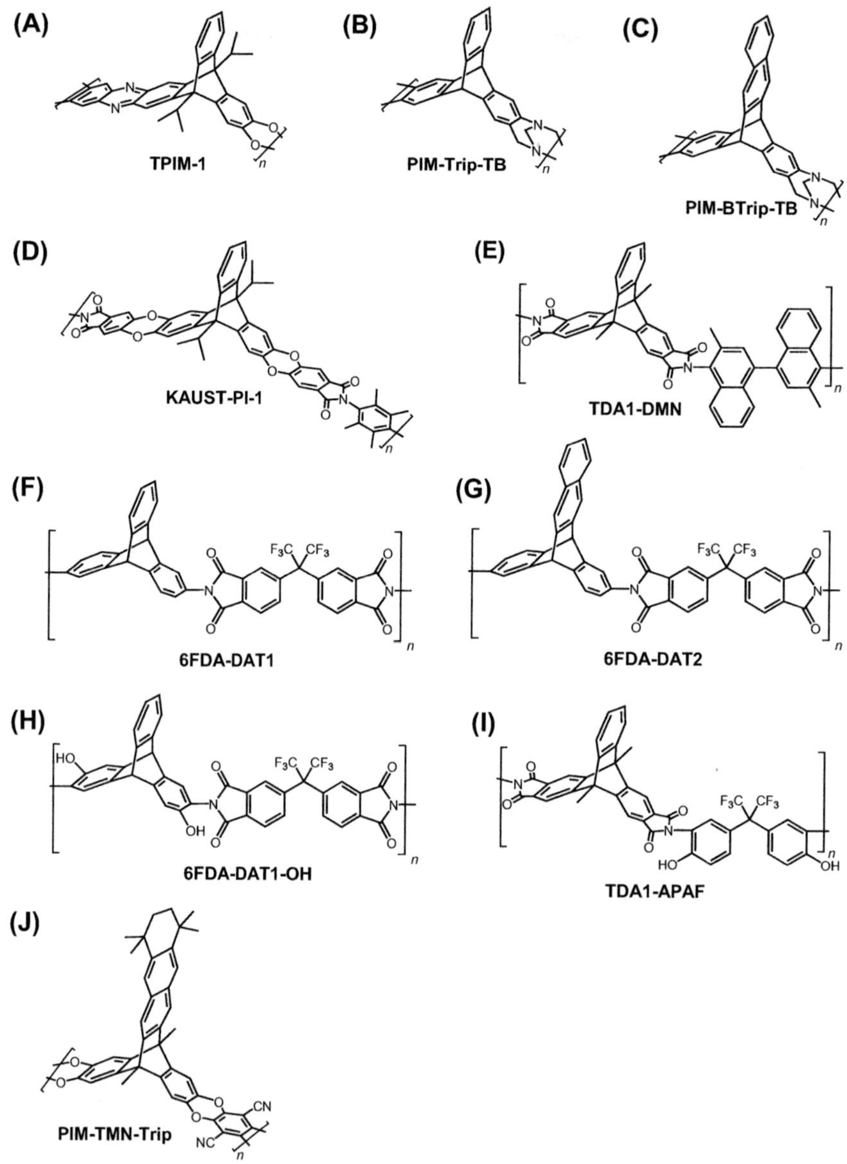

FIGURE 9.10 Polymers of intrinsic microporosity (PIMs) incorporating triptycene-based units: (A) TPIM-1, derived from an AB-type monomer [99], (B) triptycene-containing TB polymer PIM-Trip-TB [78] and (C) its benzotriptycene analogue PIM-BTrip-TB [82], (D) ultramicroporous polyimide KAUST-PI-1 [66,200–202], (E) polyimide TDA1-DMN [68], derived from a triptycene dianhydride, (F) triptycene-containing polyimide 6FDA-DAT1 and (G) its benzotriptycene analogue 6FDA-DAT2 [96,203], hydroxyl functionalized polyimides (H) 6FDA-DAT1-OH [69], (I) TDA1-APAF [204,205], and (J) ultrapermeable 2D polymer PIM-TMN-Trip [206].

3.4 Polymers of Intrinsic Microporosity Incorporating Triptycene Derivatives

Triptycene-based units (Fig. 9.2D) have been incorporated into a variety of soluble PIMs. 9,10-Bridgehead substituents can significantly affect the properties, for example, TPIM-1 (Fig. 9.10A), which has isopropyl substituents, has better gas transport properties than the equivalent polymer with linear propyl substituents [99]. Tröger's base polymers with triptycene (PIM-Trip-TB, Fig. 9.10B) [78] or benzotriptycene (PIM-BTrip-TB, Fig. 9.10C) [82] show promise for a number of important gas separations. The triptycene-based polyimide KAUST-PI-1 (Fig. 9.10D) [66,200–202] exhibits a high degree of ultramicroporosity (effective pore size <0.7 nm), which enhances the molecular-sieving behavior. Triptycene-based dianhydrides can be reacted with appropriate diamines to create polyimides such as TDA1-DMN (Fig. 9.10E) [68]. Triptycene- or benzotriptycene-based diamines may be reacted with the dianhydride 4,4′-(hexafluoroisopropylidene)diphthalic anhydride (6FDA) yielding polymers such as 6FDA-DAT1 (Fig. 9.10F) and 6FDA-DAT2 (Fig. 9.10G), respectively [96,203]. Hydroxyl-functionalized polyimides have been prepared, including 6FDA-DAT1-OH (Fig. 9.10H) [69], TDA1-APAF (Fig. 9.10I) [204,205], and other structures [207,208]. The challenge

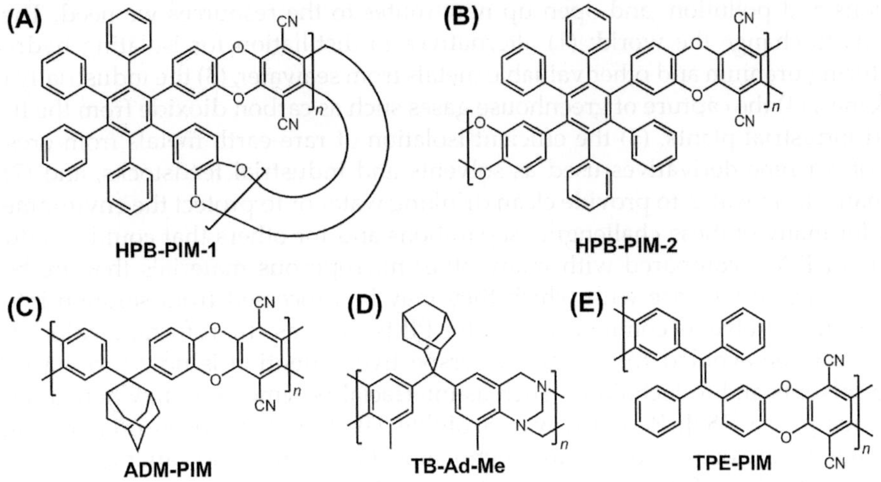

FIGURE 9.11 Polymers of intrinsic microporosity (PIMs) incorporating hexaphenylbenzene units: (A) cyclic HBP-PIM-1 and (B) HBP-PIM-2 [103,104]. PIMs incorporating adamantane units: (C) ADM-PIM [209] and (D) TB-Ad-Me [75]. A PIM incorporating tetraphenylethylene units: (E) TPE-PIM [210].

has been to develop soluble triptycene-based structures that extend in two dimensions. This was achieved with PIM-TMN-Trip (Fig. 9.10J) [206], where the triptycene bears a fused tetramethyltetrahydronaphthalene unit. The bulky substituent aids solubility in organic solvents, the equivalent benzotriptycene structure being insoluble. PIM-TMN-Trip exhibits ultrahigh gas permeability, while being more selective than other ultrapermeable polymers.

3.5 Other Polymers of Intrinsic Microporosity

Hexaphenylbenzene (Fig. 9.2G) has a propeller-like shape and a rigid structure because of steric interactions of the phenyl groups. Di-(3′,4′-dihydroxyphenyl)-tetraphenylbenzene can be utilized to synthesize PIMs by benzodioxane formation, but different regioisomers give very different products, the 1,2 regioisomer yielding predominately cyclic, oligomeric HPB-PIM-1 (Fig. 9.11A) and the 1,4 regioisomer giving high molecular weight, membrane-forming HPB-PIM-2 (Fig. 9.11B) [103]. Variants of HPB-PIM-2 with methyl, bromine, and nitrile substituents have been studied [104].

The rigid, diamondoid structure of adamantane, comprising fused cyclohexane rings, has been investigated for its ability to disrupt polymer packing and enhance gas permeability. An adamantane-containing bis(catechol) has been created and used to generate ADM-PIM (Fig. 9.11C) [209]. Adamantyl units have been incorporated into TB polymers such as TB-Ad-Me (Fig. 9.11D) [75].

The bulky, sterically hindered tetraphenylethylene group has potential as a building block for PIMs. A tetraphenyethylene-based PIM, TPE-PIM (Fig. 9.11E), and a range of copolymers have been reported [210].

In the 15 years since the synthesis of PIM-1 was first reported, researchers have generated a wide variety of polymer structures that exhibit PIM-like behavior. This has required great ingenuity and skill from chemists, but sustainability has not always been at the top of the agenda. Many potential monomers require multistep syntheses that are not easy to scale-up. Undesirable solvents, such as dimethylformamide, are frequently employed in monomer and polymer synthesis, and large volumes of solvents such as dioxane may be used in polymer purification. The potential to use renewable building blocks in monomer synthesis has not yet been explored. The challenge now is to find more sustainable and scaleable routes to PIM-like polymers.

4. POLYMER OF INTRINSIC MICROPOROSITY APPLICATIONS

The wide variety of PIM structures described above provides a valuable source of potentially useful materials for a broad range of advanced separations and other applications. Much current research is focused on the role that innovative separation materials and processes can play in helping to achieve global goals for sustainable development, in particular in relation to clean water, affordable and clean energy, responsible consumption and production, and climate action. As Sholl and Lively [211] have pointed out, the introduction of better and cheaper membranes, and of other separation processes that do not require a great deal of heat, could radically lower global energy use, reduce emissions and pollution, and open up new routes to the resources we need. They highlighted seven chemical separations to change the world: (1) alternatives to distillation for isolating hydrocarbons from crude oil, (2) ways of capturing uranium and other valuable metals from seawater, (3) the industrially important separation of alkenes from alkanes, (4) the capture of greenhouse gases such as carbon dioxide from the flue gases arising from power stations and industrial plants, (5) the efficient isolation of rare-earth metals from ores and their recycling, (6) the separation of benzene derivatives used as solvents and industrial feedstocks, and (7) the removal of salt and other contaminants from water, to provide clean drinking water or to protect the environment. PIMs are actively being investigated for many of these challenging separations and for others that could be added to the list.

A key advantage of PIMs, compared with many other microporous materials that are being investigated for advanced separations, lies in the ease with which they may be processed from solution into membranes, fibers, and coatings. However, the solvents commonly used for PIMs, such as chloroform, dichloromethane, and tetrahydrofuran, are solvents to be avoided from a green perspective. Attention is now turning to the investigation of greener alternative solvents and to techniques such as interfacial polymerization, where at least one of the monomers is in an aqueous phase [180]. Recently, water-soluble TB PIMs have been prepared, anionic character has been achieved through the use of a sulfonic acid functionalized monomer, and cationic character has been achieved through partial quaternization of a hydroxyl-functionalized version [212]. These were used to prepare TFC membranes through a layer-by-layer method using water as a solvent.

4.1 Gas Separation Membranes

Established industrial applications of gas separation membranes include hydrogen recovery in ammonia production, air separation for inert gas generation, and carbon dioxide removal in the treatment of natural gas [213,214]. To tackle separations important for the future, such as carbon dioxide capture from flue gases, new materials are needed, which offer exceptional performance and continue to work effectively under the conditions of use.

4.1.1 The Trade-off Between Permeability and Selectivity

For a membrane to be employed for large-scale gas separations, it must be productive, allowing large volumes of permeate to pass through. This may be expressed in terms of flux (the amount of permeate that passes through a certain membrane area in a given time) or permeance (flux normalized for the pressure or fugacity difference that provides the driving force for transport through the membrane). Permeance is often expressed in gas permeation units (1 gas permeation unit [GPU] $= 10^{-6}$ cm^3 [STP] cm^{-2} s^{-1} $cmHg^{-1} = 3.348 \times 10^{-10}$ mol m^{-2} s^{-1} Pa^{-1}). The permeance of a membrane depends on its effective thickness, so membrane materials are frequently compared in terms of permeability ($P =$ permeance $\times$ thickness), which may be expressed in units of barrer (1 barrer $= 10^{-10}$ cm^3 [STP] cm cm^{-2} s^{-1} $cmHg^{-1} = 3.35 \times 10^{-16}$ mol m m^{-2} s^{-1} Pa^{-1}). The transport of gases through a polymeric membrane can in many cases be understood in terms of the solution-diffusion model, in which permeation is assumed to involve (1) dissolution or sorption of the permeating species in the membrane on the feed side, (2) diffusion through the membrane, and (3) desorption at the permeate side. For this model, permeability, P, can generally be expressed as the product of a solubility or sorption coefficient, S, and a diffusion coefficient, D, thus

$$P = SD \tag{9.2}$$

As well as being permeable to a desired species, a membrane needs to be selective. For an ideal case of two gases, A and B, the selectivity, α, may be expressed as a ratio of permeabilities. Separation can arise from differences in solubility coefficient and/or from differences in diffusion coefficient, thus

$$\propto = \frac{P_A}{P_B} = \frac{S_A}{S_B}\frac{D_A}{D_B} \tag{9.3}$$

The problem for the material chemist seeking to develop new membrane materials is that attempts to increase permeability tend to lead to a reduction in selectivity and vice versa. This trade-off between permeability and selectivity was demonstrated in 1991 by Robeson [215], who for various gas pairs drew a line on a double logarithmic plot of selectivity against the permeability of the most permeable gas of the pair, indicating the upper bound of performance for the polymeric membranes that had been studied up until that time. Robeson's 1991 upper bounds for the O_2/N_2, H_2/N_2, H_2/CH_4, and CO_2/CH_4 gas pairs are shown in Fig. 9.12A–D. When gas permeation data were first reported in 2005 for two PIMs, PIM-1 and PIM-7 [178], it was found that not only were PIMs more permeable than the vast majority of polymers, apart from very high free volume polymers such as PTMSP, but also that they were sufficiently selective as to break Robeson's 1991 upper bound for several key gas pairs. In 2008,

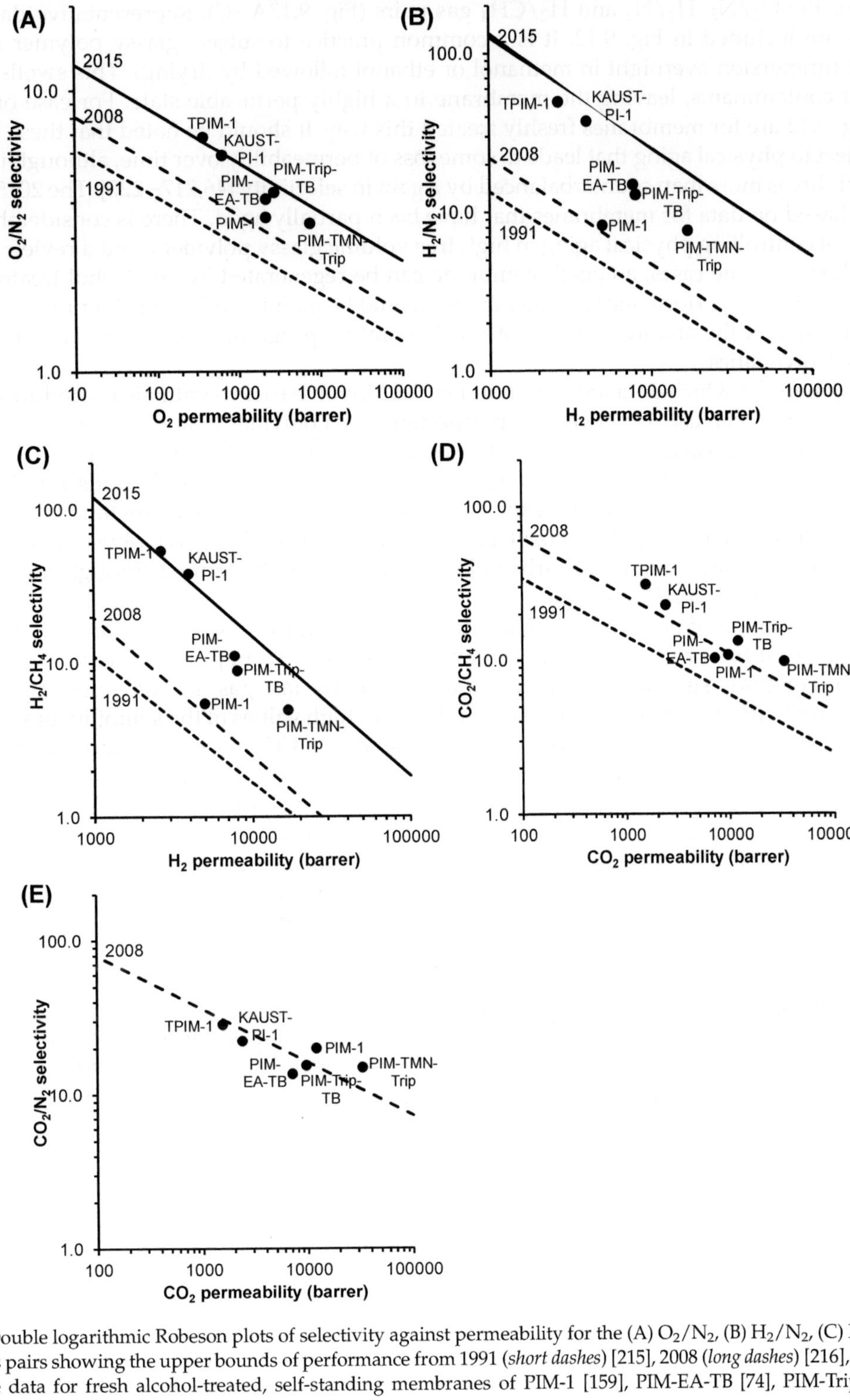

FIGURE 9.12 Double logarithmic Robeson plots of selectivity against permeability for the (A) O_2/N_2, (B) H_2/N_2, (C) H_2/CH_4, (D) CO_2/CH_4, and (E) CO_2/N_2 gas pairs showing the upper bounds of performance from 1991 (*short dashes*) [215], 2008 (*long dashes*) [216], and 2015 (*solid line*) [90] with representative data for fresh alcohol-treated, self-standing membranes of PIM-1 [159], PIM-EA-TB [74], PIM-Trip-TB [78], TPIM-1 [99], KAUST-PI-1 [200], and PIM-TMN-Trip [206].

Robeson [216] drew new upper bounds, also shown in Fig. 9.12. The CO_2/N_2 gas pair, relevant to postcombustion carbon dioxide capture, was not considered in 1991 but was included in 2008 (Fig. 9.12E).

Since 2008, the challenge has been to push the boundaries still further toward higher permeability and higher selectivity. New PIM structures, as discussed above, with relatively rigid sites of contortion such as triptycene-based units, have demonstrated that performance far beyond the 2008 upper bound can be achieved, especially for smaller gases where "molecular sieving" behavior is sought. This led to the Pinnau group [90] setting new upper bounds in 2015 for the O_2/N_2, H_2/N_2 and H_2/CH_4 gas pairs (Fig. 9.12A–C). Representative data for some high-performing PIMs are included in Fig. 9.12. It is a common practice to subject glassy polymer membranes to an alcohol treatment (immersion overnight in methanol or ethanol followed by drying). This swells the polymer and helps to flush out contaminants, leaving the membrane in a highly permeable state. For ease of comparison, the data shown in Fig. 9.12 are for membranes freshly treated this way. It should be noted that these highly permeable polymers are subject to physical aging that leads to some loss of permeability over time, although in favorable cases, the loss of permeability is more than counterbalanced by a gain in selectivity [46,217–220]. The 2015 upper bounds in Fig. 9.12A–C are based on data for membranes that have been partially aged. There is considerable research effort looking into ways of controlling physical aging in high free volume glassy polymers, and a review of the subject has been published [221]. In some cases, an aged membrane can be regenerated by an alcohol treatment as described above. In an ideal world, the performance of a membrane would be maintained for the lifetime of the plant in which it is employed, and this is the ambitious focus of a collaborative program of research, known as "SynFabFun," involving five UK Universities.

Some of the polymers for which data are shown in Fig. 9.12 involve tricky synthetic procedures that are unlikely to be scaled up for commercial use. However, the performance of a polymer that is more readily synthesized, such as PIM-1, can be tailored in a variety of ways. Controlled thermooxidative treatment of PIM-1 [170,171] leads to a cross-linked polymer structure that pushes performance toward the 2015 upper bounds for O_2/N_2, H_2/N_2 and H_2/CH_4. Selectivity may be enhanced, at the expense of permeability, by various methods of cross-linking, such as azide-based cross-linking [166,167], decarboxylation-induced cross-linking [141], physical cross-linking through interactions with polycyclic aromatic hydrocarbons [168], and cross-links formed through interactions of hydrolyzed PIM-1 with multivalent metal ions [176].

Although the upper bound of performance that can be achieved for O_2/N_2, H_2/N_2 and H_2/CH_4 has advanced dramatically since 2008, it has proved harder to achieve such substantial improvements for separations involving CO_2 (Fig. 9.12D,E). For PIMs such as PIM-1, CO_2 is a particularly fast gas, the CO_2 permeability being higher even than that of a tiny hydrogen molecule. This is linked to very high values of the solubility or sorption coefficient, S [47]. Chemical modifications that lead to enhanced interactions with CO_2, such as tetrazole modification [153], can enhance selectivity toward CO_2. However, too strong an interaction may lead to immobilizing sorption that radically decreases the diffusion coefficient, D, and hence the permeability, as has been observed for amine-PIM-1 [157,159].

In a mixed gas system, competitive sorption of one gas may have a strong impact on the permeability of another. For example, hydrocarbon vapors such as pentane, hexane, and toluene are strongly absorbed by PIM-1, reducing the permeabilities of both CO_2 and N_2, but significantly enhancing the CO_2/N_2 selectivity [222], leading to performance well above the 2008 Robeson upper bound. For practical application in postcombustion CO_2 capture, the effects of humidity and flue gas impurities must be considered [223].

PIM-based membranes have been investigated for a variety of gas separations in addition to those mentioned above. Alkene/alkane separations represent particular opportunities for membranes, to reduce the energy costs associated with conventional cryogenic distillation technologies. Promising results for propene/propane separation have been achieved for PIM-polyimides such as KAUST-PI-1 [66], for thermally treated PIM-6FDA-OH [67,190] and mixed matrix membranes (MMMs) of PIM-6FDA-OH with ZIF-8 [224], for hydrolyzed PIM-1 cross-linked with metal ions [150], and for molecular sieve membranes formed by thermal treatment of a PIM copolymer incorporating β-cyclodextrin [225]. Interesting results for ethene/ethane separation have been obtained for carbon molecular sieve membranes derived from PIM-1 [226], hydrolyzed PIM-1 with embedded boron compounds [227], PIM-6FDA-OH [189], and hydroxyl-free PIM-6FDA [191].

4.1.2 Blend Membranes

Membrane properties can be tuned by blending different polymers. The performance of the commercial polyimide Matrimid is enhanced by blending with PIM-1 [228], and the properties can be further tailored through chemical modification [229]. PIM-1/Matrimid hollow fiber membranes have been produced [230]. Hydrolyzed PIM-1 has been used to create miscible blends with Matrimid [143]. Similarly, blend membranes have been formed of PIM-1 with the polyetherimide Ultem [231,232], with partially sulfonated polyphenylenesulfone [233] and with polyethylene glycol [234], and of hydrolyzed PIM-1 with the polyamide-imide Torlon [142]. Promising results for CO_2/N_2 separation have been achieved for a blend of PIM-1 with the ether-functionalized polyphosphazene MEEK80 [235] and for blends of PIM-1 copolymers with an ionic liquid [236].

4.1.3 Mixed Matrix Membranes

Another strategy for tailoring membrane performance involves the addition of a particulate filler to a polymeric film, giving a composite or MMM [237]. This approach can be used to combine the separation performance of a crystalline or framework material with the processability of a polymer.

Porous fillers that have been combined with PIM-1 include the zeolite silicalite-1 [238,239] and metal-organic frameworks (MOFs) such as ZIF-8 [240–243], ZIF-67 [244], ZIF-71 [245], MIL-101 [246,247], UiO-66 [131,163,248,249], MFM-300(V^{III}) [250], and NUS-8 [251]. An MOF may be loaded with a task-specific ionic liquid to promote, for example, CO_2 separation [252]. Controlled thermal cross-linking of a PIM-1/MOF MMM can enhance the molecular sieving properties [172]. Although most MMMs are prepared as a simple mixture of polymer and filler, better performance may be achieved for a polymer chemically grafted to an MOF [249].

Purely organic porous materials have also been incorporated into MMMs, including cage molecules that may be crystallized in situ [253] and covalent organic frameworks [254]. Porous aromatic frameworks can help to control aging in PIM-1 and other superglassy polymers [255–257]. Hypercrosslinked polymers [258–260] are potentially cheaper and easier to scale-up than many of the other fillers that have been investigated.

Even nonporous fillers, such as silica nanoparticles [261,262], polyhedral oligomeric phenethyl-silsesquioxanes (POSS) [263], or potassium dodecahydrododecaborate particles [264], which would normally be expected to restrict gas transport, can actually increase the apparent permeability of an MMM because of their ability to disrupt the packing of a glassy polymer and create free volume. The two-dimensional material graphene, in itself a perfect gas barrier, can enhance gas transport properties at low loadings in PIM-1 [265–268]. Graphitic carbon nitride nanosheets have also been investigated [269]. The addition of graphene oxide nanosheets to PIM-1 is reported to give exceptional performance for CO_2/N_2 separation, with CO_2/N_2 selectivity up to 124 and CO_2 permeability over 6000 barrer [270]. MMMs of PIM-1 incorporating multiwall carbon nanotubes functionalized with poly(ethylene glycol) also exhibit good performance for CO_2/N_2 separation [271–273].

In addition to conventional MMMs in which discrete filler particles are dispersed in a continuous polymer phase, alternative composite structures have been explored. Submicron ZIF coatings can be grown directly on the surface of a PIM from ZnO nanocrystal precursor films [274]. Multilayer structures have been created, in which PIM-1 is sandwiched between two layers of an MOF (ZIF-8 or HKUST-1) [275]. PIM-1 can be constrained within the channels of anodic alumina membranes, enhancing selectivity [276].

4.1.4 Asymmetric and Thin Film Composite Membranes

A great deal of research on gas separation membranes has involved relatively thick (typically 40–100 μm thickness) self-standing membranes. However, for commercial application, high permeance is needed, to minimize the membrane area and thus the cost. Permeance can be enhanced by making the selective film very thin, but then a macroporous support is necessary to achieve a mechanically robust membrane. In some cases, a phase inversion process is used to create an integrally skinned asymmetric membrane, which has a dense selective skin on a more porous substructure. Alternatively, TFC membranes may be utilized, in which a selective layer of one material is coated onto a porous material of a different type. Asymmetric membranes may be produced as flat sheets or in the form of hollow fibers, which offer a high effective surface area for a relatively small module volume.

Hollow fiber membranes are commonly formed by dry-jet wet spinning, where a polymer dope is extruded through a spinneret into air and then into a coagulant water bath. The dope typically consists of the polymer dissolved in a solvent mixture that includes a relatively nonvolatile solvent such as *N*-methyl-2-pyrrolidone (NMP) and which may also include pore-forming or other additives. A double orifice spinneret is often used, which has an annulus through which the polymer dope is extruded in a cylindrical form and a central hole for a bore fluid which helps to maintain the shape and to control the coagulation. A problem encountered with polymers such as PIM-1 is the lack of a suitable nonvolatile solvent (NMP is a weak nonsolvent for PIM-1). The volatility of good solvents for PIM-1, such as tetrahydrofuran, creates problems in dry-jet wet spinning due to rapid evaporation of solvent in the air gap. However, the Lively group has demonstrated that defect-free asymmetric hollow fibers of PIM-1 can be produced by utilizing a triple orifice spinneret, with an outer annulus that enables a protective sheath of an inviscid nonsolvent to be coextruded with the dope and bore fluid [277]. PIM-1 has also been incorporated into hollow fibers as blends with other polymers such as Matrimid [230] and polyetherimide [232].

TFC membranes are produced by coating the selective layer onto the surface of a porous support. For a flat sheet TFC membrane, the support is typically an asymmetric membrane with an effective pore size in the range 10–50 nm, as may be used for ultrafiltration. Such membranes are themselves formed through a phase inversion process with a polymer such as polyacrylonitrile (PAN), coated onto a nonwoven polypropylene or polyester fabric. Sometimes a highly permeable gutter layer is used between the support and the selective layer.

When a filler is added to the selective layer of a TFC membrane, it may be referred to as a thin film nanocomposite (TFN) membrane. Many conventional MMMs have been formed with relatively large (dimensions >1 μm) filler particles, but for use in a TFN membrane, a nanoparticulate filler is necessary.

At Helmholtz-Zentrum Geesthacht, it has been demonstrated that TFN membranes incorporating functionalized multiwalled carbon nanotubes into a PIM-1 selective layer, on a PAN support, offer exceptional performance for CO_2/N_2 separation, with improved aging behavior as compared with TFC membranes without the nanotubes [271,278]. They have also shown that free volume and hence permeance can be tuned in TFC membranes with selective layers formed of copolymers of PIM-1 monomers with dibutyl anthracene maleimide [199].

Highly permeable hypercrosslinked polystyrene nanoparticles, and carbon nanoparticles derived from them, offer a cost-effective route to very high-permeance TFN membranes [260].

The Chung group has generated high performance composite hollow fiber membranes with three layers: a PIM selective layer, a cross-linked polydimethylsiloxane (PDMS) gutter layer, and a PAN support [279]. The best O_2/N_2 and CO_2/N_2 performance was obtained with a selective layer comprising a copolymer of the PIM-1 monomers with β-cyclodextrin.

Very thin (up to 30 nm thickness) selective layers of PIM-EA-TB(H_2) have been deposited onto PTMSP using the Langmuir–Blodgett technique [280]. The resulting membranes had higher CO_2 permeance but similar CO_2/N_2 selectivity to self-standing films of the PIM alone. The approach has been extended to PIM-1 and PIM-TMN-Trip, to compare the effects of 2D and 3D polymer architecture [281].

An alternative to a conventional membrane gas separation process for CO_2 capture is to use the membrane as a gas–liquid contactor, to transfer CO_2 from a flue gas into a solvent where it is chemically absorbed [282] and/or to recover CO_2 from the solvent [283].

Research on asymmetric and TFC membranes is bringing PIM-based membranes closer to industrial application. However, a number of barriers still need to be overcome before they will be accepted by the process engineering industry. Firstly, for any given application, they must be shown to operate effectively under realistic process conditions. Most data in the literature are for pure gases or binary gas mixtures, whereas real gas feeds may include moisture and other contaminants that can significantly affect the permeation properties of the membrane. Secondly, they must be shown to maintain their performance over extended periods of operation. As already discussed, high free volume glassy polymers such as PIMs are prone to physical aging, and this must be controlled if they are to be used commercially. Thirdly, issues of sustainability must be addressed with regard to the way the polymers are made, the way the membranes are fabricated, and the way the membrane modules are dealt with at the end of their useful life.

4.2 Pervaporation Membranes

Pervaporation enables the separation of volatile components from a liquid mixture and has advantages over conventional processes, such as distillation, for the separation of azeotropes, organic–organic mixtures, and thermally sensitive compounds [284]. Pervaporation is a combination of permeation and evaporation, in which a membrane is contacted with a liquid mixture and the permeate is removed as a vapor. The driving force is a chemical potential gradient arising from differences in the fugacity or partial vapor pressure of a component between the feed and permeate sides of the membrane. This is achieved by application of a vacuum or sweep gas to the permeate side of the membrane.

Transport through a pervaporation membrane can often be understood in terms of a solution-diffusion model, as outlined above for gas separation membranes. As also discussed above, the productivity of a membrane may be expressed in terms of flux, permeance, or permeability. In pervaporation studies, the productivity is frequently quoted as a mass flux, with units such as $kg\ m^{-2}\ h^{-1}$, or as a mass flux normalized for the thickness of the membrane selective layer, with units such as $\mu m\ kg\ m^{-2}\ h^{-1}$. It is also possible to quote fluxes in terms of molar amounts, or the equivalent volume of an ideal gas at standard temperature and pressure, as is often done in gas separation studies [285]. The relationship between the flux, J_i, of component i and its permeability, P_i, can be expressed as

$$J_i = \frac{P_i}{l}(f_{i,f} - f_{i,p}) \tag{9.4}$$

where l is the thickness of the membrane selective layer and $f_{i,f}$ and $f_{i,p}$ are the fugacities of component i on the feed and permeate sides of the membrane, respectively [284]. For an ideal gas, fugacity can be replaced by partial pressure. However, many vapors of interest for pervaporation show nonideal behavior, and a realistic estimate of the fugacity coefficient is needed if permeabilities are to be calculated.

The selectivity of a pervaporation membrane for one component over another may be expressed as a ratio of permeabilities, $\alpha = P_A/P_B$, as discussed above for gas separation membranes, and it has been argued that this is the best way to describe membrane performance [285,286]. Nevertheless, it is more common in pervaporation studies to quote a separation factor, β,

$$\beta = \frac{Y/(1-Y)}{X/(1-X)} \tag{9.5}$$

where Y is the mass or mole fraction of the preferentially permeating component in the permeate and X is the fraction of that component in the feed. Note that in older literature, the symbol α is often used for separation factor, but the tendency nowadays is to reserve α for a ratio of permeabilities or permeances. The overall separation factor for pervaporation, β_{pervap}, arises from two contributions, the effect of evaporation, β_{evap}, and the contribution of the membrane itself, β_{membr}, so that

$$\beta_{pervap} = \beta_{evap} \times \beta_{membr} \tag{9.6}$$

Another way to describe the separating ability is in terms of an enrichment factor, also sometimes given the symbol β, which is the ratio of concentrations of the preferentially permeating component in permeate and feed [287]. The value obtained for enrichment factor depends on the concentration units used. Care must be taken when interpreting pervaporation data to understand the basis of the results presented.

To take account of the trade-off between productivity and selectivity, membranes may be compared in terms of a pervaporation separation index, PSI,

$$\mathrm{PSI} = J_{total}(\beta - 1) \tag{9.7}$$

where J_{total} is the total flux. The PSI is zero if nothing useful happens, either because there is no flux ($J_{total} = 0$) or because there is no separation ($\beta = 1$).

Commercially, the most well-developed pervaporation membranes are hydrophilic membranes utilized, for example, for the dehydration of alcohol. However, some of the more challenging separations, such as the recovery of organics from dilute aqueous solution, and organic–organic separations, require organophilic membranes. PIM-1 is counted amongst a select group of materials suitable for the formation of organophilic pervaporation membranes, along with some other high free volume glassy polymers such as PTMSP and rubbery polymers such as PDMS. Some applications are outlined below.

4.2.1 *Phenol From Dilute Aqueous Solution*

The very first membrane data reported for PIM-1 were for the pervaporation of aqueous phenol solutions with concentrations in the range 1–5 wt.% [8]. For a self-standing PIM-1 film (40 μm thickness), the total mass flux was shown to increase from 0.20 to 0.67 kg m^{-2} h^{-1} with increasing temperature over the range 50–80°C, with separation factor, β, in the range 16–18.

4.2.2 *Ethanol From Dilute Aqueous Solution*

Pervaporation of ethanol from dilute aqueous solution with self-standing PIM-1 membranes yields a permeate that is significantly enriched in ethanol: a 10 wt.% feed yielding 50 wt.% or more ethanol in the permeate [288] and a 2.52% feed yielding 12.1% ethanol in the permeate [289]. However, higher selectivity is desired for practical application. Improved ethanol/water separation has been reported for MMMs of PIM-1 with silicalite-1 [238], but the addition of chemically modified graphene oxide did not give a significant improvement [290].

4.2.3 *Volatile Organic Compounds From Dilute Aqueous Solution*

In addition to ethanol, the separation of methanol, isopropanol, acetone, acetonitrile, acetic acid, diethyl ether, ethyl acetate, and tetrahydrofuran from 1 mol% aqueous solution by pervaporation with a PIM-1 membrane has been investigated [289]. For ethyl acetate, a very high thickness normalized mass flux (39.5 μm kg m^{-2} h^{-1}) and high separation factor (189) were reported.

4.2.4 Butanol From Dilute Aqueous Solution

The recovery from dilute aqueous solution of butanol, in particular, the 1-butanol isomer, has been the subject of recent research because of a resurgence of interest in the acetone–butanol–ethanol fermentation process for the production of biobutanol [291]. In a fermentation process, product concentrations are limited by the toxicity of the products to the microorganisms used, so selective butanol removal is desired. Butanol and water are not fully miscible, and pervaporation may be utilized to produce a concentrated permeate in which phase separation occurs and which can be further processed [292].

Self-standing PIM-1 membranes exhibit an ideal selectivity for butanol/water at 30°C, based on the mass fluxes for pure components, of $\alpha = 3.7$ [288]. It has been reported that aging of a PIM-1 membrane reduces the permeability but increases the selectivity at low butanol concentration [291]. Further research is required to understand and control aging effects under pervaporation conditions.

The use of relatively thin (1–3 μm) selective layers of PIM-1 in a TFC membrane gives high mass flux (up to 9 kg m^{-2} h^{-1}), although the thickness normalized flux decreases for thin films [293]. In a TFC membrane, the nature of the support material can have a dramatic effect on performance. PAN supports, as have been used in TFC membranes for gas separation and OSN, are not suitable for organophilic pervaporation. However, polyvinylidene fluoride (PVDF) supports with high surface porosity can give TFC membranes with good performance [293].

MMMs of PIM-1 with low loadings (up to ~1 wt.%) of alkyl-functionalized graphene-like fillers show improved separation, with values of β up to 32.9 for pervaporation of 5 wt.% aqueous butanol at 65°C, compared with $\beta = 13.5$ for pure PIM-1 under the same conditions [290]. Similar fillers can be incorporated into TFN membranes on PVDF supports, but the lateral flake size needs to be carefully controlled [294]; the addition of alkyl-functionalized graphene-like nanosheets to the selective layer can, under certain conditions, enhance both flux and separation factor. Butanol pervaporation has also been studied for MMMs on cellulose acetate microfiltration supports of PIM-1 with functionalized carbon black nanoparticles [295] and nano-fumed silica [296].

4.2.5 Dehydration of Alcohols

For dehydration of alcohols, removing the water when it is a minor component, a hydrophilic membrane is required. Base-catalyzed hydrolysis of PIM-1 yields a more hydrophilic polymer that can be blended with polyimides such as Matrimid, Torlon, and P84 [297]. The addition of the PIM to the polyimide enhances flux while maintaining a good separation. For example, a membrane of 20 wt.% hydrolyzed PIM-1 in Matrimid gave a mass flux of ~0.05 kg m^{-2} h^{-1} with a separation factor >5000 for the dehydration pervaporation of 1-butanol with 15 wt.% water at 60 °C.

4.2.6 Ethylene Glycol Separation

Industrial production of ethylene glycol generally produces a mixture from which either methanol or water needs to be removed. PIM-1 membranes have been shown to be suitable for this [298]. For example, a thickness normalized mass flux of 10.4 μm kg^{-1} m^{-2} h^{-1} and separation factor of 24.2 was obtained for the pervaporation of methanol from an 18.5 wt.% mixture with ethylene glycol at 30°C. More hydrophilic membranes formed of hydrolyzed PIM-1 show exceptional performance for the removal of water from ethylene glycol [144], with, for example, a flux of 13.7 μm kg^{-1} m^{-2} h^{-1} and separation factor of 75.9 for dehydration of a mixture of 20 wt.% water in ethylene glycol at 30°C.

4.3 Nanofiltration Membranes

Nanofiltration is a pressure-driven process in which solute molecules of molar mass in the range 200–1400 g mol^{-1} are retained by a membrane through which the solvent is able to pass. Nanofiltration of aqueous solutions is well established, but there is growing interest in nanofiltration with organic solvents (solvent-resistant or OSN) because of its potential in chemistry-related industries for the energy-efficient separation of intermediates, purification of products, and recycling of catalysts.

Fritsch et al. [169] demonstrated that cross-linked PIM-1 and PIM copolymer TFC membranes showed excellent retention of hexaphenylbenzene and of polystyrene oligomers in solvents such as *n*-heptane, outperforming the state of the art at the time.

The Topchiev Institute of Petrochemical Synthesis compared three high free volume polymers with potential for OSN, poly(4-methyl-2-pentyne), PTMSP, and PIM-1, highlighting the effects that different types of solute may have [299]. They also studied the fractional accessible volume (FAV) for PIM-1 and a range of other hydrophobic polymers, demonstrating that a threshold FAV of about 12% was necessary for liquid transport [42]. They further showed for water/ethanol mixtures that a minimum ethanol concentration was necessary for liquid transport through PIM-1 and other high free volume polymers. A PIM-1 nanofiltration membrane was applied for solvent recovery in a process for the separation of two organic solutes from a common solvent, which also utilized PIM-1 as an adsorbent [300].

Molecular simulation has been used to compare PIM-1 membranes and carbon membranes for application in ethanolamine purification [301].

For OSN in polar aprotic solvents, blend membranes have been prepared of PIM-1 with polybenzimidazole, which were stabilized by reduction of the nitriles in the PIM-1 to amine followed by treatment with HCl [302].

The Livingston group [303] generated TFC membranes of three PIMs (PIM-1, PIM-7, and PIM-8) by roll-to-roll dip coating, for application in OSN. Better coatings were obtained on a cross-linked Ultem 1000 support than on PAN supports. Of the three PIMs studied, PIM-8 showed the highest separation factor between straight and branched chain C_{16} alkanes.

OSN membranes with ultrathin selective layers of PIM-1, down to 35 nm in thickness, have also been investigated [182]. The membranes were prepared by spin-coating a PIM-1 solution onto glass or silicon, then transferring the ultrathin film onto a porous PAN or alumina support. The best permeance (18 $L\,m^{-2}\,h^{-1}\,bar^{-1}$) was obtained with a 140 nm thick PIM-1 selective layer. The performance of thinner selective layers was poorer, which may be attributed to rapid relaxation and more effective packing of the polymer in very thin films.

One of the key advantages of PIMs compared with many other microporous materials is their solution processability. However, this can also be a disadvantage when the membrane is to be used with a solvent that may swell or dissolve the polymer. To create highly cross-linked ultrathin selective layers stable to a wide range of solvents, the Livingston group [180] combined the PIM concept with the idea of generating thin films in situ by interfacial polymerization. They showed that polyarylate films, incorporating sites of contortion such as spirobisindane, could be prepared with thicknesses down to 20 nm. In this work, a spirocyclic biscatechol, or another phenol monomer, was dissolved in aqueous NaOH and the nanofilm formed at the interface with a solution of trimesoyl chloride in hexane. The membranes exhibited solvent permeances up to two orders of magnitude higher than that achieved for nanofilms without sites of contortion.

Although PIMs themselves have mostly been investigated for nanofiltration with organic solvents, carbonaceous membranes prepared by controlled carbonization of PIM-1 show potential as nanofiltration membranes for water treatment [304]. TFC membranes for nanofiltration of aqueous systems have also been prepared through layer-by-layer assembly of oppositely charged TB PIMs [212].

4.4 Adsorbents

Early work on microporous network polymers demonstrated the potential for organic polymers with high internal surface area to be used as adsorbents [9,16,18]. The advent of solution-processable PIMs opened up new possibilities to create innovative materials for pressure-swing, temperature-swing, or solvent-swing adsorption processes.

4.4.1 *Adsorption of Gases*

An understanding of the adsorption/desorption behavior of gases and vapors in PIMs is important both for their applications as gas separation membranes, where sorption or dissolution in the polymer is a key stage of membrane transport, and for their applications as solid-state adsorbents. PIMs possess a combination of the characteristics of a microporous material and of a polymer. Thus, a certain amount of adsorbate may be taken up by the effective microporosity (free volume) of the polymer, and further uptake may lead to swelling of the polymer. In a nitrogen adsorption experiment at liquid nitrogen temperature, a PIM typically shows high uptake at very low relative pressure, as is characteristic of a microporous material where there is enhanced adsorption due to multiwall interactions in pores <2 nm in dimension. After the initial large rise, the amount adsorbed shows a further steady increase as the pressure increases [7,8]. Subsequent desorption exhibits hysteresis, with the desorption curve lying above the adsorption curve down to very low relative pressures, which is unlike mesoporous materials (pore size 2–50 nm) that exhibit a hysteresis that closes at a relative pressure >0.4. The hysteresis in a PIM can be attributed to swelling of the polymer and/or to the effects of a complex network of micropores.

The adsorption of gases in glassy polymers such as PIMs can usually be analyzed in terms of a dual mode sorption model [305–310]. The dual mode model treats sorption as having two components, a Henry's law component in which the concentration of gas in the polymer is directly proportional to its partial pressure and a component that behaves as Langmuir-type sorption. Other approaches include the Guggenheim–Anderson–De Boer model [309,310] and the nonequilibrium lattice fluid model [262,311].

PIMs were included in a study of solid-state adsorbents for CO_2 capture [312]. In a rapid screening experiment, a relatively dense PIM-1 powder showed modest performance due to slow kinetics, but samples that had been precipitated with a more open morphology exhibited comparable behavior to activated carbons. The Lively group have shown that composites of PIM-1 with poly(ethylene imine) in powder or fiber form show promise for CO_2 capture [313].

The prospect of a hydrogen economy, as an alternative to one based on fossil fuels, has driven the search for efficient and safe ways to store hydrogen. PIMs represented some of the first porous organic polymers to be explored for hydrogen storage by physisorption, and there is continuing interest in the subject [17,19,21–26,177,181,239,314–322].

4.4.2 *Adsorption of Organic Vapors*

The affinity of PIM-1 for organic vapors is the basis for the first commercial product to utilize the polymer; it is employed by 3M in an end-of-life indicator for organic vapor–absorbing cartridges used in personal protection [323,324].

Adsorbent hollow fibers, which offer an attractive alternative to packed beds for removing contaminants from gas streams, are generally produced by encapsulating an adsorbent material within a polymer. The use of PIM-1 as an active binder, in the place of a conventional polymer such as polyethersulfone, significantly improves the performance of the composite fiber, as demonstrated for octane adsorption [325].

Additive manufacturing technologies, such as 3D printing, can produce complex structures that are not easily created by traditional methods. However, most of these techniques are limited in the range of materials for which they can be used. A methodology has been developed for 3D printing of solution-processable polymers, and 3D-printed PIM-1 monoliths were shown to give rapid adsorption of toluene vapor [326].

4.4.3 *Adsorption of Organic Compounds From Solution*

For sustainable engineering, efficient technologies are required for wastewater treatment in many industries. For example, the fabrication of membranes by phase inversion generates wastewater from the coagulation bath that is contaminated with aprotic solvents such as NMP or *N*,*N*-dimethylformamide. Thus, membrane fabrication lacks green credentials, even though membrane processes are generally regarded as sustainable technologies. This problem can be addressed by implementing a continuous process for treating and recycling the wastewater. PIMs and other potential adsorbents have been investigated for this process [327].

The separation and purification of complex mixtures represents an energy-intensive stage of many production processes in chemistry-related industries. A concept has been explored for separating two organic solutes from a common solvent by solvent-swing adsorption [300]. Utilizing dyes as model compounds, PIM-1 was shown selectively to adsorb a neutral solute (Oil Red O) from a mixture with an anionic solute (Remazol Brilliant Blue R) in aqueous/ethanolic solution. The adsorption process was linked to nanofiltration for solvent recovery, which also utilized PIM-1.

Chemical modification of PIM-1 (Fig. 9.4) may be used to tailor the selectivity of a PIM toward different types of organic compound. Hydrolysis of PIM-1 yields products that show increasing uptake of a cationic dye (Safranin O) and decreasing uptake of an anionic dye (Orange II), from aqueous solution, with increasing degree of carboxylation [145]. Conversely, hydroxyalkylaminoalkylamide PIMs [158,164] and amine-PIM-1 [158] show strong selectivity for the anionic dye and low affinity for the cationic dye.

4.5 Electrospun Fibrous Membranes

Electrospinning, in which a high voltage is used to draw ultrafine fibers from a polymer solution, can be used to create highly porous, fibrous membranes, suitable for use as adsorbents or for other applications. PIM-1 has been electrospun and subsequently carbonized to give high surface area electrodes [328,329]. The conditions for electrospinning PIM-1 have been explored [330]. Electrospun PIM-1 fibers have been shown to be effective adsorbents of dyes (Solvent Blue 35, Oil Red O) from ethanolic solution [331], of aniline from aqueous solution or from the air [332], and of carbendazim from methanol/water solution and phenol from aqueous solution [333].

Superhydrophobic–superoleophilic membranes that can separate oil/water mixtures and water-in-oil emulsions, and can also adsorb contaminants from oils, were achieved by electrospinning PIM-1 with polyhedral oligomeric silsesquioxane (POSS) particles [334]. Electrospun PIM-1 has also been used to support catalytically active Pd nanoparticles [335].

Amine modified electrospun PIM-1 fibers have been investigated for the removal of methyl orange from aqueous media [336]. Hydrolysis of electrospun PIM-1 fibers yields materials suitable for the selective adsorption of dyes and heavy metal ions [337]. Conditions have been found for preparing hydrolyzed PIM-1 with good solubility that can itself be electrospun [148]; the fibrous membranes were used in a simple gravity filtration system to remove a cationic dye (Methylene Blue) from aqueous solution with high efficiency by adsorption. Blends of hydrolyzed PIM-1 with hexamethylene diisocyanate have also been electrospun, subsequent thermal treatment yielding a cross-linked superhydrophobic membrane suitable for oil/water separation as well as for the adsorption of organic compounds [149]. An alternative to electrospun membranes for oil/water separation is a superhydrophobic–superoleophilic fabric prepared by an etching and dip-coating process [338].

4.6 Other Applications

Membrane-mediated enantiomer separation offers an alternative to chromatographic and other techniques. The spirobisindane unit in PIM-1 has two enantiomeric forms. Chiral (+)-PIM-1 has been prepared from the enantiomerically pure monomer, a carboxylated version was produced by acid hydrolysis, and membranes of the two polymers were shown to give high and enantioselective permeability for various racemic compounds [339]. For the case of permeation of the racemic spirobisindane monomer through a (+)-PIM-1 membrane, an enantiomeric excess of 87% was reported, suggesting that chiral PIM membranes have potential for the production of their own starting materials.

Preconcentrators enable organic compounds present in trace amounts in groundwater or other media to be concentrated up, then thermally desorbed and detected. A preconcentrator utilizing PIM-1 enabled detection of 1,1,2-trichloroethane at levels as low as 5 ppbV, and of benzene at 145 ppbV [340].

Organic solar cells require transparent conducting electrodes. It has been shown that a suitable electrode can be formed by spin-coating PIM-1 and then thermally treating it to give a graphene-like carbon nanosheet [341].

Organic light-emitting diodes (OLEDs) often incorporate an electron transport material to improve efficiency. PIM-1 has the characteristics of an n-type semiconductor and has been incorporated into a solution-processed green emitting OLED [342].

Nonaqueous redox flow batteries have a membrane that must retain redox-active molecules while allowing transport of charge-balancing ions [343–346]. Oligomeric catholytes have been tailored to match the molecular sieving properties of a PIM-1 membrane [347].

5. CONCLUDING REMARKS

The variety of polymer structures that exhibit PIM-like behavior, and the diverse applications for which they have been investigated, illustrate the wide scope for utilizing the PIM concept within energy-efficient technologies. A PIM is already used commercially in a sensor, and other applications are expected to reach the market in due course. There is a growing understanding of the relationships between macromolecular structure, free volume distribution, and the sorption, permeation, and other properties of PIM-based materials. In the field of advanced separations, there is potential to create bespoke membrane materials or adsorbents tailored for specific purposes. However, there is still work to be done to address issues of sustainability over the whole life cycle, from monomer synthesis through to product reuse or recycling. This is the challenge for the next phase of PIMs research and development.

References

[1] K.S.W. Sing, D.H. Everett, R.A.W. Haul, L. Moscou, R.A. Pierotti, J. Rouquerol, T. Siemieniewska, Reporting physisorption data for gas/solid systems with special reference to the determination of surface area and porosity (Recommendations 1984), Pure and Applied Chemistry 57 (1985) 603–619.

[2] N.B. McKeown, P.M. Budd, Exploitation of intrinsic microporosity in polymer-based materials, Macromolecules 43 (2010) 5163–5176.

[3] O.M. Ilinitch, V.B. Fenelonov, A.A. Lapkin, L.G. Okkel, V.V. Terskikh, K.I. Zamaraev, Intrinsic microporosity and gas transport in polyphenylene oxide polymers, Microporous Mesoporous Materials 31 (1999) 97–110.

[4] N.B. McKeown, Polymers of intrinsic microporosity, ISRN Materials Science (2012) 513916, 513986.

[5] N.B. McKeown, H. Li, S. Makhseed, Towards phthalocyanine network polymers for heterogeneous catalysis, Special Publications–Royal Society of Chemistry 266 (2001) 214–218.

[6] N.B. McKeown, S. Makhseed, P.M. Budd, Phthalocyanine-based nanoporous network polymers, Chemical Communications (2002) 2780–2781.

[7] P.M. Budd, B.S. Ghanem, S. Makhseed, N.B. McKeown, K.J. Msayib, C.E. Tattershall, Polymers of intrinsic microporosity (PIMs): robust, solution-processable, organic nanoporous materials, Chemical Communications (2004) 230–231.

[8] P.M. Budd, E.S. Elabas, B.S. Ghanem, S. Makhseed, N.B. McKeown, K.J. Msayib, C.E. Tattershall, D. Wang, Solution-processed, organophilic membrane derived from a polymer of intrinsic microporosity, Advanced Materials 16 (2004) 456–459.

[9] A.V. Maffei, M. Budd Peter, B. McKeown Neil, Adsorption studies of a microporous phthalocyanine network polymer, Langmuir 22 (2006) 4225–4229.

[10] H.J. Mackintosh, P.M. Budd, N.B. McKeown, Catalysis by microporous phthalocyanine and porphyrin network polymers, Journal of Materials Chemistry 18 (2008) 573–578.

[11] S. Makhseed, F. Al-Kharafi, J. Samuel, B. Ateya, Catalytic oxidation of sulfide ions using a novel microporous cobalt phthalocyanine network polymer in aqueous solution, Catalysis Communications 10 (2009) 1284–1287.

[12] G.C. Eastmond, J. Paprotny, Cyanodibenzo[1,4]dioxins: a new family of synthons for substituted dibenzo[1,4]dioxins, Chemistry Letters (1999) 479–480.

[13] G.C. Eastmond, J. Paprotny, A. Steiner, L. Swanson, Synthesis of cyanodibenzo[1,4]dioxines and their derivatives by cyano-activated fluoro displacement reactions, New Journal of Chemistry 25 (2001) 379–384.

[14] N.B. McKeown, S. Hanif, K. Msayib, C.E. Tattershall, P.M. Budd, Porphyrin-based nanoporous network polymers, Chemical Communications (2002) 2782–2783.

[15] A.R. Antonangelo, C. Grazia Bezzu, S.S. Mughal, T. Malewschik, N.B. McKeown, S. Nakagaki, A porphyrin-based microporous network polymer that acts as an efficient catalyst for cyclooctene and cyclohexane oxidation under mild conditions, Catalysis Communications 99 (2017) 100–104.

[16] P.M. Budd, B. Ghanem, K. Msayib, N.B. McKeown, C. Tattershall, A nanoporous network polymer derived from hexaazatrinaphthylene with potential as an adsorbent and catalyst support, Journal of Materials Chemistry 13 (2003) 2721–2726.

[17] N.B. McKeown, B. Gahnem, K.J. Msayib, P.M. Budd, C.E. Tattershall, K. Mahmood, S. Tan, D. Book, H.W. Langmi, A. Walton, Towards polymer-based hydrogen storage materials: engineering ultra-microporous cavities within polymers of intrinsic microporosity, Angewandte Chemie International Edition 45 (2006) 1804–1807.

[18] B.S. Ghanem, M. Hashem, K.D.M. Harris, K.J. Msayib, M. Xu, P.M. Budd, N. Chaukura, D. Book, S. Tedds, A. Walton, N.B. McKeown, Triptycene-based polymers of intrinsic microporosity: organic materials that can Be tailored for gas adsorption, Macromolecules 43 (2010) 5287–5294.

[19] B.S. Ghanem, K.J. Msayib, N.B. McKeown, K.D.M. Harris, Z. Pan, P.M. Budd, A. Butler, J. Selbie, D. Book, A. Walton, A triptycene-based polymer of intrinsic microporosity that displays enhanced surface area and hydrogen adsorption, Chemical Communications (2007) 67–69.

[20] J. Vile, M. Carta, C.G. Bezzu, N.B. McKeown, Tribenzotriquinacene-based polymers of intrinsic microporosity, Polymer Chemistry 2 (2011) 2257–2260.

[21] S. Makhseed, J. Samuel, Microporous organic polymers incorporating dicarboximide units for H_2 storage and remarkable CO_2 capture, Journal of Materials Chemistry A 1 (2013) 13004–13010.

[22] M. Budd Peter, A. Butler, J. Selbie, K. Mahmood, B. McKeown Neil, B. Ghanem, K. Msayib, D. Book, A. Walton, The potential of organic polymer-based hydrogen storage materials, Physical Chemistry Chemical Physics 9 (2007) 1802–1808.

[23] N.B. McKeown, P.M. Budd, D. Book, Microporous polymers as potential hydrogen storage materials, Macromolecular Rapid Communications 28 (2007) 995–1002.

[24] J. Germain, J.M.J. Frechet, F. Svec, Nanoporous polymers for hydrogen storage, Small 5 (2009) 1098–1111.

[25] N.B. McKeown, P.M. Budd, Polymers of intrinsic microporosity (PIMs): organic materials for membrane separations, heterogeneous catalysis and hydrogen storage, Chemical Society Reviews 35 (2006) 675–683.

[26] D. Ramimoghadam, E.M. Gray, C.J. Webb, Review of polymers of intrinsic microporosity for hydrogen storage applications, International Journal of Hydrogen Energy 41 (2016) 16944–16965.

[27] S. Chen, W. Yi, J. Duhamel, K. Heinrich, G. Bengtson, D. Fritsch, Effect of the porosity of a polymer of intrinsic microporosity (PIM) on its intrinsic fluorescence, The Journal of Physical Chemistry B 117 (2013) 5249–5260.

[28] H. Yin, Y.Z. Chua, B. Yang, C. Schick, W.J. Harrison, P.M. Budd, M. Boehning, A. Schoenhals, First clear-cut experimental evidence of a glass transition in a polymer with intrinsic microporosity: PIM-1, The Journal of Physical Chemistry Letters 9 (2018) 2003–2008.

[29] N. Konnertz, Y. Ding, W.J. Harrison, P.M. Budd, A. Schoenhals, M. Boehning, Molecular mobility of the high performance membrane polymer PIM-1 as investigated by dielectric spectroscopy, ACS Macro Letters 5 (2016) 528–532.

[30] N.B. McKeown, P.M. Budd, K. Msayib, B. Ghanem, Microporous Polymer Material, (The University of Manchester, UK). Patent US 7,690,514 B2, dated Apr 6, 2010; Patent US 8,056,732 B2, dated Nov 15, 2011; European Patent Specification EP 1,648,954 B1, date of grant Nov 5, 2014.

[31] M. Heuchel, D. Fritsch, P.M. Budd, N.B. McKeown, D. Hofmann, Atomistic packing model and free volume distribution of a polymer with intrinsic microporosity (PIM-1), Journal of Membrane Science 318 (2008) 84–99.

[32] G.S. Larsen, P. Lin, K.E. Hart, C.M. Colina, Molecular simulations of PIM-1-like polymers of intrinsic microporosity, Macromolecules 44 (2011) 6944–6951.

[33] A.G. McDermott, G.S. Larsen, P.M. Budd, C.M. Colina, J. Runt, Structural characterization of a polymer of intrinsic microporosity: X-ray scattering with interpretation enhanced by molecular dynamics simulations, Macromolecules 44 (2011) 14–16.

[34] C.G. Bezzu, M. Carta, M.-C. Ferrari, J.C. Jansen, M. Monteleone, E. Esposito, A. Fuoco, K. Hart, T.P. Liyana-Arachchi, C.M. Colina, N.B. McKeown, The synthesis, chain-packing simulation and long-term gas permeability of highly selective spirobifluorene-based polymers of intrinsic microporosity, Journal of Materials Chemistry A 6 (2018) 10507–10514.

[35] P.M. Budd, N.B. McKeown, D. Fritsch, Free volume and intrinsic microporosity in polymers, Journal of Materials Chemistry 15 (2005) 1977–1986.

[36] N.B. McKeown, P.M. Budd, K.J. Msayib, B.S. Ghanem, H.J. Kingston, C.E. Tattershall, S. Makhseed, K.J. Reynolds, D. Fritsch, Polymers of intrinsic microporosity (PIMs): bridging the void between microporous and polymeric materials, Chemistry: A European Journal 11 (2005) 2610–2620.

[37] V.P. Shantarovich, T. Suzuki, Y. Ito, K. Kondo, R.S. Yu, P.M. Budd, Y.P. Yampolskii, S.S. Berdonosov, A.A. Eliseev, Structural heterogeneity in glassy polymeric materials revealed by positron annihilation and other supplementary techniques, Physica Status Solidi C 4 (2007) 3776–3779.

[38] T. Emmler, K. Heinrich, D. Fritsch, P.M. Budd, N. Chaukura, D. Ehlers, K. Raetzke, F. Faupel, Free volume investigation of polymers of intrinsic microporosity (PIMs): PIM-1 and PIM1 copolymers incorporating ethanoanthracene units, Macromolecules 43 (2010) 6075–6084.
[39] J. Weber, J. Schmidt, A. Thomas, W. Boehlmann, Micropore analysis of polymer networks by gas sorption and ^{129}Xe NMR spectroscopy: toward a better understanding of intrinsic microporosity, Langmuir 26 (2010) 15650–15656.
[40] N. Ritter, I. Senkovska, S. Kaskel, J. Weber, Intrinsically microporous poly(imide)s: structure-porosity relationship studied by gas sorption and X-ray scattering, Macromolecules 44 (2011) 2025–2033.
[41] J. Weber, N. Du, M.D. Guiver, Influence of intermolecular interactions on the observable porosity in intrinsically microporous polymers, Macromolecules 44 (2011) 1763–1767.
[42] A. Yushkin, A. Grekhov, S. Matson, M. Bermeshev, V. Khotimsky, E. Finkelstein, P.M. Budd, V. Volkov, T.J.H. Vlugt, A. Volkov, Study of glassy polymers fractional accessible volume (FAV) by extended method of hydrostatic weighing: effect of porous structure on liquid transport, Reactive and Functional Polymers 86 (2015) 269–281.
[43] V.P. Shantarovich, T. Suzuki, Y.P. Yampol'skii, P. Budd, V.V. Gustov, I.B. Kevdina, A.V. Pastukhov, S.S. Berdonosov, V.E. Bozhevol'nov, Positronium annihilation data and actual free-volume distribution in polymers, High Energy Chemistry 41 (2007) 370–380.
[44] R. Lima de Miranda, J. Kruse, K. Raetzke, F. Faupel, D. Fritsch, V. Abetz, P.M. Budd, J.D. Selbie, N.B. McKeown, B.S. Ghanem, Unusual temperature dependence of the positron lifetime in a polymer of intrinsic microporosity, Physica Status Solidi RRL 1 (2007) 190–192.
[45] C.L. Staiger, S.J. Pas, A.J. Hill, C.J. Cornelius, Gas separation, free volume distribution, and physical aging of a highly microporous spirobisindane polymer, Chemistry of Materials 20 (2008) 2606–2608.
[46] S. Harms, K. Raetzke, F. Faupel, N. Chaukura, P.M. Budd, W. Egger, L. Ravelli, Aging and free volume in a polymer of intrinsic microporosity (PIM-1), The Journal of Adhesion 88 (2012) 608–619.
[47] P.M. Budd, N.B. McKeown, B.S. Ghanem, K.J. Msayib, D. Fritsch, L. Starannikova, N. Belov, O. Sanfirova, Y. Yampolskii, V. Shantarovich, Gas permeation parameters and other physicochemical properties of a polymer of intrinsic microporosity: polybenzodioxane PIM-1, Journal of Membrane Science 325 (2008) 851–860.
[48] E. Tocci, L. De Lorenzo, P. Bernardo, G. Clarizia, F. Bazzarelli, N.B. McKeown, M. Carta, R. Malpass-Evans, K. Friess, K. Pilnacek, M. Lanc, Y.P. Yampolskii, L. Strarannikova, V. Shantarovich, M. Mauri, J.C. Jansen, Molecular modeling and gas permeation properties of a polymer of intrinsic microporosity composed of ethanoanthracene and Troger's base units, Macromolecules 47 (2014) 7900–7916.
[49] R. Zorn, H. Yin, W. Lohstroh, W. Harrison, P.M. Budd, B.R. Pauw, M. Boehning, A. Schoenhals, Anomalies in the low frequency vibrational density of states for a polymer with intrinsic microporosity—the Boson peak of PIM-1, Physical Chemistry Chemical Physics 20 (2018) 1355–1363.
[50] A. Bondi, van der Waals volumes and radii, The Journal of Physical Chemistry 68 (1964) 441–451.
[51] D.W. van Krevelen (Ed.), Properties of Polymers: Their Correlation with Chemical Structure; Their Numerical Estimation and Prediction from Additive Group Contributions, Third Completely Revised Edition, Elsevier, Amsterdam, 1997.
[52] N. Du, G.P. Robertson, J. Song, I. Pinnau, S. Thomas, M.D. Guiver, Polymers of intrinsic microporosity containing trifluoromethyl and phenylsulfone groups as materials for membrane gas separation, Macromolecules 41 (2008) 9656–9662.
[53] W. Ogieglo, B. Ghanem, X. Ma, I. Pinnau, M. Wessling, How much do ultrathin polymers with intrinsic microporosity swell in liquids? Journal of Physical Chemistry B 120 (2016) 10403–10410.
[54] W. Ogieglo, K. Rahimi, S.B. Rauer, B. Ghanem, X. Ma, I. Pinnau, M. Wessling, How do organic vapors swell ultrathin films of polymer of intrinsic microporosity PIM-1? Journal of Physical Chemistry B 121 (2017) 7210–7220.
[55] W. Ogieglo, A. Furchner, B. Ghanem, X. Ma, I. Pinnau, M. Wessling, Mixed-penetrant sorption in ultrathin films of polymer of intrinsic microporosity PIM-1, Journal of Physical Chemistry B 121 (2017) 10190–10197.
[56] W. Ogieglo, I. Pinnau, M. Wessling, In-situ non-invasive imaging of liquid-immersed thin film composite membranes, Journal of Membrane Science 546 (2018) 206–214.
[57] W. Ogieglo, B. Ghanem, X. Ma, M. Wessling, I. Pinnau, High-pressure CO_2 sorption in polymers of intrinsic microporosity under ultrathin film confinement, ACS Applied Materials and Interfaces 10 (2018) 11369–11376.
[58] B.S. Ghanem, N.B. McKeown, P.M. Budd, J.D. Selbie, D. Fritsch, High-performance membranes from polyimides with intrinsic microporosity, Advanced Materials 20 (2008) 2766–2771.
[59] B.S. Ghanem, N.B. McKeown, P.M. Budd, N.M. Al-Harbi, D. Fritsch, K. Heinrich, L. Starannikova, A. Tokarev, Y. Yampolskii, Synthesis, characterization, and gas permeation properties of a novel group of polymers with intrinsic microporosity: PIM-polyimides, Macromolecules 42 (2009) 7881–7888.
[60] X. Ma, R. Swaidan, Y. Belmabkhout, Y. Zhu, E. Litwiller, M. Jouiad, I. Pinnau, Y. Han, Synthesis and gas transport properties of hydroxyl-functionalized polyimides with intrinsic microporosity, Macromolecules 45 (2012) 3841–3849.
[61] J. Weber, Q. Su, M. Antonietti, A. Thomas, Exploring polymers of intrinsic microporosity—microporous, soluble polyamide and polyimide, Macromolecular Rapid Communications 28 (2007) 1871–1876.
[62] X. Ma, O. Salinas, E. Litwiller, I. Pinnau, Novel spirobifluorene- and dibromospirobifluorene-based polyimides of intrinsic microporosity for gas separation applications, Macromolecules 46 (2013) 9618–9624.
[63] Y. Rogan, L. Starannikova, V. Ryzhikh, Y. Yampolskii, P. Bernardo, F. Bazzarelli, J.C. Jansen, N.B. McKeown, Synthesis and gas permeation properties of novel spirobisindane-based polyimides of intrinsic microporosity, Polymer Chemistry 4 (2013) 3813–3820.
[64] R. Swaidan, X. Ma, E. Litwiller, I. Pinnau, High pressure pure- and mixed-gas separation of CO_2/CH_4 by thermally-rearranged and carbon molecular sieve membranes derived from a polyimide of intrinsic microporosity, Journal of Membrane Science 447 (2013) 387–394.
[65] Z. Wang, D. Wang, J. Jin, Microporous polyimides with rationally designed chain structure achieving high performance for gas separation, Macromolecules 47 (2014) 7477–7483.
[66] R.J. Swaidan, B. Ghanem, R. Swaidan, E. Litwiller, I. Pinnau, Pure- and mixed-gas propylene/propane permeation properties of spiro- and triptycene-based microporous polyimides, Journal of Membrane Science 492 (2015) 116–122.
[67] R.J. Swaidan, X. Ma, E. Litwiller, I. Pinnau, Enhanced propylene/propane separation by thermal annealing of an intrinsically microporous hydroxyl-functionalized polyimide membrane, Journal of Membrane Science 495 (2015) 235–241.
[68] B. Ghanem, F. Alghunaimi, X. Ma, N. Alaslai, I. Pinnau, Synthesis and characterization of novel triptycene dianhydrides and polyimides of intrinsic microporosity based on 3,3′-dimethylnaphthidine, Polymer 101 (2016) 225–232.

[69] N. Alaslai, X. Ma, B. Ghanem, Y. Wang, F. Alghunaimi, I. Pinnau, Synthesis and characterization of a novel microporous dihydroxyl-functionalized triptycene-diamine-based polyimide for natural gas membrane separation, Macromolecular Rapid Communications 38 (2017).

[70] C.A. Terraza, L.H. Tagle, J.L. Santiago-Garcia, R.J. Canto-Acosta, M. Aguilar-Vega, R.A. Hauyon, D. Coll, P. Ortiz, G. Perez, L. Herran, B. Comesana-Gandara, N.B. McKeown, A. Tundidor-Camba, Synthesis and properties of new aromatic polyimides containing spirocyclic structures, Polymer 137 (2018) 283–292.

[71] H.B. Park, C.H. Jung, Y.M. Lee, A.J. Hill, S.J. Pas, S.T. Mudie, E. Van Wagner, B.D. Freeman, D.J. Cookson, Polymers with cavities tuned for fast selective transport of small molecules and ions, Science 318 (2007) 254–258.

[72] S. Li, H.J. Jo, S.H. Han, C.H. Park, S. Kim, P.M. Budd, Y.M. Lee, Mechanically robust thermally rearranged (TR) polymer membranes with spirobisindane for gas separation, Journal of Membrane Science 434 (2013) 137–147.

[73] H. Shamsipur, B.A. Dawood, P.M. Budd, P. Bernardo, G. Clarizia, J.C. Jansen, Thermally rearrangeable PIM-polyimides for gas separation membranes, Macromolecules 47 (2014) 5595–5606.

[74] M. Carta, R. Malpass-Evans, M. Croad, Y. Rogan, J.C. Jansen, P. Bernardo, F. Bazzarelli, N.B. McKeown, An efficient polymer molecular sieve for membrane gas separations, Science 339 (2013) 303–307.

[75] M. Carta, M. Croad, J.C. Jansen, P. Bernardo, G. Clarizia, N.B. McKeown, Synthesis of cardo-polymers using Troger's base formation, Polymer Chemistry 5 (2014) 5255–5261.

[76] M. Carta, M. Croad, K. Bugler, K.J. Msayib, N.B. McKeown, Heterogeneous organocatalysts composed of microporous polymer networks assembled by Troger's base formation, Polymer Chemistry 5 (2014) 5262–5266.

[77] M. Carta, R. Malpass-Evans, M. Croad, Y. Rogan, M. Lee, I. Rose, N.B. McKeown, The synthesis of microporous polymers using Troger's base formation, Polymer Chemistry 5 (2014) 5267–5272.

[78] M. Carta, M. Croad, R. Malpass-Evans, J.C. Jansen, P. Bernardo, G. Clarizia, K. Friess, M. Lanc, N.B. McKeown, Triptycene induced enhancement of membrane gas selectivity for microporous Troeger's base polymers, Advanced Materials 26 (2014) 3526–3531.

[79] A. Del Regno, A. Gonciaruk, L. Leay, M. Carta, M. Croad, R. Malpass-Evans, N.B. McKeown, F.R. Siperstein, Polymers of intrinsic microporosity containing Troeger base for CO_2 capture, Industrial and Engineering Chemistry Research 52 (2013) 16939–16950.

[80] E. Madrid, Y. Rong, M. Carta, N.B. McKeown, R. Malpass-Evans, G.A. Attard, T.J. Clarke, S.H. Taylor, Y.-T. Long, F. Marken, Metastable ionic diodes derived from an amine-based polymer of intrinsic microporosity, Angewandte Chemie International Edition 53 (2014) 10751–10754.

[81] E. Madrid, P. Cottis, Y. Rong, A.T. Rogers, J.M. Stone, R. Malpass-Evans, M. Carta, N.B. McKeown, F. Marken, Water desalination concept using an ionic rectifier based on a polymer of intrinsic microporosity (PIM), Journal of Materials Chemistry A 3 (2015) 15849–15853.

[82] I. Rose, M. Carta, R. Malpass-Evans, M.-C. Ferrari, P. Bernardo, G. Clarizia, J.C. Jansen, N.B. McKeown, Highly permeable benzotriptycene-based polymer of intrinsic microporosity, ACS Macro Letters 4 (2015) 912–915.

[83] M. Lee, C.G. Bezzu, M. Carta, P. Bernardo, G. Clarizia, J.C. Jansen, N.B. McKeown, Enhancing the gas permeability of Troger's base derived polyimides of intrinsic microporosity, Macromolecules 49 (2016) 4147–4154.

[84] Z. Yang, R. Guo, L. Wu, T. Xu, R. Malpass-Evans, M. Carta, B. McKeown Neil, D. Guiver Michael, Highly conductive anion-exchange membranes from microporous Troger's base polymers, Angewandte Chemie International Edition in English 55 (2016) 11499–11502.

[85] Y. Xiao, L. Zhang, L. Xu, T.-S. Chung, Molecular design of Troger's base-based polymers with intrinsic microporosity for gas separation, Journal of Membrane Science 521 (2017) 65–72.

[86] M. Carta, C.G. Bezzu, J. Vile, B.M. Kariuki, N.B. McKeown, Polymers of Intrinsic Microporosity derived from a carbocyclic analogue of Troger's base, Polymer 126 (2017) 324–329.

[87] R. Williams, L.A. Burt, E. Esposito, J.C. Jansen, E. Tocci, C. Rizzuto, M. Lanc, M. Carta, N.B. McKeown, A highly rigid and gas selective methanopentacene-based polymer of intrinsic microporosity derived from Troger's base polymerization, Journal of Materials Chemistry A 6 (2018) 5661–5667.

[88] L. Starannikova, N. Belov, V. Shantarovich, J. Zhang, J. Jin, Y. Yampolskii, Effective increase in permeability and free volume of PIM copolymers containing ethanoanthracene unit and comparison between the alternating and random copolymers, Journal of Membrane Science 548 (2018) 593–597.

[89] J. Zhou, X. Zhu, J. Hu, H. Liu, Y. Hu, J. Jiang, Mechanistic insight into highly efficient gas permeation and separation in a shape-persistent ladder polymer membrane, Physical Chemistry Chemical Physics 16 (2014) 6075–6083.

[90] R. Swaidan, B. Ghanem, I. Pinnau, Fine-tuned intrinsically ultramicroporous polymers redefine the permeability/selectivity upper bounds of membrane-based air and hydrogen separations, ACS Macro Letters 4 (2015) 947–951.

[91] Y. Zhuang, J.G. Seong, Y.S. Do, H.J. Jo, Z. Cui, J. Lee, Y.M. Lee, M.D. Guiver, Intrinsically microporous soluble polyimides incorporating Troger's base for membrane gas separation, Macromolecules 47 (2014) 3254–3262.

[92] T.M. Long, T.M. Swager, Using "internal free volume" to increase chromophore alignment, Journal of the American Chemical Society 124 (2002) 3826–3827.

[93] T.M. Long, T.M. Swager, Molecular design of free volume as a route to low-κ dielectric materials, Journal of the American Chemical Society 125 (2003) 14113–14119.

[94] N.T. Tsui, A.J. Paraskos, L. Torun, T.M. Swager, E.L. Thomas, Minimization of internal molecular free volume: a mechanism for the simultaneous enhancement of polymer stiffness, strength, and ductility, Macromolecules 39 (2006) 3350–3358.

[95] T.M. Swager, Iptycenes in the design of high performance polymers, Accounts of Chemical Research 41 (2008) 1181–1189.

[96] S.A. Sydlik, Z. Chen, T.M. Swager, Triptycene polyimides: soluble polymers with high thermal stability and low refractive indices, Macromolecules 44 (2011) 976–980.

[97] L.C.H. Moh, J.B. Goods, Y. Kim, T.M. Swager, Free volume enhanced proton exchange membranes from sulfonated triptycene poly(ether ketone), Journal of Membrane Science 549 (2018) 236–243.

[98] M. Hashem, C.G. Bezzu, B.M. Kariuki, N.B. McKeown, Enhancing the rigidity of a network polymer of intrinsic microporosity by the combined use of phthalocyanine and triptycene components, Polymer Chemistry 2 (2011) 2190–2192.

[99] B.S. Ghanem, R. Swaidan, X. Ma, E. Litwiller, I. Pinnau, Energy-efficient hydrogen separation by AB-type ladder-polymer molecular sieves, Advanced Materials 26 (2014) 6696–6700.

[100] C.G. Bezzu, M. Carta, A. Tonkins, J.C. Jansen, P. Bernardo, F. Bazzarelli, N.B. McKeown, A spirobifluorene-based polymer of intrinsic microporosity with improved performance for gas separation, Advanced Materials 24 (2012) 5930–5933.
[101] N. Du, G.P. Robertson, I. Pinnau, S. Thomas, M.D. Guiver, Copolymers of intrinsic microporosity based on 2,2′,3,3′-Tetrahydroxy-1,1′-dinaphthyl, Macromolecular Rapid Communications 30 (2009) 584–588.
[102] N. Du, G.P. Robertson, I. Pinnau, M.D. Guiver, Polymers of intrinsic microporosity with dinaphthyl and thianthrene segments, Macromolecules 43 (2010) 8580–8587.
[103] R. Short, M. Carta, C.G. Bezzu, D. Fritsch, B.M. Kariuki, N.B. McKeown, Hexaphenylbenzene-based polymers of intrinsic microporosity, Chemical Communications 47 (2011) 6822–6824.
[104] M. Carta, P. Bernardo, G. Clarizia, J.C. Jansen, N.B. McKeown, Gas permeability of hexaphenylbenzene based polymers of intrinsic microporosity, Macromolecules 47 (2014) 8320–8327.
[105] P.M. Budd, N.B. McKeown, Highly permeable polymers for gas separation membranes, Polymer Chemistry 1 (2010) 63–68.
[106] T. Masuda, E. Isobe, T. Higashimura, K. Takada, Poly[1-(trimethylsilyl)-1-propyne]: a new high polymer synthesized with transition-metal catalysts and characterized by extremely high gas permeability, Journal of the American Chemical Society 105 (1983) 7473–7474.
[107] K. Takada, H. Matsuya, T. Masuda, T. Higashimura, Gas permeability of polyacetylenes carrying substituents, Journal of Applied Polymer Science 30 (1985) 1605–1616.
[108] T. Masuda, B.Z. Tang, T. Higashimura, Ethanol-water separation by pervaporation through substituted-polyacetylene membranes, Polymer Journal 18 (1986) 565–567.
[109] T. Masuda, Y. Iguchi, B.Z. Tang, T. Higashimura, Diffusion and solution of gases in substituted polyacetylene membranes, Polymer 29 (1988) 2041–2049.
[110] B.Z. Tang, T. Masuda, T. Higashimura, Permeation of oxygen dissolved in water through substituted polyacetylene membranes, Journal of Polymer Science Part B: Polymer Physics 27 (1989) 1261–1272.
[111] T. Masuda, M. Takatsuka, B.Z. Tang, T. Higashimura, Pervaporation of organic liquid-water mixtures through substituted polyacetylene membranes, Journal of Membrane Science 49 (1990) 69–83.
[112] L.G. Toy, K. Nagai, B.D. Freeman, I. Pinnau, Z. He, T. Masuda, M. Teraguchi, Y.P. Yampolskii, Pure-gas and vapor permeation and sorption properties of poly[1-phenyl-2-[p-(trimethylsilyl)phenyl]acetylene] (PTMSDPA), Macromolecules 33 (2000) 2516–2524.
[113] K. Nagai, T. Masuda, T. Nakagawa, B.D. Freeman, I. Pinnau, Poly[1-(trimethylsilyl)-1-propyne] and related polymers: synthesis, properties and functions, Progress in Polymer Science 26 (2001) 721–798.
[114] T. Sakaguchi, K.-I. Yumoto, G. Kwak, M. Yoshikawa, T. Masuda, Pervaporation of ethanol/water and benzene/cyclohexane mixtures using novel substituted polyacetylene membranes, Polymer Bulletin 48 (2002) 271–276.
[115] M. Teraguchi, T. Masuda, Poly(diphenylacetylene) membranes with high gas permeability and remarkable chiral memory, Macromolecules 35 (2002) 1149–1151.
[116] R.D. Raharjo, H.J. Lee, B.D. Freeman, T. Sakaguchi, T. Masuda, Pure gas and vapor permeation properties of poly[1-phenyl-2-[p-(trimethylsilyl)phenyl]acetylene] (PTMSDPA) and its desilylated analog, poly[diphenylacetylene] (PDPA), Polymer 46 (2005) 6316–6324.
[117] Y. Hu, T. Sakaguchi, M. Shiotsuki, F. Sanda, T. Masuda, Synthesis and gas permeation properties of poly(diphenylacetylenes) having bulky alkyl/silyl and hydroxy groups, Journal of Membrane Science 282 (2006) 423–429.
[118] Y. Hu, M. Shiotsuki, F. Sanda, T. Masuda, Synthesis and extremely high gas permeability of polyacetylenes containing polymethylated indan/tetrahydronaphthalene moieties, Chemical Communications (2007) 4269–4270.
[119] Y. Hu, M. Shiotsuki, F. Sanda, B.D. Freeman, T. Masuda, Synthesis and properties of indan-based polyacetylenes that feature the highest gas permeability among all the existing polymers, Macromolecules 41 (2008) 8525–8532.
[120] T. Sakaguchi, Y. Hu, T. Masuda, Substituted polyacetylenes, Membrane Materials for Gas and Vapor Separation (2017) 107–142.
[121] I. Pinnau, L.G. Toy, Gas and vapor transport properties of amorphous perfluorinated copolymer membranes based on 2,2-bistrifluoromethyl-4,5-difluoro-1,3-dioxole/tetrafluoroethylene, Journal of Membrane Science 109 (1996) 125–133.
[122] A.Y. Alentiev, Y.P. Yampolskii, V.P. Shantarovich, S.M. Nemser, N.A. Plate, High transport parameters and free volume of perfluorodioxole copolymers, Journal of Membrane Science 126 (1997) 123–132.
[123] V.P. Shantarovich, I.B. Kevdina, Y.P. Yampolskii, A.Y. Alentiev, Positron annihilation lifetime study of high and low free volume glassy polymers: effects of free volume sizes on the permeability and permselectivity, Macromolecules 33 (2000) 7453–7466.
[124] A.Y. Alentiev, V.P. Shantarovich, T.C. Merkel, V.I. Bondar, B.D. Freeman, Y.P. Yampolskii, Gas and vapor sorption, permeation, and diffusion in glassy amorphous Teflon AF1600, Macromolecules 35 (2002) 9513–9522.
[125] E.S. Finkelshtein, K.L. Makovetskii, M.L. Gringolts, Y.V. Rogan, T.G. Golenko, L.E. Starannikova, Y.P. Yampolskii, V.P. Shantarovich, T. Suzuki, Addition-type polynorbornenes with $Si(CH_3)_3$ side groups: synthesis, gas permeability, and free volume, Macromolecules 39 (2006) 7022–7029.
[126] L. Starannikova, M. Pilipenko, N. Belov, Y. Yampolskii, M. Gringolts, E. Finkelshtein, Addition-type polynorbornene with $Si(CH_3)_3$ side groups: detailed study of gas permeation and thermodynamic properties, Journal of Membrane Science 323 (2008) 134–143.
[127] N. Belov, R. Nikiforov, L. Starannikova, K.R. Gmernicki, C.R. Maroon, B.K. Long, V. Shantarovich, Y. Yampolskii, A detailed investigation into the gas permeation properties of addition-type poly(5-triethoxysilyl-2-norbornene), European Polymer Journal 93 (2017) 602–611.
[128] N. Du, J. Song, G.P. Robertson, I. Pinnau, M.D. Guiver, Linear high molecular weight ladder polymer via fast polycondensation of 5,5′,6,6′-tetrahydroxy-3,3,3′,3′-tetramethylspirobisindane with 1,4-dicyanotetrafluorobenzene, Macromolecular Rapid Communications 29 (2008) 783–788.
[129] J. Song, N. Du, Y. Dai, G.P. Robertson, M.D. Guiver, S. Thomas, I. Pinnau, Linear high molecular weight ladder polymers by optimized polycondensation of tetrahydroxytetramethylspirobisindane and 1,4-dicyanotetrafluorobenzene, Macromolecules 41 (2008) 7411–7417.
[130] S. Thomas, I. Pinnau, N. Du, M.D. Guiver, Pure- and mixed-gas permeation properties of a microporous spirobisindane-based ladder polymer (PIM-1), Journal of Membrane Science 333 (2009) 125–131.
[131] M.R. Khdhayyer, E. Esposito, A. Fuoco, M. Monteleone, L. Giorno, J.C. Jansen, M.P. Attfield, P.M. Budd, Mixed matrix membranes based on UiO-66 MOFs in the polymer of intrinsic microporosity PIM-1, Separation and Purification Technology 173 (2017) 304–313.
[132] H.R. Kricheldorf, D. Fritsch, L. Vakhtangishvili, G. Schwarz, Cyclic ladder polymers by polycondensation of silylated tetrahydroxytetramethylspirobisindane with 1,4-dicyanotetrafluorobenzene, Macromolecular Chemistry and Physics 206 (2005) 2239–2247.

[133] H.R. Kricheldorf, N. Lomadze, D. Fritsch, G. Schwarz, Cyclic and telechelic ladder polymers derived from tetrahydroxytetramethylspirobisindane and 1,4-dicyanotetrafluorobenzene, Journal of Polymer Science Part A: Polymer Chemistry 44 (2006) 5344–5352.

[134] I.I. Ponomarev, I.V. Blagodatskikh, A.V. Muranov, Y.A. Volkova, D.Y. Razorenov, I.I. Ponomarev, K.M. Skupov, Dimethyl sulfoxide as a green solvent for successful precipitative polyheterocyclization based on nucleophilic aromatic substitution, resulting in high molecular weight PIM-1, Mendeleev Communications 26 (2016) 362–364.

[135] P. Zhang, X. Jiang, S. Wan, S. Dai, Advancing polymers of intrinsic microporosity by mechanochemistry, Journal of Materials Chemistry A 3 (2015) 6739–6741.

[136] P. Zhang, S. Dai, Mechanochemical synthesis of porous organic materials, Journal of Materials Chemistry A 5 (2017) 16118–16127.

[137] J. Zhang, J. Jin, R. Cooney, S. Zhang, Synthesis of polymers of intrinsic microporosity using an AB-type monomer, Polymer 57 (2015) 45–50.

[138] B. Ghanem, F. Alghunaimi, N. Alaslai, X. Ma, I. Pinnau, New phenazine-containing ladder polymer of intrinsic microporosity from a spirobisindane-based AB-type monomer, RSC Advances 6 (2016) 79625–79630.

[139] B.S. Ghanem, A facile synthesis of a novel triptycene-containing A-B monomer: precursor to polymers of intrinsic microporosity, Polymer Chemistry 3 (2012) 96–98.

[140] N. Du, G.P. Robertson, J. Song, I. Pinnau, M.D. Guiver, High-performance carboxylated polymers of intrinsic microporosity (PIMs) with tunable gas transport properties, Macromolecules 42 (2009) 6038–6043.

[141] N. Du, M.M. Dal-Cin, G.P. Robertson, M.D. Guiver, Decarboxylation-induced cross-linking of polymers of intrinsic microporosity (PIMs) for membrane gas separation, Macromolecules 45 (2012) 5134–5139.

[142] W.F. Yong, F.Y. Li, T.S. Chung, Y.W. Tong, Molecular interaction, gas transport properties and plasticization behavior of cPIM-1/Torlon blend membranes, Journal of Membrane Science 462 (2014) 119–130.

[143] W.F. Yong, T.-S. Chung, Miscible blends of carboxylated polymers of intrinsic microporosity (cPIM-1) and Matrimid, Polymer 59 (2015) 290–297.

[144] M. Chen, X. Wu, F. Soyekwo, Q. Zhang, R. Lv, A. Zhu, Q. Liu, Toward improved hydrophilicity of polymers of intrinsic microporosity for pervaporation dehydration of ethylene glycol, Separation and Purification Technology 174 (2017) 166–173.

[145] B. Satilmis, P.M. Budd, Base-catalysed hydrolysis of PIM-1: amide versus carboxylate formation, RSC Advances 4 (2014) 52189–52198.

[146] B. Santoso, P. Yanaranop, H. Kang, I.K.H. Leung, J. Jin, A critical update on the synthesis of carboxylated polymers of intrinsic microporosity (C-PIMs), Macromolecules 50 (2017) 3043–3050.

[147] W.F. Yong, T.-S. Chung, Mechanically strong and flexible hydrolyzed polymers of intrinsic microporosity (PIM-1) membranes, Journal of Polymer Science Part B: Polymer Physics 55 (2017) 344–354.

[148] B. Satilmis, P.M. Budd, T. Uyar, Systematic hydrolysis of PIM-1 and electrospinning of hydrolyzed PIM-1 ultrafine fibers for an efficient removal of dye from water, Reactive and Functional Polymers 121 (2017) 67–75.

[149] B. Satilmis, T. Uyar, Superhydrophobic hexamethylene diisocyanate modified hydrolyzed polymers of intrinsic microporosity electrospun ultrafine fibrous membrane for the adsorption of organic compounds and oil/water separation, ACS Applied Nano Materials 1 (2018) 1631–1640.

[150] K.-S. Liao, J.-Y. Lai, T.-S. Chung, Metal ion modified PIM-1 and its application for propylene/propane separation, Journal of Membrane Science 515 (2016) 36–44.

[151] P. Yanaranop, B. Santoso, R. Etzion, J. Jin, Facile conversion of nitrile to amide on polymers of intrinsic microporosity (PIM-1), Polymer 98 (2016) 244–251.

[152] C.R. Mason, L. Maynard-Atem, N.M. Al-Harbi, P.M. Budd, P. Bernardo, F. Bazzarelli, G. Clarizia, J.C. Jansen, Polymer of intrinsic microporosity incorporating thioamide functionality: preparation and gas transport properties, Macromolecules 44 (2011) 6471–6479.

[153] N. Du, H.B. Park, G.P. Robertson, M.M. Dal-Cin, T. Visser, L. Scoles, M.D. Guiver, Polymer nanosieve membranes for CO_2-capture applications, Nature Materials 10 (2011) 372–375.

[154] J.K. Moore, M.D. Guiver, N. Du, S.E. Hayes, M.S. Conradi, Molecular motions of adsorbed CO_2 on a tetrazole-functionalized PIM polymer studied with ^{13}C NMR, The Journal of Physical Chemistry C 117 (2013) 22995–22999.

[155] N. Du, G.P. Robertson, M.M. Dal-Cin, L. Scoles, M.D. Guiver, Polymers of intrinsic microporosity (PIMs) substituted with methyl tetrazole, Polymer 53 (2012) 4367–4372.

[156] J.K. Moore, R.M. Marti, M.D. Guiver, N. Du, M.S. Conradi, S.E. Hayes, CO_2 adsorption on PIMs studied with ^{13}C NMR spectroscopy, The Journal of Physical Chemistry C 122 (2018) 4403–4408.

[157] C.R. Mason, L. Maynard-Atem, K.W.J. Heard, B. Satilmis, P.M. Budd, K. Friess, M. Lanc, P. Bernardo, G. Clarizia, J.C. Jansen, Enhancement of CO_2 affinity in a polymer of intrinsic microporosity by amine modification, Macromolecules 47 (2014) 1021–1029.

[158] B. Satilmis, P.M. Budd, Selective dye adsorption by chemically-modified and thermally-treated polymers of intrinsic microporosity, Journal of Colloid and Interface Science 492 (2017) 81–91.

[159] B. Satilmis, M. Lanc, A. Fuoco, C. Rizzuto, E. Tocci, P. Bernardo, G. Clarizia, E. Esposito, M. Monteleone, M. Dendisova, K. Friess, P.M. Budd, J.C. Jansen, Temperature and pressure dependence of gas permeation in amine-modified PIM-1, Journal of Membrane Science 555 (2018) 483–496.

[160] H.A. Patel, C.T. Yavuz, Noninvasive functionalization of polymers of intrinsic microporosity for enhanced CO_2 capture, Chemical Communications 48 (2012) 9989–9991.

[161] R. Swaidan, B.S. Ghanem, E. Litwiller, I. Pinnau, Pure- and mixed-gas CO_2/CH_4 separation properties of PIM-1 and an amidoxime-functionalized PIM-1, Journal of Membrane Science 457 (2014) 95–102.

[162] Y.H. Sihn, J. Byun, H.A. Patel, W. Lee, C.T. Yavuz, Rapid extraction of uranium ions from seawater using novel porous polymeric adsorbents, RSC Advances 6 (2016) 45968–45976.

[163] Z. Wang, H. Ren, S. Zhang, F. Zhang, J. Jin, Polymers of intrinsic microporosity/metal-organic framework hybrid membranes with improved interfacial interaction for high-performance CO_2 separation, Journal of Materials Chemistry A 5 (2017) 10968–10977.

[164] B. Satilmis, M.N. Alnajrani, P.M. Budd, Hydroxyalkylaminoalkylamide PIMs: selective adsorption by ethanolamine- and diethanolamine-modified PIM-1, Macromolecules 48 (2015) 5663–5669.

[165] B.G. Kim, D. Henkensmeier, H.-J. Kim, J.H. Jang, S.W. Nam, T.-H. Lim, Sulfonation of PIM-1—towards highly oxygen permeable binders for fuel cell application, Macromolecular Research 22 (2014) 92–98.

[166] N. Du, M.M. Dal-Cin, I. Pinnau, A. Nicalek, G.P. Robertson, M.D. Guiver, Azide-based cross-linking of polymers of intrinsic microporosity (PIMs) for condensable gas separation, Macromolecular Rapid Communications 32 (2011) 631–636.
[167] M.M. Khan, G. Bengtson, S. Shishatskiy, B.N. Gacal, M. Mushfequr Rahman, S. Neumann, V. Filiz, V. Abetz, Cross-linking of polymer of intrinsic microporosity (PIM-1) via nitrene reaction and its effect on gas transport property, European Polymer Journal 49 (2013) 4157–4166.
[168] T.O. McDonald, R. Akhtar, C.H. Lau, T. Ratvijitvech, G. Cheng, R. Clowes, D.J. Adams, T. Hasell, A.I. Cooper, Using intermolecular interactions to crosslink PIM-1 and modify its gas sorption properties, Journal of Materials Chemistry A 3 (2015) 4855–4864.
[169] D. Fritsch, P. Merten, K. Heinrich, M. Lazar, M. Priske, High performance organic solvent nanofiltration membranes: development and thorough testing of thin film composite membranes made of polymers of intrinsic microporosity (PIMs), Journal of Membrane Science 401–402 (2012) 222–231.
[170] F.Y. Li, Y. Xiao, T.-S. Chung, S. Kawi, High-performance thermally self-cross-linked polymer of intrinsic microporosity (PIM-1) membranes for energy development, Macromolecules 45 (2012) 1427–1437.
[171] Q. Song, S. Cao, K. Cheetham Anthony, H. Pritchard Robyn, M. Terentjev Eugene, B. Ghalei, A. Al-Muhtaseb Shaheen, E. Sivaniah, Controlled thermal oxidative crosslinking of polymers of intrinsic microporosity towards tunable molecular sieve membranes, Nature Communications 5 (2014) 4813.
[172] Q. Song, S. Cao, R.H. Pritchard, H. Qiblawey, E.M. Terentjev, A.K. Cheetham, E. Sivaniah, Nanofiller-tuned microporous polymer molecular sieves for energy and environmental processes, Journal of Materials Chemistry A 4 (2016) 270–279.
[173] C. Liu, S.T. Wilson, D.A. Lesch, UV-Crosslinked Membranes From Polymers of Intrinsic Microporosity for Liquid Separations and Their Production, UOP LLC, USA, July 20, 2010. Patent US 7,758,751 B1.
[174] F.Y. Li, Y. Xiao, Y.K. Ong, T.-S. Chung, UV-rearranged PIM-1 polymeric membranes for advanced hydrogen purification and production, Advanced Energy Materials 2 (2012) 1456–1466.
[175] Q. Song, S. Cao, P. Zavala-Rivera, L.P. Lu, W. Li, Y. Ji, S.A. Al-Muhtaseb, A.K. Cheetham, E. Sivaniah, Photo-oxidative enhancement of polymeric molecular sieve membranes, Nature Communications 4 (2013) 1918.
[176] H. Zhao, Q. Xie, X. Ding, J. Chen, M. Hua, X. Tan, Y. Zhang, High performance post-modified polymers of intrinsic microporosity (PIM-1) membranes based on multivalent metal ions for gas separation, Journal of Membrane Science 514 (2016) 305–312.
[177] S. Makhseed, F. Ibrahim, J. Samuel, Phthalimide based polymers of intrinsic microporosity, Polymer 53 (2012) 2964–2972.
[178] P.M. Budd, K.J. Msayib, C.E. Tattershall, B.S. Ghanem, K.J. Reynolds, N.B. McKeown, D. Fritsch, Gas separation membranes from polymers of intrinsic microporosity, Journal of Membrane Science 251 (2005) 263–269.
[179] B.S. Ghanem, N.B. McKeown, P.M. Budd, D. Fritsch, Polymers of intrinsic microporosity derived from bis(phenazyl) monomers, Macromolecules 41 (2008) 1640–1646.
[180] M.F. Jimenez-Solomon, Q. Song, K.E. Jelfs, M. Munoz-Ibanez, A.G. Livingston, Polymer nanofilms with enhanced microporosity by interfacial polymerization, Nature Materials 15 (2016) 760–767.
[181] F. Ibrahim, Synthesis of linear unbranched polymers chain free of macrocyclic species and oligomers based on chloro-monomers and TTSBI, Elixir International Journal (2013) 19054–19059, 19056 pp.
[182] P. Gorgojo, S. Karan, H.C. Wong, M.F. Jimenez-Solomon, J.T. Cabral, A.G. Livingston, Ultrathin polymer films with intrinsic microporosity: anomalous solvent permeation and high flux membranes, Advanced Functional Materials 24 (2014) 4729–4737.
[183] N. Du, G.P. Robertson, I. Pinnau, M.D. Guiver, Polymers of intrinsic microporosity derived from novel disulfone-based monomers, Macromolecules 42 (2009) 6023–6030.
[184] D. Fritsch, G. Bengtson, M. Carta, N.B. McKeown, Synthesis and gas permeation properties of spirobischromane-based polymers of intrinsic microporosity, Macromolecular Chemistry and Physics 212 (2011) 1137–1146.
[185] K. Yuan, C. Liu, S. Zhang, L. Jiang, C. Liu, G. Yu, J. Wang, X. Jian, Phthalazinone-based copolymers with intrinsic microporosity (PHPIMs) and their separation performance, Journal of Membrane Science 541 (2017) 403–412.
[186] G. Bengtson, S. Neumann, V. Filiz, Membranes of polymers of intrinsic microporosity (PIM-1) modified by poly(ethylene glycol), Membranes 7 (2017), 28/21-28/21.
[187] W.A. Elmehalmey, R.A. Azzam, Y.S. Hassan, M.H. Alkordi, T.M. Madkour, Imide-based polymers of intrinsic microporosity: probing the microstructure in relation to CO_2 sorption characteristics, ACS Omega 3 (2018) 2757–2764.
[188] X. Ma, R. Swaidan, B. Teng, H. Tan, O. Salinas, E. Litwiller, Y. Han, I. Pinnau, Carbon molecular sieve gas separation membranes based on an intrinsically microporous polyimide precursor, Carbon 62 (2013) 88–96.
[189] O. Salinas, X. Ma, E. Litwiller, I. Pinnau, High-performance carbon molecular sieve membranes for ethylene/ethane separation derived from an intrinsically microporous polyimide, Journal of Membrane Science 500 (2016) 115–123.
[190] R.J. Swaidan, X. Ma, I. Pinnau, Spirobisindane-based polyimide as efficient precursor of thermally-rearranged and carbon molecular sieve membranes for enhanced propylene/propane separation, Journal of Membrane Science 520 (2016) 983–989.
[191] O. Salinas, X. Ma, Y. Wang, Y. Han, I. Pinnau, Carbon molecular sieve membrane from a microporous spirobisindane-based polyimide precursor with enhanced ethylene/ethane mixed-gas selectivity, RSC Advances (2017) 3265–3272.
[192] X. Ma, O. Salinas, E. Litwiller, I. Pinnau, Pristine and thermally-rearranged gas separation membranes from novel o-hydroxyl-functionalized spirobifluorene-based polyimides, Polymer Chemistry 5 (2014) 6914–6922.
[193] M. Carta, K.J. Msayib, P.M. Budd, N.B. McKeown, Novel spirobisindanes for use as precursors to polymers of intrinsic microporosity, Organic Letters 10 (2008) 2641–2643.
[194] M. Carta, K.J. Msayib, N.B. McKeown, Novel polymers of intrinsic microporosity (PIMs) derived from 1,1-spiro-bis(1,2,3,4-tetrahydronaphthalene)-based monomers, Tetrahedron Letters 50 (2009) 5954–5957.
[195] J. Zhang, H. Kang, J. Martin, S. Zhang, S. Thomas, T.C. Merkel, J. Jin, The enhancement of chain rigidity and gas transport performance of polymers of intrinsic microporosity via intramolecular locking of the spiro-carbon, Chemical Communications 52 (2016) 6553–6556.
[196] Y. Rogan, R. Malpass-Evans, M. Carta, M. Lee, J.C. Jansen, P. Bernardo, G. Clarizia, E. Tocci, K. Friess, M. Lanc, N.B. McKeown, A highly permeable polyimide with enhanced selectivity for membrane gas separations, Journal of Materials Chemistry A 2 (2014) 4874–4877.
[197] X. Ma, I. Pinnau, Effect of film thickness and physical aging on "intrinsic" gas permeation properties of microporous ethanoanthracene-based polyimides, Macromolecules 51 (2018) 1069–1076.

[198] M.M. Khan, G. Bengtson, S. Neumann, M.M. Rahman, V. Abetz, V. Filiz, Synthesis, characterization and gas permeation properties of anthracene maleimide-based polymers of intrinsic microporosity, RSC Advances 4 (2014) 32148–32160.

[199] M.M. Khan, V. Filiz, T. Emmler, V. Abetz, T. Koschine, K. Raetzke, F. Faupel, W. Egger, L. Ravelli, Free volume and gas permeation in anthracene maleimide-based polymers of intrinsic microporosity, Membranes 5 (2015) 214–227.

[200] B.S. Ghanem, R. Swaidan, E. Litwiller, I. Pinnau, Ultra-microporous triptycene-based polyimide membranes for high-performance gas separation, Advanced Materials 26 (2014) 3688–3692.

[201] R. Swaidan, M. Al-Saeedi, B. Ghanem, E. Litwiller, I. Pinnau, Rational design of intrinsically ultramicroporous polyimides containing bridgehead-substituted triptycene for highly selective and permeable gas separation membranes, Macromolecules 47 (2014) 5104–5114.

[202] R. Swaidan, B. Ghanem, M. Al-Saeedi, E. Litwiller, I. Pinnau, Role of intrachain rigidity in the plasticization of intrinsically microporous triptycene-based polyimide membranes in mixed-gas CO_2/CH_4 separations, Macromolecules 47 (2014) 7453–7462.

[203] F. Alghunaimi, B. Ghanem, N. Alaslai, R. Swaidan, E. Litwiller, I. Pinnau, Gas permeation and physical aging properties of iptycene diamine-based microporous polyimides, Journal of Membrane Science 490 (2015) 321–327.

[204] F. Alghunaimi, B. Ghanem, N. Alaslai, M. Mukaddam, I. Pinnau, Triptycene dimethyl-bridgehead dianhydride-based intrinsically microporous hydroxyl-functionalized polyimide for natural gas upgrading, Journal of Membrane Science 520 (2016) 240–246.

[205] F. Alghunaimi, B. Ghanem, Y. Wang, O. Salinas, N. Alaslai, I. Pinnau, Synthesis and gas permeation properties of a novel thermally-rearranged polybenzoxazole made from an intrinsically microporous hydroxyl-functionalized triptycene-based polyimide precursor, Polymer 121 (2017) 9–16.

[206] I. Rose, C.G. Bezzu, M. Carta, B. Comesana-Gandara, E. Lasseuguette, M.C. Ferrari, P. Bernardo, G. Clarizia, A. Fuoco, J.C. Jansen, K.E. Hart, T.P. Liyana-Arachchi, C.M. Colina, N.B. McKeown, Polymer ultrapermeability from the inefficient packing of 2D chains, Nature Materials 16 (2017) 932–937.

[207] R. Swaidan, B. Ghanem, E. Litwiller, I. Pinnau, Effects of hydroxyl-functionalization and sub-T_g thermal annealing on high pressure pure- and mixed-gas CO_2/CH_4 separation by polyimide membranes based on 6FDA and triptycene-containing dianhydrides, Journal of Membrane Science 475 (2015) 571–581.

[208] N. Alaslai, B. Ghanem, F. Alghunaimi, I. Pinnau, High-performance intrinsically microporous dihydroxyl-functionalized triptycene-based polyimide for natural gas separation, Polymer 91 (2016) 128–135.

[209] B. Shrimant, S.V. Shaligram, U.K. Kharul, P.P. Wadgaonkar, Synthesis, characterization, and gas permeation properties of adamantane-containing polymers of intrinsic microporosity, Journal of Polymer Science Part A: Polymer Chemistry 56 (2018) 16–24.

[210] X. Ma, I. Pinnau, A novel intrinsically microporous ladder polymer and copolymers derived from 1,1′,2,2′-tetrahydroxy-tetraphenylethylene for membrane-based gas separation, Polymer Chemistry 7 (2016) 1244–1248.

[211] S. Sholl David, P. Lively Ryan, Seven chemical separations to change the world, Nature 532 (2016) 435–437.

[212] P. Agarwal, I. Tomlinson, R.E. Hefner Jr., S. Ge, Y. Rao, T. Dikic, Thin film composite membranes from polymers of intrinsic microporosity using layer-by-layer method, Journal of Membrane Science 572 (2019) 475–479.

[213] P. Bernardo, E. Drioli, G. Golemme, Membrane gas separation: a review/state of the art, Industrial and Engineering Chemistry Research 48 (2009) 4638–4663.

[214] E. Drioli, G. Barbieri (Eds.), *Membrane Engineering for the Treatment of Gases*, vol. 1, 2011.

[215] L.M. Robeson, Correlation of separation factor versus permeability for polymeric membranes, Journal of Membrane Science 62 (1991) 165–185.

[216] L.M. Robeson, The upper bound revisited, Journal of Membrane Science 320 (2008) 390–400.

[217] P. Bernardo, F. Bazzarelli, F. Tasselli, G. Clarizia, C.R. Mason, L. Maynard-Atem, P.M. Budd, M. Lanc, K. Pilnacek, O. Vopicka, K. Friess, D. Fritsch, Y.P. Yampolskii, V. Shantarovich, J.C. Jansen, Effect of physical aging on the gas transport and sorption in PIM-1 membranes, Polymer 113 (2017) 283–294.

[218] K. Pilnacek, O. Vopicka, M. Lanc, M. Dendisova, M. Zgazar, P.M. Budd, M. Carta, R. Malpass-Evans, N.B. McKeown, K. Friess, Aging of polymers of intrinsic microporosity tracked by methanol vapour permeation, Journal of Membrane Science 520 (2016) 895–906.

[219] A.G. McDermott, P.M. Budd, N.B. McKeown, C.M. Colina, J. Runt, Physical aging of polymers of intrinsic microporosity: a SAXS/WAXS study, Journal of Materials Chemistry A 2 (2014) 11742–11752.

[220] R. Swaidan, B. Ghanem, E. Litwiller, I. Pinnau, Physical aging, plasticization and their effects on gas permeation in "rigid" polymers of intrinsic microporosity, Macromolecules 48 (2015) 6553–6561.

[221] Z.-X. Low, P.M. Budd, N.B. McKeown, D.A. Patterson, Gas permeation properties, physical aging, and its mitigation in high free volume glassy polymers, Chemical Reviews 118 (2018) 5871–5911.

[222] C.A. Scholes, J. Jin, G.W. Stevens, S.E. Kentish, Hydrocarbon solubility, permeability, and competitive sorption effects in polymer of intrinsic microporosity (PIM-1) membranes, Journal of Polymer Science Part B: Polymer Physics 54 (2016) 397–404.

[223] E. Lasseuguette, M. Carta, S. Brandani, M.-C. Ferrari, Effect of humidity and flue gas impurities on CO_2 permeation of a polymer of intrinsic microporosity for post-combustion capture, International Journal of Greenhouse Gas Control 50 (2016) 93–99.

[224] X. Ma, R.J. Swaidan, Y. Wang, C.-E. Hsiung, Y. Han, I. Pinnau, Highly compatible hydroxyl-functionalized microporous polyimide-ZIF-8 mixed matrix membranes for energy efficient propylene/propane separation, ACS Applied Nano Materials (2018) 3541–3547.

[225] J. Liu, Y. Xiao, T.-S. Chung, Flexible thermally treated 3D PIM-CD molecular sieve membranes exceeding the upper bound line for propylene/propane separation, Journal of Materials Chemistry A 5 (2017) 4583–4595.

[226] O. Salinas, X. Ma, E. Litwiller, I. Pinnau, Ethylene/ethane permeation, diffusion and gas sorption properties of carbon molecular sieve membranes derived from the prototype ladder polymer of intrinsic microporosity (PIM-1), Journal of Membrane Science 504 (2016) 133–140.

[227] K.-S. Liao, S. Japip, J.-Y. Lai, T.-S. Chung, Boron-embedded hydrolyzed PIM-1 carbon membranes for synergistic ethylene/ethane purification, Journal of Membrane Science 534 (2017) 92–99.

[228] W.F. Yong, F.Y. Li, Y.C. Xiao, P. Li, K.P. Pramoda, Y.W. Tong, T.S. Chung, Molecular engineering of PIM-1/Matrimid blend membranes for gas separation, Journal of Membrane Science 407–408 (2012) 47–57.

[229] W.F. Yong, F.Y. Li, T.-S. Chung, Y.W. Tong, Highly permeable chemically modified PIM-1/Matrimid membranes for green hydrogen purification, Journal of Materials Chemistry A 1 (2013) 13914–13925.

[230] W.F. Yong, F.Y. Li, Y.C. Xiao, T.S. Chung, Y.W. Tong, High performance PIM-1/Matrimid hollow fiber membranes for CO_2/CH_4, O_2/N_2 and CO_2/N_2 separation, Journal of Membrane Science 443 (2013) 156–169.
[231] L. Hao, P. Li, T.-S. Chung, PIM-1 as an organic filler to enhance the gas separation performance of Ultem polyetherimide, Journal of Membrane Science 453 (2014) 614–623.
[232] L. Hao, J. Zuo, T.-S. Chung, Formation of defect-free polyetherimide/PIM-1 hollow fiber membranes for gas separation, AIChE Journal 60 (2014) 3848–3858.
[233] W.F. Yong, Z.K. Lee, T.-S. Chung, M. Weber, C. Staudt, C. Maletzko, Blends of a polymer of intrinsic microporosity and partially sulfonated polyphenylenesulfone for gas separation, ChemSusChem 9 (2016) 1953–1962.
[234] M.W. Xin, G.Z. Qiu, J.L. Peng, Y. Qu, M.Z. Ai, L.L. Qing, Towards enhanced CO_2 selectivity of the PIM-1 membrane by blending with polyethylene glycol, Journal of Membrane Science 493 (2015) 147–155.
[235] A.K. Sekizkardes, V.A. Kusuma, G. Dahe, E.A. Roth, L.J. Hill, A. Marti, M. Macala, S.R. Venna, D. Hopkinson, Separation of carbon dioxide from flue gas by mixed matrix membranes using dual phase microporous polymeric constituents, Chemical Communications 52 (2016) 11768–11771.
[236] K. Halder, M.M. Khan, J. Gruenauer, S. Shishatskiy, C. Abetz, V. Filiz, V. Abetz, Blend membranes of ionic liquid and polymers of intrinsic microporosity with improved gas separation characteristics, Journal of Membrane Science 539 (2017) 368–382.
[237] R. Castro-Munoz, V. Fila, C.T. Dung, Mixed matrix membranes based on PIMs for gas permeation: principles, synthesis, and current status, Chemical Engineering Communications 204 (2017) 295–309.
[238] C.R. Mason, M.G. Buonomenna, G. Golemme, P.M. Budd, F. Galiano, A. Figoli, K. Friess, V. Hynek, New organophilic mixed matrix membranes derived from a polymer of intrinsic microporosity and silicalite-1, Polymer 54 (2013) 2222–2230.
[239] L. Zhao, D. Zhai, B. Liu, Z. Liu, C. Xu, W. Wei, Y. Chen, J. Gao, Grand Canonical Monte Carlo simulations for energy gases on PIM-1 polymer and silicalite-1, Chemical Engineering Science 68 (2012) 101–107.
[240] A.F. Bushell, M.P. Attfield, C.R. Mason, P.M. Budd, Y. Yampolskii, L. Starannikova, A. Rebrov, F. Bazzarelli, P. Bernardo, J. Carolus Jansen, M. Lanc, K. Friess, V. Shantarovich, V. Gustov, V. Isaeva, Gas permeation parameters of mixed matrix membranes based on the polymer of intrinsic microporosity PIM-1 and the zeolitic imidazolate framework ZIF-8, Journal of Membrane Science 427 (2013) 48–62.
[241] R. Semino, N.A. Ramsahye, A. Ghoufi, G. Maurin, Microscopic model of the metal-organic framework/polymer interface: a first step toward understanding the compatibility in mixed matrix membranes, ACS Applied Materials and Interfaces 8 (2016) 809–819.
[242] M. Benzaqui, R. Semino, N. Menguy, F. Carn, T. Kundu, J.-M. Guigner, N.B. McKeown, K.J. Msayib, M. Carta, R. Malpass-Evans, C. Le Guillouzer, G. Clet, N.A. Ramsahye, C. Serre, G. Maurin, N. Steunou, Toward an understanding of the microstructure and interfacial properties of PIMs/ZIF-8 mixed matrix membranes, ACS Applied Materials and Interfaces 8 (2016) 27311–27321.
[243] R. Semino, N.A. Ramsahye, A. Ghoufi, G. Maurin, Role of MOF surface defects on the microscopic structure of MOF/polymer interfaces: a computational study of the ZIF-8/PIMs systems, Microporous Mesoporous Materials 254 (2017) 184–191.
[244] X. Wu, W. Liu, H. Wu, X. Zong, L. Yang, Y. Wu, Y. Ren, C. Shi, S. Wang, Z. Jiang, Nanoporous ZIF-67 embedded polymers of intrinsic microporosity membranes with enhanced gas separation performance, Journal of Membrane Science 548 (2018) 309–318.
[245] L. Hao, K.-S. Liao, T.-S. Chung, Photo-oxidative PIM-1 based mixed matrix membranes with superior gas separation performance, Journal of Materials Chemistry A 3 (2015) 17273–17281.
[246] A.Y. Alentiev, G.N. Bondarenko, Y.V. Kostina, V.P. Shantarovich, S.N. Klyamkin, V.P. Fedin, K.A. Kovalenko, Y.P. Yampolskii, PIM-1/MIL-101 hybrid composite membrane material: transport properties and free volume, Petroleum Chemistry 54 (2014) 477–481.
[247] M. Khdhayyer, A.F. Bushell, P.M. Budd, M.P. Attfield, D. Jiang, A.D. Burrows, E. Esposito, P. Bernardo, M. Monteleone, A. Fuoco, G. Clarizia, F. Bazzarelli, A. Gordano, J.C. Jansen, Mixed matrix membranes based on MIL-101 metal-organic frameworks in polymer of intrinsic microporosity PIM-1, Separation and Purification Technology 212 (2019) 545–554.
[248] S.J.D. Smith, B.P. Ladewig, A.J. Hill, C.H. Lau, M.R. Hill, Post-synthetic Ti exchanged UiO-66 metal-organic frameworks that deliver exceptional gas permeability in mixed matrix membranes, Scientific Reports 5 (2015) 7823.
[249] N. Tien-Binh, D. Rodrigue, S. Kaliaguine, In-situ cross interface linking of PIM-1 polymer and UiO-66-NH2 for outstanding gas separation and physical aging control, Journal of Membrane Science 548 (2018) 429–438.
[250] C.G. Morris, N.M. Jacques, H.G.W. Godfrey, T. Mitra, D. Fritsch, Z. Lu, C.A. Murray, J. Potter, T.M. Cobb, F. Yuan, C.C. Tang, S. Yang, M. Schroder, Stepwise observation and quantification and mixed matrix membrane separation of CO_2 within a hydroxy-decorated porous host, Chemical Science 8 (2017) 3239–3248.
[251] Y. Cheng, S.R. Tavares, C.M. Doherty, Y. Ying, E. Sarnello, G. Maurin, M.R. Hill, T. Li, D. Zhao, Enhanced polymer crystallinity in mixed-matrix membranes induced by metal-organic framework nanosheets for efficient CO_2 capture, ACS Applied Materials and Interfaces 10 (2018) 43095–43103.
[252] J. Ma, Y. Ying, X. Guo, H. Huang, D. Liu, C. Zhong, Fabrication of mixed-matrix membrane containing metal-organic framework composite with task-specific ionic liquid for efficient CO_2 separation, Journal of Materials Chemistry A 4 (2016) 7281–7288.
[253] A.F. Bushell, P.M. Budd, M.P. Attfield, J.T.A. Jones, T. Hasell, A.I. Cooper, P. Bernardo, F. Bazzarelli, G. Clarizia, J.C. Jansen, Nanoporous organic polymer/cage composite membranes, Angewandte Chemie International Edition 52 (2013) 1253–1256.
[254] X. Wu, Z. Tian, S. Wang, D. Peng, L. Yang, Y. Wu, Q. Xin, H. Wu, Z. Jiang, Mixed matrix membranes comprising polymers of intrinsic microporosity and covalent organic framework for gas separation, Journal of Membrane Science 528 (2017) 273–283.
[255] C.H. Lau, P.T. Nguyen, M.R. Hill, A.W. Thornton, K. Konstas, C.M. Doherty, R.J. Mulder, L. Bourgeois, A.C.Y. Liu, D.J. Sprouster, J.P. Sullivan, T.J. Bastow, A.J. Hill, D.L. Gin, R.D. Noble, Ending aging in super glassy polymer membranes, Angewandte Chemie International Edition 53 (2014) 5322–5326.
[256] C.H. Lau, K. Konstas, A.W. Thornton, A.C.Y. Liu, S. Mudie, D.F. Kennedy, S.C. Howard, A.J. Hill, M.R. Hill, Gas-separation membranes loaded with porous aromatic frameworks that improve with age, Angewandte Chemie International Edition 54 (2015) 2669–2673.
[257] S.J.D. Smith, C.H. Lau, J.I. Mardel, M. Kitchin, K. Konstas, B.P. Ladewig, M.R. Hill, Physical aging in glassy mixed matrix membranes; tuning particle interaction for mechanically robust nanocomposite films, Journal of Materials Chemistry A 4 (2016) 10627–10634.
[258] T. Mitra, R.S. Bhavsar, D.J. Adams, P.M. Budd, A.I. Cooper, PIM-1 mixed matrix membranes for gas separations using cost-effective hypercrosslinked nanoparticle fillers, Chemical Communications 52 (2016) 5581–5584.

[259] D. Chen, S. Gu, Y. Fu, X. Fu, Y. Zhang, G. Yu, C. Pan, Hyper-crosslinked aromatic polymers with improved microporosity for enhanced CO_2/N_2 and CO_2/CH_4 selectivity, New Journal of Chemistry 41 (2017) 6834–6839.

[260] R.S. Bhavsar, T. Mitra, D.J. Adams, A.I. Cooper, P.M. Budd, Ultrahigh-permeance PIM-1 based thin film nanocomposite membranes on PAN supports for CO_2 separation, Journal of Membrane Science 564 (2018) 878–886.

[261] J. Ahn, W.-J. Chung, I. Pinnau, J. Song, N. Du, G.P. Robertson, M.D. Guiver, Gas transport behavior of mixed-matrix membranes composed of silica nanoparticles in a polymer of intrinsic microporosity (PIM-1), Journal of Membrane Science 346 (2010) 280–287.

[262] M.G. De Angelis, R. Gaddoni, G.C. Sarti, Gas solubility, diffusivity, permeability, and selectivity in mixed matrix membranes based on PIM-1 and fumed silica, Industrial and Engineering Chemistry Research 52 (2013) 10506–10520.

[263] N. Konnertz, Y. Ding, W.J. Harrison, P.M. Budd, A. Schoenhals, M. Boehning, Molecular mobility and gas transport properties of nanocomposites based on PIM-1 and polyhedral oligomeric phenethyl-silsesquioxanes (POSS), Journal of Membrane Science 529 (2017) 274–285.

[264] M.M. Khan, S. Shishatskiy, V. Filiz, Mixed matrix membranes of boron cosahedron and polymers of intrinsic microporosity (PIM-1) for gas separation, Membranes 8 (2018) 1/1–1/18.

[265] A. Gonciaruk, K. Althumayri, W.J. Harrison, P.M. Budd, F.R. Siperstein, PIM-1/graphene composite: a combined experimental and molecular simulation study, Microporous Mesoporous Materials 209 (2015) 126–134.

[266] K. Althumayri, W.J. Harrison, Y. Shin, J. Gardiner, C. Casiraghi, P.M. Budd, P. Bernardo, G. Clarizia, J.C. Jansen, The influence of few-layer graphene on the gas permeability of the high-free-volume polymer PIM-1, Philosophical Transactions of the Royal Society A 374 (2016), 20150031/20150031–20150031/20150039.

[267] Y. Shin, E. Prestat, K.-G. Zhou, P. Gorgojo, K. Althumayri, W. Harrison, P.M. Budd, S.J. Haigh, C. Casiraghi, Synthesis and characterization of composite membranes made of graphene and polymers of intrinsic microporosity, Carbon 102 (2016) 357–366.

[268] M. Alberto, R. Bhavsar, J.M. Luque-Alled, A. Vijayaraghavan, P.M. Budd, P. Gorgojo, Impeded physical aging in PIM-1 membranes containing graphene-like fillers, Journal of Membrane Science 563 (2018) 513–520.

[269] Z. Tian, S. Wang, Y. Wang, X. Ma, K. Cao, D. Peng, X. Wu, H. Wu, Z. Jiang, Enhanced gas separation performance of mixed matrix membranes from graphitic carbon nitride nanosheets and polymers of intrinsic microporosity, Journal of Membrane Science 514 (2016) 15–24.

[270] M. Chen, F. Soyekwo, Q. Zhang, C. Hu, A. Zhu, Q. Liu, Graphene oxide nanosheets to improve permeability and selectivity of PIM-1 membrane for carbon dioxide separation, Journal of Industrial and Engineering Chemistry 63 (2018) 296–302.

[271] M.M. Khan, V. Filiz, G. Bengtson, S. Shishatskiy, M. Rahman, V. Abetz, Functionalized carbon nanotubes mixed matrix membranes of polymers of intrinsic microporosity for gas separation, Nanoscale Research Letters 7 (2012) 1–12.

[272] M.M. Khan, V. Filiz, G. Bengtson, S. Shishatskiy, M.M. Rahman, J. Lillepaerg, V. Abetz, Enhanced gas permeability by fabricating mixed matrix membranes of functionalized multiwalled carbon nanotubes and polymers of intrinsic microporosity (PIM), Journal of Membrane Science 436 (2013) 109–120.

[273] K. Golzar, H. Modarress, S. Amjad-Iranagh, Effect of pristine and functionalized single- and multi-walled carbon nanotubes on CO_2 separation of mixed matrix membranes based on polymers of intrinsic microporosity (PIM-1): a molecular dynamics simulation study, Journal of Molecular Modeling 23 (2017) 1–25.

[274] S.M. Meckler, C. Li, W.L. Queen, T.E. Williams, J.R. Long, R. Buonsanti, D.J. Milliron, B.A. Helms, Sub-micron polymer-zeolitic imidazolate framework layered hybrids via controlled chemical transformation of naked ZnO nanocrystal films, Chemical Materials 27 (2015) 7673–7679.

[275] A. Fuoco, M.R. Khdhayyer, M.P. Attfield, E. Esposito, J.C. Jansen, P.M. Budd, Synthesis and transport properties of novel MOF/PIM-1/MOF sandwich membranes for gas separation, Membranes 7 (2017) 7/1–7/17.

[276] E. Chernova, D. Petukhov, O. Boytsova, A. Alentiev, P. Budd, Y. Yampolskii, A. Eliseev, Enhanced gas separation factors of microporous polymer constrained in the channels of anodic alumina membranes, Scientific Reports 6 (2016) 31183.

[277] M.L. Jue, V. Breedveld, R.P. Lively, Defect-free PIM-1 hollow fiber membranes, Journal of Membrane Science 530 (2017) 33–41.

[278] T. Koschine, K. Raetzke, F. Faupel, M.M. Khan, T. Emmler, V. Filiz, V. Abetz, L. Ravelli, W. Egger, Correlation of gas permeation and free volume in new and used high free volume thin film composite membranes, Journal of Polymer Science Part B: Polymer Physics 53 (2015) 213–217.

[279] C.Z. Liang, J.T. Liu, J.-Y. Lai, T.-S. Chung, High-performance multiple-layer PIM composite hollow fiber membranes for gas separation, Journal of Membrane Science 563 (2018) 93–106.

[280] J. Benito, J. Sanchez-Lainez, B. Zornoza, S. Martin, M. Carta, R. Malpass-Evans, C. Tellez, N.B. McKeown, J. Coronas, I. Gascon, Ultrathin composite polymeric membranes for CO_2/N_2 separation with minimum thickness and high CO_2 permeance, ChemSusChem 10 (2017) 4014–4017.

[281] J. Benito, J. Vidal, J. Sanchez-Lainez, B. Zornoza, C. Tellez, S. Martin, K.J. Msayib, B. Comesana-Gandara, N.B. McKeown, J. Coronas, I. Gascon, The fabrication of ultrathin films and their gas separation performance from polymers of intrinsic microporosity with two-dimensional (2D) and three-dimensional (3D) chain conformations, Journal of Colloid and Interface Science 536 (2019) 474–482.

[282] C.A. Scholes, S.E. Kentish, G.W. Stevens, D. de Montigny, Comparison of thin film composite and microporous membrane contactors for CO_2 absorption into monoethanolamine, International Journal of Greenhouse Gas Control 42 (2015) 66–74.

[283] C.A. Scholes, S.E. Kentish, G.W. Stevens, J. Jin, D. de Montigny, Thin-film composite membrane contactors for desorption of CO_2 from Monoethanolamine at elevated temperatures, Separation and Purification Technology 156 (2015) 841–847.

[284] Y.K. Ong, G.M. Shi, N.L. Le, Y.P. Tang, J. Zuo, S.P. Nunes, T.-S. Chung, Recent membrane development for pervaporation processes, Progress in Polymer Science 57 (2016) 1–31.

[285] R.W. Baker, J.G. Wijmans, Y. Huang, Permeability, permeance and selectivity: a preferred way of reporting pervaporation performance data, Journal of Membrane Science 348 (2010) 346–352.

[286] J.G. Wijmans, Process performance = membrane properties + operating conditions, Journal of Membrane Science 220 (2003) 1–3.

[287] K.W. Boeddeker, Terminology in pervaporation, Journal of Membrane Science 51 (1990) 259–272.

[288] S.V. Adymkanov, Y.P. Yampol'skii, A.M. Polyakov, P.M. Budd, K.J. Reynolds, N.B. McKeown, K.J. Msayib, Pervaporation of alcohols through highly permeable PIM-1 polymer films, Vysokomolekulyarnye Soedineniya, Seriya A Seriya B 50 (2008) 678–685.

[289] X.M. Wu, Q.G. Zhang, F. Soyekwo, Q.L. Liu, A.M. Zhu, Pervaporation removal of volatile organic compounds from aqueous solutions using the highly permeable PIM-1 membrane, AIChE Journal 62 (2016) 842–851.

[290] M. Alberto, J.M. Luque-Alled, L. Gao, M. Iliut, E. Prestat, L. Newman, S.J. Haigh, A. Vijayaraghavan, P.M. Budd, P. Gorgojo, Enhanced organophilic separations with mixed matrix membranes of polymers of intrinsic microporosity and graphene-like fillers, Journal of Membrane Science 526 (2017) 437–449.

[291] M. Zak, M. Klepic, L.C. Stastna, Z. Sedlakova, H. Vychodilova, S. Hovorka, K. Friess, A. Randova, L. Brozova, J.C. Jansen, M.R. Khdhayyer, P.M. Budd, P. Izak, Selective removal of butanol from aqueous solution by pervaporation with a PIM-1 membrane and membrane aging, Separation and Purification Technology 151 (2015) 108–114.

[292] S. van Wyk, A.G.J. van der Ham, S.R.A. Kersten, Pervaporative separation and intensification of downstream recovery of acetone-butanol-ethanol (ABE), Chemical Engineering and Processing 130 (2018) 148–159.

[293] L. Gao, M. Alberto, P. Gorgojo, G. Szekely, P.M. Budd, High-flux PIM-1/PVDF thin film composite membranes for 1-butanol/water pervaporation, Journal of Membrane Science 529 (2017) 207–214.

[294] M. Alberto, R. Bhavsar, J.M. Luque-Alled, E. Prestat, L. Gao, P.M. Budd, A. Vijayaraghavan, G. Szekely, S.M. Holmes, P. Gorgojo, Study on the formation of thin film nanocomposite (TFN) membranes of polymers of intrinsic microporosity and graphene-like fillers: effect of lateral flake size and chemical functionalization, Journal of Membrane Science 565 (2018) 390–401.

[295] Y. Lan, P. Peng, Preparation of polymer of intrinsic microporosity composite membranes and their applications for butanol recovery, Journal of Applied Polymer Science 136 (2019) 46912.

[296] Y. Lan, P. Peng, P. Chen, Preparation of polymers of intrinsic microporosity composite membranes incorporated with modified nano-fumed silica for butanol separation, Polymers for Advanced Technologies 37 (2018) 3297–3304.

[297] W.F. Yong, P. Salehian, L. Zhang, T.-S. Chung, Effects of hydrolyzed PIM-1 in polyimide-based membranes on C2-C4 alcohols dehydration via pervaporation, Journal of Membrane Science 523 (2017) 430–438.

[298] X.M. Wu, H. Guo, F. Soyekwo, Q.G. Zhang, C.X. Lin, Q.L. Liu, A.M. Zhu, Pervaporation purification of ethylene glycol using the highly permeable PIM-1 membrane, Journal of Chemical and Engineering Data 61 (2016) 579–586.

[299] S. Tsarkov, V. Khotimskiy, P.M. Budd, V. Volkov, J. Kukushkina, A. Volkov, Solvent nanofiltration through high permeability glassy polymers: effect of polymer and solute nature, Journal of Membrane Science 423–424 (2012) 65–72.

[300] T.S. Anokhina, A.A. Yushkin, P.M. Budd, A.V. Volkov, Application of PIM-1 for solvent swing adsorption and solvent recovery by nanofiltration, Separation and Purification Technology 156 (2015) 683–690.

[301] K.M. Gupta, Q. Shi, L. Sarkisov, J. Jiang, Ethanolamine purification by nanofiltration through PIM-1 and carbon membranes: a molecular simulation study, The Journal of Physical Chemistry C 121 (2017) 20539–20545.

[302] G. Ignacz, F. Fei, G. Szekely, Ion-stabilized membranes for demanding environments fabricated from polybenzimidazole and its blends with polymers of intrinsic microporosity, ACS Applied Nano Materials 1 (2018) 6349–6356.

[303] M. Cook, P.R.J. Gaffney, L.G. Peeva, A.G. Livingston, Roll-to-roll dip coating of three different PIMs for Organic Solvent Nanofiltration, Journal of Membrane Science 558 (2018) 52–63.

[304] H.J. Kim, D.-G. Kim, K. Lee, Y. Baek, Y. Yoo, Y.S. Kim, B.G. Kim, J.-C. Lee, A carbonaceous membrane based on a polymer of intrinsic microporosity (PIM-1) for water treatment, Scientific Reports 6 (2016) 36078.

[305] P. Li, T.S. Chung, D.R. Paul, Gas sorption and permeation in PIM-1, Journal of Membrane Science 432 (2013) 50–57.

[306] P. Li, T.S. Chung, D.R. Paul, Temperature dependence of gas sorption and permeation in PIM-1, Journal of Membrane Science 450 (2014) 380–388.

[307] O. Vopicka, K. Friess, V. Hynek, P. Sysel, M. Zgazar, M. Sipek, K. Pilnacek, M. Lanc, J.C. Jansen, C.R. Mason, P.M. Budd, Equilibrium and transient sorption of vapours and gases in the polymer of intrinsic microporosity PIM-1, Journal of Membrane Science 434 (2013) 148–160.

[308] O. Vopicka, M.G. De Angelis, N. Du, N. Li, M.D. Guiver, G.C. Sarti, Mixed gas sorption in glassy polymeric membranes: II. CO_2/CH_4 mixtures in a polymer of intrinsic microporosity (PIM-1), Journal of Membrane Science 459 (2014) 264–276.

[309] O. Vopicka, K. Friess, Analysis of gas sorption in glassy polymers with the GAB model: an alternative to the dual mode sorption model, Journal of Polymer Science Part B: Polymer Physics 52 (2014) 1490–1495.

[310] M. Lanc, K. Pilnacek, C.R. Mason, P.M. Budd, Y. Rogan, R. Malpass-Evans, M. Carta, B.C. Gandara, N.B. McKeown, J.C. Jansen, O. Vopicka, K. Friess, Gas sorption in polymers of intrinsic microporosity: the difference between solubility coefficients determined via time-lag and direct sorption experiments, Journal of Membrane Science 570–571 (2019) 522–536.

[311] M. Minelli, K. Friess, O. Vopicka, M.G. De Angelis, Modeling gas and vapor sorption in a polymer of intrinsic microporosity (PIM-1), Fluid Phase Equilibria 347 (2013) 35–44.

[312] E. Mangano, S. Brandani, M.C. Ferrari, H. Ahn, D. Friedrich, M.L. Lozinska, P.A. Wright, J. Kahr, R. Morris, M. Croad, N. McKeown, H. Shamsipour, P. Budd, Efficient and rapid screening of novel adsorbents for carbon capture in the UK IGSCC project, Energy Procedia 37 (2013) 40–47.

[313] S.H. Pang, M.L. Jue, J. Leisen, C.W. Jones, R.P. Lively, PIM-1 as a solution-processable "molecular basket" for CO_2 capture from dilute sources, ACS Macro Letters 4 (2015) 1415–1419.

[314] S. Makhseed, J. Samuel, A. Bumajdad, M. Abdalla, Preparation and hydrogen sorption studies on high surface area organic framework polymer, Kuwait Journal of Science and Engineering 35 (2008) 37–50.

[315] S. Makhseed, J. Samuel, A. Bumajdad, M. Hassan, Synthesis and characterization of fluoropolymers with intrinsic microporosity and their hydrogen adsorption studies, Journal of Applied Polymer Science 109 (2008) 2591–2597.

[316] J. Weber, M. Antonietti, A. Thomas, Microporous networks of high-performance polymers: elastic deformations and gas sorption properties, Macromolecules 41 (2008) 2880–2885.

[317] F. Ibrahim, Synthesis and characterization of linear polyimides with intrinsic microporosity and their hydrogen adsorption studies, Elixir International Journal (2013) 17542–17548.

[318] K. Polak-Krasna, R. Dawson, L.T. Holyfield, C.R. Bowen, A.D. Burrows, T.J. Mays, Mechanical characterisation of polymer of intrinsic microporosity PIM-1 for hydrogen storage applications, Journal of Materials Science 52 (2017) 3862–3875.

[319] K. Polak-Krasna, C. Fuhrhop, S. Rochat, A.D. Burrows, A. Georgiadis, C.R. Bowen, T.J. Mays, AFM imaging and nanoindentation of polymer of intrinsic microporosity PIM-1, International Journal of Hydrogen Energy 42 (2017) 23915–23919.

[320] S. Rochat, K. Polak-Krasna, M. Tian, L.T. Holyfield, T.J. Mays, C.R. Bowen, A.D. Burrows, Hydrogen storage in polymer-based processable microporous composites, Journal of Materials Chemistry A 5 (2017) 18752–18761.

[321] R. Kato, H. Nishide, Polymers for carrying and storing hydrogen, Polymer Journal 50 (2018) 77–82.

[322] S. Rochat, K. Polak-Krasna, M. Tian, T.J. Mays, C.R. Bowen, A.D. Burrows, Assessment of the long-term stability of the polymer of intrinsic microporosity PIM-1 for hydrogen storage applications, International Journal of Hydrogen Energy 44 (2019) 332–337.

[323] N.A. Rakow, M.S. Wendland, J.E. Trend, R.J. Poirier, D.M. Paolucci, S.P. Maki, C.S. Lyons, M.J. Swierczek, Visual indicator for trace organic volatiles, Langmuir 26 (2010) 3767–3770.

[324] J. Christopher Thomas, J.E. Trend, N.A. Rakow, M.S. Wendland, R.J. Poirier, D.M. Paolucci, Optical sensor for diverse organic vapors at ppm concentration ranges, Sensors 11 (2011) 3267–3280.

[325] C.A. Jeffs, M.W. Smith, C.A. Stone, C.G. Bezzu, K.J. Msayib, N.B. McKeown, S.P. Perera, A polymer of intrinsic microporosity as the active binder to enhance adsorption/separation properties of composite hollow fibres, Microporous Mesoporous Materials 170 (2013) 105–112.

[326] F. Zhang, Y. Ma, J. Liao, V. Breedveld, R.P. Lively, Solution-based 3D printing of polymers of intrinsic microporosity, Macromolecular Rapid Communications 39 (2018) 1800274.

[327] M. Razali, J.F. Kim, M. Attfield, P.M. Budd, E. Drioli, Y.M. Lee, G. Szekely, Sustainable wastewater treatment and recycling in membrane manufacturing, Green Chemistry 17 (2015) 5196–5205.

[328] J.S. Bonso, G.D. Kalaw, J.P. Ferraris, High surface area carbon nanofibers derived from electrospun PIM-1 for energy storage applications, Journal of Materials Chemistry A 2 (2014) 418–424.

[329] K.M. Skupov, I.I. Ponomarev, D.Y. Razorenov, V.G. Zhigalina, O.M. Zhigalina, I.I. Ponomarev, Y.A. Volkova, Y.M. Volfkovich, V.E. Sosenkin, Carbon nanofiber paper electrodes based on heterocyclic polymers for high temperature polymer electrolyte membrane fuel cell, Macromolecular Symposia 375 (2017), 1600188/1600181–1600188/1600186.

[330] E. Lasseuguette, M.-C. Ferrari, Development of microporous electrospun PIM-1 fibres, Materials Letters 177 (2016) 116–119.

[331] C. Zhang, P. Li, B. Cao, Electrospun polymer of intrinsic microporosity fibers and their use in the adsorption of contaminants from a nonaqueous system, Journal of Applied Polymer Science 133 (2016) 43475.

[332] B. Satilmis, T. Uyar, Removal of aniline from air and water by polymers of intrinsic microporosity (PIM-1) electrospun ultrafine fibers, Journal of Colloid and Interface Science 516 (2018) 317–324.

[333] Y. Pan, L. Zhang, Z. Li, L. Ma, Y. Zhang, J. Wang, J. Meng, Hierarchical porous membrane via electrospinning PIM-1 for micropollutants removal, Applied Surface Science 443 (2018) 441–451.

[334] C. Zhang, P. Li, B. Cao, Electrospun microfibrous membranes based on PIM-1/POSS with high oil wettability for separation of oil-water mixtures and cleanup of oil soluble contaminants, Industrial and Engineering Chemical Research 54 (2015) 8772–8781.

[335] K. Halder, G. Bengtson, V. Filiz, V. Abetz, Catalytically active (Pd) nanoparticles supported by electrospun PIM-1: influence of the sorption capacity of the polymer tested in the reduction of some aromatic nitro compounds, Applied Catalysis A 555 (2018) 178–188.

[336] B. Satilmis, T. Uyar, Amine modified electrospun PIM-1 ultrafine fibers for an efficient removal of methyl orange from an aqueous system, Applied Surface Science 453 (2018) 220–229.

[337] C. Zhang, P. Li, W. Huang, B. Cao, Selective adsorption and separation of organic dyes in aqueous solutions by hydrolyzed PIM-1 microfibers, Chemical Engineering Research and Design 109 (2016) 76–85.

[338] C. Zhang, P. Li, B. Cao, Fabrication of superhydrophobic-superoleophilic fabrics by an etching and dip-coating two-step method for oil-water separation, Industrial and Engineering Chemical Research 55 (2016) 5030–5035.

[339] X. Weng, J.E. Baez, M. Khiterer, M.Y. Hoe, Z. Bao, K.J. Shea, Chiral polymers of intrinsic microporosity: selective membrane permeation of enantiomers, Angewandte Chemie International Edition 54 (2015) 11214–11218.

[340] S.T. Hobson, S. Cemalovic, S.V. Patel, Preconcentration and detection of chlorinated organic compounds and benzene, Analyst 137 (2012) 1284–1289.

[341] S.-Y. Son, Y.-J. Noh, C. Bok, S. Lee, B.G. Kim, S.-I. Na, H.-I. Joh, One-step synthesis of carbon nanosheets converted from a polycyclic compound and their direct use as transparent electrodes of ITO-free organic solar cells, Nanoscale 6 (2014) 678–682.

[342] B.K. Gupta, G. Kedawat, P. Kumar, M.A. Rafiee, P. Tyagi, R. Srivastava, P.M. Ajayan, An n-type, new emerging luminescent polybenzodioxane polymer for application in solution-processed green emitting OLEDs, Journal of Materials Chemistry C 3 (2015) 2568–2574.

[343] S.E. Doris, A.L. Ward, A. Baskin, P.D. Frischmann, N. Gavvalapalli, E. Chenard, C.S. Sevov, D. Prendergast, J.S. Moore, B.A. Helms, Macromolecular design strategies for preventing active-material crossover in non-aqueous all-organic redox-flow batteries, Angewandte Chemie International Edition 56 (2017) 1595–1599.

[344] C. Li, A.L. Ward, S.E. Doris, T.A. Pascal, D. Prendergast, B.A. Helms, Polysulfide-Blocking microporous polymer membrane tailored for hybrid Li-sulfur flow batteries, Nano Letters 15 (2015) 5724–5729.

[345] S.E. Doris, A.L. Ward, P.D. Frischmann, L. Li, B.A. Helms, Understanding and controlling the chemical evolution and polysulfide-blocking ability of lithium-sulfur battery membranes cast from polymers of intrinsic microporosity, Journal of Materials Chemistry A 4 (2016) 16946–16952.

[346] A.L. Ward, S.E. Doris, L. Li, M.A. Hughes, X. Qu, K.A. Persson, B.A. Helms, Materials genomics screens for adaptive ion transport behavior by redox-switchable microporous polymer membranes in lithium-sulfur batteries, ACS Central Science 3 (2017) 399–406.

[347] K.H. Hendriks, S.G. Robinson, M.N. Braten, C.S. Sevov, B.A. Helms, M.S. Sigman, S.D. Minteer, M.S. Sanford, High-performance oligomeric catholytes for effective macromolecular separation in nonaqueous redox flow batteries, ACS Central Science 4 (2018) 189–196.

CHAPTER

10

Polymer Membranes for Sustainable Gas Separation

Elsa Lasseuguette, Maria-Chiara Ferrari

School of Engineering, The University of Edinburgh, Edinburgh, United Kingdom

OUTLINE

1. INTRODUCTION

Competing with conventional technologies such as pressure swing adsorption (PSA), chemical absorption, and cryogenic, membrane-based gas separations have played an important role in gas separation. The relatively low energy requirement of membrane processes makes them an attractive alternative to more intensive separation processes and therefore they are becoming highly relevant in any sustainable industrial development strategy.

The adoption of membrane-based gas separation is rapidly growing and now it is one of the important industrial gas separation technologies. In 2015, 77.46% of the gas separation market is based on polymeric membranes, and they are projected to witness the fastest growth rate of 8.4% from 2016 to 2024 [1]. Currently, polymeric membranes are dominating materials in this field owing to their excellent processability and easy scale up.

Sustainable Nanoscale Engineering
https://doi.org/10.1016/B978-0-12-814681-1.00010-2

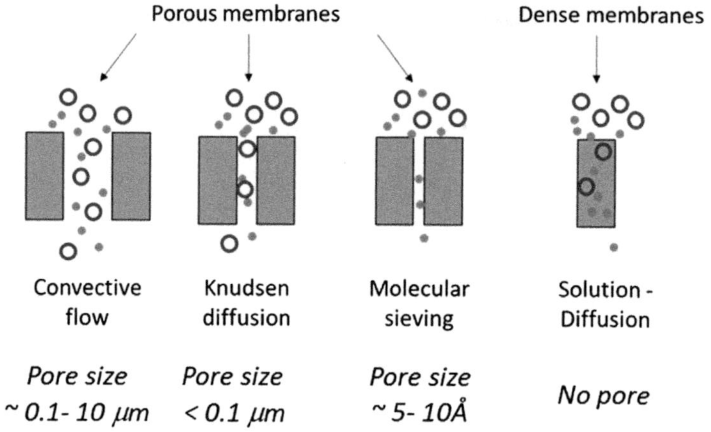

FIGURE 10.1 Transport mechanism for gases through porous and dense membranes.

Membranes have been known for over 150 years [2–4], and they have been used commercially since 1970 for gas separation [5–7]. Gas separation membranes are used in number of industrial processes, such as the production of oxygen-enriched air, separation of CO_2 and H_2O from natural gas, separation of nitrogen from air, purification of H_2, and recovery of volatile organic components (VOCs). A significant research effort has been also devoted to their potential employment for CO_2 capture.

The purpose of this chapter is to review the state of the art of polymeric membrane materials, the fundamentals of gas permeation, and their main industrial applications with particular emphasis on how nanotechnology can help deliver improved separation performances and sustainable processes.

2. TRANSPORT MECHANISM—MEMBRANE STRUCTURE

Polymeric membranes can be classified based on porosity (porous or nonporous) and their structure (asymmetric or isotropic).

The transport mechanism differs according the pore size of the polymeric membranes. Fig. 10.1 illustrates the different mechanisms of gas permeation.

If the pores are relatively large, from 0.1 to 10 μm, gases permeate through the membrane by convective flow, and no separation occurs. For pores smaller than 0.1 μm, the membrane pore radius is similar to the mean free path of the gas molecules. Transport through such pores is governed by Knudsen diffusion. The transport rate of any gas is inversely proportional to the square root of its molecular weight. These membranes are therefore not attractive for direct gas separation applications because of the small molecular weight differences.

When the pore diameter approaches the diameter of the molecules attempting to pass through the membrane, of the order 5–10 Å, the molecules are separated by molecular sieving. Transport through this kind of membrane includes both diffusion in the gas phase and diffusion of adsorbed species on the surface of the pores (surface diffusion). This phenomenon is more common in inorganic materials such as zeolites, microporous silica, or activated carbon. As pore sizes reduce, and pores disappears finally, the transport can be described by a solution-diffusion mechanism, which is the most common mechanism in the current commercial gas separation membranes, mainly based on dense polymer. In this case, each component in the gas feed dissolves in the membrane polymer at its upstream surface and then diffuses through the polymer layer along a chemical potential gradient to the opposite phase where they emerge into the downstream gas phase.

The permeability coefficient P_A of the gas molecule A, which represents the ability of the gas A to cross the membrane, can be expressed as the product of the diffusion coefficient, D, and the solubility coefficient, S, (Eq. 10.1):

$$P_A = D \times S \tag{10.1}$$

Permeability is a material property, but what is important from a process perspective is permeance (i.e., permeability divided by thickness). Most of literature on novel materials just reports test on thick, self-standing films and therefore permeability. In reality, commercial membranes will be composite structure with a very thin selective layer of high-performing polymer on a cheap support, and it is well known [8] that thin supported films have a different behavior and therefore an effort to move toward permeance for formed structures should be pursued.

The selectivity of the membrane, which is the capacity of the membrane to separate the two components A and B, can be expressed as the ratio of gas permeabilities (Eq. 10.2):

$$\alpha_{A/B} = \frac{P_A}{P_B} = \frac{D_A}{D_B} \times \frac{S_A}{S_B} \tag{10.2}$$

In literature, the so-called "ideal" selectivity is usually reported; ideal selectivity is calculated from pure gas measurements that are much more common and easy to perform but not always representative of the behavior of the materials in real conditions where a mixture of gases is present. For these reasons, in recent years, researchers have been devoting significant effort to test the materials in mixed gas conditions, calculating the mixed gas selectivity that is more closely representing the potential performance of the membrane. The solubility of the penetrant depends on its condensability and polymer–penetrant interactions. The solubility will increase as penetrant condensability, i.e., critical temperature, increases. The diffusion coefficient can be correlated to the penetrant size.

Thus, the polymer structure has a large impact on the membrane performance. Actually, if the material is below its glass transition temperature, i.e., case for glassy polymers, its polymer chains cannot rotate and are mainly fixed. Then, the effect of differences in size of the permeate gases on their relative mobility is large. Above the glass transition temperature, i.e., case for rubbery polymers, the segments of the polymer chains have sufficient thermal energy to allow rotations. Then, the effect of molecular size of the permeating gases on relative mobility is reduced. Consequently, diffusion coefficient in glassy polymers decreases more rapidly with increasing permeate size than diffusion coefficient in rubbery polymers.

Typically, glassy polymers exhibit high selectivity but low permeability, and this may again reduce over time due to the process called physical aging (loss of excess free volume [FV]), whereas most rubbery polymers show a high permeability but low selectivity.

3. UNDER HUMID CONDITIONS

It is well-known that polymer membranes exhibit a trade-off relationship between permeability and selectivity in that highly permeable polymers usually have relatively restricted selectivities. Robeson illustrated the empirical relationship in plots of log $\alpha_{A/B}$ versus log P_A with the so-called upper bound [9]. With the developments of new polymers, this upper bound is shifting toward higher permeabilities and selectivities [10] (Fig. 10.2).

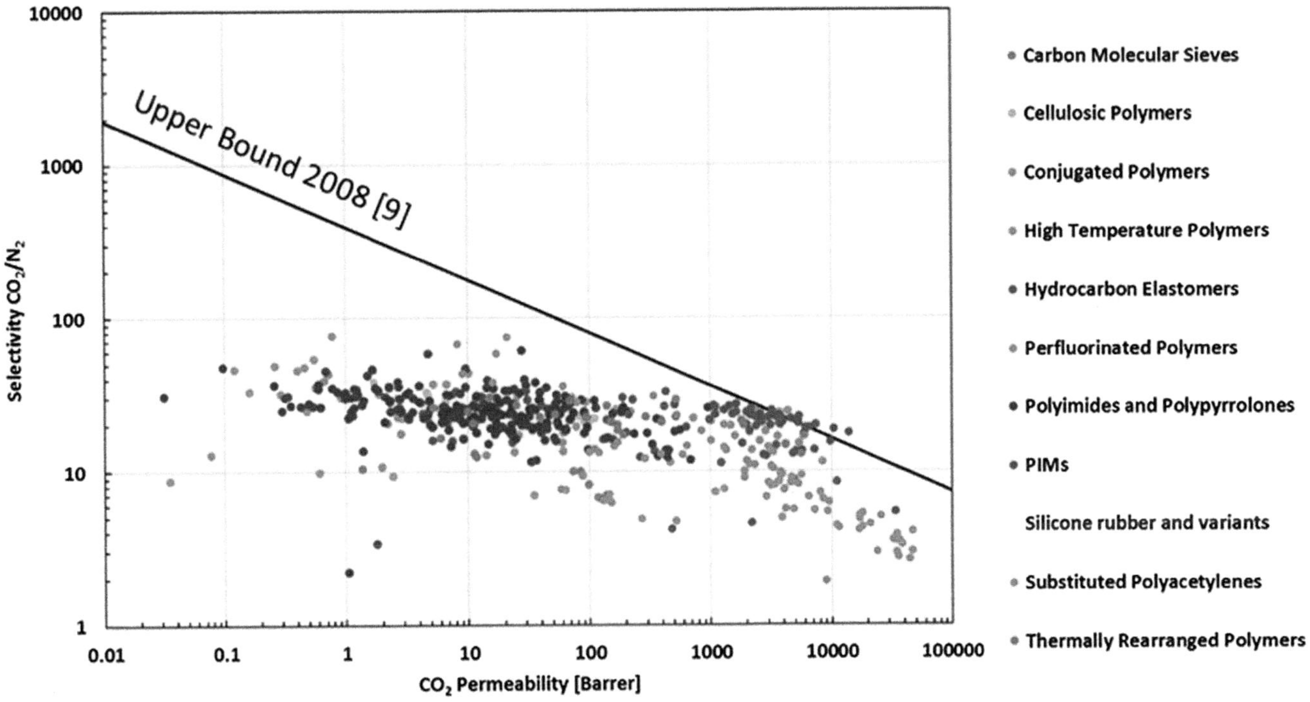

FIGURE 10.2 CO_2 permeability (barrer) versus CO_2/N_2 selectivity for polymeric membranes [11].

In recent years, polymer-based membrane materials including microporous polymers and polymer–inorganic hybrid membrane materials have demonstrated great potential for gas separation with a particular emphasis on CO_2 separation. Over 8000 research papers and reviews [12–24] on gas separation membranes have been published since 2008, leading to the proposition of a new upper bound for several separations in 2015.

In this chapter, we will focus on four classes of materials, which have become predominant in membrane research for gas separation application, namely polymers of intrinsic microporosity (PIMs), thermally rearranged (TR) polymers, mixed matrix membranes (MMMs), and facilitated transport membranes (FTMs).

3.1 Polymers With Intrinsic Microporosity

Among the many efforts to develop new polymeric structures with high performance, one recent example is the family of PIMs introduced by Budd et al. [25].

PIMs refer to a new class of rigid chain-contorted soluble microporous polymers with interconnected pores of less than 2 nm, which exhibit very high gas permeability regardless of membrane fabrication conditions [15]. Intrinsic microporosity is defined as "a continuous network of interconnected intermolecular microcavities" [26], which is formed because of the contorted shape and chain rigidity of the polymer structure. The restricted chain rotation originating from sites of contortion or spirocenters leads to an inefficiently packed matrix with high FV elements, typically above 20%. Thus, PIM membranes exhibit very high gas permeabilities, which are comparable with other high FV polymers such as Teflon AF2400 and poly(1-trimethylsily-1-propyne) (PTMSP) for most gases (Fig. 10.2).

The synthesis of the first PIMs, PIM-1 and PIM-7, are shown in Fig. 10.3.

PIM-1, the first reported high molecular weight microporous ladder polymer, and PIM-7 have demonstrated extremely high gas permeabilities with moderate ideal selectivity [25,27], especially for O_2/N_2, CO_2/N_2 pairs with values lying above the Robeson's upper bound. For instance, Budd et al. [25] reported gas permeabilities of 12,600 barrer for CO_2, 500 barrer for N_2, and 1610 barrer for O_2.

To enhance CO_2 permeation and selectivity in PIMs, recent investigations have been undertaken into (i) tuning the angle of contorted centers, (ii) introducing bulky pendant groups, or (iii) cross-linking of polymer chains.

The group of Prof N. McKeown has introduced a new class of PIMs prepared using a polymerization reaction based on the efficient formation of the bridged bicyclic diamine called Tröger's base (TB; i.e., *6H,12H*-5,11-methanodibenzo[b,f][1,5]diazocine) [28](Fig. 10.4).

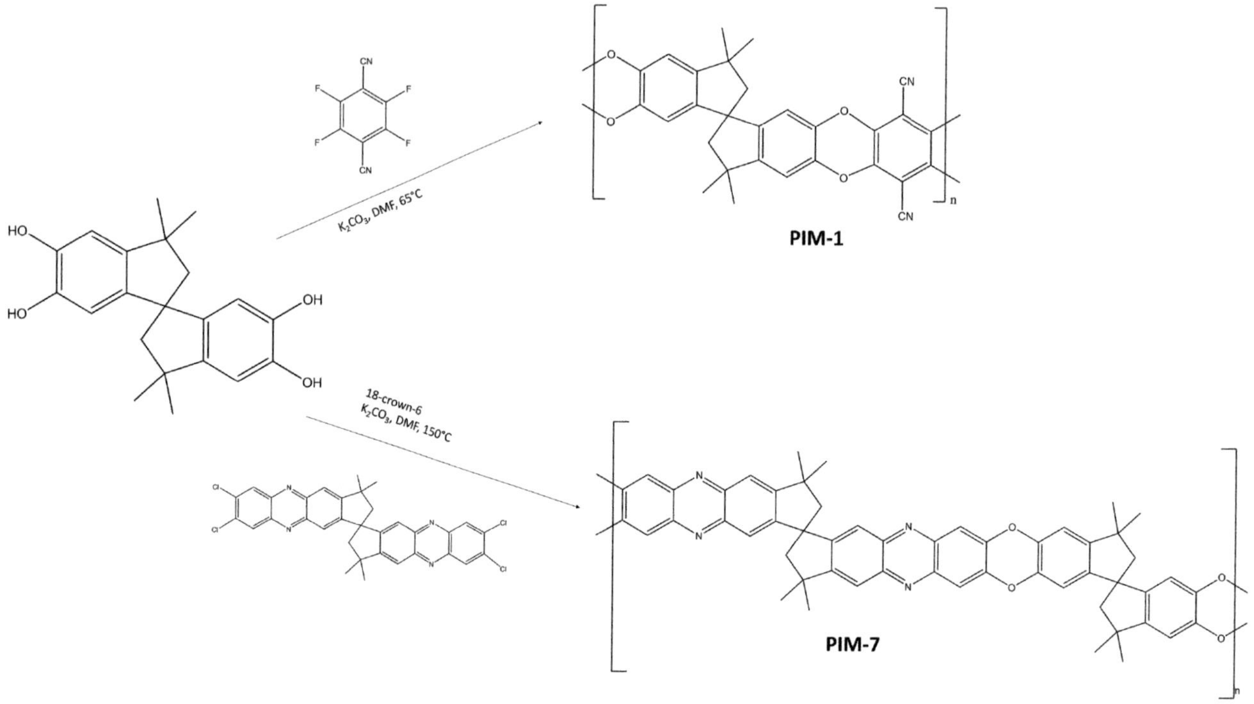

FIGURE 10.3 Synthesis and chemical structures of PIM-1 and PIM-7.

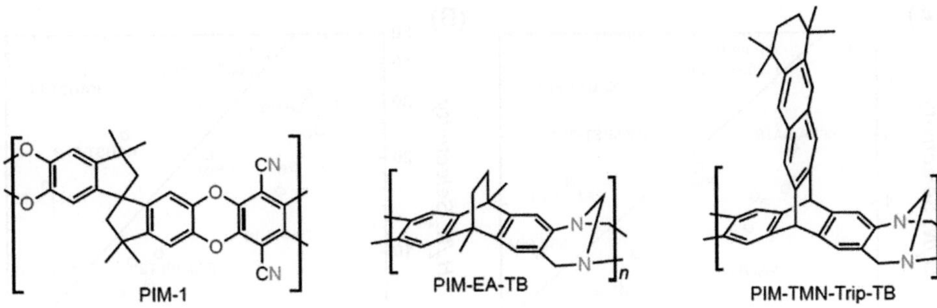

FIGURE 10.4 Molecular structure of PIM-1, PIM-EA-TB, and PIM-TMN-Trip [28,29].

PIM-EA-TB [28] demonstrates very large gas permeability ($CO_2 = 7140$ Barrer, $N_2 = 525$ Barrer, and $CH_4 = 699$ Barrer) and good selectivity ($CO_2/N_2 = 14$, $CO_2/CH_4 = 10$) so that its data lie well above the 2008 upper bounds due to exceptional diffusivity selectivity that favors the transport of gas molecules of smaller kinetic diameters. This enhanced molecular sieving performance of PIM-EA-TB was attributed to its highly rigid macromolecular structure, which does not contain the relatively flexible spirocentres or dioxin linkages found in conventional PIMs such as PIM-1. Recently, Rose et al. [29] developed an ultrapermeable PIM (PIM-TMN-Trip) that is substantially more selective than PTMSP (Fig. 10.4). They explained through molecular simulations and experimental measurement that the inefficient packing of the two-dimensional (2D) chains of PIM-TMN-Trip generates a high concentration of both small (<0.7 nm) and large (0.7–1.0 nm) micropores, the former enhancing selectivity and the latter permeability. PIM-TMN-Trip has the CO_2 permeability of 30,000 Barrer with CO_2/CH_4, CO_2/N_2, and CO_2/H_2 ideal selectivity of 15, 2, and 9, respectively.

Cross-linking by thermal or chemical treatment is another alternative approach to improve the selectivity and physical properties of PIMs. Li et al. [27] reported thermal cross-linking of PIM-1 at 300°C by the formation of triazine rings through trimerization of nitrile substituents. The thermally cross-linked PIM-1 membranes demonstrate exceptional gas separation performance that surpasses the most recent upper bounds for the important gas pairs, such as H_2/N_2, CO_2/CH_4, and CO_2/N_2. For instance, an ideal CO_2/CH_4 selectivity of 54.8, which is almost four times higher than the PIM-1 membrane, with CO_2 permeability of 4000 Barrer has been obtained.

Combining PIMs and polyimides is another way to improve the PIM performance. Actually, the introduction of spirocenters into polyimide main chains [30] induces exceptional gas permeabilities. Ghanem et al. [30] has developed several series of PIM-PI, PIM-PI1, PIM-PI3, and PIM-PI8 (Fig. 10.5). The gas permeabilities obtained for PIM-PI-8 are by far the highest ever reported for a polyimide, with a CO_2 permeability of 3700 Barrer, 2.6 times higher than a polyimide. These polymers were more selective than ultrahigh FV polymers such as PTMSP, PMP, and Teflon AF2400.

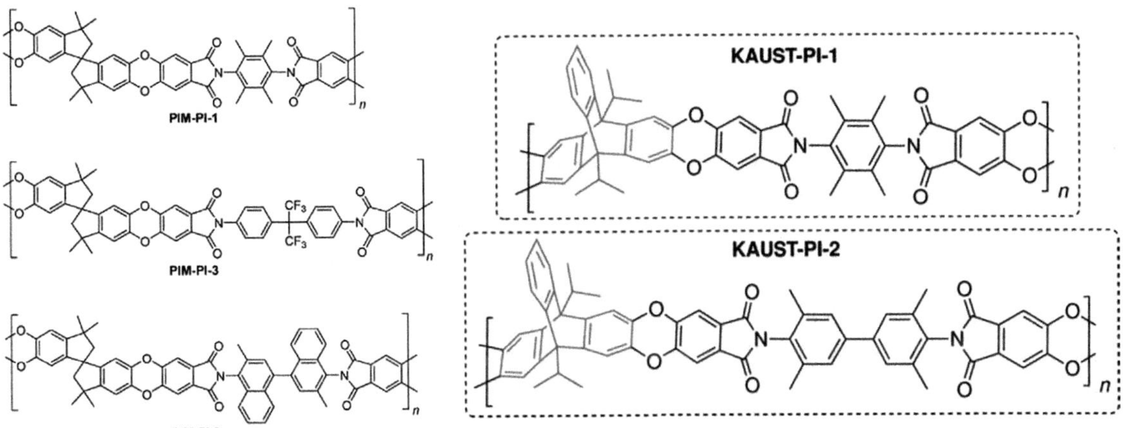

FIGURE 10.5 Modified polymers of intrinsic microporosity: PIM-PI1, PIM-PI3, and PIM-PI8 (left) and KAUST-PI (right). *Adapted from B. Ghanem, N. McKeown, P. Budd, J. Selbie and D. Fritsch, High-performance membranes from polyimides with intrinsic microporosity, Advanced Materials 20 (2008) 2766–2771; B. Ghanem, R. Swaidan, E. Litwiller, I. Pinnau, Ultra-microporous triptycene-based polyimide membranesfor high-performance gas separation, Advanced Materials 26 (2014) 3688–3692.*

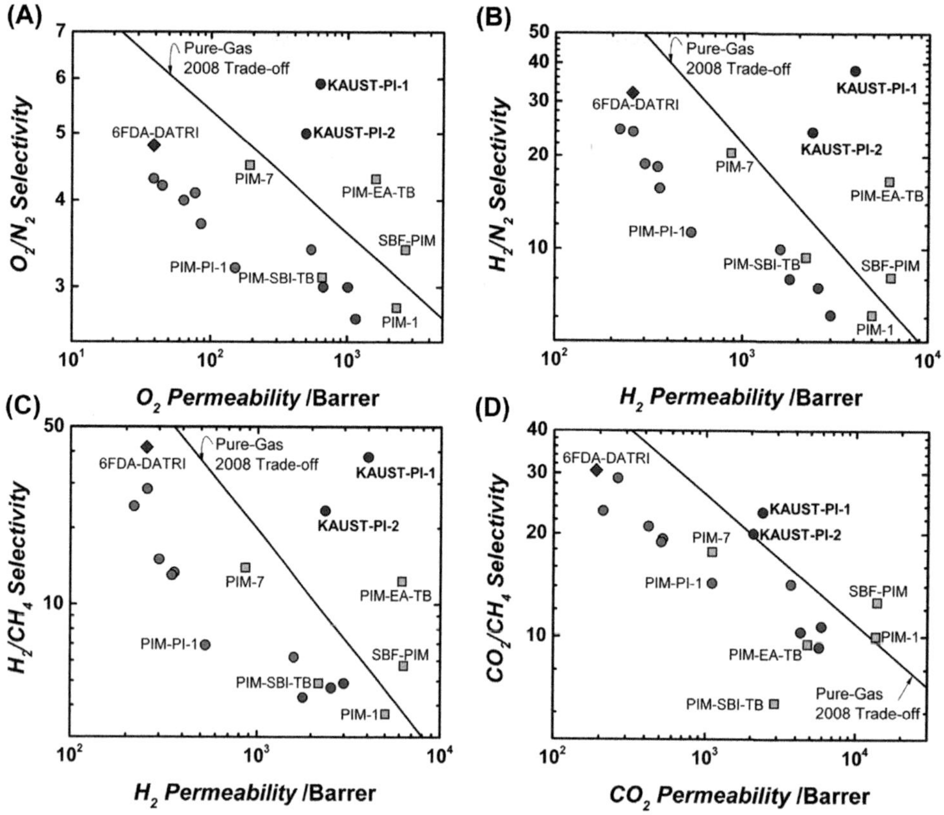

FIGURE 10.6 Separation performance for KAUST-PI [31].

However, their performances are below the current 2008 upper bound. More recently, Ghanem et al. [31] developed new PIM-PIs, called KAUST-PI (Fig. 10.5), based on the replacement of the relatively flexible spirobisindane contortion center of PIM-PI-1 with a shape-persistent 9,10-diisopropyltriptycene (TPDA) moiety. This change resulted in an extraordinary gas separation performance improvement in comparison with current PIM-polyimides (Fig. 10.6). For instance, KAUST-PI1 and KAUST-PI2 presented a CO_2 permeability of 2389 Barrer and 2071 Barrer, N_2 permeability of 107 Barrer and 98 Barrer, CH_4 permeability of 105 Barrer and 101 Barrer, and H_2 permeability of 3983 Barrer and 2368 Barrer, respectively (Fig. 10.6).

The performance is explained by the presence of small pores (<7 Å) and highly interconnected microporous structure.

3.2 Thermally Rearranged Polymer Membranes

TR polymers are considered the next generation of membrane materials because of their excellent transport properties and high thermal and chemical stability. TR polymers are aromatic polymers with heterocyclic rings, such as polybenzoxazoles (PBO) or polybenzothiazoles structures prepared via thermal rearrangement process of polyimide with hydroxyl-ortho-functional group (Fig. 10.7).

To accomplish this rearrangement, a soluble polyimide with a reactive group ortho-position to the polyimide diamine is cast as a film from common organic solvents such as N-methyl-2-pyrrolidone (NMP) or dimethylacetamide (DMAc). Next, these films are thermally transformed into TR polymers through a solid-state reaction at elevated temperatures, close to 450°C. This thermal rearrangement induces a narrow cavity size distribution and cavity size centered around 3.8 Å [32,33], which allows TR-polymer membranes to present high CO_2 permeability (2000 Barrer) and CO_2/N_2 and CO_2/CH_4 selectivities (27 and 50, respectively). The gas permeabilities of almost all TR polymer membranes are enhanced by two orders of magnitude over those of the original polymers and typical glassy polymers, confirming the presence of larger interconnected FV elements. The gas permeabilities are lower than those of PTMSP membrane, the highest gas permeable polymer, but gas selectivities toward important gas separations (e.g., CO_2/CH_4, CO_2/N_2) are 200%–300% higher than in PTMSP and also comparable with or slightly lower than in inorganic molecular sieve membranes [32].

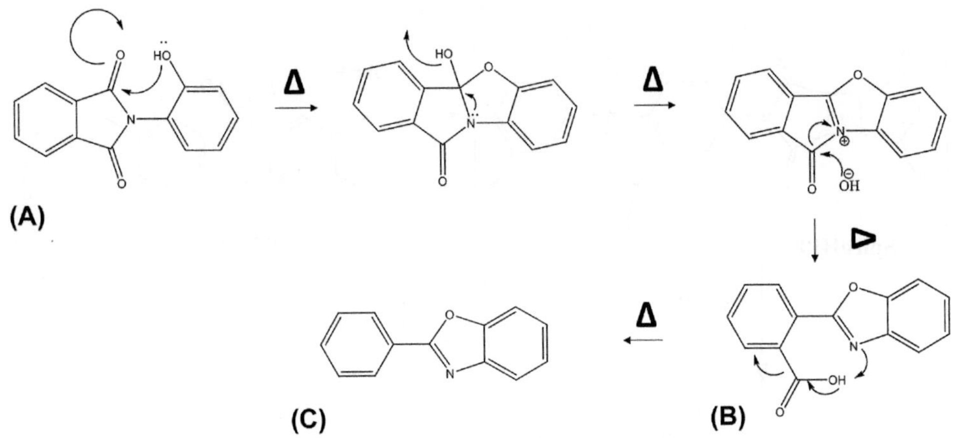

FIGURE 10.7 Proposed mechanism of thermal rearrangement from hydroxyl-containing polyimide to polybenzoxazole: (A) an imide ring with ortho-positioned hydroxyl group, (B) a carboxyl-benzoxazole ring, and (C) a benzoxazole ring.

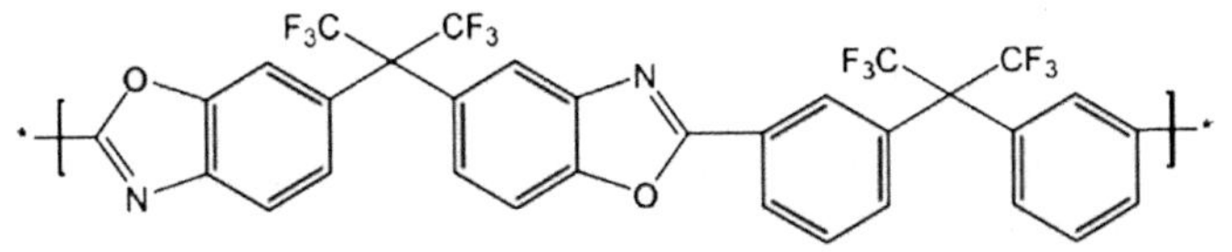
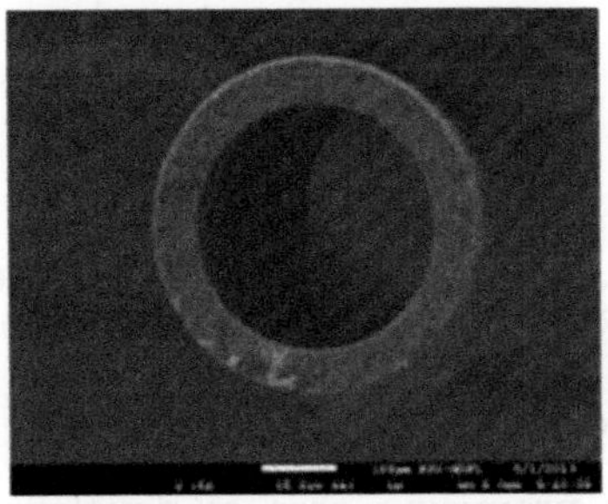

FIGURE 10.8 Chemical structure of thermally rearranged polybenzoxazole (TR-PBO) (left). SEM pictures of TR-PBO hollow fiber membranes (right). *Adapted from K. Woo, J. Lee, G. Dong, J.D.Y. Kim, W. Hung, K. Lee, G. Barbieir, E. Drioli, Y. Lee, Fabrication of thermally rearranged (TR) polybenzoxazole hollow fiber membranes with superior CO_2/N_2 separation performance, Journal of Membrane Science 490 (2015) 129–138.*

The first TR membranes have been reported by Park et al. [33] in 2007, based on PBOs. They showed excellent separation performance for CO_2/CH_4 mixtures, with high selectivity (between 20 and 40) and CO_2 permeability (between 4000 and 5500 Barrer) due to an unusual microstructure whose cavity size and distribution could be tuned by the choice of template molecules and heat treatment. They explained that the cavity size in TR polymers can be easily tuned by adding small acidic dopants (e.g., HCl and H_3PO_4) due to the basic nitrogen atoms (-C=N-) present on the heterocyclic rings (e.g., benzoxazole ring). After doping, the CO_2 permeability decreased but CO_2/CH_4 selectivity increased. However, after dedoping, the permeability and selectivity returned to their original values, indicating that the cavity size and shape can be tailored. In the same way, Scholes et al. [34] showed by varying the relative ratios of PI and PBO segments within the TR copolymer, it becomes possible to control the gas separation properties of the polymeric membrane. They have developed a series of poly(benzoxazole-co-imide) copolymers with different PBO:PI ratios. The addition of PBO moieties induced an increase of the CO_2 permeability (from 100 to 249 Barrer for 1:3 and 3:1 PBO:PI ratio, respectively) with constant selectivities (CO_2/CH_4=20, CO_2/N_2=8).

In 2015, Woo et al. [35] managed to obtain TR-polybenzoxazole (TR-PBO) hollow fiber membranes from a poly(amic acid) (HPAAc) precursor through a non–solvent-induced phase separation technique (NIPS) (Fig. 10.8), with superior pure gas permeation behavior (CO_2 permeance of 2500 GPU and CO_2/N_2 ideal selectivity of 16).

PALs measurements revealed that such an exceptional separation performance could be mainly attributed to the ideal cavity radius (3.584 Å) and ultrathin skin layer thickness (193 nm) obtained using the optimal fabrication conditions.

The major disadvantages of TR polymer membranes are the high temperatures necessary to obtain them, which are close to the decomposition temperatures of the membrane materials, and the poor mechanical properties of the membranes. One solution to overcome this problem is the development of new polyimide precursors [36–40]. For instance, Li et al. [38] have prepared mechanically robust TR polymer membranes with spirobisindane-containing hydroxyl polyimides (Fig. 10.9).

The resulting spiroTR-PBOs exhibited high FVs and superior mechanical properties in terms of elongation and tensile strength. After thermal rearrangement, spiroTR-PBO showed increased gas permeability up to 675 Barrer and high elongation at break around 20%, higher than normal TR-PBO membranes [38].

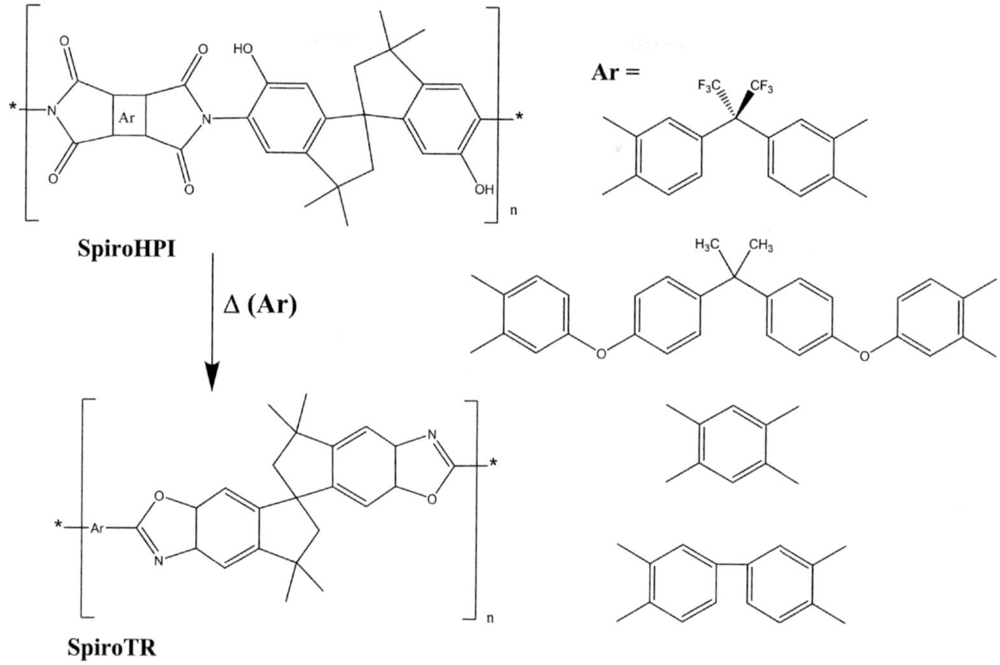

FIGURE 10.9 Chemical structure of spiroTR-PBOs. *Adapted from S. Li, J. Jin, S. Han, C.K.S. Park, P. Budd, Y. Lee, Mechanically robust thermally rearranged (TR) polymer membranes with spirobisindane for gas separation, Journal of Membrane Science 434 (2013) 137–147.*

3.3 Mixed Matrix Membranes

Another promising area of development for gas separation membranes is the use of hybrid membranes or MMMs. They are composite membranes made by combining a filler (dispersed phase) and a polymer matrix (the continuous phase) (Fig. 10.10).

By using two materials with different transport properties, such membranes have the potential to combine synergistically the easy processability of polymers and the superior gas-separation performance of filler materials.

However, the transport properties of organic/inorganic MMMs are strongly dependent on the nanoscale morphology of the membranes [41]. Fig. 10.11 shows a schematic diagram of main structures at the polymer/particle interface.

Case 1 represents an ideal morphology, without defects. Case 2 shows the detachment of polymer chains from the filler surface, causing the interface voids. It is mainly due to a poor compatibility between polymer matrix and filler. Because of the less resistance in interface voids, gas passes through interface voids instead of polymer matrix or filler, which improves gas permeability. Meanwhile, change on gas selectivity is dependent on the size of interface voids. Generally, gas selectivity decreases. Case 3 indicates that the polymer chains in direct contact with the filler surface can be rigidified compared with the bulk polymer chains. In this case, the movement of polymer chain at polymer/filler interface is restrained, which reduces gas adsorption, and then reduces gas permeation. Case 4 displays a situation in which the surface pores of the fillers has been partially sealed by the rigidified polymer chains, and therefore, gas permeability declines.

The addition of the fillers in the continuous phase plays an important role in the transport properties of MMMs. It is expected that the gas permeability increases with an increase in inorganic filler fraction, but in some cases, negative effects are also noted (Fig. 10.12).

Two possible hypotheses are formulated for the decline in permeability. One is due to the rigidification of polymeric chain, and the second is the partial blockage of the zeolite pores by polymer chains [47]. On the other hand, a higher filler loading introduces a higher risk of defects, which would induce a decrease of the selectivity.

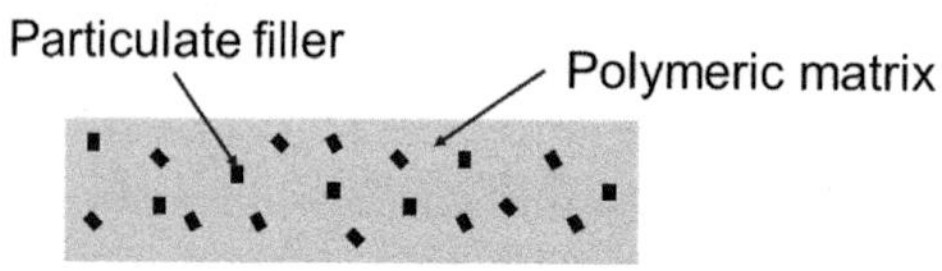

FIGURE 10.10 Mixed matrix membranes structures.

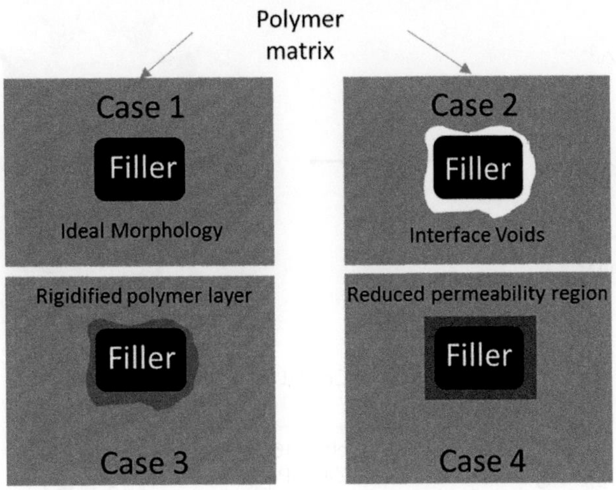

FIGURE 10.11 Nanoscale structures of mixed matrix membranes (MMMs).

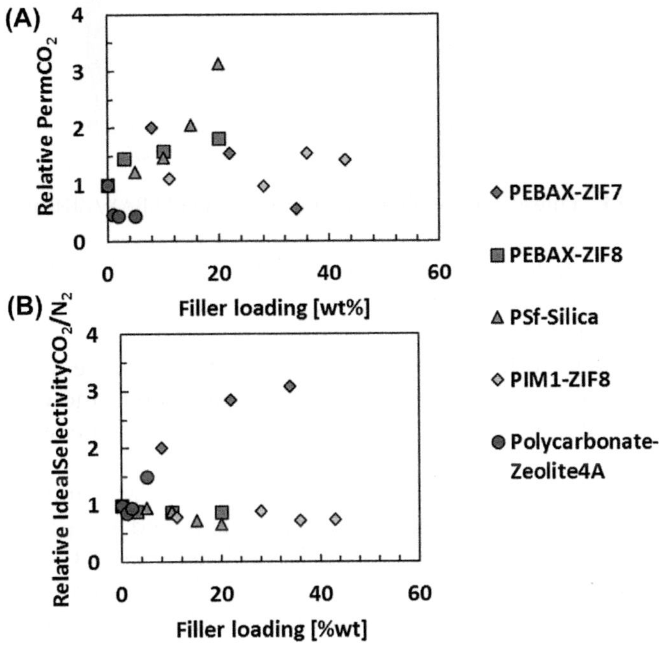

FIGURE 10.12 Effect of filler loading in the mixed matrix membrane for CO_2 permeability (A) and ideal selectivity of CO_2/N_2 (B) [42–46].

Different types of organic/inorganic additives have been used, such as POFs, zeolites, carbon nanotubes (CNTs), and graphene [48–52].

3.3.1 Porous Organic Framework

COFs (covalent organic frameworks) and MOFs (metal–organic frameworks) are porous three-dimensional crystals with extraordinarily large internal surface areas that can absorb and store enormous quantities of targeted molecules.

MOFs are chemically mutable, highly porous materials prepared from the combination of metal ions or metal oxido clusters and organic ligands. In comparison with purely inorganic zeolites, the chemical diversity of MOF structures can be used to facilitate strong interactions with the polymer bulk material. MOFs offer various advantages over zeolites (higher porosity, higher pore volume, thermal and chemical stability) and are, therefore, promising additives for MMMs [52,53]. One of the main advantages compared with zeolites comes from their organic nature that leads to a higher compatibility with the polymer matrix and therefore a lower incidence of

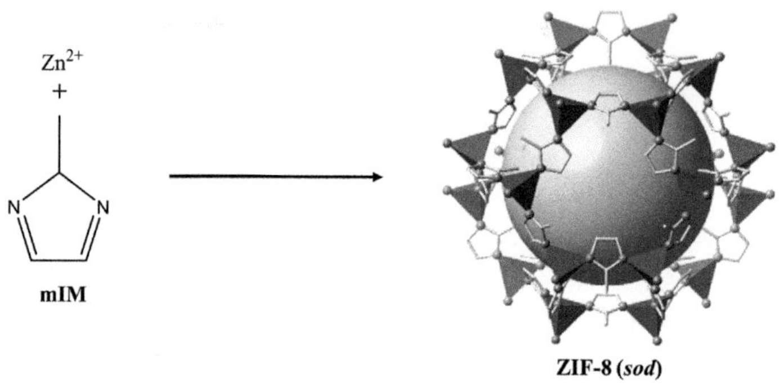

FIGURE 10.13 Crystal structure of ZIF8 [60].

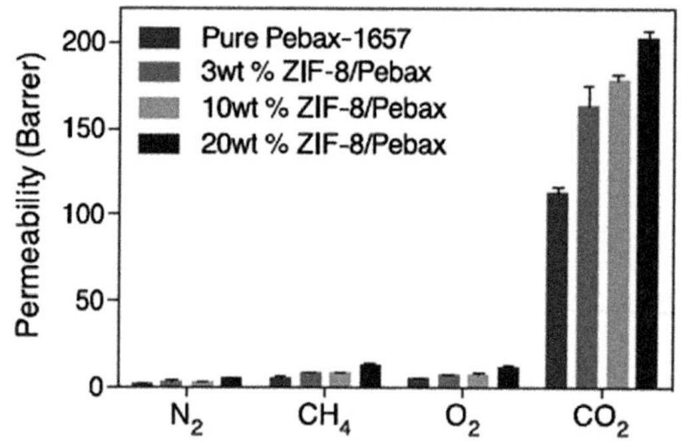

FIGURE 10.14 Pure gas permeability for pure PEBAX and PEBAX/ZIF8 membranes [42].

defects. Moreover, they are highly tailorable, leading to a very large number of possible structure to investigate [54–56]. Some of them have been used for developing inorganic membranes. These MOF membranes show promising gas separation performance [57,58].

A large body of literature is devoted to MMMs using MOFs, and the readers are referred to some of the most recent reviews [53,59]. Among these, recently ZIF-8, a zeolite imidazolate framework (Fig. 10.13), gained attention as nanoscale fillers for MMMs because of their molecular sieving effect, facile synthesis, and good compatibility with polymers.

The first ZIF-based polymer MMM using ZIF-8 as the filler phase and Matrimid as the polymer phase has been reported by Ordonez et al. [61]. Increases in ideal selectivities were obtained at the higher ZIF-8 loadings for gas pairs containing small gas molecules, such as H_2/CH_4, CO_2/CH_4, H_2/C_3H_8, and CO_2/C_3H_8, suggesting that the ZIF-8 may act as a molecular sieve for smaller gas molecules. Very recently, Sutrisna et al. [42] looked at ZIF-8/PEBAX membranes successfully combining ZIF-8 with both flat sheet and hollow fiber membranes. They noted a significant effect on the chain packing of the PEBAX phase in flat sheet membranes, which resulted in increased permeabilities (Fig. 10.14), almost certainly due to higher FV.

Recently, Liu et al. [62,63] have developed MMMs based with face-centered cubic (fcu)-MOFs incorporated into 6FDA-DAM. Adding fcu-MOFs enhanced the permeation properties of pure 6FDA-DAM membranes, and the performance gradually increased with fcu-MOF loading. However, the performance of fcu-MOF/6FDA-DAM membranes, even with high MOF loadings did not surpass the upper bound for CO_2/N_2 or N_2/CH_4, owing to the moderate performance of the pure 6FDA-DAM phase for these gas pairs. It can be expected that choosing polymers whose performances are located on the upper bound to better match the transport properties of the fcu-MOFs could further push the performance of the fcu-MOFs hybrid membranes beyond the upper bound.

Although MOF fillers have attracted considerable attention, alternative crystalline and noncrystalline organic porous materials have also been explored very recently. In contrast to MOFs, these additives have entirely organic extended frameworks with either crystalline (COFs) or amorphous structures (PAFs). They are constructed via the linkage of organic molecular building units in periodic networks via robust covalent bonds. They provided a great versatile architecture because of diversity of organic buildings units, which allowed tunable pore sizes (from 0.3 to 50 nm) and shapes [64]. The distinct advantages of POFs are their organic structures, which display excellent chemical compatibility with the organic polymer phase, and the chemical stability conferred by irreversible covalent bonding, particularly for materials such as PAFs and PIMs [52].

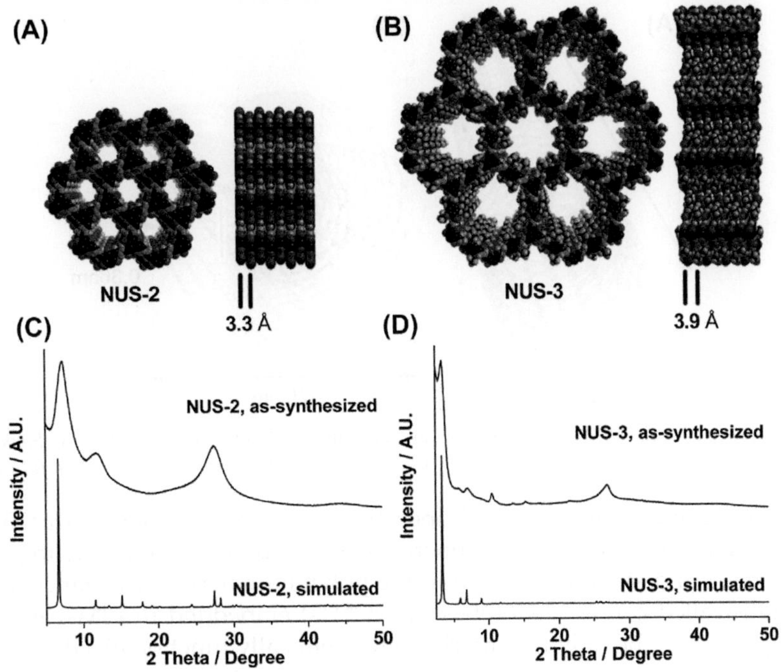

FIGURE 10.15 Simulated crystal structures of (A) NUS-2 and (B) NUS-3; PXRD patterns of (C) NUS-2 and (D) NUS-3 [65].

Only a few studies of COFs as additives to polymer matrices have been reported because of challenging synthetic procedures and, for some COFs, limited stability under humid conditions. Kang et al. [65] developed two water stable COFs (Fig. 10.15), which have been blended with poly(ether imide) and polybenzimidazole membranes.

They obtained highly homogenous membranes with an excellent H_2/CO_2 permselectivity that exceeded the 2008 Robeson upper bound. 20% of COFs within PBI induced a H_2/CO_2 electivity of 31.4, whereas for the PBI alone, the value is less than 10.

PAFs, which are amorphous COFs, can exhibit unusually high porosity and narrow pore size distributions for amorphous solids, and they present higher surface area than COFs. PAF1 has a BET of 7100 m^2g, whereas the highest reported surface area for a COF is 4120 m^2g, in COF-103. This high surface area allows PAF having high uptake capacities for hydrogen (10.7 wt% at 77K and 48 bar) and for carbon dioxide (1300 mg g^{-1} at 298K and 40 bar) [66]. Lau et al. [67,68] showed that the addition of PAF1 within super glassy polymers such as PTMSP, PIM-1, and PMP induces a drastically increase of the gas permeability, thanks to this high surface area and its extreme porosity. The CO_2 permeability increases from 25% in PTMSP to as much as 320% in PIM-1. They also demonstrated that PAF1 has the capacity to "freeze" the aging process. Super glassy polymers can be very interesting for gas separation application where a large quantity of gas needs to be treated. This interest comes from their high permeabilities and selectivity resulting of their high free fraction volume (FFV). However, they are very sensitive to physical aging due to the collapse of FFV with the time. For instance, the CO_2 permeability of PTMSP decreases from 29,800 Barrer to 8000 Barrer over 1 year. Lau et al. [67,68] showed that the presence of PAFs inside the polymer stops the aging process, with, for instance, a stability over 240 days for CO_2 permeability of PTMSP, PIM-1, and PMP polymer mixed with PAF1. Solid-state NMR analysis indicates that the antiageing mechanism is due to the intercalation of PAF within super glassy polymer chain and the local interaction of PAF with the methyl groups, which freezes the FFV. An important observation is that this mechanism is not general for nanoporous additives, as the chemical compatibility of the additive and polymer appears to be essential. For instance, ZIF8-PTMSP nanocomposites exhibit the same aging process as native PTMSP [69].

3.3.2 *Carbon Nanotubes*

CNTs (Fig. 10.16) have received great attention in recent years because of their outstanding properties, such as mechanical properties and the rapid transport for gases. It was reported that gas transport in CNTs was faster than that in any other known materials due to their inherent smoothness [70].

If CNTs were mixed with some polymers, the permeabilities of polymers may be significantly enhanced by enhancing diffusion coefficients. So far, CNTs have been used to improve the properties of some polymers. Afsari et al. [73] reviewed the use of CNTs as fillers. They showed that although CNTs have excellent separation properties,

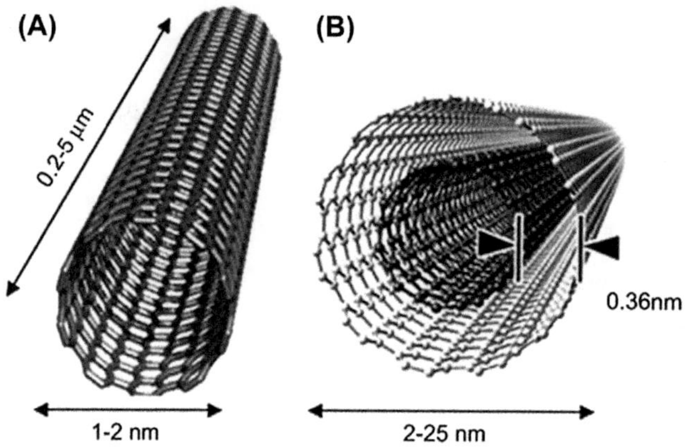

FIGURE 10.16 Single-walled (A) and multi-walled (B) carbon nanotubes. *Adapted from A. Hirsch, Funktionalisierung von einwandigen Kohlenstoffnanoröhern, Angewandte Chemie 114 (2002) 1933–1939; S. Iijima, Carbon nanotubes: past, present, and future, Physica B 323 (2002) 1–5.*

preparation of CNT-MMMs is more complicated. Proper dispersion and suitable interfacial adhesion between CNT and polymer matrix have to be guaranteed. CNT functionalization is an efficient way to prevent CNT aggregation, leading to a better dispersion and stabilization within the polymer matrix. For instance, Khan et al. [74] prepared MMM consisting of multi-walled carbon nanotubes (MWCNTs) embedded in PIM-1 matrix, via solution casting method. For a better dispersion in the polymer matrix, they chemically functionalized MWCNTs with polyethylene glycol. MMMs prepared with functionalized CNT presented better performance than the MMM with native CNT in term of dispersion, mechanical properties, and gas separation without sacrificing selectivity. With 1%wt. of filler, they obtained homogenous MMMs with a CO_2 permeability of 7813 Barrer for functionalized CNT against 6219 Barrer for native CNT and a CO_2/N_2 selectivity around 18 for the two kinds of CNT.

Brunetti et al. [75] also studied MMMs based on oxidized CNT mixed with thermally rearranged polybenzoxazole-co-imide (TR-PBOI). The addition of CNTs improved slightly the CO_2/N_2 separation performance. With 0.5%wt of CNT, they obtained a CO_2 permeability of 126 Barrer against 105 Barrer for the pure TR, whereas the N_2 permeability is similar with and without CNTs. This improvement is due to an increase of the FV and a higher degree of thermal rearrangement than the pure TR-PBOI. Another very interesting result was that the addition of CNTs did not affect the aging behavior.

3.3.3 *Graphene Oxide*

Graphene oxide (GO) possess outstanding properties for gas separation. A number of graphene-based membranes have been extensively studied, and they exhibit unique gas separation properties for various gas pairs, surpassing the typical trade-off between permeability and selectivity.

GO is a nanomaterial in the graphene-based family, which has the same carbon sheet structure as graphene and additional oxygenated functional groups attached on its planar sheet (Fig. 10.17). The presence of these functional groups, such as epoxy, hydroxyl, and carboxyl groups, improves GO compatibility with others polymers, which in turn contribute to enhance gas separation performance of the MMM.

Zhao et al. [76] developed MMM based on 1 wt% GO into polyimide. The presence of GO induced an increase of CO_2 permeability (up to 16 Barrer) and CO_2/N_2 selectivity (up to 84) 14 and 4 times higher than the neat membrane, respectively. In the same way, Zahri et al. [77] fabricated hollow fibers based on the incorporation of GO into polysulfone polymer matrix, which presents better gas separation performance than PSf hollow fibers. The presence of GO within PSf membranes increased the permeance of CO_2 as much as 14% (from 65 GPU to 74 GPU), and the selectivity of both CO_2/N_2 and CO_2/CH_4 was also enhanced as much as 158% and 74%, respectively, compared with that of neat PSf. These improvements are firstly due to the strong affinity between GO and CO_2 resultant from the

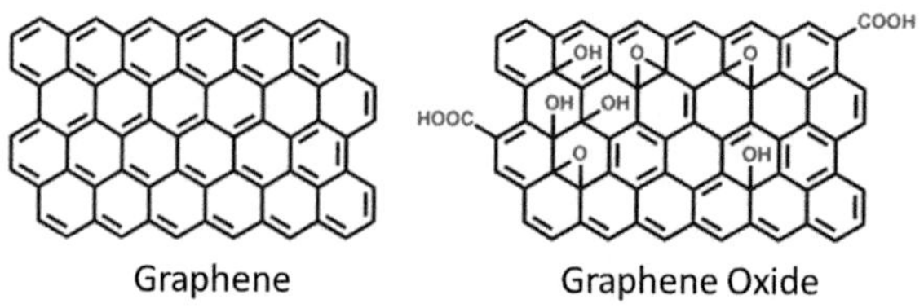

FIGURE 10.17 Chemical structure graphene and graphene oxide.

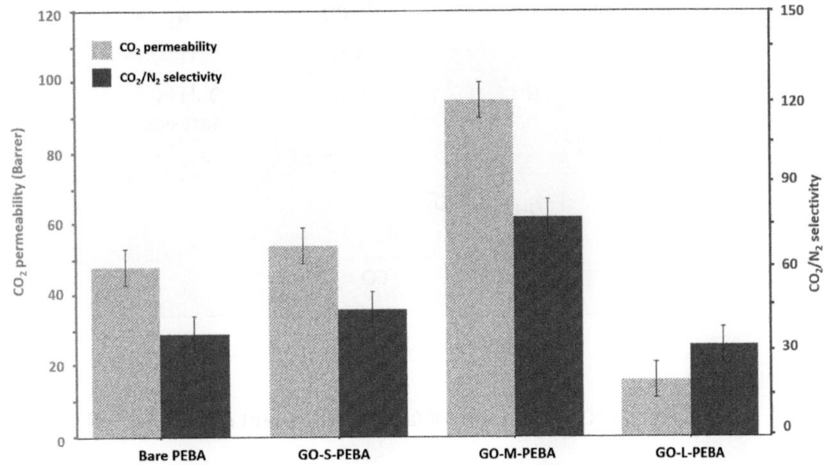

FIGURE 10.18 Effect of graphene oxide (GO) sheets lateral size on the CO_2/N_2 separation performance (1 wt% of GO) [78].

effective π–π stacking interaction, and secondly, GO acts as barrier to create tortuous path in the membrane to allow the gas with smaller molecules (CO_2) to effortlessly pass through the composite membranes, meanwhile the gas molecules with larger size (N_2 and CH_4) are hindered from passing through the membranes.

Shen et al. [78] showed that the lateral size of GO sheets has an effect on the gas separation performances. They have prepared MMMs based on polyether block amide (PEBA) mixed with GO nanosheets with different lateral sizes, 100–200 nm, 1–2 μm, and 5–10 μm, for GO-S-PEBA, GO-M-PEBA, and GO-L-PEBA respectively. GO sheets with large lateral size are beneficial for providing longer gas channel.

As shown on Fig. 10.18, the gas permeability increased dramatically when the GO sheets lateral size was in 1–2 μm. However, GO sheets with too large lateral size had a significantly negative influence on gas separation performance. The larger GO sheets lateral size can lead to poor dispersion and easier formation of an aggregated stacking structure, which is detrimental to molecular transport.

3.4 Facilitated Transport Membranes

FTMs have shown a great potential to reach performances beyond the Robeson upper bound region. FTMs involve the use of a carrier, which interacts specifically and reversibly with one of the mixture components (reactive species). In these membranes, the reactive species is transported through other mechanism in addition to the ordinary solution-diffusion, increasing its permeability. As the carrier does not interfere within the transport of the nonreactive species, selectivity also increases.

According to the mobility of carriers, FTMs can be classified into two types: mobile carriers moving freely from one side to the other and fixed-site carriers (FSC) with limited mobility bounding to the membrane matrix [79]. The transport of the target permeant through mobile carrier membranes obeys the vehicle mechanism, in which the carrier serves as a "ferryboat" plying between the two sides of membrane (Fig. 10.19A), whereas for FSC membranes, the target molecule/ion has to pass through the membrane via carrier-to-carrier hopping (Fig. 10.19B) [80].

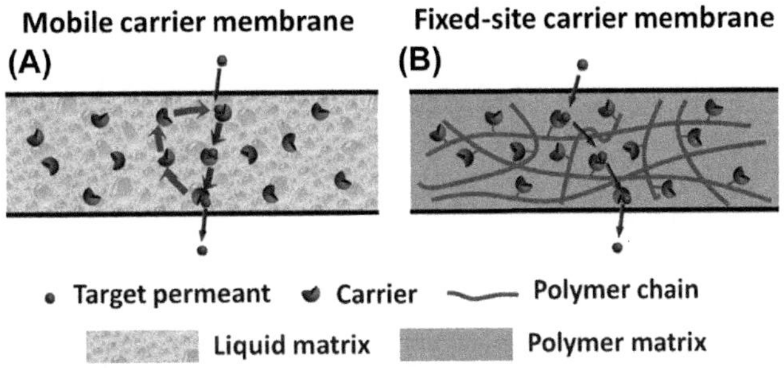

FIGURE 10.19 Schematic illustration of the carrier-mediated transport mechanisms for (A) a mobile carrier membrane and (B) a fixed-site carrier membrane. *Adapted from Y. Li, S. Wang, G. He, H. Wu, F. Pan, Z. Jiang, Facilitated transport of small molecules and ions for energy-efficient membranes, Chemical Society Reviews 44 (2015) 103–118.*

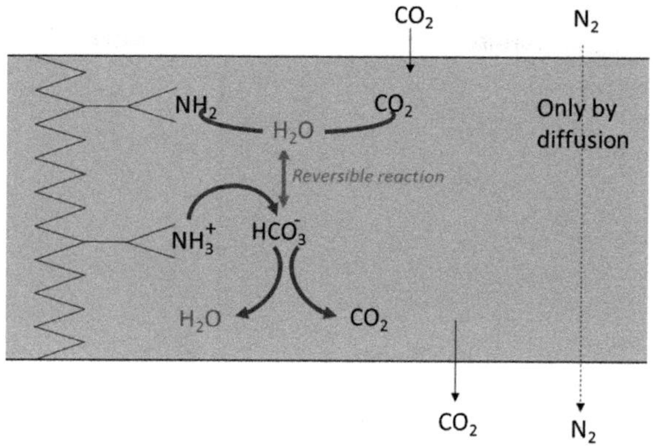

FIGURE 10.20 Mechanism of facilitated transport membranes [84].

In most industrial processes, which involve CO_2, feed streams are saturated with water vapor; hence, CO_2 permeation will be exposed to a competitive sorption of water in the membranes [81]. However, in the case of FTM, the presence of water vapor causes an enhancement of the CO_2 transport through the membrane due to the reversible reaction between the carrier and CO_2 in the feed gas. The transport of CO_2 through the membrane is carried out as carbamate and bicarbonate, and this is induced by the presence of water in the humidified flue gas streams as already mentioned [82–84]. Thus, the presence of water promotes CO_2 transport across these FTM membranes. Fig. 10.20 shows the CO_2 transport mechanism by the carrier attached to the polymer backbone, facilitated by H_2O-forming bicarbonate on the feed side interface of the membrane. CO_2 is released on the permeate side as a result of bicarbonate reacting with the protonated amine groups.

3.4.1 *Fixed-Site Carrier Facilitated Transport Membranes*

Inspired by the amine scrubbing process, in which CO_2 is absorbed by amine solvent (e.g., alkanolamine), amine groups and its derivatives are the most commonly used carriers for FTMs. The chemical structures of the most commonly used polyamine FSCs, namely polyvinylamine (PVAm), polyallylamine (PAAm) and polyethylenimine (PEI) are illustrated in Fig. 10.21.

Polyvinylamine (PVAm) has been reported as one of the best FSC membrane materials. Among the aforementioned three polyamines (PVAm, PAA, PEI), PVAm has the highest amino group concentration with the lowest steric hindrance degree, which induce a high reactivity within CO_2 [85]. However, PVAm can suffer from instability issues, due to the poor mechanical properties achieved at high humidity, which can lead to excessive matrix swelling. This problem is usually overcome by blending PVAm with a structural polymer, which can improve the mechanical properties of the matrix under wet conditions. Typically, polyvinyl alcohol (PVA) is adopted for this purpose [85,86]. Crosslinking PVAm on a microporous support, such as polysulfone, is another solution [87] to improve the mechanical properties. Hägg et al. patented [88] FTM membranes based on a polysulfone support coated with crosslinked PVAm. Their membrane exhibited high selectivity for CO_2 over N_2 and CH_4.

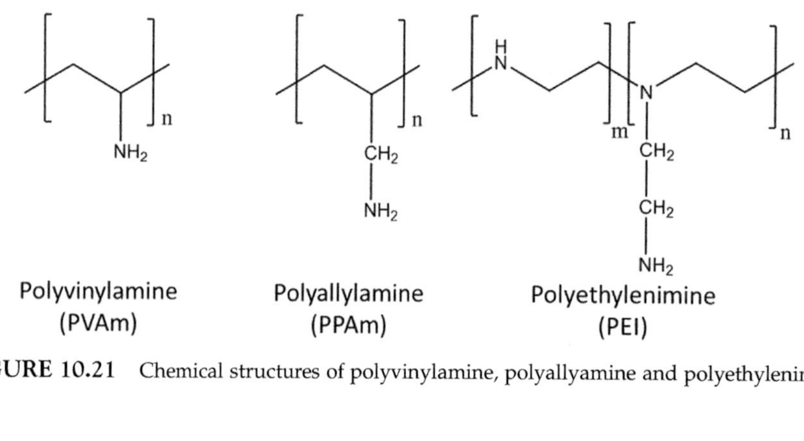

FIGURE 10.21 Chemical structures of polyvinylamine, polyallyamine and polyethylenimine.

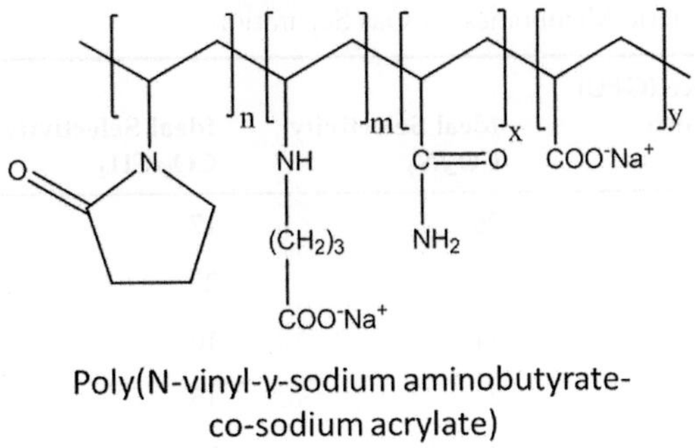

FIGURE 10.22 Chemical structure of poly(N-vinyl-γ-sodium aminobutyrate-co-sodium acrylate).

Other methods to obtain a more stable materials can rely on the use of fillers such as nanofibrillated cellulose (NFC) as a reinforcing agent in the membrane structure. Recently, Venturi et al. [89] developed mixed matrix FTMs, based on blending PVAm solution with NFC, via solvent casting procedure. The membrane containing 30% NFC showed high selectivity values of 218 and 184, for CO_2/N_2 and CO_2/CH_4 systems respectively, at 85% RH thus determining a significant improvement with respect to Robeson's 2008 upper bound. However, all films suffered from an excessive water swelling, which largely hindered the separation performances.

Others types of FSCs have been investigated. For instance, Du et al. [90] synthesized poly (N,N-dimethylaminoethyl methacrylate) by free radical bulk polymerization, and coated onto polysulfone substrates, followed by crosslinking with p-xylylene dichloride in a heptane solution. When hydrated, the membrane was shown to be more permeable to CO_2, while the membrane selectivity did not change significantly. Their membranes could demonstrate a CO_2/N_2 ideal selectivity of 50. The high permselectivity of PDMAEMA to CO_2/N_2 is attributed to the amino groups in the polymer, which interact with CO_2 favorably.

Wang et al. [91] developed FTMs with poly(N-vinyl-γ-sodium aminobutyrate-co-sodium acrylate) as FSC (Fig. 10.22).

They showed that the carriers, the secondary amine and carboxylate ions, react with CO_2 reversibly to form HCO_3^- in the humidified membrane, which facilitates the permeation of CO_2. In wet conditions, the membrane reaches a CO_2/CH_4 selectivity of 50, which stays stable after more than 160 h of operating time.

3.4.2 Mobile Carrier Facilitated Transport Membranes

Mobile carrier membranes are usually liquid membranes, of which the major ingredient is often carrier solution or carrier-bearing liquid compounds, e.g., ionic liquid. Myers et al. [92] reported the use of amino ionic liquid as CO_2 carrier. Their FTMs were based on a 1-(3-aminopropyl)-3-methylimidazolium bis(trifluoromethylsulfonyl)imide and presented high CO_2/H_2 selectivity and CO_2 permeability at elevated temperature, 15 and 1000 Barrer, respectively. Dai et al. [93] used as well ionic liquids as mobile carrier. They developed FTMs based on cocasting an ionic liquid, 1-butyl-3-methylimidazolium tetrafluoroborate ([Bmim][BF4]), with Nafion polymer. While adding 40 wt% [Bmim][BF4] to Nafion induced a CO_2 permeability increase from 1.6 to 75.4 Barrer, the presence of water could further increase the CO_2 permeability to 390 Barrer, which constitutes a 200× enhancement compared with neat Nafion. Water vapor introduced in the membranes further improved gas permeability as a "second promoter." They showed the presence of a synergy between the ionic liquid and water vapor within the Nafion matrix.

However, from the point of stability, polymeric FSC membranes are in general more favorable than membranes with mobile carriers as the carriers in FSC membrane are covalently bonded directly to the polymer matrix.

3.4.3 Conclusion

Table 10.1 Summarizes the performance of the new generation gas separation membranes discussed so far. Most of authors report permeability that is a material property, while permeance is more relevant for industrial application.

All the novel materials cited in the previous section need to be scaled up maintaining their performances, which presents several challenges. As already mentioned, most data in the literature were collected from a single gas test for self-standing membrane with a thickness of around 50–150 μm that do not represent the performance of the membrane in industrial settings (mixed gas, presence of impurities, hollow fibers modules, thin film...).

TABLE 10.1 New Generations of Polymeric Membranes for Gas Separation.

Membrane Name	CO_2 Permeance (GPU) CO_2 Permeability (Barrer)	Ideal Selectivity CO_2/N_2	Ideal Selectivity CO_2/CH_4	Form Tested	Reference
PIM1	12,600 Barrer	26	17	Film	[25,27]
	360 GPU		23	Hollow Fiber	[94]
PIM-EA-TB	7,140 Barrer	14	10	Film	[28]
PIM-TMN-Trip	30,000 Barrer	2	15	Film	[29]
KAUST-PI	2,389–2,071 Barrer	22–21	23–21	Film	[30,31]
TR-polybenzoxazoles	4,000–5,500		20–40	Film	[33]
TR-polybenzoxazoles	2,500 GPU	16		Hollow fiber by non–solvent-induced phase separation	[35]
TR-poly(benzoaxole-coimide)	655 Barrer		33	Film	[95]
ZIF8-PEBAX	200 Barrer	41		Film	[42]
MWCNT-PIM1	7,813 Barrer	18		Film	[74]
GO-PI	16 Barrer	84		Film	[76]
GO-PSf	74 GPU	44	30	Hollow Fiber	[77]
GO-PEBA	90 Barrer	85		Film	[78]
NFC-PVAm	93 Barrer	218	184	Film	[89]
PDMAEMA	80 GPU		50	Film	[90]

4. CHALLENGES IN MEMBRANE SCIENCE AND LIMITS OF TECHNOLOGY

The selection of membrane materials with good separation properties does not guarantee successful large-scale implementation in industry. Factors that influence the performance of a gas separation membrane system include the properties of the membrane (selectivity, permeability, stability) in the specific application, the physical properties and the chemical composition of the feed, and adequate engineering design including module configuration, stage cut, and flow management.

4.1 Stability Over Time

For membrane technology to become commercially viable for gas separation, membranes should possess sustainable performance over long-term operation. Physical and chemical (i.e., effect of components in the feed gas on membrane performance) aging of the membrane should be considered.

Physical aging is normally observed in glassy polymers because they are not in a thermodynamic equilibrium. The aging causes a chain relaxation accompanied by increases in density and reduction of FV, and therefore, a decrease of membrane permeability over time. In particular, glassy polymers with high FV, such as certain polyacetylenes and PIMs, are very sensitive to physical aging [96]. Therefore, significant research is focused on mitigating physical aging of such high FV polymers. Various methods have been followed to improve the aging resistance of high FV polymers, including polymer backbone modification, cross-linking, heat treatment, and mixing of nanomaterials. In his review, Low et al. [96] detailed several approaches used for preventing and/or minimizing the physical aging within PIMs. However, so far, all these approaches are not providing effective solutions for industrial applications. For instance, polymer backbone modification approach may involve complex and expensive synthesis procedures, blending of nanomaterials method is less suitable for thin film composites, and cross-linking method may produce more brittle compounds.

Methods based on in situ regeneration have been used as well with hollow fiber module configuration [97,98]. Recently, Brunetti et al. [99] developed an in situ restoring procedure for TR hollow fiber membranes using methanol. The methanol was injected directly into the aged membranes modules without any disassembly. The separation properties of restored membranes were fully recovered: CO_2 and N_2 permeance and CO_2/N_2 selectivity all reached their pristine values measured before the aging tests.

Another issue to consider is that commonly reported aging studies are conducted at constant temperature, zero stress, and under no influence from any other external conditions. Under industrial plant operating conditions such as repeated startup and shutdown or ambient temperature changes, physical aging could potentially accelerate [13]. Moreover, physical aging is also a thickness dependent, which means that thin films experience a more dramatic aging than thick films. The accelerated aging of thin films is attributed to the enhanced mobility near the surface [100] or the diffusion of FV to the surface of the film [101] that allows the polymer to reach a lower FV state more quickly than bulk samples.

Some components in the raw feed gas may alter as well the separation performance. Many polymers can be swollen, plasticized, or damaged when exposed to hydrocarbons, CO_2, or others impurities present in gas streams for CO_2 capture (such as H_2O, NOx, SOx, NH_3, H_2S...). As Scholes et al. [102] showed in his review, these impurities can have a dramatic effect on the separation performance. Actually, the presence of minor components in the membrane gas separation competes with CO_2 for separation and therefore decreases permeability, as well as degrades the membrane, altering separation performance. More recently, Lasseuguette et al. [81] studied the impact of the humidity and flue gas impurities on PIM-1 membrane performance for postcombustion capture, i.e., CO_2 permeability and CO_2/N_2 selectivity. The presence of flue gas contaminants (NOx, SOx, water) leads to a dramatic decrease of the membrane performance even after short exposure. The CO_2 permeability is reduced to just one-third of the initial value after only 5 days of exposure. The FTIR analysis suggests the presence of an acidic hydrolysis of the nitrile moieties of PIM-1. These results indicate that a full investigation including flue gas exposure tests is crucial to properly assess the potential of the material in carbon capture application.

Moreover, exposing polymeric membranes to high-pressure condensable gases, such as CO_2, leads to plasticization of the membranes and a sudden decrease in their separation performance. Plasticization is defined as an increase in the gas permeability and loss of size discrimination, causing a significant reduction of diffusivity selectivity [103]. By consequence, with pure gases, the permeability will tend to increase with increasing feed pressure. However, for gas mixtures, the opposite can be observed because the solubility of a particular gas can be significantly reduced (compared with that of the same gas in a pure state) due to competitive sorption between gas species. Therefore, the plasticization phenomenon can be suppressed [104,105].

Various strategies have been applied to reduce or suppress the CO_2 plasticization, such as polymer blending [106], thermal annealing [107], and cross-linking [108]. Recently, Isfahani et al. [109] developed a plasticization-resistant cross-linked polyurethane membrane.

4.2 Module Formation

The first consideration for the engineering design of membrane systems is the module configuration. Membrane can be assembled as flat sheets in long rolls or in the form of hollow fibers. Hollow fibers are the dominant configuration covering about 80% of the entire membrane market with spiral wound modules making up most of the rest [18]. Hollow fibers modules consist of a bundles of fibers enclosed in a pressure vessel, in which the permeate is passing either out of the fibers (bore-side feed) or into them (shell-side feed). Each module could contain up to 10^5 fibers. They are of interest for large-scale industrial applications, due to the high membrane area to module volume ratio (>1000 $m^2\,m^{-3}$), which results in high productivity per volume unit and cost efficient production. However, the method used to prepare hollow fiber ("spinnability") limits the number of materials that can be formed into high-performance membranes (Table 10.1). For instance, Jue et al. [94] investigated the fabrication of defect-free PIM-1 hollow fibers. They explained that while PIM-1 flat membranes can be easily obtained, the fabrication of defect-free PIM-1 hollow fibers can be challenging. They used THF as dope solvent, which is constraining due to its volatility. THF dopes are much more difficult to work with due to rapid solvent evaporation in the air gap after leaving the spinneret. The rapid solvent loss produced a thick (tens of micrometers) outer skin on a flattened fiber (Fig. 10.23A). To avoid this issue, they have to use a triple orifice spinneret with butanol and THF as the sheath layer, which prevents the flash vitrification type formation of the skin layer.

With 100% butanol, they obtained hollow fibers but with pinhole defects (Fig. 10.23B), whereas when THF is added, free-defect hollow fibers have been obtained (Fig. 10.23C), with a skin layer of 2.8 μm. Pinhole defects are likely caused by the rapid loss of THF into the 1-butanol sheath layer while in the air gap. Because of this solvent loss, the nascent fibers quickly phase invert after entering the water quench bath, locking into place any defects

FIGURE 10.23 PIM-1 hollow fibers (A) without triple spinneret, (B) with triple spinneret—100% butanol, (C) with triple spinneret—butanol/THF [94].

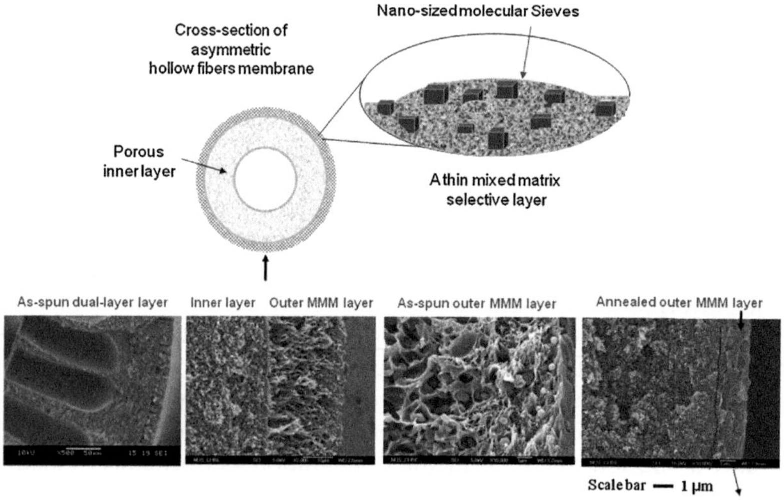

FIGURE 10.24 Schematic cross-section morphology of the dual-layer hollow fiber with a polymer/particle mixed matrix skin [115].

in the skin layer. With the addition of THF, this loss is slowed down. The separation performance of these hollow fibers was slightly higher to the ones previously reported for PIM-1 flat sheet membrane, with a relatively high CO_2 permeance of 360 GPU and CO_2/CH_4 selectivity of 23. This difference has been noted before for other glassy polymers and has been attributed to partial polymer chain alignment due to the high shear environment of the solution within the spinneret [110].

In the same way, the fabrication of MMM into hollow fibers (Fig. 10.24) with minimum defects is complex and challenging and could affect the performance of the membrane in gas separation. The performances of the hollow fibers are strongly affected by particle loading, agglomeration of particles, particle sizes, and spinning process variables (spinneret dimension, dope solution flow rate, gap height,...) [111–114].

Jiang et al. [116] investigated the fabrication of hollow fibers based on Matrimid 5218 mixed with zeolite. They showed that with the increase of the loading and the particle size, the viscosity of the dope solution is modified, which affects the particle movements. The particles may rotate causing agglomeration in the spinneret and consequently a blockage. Moreover, this agglomeration of the particles will impact the skin layer structure, with pore blockage and defects, which will contribute to lower the membrane performance.

4.3 Thin Membranes

To minimize the membrane area required for the process and the amount of specialized material needed, membrane materials must be formed into thin membranes with a selective layer 0.1–1.0 µm thick. Thin film composite membrane (TFCM) systems have been shown to be one of the best ways to achieve high permeance. Most of the TFCMs consist of a porous substrate, a gutter layer, and an ultrathin (<100 nm) top layer. Membranes with a thickness of a few tens of nanometers can be prepared using various methods, such as solution casting [117], dip and spin coating [118–120], and layer-by-layer assembly [121–123].

However, permeation properties of composite membranes with such thin selective layer are very different from thick films. The arrangement of the polymer chains in thin polymeric films differs from bulk polymers [8], resulting in different transport properties, and consequently different separation performances. For instance, Kim et al. [124] have developed thermally rearranged polybenzoxazole (TR-PBO) hollow fiber membranes using non–solvent-induced phase separation.

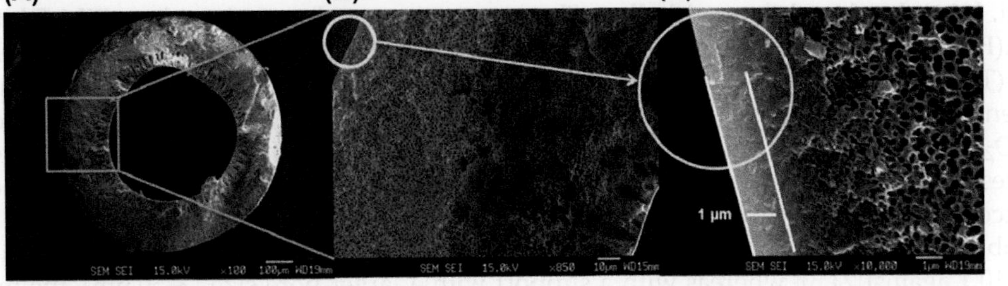

FIGURE 10.25 Cross-sectional images of thermally rearranged polybenzoxazole (TR-PBO) hollow fiber membranes (A) overall hollow fiber, (B) morphology at wall, and (C) outer skin layer [124].

They managed to obtain hollow fibers (Fig. 10.25) with up to 2 μm thin effective skin layer without defects, with CO_2 permeance of 2000 GPU and a CO_2/N_2 selectivity about 13. But these values are lower than ones from a flat sheet TR-PBO membrane casted from the same dope solution due to the aging. Actually, polymers with a rigid backbone and high FV, such as TR-PBO or PIMs, can show very promising gas separation performance in the thick dense membrane form. However, the accelerated physical aging in the thin film form results in losing their superior gas transport properties in a very short time [125]. The physical aging rate is higher in thin films than in bulk polymers, due to the higher diffusion rate of the FV to the surface. In the same way, thin films are more sensitive to plasticizing gases, such as CO_2, than thick films, even at lower pressures [126].

Furthermore, one of the challenges is to obtain this very thin layer without defects. Actually, the presence of defects on the surface of a thin layer has dramatic effects on the selectivity of the membrane. As detailed before (Section 2), for dense layer, the separation behavior is based on a solution-diffusion phenomenon. With defects, such as pinholes, the transport mechanism might change, inducing a decrease in the selectivity or even no selectivity at all. Repair techniques are usually employed to seal any defects to achieve effective dense layer thicknesses of 100 nm or less [127].

Kim et al. [128] managed to fabricate an ultrathin TR-PBOI membrane with GO scaffold, with a defect-free and ultrathin selective layer below 40 nm. Ultrathin TR-PBOI membranes were fabricated by the formation of a thin layer of precursor GO and polyimide on a porous inorganic substrate and subsequent thermal rearrangement (Fig. 10.26).

They explained that the GO scaffold inside the polymer phase not only enabled the formation of the ultrathin selective layer of TR-PBOI but also provided mechanical robustness. The resulting membrane showed remarkable gas permeance, while maintaining the gas selectivity of the pristine polymer. In particular, it had a CO_2 permeance of 1784 GPU and a CO_2/CH_4 selectivity of 32.

Finally, the asymmetric structure introduces a new challenge in the manufacturing of the modules used in industry. The influence of the support can play an important role in the separation performance hindering the transport of the gas species. This is especially important with the advent of very permeable materials: the support could become

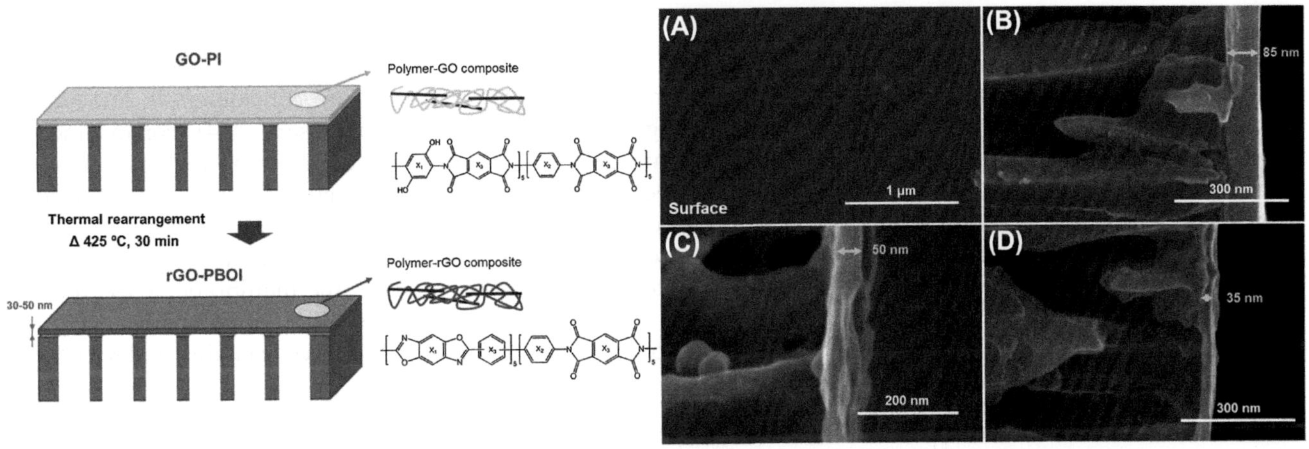

FIGURE 10.26 Schematic of the polymer–GO composite membranes (left) and SEM images of rGO–PBOI-2.0 membranes prepared from the solution in THF with a concentration of (A and B) 5.0 wt%, (C) 2.0 wt%, and (D) 1.5 wt%. (Right). *Reproduced from S. Kim, J. Hou, Y. Wang, R. Ou, G. Simon, J. Seong, Y.M. Lee, H. Wang, Highly permeable hermally rearranged polymer composite membranes with a graphene oxide scaffold for gas separation, Journal of Materials Chemistry A 6 (2018) 7668–7674.*

the limiting step in the separation needs reversing any benefit of the novel materials [129,130]. For instance, P. Bernardo et al. [131] developed flat and hollow fiber composite membranes based on PIM-EA(H_2)-TB coated on PAN. PIM-EA(H_2)-TB similar to PIM-EA(Me)-TB, it is based on the formation of the bridged bicyclic diamine called Tröger's base (TB: 6H,12H-5, 11-methanobdibenzo[b,f][1,5]diazocine). These highly rigid ladder polymers demonstrate at ambient temperature very fast gas permeability and good selectivity, surpassing the Robeson's upper bound in the case of O_2/N_2, H_2/N_2, and H_2/CH_4 gas pairs [132,133] but for self-standing membrane. Therefore, P. Bernardo et al. [131] investigated the effects of using a thin film composite instead self-standing membrane on the separation properties. According the type of support, different permeation properties were obtained. In case of flat-sheet support, the membrane prepared on native PAN support (pore size ~20 nm) had a CO_2/N_2 selectivity that is half the intrinsic one (10.1 against 21.3), whereas with a support with a larger pore size (~50 nm), the membrane present a Knudsen-like diffusion, with a diffusion higher for N_2 than CO_2. Actually, a gutter layer was required to avoid the intrusion of PIM-EA(H_2)-TB into the support pores. With the hollow fiber membranes, the support played as well an important role on the separation performance. The PAN support needed to be functionalized to enhance the compatibility with the PIM-EA(H_2)-TB coating layer and keep the high CO_2/N_2 selectivity of PIM-EA(H_2)-TB.

4.4 Sustainable Fabrication

Another limitation to these new polymers is the presence of toxic solvents used during their fabrication (polymer synthesis and membrane preparation), such as THF [94], NMP [32,33], and DMAc [33]. These toxic solvents pose severe risks for the workers' safety and the environment. For instance, during the membrane manufacturing process, the solvent contained in the dope solution is lost in the water used as the coagulation bath. Then, this mixture is sent directly to the wastewater treatment plant or to a recycling unit. Treatment of these effluents is not easy due to their toxicity, volatility, and environmental implications of the solvents used. Less toxic, or completely green, new solvents have been recently synthesized and/or experimented for the substitution of the hazard ones for the preparation of polymeric membranes mainly made of PVDF, PES, and cellulose acetate (CA) [134,135]. Conversely, no studies have been published on the use of green solvents with PIMs, probably due to the fact that PIMs can be difficult to dissolve. In the same way, in MMM application, the nanomaterials used as filler such as MOF and COF are not green to fabricate. The environmental challenges of metal–organic chemistry are complex, as they combine the environmental impacts of organic chemistry with hazards associated with metal ions, their salts, and complexes [136]. Therefore, extensive effort has been dedicated mostly to developing new "green," such as using biocompatible building blocks [137], reducing energy input [138], using water and near-supercritical solvents as the reaction medium [137,139], and avoiding bulk solvent [140].

Membrane technology, which is considered as a sustainable and energy-efficient process, should better control its environmental footprint by reducing the use of toxic solvents and nanomaterials and using "green" solvents and nanomaterials. More holistic approaches such as life-cycle assessment are required, where the whole process is evaluated in terms of environmental impact [141–143].

4.5 Process Configuration

Most of the experimental studies on the use of membrane for gas separation are considered as one stage of separation. However, to achieve high purity, a multistage system is sometimes necessary [144]. For instance, for CO_2 capture, because of the low CO_2 feed partial pressure driving force in flue gas, it is impossible for a single-stage membrane process to achieve the prescribed CO_2 recovery and purity targets even for high CO_2/N_2 selectivity materials due to pressure ratio limitations [145]. The targets can be realized in multistage configurations (Fig. 10.27).

Woo et al. [146] evaluated the separation performance of TR-PBOI hollow fiber membrane, which presents exceptional permeation properties. They observed that despite the high performance of the membrane, a second stage operation is required to achieve 90% purity. However, the disadvantage of the multistage processes is the high capital cost due to the use of a number of modules and compressors. Several studies have been devoted to optimize the configuration of multistage processes [147,148].

4.6 Conclusions

Criteria for selecting membranes for a given application are complex. Durability, mechanical integrity at the operating conditions, productivity, and separation efficiency are important stipulations that must be balanced against

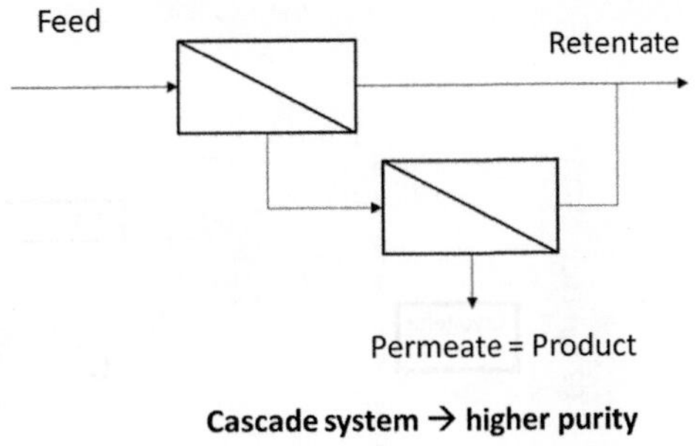

FIGURE 10.27 Multistage membranes system arrangement with permeate as product.

cost in all cases. In the absence of defects, the selectivity (or separation efficiency) is a function of the material properties at the operating conditions. The permeation rate (or productivity) is a function of the material properties and the thickness of the barrier, and the lower the thickness, the higher the productivity.

5. APPLICATIONS OF GAS SEPARATION MEMBRANES

5.1 Sustainable Solution for Gas Separation

Membranes can provide a sustainable solution for gas separation compared with traditional processes such as absorption, distillation, or cryogenic separation due to their energy efficiency (Fig. 10.28). They present a low carbon footprint; small spatial requirement do not require a thermal driving force, and no phase change is involved in the separation process.

For instance, during the cryogenic separation process, the components of gas mixtures are separated by a series of compression, refrigeration, and separation steps requiring large amount of energy (Fig. 10.29).

This low energy requirement allows the membrane technology to be a cost-effective system, in comparison with the other separation processes, and is starting to attract attention even from industries that are traditionally very conservative toward new technologies.

5.2 Current Commercial Membranes

In the following section, current commercial applications will be detailed.

The global gas separation membranes market had a value of 1720.1 million US$ in 2015 and is expected to reach 3300 million US$ by 2024 [1]. In the 1980s, the first large-scale membranes for gas separation appeared. Permea (now part of Air Products) developed hollow fibers made of polysulfone for separating hydrogen from methane.

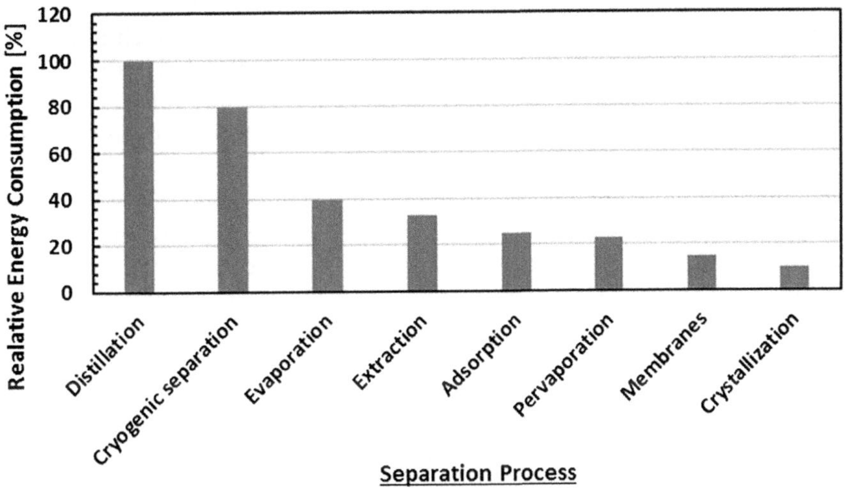

FIGURE 10.28 Relative energy consumption for different separation processes. *Adapted from U. D. o. E. Report, Materials for Separation Technology: Energy and Emission Reduction Opportunities, (2005).*

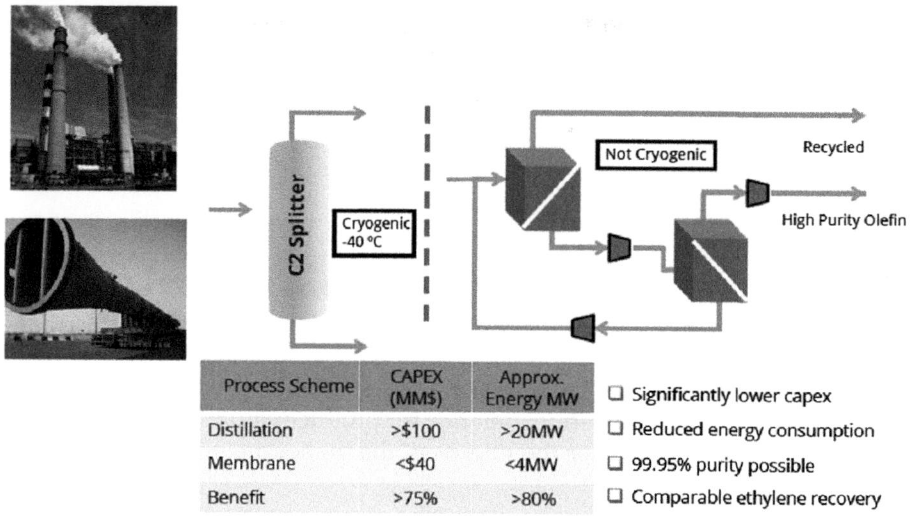

FIGURE 10.29 Comparison of cryogenic distillation with membrane technology for olefin-paraffin separation process [150].

TABLE 10.2 Current Commercial Gas Separation Membrane Systems [151].

Company	Market	Polymer	Module Type
Air Product (Prism, Cactus)	H_2, N_2, H_2O, CO_2	Polysulfone	Hollow fibers
Air Liquide (Medal)	H_2, N_2, CO_2	Polyimide, polyaramide	Hollow fibers
Generon (IGS)	N_2	Tetrabromo polycarbonate, TPX	Hollow fibers
Ube	H_2, N_2, H_2O	Polyimide	Hollow fibers
Evonik (Sepuran)	H_2, N_2, CH_4, He	Polyimide	Hollow fibers
Parker (Aquilo)	O_2, N_2	Polyphenylene oxide	Hollow fibers
Schlumberger (Cynara)	CO_2	Cellulose acetate	Hollow fibers
UOP (Separex + GMS)	H_2, CO_2	Polyimide, cellulose acetate	Spiral wounds
Membrane Technology and Research	CO_2, olefins	Fluoropolymers, polyimide	Spiral wounds
Fujifilm (Apura)	H_2, CO_2	Polydimethylsiloxane	Spiral wounds

Later, Cynara (now part of Schlumberger) and Separex (now part of UOP) manufactured a CA membrane for carbon dioxide separation from methane. The applications of membrane were expanded hereafter with nitrogen production from air, removal of hydrogen from various streams, natural gas purification, removal of VOCs from air, oxygen enrichment, and air dehydration. The current membrane gas separation industries are shown in Table 10.2.

5.2.1 *Air Separation*

The largest application of membranes for gas separation is separation of nitrogen from air. Nitrogen is traditionally produced by cryogenic distillation of air. Today, air separation is accomplished by several industrial processes, including also PSA and membranes [17]. Membrane technology is the most economical option for relatively low flow rates and purity processes. In nitrogen separation system, the retentate corresponds to the pure nitrogen, whereas the permeate is an oxygen-enriched stream ($\approx$40%–50%) with water vapor at low pressure. In particular cases, this flow can also have a value, for example, as oxygen-enriched air for combustion.

Membrane technology has a greater suitability for niche applications requiring low purity nitrogen, portability, and ease of operation. Membranes are ideal for remote areas where transport by tank cars is not economical or in specific challenging setting such as offshore platforms, aircrafts, or boats. For instance, in 2016, Evonik has developed a polyimide-based membrane Sepuran [152], which allows to obtain 99.5% N_2 purity with low air consumption, and can be used in the aerospace industry to inert the fuel tanks of planes and create a flame-resistant atmosphere. Air product has a dedicated company, Air Products Norway, which commercializes hollow fiber membranes (Prism) for nitrogen separation on-board ships and offshore installations.

5.2.2 *Air Drying*

The removal of water vapors from atmospheric air is commercially done by refrigeration or adsorption methods. The first one is not energy efficient and not effective to produce very dry air. The second needs regeneration step with heat or vacuum system. Therefore, membrane technology is very attractive because of its lightweight and of the possibility to obtain dry air with a very low dew point (~ −40°C). Almost all of the polymers have high water permeability, compared with other gases. In air-drying system, water vapors permeate through the membrane leaving air dry in the retentate.

5.2.3 *Hydrogen Separation*

Hydrogen is produced in refineries by steam-methane reforming, where methane is reacted with water to produce a mixture of hydrogen and carbon monoxide, called syngas. Syngas can be used to produce electricity, ultrapure hydrogen after a shift reaction, transportation fuels, or chemical products such as ammonia.

Currently, H_2 can be purified through one or a combination of three major processes; PSA, fractional/cryogenic distillation, or membrane separation. However, the latter method, membrane separation, is currently considered to be the most promising because of low energy consumption, continuous operation, its ease of operation, and ultimately cost effectiveness. Hydrogen recovery with membrane is currently economical if the H_2 purity required is below 99%, and the pressure loss is not a penalizing factor [153]. At present, hydrogen separation membranes are used for recovering ammonia purge gas, oxo-chemical synthesis, and refinery off-gas purification [17]. Historically, the use of Pd-based membranes was explored extensively, as they are thermally resistant and highly permeable to H_2. However, more and more polymeric membranes are used for H_2 separation. Over the past 25 years, substantial progress had been achieved in developing polymeric membranes for hydrogen purification, with polymeric membranes with improved thermal and chemical resistance (ammonia, benzene, hydrochloric acid) [154]. Typical polymers are polysulfone, polyimide, and CA (Table 10.2).

The first commercial application for hydrogen separation membranes was in the ammonia synthesis loop. In ammonia production, hydrogen and nitrogen are reacted at high pressure to form ammonia. As the conversion per pass is not 100%, the reactor is operated in recycle mode. Ammonia is continuously condensed out of the loop, and fresh synthesis gas is added.

The VaporSep-H2 unit, developed by Membrane Technology and Research (MTR), consists of a single-stage membrane system that recovers most of the hydrogen from the purge gas stream. The hydrogen permeating the membrane, almost free of methane and argon, is recycled to the synthesis gas compressor suction, pressurized, and returned to the synthesis loop. The hydrogen depleted residue, containing the purged methane and argon, is sent to reformer fuel. VaporSep-H2 systems recover more than 80% of purge gas hydrogen. Installation of such a unit will increase ammonia production by 4%−5%, without increasing gas feed to the reformer [155]. This system was installed in a Korean refinery in 2008. The unit improved the process economy and paid itself after only 1 month of operation [16].

Other hydrogen separations involve adjusting molar ratio of syngas (H_2/CO). Syngas can be produced from steam reforming of natural gas, oxidation of heavy oils, or gasification of coal or coke. Depending of the method used, the ratio H_2/CO will vary from 1 to 5 [17]. This ratio needs to be adjusted for specific synthesis application.

Prism, developed by Air Products, offers a simple method for separating H_2 from syngas to adjust the H_2:CO ratio. Membrane separation is a good fit for this application as the feed gas is treated at essentially the same pressure as the synthesis loop pressure. The hydrogen/carbon monoxide ratio can be easily adjusted to meet specific process requirements. The PRISM Membrane System can produce 85 mol% purity CO in a single stage system. For systems requiring higher purity CO, a two-stage system and a recycle compressor will improve the separation, increasing the CO product purity to 95 mol% or greater.

Refinery off-gas purification is another hydrogen-based commercial membrane application. Petroleum crude feedstock contains many different molecular weight products that must be cracked, i.e., separated before use. This process of hydrocracking relies on injecting hydrogen into the cracker to improve the reaction. The hydrogen is recycled from the hydrocracker products to the hydrocracker feed, by using polymer membranes. Membranes with H_2/CH_4 selectivities of 20−25 and very high permeability are suitable for this application. Glassy polymers such as polyimide present these characterizations and can be used for this application.

Medal membranes, developed by Air Liquid, are often used for this application. The pressurized feed gas enters the bundle from the shell side; the hydrocarbons stay under pressure while the H_2 is collected at a lower pressure from the fiber bore.

5.2.4 *Natural Gas Purification*

Natural gas is a mixture of methane, carbon dioxide, ethane, hydrogen sulphide, nitrogen, and hydrocarbons. The composition varies according the well. However, all gas fields necessitate some treatment for preventing pipeline corrosion, adjusting the heating value and dew point of the fuel. This requires the removal of acid gases, such as hydrogen sulphide and carbon dioxide, and of inert gases (N_2 and He).

Removal of acid gases from natural gases is a growing area for membrane technology. In 2010, less than 2% of the natural gas purification was realized with membrane processes, and it is estimated to grow up to 5% by 2020 [156]. Until the introduction of membrane technology, the standard removal technique of CO_2 was amine absorption. However, as with membrane technology, low purity is obtained, and amine plant is still preferred for relatively low-concentration and high flow gas streams [157].

Current commercial polymeric membranes are not selective enough to permit separation of acid gases from the natural gas with acceptable methane losses in a single-stage operation. Therefore, a two-stage process is commonly used [157].

The first membranes for natural gas purification were developed in mid-1980s by UOP (previously Grace and Separex) and by Schlumberger (ex-Cynara). They were based on CA. These CA membrane are still widely used today, but others materials gained attraction, such as polyimides (Medal from Air Liquide), polyether ether ketone (Porogen from Air Liquide) or mixed matrices membranes (yet noncommercial).

Fujifilm has developed a spiral wound membrane elements Apura. It uses spiral wound membrane modules to remove acidic gases such as CO_2 and H_2S from the natural gas feed stream. This results in an enhanced caloric product stream (known as residue or retentate) that is low in acidic gases, and a waste stream (permeate) that is lean in hydrocarbons and rich in CO_2 and H_2S.

This system has been installed in 2015 in an American company providing saleable natural gas and liquid petroleum products. After 3 months, the system showed better results than CA. The hydrocarbon recovery achieved with Apura was almost 10% higher than with CA elements. This improvement was estimated to be worth US$ 474,000 per year per bank, versus the CA elements, with a payback of less than 9 months [158]. This solution provides a lower capital expenditure (CAPEX) and OPEX for end users over alternative process solutions.

5.2.5 *Separation of Volatile Organic Components*

VOCs are hydrocarbons with high volatility. They occur naturally as well as being produced and used in many industrial processes of the oil and gas, chemical, and petrochemical industry. Typical VOC sources are crude oil, solvents, fuels, additives, alcohols, etc.

VOC emissions, caused by evaporation and other processes, can result in significant health risks. They are also known as precursor substances for ozone gas contributing to climate change. Therefore, suitable measures must be taken to minimize VOC emissions. Regulations have been implemented in many countries to minimize VOC emissions by reduction of consumption and installation of emission control systems.

The main applications in the removal of VOCs are the separation process to capture VCM (vinyl chloride monomer) from PVC production off-gas, hexane recovery in polyolefin degassing, VOC recovery in tank fill operations, vapor recovery at petrol stations, and the VOC abatement in soil decontamination.

Currently, activated carbon and thermal/catalytic oxidation are widely used for removing VOCs from gases. However, with more and more rigorous regulations on the control of VOC emissions, new efficient technologies need to be developed. Among recovery processes, gas membrane permeation offers an attractive solution due to the high affinity of elastomeric polymers for organic substances and their low-temperature process (Table 10.3).

The membranes for solvent vapor separation have to be selective for hydrocarbon vapors versus air. They also have to be chemically and mechanically stable. Polydimethylsiloxane (PDMS) is the most commonly used membrane material for this kind of separation due to its high selectivity and chemical resistance.

MTR has taken the lead in the development of membranes for this application, with the VaporSep system, based on a PDMS membrane. This system is currently used for different applications such as the separation of unreacted VCM from the polyvinylchloride production by Solvay, Oxyvinyls and the separation of hydrocarbons during the polyethylene production by Exxon, Mobil, SABIC.

VaporSep systems allow PVC producers to recover 90% to 99+% of the VCM currently lost in vent streams, providing a significant economic benefit.

TABLE 10.3 Comparison With Different Technologies for Volatile Organic Components (VOCs) Removal [16].

Technology	VOC Content	Efficiency [%]	Process Temperature (°C)	Disadvantages
Thermal oxidation	20 ppm–20% LEL[a]	95–99	371	Production of toxic gases Flow <1700 $Nm^3 h^{-1}$ can be treated
Catalytic oxidation	100–1000	90–98	149	Sensitive to impurities Flow <1700 $Nm^3 h^{-1}$ can be treated
Activated carbon	700–10,000	80–90	<54	Sensitive to moisture and ketones/esters
Membranes	<20 ppm–25% LEL	90–99	Ambient	Cost if high flow

[a]*LEL, lower explosion limit; lowest concentration in which the gas will produce fire in the presence of ignition.*

MTR has also developed with OPW Fueling Components a membrane vapor recovery system for the fuel storage tanks of retail gasoline stations, OPW Vaporsaver system. The hydrocarbon emissions are reduced by 95%–99%.

Similar small systems have been also installed in gasoline stations to minimize the release of hydrocarbons vapors to atmosphere, such as PERMEATOR system by ARID Technologies Inc. and Borsig system by Borsig Membrane Technology [159]. These systems prevent the venting of gasoline vapors (VOCs), but also the financial cost due to the associated product loss from gasoline storage tanks.

5.2.6 CO_2 Separation

The applications described above present the current commercial applications for polymeric membranes for gas separations process. Thanks to research and development progress in membrane (see above, Section 3) and in process design, many new applications for gas separation membranes are being developed and could become commercial in the next few years. Most of these future applications revolve around environmental issues and in particular separation of carbon dioxide from power plant and industrial flue gases.

Atmospheric temperature increases during the last years have been associated with the growing levels of greenhouse gas emissions. Unfortunately, this trend is expected to continue during the next decades, and CO_2 is one of the main contributors to the greenhouse gases effect. One way to reduce CO_2 emissions into the atmosphere is known as carbon capture and storage. Several approaches have been proposed to capture the CO_2 from fossil fuel power plant and industrial plant emissions, including chemical and physical absorption, adsorption, cryogenic distillation, and membrane separation. Chemical absorption with packed columns using amine solutions represents the most mature technology but is very energy intensive and suffers from various problems including flooding, foaming, and/or solvent degradation. Membrane separations present many advantages, such as compactness and easy integration in already operating facilities [13,160].

Although, numerous commercial membrane technologies are already available for natural gas sweetening, i.e., removal of CO_2 and H_2S from natural gas, this does not necessarily ensure the feasibility of membranes for flue gas treatment [161]. Actually, membranes are good for bulk separation, but generally are not an economical option for cases with either a low feed partial pressure or a high purity product requirement. Moreover, unlike well gas streams at high pressures, flue gas streams are typically at atmospheric pressure and require either costly compression or vacuum pumping to generate driving force for membrane gas separation. Despite this, many studies have shown that the membrane technology could be a cost-effective sustainable technology for CO_2 mitigation in power generation, if multistage and/or hybrid systems are using.

Merkel et al. [145] showed membrane technologies can be an attractive process for CO_2 capture. Even with a modest CO_2/N_2 selective membrane, they managed to reach DOE (Department of Energy) CO_2 capture targets by using innovative process designs. This novel process design used incoming combustion air as a sweep gas to generate driving force (Fig. 10.30).

For their analysis, they used the PolarisTM membrane with CO_2 permeance of greater than 1000 GPU and a CO_2/N_2 selectivity of 50. This permeance is tenfold higher than commercial CO_2 membranes, and the selectivity is among the highest reported for nonfacilitated transport materials. Preliminary estimates indicate this membrane process can capture 90% of CO_2 in flue gas as a sequestration-ready supercritical fluid using about 16% of plant energy at a cost as low as \$23/ton CO_2. They explained that further improvements could be possible with a higher membrane

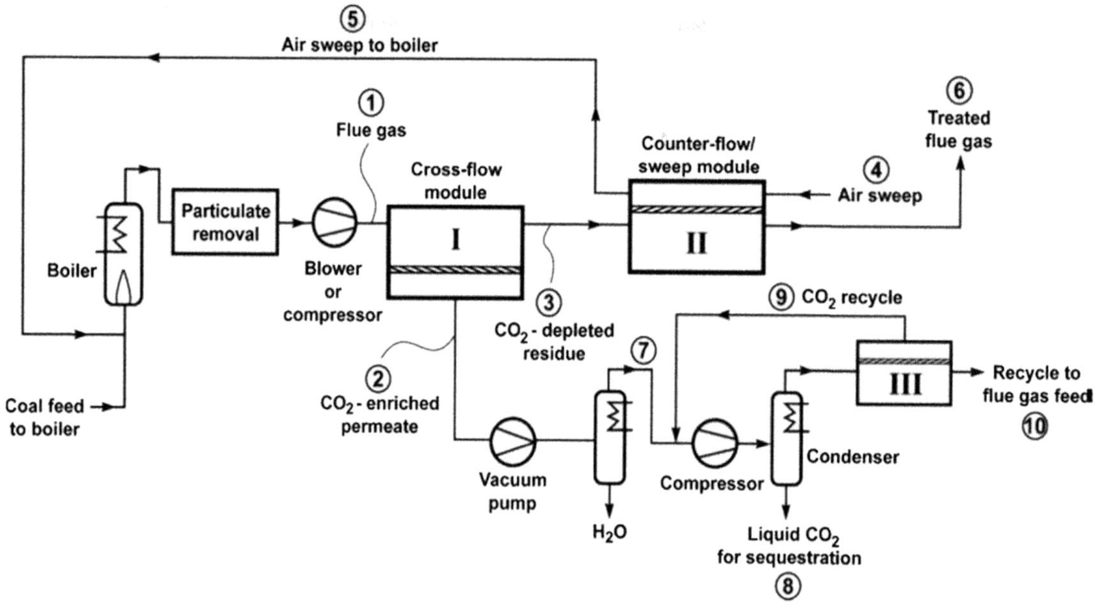

FIGURE 10.30 Simplified flow diagram of a two-step counterflow/sweep membrane process to capture and sequester CO_2 in flue gas from a coal-fired power plant [145].

FIGURE 10.31 Membrane Technology and Research (MTR) pilot system, National Carbon Capture Center (NCCC) (USA) [12].

permeance and a lower cost. In 2015, this membrane and design have been tested at the National Carbon Capture Center (NCCC) in Wilsonville (USA) (Fig. 10.31).

The field tests show stable performance with time, with a capture rate higher than 90% and a recovery higher than 65% [12].

Similarly, Bocciardo et al. [162] showed that multistage membrane separations based on PIM-1 membrane can be a competitive solution for postcombustion capture from coal-fired power plants, in alternative to amines. The overall capture performance meets the separation requirements and reduces the capture cost by more than 30%. However, the aging and the stability of the membrane with the impurities, such as H_2O, NOx, and SOx, present in the flue gas have not been taken in account.

Similar studies have been focused on the deployment of membrane in precombustion applications. In this case, the CO_2 is to be separated from hydrogen, and both CO_2 selective and hydrogen selective materials can be employed to reach the required purity and recovery. Parametric studies demonstrated that because of duty in the compression train for the CO_2, H_2 selective materials can be advantageous [14]. FTMs have also been proposed for these applications [163].

Membranes are more suitable for bulk separations and for the strict purity requirements for carbon capture, the combination of different separation processes can provide significant benefits in terms of energy requirement and costs. Air Liquide is developing a cost effective hybrid CO_2 capture process based on subambient temperature operation of a hollow fiber membrane (MEDAL) in combination with cryogenic separation. This system has been tested at the NCCC in Wilsonville (USA), and it allows reducing the CO_2 capture cost by 25% compared with the amine case [164].

FTM based on polyvinyl amine have been also tested for CO_2 capture in a pilot facility at Norcem (cement factory) in Norway [165]. The testing results indicated a 70 mol% CO_2 purity can be easily achieved in single stage. The membrane also presented a good stability by exposure to high concentration SO_2 and NO_X for a long period without significant performance change.

6. CONCLUSIONS

Over the past decade, polymeric membranes have been widely investigated for a variety of industrial gas separation applications. In today's competitive climate, membrane gas separation is now widely accepted as an economic process to produce moderate purity stream gases. To promote the practical applications of these high performance membrane materials, several major challenges should be overcome through collaboration among chemistry, materials science, and engineering. The main issues still hindering the deployment of membranes in industry are

- the performance trade-off with the development of membrane with high permeability and high selectivity,
- the stability over time (physical and chemical aging, plasticization),
- the processability of the material (spinnable and free defect thin membranes).

Latest developments in materials with high separation performances such as PIMs, TR polymers, MMMs, and FTMs are lately accompanied by an effort to investigate conditions closer to the real stream, which are encountered on industrial plants, and mitigate the traditional drawbacks of polymers. New procedures to reduce or suppress the aging have been developed such as an in situ restoring procedure. New methods to get free defect thin selective layer have been found as well. Nanotechnology is a key tool in support of the research on materials, that is, the shift to more performing materials and therefore more sustainable processes.

Although the commercialization of membranes and a large fraction of their potential applications are considered as sustainable, the availability of synthetic and manufacturing procedures mindful of sustainability and environmental impact is still lagging behind, and more efforts toward the green production of polymers should be encouraged.

References

[1] Credence Research, Gas Separation Membranes Market by Type, by Application – Growth, Future Prospects and Competitive Analysis, 2016 – 2024, Credence Research, 2017.

[2] E. Mason, From pig bladders and cracked jars to polysulfones: an historical perspective on membrane transport, Journal of Membrane Science 60 (1991) 125–145.

[3] K. Boddeker, Commentary: tracing membrane science, Journal of Membrane Science 100 (1995) 65–68.

[4] J. Nollet, Investigations on the causes for the ebullition of liquids, Journal of Membrane Science 100 (1995) 1–3.

[5] W. Koros, G. Fleming, Membrane-based gas separation, Journal of Membrane Science 83 (1993) 1–80.

[6] C. Antonson, R. Gardner, C. King, D. Ko, Analysis of gas separation by permeation in hollow fibers, Industrial and Engineering Chemistry Process Design and Development 16 (1977) 463.

[7] H. Hoehn, Heat Treatment of Membranes of Selected Polyimides, Polyesters and Polyamides, 1974. US Patent 3,822,202.

[8] B. Rowe, B. Freeman, D. Paul, Physical aging of ultrathin glassy polymer films tracked by gas permeability, Polymer 50 (2009) 5565–5575.

[9] L.M. Robeson, Correlation of separation factor versus permeability for polymeric membranes, Journal of Membrane Science 62 (1991) 165.

[10] L.M. Robeson, The upper bound revisited, Journal of Membrane Science 320 (2008) 390–400.

[11] Membrane Society of Australasia, 2017 [Online]. Available: https://www.membrane-australasia.org/polymer-gas-separation-membranes/.

[12] R. Baker, B. Freeman, T. Merkel, A membrane approach to CO_2 capture, in: CO_2 Summit III: Pathways to Carbon Capture, Utilization and Stroarge Deployment, Calabria, 2017.

[13] J. Abanades, B. Arias, A. Lyngfelt, T. Mattisson, D. Wiley, H. Li, M. Ho, E. Mangano, S. Brandani, Emerging CO_2 capture systems, International Journal of Greenhouse Gas Control 40 (2015) 126–166.

[14] T. Merkel, M. Zhou, R. Baker, Carbon dioxide capture with membranes at an IGCC power plant, Journal of Membrane Science 389 (2012) 441–450.

[15] S. Wang, X. Li, H. Wu, Z. Tian, Q. Xin, G. He, D. Peng, S. Chen, Y. Yin, Z. Jiang, M. Guiver, Advances in high permeability polymer-based membrane materials for CO_2 separations, Energy and Environmental Science 9 (2016) 1863–1890.

[16] Y. Alqaheem, A. Alomair, M. Vinoba, A. Pérez, Polymeric gas-separation membranes for petroleum refining, International Journal of Polymer Science 2017 (2017) 1–19.

[17] D. Sanders, Z. Smith, R. Guo, L. Robeson, J. McGrath, D. Paul, B. Freeman, Energy-efficient polymeric gas separation membranes for a sustainable future: a review, Polymer 54 (2013) 4729–4761.

[18] E. Drioli, G. Barbieri, Membrane Engineering for the Treatment of Gases: Volume 1: Gas-Separation Problems With Membranes, Royal Society of Chemistry, Cambridge, 2011.

[19] R. Swaidan, B. Ghanem, I. Pinnau, Fine-tuned intrinsically ultramicroporous polymers redefine the permeability/selectivity upper bounds of membrane-based air and hydrogen separations, ACS Macro Letters 4 (2015) 947–951.

[20] S. Quan, S. Li, Z. Wang, X. Yan, Z. Guo, L. Shao, A bio-inspired CO_2-philic network membrane for enhanced sustainable gas separation, Journal of Materials Chemistry A 3 (2015) 13758–13766.

[21] C. Park, E. Kim, J. Kim, Y. Lee, J. Kim, Novel semi-alicyclic polyimide membranes: synthesis,characterization, and gas separation properties, Polymer 151 (2018) 325–333.

[22] T. Zhu, X. Yang, Y. Zhang, Y. Zheng, X. He, J. Luo, Random and block copolymer membranes based on flexible etheric-aliphatic soft segments designed for CO_2/CH_4 separation, Journal of Natural Gas Science and Engineering 54 (2018) 92–101.

[23] Y. Zhuang, J. Seong, Y. Do, W. Lee, M. Lee, M. Guiver, Y. Lee, High-strength, soluble polyimide membranes incorporating Tröger's Base for gas separation, Journal of Membrane Science 504 (2016) 55–65.

[24] J. Lee, J. Jung, E. Jang, K. Lee, Y. Kang, J. Kim, CO_2-philic PBEM-g-POEM comb copolymer membranes: synthesis, characterization and CO_2/N_2 separation, Journal of Membrane Science 502 (2016) 191–201.

[25] P.M. Budd, N.B. McKeown, B.S. Ghanem, K.J. Msayib, D. Fritsch, L. Starannikova, N. Belov, O. Sanfirova, Y. Yampolskii, V. Shantarovich, Gas permeation parameters and other physicochemical properties of a polymer of intrinsic microporosity: polybenzodioxane PIM-1, Journal of Membrane Science 325 (2008) 851–860.

[26] N. McKeown, P. Budd, Polymers of intrinsic microporosity (PIMs): organic materials for membrane separations, heterogeneous catalysis and hydrogen storage, Chemical Reviews Society 35 (2006) 675–683.

[27] F.Y. Li, Y. Xiao, T.-S. Chung, S. Kawi, High-performance thermally self-cross-linked polymer of intrinsic microporosity (PIM-1) membranes for energy development, Macromolecules (2012) 1427–1437.

[28] M. Carta, R. Malpass-Evans, M. Croad, Y. Rogan, J.C. Jansen, P. Bernardo, F. Bazzarelli, N.B. McKeown, An efficient polymer molecular sieve for membrane gas separations, Science 339 (2013) 303–307.

[29] I. Rose, C. Bezzu, M. Carta, B. Comesaña-Gándara, E. Lasseuguette, M. Ferrari, P. Bernardo, G. Clarizia, A. Fuoco, J.C. Jansen, K. Hart, P. Liyana-Arachchi, M. Colina, N. McKeown, Polymer ultrapermeability from the inefficient packing of 2D chains, Nature Materials 16 (2017) 932–937.

[30] B. Ghanem, N. McKeown, P. Budd, J. Selbie, D. Fritsch, High-performance membranes from polyimides with intrinsic microporosity, Advanced Materials 20 (2008) 2766–2771.

[31] B. Ghanem, R. Swaidan, E. Litwiller, I. Pinnau, Ultra-microporous triptycene-based polyimide membranesfor high-performance gas separation, Advanced Materials 26 (2014) 3688–3692.

[32] H. Park, S. Han, C. Jung, Y. Lee, A. Hill, Thermally rearranged (TR) polymer membranes for CO_2 separation, Journal of Membrane Science 359 (2010) 11–24.

[33] H. Park, C. Jung, Y. Lee, A. Hill, S. Pas, S. Mudie, E. Van Wagner, B. Freeman, D. Cookson, Polymers with cavities tuned for fast selective transport of small molecules and ions, Science 318 (2007) 254–258.

[34] C. Scholes, C.K.S. Ribeiro, B. Freeman, Thermal rearranged poly(benzoxazole-co-imide) membranes for CO_2 separation, Journal of Membrane Science 450 (2014) 72–80.

[35] K. Woo, J. Lee, G. Dong, J.D.Y. Kim, W. Hung, K. Lee, G. Barbieir, E. Drioli, Y. Lee, Fabrication of thermally rearranged (TR) polybenzoxazole hollow fiber membranes with superior CO_2/N_2 separation performance, Journal of Membrane Science 490 (2015) 129–138.

[36] Y. Lu, H.J.L. Li, J. Song, X.G.H. Zhao, Z. Hu, T. Wang, Preparation and gas transport properties of thermally induced rigid membranes of copolyimide containing cardo moieties, Reactive and Functional Polymers 119 (2017) 134–144.

[37] B. Comesaña-Gándara, M. Calle, H. Jo, A. Hernandez, J. de la Campa, A. de Abajo, Y. Lee, Thermally rearranged polybenzoxazoles membranes with biphenyl moieties: monomer isomeric effect, Journal of Membrane Science 450 (2014) 369–379.

[38] S. Li, J. Jin, S. Han, C.K.S. Park, P. Budd, Y. Lee, Mechanically robust thermally rearranged (TR) polymer membranes with spirobisindane for gas separation, Journal of Membrane Science 434 (2013) 137–147.

[39] A. Tena, S. Rangou, S. Shishatskiy, V. Filiz, V. Abetz, Claisen thermally rearranged (CTR) polymers, Materials Science 2 (2016) 1–9.

[40] B. Diez, P. Cuadrado, A. Marcos-Fernandez, J. de la Campa, A. Tena, P. Pradanos, L. Palacio, Y. Moo Lee, C. Alvarez, A. Lozona, A. Hernandez, Thermally rearranged polybenzoxazoles made from poly(ortho-hydroxyamide)s. Characterization and evaluation as gas separation membranes, Reactive and Functional Polymers 127 (2018) 38–47.

[41] T. Chung, L. Jiang, Y. Li, S. Kulprathipanja, Mixed matrix membranes (MMMs) comprising organic polymers with dispersed inorganic fillers for gas separation, Progress in Polymer Science 32 (2007) 483–507.

[42] P.D. Sutrisna, J. Hou, H. Li, Y. Zhang, V. Chen, Improved operational stability of Pebax-based gas separation membranes with ZIF-8: a comparative study of flat sheet and composite hollow fibre membranes, Journal of Membrane Science 524 (2017) 266–279.

[43] T. Li, Y. Pan, K. Peinemann, Z. Lai, Carbon dioxide selective mixed matrix composite membrane contianing ZIF-7 nanofillers, Journal of Membrane Science 425–426 (2013) 235–242.

[44] J. Ahn, W. Chung, I. Pinnau, M. Guiver, Polysulfone/silica nanoparticle mixed-matrix membranes for gas separation, Journal of Membrane Science 314 (2008) 123–133.

[45] A. Bushell, P. Budd, M. Attfield, J. Jones, T. Hasell, A. Cooper, P. Bernardo, F. Bazzarelli, G. Clarizia, J. Jansen, Nanoporous organic polymer/cage composite membranes, Angewandte Chemie International Edition 52 (2013) 1253–1256.

[46] D. Sen, H. Kalipcilar, L. Yilmaz, Development of polycarbonate based zeolite 4A filled mixed matrix gas separation membranes, Journal of Membrane Science 303 (2007) 194–199.

[47] D. Bastani, N. Esmaeili, M. Asadollahi, Polymeric mixed matrix membranes containing zeolites as a filler for gas separation applications: a review, Journal of Industrial and Engineering Chemistry 19 (2013) 375–393.

[48] T. Gur, Permselectivity of zeolite filled polysulfone gas separation membranes, Journal of Membrane Science 93 (1994) 283–289.

[49] M.G. Süer, N. Baç, L. Yilmaz, Gas permeation characteristics of polymer-zeolite mixed matrix membranes, Journal of Membrane Science 91 (1994) 77–86.

[50] M. Hussain, A. König, Mixed-matrix membrane for gas separation: polydimethylsiloxane filled with zeolite, Chemical Engineering and Technology 35 (2012) 561–569.

[51] K. Zarshenas, A. Raisi, A. Aroujalian, Mixed matrix membrane of nano-zeolite NaX/poly (ether-blockamide) for gas separation applications, Journal of Membrane Science 510 (2016) 270–283.

[52] J. Dechnik, J. Gascon, C. Doonan, C. Janiak, C. Sumby, Mixed-matrix membranes, Angewandte Chemie International Edition 56 (2017) 9292–9310.

[53] E. Adatoz, A. Avci, S. Keskin, Opportunities and challenges of MOF-based membranes in gas separations, Separation and Purification Technology 152 (2015) 207–237.

[54] A. Uzun, S. Keskin, Site characteristics in metal organic frameworks for gas separation, Progress in Surface Science 89 (2014) 56–79.

[55] Y. Li, G. He, S. Wang, S. Yu, F. Pan, H. Wu, Z. Jiang, Recent advances in the fabrication of advanced composite membranes, Journal of Materials Chemistry A 1 (2013) 10058–10077.

[56] M. Ahmadi, E. Tas, A. Kilic, V. Kumbaraci, N. Talinli, M. Ahunbay, B. Tantekin-Ersolmaz, Highly CO_2 selective microporous metal-imidazolate framework based mixed matrix membranes, ACS Applied Materials and Interfaces 9 (2017) 35936–35946.

[57] K. Xie, Q. Fu, C. Xu, H. Lu, Q. Zhao, R. Curtain, D. Gu, P. Webley, G. Qiao, Continuous assembly of a polymer on a metal-organic framework (CAP on MOF): a 30 nm thick polymeric gas separation membrane, Energy and Environmental Science 11 (2018) 544–550.

[58] H. Maghsoudi, M. Soltanieh, Simultaneous separation of H_2S and CO_2 from CH_4 by a high silica CHA-type zeolite membrane, Journal of Membrane Science 470 (2014) 159–165.

[59] M. Vinoba, M. Bhagiyalakshmi, Y. Alqaheem, A. Alomair, A. Pérez, M. Rana, Recent progress of fillers in mixed matrix membranes for CO_2 separation: a review, Separation and Purification Technology 188 (2017) 431–450.

[60] T. Isimjan, H. Kazemian, S. Rohani, A. Ray, Photocatalytic activities of Pt/ZIF-8 loaded highly ordered TiO_2 nanotubes, Journal of Materials Chemistry 20 (2010) 10241–10245.

[61] M.J. Ordonez, J.J.B. Kenneth, J.P. Ferraris, I.H. Musselman, Molecular sieving realized with ZIF-8/Matrimid® mixed-matrix membranes, Journal of Membrane Science 361 (2010) 28–37.

[62] G. Liu, V. Chernikova, Y. Liu, K. Zhang, Y. Belmabkhout, O. Shekhah, C. Zhang, S. Yi, M. Eddaoudi, W. Koros, Mixed matrix formulations with MOF molecular sieving for key energy-intensive separations, Nature Materials 17 (2018) 283–289.

[63] Y. Liu, G. Liu, C. Zhang, W. Qiu, S. Yi, V. Chernikova, Z. Chen, Y. Belmabkhout, O. Shekkah, M. Eddaoudi, W. Koros, Enhanced CO_2/CH_4 separation performance of a mixed matrix membrane based on tailored MOF-polymer formulations, Advanced Science 5 (2018) 1800981–1800986.

[64] X. Gao, X. Zou, H. Ma, S. Meng, G. Zhu, Highly selective and permeable porous organic framework membrane for CO_2 capture, Advanced Materials 26 (2014) 3644–3648.

[65] Z. Kang, Y. Peng, Y. Qian, D. Yuan, M. Addicoat, T. Heine, Z. Hu, L. Tee, Z. Guo, D. Zhoa, Mixed matrix membranes (MMMs) comprising exfoliated 2D covalent organic frameworks (COFs) for efficient CO_2 separation, Chemistry of Materials 28 (2016) 1277–1285.

[66] T. Ben, H. Ren, S.C.D. Ma, J. Lan, X. Jing, W. Wang, J. Xu, F. Deng, J. Simmons, S. Qiu, G. Zhu, Targeted synthesis of a porous aromatic framework with high stability and exceptionally high surface area, Angewandte Chemie International Edition 48 (2009) 9457–9460.

[67] C.H. Lau, K. Konstas, C.M. Doherty, M. Carta, R.-M. Evans, E. Lasseuguette, M.C. Ferrari, N.B. McKeown, M.R. Hill, Tailoring Compatibility in Ultrapermeable Blends to Switch off Plasticization and Physical Aging, 2018. Unpublished Data.

[68] S. Smith, C. Lau, J. Mardel, M. Kitchin, K. Konstas, B. Ladewig, M. Hill, Physical aging in glassy mixed matrix membranes; tuning particle interaction for mechanically robust nanocomposite films, Journal of Materials Chemistry A 4 (4) (2016) 10627–10634, 2016.

[69] K. De Sitter, P. Winberg, J. D'Haen, C. Dotremont, R. Leysen, J. Martens, S. Mullens, F. Maurer, I. Vankelecom, Silica filled poly(1-trimethylsilyl-1-propyne) nanocomposite membranes: relation between the transport of gases and structural characteristics, Journal of Membrane Science 278 (2006) 83–91.

[70] A. Skoulidas, D. Ackerman, J. Johnson, D. Sholl, Rapid transport of gases in carbon nanotubes, Physical Review Letters 89 (2002) 1–4.

[71] A. Hirsch, Funktionalisierung von einwandigen Kohlenstoffnanoröhern, Angewandte Chemie 114 (2002) 1933–1939.

[72] S. Iijima, Carbon nanotubes: past, present, and future, Physica B 323 (2002) 1–5.

[73] M. Nejad, M. Asghari, M. Afsari, Investigation of carbon nanotubes in mixed matrix membranes for gas separation: a review, ChemBioEng Reviews 3 (2016) 276–298.

[74] M. Khan, V. Filiz, G. Bengtson, S. Shishatskiy, M. Rahman, J. Lillepaerg, V. Abetz, Enhanced gas permeability by fabricating mixed matrix membranes of functionalized multiwalled carbon nanotubes and polymers of intrinsic microporosity (PIM), Journal of Membrane Science 436 (2013) 109–120.

[75] A. Brunetti, M. Cersosimo, J. Kim, G. Dong, E. Fontananova, Y. Lee, E. Drioli, G. Barbieri, Thermally rearranged mixed matrix membranes for CO_2 separation: an aging study, International Journal of Greenhouse Gas Control 61 (2017) 16–26.

[76] L. Zhao, C. Cheng, Y. Chen, T. Wang, C. Du, L. Wu, Enhancement on the permeation performance of polyimide mixed matrix membranes by incorporation of graphene oxide with different oxidation degrees, Polymers for Advanced Technologies 26 (2015) 330–337.

[77] K. Zahri, K. Wong, P. Goh, A. Ismail, Graphene oxide/polysulfone hollow fiber mixed matrix membranes for gas separation, RSC Advances 6 (2016) 89130–89139.

[78] J. Shen, M. Zhang, G. Liu, K. Guan, W. Jin, Size effects of graphene oxide on mixed matrix membranes for CO_2 separation, AIChE Journal 62 (2016) 2843–2852.

[79] A. Figoli, W. Sager, M. Mulder, Facilitated oxygen transport in liquid membranes: review and new concepts, Journal of Membrane Science 181 (2001) 97–110.

[80] Y. Li, S. Wang, G. He, H. Wu, F. Pan, Z. Jiang, Facilitated transport of small molecules and ions for energy-efficient membranes, Chemical Society Reviews 44 (2015) 103–118.

[81] E. Lasseuguette, M. Carta, S. Brandani, M.-C. Ferrari, Effect of humidity and flue gas impurities on CO_2 permeation of apolymer of intrinsic microporosity for post-combustion capture, International Journal of Greenhouse Gas Control 50 (2016) 93–99.

[82] M. Yoshikawa, T. Ezaki, K. Sanui, N. Ogata, Selective permeation of carbon dioxide through synthetic polymer membranes having pyridine moiety as a fixed carrier, Journal of Applied Polymer Science 35 (1988) 145–154.

[83] H. Matsuyama, M. Teramoto, H. Sakakura, Selective permeation of CO_2 through poly 2-(N,N-dimethyl)aminoethyl methacrylate membrane prepared by plasma-graft polymerization technique, Journal of Membrane Science 114 (1996) 193–200.

[84] T.-J. Kim, B. Li, M.-B. Hägg, Novel fixed-site–carrier polyvinylamine membrane for carbon dioxide capture, Journal of Polymer Science Part B: Polymer Physics 42 (2004) 4326–4336.

[85] L. Deng, T. Kim, M. Hägg, Facilitated transport of CO_2 in novel PVAm/PVA blend membrane, Journal of Membrane Science 154–163 (2009) 340.

[86] L. Deng, M. Hagg, Fabrication and evaluation of a blend facilitated transport membrane for CO_2/CH_4 separation, Industrial and Engineering Chemistry Research (2015) 11139–11150.

[87] M. Sandru, S.H. Haukebø, M.-B. Hägg, Composite hollow fiber membranes for CO_2 capture, Journal of Membrane Science 346 (2010) 172–186.

[88] M. Hagg, T. Kim, B. Li, Membrane for Separating CO_2 and Process for the Production Thereof, September 29, 2005. Patent WO2005089907 A1.

[89] D. Venturi, D. Grupkovic, L. Sisti, M. Baschetti, Effect of humidity and nanocellulose content on Polyvinylaminenanocellulose hybrid membranes for CO_2 capture, Journal of Membrane Science 548 (2018) 263–274.

[90] J. Du, L. Liu, A. Chakma, X. Feng, A study of gas transport through interfacially formed poly(N,N-dimethylaminoethyl methacrylate) membranes, Chemical Engineering Journal 156 (2010) 33–39.

[91] Z. Wang, C. Yi, Y. Zhang, J. Wang, S. Wang, CO_2-Facilitated transport through poly(N-vinyl-sodium aminobutyrate-co-sodium acrylate)/polysulfone composite membranes, Journal of Applied Polymer Science 100 (2006) 275–282.

[92] C. Myers, H. Pennline, D. Luebke, J. Ilconich, J.K. Dixon, E.J. Maginn, J.F. Brennecke, High temperature separation of carbon dioxide/hydrogen mixtures using facilitated supported ionic liquid membranes, Journal of Membrane Science 322 (2008) 28–31.

[93] Z. Dai, L. Ansaloni, J. Ryan, R. Spontak, L. Deng, Nafion/IL hybrid membranes with tuned nanostructure for enhanced CO_2 spearation: effects of ionic liquid and water vapor, Green Chemistry 20 (2018) 1391–1404.

[94] M. Jue, V. Breedveld, R. Lively, Defect-free PIM-1 hollow fiber membranes, Journal of Membrane Science 530 (2017) 33–41.

[95] M. Calle, H. Jin Jo, C. Doherty, A. Hill, Y. Lee, Cross-linked thermally rearranged poly(benzoxazole-co-imide) membranes prepared from ortho-hydroxycopolyimides containing pendant carboxyl groups and gas separation properties, Macromolecules 48 (2015) 2603–2613.

[96] Z. Low, P. Budd, N. McKeown, D. Patterson, Gas permeation properties, physical aging, and its mitigation in high free volume glassy polymers, Chemical Reviews 118 (12) (2018) 5871–5911.

[97] C. Jones, W. Koros, Carbon molecular sive gas separation membranes-II. Regeneration following organic exposure, Carbon 32 (1994) 1427–1432.

[98] H. Xuezhong, M. Hagg, Hollow fiber carbon membranes: from materials to application, Chemical Engineering Journal 2015–2016 (2013) 440–448.

[99] A. Brunetti, M. Cersosimo, G. Dong, K. Woo, J. Lee, J. Kim, Y. Lee, E. Drioli, G. Barbieri, In situ restoring of aged thermally rearranged gas separation, Journal of Membrane Science 520 (2016) 671–678.

[100] B. Rowe, B. Freeman, D. Paul, Membrane Engineering for the Treatment of Gases, Royal Society of Chemistry, Croydon, UK, 2011.

[101] Y. Huang, D. Paul, Physical aging of thin glassy polymer films monitored by optical properties, Macromolecules 39 (2006) 1554–1559.

[102] A. Scholes, S. Kentish, G. Stevens, Effects of minor components in carbon dioxide capture using polymeric gas separation membranes, Separation and Purification Reviews 38 (2009) 1–44.

[103] R. Baker, Membrane Technology and Applications, John Wiley, West Sussex, 2000.

[104] S. Lee, M. Binns, J. Lee, J. Moon, J. Yeo, Y. Yeo, Y. Lee, J. Kim, Membrane separation process for CO_2 capture from mixed gases using TR and XTR hollow fiber membranes: process modeling and experiments, Journal of Membrane Science 541 (2017) 224–234.

[105] B. Stroy, W. Koros, Comparison of three models for permeation of CO_2/CH_4 mixtures in poly(phenylene oxide), Journal of Polymer Science Part B Polymer Physics 27 (1989) 1927–1948.

[106] S. Japip, H. Wang, Y. Xiao, T. Chung, HighlHighly permeable zeolitic imidazolate framework (ZIF)-71 nano-particles enhanced polyimide membranes for gas separation, Journal of Membrane Science 467 (2014) 162–174.

[107] R. Guo, D. Sanders, Z. Smith, B. Freeman, D. Paul, J. McGrath, Synthesis and characterization of thermally rearranged (TR) polymers: effect of glass transition temperature of aromatic poly(hydroxyimide) precursors on TR process and gas permeation properties, Journal of Materials Chemistry A 19 (2013) 6063–6072.

[108] B. Kraftschik, W. Koros, Cross-linkable polyimide membranes for improved plasticisation resistance and permselectivity in sour gas separations, Macromolecules 46 (2013) 6908–6921.

[109] A. Pournaghshband Isfahani, B. Ghalei, K. Wakimoto, R. Bagheri, E. Sivaniah, M. Sadeghi, Plasticization resistant crosslinked polyurethane gas separation membranes, Journal of Materials Chemistry A 4 (2016) 17431–17439.

[110] T. Chung, W. Lin, H. Rohitkumar, The effect of shear rates on gas separation performance of 6FDA-durene polyimide hollow fibers, Journal of Membrane Science 167 (2000) 55–66.

[111] M. Mubashir, Y. Fong, C. Leng, L. Keong, Issues and current trends of hollow-fiber mixed-matrix membranes for CO_2 separation from N_2 and CH_4, Chemical Engineering and Technology 41 (2018) 235–252.

[112] H. Zhu, X. Jie, L. Wang, G. Kang, D. Liu, Y. Cao, Effect of MIL-53 on phase inversion and gas separation performance of mixed matrix hollow fiber membranes, RSC Advances 6 (2016) 69124–69134.

[113] A. Zulhairun, B. Ng, A. Ismail, R. Murali, M. Abdullah, Production of mixed matrix hollow fiber membrane for CO_2/CH_4 separation, Separation and Purification Technology 137 (2014) 1–12.

[114] H. Zhu, X. Jie, Y. Cao, Fabrication of functionalized MOFs incorporated mixed matrix hollow fiber membrane for gas separation, Journal of Chemistry (2017) 1–10.

[115] N. Peng, N. Widjojo, P.T.M. Sukitpaneenit, G. Lipscomb, T. Chung, J. Lai, Evolution of polymeric hollow fibers as sustainable technologies: past, present and future, Progress in Polymer Science 37 (2012) 1401–1424.

[116] Y. Li, T. Chung, Z. Huang, S. Kulprathipanja, Dual-layer polyethersulfone (PES)/BTDA-TDI/MDI co-polyimide (P84) hollow fiber membranes with a submicron PES–zeolite beta mixed matrix dense-selective layer for gas separation, Journal of Membrane Science 277 (2006) 28–37.
[117] W. Yave, A. Car, J. Wind, K. Peinemann, Nanometric thin film membranes manufactured on square meter scale: ultra-thin films for CO_2 capture, Nanotechnology 21 (2010) 395301–395308.
[118] R. Vendamme, S. Onoue, A. Nakao, T. Kunitake, Robust free-standing nanomembranes of organic/inorganic interpenetrating networks, Nature Materials 5 (2006) 494–501.
[119] H. Watanabe, E. Muto, T. Ohzono, A. Nakao, T. Kunitake, Giant nanomembrane of covalently-hybridized epoxy resin and silica, Journal of Materials Chemistry 19 (2009) 2425–2431.
[120] E. Lasseuguette, J. Rouch, J. Remigy, Hollow-fiber coating: application to preparation of composite hollow-fiber membrane for gas separation, Industrial and Engineering Chemistry Research 52 (2013) 13146–13158.
[121] S. Markutsya, C. Jiang, Y. Pikus, V. Tsukruk, Freely suspended layer-by-layer nanomembranes: testing micromechanical properties, Advanced Functional Materials 15 (2005) 771–780.
[122] D. Kulkarni, I. Choi, S. Singamaneni, V. Tsukruk, Graphene oxide–polyelectrolyte nanomembranes, ACS Nano 4 (2010) 4667–4676.
[123] C. Nardin, M. Winterhalter, W. Meier, Giant free-standing ABA triblock copolymer membranes, Langmuir 16 (2000) 7708–7712.
[124] S. Kim, S. Han, Y. Lee, Thermally rearranged (TR) polybenzoxazole hollow fiber membranes for CO_2 capture, Journal of Membrane Science 403–404 (2012) 169–178.
[125] S. Harms, K. Rätzke, F. Faupel, N. Chaukura, P. Budd, W. Egger, L. Ravelli, Aging and free volume in a polymer of intrinsic microporosity (PIM-1), The Journal of Adhesion 88 (2012) 608–619.
[126] N. Horn, D. Paul, Carbon dioxide plasticization and conditioning effects in thick vs. thin glassy polymer films, Polymer 52 (2011) 1619–1627.
[127] J. Henis, M. Tripodi, The developing technology of gas separating membranes, Science 4592 (1983) 11–17.
[128] S. Kim, J. Hou, Y. Wang, R. Ou, G. Simon, J. Seong, Y.M. Lee, H. Wang, Highly permeable thermally rearranged polymer composite membranes with a graphene oxide scaffold for gas separation, Journal of Materials Chemistry A 6 (2018) 7668–7674.
[129] Z. Tong, V. Vakharia, M. Gasda, W. Ho, Water vapor and CO_2 transport through amine-containing facilitated transport membranes, Reactive and Functional Polymers 86 (2015) 111–116.
[130] W. Koros, R. Mahajan, Pushing the limits on possibilities for large scale gas separation: which strategies? Journal of Membrane Science 175 (2000) 181–196.
[131] P. Bernardo, V. Scorzafave, G. Clarizia, E. Tocci, J. Jansen, A. Borgogno, R. Malpass-Evans, N. McKeown, M. Carta, F. Tasselli, Thin film composite membranes based on a polymer of intrinsic microporosity derived from Tröger's base: a combined experimental and computational investigation of the role of residual casting solvent, Journal of Membrane Science 569 (2019) 17–31.
[132] M. Carta, M. Croad, R. Malpass-Evans, J.C. Jansen, P. Bernardo, G. Clarizia, K. Friess, M. Lanč, N. McKeown, Triptycene induced enhancement of membrane gas selectivity for microporous tröger's base polymers, Advanced Materials 26 (2014) 3526–3531.
[133] E. Tocci, L.D. Lorenzo, P. Bernardo, G. Clarizia, F. Bazzarelli, N.B. Mckeown, M. Carta, R. Malpass-Evans, K. Friess, K. Pilnáček, M. Lanč, Y.P. Yampolskii, L. Strarannikova, V. Shantarovich, M. Mauri, J.C. Jansen, Molecular modeling and gas permeation properties of a polymer of intrinsic microporosity composed of ethanoanthracene and Tröger's base units, Macromolecules 47 (2014) 7900–7916.
[134] Y. Medina-Gonzalez, P. Aimar, J. Lahitte, J.C. Remigy, Towards green membranes: preparation of cellulose acetate ultrafiltration membranes using methyl lactate as a biosolvent, International Journal of Sustainable Engineering 4 (2011) 75–83.
[135] A. Figoli, T. Marino, S. Simone, E. Di Nicolò, X. Li, T. He, S. Tornaghi, E. Drioli, Towards non-toxic solvents for membrane preparation: a review, Green Chemistry 16 (2014) 4034–4059.
[136] P. Julien, C. Mottillo, T. Friščić, Metal–organic frameworks meet scalable and sustainable synthesis, Green Chemistry 19 (2017) 2729–2747.
[137] K. Matsuyama, N. Hayashi, M. Yokomizo, T. Kato, K. Ohara, T. Okuyama, Supercritical carbon dioxide-assisted drug loading and release from biocompatible porous metal–organic frameworks, Journal of Materials Chemistry B 43 (2014) 7551–7558.
[138] J. Klinowski, F. Almeida Paz, P. Silva, J. Rocha, Microwave-assisted synthesis of metal–organic frameworks, Dalton Transactions 40 (2011) 321–330.
[139] N. Stock, S. Biswas, Fundamentals of green chemistry: efficiency in reaction design, Chemical Society Reviews 41 (2012) 1437–1451.
[140] C. Mottillo, T. Friščić, Advances in solid-state transformations of coordination bonds: from the ball mill to the aging chamber, Molecules 22 (2017) 144.
[141] A. Martins, N. Caetano, T. Mata, LCA for membrane processes, in: Sustainable Membrane Technology for Water and Wastewater Treatment. Green Chemistry and Sustainable Technology, Springer, Singapore, 2017.
[142] A. Schreiber, J. Marx, P. Zapp, Environmental assessment of a membrane-based air separation for a coal-fired oxyfuel power plant, Journal of Membrane Science 440 (2013) 122–133.
[143] S. Troy, A. Schreober, P. Zapp, Life cycle assessment of membrane carbon capture and storage, Celan Technology Environemental Policy 18 (2016) 1641–1654.
[144] L. Zhao, E. Riensche, R. Memnzer, L. Blum, D. Stolten, A parametric study of CO_2/N_2 gas separation membrane processes for post-combustion capture, Journal of Membrane Science 325 (2008) 284–294.
[145] T. Merkel, H. Lin, X. Wei, R. Baker, Power plant post-combustion carbon dioxide capture: an opportunity for membranes, Journal of Membrane Science 359 (2010) 126–129.
[146] K. Woo, G. Dong, J. Lee, J. Kim, Y. Do, W. Lee, H. Lee, Y. Lee, Ternary mixed-gas separation for flue gas CO_2 capture using high performance thermally rearranged (TR) hollow fiber membranes, Journal of Membrane Science 510 (2016) 472–480.
[147] P. Gabrielli, M. Gazzani, M. Mazzotti, On the optimal design of membrane-based gas separation processes, Journal of Membrane Science 526 (2017) 118–130.
[148] A. Arias, M. Mussati, P. Mores, N. Scenna, J. Caballero, S. Mussati, Optimization of multi-stage membrane systems for CO_2 capture from flue gas, International Journal of Greenhouse Gas Control 53 (2016) 371–390.
[149] U. D. o. E., Materials for Separation Technology: Energy and Emission Reduction Opportunities, 2005.
[150] S. Duong, White Paper: Cost Effective Alternative to Distillation for Olefin Purification and Extraction, [Online]. Available: https://www.imtexmembranes.com/cost-effective-alternative-to-distillation-for-olefin-purification-and-extraction.

[151] P.S. Puri, Commercial applications of membranes in gas separations, in: Membrane Engineering for the Treatment of Gases, Gas-Separation Problems With Membranes, vol. 1, Royal Society of Chemistry, 2011, pp. 215–243.
[152] Evonik [Online]. Available: http://www.sepuran.com/product/sepuran/en/nitrogen-generation/pages/sepuran-n-two.aspx, 2017.
[153] L.-M. Sun, J.-Y. Thonneleir, Permeation gazeuse, Techniques de l'Ingenieur J2810 (2000).
[154] N.W. Ockwig, T.M. Nenoff, Membranes for hydrogen separation, Chemical Reviews 107 (2007) 4078–4110.
[155] MTR Membrane and Technology Research, 2017 [Online]. Available: http://www.mtrinc.com/.
[156] R.W. Baker, Future directions of membrane gas separation technology, Industrial and Engineering Chemistry Research 41 (2002) 1393–1411.
[157] R. Baker, K. Lokhandwala, Natural gas processing with membranes: an overview, Industrial and Engineering Chemistry Research 47 (2008) 2109–2112.
[158] J. Sabey, O. Gerrits [Online]. Available: https://www.hydrocarbonengineering.com, 01 05 2015 https://www.hydrocarbonengineering.com/special-reports/01052015/membrane-technology-hydrocarbon-opex-082/.
[159] ARID Technologies, 2017 [Online]. Available: http://www.aridtech.com.
[160] M.T. Ho, G.W. Allinson, D.E. Wiley, Reducing the cost of CO_2 capture from flue gases using membrane technology, Industrial and Engineering Chemistry Research 47 (2008) 1562–1568.
[161] Twister, March 2017 [Online]. Available: https://www.twisterbv.com/praesent-at-nulla-lorem-ut-erat-nulla-laoreet-ut-est-vitae-feugiat-interdum-urna-3/.
[162] M.-C. Ferrari, D. Bocciardo, S. Brandani, Integration of multi-stage membrane carbon capture processes to coal-fired power plants using highly permeable polymers, Green Energy and Environment 1 (2016) 211–221.
[163] V. Vakharia, R. Ramasubramanian, W. Ho, An experimental and modeling study of CO_2-selective membranes for IGCC syngas purification, Journal of Membrane Science 488 (2015) 56–66.
[164] American Air Liquide Inc., CO_2 Capture by Cold Membrane Operation with Actual Power Plant Flue Gas Field Test Summary Report, 2017.
[165] M. Hägg, A. Lindbråthen, X. He, S. Nodeland, T. Cantero, Pilot demonstration-reporting on CO_2 capture from a cement plant using hollow fiber process, Energy Procedia 114 (2017) 6150–6165.

CHAPTER

11

Block Copolymer Membranes

Suzana P. Nunes

King Abdullah University of Science and Technology (KAUST), Biological and Environmental Science and Engineering (BESE), Advanced Membranes and Porous Materials Center, Thuwal, Saudi Arabia

OUTLINE

1. BLOCK COPOLYMER SELF-ASSEMBLY AND MORPHOLOGY

1.1 Block Copolymer Equilibrium Morphology

Block copolymers have naturally attracted the interest of membrane scientists for decades. They are constituted by a series of monomeric unities of the same composition, adjacently placed in the polymer molecule, and covalently bonded to another series of chemically distinct monomeric unities. The current methods available for the preparation of block copolymers are discussed in an excellent review by Polymeropoulos et al. [1]. The preparation of well-defined block copolymers with low polydispersity has been made possible with the advent of the living anionic polymerization in the 50s. More recently, new polymerization techniques were introduced, such as atom transfer radical polymerization, reversible addition-fragmentation chain transfer polymerization (RAFT), ring-opening polymerization, and click chemistry, expanding the number of possible architectures and chemical compositions.

Sustainable Nanoscale Engineering
https://doi.org/10.1016/B978-0-12-814681-1.00011-4

The thermodynamic interaction between different blocks guides the degree of segregation in the melt or in solution and is the main reason for the versatility of morphologies seen in these materials. When applied to membranes, block copolymers offer the possibility of precisely tuning them at the nanoscale level and combining materials with different properties in one single membrane. Excellent reviews on the block copolymer self-assembly in the solid state have been published by Bates et al. [2–4], discussing in detail the morphology as a function of the block–block Flory–Huggins thermodynamic interaction parameter, χ_{AB}, the degree of polymerization, N, and the size fraction of each block $f_A = N_A/N$. For the self-assembly to take place and lead to an ordered arrangement as spheres, cylinders, or lamellae, a minimum value of $\chi_{AB}N < 10.5$, as shown in Fig. 11.1. For block copolymer dense films or porous membranes prepared by sacrificial etching of one of the blocks after the self-assembly, the order–disorder phase diagram is fundamental. These are equilibrium or metastable morphologies, and their prediction is valid only for systems enabled to anneal long enough to delete any history of solvent influence or nonequilibrium condition. For the membrane application, effective paths for the transport and high selectivity would be preferentially provided by cylindrical or gyroid morphologies.

1.2 Self-Assembly of Block Copolymers in Solution

The polymeric membrane manufacture mostly includes a step of solution casting. The self-assembly of block copolymers in solution leads to an ordered morphology, which depends not only on the interaction between the block but also the interaction between each block and the solvents and copolymer concentration. The large variety of morphologies of block copolymers in solution could be explained and roughly predicted by using the simplified concept of packing factors, proposed initially for surfactants, as in Fig. 11.2 [5], but can be extended to copolymers in solution.

The form of the assemblies is guided by the packing parameter, p, which is given by the relation between the parameters v, l, and a (Fig. 11.2). These parameters are directly related to the size of the blocks, their relative fraction, and the thermodynamic interaction between the blocks themselves and between the solvent and each of the blocks. In a diluted solution of block copolymers in a highly selective solvent, micelles ($p < 1/3$) are favored. In concentrated solutions of block copolymers with strong repulsion between blocks and relatively similar interactions with the solvent, bilayers might be favored. Fig. 11.2 is, however, a simplified picture of the large number of possible morphologies in copolymer solutions. The complexity of these morphologies increases by increasing the number of distinct blocks per molecule and the number of solvents in the solution. Another important factor is the copolymer concentration. Each of the assemblies depicted in Fig. 11.2 can themselves arrange in ordered systems. In this way, micelles can organize themselves in cubic or hexagonal arrangements, the cylinders can also assume a hexagonal order, bilayers can arrange themselves in lamellar arrangements, etc. The influence of the form of self-assembly on the membrane morphology will be discussed later in Section 5.2.2.

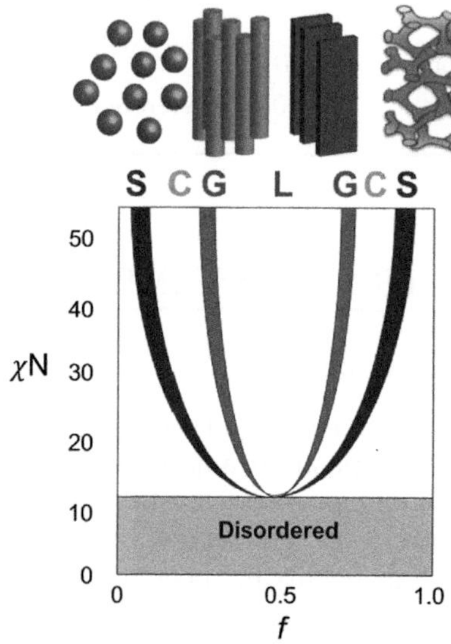

FIGURE 11.1 Diblock copolymer morphology in the melt: spheres, cylinders, lamellae, and gyroid. *Adapted from F.S. Bates, G.H. Fredrickson, Block copolymers designer soft materials, Physics Today 52 (1999) 32.*

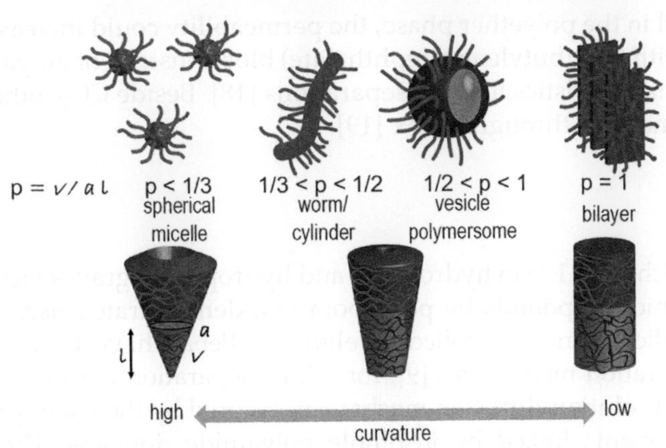

FIGURE 11.2 Diblock copolymer morphologies in solution, where p is the packing parameter, v is the volume of the solvophobic block, l is its length, and a is its contact area with the solvophilic block. *Adapted from J.N. Israelachvili, Intermolecular and Surface Forces, 2011; S.P. Nunes, Block copolymer membranes for aqueous solution applications, Macromolecules 49 (2016) 2905–2916.*

2. BLOCK COPOLYMERS FOR DENSE MEMBRANE PREPARATION

The most explored block copolymers for membrane fabrication are those commercialized as Pebax by Arkema and Nexar by Kraton (Scheme 11.1). Pebax is a polyether-*b*-polyamide copolymer with different degrees of hydrophilicity. Nexar is a pentablock of *tert*-styrene, hydrogenated isoprene, and randomly sulfonated styrene. Both copolymers have been explored for different membrane applications, such as gas separation [7], ultra- and nanofiltration [8,9], pervaporation [10], fuel cell, and batteries.

2.1 Gas Separation

The preferential interaction of the polyether block with CO_2 in Pebax favors its transport, while the polyamide limits the plasticization and swelling and provides a high mechanical stability. Pebax is hardly soluble in single solvents, due to the polyamide block, and this delayed its use for membrane manufacture. In the first reports [7], Pebax was dissolved in n-butanol or 1,1,2-trichloroethane, by heating close to the solvent boiling point. Solvent mixtures in many cases successfully dissolve block copolymers in concentrations useful for membrane preparation. Water/ethanol was later successfully applied as a convenient solvent for the hydrophilic Pebax grade [11]. Poly(ethylene oxide) alone or similar polyethers do not lead to good membranes for CO_2 separation. One issue is the plasticization as the CO_2 transport proceeds, and consequent reduction of the selectivity in real operation. Another issue is the tendency to crystallize, leading to a low gas permeation. The use of a copolymer of polyether and polyamide has secured CO_2 permeability of 73 Barrer for the pure copolymers and CO_2/N_2 and CO_2/CH_4 selectivity values [12] of 15.6, suitable for industrial applications [13,14]. By adding poly(ethylene glycol) (PEG) [12] or other components such as ionic liquids [15], oligomers of ionic liquid silica [11], and metal–organic

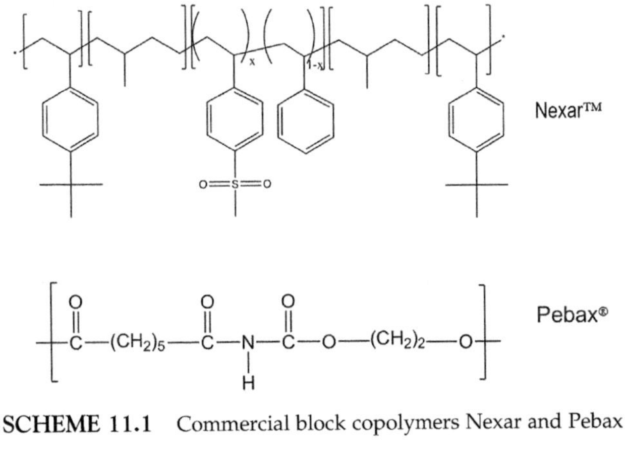

SCHEME 11.1 Commercial block copolymers Nexar and Pebax

frameworks [16,17] dispersed in the polyether phase, the permeability could increase above 150 Barrer. Polyactive is a commercial copolymer with poly(butylene terephthalate) block, instead of polyamide, which also has excellent permeation and selectivity characteristics for CO_2 separations [18]. Beside CO_2, other acid gases, such as H_2S and SO_2, have a preferential permeation through Pebax [19].

2.2 Liquid Separation

The availability of Pebax (Scheme 11.1) in hydrophilic and hydrophobic grades facilitates its application for dehydration and removal of organic compounds by pervaporation, demonstrated decades ago [10,20–22]. The nanostructuration, high hydrophilicity, and controlled swelling of Pebax have been responsible for the successful performance of dense ultrafiltration membranes [9], for which separation does not rely on well-defined pore size but rather on a combination of solution-diffusion mechanism favored by the water polyether affinity and high flexibility of the hydrophilic segments linked by insoluble polyamide domains. The hydrophilicity also helps in decreasing the fouling susceptibility. Cross-linking strategies and the incorporation of fillers and inorganic phases could decrease the molecular weight cut off of these membranes for application in nanofiltration in aqueous and organic medium or in bioelectrochemical devices [8,23,24].

The copolymer has been mainly explored for the preparation of dense layers on porous supports for different membrane applications. Geise et al. [25] were the first to report on Nexar application as a membrane, investigating it for desalination. Thong et al. [26] demonstrated the application of Nexar nanofiltration membranes for removing heavy metal ions from wastewater with water permeation of 2.4 $L\ m^{-2}\ h^{-1}\ bar^{-1}$. The membrane was prepared using a polyethyleneimine gutter layer between a porous support and the copolymer layer. The reported solute rejection corresponds to that of a membrane with 0.5 nm "pore" diameter; however, like for Pebax nanofiltration membranes, a simple size-sieving in this separation range is not necessarily the dominant acting transport mechanism [27]. Nexar was used for alcohol dehydration via pervaporation as a single layer [28] or as the polyanion in a layer-by-layer membrane on hydrolyzed polyacrylonitrile [29]. The morphology is important. The cyclic hydration of Nexar films promotes a transition to a more interconnected network of sulfonated domains [30] that favors the water transport. By choosing the right solvents to cast Nexar films, the water transport can be several times increased [31].

Besides the use of commercial block copolymers, the development of membranes for liquid separation, based on lab-synthesized copolymers, has been reported by different groups. For instance, one of the approaches explored multiblock copolymers with poly(dimethyl siloxane) grafting for bioethanol recovery as a sustainable membrane process [32].

2.3 Fuel Cell and Batteries

Because of the sulfonated blocks of Nexar, it has been considered for proton transport and required for a fuel cell, vanadium flow battery, or ionomer actuators [33]. As for water transport, the morphology is very important for the proton conductivity. Although this can be controlled, the stability under oxidative conditions is also mandatory for fuel cell applications.

The synthesis and testing of cation-exchange copolymer membranes with protons conductive blocks for fuel cell have been proposed by Taeger et al. [34] as sulfonated aramides and by Alvarez-Gallego et al. [35] as sulfonated polynaphthalimides with benzimidazole pendant groups. The influence of the morphology on proton-conductive membranes based on block copolymers, as well as for application in lithium and redox flow batteries, actuators, and electrodialysis, has been revised by Shin et al. [36]. More recent developments based on fluorinated block copolymers for anion exchange membranes with high free volume for hydroxide ion transport have been reported [37].

3. BLOCK COPOLYMERS FOR SURFACE MODIFICATION OF MEMBRANES FOR WATER-BASED SEPARATIONS

The membrane surface functionalization with PEG or zwitterionic segments is a common strategy to minimize fouling during the filtration of aqueous streams. McCloskey et al. [38] adopted a polydopamine coating layer as a platform for further functionalization and incorporation of PEG segments on membrane surfaces. This grafting strategy has been later used by many groups leading to covalently bonded hydrophilic segments [39,40]. An interesting approach for membrane modification was proposed by Xie et al. [41]. Membranes were prepared from copolymers of plain polysulfone and azidomethylated polysulfone. A postfunctionalization by click reaction with propargyl alcohol attached hydrophilic 1,2,3-triazole groups to the membrane. The click reaction strategy could be in principle used for a large variety of polymers and copolymer membranes. Instead of grafting or postfunctionalization, linear block copolymers can be used as additives. In this case, the hydrophilic segment is normally distributed on the surface, while the second block strongly interacts with the membrane, physically anchoring the chain to the surface. A recent example is the use of poly(ethylene oxide)-*b*-poly(methyl methacrylate) as additive for poly(vinylidene fluoride) (PVDF) ultrafiltration membranes [42]. Poly(methyl methacrylate) and PVDF are miscible and have been previously used as homopolymer blend for membrane preparation [43]. When the block copolymer is used, the acrylate segment is miscible with the PVDF matrix, but the poly(ethylene oxide) block tends to migrate to the membrane–water interface. A different approach for modification of PVDF membranes with sustainable antibiofouling protection was reported, consisting of the covalent immobilization of silver nanoparticles, mediated by a thiol-end functional amphiphilic block copolymer linker [44].

The nanostructuration of alternating hydrophobic and hydrophilic groups is believed to be even more effective in mitigating fouling. Self-assembled copolymers with PEG and fluoroblocks are proposed as fouling release coatings [45–48]. Oligomeric zwitterionic building blocks have been recently integrated into an interfacially polymerized polyamide membrane with successful increase of fouling resistance [49]. More recently, copolymers containing peptides blocks are being explored as antifouling coatings. Zwitterionic peptides [50] or zwitterionic block copolymers based on glutamic acid and lysine [51] have been recently reported for this purpose.

4. BLOCK COPOLYMERS FOR POROUS ULTRA- AND NANOFILTRATION MEMBRANES

The variety of morphologies that can be achieved with block copolymer has attracted the imagination of membrane scientists for long, and different approaches have been tested for the preparation of porous materials and membranes. They could be summarized as two main strategies: (1) the preparation of dense films, followed by etching or partial solubilization and (2) self-assembly and non–solvent-induced phase separation (SNIPS).

4.1 Preparation by Etching of Dense Films

The pioneer approaches on porous block copolymer membranes were based on a selective etching of one of the blocks, for instance, by exposing polyisoprene-*b*-poly(4-vinylphenyl-dimethyl-2-propoxysilane) or polystyrene-b-polyisoprene copolymers to ozone. The ozonolysis etches out the isoprene blocks [52–54]. Plasma etching, reactive-ion etching, and UV-induced degradation are other applied strategies, besides ozonolysis [55–57]. The sacrifice of poly(lactic acid) (PLA) blocks, by etching with alkaline solutions, has been explored in different ways by the Hillmyer's group [58–62]. The etching has been recently applied to membranes prepared from copolymers with robust norbornene blocks bearing amines and side chains of polylactide and liquid crystal forming blocks [63]. Particularly, interesting approaches were proposed, combining a polymerization-induced phase separation and in situ block copolymer formation, as well as cross-linking strategies, followed by selective etching of segments to form pores [62,64,65]. The combination of block copolymer self-assembly, interfacial polymerization, and selective etching is also a promising approach [61]. The main focus of this review is, however, the block copolymer membranes prepared by non–solvent-induced phase separation as discussed below.

4.2 Preparation by Self-Assembly and Non–Solvent-Induced Phase Separation

4.2.1 *First Reports and Basic Principles*

The combination of the block copolymer self-assembly and the non–solvent-induced phase separation process, leading to an isoporous membrane without an etching step or selective dissolution of a pore-forming additive, was first demonstrated by Peinemann in the frame of the European project COMPOSE—"Multicomponent nanostructured materials for separation membranes" (2004–07) [66]. The process consists of casting a copolymer solution, followed by a short partial solvent evaporation and immersion in a nonsolvent, normally water, and is represented in Fig. 11.3.

The first results [66] were reported with polystyrene-*b*-poly(4vinyl pyridine) (PS-*b*-P4VP) (Scheme 11.2). Peinemann chose the same optimized conditions and solvents he previously used for the preparation of polyimide dense asymmetric membranes, using a mixture of dimethylformamide (DMF) and tetrahydrofuran (THF) as solvent.

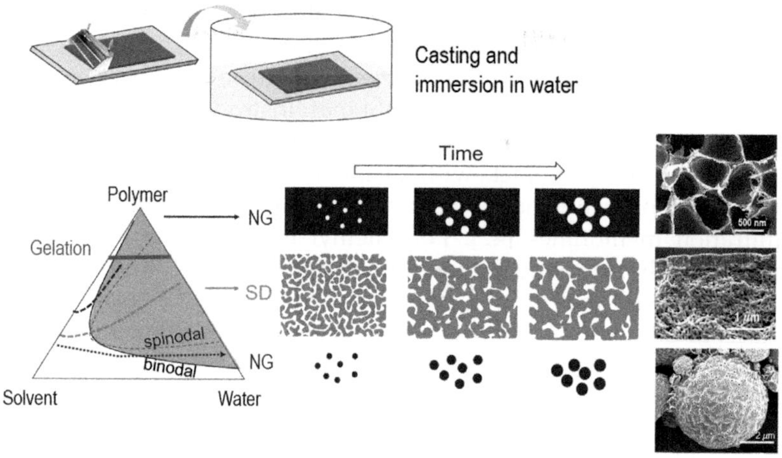

FIGURE 11.3 Solution casting and immersion in water and the consequent phase separation process. Depending on the water–solvent exchange path, the phase separation may follow the mechanisms of spinodal decomposition (SD) or nucleation and growth (NG). At low polymer concentration, NG leads to the formation of particles, whereas at high concentration, polymer-lean nuclei may form a honeycomb network of cavities. Asymmetric membranes with a gradient of pores from the top to the bottom and a highly ordered isoporous layer on the top can be favored by SD. A gelation interrupts the phase separation, immobilizing the developing morphology and trapping it in a nonequilibrium thermodynamic state. *Images adapted from S.P. Nunes, Block copolymer membranes for aqueous solution applications, Macromolecules 49 (2016) 2905–2916; S.P. Nunes, R. Sougrat, B. Hooghan, D.H. Anjum, A.R. Behzad, L. Zhao, N. Pradeep, I. Pinnau, U. Vainio, K.V. Peinemann, Ultraporous films with uniform nanochannels by block copolymer micelles assembly, Macromolecules 43 (2010) 8079–8085; H. Yu, X. Qiu, S.P. Nunes, K.V. Peinemann, Biomimetic block copolymer particles with gated nanopores and ultrahigh protein sorption capacity, Nature Communications 5 (2014) 4110; S.M. Chisca, Valentina; Sougrat, R.; Behzad, A.; Nunes, S.P., Artificial 3D Hierarchical and Isotropic Porous Polymeric Materials, vol. 4, 2018, pp. eaat0713.*

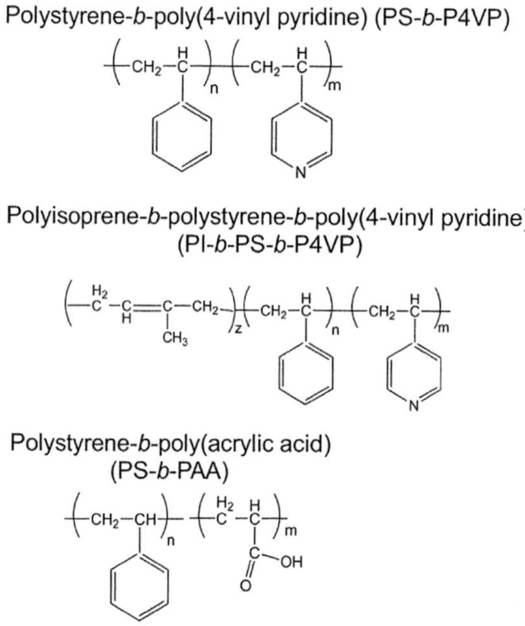

SCHEME 11.2 Block copolymers commonly used for self-assembly and non–solvent-induced phase separation (SNIPS) membrane preparation.

A membrane with densely distributed pores with the same size was obtained, as shown in Fig. 11.4. This morphology had not been seen before in membranes prepared by phase inversion, without a postetching step. However, at that time, the reproducibility with different batches of the copolymer was a critical issue. Only years later [67], after the better understanding of the formation mechanism and the influence of the copolymer self-assembly, forming micelles that themselves assume a regular order in the casting solution, isoporous membranes could be properly reproduced and explored in different approaches.

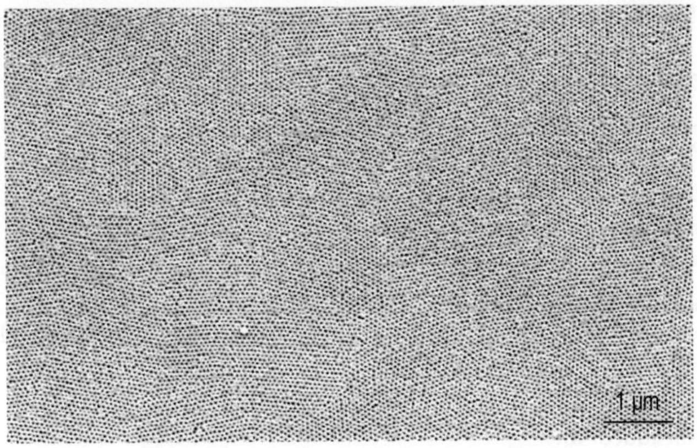

FIGURE 11.4 Scanning electron microscopy (SEM) image of the surface of a polystyrene-*b*-poly(4vinyl pyridine) (PS-*b*-P4VP) isoporous membrane, prepared by self-assembly and non−solvent-induced phase separation (SNIPS).

Since these initial reports [67], SNIPS is developing as a membrane field itself, investigated by different groups. Table 11.1 summarizes the main achievements in the last decade on the SNIPS process. The combination of block copolymers and classical polymer solution thermodynamic is an emerging field with broad opportunities for developing membranes for aqueous and solvent-based separations with strict size-sieving selectivity, allied to a variety of possible functionalization strategies for separations stimulated by chemical affinity. Stimuli-responsive gates could extend the application as dynamic systems in the biomedical sector and sensors industry.

The membrane formation by SNIPS follows in part the phase inversion mechanism widely used for commercial membranes [6,67], summarized in Fig. 11.3. When the polymer solution is cast and immersed in water, the solvent−nonsolvent exchange brings the system to a thermodynamic instability and leads to a progressive separation in a polymer-lean and polymer-concentrated phases. The initial polymer concentration in solution is the main parameter to decide if the result will be a self-standing film (high concentration), useful for membrane application, or dispersed isolated particles (low concentration). There are two classical mechanisms of phase separation: spinodal decomposition (SD) and nucleation and growth (NG). SD dominates if the solvent−nonsolvent exchange is fast and the thermodynamic conditions rapidly deteriorate to favor the separation, "quenching" the system into the unstable region of the phase diagram. Even a relatively slow exchange can favor SD if the paths into the two-phase area of the phase diagram do not cross a metastable condition (red line in Fig. 11.3). On the other hand, NG is the favored mechanism if the system goes through a metastable condition for some time, before fully separating in two phases. SD starts with a local and continues concentration oscillation. The polymer concentration difference between the two developing phases smoothly increases with the time. If the initial polymer concentration is high enough, a highly viscous and continuous polymer matrix is formed and the separation is kinetically trapped. In the case of NG polymer lean (or polymer concentrated, depending on the total polymer content in the system), nuclei are formed. The polymer concentration in the nuclei remains practically constant as the separation proceeds; the nuclei grow by in size, by transport of polymer and some solvent molecules from the continuous phase. Homopolymer and block copolymer solutions follow these mechanisms of "macrophase" separation.

What really makes the difference when block copolymers are used is that, in addition to the SD or NG "macrophase" separation, they have the ability to self-assembly in solution in the nanoscale. In diluted solutions, they assemble into individual micelles, rods, or vesicles, as in Fig. 11.2. If the concentration is high enough, these separated unities will themselves arrange in a supramolecular assembly, forming an ordered network. This micelle assembly highly influences the final membrane morphology. The first evidences for that were proposed and demonstrated by Nunes et al. [67]. Details on the exact micelle assembly and a complete understanding of all the stages leading to the final pore formation are still a topic of discussion and investigation. A comprehensive description of the pore formation is complex and might need to consider additional aspects, such as surface phenomena and their influence on the late stages of the pore formation [6].

Fig. 11.3 shows three different block copolymer structures, which are tuned by navigating the phase diagram with different paths. The middle one is the most commonly reported asymmetric isoporous membrane, mainly following the SD phase separation mechanism. Highly interconnected pores are formed, with a gradient of growing pore sizes from the top to the bottom of the membrane. Only the top layer (<500 nm) is isoporous (see surface image in Fig. 11.4) with dominant influence of the block copolymer self-assembly. If the initial copolymer concentration is low, a continuous film will not be formed, but the NG mechanism can promote the formation of porous particles

TABLE 11.1 Milestones in Block Copolymer Membranes Prepared by Self-Assembly and Non–Solvent-Induced Phase Separation (SNIPS).

Date	Milestone	Reference
2007	First isoporous SNIPS membranes (PS-*b*-P4VP in DMF/THF casting solution)	[66]
2010	Reproducible PS-*b*-P4VP membrane with added copper acetate Report on the influence of micelle formation in solution on the membrane morphology; use of SAXS, FIB, and cryo TEM and cryo SEM to demonstrate it	[67]
2011	pH responsive isoporous membranes (pore size switching from 20 to 1 nm) Block copolymer membranes prepared on nonwoven to improve stability and facilitate the scale-up in machines ("seed fund" initiative at KAUST)	[70]
2011	Proposition of more efficient solvent mixtures, including dioxane for better membranes reproducibility without added salts	[71]
2011	Isoporous triblock membranes based on PI-*b*-PS-*b*-P4VP	[72]
2012	SAXS demonstration that the self-assembly order in solution correlates with the final membrane morphology	[73]
2013	Block copolymer membranes fabrication in machines	[74]
2013	First isoporous hollow fiber membranes	[74–76]
2013	Catalytic isoporous membranes with gold nanoparticles	[75]
2013	Influence of complexing additives with strong H-bond formation and the relevance of micelle rigidity; rheological investigation and correlation with morphology	[77]
2013	Use of dissipative particle dynamic combined to SAXS to demonstrate the isoporous membrane formation	[78]
2013	Spinout TeraPore founded to manufacture block copolymer membranes	Teraporetech.com
2014	First dual layer hollow fiber membranes with homopolymer as internal layer and isoporous block copolymer outer layer	[79]
2014	PS-*b*-PEO isoporous membranes and the influence of molecular weight on the pore size	[80]
2014	RAFT copolymerization applied to the preparation of isoporous triblock membranes	[81]
2014	Pore tuning between 3 and 20 nm with electroless gold deposition	[82]
2014	Anfiro start-up (anfiro.com)	
2015	Time-resolved GISAXS applied to follow the membrane formation	[83]
2015	Asymmetric isoporous membranes used as template for inorganic porous materials	[84,85]
2015	Ionic liquids used to drive the block copolymer self-assembly and the membrane morphology	[86]
2015	Membranes based on polysulfone copolymers with azide groups for further click chemistry postmodification	[41]
2015	Used of mixtures of PS-*b*-P4VP of different molecular weights to tune the pore size down to 20 nm	[87]
2015	Use of glycerol and pH protonation to reduce the pore size to 5 nm	[85]
2015	Blends of PS-*b*-P4VP and PS-*b*-PAA copolymers to obtain membranes with pore size in the range of nanofiltration (ca. 1.5 nm)	[88,89]
2016	Photoresponsive membranes based on anthracene blocks	[90]
2016	Block copolymer membranes without polystyrene block: polysulfone and acrylic acid blocks	[91,92]
2016	Semiempirical trendline solvent selection proposed for block copolymer membranes	[93]
2018	Membranes prepared from diverse block compositions (e.g., PS-*b*-PSOH-*b*-PS, thiol-modified SBS)	[94,95]
2018	3D isotropic isoporous films prepared from block copolymers	[69,96]
2018	Membranes prepared from copolymers with polypeptide blocks	[97]

with internal order [68]. Here, again the block copolymer self-assembly also plays an important role, and the order is influenced by the minimization of the energy in a confined state. The system depicted on the top right of Fig. 11.3 was most recently reported and will be discussed at the end of this chapter [69]. In this case, a metastable stage develops with the formation of polymer-lean nuclei. The morphology is immobilized, as the nuclei grow surrounded by a concentrated polymer matrix. The contribution of the block copolymer self-assembly in this case is the formation of a highly ordered isoporous morphology in the walls interconnecting the micrometer-sized cavities.

4.2.2 *Solvent Optimization and Influence of the Block Copolymer Order in Solution on the Membrane Morphology*

Choosing the right solvent mixture is essential to guide the final membrane morphology, as shown in Fig. 11.5. DMF/THF luckily worked for the first reported isoporous PS-*b*-P4VP membrane, but this is not the most convenient solvent mixture for this system. Nunes et al. [71] early demonstrated that by adding dioxane to the mixture, a much more regular and ordered membrane morphology could be reproduced. This could be explained by correlating it with the order observed in the binary and ternary DMF/THF/dioxane solutions, as visualized by cryoscanning electron microscopy [71,83] in Fig. 11.5 and by small-angle scattering methods [83,98]. A semiempirical method for identifying convenient solvent mixtures for block copolymers leading to isoporous membranes was proposed by Sutisna et al. [93]. They plotted a ratio of Flory–Huggins interaction parameters as a function of the blocks volume ratios for a number of investigated systems that were successful or unsuccessful in leading to isopores, as shown in Fig. 11.6. This is an excellent guide for optimization, particularly when new block copolymer compositions start to be tested, but it is just the first step. Once the best solvent mixture has been identified, the adjustment of the polymer concentration is the next essential step to guarantee a stable self-assembly.

The self-assembly in the casting solution of block copolymer isoporous membranes has been confirmed in detailed by small-angle X-ray scattering static [73,98,99] and time-resolved experiments, pioneered by KAUST and Cornell groups [83]. Experiments have been later conducted by different groups. Even if in detail the interpretations might differ, the proposed relevance of the self-assembly in solution has been confirmed by all groups [88,99–102]. By comparing a PS-*b*-P4VP diblock and the corresponding triblock copolymer with isoprene as the third block, it was observed that hexagonal or body-centered cubic self-assembly in solution was correlated to the same

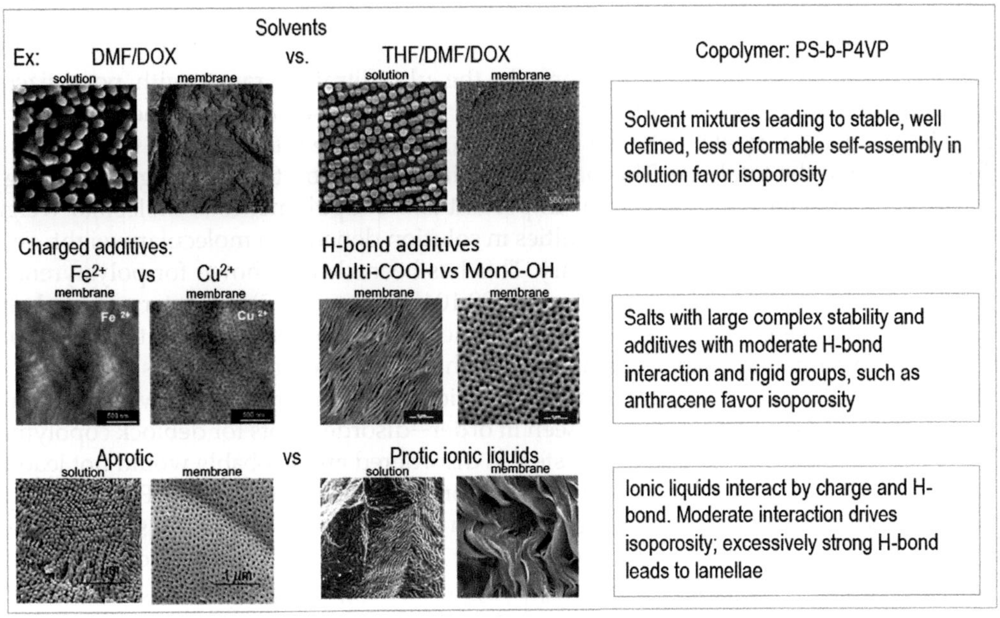

FIGURE 11.5 Influence of solvent and additives on the self-assembly and non–solvent-induced phase separation (SNIPS) block copolymer membrane morphology. *Images adapted from S.P. Nunes, A.R. Behzad, B. Hooghan, R. Sougrat, M. Karunakaran, N. Pradeep, U. Vainio, K.V. Peinemann, Switchable pH-responsive polymeric membranes prepared via block copolymer micelle assembly, ACS Nano 5 (2011) 3516–3522; S.P. Nunes, M. Karunakaran, N. Pradeep, A.R. Behzad, B. Hooghan, R. Sougrat, H. He, K.V. Peinemann, From micelle supramolecular assemblies in selective solvents to isoporous membranes, Langmuir 27 (2011) 10184–10190; P. Madhavan, K.V. Peinemann, S.P. Nunes, Complexation-tailored morphology of asymmetric block copolymer membranes, ACS Applied Materials and Interfaces 5 (2013) 7152–7159; P. Madhavan, R. Sougrat, A.R. Behzad, K.-V. Peinemann, S.P. Nunes, Ionic liquids as self-assembly guide for the formation of nanostructured block copolymer membranes, Journal of Membrane Science 492 (2015) 568–577.*

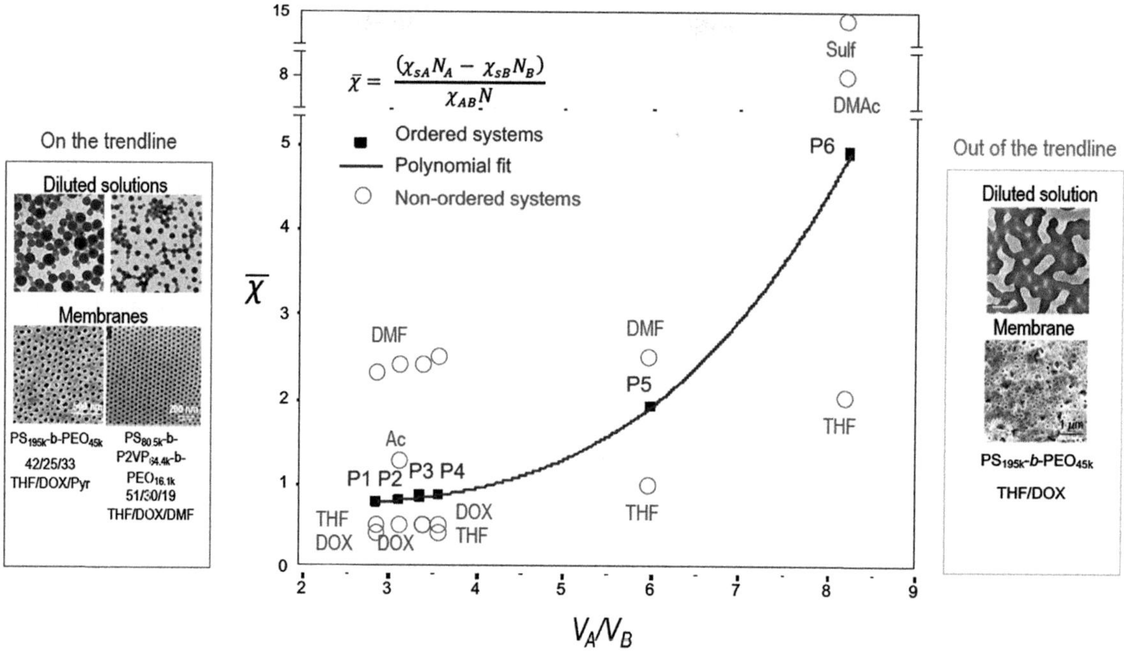

FIGURE 11.6 Trendline guide for solvent mixtures screening in the membrane preparation. The red (black in print version) curve represents the compositions that led to isoporous morphology, whereas the green dots are relative to nonuniform. The TEM microscopy of micelles formed in the diluted solutions and the corresponding membranes illustrate the correlation of the trendline and the obtained morphologies. *Adapted from B. Sutisna, G. Polymeropoulos, V. Musteata, K.-V. Peinemann, A. Avgeropoulos, D.-M. Smilgies, N. Hadjichristidis, S.P. Nunes, Design of block copolymer membranes using segregation strength trend lines, Molecular Systems Design and Engineering 1 (2016) 278–289.*

kind of pore arrangement in the membrane pore distribution [73]. Therefore, not only the presence of order in solution but also its kind is essential.

4.2.3 Tailoring Pore Size

Most membranes prepared by SNIPS have pores in the ultrafiltration range with pore sizes above 10 nm. Preparing nanofiltration membranes by SNIPS is challenging. The self-assembly in the melt and in solution is highly dependent on the interaction parameter between blocks, χ_{AB}, and on the ratio between the degrees of polymerization of each block relative to the total one. This has a direct implication on the resulting morphology of the SNIPS membranes. The other parameter that is at least as important is the copolymer total molecular weight. The dimensions of micelles, cylinders, and other assembled unities in solution depend on molecular weight, and the pore sizes of the resulting membranes are also expected to vary. This has been clearly shown for polystyrene–poly(ethylene oxide) (PS-*b*-PEO) [103] and PI-*b*-PS-*b*-P4VP membranes [80,104] (Scheme 11.2) and later reported also for mixtures of PS-*b*-P4VP [87] of different molecular weights (Fig. 11.7). However, a hypothetical strategy of reducing the molecular weight enough to obtain nanofiltration membranes with well-defined round pores of size far below 10 nm is restricted by the fact that the block segregations is a function of the interaction between block and also of the polymerization degree, N. This can be more easily seen in order–disorder plots for deblock copolymer melts. Below a certain polymerization degree, χ_{AB} N < 0.5, the system is disordered and probably would not lead to an isoporous membrane formation. As mentioned, when the membrane pores are constituted by P4VP or polyacrylic acid (PAA), a change in pH can cause a drastic change of pore size down to 1–3.5 nm, but this is also associated with a strong undesirable permeation decrease even below 1 L m^{-2} h^{-1} bar^{-1}. Gu et al. [85] added glycerol to the casting solution of isoporous membranes. This increased the pore siz, and by controlling the pH, the blocks on the wall expanded but did not close the pores completely. About 5 nm was left open for a permeation of 15 L m^{-2} h^{-1} bar^{-1}. The most effective demonstrated strategy to prepare isoporous nanofiltration membranes so far has been blending PS-*b*-P4VP copolymers with PS-*b*-PAA of much lower molecular weight. A high density of pores as small as 1.5 nm, rejection of 600 g mol^{-1}, and a high permeation of 432 15 L m^{-2} h^{-1} bar^{-1} could then be obtained. Other approaches for reducing the pore sizes are based on postmodification [6] by atomic layer deposition [98] and electroless gold deposition [82].

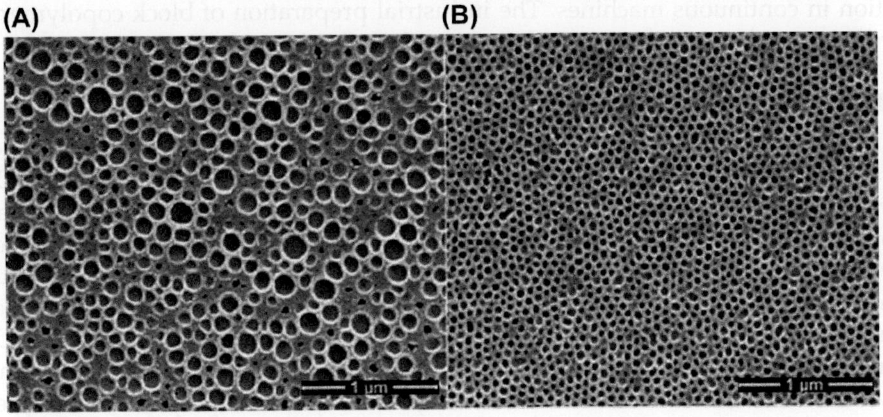

FIGURE 11.7 Block copolymer membranes prepared from 22 wt% (A) PS_{227K}-*b*-PEO_{61K} and (B) PS_{138K}-*b*-PEO_{18K} solutions in tetrahydrofuran (THF)/DMAc/sulfolane. *Adapted from M. Karunakaran, S.P. Nunes, X. Qiu, H. Yu, K.-V. Peinemann, Isoporous PS-b-PEO ultrafiltration membranes via self-assembly and water-induced phase separation, Journal of Membrane Science 453 (2014) 471–477.*

4.2.4 *Influence of Additives and Complexing Agents*

Having in mind the relevance of the self-assembly in solution, and the need to stabilize it, salts were added in the casting solution of PS-*b*-P4VP. A high reproducibility was achieved with the same order depicted in Fig. 11.4 [67,70]. The kind of cation and its ability to complex with P4VP strongly influence the final regularity of the membrane, Cu^{2+} being the most effective among those tested, being also the one with the highest metal–pyridine complex stability constant [70]. The complexation by H-bond can stabilize the self-assembly in solution as well [77]. It has been demonstrated that excessively strong H-bond interactions, as provided by mellitic acid, which has six carboxylic acid groups per molecule, or 1,3,5-tris(4′carboxyl[1,1′biphenyl]-4-yl) benzene, with three carboxylic groups, can cross-link into a lamellar morphology, unsuitable for membranes. Surprisingly, the rigid monofunctionalized 9-anthracenemethanol molecules were the most effective additives in a series of complexing agents. This indicated that increasing the rigidity of the micelles is of primary relevance. A low deformability helps to reach the isoporosity. A combination of electrostatic forces and H-bond interactions is provided by the addition of ionic liquids to a block copolymer solution. By changing the ionic liquid composition, the membrane morphology can be guided or completely modified [86]. By adding protic ionic liquids to a PS-*b*-P4VP solution, a lamellar morphology was achieved, while with aprotic analogues, with less stronger interactions, membranes with circular pores were favored. Fig. 11.5 summarizes the effect of different additives and solvents.

4.2.5 *Beyond Size: Separation by Charge and Stimuli Response*

The isoporosity characteristic of the block copolymer membranes, as shown in Fig. 11.4, provides better size selectivity than commercial homopolymer membranes in the ultrafiltration range. This is reflected by the ability of separating macromolecules of similar molecular weights. Besides this advantage in size-sieving, the separation ability of PS-*b*-P4VP membranes can be additionally tuned by the pH. The membranes are stimuli-responsive [70]. At low pH, the pyridine groups are protonated, becoming positively charged. The hydrophilic segments, which constitute the pore wall, stretch reducing the pore size. The process is reversible. The pore reduction or enlargement can be tuned by controlling the pH. In addition, the surface charge might contribute to the separation. This is particularly useful in the case of proteins, which can be preferentially positive or negative depending also on pH. PAA blocks respond to pH in an opposite way to pyridine. Isoporous membranes and particles based on PS-*b*-PAA have their pores reduced and the surface turning positive at high pH [68,105,106]. Their capacity of binding metals also changes with pH and has been explored for selective ion transport [107]. The response to light has been demonstrated for block copolymer membranes with anthracene pending groups [90].

4.2.6 *From Lab to Technical Manufacture*

Block copolymer membranes based on PS-*b*-P4VP or similar di- or triblock copolymers have been progressing since the first reports [66,67]. Their manufacture has been demonstrated as flat sheet or hollow fiber [75,76,79,108,109] in the lab. The scale-up of flat-sheet block copolymer membranes in semitechnical machines have been demonstrated at KAUST [74], being a topic of a "seed fund" initiative in 2010. As for most commercial polymeric membranes, a nonwoven is commonly used to increase the mechanical stability of the membranes and allow the preparation in continuous machines. The industrial preparation of block copolymer membranes is now being explored by Terapore (https://teraporetech.com) and Anfiro (www.anfiro.com).

4.2.7 Exploring Different Chemistries

PS-*b*-P4VP and PI-*b*-PS-*b*-P4VP copolymer membranes have been the most explored ones and have been serving as a platform for further functionalization and modifications targeting different applications. By incorporating additives such as silver [110], gold [75], TiO_2 [111], metal–organic frameworks [112], and carbon nanotubes [113], functions such as biofouling resistance, catalytic activity, drug delivery, and sensing have been added to the membranes. The copolymers with pyridine or PEO blocks are, however, just a few of innumerous others with great potential for membranes. Probably, the big chances for block copolymer membranes still to be explored are the exploration of new chemistries and functionalizations. One important aspect is the potential advantage that would bring the substitution of the polystyrene block by other segments more frequently used for stable membranes. Examples could be polysulfone, polyacrylonitrile, or fluorinated backbones. Polysulfone block copolymers were prepared by Xie et al. [41,91,92] with the purpose of membrane preparation. In the first report, copolymers with azide groups for the postfunctionalization by click chemistry [41] were prepared. More recently postfunctionalization by click chemistry has been applied to PS-*b*-P4VP [114]. In a different approach, a combination of polycondensation and RAFT copolymerization was applied to prepare poly(t-butyl acrylate)-*b*-polysulfone-*b*-poly(t-butyl acrylate) [91]. The hydrolysis of the acrylate leads to the formation of poly(acrylic acid) blocks. The chelation with salts guided the formation of an ordered structure [92].

Anionic copolymerization is one of the most effective methods to obtain macromolecules with minimum polydispersity and well-defined blocks. Different copolymer systems (Scheme 11.3) were recently explored by the Nunes's group in collaboration with Hadjichristidis' group at KAUST. The first of them is *poly(styrene-b-tert-butoxystyrene-b-styrene)* [94] and clearly demonstrates the relevance of the pore functionalization. This copolymer offers a large density of hydroxy groups, which are effective hydrogen bonds promoters. Small H-bond promoters such as pyridine and imidazole were used as additives. The nanopores were small enough to retain 97% of PEG molecules with 10 kg mol^{-1} molecular weight and 42% of PEG with 3 kg mol^{-1} but reject only 32% cytrochrome c (12.4 kg mol^{-1}) and 25% of lysozyme (14.3 kg mol^{-1}). The Stokes radii of these compound are similar, around 1.5–2.8 nm. PEG 3 kg mol^{-1} is the lowest one. The permeant dimensions are similar to the pore size. While permeating, a strong contact with the pore walls is promoted. Probably, the OH groups on the pore walls preferentially facilitate the transport of the proteins with the right balance of reversible H-bonds formation.

Another recent example is a membrane prepared from polystyrene-*b*-polybutadiene-*b*-polystyrene (SBS) prepared also by anionic polymerization [95]. SBS is commercially available in large scale and low cost with a block ratio different than that used for the membrane preparation. This is important to mention because it indicates that the cost of block copolymers can be considerably reduced if produced in the technical scale and should not be seen as an obstacle for membrane manufacture. The approach applied to the SBS membrane took advantage of the butadiene double bond, reacting it with a thiol molecule. Hydrophilic channels could then be obtained. The strategy could be in principle applied to any copolymer with double bonds and at the same time to other thiol molecules, working as a platform for tuning functionalization.

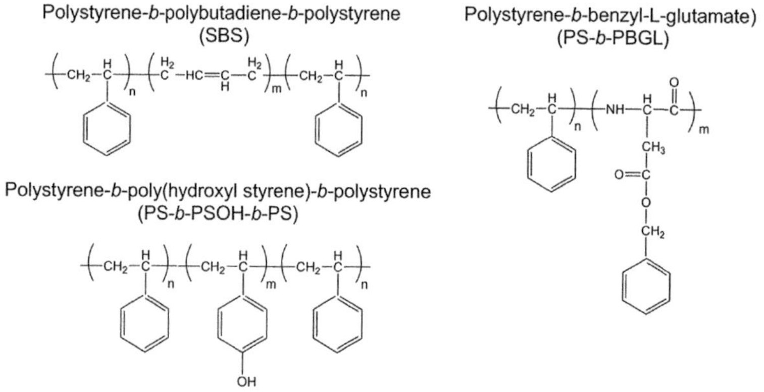

SCHEME 11.3 Block copolymers recently used for self-assembly and non–solvent-induced phase separation (SNIPS) membranes [94,95,97].

Finally, membranes were prepared from poly(styrene-*b*-γ-benzyl-L-glutamate) casting solutions [97]. The peptide block has a helicoidal conformation, guiding the formation of nanochannels and providing H-bonds.

5. BLOCK COPOLYMERS FOR BIOMIMETIC 3D HIERARCHICAL AND ISOTROPIC STRUCTURES

Most porous films prepared by SNIPS are asymmetric membranes, only the thin (<500 nm) top layer being isoporous. Fully isoporous 3D hierarchical structures [69,96] have been now demonstrated, mimicking the complex morphology of biological systems with the perspective of applications also for separations and selective transport. In this case, solutions of copolymers with both hydrophobic blocks, such as polystyrene-b-poly(t-butyl acrylate) (PS-*b*-PtBA), were prepared in DMF solvent mixtures. A phase separation by NG takes place with the polymer-lean nuclei developing to cavities of micrometer size. A high density of isopores with 15 nm diameter is formed, interconnecting all cavities, as shown in Fig. 11.8. The hydrophobicity can be controlled by hydrolysis of PS-*b*-PtBA (Scheme 11.4). This morphology has potential application for separations, as well as for catalysis and as biomedical scaffolds. Analogous systems are being now investigated and tested for different applications.

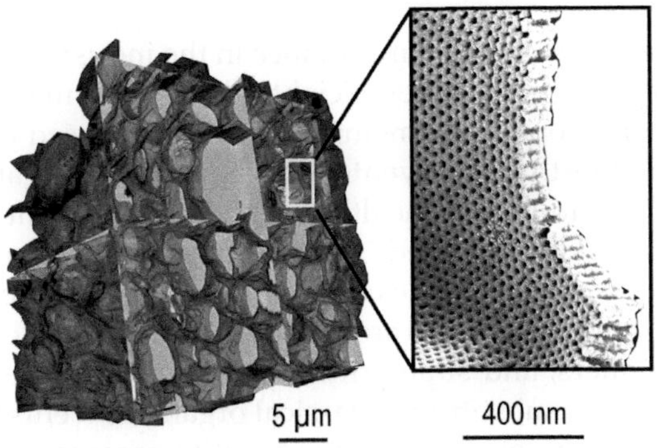

FIGURE 11.8 3D hierarchical porous structure prepared by block copolymer self-assembly and nucleation and growth: serial block face SEM image on the left, showing the cross-section of a porous isotropic film with micrometer-size cavities and a high SEM magnification of the wall between cavities. *Adapted from S.M. Chisca, Valentina; Sougrat, R.; Behzad, A.; Nunes, S.P., Artificial 3D Hierarchical and Isotropic Porous Polymeric Materials, vol. 4, 2018, pp. eaat0713.*

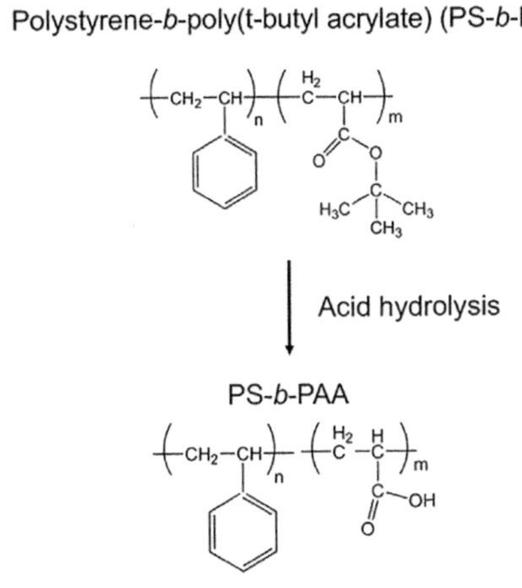

SCHEME 11.4 Block copolymer for 3D hierarchical isotropic structures: PS-*b*-PtBA and its hydrolysis to PS-b-PAA for carboxylation and hydrophilization[69].

6. SUSTAINABILITY OF BLOCK COPOLYMER MEMBRANES

The membrane technology is seen as a competitive energy-saving process for many applications such as seawater desalination. It has the potential to contribute even more to sustainability in terms of process intensification in the chemical and pharmaceutical industry, enabling separations at lower temperatures and with lower CO_2 footprint or enabling the use of renewable feedstocks in a more effective way. These aspects address some of the principles of Green Chemistry, which have been recently reviewed [29,115]. Commercial block copolymer membranes have been contributing to sustainable separations for decades. This is the case of copolymers such as Pebax or analogous dense membranes, which are applied in large industrial scale for CO_2 separation, for instance, to reduce the impact of coal plants to the environment [13,116]. Another sustainable process demonstrated decades ago using Pebax, however, with a hydrophobic grade, is the in situ recovery and enrichment of organic products from fermentation broths by pervaporation or perstraction, aiming at the production of oxychemicals by continuous bioprocessing [117]. Recently, membranes based on another hydrophobic copolymer, polystyrene-*b*-polydimethylsiloxane-*b*-polystyrene, have been applied to remove organic inhibitors (furfural and acetic acid) of lignocellulosic hydrolysate from fermentation broths by pervaporation, leading to a much more effective production of bioethanol [118]. Poly(butylnorbornene)-*b*-poly (hydroxyhexafluoroisopropyl norbornene) copolymer membranes are under investigation for the separation of biobutanol from water [119]. The nanostructure of the block copolymer provides a better control of the swelling and separation characteristics. Block copolymer membranes have been considered for renewable energy conversion, such as fuel cell, as a possibility of providing effective paths for proton exchange (copolymers of sulfonated polyimide with benzimidazole pendant groups) [35] or anion exchange (copolymers based on polymerized ionic liquids) [120].

Although the membrane technology is gaining importance in the industry for more sustainable separations, the membrane manufacture itself needs to be more beneficial to the environment. This is valid for block copolymer membranes, as well as for classical homopolymer membranes currently used in large scale. Most of the membrane manufacture is solvent-based. The most used solvents for membrane fabrication, such as DMF and tetrahydrofuran, are considered relatively toxic to human health and the environment [121–123]. The solvent waste and the emission of volatile organics should be avoided. Alternative greener solvents should substitute the currently applied ones or when possible nonsolvent processes for membrane preparation should be introduced. A number of sustainable solvents derived from plant biomass are now under consideration [123]. Besides those, ionic liquids, deep eutectic solvents, liquid polymers, and supercritical CO_2 are other alternatives. Our group has dedicated efforts in the preparation of membranes substituting classical organic solvents by ionic liquids [124–129]. In addition to the environmental advantage, by eliminating the volatile organic emission, we have observed that the phase separation process during the membrane fabrication follows different paths, leading to high porosity and permeance and even to a shift of the molecular weight cutoff of membranes to lower values, extending the range of applications of polyethersulfone, cellulose, cellulose acetate, and other polymers. Ionic liquids should be recycled and reused in the membrane manufacturing process. For block copolymers membranes, the use of green solvents is still in an early stage. We demonstrated that the addition of small amounts of ionic liquids [86] strongly influences the self-assembly. Aprotic ionic liquids could induce the self-assembly and order, promoting the formation of hexagonal spherical pores, while protic ionic liquids with stronger interaction with the copolymer led to a lamellar structure. The use of ionic liquids as single solvent for block copolymer membrane synthesis shall proceed in the near future. There is a variety of ionic liquids as potential solvents, and we anticipate that by slightly changing the chemical composition, a corresponding change of the pore morphology would be observed. The theoretical prediction of the best ionic liquid for the membrane manufacture might require methods other than the evaluation of the Flory–Huggins parameters, as the electrostatic interaction can be stronger that supported by this model. Besides ionic liquids, other green solvents have been seldomly tested for block copolymers.

Membranes have been prepared from a large number of polymeric materials. This includes many polymers synthesized from fossil fuels, such as polysulfone and polyethersulfone, PVDF, polyamide, and polyacrylonitrile. The majority of membranes prepared from block copolymers contains polystyrene blocks. There is a growing concern with the use of nondegradable plastics and their damage to the environment. The use of renewable feedstocks for the production of polymeric materials and membrane manufacture should be more investigated. Cellulose and its derivatives were among the first materials used for membrane fabrication. The cellulose membrane manufacture itself needed to rely on aggressive processes of solubilization. Now cellulose can be dissolved in ionic liquids, making its manufacture into membranes much more sustainable [127,129]. Block or graft copolymers [130] based on cellulose should be explored for functionalized membrane fabrication in the future. The synthesis of biobased plastics is developing [131,132]. Polylactides are being prepared by fermentation of sugar to lactic acid. Polylactides or PLA have constituted block copolymers reported [133] in different approaches of membrane preparation [60,134,135]. More copolymers are being prepared from renewable sources opening new perspectives for membrane preparation. For instance, block copolymers synthesized from rosin-based monomers and norbornene were newly reported with high toughness [136]. In addition to the sustainability factor, block copolymers with segments mimicking or prepared from biological systems can lead to morphologies not obtained before in membranes, which could promote transport by hydrogen bond or by forming nanochannels due to helicoidal structures. For instance, we recently report membranes prepared from block copolymers of polystyrene and a polypeptide [137]. Finally, biodegradable polymers could be used for membrane fabrication, as long as the time of degradation could be well-tailored for the application. A complete analysis of the sustainability of block copolymer membranes manufacture is, however, complex and needs to evaluate the copolymer synthesis itself, solvent used for membrane preparation. How fast the manufacturing processes will be substituted by more sustainable ones will depend on cost, regulations, and the advantages the resulting membranes will bring.

7. CONCLUSION: ADVANTAGES, DRAWBACKS, AND PERSPECTIVES

The main advantages of block copolymers for membrane fabrication are the precise pore morphology, much better defined than regular homopolymers can provide, high porosity, and rich possibilities of functionalization. The SNIPS method leads to isoporous selective layers in asymmetric membranes practically in one stage, without etching. The resulting membranes have the potential to effectively enable separations for instance of biological systems in a much more effective way than done by other classes of membranes. Drawbacks for immediate commercialization at this moment are still the relatively high cost of the copolymers optimized for isoporous membranes and the fact that the morphology is more sensitive to minor variations of molecular weight than homopolymer membranes. The first step in the scale-up should, therefore, be that of the copolymer synthesis. Block copolymers such as styrene-butadiene-styrene are available in large scale, and this should be feasible. Other aspects of scale-up do not differ from homopolymer membranes, and strategies such as the use of nonwoven supports in continuous machines have been long being demonstrated [74].

Aging aspects are expected to be similar to those observed for homopolymers of analogous chemical composition. The majority of block copolymer membranes reported so far contain polystyrene blocks. To extend the membrane application, copolymers with blocks more stable than polystyrene in terms of temperature, and solvent resistance should be investigated. Blocks with compositions analogous to those currently used for homopolymer membranes, such as polysulfone and PVDF, would give more robust systems. However, the methods of polymerization should also guarantee low polydispersity and well-defined blocks. Cross-linking strategies would be important to secure solvent resistance. Cross-linking examples have been reported [138–140] for membrane systems requiring block etching for pore formation and for microgels [141]. Pyrolysis of block copolymer films [142,143] could be extended to membranes.

Block copolymer membranes such as those based on Pebax or similar ones are already being used in large-scale sustainable process of separation. The process of membrane manufacture needs to be more sustainable, as it is the case of homopolymeric membranes. The minimization of solvents and substitution for greener alternatives are already being investigated. New polymers and copolymers from renewable sources will be explored in the future.

Acknowledgments

The Nunes's work on block copolymer membranes has been supported by the King Abdullah University of Science and Technology (KAUST). This chapter is a summary of the research on block copolymer membranes with selected examples from different groups in the field. The author thanks all coauthors and collaborators of the Nanostructured Polymeric Lab, Klaus Peinemann, Nikos Hadjichristidis, Victor Calo, their coworkers, and the core lab at KAUST. Special thanks also to Detlef Smilgies and Uli Wiesner's group at Cornell and the synchrotron scientists at the LNLS in Brazil and DESY, Germany.

References

[1] G. Polymeropoulos, G. Zapsas, K. Ntetsikas, P. Bilalis, Y. Gnanou, N. Hadjichristidis, 50th anniversary perspective: polymers with complex architectures, Macromolecules 50 (2017) 1253–1290.

[2] F.S. Bates, G.H. Fredrickson, Block copolymers designer soft materials, Physics Today 52 (1999) 32.

[3] F.S. Bates, M.A. Hillmyer, T.P. Lodge, C.M. Bates, K.T. Delaney, G.H. Fredrickson, Multiblock polymers: panacea or pandora's box? Science 434 (2012) 440.

[4] C.M. Bates, F.S. Bates, 50th anniversary perspective: block polymers—pure potential, Macromolecules 50 (2016) 3–22.
[5] J.N. Israelachvili, Intermolecular and Surface Forces, 2011.
[6] S.P. Nunes, Block copolymer membranes for aqueous solution applications, Macromolecules 49 (2016) 2905–2916.
[7] I.P. Blume, Ingo, Composite Membrane, Method of Preparation and Use, MTR, US, 1987.
[8] M.L.N. Sforça, S.P. Nunes, K.V. Peinemann, Composite nanofiltration membranes prepared by in situ polycondensation of amines in a poly(ethylene oxide-b-amide) layer, Journal of Membrane Science 135 (1997) 179–186.
[9] S.P. Nunes, M.L. Sforça, K.V. Peinemann, Dense hydrophilic composite membranes for ultrafiltration, Journal of Membrane Science 106 (1995) 49–56.
[10] K.W.B. Boeddeker, G. Bengtson, H. Pingel, Pervaporation of isomeric butanols, Journal of Membrane Science 54 (1990) 1–12.
[11] S. Shishatskiy, J.R. Pauls, S.P. Nunes, K.-V. Peinemann, Quaternary ammonium membrane materials for CO_2 separation, Journal of Membrane Science 359 (2010) 44–53.
[12] A. Car, C. Stropnik, W. Yave, K.-V. Peinemann, PEG modified poly(amide-b-ethylene oxide) membranes for CO_2 separation, Journal of Membrane Science 307 (2008) 88–95.
[13] T.C. Merkel, H. Lin, X. Wei, R. Baker, Power plant post-combustion carbon dioxide capture: an opportunity for membranes, Journal of Membrane Science 359 (2010) 126–139.
[14] R.W. Baker, B.T. Low, Gas separation membrane materials: a perspective, Macromolecules 47 (2014) 6999–7013.
[15] W. Fam, J. Mansouri, H. Li, V. Chen, Improving CO_2 separation performance of thin film composite hollow fiber with Pebax®1657/ionic liquid gel membranes, Journal of Membrane Science 537 (2017) 54–68.
[16] T. Khosravi, M. Omidkhah, S. Kaliaguine, D. Rodrigue, Amine-functionalized CuBTC/poly(ether-b-amide-6) (Pebax® MH 1657) mixed matrix membranes for CO_2/CH_4 separation, The Canadian Journal of Chemical Engineering 95 (2017) 2024–2033.
[17] P.D. Sutrisna, J. Hou, M.Y. Zulkifli, H. Li, Y. Zhang, W. Liang, D.M. D'Alessandro, V. Chen, Surface functionalized UiO-66/Pebax-based ultrathin composite hollow fiber gas separation membranes, Journal of Materials Chemistry A 6 (2018) 918–931.
[18] W. Yave, A. Car, S.S. Funari, S.P. Nunes, K.-V. Peinemann, CO_2-philic polymer membrane with extremely high separation performance, Macromolecules 43 (2010) 326–333.
[19] X. Ren, J. Ren, M. Deng, Poly(amide-6-b-ethylene oxide) membranes for sour gas separation, Separation and Purification Technology 89 (2012) 1–8.
[20] K.W. Boeddeker, Pervaporation of Phenols in Germany, 1989.
[21] K.W. Boddeker, G. Bengtson, E. Bode, Pervaporation of low volatility aromatics from water, Journal of Membrane Science 53 (1990) 143–158.
[22] M.L. Sforça, I.V.P. Yoshida, C.P. Borges, S.P. Nunes, Hybrid membranes based on SiO_2/polyether-b-polyamide: morphology and applications, Journal of Applied Polymer Science 82 (2001) 178–185.
[23] J. Aburabie, K.-V. Peinemann, Crosslinked poly(ether block amide) composite membranes for organic solvent nanofiltration applications, Journal of Membrane Science 523 (2017) 264–272.
[24] C. Cuevas, D. Kim, K.P. Katuri, P. Saikaly, S.P. Nunes, Electrochemically active polymeric hollow fibers based on poly(ether- b -amide)/carbon nanotubes, Journal of Membrane Science 545 (2018) 323–328.
[25] G.M. Geise, B.D. Freeman, D.R. Paul, Characterization of a sulfonated pentablock copolymer for desalination applications, Polymer 51 (2010) 5815–5822.
[26] Z. Thong, G. Han, Y. Cui, J. Gao, T.S. Chung, S.Y. Chan, S. Wei, Novel nanofiltration membranes consisting of a sulfonated pentablock copolymer rejection layer for heavy metal removal, Environmental Science Technology 48 (2014) 13880–13887.
[27] P. Marchetti, A.G. Livingston, Predictive membrane transport models for organic solvent nanofiltration: how complex do we need to be? Journal of Membrane Science 476 (2015) 530–553.
[28] J. Zuo, G.M. Shi, S. Wei, T.S. Chung, The development of novel Nexar block copolymer/Ultem composite membranes for C2-C4 alcohols dehydration via pervaporation, ACS Applied Materials and Interfaces 6 (2014) 13874–13883.
[29] G.M. Shi, J. Zuo, S.H. Tang, S. Wei, T.S. Chung, Layer-by-layer (LbL) polyelectrolyte membrane with Nexar™ polymer as a polyanion for pervaporation dehydration of ethanol, Separation and Purification Technology 140 (2015) 13–22.
[30] P.V. Truong, S. Shingleton, M. Kammoun, R.L. Black, M. Charendoff, C. Willis, H. Ardebili, G.E. Stein, Structure and properties of sulfonated pentablock terpolymer films as a function of wet–dry cycles, Macromolecules 51 (2018) 2203–2215.
[31] F.H.V.H. Akhtar, H. Villalobos, L. Francisco, R. Shevate, M. Kumar, S.P. Nunes, U. Schwingenschloegl, K.-V. Peinemann, Highways for Water Molecules: Interplay between Nanostructure and Water Vapor Transport in Block Copolymer Membranes, 2018.
[32] S. Vijay Kumar, C. Arnal-Herault, M. Wang, J.r.m. Babin, A. Jonquieres, Multiblock copolymer grafting for butanol biofuel recovery by a sustainable membrane process, ACS Applied Materials and Interfaces 8 (2016) 16262–16272.
[33] F. Huang, T.D. Largier, W. Zheng, C.J. Cornelius, Pentablock copolymer morphology dependent transport and its impact upon film swelling, proton conductivity, hydrogen fuel cell operation, vanadium flow battery function, and electroactive actuator performance, Journal of Membrane Science 545 (2018) 1–10.
[34] A. Taeger, C. Vogel, D. Lehmann, D. Jehnichen, H. Komber, J. Meier-Haack, N.A. Ochoa, S.P. Nunes, K.V. Peinemann, Ion exchange membranes derived from sulfonated polyaramides, Reactive and Functional Polymers 57 (2003) 77–92.
[35] Y. Álvarez-Gallego, B. Ruffmann, V. Silva, H. Silva, A.E. Lozano, J.G. de la Campa, S.P. Nunes, J. de Abajo, Sulfonated polynaphthalimides with benzimidazole pendant groups, Polymer 49 (2008) 3875–3883.
[36] D.W. Shin, M.D. Guiver, Y.M. Lee, Hydrocarbon-based polymer electrolyte membranes: importance of morphology on ion transport and membrane stability, Chemical Reviews 117 (2017) 4759–4805.
[37] K. Zhang, S. Gong, B. Zhao, Y. Liu, N.A. Qaisrani, L. Li, F. Zhang, G. He, Bent-twisted block copolymer anion exchange membrane with improved conductivity, Journal of Membrane Science 550 (2018) 59–71.
[38] B.D. McCloskey, H.B. Park, H. Ju, B.W. Rowe, D.J. Miller, B.D. Freeman, A bioinspired fouling-resistant surface modification for water purification membranes, Journal of Membrane Science 413–414 (2012) 82–90.
[39] H.-C. Yang, J. Luo, Y. Lv, P. Shen, Z.-K. Xu, Surface engineering of polymer membranes via mussel-inspired chemistry, Journal of Membrane Science 483 (2015) 42–59.

[40] N.L. Le, M. Ulbricht, S.P. Nunes, How do polyethylene glycol and poly(sulfobetaine) hydrogel layers on ultrafiltration membranes minimize fouling and stay stable in cleaning chemicals? Industrial and Engineering Chemistry Research 56 (2017) 6785–6795.

[41] Y. Xie, R. Tayouo, S.P. Nunes, Low fouling polysulfone ultrafiltration membrane via click chemistry, Journal of Applied Polymer Science 132 (2015).

[42] J. Meyer, M. Ulbricht, Poly(ethylene oxide)- block -poly(methyl methacrylate) diblock copolymers as functional additive for poly(vinylidene fluoride) ultrafiltration membranes with tailored separation performance, Journal of Membrane Science 545 (2018) 301–311.

[43] S.P. Nunes, K.V. Peinemann, Ultrafiltration membranes from PVDF/PMMA blends, Journal of Membrane Science 73 (1992) 25–35.

[44] S.Y. Park, J.W. Chung, Y.K. Chae, S.-Y. Kwak, Amphiphilic thiol functional linker mediated sustainable anti-biofouling ultrafiltration nanocomposite comprising a silver nanoparticles and poly (vinylidene fluoride) membrane, ACS Applied Materials and Interfaces 5 (2013) 10705–10714.

[45] C.J. Weinman, N. Gunari, S. Krishnan, R. Dong, M.Y. Paik, K.E. Sohn, G.C. Walker, E.J. Kramer, D.A. Fischer, C.K. Ober, Protein adsorption resistance of anti-biofouling block copolymers containing amphiphilic side chains, Soft Matter 6 (2010).

[46] B.H. Tan, H. Hussain, K.C. Chaw, G.H. Dickinson, C.S. Gudipati, W.R. Birch, S.L.M. Teo, C. He, Y. Liu, T.P. Davis, Barnacle repellent nanostructured surfaces formed by the self-assembly of amphiphilic block copolymers, Polymer Chemistry 1 (2010) 276–279.

[47] M. Lejars, A. Margaillan, C. Bressy, Fouling release coatings: a nontoxic alternative to biocidal antifouling coatings, Chemical Reviews 112 (2012) 4347–4390.

[48] S. Krishnan, N. Wang, C. Ober, J.A. Finlay, M. Callow, J. Callow, A. Hexemer, K. Sohn, E. Kramer, D. Fischer, Comparison of the fouling release properties of hydrophobic fluorinated and hydrophilic PEGylated block copolymer surfaces, Biomacromolecules 7 (2006) 1449–1462.

[49] P.H.H. Duong, K. Daumann, P.-Y. Hong, M. Ulbricht, S.P. Nunes, Interfacial polymerization of zwitterionic building blocks for high-flux nanofiltration membranes, Langmuir 35 (5) (2019) 1284–1293.

[50] J.A. Walker, K.J. Robinson, C. Munro, T. Gengenbach, D.A. Muller, P.R. Young, L.H.L. Lua, S.R. Corrie, Antibody-binding, antifouling surface coatings based on recombinant expression of zwitterionic EK peptides, Langmuir 35 (5) (2019) 1266–1272.

[51] M. Piatkovsky, H. Acar, A.B. Marciel, M. Tirrell, M. Herzberg, A zwitterionic block-copolymer, based on glutamic acid and lysine, reduces the biofouling of UF and RO membranes, Journal of Membrane Science 549 (2018) 507–514.

[52] J.S. Lee, A. Hirao, S. Nakahama, Polymerization of monomers containing functional silyl groups. 5. Synthesis of new porous membranes with functional groups, Macromolecules 21 (1988) 274–276.

[53] J.S. Lee, A. Hirao, S. Nakahama, Polymerization of monomers containing functional silyl groups. 7. Porous membranes with controlled microstructures, Macromolecules 22 (1989) 2602–2606.

[54] S.Y. Chen, Y. Huang, R.C.C. Tsiang, Ozonolysis efficiency of PS-b-PI block copolymers for forming nanoporous polystyrene, Journal of Polymer Science Part A: Polymer Chemistry 46 (2008) 1964–1973.

[55] M. Park, C. Harrison, P.M. Chaikin, R.A. Register, D.H. Adamson, Block copolymer lithography: periodic arrays of ~ 1011 holes in 1 square centimeter, Science 276 (1997) 1401–1404.

[56] S. Park, J.Y. Wang, B. Kim, J. Xu, T.P. Russell, A simple route to highly oriented and ordered nanoporous block copolymer templates, ACS Nano 2 (2008) 766–772.

[57] T. Thurn-Albrecht, R. Steiner, J. DeRouchey, C.M. Stafford, E. Huang, M. Bal, M. Tuominen, C.J. Hawker, T.P. Russell, Nanoscopic templates from oriented block copolymer films, Advanced Materials 12 (2000) 787–791.

[58] W.A. Phillip, M.A. Hillmyer, E. Cussler, Cylinder orientation mechanism in block copolymer thin films upon solvent evaporation, Macromolecules 43 (2010) 7763–7770.

[59] L. Chen, W.A. Phillip, E. Cussler, M.A. Hillmyer, Robust nanoporous membranes templated by a doubly reactive block copolymer, Journal of the American Chemical Society 129 (2007) 13786–13787.

[60] E.A. Jackson, M.A. Hillmyer, Nanoporous membranes derived from block copolymers: from drug delivery to water filtration, ACS Nano 4 (2010) 3548–3553.

[61] M. Seo, D. Moll, C. Silvis, A. Roy, S. Querelle, M.A. Hillmyer, Interfacial polymerization of reactive block polymers for the preparation of composite ultrafiltration membranes, Industrial and Engineering Chemistry Research 53 (2014) 18575–18579.

[62] M. Seo, M.A. Hillmyer, Reticulated nanoporous polymers by controlled polymerization-induced microphase separation, Science 336 (2012) 1422–1425.

[63] D. Ndaya, R. Bosire, L. Mahajan, S. Huh, R. Kasi, Synthesis of ordered, functional, robust nanoporous membranes from liquid crystalline brush-like triblock copolymers, Polymer Chemistry 9 (2018) 1404–1411.

[64] L.D. McIntosh, M.W. Schulze, M.T. Irwin, M.A. Hillmyer, T.P. Lodge, Evolution of morphology, modulus, and conductivity in polymer electrolytes prepared via polymerization-induced phase separation, Macromolecules 48 (2015) 1418–1428.

[65] S.A. Saba, M.P. Mousavi, P. Bühlmann, M.A. Hillmyer, Hierarchically porous polymer monoliths by combining controlled macro-and microphase separation, Journal of the American Chemical Society 137 (2015) 8896–8899.

[66] K.V. Peinemann, V. Abetz, P.F.W. Simon, Asymmetric superstructure formed in a block copolymer via phase separation, Nature Materials 6 (2007) 992–996.

[67] S.P. Nunes, R. Sougrat, B. Hooghan, D.H. Anjum, A.R. Behzad, L. Zhao, N. Pradeep, I. Pinnau, U. Vainio, K.V. Peinemann, Ultraporous films with uniform nanochannels by block copolymer micelles assembly, Macromolecules 43 (2010) 8079–8085.

[68] H. Yu, X. Qiu, S.P. Nunes, K.V. Peinemann, Biomimetic block copolymer particles with gated nanopores and ultrahigh protein sorption capacity, Nature Communications 5 (2014) 4110.

[69] S. Chisca, V.-E. Musteata, R. Sougrat, A. Behzad, S.P. Nunes, Artificial 3D Hierarchical and Isotropic Porous Polymeric Materials, vol. 4, 2018, p. eaat0713.

[70] S.P. Nunes, A.R. Behzad, B. Hooghan, R. Sougrat, M. Karunakaran, N. Pradeep, U. Vainio, K.V. Peinemann, Switchable pH-responsive polymeric membranes prepared via block copolymer micelle assembly, ACS Nano 5 (2011) 3516–3522.

[71] S.P. Nunes, M. Karunakaran, N. Pradeep, A.R. Behzad, B. Hooghan, R. Sougrat, H. He, K.V. Peinemann, From micelle supramolecular assemblies in selective solvents to isoporous membranes, Langmuir 27 (2011) 10184–10190.

[72] W.A. Phillip, R. Mika Dorin, J. Werner, E.M.V. Hoek, U. Wiesner, M. Elimelech, Tuning structure and properties of graded triblock terpolymer-based mesoporous and hybrid films, Nano Letters 11 (2011) 2892–2900.
[73] R.M. Dorin, D.S. Marques, H. Sai, U. Vainio, W.A. Phillip, K.V. Peinemann, S.P. Nunes, U. Wiesner, Solution small-angle X-ray scattering as a screening and predictive tool in the fabrication of asymmetric block copolymer membranes, ACS Macro Letters 1 (2012) 614–617.
[74] S.P. Nunes, A.R. Behzad, K.-V. Peinemann, Self-assembled block copolymer membranes: from basic research to large-scale manufacturing, Journal of Materials Research 28 (2013) 2661–2665.
[75] R. Hilke, N. Pradeep, P. Madhavan, U. Vainio, A.R. Behzad, R. Sougrat, S.P. Nunes, K.-V. Peinemann, Block copolymer hollow fiber membranes with catalytic activity and pH-response, ACS Applied Materials and Interfaces 5 (2013) 7001–7006.
[76] M. Radjabian, J. Koll, K. Buhr, U.A. Handge, V. Abetz, Hollow fibre spinning of block copolymers: influence of spinning conditions on morphological properties, Polymer (2013).
[77] P. Madhavan, K.V. Peinemann, S.P. Nunes, Complexation-tailored morphology of asymmetric block copolymer membranes, ACS Applied Materials and Interfaces 5 (2013) 7152–7159.
[78] D.S. Marques, Micellar self-assembly of block copolymers for fabrication of nanostructured membranes, King Abdullah University of Science and Technology (2013). PhD Thesis.
[79] R. Hilke, N. Pradeep, A.R. Behzad, S.P. Nunes, K.-V. Peinemann, Block copolymer/homopolymer dual-layer hollow fiber membranes, Journal of Membrane Science 472 (2014) 39–44.
[80] M. Karunakaran, S.P. Nunes, X. Qiu, H. Yu, K.-V. Peinemann, Isoporous PS-b-PEO ultrafiltration membranes via self-assembly and water-induced phase separation, Journal of Membrane Science 453 (2014) 471–477.
[81] R.A. Mulvenna, J.L. Weidman, B. Jing, J.A. Pople, Y. Zhu, B.W. Boudouris, W.A. Phillip, Tunable nanoporous membranes with chemically-tailored pore walls from triblock polymer templates, Journal of Membrane Science 470 (2014) 246–256.
[82] H. Yu, X. Qiu, S.P. Nunes, K.V. Peinemann, Self-assembled isoporous block copolymer membranes with tuned pore sizes, Angewandte Chemie 126 (2014) 10236–10240.
[83] D.S. Marques, R.M. Dorin, U. Wiesner, D.-M. Smilgies, A.R. Behzad, U. Vainio, K.-V. Peinemann, S.P. Nunes, Time-resolved GISAXS and cryo-microscopy characterization of block copolymer membrane formation, Polymer 55 (2014) 1327–1332.
[84] Y. Gu, J.G. Werner, R.M. Dorin, S.W. Robbins, U. Wiesner, Graded porous inorganic materials derived from self-assembled block copolymer templates, Nanoscale 7 (2015) 5826–5834.
[85] Y. Gu, U. Wiesner, Tailoring pore size of graded mesoporous block copolymer membranes: moving from ultrafiltration toward nanofiltration, Macromolecules 48 (2015) 6153–6159.
[86] P. Madhavan, R. Sougrat, A.R. Behzad, K.-V. Peinemann, S.P. Nunes, Ionic liquids as self-assembly guide for the formation of nanostructured block copolymer membranes, Journal of Membrane Science 492 (2015) 568–577.
[87] M. Radjabian, V. Abetz, Tailored pore sizes in integral asymmetric membranes formed by blends of block copolymers, Advanced Materials 27 (2015) 352–355.
[88] H. Yu, X. Qiu, N. Moreno, Z. Ma, V.M. Calo, S.P. Nunes, K.V. Peinemann, Self-assembled asymmetric block copolymer membranes: bridging the gap from ultra-to nanofiltration, Angewandte Chemie International Edition 54 (2015) 13937–13941.
[89] N. Moreno, S.P. Nunes, K.-V. Peinemann, V.M. Calo, Topology and shape control for assemblies of block copolymer blends in solution, Macromolecules 48 (2015) 8036–8044.
[90] P. Madhavan, B. Sutisna, R. Sougrat, S.P. Nunes, Photoresponsive nanostructured membranes, RSC Advances 6 (2016) 75594–75601.
[91] Y. Xie, N. Moreno, V.M. Calo, H. Cheng, P.-Y. Hong, R. Sougrat, A.R. Behzad, R. Tayouo, S.P. Nunes, Synthesis of highly porous poly(tert-butyl acrylate)-b-polysulfone-b-poly(tert-butyl acrylate) asymmetric membranes, Polymer Chemistry 7 (2016) 3076–3089.
[92] Y. Xie, B. Sutisna, S.P. Nunes, Membranes prepared by self-assembly and chelation assisted phase inversion, Chemical Communications (Cambridge) 53 (2017) 6609–6612.
[93] B. Sutisna, G. Polymeropoulos, V. Musteata, K.-V. Peinemann, A. Avgeropoulos, D.-M. Smilgies, N. Hadjichristidis, S.P. Nunes, Design of block copolymer membranes using segregation strength trend lines, Molecular Systems Design and Engineering 1 (2016) 278–289.
[94] B. Sutisna, G. Polymeropoulos, E. Mygiakis, V. Musteata, K.V. Peinemann, D.M. Smilgies, N. Hadjichristidis, S.P. Nunes, Artificial membranes with selective nanochannels for protein transport, Polymer Chemistry 7 (2016) 6189–6201.
[95] B. Sutisna, G. Polymeropoulos, V. Musteata, R. Sougrat, D.M. Smilgies, K.V. Peinemann, N. Hadjichristidis, S.P. Nunes, Functionalized nanochannels from self-assembled and photomodified poly(styrene-b-butadiene-b-styrene), Small 14 (2018) e1701885.
[96] V.-E. Musteata, S. Chisca, F. Meneau, D.-M. Smilgies, S.P. Nunes, Carboxyl-functionalized nanochannels based on block copolymer hierarchical structures, Faraday Discussions (2018).
[97] B. Sutisna, P. Bilalis, V. Musteata, D. Smilgies, K.-V. Peinemann, N. Hadjichristidis, S.P. Nunes, Self-assembled membranes with feather-like and lamellar morphologies containing α-helical polypeptides, Macromolecules 51 (2018) 8174–8187.
[98] D.S. Marques, U. Vainio, N.M. Chaparro, V.M. Calo, A.R. Bezahd, J.W. Pitera, K.-V. Peinemann, S.P. Nunes, Self-assembly in casting solutions of block copolymer membranes, Soft Matter 9 (2013) 5557–5564.
[99] V. Musteata, B. Sutisna, G. Polymeropoulos, A. Avgeropoulos, F. Meneau, K.-V. Peinemann, N. Hadjichristidis, S.P. Nunes, Self-assembly of polystyrene- b -poly(2-vinylpyridine)- b -poly(ethylene oxide) triblock terpolymers, European Polymer Journal 100 (2018) 121–131.
[100] C. Stegelmeier, A. Exner, S. Hauschild, V. Filiz, J. Perlich, S.V. Roth, V. Abetz, S. Förster, Evaporation-induced block copolymer self-assembly into membranes studied by in situ synchrotron SAXS, Macromolecules 48 (2015) 1524–1530.
[101] S. Dami, C. Abetz, B. Fischer, M. Radjabian, P. Georgopanos, V. Abetz, A correlation between structural features of an amphiphilic diblock copolymer in solution and the structure of the porous surface in an integral asymmetric membrane, Polymer 126 (2017) 376–385.
[102] Y. Gu, R.M. Dorin, K.W. Tan, D.-M. Smilgies, U. Wiesner, In situ study of evaporation-induced surface structure evolution in asymmetric triblock terpolymer membranes, Macromolecules 49 (2016) 4195–4201.
[103] L. Guo, Z. Zhong, Y. Wang, Atomic layer deposition on block copolymer membranes with gyroidal nanopores toward periodically nanostructured vapor sensors: nanotubes versus nanorods, Advanced Materials Interfaces 3 (2016).
[104] R.M. Dorin, W.A. Phillip, H. Sai, J. Werner, M. Elimelech, U. Wiesner, Designing block copolymer architectures for targeted membrane performance, Polymer 55 (2014) 347–353.

[105] X. Qiu, H. Yu, M. Karunakaran, N. Pradeep, S.P. Nunes, K.-V. Peinemann, Selective separation of similarly sized proteins with tunable nanoporous block copolymer membranes, ACS Nano 7 (2012) 768–776.
[106] J.L. Weidman, R.A. Mulvenna, B.W. Boudouris, W.A. Phillip, Nanostructured membranes from triblock polymer precursors as high capacity copper adsorbents, Langmuir 31 (2015) 11113–11123.
[107] S. Qu, T. Dilenschneider, W.A. Phillip, Preparation of chemically-tailored copolymer membranes with tunable ion transport properties, ACS Applied Materials and Interfaces 7 (2015) 19746–19754.
[108] K. Sankhala, J. Koll, M. Radjabian, U.A. Handge, V. Abetz, A pathway to fabricate hollow fiber membranes with isoporous inner surface, Advanced Materials Interfaces (2017) 4.
[109] M. Radjabian, J. Koll, K. Buhr, U. Vainio, C. Abetz, U.A. Handge, V. Abetz, Tailoring the morphology of self-assembled block copolymer hollow fiber membranes, Polymer 55 (2014) 2986–2997.
[110] P. Madhavan, P.Y. Hong, R. Sougrat, S.P. Nunes, Silver-enhanced block copolymer membranes with biocidal activity, ACS Applied Materials and Interfaces 6 (2014) 18497–18501.
[111] Y. Gu, R.M. Dorin, U. Wiesner, Asymmetric organic–inorganic hybrid membrane formation via block copolymer–nanoparticle co-assembly, Nano Letters 13 (2013) 5323–5328.
[112] H. Yu, X. Qiu, P. Neelakanda, L. Deng, N.M. Khashab, S.P. Nunes, K.V. Peinemann, Hollow ZIF-8 nanoworms from block copolymer templates, Scientific Reports 5 (2015) 15275.
[113] R. Shevate, M.A. Haque, F.H. Akhtar, L.F. Villalobos, T. Wu, K.V. Peinemann, Embedding 1D conducting channels into 3D isoporous polymer films for high-performance humidity sensing, Angewandte Chemie International Edition 57 (2018) 11218–11222.
[114] C. Höhme, V. Filiz, C. Abetz, P. Georgopanos, N. Scharnagl, V. Abetz, Postfunctionalization of nanoporous block copolymer membranes via click reaction on polydopamine for liquid phase separation, ACS Applied Nano Materials 1 (2018) 3124–3136.
[115] H.C. Erythropel, J.B. Zimmerman, T.M. de Winter, L. Petitjean, F. Melnikov, C.H. Lam, A.W. Lounsbury, K.E. Mellor, N.Z. Janković, Q. Tu, The Green ChemisTREE: 20 years after taking root with the 12 principles, Green Chemistry 20 (2018) 1929–1961.
[116] T.C. Merkel, M. Zhou, R.W. Baker, Carbon dioxide capture with membranes at an IGCC power plant, Journal of Membrane Science 389 (2012) 441–450.
[117] K.W. Boeddeker, G. Bengtson, Selective pervaporation of organics from water, in: GKSS-Forschungszentrum Geesthacht G.m.b.H., Geesthacht-Tesperhude (Germany, F.R.), 1989, p. 26. Medium: X; Size.
[118] D.R. Greer, T.P. Basso, A.B. Ibanez, S. Bauer, J.M. Skerker, A.E. Ozcam, D. Leon, C. Shin, A.P. Arkin, N.P. Balsara, Fermentation of hydrolysate detoxified by pervaporation through block copolymer membranes, Green Chemistry 16 (2014) 4206–4213.
[119] C. Ye, T. Takigawa, O. Burtovvy, L. Langsdorf, D. Jablonski, A. Bell, B.D. Vogt, Impact of nanostructure on mechanical properties of norbornene-based block copolymers under simulated operating conditions for bio-butanol membranes, ACS Applied Materials and Interfaces 7 (2015) 11765–11774.
[120] J.R. Nykaza, R. Benjamin, K.M. Meek, Y.A. Elabd, Polymerized ionic liquid diblock copolymer as an ionomer and anion exchange membrane for alkaline fuel cells, Chemical Engineering Science 154 (2016) 119–127.
[121] L. Cseri, M. Razali, P. Pogany, G. Szekely, Organic solvents in sustainable synthesis and engineering, in: Green Chemistry, Elsevier, 2018, pp. 513–553.
[122] A. Figoli, T. Marino, S. Simone, E. Di Nicolo, X.-M. Li, T. He, S. Tornaghi, E. Drioli, Towards non-toxic solvents for membrane preparation: a review, Green Chemistry 16 (2014) 4034–4059.
[123] C.M. Alder, J.D. Hayler, R.K. Henderson, A.M. Redman, L. Shukla, L.E. Shuster, H.F. Sneddon, Updating and further expanding GSK's solvent sustainability guide, Green Chemistry 18 (2016) 3879–3890.
[124] D. Kim, O.R. Salazar, S.P. Nunes, Membrane manufacture for peptide separation, Green Chemistry 18 (2016) 5151–5159.
[125] D. Kim, H. Vovusha, U. Schwingenschlögl, S.P. Nunes, Polyethersulfone flat sheet and hollow fiber membranes from solutions in ionic liquids, Journal of Membrane Science 539 (2017) 161–171.
[126] D. Kim, S.P. Nunes, Poly (ether imide sulfone) membranes from solutions in ionic liquids, Industrial and Engineering Chemistry Research 56 (2017) 14914–14922.
[127] D. Kim, S. Livazovic, G. Falca, S.P. Nunes, Oil-water separation using membranes manufactured from cellulose/ionic liquid solutions, ACS Sustainable Chemistry and Engineering 7 (2019) 5649–5659.
[128] D. Kim, N.L. Le, S.P. Nunes, The effects of a co-solvent on fabrication of cellulose acetate membranes from solutions in 1-ethyl-3-methylimidazolium acetate, Journal of Membrane Science 520 (2016) 540–549.
[129] S. Livazovic, Z. Li, A.R. Behzad, K.-V. Peinemann, S.P. Nunes, Cellulose multilayer membranes manufacture with ionic liquid, Journal of Membrane Science 490 (2015) 282–293.
[130] H. Kang, R. Liu, Y. Huang, Cellulose derivatives and graft copolymers as blocks for functional materials, Polymer International 62 (2013) 338–344.
[131] M.A. Hillmyer, The promise of plastics from plants, Science 358 (2017) 868–870.
[132] D.K. Schneiderman, M.A. Hillmyer, 50th anniversary perspective: there is a great future in sustainable polymers, Macromolecules 50 (2017) 3733–3749.
[133] M.A. Hillmyer, W.B. Tolman, Aliphatic polyester block polymers: renewable, degradable, and sustainable, Accounts of Chemical Research 47 (2014) 2390–2396.
[134] W.A. Phillip, J. Rzayev, M.A. Hillmyer, E. Cussler, Gas and water liquid transport through nanoporous block copolymer membranes, Journal of Membrane Science 286 (2006) 144–152.
[135] L. Zhu, F. Liu, X. Yu, L. Xue, Poly (lactic acid) hemodialysis membranes with poly (lactic acid)-block-poly (2-hydroxyethyl methacrylate) copolymer as additive: preparation, characterization, and performance, ACS Applied Materials and Interfaces 7 (2015) 17748–17755.
[136] M.A. Rahman, H.N. Lokupitiya, M.S. Ganewatta, L. Yuan, M. Stefik, C. Tang, Designing block copolymer architectures toward tough bioplastics from natural rosin, Macromolecules 50 (2017) 2069–2077.
[137] B. Sutisna, P. Bilalis, V. Musteata, D.-M. Smilgies, K.-V. Peinemann, N. Hadjichristidis, S.P. Nunes, Self-assembled membranes with feather-like and lamellar morphologies containing α-helical polypeptides, Macromolecules 51 (2018) 8174–8187.

[138] M. Seo, Robust mesoporous polymers derived from cross-linked block polymer precursors, in: Submicron Porous Materials, Springer, 2017, pp. 53–79.

[139] T. Vidil, N. Hampu, M.A. Hillmyer, Nanoporous thermosets with percolating pores from block polymers chemically fixed above the order–disorder transition, ACS Central Science 3 (2017) 1114–1120.

[140] S.A. Chopade, J.G. Au, Z. Li, P.W. Schmidt, M.A. Hillmyer, T.P. Lodge, Robust polymer electrolyte membranes with high ambient-temperature lithium-ion conductivity via polymerization-induced microphase separation, ACS Applied Materials and Interfaces 9 (2017) 14561–14565.

[141] X. Wang, J.L. Davis, B.M. Aden, B.S. Lokitz, S.M. Kilbey, Versatile synthesis of amine-reactive microgels by self-assembly of azlactone-containing block copolymers, Macromolecules 51 (2018) 3691–3701.

[142] Q. Zhang, F. Matsuoka, H.S. Suh, P.A. Beaucage, S. Xiong, D.M. Smilgies, K.W. Tan, J.G. Werner, P.F. Nealey, U.B. Wiesner, Pathways to mesoporous resin/carbon thin films with alternating gyroid morphology, ACS Nano 12 (2018) 347–358.

[143] T. Kowalewski, N.V. Tsarevsky, K. Matyjaszewski, Nanostructured carbon arrays from block copolymers of polyacrylonitrile, Journal of the American Chemical Society 124 (2002) 10632–10633.

CHAPTER

12

Imprinted Materials: From Green Chemistry to Sustainable Engineering

Rüstem Keçili[1], *Ecevit Yılmaz*[2], *Arzu Ersöz*[3,4], *Rıdvan Say*[4]

[1]Yunus Emre Vocational School of Health Services, Department of Medical Services and Techniques, Anadolu University, Eskisehir, Turkey; [2]YLS Consulting AB, Växtskyddsvägen, Alnarp, Sweden; [3]Faculty of Science, Chemistry Department, Eskisehir Technical University, Eskisehir, Turkey; [4]Bionkit Ltd., Technology Park, Eskisehir, Turkey

OUTLINE

1. INTRODUCTION

Green chemistry refers to the design and application of chemical processes and products that eliminate or reduce the generation and use of chemicals hazardous to environment and health. The term "green chemistry" is widely used as a synonym of the term "sustainable chemistry." However, these terms are not synonymous. Sustainable chemistry is a broader term that considers financial, social, and environmental impacts related to production and applications of products and chemicals (Fig. 12.1).

Green chemistry is the core for innovative approaches and new strategies for the development of sustainable chemistry.

Sustainable Nanoscale Engineering
https://doi.org/10.1016/B978-0-12-814681-1.00012-6

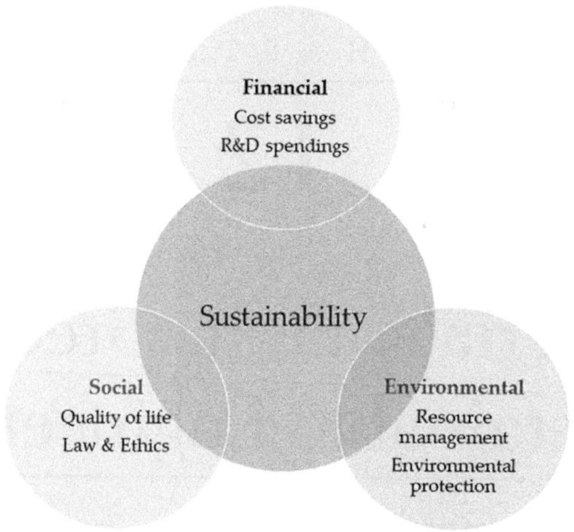

FIGURE 12.1 The aspects of sustainability. The term sustainability is a balance between financial, social, and environmental factors.

Twelve key principles of green chemistry were proposed by Anastas and Warner in 1998 [1].

1. *Waste prevention:* Design and application of synthesis techniques to prevent waste components.
2. *Atom economy:* Design and application of synthesis techniques to increase the incorporation of all starting materials to obtain high-yield product.
3. *The use of less hazardous chemicals:* Design and application of synthesis techniques to reduce the environmental pollution by generating compounds with lower toxicity.
4. *The use of safer chemicals:* Design chemical products that exhibit minimum toxic effects to the environment.
5. *The use of safer solvents and process conditions:* Avoid using toxic solvents and chemicals.
6. *Energy efficiency:* Conduct all processes at mild conditions such as room temperature, lower pressure, etc.
7. *The use of renewable feedstocks:* Use of renewable starting materials also called as feedstocks such as materials obtained from agricultural products or waste.
8. *Reduce derivatives:* Avoid using any protective or derivatization agents if possible.
9. *Catalyst efficiency:* Use green catalysts to reduce the waste.
10. *Degradable design:* Design degradable chemicals and products that are not potentially hazardous to environment.
11. *Real-time analysis:* Real-time analysis, control, and monitoring of the compounds during the process to eliminate and minimize undesired products.
12. *Minimize the potential hazards and accidents:* Choose appropriate chemicals to reduce their potential hazards and minimize potential risks for the accidents such as fires, explosions, and releases of chemicals into the environment.
13. *Minimization of synthetic steps:* Use of ecoeffective synthetic routes by minimization of synthetic steps and isolations.

Solvents play a crucial role in designing more sustainable synthesis. The use of safe and environmentally friendly solvents such as water, supercritical CO_2, ionic liquids, and isopropanol in the industrial processes is a greener route to reduce environmental impacts. In pharmaceutical industry, GlaxoSmithKline and Pfizer reported the solvent selection guides for their employees who work on drug design and development [2,3] (Fig. 12.2).

On the other hand, green engineering refers to industrial applications of green chemistry. Therefore, the main aim of green engineering is the design and application of industrial processes that reduces the generation of waste components. Energy efficiency is another aim of green engineering as the higher energy consumption leads to various environmental impacts such as global warming. Therefore, all industrial processes should be carried out at mild conditions such as room temperature, lower pressure, etc.

The industrial awareness of green chemistry is already in practice, and the near future will show an enhanced role of the green approaches to prevent environmental pollution and the design of safer, greener, and cleaner industrial processes.

The following sections start with the main principles of molecular imprinting technology. Then, green aspects of this technology are demonstrated. The determination of the most appropriate green porogenic solvents (i.e., supercritical carbon dioxide [$scCO_2$], ionic liquids, etc), the use of green functional monomers, optimization of elution conditions, maximizing lifetime and reusability of molecularly imprinted polymers (MIPs), computational design, and

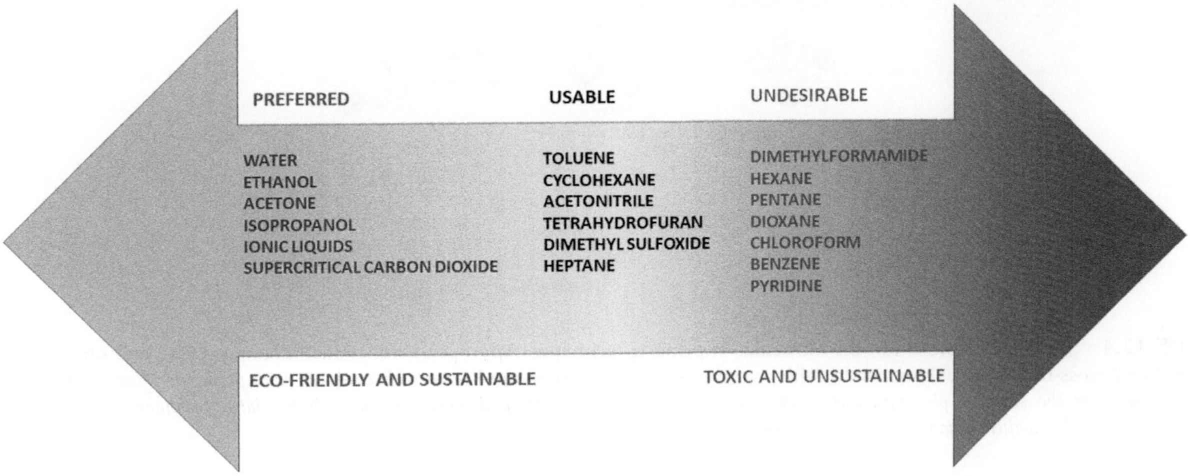

FIGURE 12.2 Solvent selection guide for industrial processes. *Reversible arrow* demonstrates the preferred, useable, and undesirable solvents [2,3].

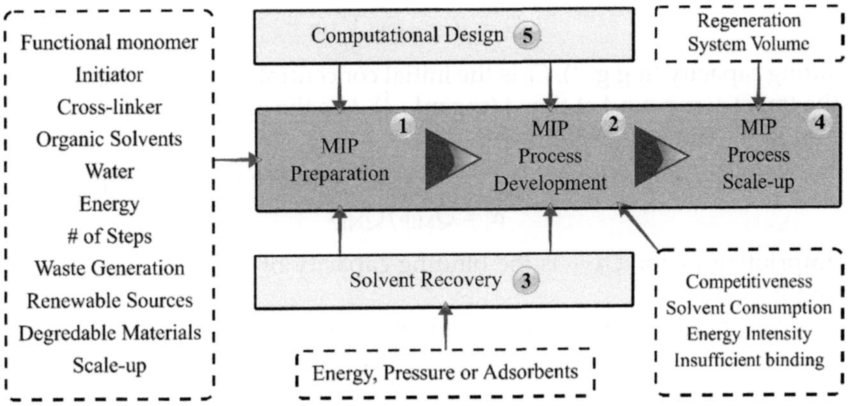

FIGURE 12.3 Aspects of molecular imprinting approach: from molecularly imprinted polymer (MIP) preparation to MIP process development until scale up of MIP process as well as supporting computational design and solvent recovery techniques. Numbers in circles demonstrate the sections of the chapter, while the dashed boxes list the parameters affecting the greenness and sustainability of MIPs in the respective sections.

minimizing waste generation during MIP preparation process are described in the further sections. In addition, various reported studies on green approaches to design and development of selective MIPs in micro- and nanoscale applications are presented (Fig. 12.3).

2. PREPARATION OF IMPRINTED POLYMERS

MIPs are tailor-made materials bearing selective recognition sites with great binding affinity toward the target compound. In the synthesis procedure for the selective MIPs, the polymerization of suitable functional monomers with a cross-linker in the existence of the target compound that is also called template is carried out. The schematic representation of a typical molecular imprinting process is shown in Fig. 12.4. Because of their excellent binding behavior toward the target compound, MIPs have found wide application areas where high selectivity and binding affinity toward the target compound is essential, i.e., solid-phase extraction (SPE) [4], biosensor platforms [5], electrochemical sensors [6], and metal catalysis [7] applications, etc.

MIPs are facile to design and prepare, robust at harsh operating conditions (high temperature, pH values [common limits vary between 1 and 10] and pressure, organic solvents), cost-effective, and reusable artificial materials. In addition, MIPs can be successfully used as tailor-made "plastic receptors" especially when natural receptors are not available [8].

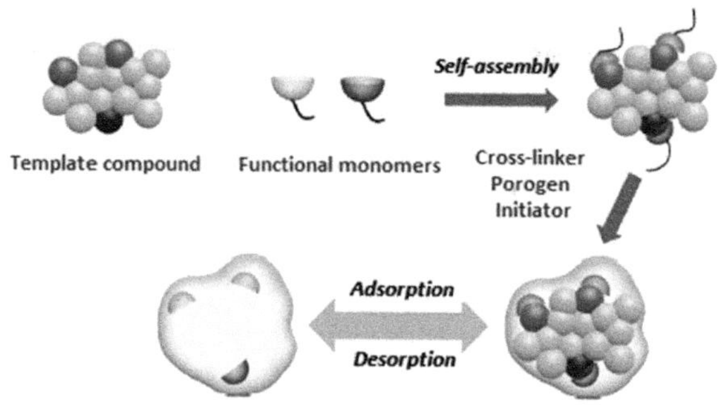

FIGURE 12.4 Schematic depiction of the molecular imprinting technique. Appropriate functional monomers (red and blue [black in print versions]) and cross-linker are polymerized in the presence of template compound (gray). *Reproduced with permission from T. Kubo, K. Otsuka, Recent progress for the selective pharmaceutical analyses using molecularly imprinted adsorbents and their related techniques: a review, Journal of Pharmaceutical and Biomedical Analysis 130 (2016) 68–80.*

Various parameters that are commonly used to quantify the obtained results from molecular imprinting studies are briefly described in the following:

The binding capacity of the MIP or NIP (Q) is calculated by using the following formula.

$$Q = (C_0 - C_t)\ V/m \tag{12.1}$$

where Q refers the binding capacity (mg g^{-1}), C_0 is the initial concentration of the target compound (mg mL^{-1}), C_t is the concentration of the target compound at time t (mg mL^{-1}), V is the solution volume (mL), and m is the weight of the MIP or NIP (g) used in the binding experiments.

On the other hand, the imprinting factor (α) is determined using Eq. (12.2):

$$\alpha = Q_{MIP}/Q_{NIP} \tag{12.2}$$

where α defines the imprinting factor, Q_{MIP} is the binding capacity of the MIP (mg g^{-1}), and Q_{NIP} is the binding capacity of the NIP (mg g^{-1}).

Distribution coefficient (K_d) is calculated using the following equation:

$$K_d = (C_i - C_f)\ V/C_f m \tag{12.3}$$

where K_d is distribution coefficient and C_i and C_f are initial and final concentration of the target compound, respectively. V is the solution volume (mL) and m is the weight of the MIP or NIP (g).

On the other hand, selectivity coefficient (k) for the target compound in the existence of other competitor compounds can be determined from Eq. (12.4).

$$k = K_{d\ target\ compound}/K_{d\ competitor} \tag{12.4}$$

Relative selectivity coefficient (k') is calculated by using the following formula:

$$k' = k_{MIP}/k_{NIP} \tag{12.5}$$

2.1 Toward Green Chemistry and Engineering of Molecularly Imprinted Polymers

2.1.1 Green Monomers

The synthesis of monomers by using renewable materials is crucial for the green aspects of polymers [10,11]. Various monomers were successfully synthesized by applying green approaches. For example, Riaz et al. synthesized aromatic monomers by solvothermal liquefaction of lignin, which is the most naturally abundant and crucial biopolymeric structure that exists in the cell wall of plants [12]. In another interesting study conducted by Lu and colleagues [13], biosynthesis of ω-hydroxyfatty acids as monomers was carried out using a strain of the diploid yeast *Candida tropicalis*. Cruz-Aguilar and coworkers reported the synthesis method for allylic (2-(((allyloxy)carbonyl)oxy) ethyl (bis(2-(((allyloxy)carbonyl)oxy)ethyl) terephthalate and 2-hydroxyethyl) terephthalate) and acrylic (bis(2-(acryloyloxy)ethyl) terephthalate) bifunctional cross-linking monomers, which are derived from PET waste [14]. In an important study carried out by Jaswal and Gaur [15], various bio-based monomers including

TABLE 12.1 Various Applications of Molecularly Imprinted Polymer (MIPs) Using Ionic Liquids as Green Functional Monomers.

Ionic Liquid as Functional Monomer	Template	Target Compound	Binding Capacity (mg g^{-1})	Reference
1-vinyl-3-butylimidazolium chloride	Lysozyme	Lysozyme	763.36	[24]
1-(α-allylacetate)-3-N-(3-aminopropyl)-imidazolium chloride	Lysozyme	Lysozyme	213.7	[25]
Mono-6A-deoxy-6-(1-vinylimidazolium)-b-cyclodextrin tosylate	C-terminus of Cytochrome c, residues 96–104	Cytochrome c	67.6	[26]
1-vinyl-3-butylimidazolium tetrafluoroborate	6-benzylaminopurine	6-benzylaminopurine	49.55	[27]
1-vinyl-3-butylimidazolium chloride	Benzoic acid	Salicylic acid	29.75	[28]
1-ally-3-vinylimidazolium chloride	3,4-dihydroxybenzenepropanoic acid	3,4-dihydroxybenzenepropanoic acid	0.22	[29]

methacrylated guaiacol, methacrylated eugenol using methacrylic anhydride in the existence of catalyst 4-dimethylaminopyridine. The authors reported that these green monomers are great alternatives to petroleum-based monomers as they are cheap and environmentally friendly and their volatility is quite low. Lima et al. proposed a facile approach for the synthesis of green monomers sobrerol methacrylate and sobrerol acrylate [16]. In this study, sobrerol was synthesized by applying a green strategy that involved the hydration of α-pinene oxide in the presence of CO_2. The results indicated that the synthesized green monomers have low volatility and green alternative to conventional styrene monomer.

In a study conducted by Yao and colleagues [17], green phenolic monomers were successfully synthesized by using alkali lignin material, which is a by-product of paper and pulp industry. Hayama et al. reported the synthesis of glycomonomer and various functional monomers by using 4-chloromethylstyrene as the starting material [18]. In another study [19], the synthesis of allyl saccharide/vinyl copolymers was carried out using renewable feedstocks such as α,α α,α'-trehalose and D-glucose to obtain green monomers. On the other hand, ionic liquids as green functional monomers were extensively used for the preparation of MIPs [20–22]. The attention on the use of environmentally friendly ionic liquids in molecular imprinting process emanates from the unique features of these compounds such as high thermal stability, high viscosity, and low vapor pressure and miscibility with a broad range of organic solvents. Marcelo and colleagues used green monomer itaconic acid for the preparation of MIPs by using supercritical fluid technology [23]. The prepared green MIPs were successfully used for the drug delivery applications (Table 12.1).

The raw material usage and the number of ingredients in the synthesis of MIPs can be minimized by initiator-free synthesis. For example, the research group of Haupt reported the self-initiated synthesis of MIPs toward various compounds including trypsin, propranolol, and testosterone [30].

2.1.2 *Green Solvents for Porogens*

Porogens play a crucial role in the preparation of MIPs. All components (i.e., functional monomer, template compound, etc) should be dissolved in porogenic solvent, and functional monomer–template preorganized complex is formed through noncovalent or covalent interactions. Thus, the choice of an appropriate porogen affects the formation of functional monomer–template preorganized complex. Selection of the right solvent can enhance performance and minimize the environmental burden of chemical manufacturing [31]. Green porogenic solvents such as water [32,33] and ionic liquids [34] were successfully used for the preparation of MIPs. For example, 1-butyl-3-methylimidazolium tetrafluoroborate as the green porogenic solvent was used for the preparation of MIPs toward testosterone [34].

$scCO_2$ is another green alternative to replace conventional organic solvents widely used in polymerization reactions as porogens. $scCO_2$ has many superiorities such as being nonhazardous, low cost, and inert. In addition, the removal of $scCO_2$ from reaction medium can easily be carried out. The green solvent $scCO_2$ was successfully used for the preparation MIPs. For example, Rebocho et al. developed and MIP-based electrochemical sensor for the sensitive detection of bisphenol A (BPA) [35]. For this purpose, the functional monomer ferrocenylmethyl methacrylate and cross-linker ethylene glycol dimethacrylate (EGDMA) were copolymerized in the green porogenic solvent $scCO_2$. The obtained results showed that the developed MIP-based electrochemical sensor displayed great recognition behavior toward the target compound BPA in the concentration range between 4.7 and 8.0 nM. The limit of detection and limit of quantification values were achieved as 3.2 and 4.7 nM, respectively.

In another study reported by Viveiros and colleagues [36], MIPs were prepared for the removal of pharmaceutical impurity acetamide. In their study, itaconic acid and 2-hydroxyethyl methacrylate were chosen as the functional monomers for the preparation of MIPs in green solvent scCO2. The obtained results confirmed that the prepared MIPs exhibited great affinity and selectivity toward the target compound acetamide in the presence of other competing compounds such as benzamide and trimethylacetamide.

Ye et al. reported the preparation of imprinted nanoparticles in $scCO_2$ [37]. They used methacrylic acid (MAA), propranolol, and divinylbenzene (DVB) as the functional monomer, template compound, and cross-linker, respectively. The prepared imprinted nanoparticles were successfully used for the binding of propranolol in aqueous solutions and organic solvents.

In another interesting study [38], Soares da Silva and coworkers prepared molecularly imprinted composite membranes for the removal of BPA from aqueous solutions. In their study, firstly, the functional monomer MAA and the cross-linker EGDMA were polymerized in the presence of template compound BPA using $scCO_2$ as the green solvent. Then, $scCO_2$-assisted MIP-based membranes were produced by phase-inversion technique. The highest binding capacity for the target BPA was achieved as 625 $\mu mol\ g^{-1}$.

Ferreira and colleagues reported the preparation of selective MIPs in $scCO_2$ for the removal of oxidized products from diesel samples [39]. For this purpose, model substrate dibenzothiophene sulfone was oxidized using various acid catalysts and also phase catalysts including ionic liquids trihexyl(tetradecyl)phosphoniumchloride and Aliquat. Then, the oxidation products dibenzothiophene sulfone and dibenzothiophene sulfoxide were successfully removed from diesel samples by using the prepared MIPs.

Lourenço et al. developed a molecularly imprinted CdTe quantum dot-based ultrasensitive sensor toward BPA [40]. The green imprinting process was carried by using MAA, EGDMA, BPA, and $scCO_2$ as the functional monomer, cross-linker, template and green solvent, respectively. The prepared imprinted nanosensor showed great recognition behavior toward the target compound BPA, and the limit of detection was achieved as 4 nM.

Another study on the use of green solvent $scCO_2$ for the preparation of MIPs was conducted by Lee and colleagues [41]. In their study, $scCO_2$-assisted synthesis of MIPs was carried out for the removal of carbamazepine from aqueous solutions. The prepared MIPs displayed high affinity and selectivity for carbamazepine in the presence of other competing compounds including 2,4-dichlorophenoxyacetic acid, aspirin, and acetaminophen. The maximum binding capacity was obtained as 101.61 $\mu mol\ g^{-1}$.

On the other hand, carbon capture, utilization, and storage are crucial and challenging issues [42]. In a recent attempt toward sustainable MIP synthesis and application [43], a novel MIP toward CO_2 was prepared. For this purpose, an amino-terminated hyperbranched polymer was synthesized by the reaction between and methyl acrylate and amine groups. Subsequently, CO_2 imprinted hyperbranched polymer was prepared by using the cross-linker glutaraldehyde and green solvent water. The prepared MIPs showed high adsorption behavior toward the target compound CO_2 with a maximum adsorption capacity of 7.65 $mmol\ g^{-1}$.

In another study conducted by Zhao et al. [44], selective MIPs were developed for the efficient separation of CO_2 from flue gas samples. In this study, acrylamide (AAm), ethanedioic acid, and EGDMA were used as the functional monomer, template compound, and cross-linker, respectively. The maximum CO_2 adsorption capacity was achieved as 0.57 $mmol\ g^{-1}$.

Nabavi and coworkers designed and prepared highly selective MIPs for the capture of CO_2 [45]. For this purpose, AAm and EGDMA were chosen as the functional monomer and cross-linker, respectively. Selective MIPs toward CO_2 were prepared by using oil-in-oil suspension polymerization approach. The prepared MIPs were successfully applied for the capture of CO_2 molecules. The obtained maximum capture of CO_2 was 1.1 $mmol\ g^{-1}$.

On the other hand, the long-term reusability and stability are another green aspects of MIPs. MIPs are mostly thermally stable in the temperature range from room temperature to 250°C [46]. Thermal stability behavior of the MIPs was successfully determined by thermogravimetric analyses [47,48]. The obtained results confirmed that the use of MIPs at high temperature values does not affect the polymeric structure and damage the MIPs in any way. Thus, it should be noted that the abundant existence of MIPs in the environment should be avoided as these materials are not easily degraded under ambient conditions. Furthermore, long-term reusability and stability of MIPs is crucial in green chemistry perspective as these features entail minimization of solid waste.

2.1.3 *Green Polymers and Polymerization Techniques*

MIPs can be prepared by applying various methods such as bulk polymerization, suspension polymerization, emulsion polymerization, and precipitation polymerization. These techniques are briefly explained in the following sections.

2.1.3.1 Bulk Polymerization

In the bulk polymerization, functional monomers directly participate in polymerization process. The initiator, functional monomers, and cross-linker are dissolved in an appropriate solvent. Then, dissolved oxygen, which negatively effects the polymerization yield in the solution, is removed to obtain an inert atmosphere. Finally, polymerization is initiated by applying heat or UV light and completed after 18–24 h. Then, the obtained monolithic polymer is crushed, ground, and sieved. The polymeric particles obtained by bulk polymerization have irregular shapes. This polymerization technique has the following advantages:

- Facile
- Cheap
- Fast process

However, bulk polymerization has some drawbacks such as the following:

- Viscosity increases during the polymerization that leads to difficulties in stirring of the polymer solution and heat transfer.
- The unreacted functional monomers should be removed from the reaction medium by applying a washing step.
- After polymerization, crushing, grinding, and sieving steps should be applied to obtain desired particle size distribution.
- The grinding and sieving process may lead to loss of polymeric particles.

Because of the disadvantages mentioned above, bulk polymerization is not preferred for the synthesis of polymers in large scale. Schematic depiction of the MIP preparation by using bulk polymerization is given in Fig. 12.5.

2.1.3.2 Suspension (Bead) Polymerization

Suspension or bead polymerization can overcome the drawbacks of the bulk polymerization. There are two phases in this technique. One of these phases is the solvent such as water containing additives (polyvinyl alcohol [PVA], etc.). This phase is also called as dispersing medium. The other phase is organic phase that consists of the functional monomer, cross-linker, template compound, and initiator. The polymerization takes place in liquid monomer droplets. During the polymerization process, each monomer droplet slowly converts to solid spherical polymer particles. The surfactants such as PVA help to reduce viscosity and prevent agglomeration of spherical particles by forming a protective layer on the surface of the particles. The particle size distribution of the obtained spherical beads varies from 50 to 500 μm depending on the polymerization conditions.

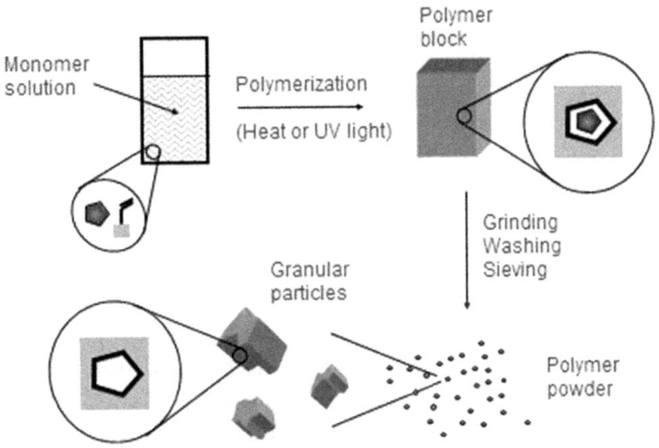

FIGURE 12.5 Schematic depiction of the molecularly imprinted polymer (MIP) preparation by using bulk polymerization. Polymer block is obtained by bulk polymerization. Grinding, washing, and sieving steps are applied to obtain polymer powder.

Suspension polymerization has some advantages over bulk polymerization such as low viscosity due to suspension and temperature control by continuous stirring.

2.1.3.3 Emulsion Polymerization

Emulsion polymerization technique is commonly applied for the preparation of polymeric latex particles [49]. In this technique, a small amount of water-insoluble monomer is dispersed in water phase, and water-soluble initiators are added into the solution. An appropriate surfactant that shows hydrophobic and hydrophilic feature is used for the stabilization of the emulsion of the monomer in water phase. Surfactant molecules come together and form aggregates also called "micelles." Then, free radical initiator molecules in water phase migrate into the micelles, and polymerization is initiated. The polymerization kinetics consists of three stages. In the first stage, monomers diffuse from droplet into micelles and polymerization starts to form polymeric particles. Second stage corresponds to growth of polymeric particles by propagation. During this stage, the monomer concentration inside the polymeric particles and number of polymeric particles remain almost constant. In the final stage, the rate of the polymerization decreases, and no monomer droplets are observed. The remaining monomers in the particles are polymerized. The size distribution of the obtained latex particles varies from 50 to 700 nm depending on the polymerization conditions.

2.1.3.4 Precipitation Polymerization

Precipitation polymerization is used for the preparation of polymeric nanoparticles. In this technique, highly diluted monomer solution is used (ca 2% v/v). The agglomeration of the polymeric chains does not occur during the polymerization process. Monomers and oligomers continue to participate in the structure of the polymeric chains. Finally, participation occurs when they become insoluble in the reaction media due to the size of polymeric particles. The precipitation polymerization approach is facile and fast compared with other polymerization approaches [50]. In spite of these advantages, this approach has some limitations:

- The polymerization in highly diluted monomer solution may lead to weak interactions between the monomer and target compound.
- It is difficult to control the size of the nanoparticles.

Both suspension and precipitation polymerization approaches lead to large production of the liquid waste. Therefore, the bulk polymerization technique is relatively greener compared with suspension and precipitation polymerization techniques.

2.1.3.5 Iniferter Polymerization

This alternative approach was developed to overcome the drawbacks of precipitation polymerization. The size of the particles during the polymerization can be controlled in this polymerization technique. For this purpose, a chemical agent also called "iniferter" is added into the polymerization medium. It behaves as an initiator, chain transfer agent, and terminator during the polymerization process that provides the exact controllability of the polymerization kinetics. The yield of the obtained nanoparticles is affected by various parameters such as iniferter concentration, reaction time, and type of chain terminator agent. The considerable disadvantage of the iniferter polymerization approach is the higher complexity of the reaction medium, which is an additional factor that has to be optimized.

The iniferter polymerization approach was successfully applied for the preparation of MIPs. In a reported study by Garcia-Soto et al. [51], testosterone-selective MIPs were prepared by using photoiniferter polymerization technique under visible light. The obtained results indicated that prepared MIPs exhibited high binding affinity toward the target compound testosterone.

2.1.3.6 Reversible Addition Fragmentation Chain Transfer Polymerization

Free radical polymerization is the most extensively applied approach for the preparation of MIPs as it has various advantages (i.e., simplicity and rapidity in preparation, no requirement for expensive and sophisticated instrumentation, etc.). However, this technique has some drawbacks. For example, it is not easy to control the rate of chain propagation during the free-radical polymerization process, and the prepared polymers usually have a broad particle size distribution. To overcome these drawbacks, reversible addition fragmentation chain transfer (RAFT) polymerization technique has been proposed as a controlled radical polymerization approach [52]. RAFT polymerization technique has various advantages as briefly described in the following:

- A broad range of functional monomers can be used, and polymerization process takes place under mild polymerization conditions.
- The functional groups can easily be introduced into the polymer chain by adjusting the RAFT agent structure

❖ RAFT polymerization technique displays great versatility in preparation of well-defined complex architecture polymers such as graft and block polymers.

Surface-initiated RAFT polymerization was successfully applied for the preparation of core–shell MIPs [53–56]. Various applications of RAFT polymerization technique (i.e., degradable polymers, self-plasticized polymers, and miscellaneous green materials, etc.) were reported in a comprehensive review by Tsarevsky and Matyjaszewski [57] and Semsarilar and Perrier [58]. Phommalysack-Lovan et al. reported the application of photoinduced electron/energy transfer (PET)-RAFT approach for the green preparation of polymers [59]. In another study carried out by West and coworkers [60], polymerization of methyl methacrylate (MMA) by using RAFT polymerization technique was reported. In this study, poly(ethylene glycol) was chosen as the green solvent in the polymerization process.

3. COMPUTATIONAL DESIGN OF MOLECULARLY IMPRINTED POLYMERS

The design and preparation of MIPs toward target compound/s require so much effort, suitable instrumentation, and time-consuming work. Large volumes of solvents and significant amounts of chemicals are used in the production of these polymers, which also lead to generation of significant quantities of waste. The implementation of molecular modeling approaches in the design and production processes of MIPs can overcome these drawbacks. Suitable computations and simulation approaches ensure the facile determination of the nature of chemical interactions between the polymerization reagents (i.e., functional monomer, template compound, and cross-linker). Therefore, the best configurations of the required polymerization reagents can easily be determined for the preparation of selective MIPs. This allows for a considerable decrease in the amounts of reagents and solvents and obtaining the high selective MIPs toward the target compounds [61–64]. In the modern computational tools that provide the rational design of MIPs by screening the potential functional monomers before the polymerization process, the classical trial and error technique is eliminated [65]. In addition, waste is not generated, which is one of the green chemistry requirements.

The number of the reported studies in this field is gradually increasing. Various approaches such as density functional theory and molecular dynamics simulations are applied for the rational design of MIPs and the investigation of the interactions between the functional monomer and target compound [66–68]. In these computational approaches, different computer softwares including Gaussian [69], Avogadro [70], and Studio [66] are commonly used for the calculation of energy, optimization of structural geometry, and molecular dynamics simulations. The most widely applied computational approach for the determination of appropriate functional monomer in the MIP preparation process is density functional theory. On the other hand, molecular dynamic simulations are usually applied for the validation of the chemical interactions between the functional monomer and target compound. The use of computational methods was also extended to screening and determination of appropriate porogenic solvents for the preparation of MIPs [69,71].

Dineiro and colleagues applied Gaussian computational approach for the rational design of MIPs for homovanillic acid. In their study, the density functional theory methodology was used for rational design of homovanillic acid–selective MIPs that provide the determination of the most appropriate functional monomer and porogenic solvent for the synthesis of MIPs [72]. In this methodology, a comparison study on the stabilization energies of the prepolymerization adducts between the functional monomer and target compound is carried out. The same research group also proposed another alternative approach, the prediction based on the density functional theory and experimental investigation for the synthesis of MIPs toward homovanillic acid. The obtained results indicated that trifluoromethacrylic acid and toluene were the most appropriate functional monomer and porogenic solvent, respectively [73].

On the other hand, SYBYL, which is a user-friendly software, is generally used in drug design applications. The group of Pilestsky successfully applied this software to determine the affinity interactions between molecules involved in preparation of MIPs. In 2001, they reported the application of SYBYL in the design and preparation of selective MIPs toward ephedrine [74].

4. APPLICATIONS OF IMPRINTED POLYMERS

4.1 Solid-Phase Extraction With Molecularly Imprinted Polymers

SPE is an efficient tool for the sample preparation in environmental chemistry. SPE was first applied in 1940s [75], and the rapid progresses in various SPE applications were initiated in 1970s. It is now successfully applied in different field of science especially in environmental analysis. Conventional SPE materials such as silica-based [76,77], carbon-based [78,79] and clay-based [80] resins were used in various SPE examples. Many adsorbents for SPE applications now commercially exist in different format on the market such as SPE tubes and pipet tip formats such as Oasis HLB (produced by Waters), OMIX (produced by Agilent), and MonoTips (produced by GL Sciences), etc.

The basic principles of SPE are similar to liquid–liquid extraction. Both methods involve the distribution of dissolved species between two phases. However, SPE involves the dispersion of the analyte between a liquid (sample medium) and a solid (adsorbent) phase instead of the two liquid phases that are not mixed together as in liquid-liquid extraction. This technique allows the enrichment and purification of the analytes on a solid adsorbent through adsorption from the solution. Basically, this method involves passing a liquid through a column holding analytes, such as a cartridge, resin, or a disc, and then recovering the captured analytes with a suitable solvent. A typical SPE process composed of four different stages including conditioning, sample loading, washing, and eluting is schematically demonstrated in Fig. 12.6.

Conditioning step: The aim of the conditioning step is to condition the solid adsorbent in the SPE column using a suitable solvent that is the same as the solvent of the sample. In another words, this step aims to activate the chromatographic SPE adsorbent to allow appropriate phase interface with the sample, which will be applied to the SPE column. Acetonitrile (MeCN), methanol (MeOH), ethanol (EtOH), dichloromethane, and tetrahydrofuran are the most widely used organic solvents in the conditioning step. These solvents function in a surfactant capacity that allows the interaction of aqueous sample having high polarity with the hydrophobic adsorbent. This conditioning step is very crucial as it ensures that the adsorbent material is wetted and the functional groups are solvated. In addition, it removes impurities that may initially exist on the solid adsorbent. In fact, this step removes the air in the column to fill the cavities with solvent. The choice of conditioner depends on the nature of the solvent and the solid adsorbent. Between the treatment and sample treatment steps, the solid adsorbent should not be allowed to dry. Otherwise, the analytes do not efficiently adsorb to the column, and low recovery is achieved. If the adsorbent remains dry for more than a few minutes, it must be reconditioned.

Sample loading step: In the sample loading step, the sample is passed through the SPE column filled with selective MIP particles as the adsorbent. Depending on the system used, sample volumes can range from 1 mL to 1 L. The sample can be sent to the column by gravity, pumping, vacuum, or an automatic system. One of the most important issues of this step is the flow rate of the sample in the column.

The importance of a sufficiently slow flow rate is related to the interaction time of the target compound/s with the MIP in the column. The flow rate of the sample in the column should be slow enough to allow the efficient adsorption of the target compound/s to the adsorbent, and it should be high enough to avoid excess time. During this step, the target compound/s is concentrated on the solid adsorbent in the column.

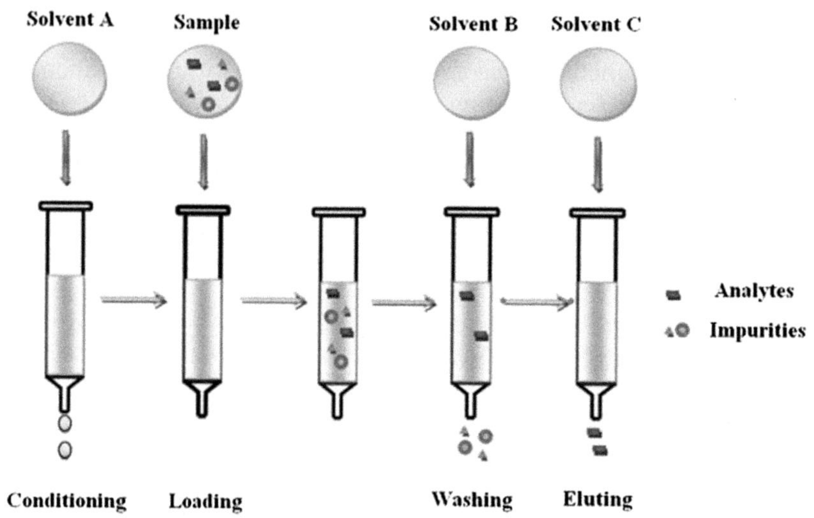

FIGURE 12.6 Solid-phase extraction (SPE) process using molecularly imprinted polymer (MIPs). Solvent A, B, and C represent the appropriate solvents for the conditioning, washing, and elution step of the SPE process, respectively. *Reproduced with permission from J. Płotka-Wasylka, N. Szczepańska, M. de la Guardia, J. Namieśnik, Miniaturized solid-phase extraction techniques, TRAC Trends in Analytical Chemistry 73 (2015) 19–38.*

TABLE 12.2 Some Commercially Available Molecularly Imprinted Polymer (MIP)—Based Solid-Phase Extraction (SPE) Products In the Market.

Company	SPE Product	Application
AFFINISEP	AFFINIMIP SPE FumoZON	SPE of fumonisins and zearalenone from food samples
	AFFINIMIP SPE Deoxynivalenol	SPE of deoxynivalenol from food samples
	AFFINIMIP SPE Multimyco	SPE of fumonisins, ochratoxinA, zearalenone aflatoxins, and deoxynivalenol from cereals
	AFFINIMIP SPE Metanephrines	SPE of metanephrines from biological matrices
	AFFINIMIP SPE Glyphosate – AMPA	SPE of glyphosate and aminomethylphosphonic acid (AMPA) from water
	AFFINIMIP SPE Picolinic herbicides	SPE of picloram, clopyralid, and aminopyralid from environmental matrices
Biotage	AFFINILUTE MIP – Clenbuterol	SPE of clenbuterol from biological matrices
	AFFINILUTE MIP – β-Agonist	SPE of β-agonist from biological matrices
	AFFINILUTE MIP-Triazine	SPE of triazines from environmental matrices
	AFFINILUTE MIP β-Blocker	SPE of β-blocker from biological matrices
	AFFINILUTE MIP-Amphetamine	SPE of amphetamine from biological matrices
	AFFINILUTE MIP-PAH	SPE of polyaromatic hydrocarbons from organic matrices
Sigma Aldrich	SupelMIP SPE – Patulin	SPE of patulin from fruit and fruit products
	SupelMIP SPE – Aminoglycosides	SPE of aminoglycoside antibiotics from tissue samples and cell culture
	SupelMIP SPE – Bisphenol A (BPA)	SPE of BPA from environmental food samples
	SupelMIP SPE – Riboflavin (vitamin B2)	SPE of riboflavin from milk and other aqueous samples

Washing step: After the loading of the sample to the MIP in the column, a washing step is required. The washing step is applied for the removal of undesired impurities from the adsorbent while allowing to leave target compound/s on the MIP. The washing of the SPE column is commonly carried out with an appropriate solvent which has higher elution strength than the solvent in which the sample was dissolved and loaded to the SPE column. In most of the cases, the solvent that is used in the washing step is a weaker type of the solvent.

Elution step: The final step of a typical SPE process is elution of the target compound/s from the MIP by using an appropriate elution solvent. The elution step sometimes can be composed of more than one step, where the retentive interactions between the functional groups exist on the MIP and the target compound/s are disrupted.

Commercial MIP-based SPE products are marketed today by Affinisep (Petit-Couronne, France), Biotage (Lund, Sweden), and Sigma–Aldrich (St. Louis, USA) [82–84] (Table 12.2).

4.2 Pharmaceutical Manufacturing With Molecularly Imprinted Polymers

Environmental pollution is one of the significant international concerns today. Various emerging pollutants in environment such as pharmaceutical compounds display harmful impacts on plants, animals, and humans. Therefore, the design and synthesis of effective materials for the sensitive detection and removal of these environmental pollutants are very important.

The research group of Sellergren reported to the first application of SPE based on selective MIPs [85]. In their work, selective MIP was designed and synthesized for the SPE of target drug compound pentamidine that is used for the treatment of AIDS disease. In the MIP synthesis, 2-propanol and water were used as green solvents. After this report on the successful use of MIPs for SPE, scientists put so much effort on the design and synthesis of new MIPs for the selective extraction of many different compounds from various matrices [86–97].

In another important research, selective MIP spheres were designed and prepared for the efficient extraction of an antimicrobial veterinary drug enrofloxacin [98]. In this report, the authors have chosen MAA and 2-hydroxyethyl methacrylate as the functional monomer and hydrophilic comonomer, respectively (Fig. 12.7). Acetonitrile was chosen as the porogen, which is a useable solvent in terms of its greenness. The results confirmed that the prepared MIP

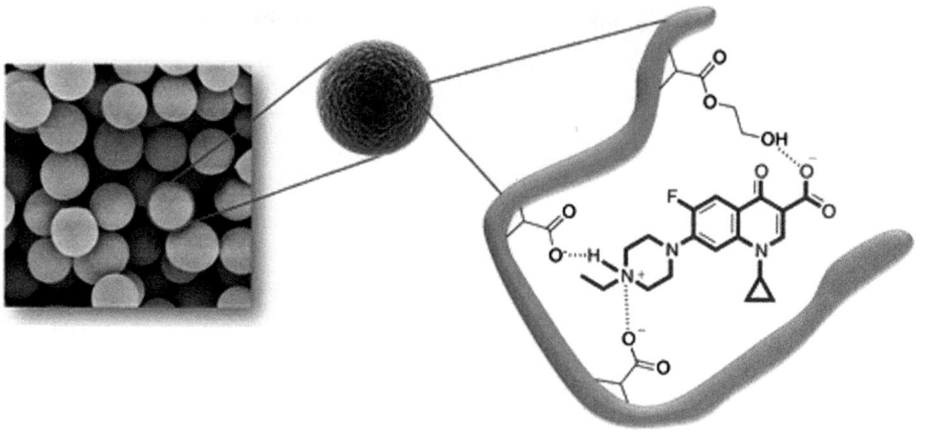

FIGURE 12.7 The SEM image and the schematic depiction of the enrofloxacin imprinted spheres for enrofloxacin. The particle size of the enrofloxacin imprinted spheres is around 3 μm. *Reproduced with permission from E. Benito-Peña, F.Navarro-Villoslada, S. Carrasco, S. Jockusch, M. F. Ottaviani, M.C. Moreno-Bondi, Experimental mixture design as a tool for the synthesis of antimicrobial selective molecularly imprinted monodisperse microbeads, ACS Applied Materials and Interfaces 7 (2015) 10966–10976.*

spheres showed great selectivity and binding affinity toward the target drug compound enrofloxacin. The obtained affinity constant (K_a) value was 7558 M^{-1}.

Dai et al. prepared MIPs for the selective removal of a nonsteroidal antiinflammatory drug diclofenac from contaminated river water samples [99]. The functional monomer 2-vinyl pyridine (2-VP) and cross-linker EGDMA were preferred for the synthesis of MIPs toward diclofenac. The binding behavior of the MIPs toward diclofenac was compared with activated carbon, and the results showed that MIPs exhibited higher binding toward diclofenac with a binding capacity of 324.8 mg g^{-1}. The reusability of the MIPs was also investigated, and MIPs maintained their high binding behavior toward diclofenac after 12 cycles without considerable loss in their binding performance.

Same research group prepared a novel MIP using multitemplates for the removal of naproxen, diclofenac, ibuprofen, ketoprofen, and clofibric acid from lake water samples [100]. 2-VP was chosen as the functional monomer. The binding behavior of the imprinted polymer toward target pharmaceutical compounds was investigated in detail. The results confirmed that the prepared MIP displayed excellent selectivity and affinity for the five target compounds. The maximum binding capacities of the prepared MIP toward naproxen, diclofenac, ibuprofen, ketoprofen, and clofibric acid were 60.7, 61.3, 60.7, 48.7, and 52 μg g^{-1}, respectively.

In an important study carried out by Guo et al. [101], MIPs were prepared for the online SPE of herbicide chlorsulfuron in surface water samples. In their study, vinylimidazolium ionic liquid as the green functional monomer was used for the preparation of chlorsulfuron-selective MIPs. The results showed that the prepared MIPs exhibited high affinity and selectivity toward the target compound chlorsulfuron. The binding equilibrium of chlorsulfuron was obtained within 5 min. The prepared MIPs displayed higher binding in water than in methanol. This result could be explained by the hydrophobic interactions between MIPs and chlorsulfuron. The maximum binding capacity of MIPs in water was found as approximately 28 mg g^{-1}.

In another study [102], synephrine, which is a bioactive alkaloid naturally found in citrus plants, was successfully separated from water–methanol mixture using selective MIPs. For this purpose, various ionic liquids such as 1-vinyl-3-carboxypentylimidazolium bromide, 1-vinyl-3-carboxymethylimidazolium bromide, 1-vinyl-3-carboxybutylimidazolium bromide, and 1-vinyl-3-carboxyethylimidazolium bromide as green functional monomers were chosen for the preparation of different MIPs toward synephrine. The best results were achieved by using the MIP having the functional monomer 1-vinyl-3-carboxybutylimidazolium bromide. The maximum binding capacity of the MIPs was obtained as approximately 74 μmol g^{-1}.

On the other hand, magnetic MIPs can efficiently be used in many applications. Magnetic nanoparticles are synthesized by using various magnetic materials such as manganese, iron, cobalt, their oxides (i.e., Fe_2O_3 and Fe_3O_4, etc.), and derivatives (i.e., $CoFe_2O_4$ and $MnFe_2O_4$, etc.) due to their great magnetic features and biological suitability. Typically, magnetic MIPs are directly dispersed in the sample solutions to quickly extract target compound/s as they can be readily recovered by a magnet, which overcomes problems with conventional approaches of using adsorbent materials, time consumption, and large volume of samples.

Chen and coworkers prepared magnetic MIPs toward sulfonamides and their acetylated metabolites [103]. The magnetic MIPs were synthesized by using MAA and EGDMA as the functional monomer and cross-linker, respectively. Green solvent water and dimethyl sulfoxide were used for the preparation of magnetic MIPs. The extraction of four sulfonamides sulfadiazine, sulfamonomethoxine, sulfamethoxydiazine, sulfaquinoxaline, and their acetylated metabolites from lake and river water samples was successfully achieved. The maximum binding capacity of the magnetic MIPs was achieved as 23.3 μmol g^{-1}. The optimum conditions for the extraction procedure were also determined. The results confirmed that the highest extraction was achieved using 30% aqueous solutions of methanol as the washing solvent and methanol/acetic acid (95:5, v:v) as the elution solvent. The detection of the target compounds was performed by LC-MS, and the limit of detection values varied in the range between 0.38 and 1.32 ng L^{-1}.

In another study, Zhang and coworkers developed magnetic MIPs for the separation of hesperitin from wastewater samples [104]. For this purpose, the 2-VP (functional monomer) and EGDMA (cross-linker) were chosen for the synthesis of magnetic MIPs toward hesperitin. The prepared MIPs showed high selectivity for hesperitin in the existence of other competing compounds such as apigenin, rutin, and chlorogenic acid. The maximum binding toward hesperitin was found as 64.7 μmol g^{-1}.

The separation of chiral compounds is also crucial and challenging task in pharmaceutical industry as these compounds exhibit different toxicological and pharmacological features [105]. Szekely and his research group developed MIPs for selective separation of chiral amines [106]. For this purpose, they synthesized different crown ether derivatives that were used as functional monomer anchors for the preparation of enantioselective MIPs toward amines. The controlled release and enantioseparation performance of the prepared MIPs were investigated in details using racemic mixture of 1-(1-naphthyl)ethylamine hydrogen perchlorate. The results showed that incorporation of pyridine moiety into the MIP increased the binding capacity up to 60%. In addition, the prepared selective MIPs released the drug in a more sustained way than their corresponding NIPs.

On the other hand, Szekely et al. also prepared MIP/organic solvent nanofiltration (OSN) membrane composites for the selective removal of pharmaceutical compound roxithromycin in the presence of 2-aminopyrimidine and N,N-dimethylaminopyridine, which are building block and catalyst, respectively [107]. In this study, polybenzimidazole was used as the polymeric membrane for the preparation of MIP/membrane composite for the first time. MIP/membrane composite showed fourfold higher flux over its corresponding nonimprinted membrane.

In another interesting study conducted by Ansell et al. [108], the separation of (±)-ephedrine enantiomers was successfully carried out by using MIP-based chiral stationary phases in supercritical fluid chromatography. The optimized mobile phase was $scCO_2$ with a modifier composed of methanol/isopropylamine/water (93:5:2 [v/v/v]).

The imprinting of macromolecules or large biomolecules is a challenging process. One of the efficient techniques for the development of imprinted materials toward these compounds is epitope imprinting. In this technique, a short fragment (epitope) of the target compound is chosen as the template, and the target macromolecule can efficiently be captured by recognizing the epitope of the target molecule [109]. The group of Sellergren has developed MIPs for the extraction of lipopolysaccharide endotoxins. In their study, polymerizable urea and bisimidazolium functional monomers and template compounds, phenylphosphonate and phosphatidic acid, were used for the synthesis of 12 different MIPs toward target endotoxin. The obtained results confirmed that the prepared MIP having both urea and imidazolium functional monomers showed highest binding affinity toward target endotoxin.

Potentially genotoxic impurities (GTIs) are compounds that may potentially exist in the final pharmaceutical compounds as a result of the production process. GTIs have the potential to interact DNA and may lead to genetic mutations and structural rearrangements result from chromosomal breaks in humans. Thus, exposure to even trace level of GTIs exists in the final pharmaceutical compound is a crucial concern [110,111]. Therefore, selective recognition and removal of GTIs from active pharmaceutical ingredients (APIs) using innovative and green technologies is a significant issue for the pharmaceutical industry.

The limits of GTI concentration in pharmaceutical compounds derived from the toxicological concern values are determined from the estimated daily dose using the following formula:

$$\text{Concentration limit (ppm)} = \text{TTC}\left(\mu\text{g day}^{-1}\right)/\text{dose}\left(\text{g day}^{-1}\right)$$

Viveiros and colleagues proposed a green strategy for the development of selective MIPs toward potential GTI acetamide [112]. For this purpose, silica beads were first functionalized with 3-(trimethoxysilyl)propyl methacrylate. Then, MIP layer was prepared on the surface of modified silica beads using $scCO_2$ as the green solvent (Fig. 12.8). Modification of silica beads were also carried out by using microwave plasma technique. The prepared MIPs were

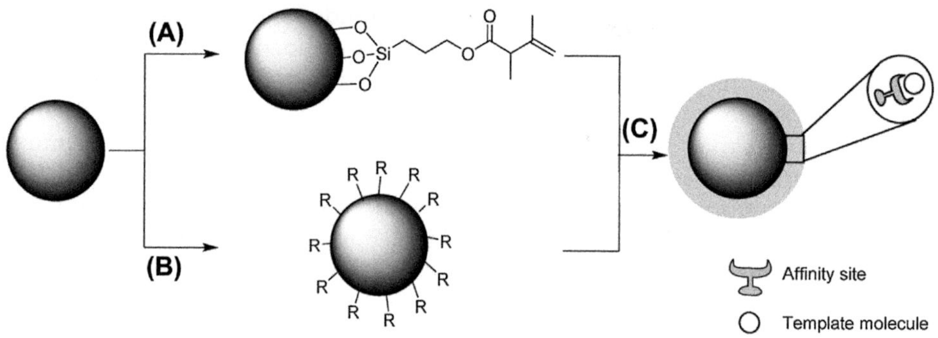

FIGURE 12.8 Schematic depiction of the preparation of silica beads coated with molecularly imprinted polymer (MIP) layer. (A) Modification of silica beads with 3-(trimethoxysilyl)propyl methacrylate in supercritical CO_2. (B) Modification of silica beads by using plasma. (C) Silica beads having MIP layer prepared in supercritical CO_2. *Reproduced with permission from R. Viveiros, F.M. Dias, L.B. Maia, W. Heggie, T. Casimiro, Green strategy to produce large core–shell affinity beads for gravity-driven API purification processes, Journal of Industrial and Engineering Chemistry 54 (2017) 341–349.*

successfully applied for the extraction of acetamide from beclomethasone dipropionate, which is the model API. The results showed that 100% of acetamide was removed by using selective MIPs with only very little loss of API (0.37%).

Same research group developed methacrylamide- and acrylate-based MIPs using $scCO_2$ as the green solvent [113]. The prepared methacrylamide-based MIPs exhibited the best binding performance and selectivity toward acetamide in the presence of other competing compounds including pivalamide and benzamide. The binding capacity of the prepared MIPs was achieved as 2.21 mmol g^{-1}.

In another study conducted by Esteves and coworkers, MIPs were prepared for the selective removal of potential GTI 4-dimethylaminopyridine from the API mometasone furoate [114]. The prepared MIPs exhibited excellent binding performance toward target GTI 4-dimethylaminopyridine even in a typical industrial case study where low amount of GTI exist in high amount of API solution. The prepared selective MIPs could successfully remove 93% of GTI 4-dimethylaminopyridine with a loss of 12% API mometasone furoate. The authors also proposed a hybrid approach composed of OSN technique and MIPs for the removal of target GTI. They could remove 99.7% of GTI with a loss of 8% API applying hybrid approach.

Szekely and his colleagues [115] prepared selective MIPs for the removal of potentially GTI 1,3-diisopropylurea (IPU) from the model APIs mometasone furoate, keppra, and roxithromycin. The synthesized MIPs showed high binding toward 1,3-diisopropylurea (80% binding), while the nonspecific API binding to the MIP was 15%.

In another research performed by the same research group, MIPs were effectively used for the removal of potential GTIs methyl tosylate and acetamide from crude drug samples halobetasole propionate and etodolac [116]. The prepared acetamide imprinted polymer showed 100% acetamide removal with only 14% binding to model API etodolac. On the other hand, the prepared methyl tosylate imprinted polymer exhibited ca. 45% methyl tosylate removal, while 15% API halobetasole propionate was nonspecifically bound to the MIP.

OSN and molecular imprinting approach combined for the removal of GTI IPU from mometasone furoate that is a model API was reported by the same group [117]. In the first stage, OSN was applied for the separation of IPU and then the remaining IPU retained in the solvent was selectively extracted from API solutions by using MIPs. 90% IPU was successfully separated by applying OSN, and the prepared MIPs successfully removed 83% of the remaining IPU with the loss of 2.5% of mometasone furoate.

In a study conducted by Keçili et al. [118] MIP libraries were screened for the fast identification of a hit polymer for the selective removal of potentially GTIs aminopyridines from model APIs tenoxicam and piroxicam. The obtained results from the screening experiments showed that a nornicotine-imprinted polymer exhibited high affinity toward a range of target aminopyridines. The identified MIP successfully removed 2-aminopyridine, 3-aminopyridine, and 5-amino-2-methylpyridine with a little loss of API.

Del Blanco and his coworkers developed MIP-based membranes for selective separation of 4,4′-methylenedianiline [119]. For this purpose, they prepared various membranes such as poly(acrylic acid-co-acrylonitrile), poly(itaconic acid-co-acrylonitrile), and poly(methacrylic acid-co-acrylonitrile). The obtained results showed that the prepared imprinted poly(acrylic acid-co-acrylonitrile) membranes showed the maximum toward the target GTI 4,4′-methylenedianiline (2.6 $\mu mol/g_{memb}$).

Apart from the studies published above, regarding the commercialization of MIPs in the pharmaceutical applications, a number of studies were reported in recently published patent applications by various companies [120–123].

Phenolic compounds are among the most crucial pollutants existing in environmental samples such as natural water, wastewater, river, and lake water. These compounds that exhibit high toxicity can be generated from the production processes of various industries such as drugs, textiles, papers, dyes, and refineries [124,125]. The release of the generated phenolic compounds causes the pollution of environment. In 2014, Environmental Protection Agency in United States designated phenolic compounds as priority pollutants [126]. The long-term exposure of phenolic compounds may cause serious health problems such as vertigo, muscle weakness, and respiratory diseases. Thus, efficient and selective removal of these pollutants from environmental samples is required.

In a reported study [127], Meng et al. prepared MIPs for the efficient removal of α-estradiol from contaminated lake water samples. Selective MIPs were prepared by using the functional monomer acrylamide (AAm) and cross-linker trimethylolpropane trimethacrylate. The prepared MIPs exhibited outstanding selectivity and affinity toward α-estradiol. The reusability of the MIPs was also investigated, and MIPs maintained their binding capacity for target phenolic compound α-estradiol after five cycles without significant loss in their binding performance.

Zhang and coworkers developed a magnetic MIP-based MWCNT composite material for the extraction of BPA from water samples [128]. For this purpose, MAA was chosen as the functional monomer for the preparation of magnetic composite selective to BPA (Fig. 12.9). The obtained results from the binding studies of BPA to the magnetic MIP-based composite in batch mode indicated that the prepared magnetic MIP-based multiwalled carbon nanotube composite showed high affinity and selectivity toward BPA. The maximum binding capacity of the imprinted composite material was obtained as 49.3 $\mu mol\ g^{-1}$.

In another study, an MIP-based membrane having chitosan was prepared for the effective removal of 4-nitrophenol from water [129]. The study was also aimed to design and prepare a biocompatible MIP-based material using a cheap and green biopolymer chitosan. 4-[(4-hydroxy)phenylazo] benzenesulfonic acid, polyetylene glycol, and glutaraldehyde were chosen as the ligand, porogen, and cross-linker, respectively. The imprinted membrane

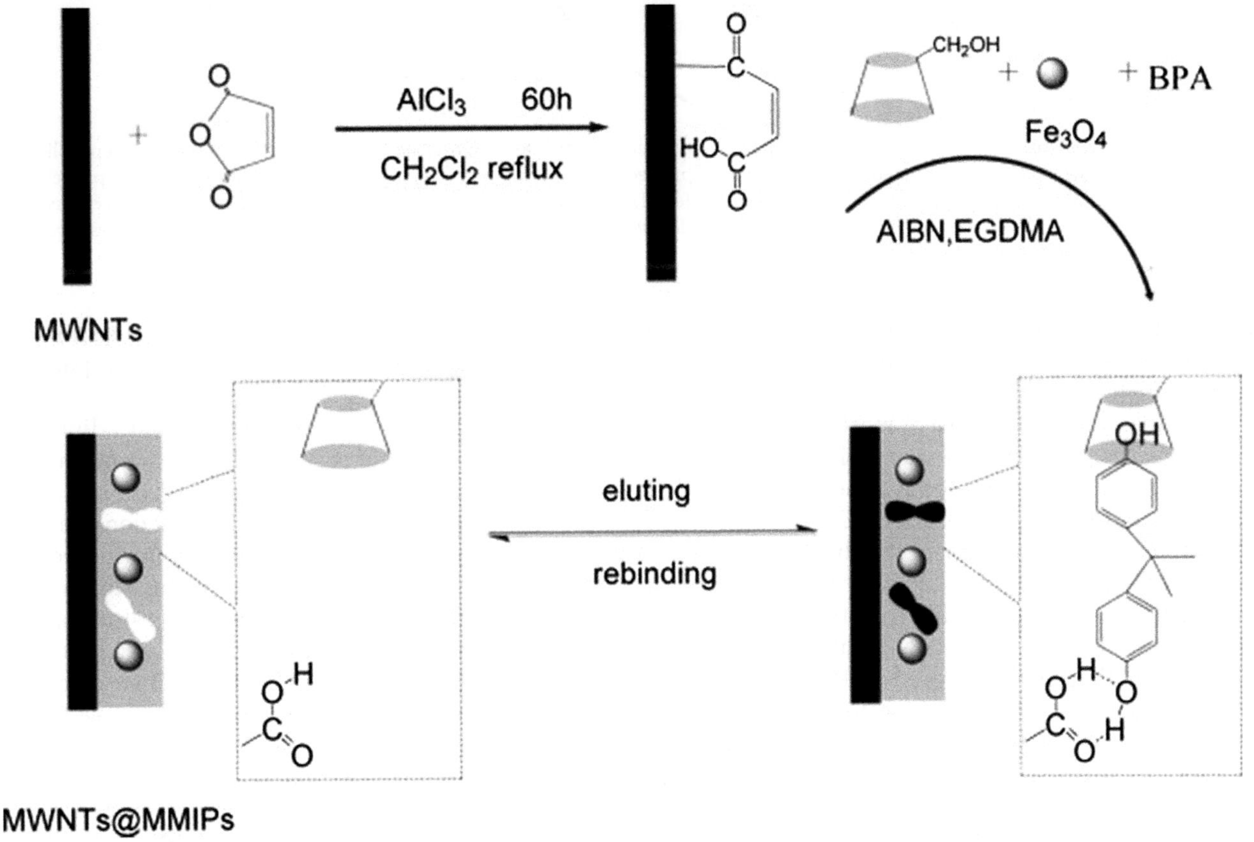

FIGURE 12.9 Schematic depiction of the preparation of molecularly imprinted polymer (MIP)–based MWCNTs composite toward BPA. BPA contains two interaction sites including benzene ring and hydroxyl group, which is available to form hydrogen bond with MAA. Considering β-CD owning a cavity that could hold benzene ring, BPA can be situated within the cavity after the polymerization. *Reproduced with permission from Z. Zang, X. Chen, W. Rao, H. Chen, R. Cai, Synthesis and properties of magnetic molecularly imprinted polymers based on multiwalled carbon nanotubes for magnetic extraction of bisphenol A from water, Journal of Chromatography B 965 (2014) 190–196.*

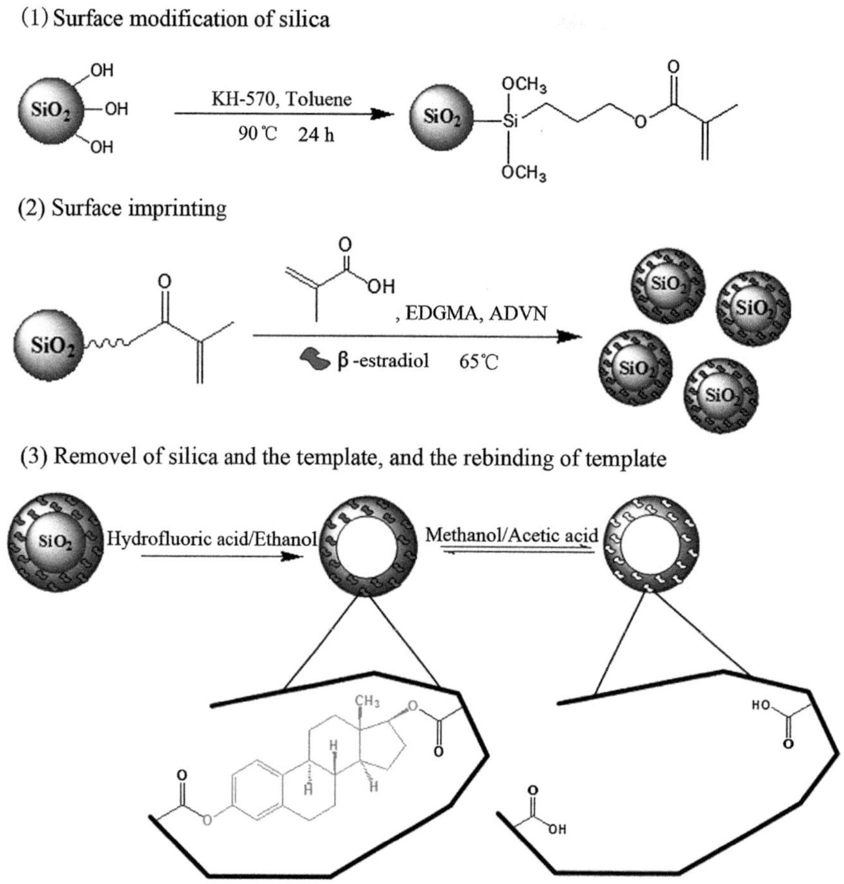

FIGURE 12.10 Schematic depiction of the preparation of molecularly imprinted hollow beads for β-estradiol. In the first step, colloidal SiO_2 was dispersed in toluene. In the second step, β-estradiol was imprinted on the surface of SiO_2 particles. Finally, removal of silica particles and the template was carried out by using HF/EtOH (2/98, v/v) and MeOH/CH_3COOH (80/20, v/v), respectively. *Reproduced with permission from W. Chen, M. Xue, F. Xue, X. Mu, Z. Xu, Z. Meng, G. Zhu, K.J. Shea, Molecularly imprinted hollow spheres for the solid phase extraction of estrogens, Talanta 140 (2015) 68–72.*

was effectively used for the removal of 4-nitrophenol from environmental water samples. The results showed that the highest 4-nitrophenol binding was obtained in 8 h. The obtained maximum binding capacity was 723.25 μmol g^{-1}.

The research group of Shea prepared molecularly imprinted hollow beads for the removal of β-estradiol from tap water samples [130]. In their study, SiO_2 nanoparticles were used as the sacrificial support. After surface modification with 3-(trimethoxysilyl)propyl methacrylate, MIP layer toward β-estradiol was synthesized on the surface of the SiO_2 nanoparticles using MAA, EGDMA, and 2,2′-azobis-(2,4-dimethylvaleronitrile) as the functional monomer, cross-linker and initiator, respectively (Fig. 12.10). The results confirmed that the synthesized imprinted hollow beads showed high binding behavior toward target β-estradiol in the presence of other competing estrogenic compounds such as ethinyl estradiol, diethylstilbestrolum, estrone, and ethisterone. The maximum β-estradiol binding was obtained in very short time (15 min), and the maximum binding capacity was calculated as 44.5 μmol g^{-1}.

Soares da Silva et al. prepared MIP-based poly(methyl methacrylate) membranes for the selective separation of BPA from aqueous solutions [38]. In this study, MIPs were synthesized by using the EGDMA and MAA as the cross-linker and functional monomer, respectively. The polymerization was carried out in the green solvent supercritical CO_2. Then, MIPs/membrane composite toward BPA was prepared by using phase inversion technique. The MIP-based composite membranes prepared by using a green approach were successfully applied for the removal of BPA from aqueous solutions. The binding capacity of the imprinted membranes toward BPA was found as 625 μmol g^{-1}.

In another interesting study conducted by Shen and colleagues [131], SiO_2 particles having MIP layer were prepared for the selective removal of tetrabromobisphenol A (TBBPA) from river water samples. In their study, tetrachlorobisphenol A was used as the dummy template for the preparation of MIP toward TBBPA (Fig. 12.11). The

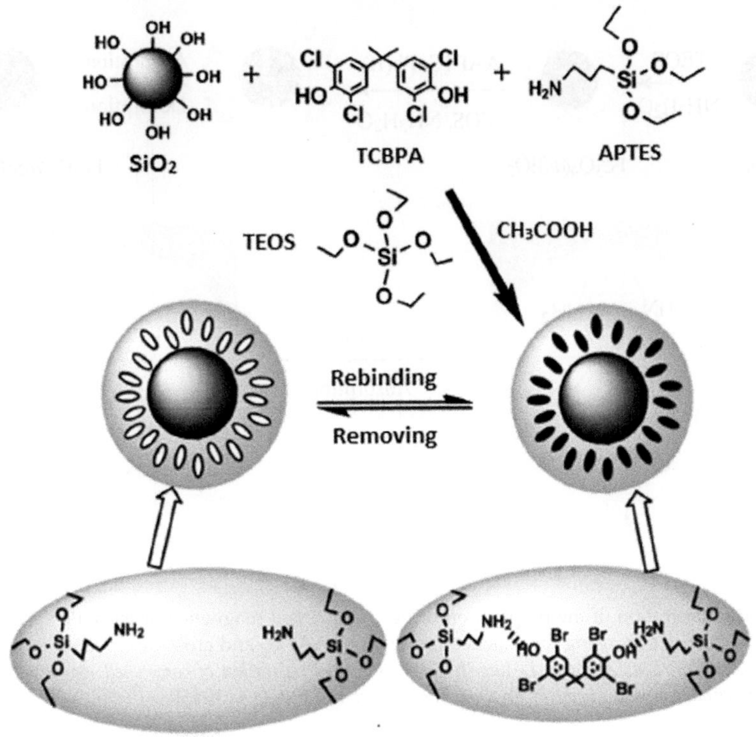

FIGURE 12.11 Schematic depiction of the preparation of SiO_2/MIP composite toward tetrabromobisphenol A (TBBPA). Dummy template imprinted polymer was prepared on the surface of SiO_2 particles using the functional monomer 3-aminopropyltriethoxysilane (APTES) and cross-linker tetraethoxysilane (TEOS) by applying sol–gel process. *Reproduced with permission from W. Shen, G. Xu, F. Wei, J. Yang, Z. Cai, Q. Hu, Preparation and application of imprinted polymer for tetrabromobisphenol A using tetrachlorobisphenol A as the dummy template, Analytical Methods 5 (2013) 5208–5214.*

prepared imprinted SiO_2 particles displayed fast binding kinetics (20 min) and high binding capacity (230 μmol g^{-1}) toward the target compound TBBPA.

Selective extraction of valuable phenolic compounds from natural sources (i.e., plant materials and agricultural wastes) is also crucial process. The group of Szekely has reported the use of MIP/nanofiltration hybrid process for the selective extraction of phenolic antioxidant oleuropein from olive leaves using green solvents [132]. The developed sustainable process has been successfully applied for the continuous extraction of oleuropein from olive leaf extracts. The results showed that oleuropein was selectively recovered from olive leaf extracts with a purity of 99.7%. The binding capacity of the prepared selective MIP was obtained as 1.75 g kg^{-1}. Sustainability assessment of the developed green process was also investigated. 97.5% of the green solvent ethyl acetate was successfully recycled as well as decreasing the carbon footprint by 44.5%.

In another interesting work reported by Zhi and colleagues [133], an MIP was prepared for the selective recognition and extraction of gossypol, a polyphenolic compound that is isolated from cotton plants. For this purpose, the functional monomer dimethylaminoethyl methacrylate was used for the synthesis of gossypol imprinted polymers. The results indicated that the prepared MIP showed great affinity toward gossypol with an excellent binding capacity of 632 mg g^{-1}.

4.3 Water Treatment With Molecularly Imprinted Polymers

To address the increasing demand for clean water due to the rapidly increasing social and economic growth in the 21st century, materials and processes for water treatment are being rapidly developed to safeguard water security.

Ion imprinted polymers (IIPs) as selective SPE adsorbents for the target ions were first introduced by Nishide and coworkers in 1976 [134]. Today, so many IIPs were synthesized and successfully applied for the selective SPE of target metal ions from environmental samples.

For example, the research group of Zhang reported the design and synthesis of ion-imprinted magnetic nanocomposites for the selective SPE of Pb^{2+} from environmental samples [135]. The functional monomer

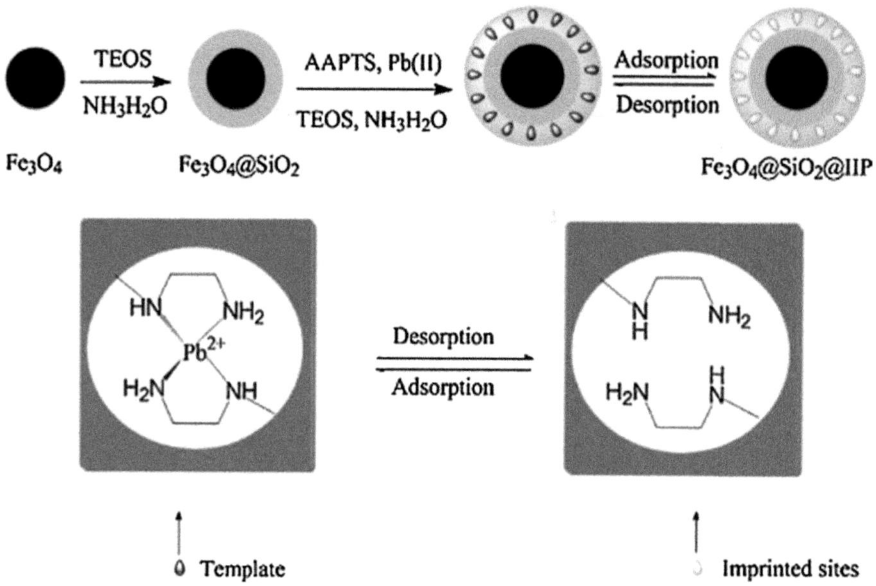

FIGURE 12.12 Schematic demonstration of the preparation of Pb^{2+} imprinted magnetic particles. Pb^{2+}-selective MIP was prepared on the surface of magnetic $Fe_3O_4@SiO_2$ particles by using the functional monomer AAPTS and cross-linker tetraethoxysilane (TEOS). *Reproduced with permission from M. Zhang, Z. Zhang, Y. Liu, X. Yang, L. Luo, J. Chen, S. Yao, Preparation of core–shell magnetic ion-imprinted polymer for selective extraction of Pb(II) from environmental samples, Chemical Engineering Journal 178 (2011) 443–450.*

3-(2-aminoethylamino)propyltrimethoxysilane (AAPTS) and cross-linker TEOS were used for the synthesis of Pb^{2+} imprinted magnetic adsorbent (Fig. 12.12). The effects of different factors such as pH and sample volume on the binding of target ion were also investigated. The binding capacity of the nanocomposites for Pb^{2+} ions was obtained as to be 19.61 mg g^{-1}.

Yin and colleagues reported the use of photocatalytic magnetic IIPs ($Fe_3O_4@SiO_2@TiO_2$) for the removal of Ag^+ ions from wastewater samples [136]. In their study, sol–gel technique was used for the preparation of Ag^+-imprinted magnetic polymer that exhibits photocatalytic activity. The prepared IIPs showed excellent affinity and selectivity toward Ag^+ ions with a maximum binding capacity of 35.5 mg g^{-1}. The obtained selectivity factors were 68.11, 10.62, 27.83, and 13.3 for Ag^+/Ni^{2+}, Ag^+/Li^+, Ag^+/Co^{2+}, and Ag^+/Cu^{2+} binary systems, respectively. In addition, the bound Ag^+ could be reduced to elemental Ag by TiO_2 in the structure of magnetic IIP. The reusability of the IIPs was also investigated. The obtained results confirmed that the IIPs maintained their binding behavior for Ag^+ ions after four cycles without significant loss in their binding performance, which indicates their promising reusability and stability. From the practical point of view, the design and preparation of the reusable IIP-based photocatalysts is a convenient, efficient, and ecofriendly approach for the removal of Ag^+ ions from wastewater samples.

In another study published by Luo and coworkers [137], novel IIPs were used for the selective separation of Pb^{2+} ions from lake water samples. For this purpose, 4-vinylbenzo-18-crown-6 and EGDMA were used as the functional monomer and cross-linker, respectively, for the preparation of IIP toward Pb^{2+} ions. The schematic depiction of the preparation of Pb^{2+} imprinted polymer is shown in Fig. 12.13. The obtained results confirmed that the IIPs displayed excellent recognition ability and selectivity toward Pb^{2+} ions in the presence of Zn^{2+}, Co^{2+}, Ni^{2+}, and Cd^{2+} with the selectivity coefficients of 617.8, 500.6, 52.3, and 201.15 for Pb^{2+}/Zn^{2+}, Pb^{2+}/Co^{2+}, Pb^{2+}/Ni^{2+}, and Pb^{2+}/Cd^{2+} binary systems, respectively. The maximum binding capacities of IIP and its corresponding NIP toward target Pb^{2+} ions were 27.95 and 13.54 mg g^{-1}, respectively.

Extraction of various rare-earth metals (i.e., Nd^{3+}) from environmental samples is crucial process because these metals are extensively used as microadditives in the production of functional materials such as secondary batteries, high-temperature superconductors, and laser materials, etc. [138,139]. In an interesting study, IIPs toward Nd^{3+} were developed by Dolak et al. [140]. For this purpose, EGDMA and N-methacryloylamido folic acid were chosen as the cross-linker and functional monomer, respectively. The effects of various factors such as initial metal concentration, pH and time on the extraction of Nd^{3+} from mineral bastnasite samples were investigated. Extraction of Nd^{3+} from mineral bastnasite samples was performed using aqueous solutions of the real sample, which is important for the greenness of the extraction process. The maximum extraction of Nd^{3+} was achieved in 30 min.

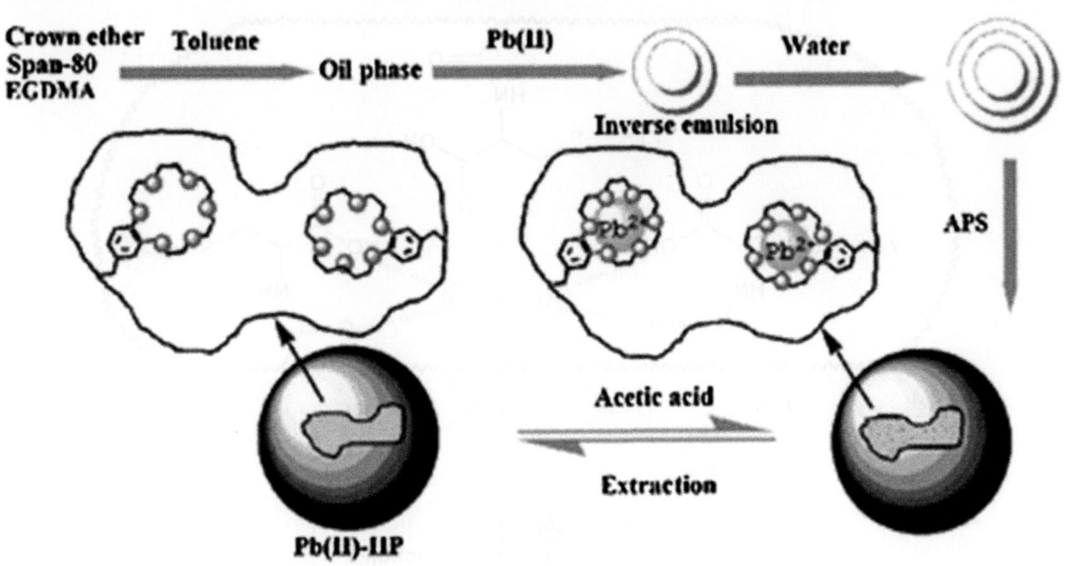

FIGURE 12.13 Schematic demonstration of the synthesis of Pb^{2+} imprinted polymers. The functional monomer 4-vinylbenzo-18-crown-6, Span-80, and cross-linker EGDMA were dissolved in toluene to form oil phase. Then, IIPs toward Pb^{2+} were prepared by using inverse emulsion polymerization technique. *Reproduced with permission from X. Luo, L. Liu, F. Deng, S. Luo, Novel ion-imprinted polymer using crown ether as a functional monomer for selective removal of Pb(II) ions in real environmental water samples, Journal of Materials Chemistry A 1 (2013) 8280–8286.*

The prepared IIPs displayed great selectivity for Nd^{3+} ions in the existence of Ce^{3+}, La^{3+}, and Eu^{3+}, and the maximum capacity was 14.6 mg g^{-1}. On the other hand, the reusability of the prepared IIPs was also investigated. The results showed that the IIPs maintained their binding behavior toward Nd^{3+} after 12 cycles without significant loss in their binding performance, which is crucial for the sustainability aspect of the prepared IIPs.

Xi and colleagues synthesized IIPs for the selective SPE of Cd^{2+} ions from wastewater [141]. For this purpose, MAA and dithizone were chosen as the functional monomer and complexing agent, respectively. The results showed that the synthesized IIPs displayed excellent affinity toward their target Cd^{2+} ions. The maximum Cd^{2+}-binding capacity was achieved as 46.5 mg g^{-1}, and the maximum removal of Cd^{2+} ions from wastewater was achieved in 20 min.

Roy et al. [142] developed molecularly imprinted membranes having cysteine modified-ZnS nanoparticles for the selective separation of As^{3+} and As^{5+} ions from contaminated drinking water. The schematic depiction of the interaction of functional monomers and target ion is given in Fig. 12.14. The prepared imprinted membrane was successfully used for the efficient removal of target ions of As^{3+} and As^{5+}. The maximum binding capacity of the imprinted membrane for As^{3+} and As^{5+} was obtained as 151 and 130 mg g^{-1}, respectively. Reusability experiments were also performed, and the prepared imprinted membranes maintained their high binding capacity toward target ions after 25 binding-elution cycles.

The research group of Szekely successfully applied membrane technology as a green and sustainable process for the efficient treatment of wastewater samples [143]. For this purpose, various materials (i.e., MIPs, graphene, charcoal, zeolites, polymers with intrinsic microporosity, metal organic frameworks, and commercial resins) were thoroughly screened using membrane fabrication for the efficient removal of N,N-dimethylformamide and N-methyl-2-pyrrolidone from contaminated water samples. The achieved results indicated that most of the materials exhibited feasible removal performance to treat the wastewater samples. Particularly, MIPs having phenylboronic acid functional groups displayed superior removal performance toward the target contaminants N-methyl-2-pyrrolidone and N,N-dimethylformamide capable of high binding (ca. 502 and 677 mmol kg^{-1}, respectively).

In a study reported by Zheng et al. [144], silica-based IIPs were developed for the efficient separation of Dy (III) ions from NdFeB magnets in an acidic system. In their study, the maximum binding capacity was achieved as 22.33 mg g^{-1} at pH 2.0.

Dye compounds are widely used in various industries such as paper, food, cosmetic, and textile [145,146], and these compounds have the potential to generate large volumes of contaminated effluents in the environmental samples. The release of these effluents into the environment is one of the serious global environmental problems

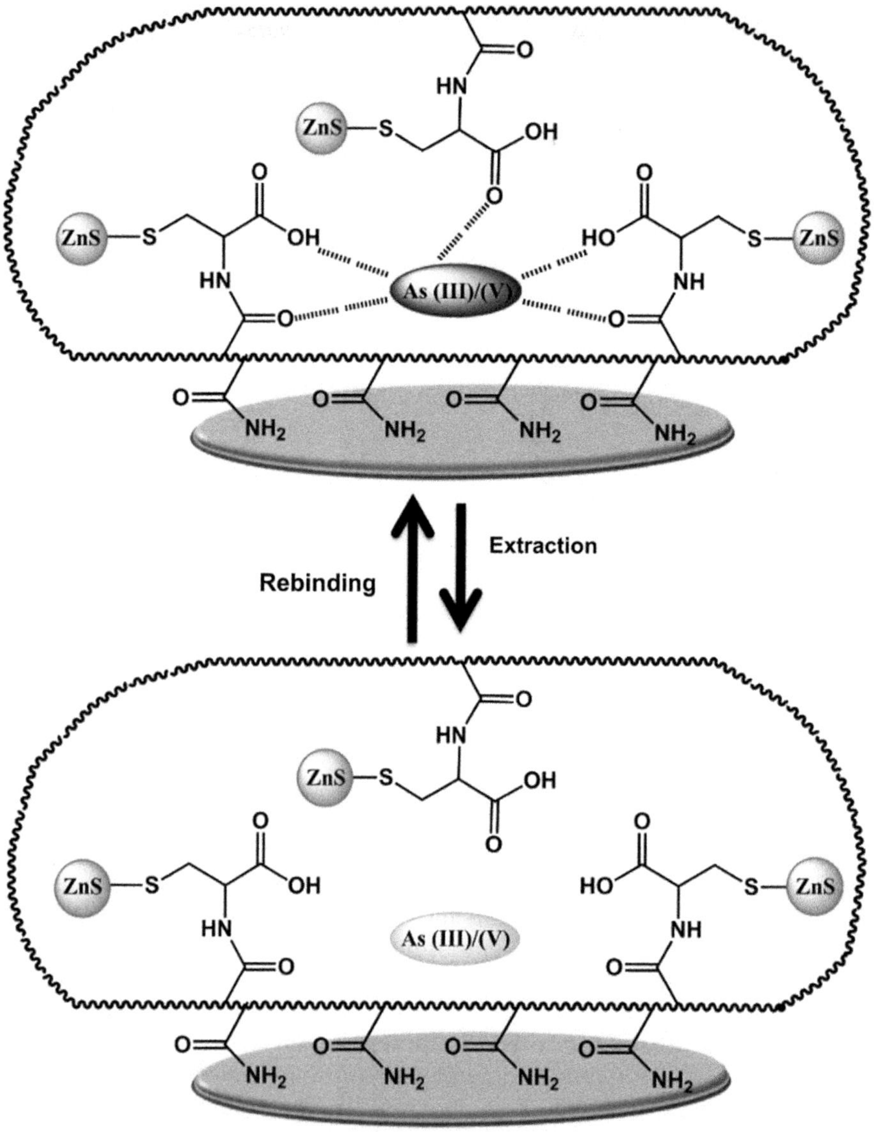

FIGURE 12.14 Schematic depiction of the interaction between the functional monomers present in the imprinted membrane and target ion. *Reproduced with permission from E. Roy, S. Patra, R. Madhuri, P.K. Sharma, A single solution for arsenite and arsenate removal from drinking water using cysteine@ZnS:TiO$_2$ nanoparticle modified molecularly imprinted biofouling-resistant filtration membrane, Chemical Engineering Journal 304 (2016) 259–270.*

that need to be solved. These environmental pollutants must be efficiently analyzed and removed from the environmental samples, which is a challenging and crucial process. Thus, simple, effective, cheap, and environmentally friendly techniques are required for the sensitive analysis of these dye contaminants in environmental samples.

Chemical and biological techniques can be applied for the removal of dye pollutants from environmental samples such as contaminated effluents. In biological techniques, various microorganisms (i.e., fungi, bacteria, algae, etc.) have been used [147–149]. Although this technique is cheap and environmentally friendly, it is challenging and not efficient as the biological approach takes much time and cannot completely degrade dye compounds. In addition, the adsorbents in biological processes are not reusable.

Traditional adsorbents (i.e., activated carbon) have been extensively used for the extraction of dye pollutants from environmental samples as they show high extraction capacity [150–153]. However, activated carbon has various disadvantages such as regeneration problems and high cost. Imprinted materials can overcome these drawbacks of conventional techniques. Melvin and colleagues synthesized MIP-based polysulfone membranes embedded with TiO_2 nanoparticles and effectively applied for the removal of methylene blue dye from aqueous solutions [154]. The imprinted composite membrane displayed high binding ability and selectivity toward methylene blue in the existence of methylene orange dye. The selectivity factor was achieved as 2.0. Another research on

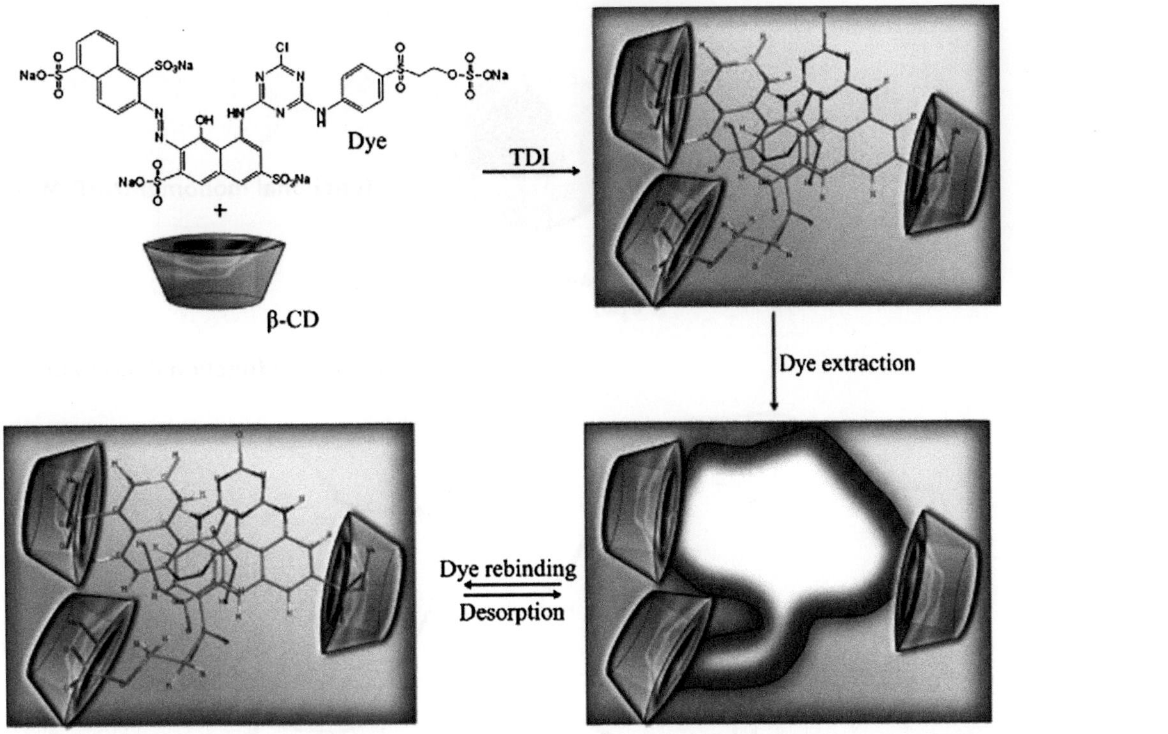

FIGURE 12.15 Schematic demonstration of the synthesis of β-cyclodextrin-based MIPs for Remazol Red 3BS. The binding of dye molecules onto β-cylcodextrin was achieved through three sulfonate groups. *Reproduced with permission from G.Z. Kyzas, N.K. Lazaridis, D.N. Bikiaris, Optimization of chitosan and β-cyclodextrin molecularly imprinted polymer synthesis for dye adsorption, Carbohydrate Polymers 91 (2013) 198–208.*

the removal of methylene blue dye from aqueous solutions was carried out by Asman et al. [155]. They synthesized selective-MIPs using MAA as the functional monomer. The results indicated that the prepared MIP effectively removed methylene blue from aqueous solutions. The maximum binding of methylene blue was obtained at pH 5.0 in 30 s.

The group of Kyzas synthesized β-cyclodextrin-based MIPs having chitosan for the removal of dye pollutant Remazol Red 3BS from water samples [156]. Fig. 12.15 shows the schematic demonstration of the synthesis of MIPs for Remazol Red 3BS. The results confirmed that MIPs showed high binding behavior toward Remazol Red 3BS, and the achieved maximum binding capacity was 35 mg g^{-1}.

Al-Degs et al. synthesized MIPs for Cibacron reactive red dye [157]. In their study, MMA and MAA were chosen as functional monomers for the synthesis of MIPs toward Cibacron reactive red dye. The removal of Cibacron reactive red dye by the synthesized two different polymers (MIP 1 and MIP 2) was 79.2% and 38.2% for MIP 1 and MIP 2, respectively. In another study, Gao et al. synthesized molecularly imprinted SiO_2 particles toward sunset yellow dye. Acryloyloxyethyl trimethyl ammonium chloride was chosen as the functional monomer [158]. The synthesized imprinted SiO_2 particles showed high extraction performance for sunset yellow dye. The achieved maximum extraction capacity was 416 mg g^{-1}.

Li and colleagues synthesized magnetic MIPs for the removal of Rhodamine B (RhB) dye from water samples [159]. For this purpose, MIP layer was coated on the surface of magnetic silica nanoparticles (Fe_3O_4@SiO_2) bearing fluorophore nitrobenzoxadiazole as schematically demonstrated in Fig. 12.16. The obtained results showed the synthesized RhB-imprinted silica nanoparticles were effectively applied for the selective separation of RhB from aqueous solutions. The maximum binding capacity was achieved as 29.64 mg g^{-1} (Table 12.3).

4.4 Sustainable Catalysis With Molecularly Imprinted Polymers

Many biochemical and chemical catalyses take place under mild conditions by using enzymes. Enzymes as natural catalysts exhibit excellent affinity and regioselectivity toward their substrates with a high degree of reaction rates [194]. However, enzymes have some limitations such as low stability under harsh conditions (e.g., organic solvents, high pH, temperature, and pressure, etc.), high cost, and difficulties in separation of the

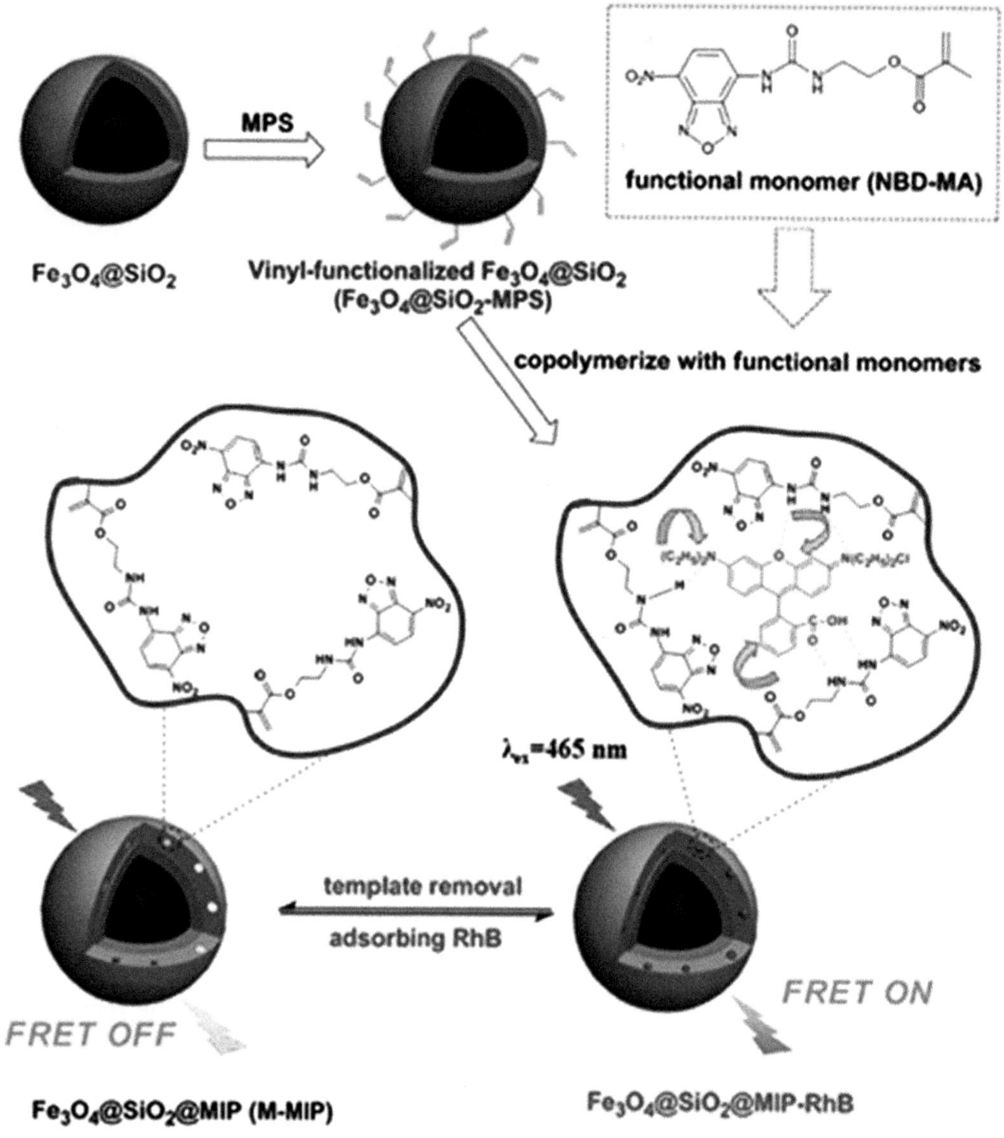

FIGURE 12.16 Schematic depiction of the synthesis of magnetic molecularly imprinted polymers (MIPs) toward Rhodamine B and fluorescence resonance energy transfer (FRET) process between the functional monomers that act as the donor and the target dye that acts as the acceptor because of their well-matched adsorption and emission bands. *Reproduced with permission from H. Li, N. Li, J. Jiang, D. Chen, Q. Xu, H. Li, J. He, J. Lu, Molecularly imprinted magnetic microparticles for the simultaneous detection and extraction of Rhodamine B, Sensors and Actuators B: Chemical 246 (2017) 286–292.*

enzyme from the reaction medium. Thus, the design and preparation of efficient and robust artificial catalysts by using green approaches that can imitate the catalytic features of enzymes has received great attention from the researchers. Although artificial equivalents cannot compete with enzymes in terms of regioselectivity, affinity, and turnover numbers, considerable progresses were achieved in the design and preparation of tailor-made artificial catalysts in the past decade [195–198]. MIP-based catalysts are another interesting application of MIPs. For the preparation of MIP-based artificial catalysts, appropriate functional monomers are incorporated in the polymeric structure by choosing the substrate of desired catalyst or the transition state analogue of the reaction as the template compound. After removal of the template from the polymeric network, the obtained polymer behaves as an artificial catalyst toward the desired chemical or biochemical reaction. MIP-based artificial catalysts are easy to prepare, inexpensive, stable at harsh operating conditions, reusable, and capable of selective recognition of the target compound. Some examples reported in the literature are given in the following sections.

TABLE 12.3 Environmental Applications of Various Imprinted Materials.

	Reference	Composition of Imprinted Material	Target Compound	Binding Capacity ($mg\ g^{-1}$)	Sample
Pharmaceutical compounds	[160]	MIP layer–coated ZnS	Ciprofloxacin	16.94	Aqueous solutions
	[161]	MIP having functional monomer MAA	Pseudoephedrine	47.6	Aqueous solutions
	[162]	MIP fiber having functional monomer AA	Abacavir	149	Surface water, wastewater
	[163]	Carbon nanotube–MIP composite prepared by using functional monomer AAm	Ribavirin	10.5	Aqueous solutions
	[164]	MIPs having 2-VP as the functional monomer	Clofibric acid Carbamazepine	82.2 74.2	Tap water, lake and river water samples
	[165]	MIP-based magnetic silica gel prepared by sol–gel technique	Enrofloxacin	45.74	Aqueous solutions
	[166]	MIPs having functional monomer MAA	Amikacin	60.6	Aqueous solutions
	[167]	Carbon nanoparticles covalently grafted with MIP nanoparticles having MAA as the functional monomer	Ofloxacin	41	Aqueous solutions
Metal ions	[168]	Magnetic IIP having functional monomer 1-vinylimidazole	Co(II)	23.1	Industrial wastewater
	[169]	IIP/silica particles having functional monomer 3-(γ-aminoethylamino) propyltrimethoxysilane	Cd(II)	40	Mine wastewater
	[170]	IIP having functional monomer 4-VP	(UO_2)(II)	98.5	Nuclear power reactor effluents
	[171]	Chitosan/IIP composite having dithiocarbamate	Sr(II)	86.7	Aqueous solutions
	[172]	Graphene–SiO_2–IIP nanocomposite having functional monomer AAm	Ni(II)	81.73	Aqueous solutions
	[173]	IIP having functional monomer 4-VP	Cr(VI)	286.6	Aqueous solutions
	[174]	V(IV)-imprinted magnetic nanoparticles having functional monomer 3-(triethoxysilyl)propyl methacrylate	V(IV)	24	Crude oil
	[175]	IIP having functional monomer MAA	Nd(III)	9	Tap water and river water
	[176]	IIP having functional monomer MAA	Tl(I)	18.3	Tap water, river water, groundwater, wastewater, and dam water
	[177]	IIP having functional monomer MAA	Cu(II)	38.8	Mineral water and river water
	[178]	Ce(III) imprinted polymer prepared by using sacrificial support attapulgite	Ce(III)	130	River sediment
	[179]	IIP having functional monomer MAA	Mo(VI)	3.8	Lake water and tap water
	[180]	Pb(II) imprinted polymer–gelatin composite grafted with 8-hydroxyquinoline	Pb(II)	235.7	Aqueous solutions
	[181]	Cd(II)-imprinted polymers having functional monomer 3-allylrhodanine	Cd(II)	65.5	Wastewater
	[182]	Ni(II) imprinted SBA-15 and MCM-41 mesoporous organosilicas	Ni(II)	22.9 and 22.8	Wastewater

Continued

TABLE 12.3 Environmental Applications of Various Imprinted Materials.—cont'd

	Reference	Composition of Imprinted Material	Target Compound	Binding Capacity (mg g^{-1})	Sample
	[183]	IIP having functional monomer MAA	Fe(III)	40.41	Mineral water and river water
Phenolic compounds	[184]	Hydrophilic MIP having functional monomers AAm and glycidyl methacrylate	2,4-dinitrophenol	138.9	Wastewater
	[185]	Magnetic MIP nanoparticles having functional monomer 3-aminopropyltriethoxysilane	17β-estradiol	16.9	Lake water
	[186]	Magnetic MIP nanoparticles having functional monomer MAA	Dienestrol	4.68	Seawater
	[187]	Magnetic MIP nanoparticles prepared by using bisphenol F as the dummy template	Bisphenol A	450.6	Tap water Drinking water River water
	[188]	MIP particles prepared by using dummy template sulfadiazine	Phenol	6.09	Wastewater
	[189]	Halloysite nanotube–MIP composite having functional monomer MAA	2,4,6-Trichlorophenol	246.73	Environmental water samples
Dye compounds	[190]	MIP/chitosan-bearing Ti (IV)	Active brilliant red X-3B	161.1	Aqueous solutions
	[191]	MIP synthesized by using functional monomer MAA	Malachite Green	3.5	Seawater
	[47]	MIP synthesized by using functional monomer 1-vinylimidazole	Acid Green 16	6.90	Textile effluent and tap water
	[192]	Molecularly imprinted membrane synthesized by using functional monomer MAA	Rhodamine B	100.1	Lake water
	[193]	MIP-chitosan-TiO_2 nanocomposite using epichlorohydrin as the cross-linker	Rose bengal	79.4	Wastewater

2-VP, 2-vinyl pyridine; *AAm*, acrylamide; *MAA*, methacrylic acid; *MIP*, molecularly imprinted polymer.

4.4.1 Molecularly Imprinted Polymer–Assisted Hydrolysis

The hydrolysis reactions are important in all biological systems. The preparation of MIP-based catalysts for the hydrolysis reactions has gained great interest from the researchers who work on the design and construction of robust artificial catalysts. One of the most extensively studied enzymes is chymotrypsin with well-known catalytic mechanism [199], which has been chosen as a model enzyme to design and prepare robust artificial catalysts based on molecular imprinting approach [200].

In an interesting study [201], Mathew et al. developed an MIP catalyst as chymotrypsin mimic using a transition state analogue (TSA) phenyl-1-(N-benzyloxycarbonyl amino)-2-(phenyl)ethyl phosphonate as the template compound. The results showed that chymotrypsin mimic MIP catalyst exhibited high peptidase activity for the hydrolysis of dipeptides.

In another reported research [202], a clay-/MIP-based artificial nanoenzyme was prepared for the selective hydrolysis of paraoxon. In their study, after modification of the organosmectite with [2-(methacryloyloxy)ethyl] dimethyl hexadecyl ammonium bromide (MOEDMAC16), the metal chelate functional monomer methacryloyl amidohistidine-Cu^{2+} (MAH-Cu^{2+}) was used for the preparation of paraoxon imprinted-composite catalyst. The MIP-based composite catalyst exhibited high hydrolytic activity toward paraoxon.

Markowitz et al. prepared MIP-based silica nanocomposites for the hydrolysis of substrates of chymotrypsin and trypsin [203]. For this purpose, a TSA of α-chymotrypsin was chosen as the template to prepare α-chymotrypsin-like nanocatalyst as schematically shown in Fig. 12.17. The functional silane groups modified with the amino acids, which are found in the catalytic center of the α-chymotrypsin, were used in the preparation of silica nanoparticles.

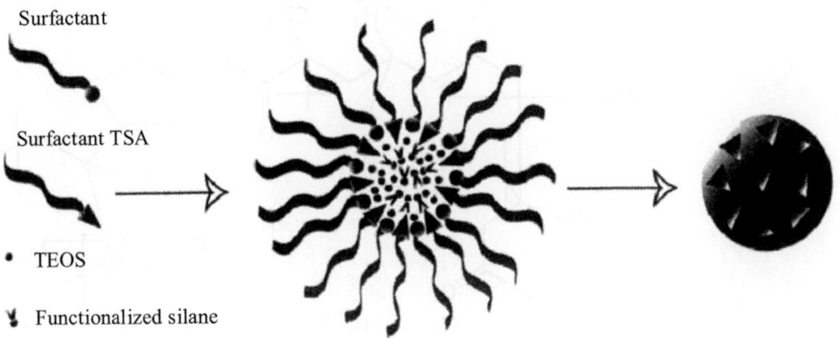

FIGURE 12.17 Schematic demonstration of the preparation of molecularly imprinted polymer (MIP)–based catalytic silica nanoparticles. After the formation of a water-in-oil microemulsion by mixing a nonionic surfactant and the surfactant TSA with cyclohexane, water and ammoniated ethanol, tetraethoxysilane (TEOS) and functionalized silanes are added to initiate the particle formation and surface imprinting. *Reproduced with permission from M.A. Markowitz, P.R. Kust, G. Deng, P.E. Schoen, J.S. Dordick, D.S. Clark, B.P. Gaber, Catalytic silica particles via templated-directed molecular imprinting, Langmuir 16 (2000) 1759–1765.*

The activity of the prepared MIP-based silica nanocomposites was performed by monitoring the hydrolysis of the substrates succinyl-Ala-Ala-Pro-Phe-*p*-nitroanilide and benzoyl-DL-arginine-*p*-nitroanilide. The results confirmed that the prepared nanocatalyst showed high enantioselective hydrolytic activity toward substrates.

Keçili et al. prepared magnetic MIP nanoparticles that display hydrolytic activity toward p-nitrophenyl palmitate (p-NPP) [204]. Trimethoxysilyl propyl methacrylate was used for the modification of the surface of the magnetic Fe_2O_3 nanoparticles. In this study, methacryloyl amidohistidine (MAH), methacryloyl amidoserine (MASe), methacryloyl amidoglutamic acid (MAGA), and pNPP were chosen as functional monomers and template, respectively (Fig. 12.18). The prepared magnetic MIP-based nanoenzyme showed excellent activity for the p-NPP hydrolysis and was also successfully applied for the esterification reactions. The conversion of jasmonic acid to methyl jasmonate and cis-jasmone by using magnetic MIP-based nanoenzyme was obtained as 56.71% and 43.3%, respectively. In addition, the highest conversion value (100%) was achieved for the transesterification of oleic acid to methyl oleate by using catalytic MIP nanoparticles.

In a study reported by Wang and coworkers [205], a substrate imprinted lipase nanogel was prepared for the efficient catalysis of the transesterification reactions. For this purpose, imprinted nanogel was prepared by using acrylamide and palmitic acid as the functional monomer and template compound, respectively. The obtained results indicated that one-step synthesis of chloramphenicol palmitate by using the prepared palmitic acid imprinted nanogel gave a yield of approximately 99% in 12 h.

4.4.2 MIP-Assisted Photocatalysis

MIPs have also successfully been applied for the selective photodegradation of various polutants in environmental samples. For example, Luo and co-workers developed TiO_2/WO_3/MIP-based composite nanocatalyst for the degradation of 2-nitrophenol and 4-nitrophenol [206]. One-step sol-gel technique was used for the preparation of composite nanocatalyst by using tetrabutyl orthotitanate as titanium source which is also the functional monomer precursor. The photocatalytic activity of the TiO_2/WO_3/MIP-based composite nanocatalyst toward the target compounds is more than 2 times than that of control catalyst (NIP).

Arabzadeh et al. prepared TiO_2/MIP-based composite nanocatalyst for selective degradation of tartrazine [207]. For this purpose, AAm and N,N′-methylene-bis-acrylamide were chosen as the functional monomer and cross-linker, respectively. The characterization studies of the nanocatalyst were carried out by using FT-IR, SEM, TEM and thermal gravimetric analyses. The results obtained from the photocatalytic activity of the prepared TiO_2/MIP-based composite nanocatalyst toward target tartrazine showed that the nanocomposite catalyst exhibits high activity for photodegradation of tartrazine under UV light.

Coelho de Escobar and co-workers prepared TiO_2/MIP photocatalyst for selective degradation of diclofenac which is a pharmaceutical pollutant in environmental samples [208]. MAA was used as the functional monomer for the preparation of MIP-based photocatalyst. The prepared TiO_2/MIP photocatalyst showed excellent selectivity toward the target compound diclofenac in the presence of other pharmaceutical pollutants such as paracetamol and fluoxetine. The reusability experiments showed that the prepared TiO_2/MIP photocatalyst maintained its photodegradation activity toward diclofenac after six cycles.

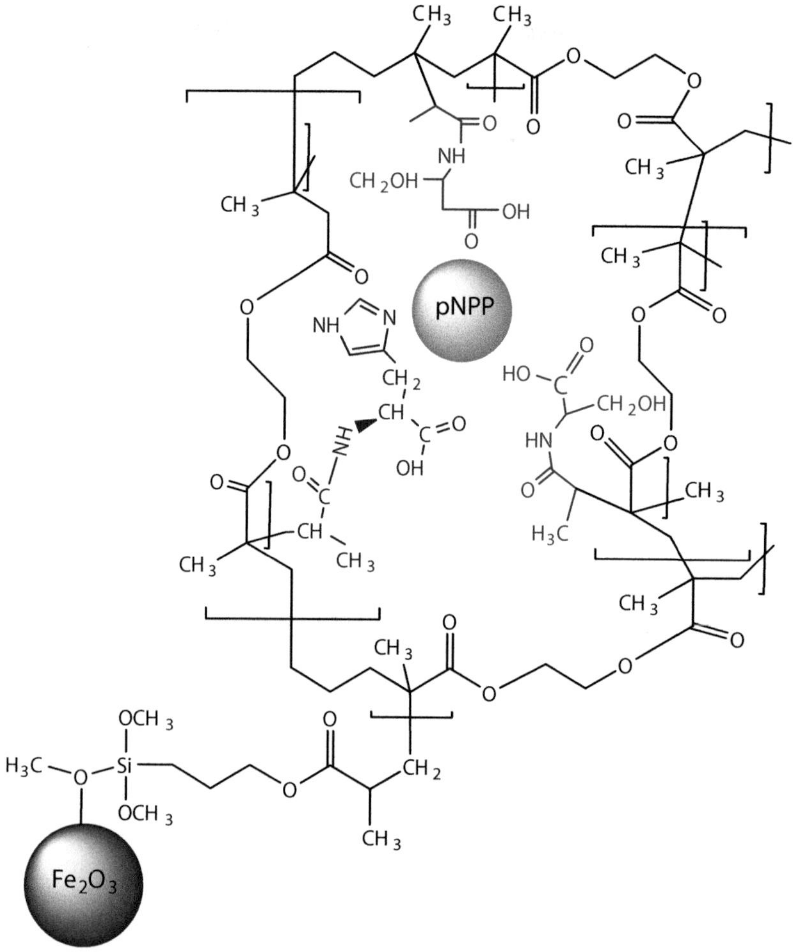

FIGURE 12.18 Schematic illustration of the magnetic nanocomposite catalyst toward p-nitrophenyl palmitate (pNPP). After the modification of magnetic Fe_2O_3 nanoparticles with trimethoxysilyl propyl methacrylate, MIP layer on the surface of nanoparticles was prepared by using MAH, MASe, and MAGA as functional monomers. *Reproduced with permission from R. Keçili, A. Atılır Özcan, A. Ersöz, D. Hür, A. Denizli, R. Say, Superparamagnetic nanotraps containing MIP based mimic lipase for biotransformations uses, Journal of Nanoparticle Research 13(5) (2011) 2073–2079.*

In another study, a molecularly imprinted TiO_2 photocatalyst conjugated with thiol groups for the removal of 2,4-dinitrophenol (2,4-DNP) from wastewater was developed by Zhou et al. [209]. MIP-based TiO_2 photocatalyst was prepared in water as a green solvent using o-phenylenediamine as the functional monomer. The results confirmed that the prepared MIP-based green photocatalyst displayed excellent selectivity and degradation activity toward 2,4-DNP in wastewater.

4.4.3 MIP-Assisted Elimination Reactions

The first attempts in the design and preparation of MIPs in elimination reactions were carried out by Beach and Shea [210]. They investigated the dehydrofluorination of 4-fluoro-4-nitrophenylbutan-2-one using MIP-based catalyst. For this purpose, the functional monomer N-(2-aminoethyl)-methacrylamide and the template compound benzylmalonic acid were used for the preparation of MIP-based catalyst. The prepared catalyst showed high catalytic activity toward the β-elimation of HF with a rate enhancement of 12.8-fold.

Müller and colleagues developed a MIP-based catalyst using the functional monomer MAA and template N-isopropylbenzyl amine for the catalysis of β-elimation reaction [211]. In this study, the catalysis of dehydrofluorination was carried out via an oriented −COOH functional group and 2.4-fold rate enhancement was obtained.

In a study reported by Bonomi et al. [212], molecularly imprinted nanogels were prepared for the catalysis of Kemp elimination reactions. The functional monomer 4-VP and template compound 5-nitro indole were used for the preparation of imprinted nanogels with catalytic activity. The results showed that the prepared 5-nitro indole imprinted nanogels exhibited high catalytic activity toward the substrate 1,2-benzisoxazole. Substrate selectivity of the prepared catalytic nanogels was also investigated using 5-Cl-benzisoxazole which is a substrate analogue. The catalytic nanogels displayed lower affinity toward 5-Cl-benzisoxazole compared to the substrate 1,2-benzisoxazole.

In another interesting study, Mosbach and his research group prepared catalytic MIPs toward other elimination reactions [213]. In their study, it was aimed to design and prepare MIPs for the catalytic decarboxylation of benzisoxazoles. Catalytic MIPs were prepared by using covalent imprinting approach. The prepared catalytic MIPs exhibited high catalytic activity toward 5-nitro-3-carboxybenzisoxazole with a rate enhancement of 11-fold.

4.4.4 *Molecularly Imprinted Polymer–Assisted C–C Bond Formations*

A few studies have been reported on the use of MIPs in C–C bond formation reactions. These reactions are one of the most challenging transformations because they proceed via unfavorable catalytic routes for entropic reasons.

Klaus Mosbach and his research group developed the first MIP-based artificial catalyst toward C–C bond formation [214]. For this purpose, styrene-DVB copolymer–based MIP was prepared using a TSA as the template compound. The synthesized MIP exhibited high aldolase activity for the C–C bond formation reaction between acetophenone and benzaldehyde.

In another interesting study reported by the group of Resmini [215], the first example of a novel MIP nanogel that shows type I aldolase activity was prepared by using radical polymerization technique. A diketone that is the TSA and proline derivative were used as template compound and functional monomer, respectively. The prepared MIP-based nanogels displayed high catalytic activity in C–C bond formation reactions. The catalytic rate constant (K_{cat}) was obtained as 2.6×10^{-3} min $^{-1}$.

Diels–Alder reaction, which is one of the most crucial reaction in the formation of C–C bonds, is due to its high stereoselectivity. Kirsch et al. prepared MIPs for the catalysis of a Diels–Alder reaction [216]. For this purpose, MAA-DVB copolymer–based MIP using a TSA as the template was prepared. The prepared MIP was successfully used for the catalysis of cycloaddition reaction of *N,N*-dimethyl acrylamide and 1,3-butadiene carbamic acid benzyl ester. An enhancement rate of 20-fold was achieved by using MIP-based catalyst compared with the control solution reaction.

5. CONCLUSIONS AND FUTURE TRENDS

This chapter highlights a comprehensive overview of the green aspects of imprinted materials as selective adsorbents in the recognition and extraction of various compounds such as metal ions, pharmaceutical compounds, phenolic compounds, and dye compounds from environmental samples and the purification of valuable compounds from natural sources. The design and methodologies for the preparation of these materials, as well as their potential in design and developing sustainable separation processes and biomimetic catalyses were critically discussed. Industrial case studies were demonstrated and the future perspective of imprinting technology was evaluated. A number of significant studies were reported in the rational design and preparation of imprinted materials, where the careful selection of polymerization agents (safer chemicals) is needed for the minimization of waste components, which is one of the aspects of green chemistry. The reported studies showed that imprinted materials having excellent selectivity and binding affinity for the desired compounds are promising and efficient materials for the selective recognition and separation of target compounds from complex samples. In addition, these engineered materials can be sucessfully used for large-scale extraction processes and catalytic reactions, which are not catalyzed by natural catalysts such as degradation of environmental pollutants.

The rapid growth of environmental awareness and strict regulatory rules for the use of reagents, solvents, and financial competitiveness are crucial and challenging issues for researchers to explore greener approaches in the industrial processes, decreasing waste generation while improving the effectivity of the processes that can be obtained by the use of the green chemistry principles. The molecular imprinting technology has much to gain in the application of these environmentally friendly tools because new green functional monomers (i.e., ionic liquids) and porogens such as supercritical CO_2 combined with computational approaches can successfully optimize the process and provide a facile scale-up that is crucial for the industrial applications of MIPs.

In spite of the many successful examples of imprinted materials in the literature, the green technologies applied for the preparation of new materials are still under development, and more efforts are needed to prepare environmentally friendly, low-cost, selective, and reusable imprinted materials. There are still so many opportunities to be explored in green aspects of molecular imprinting technique that include the search for greener functional monomers, solvents, cross-linkers, and initiators.

References

[1] P.T. Anastas, J.C. Warner, Green Chemistry: Theory and Practice, Oxford University Press: Oxford University Press, New York, USA, 1998.

[2] C. Jiménez-González, A.D. Curzons, D.J.C. Constable, V.L. Cunningham, Expanding GSK's solvent selection guide-application of life cycle assessment to enhance solvent selections, Clean Technologies and Environmental Policy 7 (2004) 42–50.

[3] K. Kim Alfonsi, J. Colberg, P.J. Dunn, T. Fevig, S. Jennings, T.A. Johnson, H.P. Kleine, C. Knight, M.A. Nagy, D.A. Perry, M. Stefaniak, Green chemistry tools to influence a medicinal chemistry and research chemistry based organisation, Green Chemistry 10 (2008) 31–36.

[4] F.G. Tamayo, E. Turiel, A. Martin-Esteban, Molecularly imprinted polymers for solid-phase extraction and solid-phase microextraction: recent developments and future trends, Journal of Chromatography A 1152 (2007) 32–40.

[5] K. Yano, I. Karube, Molecularly imprinted polymers for biosensor applications, Trends in Analytical Chemistry 18 (3) (1999) 199–204.

[6] D. Antuna-Jimenez, G. Diaz-Diaz, M. Carmen Blanco-Lopez, M.J. Lobo-Castanon, A.J. Miranda-Ordieres, P. Tunon-Blanco, Molecularly imprinted electrochemical sensors: past, present, and future, in: S. Li, S.A. Piletsky (Eds.), Chapter 1, Molecularly Imprinted Sensors, Elsevier Science, Amsterdam, Netherlands, 2012.

[7] G. Szekely, C. Didaskalou, Biomimics of metalloenzymes via imprinting, in: S. Li, S.A. Piletsky (Eds.), Chapter 7, Molecularly Imprinted Catalysts: Principles, Syntheses, and Applications, Elsevier Science, Amsterdam, Netherlands, 2016.

[8] J. Wackerlig, R. Schirhagl, Applications of molecularly imprinted polymer nanoparticles and their advances toward industrial use: a review, Analytical Chemistry 88 (2016) 250–261.

[9] T. Kubo, K. Otsuka, Recent progress for the selective pharmaceutical analyses using molecularly imprinted adsorbents and their related techniques: a review, Journal of Pharmaceutical and Biomedical Analysis 130 (2016) 68–80.

[10] M.A.R. Meier, J.O. Metzger, U.S. Schubert, Plant oil renewable resources as green alternatives in polymer science, Chemical Society Reviews 36 (2007) 1788–1802.

[11] S. Molina-Gutiérrez, V. Ladmiral, R. Bongiovanni, S. Caillol, P. Lacroix-Desmazes, Radical polymerization of biobased monomers in aqueous dispersed media, Green Chemistry 21 (2019) 36–53.

[12] A. Riaz, D. Verma, H. Zeb, J.H. Lee, J.C. Kim, S.K. Kwak, J. Kim, Solvothermal liquefaction of alkali lignin to obtain a high yield of aromatic monomers while suppressing solvent consumption, Green Chemistry 20 (2018) 4957–4974.

[13] W. Lu, J.E. Ness, W. Xie, X. Zhang, J. Minshull, R.A. Gross, Biosynthesis of monomers for plastics from renewable oils, Journal of the American Chemical Society 132 (2010) 15451–15455.

[14] A. Cruz-Aguilar, A.M. Herrera-González, R.A. Vázquez-García1, D. Navarro-Rodríguez, J. Coreño, Synthesis of acrylic and allylic bifunctional crosslinking monomers derived from PET waste, IOP Conference Series: Materials Science and Engineering 45 (2013) 012007.

[15] S. Jaswal, B. Gaur, Green methacrylated lignin model compounds as reactive monomers with low VOC emission for thermosetting resins, Green Processing and Synthesis 4 (2015) 191–202.

[16] M.S. Lima, C.S.M.F. Costa, J.F.J. Coelho, A.C. Fonseca, A.C. Serra, A simple strategy toward the substitution of styrene by sobrerol-based monomers in unsaturated polyester resins, Green Chemistry 20 (2018) 4880–4890.

[17] B. Yao, P. Kolla, R. Koodali, S. Balaranjan, S. Shrestha, A. Smirnova, Laccase–natural mediator systems for "green" synthesis of phenolic monomers from alkali lignin, Sustainable Energy and Fuels 1 (2017) 1573–1579.

[18] R. Hayama, T. Koyama, T. Matsushita, K. Hatano, K. Matsuoka, Preparation of functional monomers as precursors of bioprobes from a common styrene derivative and polymer synthesis, Molecules 23 (2018) 2875.

[19] A. Papacchini1, M.R. Telaretti Leggieri, L. Zucchini, M.A. Ortenzi, F. Ridi, D. Giomi, A. Salvini, Modified α,α'-trehalose and D-glucose: green monomers for the synthesis of vinyl copolymers, Royal Society Open Science 5 (2018) 171313.

[20] Y. Yuan, S. Liang, H. Yan, Z. Ma, Y. Liu, Ionic liquid-molecularly imprinted polymers for pipettetip solid-phase extraction of (Z)-3-(chloromethylene)-6-flourothiochroman-4-one in urine, Journal of Chromatography A 1408 (2015) 49–55.

[21] Y. Wang, M. Han, G. Liu, X. Hou, Y. Huang, K. Wu, C. Li, Molecularly imprinted electrochemical sensing interface based on in-situ-polymerization of amino-functionalized ionic liquid for specific recognition of bovine serum albumin, Biosensors and Bioelectronics 74 (2015) 792–798.

[22] F. Qiao, M. Gao, H. Yan, Molecularly imprinted ionic liquid magnetic microspheres for the rapid isolation of organochlorine pesticides in environmental water, Journal of Separation Science 39 (2016) 1310–1315.

[23] G. Marcelo, I.C. Ferreira, R. Viveiros, T. Casimiro, Development of itaconic acid-based molecular imprinted polymers using supercritical fluid technology for pH-triggered drug delivery, International Journal of Pharmaceutics 542 (2018) 125–131.

[24] S. Yuan, Q. Deng, G. Fang, J. Wu, W. Li, S. Wang, Protein imprinted ionic liquid polymer on the surface of multiwall carbon nanotubes with high binding capacity for lysozyme, Journal of Chromatography B 960 (2014) 239–246.

[25] W. Xu, Q. Dai, Y. Wang, X. Hu, P. Xu, R. Ni, J. Meng, Creating magnetic ionic liquid-molecularly imprinted polymers for selective extraction of lysozyme, RSC Advances 8 (2018) 21850–21856.

[26] X. Zhang, N. Zhang, C. Du, P. Guan, X. Gao, C. Wang, Y. Du, S. Ding, X. Hu, Preparation of magnetic epitope imprinted polymer microspheres using cyclodextrin-based ionic liquids as functional monomer for highly selective and effective enrichment of cytochrome c, Chemical Engineering Journal 317 (2017) 988–998.

[27] X. Zhu, Y. Zeng, Z. Zhang, Y. Yang, Y. Zhai, H. Wang, L. Liu, J. Hu, L. Li, A new composite of graphene and molecularly imprinted polymer based on ionic liquids as functional monomer and cross-linker for electrochemical sensing 6-benzylaminopurine, Biosensors and Bioelectronics 108 (2018) 38–45.

[28] H. Xiang, M. Peng, H. Li, S. Peng, S. Shi, High-capacity hollow porous dummy molecular imprinted polymers using ionic liquid as functional monomer for selective recognition of salicylic acid, Journal of Pharmaceutical and Biomedical Analysis 133 (2017) 75–81.

[29] L. Chen, X. Huang, Preparation and application of a poly (ionic liquid)-based molecularly imprinted polymer for multiple monolithic fiber solid-phase microextraction of phenolic acids in fruit juice and beer samples, Analyst 142 (2017) 4039–4047.

[30] M. Panagiotopoulou, S. Beyazit, S. Nestora, K. Haupt, B.T.S. Bui, Initiator-free synthesis of molecularly imprinted polymers by polymerization of self-initiated monomers, Polymer 66 (2015) 43–51.

[31] L. Cseri, M. Razali, P. Pogany, G. Szekely, Organic solvents in sustainable synthesis and engineering, in: B. Török, T. Dransfield (Eds.), Chapter 3.15, Green Chemistry, Elsevier Science, Amsterdam, Netherlands, 2018.

[32] M. Wang, X. Chang, X. Wu, H. Yan, F. Qiao, Water-compatible dummy molecularly imprinted resin prepared in aqueous solution for green miniaturized solid-phase extraction of plant growth regulators, Journal of Chromatography A 1458 (2016) 9–17.
[33] X. Wang, L. Tang, F. Yang, L. Ying, Y. Huang, Z. Liu, Green synthesis of watercompatible and thermos-responsive molecularly imprinted nanoparticles, European Polymer Journal 92 (2017) 174–184.
[34] C. He, Y. Long, J. Pan, K. Li, F. Liu, Molecularly imprinted silica prepared with immiscible ionic liquid as solvent and porogen for selective recognition of testosterone, Talanta 74 (2008) 1126–1131.
[35] S. Rebocho, C.M. Cordas, R. Viveiros, T. Casimiro, Development of a ferrocenyl-based MIP in supercritical carbon dioxide: towards an electrochemical sensor for bisphenol A, The Journal of Supercritical Fluids 135 (2018) 98–104.
[36] R. Viveiros, K. Karim, S.A. Piletsky, W. Heggie, T. Casimiro, Development of a molecularly imprinted polymer for a pharmaceutical impurity in supercritical CO_2: rational design using computational approach, Journal of Cleaner Production 168 (2017) 1025–1031.
[37] L. Ye, K. Yoshimatsu, D. Kolodziej, J. Da Cruz Francisco, E.S. Dey, Preparation of molecularly imprinted polymers in supercritical carbon dioxide, Journal of Applied Polymer Science 102 (2006) 2863–2867.
[38] M.S. da Silva, R. Viveiros, M.B. Coelho, A. Aguiar-Ricardo, T. Casimiro, Supercritical CO_2-assistedpreparationofaPMMAcomposite membrane for bisphenol A recognition in aqueous environment, Chemical Engineering Science 68 (2012) 94–100.
[39] J.P. Ferreira, R. Viveiros, A. Lourenço, M. Soares da Silva, A. Rosatella, T. Casimiro, C.A.M. Afonso, Integrated desulfurization of diesel by combination of metal-free oxidation and product removal by molecularly imprinted polymers, RSC Advances 4 (2014) 54948–54952.
[40] A. Lourenço, R. Viveiros, A. Mouro, J.C. Lima, V.D.B. Bonifacio, T. Casimiro, Supercritical CO_2-assisted synthesis of an ultrasensitive amphibious quantum dot molecularly imprinted sensor, RSC Advances 4 (2014) 63338–63341.
[41] J.-C. Lee, C.-R. Kim, H.-S. Byun, Synthesis and adsorption properties of carbamazepine imprinted polymer by dispersion polymerization in supercritical carbon dioxide, Korean Journal of Chemical Engineering 31 (12) (2014) 2266–2273.
[42] J. Leclaire, D.J. Heldebrant, A call to (green) arms: a rallying cry for green chemistry and engineering for CO_2 capture, utilisation and storage, Green Chemistry 20 (2018) 5058–5081.
[43] H. He, L. Zhuang, S. Chen, H. Liua, Q. Li, Structure design of a hyperbranched polyamine adsorbent for CO_2 adsorption, Green Chemistry 18 (2016) 5859–5869.
[44] Y. Zhao, Y. Shen, G. Ma, R. Hao, Adsorption separation of carbon dioxide from flue gas by a molecularly imprinted adsorbent, Environmental Science and Technology 48 (2014) 1601–1608.
[45] S.A. Nabavi, G.T. Vladisavljevic, Y. Zhu, V. Manovic, Synthesis of size-tunable CO_2-philic imprinted polymeric particles (MIPs) for low-pressure CO_2 capture using oil-in-oil suspension polymerization, Environmental Science and Technology 51 (2017) 11476–11483.
[46] L.M. Madikizela, N.T. Tavengwa, H. Tutu, L. Chimuka, Green aspects in molecular imprinting technology: from design to environmental applications, Trends in Analytical Chemistry 17 (2018) 14–22.
[47] M.V. Foguel, N.T.B. Pedro, A. Wong, S. Khan, M.V.B. Zanoni, M.P.T. Sotomayor, Synthesis and evaluation of a molecularly imprinted polymer for selective adsorption and quantification of Acid Green 16 textile dye in water samples, Talanta 170 (2017) 244–251.
[48] A.L.M. Ruela, E.C. Figueiredo, G.R. Pereira, Molecularly imprinted polymers as nicotine transdermal delivery systems, Chemical Engineering Journal 248 (2014) 1–8.
[49] J. Qiu, B. Charleux, K. Matyjaszewski, Controlled/living radical polymerization in aqueous media. Homogeneous and heterogeneous systems, Progress in Polymer Science 26 (2001) 2083–2134.
[50] A. Poma, A.P.F. Turner, S.A. Piletsky, Advances in the manufacture of MIP nanoparticles, Trends in Biotechnology 28 (2010) 629–637.
[51] M.J. Garcia-Soto, K. Haupt, C. Gonzato, Synthesis of molecularly imprinted polymers by photo-iniferter polymerization under visible light, Polymer Chemistry 8 (2017) 4830–4834.
[52] G. Moad, E. Rizzardo, S.H. Thang, Living radical polymerization by the RAFT process- a second update, Australian Journal of Chemistry 62 (2009) 1402–1472.
[53] L. Sheng, Y. Jin, Y. He, Y. Huang, L. Yan, R. Zhao, Well-defined magnetic surface imprinted nanoparticles for selective enrichment of 2,4-dichlorophenoxyacetic acid in real samples, Talanta 174 (2017) 725–732.
[54] L. Wan, Z. Chen, C. Huang, X. Shen, Core-shell molecularly imprinted particles, Trends in Analytical Chemistry 95 (2017) 110–121.
[55] Y. Li, X. Li, J. Chu, C. Dong, J.Q. Yuan, Synthesis of core-shell magnetic molecular imprinted polymer by the surface RAFT polymerization for the fast and selective removal of endocrine disrupting chemicals from aqueous solutions, Environmental Pollution 158 (2010) 2317–2323.
[56] X. Xie, L. Chen, X. Pan, S. Wang, Synthesis of magnetic molecularly imprinted polymers by reversible addition fragmentation chain transfer strategy and its application in the Sudan dyes residue analysis, Journal of Chromatography A 1405 (2015) 32–39.
[57] N.V. Tsarevsky, K. Matyjaszewski, "Green" atom transfer radical polymerization: from process design to preparation of well-defined environmentally friendly polymeric materials, Chemical Reviews 107 (2007) 2270–2299.
[58] M. Semsarilar, S. Perrier, 'Green' reversible addition-fragmentation chain-transfer (RAFT) polymerization, Nature Chemistry 2 (2010) 811–820.
[59] J. Phommalysack-Lovan, Y. Chu, C. Boyer, J. Xu, PET-RAFT polymerisation: towards green and precision polymer manufacturing, Chemical Communications 54 (2018) 6591–6606.
[60] A.G. West, C. Barner-Kowollik, S. Perrier, Poly(ethylene glycol) as a 'green solvent' for the RAFT polymerization of methyl methacrylate, Polymer 51 (2010) 3836–3842.
[61] A. Azimi, M. Javanbakht, Computational prediction and experimental selectivity coefficients for hydroxyzine and cetirizine molecularly imprinted polymer based potentiometric sensors, Analytica Chimica Acta 812 (2014) 184–190.
[62] N.T. Abdel Ghani, R.M. El Nashar, F.M. Abdel-Haleem, A. Madbouly, Computational design, synthesis and application of a new selective molecularly imprinted polymer for electrochemical detection, Electroanalysis 28 (2016) 1530–1538.
[63] M.M. Cascant, C. Breil, S. Garrigues, M. de la Guardia, A.S. Fabiano-Tixier, F. Chemat, A green analytical chemistry approach for lipid extraction:computation methods in the selection of green solvents as alternative to hexane, Analytical and Bioanalytical Chemistry 409 (2017) 3527–3539.
[64] I.A. Nicholls, H.S. Andersson, C. Charlton, H. Henschel, B.C.G. Karlsson, J.G. Karlsson, J. O'Mahony, A.M. Rosengren, K.J. Rosengren, S. Wikman, Theoretical and computational strategies for rational molecularly imprinted polymer design, Biosensors and Bioelectronics 25 (2009) 543–552.

[65] M. Tabandeh, S. Ghassamipour, H. Aqababa, M. Tabatabaei, M. Hasheminejad, Computational design and synthesis of molecular imprinted polymers for selective extraction of allopurinol from human plasma, Journal of Chromatography B 898 (2012) 24–31.
[66] L.M. Madikizela, P.S. Mdluli, L. Chimuka, Experimental and theoretical study of molecular interactions between 2-vinyl pyridine and acidic pharmaceuticals used as multi-template molecules in molecularly imprinted polymer, Reactive and Functional Polymers 103 (2016) 33–43.
[67] N. Martins, E.P. Carreiro, A. Locati, J.P.P. Ramalho, M.J. Cabrita, A.J. Burke, R. Garcia, Design and development of molecularly imprinted polymers for the selective extraction of deltamethrin in olive oil: an integrated computational-assisted approach, Journal of Chromatography A 1409 (2015) 1–10.
[68] B.B. Prasad, G. Rai, Study on monomer suitability toward the template in molecularly imprinted polymer: an ab initio approach, Spectrochimica Acta 88 (2012) 82–89.
[69] F. Ahmadi, H. Rezaei, R. Tahvilian, Computational-aided design of molecularly imprinted polymer for selective extraction of methadone from plasma and saliva and determination by gas chromatography, Journal of Chromatography A 1270 (2012) 9–19.
[70] N. Karimian, M.B. Gholivand, F. Taherkhani, Computational design and development of a novel voltammetric sensor for minoxidil detection based on electropolymerized molecularly imprinted polymer, Journal of Electroanalytical Chemistry 740 (2015) 45–52.
[71] F. Meier, B. Schott, D. Riedel, B. Mizaikoff, Computational and experimental study on the influence of the porogen on the selectivity of 4-nitrophenol molecularly imprinted polymers, Analytica Chimica Acta 744 (2012) 68–74.
[72] Y. Dineiro, M.I. Menéndez, M.C. Blanco-López, M.J. Lobo-Castanon, A.J. Miranda-Ordieres, P. Tunon-Blanco, Computational approach to the rational design of molecularly imprinted polymers for voltammetric sensing of homovanillic acid, Analytical Chemistry 77 (2005) 6741–6746.
[73] Y. Dineiro, M.I. Menéndez, M.C. Blanco-López, M.J. Lobo-Castanon, A.J. Miranda-Ordieres, P. Tunon-Blanco, Computational predictions and experimental affinity distributions for a homovanillic acid molecularly imprinted polymer, Biosensors and Bioelectronics 22 (2006) 364–371.
[74] S.A. Piletsky, K. Karim, E.V. Piletska, A.P.F. Turner, C.J. Day, K.W. Freebairn, C. Legge, Recognition of ephedrine enantiomers by molecularly imprinted polymers designed using a computational approach, Analyst 126 (2001) 1826–1830.
[75] I. Liska, Fifty years of solid-phase extraction in water analysis–historical development and overview, Journal of Chromatography A 885 (1–2) (2000) 3–16.
[76] J. Cazes, Encyclopedia of Chromatography, CRC Press, Germany, 2009.
[77] B.Y. Spivakov, G.L. Malofeeva, O.M. Petruskhin, Solid-phase extraction on alkyl-bonded silica gels in inorganic analysis, Analytical Science 22 (2006) 503–519.
[78] K. Pyrzynska, Application of carbon sorbents for the concentration and separation of metal ions, Analytical Science 23 (2007) 631–637.
[79] J.A. Rodríguez, K.A. Escamilla-Lara, A. Guevara-Lara, J.M. Miranda, M.E. Páez-Hernández, Application of an activated carbon-based support for magnetic solid phase extraction followed by spectrophotometric determination of tartrazine in commercial beverages, International Journal of Analytical Chemistry (2015) 291827.
[80] M.G. Valdes, A.I. Perez-Cordoves, M.E. Diaz-Garcia, Zeolites and zeolite-based materials in analytical chemistry, TRAC Trends in Analytical Chemistry 25 (2006) 24–40.
[81] J. Płotka-Wasylka, N. Szczepańska, M. de la Guardia, J. Namieśnik, Miniaturized solid-phase extraction techniques, TRAC Trends in Analytical Chemistry 73 (2015) 19–38.
[82] www.biotage.com.
[83] www.affinisep.com.
[84] www.sigma-aldrich.com.
[85] B. Sellergren, Direct drug determination by selective sample enrichment on an imprinted polymer, Analytical Chemistry 66 (1994) 1578–1582.
[86] E. Caro, M. Marce, P.A.G. Cormack, D.C. Sherrington, F. Borrull, On-line solid-phase extraction with molecularly imprinted polymers to selectively extract substituted 4-chlorophenols and 4-nitrophenol from water, Journal of Chromatography A 995 (2003) 233–238.
[87] F. Chapuis, V. Pichon, F. Lanza, B. Sellergren, M.C. Hennion, Retention mechanism of analytes in the solid-phase extraction process using molecularly imprinted polymers: application to the extraction of triazines from complex matrices, Journal of Chromatography B 804 (2004) 93–101.
[88] J.P. Lai, R. Niessner, D. Knopp, Benzo[a]pyrene imprinted polymers: synthesis, characterization and SPE application in water and coffee samples, Analytica Chimica Acta 522 (2004) 137–144.
[89] O. Brüggemann, A. Visnjevski, R. Burch, P. Patel, Selective extraction of antioxidants with molecularly imprinted polymers, Analytica Chimica Acta 504 (2004) 81–88.
[90] F. Puoci, A. Scoma, G. Cirillo, L. Bertin, F. Fava, N. Picci, Selective extraction and purification of gallic acid from actual site olive mill wastewaters by means of molecularly imprinted microparticles, Chemical Engineering Journal 198–199 (2012) 529–535.
[91] J. Xin, X. Qiao, Z. Xu, J. Zhou, Molecularly imprinted polymer as sorbent for solid-phase extraction coupling to gas chromatography for the simultaneous determination of trichlorfon and monocrotophos residues in vegetables, Food Analytical Methods 6 (2013) 274–281.
[92] Y. Li, C. Zheng, X. Sun, B. Ouyang, P. Ni, Y. Zhang, Identification of 3-chloro-1,2-propandiol using molecularly imprinted composite solid-phase extraction materials, Analytical and Bioanalytical Chemistry 406 (2014) 6319–6327.
[93] X. He, X. Mei, J. Wang, Z. Lian, L. Tan, W. Wu, Determination of diethylstilbestrol in seawater by molecularly imprinted solid-phase extraction coupled with high-performance liquid chromatography, Marine Pollution Bulletin 102 (2016) 142–147.
[94] S. Han, X. Li, Y. Wang, S. Chen, Multifunctional imprinted polymers based on CdTe/CdS and magnetic graphene oxide for selective recognition and separation of p-t-octylphenol, Chemical Engineering Journal 271 (2015) 87–95.
[95] E.C. Morais, R. Brambilla, G.G. Correa, V. Dalmoro, J.H.Z. Dos Santos, Imprinted silicas for paracetamol preconcentration prepared by the sol–gel process, Journal of Sol-Gel Science and Technology 83 (2017) 90–99.
[96] I. Vasconcelos, C. Fernandes, Magnetic solid phase extraction for determination of drugs in biological matrices, Trends in Analytical Chemistry 89 (2017) 41–52.
[97] S. Ansari, M. Karimi, Synthesis and application of molecularly imprinted polymer for highly selective solid phase extraction trace amount of sotalol from human urine samples: optimization by central composite design (CCD), Medicinal Chemistry Research (2017) 1–14.

[98] E. Benito-Peña, F. Navarro-Villoslada, S. Carrasco, S. Jockusch, M.F. Ottaviani, M.C. Moreno-Bondi, Experimental mixture design as a tool for the synthesis of antimicrobial selective molecularly imprinted monodisperse microbeads, ACS Applied Materials and Interfaces 7 (2015) 10966–10976.

[99] C.-M. Dai, S.-U. Geissen, Y.-L. Zhang, Y.-J. Zhang, X.-F. Zhou, Selective removal of diclofenac from contaminated water using molecularly imprinted polymer microspheres, Environmental Pollution 159 (2011) 1660–1666.

[100] Y.-P. Duan, C.-M. Dai, Y.-L. Zhang, L. Chen, Selective trace enrichment of acidic pharmaceuticals in real water and sediment samples based on solid-phase extraction using multi-templates molecularly imprinted polymers, Analytica Chimica Acta 758 (2013) 93–100.

[101] L. Guo, Q. Deng, G. Fang, W. Gao, S. Wang, Preparation and evaluation of molecularly imprinted ionic liquids polymer as sorbent for on-line solid-phase extraction of chlorsulfuron in environmental water samples, Journal of Chromatography A 1218 (2011) 6271–6277.

[102] J.-P. Fan, Z.-Y. Tian, S. Tong, X.-H. Zhang, Y.-L. Xie, R. Xu, Y. Qin, L. Li, J.-H. Zhu, X.-K. Ouyang, A novel molecularly imprinted polymer of the specific ionic liquid monomer for selective separation of synephrine from methanol-water media, Food Chemistry 141 (2013) 3578–3585.

[103] H. Chen, Y. Zhang, B. Gao, Y. Xu, Q. Zhao, J. Hou, G. Li, H. Wang, L. Ding, J. Ding, C. Zhao, Fast determination of sulfonamides and their acetylated metabolites from environmental water based on magnetic molecularly imprinted polymers, Environmental Science and Pollution Research International 20 (2013) 8567–8578.

[104] X. Zhang, F. Shia, J. Jiang, A new simple extraction method for the extraction of hesperitin from wastewater, New Journal of Chemistry 40 (2016) 7290–7293.

[105] S. Yang, Y. Wang, Y. Jiang, S. Li, W. Liu, Molecularly imprinted polymers for the identification and separation of chiral drugs and biomolecules, Polymers 8 (6) (2016) 216.

[106] J. Kupai, E. Rojik, P. Huszthy, G. Szekely, Role of chirality and macroring in imprinted polymers with enantiodiscriminative power, ACS Applied Materials and Interfaces 7 (2015) 9516–9525.

[107] G. Szekely, I.B. Valtcheva, J.F. Kim, A.G. Livingston, Molecularly imprinted organic solvent nanofiltration membranes-Revealing molecular recognition and solute rejection behaviour, Reactive and Functional Polymers 86 (2015) 215–224.

[108] R.J. Ansell, J.K.L. Kuah, D. Wang, C.E. Jackson, K.D. Bartle, A.A. Clifford, Imprinted polymers for chiral resolution of (±)-ephedrine, 4: packed column supercritical fluid chromatography using molecularly imprinted chiral stationary phases, Journal of Chromatography A 1264 (2012) 117–123.

[109] R. Sulc, G. Szekely, S. Shinde, C. Wierzbicka, F. Vilela, D. Bauer, B. Sellergren, Phospholipid imprinted polymers as selective endotoxin scavengers, Scientific Reports 7 (2017) 44299.

[110] K. Kondo, A. Watanabe, Y. Iwanaga, I. Abe, H. Tanaka, M.H. Nagaoka, H. Akiyama, T. Maitani, Determination of genotoxic phenylhydrazine agaritine in mushrooms using liquid chromatography–electrospray ionization tandem mass spectrometry, Food Additives and Contaminants 23 (2006) 1179–1186.

[111] B. Leblanc, C. Charuel, W. Ku, R. Ogilvie, Acceptability of low levels of genotoxic impurities in new drug substances, International Journal of Pharmaceutical Medicine 18 (2004) 215–220.

[112] R. Viveiros, F.M. Dias, L.B. Maia, W. Heggie, T. Casimiro, Green strategy to produce large core–shell affinity beads for gravity-driven API purification processes, Journal of Industrial and Engineering Chemistry 54 (2017) 341–349.

[113] R. Viveiros, M.I. Lopes, W. Heggie, T. Casimiro, Green approach on the development of lock-and-key polymers for API purification, Chemical Engineering Journal 308 (2017) 229–239.

[114] T. Esteves, R. Viveiros, J. Bandarra, W. Heggie, T. Casimiro, F.C. Ferreira, Molecularly imprinted polymer strategies for removal of a genotoxic impurity, 4-dimethylaminopyridine, from an active pharmaceutical ingredient post-reaction stream, Separation and Purification Technology 163 (2016) 206–214.

[115] G. Székely, J. Bandarra, W. Heggie, F.C. Ferreira, B. Sellergren, Design, preparation and characterization of novel molecularly imprinted polymers for removal of potentially genotoxic 1, 3-diisopropylurea from API solutions, Separation and Purification Technology 86 (2012) 190–198.

[116] G. Székely, E. Fritz, J. Bandarra, W. Heggie, B. Sellergren, Removal of potentially genotoxic acetamide and arylsulfonate impurities from crude drugs by molecular imprinting, Journal of Chromatography A 1240 (2012) 52–58.

[117] G. Székely, J. Bandarra, W. Heggie, B. Sellergren, F.C. Ferreira, A hybrid approach to reach stringent low genotoxic impurity contents in active pharmaceutical ingredients: combining molecularly imprinted polymers and organic solvent nanofiltration for removal of 1,3-diisopropylurea, Separation and Purification Technology 86 (2012) 79–87.

[118] R. Kecili, J. Billing, D. Nivhede, B. Sellergren, A. Rees, E. Yilmaz, Fast identification of selective resins for removal of genotoxic aminopyridine impurities via screening of molecularly imprinted polymer libraries, Journal of Chromatography A 1339 (2014) 65–72.

[119] S.G. Del Blanco, L. Donato, E. Drioli, Development of molecularly imprinted membranes for selective recognition of primary amines in organic medium, Separation and Purification Technology 87 (2012) 40–46.

[120] J.P. Farr, M.M. Stuckey, M.J. Petrin, W.P. Sibert, Inventors; DeCaf Company LLC, Assignee, Fiber Based Molecularly Imprinted Polymers for the Removal of a Significant Fraction of Target Imprintable Entities, US20160199752A1, July 14, 2016.

[121] R.H. Knop, T. Griffin, V. Perez-Luna, W.B. Leslie, E. Johnson, Inventors; MedTech Detect LLC, Assignee, Article and Method for Detection of a Drug in a Liquid Medium US201414190960, February 26, 2014.

[122] S. Jönsson, E. Yilmaz, S. Kronauer, Inventors; Biotage AB, Assignee, Porous Polymeric Separation Material, EP20120128842, March 28, 2012.

[123] B. Sellergren, A. Hall, K. Chenon, R. Karmalkar, Inventors; MIP Technologies AB, Assignee, Functional Monomers for Molecular Recognition and Catalysis, US20050159570A1, July 21, 2005.

[124] S. Mohammadi, A. Kargari, H. Sanaeepur, K. Abbassian, A. Najafi, E. Mofarrah, Phenol removal from industrial wastewaters: a short review, Desalination and Water Treatment 53 (2015) 2215–2234.

[125] X. Sun, C. Wang, Y. Li, W. Wang, J. We, Treatment of phenolic wastewater by combined UF and NF/RO processes, Desalination 355 (2015) 68–74.

[126] Emergency Planning and Community Right-to-Know Act (EPCRA) Section 313 Chemical List for Reporting Year 2014. Available from: http://www.epa.gov/toxics-release-inventory-triprogram/tri-chemical-list-ry-2014-including-toxic-chemicalcategories.

[127] Z. Meng, W. Chen, A. Mulchandani, Removal of estrogenic pollutants from contaminated water using molecularly imprinted polymers, Environmental Science and Technology 39 (2005) 8958–8962.

[128] Z. Zang, X. Chen, W. Rao, H. Chen, R. Cai, Synthesis and properties of magnetic molecularly imprinted polymers based on multiwalled carbon nanotubes for magnetic extraction of bisphenol A from water, Journal of Chromatography B 965 (2014) 190–196.
[129] M.P. Di Bello, L. Mergola, S. Scorrano, R. Del Sole, Towards a new strategy of a chitosan-based molecularly imprinted membrane for removal of 4-nitrophenol in real water samples, Polymer International 66 (2017) 1055–1063.
[130] W. Chen, M. Xue, F. Xue, X. Mu, Z. Xu, Z. Meng, G. Zhu, K.J. Shea, Molecularly imprinted hollow spheres for the solid phase extraction of estrogens, Talanta 140 (2015) 68–72.
[131] W. Shen, G. Xu, F. Wei, J. Yang, Z. Cai, Q. Hu, Preparation and application of imprinted polymer for tetrabromobisphenol A using tetrachlorobisphenol A as the dummy template, Analytical Methods 5 (2013) 5208–5214.
[132] C. Didaskalou, C.P. Fonte, S. Büyüktiryaki, R. Keçili, G. Szekely, Valorisation of agricultural waste with adsorption/nanofiltration hybrid process: from materials to sustainable process design, Green Chemistry 19 (2017) 3116–3125.
[133] K. Zhi, L. Wang, Y. Zhang, X. Zhang, L. Zhang, L. Liu, J. Yao, W. Xiang, Preparation and evaluation of molecularly imprinted polymer for selective recognition and adsorption of gossypol, Journal of Molecular Recognition 31 (2018) e2627.
[134] H. Nishide, J. Deguchi, E. Tsuchida, Selective adsorption of metal ions on crosslinked poly(vinylpyridine) resin prepared with a metal ions as a template, Chemistry Letters 5 (1976) 169–176.
[135] M. Zhang, Z. Zhang, Y. Liu, X. Yang, L. Luo, J. Chen, S. Yao, Preparation of core–shell magnetic ion-imprinted polymer for selective extraction of Pb(II) from environmental samples, Chemical Engineering Journal 178 (2011) 443–450.
[136] X. Yin, J. Long, Y. Xi, X. Luo, Recovery of silver from wastewater using a new magnetic photocatalytic ion-imprinted polymer, ACS Sustainable Chemistry and Engineering 5 (2017) 2090–2097.
[137] X. Luo, L. Liu, F. Deng, S. Luo, Novel ion-imprinted polymer using crown ether as a functional monomer for selective removal of Pb(II) ions in real environmental water samples, Journal of Materials Chemistry A 1 (2013) 8280–8286.
[138] C.A. Morais, V.S.T. Ciminelli, Process development for the recovery of high-grade lanthanum by solvent extraction, Hydrometallurgy 73 (2004) 237–244.
[139] P. Maestro, D. Huguenin, Industrial applications of rare earths: which way for the end of the century, Journal of Alloys and Compounds 225 (1995) 520–528.
[140] İ. Dolak, R. Keçili, D. Hür, A. Ersöz, R. Say, Ion-imprinted polymers for selective recognition of neodymium(III) in environmental samples, Industrial and Engineering Chemistry Research 54 (2015) 5328–5335.
[141] Y. Xi, Y. Luo, J. Luo, X. Luo, Removal of cadmium(II) from wastewater using novel cadmium ion-imprinted polymers, Journal of Chemical and Engineering Data 60 (2015) 3253–3261.
[142] E. Roy, S. Patra, R. Madhuri, P.K. Sharma, A single solution for arsenite and arsenate removal from drinking water using cysteine@ZnS:TiO_2 nanoparticle modified molecularly imprinted biofouling-resistant filtration membrane, Chemical Engineering Journal 304 (2016) 259–270.
[143] M. Razali, J.F. Kim, M. Attfield, P.M. Budd, E. Drioli, Y.M. Lee, G. Szekely, Sustainable wastewater treatment and recycling in membrane manufacturing, Green Chemistry 17 (2015) 5196–5205.
[144] X. Zheng, E. Liu, F. Zhang, Y. Yan, J. Pan, Efficient adsorption and separation of dysprosium from NdFeB magnets in an acidic system by ion imprinted mesoporous silica sealed in a dialysis bag, Green Chemistry 18 (2016) 5031–5040.
[145] K. Hunger, Industrial Dyes: Chemistry, Properties, Applications, Wiley-VCH, Weinheim, 2003.
[146] H. Kusic, N. Koprivanac, A.L. Bozic, Environmental aspects on the photodegradation of reactive triazine dyes in aqueous media, Journal of Photochemistry and Photobiology A 252 (2013) 131–144.
[147] L. Tan, S.X. Ning, X.W. Zhang, S.N. Shi, Aerobic decolorization and degradation of azo dyes by growing cells of a newly isolated yeast Candida tropicalis TL-F1, Bioresource Technology 138 (2013) 307–313.
[148] A. Pandey, P. Singh, L. Iyengar, Bacterial decolorization and degradation of azo dyes, International Biodeterioration and Biodegradation 59 (2007) 73–84.
[149] H. Ali, Biodegradation of synthetic dyes-a review, Water Air and Soil Pollution 213 (2010) 251–273.
[150] R. Hazzaa, M. Hussein, Adsorption of cationic dye from aqueous solution onto activated carbon prepared from olive stones, Environmental Technology and Innovation 4 (2015) 36–51.
[151] J.E. Aguiar, J.C.A. De Oliveira, P.F.G. Silvino, J.A. Neto, I.J. Silva, S.M.P. Lucena, Correlation between PSD and adsorption of anionic dyes with different molecular weigths on activated carbon, Colloids and Surfaces A: Physicochemical and Engineering Aspects 496 (2016) 125–131.
[152] T.L. Silva, A. Ronix, O. Pezoti, L.S. Souza, P.K.T. Leandro, K.C. Bedin, K.K. Beltrame, A.L. Cazetta, V.C. Almeida, Mesoporous activated carbon from industrial laundry sewage sludge: adsorption studies of reactive dye Remazol Brilliant Blue R, Chemical Engineering Journal 303 (1) (2016) 467–476.
[153] A. Regti, M.R. Laamari, S.-E. Stiriba, M. El Haddad, Use of response factorial design for process optimization of basic dye adsorption onto activated carbon derived from Persea species, Microchemical Journal 130 (2017) 129–136.
[154] H.K. Melvin Ng, C.P. Leo, A.Z. Abdullah, Selective removal of dyes by molecular imprinted TiO_2 nanoparticles in polysulfone ultrafiltration membrane, Journal of Environmental Chemical Engineering 5 (2017) 3991–3998.
[155] S. Asman, N.A. Yusof, A.H. Abdullah, M.J. Haron, Synthesis and characterization of a molecularly imprinted polymer for methylene blue, Asian Journal of Chemistry 23 (2011) 4786–4794.
[156] G.Z. Kyzas, N.K. Lazaridis, D.N. Bikiaris, Optimization of chitosan and β-cyclodextrin molecularly imprinted polymer synthesis for dye adsorption, Carbohydrate Polymers 91 (2013) 198–208.
[157] Y.S. Al-Degs, A.S. Abu-surrah, K.A. Ibrahim, Preparation of highly selective solid-phase extractants for Cibacron reactive dyes using molecularly imprinted polymers, Analytical and Bioanalytical Chemistry 393 (2009) 1055–1062.
[158] B. Gao, T. Chen, Z. Qiao, L. Fang, Preparation of acid dye molecule surface-imprinted material for effective removal of acid dyes from water and study on its molecule recognition performance, Separation Science and Technology 50 (2015) 1108–1119.
[159] H. Li, N. Li, J. Jiang, D. Chen, Q. Xu, H. Li, J. He, J. Lu, Molecularly imprinted magnetic microparticles for the simultaneous detection and extraction of Rhodamine B, Sensors and Actuators B: Chemical 246 (2017) 286–292.
[160] X. Zhang, X. Gao, P. Huo, Y. Yan, Selective adsorption of micro ciprofloxacin by molecularly imprinted functionalized polymers appended onto ZnS, Environmental Technology 33 (2012) 2019–2025.

[161] N. Arabzadeh, M. Abdouss, Synthesis and characterization of molecularly imprinted polymers for selective solid phase extraction of pseudoephedrine, Colloid Journal 72 (2010) 446–455.
[162] Z. Terzopoulou, M. Papageorgiou, G.Z. Kyzas, D.N. Bikiaris, D.A. Lambropoulou, Preparation of molecularly imprinted solid-phase microextraction fiber for the selective removal and extraction of the antiviral drug abacavir in environmental and biological matrices, Analytica Chimica Acta 913 (2016) 63–75.
[163] L. Xu, Z.-F. Xu, Molecularly imprinted polymer based on multiwalled carbon nanotubes for ribavirin recognition, Journal of Polymer Research 19 (2012) 9942.
[164] C.M. Dai, J. Zhang, Y.L. Zhang, X.F. Zhou, Y.P. Duan, S.G. Liu, Removal of carbamazepine and clofibric acid from water using double templates-molecularly imprinted polymers, Environmental Science and Pollution Research 20 (2013) 5492–5501.
[165] L. An, J. Wang, Z. Pang, R. Xi, Magnetic molecularly imprinted silica gel for enrofloxacin recognition, Colloids and Surfaces A: Physicochemical and Engineering Aspects 452 (2014) 125–128.
[166] S.K. Bajpai, S. Jhariya, Selective removal of amikacin from simulated polluted water using molecularly imprinting polymer (MIP), Journal of Macromolecular Science Part A 52 (2015) 901–911.
[167] F. Tan, D. Sun, J. Gao, Q. Zhao, X. Wang, F. Teng, X. Quan, J. Chen, Preparation of molecularly imprinted polymer nanoparticles for selective removal of fluoroquinolone antibiotics in aqueous solution, Journal of Hazardous Materials 244–245 (2013) 750–757.
[168] R. Kang, L. Qiu, L. Fang, R. Yu, Y. Chen, X. Lu, X. Luo, A novel magnetic and hydrophilic ion-imprinted polymer as a selective sorbent for the removal of cobalt ions from industrial wastewater, Journal of Environmental Chemical Engineering 4 (2016) 2268–2277.
[169] W. Li, R. He, L. Tan, S. Xu, C. Kang, C. Wei, Y. Tang, One-step synthesis of periodic ion imprinted mesoporous silica particles for highly specific removal of Cd^{2+} from mine wastewater, Journal of Sol-Gel Science and Technology 78 (2016) 632–640.
[170] C.R. Preeatha, J.M. Gladis, T.P. Rao, Removal of Toxic Uranium from synthetic nuclear power reactor effluents using uranyl ion imprinted polymer particles, Environmental Science and Technology 40 (2006) 3070–3074.
[171] F. Liu, Y. Liu, L. Ni, X. Meng, Z. Hu, G. Zhong, M. Meng, Y. Wang, J. Han, Efficient static and dynamic removal of Sr(II) from aqueous solution using chitosan ion-imprinted polymer functionalized with dithiocarbamate, Journal of Environmental Chemical Engineering 3 (2015) 1061–1071.
[172] Y. Liu, X. Meng, Z. Liu, M. Meng, F. Jiang, M. Luo, L. Ni, J. Qiu, F. Liu, G. Zhong, Preparation of a two-dimensional ion-imprinted polymer based on a graphene oxide/SiO_2 composite for the selective adsorption of nickel ions, Langmuir 31 (2015) 8841–8851.
[173] D. Kong, F. Zhang, K. Wang, Z. Ren, W. Zhang, Fast removal of Cr(VI) from aqueous solution using Cr(VI)-imprinted polymer particles, Industrial and Engineering Chemistry Research 53 (2014) 4434–4441.
[174] A. Aliakbari, M.M. Amini, K. Mehrani, H.R.M. Zadeh, Magnetic ion imprinted polymer nanoparticles for the preconcentration of vanadium(IV) ions, Microchimica Acta 181 (2014) 1931–1938.
[175] M. Moussa, M.M. Ndiaye, T. Pinta, V. Pichon, T. Vercoute, N. Delaunay, Selective solid phase extraction of lanthanides from tap and river waters with ion imprinted polymers, Analytica Chimica Acta 963 (2017) 44–52.
[176] M. Fayazi, M. Ghanei-Motlagh, M.A. Taher, R. Ghanei-Motlagh, M.R. Salavati, Synthesis and application of a novel nanostructured ion-imprinted polymer for the preconcentration and determination of thallium(I) ions in water samples, Journal of Hazardous Materials 309 (2016) 27–36.
[177] M. Roushani, S. Abbasi, H. Khani, Synthesis and application of ion-imprinted polymer nanoparticles for the extraction and preconcentration of copper ions in environmental water samples, Environmental Monitoring and Assessment 187 (2015) 219.
[178] J. Pan, X. Zou, C. Li, Y. Liu, Y. Yan, J. Han, Synthesis and applications of Ce(III)-imprinted polymer based on attapulgite as the sacrificial support material for selective separation of cerium(III) ions, Microchimica Acta 1–2 (2010) 151–160.
[179] F. Ardestani, M.H. Hosseini, M. Taghizadeh, M.R. Pourjavid, M. Rezaee, Synthesis and characterization of nanopore Mo(VI)-imprinted polymer and its application as solid phase for extraction, separation and preconcentration of molybdenum ions from water samples, Journal of the Brazilian Chemical Society 27 (2016) 1279–1289.
[180] K. Huang, B. Li, F. Zhou, S. Mei, Y. Zhou, T. Jing, Selective solid-phase extraction of lead ions in water samples using three-dimensional ion-imprinted polymers, Analytical Chemistry 88 (2016) 6820–6826.
[181] X. Luo, Y. Xi, H. Yu, X. Yin, S. Luo, Capturing cadmium(II) ion from wastewater containing solid particles and floccules using ion-imprinted polymers with broom effect, Industrial and Engineering Chemistry Research 56 (9) (2017) 2350–2358.
[182] R. He, Z. Wang, L. Tan, Y. Zhong, W. Li, D. Xing, C. Wei, Y. Tang, Design and fabrication of highly ordered ion imprinted SBA-15 and MCM-41 mesoporous organosilicas for efficient removal of Ni^{2+} from different properties of wastewaters, Microporous and Mesoporous Materials 257 (2018) 212–221.
[183] M. Roushani, T.M. Beygi, Z. Saedi, Synthesis and application of ion-imprinted polymer for extraction and pre-concentration of iron ions in environmental water and food samples, Spectrochimica Acta Part A: Molecular and Biomolecular Spectroscopy 153 (2016) 637–644.
[184] M. Fang, F. Lei, J. Zhou, Y.-N. Wu, Z.-Y. Gong, Rapid, simple and selective determination of 2,4-dinitrophenol by molecularly imprinted spin column extraction coupled with fluorescence detection, Chinese Chemical Letters 25 (2014) 1492–1494.
[185] R. Gao, Y. Hao, L. Zhang, X. Cui, D. Liu, Y. Tang, Core–shell nano-sized magnetic molecularly imprinted solid phase extractant coupled with HPLC for the selective isolation and determination of 17β-estradiol in a lake water sample, Analytical Methods 6 (2014) 9791–9799.
[186] X.-P. He, Z.-R. Lian, L.-J. Tan, J.-T. Wang, Preparation and characterization of magnetic molecularly imprinted polymers for selective trace extraction of dienestrol in seawater, Journal of Chromatography A 1469 (2016) 8–16.
[187] Z. Lin, W. Cheng, Y. Li, Z. Liu, X. Chen, C. Huang, A novel superparamagnetic surface molecularly imprinted nanoparticle adopting dummy template: an efficient solid-phase extraction adsorbent for bisphenol A, Analytica Chimica Acta 720 (2012) 71–76.
[188] Y. Huang, Y. Xu, Q. He, B. Du, Y. Cao, Preparation and characteristics of a dummy molecularly imprinted polymer for phenol, Journal of Applied Polymer Science 128 (2013) 3256–3262.
[189] J. Pan, H. Yao, L. Xu, H. Ou, P. Huo, X. Li, Y. Yan, Selective recognition of 2,4,6-trichlorophenol by molecularly imprinted polymers based on magnetic halloysite nanotubes composites, The Journal of Physical Chemistry C 115 (13) (2011) 5440–5449.
[190] H. Deng, Z. Wei, X. Wang, Enhanced adsorption of active brilliant red X-3B dye on chitosan molecularly imprinted polymer functionalized with Ti(IV) as Lewis acid, Carbohydrate Polymers 157 (2017) 1190–1197.

[191] Z. Lian, J. Wang, Molecularly imprinted polymer for selective extraction of malachite green from seawater and seafood coupled with high-performance liquid chromatographic determination, Marine Pollution Bulletin 64 (2012) 2656–2662.
[192] L. Lu, X. Yue, F. Lin, F. Huang, B. Zhang, Z. Lin, Template-synthesized ultra-thin molecularly imprinted polymers membrane for the selective preconcentration of dyes, Journal of Materials Chemistry A 3 (2015) 10959–10968.
[193] M.A. Ahmed, N.M. Abdelbar, A.A. Mohamed, Molecular imprinted chitosan-TiO_2 nanocomposite for the selectiveremoval of Rose Bengal from wastewater, International Journal of Biological Macromolecules 107 (2018) 1046–1053.
[194] N. Gurung, S. Ray, S. Bose, V. Rai, A broader view: microbial enzymes and their relevance in industries, medicine and beyond, BioMed Research International 2013 (2013) 329121.
[195] W.B. Motherwell, M.J. Bingham, Y. Six, Recent progress in the design and synthesis of artificial Enzymes, Tetrahedron 57 (2001) 4663–4686.
[196] G. Wulff, Enzyme-like catalysis by molecularly imprinted polymers, Chemical Reviews 102 (2002) 1–27.
[197] K. Johnsson, R.K. Allemann, H. Widmer, S.A. Benner, Synthesis, structure and activity of artificial, rationally designed catalytic polypeptides, Nature 365 (1993) 530–532.
[198] P. Wirsching, J.A. Ashley, S.J. Benkovic, K.D. Janda, R.A. Lerner, An unexpectedly efficient catalytic antibody operating by ping-pong and induced fit mechanisms, Science 252 (1991) 680–685.
[199] D. Voet, J.G. Voet, Biochemistry, John Wiley & Sons, 1995.
[200] C. Alexander, L. Davidson, W. Hayes, Imprinted polymers: artificial molecular recognition materials with applications in synthesis and catalysis, Tetrahedron 59 (2003) 2025–2057.
[201] D. Mathew, B. Thomas, K.S. Devaky, Biomimetic recognition and peptidase activities of transition state analogue imprinted chymotrypsin mimics, Reactive and Functional Polymers 124 (2018) 121–128.
[202] M. Erdem, R. Say, A. Ersöz, A. Denizli, H. Türk, Imprinted polymer/organo-smectite nanocomposites for paraoxon hydrolysis, Applied Clay Science 47 (2010) 223–228.
[203] M.A. Markowitz, P.R. Kust, G. Deng, P.E. Schoen, J.S. Dordick, D.S. Clark, B.P. Gaber, Catalytic silica particles via templated-directed molecular imprinting, Langmuir 16 (2000) 1759–1765.
[204] R. Keçili, A. Atılır Özcan, A. Ersöz, D. Hür, A. Denizli, R. Say, Superparamagnetic nanotraps containing MIP based mimic lipase for biotransformations uses, Journal of Nanoparticle Research 13 (5) (2011) 2073–2079.
[205] R. Wang, Y. Zhang, J. Huang, D. Lu, J. Ge, Z. Liu, Substrate imprinted lipase nanogel for one-step synthesis of chloramphenicol palmitate, Green Chemistry 15 (2013) 1155–1158.
[206] X. Luo, F. Deng, L. Min, S. Luo, B. Guo, G. Zeng, C. Au, Facile one step synthesis of inorganic-framework molecularly imprinted TiO_2/WO_3 nanocomposite and its molecular recognitive photocatalytic degradation of target contaminant, Environmental Science and Technology 47 (2013) 7404–7412.
[207] N. Arabzadeh, A. Khosravi, A. Mohammadi, N.M. Mahmoodi, Enhanced photodegradation of hazardous tartrazine by composite of nano-molecularly imprinted polymer-nanophotocatalyst with high efficiency, Desalination and Water Treatment 57 (2016) 3142–3151.
[208] C. Coelho de Escobar, Y.P. Moreno Ruiz, J.H. Zimnoch dos Santos, L. Ye, Molecularly imprinted TiO_2 photocatalysts for degradation of diclofenac in water, Colloids and Surfaces A 538 (2018) 729–738.
[209] X. Zhou, C. Lai, D. Huang, G. Zeng, L. Chen, L. Qin, P. Xu, M. Chenga, C. Huang, C. Zhang, C. Zhou, Preparation of water-compatible molecularly imprinted thiol-functionalized activated titanium dioxide: selective adsorption and efficient photodegradation of 2, 4-dinitrophenol in aqueous solution, Journal of Hazardous Materials 346 (2018) 113–123.
[210] J.V. Beach, K. Shea, Designed catalysts. A synthetic network polymer that catalyses the dehydrofluorination of 4-fluoro-4(p-nitrophenyl) butan-2-one, Journal of the American Chemical Society 116 (1994) 379–380.
[211] R. Müller, L.I. Andersson, K. Mosbach, Molecularly imprinted polymers facilitating a β-elimination reaction, Macromolecular Chemical Rapid Communications 14 (1993) 637–641.
[212] P. Bonomi, A. Servant, M. Resmini, Modulation of imprinting efficiency in nanogels with catalytic activity in the Kemp elimination, Journal of Molecular Recognition 25 (2012) 352–360.
[213] X.-C. Liu, K. Mosbach, Catalysis of benzisoxazole isomerization by molecularly imprinted polymers, Macromolecular Rapid Communications 19 (1998) 671–674.
[214] J. Matsui, I.A. Nicholls, I. Karube, K. Mosbach, Carbon-carbon bond formation using substrate selective catalytic polymers prepared by molecular imprinting: an artificial class II aldolase, Journal of Organic Chemistry 61 (1996) 5414–5417.
[215] D. Carboni, K. Flavin, A. Servant, V. Gouverneur, M. Resmini, The first example of molecularly ımprinted nanogels with aldolase type I activity, Chemistry: A European Journal 14 (2008) 7059–7065.
[216] N. Kirsch, J. Hedin-Dahlström, H. Henschel, M.J. Whitcombe, S. Wikman, I.A. Nicholls, Molecularly imprinted polymer catalysis of a Diels-Alder reaction, Journal of Molecular Catalysis B: Enzymatic 58 (2009) 110–117.

CHAPTER

13

Fundamental Theory and Molecular Design of Thermoresponsive Polymers Expandable to Sustainable and Smart Materials

Kenta Kokado, Masami Naya, Kazuki Sada

Department of Chemistry, Faculty of Science, Hokkaido University, Sapporo, Hokkaido, Japan

OUTLINE

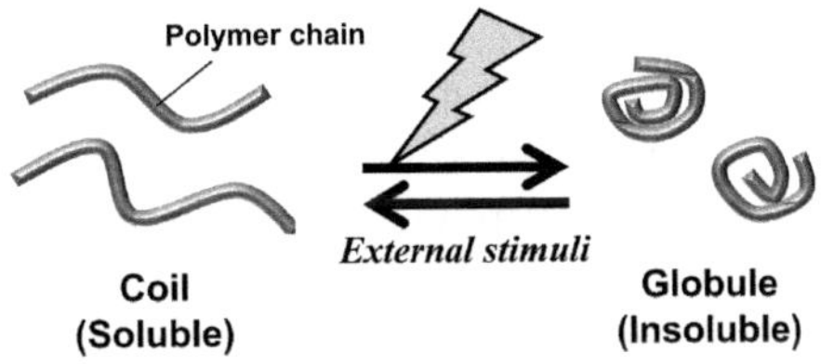

Sustainable Nanoscale Engineering
https://doi.org/10.1016/B978-0-12-814681-1.00013-8

1. INTRODUCTION

Our modern civilization has prospered especially over the last 100 years. However, considering the finite natural resources of the earth, we must immediately start constructing a sustainable society in which the resources have to be utilized in the most efficient ways possible. One of the promising solutions to improve the efficiency of materials is to reuse and recycle; thus, consumption of the natural resources necessary for preparation of the materials should be reduced. For high reusability or recyclability, multiple uses of materials are essential because it will reduce the necessary amount of resources. Additionally, sustainable materials should show their functions only at the desired time, and they should be inactive or dormant at other time, as active materials would be easily deactivated constantly and decomposed by the surrounding environment. In this context, chemists have paid significant attention to switchable and intelligent matters that offer a switching property to other functional materials; hence, the function of the functional materials can be controlled through the attached switchable matters. Among the switchable matters, stimuli-responsive polymers are the most important class because they can exhibit large conformational changes in their solution between the coil (dissolved) and globule (aggregated) states, in response to external stimulating fields such as temperature, light, electric and magnetic fields, and ultrasound, without undergoing any chemical composition change [1,2]. In this chapter, the fundamental understanding and design of stimuli-responsive polymers is discussed by considering the thermoresponsivity mechanism of the polymers.

2. UNDERSTANDING THEMORESPONSIVE POLYMERS

Heating, thus temperature elevation, accelerates molecular motions. As a result of heating, a substance usually becomes easier to dissolve in a solvent, from which one can imagine recrystallization in chemical experiments. This phenomenon is defined as the upper critical solution temperature (UCST)−type phase transition, in which the UCST denotes the solvent, and solutes are phase-separated under the critical temperature. The opposite phenomenon is defined as the lower critical solution temperature (LCST)−type phase transition. Only limited examples of a low molecular weight compound combination can exhibit the LCST-type phase transition; however, the phenomenon is more common in the aqueous solution of polymers. Certain types of polymers in water precipitate on heating, while the LCST-type phase transition of low molecular weight compounds only show a liquid−liquid phase separation.

From the viewpoint of thermodynamics, the elevation of temperature means the increasing contribution of the entropic term in the Gibbs free energy, usually leading to a more dissipative state. When two substances are mixed, the mixture has two states, i.e., the homogeneously mixed phase state and phase separated (two phases) state. The state is defined by the applied pressure, temperature, or composition of the substances. Under a constant pressure, the phase diagram can be depicted with the composition on the horizontal axis and the temperature on the vertical axis. In the case of being miscible at a lower temperature and immiscible at a higher temperature, it produces a curve that is convex downward. The minimum point is defined as the LCST (Fig. 13.1A), as it is the lowest temperature to cause the phase transition. In the opposite case, the curve becomes convex upwards, and the maximum point is defined as the UCST (Fig. 13.1B), as it is the highest temperature to cause the phase transition. At the boundary of the two states, the Gibbs free energy of both states becomes equal, thus the high and low relationship of the Gibbs free energy of both states changes when the temperature passes over the boundary (Fig. 13.1C). As a result, the state drastically changes, and the temperature is defined as the cloud point.

For low molecular weight compounds, the UCST-type phase transition is widely known, intuitively as recrystallization as mentioned above (Fig. 13.1D). In this case, at a lower temperature, the two substances are phase separated, but at a higher temperature, they are homogeneously mixed due to the increased contribution of the entropy term. On another front, the LCST-type phase transition is rarely observed for low molecular weight compounds, including water-triethylamine, water-nicotine [3], or water−ionic liquid [4]. In these cases, the two components form a relatively highly ordered structure via hydrogen bonding and other intermolecular interactions at the lower temperature, which is cleaved at the higher temperature, resulting in the observed phase separation (Fig. 13.1D). The two components must have strong intermolecular interactions to cause the LCST-type phase transition that forms a highly ordered structure in the homogeneously mixed state.

The molecular structure of a polymer is quite different from that of a low molecular weight compound. Therefore, a polymer is stated differently from a low molecular weight compound in thermodynamics by the Flory−Huggins

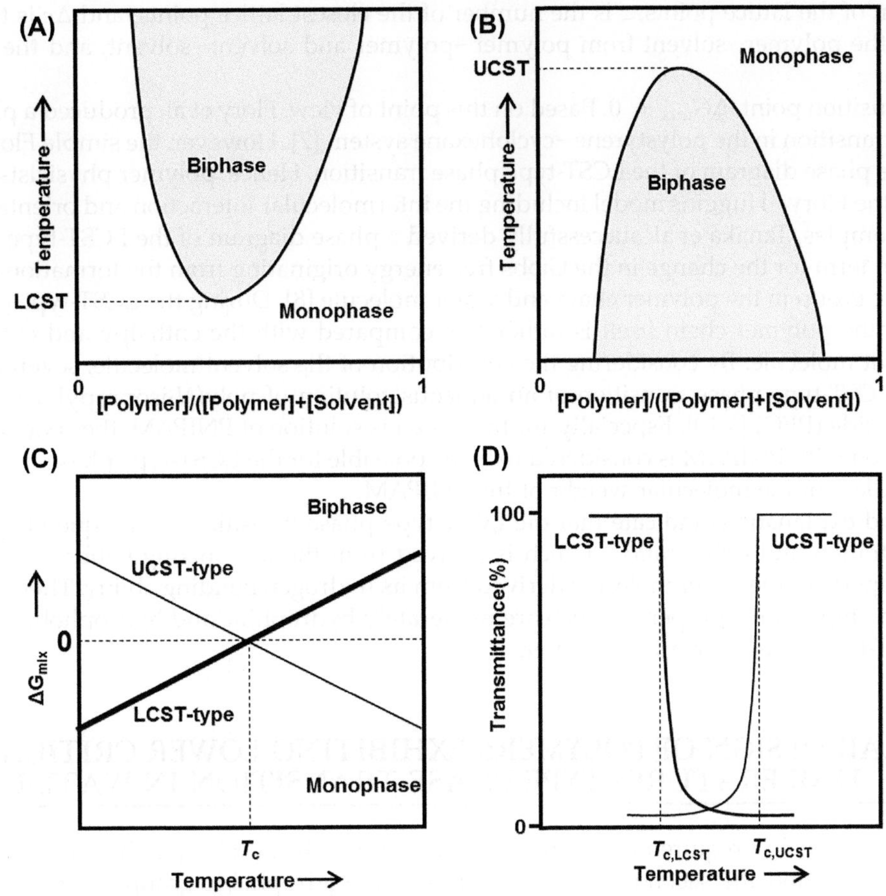

FIGURE 13.1 Representative image of phase diagram for (A) lower critical solution temperature (LCST)– and (B) upper critical solution temperature (UCST)–type phase transition of polymers. Horizontal axis: composition ration of polymer and solvent. Vertical axis: temperature. (C) Mixing Gibbs free energy change versus temperature for the LCST- and UCST-type phase transition. (D) Typical transmittance change to determine LCST- or UCST-type phase transition.

theory [5,6], which is hypothesized in a lattice model with a similar volume of the monomer unit in the polymer chain and the solvent molecule. The mixing entropy change in the solvent (N_0 molecules) and polymer (N_1 chains) can be described as follows:

$$\Delta S_{\text{mix, polym}} = -k_b \left(N_0 \ln \phi_0 + N_1 \ln \phi_1\right) \tag{13.1}$$

where k_b is the Boltzmann constant and φ_0 and φ_1 are the volume fractions of the solvent and polymer, respectively. In the case of a low molecular weight compound such as a monomer, the mixing entropy change of the solvent (N_0 molecules) and the monomer (nN_1 molecules) can be described as follows:

$$\Delta S_{\text{mix, monom}} = -k_b \left(N_0 \ln \phi_0 + nN_1 \ln \phi_1\right) \tag{13.2}$$

where n is the number of monomer units in a polymer chain. From the difference in formulas (13.1) and (13.2), the polymerization decreases the mixing entropy as follows:

$$\Delta S_{\text{mix, monom}} - \Delta S_{\text{mix, polym}} = -k_b(n-1)N_1 \ln \phi_1 \tag{13.3}$$

This formula indicates that the mixing entropy of a polymer solution is much lower than that of a solution of a low molecular weight compound. By combining formula (13.1) with the mixing enthalpy of the polymer solution, the mixing Gibbs free energy can be described as follows:

$$\Delta G_{\text{mix}} = -k_b \left(N_0 \ln \phi_0 + N_1 \ln \phi_1\right) + \Omega z \Delta\varepsilon \phi_0 \phi_1 \tag{13.4}$$

where Ω is the sum of the lattice points, z is the number of the closest lattice points, and $\Delta\varepsilon$ is the difference in the contact energy of the polymer–solvent from polymer–polymer and solvent–solvent, and the volume change on mixing is ignored.

At the phase transition point, $\Delta G_{\text{mix}} = 0$. Based on this point of view, Flory et al. produced a phase diagram of the UCST-type phase transition in the polystyrene–cyclohexane system [7]. However, the simple Flory–Huggins model cannot produce the phase diagram of the LCST-type phase transition. Hence, polymer physicists have tried to add a correction term to the Flory–Huggins model including the intermolecular interaction and orientation of the polymer and solvent. As examples, Tanaka et al. successfully derived a phase diagram of the LCST-type phase transition by adding a correction term for the change in the Gibbs free energy originating from the formation of a cluster consisting of a hydrophilic group in the polymer chain and water molecule [8]. During the LCST-type phase transition, the entropy change in the polymer chain itself is rather low compared with the enthalpy and entropy change in the surrounding solvent molecule. By considering the contribution of the solvent molecule, several models have been proposed for the LCST-type phase transition of an aqueous solution of poly(*N*-isopropyl acrylamide) (PNIPAM) and polyethylene oxide (PEO) [9,10]. Especially, for the aqueous solution of PNIPAM, the cooperativity of the water molecule coordinated with PNIPAM is considered to be responsible for the LCST-type phase transition at a constant temperature regardless of the molecular weight of the PNIPAM.

These results and explanations indicate that the LCST-type phase transition in an aqueous polymer solution is caused by the hydration and dehydration, which is derived from the low mixing Gibbs entropy of the polymer and associative property of the water molecule derived from its hydrogen bonding ability. Therefore, it is reasonable that polymers with amphiphilic property, which are moderately hydrophilic and hydrophobic, have an inclination to exhibit the LCST-type phase transition in water.

3. MOLECULAR DESIGN OF POLYMERS EXHIBITING LOWER CRITICAL SOLUTION TEMPERATURE–TYPE PHASE TRANSITION IN WATER

As indicated above, the first requirement for polymers exhibiting the LCST-type phase transition in water is having hydrophilic groups, and the second requirement is having hydrophobic groups (Fig. 13.2). The hydrophilic groups associate with themselves when the polymer aggregates, and they should also determine the orientation of the water molecules and/or hydration/dehydration balance.

Most research studies of the LCST-type phase transition in water have dealt with PNIPAM, thus one can consider that the LCST-type phase transition is a particular phenomenon for PNIPAM. However, most amphiphilic polymers principally exhibit the LCST-type phase transition in water, and various kinds of amphiphilic polymers have been actually reported to exhibit it. The typical examples are listed in Table 13.1. The cloud point (T_c) is defined as the

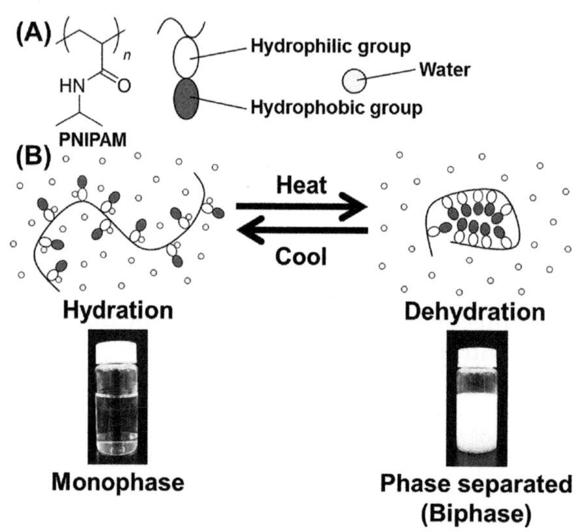

FIGURE 13.2 (A) The structural requirement for polymers exhibiting lower critical solution temperature (LCST)–type phase transition in water, and the chemical structure of PNIPAM as a typical example. (B) Schematic image and photograph of LCST-type phase transition of aqueous polymer solution.

TABLE 13.1 Polymers With Lower Critical Solution Temperature (LCST)–Type Phase Transition in Water.

Polymers		LCST or T_c	References
Poly(*N*-isopropyl acrylamide)	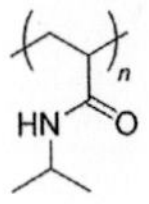	32	[11]
Poly(*N*-*n*-propyl acrylamide)	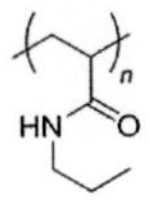	23	[13,14]
Poly(*N*-cyclopropyl acrylamide)	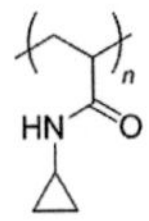	49	[14]
Poly(*N*,*N*-diethyl acrylamide) DEAM	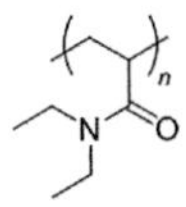	32	[15]
Poly(*N*-vinylisobutylamide)	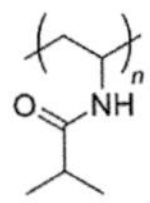	39	[16]
Poly(ethylene oxide)	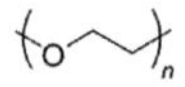	99–176	[17]
Poly(propylene oxide)	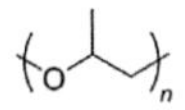	0–50	[18]
Poly(oligo(ethyleneglycol) methacrylate)OEGMA	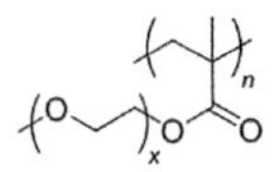	26 ($x = 2$) 52($x = 3$)	[19]
Poly(styrene) containing oligo(oxyethylene)	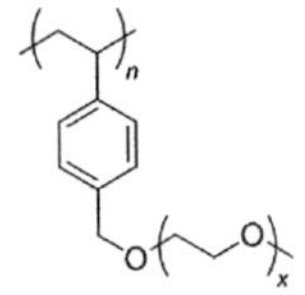	13 ($x = 3$) 39 ($x = 4$) 55 ($x = 4$)	[20]
Poly(phosphazene) containing oligo(oxyethylene)	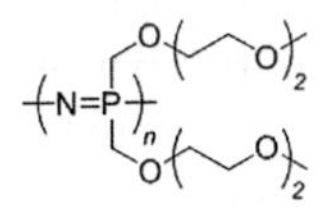	65	[21]
Poly(norbornene) containing oligo(oxyethylene)	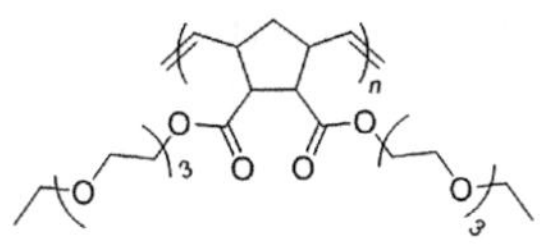	26	[22]
Poly(lactide) containing oligo(oxyethylene)	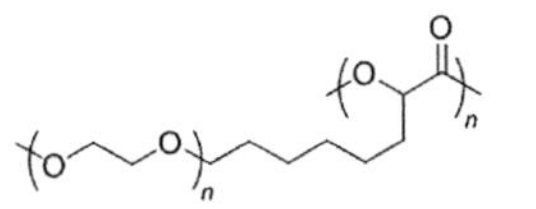	17 ($x = 3$) 37 ($x = 4$)	[23]
Poly(*N*-vinyl pyrrolidone)	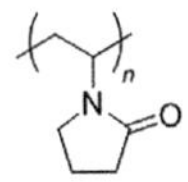	30	[24]
Polyi(*N*-vinyl caprolactam)	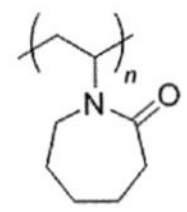	31	[24,25]
Poly(2-ethyl-2-oxazoline)	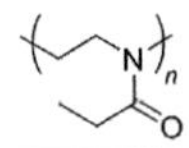	62–65	[26]

Continued

TABLE 13.1 Polymers With Lower Critical Solution Temperature (LCST)–Type Phase Transition in Water.—cont'd

Polymers		LCST or T_c	References
Poly(2-ethyl-2-oxazine)	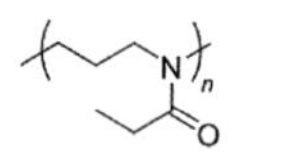	56	[27]
Poly(methyl vinyl ether)	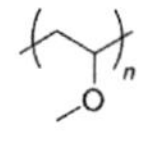	35	[28]
Poly(2-methoxyethyl vinyl ether)	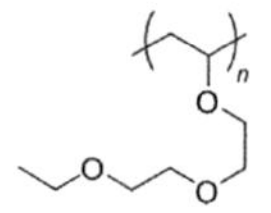	41	[29]
Poly(N N-dimethylaminoethyl methacrylate)	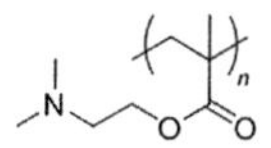	78	[30]
Poly(2-hydroxypropyl acrylate)	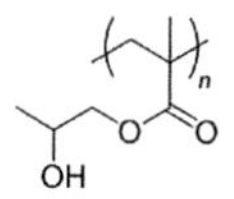	37	[31]
Poly(vinyl alcohol-*r*-vinyl acetal)	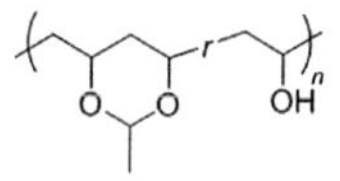	17–41	[32]
Poly(2-ethoxy-2-oxo-1,3,2-dioxaphospholane)	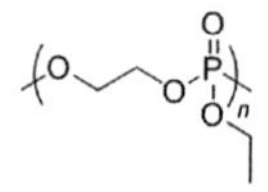	38	[33]
Elastin-like polypeptide	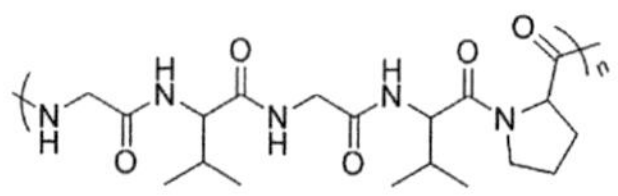	27–10	[34]
2-Hydioxy-3-butoxypropyl starches	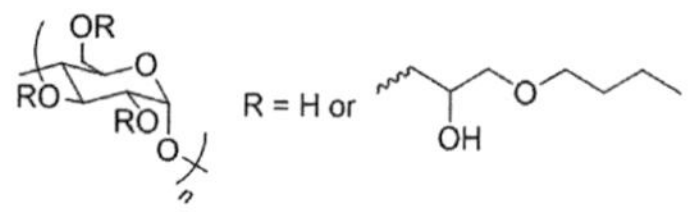	4.5–32.5	[35]

temperature at which a polymer solution with a given composition (φ) undergoes phase separation, while the LCST is the minimum of the $T_c(\varphi)$ function (Fig. 13.1D). PNIPAM is the best-known polymer exhibiting the LCST-type phase transition in water (see also review [34]), and it was vigorously investigated since the first report of its LCST-type phase transition in water in 1967 [11]. PNIPAM contains a hydrophilic amide group and hydrophobic isopropyl group; thus, it comprises an amphiphilic molecular structure. Moreover, the LCST of PNIPAM is observed at 32°C in water, which is close to the physiological condition; thus, many studies of PNIPAM have targeted biomedical applications [12]. The effect of the hydrophobic isopropyl group on the PNIPAM has been investigated, and its replacement by an *n*-propyl or cyclopropyl group drastically affected the T_c of the LCST-type phase transition [13,14]. Poly(*N*,*N*-diethylacrylamide) exhibited the LCST-type phase transition, and its T_c was the same as that of PNIPAM [15]. Poly(*N*-vinyl isobutylamide) exhibits the LCST-type phase transition, and T_c was observed at a higher temperature compared with that of PNIPAM [16]. These examples clearly show that even a slight difference in the molecular structure can shift T_c.

PEO is another important source to exhibit the LCST-type phase transition in water [17], but the research examples of its thermoresponsivity including applications is far less than those of PNIPAM due to the requirement of extreme conditions above the water boiling point and strong molecular weight dependence (e.g., $T_c = 163°C$ for PEO with $M_n = 2290$). Compared with PEO, poly(propylene oxide) (PPO) is more hydrophobic and exhibits the LCST-type phase transition at a temperature lower than 100°C, while the strong molecular weight dependence still remains [18]. Instead of its use as the main chain, the employment of PEO as a side group successfully provided various polymers with the LCST-type phase transition, such as poly(methacrylate) [19], poly(styrene) [20], poly(phosphazene) [21], poly(norbornene) [22], and poly(lactide) [23]. The elongation of the PEO chain typically resulted in elevation of the LCST; therefore, in these polymers, the PEO chain functions as the hydrophilic group and the main chain as the hydrophobic group.

As other polymers with the amide side chain, poly(*N*-vinylpyrrolidone) and poly(*N*-vinyl caprolactam) (PVCL) exhibit the LCST-type phase transition [24,25]. As examples, the T_c of PVCL was observed at around 30°C. Amide polymers with a nitrogen atom in the main chain, such as poly(oxazoline) and poly(oxazine), exhibit the LCST-type phase transition with the T_c of around 60°C [26,27]. These polymers have a hydrophilic amide group analogous to PNIPAM, although the main chain structure is different from PNIPAM. As an analogue of PEO and PPO, poly(vinyl ether)s also exhibit the LCST-type phase transition. As examples, the LCSTs of poly(methyl vinyl ether) and poly(2-methoxyethyl vinyl ether) are 35 and 41°C, respectively [28,29]. These polymers have a hydrophilic ether group analogous to PEO and PPO, though the main chain consists of only carbon atoms.

For other synthetic polymers, poly(*N*,*N*-dimethylaminoethyl methacrylate) (PDMAEMA), tethering as amino group, they exhibit an LCST-type phase transition with the T_c of 78°C [30], and poly(2-hydroxypropyl acrylate), having a hydroxy group, exhibits an LCST-type phase transition with T_c of 37°C [31]. Poly(vinyl alcohol) (PVA) possesses too high a hydrophilicity to exhibit a thermoresponsivity, similar to PEO, whereas partial acetylation of the PVA provides polymers exhibiting an LCST-type phase transition with a variable T_c dependent on the acetylation degree between 28 and 41°C[32].

Biocompatible polymers exhibiting the LCST-type phase transition have recently been reported. Poly(phosphoester) is biodegradable under ambient conditions, and it exhibits an LCST-type phase transition [33]. The elastin-like polypeptide, derived from the elastic protein in connective tissue, exhibits an LCST-type phase transition with the T_c of 27°C, and alteration of the amino acid residue can elevate the T_c above 40°C [35]. Among carbohydrate derivatives, methylcellulose exhibits an LCST-type sol–gel transition [36], and partial 2-hydroxy-3-butoxypropylation of amylose leads to the LCST-type phase transition in which the T_c can be lowered by increasing the 2-hydroxy-3-butoxypropylation degree[37].

Ritter et al. reported that a polymer bearing adamantyl moieties exhibited LCST-type phase transition in the presence of methylated β-cyclodextrins (β-CD) in water due to the dissociation of β-CD from the adamantyl group in the polymer chain on heating and the subsequent association of adamantyl groups through a hydrophobic interaction [38,39].

As reviewed in this section, many polymers exhibiting an LCST-type phase transition in water have been reported, other than PNIPAM, and they show various T_c values. In all the polymers introduced here, both a hydrophilic group and hydrophobic group are found, in which the hydrophilic group strongly interacts with water molecules, and the hydrophobic group associates with itself. Such an amphiphilic molecular structure is a characteristic of thermoresponsive polymers in water. Increasing the hydrophilic group content elevates the T_c and vice versa. These facts mean that the T_c can be controlled by copolymerization or polymer modification.

4. MOLECULAR DESIGN OF POLYMERS EXHIBITING LOWER CRITICAL SOLUTION TEMPERATURE–TYPE PHASE TRANSITION IN ORGANIC SOLVENTS

As already mentioned, the water molecules can form hydrogen bonding with each other, and this intermolecular interaction is essentially responsible for the LCST-type phase transition of the aqueous polymer solution. Compared with water, organic solvent molecules generally possess a weaker interaction with each other; thus, the LCST-type phase transition is scarcely observed by simply dissolving a polymer in an organic solvent. Polymers often exhibit LCST-type phase transition under extremely high temperature and pressure condition even in nonaqueous condition [40,41], due to free volume effect; however, the examples of polymers exhibiting LCST-type phase transition under nonaqueous ambient condition are very few. As a limited example of the LCST-phase transition under nonaqueous and ambient conditions, poly(7-methacryloyloxycoumarin) and its derivatives exhibited LCST-type phase transition in halogenated solvents [42]. The dimerization of the coumarin moiety in the polymer chain readily changed T_c to lower temperature. Poly(benzyl methacrylate) in an ionic liquid [43] and poly(vinyl ether) tethering imidazolium in chloroform (Fig. 13.3) [44] were reported to exhibit an LCST-type phase transition. In these cases, the interaction of the polymer ionic liquid and salt moiety on the polymer-solvent should have a particular interaction resulting in the observed LCST-type phase transition.

In this context, the author and coworkers reported a versatile LCST-type phase transition in an organic solvent using ternary systems consisting of a polymer, an "effector" (a small molecule that governs the solubility of the polymer), and a solvent, in which the three components should meet the following criteria; (1). A polymer having strong self-associating functional groups to provide an attractive interaction between the polymer chains. (2). An effector interacting with the polymer chains by a relatively weak intermolecular interaction. (3). A solvent inert to the intermolecular interaction. Until now, the charge-transfer (CT) interaction and hydrogen bonding were successfully

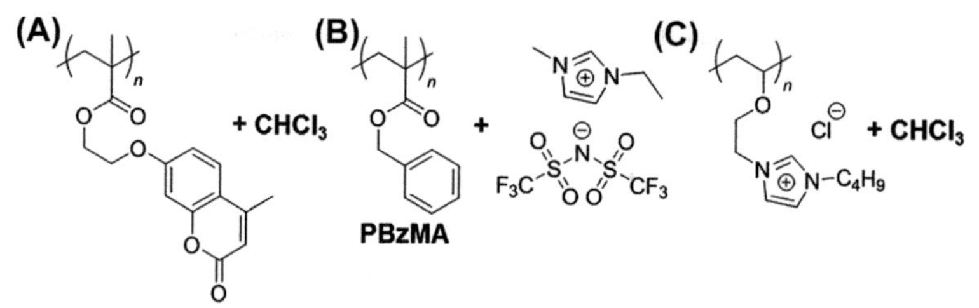

FIGURE 13.3 The polymer solution systems exhibiting lower critical solution temperature (LCST)–type phase transition in organic solvents. *From (A) E. Sato, Y. Masuda, J. Kadota, T. Nishiyama, H. Horibe, European Polymer Journal 69 (2015) 605–615. (B) T. Ueki, M. Watanabe, Langmuir 23 (2007) 988–990. (C) K. Seno, S. Kanaoka, S. Aoshima, Journal of Polymer Science Part A: Polymer Chemistry 46 (2008)5724–5733.*

utilized to cause the LCST-type phase transition in organic solvents [45]. In the case of the CT interaction, poly(acrylate) tethering the pyrenyl group (PPMA) was employed as the polymer and pyromellitic diimide (PDI), naphthalene diimide, and mellitic triimide with long alkyl chains as the effector because electron-rich (donor) and electron-deficient (acceptor) aromatic rings usually form the CT interaction. PPMA was insoluble in various organic solvents, such as 1,2-dichloroethane, ethyl acetate, or toluene, due to the strong $\pi-\pi$ stacking of the pyrene group; however, it was dissolved in the presence of the effector. Furthermore, it exhibited an LCST-type phase transition. The CT interaction between PPMA and the effector first dissolved the PPMA, and the CT interaction was cleaved upon heating that caused the phase separation. Thanks to the CT interaction, the association constant (K_A) between the PPMA and effector and other thermodynamic parameters were determined by UV-Vis spectroscopy. T_c was readily controlled by changing the concentration of the effectors or K_A dependent on the effector, and the comparison of the CT complex and PPMA monomer unit suggested that around a 10% association degree (the ratio of associating pyrene units per all pyrene units) is sufficient to dissolve the PPMA regardless of the employed effectors. In this sense, the authors of this work prepared a gel by chemically cross-linking above the polymer [46]. The PPMA gel was not swollen in organic solvents, such as toluene, but mixing of the effectors drastically increased the swelling degree. Furthermore, the swollen gel with the aid of the effector showed a volume phase transition temperature, thus the gel collapsed at a higher temperature through disassembly of the CT complex.

By using hydrogen bonding, the LCST-type phase transition in organic solvents was also achieved [47]. PBPU, a poly(acrylate) tethering ureido group on the side chain, is not soluble in organic solvents other than alcohols and amides due to the strong hydrogen bonding of the ureido groups. However, after the addition of effectors, such as alcohols or amides with long alkyl chains, the PBPU was dissolved in the various organic solvents, and it showed an LCST-type phase transition on heating. Hydrogen bonding between PBPU and the effector first dissolved the PBPU, and the hydrogen bonding was cleaved on heating to cause the phase separation. The T_c was also tuned by changing the concentration of the effector. In comparison to the LCST-type phase transition in water, the effector can be regarded as a water molecule and is designated to interact with the polymer chain due to its relatively higher K_A compared with the water molecule. These results will be of importance both in understanding the principle and practical design of materials, thus they will allow more latitude for application of the LCST-type phase transition of polymer solutions.

5. MOLECULAR DESIGN OF POLYMERS EXHIBITING UPPER CRITICAL SOLUTION TEMPERATURE–TYPE PHASE TRANSITION IN WATER

As mentioned above, UCST-type phase transition of polymer solution is the opposite thermoresponsivity to LCST-type; thus, the polymer is insoluble at lower temperature than T_c and soluble at higher temperature. The mixing Gibbs energy (ΔG_{mix}) is positive at lower temperature; thus, the requirement for polymers exhibiting the UCST-phase transition is stronger polymer–polymer interaction at lower temperature. In the case of water, the water molecule can form hydrogen bonding with polymer chain; therefore, strong ionic interaction or multiple hydrogen bonding is the necessary condition for polymers exhibiting UCST-type phase transition in water (Fig. 13.4).

The typical examples of polymers exhibiting UCST-type phase transition in water are listed in Table 13.2. To utilize ionic interaction as an attractive force between polymer chains, zwitter ionic structure is one of the most reasonable solution because the opposite charge is necessary to produce the attractive force between the polymer chains. As the polyzwitterion, poly(*N*-(3-sulfopropyl)-*N*-methacryloxyethyl-*N*,*N*-dimethyl ammonium) (PSPE) is often

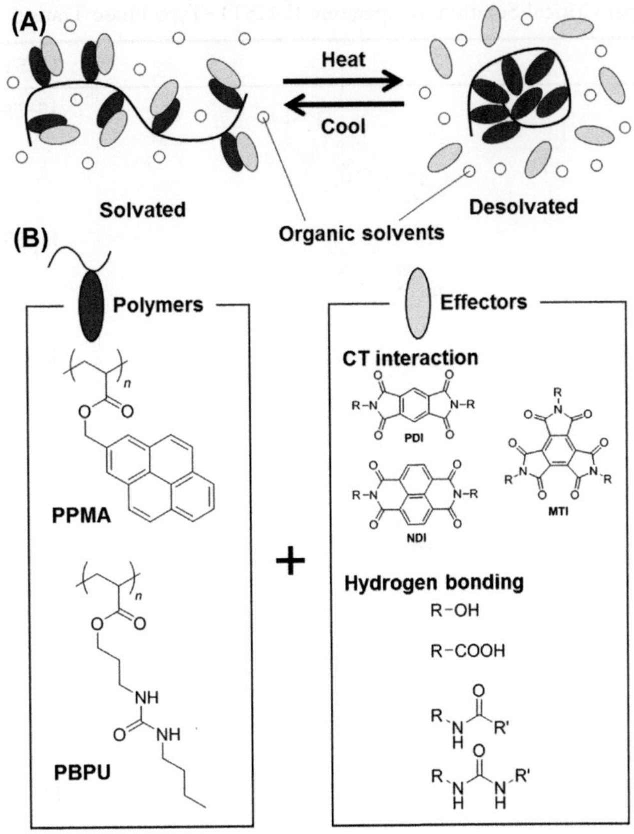

FIGURE 13.4 (A) Schematic image of polymers exhibiting lower critical solution temperature (LCST)—phase transition mediated by effectors in organic solvents. (B) Chemical structures of polymers and effectors for LCST-phase transition in organic solvents.

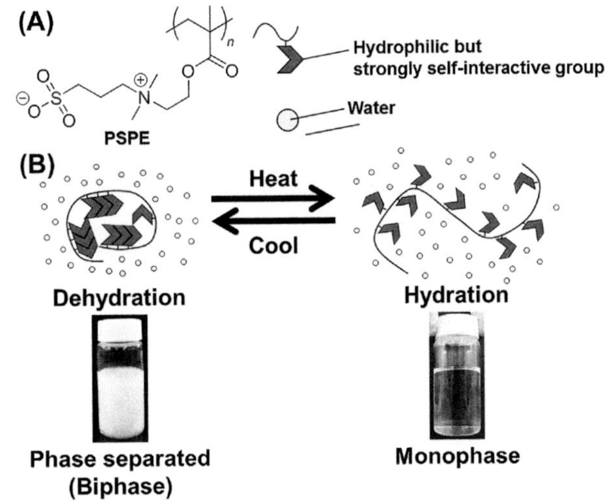

FIGURE 13.5 (A) The structural requirement for polymers exhibiting upper critical solution temperature (UCST)—type phase transition in water, and the chemical structure of PSPE as a typical example. (B) Schematic image and photograph of UCST-type phase transition of aqueous polymer solution.

employed as a polymer exhibiting UCST-type phase transition in water. Schulz et al. reported the UCST-type phase transition of PSPE in water [48]. The addition of salts predominantly resulted in lowering T_c, and "soft" anions and cations showed larger effect on the shift of T_c. For anions, the enhancement of solubility are $SCN^- > Cl^- > CO_3^{2-}$, $I^- > Br^- > Cl^-$, and for cations, $Ag^+ > NBu_4^+ >> K^+ = Na^+ > Li^+$, $Ba^{2+} = Ca^{2+} > Mg^{2+}$. Unperturbed dimension of PSPE was determined by Huglin et al. [49] Laschewski et al. reported the synthesis and UCST-type phase transition of

TABLE 13.2 Polymers With Upper Critical Solution Temperature (UCST)–Type Phase Transition in Water.

Polymers		UCST or T_c	References
Poly(*N*-(3-sulphopropyl)-*N*-methacryloxyethyl–*N*, *N*- dimethyl ammonium)	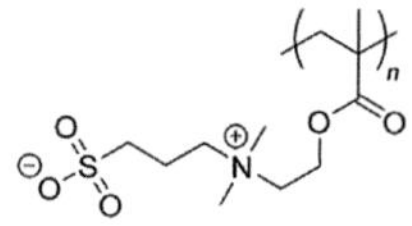	15–55	[48]
Poly(1-methyl-3-vinyloxyethyl-imidazolium tetrafluoroborate)	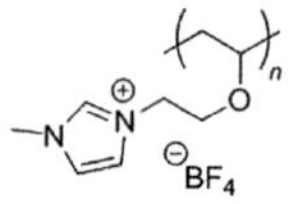	15–55	[53]
Polyethylene oxide-*b*-2-vinylpyridinium peroxydisulfate)	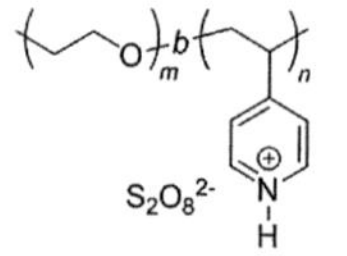	40–70	[54]
Poly(N-acryloyl glycinamide)	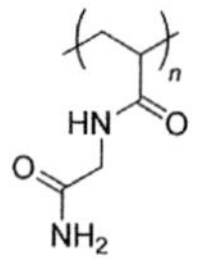	20–25	[55,56]
Poly(*N*-acryloyl glycinamide-*r*-*N*-acetyl acrylamide)	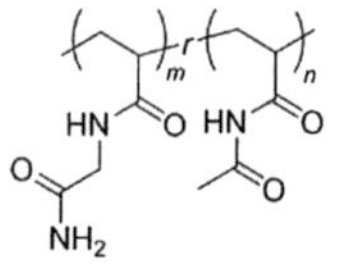	15–20	[55]
Poly(*N*-acryloylasparaginamide)	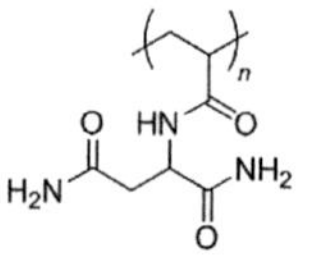	20–30	[57]
Poly(*N*-isopropyl methacrylamide)	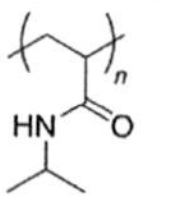	45	[63]
Poly(6-(acryloylmethyl)uracil)	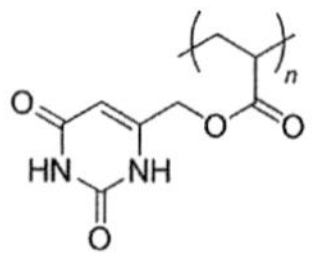	60	[65]
Ureidated poly(allylamine)	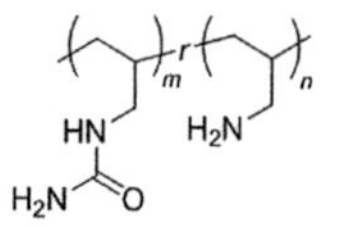	10–60	[66]
Poly(*N*-isopropyl acrylamide-*r*-acrylic acid)	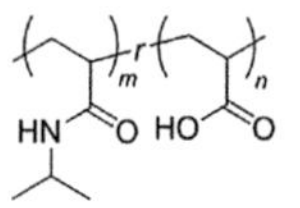	10–90	[67]
Poly(*N*,*N*-diethylacrylamide-*r*-acrylic acid)	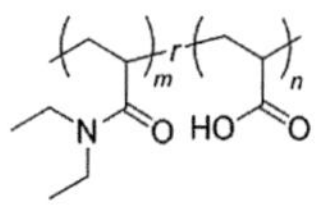	10–60	[68]
Poly(acrylonitrile-*r*-acrylamide)	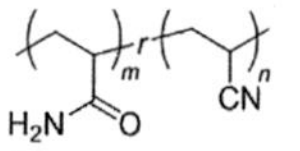	5–60	[70]

polyzwitterions similar with PSPE, and they also demonstrated the soap property of the polymer, which is influenced by the addition of salts, such as surface tension as well as X-ray powder diffraction pattern [50,51]. The effect of salt addition was also investigated by Bendejacq et al. by using PSPE and a related polymer, and they found the threshold for the amount of salt [52]. At larger salt concentration than the threshold, the addition of salts turned to enhance the solubility of the polymer, thus lowering T_c, due to the screening of the attractive force. On the contrary, at smaller concentration than the threshold, partial hydrolysis of the zwitter ionic moiety resulted in a specific interaction between the polymer chain and salts, revealed by a zeta potential measurement.

Few polymers with one kind of charge have been reported to exhibit UCST-type phase transition. Aoshima et al. reported poly(vinyl ether) derivative tethering alkylimidazolium tetrafluoroborate on the side chain exhibiting UCST-type phase transition in water [53]. Chen et al. reported that a block copolymer poly(ethylene oxide-*b*-2-vinylpyridinium peroxydisulfate) exhibited UCST-type phase transition in water with pH 1.3 [54]. By reducing the peroxydisulfate, the interaction between the polymer chains is weakened, and the polymer turned to be soluble at room temperature.

Multiple hydrogen bonding is another solution to reinforce the polymer–polymer interactions in water. From this point of view, the derivatives of amino acids can be a promising candidate to achieve polymers with UCST-type phase transition. In 2012, Agarwal et al. reported that poly(*N*-acryloyl glycinamide) (PNAGA) and poly(*N*-acryloyl glycinamide-*r*-*N*-acetyl acrylamide) exhibited UCST-type phase transition [55]. The synthesis of PNAGA was originally reported by Haas et al. [56], and they also found the gelation behavior of PNAGA. Agarwal et al. also reported that the complete removal of ionic species (acrylic acid [AA] and carboxymethyl acrylate in this case) was responsible for the finding of the UCST-type phase transition in PNAGA [57]. Copolymerization with hydrophobic monomers such as butyl acrylate (BA) and styrene led to increase of T_c. The same group reported the controlled radical polymerization, reversible addition–fragmentation transfer (RAFT) polymerization of NAGA, and the end groups exerted a strong effect on T_c [58]. The polymer obtained from nonionic- and ionic-type RAFT agent resulted in the clearly opposite response to ionic strength of the solution. The lengthening of polymer chain caused the lowering of T_c. Lutz et al. demonstrated UCST-type gelation behavior of PNAGA [59]. The same group reported that poly(*N*-acryloylasparaginamide) exhibited UCST-type phase transition in pure water and physiological conditions [60]. In this case, the lengthening of polymer chain caused the elevation of T_c.

Apart from amino acids, amides, ureides, and carboxylic acids are also promising candidates to achieve polymers with UCST-type phase transition. Poly(methacrylamides) are well-known polymers exhibiting UCST-type phase transition in water [61–64]. For example, Netopilík et al. reported the UCST-type phase transition of poly(*N*-isopropyl methacrylamide) in water [63]. By using an uracil derivative, Aoki et al. reported UCST-type phase transition of poly(6-(acryloylmethyl)uracil) in water [65]. Maruyama et al. synthesized partial side chain ureidation for poly(allylamine) and poly(L-ornithine) and reported UCST-type phase transition in water and various salts aqueous solution [66]. By using AA, some copolymers can show UCST-type phase transition [67,68]. Bokias et al. reported UCST-type phase transition of poly(NIPAM-*r*-AA) in the case of 10% NIPAM ratio [67]. Fang et al. reported UCST-type phase transition in water of poly(*N*,*N*-diethylacrylamide (DEA)-*r*-AA) in the case of 6% and 11% DEA ratio [68]. In the both cases, increase of DEA feed resulted in the change of thermoresponsivity to LCST-type phase transition in water. Partially substituted PVA is also useful to obtain thermoresponsive polymers. Wang et al. reported a graft polymer, poly(*p*-dioxanone) (PPDO)-grafted PVA exhibiting UCST-type phase transition in water in the case of longer PPDO length [69]. With shorter PPDO chain, the graft polymer exhibited LCST-type phase transition in water. As other unit structure, Agarwal et al. reported UCST-type phase transition of poly(acrylonitrile-*r*-acrylamide) in water, and increase of acrylonitrile content resulted in elevation of T_c[70].

Combination of two monomers that are, respectively, responsible for LCST- and UCST-type phase transition leads to occurrence of the both thermoresponsivity in one scan of heating or cooling. The typical examples of polymers exhibiting LCST- and UCST-type phase transitions in water are listed in Table 13.3. For examples, Laschewsky et al. reported that poly(NIPAM-*b*-SPE) exhibited insoluble–soluble–insoluble phase transition on heating and cooling, thus UCST-type phase transition at lower T_c and LCST-type at higher T_c in water [71]. Tenhu et al. reported the salt effect on the thermoresponsivity of poly(NIPAM-*b*-SPE), and the increase of salt concentration decreased the T_c for the both LCST- and UCST-type phase transition, thus the salt addition showed opposite effects on the two thermoresponsivity [72]. Chang et al. reported the synthesis of a block copolymer consisting of NIPAM and sulfobetaine methacrylate (SBMA), poly(NIPAM-*b*-SBMA), exhibiting UCST-type phase transition at lower T_c and LCST-type at higher T_c in water [73]. They additionally fabricated a surface plasmon resonance sensor covered with

TABLE 13.3 Polymers With Lower Critical Solution Temperature (LCST)– and Upper Critical Solution Temperature (UCST)–Type Phase Transitions in Water.

Polymers		UCST or T_c	LCST or T_c	References
P(NIPAM-*b*-SPE)	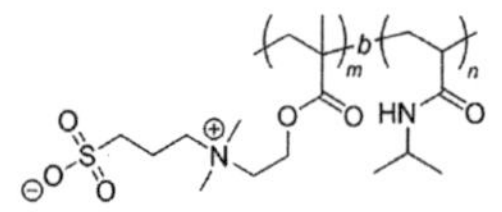	5–15	30–35	[72,73]
P(DEA-*b*-SPE)	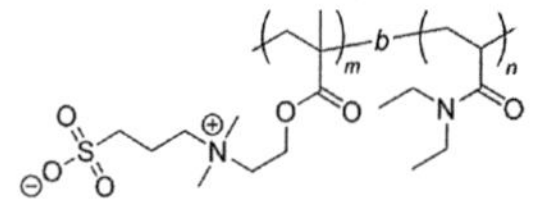	10–15	45–55	[74]
Partially ionized P((OEGMA-*r*-MEO$_2$MA)-*b*-DMAPMA)	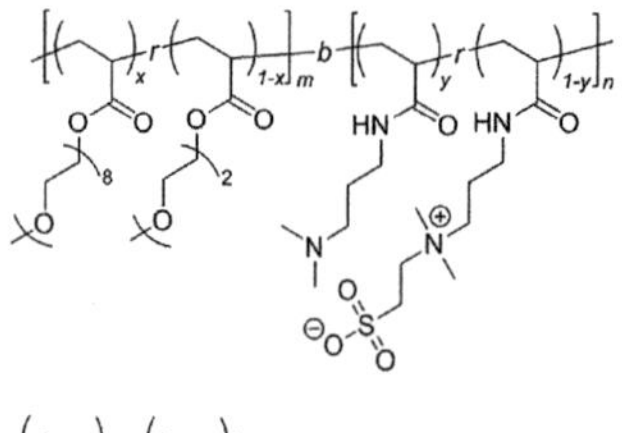	20–30	50–60	[75]
P(MOVE-*b*-[EtIm][BF$_4$])	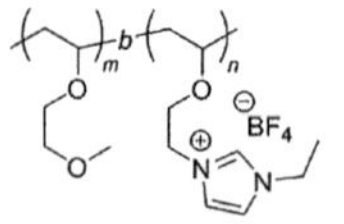	5–15	70–80	[76]
P(HEMA-*b*-MAPTAC)	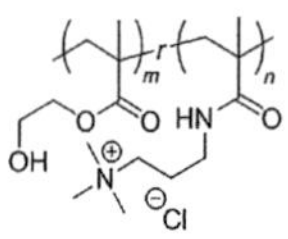	70–80	20–30	[77]
P(NVA-*r*-AA)	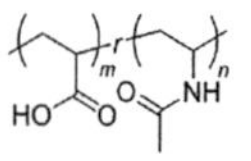	80–95	30–40	[78]
P(A-Pro-OMe-*b*-A-Hyp-OH)	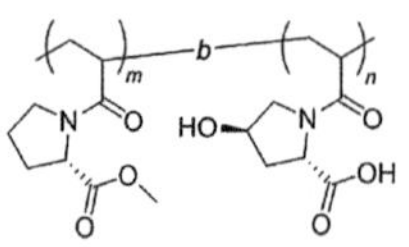	40–50	15–25	[79]

poly(NIPAM-*b*-SBMA) and found extremely low protein absorption and high anticoagulant activity of the polymer. Maeda et al. reported that poly(DEA-b-SPE) exhibited UCST-type phase transition at lower T_c and LCST-type at higher T_c in water [74]. You et al. reported the synthesis of a block copolymer consisting of 2-(2-methoxyethoxy)ethyl methacrylate (MEO2MA), oligo(ethylene glycol)methacrylate (OEGMA), 3-(*N*,*N*-dimethylamino)propyl methacrylate (DMAPMA), and poly((OEGMA-*r*-MEO2MA)-*b*-DMAPMA), and partial ionization of DMAPMA moiety with 1,3-propanesultone resulted in the formation of zwitter ionic moiety, leading to UCST-type phase transition at lower T_c, and LCST-type at higher T_c in water [75]. The T_c of UCST-type phase transition was tuned by the ionization ratio, and the T_c of LCST-type phase transition was tuned by the ratio of OEGMA moiety. Aoshima and Winnik et al. reported that a block copolymer derived from 2-methoxyethyl vinyl ether and 1-(2-vinyloxyethyl)-3-ethylimidasolium tetrafluoroborate exhibited UCST-type phase transition at lower T_c and LCST-type at higher T_c in water [76].

For the combination with the opposite thermoresponsivity, i.e., soluble–insoluble–soluble phase transition on heating and cooling. Stöver et al. reported that random copolymers derived from mainly 2-hydroxyethyl methacrylate (HEMA) and small amount of ionic monomers such as methacrylaminopropyl trimethylammonium chloride (MAPTAC) or methacryloyloxyethyl trimethylammonium chloride (MOETAC) exhibited LCST-type phase transition at lower T_c and UCST-type at higher T_c in water [77]. The addition of salt always decreased the solubility, thus the T_c for LCST-type phase transition decreased and the T_c for UCST-type phase transition increased on the addition of salt. Mori et al. reported that a random copolymer derived from *N*-vinyl acetamide (NVA) and AA, poly(NVA-*r*-AA), exhibited LCST-type phase transition at lower T_c and UCST-type at higher T_c in water [78]. The salt addition decreased the solubility of the polymer, but the addition of urea increased the solubility. Mori et al. reported the synthesis of a block copolymer derived from *N*-acryloyl-L-proline methyl ester (A-Pro-OMe) and *N*-acryloyl-4-*trans*-hydroxyl-L-proline (A-Hyp-OH), poly(A-Pro-OMe-*b*-A-Hyp-OH), exhibiting LCST-type phase transition at lower T_c and UCST-type at higher T_c in water [79]. Interestingly, the random copolymer analogue was always soluble at any temperature, indicating the importance of the monomer sequence for the thermoresponsivity.

PDMAEMA is a well-known polymer exhibiting LCST-type phase transition in water, but Müller et al. reported that the addition of hexacyanocobaltate(III) to PDMAEMA aqueous solution newly induced UCST-type phase transition, thus resulting in LCST-type phase transition at lower T_c and UCST-type at higher T_c in water [80]. Additionally, photoirradiation induces the valence switch of cobalt, causing the decrease of the T_c for the both UCST- and LCST-type phase transition. Hoogenboom et al. reported that a block copolymer, poly(OEGMA-*b*-DMAEMA), exhibited LCST-type phase transition at lower T_c and UCST-type at higher T_c in water in the presence of hexacyanocobaltate(III) or NaCl[81].

At extremely high temperature, PEO exhibits UCST-type phase transition in water, while it exhibits LCST-type phase transition at lower temperature [17,82–85]. The phenomenon was firstly reported in 1957 [17,82], and it has been thoroughly investigated since then through thermooptical analyzing system [83], theoretical modeling [84], and the end group manipulation [85]. A similar phase diagram was obtained by partial butylation of PVA reported by Imai et al.[86].

6. MOLECULAR DESIGN OF POLYMERS EXHIBITING UPPER CRITICAL SOLUTION TEMPERATURE–TYPE PHASE TRANSITION IN ORGANIC SOLVENTS

In nonaqueous condition, the interaction between solvent molecules is relatively small, thus a mild polymer–polymer interaction is enough to exhibit UCST-type phase transition, differently from the aqueous condition. Therefore, van der Waals interaction or π–π interaction can be employed as the polymer–polymer interaction for UCST-type phase transition. Aoshima et al. reported that polymers derived from various vinyl ether derivatives having alkyl, cholesteryl, or biphenyl pendants exhibited UCST-type phase transition in organic solvents [87]. Amongst the synthesized polymers, poly(dodecyl vinyl ether) exhibited UCST-type phase transition in various organic solvents such as hexane, toluene, chloroform, THF, dichloromethane, and diethyl ether, due to the crystalline nature derived from the long alkyl chain, while shorter alkyl chain did not show any crystallization peak in DSC measurement. The same group reported the synthesis of polymers derived from fluorine-containing vinyl ethers, and random copolymerization with isobutyl vinyl ether resulted in polymers exhibiting UCST-type phase transition in organic solvents, in which increasing the ratio of fluorine-containing monomers shifted the responding organic solvents to less polar [88]. Jäkle et al. reported the synthesis of a polymer derived from triisopropylphenylvinylbiphenyl-borinic acid, and addition of trace water in polymer solution in DMSO resulted in UCST-type phase transition [89]. In ionic liquid such as ethylmethyl imidazolium bis(trifluoromethylsulfonyl)imide ([C2mim][TFSI]), a block copolymer derived from benzyl methacrylate (BzMA), NIPAM, and AAM, P(BzMA-*b*-(NIPAM-*r*-AAM), exhibited UCST- and LCST-type phase transition, thus insoluble–soluble–insoluble phase transition on heating and cooling [90]. By using PBPU and strongly interacting effectors such as carboxylic acids, halogen anions, ureides, and multiple alcohol, the authors of this work reported the UCST-type phase transition in organic solvents [47]. Moreover, mixing with effectors for LCST- and UCST-type phase transitions, insoluble–soluble–insoluble phase transition was achieved in a wide range of temperature.

7. THE APPLICATION OF THERMORESPONSIVE POLYMERS

7.1 Drug Delivery System

Synthetic polymers have attracted much attention in application for drug delivery as a therapeutic agent. Because of their high stability with long circulation time and improved tissue targeting, synthetic polymers are employed as various drug delivery systems (DDSs), and thermoresponsive polymers have been expected as one of the best platforms of DDS. For examples, Kim et al. reported the preparation of DDS consisting of poly(NIPAM-*r*-BMA-*r*-AA) containing human calcitonin (hCT) via precipitation of the polymers at the interface of cold aqueous solution droplets and warm oil phase and subsequent drying process [91]. The contained hCT was quantitatively released in 60 min at 37°C, pH 7.4. Milašinović et al. reported the preparation of hydrogel of PNIPAM cross-linked by itaconic acid, which contains a model protein, lipase [92]. More than 80% of lipase was released after 1 day at 37°C.

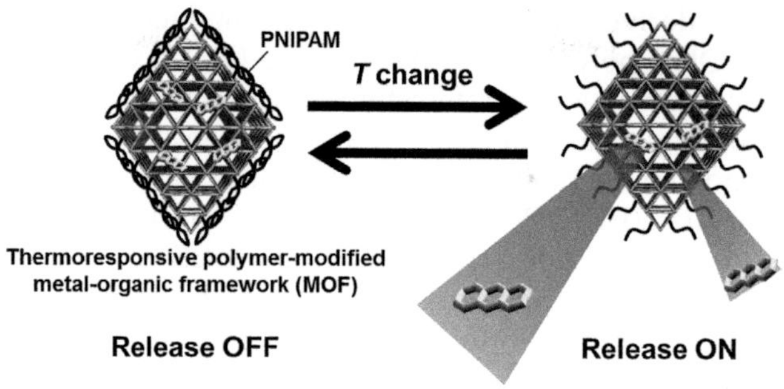

FIGURE 13.6 Thermoresponsive polymer (PNIPAM)–modified metal–organic framework (MOF). *From S. Nagata, K. Kokado, K. Sada, Chemical Communications 51 (2015) 8614–8617.*

Cheng and Lin et al. reported thermally controlled drug release of poly(NIPAM-*r*-MAA) coating microsphere of $NaYF_4{:}Yb^{3+}/Er^{3+}$ [93]. The included doxorubicin was readily released in 5 h under 37°C, pH 5.0, and the presence of $NaYF_4{:}Yb^{3+}/Er^{3+}$ allows fluorescence microscopic observation. The authors of this work reported the coverage of metal–organic framework with PNIPAM via activated ester amidation [94]. The included small molecules such as resorufine, caffeine, or ibuprofen were readily released at lower temperature, and ON–OFF switching of the release was achieved by changing the temperature (Fig. 13.6). Injectable hydrogel is also a useful material for DDS. Stayton et al. reported the synthesis of copolymer of NIPAM and propylacrylic acid (PAA), poly(NIPAM-*r*-PAA), and the included vascular endothelial growth factor was released in 5 days under 37°C pH 7.0 [95]. Wagner et al. reported preparation of injectable hydrogel derived from NIPAM, HEMA, biodegradable polylactide methacrylate (MAPLA), and methacryloxy *N*-hydroxysuccinimide (NHSMA) [96]. The attached protein, bovine serum albumin (BSA), was sequentially released in dozens of days.

7.2 Tissue Engineering

Tissue engineering is well known to be an integrated research field of engineering and the life sciences to fabricate biological substitutes to repair or improve tissue function. Tissue engineering targets the replacement of biological damaged tissue or generation of repaired organs for a wide range of medical conditions. For this purpose, thermoresponsive polymers have been generally employed in tissue engineering mainly for the cell culture and injectable gels for the scaffold. In cell culture, thermoresponsive property of the polymers is used to switch the cell attachment and detachment from the surface of cell culture substratum (Fig. 13.7). In 1990, Takezawa et al. firstly reported the culture of human dermal fibroblasts by conjugating it with collagen mixed with PNIPAM [97]. At 37°C, above T_c of PNIPAM, monolayered fibroblasts was cultured on the substratum containing the PNIPAM, and it was completely detached from the substratum by simply cooling the temperature to less than T_c, without the use of any conventional detaching agents such as trypsin or EDTA. Okano et al. reported the cell culture growth of bovine endothelial cells and rat hepatocytes on plasma irradiation induced PNIPAM-grafted polystyrene dish [98]. As the result, cell sheet was obtained in each case, without concomitant of PNIPAM. The same group reported culture of bovine carotid artery endothelial cell on PNIPAM-grafted polystyrene dish, and the thinner layer of PNIPAM was found to be effective in the formation of cell sheet [99]. Alexander et al. reported the use of grafted copolymer derived from NIPAM

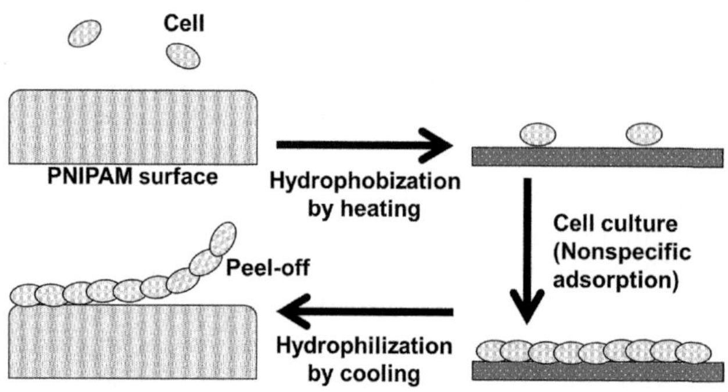

FIGURE 13.7 Schematic image of cell culture on thermoresponsive polymer (PNIPAM). The cells are attached via nonspecific adsorption after heating to body temperature. After cultured, the substratum is cooled to peel off the sheet-shaped cell due to hydrophilization of the substratum.

and *N-t*-butylacrylamide (NtBAM), poly(NIPAM-*r*-NtBAM), on a polystyrene dish and examined attachment of Gram-negative motile bacterium (*Salmonella typhimurium*) and Gram-positive nonmotile species (*Bacillus cereus*) as well as proteins such as BSA and cytochrome c [100]. Canavan reported the fabrication of PNIPAM surface with sol–gel reaction of tetraethylorthosilicate, and the cell such as bovine aortic endothelial cell was cultured on the surface [101].

7.3 Thermoresponsive Catalyst

The use of soluble polymers to recover catalysts and ligands is one of the most effective ways to achieve a sustainable society, as catalysts often contain rare, expensive, and environment-poisoning chemical species such as transition metals. From this viewpoint, catalysts linked with soluble long polymer chain have been paid much attention, and the employment of thermoresponsive polymer has been also explored to add smartness to catalysts. Bergbreiter reported the synthesis of telechelic polymer of PEO-*b*-PPO-*b*-PEO or having phosphines on the chain end coordinating to Rh(I) ion and examined the catalytic hydrogenation of allyl alcohol in water [102] (Type 1 in Fig. 13.8). PEO-*b*-PPO-*b*-PEO intrinsically shows LCST-type phase transition with T_c of around 15°C in water. As the result, the hydrogen uptake was increased at 0°C, but the reaction was completely suppressed to less than 10^{-2} at 40°C. ON–OFF switching was achieved by intermittently changing the temperature [103]. By using PNIPAM, the solubility control of the substrate for hydrogenation of nitro group was reported by the same group [104,105]. Use of NIPAM allows the separation of the catalyst as a powder, which is an important property for recyclability of catalysts [106]. For other examples, Francis reported the preparation of PNIPAM-tethered cellulase and its catalytic activity for cellulose degradation and the recyclability [107]. Datta et al. reported the preparation of PNIPAM-tethered β-glycosidase and its hydrolysis activity for cellobiose degradation and the recyclability [108].

Phase transfer catalysts are another way to regulate the catalysis, and the switching of hydrophilicity of thermoresponsive polymers is suitable means for phase transfer (Fig. 13.8, Type 2). Jin et al. synthesized a series of PEO-substituted triphenylphosphines, $Ph_{3-m}P[Ph\text{-}p\text{-}(OCH_2CH_2)_nOH]_m$ (PEO-TPP), and examined phase-transfer catalysis for rhodium-catalyzed two-phase hydroformylation of olefins [109]. As the result, at higher temperature than T_c, the catalysts feasibly transferred to the organic phase, and the reaction was effectively promoted. The same group also reported the synthesis of octyl-PEO-phenylene-phosphite and rhodium complex and carried out hydroformylation of styrene, resulting in high styrene conversion and high selectivity of branched structure [110]. Stark et al. reported the preparation of C/Co@PNIPAM nanoparticle, followed by coordination with Pd(0) via diphenyl phosphine moiety in the polymer chain [111]. The catalyst particle could show phase transfer ability

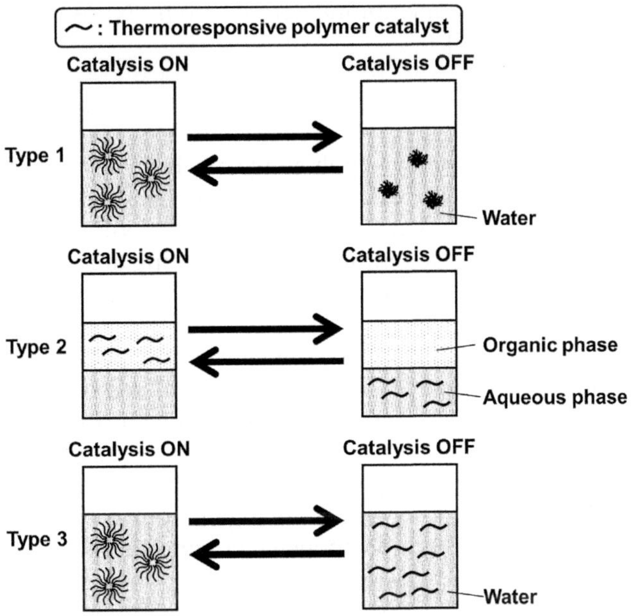

FIGURE 13.8 Schematic image of thermoresponsive catalysis systems using thermoresponsive polymers. Type 1: The reaction is catalyzed when the catalyst forms a micelle or is dissolved, and it is shut off when the catalyst is collapsed. Type 2: The reaction is catalyzed when the catalyst is dissolved in the organic phase, and it is shut off when the catalyst is dissolved in the aqueous phase. Type 3: The reaction is catalyzed when the catalyst forms a micelle, and it is shut off when the catalyst is dissolved.

across the T_c as well as magnetic property enabling the collection by an external magnet and effectively catalyzed Suzuki-Miyaura coupling with good recyclability.

Above T_c, formation of micelle provides a suitable nanoreactor for asymmetric reaction (Fig. 13.8, Type 3). Suzuki et al. reported the synthesis of copolymer consisting of NIPAM, PEG monomethyl ether acrylate, and *O*-acryloyl-*trans*-4-hydroxy-L-proline (ProlA) for catalysis in asymmetric aldol reaction in water [112]. As the result, high temperature than T_c provided high yield, high *anti/syn* ratio, and high *ee*, probably due to the formation of pseudo-hydrophobic environment effective for controlling the stereoselectivity in proline-catalyzed aldol reactions. Similar tendency was reported by O'Reilly and Monteiro et al., who reported the synthesis of a copolymer derived from *N,N*-dimethyl acrylamide, NIPAM, BA, and ProlA, PDMA-*b*-P(NIPAM-*r*-BA-*r*-ProlA), and applied it to the same asymmetric aldol reaction in water [113]. The elevation of temperature across T_c formed micelle-like nanoreactor, which efficiently promoted the reaction with high *anti/syn* ratio and high *ee*. Zhang et al. reported the preparation of PNIPAM-stabilized palladium nanoparticle catalyst for the hydrodechlorination of 4-chlorophenol and the reduction of 4-nitrophenol (4-NP) in water, and good recyclability was confirmed[114].

7.4 Thermoresponsive Separation

The thermoresponsivity of these polymers can also switch their affinity with other molecules, which is applicable to separation techniques such as chromatography and membranes. For examples, Kanazawa et al. reported that the PNIPAM-modified silica showed temperature-dependent separation profile for mixed steroid compounds when it was utilized as the stationary phase of column chromatography [115]. As the temperature elevated, the hydrophobicity gradually increased, and the difference in retention time became larger enough to separate the peaks. Okano et al. also reported the higher stability of the grafted polymer brush on silica surface than hydrogel [116]. Aoyagi et al. reported the utilization of inductive heating via altering magnetic field to control the separation profile. These features are also reviewed by Okano et al. [117,118] and Kanazawa et al.[119].

Not only column chromatography but also membrane can also serve a substance separation system based on filtration. Kim et al. reported the thermoresponsive hydrogel membrane derived from PNIPAM copolymer exhibiting various permeation profiles depending on the applied temperature and molecular size [120]. Nonaka et al. reported the PVA-*graft*-PNIPAM membrane showing thermoresponsive permeation of glucose and insulin [121]. Aminabhavi et al. reported PNIPAM-sodium alginate semiinterpenetrating membrane undergoing water and ethanol separation [122]. Song et al. reported PMMA/PNIPAM electrospun nanofibrous mats separating BSA depending on the applied temperature [123]. Tripathi et al. reported the fabrication of hollowed membrane derived from PNIPAM, showing thermosresponsive permeation behavior [124]. Wang et al. reported the fabrication of robust thermoresponsive polymer composite membrane derived from PNIPAM and elastic polyurethane showing thermoresponsive oil—water separation [125].

7.5 Sensory Application

Several sensory applications of thermoresponsive polymers have been also reported. For examples, Maeda et al. reported single-stranded DNA-grafted PNIPAM exhibiting thermoresponsive formation of colloidal nanoparticles, also depending on $MgCl_2$ concentration [126]. Wanless et al. reported PNIPAM brush surface showing thermoresponsive change in thickness or contact angle depending on surrounding anion [127]. Advincula et al. reported the fabrication of freestanding polymer membrane consisting of polysaccharide and grafted PNIPAM exhibiting surfactant-induced bending behavior [128].

By tethering solvatochromic dyes on thermoresponsive polymers, the polymer can become a sensor for temperature and other surrounding conditions such as pH. Hoogenboom et al. reported the preparation of OEGMA-based polymer tethering 4-amino-4′-nitro-azobenzene, and the polymer showed color change responding to temperature and pH [129]. Zhang et al. reported the preparation of comb-like poly(methacrylate)s tethering dendritic OEG side groups and 4-amino-4′-nitro-azobenzene, and the polymers showed color change responding to temperature and pH [130,131]. Kim et al. reported the use of fluorogenic π-conjugated polymer, poly(diacetylene), for a temperature indicator in a microfluidic device [132]. The thermoresponsive polymers tethering solvatochromic dyes for sensory applications are summarized by Schubert and Hoogenboom [133].

7.6 Oil Recovery

For the extraction of residual oil resources from depleted oil reservoirs, oil recovery has been recognized as a requisite technique. From a viewpoint of sustainability, it can enlarge the oil output from an already existing oil well, and it can diminish the need for additional drilling or another new oil well. Several researches have demonstrated that the thermoresponsive polymer can be one of the promising candidates for the oil recovery materials. For examples, Feng et al. reported thermos-thickening and salt-thickening behavior of a graft copolymer, in which the side chain consists of acrylamide and diacetone acrylamide [134]. Similar behavior was recently reported by Guo et al., in which a graft copolymer in which the side chain consists of *N,N*-dimethyl acrylamide and diacetone acrylamide was employed [135]. These examples exhibit the potential application of thermosresponsive polymers for oil recovery.

As another example, extraction of oil from oil sand would also contribute to reduction of natural resource consumption. Duhamel et al. reported the extraction of oil from oil sand by using thermoresponsive polymeric surfactant, which consists of PEO-*b*-PMEO2MA [136]. At higher temperature, the surfactant extracts oil from oil sand, which was released at lower temperature.

8. OTHER STIMULI

Other stimuli than heat can be applied to trigger a change in polymer conformation from coil to globule and vice versa, such as pH [137,138], light [139,140], redox [141,142], and enzymatic reaction [143,144]. As the change triggered by these stimuli involves the change in chemical potential of the system, for examples derived from the change in chemical composition of the polymer chain, the coil-globule change is not a "phase transition." Nonetheless, the control of polymer conformation by applying these stimuli will be a promising tool to obtain various smart and sustainable materials.

9. POLYMER SYNTHESIS

As reviewed in this chapter, thermoresponsive polymers with various molecular structures have been reported. Although most of thermoresponsive polymers are derived from fossil fuels and following the petrochemical industry and organic synthesis, from the viewpoint of sustainability, the supply of their monomers via biosynthetic process will be of importance. Production of polymers via biosynthetic process is summarized in several reviews [145–147]. The synthetic method of polymers from renewable starting materials has also become one of the attractive topics in polymer chemistry [148]. In laboratory scale, thermoresponsive polymers such as PNIPAM are usually synthesized by radical polymerization, and the improvement of polymerization has been also explored recently. Atom-transfer radical polymerization (ATRP) is the most frequently used controlled radical polymerization using metal complex catalysis, and several attempts to make them more eco-friendly have been reported. For examples, Matyjaszewski et al. invented activators regenerated by electron transfer (ARGET) ATRP, which enabled reduction of metal amount to ppm order and use of eco-friendly solvents such as water [149,150]. Hawker et al. reported metal-free ATRP mediated by light irradiation [151].

The concept of recycle and life cycle assessment of polymer materials are also mentioned in several reviews [152–154]. Additionally, the synthetic process of monomers for thermoresponsive polymer has been improved from the viewpoint of green chemistry. In general synthesis, NIPAM is synthesized from acrylonitrile and isopropanol via Ritter reaction [155] in the presence of recoverable zeolite [156]. The first ingredient, acrylonitrile, is one of the most important chemicals for chemical industry, which is produced more than 6 Mt/y, and is synthesized from propylene treated in the presence of ammonia and oxygen, so-called Sohio process or ammoxidation, obtaining hydrogen cyanide and acetonitrile as by-product. The production of acrylonitrile from biomass-related compounds such as lactic acid, glycerol, glutamic acid, or sugars has been recently reported [157–159]. The second ingredient, isopropanol, is also synthesized from propylene treated with strong acid, followed by hydrolysis. Hanai et al. reported the production of isopropanol from sugars by metabolic process of bacteria such as *Escherichia coli* [160,161].

At the present moment, green chemistry concepts are slightly utilized in the researches of thermoresponsive polymers; however, polymers derived from biosynthetic and recycled processes will be used in the near future.

10. CONCLUSION AND PERSPECTIVES

The LCST- and UCST-type phase transition of a polymer solution can induce aggregation of the solid on heating and cooling. These phenomena not only enable changes in the turbidity of the polymer solution but also cause a large volumetric and affinity change in response to external stimuli such as a temperature change. Therefore, applying such polymers on other materials easily produces a switching property to other materials. Indeed, DDSs, tissue engineering, catalysts, sensing devices, and chromatography, having a thermoresponsivity, have already been explored. By utilizing the fascinating property of thermoresponsive polymers, intelligent and smart ability can be imparted to various objects such as adsorbents, fibers, plastics, films, coatings, and composites to provide next-generation materials. Furthermore, based on the knowledge of thermoresponsive polymers, polymers responding to other stimuli, such as pH, light, redox, or enzyme, have also been investigated. The development of these stimuli responsive polymers will lead to promising materials to achieve smart and sustainable products.

References

[1] D. Roy, J.N. Cambre, B.S. Sumerlin, Future perspectives and recent advances in stimuli-responsive materials. Progress in Polymer Science 35 (2010) 278–301.

[2] C. Weber, R. Hoogenboom, S. Schubert, Ulrich, Temperature responsive bio-compatible polymers based on poly(ethylene oxide) and poly(2-oxazoline)s. Progress in Polymer Science 37 (2012) 686–714.

[3] D. Langbridge, A.S.C. Lawrence, R. Stenson, Solubility in soap solutions. VI. The effect of soap upon upper and lower consolute temperatures in the nicotine-water and triethylamine-water systems. Journal of Colloid Science 11 (1956) 585–592.

[4] Y. Kohno, S. Saita, Y. Men, J. Yuan, H. Ohno, Thermoresponsive polyelectrolytes derived from ionic liquids. Polymer Chemistry 6 (2015) 2163–2178.

[5] P.J. Flory, Thermodynamics of High Polymer Solutions. The Journal of Chemical Physics 10 (1942) 51–61.

[6] M.L. Huggins, Some Properties of Solutions of Long-chain Compounds. The Journal of Physical Chemistry 46 (1942) 151–158.

[7] A.R. Shultz, P.J. Flory, Phase Equilibria in Polymer—Solvent Systems. Journal of the American Chemical Society 74 (1952) 4760–4767.

[8] A. Matsuyama, F. Tanaka, Theory of solvation-induced reentrant phase separation in polymer solutions, Physical Review Letters 65 (1990) 341–344.

[9] S. Bekiranov, R. Bruinsma, P. Pincus, Solution behavior of polyethylene oxide in water as a function of temperature and pressure Physical Review E 55 (1997) 577–585.

[10] Y. Okada, F. Tanaka, Cooperative Hydration, Chain Collapse, and Flat LCST Behavior in Aqueous Poly(N-isopropylacrylamide) Solutions. Macromolecules 38 (2005) 4465–4471.

[11] J.S. Scarpa, D.D. Mueller, I.M. Klotz, Slow hydrogen-deuterium exchange in a non-.alpha.-helical polyamide. Journal of the American Chemical Society 89 (1967) 6024–6030.

[12] E.S. Gil, S.M. Hudson, Stimuli-reponsive polymers and their bioconjugates. Progress in Polymer Science 29 (2004) 1173–1222.

[13] D. Ito, K. Kubota, Solution Properties and Thermal Behavior of Poly(N-n-propylacrylamide) in Water, Macromolecules 30 (1997) 7828–7834.

[14] Y. Maeda, T. Nakamura, I. Ikeda, Changes in the Hydration States of Poly(N-n-propylmethacrylamide) and Poly(N-isopropylmethacrylamide) during Their Phase Transitions in Water Observed by FTIR Spectroscopy. Macromolecules 34 (2001) 8246–8251.

[15] I. Idziak, D. Avoce, D. Lessard, D. Gravel, X.X. Zhu, Thermosensitivity of Aqueous Solutions of Poly(N,N-diethylacrylamide). Macromolecules 32 (1999) 1260–1263.

[16] K. Suwa, K. Morishita, A. Kishida, M. Akashi, Synthesis and functionalities of poly(N-vinylalkylamide). V. Control of a lower critical solution temperature of poly(N-vinylalkylamide). Journal of Polymer Science Part A: Polymer Chemistry 35 (1997) 3087–3094.

[17] S. Saeki, N. Kuwahara, M. Nakata, M. Kaneko, Upper and lower critical solution temperatures in poly (ethylene glycol) solutions, Polymer 17 (1976) 685–689.

[18] P. Firman, M. Kahlweit, Phase behavior of the ternary system H2O-oil-polypropyleneglycol (PPG). Colloid and Polymer Science 264 (1986) 936–942.

[19] S. Han, M. Hagiwara, T. Ishizone, Synthesis of Thermally Sensitive Water-Soluble Polymethacrylates by Living Anionic Polymerizations of Oligo(ethylene glycol) Methyl Ether Methacrylates, Macromolecules 36 (2003) 8312–8319.

[20] B. Zhao, D. Li, F. Hua, D.R. Green, Synthesis of Thermosensitive Water-Soluble Polystyrenics with Pendant Methoxyoligo(ethylene glycol) Groups by Nitroxide-Mediated Radical Polymerization, Macromolecules 38 (2005) 9509–9517.

[21] H.R. Allcock, G.K. Dudley, Lower Critical Solubility Temperature Study of Alkyl Ether Based Polyphosphazenes, Macromolecules 29 (1996) 1313–1319.

[22] T. Bauer, C.J. Slugovc, The thermo responsive behavior of glycol functionalized ring opening metathesis polymers, Polymer Science Part A: Polymer Chemistry 48 (2010) 2098–2108.

[23] X. Jiang, M.R. Smith, G.L. Baker, Water-Soluble Thermoresponsive Polylactides, Macromolecules 41 (2008) 318–324.

[24] Y. Maeda, T. Nakamura, I. Ikeda, Hydration and Phase Behavior of Poly(N-vinylcaprolactam) and Poly(N-vinylpyrrolidone) in Water, Macromolecules 32 (2002) 217–222.

[25] A.C.W. Lau, C. Wu, Thermally Sensitive and Biocompatible Poly(N-vinylcaprolactam): Synthesis and Characterization of High Molar Mass Linear Chains, Macromolecules 32 (1999) 581–584.

[26] D. Christova, R. Velichkova, W. Loos, E.J. Goethals, F. Du Prez, New thermo-responsive polymer materials based on poly(2-ethyl-2-oxazoline) segments, Polymer 44 (2003) 2255–2261.

[27] M.M. Bloksma, R.M. Paulus, H.P.C. Van Kuringen, F. van der Woerdt, H.M.L. Lambermont-Thijs, U.S. Schubert, R. Hoogenboom, Thermoresponsive Poly(2-oxazine)s, Macromolecular Rapid Communications 33 (2012) 92–96.

[28] Y. Maeda, IR Spectroscopic Study on the Hydration and the Phase Transition of Poly(vinyl methyl ether) in Water, Langmuir 17 (2001) 1737–1742.

[29] Y. Matsuda, Y. Miyazaki, S. Sugihara, S. Aoshima, K. Saito, T.J. Sato, Phase separation behavior of aqueous solutions of a thermoresponsive polymer, Polymer Science Part B: Polymer Physics 43 (2005) 2937–2949.

[30] F.A. Plamper, M. Ruppel, A. Schmalz, O. Borisov, M. Ballauff, A.H.E. Müller, Tuning the Thermoresponsive Properties of Weak Polyelectrolytes: Aqueous Solutions of Star-Shaped and Linear Poly(N,N-dimethylaminoethyl Methacrylate), Macromolecules 40 (2007) 8361–8366.

[31] C.D. Vo, J. Rosselgong, S.P. Armes, N. Tirelli, Stimulus-responsive polymers based on 2-hydroxypropyl acrylate prepared by RAFT polymerization, Journal of Polymer Science Part A: Polymer Chemistry 48 (2010) 2032–2043.

[32] D. Christova, S. Ivanova, G. Ivanova, Water-soluble temperature-responsive poly(vinyl alcohol-co-vinyl acetal)s, Polymer Bulletin 50 (2003) 367–372.

[33] Y. Iwasaki, C. Wachiralarpphaithoon, K. Akiyoshi, Novel Thermoresponsive Polymers Having Biodegradable Phosphoester Backbones, Macromolecules 40 (2007) 8136–8138.

[34] A. Halperin, M. Kröger, F.M. Winnik, Poly(N-isopropylacrylamide) Phase Diagrams: Fifty Years of Research. Angewandte Chemie International Edition 54 (2015) 15342–15367.

[35] D.E. Meyer, B.C. Shin, G.A. Kong, M.W. Dewhirst, A. Chilkoti, Drug targeting using thermally responsive polymers and local hyperthermia, Journal of Controlled Release 74 (2001) 213–224.

[36] E. Heymann, Studies on sol-gel transformations. I. The inverse sol-gel transformation of methylcellulose in water, Transactions of the Faraday Society 31 (1935) 846–864.

[37] B. Ju, D. Yan, S. Zhang, Micelles self-assembled from thermoresponsive 2-hydroxy-3-butoxypropyl starches for drug delivery, Carbohydrate Polymers 87 (2012) 1404–1409.

[38] S. Schmitz, H. Ritter, Unusual Solubility Properties of Polymethacrylamides as a Result of Supramolecular Interactions with Cyclodextrin, Angewandte Chemie International Edition 44 (2005) 5658–5661.

[39] O. Kretschmann, C. Steffens, H. Ritter, Cyclodextrin Complexes of Polymers Bearing Adamantyl Groups: Host–Guest Interactions and the Effect of Spacers on Water Solubility, Angewandte Chemie International Edition 46 (2007) 2708–2711.

[40] J.M.G. Cowie, A. Maconnachie, R.J. Ranson, Phase Equilibria in Cellulose Acetate-Acetone Solutions. The Effect of the Degree of Substitution and Molecular Weight on Upper and Lower Critical Solution Temperatures, Macromolecules 4 (1971) 57–61.

[41] W. Burchard, Thermodynamic solution properties of three polymers exhibiting lower critical solution temperature, Polymer 10 (1969) 467–475.

[42] E. Sato, Y. Masuda, J. Kadota, T. Nishiyama, H. Horibe, Dual stimuli-responsive homopolymers: Thermo- and photo-responsive properties of coumarin-containing polymers in organic solvents, European Polymer Journal 69 (2015) 605–615.

[43] T. Ueki, M. Watanabe, Lower Critical Solution Temperature Behavior of Linear Polymers in Ionic Liquids and the Corresponding Volume Phase Transition of Polymer Gels, Langmuir 23 (2007) 988–990.

[44] K. Seno, S. Kanaoka, S. Aoshima, Synthesis and LCST-type phase separation behavior in organic solvents of poly(vinyl ethers) with pendant imidazolium or pyridinium salts, Journal of Polymer Science Part A: Polymer Chemistry 46 (2008) 5724–5733.

[45] S. Amemori, K. Kokado, K. Sada, Polymer Phase-Transition Behavior Driven by a Charge-Transfer Interaction. Angewandte Chemie International Edition 52 (2013) 4174–4178.

[46] D.H. Gharib, S. Amemori, M. Naya, K. Kokado, K. Sada, Gel thermoresponsiveness driven by switching of the charge-transfer interaction, RSC Advances 5 (2015) 89319–89322.

[47] S. Amemori, K. Kokado, K. Sada, Fundamental Molecular Design for Precise Control of Thermoresponsiveness of Organic Polymers by Using Ternary Systems, Journal of the American Chemical Society 134 (2012) 8344–8347.

[48] D.N. Schulz, D.G. Peiffer, P.K. Agarwal, J. Larabee, J.J. Kaladas, L. Soni, B. Handwerker, R.T. Garner, Phase behaviour and solution properties of sulphobetaine polymers, Polymer 27 (1986) 1734–1742.

[49] M.B. Huglin, M.A. Radwan, Unperturbed dimensions of a zwitterionic polymethacrylate, Polymer International 26 (1991) 97–104.

[50] A. Laschewsky, I. Zerbe, Polymerizable and polymeric zwitterionic surfactants: 1. Synthesis and bulk properties, Polymer 32 (1991) 2070–2080.

[51] P. Kiiberle, A. Lawhewsky, T.D. Lomax, Interactions of a zwitterionic polysoap and its cationic analog with inorganic salts, Macromolecular Chemical Rapid Communications 12 (1991) 427–433.

[52] P. Mary, D.D. Bendejacq, M.-P. Labeau, P. Dupuis, Reconciling Low- and High-Salt Solution Behavior of Sulfobetaine Polyzwitterions, The Journal of Physical Chemistry B 111 (2007) 7767–7777.

[53] H. Yoshimitsu, A. Kanazawa, S. Kanaoka, S. Aoshima, Well-Defined Polymeric Ionic Liquids with an Upper Critical Solution Temperature in Water, Macromolecules 45 (2012) 9427–9434.

[54] X. Jia, D. Chen, M. Jiang, Preparation of PEO-*b*-P2VPH$^+$-$S_2O_8^{2-}$ micelles in water and their reversible UCST and redox-responsive behavior, Chemical Communications (2006) 1736–1738.

[55] J. Seuring, S. Agarwal, Non-Ionic Homo- and Copolymers with H-Donor and H-Acceptor Units with an UCST in Water, Macromolecular Chemistry and Physics 211 (2010) 2109–2117.

[56] H.C. Haas, N.W. Schuler, Thermally reversible homopolymer gel systems, Journal of Polymer Science Part C: Polymer Letters 2 (1964) 1095–1096.

[57] J. Seuring, F.M. Bayer, K. Huber, S. Agarwal, Upper Critical Solution Temperature of Poly(N-acryloyl glycinamide) in Water: A Concealed Property, Macromolecules 45 (2012) 374–384.

[58] F. Liu, J. Seuring, S. Agarwal, Controlled radical polymerization of N-acryloylglycinamide and UCST-type phase transition of the polymers, Journal of Polymer Science Part A: Polymer Chemistry 50 (2012) 4920–4928.

[59] S. Glatzel, A. Laschewsky, J.F. Lutz, Well-Defined Uncharged Polymers with a Sharp UCST in Water and in Physiological Milieu, Macromolecules 44 (2011) 413–415.

[60] S. Glatzel, N. Badi, M. Päch, A. Laschewsky, J.F. Lutz, Well-defined synthetic polymers with a protein-like gelation behavior in water, Chemical Communications 46 (2010) 4517–4519.

[61] A. Silberberg, J. Eliassaf, A. Katchalsky, Temperature-dependence of light scattering and intrinsic viscosity of hydrogen bonding polymers, Journal of Polymer Science 23 (1957) 259–284.

[62] J. Eliassaf, The effect of hydrogen bonding addends on the dilute solution viscosity of poly(methacrylamide), Journal of Applied Polymer Science 111 (1960) 372.

[63] M. Netopilík, M. Bohdanecw, V. Chytrý, K. Ulbrich, Cloud point of poly(N-isopropylmethacrylamide) solutions in water: is it really a point?, Macromolecular Rapid Communications 18 (1997) 107–111.

[64] M. Kano, E. Kokufuta, On the Temperature-Responsive Polymers and Gels Based on N-Propylacrylamides and N-Propylmethacrylamides, Langmuir 25 (2009) 8649–8655.

[65] T. Aoki, K. Nakamura, K. Sanui, A. Kikuchi, T. Okano, Y. Sakurai, N. Ogata, Adenosine-Induced Changes of the Phase Transition of Poly(6-(acryloyloxymethyl)uracil) Aqueous Solution, Polymer Journal 31 (1999) 1185–1188.

[66] N. Shimada, H. Ino, K. Maie, M. Nakayama, A. Kano, A. Maruyama, Ureido-Derivatized Polymers Based on Both Poly(allylurea) and Poly(l-citrulline) Exhibit UCST-Type Phase Transition Behavior under Physiologically Relevant Conditions, Biomacromolecules 12 (2011) 3418–3422.

[67] G. Bokias, G. Staikos, I. Iliopoulos, Solution properties and phase behaviour of copolymers of acrylic acid with N-isopropylacrylamide: the importance of the intrachain hydrogen bonding, Polymer 41 (2000) 7399–7405.

[68] J. Fang, F. Bian, W. Shen, A study on solution properties of poly(N,N-diethylacrylamide-co-acrylic acid), Journal of Applied Polymer Science 110 (2008) 3373–3378.

[69] G. Wu, S.-C. Chen, Q. Zhan, Y.-Z. Wang, Well-Defined Amphiphilic Biodegradable Comb-Like Graft Copolymers: Their Unique Architecture-Determined LCST and UCST Thermoresponsivity Macromolecules 44 (2011) 999–1008.

[70] J. Seuring, S. Agarwal, First Example of a Universal and Cost-Effective Approach: Polymers with Tunable Upper Critical Solution Temperature in Water and Electrolyte Solution, Macromolecules 45 (2012) 3910–3918.

[71] M. Arotçaréna, B. Heise, S. Ishaya, A. Laschewsky, Switching the Inside and the Outside of Aggregates of Water-Soluble Block Copolymers with Double Thermoresponsivity, Journal of the American Chemical Society 124 (2002) 3787–3793.

[72] J. Virtanen, M. Arotçaréna, B. Heise, S. Ishaya, A. Laschewsky, H. Tenhu, Dissolution and Aggregation of a Poly(NIPA-block-sulfobetaine) Copolymer in Water and Saline Aqueous Solutions, Langmuir 18 (2002) 5360–5365.

[73] Y. Chang, W.-H. Chen, W. Yandi, Y.-J. Shih, W.-L. Chu, Y.-L. Liu, C.-W. Chu, R.-C. Ruaan, A. Higuchi, Dual-Thermoresponsive Phase Behavior of Blood Compatible Zwitterionic Copolymers Containing Nonionic Poly(N-isopropyl acrylamide), Biomacromolecules 10 (2009) 2092–2100.

[74] Y. Maeda, H. Mochiduki, I. Ikeda, Hydration Changes during Thermosensitive Association of a Block Copolymer Consisting of LCST and UCST Blocks, Macromolecualar Rapid Communications 25 (2004) 1330–1334.

[75] H.-Y. Tian, J.-J. Yan, D. Wang, C. Gu, Y.-Z. You, X.-S. Chen, Synthesis of Thermo-Responsive Polymers With Both Tunable UCST and LCST, Macromolecular Rapid Communications 32 (2011) 660–664.

[76] H. Yoshimitsu, E. Korchagina, A. Kanazawa, S. Kanaoka, F.M. Winnik, S. Aoshima, Shape-switching self-assembly of new diblock copolymers with UCST-type and LCST-type segments in water, Polymer Chemistry 7 (2016) 2062–2068.

[77] R. Longenecker, T. Mu, M. Hanna, N.A.D. Burke, H.D.H. Stöver, Thermally Responsive 2-Hydroxyethyl Methacrylate Polymers: Soluble–Insoluble and Soluble–Insoluble–Soluble Transitions, Macromolecules 44 (2011) 8962–8971.

[78] T. Mori, M. Nakashima, Y. Fukuda, K. Minagawa, M. Tanaka, Y. Maeda, Soluble–Insoluble–Soluble Transitions of Aqueous Poly(N-vinylacetamide-co-acrylic acid) Solutions, Langmuir 22 (2006) 4336–4342.

[79] H. Mori, I. Kato, S. Saito, T. Endo, Proline-Based Block Copolymers Displaying Upper and Lower Critical Solution Temperatures, Macromolecules 43 (2010) 1289–1298.

[80] F.A. Plamper, A. Schmalz, M. Ballauff, A.H.E. Muller, Tuning the Thermoresponsiveness of Weak Polyelectrolytes by pH and Light: Lower and Upper Critical-Solution Temperature of Poly(N,N-dimethylaminoethyl methacrylate), Journal of the American Chemical Society 129 (2007) 14538–14539.

[81] Q. Zhang, J.-D. Hong, R. Hoogenboom, A triple thermoresponsive schizophrenic diblock copolymer, Polymer Chemistry 4 (2013) 4322–4325.

[82] G.N. Malcolm, J.S. Rowlinson, The thermodynamic properties of aqueous solutions of polyethylene glycol, polypropylene glycol and dioxane, Transactions of the Faraday Society 53 (1957) 921–931.

[83] Y.C. Bae, S.M. Lambert, D.S. Soane, J.M. Prausnitz, Cloud-point curves of polymer solutions from thermooptical measurements, Macromolecules 24 (1991) 4403–4407.

[84] E.E. Dormidontova, Role of Competitive PEO–Water and Water–Water Hydrogen Bonding in Aqueous Solution PEO Behavior, Macromolecules 35 (2002) 987–1001.

[85] E.E. Dormidontova, Influence of End Groups on Phase Behavior and Properties of PEO in Aqueous Solutions, Macromolecules 37 (2004) 7747–7761.

[86] T. Shiomi, K. Imai, C. Watanabe, M. Miya, Thermodynamic and conformational properties of partially butyralized poly(vinyl alcohol) in aqueous solution, Journal of Polymer Science Polymer Physics Edition 22 (1984) 1305–1312.

[87] K. Seno, A. Date, S. Kanaoka, S. Aoshima, Synthesis and solution properties of poly(vinyl ether)s with long alkyl chain, biphenyl, and cholesteryl pendants, Journal of Polymer Science Part A: Polymer Chemistry 46 (2008) 4392–4406.

[88] H. Shimomoto, D. Fukami, S. Kanaoka, S. Aoshima, Fluorine-containing vinyl ether polymers: Living cationic polymerization in fluorinated solvents as new media and unique solubility characteristics in organic solvents, Journal of Polymer Science Part A: Polymer Chemistry 49 (2011) 1174–1182.

[89] W.-M. Wan, F. Cheng, F. Jäkle, A Borinic Acid Polymer with Fluoride Ion- and Thermo-responsive Properties that are Tunable over a Wide Temperature Range, Angewandte Chemie International Edition 53 (2014) 8934–8938.

[90] T. Ueki, M. Watanabe, T.P. Lodge, Doubly Thermosensitive Self-Assembly of Diblock Copolymers in Ionic Liquids, Macromolecules 42 (2009) 1315–1320.

[91] A. Serres, M. Baudys, S.W. Kim, Temperature and pH-sensitive Polymers for Human Calcitonin Delivery. Pharmaceutical Research 13 (1996) 196–201.

[92] N. Milašinović, M. Kalagasidis Krušić, Z. Knežević-Jugović, J. Filipović, Hydrogels of N-isopropylacrylamide copolymers with controlled release of a model protein, International Journal of Pharmaceuticals 383 (2010) 53–61.

[93] Y. Dai, P. Ma, Z. Cheng, X. Kang, X. Zhang, Z. Hou, C. Li, D. Yang, X. Zhai, J. Lin, Up-Conversion Cell Imaging and pH-Induced Thermally Controlled Drug Release from $NaYF_4$:Yb^{3+}/Er^{3+}@Hydrogel Core–Shell Hybrid Microspheres, ACS Nano 6 (2012) 3327–3338.

[94] S. Nagata, K. Kokado, K. Sada, Metal–organic framework tethering PNIPAM for ON–OFF controlled release in solution, Chemical Communications 51 (2015) 8614–8617.
[95] J.C. Garbern, A.S. Hoffman, P.S. Stayton, Injectable pH- and Temperature-Responsive Poly(N-isopropylacrylamide-co-propylacrylic acid) Copolymers for Delivery of Angiogenic Growth Factors, Biomacromolecules 11 (2010) 1833–1839.
[96] D.J. Overstreet, R. Huynh, K. Jarbo, R.Y. McLemore, B.L. Vernon, *In situ* forming, resorbable graft copolymer hydrogels providing controlled drug release, Journal of Biomedical Materials Research Part A 101A (2013) 1437–1446.
[97] T. Takezawa, Y. Mori, K. Yoshizato, Cell Culture on a Thermo-Responsive Polymer Surface. Biotechnology 8 (1990) 854–856.
[98] T. Okano, N. Yamada, H. Sakai, Y. Sakurai, A novel recovery system for cultured cells using plasma-treated polystyrene dishes grafted with poly(N-isopropylacrylamide), Journal of Biomedical Materials Research 27 (1993) 1243–1251.
[99] Y. Akiyama, A. Kikuchi, M. Yamato, T. Okano, Ultrathin Poly(N-isopropylacrylamide) Grafted Layer on Polystyrene Surfaces for Cell Adhesion/Detachment Control, Langmuir 20 (2004) 5506–5511.
[100] D. Cunliffe, C. de las Heras Alarcón, V. Peters, J.R. Smith, C. Alexander, Thermoresponsive Surface-Grafted Poly(N–isopropylacrylamide) Copolymers: Effect of Phase Transitions on Protein and Bacterial Attachment, Langmuir 19 (2003) 2888–2899.
[101] J.A. Reed, A.E. Lucero, S. Hu, L.K. Ista, M.T. Bore, G.P. Lopez, H.E. Canavan, A Low-Cost, Rapid Deposition Method for "Smart" Films: Applications in Mammalian Cell Release, ACS Applied Materials and Interfaces 2 (2010) 1048–1051.
[102] D.E. Bergbreiter, L. Zhang, V.M. Mariagnanam, Smart ligands that regulate homogeneously catalyzed reactions, Journal of the American Chemical Society 115 (1993) 9295–9296.
[103] D.E. Bergbreiter, V.M. Mariagnanam, L. Zhang, Polymer ligands that can regulate reaction temperature in "smart" catalysts, Advanced Materials 7 (1995) 69–71.
[104] D.E. Bergbreiter, J.W. Caraway, Thermoresponsive Polymer-Bound Substrates, Journal of the American Chemical Society 118 (1996) 6092–6093.
[105] D.E. Bergbreiter, B. Case, Y.-S. Liu, J.W. Caraway, Poly(N-isopropylacrylamide) Soluble Polymer Supports in Catalysis and Synthesis, Macromolecules 31 (1998) 6053–6062.
[106] D.E. Bergbreiter, Y.-S. Liu, Water-soluble polymer-bound, recoverable palladium(0)-phosphine catalysts, Tetrahedron Letters 38 (1997) 7843–7846.
[107] K.J. Mackenzie, M.B. Francis, Recyclable Thermoresponsive Polymer–Cellulase Bioconjugates for Biomass Depolymerization, Journal of the American Chemical Society 135 (2013) 293–300.
[108] I. Mukherjee, S.K. Sinha, S. Datta, P. De, Recyclable Thermoresponsive Polymer–β-Glucosidase Conjugate with Intact Hydrolysis Activity, Biomacromolecules 19 (2018) 2286–2293.
[109] X. Zheng, J. Jiang, X. Lium, Z. Jin, Thermoregulated phase transfer ligands and catalysis. III. Aqueous/organic two-phase hydroformylation of higher olefins by thermoregulated phase-transfer catalysis, Catalysis Today 44 (1998) 175–182.
[110] R. Chen, X. Liu, Z. Jin, Thermoregulated phase-transfer ligands and catalysis. Part VI. Two-phase hydroformylation of styrene catalyzed by the thermoregulated phase-transfer catalyst OPGPP/Rh, Journal of Organometallic Chemistry 571 (1998) 201–204.
[111] M. Zeltner, A. Schätz, M.L. Hefti, W.J. Stark, Magnetothermally responsive C/Co@PNIPAM-nanoparticles enable preparation of self-separating phase-switching palladium catalysts, Journal of Materials Chemistry 21 (2011) 2991–2996.
[112] N. Suzuki, T. Inoue, T. Asada, R. Akebi, G. Kobayashi, M. Rikukawa, Y. Masuyama, M. Ogasawara, T. Takahashi, H.T. Thang, Asymmetric Aldol Reaction on Water Using an Organocatalyst Tethered on a Thermoresponsive Block Copolymer, Chemistry Letters 42 (2013) 1493–1495.
[113] H.A. Zayas, A. Lu, D. Valade, F. Amir, Z. Jia, R.K. O'Reilly, M.J. Monteiro, Thermoresponsive Polymer-Supported l-Proline Micelle Catalysts for the Direct Asymmetric Aldol Reaction in Water, ACS Macro Letters 2 (2013) 327–331.
[114] Q. Wang, J. Wang, D. Wang, M. Turhong, M. Zhang, Recyclable and effective Pd/poly(N-isopropylacrylamide) catalyst for hydrodechlorination of 4-chlorophenol in aqueous solution, Chemical Engineering Journal 280 (2015) 158–164.
[115] H. Kanazawa, K. Yamamoto, Y. Matsushima, N. Takai, A. Kikuchi, Y. Sakurai, T. Okano, Temperature-Responsive Chromatography Using Poly(N-isopropylacrylamide)-Modified Silica, Analytical Chemistry 68 (1996) 100–105.
[116] K. Nagase, J. Kobayashi, A. Kikuchi, Y. Akiyama, H. Kanazawa, T. Okano, High Stability of Thermoresponsive Polymer-Brush-Grafted Silica Beads as Chromatography Matrices, ACS Applied Materials and Interfaces 4 (2012) 1998–2008.
[117] A. Kikuchi, T. Okano, Temperature-responsive, polymer-modified surfaces for green chromatography, Macromolecular Symposia 207 (2004) 217–227.
[118] K. Nagase, T. Okano, Thermoresponsive-polymer-based materials for temperature-modulated bioanalysis and bioseparations, Journal of Materials Chemistry Part B 4 (2016) 6381–6397.
[119] E. Ayano, H. Kanazawa, Aqueous chromatography system using temperature-responsive polymer-modified stationary phases, Journal of Separation Science 29 (2006) 738–749.
[120] H. Feil, Y.H. Bae, J. Feijen, S.W. Kim, Molecular separation by thermosensitive hydrogel membranes, Journal of Membrane Science 64 (1991) 283–294.
[121] T. Nonaka, T. Ogata, S. Kurihara, Preparation of poly(vinyl alcohol)-graft-N-isopropylacrylamide copolymer membranes and permeation of solutes through the membranes, Journal of Applied Polymer Science 52 (1994) 951–957.
[122] S.B. Teli, G.S. Gokavi, T.M. Aminabhavi, Novel sodium alginate-poly(N-isopropylacrylamide) semi-interpenetrating polymer network membranes for pervaporation separation of water + ethanol mixtures, Separation and Purification Technology 56 (2007) 150–157.
[123] B.P. Tripathi, N.C. Dubey, M. Stamm, Hollow Microgel Based Ultrathin Thermoresponsive Membranes for Separation, Synthesis, and Catalytic Applications, ACS Applied Materials and Interfaces 6 (2014) 17702–17712.
[124] F. Song, X.-L. Wang, Y.-Z. Wang, Fabrication of novel thermo-responsive electrospun nanofibrous mats and their application in bioseparation, European Polymer Journal 47 (2011) 1885–1892.
[125] R. Ou, J. Wei, L. Jiang, G.P. Simon, H. Wang, Robust Thermoresponsive Polymer Composite Membrane with Switchable Superhydrophilicity and Superhydrophobicity for Efficient Oil–Water Separation, Environmental Science and Technology 50 (2016) 906–914.
[126] T. Mori, M. Maeda, Temperature-Responsive Formation of Colloidal Nanoparticles from Poly(N-isopropylacrylamide) Grafted with Single-Stranded DNA, Langmuir 20 (2004) 313–319.

[127] T. Fujie, J.Y. Park, A. Murata, N.C. Estillore, M.C.R. Tria, S. Takeoka, R.C. Advincula, Hydrodynamic Transformation of a Freestanding Polymer Nanosheet Induced by a Thermoresponsive Surface, ACS Applied Materials and Interfaces 1 (2009) 1404–1413.
[128] B.A. Humphreys, J.D. Willott, T.J. Murdoch, G.B. Webber, E.J. Wanless, Specific ion modulated thermoresponse of poly(N-isopropylacrylamide) brushes, Physical Chemistry Chemical Physics 18 (2016) 6037–6046.
[129] C. Pietsch, R. Hoogenboom, U.S. Schubert, Soluble Polymeric Dual Sensor for Temperature and pH Value, Angewandte Chemie International Edition 48 (2009) 5653–5656.
[130] L. Liu, W. Li, K. Liu, J. Yan, G. Hu, A. Zhang, Comblike Thermoresponsive Polymers with Sharp Transitions: Synthesis, Characterization, and Their Use as Sensitive Colorimetric Sensors, Macromolecules 44 (2011) 8614–8621.
[131] L. Liu, W. Li, J. Yan, A. Zhang, Thermoresponsive dendronized polymeric sensors. Journal of Polymer Science Part A: Polymer Chemistry 52 (2014) 1706–1713.
[132] S. Ryu, I. Yoo, S. Song, B. Yoon, J.-M. Kim, A Thermoresponsive Fluorogenic Conjugated Polymer for a Temperature Sensor in Microfluidic Devices, Journal of the American Chemical Society 131 (2009) 3800–3801.
[133] C. Pietsch, R. Hoogenboom, U.S. Schubert, Aqueous polymeric sensors based on temperature-induced polymer phase transitions and solvatochromic dyes, Chemical Communications 47 (2011) 8750–8765.
[134] Y. Wang, Y. Feng, B. Wang, Z. Lu, A novel thermoviscosifying water-soluble polymer: Synthesis and aqueous solution properties, Journal of Applied Polymer Science 116 (2010) 3516–3524.
[135] Y. Guo, R. Song, R. Feng, G. Dai, Y. Liang, D. Pu, X. Zhang, Z. Ye, Thermoresponsive behavior of graft copolymers based on poly(*N*,*N*-dimethylacrylamide-co-diacetoneacrylamide) side chains, Journal of Applied Polymer Science 136 (2019) 47051.
[136] B. Yang, J. Duhamel, Extraction of Oil from Oil Sands Using Thermoresponsive Polymeric Surfactants, ACS Applied Materials and Interfaces 7 (2015) 5879–5889.
[137] D. Schmaljohann, Thermo- and pH-responsive polymers in drug delivery, Advanced Drug Delivery Reviews 58 (2006) 1655–1670.
[138] S. Dai, P. Ravi, K.C. Tam, pH-Responsive polymers: synthesis, properties and applications, Soft Matter 4 (2008) 435–449.
[139] M. Irie, Photoresponsive polymers, Advanced Polymer Science 94 (1990) 27–67.
[140] Y. Huang, R. Dong, X. Zhu, D. Yan, Photo-responsive polymeric micelles, Soft Matter 10 (2014) 6121–6138.
[141] M. Huo, J. Yuan, L. Tao, Y. Wei, Redox-responsive polymers for drug delivery: from molecular design to applications, Polymer Chemistry 5 (2014) 1519–1528.
[142] X. Ma, H. Tian, Stimuli-Responsive Supramolecular Polymers in Aqueous Solution, Accounts of Chemical Research 47 (2014) 1971–1981.
[143] J. Hu, G. Zhang, S. Liu, Enzyme-responsive polymeric assemblies, nanoparticles and hydrogels, Chemical Society Reviews 41 (2012) 5933–5949.
[144] M. Zelzer, S.J. Todd, A.R. Hirst, T.O. McDonald, R.V. Ulijn, Enzyme responsive materials: design strategies and future developments, Biomaterials Science 1 (2013) 11–39.
[145] M. Tabone, J.J. Cregg, E.J. Beckman, A.E. Landis, Sustainability Metrics: Life Cycle Assessment and Green Design in Polymers, Environmental Science and Technology 44 (2010) 8264–8269.
[146] C.R. Álvarez-Chávez, S. Edwards, R. Moure-Eraso, K. Geiser, Sustainability of bio-based plastics: general comparative analysis and recommendations for improvement, Journal of Cleaner Production 23 (2012) 47–56.
[147] T.A. Hottle, M.M. Bilec, A.E. Landis, Sustainability assessments of bio-based polymers, Polymer Degradation and Stability 98 (2013) 1898–1907.
[148] S. Kobayashi, Green polymer chemistry: new methods of polymer synthesis using renewable starting material, Structural Chemistry 28 (2017) 461–474.
[149] K. Matyjaszewski, W. Jakubowski, K. Min, W. Tang, J. Huang, W.A. Braunecker, N.V. Tsarevsky, Diminishing catalyst concentration in atom transfer radical polymerization with reducing agents, Proceedings of the National Academy of Sciences of the United States of America 103 (2006) 15309–15314.
[150] W. Jakubowski, K. Matyjaszewski, Activators Regenerated by Electron Transfer for Atom-Transfer Radical Polymerization of (Meth)acrylates and Related Block Copolymers, Angewandte Chemie International Edition 45 (2006) 4482–4486.
[151] E.H. Discekici, A. Anastasaki, J. Read de Alaniz, C.J. Hawker, Evolution and Future Directions of Metal-Free Atom Transfer Radical Polymerization, Macromolecules 51 (2018) 7421–7434.
[152] N. Hernández, R.C. Williams, E.W. Cochran, The battle for the "green" polymer. Different approaches for biopolymer synthesis: bioadvantaged vs. bioreplacement, Organic and Biomolecular Chemistry 12 (2014) 2834–2849.
[153] P.F.H. Harmsen, M.M. Hackmann, H.L. Bos, Green building blocks for bio-based plastics, Biofuels, Bioproducts and Biorefining 8 (2014) 306–324.
[154] A. Gandini, T.M. Lacerda, From monomers to polymers from renewable resources: Recent advances, Progress in Polymer Science 48 (2015) 1–39.
[155] J.J. Ritter, P.P. Minieri, A New Reaction of Nitriles. I. Amides from Alkenes and Mononitriles, Journal of the American Chemical Society 70 (1948) 4045–4548.
[156] S.-C. Wu, J.B. Wang, T.-C. Tsai, Catalysis of Zeolite in the Amidation of Alcohols with Nitriles for the Synthesis of N-Alkylacrylamides. Topics in Catalysis 53 (2010) 1419–1429.
[157] D. Mack, S. Schätzle, Y. Traa, E. Klemm, Synthesis of Acrylonitrile from Renewable Lactic Acid, ChemSusChem (2019), https://doi.org/10.1002/cssc.201802914.
[158] R.K. Grasselli, F. Trifirò, Acrylonitrile from Biomass: Still Far from Being a Sustainable Process, Topics in Catalysis 59 (2016) 1651–1658.
[159] J. van Haveren, E.L. Scott, J. Sanders, Bulk chemicals from biomass, Biofuels, Bioproducts and Biorefining 2 (2008) 41–57.
[160] K. Inokuma, J.C. Liao, M. Okamoto, T. Hanai, Improvement of isopropanol production by metabolically engineered Escherichia coli using gas stripping, Journal of Bioscience and Bioengineering 110 (2010) 696–701.
[161] Y. Hirokawa, Y. Dempo, E. Fukusaki, T. Hanai, Metabolic engineering for isopropanol production by an engineered cyanobacterium, *Synechococcus elongatus* PCC 7942, under photosynthetic conditions, Journal of Bioscience Bioengineering 123 (2017) 39–45.

CHAPTER

14

Nanostructured Membranes for Enhanced Forward Osmosis and Pressure-Retarded Osmosis

Chun Feng Wan, Yue Cui, Wen Xiao Gai, Zhen Lei Cheng, Tai-Shung Chung

Department of Chemical and Biomolecular Engineering, National University of Singapore, Singapore

OUTLINE

List of Abbreviations

A Pure water permeability
AL-DS Active layer facing the draw solution
AL-FS Active layer facing the feed solution
AQP Aquaporin
B Salt permeability
BDE Bond dissociation energy
BSA Bovine serum albumin
CA Cellulose acetate
CS Chitosan
CNT Carbon nanotubes
CQD Carbon quantum dot
CRR Chemical reduction route
DI Deionized
ECP External concentration polarization
FO Forward osmosis
fCNT Functionalized multi-walled carbon nanotube
gMH $g\ m^{-2}\ hr^{-1}$
GA Glutaraldehyde
GO Graphene oxide
ICP Internal concentration polarization
LbL Layer-by-layer
LDH Layered double hydroxide

Sustainable Nanoscale Engineering
https://doi.org/10.1016/B978-0-12-814681-1.00014-X

LMH L m^{-2} hr^{-1}
LCA Life cycle assessment
LTA Linde type A zeolite
MOF Metal–organic framework
MPC Methacryloyloxyethyl phosphorylcholine
MPD m-Phenylenediamine
MWCNT Multi-walled carbon nanotubes
MXDA m-Xylylenediamine
NF Nanofiltration
NP Nanoparticle
NR Nanorod
PAA Poly(acrylic acid)
PAH Poly(allylamine hydrochloride)
PAN Polyacrylonitrile
PCTE Polycarbonate tracked-etched
PDA Polydopamine
PEI Polyetherimide
PES Polyethersulfone
PLL Poly-L-lysine
POM Polyoxometalates
PRO Pressure-retarded osmosis
PSf Polysulfone
PSS Poly(sodium 4-styrene sulfonate)
PVDF Polyvinylidene fluoride
S Structural parameter
TFC Thin-film composite
TFN Thin film nanocomposite
TMC Trimesoyl chloride
TSC Trisodium citrate
USR Ultrasonication route
ΔP Hydraulic pressure difference between the draw and feed solutions
ε Porosity
λ Wall thickness of the hollow fiber membranes
τ Tortuosity

1. INTRODUCTION

Forward osmosis (FO) and pressure-retarded osmosis (PRO) have received increasing attentions for clean water and renewable energy productions [1–5]. Both technologies exploit the naturally and osmotically induced water transport across a semipermeable membrane. In FO (Fig. 14.1A), the mixing is spontaneously driven by the salinity gradient between a high-salinity draw solution and a low-salinity feed solution across a semipermeable membrane that allows water to flow through but rejects the ions and other impurities [6,7]. FO consumes less energy to transport water across the membrane than pressure-driven membrane processes [1–3]. The product of FO is a mixture of the draw and feed solutions. FO requires a secondary separation process to produce clean water and regenerate the draw solution. As novel draw solutes become available, clean water can be separated from the draw solutes by a magnetic field or low-temperature thermal processes that are more energy-efficient than conventional pressure-driven processes [8–10]. In addition, fertilizers or sugars can be used as the draw solutes to collect freshwater from wastewater in an FO process for irrigation and drinking purposes, respectively [11,12]. FO membranes also have low fouling propensity. For example, alginate fouling in FO is almost fully reversible, with more than 98% recovery of water flux after a simple water rinse without any chemical cleaning reagents [13]. These advantages make FO a sustainable technology with lower energy and chemical consumptions.

PRO is the predominant technology to harvest the osmotic energy from a salinity gradient. PRO (Fig. 14.1B) employs a hydraulic pressure on the high-salinity draw solution side. When the feed solution permeates through the membrane, the permeate is instantly pressurized and can be utilized to generate energy via a turbine or an energy-recovery device [14–16]. As osmotic energy is generated from solutions with different salinities instead of conventional fuels, osmotic energy is a cleaner and more sustainable form of energy. The global potential of osmotic energy at the estuaries where rivers meet the seas is 2.6 TW, which is higher than the current global electricity consumption [17]. Recent researches have explored the integration of PRO with desalination technologies to use osmotic energy to lower the energy dependence of desalination on fossil fuel and reduce the CO_2 footprint [18].

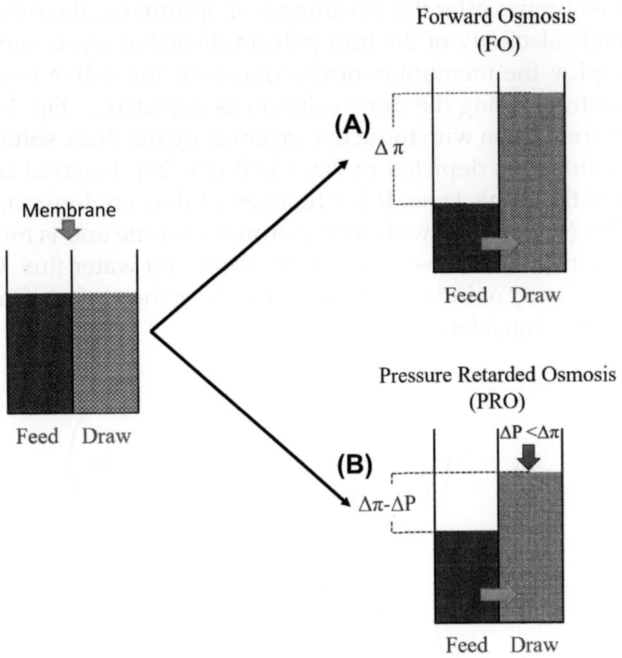

FIGURE 14.1 Steady states of (A) forward osmosis and (B) pressure-retarded osmosis.

Membranes are the hearts of both FO and PRO technologies. As shown in Fig. 14.2, most of the FO and PRO membranes are composite membranes composed of a substrate and an active layer formed by interfacial polymerization, coating, crosslinking, or layer-by-layer (LbL) assembly [19,20]. The substrate is usually prepared via non—solvent-induced phase inversion, where a homogeneous polymer solution is transformed into a polymer-rich phase and polymer-lean phase. The polymer-rich phase forms the polymer matrix of the substrate, while the polymer-lean phase forms the pores of the substrate [21]. Thin-film composite (TFC) membranes fabricated by interfacial polymerization are most widely investigated and employed in FO and PRO applications. Typically, during the interfacial polymerization, the membrane substrate first contacts with an aqueous diamine monomer (such as m-phenylenediamine, MPD) solution for a certain period. The extra solution on the membrane surface is then removed by drying, purging, or wiping. Subsequently, an organic solution containing acyl chloride (such as trimesoyl chloride, TMC) is brought to contact with the membrane substrate, react with the diamine absorbed on the membrane substrate, and form the polyamide active layer on the membrane surface. The self-terminating nature of the interfacial polymerization reaction limits the thickness of the polyamide active layer in the range of 100

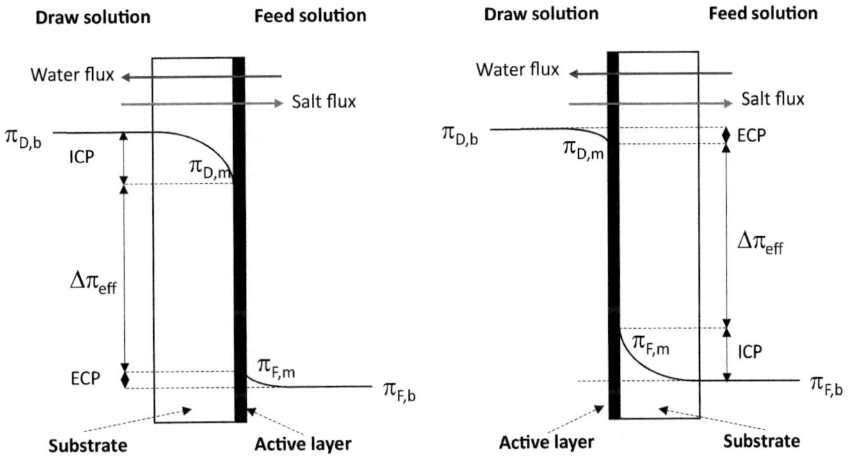

FIGURE 14.2 Concentration profiles of membranes under (left) pressure-retarded osmosis (PRO) (active layer facing the feed solution, AL-FS) and (right) forward osmosis (FO) (active layer facing the draw solution, AL-DS) modes.

nanometers. Therefore, TFC membranes offer the advantages of optimizing the robustness and structure of the substrates and the permeability and selectivity of the thin polyamide active layers separately [22–24].

FO applications usually employ the membrane orientation with the active layer facing the feed solution (i.e., AL-FS orientation) and the substrate facing the draw solution as depicted in Fig. 14.2A, whereas PRO applications usually employ the membrane orientation with the active layer facing the draw solution (i.e., AL-DS orientation) and the substrate facing the feed solution as depicted in Fig. 14.2B [25–28]. External concentration polarization (ECP) occurs on the external surface of the active layer. It is a function of flow configuration and operation conditions. Internal concentration polarization (ICP) occurs within the porous substrate and is much more significant than ECP. It can cause dramatic decreases in the effective osmotic driving force and water flux. Unlike ECP, ICP cannot be eliminated by increasing cross-flow velocity or inducing turbulence along the surface. The water flux can be expressed as follows when only the ICP effect is considered [3,5].

For the AL-FS configuration

$$J_w = A\left(\frac{\pi_{D,b}\exp\left(-\frac{J_wS}{D}\right) - \pi_{F,b}}{1 - \frac{B}{J_w}\left(\exp\left(-\frac{J_wS}{D}\right) - 1\right)} - \Delta P\right)$$

For the AL-DS configuration

$$J_w = A\left(\frac{\pi_{D,b} - \pi_{F,b}\exp\left(\frac{J_wS}{D}\right)}{1 + \frac{B}{J_w}\left(\exp\left(\frac{J_wS}{D}\right) - 1\right)} - \Delta P\right)$$

where A and B are the pure water permeability and salt permeability, respectively. $\pi_{D,b}$ and $\pi_{F,b}$ are the bulk osmotic pressures of the draw and feed solutions, respectively. ΔP is the hydraulic pressure difference between the draw and feed solutions, which is equal to zero in FO operations. S is the structural parameter of the membrane substrate that is defined as follows.

$$S = \frac{\tau\lambda}{\varepsilon}$$

where τ, ε, and λ are the tortuosity, porosity, and wall thickness of the hollow fiber membranes, respectively.

The membrane orientation also has significant impacts on membrane fouling during FO and PRO operations. Membrane fouling in the AL-FS orientation occurs by deposition of foulants from the feed solution onto the surface of the active layer and formation of cake layers. Membrane fouling in the AL-DS orientation is more complicated because the porous substrate is facing the feed solution. Although larger foulants are deposited onto the outer surface of the porous substrate, smaller foulants may be dragged into the porous substrate with the aid of water convection and then accumulated underneath the active layer and inside the pores [19,29].

The desirable FO and PRO membranes must have high pure water permeability, good hydrophilicity, excellent salt rejection, open pore structure, low structural parameter, and stable long-term performance [5–7]. Moreover, PRO membranes need to have strong mechanical properties for the high-pressure operation [14–16]. Applications of nanoparticles (NPs) in engineering osmosis have been extensively investigated, owing to their unique properties to improve the physicochemical properties of the membranes [30–33]. Generally, incorporating NPs into polymer dopes would not only enhance the solvent exchange rate between the solvent and the nonsolvent but also increase the pore connectivity, wettability, hydrophilicity, and mechanical strength of the substrates. Some NPs can be also directly deposited onto the substrates and act as the active layer [30–33]. Altering the surface hydrophilicity and substrate structure may change the transport of water and salts across the membrane, alleviate ECP and ICP effects, and lead to improving the membrane performance [34–37]. In addition, the substrate's hydrophilicity and pore structure play an important role in determining the formation of a thin and defect-free polyamide active layer [38–41]. Therefore, NPs are popular additives to improve the permeability, rejection, and antifouling properties of the polyamide layer. In this chapter, we will discuss the roles of various NPs in the fabrication of FO and PRO membranes and bring insights to future membrane researches.

2. GRAPHENE OXIDE AND ITS DERIVATIVES

Graphene oxide (GO), whose structure is shown in Fig. 14.3A, is a promising nanomaterial because it possesses a high surface-to-volume ratio and many hydrophilic functional groups, such as carboxyl, epoxy, and hydroxyl groups [42,45,46]. It also has great tensile strength, Young's modulus, and ability to absorb pressure. Therefore, GO offers great potential to improve the structural, mechanical, hydrophilicity, and antifouling properties of TFC membranes. Research efforts have been contributed to the preparation of substrates with GO-based fillers. Park et al. incorporated GO nanosheets in polysulfone (PSf) substrates to obtain PSf/GO composite substrates for TFC FO membranes. The incorporation of GO resulted in TFC membranes with an increased membrane hydrophilicity and a reduced membrane structural parameter. At the optimal addition of 0.25wt% GO, the incorporation of GO also promoted the effective formation of the polyamide layer. The pure water permeability was increased from 0.91 $L\,m^{-2}\,h^{-1}$ (LMH) of the unmodified one to 1.76 LMH of the GO modified one, while their corresponding salt rejections were increased from 97.0% to 98.7% using a 1000 mg L^{-1} NaCl solution as the feed. The enhanced properties arose from the increased water flux and reduced reverse salt flux. However, a high GO loading of above 0.5wt% caused a significant decline in FO water flux because poor GO dispersion in dope solutions deteriorated the structural properties of the substrate layer [47]. Lim et al. also incorporated GO in the casting solution of a TFC FO membrane to reduce the ICP effects [48].

Shen et al. incorporated GO in the polyamide layer to form thin film nanocomposite (TFN) membranes using an MPD/GO mixture as the aqueous solution in interfacial polymerization. The GO incorporated layer became thinner, smoother, and more hydrophilic because the horizontal deposit of GO nanosheets on the membrane surface retarded the diffusion of MPD into the organic phase. The resultant membrane exhibited a higher pure water permeability, a higher salt rejection, a lower salt diffusivity, and a lower structural parameter up to 200 ppm of GO. By using 200 ppm GO, the resultant membrane possessed a pure water permeability of 1.66 LMH bar^{-1}, a salt permeability of 0.24 LMH, and a structural parameter of 0.104 mm. At higher GO concentrations, defects were formed on the polyamide layer, and the salt rejection was compromised. Moreover, the GO-incorporated TFN membrane had a more reversible alginate membrane fouling [49]. Lai et al. employed a similar approach to fabricate the TFN membranes by using a PIP/GO mixture solution as the aqueous solution for the interfacial polymerization and obtained a 31.4% higher water flux than the pristine ones [50]. GO/Fe_3O_4 nanohybrids were also employed in the in situ interfacial polymerization to fabricate TFN-MMGO/Fe_3O_4 membranes, which achieved a water flux of 117.4% higher than the pristine TFC membrane. Using 1 M NaCl and deionized (DI) water as the feed pair, the TFN-MMGO/Fe_3O_4 membrane can achieve a water flux of 62 LMH under the PRO mode and 55 LMH under the FO mode, respectively [51]. Polydopamine (PDA)-induced GO and GO-silver NPs, thanks to their remarkable bactericidal properties, were successfully attached to the polyamide layer of the TFC membranes to mitigate the FO biofouling [52,53].

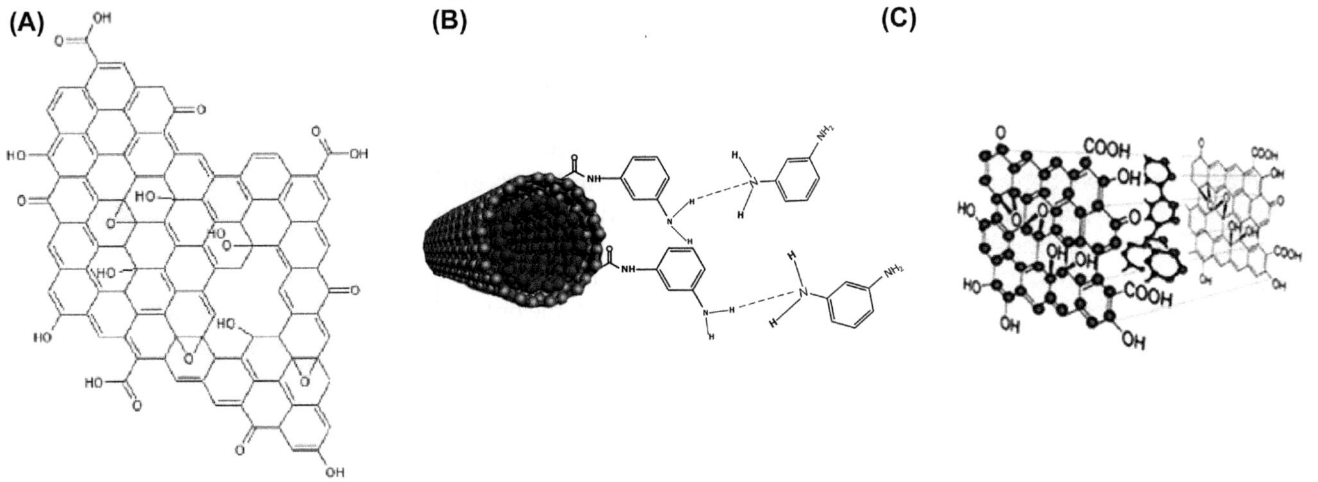

FIGURE 14.3 Chemical structures of (A) graphene oxide, (B) carbon nanotube, and (C) carbon quantum dot. *Copyright obtained from V.B. Mohan, K.-T. Lau, D. Hui, D. Bhattacharyya, Graphene-based materials and their composites: a review on production, applications and product limitations, Composites Part B 142 (2018) 200–220; M. Tian, R. Wong, K, Goh, Y. Liao, A.G. Fane, Synthesis and characterization of high-performance novel thin film nanocomposite PRO membranes with tiered nanofiber support reinforced by functionalized carbon nanotubes, Journal of Membrane Science 486 (2015) 151–160; D.L. Zhao, T.-S. Chung, Applications of carbon quantum dots (CQDs) in membrane technologies: a review, Water Research 147 (2018) 43–49.*

Other than applying GO and its derivatives to fabricate TFN membranes and to modify the polyamide active layers, these NPs can be crosslinked, filtered, and deposited by means of LbL to form the selective layer of the FO membrane. Jin et al. fabricated polyamide-crosslinked GO membranes for FO. First, intracrosslinked GO aggregates by m-xylylenediamine (MXDA) were filtered through a poly(allylamine hydrochloride) (PAH)-treated polyethersulfone (PES) flat-sheet membrane via vacuum filtration at room temperature, to form an MXDA-GO thin film on top of the PES substrate, and was then heated in an oven for further crosslinking. Second, the MXDA-GO membrane was contacted with a TMC hexane solution to intercrosslink the GO aggregates and seal the large gaps between them. The membrane with an optimal loading of 2 mL GO yielded an FO water flux of 13.2 LMH and a PRO water flux of 18.8 LMH using 0.25 M trisodium citrate (TSC) as the draw solution and DI water as the feed solution. However, the polyamide-crosslinked GO membrane showed a high solute flux, due to the formation of large GO aggregates that potentially increased heterogeneity and defects of the PA-GO membrane [54]. Yang et al. used a vacuum-assisted filtration to filter the GO laminates onto membrane substrates, chemically reduced the GO layer, and then dipped the membranes into a PDA solution to increase their hydrophilicity. As a result, the PDA-coated reduced GO membranes achieved an outstanding water flux of 36.6 LMH with a reverse solute flux of 0.042 mol m^{-2} h^{-1} using 1 M dextrose as the draw solution and 0.1 M NaCl as the feed [55]. Salehi et al. LbL assembled the positive chitosan (CS) and negative GO nanosheets via electrostatic interaction to form the active layer on a porous PES substrate. The LbL FO membranes coated by CS/GO bilayers showed water fluxes of 2–4 times higher than the TFC counterparts [56].

3. CARBON NANOTUBES

Carbon nanotubes (Fig. 14.3B), with superior separation capability as well as excellent physical properties including high tensile strength, can be used as potential fillers in the fabrication of nanocomposite FO membranes. Cellulose acetate (CA) membranes comprising functionalized multi-walled carbon nanotubes (f-MWCNTs) possessed superior fouling resistance to alginate in FO tests. The 1% f-MWCNTs-CA membrane had a flux decline of 57% less than the pristine CA membrane because of the enhanced electrostatic repulsion between the fCNT-CA membrane and the alginate foulant [58]. The TFC membranes made from the substrates embedded with MWCNTs exhibited a better separation performance due to the increased porosity of the supports, reduced ICP effects, and the smoother polyamide surface. The addition of MWCNTs up to 2wt% not only improved both the water permeability and salt rejection but also significantly enhanced tensile strength of supports [59]. Amine f-MWCNTs were also explored as additives in the aqueous MPD solution for interfacial polymerization. As shown in Fig. 14.4, with f-MWCNT loadings of 0.01–0.1wt%, the resultant TFN membranes showed a higher surface roughness (R_a). The membranes also had enhanced hydrophilicity and higher water permeability. The TFN membranes consisting of f-MWCNTs can generate a water flux of 40 LMH under the FO mode and a water flux of 95.7 LMH under the PRO mode using 2 M NaCl as the draw and 10 mM NaCl as the feed [57].

The superior mechanical properties of carbon nanotubes have sparked great interests of using them as additives in the substrates of PRO membranes. Son et al. adopted a similar approach by blending carbon nanotubes in the substrate of a PES TFC membrane as shown in Fig. 14.5. The polyamide selective layer was then etched by NaOCl. Because of the increased porosity, the more hydrophilic substrate, and the chemical etching, the water flux and the maximum power density of the TFN membrane were 87% and 110% greater than those of the pristine TFC membrane, respectively, using 0.5 M NaCl as the draw solution and DI water as the feed [60]. Tian et al. fabricated a novel TFC PRO membrane consisting of electrospun polyetherimide (PEI) nanofibers reinforced by f-CNTs and a thin film polyamide selective layer. The f-CNTs increased the mechanical strength of PEI nanofibers so that the membrane could withstand the high pressure in PRO operations. As a result, the optimized membrane can withstand a pressure up to 24 bar and generate a peak power density of 17.3 W m^{-2} at 16.9 bar using 1.0 M NaCl as the draw solution and DI water as the feed, while the TFC membrane without f-CNTs can only survive 7 bar [43].

4. CARBON QUANTUM DOTS

As depicted in Fig. 14.3C, carbon quantum dots (CQDs) possess hydrophilic hydroxyl and carboxyl groups on the edge of the nanosheets and hydrophobic aromatic rings on the plane of the nanosheets. The desirable features of CQDs, such as excellent hydrophilicity, low toxicity, environmental friendliness, low cost, and high water solubility,

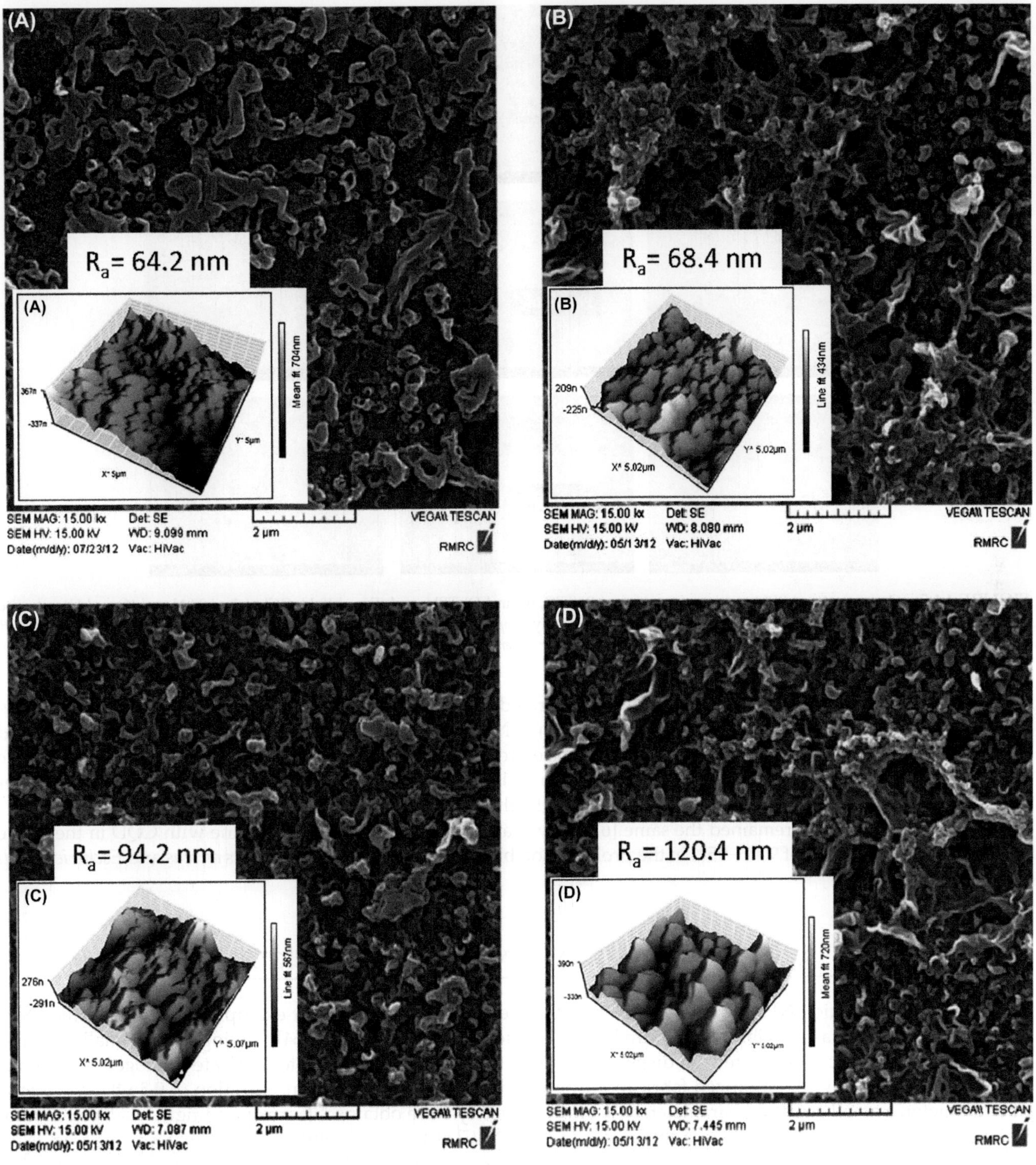

FIGURE 14.4 SEM morphologies and AFM analyses of the top surfaces of (A) a pristine TFC membrane, TFN membranes with (B) 0.01 wt/vol % (C) 0.05 wt/vol%, and (D) 0.1 wt/vol% functionalized multi-walled carbon nanotube (f-MWCNT) loadings. *Copyright obtained from M. Amini, M. Jahanshahi, A. Rahimpour, Synthesis of novel thin film nanocomposite (TFN) forward osmosis membranes using functionalized multi-walled carbon nanotubes, Journal of Membrane Science 435 (2013) 233–241.*

make them promising additives in FO and PRO membranes [44,61,62]. CQDs NPs can be covalently bonded to the polyamide chains of TFC FO membranes. The optimized TFC membrane with a CQD loading of 0.05wt% possessed a hydrophilic and neutrally charged membrane surface. It had an enhanced water flux of 12.9 LMH and a comparable reverse salt flux of 1.41 g m^{-2} h^{-1} (gMH) under FO tests using DI water and 0.5 M $MgCl_2$ as the feed and draw

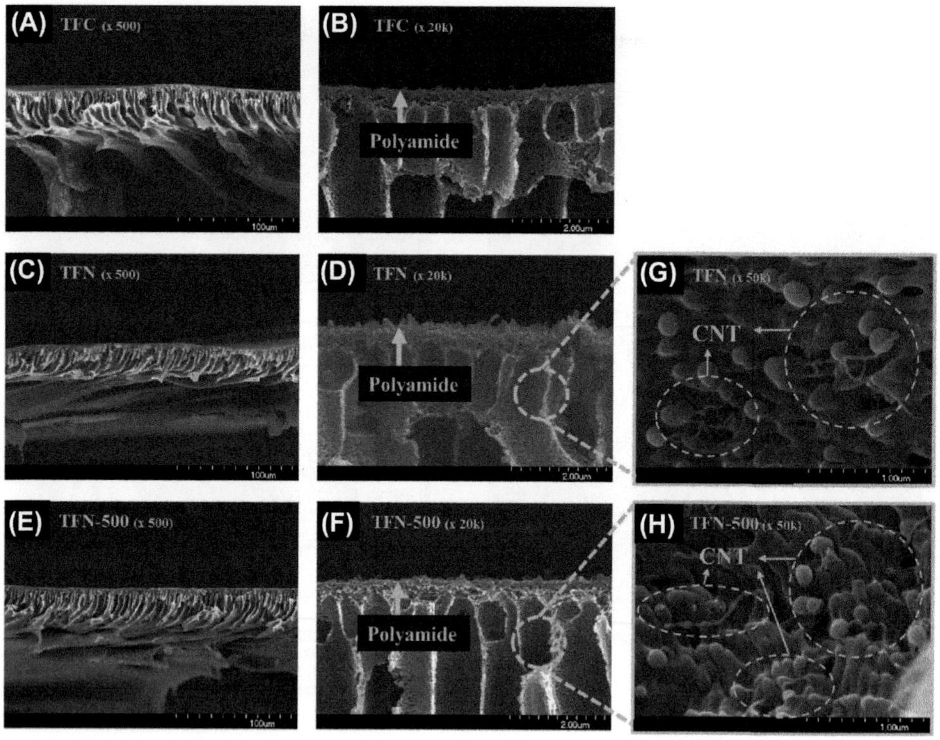

FIGURE 14.5 Cross-sectional SEM images of the (A and B) TFC, (C and D) TFN, and (E and F) NaOCl-etched TFN and (G and H) enlarged cross-section of TFN and NaOCl-etched TFN membranes, respectively. *Copyright obtained from M. Son, et al. Thin-film nanocomposite membrane with CNT positioning in support layer for energy harvesting from saline water, Chemical Engineering Journal 284 (2016) 68–77.*

solutions, respectively [61]. CQDs can also be incorporated into the polyamide selective layers via interfacial polymerization to form TFN PRO membranes. The addition of Na^+-functionalized CQDs (Na^+-CQD) not only increased the hydrophilic oxygen-containing groups and surface area of the polyamide layer but also made the polyamide network looser and thinner. With the optimal addition of 1wt% Na^+-CQD in the polyamide active layer, the water flux at 23 bar was increased from 44.5 LMH to 53.5 LMH, the power density was increased from 28.4 to 34.2 W m^{-2}, while the reverse salt flux remained the same [62]. It was also found that a TFN membrane with CQD in the active layer showed a reduction of 90%–95% in the proliferation biofouling tests of *Staphylococcus aureus* and *Escherichia coli* over 24 h [63].

5. METAL AND METAL OXIDE NANOPARTICLESS

Metal and metal oxide NPs are often added into the membrane substrates. For example, amorphous silica NPs were embedded into a nanofibrous polyvinylidene fluoride (PVDF) substrate [64]. Owing to the small size and high surface area, silica NPs were well dispersed in dope solutions and allowed to fabricate homogenous electrospun nanofiber mats. With optimized fabrication conditions and an optimized concentration of 0.5wt%, a reduced fiber diameter, a small structure parameter, and a low tortuosity were obtained. Correspondingly, a salt rejection of 99.7% and a water flux of 83 LMH were obtained using 2 M NaCl as the draw solution and DI water as the feed. With a loading of 0.2wt%, the TFN membrane showed an FO water flux of 31 LMH and a reverse salt flux of 6.25 gMH, whereas the FO water flux and reverse salt flux of the pristine TFC membrane were 18 LMH and 8.45 gMH, respectively. Emadzadeh et al. also attempted to embed commercial TiO_2 NPs into membrane substrates, followed by depositing the polyamide layer on top of the substrates via interfacial polymerization [65,66]. The water flux was increased with the addition of TiO_2 NPs in membrane substrates. When 0.5wt% TiO_2 NPs were added into the porous substrate, the water fluxes were approximately 71.5%–90% higher than those obtained by the pristine TFC membranes using 10 mM NaCl and 2 M NaCl as feed and draw solutions, respectively. The flux improvements were attributed to the porous substrate structure, improved hydrophilicity, small S value, and hence a relatively low ICP [67].

The addition of NPs into the substrate would also help to improve the antifouling properties of the membrane, owing to the improved hydrophilicity or the antiseptic properties of the NPs. It was reported that even when the TiO_2 concentration was as low as 0.5% in the PSf substrate, the biofouling of bovine serum albumin (BSA) on the substrate was significantly reduced. A hydrolayer was formed on the pore surface of the substrate, preventing biofoulants from adsorption and thus lowering the transport resistance of water. Similarly, silver (Ag) NPs were firstly deposited on the membrane surface by a photoinduced growth approach, and then TiO_2 NPs were coated on top of the Ag NPs via the charge-driven self-assembly [68]. It was found that the bacterial growth rate on the modified membrane was 11 times lower than that on the pristine membrane. Meanwhile, the water flux recovery of the modified membrane was significantly higher than the pristine one by simply rinsing with DI water. The improved antifouling properties can be attributed to the antibacterial properties and the ability to decompose the organic matter of Ag-TiO_2 NPs. Lu et al. blended layered double hydroxide NPs (LDH-NPs) of Mg and Al into the porous PSf substrate [69]. Consequently, the modified membranes exhibited a higher water permeability without increasing the reverse salt flux. The highest water flux obtained at the LDH-NPs concentration of 2wt% was 42.5% higher than that obtained by the pristine membrane. Rastgar et al. incorporated ZnO and ZnO-SiO core–shell NPs of different surface characteristics into the PSf substrate to reduce ICP. The substrate morphology transformed from a sponge-like into a finger-like structure with the addition of NPs, and the permeate flux was enhanced up to 117% [30].

In PRO applications, NPs were also specifically incorporated into the membrane substrates to alleviate membrane fouling. For example, Ag was incorporated into polyacrylonitrile (PAN) to form a PAN-Ag nanocomposite substrate, on which a crosslinked LbL selective layer was deposited [70]. The membrane performance was optimized at the Ag concentration of 0.02wt% in the substrate. The water permeability was 24% higher than the pristine membrane. The nanocomposite membranes showed an excellent antibiofouling performance when the Ag concentration reached 0.10wt%. In PRO tests, the water flux maintained at 88.2% when the membrane was tested by a feed solution containing *Comamonas testosteroni* I2. In another study, the TiO_2 NPs was coated on the substrate of a PRO membrane via sol–gel-derived spray coating method [71]. Subsequently, a thin polyamide layer was deposited on the coated substrate via interfacial polymerization. The NPs enhanced the substrate hydrophilicity and imparted a negative charge to the membrane surface. At the optimum TiO_2 NP concentration of 0.1wt%, the water flux was increased by 25% and the reverse salt flux was reduced by 50%, as compared with the pristine TFC membrane. In the subsequent fouling tests with humic acid, the modified PRO membrane showed a flux reduction of 32% lower than a commercial TFC membrane.

The applications of metal and metal oxide NPs as additives in the polyamide active layer to fabricate TFN membranes were also widely investigated. The impact of SiO_2 NPs on the morphology and performance of TFC FO membranes was investigated [72]. The SiO_2 was added into the polyamide layer with a concentration varying from 0.01 to 0.1wt%. The best performance was achieved when 0.1 wt/v% SiO_2 was added into the polyamide layer. Water fluxes of approximately 36 LMH under the PRO mode and 23 LMH under the FO mode were obtained using 10 mM NaCl and 1 M NaCl as feed and draw solutions, respectively, doubling the water fluxes obtained by the pristine TFC membrane. Meanwhile, the reverse salt flux decreased when the SiO_2 concentration increased and reached the lowest when the SiO_2 concentration increased to 0.1%. Another TFN membrane was fabricated by dispersing silica NPs functionalized by quaternary ammonium groups and subsequently coated using superhydrophilic wheel polyoxometalates (POM) in the MPD solution [73]. TiO_2 NPs were another type of widely applied NPs in water applications. For example, it was found that the surface morphology gradually evolved into a "leaf-like" structure, and with 0.05% TiO_2 NPs in the polyamide layer, the water flux reached 26.4 LMH under the FO mode when 0.5 M and 10 mM NaCl solutions were adopted as draw and feed solutions, respectively. More importantly, the TiO_2 NP incorporated membrane exhibited strong antifouling properties toward organic foulants. The water flux was able to be fully recovered by a simple rinsing process [74]. Similarly, the surface of the TFC membrane was mineralized by silver chloride (AgCl) in a soaking process as shown in Fig. 14.6 [75]. The modified membrane surface became smoother, more hydrophilic, and more negatively charged. As a result, it exhibited a higher water flux, higher salt rejection, lower flux decline in the BSA solution, and higher water flux recovery after simple rinsing. Aside TFN membranes, Ag nanocomposite membranes can be fabricated via LbL assembly on a PAN substrate for FO applications [76]. The resultant membrane displayed significant antibacterial properties against both Gram-positive *Bacillus subtilis* and Gram-negative *E. coli*. Recent work employed both passive and active strategies to mitigate membrane biofouling of TFC FO membranes [77]. The TFC membrane was first coated with PDA to enhance the passive antiadhesive property. Subsequently, Ag NPs were grown in situ on the surfaces of the active layer and the substrate layer to enhance the antibacterial property. During fouling tests with *Pseudomonas aeruginosa* over a total permeation volume of 250 mL, the new membrane exhibited 21.9% and 8.5% water flux declines in PRO and FO modes, respectively. In comparison, the unmodified TFC

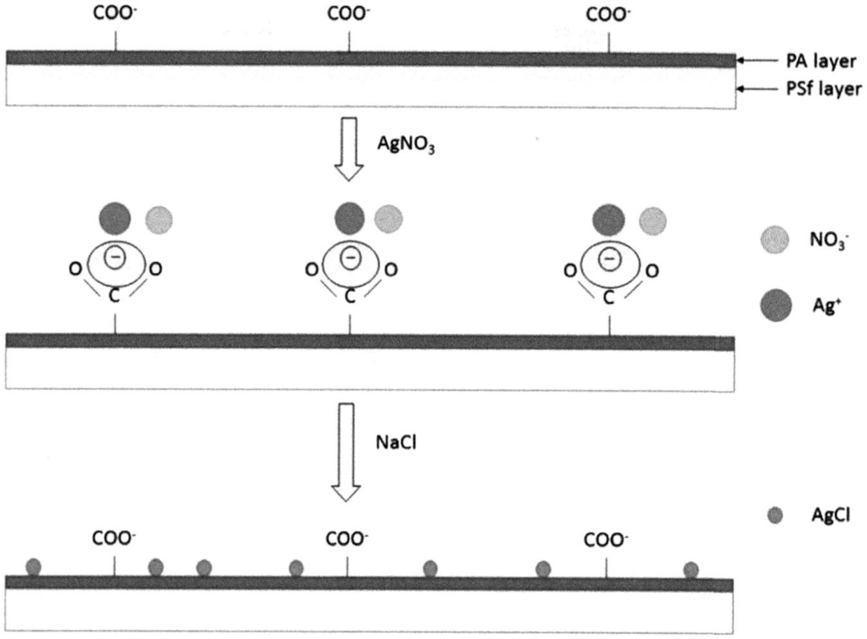

FIGURE 14.6 Concept of depositing AgCl particles on the surface of thin-film composite (TFC) polyamide membranes during surface mineralization. *Copyright obtained from H. Jin, et al., Synthesis of AgCl mineralized thin film composite polyamide membranes to enhance performance and antifouling properties in forward osmosis, Industrial and Engineering Chemistry Research 56 (2017) 1064–1073.*

membrane showed 56.4% and 56.6% declines in PRO and FO modes, respectively. Interestingly, metal oxide NPs can be washed from the active layer and/or the substrate layer and hence form nanovoids. The nanovoids offered additional channels for water transport. However, excessive nanovoids caused deterioration of the membrane integrity [78]. For example, ZnO NPs and nanorods with different nanostructures were added to PES substrates and then dissolved by 1 M HCl solutions before depositing the polyamide layer by interfacial polymerization. The resultant TFC membranes with interconnected porous substrates exhibited significantly lower S parameters. With increasing concentration of the nanovoids, the FO water flux was increased at the sacrifice of the salt rejection [79].

6. ZEOLITES

Zeolites are nanoporous aluminosilicate materials that accommodate alkali and alkali-earth metals. Zeolites have a 3D tetrahedral framework structure in which each oxygen atom is shared by two tetrahedra forming specific intracrystalline cavities and channels with dimensions of typically <1.0 nm [80]. The intrinsically unique pore size between the diameter of hydrated salts and water molecules makes zeolites capable of providing a low-resistance pathway for water molecules, rejecting hydrated salt ions, and therefore offering superior size and shape selectivity [81,82]. The porous structure of zeolites with many cavities and channels significantly increases the surface areas of zeolites [83]. In addition, by varying the silicon and aluminum compositions during syntheses, zeolite can be modified to be highly hydrophilic and highly permeable to water [84,85]. In 2001, Lin et al. used molecular dynamic simulations to show that thin zeolite membranes could be potentially used in reverse osmosis separations [86]. Since then, zeolites have been extensively studied in the field of engineering osmosis. The typical method employed in these studies was either to distribute zeolite NPs into an organic solution during the phase inversion of membranes or to fabricate the active layer via interfacial polymerization.

Ma et al. fabricated polysulfone-nanocomposite (PSfN) substrates by incorporating zeolite NPs into PSf to reduce the ICP effects in FO operations [87]. The PSfN0.5 substrate, which had a zeolite loading of 0.5wt%, showed a higher surface porosity, a lower water contact angle, and a thinner substrate thickness. In addition, the porous zeolite NPs provided additional water pathways, improved pore connectivity, and reduced substrate's tortuosity [88]. Therefore, the PSf substrate in the pristine TFC membrane greatly reduced its structure parameter from 0.96 to 0.34 mm for the PSfN substrate in the PSfN0.5-TFC membrane, while the latter has an FO water flux two times higher than the former. In addition, the PSfN0.5-TFC membrane had a water flux of 80 LMH under the PRO mode and 40 LMH under the FO mode when using 2 M NaCl and 10 mM NaCl as the feed pair. Nanostructured zeolites (i.e., clinoptilolite) of 0–0.6wt% were also incorporated into a PES substrate and proved to be effective in reducing the ICP effects [89].

In addition to modifications of the substrates of TFC membranes, porous NPs have been incorporated into the polyamide layer as an alternative way to improve the FO performance. For example, zeolite-polyamide/PSf TFN membranes were prepared via interfacial polymerization [90]. The addition of NaY zeolite NPs changed the intrinsic separation properties of the polyamide selective layer significantly. The water permeability was enhanced when the zeolite loading was within the range of 0.02–0.1 wt/v% in the TMC/hexane solution because the porous zeolite provided additional water pathways. However, when the zeolite loading increased from 0.1 wt/v%, the water permeability of the TFN membranes decreased due to the formation of a thicker polyamide layer. The water permeability of the most permeable TFN membrane with a zeolite loading of 0.1 wt/v% was approximately 80% higher than that of the pristine TFC membrane. It had water fluxes of 30.7 LMH under the PRO mode and 14.6 LMH under the FO mode when using 1 M NaCl as the draw solution and 10 mM NaCl as the feed solution.

As predicted by the Maxwell mixing model, the water permeability of a zeolite/polymer nanocomposite membrane would increase with an increase in zeolite loading [91]. However, the novel zeolite/polymer nanocomposite membranes could not achieve a high zeolite loading because of the poor compatibility between the zeolite and polymer. Therefore, Kang et al. presented a highly efficient LbL assembly approach to increase the zeolite loading in nanocomposite membranes [92]. The negatively charged Linde type A (LTA) zeolite NPs were sandwiched between the negatively charged poly(acrylic acid) (PAA) layer and the positively charged PEI layer, forming a PEI–LTA–PAA trilayer on top of a PDA-treated PAN substrate. Because of electrostatic interactions, the compatibility between the zeolite and polymers was drastically improved. Within the zeolite loadings of 30–60wt%, the water permeability of the PEI–LTA–PAA trilayer nanocomposite membrane was enhanced more than twofold without compromising membrane selectivity, as compared with a control PEI–PAA bilayer membrane. As a result, the water flux of the two-trilayer membrane was 17.3 LMH, doubling the water flux of the PEI–PAA bilayer membrane, while the reverse solute fluxes of the two membranes were similar using 1 M $MgCl_2$ and DI water as the feed pair under the FO mode.

7. METAL–ORGANIC FRAMEWORKS

Metal–organic frameworks (MOFs), a family of novel highly crystalline and highly porous nanomaterials, have also been extensively explored to enhance membranes performance. In comparison with hydrophilic inorganic porous nanomaterials, porous MOFs materials have an organic and flexible backbone in addition to high surface area and pore volume [93]. Because of the presence of both metal ions/clusters and organic ligands, MOFs have three-dimensional networks with better compatibility with polymeric matrices. Hence, it may minimize the formation of nonselective voids between MOFs and polymer matrices [94]. Moreover, the tuneable chemical structure and physical characteristics of MOFs, such as pore size/shape, surface properties, and particles size, make them great additives in membranes for water treatment [95].

Lee et al. utilized MOFs as a removable filler to fabricate MOF-based porous matrix membranes, aiming to control ICP inside the substrate of FO membranes by reducing the structure parameter [96]. In their work, three kinds of MOF particles with different water stability (HKUST-1, MIL-53, Fe-BTC) were added into PAN polymer solutions to prepare MOF-based porous matrix substrates via phase inversion. Then a LbL method was employed to deposit a thin selective layer on top of the porous matrix substrates. As shown in Fig. 14.7, this MOF-based porous matrix strategy leads to form macropores in the porous matrix substrates, thus enhancing membrane porosity and decreasing membrane tortuosity at the same time. For MOF particles with a lower bond dissociation energy (BDE) between the metal ions and organic linkers, the corresponding MOF-based porous matrix membranes showed a higher membrane bulk porosity. As a result, the mass transfer coefficient of water inside the porous matrix substrates would be improved. PMM_{C300}, with the smallest BDE, showed the highest water flux of 132 LMH using 3.0 M $MgCl_2$ and DI water as the feed pair [96]. Similarly, Zirehpour et al. also embedded a hydrophilic copper-based MOF (Cu-BTC) into a cellulosic membrane matrix to improve membrane morphology and performance for FO applications by mitigating the ICP effects [94]. The MOF loading was adjusted to modify the overall porosity, pore interconnectivity, and hydrophilicity of the cellulosic membranes. As a result, the pure water permeability of the modified membrane was enhanced by 72% compared with the unmodified one, and the structural parameter

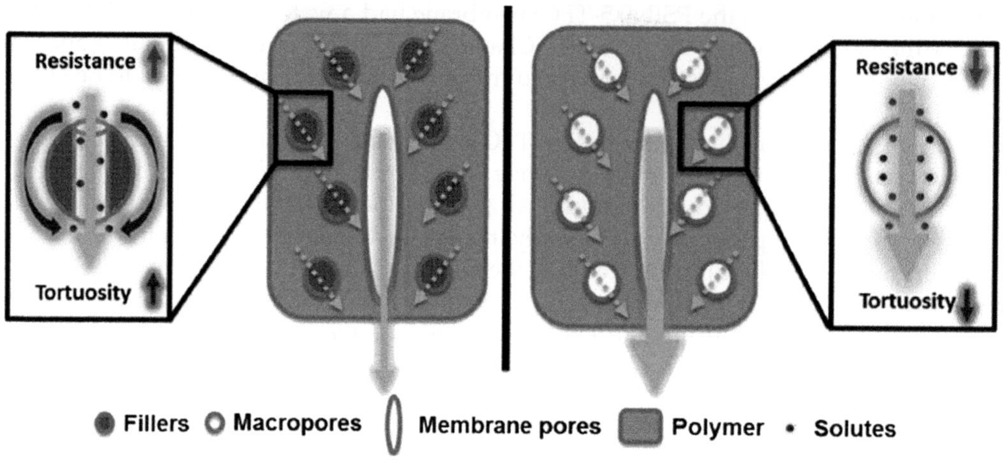

FIGURE 14.7 Schematic diagram of water transport across the macropores inside the metal–organic framework (MOF)-based polymer matrix, which provide additional pathways for water molecules to pass through the support layer with lower resistance and shorter tortuosity. *Copyright obtained from J.-Y. Lee, Q. She, F. Huo, C.Y. Tang, Metal-organic framework-based porous matrix membranes for improving mass transfer in forward osmosis membranes, Journal of Membrane Science 492 (2015) 392–399.*

of the modified membrane was reduced to 136 mm. Because of the hydrophilic surface and highly porous structure, Cu-BTC particles could absorb water and provide additional water transport pathways through the membrane. Consequently, compared with the unmodified membrane, the water flux of the modified membrane showed an enhancement about 180% without deteriorating the salt rejection. The water flux of the modified membrane was above 30 LMH using a 2 M NaCl draw solution. In addition, its antifouling properties were considerably enhanced after the modification due to the improved membrane hydrophilicity. Thus, its water flux was more easily recovered through physical cleaning.

In addition to the modification of FO membrane substrates, MOFs are also utilized to modify the polyamide layer of FO TFC membranes. Ma et al. and Zirehpour et al. incorporated Zirconiumv (IV)-carboxylate MOF UiO-66 NPs and Ag-MOF into polyamide layers by dissolving them in the organic phase through interfacial polymerization, respectively [97,98]. In comparison with the unmodified TFC membranes, the incorporation of MOFs NPs influences the active layer in terms of hydrophilicity and transport properties without deteriorating the selectivity. The intrinsic separation properties of the modified membranes showed great enhancements because of the molecular sieving and hydrophilic nature of these MOF particles. The water permeability of the modified TFC membranes was enhanced by 52% and 129% after the incorporation of Zr-MOFs and Ag-MOFs, respectively, while the salt rejection was maintained at a high level (95%). The modified membranes with 0.1wt% UiO-66 presented the maximum water flux of 51.3 LMH and 27 LMH under PRO and FO modes, respectively, using 2 M NaCl as the draw solution and DI water as the feed. Under the same testing conditions, the modified membranes with 0.04% Ag-MOF showed the highest water flux of 82 LMH under the PRO mode. Zirehpour et al. also utilized the antibacterial activity of Ag-based MOF to control biofouling in FO [99]. The amount of viable bacteria attached onto the membrane incorporated with MOFs was 90% less than the unmodified one. The biofilm formation was also significantly suppressed because of the enhanced antiadhesive and antimicrobial properties of MOFs. During the biofouling experiments using *E. coli*, the water flux of the MOF-modified membranes decreased only 8% after 24 h, while the water flux of the control decreased 21% after 24 h.

8. POLYELECTROLYTES

Polyelectrolyte membranes assembled via the LbL process offer good water flux and rejection toward divalent ions accompanied with decent solvent resistance and thermal stability [31,100,101]. This assembly technique was first proposed by Decher [102]. In principle, LbL membranes are fabricated by alternatively soaking the substrate to positively and negatively charged polyelectrolytes. A typical polyelectrolyte pair would be PAH and poly(sodium 4-styrene sulfonate) (PSS). Before soaking, the membrane substrate may need some chemical treatments to yield negatively charged hydrophilic functional groups for polycation attraction to initialize the LbL assembly [103–105]. Growth of nanometer-thick polycation/polyanion layers with nanofiltration (NF)-like separation performance is achieved via electrostatic interaction. Posttreatments such as chemical or UV crosslinking can tighten polyelectrolyte layers and further enhance membranes' stability. Therefore, the LbL deposition of polyelectrolytes provides a high degree of control to fabricate ultrathin selective films with good stability, precise thickness, and tuneable surface morphology for engineering osmosis.

The first study of fabricating polyelectrolyte PAH/PSS multilayer FO membranes via LbL assembly was reported in 2011 [100]. In the FO mode, a remarkable flux of 28.7 LMH was obtained by using 1 M $MgCl_2$ and DI water as draw and feed solutions, respectively. However, the reverse solute flux was severe. Assembling a larger number of polyelectrolyte bilayers improved the solute rejection but inevitably reduced the water permeability [103–105]. Chemical crosslinking modifications enhanced the separation performance as well as the long-term stability [106–108]. Comparing with the noncrosslinked membranes, the glutaraldehyde (GA)-crosslinked PAH/PSS FO membrane exhibited better $MgCl_2$ rejections. More importantly, it was possible to reach a water flux of around 100 LMH using DI water and 2–3 M $MgCl_2$ as the feed pair under the PRO mode [104]. A subsequent exposure of the GA-crosslinked PAH/PSS FO membrane to UV irradiation under vacuum further elevated its rejections toward monovalent ions [105]. Alternating layers of PAA and PEI were deposited on an electrospun PVDF nanofiber supported to form a TFC membrane and achieved a pure water permeability of 4.12 $L\ m^{-2}\ h^{-1}\ bar^{-1}$ and FO water flux of 45.2 $L\ m^{-2}\ h^{-1}$ using 2M NaCl solution and DI water as the feed pair [108]. Other fabrication parameters such as pH, salt polyelectrolyte concentration, and substrate characteristics have also been investigated to tailor the membrane surface with enhanced FO performances [106]. The aforementioned studies of polyelectrolyte FO membranes were all based on flat-sheet membranes for easy LbL assembly. Later researches were extended to hollow fiber membranes [109,110]. Polyelectrolytes were LbL assembled to form an NF-like second skin on the back surface of TFC membrane substrates. Comparing with single-skinned membranes, double-skinned LbL assembly polyelectrolyte FO membranes had much better antifouling performance because the LbL second skin can greatly reduce the foulant clogging from the back surface of porous supports and minimize fouling within the porous substrates [109–111].

9. AQUAPORINS

Nature offers the most efficient way for water transport across membranes with the aid of aquaporins (AQPs) under an osmotic pressure gradient. AQPs are water channel proteins typically bound in phospholipid cell membranes, exhibiting an ultrahigh water permeability up to 109 water molecules per second per channel [112]. AQPs can quickly deliver water across the cell wall at a sufficiently low energy. In 1992, Agre and his coworks from Johns Hopkins University discovered aquaporin-1 in red blood cells and won the Nobel Prize [113]. Later, more than 450 members within the AQP family have been identified [114]. Kumar et al. were the pioneers incorporating aquaporin Z (AqpZ) from *E. coli* bacterial cells into a polymeric membrane using an amphiphilic triblock polymer as vesicles [115]. The resultant protein–polymer membrane had over an order of magnitude rise in water permeability than the unmodified one with theoretically perfect rejections toward glucose, glycerol, salt, and urea. These pioneering works sparked a wave of studies using AQPs in FO and PRO membranes.

There are two types of biomimetic membranes; namely, (1) membranes with an AQP-based active layer on a porous support in Fig. 14.8A [117–120] and (2) AQP-based TFN membranes in Fig. 14.8B [121–123]. The structure of AQP-based active layer membranes is analogous to that of biological cellular membranes, where AQPs embedded in the lipid bilayer regulate the water transport and solute rejection. This allows the potential synthesis of biomimetic membranes with an ultrafast water permeability and a relatively high salt rejection. Nonetheless, the key hurdles in this design are (1) defects formed because of the thin and delicate self-assembled amphiphilic matrix, (2) a low surface coverage of AQPs on the porous substrate, (3) potential delamination at the interface between the rejection layer and the substrate, (4) maintenance of the AQP activity, and (5) difficulties of scaling up for industrial applications [117–120]. Significant progresses have been made during the recent years. Chung's group applied vesicles of an ABA block copolymer with methacrylate head groups containing AqpZ onto acrylate functionalized gold-coated polycarbonate tracked-etched (PCTE) substrates under pressure, followed by UV crosslinking to form a covalent linkage between functional groups of ABA and the acrylate residues of PCTE. This created an ultrathin selective layer on the substrate with an area of 9.6 mm^2, which displayed a water flux of 18 LMH at an AqpZ to ABA molar ratio of 1:400 using 0.5 M NaCl as the draw solution and ultrapure water as the feed solution [117]. To further improve the membrane mechanical strength and seal any defects between vesicles and substrate pores, AqpZ-embedded block copolymer vesicles were UV-crosslinked with both acrylate and disulphide groups, immobilized on gold-coated PCTE substrates via covalent binding, and then encapsulated by a PDA-histidine coating [118].

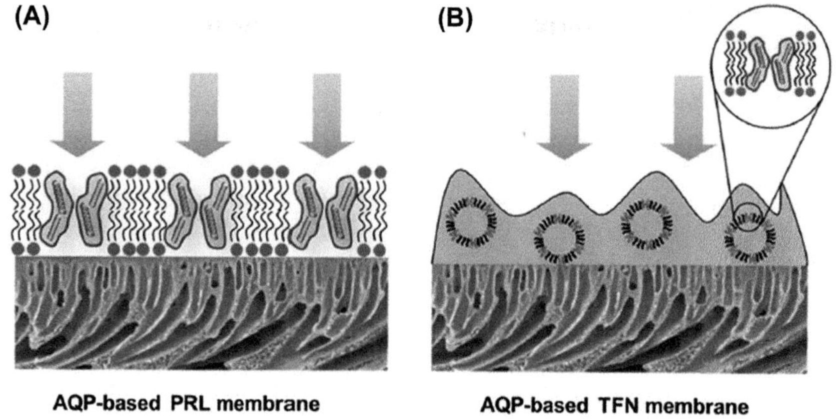

FIGURE 14.8 Schematic diagrams of aquaporin (AQP)-based desalination membranes: (A) AQP-based active layer on a porous substrate; (B) AQP-based TFN membranes on a porous substrate. *Copyright obtained from Z. Yang, X.H. Ma, C.Y. Tang, Recent development of novel membranes for desalination, Desalination 434 (2018) 37–59.*

By sealing the defects, the water flux was doubled to 40 LMH under the same FO testing conditions. However, a decrease of 25% in NaCl rejection between its FO and NF tests suggested the potential membrane deformation under applied hydraulic pressures. To overcome this, Xie et al. reported an in situ "surface imprinting" polymerization strategy, which immobilized AqpZ-containing polymer vesicles on a porous membrane support [119]. This membrane could maintain the transport characteristics under 5 bar but showed a low FO water flux. Ding et al. deposited an AqpZ embedded bilayer with positively charged phospholipid on a PDA-coated PSf substrate [120]. A water flux of 23 LMH and a reverse salt flux of 3.1 gMH were obtained with a relatively large membrane area of 36 cm^2 by using 2 M $MgCl_2$ and DI water as draw and feed solutions, respectively, under the FO mode.

Comparing with AQP-based active layer membranes, AQP-based TFN membranes can be fabricated in a larger membrane area with enhanced membrane stability. The dense salt-rejecting matrix, such as polyelectrolyte and polyamide, not only ensures an adequate salt rejection but also functions as a protection shield of AQPs to enhance the membrane's physical and chemical tolerances. However, the ultimate water transport rate in the AQP-based TFN design is limited by the dense salt-rejection matrix, while the AQP-based active layer design may take full advantages of the high permeability of AQPs. One strategy of fabricating AQP-based TFN membranes is to use the LbL polyelectrolyte assembly approach [122]. The AqpZ-containing vesicles were sandwiched between the polyelectrolyte PAH/PSS-PAA multilayers on top of a PAN substrate. To further enhance the stability and deposition of AqpZ-containing vesicles, the vesicles were protected by poly-L-lysine (PLL) and incorporated with magnetic NPs. Under a strong magnetic force, the adsorption and coverage of PLL-encapsulated magnetic proteoliposomes on the membrane surface were greatly improved. The prepared membrane with 2% magnetic proteoliposomes showed a water flux of 83.5 LMH against DI water feed when using 1.5 M $MgCl_2$ as the draw solution under the PRO mode, 70% higher than the control membrane. Another approach was to immobilize the AqpZ-containing vesicles on the membrane substrate and subsequently coat a polyamide layer by a non–gas-assisted interfacial polymerization process [123,124]. With an optimized AQP-to-lipid ratio of 0.01, the AQP-based hollow fiber membrane showed a remarkable water flux of 55.2 LMH and a low reverse salt flux of 4.5 gMH when being tested using 0.5 M NaCl and DI water as the feed pair under the PRO mode [123]. The structure of hollow fiber substrates was then optimized to achieve a lower structural parameter [124]. As a result, the water flux was elevated to 64.7 LMH, and the reverse salt flux was slightly increased to 8.3 gMH under the same FO testing conditions.

An important consideration of biomimetic membranes is their stability in real applications over a prolonged operation duration. In this case, the dense polymeric matrix of AQP-based TFN membranes can provide better protection to AQPs against physical and chemical attacks. For instance, Jensen designed the utilization of AQP channels in PRO processes for osmotic power generation [125]. Some promising progresses have also been made toward commercialization of biomimetic membranes by AQP A/S and its subsidiary Aquaporin Asia Pte. Ltd. [126–128]. The AQP-based TFN FO flat-sheet membrane is now commercially available, which could achieve water fluxes of 14 LMH and 8.8 LMH with low reverse salt fluxes of 4.6 gMH and 4 gMH under PRO and FO modes, respectively, using 1.0 M NaCl as the draw solution and DI water as the feed solution [128]. To investigate the stability of the commercial biomimetic membrane, Li et al. tested the membrane against a wide range of chemical cleaning agents [127]. It was found that the membrane possessed a stable salt rejection up to 99% with an increase in water permeability possibly due to the change of membrane structure and the hydrophilicity increase in the surrounding area of AQPs. A stable water flux of 4.6 LMH and a high salt rejection of 98.5% were maintained in a 15-day test using seawater as the draw solution and secondary wastewater effluent as the feed solution.

10. ZWITTERIONS

Zwitterionic polymers possess both cationic and anionic groups in the same monomer unit while maintaining overall charge neutrality. On one hand, the localized charges facilitate the strong bonding between zwitterionic polymers and water molecules. On the other hand, the electrical neutrality of the zwitterionic polymers could reduce the electrostatic interactions between the membrane surface and many foulants in the feed streams [129–131]. Therefore, zwitterionic polymers cannot only enhance water transport across the membrane but also improve antifouling properties of the membranes [132–134]. So far, different types of zwitterionic polymers have been incorporated into FO and PRO membranes by means of interfacial polymerization [129–131] and surface modification [132–134].

Yu et al. introduced polyzwitterions (e.g., poly [2-(methacryloyloxy)ethyl] dimethyl-(3-sulfopropyl) ammonium hydroxide) into the polyamide selective layer of FO membranes via a click chemistry approach by forming covalent bonds between the azide and alkynyl groups [114]. Comparing with the unmodified membrane, the modified membrane exhibited a negligible water flux decline during FO fouling tests and full water flux recovery after cleaning. Moreover, Park et al. explored the modification of the polyamide layer using another polyzwitterion (3-(trimethoxysilyl)-propyldimethyloctadecyl ammonium chloride) via the reaction with the -NH- groups of the polyamide layer [115]. This modification improved the antibacterial function of TFC membranes and inhibited the biofilm formation on membrane surface during FO fouling tests.

In PRO applications, zwitterionic polymers are attached to the back surface of the substrates of TFC membranes to mitigate fouling from the feed solution. Cai et al. grafted zwitterionic copolymers with thiol functionalities onto the PDA pretreated outer surface of TFC hollow fiber membranes via Michael addition, as shown in Fig. 14.9 [132]. The grafted membranes were effective in reducing bacterial adhesion and protein adsorption. A substantial improvement of water flux recovery up to 98% was achieved after backwashing and hydraulic pressure impulsion. A similar

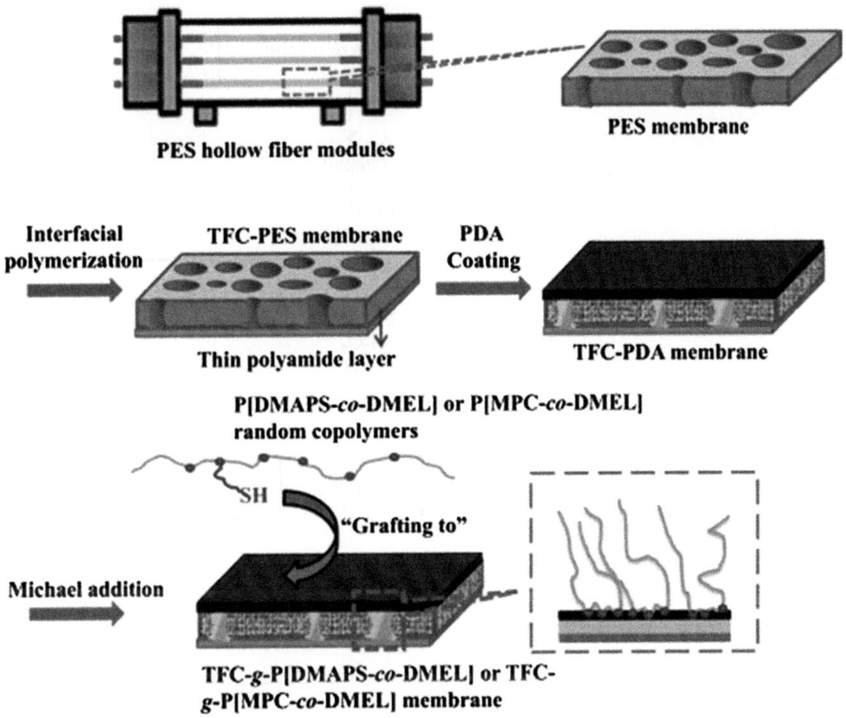

FIGURE 14.9 Schematic procedures to graft zwitterionic polymers on the back surface of the substrate of thin-film composite (TFC) membranes. *Copyright obtained from D.L. Zhao, S. Das, T.-S. Chung, Carbon quantum dots grafted antifouling membranes for osmotic power generation via pressure-retarded osmosis process, Environmental Science and Technology 51 (2017) 14016–14023.*

approach was taken by Le et al. by grafting zwitterionic poly(sulfobetaine) onto the lumen side of hollow fiber membranes to increase the fouling resistance [133]. To streamline the synthesis and coating procedures, 2-methacryloyloxyethyl phosphorylcholine (MPC) was modified and grafted onto the PDA-coated PES hollow fiber substrate. Comparing with the pristine PES and PES-PDA substrates, the MPC-modified substrate exhibited high resistance to protein adsorption and bacteria adhesion [134].

11. SUSTAINABILITY OF NANOSTRUCTURED MEMBRANES

Because large amounts of NPs are incorporated into the membranes for FO and PRO application, more researches are concerned about the sustainability, toxicity, and environmental impacts of NPs in cradle-to-gate and cradle-to-grave analyses [135,136]. A life cycle assessment (LCA) focusing on the energy use, blue water footprint, human toxicity, and ecotoxicity was conducted on graphene production by ultrasonication and chemical reduction methods. The cradle-to-gate life cycles of both routes are shown in Fig. 14.10. The study revealed that the ultrasonication route (USR) had lower energy and water usages but higher human and ecotoxicity impacts, compared to the chemical reduction route (CRR). Solvent recovery was identified to be crucial to reduce the blue water prints of both methods [136]. Biomolecules, such as L-alanine, were employed in a one-step economical and green method for chemical reduction of GO without using alkaline or additional chemicals [137]. For Ag, copper, and zinc NPs, LCAs showed that the high energy consumptions to produce the pure metals were the main cause of environmental impacts. Moreover, the energy consumptions and environmental footprints were largely determined by the weight fraction of NPs in the products [138]. Progresses have been made in developing green synthesis routes of NPs at ambient temperatures and neutral pH conditions with reduced solvent and energy consumptions [139,140].

Green chemistry routes of membrane syntheses have also been proposed to promote sustainability of membrane technologies [141]. Biopolymers derived from animals, vegetables, and bacterial fermentation products are promising candidates to replace conventional fossil-based polymers in membrane fabrication to reduce plastic wastes when the membranes are disposed [142]. Such biopolymers offer the advantages of biodegradability, low toxicity, biocompatibility, and possibility to recycle them. Great challenges and opportunities lie in the reduction of solvent consumption, promotion of solvent recycling, and exploration of "green" solvents [143]. For example, PolarClean

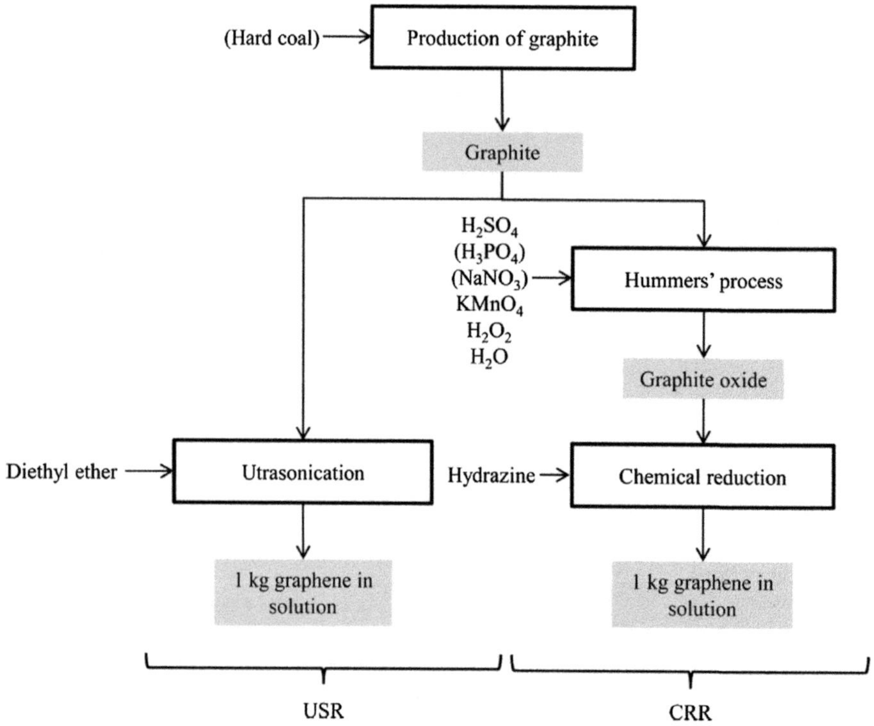

FIGURE 14.10 Flowchart describing the cradle-to-gate life cycle of graphene produced via the ultrasonication route (USR) and the chemical reduction route (CRR), respectively. *Copyright obtained from R. Arvidsson, D. Kushnir, B.A. Sanden, S. Molander, Prospective life cycle assessment of graphene production by ultrasonication and chemical reduction, Environmental Science and Technology 48 (2014) 4529–4536.*

with a high solubility of conventional polymers, low toxicity, and good biodegradability is used as a green solvent to prepare PES, PSf, PVDF, and CA membranes via non–solvent-induced phase separation [144]. Results showed that the performances of the prepared membranes were competitive with the current state-of-the-art membranes in ultrafiltration and NF.

12. SUMMARIES AND OUTLOOKS

Membrane technologies for FO and PRO have significantly advanced in the last decades. To further improve the membrane performance, a wide range of nanomaterials, such as graphene-based materials, carbon nanotubes, zeolites, MOFs, and AQPs, have been utilized to modify the membrane substrates, form the entire TFN membranes, or only form the active layer. These nanomaterials have shown effectiveness in achieving favorable membrane structures and enhancing desirable physicochemical properties, such as transport and antifouling characteristics of FO and PRO membranes. In one example, NPs can increase substrate porosity, reduce substrate tortuosity, and therefore reduce the ICP effects within the substrates. In another example, NPs can be used to fabricate TFN membranes with enhanced hydrophilicity, improved morphology, and additional water pathways in the polyamide layer. Fig. 14.11 summarizes the commercial potential, membrane enhancement, and cost of membrane production of various nanostructured membranes discussed in this chapter.

In terms of material and membrane research, more researches are required to synthesize environmental friendly, low-cost, and easy-to-produce NPs. The release of NPs from TFN membranes or substrates needs to be closely monitored to prevent their contamination in the feed and draw solutions. Moreover, most of the current researches are based on flat-sheet membranes. In contrast to flat-sheet membranes, hollow fiber membranes are self-supporting, which eliminates the need of spacers and offers a denser packing. The recent development of TFC hollow fiber membranes provides an interesting alternative configuration to study the effects of NPs in FO and PRO membranes. TFN membranes can be readily fabricated via interfacial polymerization on the inner surface of commercial-size hollow fiber modules. The bottlenecks of fabricating TFN modules are how to (1) disperse NPs in the interfacial polymerization solutions and (2) ensure a uniform distribution and coating of the solutions and NPs on the inner surfaces of membranes. Fouling is a great challenge to FO and PRO operations. Especially in PRO, the ICP effects within the substrates accelerate fouling and scaling. Although various NPs have

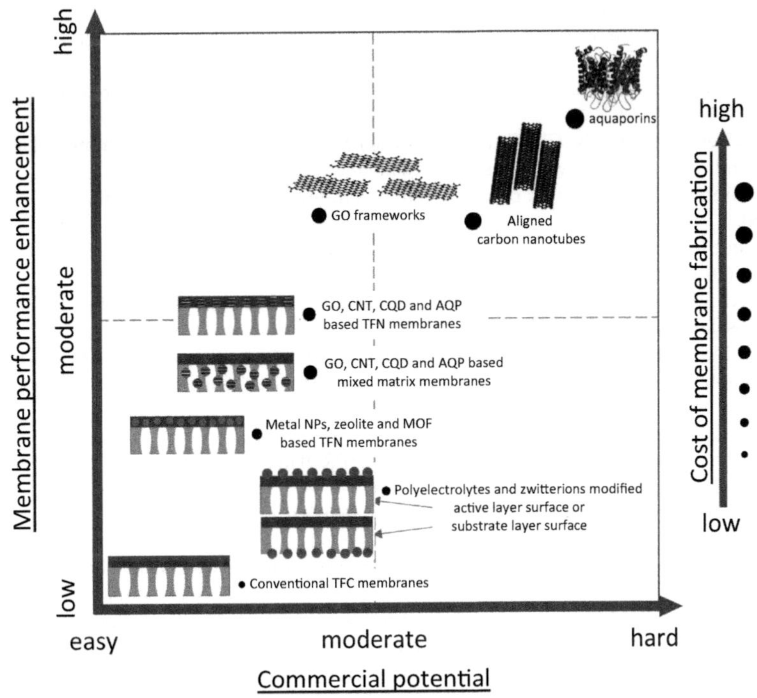

FIGURE 14.11 The commercial potential membrane performance enhancement and cost of various membranes. (A similar method to [116,145] is used to assess the membranes.).

demonstrated superior antifouling properties against big foulants, the abilities of these NPs to prevent the penetration of smaller molecules remain unstudied. It is therefore imperative to study the antifouling effects of NPs with the actual feeds, such as municipal and industrial wastewater, and screen the suitable NPs for the target applications.

In the aspect of membrane scale-up and operation, the majority of NP applications in FO and PRO membranes have been demonstrated in small membrane coupons, with the exception of AQP-based TFN membranes that have been fabricated into commercial size modules. Fabrication of large-scale membrane modules are important to evaluate whether the membranes can be produced in a large area with minimal defects, high uniformity, and the flexibility to roll up into a spiral wound module or assemble into a hollow fiber module. Fabrication of large-scale membrane modules, more importantly, is the stepping stone for pilot and long-term tests in simulated and realistic conditions. These relevant tests are essential to assess their stability and robustness, reveal potential issues for future operations, and evaluate the economic feasibility of the technology. As more information and data about the nanostructured FO and PRO become available, detailed LCA and technoeconomic analyses need to be conducted to assess the feasibility of commercial production of the new membranes. Environmental safety and health have to be addressed by developing quality check and risk assessment protocols, to ensure a positive acceptance and adoption of the new membrane technologies.

References

[1] T.-S. Chung, X. Li, R.C. Ong, Q.C. Ge, H.L. Wang, G. Han, Emerging forward osmosis (FO) technologies and challenges ahead for clean water and clean energy applications, Current Opinion in Chemical Engineering 1 (2012) 246–257.

[2] B. Van der Bruggen, P. Luis, Forward osmosis: understanding the hype, Reviews in Chemical Engineering 31 (2015) 1–12.

[3] T.Y. Cath, A.E. Childress, M. Elimelech, Forward osmosis: principles, applications, and recent developments, Journal of Membrane Science 281 (2006) 70–87.

[4] T. Thorsen, T. Holt, The potential for power production from salinity gradients by pressure retarded osmosis, Journal of Membrane Science 335 (2009) 103–110.

[5] A. Achilli, T.Y. Cath, A.E. Childress, Power generation with pressure retarded osmosis: an experimental and theoretical investigation, Journal of Membrane Science 343 (2009) 42–52.

[6] K. Lutchmiah, A.R.D. Verliefde, K. Roest, L.C. Rietveld, E.R. Cornelissen, Forward osmosis for application in wastewater treatment: a review, Water Research 58 (2014) 179–197.

[7] T.-S. Chung, L. Luo, C.F. Wan, Y. Cui, G. Amy, What is next for forward osmosis (FO) and pressure retarded osmosis (PRO), Separation and Purification Technology 156 (2015) 856–860.

[8] Q.C. Ge, M. Ling, T.S. Chung, Draw solutions for forward osmosis processes: developments, challenges, and prospects for the future, Journal of Membrane Science 442 (2013) 225–237.

[9] J.R. McCutcheon, R.L. McGinnis, M. Elimelech, A novel ammonia-carbon dioxide forward (direct) osmosis desalination process, Desalination 174 (2005) 1–11.

[10] S. Zhang, P. Wang, X. Fu, T.S. Chung, Sustainable water recovery from oily wastewater via forward osmosis-membrane distillation (FO-MD), Water Research 52 (2014) 112–121.

[11] J.E. Kim, S. Phuntsho, L. Chekli, S. Hong, N. Ghaffour, T. Leiknes, J.Y. Choi, H.K. Shon, Environmental and economic impacts of fertilizer drawn forward osmosis and nanofiltration hybrid system, Desalination 416 (2017) 76–85.

[12] K. Stache, Apparatus for Transforming Seawater, Brackish Water, Polluted Water or the Like Into a Nutrious Drink by Means of Osmosis, 1989. US Patent 4,879,030.

[13] B. Mi, M. Elimelech, Organic fouling of forward osmosis membranes: fouling reversibility and cleaning without chemical reagents, Journal of Membrane Science 348 (2010) 337–345.

[14] S.E. Skilhagen, J.E. Dugstad, R.J. Aaberg, Osmotic power – power production based on the osmotic pressure difference between waters with varying salt gradients, Desalination 220 (2008) 476–482.

[15] B.E. Logan, M. Elimelech, Membrane-based processes for sustainable power generation using water, Nature 488 (2012) 313–319.

[16] C.F. Wan, T.S. Chung, Osmotic power generation by pressure retarded osmosis using seawater brine as the draw solution and wastewater brine as the feed, Journal of Membrane Science 479 (2015) 148–158.

[17] R.J. Aaberg, Osmotic power, Refocus 4 (2003) 48–50.

[18] C.F. Wan, T.S. Chung, Energy recovery by pressure retarded osmosis (PRO) in SWRO–PRO integrated processes, Applied Energy 162 (2016) 687–698.

[19] G. Han, S. Zhang, X. Li, T.-S. Chung, Progress in pressure retarded osmosis (PRO) membranes for osmotic power generation, Progress in Polymer Science 51 (2015) 1–27.

[20] Y.-N. Wang, L. Goh, X. Li, L. Setiawan, R. Wang, Membranes and processes for forward osmosis-based desalination: recent advances and future prospects, Desalination 434 (2018) 81–99.

[21] N. Peng, N. Widjojo, P. Sukitpaneenit, M.M. Teoh, G.G. Lipscomb, T.S. Chung, J.-Y. Lai, Evolution of polymeric hollow fibers as sustainable technologies: past, present and future, Progress in Polymer Science 37 (2012) 1401–1424.

[22] Y. Cui, X.Y. Liu, T.-S. Chung, Enhanced osmotic energy generation from salinity gradients by modifying thin film composite membranes, Chemical Engineering Journal 242 (2014) 195–203.

[23] S. Zhang, P. Sukitpaneenit, T.S. Chung, Design of robust hollow fiber membranes with high power density for osmotic energy production, Chemical Engineering Journal 241 (2014) 457–465.

[24] Z.L. Cheng, X. Li, Y.D. Liu, T.-S. Chung, Robust outer-selective thin-film composite polyethersulfone hollow fiber membranes with low reverse salt flux for renewable salinity-gradient energy generation, Journal of Membrane Science 506 (2016) 119–129.
[25] Y. Cui, Q.C. Ge, X.Y. Liu, T.-S. Chung, Novel forward osmosis process to effectively remove heavy metal ions, Journal of Membrane Science 467 (2014) 188–194.
[26] K.Y. Wang, T.-S. Chung, G. Amy, Developing thin-film-composite forward osmosis membranes based on the PES/SPSf substrate through interfacial polymerization, AIChE Journal 58 (2012) 770–781.
[27] X. Li, S. Zhang, F.J. Fu, T.-S. Chung, Deformation and reinforcement of thin-film composite (TFC) polyamide-imide (PAI) membranes for osmotic power generation, Journal of Membrane Science 434 (2013) 204–217.
[28] D.I. Kim, J. Kim, S. Hong, Changing membrane orientation in pressure retarded osmosis for sustainable power generation with low fouling, Desalination 389 (2016) 197–206.
[29] Q. She, R. Wang, A.G. Fane, C.Y. Tang, Membrane fouling in osmotically driven membrane processes: a review, Journal of Membrane Science 499 (2016) 201–233.
[30] M. Rastgar, A. Shakeri, A. Bozorg, H. Salehi, V. Saadattalab, Impact of nanoparticles surface characteristics on pore structure and performance of forward osmosis membranes, Desalination 421 (2017) 179–189.
[31] D. Li, Y. Yan, H. Wang, Recent advances in polymer and polymer composite membranes for reverse and forward osmosis processes, Progress in Polymer Science 61 (2016) 104–155.
[32] W. Xu, Q. Chen, Q.C. Ge, Recent advances in forward osmosis (FO) membrane: chemical modifications on membranes for FO processes, Desalination 419 (2017) 101–116.
[33] W.A. Suwaileh, D.J. Johnson, S. Sarp, N. Hilal, Advances in forward osmosis membranes: altering the sub-layer structure via recent fabrication and chemical modification approaches, Desalination 436 (2018) 176–201.
[34] N.Y. Yip, A. Tiraferri, W.A. Phillip, J.D. Schiffman, L.A. Hoover, Y.C. Kim, M. Elimelech, Thin-film composite pressure retarded osmosis membranes for sustainable power generation from salinity gradients, Environmental Science and Technology 45 (2011) 4360–4369.
[35] T.Y. Cath, M. Elimelech, J.R. McCutcheon, R.L. McGinnis, A. Achilli, D. Anastasio, A.R. Brady, A.E. Childress, I.V. Farr, N.T. Hancock, J. Lampi, L.D. Nghiem, M. Xie, N.Y. Yip, Standard methodology for evaluating membrane performance in osmotically driven membrane processes, Desalination 312 (2013) 31–38.
[36] S. Zhao, L. Zou, Relating solution physicochemical properties to internal concentration polarization in forward osmosis, Journal of Membrane Science 379 (2011) 459–467.
[37] K.L. Lee, R.W. Baker, H.K. Lonsdale, Membranes for power generation by pressure-retarded osmosis, Journal of Membrane Science 8 (1981) 141–171.
[38] J. Ren, M.R. Chowdhury, J. Qi, L. Xia, B.D. Huey, J.R. McCutcheon, Relating osmotic performance of thin film composite hollow fiber membranes to support layer surface pore size, Journal of Membrane Science 540 (2017) 344–353.
[39] C.F. Wan, T. Yang, W.X. Gai, Y.D. Li, T.-S. Chung, Thin-film composite hollow fiber membrane with inorganic salt additives for high mechanical strength and high power density for pressure-retarded osmosis, Journal of Membrane Science 555 (2018) 388–397.
[40] A.K. Ghosh, E.M.V. Hoek, Impacts of support membrane structure and chemistry on polyamide-polysulfone interfacial composite membranes, Journal of Membrane Science 336 (2009) 140–148.
[41] M.L. Lind, A.K. Ghosh, A. Jawor, X. Huang, W. Hou, Y. Yang, E.M.V. Hoek, Influence of zeolite crystal size on zeolite-polyamide thin film nanocomposite membranes, Langmuir 25 (2009) 10139.
[42] V.B. Mohan, K.-T. Lau, D. Hui, D. Bhattacharyya, Graphene-based materials and their composites: a review on production, applications and product limitations, Composites Part B 142 (2018) 200–220.
[43] M. Tian, R. Wong, K. Goh, Y. Liao, A.G. Fane, Synthesis and characterization of high-performance novel thin film nanocomposite PRO membranes with tiered nanofiber support reinforced by functionalized carbon nanotubes, Journal of Membrane Science 486 (2015) 151–160.
[44] D.L. Zhao, T.-S. Chung, Applications of carbon quantum dots (CQDs) in membrane technologies: a review, Water Research 147 (2018) 43–49.
[45] N. Song, X. Gao, Z. Ma, X. Wang, Y. Wei, C. Gao, A review of graphene-based separation membrane: materials, characteristics, preparation and applications, Desalination 437 (2018) 59–72.
[46] H.M. Hegab, L. Zou, Graphene oxide-assisted membranes: fabrication and potential applications in desalination and water purification, Journal of Membrane Science 484 (2015) 95–106.
[47] M.J. Park, S. Phuntsho, T. He, G.M. Nisola, L.D. Tijing, X.-M. Li, G. Chen, W.-J. Chung, H.K. Shon, Graphene oxide incorporated polysulfone substrate for the fabrication of flat-sheet thin-film composite forward osmosis membranes, J. Membr. Sci. 493 (2015) 496–507.
[48] S. Lim, M.J. Park, S. Phuntsho, L.D. Tijing, G.M. Nisola, W.-G. Shim, W.-J. Chung, H.K. Shon, Dual-layered nanocomposite substrate membrane based on polysulfone/graphene oxide for mitigating internal concentration polarization in forward osmosis, Polymer 110 (2017) 36–48.
[49] L. Shen, S. Xiong, Y. Wang, Graphene oxide incorporated thin-film composite membranes for forward osmosis applications, Chemical Engineering Science 143 (2016) 194–205.
[50] G.S. Lai, Tailor-made thin film nanocomposite membrane incorporated with graphene oxide using novel interfacial polymerization technique for enhanced water separation, Chemical Engineering Journal 344 (2018) 524–534.
[51] M. Rastgar, A. Shakeri, A. Bozorg, H. Salehi, V. Saadattalab, Highly-efficient forward osmosis membrane tailored by magnetically responsive graphene oxide/Fe_3O_4 nanohybrid, Applied Surface Science 441 (2018) 923–935.
[52] H.M. Hegab, A. ElMekawy, T.G. Barclay, A. Michelmore, L. Zou, C.P. Saint, M. Ginic-Markovic, Effective in-situ chemical surface modification of forward osmosis membranes with polydopamine-induced graphene oxide for biofouling mitigation, Desalination 385 (2016) 126–137.
[53] A.F. Faria, C. Liu, M. Xie, F. Perreault, L.D. Nghiem, J. Ma, M. Elimelech, Thin-film composite forward osmosis membranes functionalized with graphene oxide-silver nanocomposites for biofouling control, Journal of Membrane Science 525 (2017) 146–156.
[54] L. Jin, Z. Wang, S. Zheng, B. Mi, Polyamide-crosslinked graphene oxide membrane for forward osmosis, Journal of Membrane Science 545 (2018) 11–18.
[55] E. Yang, C.-M. Kim, J.-H. Song, H. Ki, M.-H. Ham, I.S. Kim, Enhanced desalination performance of forward osmosis membranes based on reduced graphene oxide laminates coated with hydrophilic polydopamine, Carbon 117 (2017) 293–300.

[56] H. Salehi, M. Rastgar, A. Shakeri, Anti-fouling and high water permeable forward osmosis membrane fabricated via layer by layer assembly of chitosan/graphene oxide, Applied Surface Science 413 (2017) 99–108.
[57] M. Amini, M. Jahanshahi, A. Rahimpour, Synthesis of novel thin film nanocomposite (TFN) forward osmosis membranes using functionalized multi-walled carbon nanotubes, Journal of Membrane Science 435 (2013) 233–241.
[58] H.G. Choi, Alginate fouling reduction of functionalized carbon nanotube blended cellulose acetate membrane in forward osmosis, Chemosphere 136 (2015) 204–210.
[59] Y. Wang, R. Ou, Q. Ge, H. Wang, T. Xu, Preparation of polyether-sulfone/carbon nanotube substrate for high-performance forward osmosis membrane, Desalination 330 (2013) 70–78.
[60] M. Son, H. Park, L. Liu, H. Choi, J.H. Kim, H. Choi, Thin-film nanocomposite membrane with CNT positioning in support layer for energy harvesting from saline water, Chemical Engineering Journal 284 (2016) 68–77.
[61] S. Xu, F. Li, B. Su, M.Z. Hu, X. Gao, C. Gao, Novel graphene quantum dots (GQDs)-incorporated thin film composite (TFC) membranes for forward osmosis (FO) desalination, Desalination 451 (2019) 219–230.
[62] W.X. Gai, D.L. Zhao, T.S. Chung, Novel thin film composite hollow fiber membranes incorporated with carbon quantum dots for osmotic power generation, Journal of Membrane Science 551 (2018) 94–102.
[63] S.F. Seyedpour, A. Rahimpour, A.A. Shamsabadi, M. Soroush, Improved performance and antifouling propertiesof thin-film composite polyamide membranesmodified with nano-sized bactericidal graphenequantum dots for forward osmosis, Chemical Engineering Research and Design 139 (2018) 321–334.
[64] M. Obaid, Z.K. Ghouri, O.A. Fadali, K.A. Khalil, A.A. Almajid, N.A.M. Barakat, Amorphous SiO2 NP-incorporated poly(vinylidene fluoride) electrospun nanofiber membrane for high flux forward osmosis desalination, ACS Applied Materials and Interfaces 8 (2016) 4561–4574.
[65] D. Emadzadeh, W.J. Lau, T. Matsuura, M. Rahbari-Sisakht, A.F. Ismail, A novel thin film composite forward osmosis membrane prepared from PSf-TiO_2 nanocomposite substrate for water desalination, Chemical Engineering Journal 237 (2014) 70–80.
[66] D. Emadzadeh, W.J. Lau, T. Matsuura, A.F. Ismail, M. Rahbari-Sisakht, Synthesis and characterization of thin film nanocomposite forward osmosis membrane with hydrophilic nanocomposite support to reduce internal concentration polarization, Journal of Membrane Science 449 (2014) 74–85.
[67] D. Emadzadeh, W.J. Lau, A.F. Ismail, Synthesis of thin film nanocomposite forward osmosis membrane with enhancement in water flux without sacrificing salt rejection, Desalination 330 (2013) 90–99.
[68] A. Nguyen, L. Zou, C. Priest, Evaluating the antifouling effects of silver nanoparticles regenerated by TiO_2 on forward osmosis membrane, Journal of Membrane Science 454 (2014) 264–271.
[69] P. Lu, S. Liang, L. Qiu, Y. Gao, Q. Wang, Thin film nanocomposite forward osmosis membranes based on layered double hydroxide nanoparticles blended substrates, Journal of Membrane Science 504 (2016) 196–205.
[70] X. Liu, L.-X. Foo, Y. Li, J.-Y. Lee, B. Cao, C.Y. Tang, Fabrication and characterization of nanocomposite pressure retarded osmosis (PRO) membranes with excellent anti-biofouling property and enhanced water permeability, Desalination 389 (2016) 137–148.
[71] J. Kim, D. Suh, C. Kim, Y. Baek, B. Lee, H.J. Kim, J.-C. Lee, J. Yoon, A high-performance and fouling resistant thin-film composite membrane prepared via coating TiO2 nanoparticles by sol-gelderived spray method for PRO applications, Desalination 397 (2016) 157–164.
[72] N. Niksefat, M. Jahanshahi, A. Rahimpour, The effect of SiO_2 nanoparticles on morphology and performance of thin film composite membranes for forward osmosis application, Desalination 343 (2014) 140–146.
[73] A. Shakeri, H. Salehi, F. Ghorbani, M. Amini, H. Naslhajian, Polyoxometalate based thin film nanocomposite forward osmosis membrane: superhydrophilic, anti-fouling, and high water permeable, Journal of Colloid and Interface Science 536 (2019) 328–338.
[74] D. Emadzadeh, M. Ghanbari, W.J. Lau, M. Rahbari-Sisakht, T. Matsuura, A.F. Ismail, B. Kruczek, Solvothermal synthesis of nanoporous TiO2: the impact on thin-film composite membranes for engineered osmosis application, Nanotechnology 27 (2016) 345702.
[75] H. Jin, F. Rivers, H. Yin, T. Lai, P. Cay-Durgun, A. Khosravi, M.L. Lind, P. Yu, Synthesis of AgCl mineralized thin film composite polyamide membranes to enhance performance and antifouling properties in forward osmosis, Industrial and Engineering Chemistry Research 56 (2017) 1064–1073.
[76] X. Liu, S. Qi, Y. Li, L. Yang, B. Cao, C.Y. Tang, Synthesis and characterization of novel antibacterial silver nanocomposite nanofiltration and forward osmosis membranes based on layer-by-layer assembly, Water Research 47 (2013) 3081–3092.
[77] L. Qi, Y. Hu, Z. Liu, X. An, E. Bar-Zeev, Anti-biofouling performance of thin-film composite forward-osmosis membranes containing passive and active moieties, Environmental Science and Technology 52 (2018) 9684–9693.
[78] Z. Yang, X. Huang, X.-H. Ma, Z.-W. Zhou, H. Guo, Z. Yao, S.-P. Feng, C.Y. Tang, Fabrication of a novel and green thin-film composite membrane containing nanovoids for water purification, Journal of Membrane Science 570-571 (2019) 314–321.
[79] M. Rastgar, A. Bozorg, A. Shakeri, Novel dimensionally controlled nanopore forming template in forward osmosis membranes, Environmental Science and Technology 52 (2018) 2704–2716.
[80] Y.H. Teow, A.W. Mohammad, New generation nanomaterials for water desalination: a review, Desalination 451 (2019) 2–17.
[81] M.C. Duke, J. O'Brien-Abraham, N. Milne, B. Zhu, J.Y.S. Lin, J.C. Diniz da Costa, Seawater desalination performance of MFI type membranes made by secondary growth, Separation and Purification Technology 68 (2009) 343–350.
[82] G.-R. Xu, J.-N. Wang, C.-J. Li, Strategies for improving the performance of the polyamide thin film composite (PA-TFC) reverse osmosis (RO) membranes: surface modifications and nanoparticles incorporations, Desalination 328 (2013) 83–100.
[83] M. Lenarda, M. Da Ros, M. Casagrande, L. Storaro, R. Ganzerla, Post-synthetic thermal and chemical treatments of H-BEA zeolite: effects on the catalytic activity, Inorganica Chimica Acta 349 (2003) 195–202.
[84] T. Kawai, K. Tsutsumi, Evaluation of hydrophilic-hydrophobic character of zeolites by measurements of their immersional heats in water, Colloid and Polymer Science 270 (1992) 711–715.
[85] J.M. Castillo, J. Silvestre-Albero, F. Rodriguez-Reinoso, T.J.H. Vlugt, S. Calero, Water adsorption in hydrophilic zeolites: experiment and simulation, Physical Chemistry Chemical Physics 15 (2013) 17374.
[86] J. Lin, S. Murad, A computer simulation study of the separation of aqueous solutions using thin zeolite membranes, Molecular Physics 99 (2001) 1175–1181.

[87] N. Ma, J. Wei, S. Qi, Y. Zhao, Y. Gao, C.Y. Tang, Nanocomposite substrates for controlling internal concentration polarization in forward osmosis membranes, Journal of Membrane Science 441 (2013) 54–62.
[88] W. Li, Y. Gao, C.Y. Tang, Network modelling for studying the effect of support structure on internal concentration polarization during forward osmosis: model development and theoretical analysis with FEM, Journal of Membrane Science 379 (2011) 307–321.
[89] T.M. Salehi, M. Peyravi, M. Jahanshahi, W.-J. Lau, A.S. Rad, Impacts of zeolite nanoparticles on substrate properties of thin film nanocomposite membranes for engineered osmosis, Journal of Nanoparticle Research 20 (2018) 113 1–11315.
[90] N. Ma, J. Wei, R. Liao, C.Y. Tang, Zeolite-polyamide thin film nanocomposite membranes: towards enhanced performance for forward osmosis, Journal of Membrane Science 405–406 (2012) 149–157.
[91] M.M. Pendergast, E.M.V. Hoek, A review of water treatment membrane nanotechnologies, Environmental Science and Technology 4 (2011) 1946.
[92] Y. Kang, L. Emdadi, M.J. Lee, D. Liu, B. Mi, Layer-by-layer assembly of zeolite/polyelectrolyte nanocomposite membranes with high zeolite loading, Environmental Science and Technology Letters 1 (2014) 504–509.
[93] Y. Ji, W. Qian, Y. Yu, Q. An, L. Liu, Y. Zhou, C. Gao, Recent developments in nanofiltration membranes based on nanomaterials, Chinese Journal of Chemical Engineering 25 (2017) 1639–1652.
[94] A. Zirehpour, A. Rahimpour, S. Khoshhal, M.D. Firouzjaei, A.A. Ghoreyshi, The impact of MOF feasibility to improve the desalination performance and antifouling properties of FO membranes, RSC Advances 6 (2016) 70174–70185.
[95] X. Li, A. Sotto, J. Li, B. Van der Bruggen, Progress and perspectives for synthesis of sustainable antifouling composite membranes containing in situ generated nanoparticles, Journal of Membrane Science 524 (2017) 502–528.
[96] J.-Y. Lee, Q. She, F. Huo, C.Y. Tang, Metal-organic framework-based porous matrix membranes for improving mass transfer in forward osmosis membranes, Journal of Membrane Science 492 (2015) 392–399.
[97] D. Ma, S.B. Peh, G. Han, S.B. Chen, Thin-film nanocomposite (TFN) membranes incorporated with super-hydrophilic metal-organic framework (MOF) UiO-66: toward enhancement of water flux and salt rejection, ACS Applied Materials and Interfaces 9 (2017) 7523–7534.
[98] A. Zirehpour, A. Rahimpour, M. Ulbricht, Nano-sized metal organic framework to improve the structural properties and desalination performance of thin film composite forward osmosis membrane, Journal of Membrane Science 531 (2017) 59–67.
[99] A. Zirehpour, A. Rahimpour, A.A. Shamsabadi, M. Sharifian Gh, M. Soroush, Mitigation of thin-film composite membrane biofouling via immobilizing nano-sized biocidal reservoirs in the membrane active layer, Environmental Science and Technology 51 (2017) 5511–5522.
[100] Q. Saren, C.Q. Qiu, C.Y. Tang, Synthesis and characterization of novel forward osmosis membranes based on layer-by-layer assembly, Environmental Science and Technology 45 (2011) 5201–5208.
[101] M.L. Bruening, D.M. Dotzauer, P. Jain, L. Ouyang, G.L. Baker, Creation of functional membranes using polyelectrolyte multilayers and polymer brushes, Langmuir 24 (2008) 7663–7673.
[102] G. Decher, Fuzzy nanoassemblies: toward layered polymeric multicomposites, Science 277 (1997) 1232–1237.
[103] S. Qi, W. Li, Y. Zhao, N. Ma, J. Wei, T.W. Chin, C.Y. Tang, Influence of the properties of layer-by-layer active layers on forward osmosis performance, Journal of Membrane Science 423–2424 (2012) 536–542.
[104] C.Q. Qiu, S.R. Qi, C.Y. Tang, Synthesis of high flux forward osmosis membranes by chemically crosslinked layer-by-layer polyelectrolytes, Journal of Membrane Science 381 (2011) 74–80.
[105] P.H.H. Duong, J. Zuo, T.-S. Chung, Highly crosslinked layer-by-layer polyelectrolyte FO membranes: understanding effects of salt concentration and deposition time on FO performance, Journal of Membrane Science 427 (2013) 411–421.
[106] Y. Cui, H. Wang, H. Wang, T.-S. Chung, Micro-morphology and formation of layer-by-layer membranes and their performance in osmotically driven processes, Chemical Engineering Science 101 (2013) 13–26.
[107] F. Wang, J. Feng, C. Gao, Manipulating the properties of coacervated polyelectrolyte microcapsules by chemical crosslinking, Colloid and Polymer Science 286 (2008) 951–957.
[108] R.R. Gonzales, M.J. Park, L. Tijing, D.S. Han, S. Phuntsho, H.K. Shon, Modification of nanofiber support layer for thin film composite forward osmosis membranes via layer-by-layer polyelectrolyte deposition, Membranes 8 (2018) 70.
[109] C. Liu, W. Fang, S. Chou, L. Shi, A.G. Fane, R. Wang, Fabrication of layer-by layer assembled FO hollow fiber membranes and their performances using low concentration draw solutions, Desalination 308 (2013) 147–153.
[110] W. Fang, C. Liu, L. Shi, R. Wang, Composite forward osmosis hollow fiber membranes: integration of RO- and NF-like selective layers for enhanced organic fouling resistance, Journal of Membrane Science 492 (2015) 147–155.
[111] S. Qi, C.Q. Qiu, Y. Zhao, C.Y. Tang, Double-skinned forward osmosis membranes based on layer-by-layer assembly-FO performance and fouling behavior, Journal of Membrane Science 405–406 (2012) 20–29.
[112] M.Ø. Jensen, O.G. Mouritsen, Single-channel water permeabilities of *Escherichia coli* aquaporins AqpZ and GlpF, Biophysical Journal 90 (2006) 2270–2284.
[113] G.M. Preston, T.P. Carroll, W.B. Guggino, P. Agre, Appearance of water channels in *Xenopus oocytes* expressing red cell CHIP28 protein, Science 256 (1992) 385–387.
[114] S. Sasaki, Introduction for special issue for aquaporin, Pflugers Archiv: European Journal of Physiology 456 (2008) 647–649.
[115] M. Kumar, M. Grzelakowski, J. Zilles, M. Clark, W. Meier, Highly permeable polymeric membranes based on the incorporation of the functional water channel protein aquaporin Z, Proceedings of the National Academy of Sciences of the United States of America 104 (2007) 20719–20724.
[116] Z. Yang, X.H. Ma, C.Y. Tang, Recent development of novel membranes for desalination, Desalination 434 (2018) 37–59.
[117] H.L. Wang, T.S. Chung, Y.W. Tong, K. Jeyaseelan, A. Armugam, Z. Chen, M. Hong, W. Meier, Highly permeable and selective pore-spanning biomimetic membrane embedded with aquaporin Z, Small 8 (2012) 1185–1190.
[118] H.L. Wang, T.S. Chung, T.W. Tong, K. Jeyaseelan, A. Armugam, H.H.P. Duong, F. Fu, H. Seah, J. Yang, M. Hong, Mechanically robust and highly permeable AquaporinZ biomimetic membranes, Journal of Membrane Science 434 (2013) 130–136.
[119] W. Xie, F. He, B. Wang, T.S. Chung, K. Jeyaseelan, A. Armugam, Y.W. Tong, An aquaporin-based vesicle-embedded polymeric membrane for low energy water filtration, Journal of Materials Chemistry A 26 (2013) 7592–7600.
[120] W. Ding, J. Cai, Z. Yu, Q. Wang, Z. Xu, Z. Wang, C. Gao, Fabrication of an aquaporin-based forward osmosis membrane through covalent bonding of a lipid bilayer to a microporous support, Journal of Materials Chemistry A 40 (2015) 20118–20126.

[121] C. Tang, Z. Wang, I. Petrinić, A.G. Fane, C. Hélix-Nielsen, Biomimetic aquaporin membranes coming of age, Desalination 368 (2015) 89–105.
[122] G. Sun, T.S. Chung, N. Chen, X. Lu, Q. Zhao, Highly permeable aquaporin-embedded biomimetic membranes featuring a magnetic-aided approach, RSC Advances 3 (2013) 9178–9184.
[123] X. Li, S. Chou, R. Wang, L. Shi, W. Fang, G. Chaitra, C.Y. Tang, J. Torres, X. Hu, A.G. Fane, Nature gives the best solution for desalination: aquaporin-based hollow fiber composite membrane with superior performance, Journal of Membrane Science 494 (2015) 68–77.
[124] X. Li, C.H. Loh, R. Wang, W. Widjajanti, J. Torres, Fabrication of a robust high-performance FO membrane by optimizing substrate structure and incorporating aquaporin into selective layer, Journal of Membrane Science 525 (2017) 257–268.
[125] P.H. Jensen, Biomimetic Water Membrane Comprising Aquaporins Used in the Production of Salinity Power, 2011. EP 1,937,395 B1.
[126] H.T. Madsen, N. Bajraktari, C. Hélix-Nielsen, B. Van der Bruggen, E.G. Søgaard, Use of biomimetic forward osmosis membrane for trace organics removal, Journal of Membrane Science 476 (2015) 469–474.
[127] Z. Li, R.V. Linares, S. Bucs, L. Fortunato, C. Hélix-Nielsen, J.S. Vrouwenvelder, N. Ghaffour, T. Leiknes, G. Amy, Aquaporin based biomimetic membrane in forward osmosis: chemical cleaning resistance and practical operation, Desalination 420 (2017) 208–215.
[128] L. Xia, M.F. Andersen, C. Hélix-Nielsen, J.R. McCutcheon, Novel commercial aquaporin flat-sheet membrane for forward osmosis, Industrial and Engineering Chemistry Research 56 (2017) 11919–11925.
[129] J.S. Liu, Y. Zhan, T.W. Xu, G.Q. Shao, Preparation and characterizations of novel zwitterionic membranes, Journal of Membrane Science 325 (2008) 495–502.
[130] H.Y. Yu, Y. Kang, Y. Liu, B. Mi, Grafting polyzwitterions onto polyamide by click chemistry and nucleophilic substitution on nitrogen: a novel approach to enhance membrane fouling resistance, Journal of Membrane Science 449 (2014) 50–57.
[131] K.H. Park, S.H. Yu, H.S. Kim, H.D. Park, Inhibition of biofouling by modification of forward osmosis membrane using quaternary ammonium cation, Water Science and Technology 72 (2015) 738–745.
[132] T. Cai, X. Li, C. Wan, T.-S. Chung, Zwitterionic polymers grafted poly (ether sulfone) hollow fiber membranes and their antifouling behaviors for osmotic power generation, Journal of Membrane Science 497 (2016) 142–152.
[133] N.L. Le, M. Quilitzsch, H. Cheng, P.Y. Hong, M. Ulbricht, S.P. Nunes, T.-S. Chung, Hollow fiber membrane lumen modified by polyzwitterionic grafting, Journal of Membrane Science 522 (2017) 1–11.
[134] D.L. Zhao, S. Das, T.-S. Chung, Carbon quantum dots grafted antifouling membranes for osmotic power generation via pressure-retarded osmosis process, Environmental Science and Technology 51 (2017) 14016–14023.
[135] D.M. Mitrano, S. Motellier, S. Clavaguera, B. Nowack, Review of nanomaterial aging and transformations through the life cycle of nano-enhanced products, Environment International 77 (2015) 132–147.
[136] R. Arvidsson, D. Kushnir, B.A. Sanden, S. Molander, Prospective life cycle assessment of graphene production by ultrasonication and chemical reduction, Environmental Science and Technology 48 (2014) 4529–4536.
[137] J. Wang, E.C. Salihi, L. Siller, Green reduction of graphene oxide using alanine, Materials Science and Engineering C 72 (2017) 1–6.
[138] M. Slotte, R. Zevenhoven, Energy requirements and life cycle assessment of production and product integration of silver, copper and zinc nanoparticles, Journal of Cleaner Production 148 (2017) 948–957.
[139] K. Parveen, V. Banse, L. Ledwani, Green synthesis of nanoparticles: their advantages and disadvantages, AIP Conference Proceedings 1724 (2016) 1–7.
[140] X. Meng, F.-S. Xiao, Green routes for synthesis of zeolites, Chemical Reviews 114 (2014) 1521–1543.
[141] R.A. Sheldon, The E factor 25 years on: the rise of green chemistry and sustainability, Green Chemistry 19 (2017) 18–43.
[142] F. Galiano, K. Briceño, T. Marino, A. Molino, K.V. Christensen, A. Figoli, Advances in biopolymer-based membrane preparation and applications, Journal of Membrane Science 564 (2018) 562–586.
[143] G. Szekely, M.F. Jimenez-Solomon, P. Marchetti, J.F. Kim, A.G. Livingston, Sustainability assessment of organic solvent nanofiltration: from fabrication to application, Green Chemistry 16 (2014) 4431–4606.
[144] H.H. Wang, J.T. Jung, J.F. Kim, S. Kim, E. Drioli, Y.M. Lee, A novel green solvent alternative for polymeric membrane preparation via nonsolvent-induced phase separation (NIPS), Journal of Membrane Science 574 (2019) 44–54.
[145] M.G. Buonomenna, Nano-enhanced reverse osmosis membranes, Desalination 314 (2013) 73–88.

C H A P T E R

15

Sustainable Approaches for Materials Engineering With Supercritical Carbon Dioxide

Ikuo Ushiki[1], *Kiyoshi Matsuyama*[2], *Richard L. Smith, Jr.*[3]

[1]Hiroshima University, Hiroshima, Hiroshima-ken, Japan; [2]Fukuoka Institute of Technology, Fukuoka, Fukuoka-ken, Japan; [3]Tohoku University, Sendai, Miyagi-ken, Japan

O U T L I N E

1. REDUCING MATERIAL INTENSITY WITH SUPERCRITICAL CARBON DIOXIDE

The world population on planet Earth is expected to reach 9.6 billion (10^9) people in 2050, and if present consumption and production patterns are maintained, humankind will require the equivalent of three planets of natural resources to supply all of society's needs [1]. Fig.15.1 shows the material intensity in various regions of the world in 2011, where it can be seen that the Asia–Pacific region is driving material intensity.

Although material intensity is projected to decrease by 2060, material consumption is expected to increase from 79 Gt in 2011 to 167 Gt in 2060, corresponding to a usage of 35 kg per capita per day in 2011 to 45 kg $capita^{-1}$ day^{-1} in 2060 [2].

Intuitively, population growth might seem to be the major factor in material consumption, but on the contrary, the main drivers for material consumption are growing affluence and material intensity [3]. Affluence is the human desire to have far more than one needs to live in a social setting, while material intensity is the use of far more resources and energy than are necessary to produce a product in an economic setting or in other words using processes that are unsustainable. The desire for affluence can be addressed with philosophy (e.g., UN Good Life Goals, Ref. [4]), while the need for material intensity can be reduced by using sustainable methods that are more efficient than present methods. Ideally, sustainable methods should be decoupled from raw economics so that their

Sustainable Nanoscale Engineering
https://doi.org/10.1016/B978-0-12-814681-1.00015-1

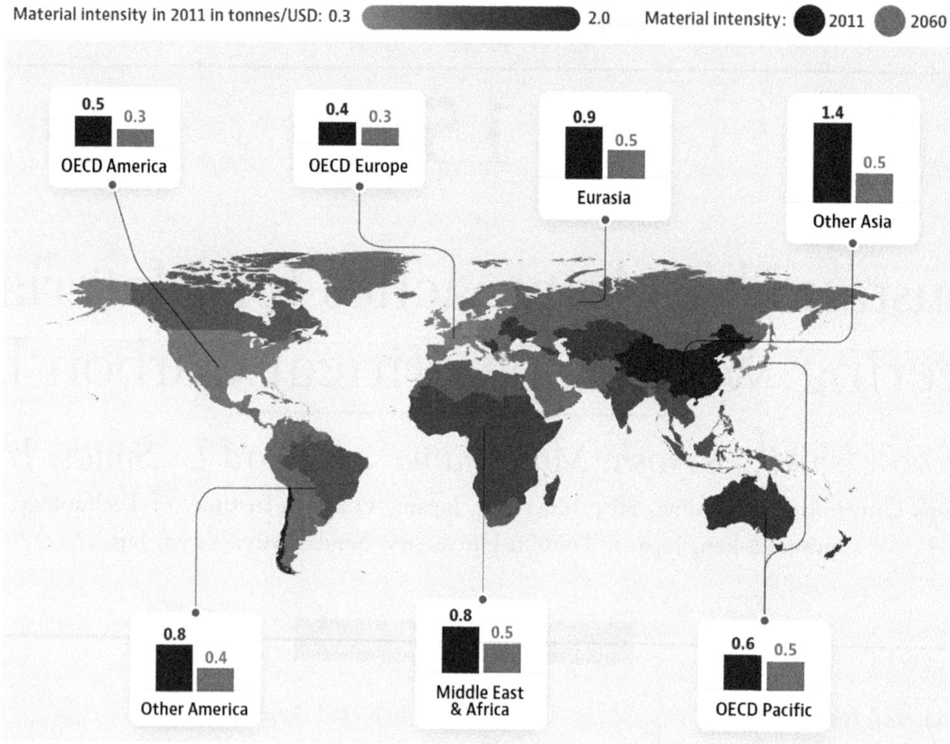

FIGURE 15.1 Material intensity (MI) in regions of the world that is defined as a ratio of domestic materials consumption (DMC) to gross domestic product (GDP) or MI = DMC/GDP. *Source OCED, Global Materials Resource Outlook to 2060, 2018. https://www.oecd.org/environment/waste/highlights-global-material-resources-outlook-to-2060.pdf.*

benefit to society and to the environment can be realized. It is evident that consumption and production patterns will require radical restructuring in the near future to allow decoupling of economic growth from resource use and environmental degradation so as to obtain a form of sustainability.

In 2000, the United Nations proposed a set of millennium development goals that lead to the following accomplishments over a 15 year period (2000–15) compared with data in 1990: (i) 1 billion people were removed from extreme poverty, (ii) child mortality dropped by more than 50%, (iii) children out of school decreased by more than 50%, and (iv) HIV/AIDS infections reduced by almost 40% compared with 2000 data. As the result of a United Nations Conference on Sustainable Development in Rio de Janeiro in 2012, a new set of 17 sustainable development goals (SDGs) were formulated [1] for the target year of 2030. The 17 SDGs include new areas related to climate change, economic inequality, innovation, and sustainable consumption (Table 15.1). SDG#12 pertains to the objectives of this chapter.

TABLE 15.1 United Nations Sustainable Development Goals (SDGs) Adopted in 2013 With Targets Being Set for 2030 [1].

(1) No poverty	(10) Reduced inequalities
(2) Zero hunger	(11) Sustainable cities and communities
(3) Good health and well-being	(12) Responsible consumption and production
(4) Quality education	(13) Climate action
(5) Gender equality	(14) Life below water
(6) Clean water and sanitation	(15) Life on land
(7) Affordable and clean energy	(16) Peace, justice, and strong institutions
(8) Decent work and economic growth	(17) Partnerships for the goals
(9) Industry, innovation, and infrastructure	

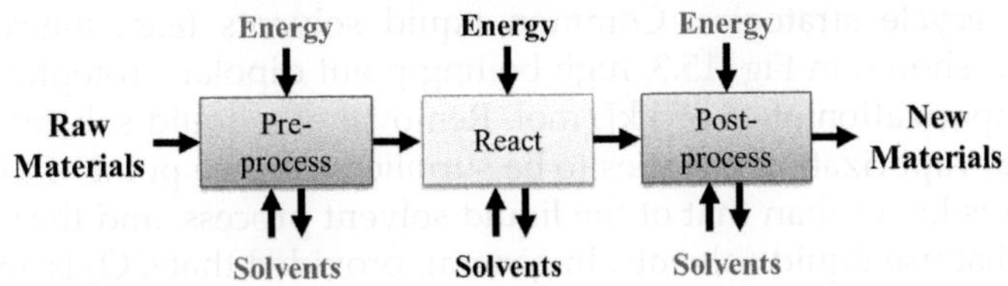

FIGURE 15.2 Typical block diagram process for materials synthesis.

UN SDG#12 addresses consumption and production including material intensity. In a typical process for synthesizing new materials, raw materials (precursors) are preprocessed in any number of steps before reaction (Fig. 15.2). After reaction, products are postprocessed to remove unreacted precursors, by-products, and solvents. Each step requires a different combination of conditions, solvents, and separations all of which require energy and make large-scale production of materials energy and material intensive.

Many researchers, including the authors, report on one-pot reactions ("React" in Fig. 15.2) to synthesize new materials or apply green catalysts but neglect discussion and analysis of preprocessing steps of raw materials and postprocessing steps of the product that often can require far more energy and solvents than in the reaction step. By using carbon dioxide in its supercritical state (see Section 15.2), the number of solvents and processing steps for producing a material can be reduced, solvent recycle can be simplified, and the energy required for separation of product and solvents can be lowered, and therefore, there can a be a reduction in material intensity through the use of CO_2 as solvent or processing aid. For example, Fig. 15.3 shows a comparison between deposition of a precursor onto a substrate with a liquid solvent and with CO_2.

In the liquid solvent (molecular, ionic, deep eutectic, etc.) process (Fig. 15.3), the precursor is dissolved into the liquid solvent that is then contacted with the substrate. After adsorption of the precursor onto the substrate occurs, the liquid solvent must be removed by one or more washing steps that generate waste liquid solvent with contaminants that must be recycled or treated.

In the CO_2 process (Fig. 15.3), CO_2 must be compressed to the required conditions of a dense gas so that sufficient precursor can be dissolved into CO_2 or CO_2 can be used in the liquid state ($P^{sat}_{CO_2}$ = 6.43 MPa at 25°C). Then, just as in the liquid solvent process, the precursor is allowed to adsorb onto the substrate. After adsorption of the precursor is completed, however, removal of CO_2 from the precursor is easily accomplished by one of many ways with depressurization of the vessel being commonly used. Removal of CO_2 from the substrate is complete because CO_2 is a gas at ambient conditions, and recycle of the CO_2 is simple because CO_2 is a highly selective solvent such that it is unlikely that spent CO_2 contains substrate contaminants.

In the deposition processes shown in Fig. 15.3, the CO_2 process uses fewer organic solvents and it requires far fewer steps and energy than the liquid solvent process. The CO_2 process allows simple recycle strategies, whereas the liquid solvent process generates mixed solvents and contaminants associated with substrate dissolution, all of

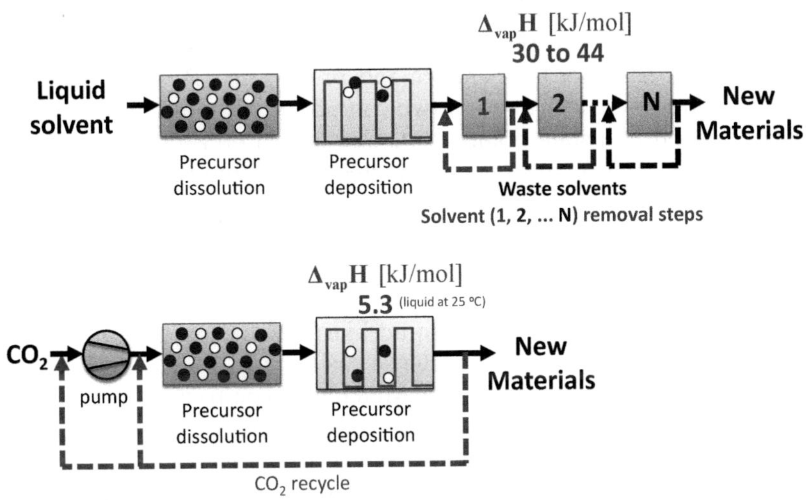

FIGURE 15.3 Precursor deposition onto a substrate for making new materials with a liquid solvent and with CO_2 showing the lack of solvent removal steps for the CO_2 process.

which require specialized recycle strategies. Common liquid solvents (e.g., *n*-hexane, methanol, water) have enthalpies of vaporization as shown in Fig. 15.3; high boiling point dipolar protophilic solvents (e.g., DMF, DMA, NMP) have enthalpies of vaporization of 45–55 kJ/mol. Removal of a liquid solvent from a substrate and recycle of the waste solvent requires vaporization energies to be supplied at some point of the process. Thus, the material intensity of the CO_2 process is lower than that of the liquid solvent process, and the CO_2 process is closer to being sustainable than processes that use liquid solvents. In general, provided that CO_2 is applicable to a given precursor and substrate according to solubility and compatibility constraints, CO_2 processes tend to be greener than those material processes that use liquid solvents.

There are notable performance benefits of using CO_2 in its dense gas state or supercritical state that is referred to as supercritical carbon dioxide ($scCO_2$) for materials processing. Some of the performance benefits of using $scCO_2$ materials processing include deep precursor penetration into substrate mesopore structures, highly efficient adsorption steps (short times), control or adjustment of transport properties, and the maintaining of substrate integrity during processing steps especially for delicate structures. There are a number of performance benefits for processing materials with $scCO_2$ demonstrated in this chapter, which are difficult or impossible to be achieved by other methods.

In the evaluation of reaction solvents for sustainability, $scCO_2$ has favorable solvent life cycle assessment in terms of ease of use, reusability, health and safety, environmental impact, and cost [5]. In all fairness to the many available approaches for design and synthesis of functional materials, application of $scCO_2$ as a solvent or processing fluid does require the use of a pressurized system. Nevertheless, the merits of using $scCO_2$ as a solvent far outweigh its requirement for a pressurized system, with one commercial example of a commonly used substance being offered to the reader, which may be of some surprise:

> Making domestic hot water with $scCO_2$ heating systems is twice as efficient as using natural gas heating and four times more efficient than using electrical heating and it is even more economical that those methods (see Section 15.2).

In this chapter, our objective is to introduce several methods for using $scCO_2$ as a design aid, a processing fluid, or a synthesis solvent in materials engineering. Carbon dioxide is an environmentally friendly fluid that can be used to replace hazardous or organic solvents and to reduce the amounts of solvents being used, thus use of $scCO_2$ offers the possibility of improving the sustainability of a process. Carbon dioxide in its supercritical state is a highly versatile solvent for material processing, and its application allows low-energy production and design of high-performance functional materials essential for developing green chemical processes [6]. Carbon dioxide can be used to reduce material intensity in applications such as (i) catalyst synthesis, (ii) membrane formation, (iii) metal–organic framework (MOF) processing and synthesis, (iv) cellulose nanofibril (CNF) modification, and (v) aerogel drying. In the next section, some of the key properties of $scCO_2$ are introduced so that the reader can become familiar with aspects of using $scCO_2$ as a solvent and then several applications studied by the authors will be highlighted.

2. CHOOSING PATHS ON THE CO_2 PHASE DIAGRAM

The physical state (solid [s], liquid [l], vapor [v], or gas) of a pure substance is described by a phase diagram (Fig. 15.4) in terms of conditions of either temperature and pressure (T, P) or temperature and density (T, ρ). A single-phase state (s, l, or v) requires specification of two conditions to determine its location on a phase diagram. Fig. 15.4A shows the specification of a state point at 0°C and 6 MPa on a P-T phase diagram. Phases that are in equilibrium with each other (s-v, s-l, l-v) have one less degree of freedom (Gibbs phase rule) and are specified by one condition and the phase boundary. Phases that are in equilibrium appear as a curve on a P-T diagram (Fig. 15.4A) or as a surface on a P-ρ-T diagram.

The critical point (T_c, P_c, ρ_c) is the terminal point on the vapor pressure (l-v) curve of a pure substance, and for CO_2, the critical point is $T_c = 30.9782°C$, $P_c = 7.3773$ MPa, and $\rho_c = 467.6\ kg/m^3$. According to the value of ρ_c of CO_2, its density is neither gas-like ($\rho \cong 1\ kg/m^3$) nor liquid-like ($\rho \cong 1000\ kg/m^3$) and intermediate to those values so that CO_2 in its critical state is like a homogeneous dense gas (l = v).

A change in state of CO_2 from 50°C, 0.1 MPa to 0°C, 6 MPa for two paths is shown in Fig. 15.4. In Path 1, vapor at 50°C is isothermally compressed to 6 MPa and then isobarically cooled at 6 MPa to 0°C. In Path 2, vapor at 50°C is isothermally compressed to 8 MPa and then it is isobarically cooled at 8 MPa to 0°C, and finally, it is isothermally decompressed at 0°C to 6 MPa. Although the final and initial states are identical for the two paths, Path 1 crosses an l-v phase boundary so that condensation occurs, while Path 2 avoids any phase boundary, so that the phase remains homogeneous throughout the change in conditions. On the P-ρ-T diagram (Fig. 15.4B), the change in states

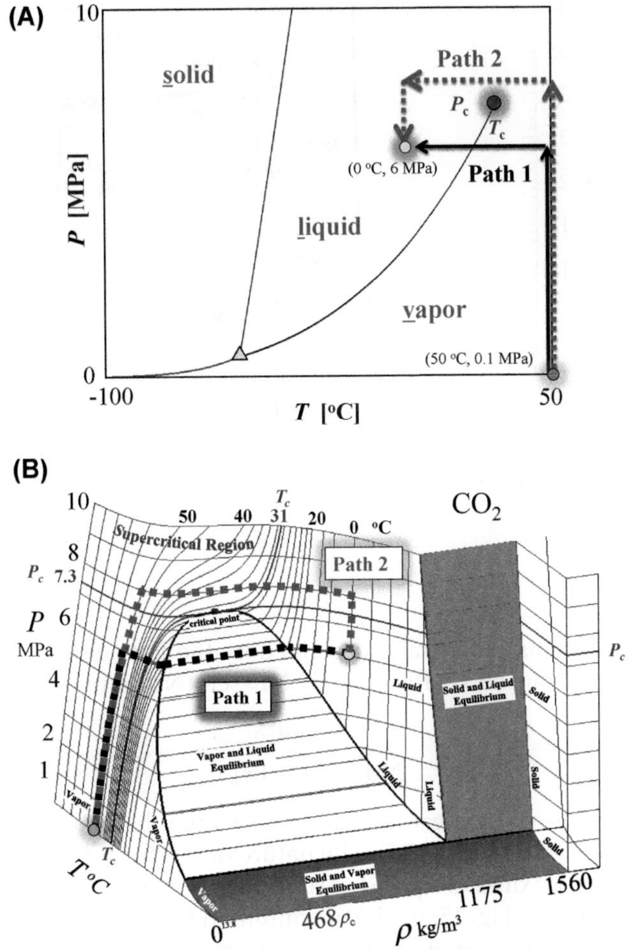

FIGURE 15.4 Phase diagrams of carbon dioxide: (a) pressure–temperature (P–T) diagram and (b) pressure–density–temperature (P–ρ–T) diagram. Path 1 undergoes a phase transition from gas to liquid, while Path 2 remains as a homogeneous phase even though the two paths begin and end at identical conditions. Calculated lines are drawn to scale with methods [7] and programs [8] described in those texts.

can be seen as conditions that move along the surface. From the point of view of processing materials, Path 2 allows homogeneous impregnation or deposition of substances or additives into a substrate, while the reverse of Path 2 allows homogeneous removal of substances, additives, or solvents common in drying or desolventation steps.

When CO_2 is at a temperature above T_c (e.g., $T > 31°C$), it does not condense into a liquid phase when compressed, and so, it is frequently referred to as being in a *fluid state*. When CO_2 is at a temperature above T_c (e.g., $T > 31°C$) and is at a pressure above P_c (e.g., $P > 7.3$ MPa) or is at a density that is close to ρ_c, then its condition is referred to as *supercritical* and the terminology *supercritical fluid* or $scCO_2$ is used. Some of the most attractive features of using $scCO_2$ as a solvent are that conditions and state changes can be chosen to obtain a desired range of transport properties and that the homogeneous fluid state can be chosen according to the requirements of the processing step.

3. HEAT AND MASS TRANSFER CHARACTERISTICS OF CO_2

$scCO_2$ (40°C, 10 MPa) has a liquid-like density ($\rho = 628.6$ kg/m^3), a gas-like viscosity ($\mu = 4.93 \times 10^{-5}$ mPa s), a liquid-like thermal conductivity ($k = 0.0684$ W m^{-1} K^{-1}) and a gas-like self-diffusion coefficient ($D = 3.28 \times 10^{-8}$ m^2 s^{-1}) so that its transport properties are highly favorable for promoting heat and mass transfer in materials engineering [7]. Conditions of CO_2 can be chosen so that they allow a reduction in the number of processing steps and processing times (Fig. 15.3), provide efficient heat transfer of materials, or promote mass transfer in materials, all of which contribute to making a process more sustainable than when using liquid solvents.

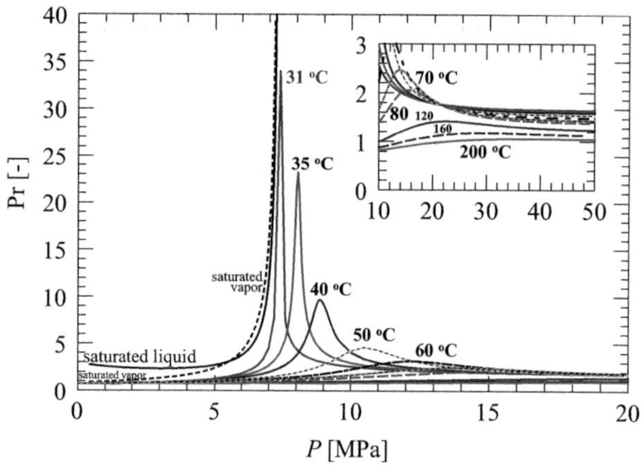

FIGURE 15.5 Prandtl (Pr) numbers of carbon dioxide for the saturation region and at selected supercritical isotherms versus pressure. Inset: enlargement of P versus Pr number region. All lines were calculated with methods in Ref. [7], and the Span–Wagner equation of state for CO_2 [11] and programs from Ref. [8].

3.1 Heat Transfer

Heat transfer in engineering systems is generally described by the Reynolds (Re) and the Prandtl (Pr) numbers that are used to estimate local heat transfer coefficients via Nusselt (Nu) number correlations:

$$\mathrm{Pr} = C_p \mu / k \tag{15.1}$$

$$\mathrm{Nu} = f(\mathrm{Re}, \mathrm{Pr}) \tag{15.2}$$

At the critical point, the constant pressure heat capacity, C_p, diverges because of the order of the phase transition [9,10], so that the Pr number also diverges (Fig. 15.5). The nature of the strong divergence in C_p persists throughout the supercritical region and causes maxima in C_p and corresponding maxima in the Pr number even as conditions become removed from the critical point (Fig. 15.5). The maximum in C_p is known as the *pseudocritical* point (T_{pc}, P_{pc}) and the locus of T_{pc} and P_{pc} forms the pseudocritical line that can cause enhancement of heat transfer at low heat flux or deterioration of heat transfer at high heat flux [12].

In the supercritical region of CO_2, heat transfer is generally enhanced, and now it is a common practice to transfer heat over a range of temperatures at a constant pressure above P_c (Fig. 15.5, inset), which is referred to as a *transcritical* path, meaning that there is no phase change during energy transport. For example, energy devices (refrigerators, water heaters) based on $scCO_2$ and transcritical paths have been widely sold in Japan over the past 10 years; typical devices for making hot water are four times more efficient than natural gas heating and use 66% less electricity than electrical heating of water [7,13–15]. In materials engineering, operations such as drying or desolventation could take advantage of transcritical paths to improve the efficiency of heat transfer and to improve fabrication methods.

3.2 Mass Transfer

Mass transfer in engineering systems is generally described by the Reynolds (Re) and the Schmidt (Sc) numbers that are used to estimate local mass transfer coefficients via Sherwood (Sh) number correlations:

$$\mathrm{Sc} = \mu / (\rho D) \tag{15.3}$$

$$\mathrm{Sh} = f(\mathrm{Re}, \mathrm{Sc}) \tag{15.4}$$

At 25°C, the Schmidt number of the saturated liquid is $\mathrm{Sc}^L = 3.428$ and that of the saturated vapor is $\mathrm{Sc}^V = 1.136$, while, at 31°C and 10 MPa, Sc = 3.687, which is higher than the liquid value (Fig. 15.5).

A value of the Sc number greater than one means that mass transfer occurs by the mode of convection (fast) rather than by diffusion (slow), so that the chosen conditions can be used to promote mass transfer by convection. As shown in Fig. 15.6, at conditions typical for $scCO_2$, the Sc number varies from 2.389 (40°C, 10 MPa) to 9.735 (40°C, 50 MPa), and the variation is approximately linear (Fig. 15.6), $\mathrm{Pr} = 0.1696 \cdot p[\mathrm{MPa}] + 1.2191$ except in the

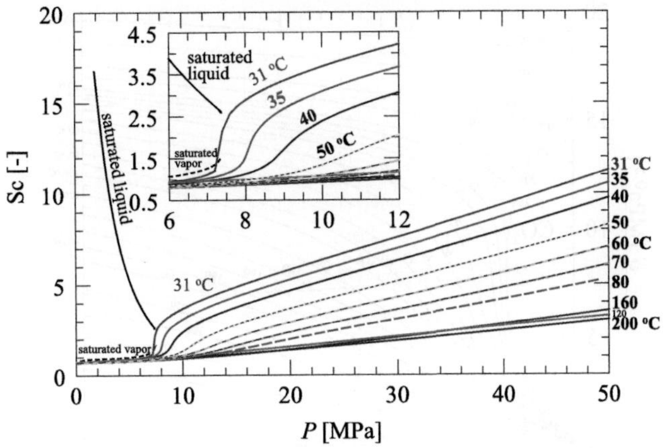

FIGURE 15.6 Schmidt (Sc) numbers of carbon dioxide for the saturation region and selected supercritical isotherms versus pressure. Inset: enlargement of P versus Sc number region. All lines were calculated with methods in Ref. [7], and the Span–Wagner equation of state for CO_2 [11] and programs from Ref. [8]. Self-diffusion coefficients for the Sc number were calculated as a function of T and ρ from the correlation given in Ref. [16].

region near the critical point (Fig. 15.6, inset) Above pressures of around 10 MPa, the Sc number is directly proportional to pressure at a given temperature (Fig. 15.6), thus estimation of mass transfer characteristics is simple and the pressure of $scCO_2$ can be used to control mass transport in a given process or protocol. However, Fig. 15.6 is based on the self-diffusion coefficient of CO_2 and typically, pairwise (1–2) diffusion coefficients are needed for which both experimental methods and predictive equations are available [17–25].

4. SOLUBILITY CHARACTERISTICS OF CO_2

Carbon dioxide (O=C=O) has two symmetrically opposed permanent electrical dipoles, so that the net dipole moment is zero, making it a nonpolar fluid. Nevertheless, CO_2 has a permanent electrical quadrupole moment, and its resonance structure makes it a weak Lewis acid so that it reacts with nucleophiles such as bases or basic groups in polymers or materials.

The physical chemistry of CO_2 is of interest in materials engineering because CO_2 can be used as a processing fluid to impart desired characteristics in a substrate. The energy of complete vaporization of a saturated liquid to the ideal gas state under isothermal conditions relative to its molar liquid volume is known as the *cohesive energy density*, *c*, and is defined as [26]

$$c = \frac{\Delta \underline{U}^{vap}}{\underline{V}^L} \tag{15.5}$$

where $\Delta \underline{U}^{vap}$ is the molar internal energy of vaporization and $\underline{V}^L$ is the molar volume of the liquid. The square root of this quantity is known as the *Hildebrand solubility parameter* and is given the symbol, δ:

$$\delta \equiv \sqrt{c} = \sqrt{\frac{\Delta \underline{U}^{vap}}{\underline{V}^L}} \approx \sqrt{\frac{\Delta \underline{H}^{vap} - RT}{\underline{V}^L}} \tag{15.6}$$

Eq. (15.6) is typically used to determine δ at 25°C from an available enthalpy of vaporization $\Delta \underline{H}^{vap}$ and a liquid density; δ has units of $MPa^{1/2}$. Solubility parameters of two substances that are close to each other tend to form homogeneous solutions [26] and to have activity coefficients that are close to one. Hansen [27] has expanded the concept of solubility parameter theory to break the contributions of vaporization into dispersion (δ_d), polar (δ_p), and hydrogen bonding (δ_h) forces that are now referred to as Hansen solubility parameters (HSPs). For CO_2 at 25°C, $\delta_d = 15.6$ $MPa^{1/2}$, $\delta_p = 5.2$ $MPa^{1/2}$, and $\delta_h = 5.8$ $MPa^{1/2}$, or these are combined ($\delta^2 = \delta_d^2 + \delta_p^2 + \delta_h^2$), $\delta = 17.4$ $MPa^{1/2}$ as estimated by low-pressure gas solubility data in a large number of solvents [27]. HSP theory is well developed and is broadly applicable to industrial systems (asphalts, biological materials, coatings, crude oil, paints, pigments, polymers, surfaces) [27].

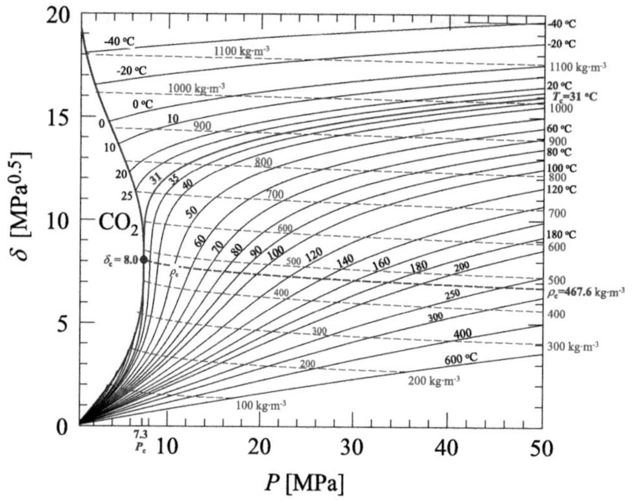

FIGURE 15.7 Calculated solubility parameter of carbon dioxide based on Eq. (15.7) and the Span–Wagner equation of state [11] using the text and programs in Refs. [7,8].

For a supercritical fluid, relations such as Eq. (15.6) are no longer valid as there is no vaporization in the supercritical region, and properties vary greatly with changes in temperature or pressure. For a pure fluid, a possible definition for solubility parameter is [6,28]

$$\delta = \left(\frac{\underline{U}^{ig} - \underline{U}}{\underline{V}}\right)^{1/2} \tag{15.7}$$

where ig refers to the ideal gas state. Fig. 15.7 shows a plot of Eq. (15.7) over a range of temperatures and pressures. It can be seen that over the workable range of conditions (T, 25–200°C; P, 0.1 –50 MPa), δ has a maximum value of 16.5 MPa$^{1/2}$.

Thus, as a guide for CO_2 interactions with a solvent or substrate, Fig. 15.7 provides the qualitative behavior of the solubility parameter that is useful for considering conditions for dissolution of solvents or additives into the supercritical phase or for the dissolution of the supercritical phase into a solid or liquid substrate.

5. REQUIRED AND DESIRABLE PHYSICAL PROPERTY DATA FOR SELECTED APPLICATIONS

5.1 Overview for Selected Applications

In the synthesis of materials using $scCO_2$ as a processing aid, certain physical properties are needed to assess the conditions and starting materials to be used. For selected applications in this chapter, Table 15.2 provides an overview. In the application of metal deposition, precursor solubility and reactivity are required in the CO_2 phase including the presence of cosolvents. Precursor adsorption kinetics, isotherms, diffusion coefficients, and other data to assess penetration of the precursors into the porous solids would be desirable, but few data are available [29–31]. Molecular simulations will become increasingly important to obtain the necessary physical properties required in materials engineering [32–34]. As an example, competitive adsorption has to be considered in multiprecursor or precursor-cosolvent systems or in porous solids. In nanocellular foaming applications, properties of the polymer and properties of the polymer solution with CO_2 are required. In drying applications, stability of the MOFs or CNF composite aerogels in the chosen solvent is needed, and for MOFs, solvent reactivity can be important as there can be by-products formed in the synthesis.

5.2 Solubility of Metal Complexes in CO_2

As metal deposition with $scCO_2$ can be operated over wide ranges of temperature and pressure, measurement and theoretical modeling of solubilities of metal complexes in $scCO_2$ are necessary for the process design. Haruki et al. [35–37] measured solubilities of various acetylacetonate-type metal complexes in $scCO_2$ using in situ

TABLE 15.2 Overview of Required and Desirable Physical Property Data for Selected Applications in This Chapter.

Application	Phase	Required Data	Desirable Data
Metal deposition	Supercritical carbon dioxide ($scCO_2$)	Precursor solubility Precursor reactivity	Precursor adsorption isotherm onto substrate in $scCO_2$ Precursor diffusivity in $scCO_2$ and inside substrate pore
Nanocellular foaming	Polymer	CO_2 sorption isotherm CO_2 solubility T_g, w_c, ρ, M_n, M_w CO_2–polymer-specific interactions	CO_2 diffusivity in polymer CO_2–polymer rheology
Metal–organic framework (MOF) drying	MOF	Stability in solvent Solvent reactivity	Solvent 1 and solvent 2 diffusivity Solvent sh number
	$scCO_2$	Solvent solubility	Solvent/linker diffusivity Solvent/linker Sh numbers
CNF aerogel drying	Aerogel	Stability in solvent	Solvent solubility Solvent diffusivity in water Solvent Sh number
	$scCO_2$	Solvent solubility	Solvent Sh number

UV-vis spectrometry. Aschenbrenner et al. [38] used a dynamic-gravimetric method to measure the solubility of metal complexes in $scCO_2$.

Solubilities of metal complexes are summarized in reviews [39,40] and in a text [41]. Furthermore, several methods for modeling solubilities of metal complexes in $scCO_2$ have been proposed, including Chrastil model [42] and regular solution theories [43,44]. Ushiki et al. [45] reported thermodynamic modeling of solubilities of acetylacetonate-type metal complexes in $scCO_2$ using the PC-SAFT (perturbed-chain statistical associating fluid theory) equation of state [46]; they obtained relatively good predictions of the solubilities using the PC-SAFT equation of state via generalizations of the pure component PC-SAFT parameters of the metal complexes while considering chemical structure and physical properties of the complexes.

6. SELECTED APPLICATIONS

6.1 Metal Nanoparticle Deposition Onto Porous Substrates

Metal deposition into solid substrates using $scCO_2$ was pioneered by Watkins and McCarthy [47], who studied the synthesis of polymer/metal nanocomposites. The basic technique of metal deposition using $scCO_2$ was further developed by others [48–54] and is widely applicable to many types of metal nanoparticles and substrates [55]. Metal nanoparticle deposition onto substrates using $scCO_2$ has three basic steps (Fig. 15.8).

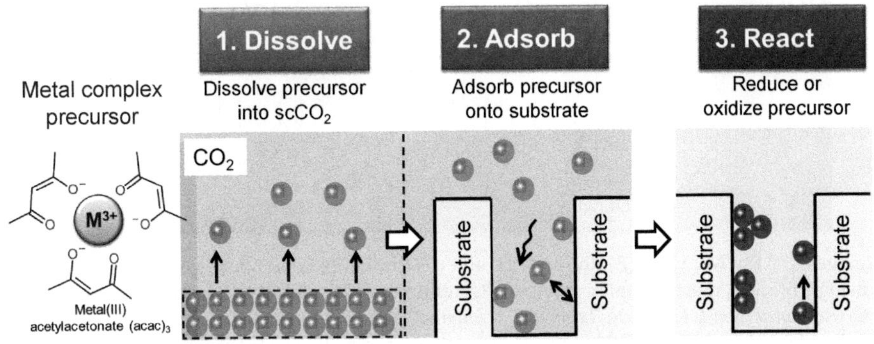

FIGURE 15.8 Three basic steps for deposition of metal nanoparticles onto substrates using supercritical carbon dioxide ($scCO_2$) as the solvent and a metal complex precursor.

TABLE 15.3 Characterization of Mesoporous Silicas and Deposition Results for $Rh(acac)_3$ Metal Complex Precursor With Supercritical carbon dioxide ($scCO_2$) (60°C, 15 MPa, 18 h adsorption Time, Ex Situ Activation via 3 h calcination in Air at 400°C).

Characterization	MCM-41	MSU-H	HMS
Specific surface area ($m^2\ g^{-1}$)	1011	585	759
Mean pore diameter (nm)	2.7	8.2	3.2
Pore volume ($cm^3\ g^{-1}$)	1.15	0.80	1.60
Pore structure	2D hexagonal	2D hexagonal	3D disordered
Mean particle diameter (μm)	30	35	25
Deposition results			
Rh mean particle diameter (nm)	2.4	4.4	12.9
Rh loading onto substrate (wt%)	4.9	5.4	7.2

Adapted from I. Ushiki, N. Takahashi, T. Shimizu, Y. Sato, M. Ota, R.L. Smith, H. Inomata, Adsorption equilibria of rhodium acetylacetonate with MCM-41, MSU-H, and HMS silica substrates in supercritical carbon dioxide for preparing catalytic mesoporous materials, Journal of Supercritical Fluids 120 (2017) 240–248. https://dx.doi.org/10.1016/j.supflu.2016.05.032, Copyright Elsevier, 2017.

As shown in Fig. 15.8, in step 1, the metal complex precursor is dissolved into $scCO_2$ to form a homogeneous solution at appropriate conditions (e.g., 60°C, 15 MPa). If the solubility of the precursor in $scCO_2$ is too low for the application, its solubility can be increased by adding a cosolvent such as ethanol or acetone in low concentrations (e.g., 5%). In step 2 (Fig. 15.8), the solution is contacted with the substrate in a pressure vessel, and adsorption of the precursor onto the substrate occurs. In step 2, the role of $scCO_2$ acts to increase mass transfer and to promote diffusion and penetration of the precursor into the substrate. In step 3 (Fig. 15.8), the substrate is either removed from the pressure vessel and subjected to reducing or oxidizing conditions at atmospheric conditions (ex situ activation) or it is subjected to a reducing or oxidizing gas mixture in the pressure vessel (in situ activation).

Ushiki et al. [31] studied rhodium deposition onto several types of mesoporous silicas using $scCO_2$ for the purpose of preparing catalysts similar to those synthesized with cobalt [56]. Properties of the mesoporous silicas are shown in Table 15.3, where it can be seen that the substrates have a wide range of surface areas, pore diameters, and pore volumes and that there are two different types of pore structures in the materials.

With the basic steps (Fig. 15.8), Rh deposition with $scCO_2$ could be realized as shown in Table 15.3 and Fig. 15.9. Rh could be deposited into porous structures regardless of the different pore diameters or pore structures of the substrates (Table 15.3), and the Rh particles are uniformly dispersed within the structures and have nanosize scale (Fig. 15.9). Thus, this example illustrates that $scCO_2$ can readily be applied to deposit or impregnate metals into porous substrates that have relatively different characteristics.

With proper choice of metal complex precursors, conditions, and gas mixtures, it is possible to combine the three steps shown in Fig. 15.8 into a sequence of processes [50] in a single vessel that is known as supercritical fluid reactive deposition (SFRD) [57]. As shown in Fig. 15.10, the metal deposited into the substrate is also active for promoting metal reduction.

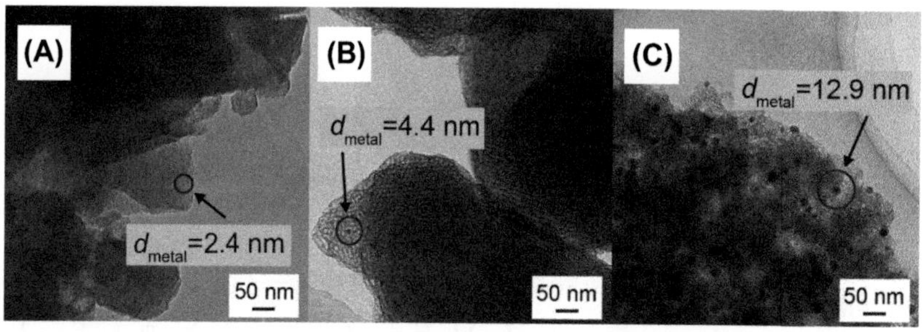

FIGURE 15.9 TEM images: (A) Rh/MCM-41, (B) Rh/MSU-H, and (C) Rh/HMS, for Rh deposition onto mesoporous silica with supercritical carbon dioxide ($scCO_2$) using $Rh(acac)_3$ metal complex precursor. Conditions: 60°C, 15 MPa, 18 h adsorption time, and ex situ activation via 3 h calcination in air at 400°C. *Reprinted from I. Ushiki, N. Takahashi, T. Shimizu, Y. Sato, M. Ota, R.L. Smith, H. Inomata, Adsorption equilibria of rhodium acetylacetonate with MCM-41, MSU-H, and HMS silica substrates in supercritical carbon dioxide for preparing catalytic mesoporous materials, Journal of Supercritical Fluids 120 (2017) 240–248. https://dx.doi.org/10.1016/j.supflu.2016.05.032, copyright Elsevier, 2017.*

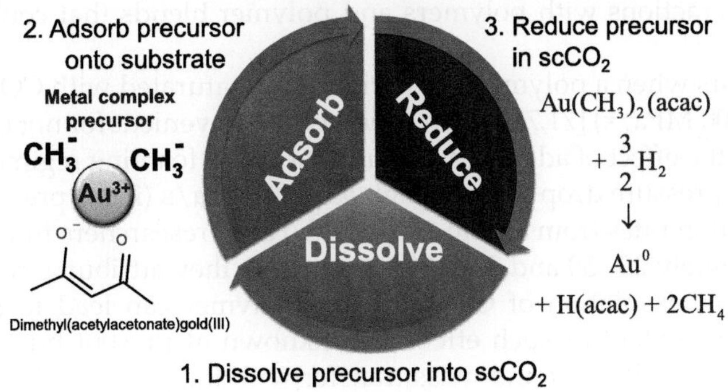

FIGURE 15.10 Supercritical fluid reactive deposition (SFRD) for a typical metal precursor that combines three steps into a sequence of processes.

Müller and Türk [50] show examples for Au, Ag, and bimetallic AuAg nanoparticles for nanopowder substrates ($d < 50$ nm), titanium dioxide (TiO_2), aluminum oxide (γ-Al_2O_3), β-cyclodextrin (β-CD), and Black Pearls carbon black (BP-2000) with deposited particle (x_{50} sizes, loadings) of Au onto substrates being TiO_2 (20.8 nm, 3.2 wt%), γ-Al_2O_3 (11.3 nm, 3.6 wt%), β-CD (25.5 nm, 0.4 wt%), and BP-2000 (2.4 nm, 3.3 wt%). Thus, the technique is applicable to many types of metal complexes, substrates, mixed-metals, sequential depositions, including the design of bifunctional catalysts [58].

For a more practical example of SFRD, Morère et al. [59] evaluated two methodologies for dispersing Ru nanoparticles into mesoporous silica, namely, simple metal complex deposition and metal complex reactive deposition for the purpose of making hydrogenation catalysts. The catalytic materials synthesized with SFRD allowed partial hydrogenation of hydrocarbons, thus demonstrating their utility. A distinct advantage of using $scCO_2$ in the deposition methodologies is that nanoparticles can be uniformly dispersed throughout the pore sizes of silica substrates.

There are many possibilities for using $scCO_2$ for materials production with deposition or reactive deposition methodologies. The reader is referred to reviews [55] and critical opinions on the topic [57].

6.2 Nanocellular Foaming of Polymers

The foaming of polymers with $scCO_2$ is an active area for both researchers and industry with an endless variety of blends, high-performance materials, composites, renewable materials, and biopolymers gaining attention. Research needs for foaming polymers with $scCO_2$ were presented in 2018 by Di Maio and Kiran [60]. Some of the key challenges of foaming polymers with $scCO_2$ highlighted in that work [60] were

1. Extruder designs for handling high levels of CO_2.
2. All-in-one instruments for measuring CO_2 sorption for many types of polymers.
3. Methods for nanocellular foaming of polymers and commodity plastics.
4. Methods for estimating the level of CO_2 in a polymer matrix.
5. Methods for estimating conditions for target polymer foam characteristics.
6. Methods for controlling expansion/shrinkage of rubbery polymers/elastomers.
7. Theories for understanding nucleation mechanisms and particle interactions.

Reviews regarding $scCO_2$ and polymer foaming are available for extruder design [61], technological processing [62], and biomedical and tissue engineering application [63]. Limitations in the technological foaming of polymers (e.g., polystyrene) with $scCO_2$ and present extrusion technology seems to be (i) the inability of being able to form large blocks of material and (ii) the minimum foam density that can be obtained (30 kg m^{-3}), which is much larger than foams manufactured with thermally induced phase separation techniques (10 kg m^{-3}) [62].

Nanoporous polymeric materials [64], nanocomposites [65], and nanoporous polymeric materials created with $scCO_2$ and $scCO_2$-expanded liquids [66] have merited reviews and are considered to be highly active research areas. For example, aspects of block copolymer nanocellular and microcellular foams were studied by Pinto et al. [67], who found that nucleation and foaming mechanisms were different in regions of the polymer blends that contained CO_2-philic polymers. The early work of Kazarian [68,69] describes many of the phenomenological effects being reported today for $scCO_2$ interactions with polymers and polymer blends that could be favorably explored with spectroscopy [70].

Nanocellular foaming occurs when a polymeric material that is saturated with CO_2 at high pressures is subjected to high pressure drop rates (500 MPa/s) [71,72]. Such a method is convenient for not only studying nucleation mechanisms but also for exploring the effect of additives on nucleation and foaming regimes. Ono et al. [73] developed an apparatus that could achieve pressure drop rates as high as 2000 MPa/s (i.e., a pressure drop of 100 MPa in 50 ms) and could achieve cell nucleation rates from 10^{12} to 10^{14} cm^{-3}. Those researchers found that there were two regimes in the cell nucleation rates, namely 20–50 and 50–100 MPa, which they attributed to the Joule–Thomson inversion curve to some extent. Namely, exsolution of CO_2 from the polymer can lead to temperature increases at high pressures (c.50 –100 MPa). The extent of such effects is unknown at present but emphasizes the importance of the physical properties of CO_2 and its heat and mass transfer with the polymer during the expansion process.

6.3 Metal Deposition Into Metal Organic Frameworks

MOFs are crystalline solids made up of metal ions and organic linkers that can be engineered to have one-, two-, and three-dimensional (3D) architectures with characteristics such as defined pore sizes and pore volumes, extremely large specific surface areas, and hierarchically ordered cage sizes (Fig. 15.11). MOFs are recognized as a distinctive new class of crystalline nanoporous materials that have potential applications in gas storage, separations, catalysis, biomedicine, and chemical sensing [77].

Fig. 15.11 shows an image of an MOF that is denoted as MIL-101 (Materials of Institute Lavoisier) that has zeotype (zeolite-like) cubic structure, an extremely large cell volume (c.700,000 Å^3), large pore sizes and cavities (c.30 Å), and a surface area of almost 6000 m^2 g^{-1} [78]. Unlike zeolites, MOFs can be completely tailor-designed in structure so that research activity on their synthesis for applications in science and industry is very high. Fig. 15.12 shows the basic steps to synthesize an MOF with solvothermal methods. Choice of the conditions (e.g., precursors, linkers, concentrations, reaction time, temperature, solvent, pH, modulators [additives]) is highly critical in step 1, and it is probably the reason why many protocols are named according to the origin (location) of the discovery (e.g., MIL-101).

Historically, step 3 (Fig. 15.12) was not used until it was realized that structural collapse could be avoided by replacing the high-boiling solvent with a low-boiling solvent [80]. Steps 2 to 4 are generally referred to as "activation," and in these steps, solvent exchange, capillary forces, residual reaction solvent or linker, or removal of the final solvent can cause collapse of the structure [79]. In 2009, it was realized that supercritical fluid drying with CO_2 could be used to remove the low-boiling solvent [81–83], and thus, efficient protocols could be developed for MOF synthesis. Nowadays, drying with $scCO_2$ is commonly used in the preparation of MOFs.

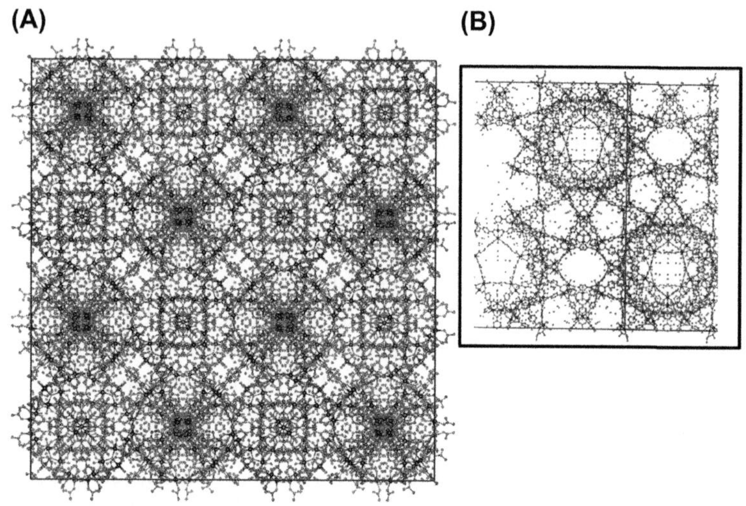

FIGURE 15.11 Image of a metal–organic framework (MOF): (A) MIL-101 (Cr), based on Ref. [74], and (B) rotated image showing cavities and pentagonal (c.12 Å openings) and hexagonal cages (c.14.5 Å × 16 Å) free apertures. Crystallographic information file (CIF) was obtained from Ref. [74], and images were generated with VESTA 3 software (version 3.4.6) Refs. [75,76].

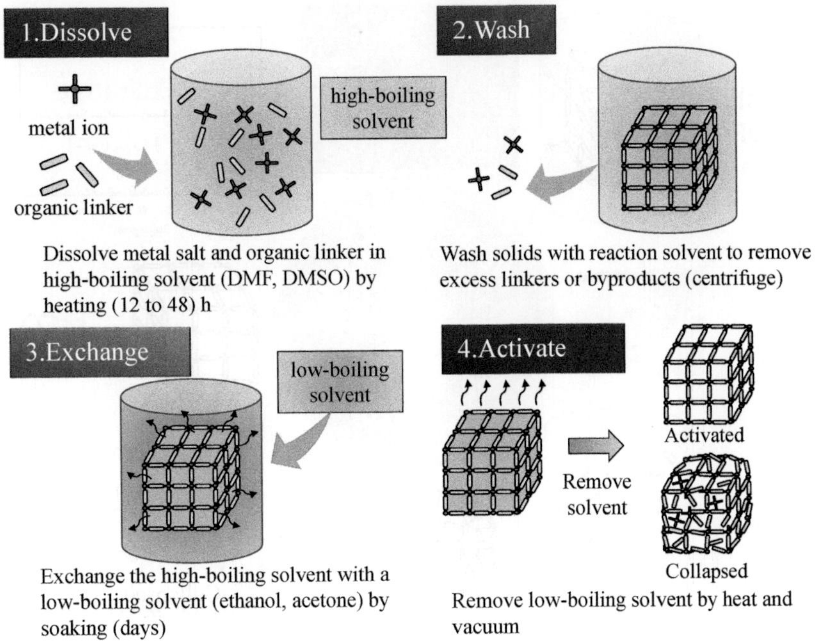

FIGURE 15.12 General steps used for synthesizing metal–organic frameworks (MOFs) with solvothermal techniques based on best practices given in Ref. [79].

Supercritical fluid deposition can be applied to MOF materials to synthesize catalysts, and one great advantage of $scCO_2$ is that it interacts with structures as a gas and is especially favorable for penetrating MOFs and transporting additives such as metal complexes into their structures. Fig. 15.13 shows TEM images [84,85] of the MOF, MIL-101 (Cr). The MIL-101 (Cr) MOFs that were used as substrates were synthesized according to procedures in the literature [86]. In Fig. 15.13A, the pristine MIL-101 (Cr) structure is shown, while in Fig. 15.13B, deposition of Pt by $scCO_2$ is shown. For comparison, deposition of Pt by a conventional method is shown (Fig. 15.13C). Two important and general findings can be revealed by the images in Fig. 15.13 regarding the $scCO_2$ deposition on MIL-101 (Cr): (i) Pt deposition and impregnation into the MOF structure seems to be isotropic and (ii) Pt particles are not agglomerated. On the other hand, by conventional deposition (impregnation) techniques (Fig. 15.13C), Pt particles are agglomerated

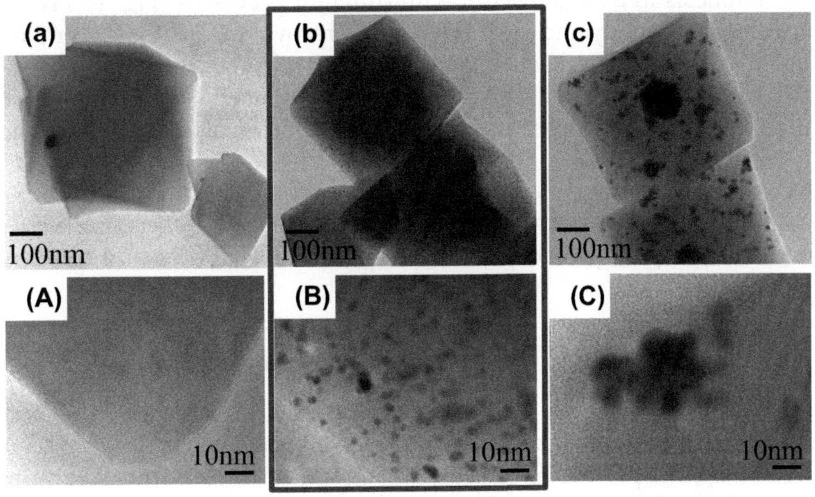

FIGURE 15.13 TEM images of MIL-101 (Cr): (a) pristine MIL-101 (Cr); (b) Pt@MIL-101 (Cr) prepared by supercritical carbon dioxide ($scCO_2$) deposition; and (c) Pt@MIL-101 (Cr) prepared by conventional methods. Enlargements of the respective images are shown in (A, B, and C). *Images taken from K. Matsuyama, Supercritical fluid processing for metal–organic frameworks, porous coordination polymers, and covalent organic frameworks, The Journal of Supercritical Fluids 134 (2018) 197–203. https://dx.doi.org/10.1016/j.supflu.2017.12.004, K. Matsuyama, M. Motomura, T. Kato, T. Okuyama, H. Muto, Catalytically active Pt nanoparticles immobilized inside the pores of metal organic framework using supercritical CO_2 solutions, Microporous and Mesoporous Materials 225 (2016) 26–32. https://dx.doi.org/10.1016/j.micromeso.2015.12.005. Copyright, Elsevier, 2016, 2018.*

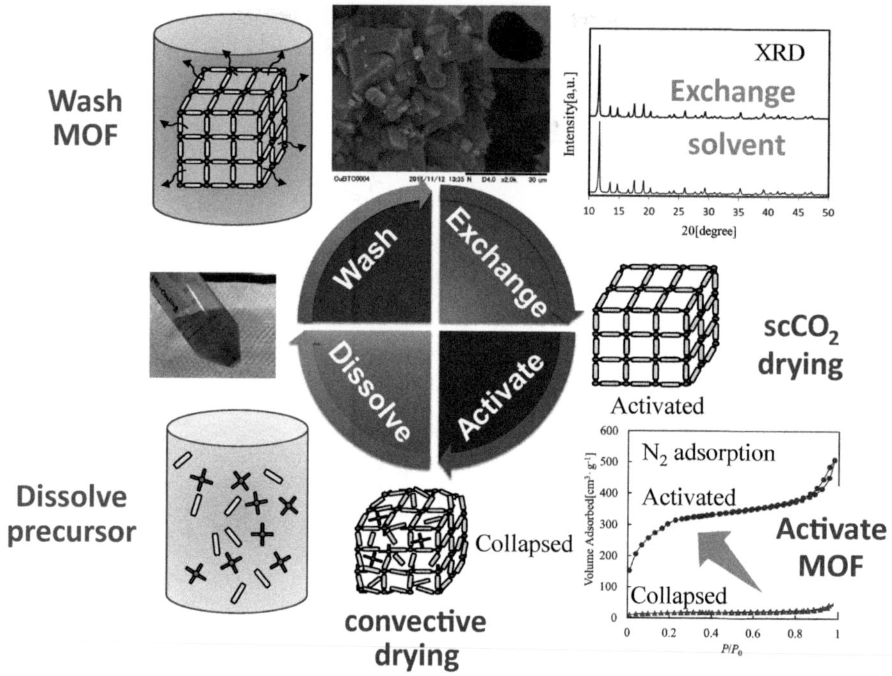

FIGURE 15.14 Schematic of metal–organic framework (MOF) synthesis in a single pressure vessel.

and are not uniformly distributed in the MOF structure. Thus, a clear advantage can be seen for using $scCO_2$ for metal deposition into MOFs.

According to physical properties of MIL-101 (Cr) shown in Fig. 15.13, they are almost the same ($S_{BET} = 3400\ m^2\ g^{-1}$, pore volume $= 1.7\ cm^3\ g^{-1}$, average pore $d = 2$ nm), but the catalytic activities (conversion, X) for hydrogen generation from NH_3BH_3 for the $scCO_2$ prepared catalyst ($X = 60\%$) were found to be about five times higher than for the conventionally prepared catalyst ($X = 12\%$).

Researchers have also been active in developing protocols for MOFs that do not use high-boiling solvents and essentially avoid steps 2–3 in Fig. 15.12 by using $scCO_2$ as the solvent for precursors and linkers that are soluble in $scCO_2$ [87]. In other words, similar to the strategy of Müller and Türk [50], the entire protocol could be performed in a single vessel as shown schematically by the process flow in Fig. 15.14.

Combination of MOF synthesis in a single vessel with different processes (Fig. 15.14) is attractive because each process step can be performed efficiently and with low-energy recovery steps for solvent aids so that sustainability can be more closely approached. Although such a combination of precursors and linkers might seem unlikely, López-Periago et al. [87] prepared Cu-based, 1D MOFs, using Cu(II) acetylacetonate and two linear linkers, bidentate 4,4′-bipyridine and 4,4′-trimethylenedipyridine, using $scCO_2$ at 60°C and 20 MPa and achieved yields of MOFs that were close to 100%.

Supercritical reactive crystallization has been used to prepare other MOFs and porous coordination polymers (PCPs) [88], and ionic liquids have also been shown to be effective for synthesizing MOFs [89–91] so that there are many opportunities for developing protocols for synthesis of MOFs, PCPs, and covalent organic frameworks.

6.4 Drying of Ag Nanoparticle–Cellulose Nanofibrils Composite Aerogels

Protocols for preparing composite aerogels from cellulose nanocrystals and CNFs have been developed for many different applications and involve acid or enzymatic preprocess steps, mechanical steps, and posttreatment steps [92]. After initial preprocessing, the protocol can be broken down into several steps as shown in Fig. 15.15.

The main processing challenge in aerogel synthesis protocols besides the critical challenge of starting materials and conditions is similar to that for MOFs discussed in the previous section, namely, avoiding structural collapse or deformation during the solvent removal step. Commonly, step 3 (Fig. 15.15) is carried out by $scCO_2$ drying or by freeze-drying (ice sublimation) [92]. Thus, $scCO_2$ has broad application in processing materials that have delicate structures.

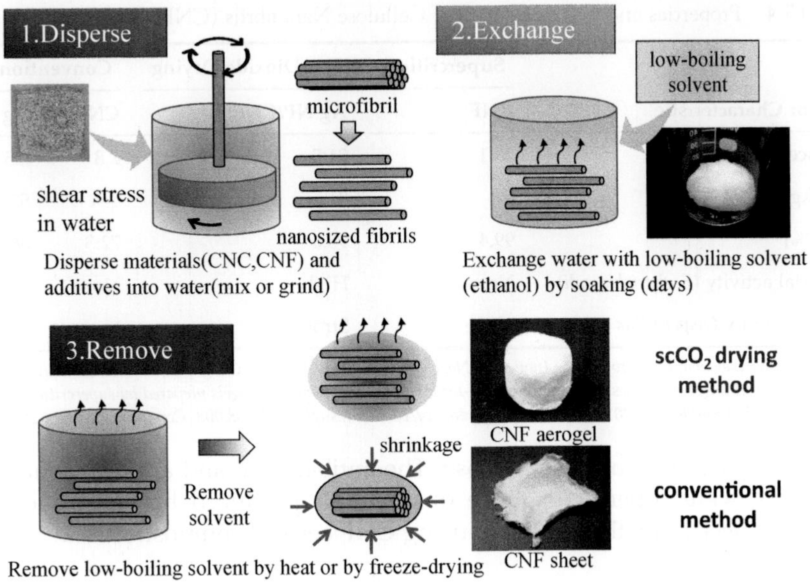

FIGURE 15.15 Process steps for preparing cellulose nanocrystals (CNCs) and cellulose nanofibrils (CNFs) composite aerogels.

In the process shown in Fig. 15.15, CNF aerogels can be enhanced to have special properties by using additives during the dispersion step or by using deposition techniques in a similar way as those outlined for MOFs (Section 15.6.3) with an example being to load the aerogel with compounds such as pharmaceuticals or bioactive compounds.

Matsuyama et al. [93] prepared CNF hydrogel solutions by ultrafine friction grinding. The hydrogel solution was soaked in ethanol to remove water. Then a portion of the material was mixed with Ag nanoparticles (Ag NPs) that had been dispersed into ethylene glycol by ball milling. The samples were subjected to either $scCO_2$ drying or conventional drying by heating at 100°C (Fig. 15.16) to remove solvent. As shown in Fig. 15.16, the physical appearance of the materials produced is quite different; composites formed with $scCO_2$ drying maintained their structural features regardless of the addition of Ag NPs, whereas conventional drying gave different distorted forms.

Quantitative properties and qualitative characteristics of the prepared materials are given in Table 15.4. As shown in Table 15.4, materials prepared with $scCO_2$ drying had aerogel-like properties that did not vary much regardless of the absence or presence of Ag NPs. On the other hand, materials prepared with conventional drying had low surface

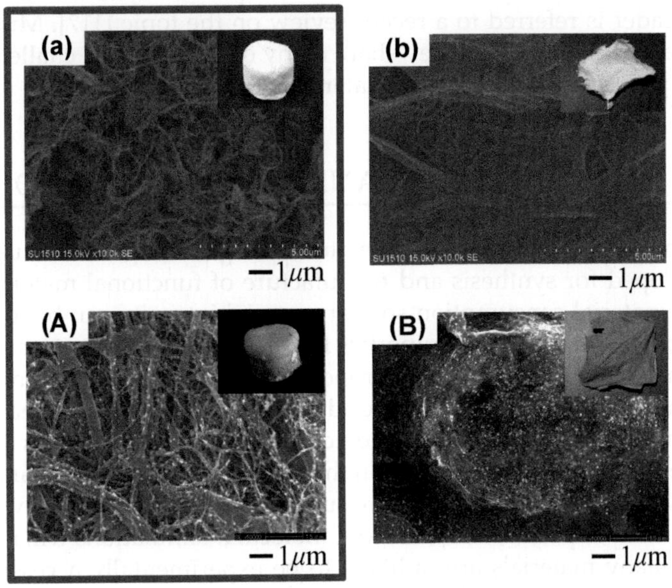

FIGURE 15.16 SEM images of cellulose nanofibrils (CNF) materials [93]: (a) sponge-like CNF aerogel prepared by supercritical carbon dioxide ($scCO_2$) drying, (b) sheet-like CNF structure prepared by conventional drying, with Ag NPs; (A) sponge-like Ag NP/CNF composite aerogel prepared by $scCO_2$ drying and (B) sheet-like Ag NP/CNF composite prepared by conventional drying. *Copyright Elsevier, 2019.*

TABLE 15.4 Properties and Characteristics of Cellulose Nanofibrils (CNF) materials.

	Supercritical Carbon Dioxide Drying		Conventional Drying	
Property or Characteristic	**CNF**	**Ag NP/CNF**	**CNF**	**Ag NP/CNF**
BET surface area [$m^2\ g^{-1}$]	33.1	31.5	7.8	6.5
Density [$kg\ m^{-3}$]	9	21	444	538
Porosity [%]	99.4	98.6	72.3	66.4
Antibacterial activity (*Escherichia coli*)	None	High	None	Low
Antifungal activity (*Aspergillus niger*)	None	Strong	None	Weak

Adapted from K. Matsuyama, K. Morotomi, S. Inoue, M. Nakashima, H. Nakashima, T. Okuyama, T. Kato, H. Muto, H. Sugiyama, Antibacterial and antifungal properties of Ag nanoparticle-loaded cellulose nanofiber aerogels prepared by supercritical CO_2 drying, Journal of Supercritical Fluids 143 (2019) 1–7. https://dx.doi.org/10.1016/j.supflu.2018.08.008. Copyright Elsevier, 2019.

areas and high densities typical of structural collapse. The antibacterial and antifungal activities of materials produced with $scCO_2$ drying were superior to those of materials produced by conventional drying (Table 15.4). Thus, $scCO_2$ drying is an effective method for producing CNF aerogel composites with special characteristics.

6.5 Other Materials and Processes

There are many other interesting materials that have been synthesized or processed with $scCO_2$ that can only be touched on in this section. Cotton can be impregnated with nanoparticles using $scCO_2$ to give it special characteristics [94,95] such as the ability to accept Cu in electroless plating [96] or to give cotton antibacterial properties [97]. Exfoliation of graphene [98–103] and inorganic materials [104] can be promoted with $scCO_2$. Nondestructive extraction of 3D Ag nanostructures from nanoseeds is possible with $scCO_2$ as a processing fluid [105]. Carbon dioxide can be used as a dispersant aid for production of polypropylene/nano-$CaCO_3$ composites [106] or for synthesis of Si/graphene composites for Li-ion battery anodes [107]. $scCO_2$ has application to the synthesis of templates or their removal [108,109], the design chiral molecular recognition polymeric materials [110], the processing and design of aerogels coatings, exfoliation, intercalation, composites, and many other areas as detailed in a review on materials processing [111]. The use of $scCO_2$ in biphasic systems allows for cleaner manufacture of chemicals [112], which could be readily extended to the synthesis of materials and to material processing. Finally, many of the principles introduced in this chapter overlap with methods being developed for application of $scCO_2$ to food [113] and pharmaceutical [114–116] processing, and although these areas are of great interest, their discussion is beyond the scope of this chapter. There are many applications of nanoparticle synthesis that use superfluids other than $scCO_2$ and the interested reader is referred to a recent review on the topic [117]. Much can be learned from techniques being applied in related fields because they share many of the common challenges of materials engineering for developing sustainable processes that have low material intensity.

7. CONCLUSIONS AND FUTURE OUTLOOK

In conclusion, $scCO_2$ is highly versatile solvent in materials engineering, and its use will allow sustainable and green processes to be developed for synthesis and manufacture of functional materials. Especially, $scCO_2$ can be used to reduce energy and material consumption so that material intensity can be lowered. The key to successful application of $scCO_2$ is to exploit its favorable properties for both heat and mass transfer. Applications in materials engineering that take advantage of the mass transfer properties of $scCO_2$ (viscosity reduction, diffusivity) are widely demonstrated in the literature; however, those that take advantage of the favorable heat transfer properties of $scCO_2$, especially in the transcritical region, are much less studied beyond the use of CO_2 as a working fluid in energy systems. Development of materials engineering systems in the future will need critical analysis of both preprocess and postprocess energy and solvent consumption to achieve sustainable production of new materials. Moreover, to truly understand the behavior of many experimental protocols, molecular simulations will need to be vigorously pursued as available properties for many materials are unlikely to be experimentally accessible. The authors expect that $scCO_2$ will play an important role in materials engineering and look forward to the excitement of new protocols, methodologies, and applications that will be discovered in the near future.

References

[1] UN, Sustainable Development Goals, June 1, 2019. https://www.unenvironment.org/explore-topics/sustainable-development-goals/why-do-sustainable-development-goals-matter.

[2] OECD, Global Materials Resource Outlook to 2060, June 11, 2019. https://www.oecd.org/environment/waste/highlights-global-material-resources-outlook-to-2060.pdf.

[3] UN, Recent Trends in Material Flows and Resource Productivity in Asia and the Pacific, June 6, 2019. http://wedocs.unep.org/handle/20.500.11822/9318.

[4] GoodLife Goals, June 6, 2019. https://sdghub.com/goodlifegoals/.

[5] J.H. Clark, S.J. Tavener, Alternative solvents: shades of green, Organic Process Research and Development 11 (2007) 149–155. https://dx.doi.org/10.1021/op060160g.

[6] H. Machida, M. Takesue, R.L. Smith, Green chemical processes with supercritical fluids: properties, materials, separations and energy, Journal of Supercritical Fluids 60 (2011) 2–15. https://dx.doi.org/10.1016/j.supflu.2011.04.016.

[7] R.L. Smith Jr., H. Inomata, C. Peters, Introduction to Supercritical Fluids, June 6, 2019. https://www.elsevier.com/books/introduction-to-supercritical-fluids/smith/978-0-444-52215-3.

[8] R.L. Smith Jr., H. Inomata, C. Peters, Introduction to Supercritical Fluids – A Spreadsheet-Based Approach – Excel programs, June 6, 2019. https://booksite.elsevier.com/9780444522153/.

[9] C. Domb, The Critical Point: A Historical Introduction to the Modern Theory of Critical Phenomena, Taylor & Francis, London, 1996.

[10] J.M.H. Levelt Sengers, G. Morrison, G. Nielson, R.F. Chang, C.M. Everhart, Thermodynamic behavior of supercritical fluid mixtures, International Journal of Thermophysics 7 (1986) 231–243. https://dx.doi.org/10.1007/BF00500151.

[11] R. Span, W. Wagner, A new equation of state for carbon dioxide covering the fluid region from the triple-point temperature to 1100 K at pressures up to 800 MPa, Journal of Physical and Chemical Reference Data 25 (1996) 1509–1596. https://dx.doi.org/10.1063/1.555991.

[12] X. Cheng, T. Schulenberg, Heat Transfer at Supercritical Pressures – Literature Review and Application to an HPLWR, Forschungszentrum Karlsruhe GmbH, Karlsruhe, 2001. http://digbib.ubka.uni.

[13] EcoCute Models, June 6, 2019. http://www.r744.com/articles/939/new_ecocute_models_for_cold_climates.

[14] Wikipedia – EcoCute, June 6, 2019. https://en.wikipedia.org/wiki/EcoCute.

[15] EcoCute News, June 6, 2019. http://www.r744.com/news.

[16] H. Liu, E.A. Macedo, Accurate correlations for the self-diffusion coefficients of CO_2, CH_4, C_2H_4, H_2O, and D_2O over wide ranges of temperature and pressure, The Journal of Supercritical Fluids 8 (1995) 310–317. https://dx.doi.org/10.1016/0896-8446(95)90006-3.

[17] T. Funazukuri, C.Y. Kong, S. Kagei, Binary diffusion coefficients in supercritical fluids: recent progress in measurements and correlations for binary diffusion coefficients, Journal of Supercritical Fluids 38 (2006) 201–210. https://dx.doi.org/10.1016/j.supflu.2006.02.016.

[18] T. Funazukuri, C.Y. Kong, S. Kagei, Predictive correlation of binary diffusion and self-diffusion coefficients under supercritical and liquid conditions, Journal of Supercritical Fluids 46 (2008) 280–284. https://dx.doi.org/10.1016/j.supflu.2008.03.004.

[19] C.Y. Kong, K. Sone, T. Sako, T. Funazukuri, S. Kagei, Solubility determination of organometallic complexes in supercritical carbon dioxide by chromatographic impulse response method, Fluid Phase Equilibria 302 (2011) 347–353. https://dx.doi.org/10.1016/j.fluid.2010.09.034.

[20] C.Y. Kong, T. Funazukuri, S. Kagei, G. Wang, F. Lu, T. Sako, Applications of the chromatographic impulse response method in supercritical fluid chromatography, Journal of Chromatography A 1250 (2012) 141–156. https://dx.doi.org/10.1016/j.chroma.2012.04.033.

[21] C.Y. Kong, T. Siratori, G. Wang, T. Sako, T. Funazukuri, Binary diffusion coefficients of platinum(II) acetylacetonate in supercritical carbon dioxide, Journal of Chemical and Engineering Data 58 (2013) 2919–2924. https://dx.doi.org/10.1021/je400313n.

[22] C.Y. Kong, K. Watanabe, T. Funazukuri, Determination and correlation of infinite dilution binary diffusion coefficients for aluminum acetylacetonate in supercritical and liquid fluids, Fluid Phase Equilibria 420 (2015) 83–88. https://dx.doi.org/10.1016/j.fluid.2016.01.042.

[23] C.Y. Kong, Y. Yakumaru, T. Funazukuri, Measurement and correlation of binary diffusion coefficients of lithium acetylacetonate in supercritical carbon dioxide and in liquid ethanol, Journal of Supercritical Fluids 104 (2015) 265–271. https://dx.doi.org/10.1016/j.supflu.2015.07.003.

[24] C.Y. Kong, K. Watanabe, T. Funazukuri, Measurement and correlation of the diffusion coefficients of chromium(III) acetylacetonate at infinite dilution in supercritical carbon dioxide and in liquid ethanol, Journal of Chemical Thermodynamics 105 (2017) 86–93. https://dx.doi.org/10.1016/j.jct.2016.10.012.

[25] T. Funazukuri, Concerning the determination and predictive correlation of diffusion coefficients in supercritical fluids and their mixtures, Journal of Supercritical Fluids 134 (2018) 28–32. https://dx.doi.org/10.1016/j.supflu.2017.11.035.

[26] J.M. Prausnitz, R.N. Lichtenthaler, E.G. de Azevedo, Molecular Thermodynamics of Fluid-Phase Equilibria, Prentice-Hall, Englewood Cliffs, 1986.

[27] C.M. Hansen, Hansen Solubility Parameters – A User's Handbook, CRC Press Taylor & Francis Group LLC, Boca Raton, 2007.

[28] S.R. Allada, Solubility parameters of supercritical fluids, Industrial and Engineering Chemistry Process Design and Development 23 (1984) 344–348. https://dx.doi.org/10.1021/i200025a028.

[29] I. Ushiki, M. Koike, T. Shimizu, Y. Sato, S. Takishima, H. Inomata, Measurement and modeling of adsorption equilibria of cobalt (III) acetylacetonate on MCM-41 mesoporous silica in the presence of supercritical carbon dioxide with methanol co-solvent, Journal of Supercritical Fluids 140 (2018) 329–335. https://dx.doi.org/10.1016/j.supflu.2018.06.015.

[30] I. Ushiki, N. Takahashi, M. Koike, Y. Sato, S. Takishima, H. Inomata, Adsorption kinetics of rhodium (III) acetylacetonate onto mesoporous silica adsorbents in the presence of supercritical carbon dioxide, Journal of Supercritical Fluids 135 (2018) 137–144. https://dx.doi.org/10.1016/j.supflu.2018.01.017.

[31] I. Ushiki, N. Takahashi, T. Shimizu, Y. Sato, M. Ota, R.L. Smith, H. Inomata, Adsorption equilibria of rhodium acetylacetonate with MCM-41, MSU-H, and HMS silica substrates in supercritical carbon dioxide for preparing catalytic mesoporous materials, Journal of Supercritical Fluids 120 (2017) 240–248. https://dx.doi.org/10.1016/j.supflu.2016.05.032.

[32] W. Jeong, D.W. Lim, S. Kim, A. Harale, M. Yoon, M.P. Suh, J. Kim, Modeling adsorption properties of structurally deformed metal–organic frameworks using structure–property map, Proceedings of the National Academy of Sciences of the United States of America 114 (2017) 7923–7928. https://dx.doi.org/10.1073/pnas.1706330114.

[33] H. Tanaka, M.T. Miyahara, Free energy calculations for adsorption-induced deformation of flexible metal–organic frameworks, Current Opinion in Chemical Engineering (2019) 19–25. https://dx.doi.org/10.1016/j.coche.2019.01.001.

[34] M. Yoneya, S. Tsuzuki, M. Aoyagi, Simulation of metal-organic framework self-assembly, Physical Chemistry Chemical Physics 17 (2015) 8649–8652. https://dx.doi.org/10.1039/c5cp00379b.

[35] M. Haruki, F. Kobayashi, K. Kishimoto, S.-I. Kihara, S. Takishima, Measurement of the solubility of metal complexes in supercritical carbon dioxide using a UV–vis spectrometer, Fluid Phase Equilibria 280 (2009) 49–55. https://dx.doi.org/https://doi.org/10.1016/j.fluid.2009.03.010.

[36] M. Haruki, F. Kobayashi, M. Okamoto, S.-I. Kihara, S. Takishima, Solubility of β-diketonate complexes for cobalt(III) and chromium(III) in supercritical carbon dioxide, Fluid Phase Equilibria 297 (2010) 155–161. https://dx.doi.org/https://doi.org/10.1016/j.fluid.2010.02.028.

[37] M. Haruki, F. Kobayashi, S.-i. Kihara, S. Takishima, Solubility of β-diketonate complexes of copper(II) and cobalt(II) in supercritical carbon dioxide, Journal of Chemical and Engineering Data 56 (2011) 2230–2235. https://dx.doi.org/10.1021/je101255v.

[38] O. Aschenbrenner, S. Kemper, N. Dahmen, K. Schaber, E. Dinjus, Solubility of β-diketonates, cyclopentadienyls, and cyclooctadiene complexes with various metals in supercritical carbon dioxide, Journal of Supercritical Fluids 41 (2007) 179–186. https://dx.doi.org/10.1016/j.supflu.2006.10.011.

[39] Ž. Knez, D. Cör, M. Knez Hrnčič, Solubility of solids in sub- and supercritical fluids: a review 2010–2017, Journal of Chemical and Engineering Data 63 (2018) 860–884. https://dx.doi.org/10.1021/acs.jced.7b00778.

[40] W.H. Teoh, R. Mammucari, N.R. Foster, Solubility of organometallic complexes in supercritical carbon dioxide: a review, Journal of Organometallic Chemistry 724 (2013) 102–116. https://dx.doi.org/10.1016/j.jorganchem.2012.10.005.

[41] C. Erkey, Chapter 3 – Thermodynamics of mixtures of metal complexes with supercritical fluids, in: C. Erkey (Ed.), Supercritical Fluid Science and Technology, Elsevier, 2011, pp. 21–40.

[42] W.C. Andersen, R.E. Sievers, A.F. Lagalante, T.J. Bruno, Solubilities of cerium(IV), Terbium(III), and Iron(III) β-diketonates in supercritical carbon dioxide, Journal of Chemical and Engineering Data 46 (2001) 1045–1049. https://dx.doi.org/10.1021/je000257c.

[43] A.F. Lagalante, B.N. Hansen, T.J. Bruno, R.E. Sievers, Solubilities of copper(II) and chromium(III) .beta.-Diketonates in supercritical carbon dioxide, Inorganic Chemistry 34 (1995) 5781–5785. https://dx.doi.org/10.1021/ic00127a015.

[44] M. Haruki, K. Kishimoto, F. Kobayashi, S.-I. Kihara, S. Takishima, A new correlation and prediction method for the solubility of metal complexes in supercritical carbon dioxide using regular solution theory with the COSMO-RS method, Journal of Chemical Engineering of Japan 42 (2009) 309–318. https://dx.doi.org/10.1252/jcej.08we280.

[45] I. Ushiki, Y. Sato, S. Takishima, H. Inomata, Thermodynamic modeling of the solubility of acetylacetonate-type metal precursors in supercritical carbon dioxide using the PC-SAFT equation of state, Journal of Chemical Engineering of Japan 52 (2019) 243–252. https://dx.doi.org/10.1252/jcej.18we194.

[46] J. Gross, G. Sadowski, Perturbed-chain SAFT: an equation of state based on a perturbation theory for chain molecules, Industrial and Engineering Chemistry Research 40 (2001) 1244–1260. https://dx.doi.org/10.1021/ie0003887.

[47] J.J. Watkins, T.J. McCarthy, Polymer/metal nanocomposite synthesis in supercritical CO_2, Chemistry of Materials 7 (1995) 1991–1994. https://dx.doi.org/10.1021/cm00059a001.

[48] C. Erkey, Supercritical Fluids and Organometallic Compounds: From Recovery of Trace Metals to Synthesis of Nanostructured Materials, Elsevier, 2011.

[49] J. Morère, M.J. Tenorio, M.J. Torralvo, C. Pando, J.A.R. Renuncio, A. Cabañas, Deposition of Pd into mesoporous silica SBA-15 using supercritical carbon dioxide, The Journal of Supercritical Fluids 56 (2011) 213–222. https://dx.doi.org/10.1016/j.supflu.2010.12.012.

[50] S. Müller, M. Türk, Production of supported gold and gold–silver nanoparticles by supercritical fluid reactive deposition: effect of substrate properties, The Journal of Supercritical Fluids 96 (2015) 287–297. https://dx.doi.org/10.1016/j.supflu.2014.09.035.

[51] E. Kondoh, Deposition of Ru thin films from supercritical carbon dioxide fluids, Japanese Journal of Applied Physics 44 (2005) 5799–5802. https://dx.doi.org/10.1143/jjap.44.5799.

[52] K. Jung, T. Momose, Y. Shimogaki, Strontium ruthenium oxide deposition in supercritical carbon dioxide using a closed reactor system, Journal of Supercritical Fluids 79 (2013) 244–250. https://dx.doi.org/10.1016/j.supflu.2013.03.008.

[53] E. Kondoh, J. Fukuda, Deposition kinetics and narrow-gap-filling in Cu thin film growth from supercritical carbon dioxide fluids, Journal of Supercritical Fluids 44 (2008) 466–474. https://dx.doi.org/10.1016/j.supflu.2007.12.004.

[54] E. Kondoh, T. Ueno, S. Kurita, M. Watanabe, S. Yamamoto, T. Suemasu, Supercritical CO_2 reactor for wafer-scale thin film deposition: reactor concept, numerical results, and Cu deposition, Journal of Supercritical Fluids 104 (2015) 49–53. https://dx.doi.org/10.1016/j.supflu.2015.05.021.

[55] S.E. Bozbağ, C. Erkey, Supercritical deposition: current status and perspectives for the preparation of supported metal nanostructures, The Journal of Supercritical Fluids 96 (2015) 298–312. https://dx.doi.org/10.1016/j.supflu.2014.09.036.

[56] T. Shimizu, I. Ushiki, M. Ota, Y. Sato, N. Koizumi, H. Inomata, Preparation of mesoporous silica supported cobalt catalysts using supercritical fluids for Fischer-Tropsch synthesis, Chemical Engineering Research and Design 95 (2015) 64–68. https://dx.doi.org/10.1016/j.cherd.2015.01.005.

[57] M. Türk, C. Erkey, Synthesis of supported nanoparticles in supercritical fluids by supercritical fluid reactive deposition: current state, further perspectives and needs, The Journal of Supercritical Fluids 134 (2018) 176–183. https://dx.doi.org/10.1016/j.supflu.2017.12.010.

[58] Z. Fang, R.L. Smith Jr., H. Li, Production of Biofuels and Chemicals with Bifunctional Catalysts, Springer-Nature, Singapore, 2017. https://www.springer.com/gb/book/9789811051364.

[59] J. Morère, M.J. Torralvo, C. Pando, J.A.R. Renuncio, A. Cabañas, Supercritical fluid deposition of Ru nanoparticles onto SiO_2 SBA-15 as a sustainable method to prepare selective hydrogenation catalysts, RSC Advances 5 (2015) 38880–38891. https://dx.doi.org/10.1039/C5RA04969E.

[60] E. Di Maio, E. Kiran, Foaming of polymers with supercritical fluids and perspectives on the current knowledge gaps and challenges, The Journal of Supercritical Fluids 134 (2018) 157–166. https://dx.doi.org/https://doi.org/10.1016/j.supflu.2017.11.013.

[61] M. Sauceau, J. Fages, A. Common, C. Nikitine, E. Rodier, New challenges in polymer foaming: a review of extrusion processes assisted by supercritical carbon dioxide, Progress in Polymer Science 36 (2011) 749–766. https://dx.doi.org//10.1016/j.progpolymsci.2010.12.004.

[62] L.J.M. Jacobs, M.F. Kemmere, J.T.F. Keurentjes, Sustainable polymer foaming using high pressure carbon dioxide: a review on fundamentals, processes and applications, Green Chemistry 10 (2008) 731–738. https://dx.doi.org/10.1039/B801895B.

[63] A.R.C. Duarte, J.F. Mano, R.L. Reis, Supercritical fluids in biomedical and tissue engineering applications: a review, International Materials Reviews 54 (2009) 214–222. https://dx.doi.org/10.1179/174328009X411181.

[64] B. Notario, J. Pinto, M.A. Rodriguez-Perez, Nanoporous polymeric materials: a new class of materials with enhanced properties, Progress in Materials Science 78–79 (2016) 93–139. https://dx.doi.org/10.1016/j.pmatsci.2016.02.002.

[65] A. Fraleoni-Morgera, M. Chhikara, Polymer-Based Nano-Composites for Thermal Insulation, Advanced Engineering Materials, 2019. https://dx.doi.org/10.1002/adem.201801162.

[66] A. Zhang, Q. Zhang, H. Bai, L. Li, J. Li, Polymeric nanoporous materials fabricated with supercritical CO_2 and CO_2-expanded liquids, Chemical Society Reviews 43 (2014) 6938–6953. https://dx.doi.org/10.1039/c4cs00100a.

[67] J. Pinto, J.A. Reglero-Ruiz, M. Dumon, M.A. Rodriguez-Perez, Temperature influence and CO_2 transport in foaming processes of poly(methyl methacrylate)-block copolymer nanocellular and microcellular foams, Journal of Supercritical Fluids 94 (2014) 198–205. https://dx.doi.org/10.1016/j.supflu.2014.07.021.

[68] S.G. Kazarian, Polymer processing with supercritical fluids, Polymer Science – Series C 42 (2000) 78–101. https://www.imperial.ac.uk/media/imperial-college/research-centres-and-groups/vibrational-spectroscopy-and-chemical-imaging/reviewscf.pdf.

[69] S.G. Kazarian, M.F. Vincent, F.V. Bright, C.L. Liotta, C.A. Eckert, Specific intermolecular interaction of carbon dioxide with polymers, Journal of the American Chemical Society 118 (1996) 1729–1736. https://dx.doi.org/10.1021/ja950416q.

[70] A.V. Ewing, S.G. Kazarian, Current trends and opportunities for the applications of in situ vibrational spectroscopy to investigate the supercritical fluid processing of polymers, The Journal of Supercritical Fluids 134 (2018) 88–95. https://dx.doi.org/10.1016/j.supflu.2017.12.011.

[71] D. Tammaro, V. Contaldi, M.G.P. Carbone, E. Di Maio, S. Iannace, A novel lab-scale batch foaming equipment: the mini-batch, Journal of Cellular Plastics 52 (2015) 533–543. https://dx.doi.org/10.1177/0021955X15584654.

[72] D. Tammaro, A. Astarita, E. Di Maio, S. Iannace, Polystyrene foaming at high pressure drop rates, Industrial & Engineering Chemistry Research 55 (2016) 5696–5701. https://dx.doi.org/10.1021/acs.iecr.5b04911.

[73] T. Ono, X. Wu, T. Furuya, S. Yoda, Development of a rapid decompression system for nanocellular foaming at 100 MPa, The Journal of Supercritical Fluids 149 (2019) 26–33. https://dx.doi.org/10.1016/j.supflu.2019.03.015.

[74] O.I. Lebedev, F. Millange, C. Serre, G. Van Tendeloo, G. Férey, First direct imaging of giant pores of the Metal–Organic framework MIL-101, Chemistry of Materials 17 (2005) 6525–6527. https://dx.doi.org/10.1021/cm051870o.

[75] K. Momma, F. Izumi, VESTA 3 for three-dimensional visualization of crystal, volumetric and morphology data, Journal of Applied Crystallography 44 (2011) 1272–1276. https://dx.doi.org/10.1107/S0021889811038970.

[76] K. Momma, Visualization for Electronic and Structural Analysis, June 6, 2019. https://jp-minerals.org/soft/en/.

[77] H. Furukawa, K.E. Cordova, M. O'Keeffe, O.M. Yaghi, The chemistry and applications of metal-organic frameworks, Science 341 (2013) 1230444. https://dx.doi.org/10.1126/science.1230444.

[78] C. Férey, C. Mellot-Draznieks, C. Serre, F. Millange, J. Dutour, S. Surblé, I. Margiolaki, Chemistry: a chromium terephthalate-based solid with unusually large pore volumes and surface area, Science 309 (2005) 2040–2042. https://dx.doi.org/10.1126/science.1116275.

[79] A.J. Howarth, A.W. Peters, N.A. Vermeulen, T.C. Wang, J.T. Hupp, O.K. Farha, Best practices for the synthesis, activation, and characterization of metal–organic frameworks, Chemistry of Materials 29 (2017) 26–39. https://dx.doi.org/10.1021/acs.chemmater.6b02626.

[80] H. Li, M. Eddaoudi, M. O'Keeffe, O.M. Yaghi, Design and synthesis of an exceptionally stable and highly porous metal- organic framework, Nature 402 (1999) 276–279. https://dx.doi.org/10.1038/46248.

[81] O.K. Farha, I. Eryazici, N.C. Jeong, B.G. Hauser, C.E. Wilmer, A.A. Sarjeant, R.Q. Snurr, S.T. Nguyen, A.O. Yazaydin, J.T. Hupp, Metal-organic framework materials with ultrahigh surface areas: is the sky the limit? Journal of the American Chemical Society 134 (2012) 15016–15021. https://dx.doi.org/10.1021/ja3055639.

[82] O.K. Farha, J.T. Hupp, Rational design, synthesis, purification, and activation of Metal–Organic framework materials, Accounts of Chemical Research 43 (2010) 1166–1175. https://dx.doi.org/10.1021/ar1000617.

[83] A.P. Nelson, O.K. Farha, K.L. Mulfort, J.T. Hupp, Supercritical processing as a route to high internal surface areas and permanent microporosity in Metal–Organic framework materials, Journal of the American Chemical Society 131 (2009) 458–460. https://dx.doi.org/10.1021/ja808853q.

[84] K. Matsuyama, Supercritical fluid processing for metal–organic frameworks, porous coordination polymers, and covalent organic frameworks, The Journal of Supercritical Fluids 134 (2018) 197–203. https://dx.doi.org/10.1016/j.supflu.2017.12.004.

[85] K. Matsuyama, M. Motomura, T. Kato, T. Okuyama, H. Muto, Catalytically active Pt nanoparticles immobilized inside the pores of metal organic framework using supercritical CO_2 solutions, Microporous and Mesoporous Materials 225 (2016) 26–32. https://dx.doi.org/10.1016/j.micromeso.2015.12.005.

[86] P. Horcajada, C. Serre, M. Vallet-Regí, M. Sebban, F. Taulelle, G. Férey, Metal–organic frameworks as efficient materials for drug delivery, Angewandte Chemie International Edition 45 (2006) 5974–5978. https://dx.doi.org/10.1002/anie.200601878.

[87] A. López-Periago, O. Vallcorba, C. Frontera, C. Domingo, J.A. Ayllón, Exploring a novel preparation method of 1D metal organic frameworks based on supercritical CO_2, Dalton Transactions 44 (2015) 7548–7553. https://dx.doi.org/10.1039/C5DT00521C.

[88] A.M. López-Periago, N. Portoles-Gil, P. López-Domínguez, J. Fraile, J. Saurina, N. Aliaga-Alcalde, G. Tobias, J.A. Ayllón, C. Domingo, Metal–organic frameworks precipitated by reactive crystallization in supercritical CO_2, Crystal Growth and Design 17 (2017) 2864–2872. https://dx.doi.org/10.1021/acs.cgd.7b00378.

[89] A. López-Periago, P. López-Domínguez, J. Pérez Barrio, G. Tobias, C. Domingo, Binary supercritical CO_2 solvent mixtures for the synthesis of 3D metal-organic frameworks, Microporous and Mesoporous Materials 234 (2016) 155–161. https://dx.doi.org/10.1016/j.micromeso.2016.07.014.

[90] X. Sang, J. Zhang, J. Xiang, J. Cui, L. Zheng, J. Zhang, Z. Wu, Z. Li, G. Mo, Y. Xu, J. Song, C. Liu, X. Tan, T. Luo, B. Zhang, B. Han, Ionic liquid accelerates the crystallization of Zr-based metal–organic frameworks, Nature Communications 8 (2017) 175. https://dx.doi.org/10.1038/s41467-017-00226-y.

[91] Y. Zhao, J. Zhang, B. Han, J. Song, J. Li, Q. Wang, Metal-organic framework nanospheres with well-ordered mesopores synthesized in an ionic liquid/CO_2/surfactant system, Angewandte Chemie – International Edition 50 (2011) 636–639. https://dx.doi.org/10.1002/anie.201005314.

[92] N. Lavoine, L. Bergström, Nanocellulose-based foams and aerogels: processing, properties, and applications, Journal of Materials Chemistry A 5 (2017) 16105–16117. https://dx.doi.org/10.1039/C7TA02807E.

[93] K. Matsuyama, K. Morotomi, S. Inoue, M. Nakashima, H. Nakashima, T. Okuyama, T. Kato, H. Muto, H. Sugiyama, Antibacterial and antifungal properties of Ag nanoparticle-loaded cellulose nanofiber aerogels prepared by supercritical CO_2 drying, Journal of Supercritical Fluids 143 (2019) 1–7. https://dx.doi.org/10.1016/j.supflu.2018.08.008.

[94] S. Katayama, S. Sameshima, S. Yonezawa, Y. Iwai, Development of a method for injection of nanoparticles into cotton fibers by use of supercritical carbon dioxide, Kagaku Kogaku Ronbunshu 41 (2015) 200–206. https://dx.doi.org/10.1252/kakoronbunshu.41.200.

[95] S. Katayama, L. Zhao, S. Yonezawa, Y. Iwai, Modification of the surface of cotton with supercritical carbon dioxide and water to support nanoparticles, Journal of Supercritical Fluids 61 (2012) 199–205. https://dx.doi.org/10.1016/j.supflu.2011.10.008.

[96] Y. Iwai, S. Sameshima, S. Yonezawa, S. Katayama, Fabrication of conductive cotton by electroless plating method with supercritical carbon dioxide, Journal of Supercritical Fluids 100 (2015) 46–51. https://dx.doi.org/10.1016/j.supflu.2015.02.027.

[97] Y. Iwai, S. Sameshima, S. Yonezawa, S. Katayama, Injection of silver nano-particles in cotton fibers by using supercritical carbon dioxide, in: 10th Asian Control Conference: Emerging Control Techniques for a Sustainable World, ASCC 2015, 2015.

[98] Z. Chen, H. Miao, J. Wu, Y. Tang, W. Yang, L. Hou, F. Yang, X. Tian, L. Zhang, Y. Li, Scalable production of hydrophilic graphene nanosheets via in situ ball-milling-assisted supercritical CO_2 exfoliation, Industrial and Engineering Chemistry Research 56 (2017) 6939–6944. https://dx.doi.org/10.1021/acs.iecr.7b00594.

[99] Y. Gai, W. Wang, D. Xiao, H. Tan, M. Lin, Y. Zhao, Exfoliation of graphite into graphene by a rotor-stator in supercritical CO_2: experiment and simulation, Industrial and Engineering Chemistry Research 57 (2018) 8220–8229. https://dx.doi.org/10.1021/acs.iecr.8b01726.

[100] Y. Gai, W. Wang, D. Xiao, Y. Zhao, Ultrasound coupled with supercritical carbon dioxide for exfoliation of graphene: simulation and experiment, Ultrasonics Sonochemistry 41 (2018) 181–188. https://dx.doi.org/10.1016/j.ultsonch.2017.09.007.

[101] S.M. Hamidinejad, R.K.M. Chu, B. Zhao, C.B. Park, T. Filleter, Enhanced thermal conductivity of graphene nanoplatelet-polymer nanocomposites fabricated via supercritical fluid-assisted in situ exfoliation, ACS Applied Materials and Interfaces 10 (2018) 1225–1236. https://dx.doi.org/10.1021/acsami.7b15170.

[102] L. Li, J. Xu, G. Li, X. Jia, Y. Li, F. Yang, L. Zhang, C. Xu, J. Gao, Y. Liu, Z. Fang, Preparation of graphene nanosheets by shear-assisted supercritical CO_2 exfoliation, Chemical Engineering Journal 284 (2016) 78–84. https://dx.doi.org/10.1016/j.cej.2015.08.077.

[103] Y. Wang, Z. Chen, Z. Wu, Y. Li, W. Yang, Y. Li, High-efficiency production of graphene by supercritical CO_2 exfoliation with rapid expansion, Langmuir 34 (2018) 7797–7804. https://dx.doi.org/10.1021/acs.langmuir.8b01030.

[104] A.D. Yapryntsev, K.B. Ustinovich, A.A. Rodina, V.A. Lebedev, O.I. Pokrovskiy, K.E. Yorov, A.V. Gavrikov, A.E. Baranchikov, V.K. Ivanov, Exfoliation of layered yttrium hydroxide by rapid expansion of supercritical suspensions, Journal of Supercritical Fluids 150 (2019) 40–48. https://dx.doi.org/10.1016/j.supflu.2019.04.012.

[105] N. Taguchi, N. Takeyasu, S. Kawata, Extraction of three-dimensional silver nanostructures with supercritical fluid, Applied Physics Express 11 (2018). https://dx.doi.org/10.7567/APEX.11.025201.

[106] S. Chen, X. Wang, X. Ma, K. Wang, Morphology and properties of polypropylene/nano-$CaCO_3$ composites prepared by supercritical carbon dioxide-assisted extrusion, Journal of Materials Science 51 (2016) 708–718. https://dx.doi.org/10.1007/s10853-015-9272-x.

[107] S.H. Lee, S. Park, M. Kim, D. Yoon, C. Chanthad, M. Cho, J. Kim, J.H. Park, Y. Lee, Supercritical carbon dioxide-assisted process for well-dispersed silicon/graphene composite as a Li ion battery anode, Scientific Reports 6 (2016) 32011. https://dx.doi.org/10.1038/srep32011. https://www.nature.com/articles/srep32011#supplementary-information.

[108] I. Ushiki, S. Teratani, M. Ota, Y. Sato, H. Inomata, Extraction of template agents from porous silica using supercritical carbon dioxide-entrainer method, Kagaku Kogaku Ronbunshu 37 (2011) 512–517. https://dx.doi.org/10.1252/kakoronbunshu.37.512.

[109] E. Prouzet, A. Kacheff, G. Aubert, A. Bentaleb, R. Backov, C. Aymonier, Toward a sustainable preparation of tunable mesoporous silica, The Journal of Supercritical Fluids 143 (2019) 139–145. https://dx.doi.org/10.1016/j.supflu.2018.07.008.

[110] M. Soares da Silva, E.R. Vão, M. Temtem, L. Mafra, J. Caldeira, A. Aguiar-Ricardo, T. Casimiro, Clean synthesis of molecular recognition polymeric materials with chiral sensing capability using supercritical fluid technology. Application as HPLC stationary phases, Biosensors and Bioelectronics 25 (2010) 1742–1747. https://dx.doi.org/10.1016/j.bios.2009.12.023.

[111] X. Zhang, S. Heinonen, E. Levänen, Applications of supercritical carbon dioxide in materials processing and synthesis, RSC Advances 4 (2014) 61137–61152. https://dx.doi.org/10.1039/C4RA10662H.

[112] Y. Medina-Gonzalez, S. Camy, J.-S. Condoret, $ScCO_2$/Green solvents: biphasic promising systems for cleaner chemicals manufacturing, ACS Sustainable Chemistry and Engineering 2 (2014) 2623–2636. https://dx.doi.org/10.1021/sc5004314.

[113] F. Temelli, Perspectives on the use of supercritical particle formation technologies for food ingredients, Journal of Supercritical Fluids 134 (2018) 244–251. https://dx.doi.org/10.1016/j.supflu.2017.11.010.

[114] L. Baldino, E. Reverchon, Challenges in the production of pharmaceutical and food related compounds by SC-CO_2 processing of vegetable matter, Journal of Supercritical Fluids 134 (2018) 269–273. https://dx.doi.org/10.1016/j.supflu.2017.11.034.

[115] P. Trucillo, R. Campardelli, E. Reverchon, A versatile supercritical assisted process for the one-shot production of liposomes, Journal of Supercritical Fluids (2019) 136–143. https://dx.doi.org/10.1016/j.supflu.2019.01.015.

[116] K. Mishima, M. Honjo, T. Sharmin, S. Ito, R. Kawakami, T. Kato, M. Misumi, T. Suetsugu, H. Orii, H. Kawano, K. Irie, K. Sano, K. Mishima, T. Harada, M. Ouchi, Gas-saturated solution process to obtain microcomposite particles of alpha lipoic acid/hydrogenated colza oil in supercritical carbon dioxide, Pharmaceutical Development and Technology 21 (2016) 737–748. https://dx.doi.org/10.3109/10837450.2015.1049707.

[117] M.K.M. Lane, J.B. Zimmerman, Controlling metal oxide nanoparticle size and shape with supercritical fluid synthesis, Green Chemistry 21 (2019) 3769–3781. https://doi.org/10.1039/C9GC01619H.

CHAPTER

16

Rational Design of Continuous Flow Processes for Synthesis of Functional Molecules

Alexei Lapkin[1,2]

[1]Department of Chemical Engineering and Biotechnology, University of Cambridge, Cambridge, United Kingdom; [2]Cambridge Centre for Advanced Research and Education in Singapore Ltd, Singapore

OUTLINE

1. INTRODUCTION

Continuous chemical processes traditionally were the domain of large-scale manufacture in petrochemical and bulk material industries, whereas manufacture of functional molecules and materials has, until recently, been almost exclusively done in batch or fed-batch type reactors. This stems from the characteristics of the distinctively different manufacturing paradigms: the conventional continuous chemical processes are highly optimized for a specific reaction–separation sequence, located within large facilities to benefit from shared services, and are highly material- and energy-integrated. This allows to fully exploit the economy of scale. A diverse range of reactors and separators can be found in the specific applications in petrochemical and bulk material industries. These are typically classified into packed-bed reactors for gas–solid reactions, packed-bed reactors with intercoolers or multitubular packed-bed reactors for reactions with large heat effects, fluidized bed or pneumatic reactors for gas–solid, liquid–solid, and gas–liquid–solid reactions where small solid particles must be used, and bubble columns and trickle-bed reactors for gas–liquid and gas–liquid–solid reactions; see Fig. 16.1 for schematic illustration of the most common types of large-scale continuous reactors used in petrochemical and bulk materials industries.

The different variations of process technology used in large-scale continuous chemical manufacture cover a very broad range of operating conditions: temperature in different continuous processes range from cryogenic to 1000°C, whereas pressures range from subatmospheric to 100s of bar, and residence times range from subseconds to hours, although the long overall residence time is frequently attained through recycling.

https://doi.org/10.1016/B978-0-12-814681-1.00016-3

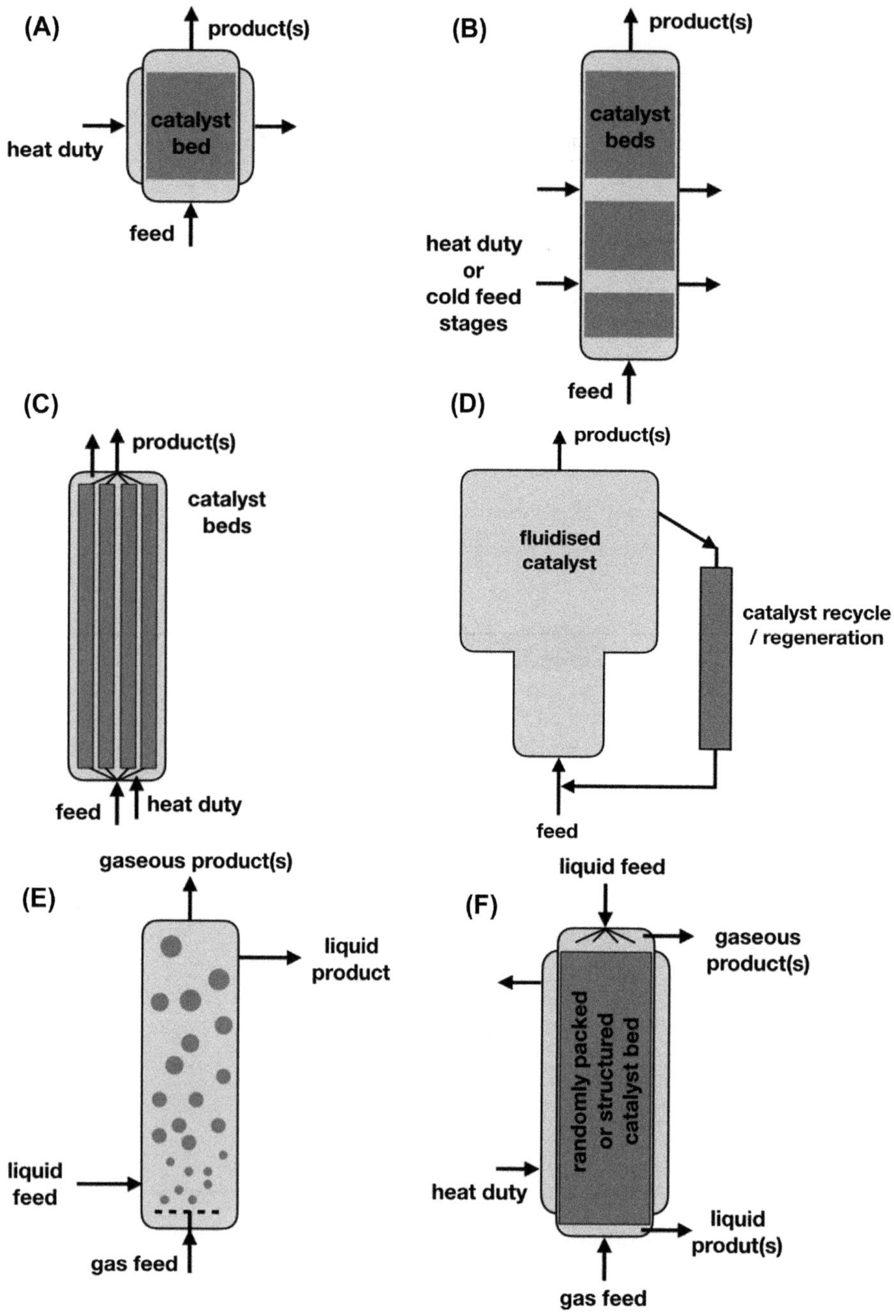

FIGURE 16.1 Schematic illustration of the reactor types used in large-scale petrochemical and material manufacturing: (A) a jacketed packed-bed reactor with a single catalyst bed, suitable for reactions with low heats of reactions, (B) multibed reactors with interstage cooling, suitable for reactions with high exotherms; note the size of beds is progressively increasing as high conversion requires longer residence times, (C) multitubular packed-bed reactors for highly exo- or endothermic reactions; the reactor of choice for steam reforming, for example, (D) a fluidized bed catalyst, suitable for reactions with short residence times, requiring small catalyst particles and good heat transfer; see, for example, fluid catalytic cracking, characterized by rapid catalyst deactivation, (E) bubble columns for continuous gas–liquid reactions with moderate heat effects, and (F) jacketed trickle-bed reactors with either random or structured catalytic packing; random packed trickle–bed reactors most frequently are operated in co-flow regime, with both gas and liquid streams flowing downward, whereas structured packing trickle beds allow for the countercurrent operation, characterized by enhanced mass transfer.

The large variety of continuous reactor designs in bulk chemical and material manufacturing is complemented by a set of commonly used separation processes: distillation, cryogenic distillation, adsorption and absorption, gas and liquid separations by membranes, phase splitting (for example, mixer-settler type devices), continuous crystallization, and so on.

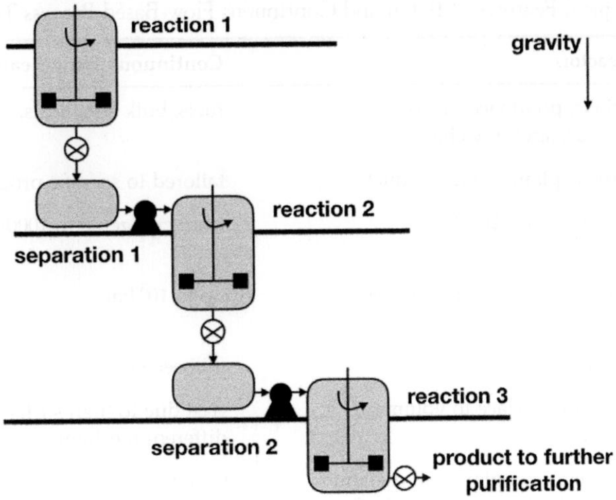

FIGURE 16.2 Schematic illustration of a generic vertically organized sequential batch processing plant, based on stirred tank reactors.

Quite opposite to bulk chemical manufacture, the manufacture of *functional* molecules and materials is done predominantly in multiuse *batch* plants, frequently organized vertically for the multistep syntheses, such that the final product is collected at the bottom of a process plant, see Fig. 16.2. Industrial-scale batch operations range from 10s of liters scale, in the case of pharmaceutical active ingredients or custom manufacture of chemicals, to close to 100 cubic meters, for example, in the case of emulsion polymerization.

A stirred vessel operated in batch regime is one of the most versatile process technologies, and for good reasons. Stirred tanks are conceptually identical to round bottom flasks all chemists are familiar with through their undergraduate training and early-stage R&D work; hence, the concept of scaling to a large vessel is very simple, does not require complex engineering, and as a result rather fast. Because a variety of different products could be made in the same vessel with minor modifications, or with adjustment of reaction time, temperature, and the stirrer speed, most operators would opt for running a suboptimal process, instead of investing into an optimal, but perhaps more product-specific hardware.

The drawbacks of batch stirred tank process technology are many: industrial-scale stirred tank reactors are frequently limited to relatively low pressures of 10–30 bar due to strong correlation of pressure–size–risk–cost. As a result, for many reactions run in organic solvents, operating temperatures are limited by the vapor pressure of the solvent. Within a stirred tank reactor, molecules experience all possible concentrations and reaction times. This has a consequence on selectivity in the case of reactions with complex kinetic schemes. This is made worse by the fact that stirred tank reactors are usually poorly mixed: the areas near impellers and baffles experience significantly different sheer rates and mixing efficiencies compared with areas away from the impeller. This is particularly important at scale. Finally, the rates of heat transfer in large-scale stirred tank reactors are usually rather poor, thus many reactions are artificially slowed-down by reducing reaction temperature or reactant's concentration to ensure the rate of reaction is commensurable with the maximum heat flux through the available cooling system. Table 16.1 gives an overview of typical features of batch and continuous flow processes.

Given hindsight, it seems obvious and sensible to combine the benefits of flexibility of the stirred tank reactor technology with the benefits of task-optimized and broad operating conditions of the continuous process technology. Over the last three decades, the universe of chemical processing techniques has significantly expanded with the increasingly wide adoption of *process intensification* (PI) methods [1–4]: combining reactions with separations to take advantage of synergy between thermodynamics and kinetics of processes, reducing length scales for mass and heat transfer to shift reactions toward kinetic control, and so on. The concept of PI allows, to some extent, to merge the two drastically different manufacturing paradigms—batch and continuous processes—as PI often allows to shrink process size by significantly enhancing efficiency of the individual steps, and thus small versions of highly efficient processes become available for bespoke manufacture of functional products [1–8]. More recently, the synthetic chemistry community has picked up on the same idea, giving it their own term "flow chemistry" [9–21]. This chapter develops the idea of transformation of continuous processes to manufacture of functional molecules and materials.

The overall process design of chemical plants is a complex process, which has been extensively researched and is the foundation of the discipline of chemical process design [22,23]. Here, we only note some basic ideas for

TABLE 16.1 A Comparison of Typical Features of Batch and Continuous Flow-Based Process Technologies.

Features	Batch Reactors	Continuous Flow Reactors
Typical chemical industry sector	formulations, pharmaceuticals, agrichemicals, specialty chemicals	fuels, bulk chemicals, intermediates
Flexibility of use	multipurpose plants; multiproduct	tailored to specific product
Range of operating temperatures	from cryogenic to ∽200°C	from cryogenic to 1000°C
Range of operating pressures	up to 10 bar (for over 10 m^3 vessels)	up to 10^3 bar
Residence time	minutes—hours	milliseconds—minutes
Heat transfer scalability	*poor* due to low surface to volume ratio at scale	*good* due to high surface to volume ratio and possible different mechanisms (staged injection, intercoolers, etc.)
Mixing	*very poor* due to complex fluid dynamics and significant inhomogeneities in flow	Can be *well controlled* through multiple available mechanisms; available mixing times in the milliseconds range
Residence time distribution	molecules experience all concentrations and reaction times	May approach *plug flow* approximation of residence time distribution
Control	*dynamic* control over the reaction progress; scheduling	maintaining *steady state*, response to disturbances, and long-term dynamics (e.g., catalyst deactivation)
Process integration	typically *not integrated*	frequently heat and/or mass *integrated* in complex facilities
Multiphase processes	liquid phase; slurries, gas—liquid, liquid—liquid, and three phases are known	particulate flow is frequently used in *large-scale* continuous flow systems; gas—liquid and liquid—liquid systems also known; particulate flow is problematic for small-scale continuous flow systems

integrated process design. A chemical process is a hierarchy of *functions* or *operations*, such as reaction, separation, recycle, and control, and thus could be tackled in a systematic way through a hierarchical process design methodology [24,25]. Optimization of integrated *continuous* plants is done for the complete flowsheets, including the possible options of different separation technologies, generating hybrid reaction—separation unit operations, checking different orders of unit operations and different configurations of the recycle streams. Such plant-wide optimization problems are rather complex, yet feasible to solve, using recent advances in the optimization methods [26]. Finally, new methods of process synthesis, such as function-based modeling [27] and new optimization approaches for synthesis of large process systems [28], are particularly exciting developments in this traditional field, opening new opportunities for computer-aided process design within the context of Industry 4.0 [29].

Today, there is a significant overlap between research groups in synthetic chemistry and process engineering, working in the area of developing new reactions and new functional molecules. This is evidenced by the significant increase in the number of interdisciplinary publications and the emergence of new scientific journals specifically addressing this interface: ACS Sustainable Chemistry and Engineering, RSC Reaction Chemistry and Engineering, and RSC Molecular Engineering are the most notable and successful examples. Yet up till now there is no widely acceptable generic methodology for design of continuous flow processes for manufacture of *functional* products. In this chapter, we bring together some emerging elements of what is likely to become a transformative methodology for rational design of continuous flow processes, compliant with aspirations for Industry 4.0.

1.1 Part 1. The Desired Characteristics of Manufacturing Functional Products in Flow

Manufacture of functional molecules is characterized by relatively small volumes, requirement for high product purity or narrow specification of the end product with respect of performance, or impurities content, and short life time of a product. As a result, product development must be fast, and manufacturing scale-up must suit multiple possible throughput requirements. An ideal manufacturing process would offer tight control over product quality, quick response to fluctuations in market demand, and be easily implemented with low capital cost.

2. PRODUCT QUALITY IN FLOW PROCESSES

The topic of tight control of operating conditions in continuous flow reactors, which is linked with the improved product quality, has been extensively investigated and reviewed in recent years [9,10,14,17–20,30–32]. There are six key concepts that must be considered with regard to control of product quality: (i) control of mean residence time and (ii) of residence time distribution, (iii) control of the concentration distribution inside a reactor, (iv) control of temperature, (v) control of sheer, and (vi) tailoring of residence time to the desired mixing time. Here, we will briefly discuss these six concepts.

In a continuous flow reactor with a reactor volume V [m^3] and feed molar flow rate F_{feed} [mol s^{-1}], the *space time*, t [s], is given by Eq. (16.1).

$$t = \frac{V}{F} \tag{16.1}$$

This is sometime reported in the literature in place of *residence time*, which is, generally, incorrect. Mean residence time, τ [s], more specifically, mean residence time of exit distribution (or also called age distribution) $E(t)$, has a meaning of a mean time an element of fluid spends within a reactor. It is calculated based on exit distribution $E(t)$ as shown in Eq. (16.2), where exit distribution is calculated according to Eq. (16.3) [33]. Eq. (16.2) is a first moment of exit distribution $E(t)$.

$$\tau = \int_{t_1}^{t_2} tE(t)dt \tag{16.2}$$

$$E(t) = \frac{C(t)}{\int_0^\infty C(t)dt} \tag{16.3}$$

where *C(t)* is the concentration at the exit of a reactor at time t. The concept of exit distribution and residence time is best illustrated on an example of a plug flow tubular reactor, see Fig. 16.3.

In some circumstances, the solution of Eq. (16.2) is identical to Eq. (16.1), for example, in a perfectly mixed continuous stirred tank reactor, but not as a general rule. Because calculating exit distributions from theory for an arbitrary reactor is not straightforward, evaluation of mean residence time should be done experimentally from the measured concentrations of a tracer as shown in the annotation to Fig. 16.3. In many cases, especially when packed reactors are used, residence time distribution curves exhibit complex behavior, due to presence of stagnant zones or channeling of fluid within the bed. In such cases, approximating residence time by the reactor' space time introduces a very large error and should be avoided.

Most often, reactants are fed into a rector not premixed. In this case, distribution of residence time within a reactor necessarily influences the distribution of concentrations. As a result, for complex kinetic schemes, reaction yield and selectivity are dependent on the residence time distribution. This is one of the key design criteria for processes with tight specification of the product.

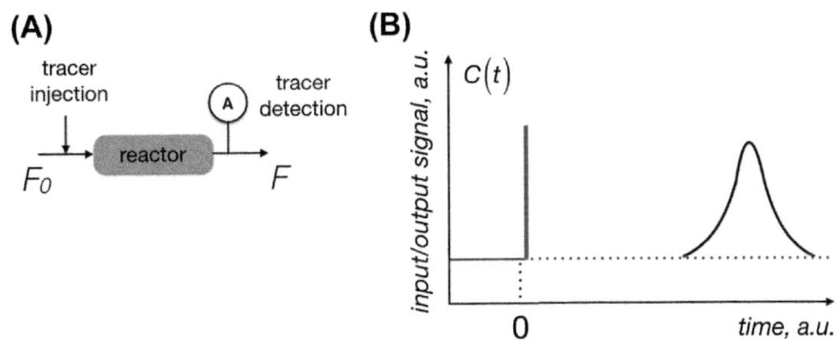

FIGURE 16.3 An illustration of the concept of residence time and exit distribution function. (A) A small amount of a tracer that does not disturb the flow and is not interacting with the reactor is injected in a pulse at the inlet of the reactor and tracer is detected at the exit of the reactor. (B) The tracer input is represented by a delta function at time $t = 0$; its response is measured at the reactor's exit and exhibits a characteristic broadening of the peak due to imperfect flow.

A more direct way of adjusting concentrations in flow reactors is staged input or withdrawal of reagents. This is frequently used to control rate of very fast or highly exothermic reactions in staged reactors with intercooling, Fig. 16.1B. Intensification of reactors by combining several functions into a single device (multifunctional reactors [34,35]) allows to combine concentration distribution and temperature control, thus presenting new degrees of freedom for design of chemical reactors. New manufacturing techniques, such as diffusion bonding and 3D printing, allow to produce complex reactor architectures, with most imaginative internal components. As an example, reactor heat exchangers with staged injection of reactants were developed for highly exothermic three-phase oxidation [36,37] and hydrogenation reactions [38]. To achieve good control of temperature inside a reactor operating under kinetic control of a fast, exothermic reaction, it is necessary to significantly reduce width of the reactor to increase surface to volume ratio. In this case, it will be feasible to avoid development of radial and axial temperature profiles within the catalyst bed. This can be done by reducing the reactor dimensions to the so-called mesoscale (mm scale), and then by combining multiple functionalities within structured mm scale reactors. This technology has found commercial applications in compact gas-to-liquid (GTL) processes, converting flare gas into liquid fuel (commercialized by CompactGTL Ltd.).

Control of sheer and tailoring residence time to mixing time is particularly important in the applications of continuous processes to synthesis of nanostructured materials. In many cases, when nanomaterials are produced through very fast precipitation reactions, distribution of residence times and long residence times in a reactor result in highly polydisperse materials, and in irreproducible results between manufacturing campaigns. This is due to the fact that the rates of particle formation and primary crystalline particle growth in the precipitation reactions are extremely fast, in the order of 1 ms, whereas residence times in most industrial-scale reactors is several orders of magnitude longer. As a result, secondary processes, such as particle growth through aggregation will take place. It was shown in the earlier literature that if mixing time in a reactor was commensurable with the characteristic particle growth time, the obtained particle size distribution would be significantly improved [39]. This can be achieved in a rotating packed-bed reactor (Hi-g technology), which has been commercialized for production of controlled-size nanomaterials [40]. Additional effect that also controls the formation of nanoparticles is sheer. A recent breakthrough in the synthesis of nanomaterials is the ability to precisely control nanoparticles' shape by tight control of sheer during synthesis, while also ensuring the mixing time is close to the particle growth characteristic time [41,42]. Reactor geometry may also be used for control of nanoparticle size distribution in continuous synthesis in microreactors, as recently shown in the example of synthesis of Ag nanoparticles in a helical flow reactor [43].

3. RESPONSE TO MARKET DEMAND

The ability to rapidly change production throughput is a significant challenge for chemical manufacturing. None of the traditional chemical manufacturing methods are particularly good in this respect: the large-scale continuous processes would typically have a relatively narrow range of capacity, when a plant operates under optimal conditions; similarly, capacity of batch operations is determined by the size and number of stirred vessels that are available. In both cases, the problem of significantly larger demand is solved by building new capacity, whereas the problem of the significantly lower demand results in spare capacity. There appears to be no "silver bullet" in transition from conventional stirred tank reactor technology to continuous flow technology with respect of the flexibility in manufacture. However, there is a growing demand for such flexibility as new business models, especially the "Manufacture on Demand" paradigm, will offer better long-term financial returns and stability to the industry.

The likely pathway to the new type of continuous manufacturing processes is through further development of the PI concepts, in combination with new modeling tools and new methods to fabricate components for chemical plants. Let us consider why this may be an appropriate development direction.

PI methods and tools aim to significantly increase space time yield (STY) of a process. STY is a measure of throughput per unit volume of a reactor and is typically given in the units of (product amount, say in kg or mol) × (reactor volume, $m^3)^{-1}$ × (unit of time, $s)^{-1}$. A significant increase in STY without the increase in capital cost of producing a new device will result in a significantly larger range of potential capacities that could be accessed with the same capital investment. There are good examples of such developments in PI methods. In the BASF' BASIL process, the STY of a specific process was increased from 8 to 690,000 kg m^{-3} h^{-1} through introduction of a recyclable acid scavenger that also acted as a catalyst [44]. This example links to our further discussion of the topic of Molecular Context in Section 2 of this chapter, because the key innovation in the BASIL process is in using a new functional molecule as a recyclable reagent, and not only in the reactor technology.

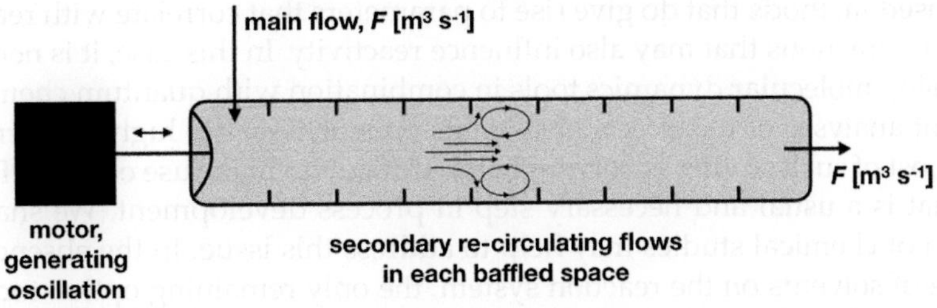

FIGURE 16.4 Schematic illustration of a principle of oscillatory baffle reactor. Oscillations are generated by a piston pump and superimposed on the main flow of reactants through the reactor. The total flow rate dictates the reactor space time, whereas oscillations generate secondary flows which promote agitation and mixing.

One of the success stories of PI through reactor technology is the development of an oscillatory baffle reactor (OBR), see Fig. 16.4. The OBR technology was developed from fundamental research on hydrodynamics and rheology in the group of Prof. Malcolm Mackley [45]. The OBR allows to combine enhanced mixing, typically achieved through faster flow rates and, hence, short residence times, with the requirement for many processes to have long residence time [7,46]. Coupled with the ability to handle multiphase flows and very different scales, as well as low energy demand, the technology has been adopted in several types of chemical manufacturing, from continuous crystallization to reactions in slurries.

In another example of simultaneous chemistry process development, at least two groups have successfully developed new process chemistries and reactor technologies for continuous manufacture of artemisinin-based antimalarial drugs [47–50]. These processes are based on the ability to handle solids in flow at mm scale, or to replace solid reagents with the soluble alternatives, and in the ability to achieve fast removal of heat from reactions in the structured mesoscale reactors.

4. CAPITAL COST AND EASE OF IMPLEMENTATION

One of key aspects of introduction of new technologies is the question about new capital expenditure. Chemical industry today is a mature industry with significant depreciated capital investment. There is strong resistance in the traditional chemical manufacturing to invest in new technologies when old ones still work well. With respect of continuous processes, the widely accepted advantages of implementing continuous technology (relevant to this subsection) is in accelerating implementation and increased throughput [51]. The increased throughput is already discussed above within the context of PI. The accelerated implementation should come from modularization and standardization of the new types of equipment, which would also enable a similar type of multiproduct use of the *continuous* equipment that is characteristic of the stirred tank technology. This is critical to ensure economic viability of "manufacture on demand" and product customization paradigms of chemical products and is a similar requirement to that in other manufacturing sectors where the new manufacturing paradigms are already a reality, such as automotive, clothing, etc. In these sectors, lowering the cost of making one item has been achieved through adopting new digital manufacturing techniques. There is no reason to think that chemical manufacturing is incapable of reaching a similar low cost of production for small runs (batches).

4.1 Part 2. The Concept of Molecular Context

Until now, the discussion of continuous processes was mainly restricted to technology. However, already the example of PI by BASF involved changing the reaction by introducing a new recyclable reactant—an ionic liquid—which dramatically changed the *molecular context* of the reaction, enabling a much faster rate of the reaction. This example signifies the importance of understanding molecular context in process development. To further elaborate on this point, let us consider the role of solvents in a reaction, and then the role of supports in heterogeneous catalysis.

It is well understood that in many chemical reactions a solvent is not simply a medium that allows to bring molecules of reactants together, or a convenient heat conduction medium, but a solvent may play an active role in a reaction. In the now well-advanced field of computer-aided molecular design (CAMD), significant effort was dedicated to developing models that incorporate solvent effects on reactivity, as well as on solubility [52–56].

The thermodynamics-based methods that do give rise to parameters that correlate with reactivity do not account for more complex dynamic interactions that may also influence reactivity. In this case, it is necessary to explicitly calculate solvent dynamics using molecular dynamics tools in combination with quantum chemistry calculations of reactivity [57]. These types of analyses of the effects of solvents on reactivity are highly informative. At the same time, because computational cost of such studies is very high, it is difficult to make use of MD-DFT work for the purpose of screening of solvents that is a usual and necessary step in process development. We shall see in Section 3 of this chapter how digitization of chemical studies may help to address this issue. In the absence of detailed mechanistic knowledge of the effects of solvents on the reaction system, the only remaining option today is to exploit empirical methods of study, such as the kinetic profile methods [58], which allow to determine empirical relationships between components of a reacting system.

In the case of heterogeneous catalysis, the role of a support is reasonably well understood, and different supports are classified into those that are *inert* during a reaction and their sole purpose is to disperse an active catalyst and to simplify the process (for example, by attaching active sites to a structure that is easily removed from a reactor, or that can be packed into a reactor, and so on), and those supports where support plays an *active role* during a reaction (for example by facilitating spillover of activated reactant from a catalytic site, or by modulating activity of a catalyst through electronic interactions, and so on). In the latter case, the presence of a support with a specific active site changes the molecular context of the site—the two systems (with and without a support) are chemically not identical.

Most interestingly, the two examples of molecular context we chose for this chapter—the effect of solvents on reactivity and the effect of catalyst support on activity of a catalytic site—are also linked in many cases of the liquid phase catalytic reactions. Recent studies by NMR relaxometry methods allow to identify the species adsorbed on the surface, as well as the differences in diffusion coefficients of solvent and/or reactant molecules in the bulk fluid and inside the pores of a catalyst support. As a result, it is now possible to reveal the effects of solvents on reactivity in heterogeneous catalysts, which can also be modulated by varying the support materials [59].

In all three examples of the importance of molecular context for process development, the common thread is the difficulty of unraveling the underlying chemical or physical interactions. The methods used to investigate such phenomena are difficult and frequently not accessible to a typical academic or industrial laboratory. To try to develop the concept of molecular context into a practically useful methodology, we have suggested to assemble description of multiple phenomena into a structured knowledge base that can be used for generation of more accurate process models with explicit description of molecular context [60]. A possible way of describing "context" is by considering six main parameters and their associated information: substance, structure, space, time, energy, and information [60]. Let us expand on the original publication and describe these six parameters for the case of molecular context in more detail, see Fig. 16.5.

"Substance" is the information about all species that are found in the same space and interacting. "Structure" corresponds to molecular structure, taking into account the other species present, and not only of the main compound, because structure maybe affected by the presence of other molecules. "Space" corresponds to molecular volume within the specific context of composition, energy, and time. "Time" is important only in cases when one is considering time-dependent phenomena. This could be the dependence of structuring of surfactants in some emulsion systems on their processing time or time elapsed because mixing the components of an emulsion or specific conformation of an enzyme that depends on the stage of a chemical reaction it catalyzes, and so on. "Energy" is the energy *field* in which molecules are found. This does not necessarily mean only temperature, as quite a few molecular systems may be sensitive to light, electric and/or magnetic field, and behavior of particles or large

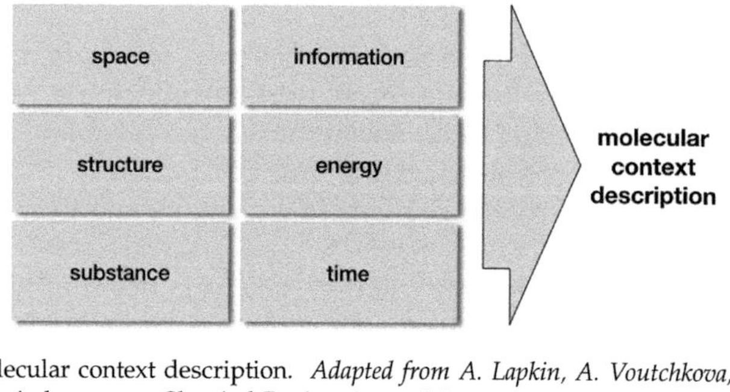

FIGURE 16.5 A six-parameter molecular context description. *Adapted from A. Lapkin, A. Voutchkova, P. Anastas, A conceptual framework for description of complexity in intensive chemical processes, Chemical Engineering and Processing: Process Intensification 50 (2011) 1027–1034.*

molecules can be affected by the presence or absence of gravity. Finally, "information" relates to entropy of the system: number of possible conformations and their distribution, as well as the dependence of conformation distribution on the neighboring molecules, other possible states of the molecular systems. The information-rich molecular systems typically are large molecules where entropic effects could be highly significant for the overall behavior of the system.

A systematic description of molecular context has been proposed as a means of identifying potential functional relationships between species and their properties, which could be used in rational design of processes. Thus, design of active molecular systems—systems that can be activated by an external stimulus to perform a certain action—might be easier accomplished if the potential for interactions could be systematically evaluated. Examples of such systems are switchable solvents, where a significant change in solvation can be induced by a modest change of temperature or composition [61–63]. There are, however, natural limitations to this approach. The first limitation stems from very high computational cost of generating the necessary data, which may never allow to fully exploit the idea of screening for potential useful functions in a complex molecular system. The second limitation stems from the fact that to compute properties we need to know what we want to compute in the first place, whereas to date, serendipity plays a significant role in discovery of useful new molecular systems. As a result, the more plausible path toward design of molecular systems is through a hybrid approach, making use of data generated by both, rigorous computational tools based on first principles or their reasonable approximations, as well as models generated by machine learning and artificial intelligence tools. We shall now discuss the emerging opportunities for design of chemical processes using the new tools of digital molecular technologies (DMTs).

4.2 Part 3. Digital Molecular Technology

To date, chemistry and chemical process development are firmly grounded into "analogue" world. The way we represent information is completely analogue. Consider the representation of chemical information shown in Fig. 16.6. The information in Fig. 16.6A is the way how any chemist would expect the information about a reaction

(A)

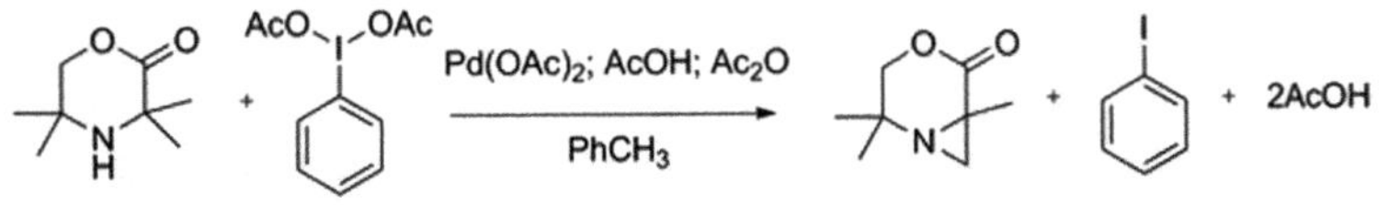

Compound	Amount fed [mol s-1]
3,3,5,5-tetramethylmorpholin-2-one	8.3 x 10^{-7}
(diacetoxyiodo)benzene	8.3 x 10^{-7}
acetic acid	8.3 x 10^{-6}
Palladium(II) acetate	4.2 x 10^{-9}
Acetic anhydride	1.7 x 10^{-6}
Toluene	1.5 x 10^{-4}
Property	**Value**
Reaction temperature	393 K
Reaction Pressure	6x10^{6} Pa
Yield	0.90

Residence time [sec]	Conversion
60	0.06
120	0.14
180	0.20
240	0.32
300	0.40
360	0.52
420	0.70
480	0.90
540	1.00
600	1.00

(B)


```
RInChI=0.03.1S/C10H11IO4/c1-8(12)14-11(15-9(2)13)10-6-4-3-5-7-10/h3-7H,1-2H3!
C8H15NO2/c1-7(2)5-11-6(10)8(3,4)9-7/h9H,5H2,1-4H3<>C2H4O2/c1-2(3)4/h1H3,(H,
3,4)!C6H5I/c7-6-4-2-1-3-5-6/h1-5H!C8H13NO2/c1-7(2)5-11-6(10)8(3)4-9(7)8/
h4-5H2,1-3H3<>2C2H4O2.Pd/c2*1-2(3)4;/h2*1H3,(H,3,4);/q;;+2/p-2!C2H4O2/
c1-2(3)4/h1H3,(H,3,4)!C4H6O3/c1-3(5)7-4(2)6/h1-2H3!C7H8/c1-7-5-3-2-4-6-7/h2-6H,
1H3/d+

ProcAuxInfo=$$PAI0.01|1:2|1;1|2;1;1|393|6E5|
60:0.06;120:0.14;180:0.20;240:0.32;300:0.40;360:0.52;420:0.70;480:0.90;540:1.0;600
:1.0|?;?;0.90|8.3E-7;8.3E-7|0;0;0|4.2E-9:m:f;8.3E-6:m:f;1.7E-6:m:f;1.5E-4:m:f|1E-5
```


FIGURE 16.6 Analogue versus digital chemical data. (A) Analogue representation of a specific chemical reaction, adapted from our previous research [65,66]. (B) Digital representation of the same information. The RInChI string is based on the standard extension proposed in [64].

to be presented. This representation shows all reactants, reagents, and products, and the tables provide all the information necessary to reproduce the experiment: all quantities and reaction conditions, as well as conversion data as a function of residence time, which unequivocally identifies this reaction as being done in flow. This is an entirely analogue representation of chemical information and it cannot be directly used by *algorithms*. The representation shown in Fig. 16.6B is a machine-readable information which is identical in the information content to the analogue information in Fig. 16.6A: it also identifies all molecules involved and contains all the information given in the tables, but this representation can now be passed on between algorithms, because the rules of decoding this string are unique and are maintained by a specified standard.

This specific example is using one possible way of codifying chemical information, an extension to the RInChI standard [64]. There are alternative methodologies being developed in parallel, but the main idea behind these efforts is to ensure that we can uniquely identify molecules, reactions, and reaction conditions, such that representation of experimental facts is transformed into the digital world. This is the first and necessary step toward digitization of chemistry and chemical process development.

So, what is DMT? Current use of digital technologies in the fields of chemistry and process engineering are very limited. We have a number of ways to identify molecules (CAS number, Reaxys ID, SMILE string, InChI string, etc), which we use for ordering chemicals, checking stocks and prices, managing chemical inventories (frequently also using bar codes in combination with CAS numbers to add more information to the digital record), and searching for chemical information in databases, such as Reaxys or SciFinder. However, as soon as we start working on developing chemical syntheses or understanding of reaction mechanisms, we move entirely into the analogue world. The digital representation of molecules that exists in databases has no connection to the digital representation of molecules in quantum computational models. The interpretation of mechanisms, even if based on results of calculations done within digital representation of molecules (for example, in DFT models) is done entirely using tools of analogue representation: the old models of molecular orbitals, and so on. There are only few pioneering publications on algorithmic use of chemical data, for example, for design of catalysts [67].

Laboratory work in chemistry is entirely analogue and it is still extremely rare when researchers would make use of digital tools, such as advanced design of experiment methodologies (we shall provide references to all mentioned tools further in this section as we assemble the concept of DMT). Most of laboratory work done to date is manual and outcomes of experiments are recorded by hand in lab journals. These outcomes may eventually be represented in a way that is amenable to interpretation by fellow researchers: 2D or 3D plots, or sets of data in tables. In very few fields of chemical research, tools of statistical multivariate analysis are used, for example, linear or nonlinear dimensionality reduction techniques in analytical chemistry. To use these tools, the manually collected data need to be converted into the digital representation that can be shared with algorithms and results of algorithms again interpreted as 2D or 3D plots we are used to see.

In process development, the use of algorithms is more advanced: chemical engineers are used to dealing with large sets of data describing thermodynamic, kinetic, mass, and heat transfer aspects of a process, to use optimization algorithms. The field of multiscale modeling is also developing very rapidly. CAMD and computer-aided process design are both reasonably well developed. Most tools, however, use information in a specific format and it is a painful process that all chemical engineers have to go through, which is to ensure that data used at the start of process development is of good quality, reliable, and is in the right format for the tools to be used.

In chemical manufacture, some industry sectors are much more advanced than others with respect of using digital technologies. There are a number of examples of sophisticated "smart" plants that use real-time process analytics, real-time control, and optimization algorithm, anomaly, and fault detection algorithms. These examples are, unfortunately, very few. Yet, it is well understood that real-time process optimization, which is only feasible with the use of advanced digital tools, gives significant and rapid economic return through reducing waste and improving product quality. It is also true that today's chemical manufacture is largely based on very old processing paradigms: either large integrated continuous processes in bulk chemical manufacture or processes based on flexible stirred tank reactor technology, used in the multistep syntheses of functional molecules and materials at the much smaller scales, compared with the continuous bulk chemical manufacturing processes. Let us try to assemble the different aspects of the chemical development and manufacture into a concept of DMT, see Fig. 16.7. To support this vision, we shall use specific examples and methodologies, which are pioneering different elements of the overall picture.

The cornerstone of any digital technology is data. Therefore, the common theme through all aspects of DMT is *ontological* representation of data and methods of working with the data [68–71]. In the absence of a common representation of data and methods, the algorithmic use of data, data sharing, and algorithmic experimental platforms are impossible. This is one of the most urgent aspects of digital technologies that requires rapid development and adoption.

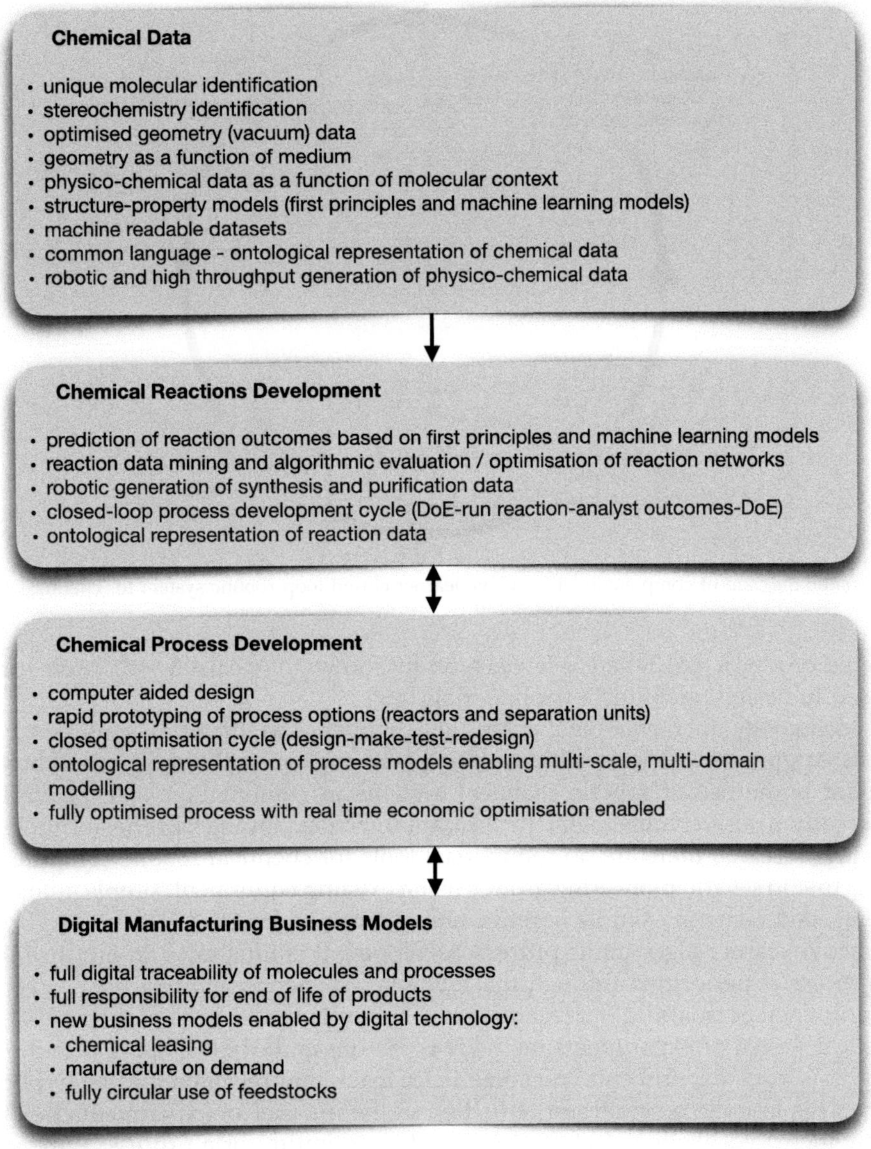

FIGURE 16.7 Elements of Digital Molecular Technology.

Algorithmic research in chemistry is experiencing a real boom, with a very rapidly expanding literature on machine learning, data mining, and artificial intelligence methods. One premise of this area of work is the ability to deal with complete knowledge of chemistry using data mining algorithms. This has led to the development of a set of methods based on network theory and graph tools [72–78]. These methods allow to remove scientist's bias in developing process routes, as well as posing a number of new types of research questions—how to fit a synthesis route to a particular supply chain scenario, how to build into synthesis some redundancy by using more of "hub" molecules, how to optimize reaction networks for minimum overall energy and mass intensities, and so on. These methods are already very useful and are being adopted by industry, but are not yet widely accessible.

Another set of questions in algorithmic chemistry is whether chemical syntheses could be designed by machine learning or artificial intelligence algorithms, trained on past chemical data. There are some spectacular success stories in this emerging field already [79–83], but there are also some more sobering views on the generality of this approach in chemistry [84]. Regardless of the likely limitations, it is already quite obvious that there exists a tremendous opportunity to speed up development of new chemical entities, new reactions, and materials. In the area of materials and formulated products, machine learning methods based on data (high throughput and robotic data collection) are particularly useful. Thus, relatively simple classification algorithms can describe complex

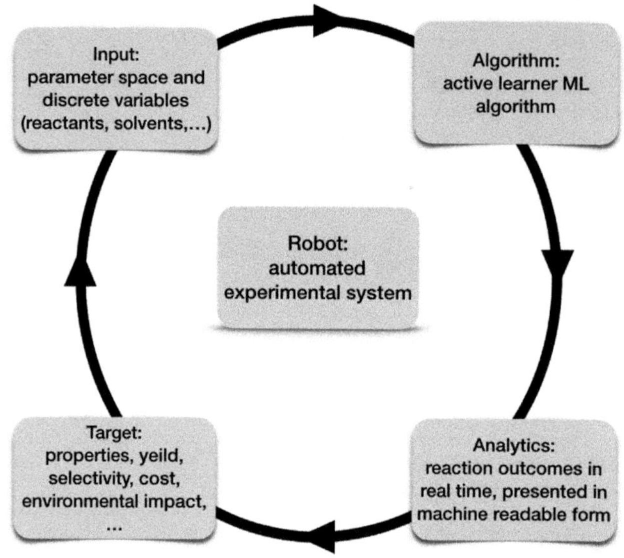

FIGURE 16.8 A schematic diagram of components of an active learner closed loop robotic system for chemical reaction development and optimization.

behavior of formulated products [85], whereas it was recently shown that learning structure–property relationships from data can be used to design new functional materials [86].

The supporting technology to developing chemical reactions and material synthesis is robotic and high-throughput synthesis equipment. HT is a well-developed area and we will not discuss it here. The innovation in chemical development is the use of robotic chemical systems in combination with active learning algorithms. Such systems are already in use commercially, for example, for developing libraries of molecules for bioactivity screening [87]. In wider research there are no commercial tools for chemical robotic equipment yet. All published examples are put together in several pioneering groups using existing equipment, simple automation tools, available analytical instruments, and relatively simple optimization algorithms [66,88–97].

The principle of active learner algorithmic process development is illustrated in Fig. 16.8. In such systems, the optimization of a process is performed through the following sequence: hypothesis → experimental design → experiment → experimental evaluation → reevaluation of the hypothesis → experimental design. The difference from classical statistical design of experiments methods is that instead of developing a grid of experiments based on one initial hypothesis, each experimental outcome is fed back into reevaluating the hypothesis and designing the next experiment. This leads to a significant reduction in the number of experiments that must be undertaken and in a significantly increased likelihood of finding a global optimum for a process. Some systems are based on one-by-one sequence of experiment-analysis-DoE, whereas some are adapted to batch-sequential workflow, when a series of experiments are designed and performed.

All elements of this system are essential. The easiest application of the robotic experimental systems is in optimization of continuous variables, such as process conditions (temperature, pressure, flow rates, etc), for a fixed chemical recipe, when all discrete variables (selection of solvent, additives, catalyst, etc.) are fixed. In this case, there is no need to set up a robotic system for change of discrete inputs. In the case when discrete variables are also included in the optimization, more complex systems with liquid and solid handling robots are required, which significantly increases the cost. This also radically changes the approach to design of experiments, see below.

The experiment is done in a system that can be easily controlled, filled, and cleaned. For this reason, automated workflow is relatively easy to do in flow, when variation of composition and reaction conditions can be easily set up and there is little, if any, requirement for cleaning of the system between the experiments. Experiments could be done in continuous flow, or in dispersed flow, when reactants are dispensed into droplets within a continuous phase, such that each droplet is an individual "reactor." Experiments can also be done in batch, but the complexity and cost increase dramatically with the increase in the difficulty of experimental conditions: high pressure, evaporating solvents, presence of gases, or need for inert atmosphere are very difficult to achieve in robotic batch systems at a moderate cost.

The reaction outcomes must be evaluated to generate information for decision-making. What is typically measured are concentrations (by fast spectroscopic methods such as UV, IR, Raman, or low field NMR, by mass spec, or slower methods such as GC or HPLC) and particle size (by in line light scattering techniques). It is also possible to measure pH and conductivity. In some cases, still images or videos are recorded for image analysis. The desired outcome of analytical data collection is to have sufficient information about the process to decide on the next experiment using an algorithm. As a result, the aim of the analytical part of the robotic system is to provide data on concentrations, structure, and performance, preferably within a time commensurable to the time of an experiment, or even faster. Analytical part is frequently the bottleneck in the robotic workflow, as many methods are much slower than the experiment time. This constraint is particularly problematic for high-throughput batch-sequential workflows, when a large number of samples need to be analyzed as quickly as possible to set the next batch of experiments and avoid equipment being idle. There are some very promising advances in high-throughput analytical methods. Thus, a team at Merck has managed to significantly increase the throughput of HPLC analysis [98], whereas AstraZeneca codeveloped with Waters and Labcyte an acoustic mass spectroscopy method to achieve sample throughputs of up to 10,000 data points per hour [99].

The final component of the robotic chemical system is the decision algorithms, see Fig. 16.8. In the simple realization of self-optimization even conventional gradient-based optimization algorithms could be used. However, these algorithms usually are data-hungry and may lead to large numbers of experiments. A more promising approach, which has seen some excellent results, is the use of active learning ML algorithms. Such algorithms could be tuned to either rapidly find an optimal solution or to develop a good statistical model of the process, the two extremes that would lead to data-lean or data-rich solutions. Some early examples of machine learning algorithms designed specifically for self-optimization of chemical processes are Bayesian algorithms MOAL (multiobjective active learner) [94] and TSEMO (Thompson Sampling Efficient Multiobjective Optimization) [97]. The two algorithms have been applied for such problems as optimization of recipes for semibatch polymerization [96], optimization of synthetic chemistry processes in flow [66,89], as well as optimization of process flowsheets with regard of two independent simultaneous targets of lowest cost and lowest environmental impact [100]. As the field of machine learning is expanding to new areas, including chemistry and chemical processing, there will be new methods and tools developed for different types of problems. This is an area of chemical engineering that has emerged very rapidly and which is beginning to make an impact in both academia and industry.

Arguably, one of the greatest impacts of digital technology that we should see in the near future is in the way how molecules are manufactured. Digital technologies are rapidly transforming pretty much every industry, from medical diagnostics and legal services to logistics, navigation, distribution of content, financial services, and manufacture of custom products. It is very difficult to speculate what effect digital technologies would have on the way how molecules are manufactured. However, it is more than likely that several sectors of chemical industries, such as manufacture of advanced functional products (pharmaceuticals, agrochemicals, etc.), will be transformed to "manufacture on demand" paradigm, with the aid of both, the new robotic processing tools and the new digital chemical information tools. Such concepts of managing chemicals as "chemical leasing" [101], which are aimed to developing a more sustainable use of chemicals, should be much easier to implement in the fully digitalized chemical supply chain, see Parts 5 and 6 of this chapter.

4.3 Part 4. Toward Rational Design of Flow Processes

The tools of digital chemistry will ultimately allow rational design of processes. An example of such an approach was recently demonstrated for an sp3 C—H activation reaction [65]. Based on an accepted and computationally verified reaction mechanism, the initial kinetic model and its parameters were identified from DFT calculations, and later fine-tuned against experimental data. The extrapolating (predicting) ability of the developed model was tested by first predicting, and then confirming, the reaction behavior in a new-to-model process arrangement—fed-batch compared with batch that was used in model development. As the model was confirmed to be predictive, it was then used to optimize reaction conditions for a flow process first in silico, and then confirmed experimentally. In this case, the highly idealized workflow from a fully mechanistic model to in silico process design was shown, see Fig. 16.9.

To further translate models developed for the individual reaction and separation steps, it is necessary to find a way of evaluating the different potential combinations of discrete choices that comprise a complete multistep process: it is clear, for example, that a choice of a solvent in one reaction step may affect how the following separation is being done and so on. This problem is, in principle, well-understood in the domain of process systems engineering as a problem of *superstructure optimization*. Recently, new approaches that significantly expand the decision space and allow rational design of complex processes have emerged. One of such methods is based on functional description of a process [27]. Each step of a process is first simplified to a functional model, shown schematically in Fig. 16.10. Fig. 16.10a shows a control volume with fluxes of heat and mass that are required to accomplish the

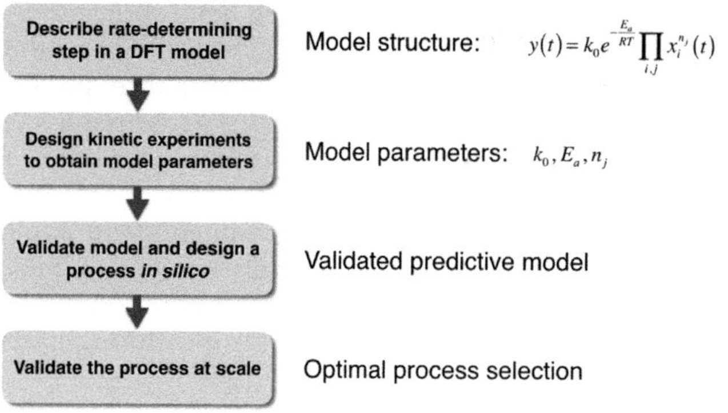

FIGURE 16.9 An idealized workflow for rational process design based on mechanistic reaction understanding.

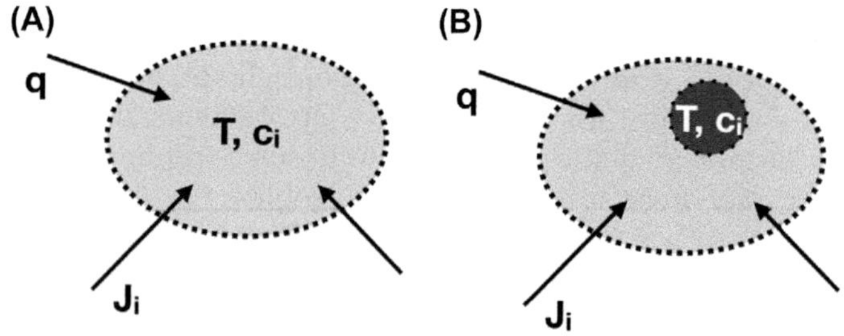

FIGURE 16.10 Formulation of a process based on a controlled volume of constant properties and a balance of mass and energy fluxes required for optimal state variable trajectory to the desired optimal solution.

desired task, whereas Fig. 16.10b shows a more complex formulations, allowing for multiple domains of constant conditions within a control volume, for example, in the case of a multiphase reaction. The model is then optimized with respect of target criteria, such as highest yield at desired cost, or others, using input/output fluxes as optimization variables. Analysis of solutions can lead to nonintuitive process designs. This methodology can then be extended to the development of multistep processes and optimization of a complete process superstructure [114]. This approach is agnostic to whether the target process is based on batch, fed-batch, flow, or a combination of all types of processes, which is potentially more useful for the future rational process design methodologies.

4.4 Part 5. Environmental Impacts of Continuous Flow Processes

Let us now evaluate some advantages and disadvantages of continuous flow syntheses of functional molecules and materials with respect of their environmental and human health impacts, compared with batch processes. This forms a part of a sustainability assessment—the environmental pillar of sustainability. In the previous sections of this chapter, we introduced key features of continuous manufacturing of functional molecules/materials and outlined the emerging methodology for rational design of processes, largely based on the introduction of DMTs. Three key conditions for the introduction of the new development methods and of new products are (i) improved performance (new materials with new functions, speed to market) at (ii) reduced cost (cost-efficient manufacture, cost of raw materials) and (iii) reduced environmental impact. Clearly, without (i) and (ii) new process would not be viable, but condition (iii) must also be satisfied. We argued [102,103] that environmental impact must be evaluated according to life cycle assessment (LCA) methodology [104] as opposed to the indicator-based methods that do not include into evaluation the impacts of the supply chain or the impacts of the product/process end-of-life. The abridged environmental evaluation, performed within gate-to-gate system boundaries (only including the process under consideration and not including supply chain and end-of-life) is much cheaper and faster to perform, but such evaluations would automatically miss any impacts of such changes as replacement of solvents or reduction in the use of specific stoichiometric reagents, as such impacts are mainly associated with the processes that take place upstream in the supply chain. Let us illustrate this with a recent example of a LCA evaluation of a proposed continuous flow synthesis based on one of key pharmaceutical transformations.

Buchwald–Hartwig amination is one of key transformations used in a large number of pharmaceutical syntheses. It is one of the more difficult transformations to be performed in flow, due to the formation of insoluble by-products.

The development of a potentially viable continuous flow Buchwald–Hartwig amination reaction is described in [105] and details of LCA are described in [106]. To evaluate environmental impact, pharmaceutical industry would typically use process mass intensity (PMI) indicator, which takes into account actual masses of all reagents and wastes generated [107]. PMI, however, does not account for impacts that originate in the manufacture of the molecules that are used in the given process. As was shown in [106], the PMI of the new flow process based on a new reactor concept and a new catalyst was significantly lower than the PMI of the original batch process. However, results of full LCA had shown that due to the more complex manufacture of the catalyst used in the flow process, the advantage of the new process in terms of the environmental impact was marginal in the most optimistic process scenario. This stems from the frequently observed increased inventory of solvents when a process is transformed from a batch to a continuous version: small-scale continuous reactors do not cope well with the presence of particles [11]; hence, to prevent the potential precipitation of reactants, by-products or products lower concentrations are used. This has been noted in a number of studies on continuous flow processes [12,48]. To resolve this issue, a potential flow process would have to include solvent recovery/recycle or incineration with energy recovery. Both solutions will reduce life cycle impacts from the solvent inventory.

Evaluation of life cycle impacts also allows to pick up more potential problems, than just a few preselected issues, such as mass of waste or very commonly used greenhouse gas emissions indicator (global warming potential). In a recent study [108], LCA of a continuous supercritical synthesis of TiO_2 nanoparticles revealed that although the new process is characterized by reduced impact in the climate change category, but not in the human health category. This study also highlights a number of methodological issues in LCA which, if not taken into account, may lead to erroneous conclusions. These relate to the quality of the data used in LCA models and the underlying assumptions of the statistical significance of the data in life cycle inventories.

To date, there are very few studies of environmental impacts of continuous flow processes performed on the basis of LCA. Results obtained on the basis of abridged studies, which include only one or two indicators and not involve the upstream and end-of-life, are not generally sufficient to argue any environmental benefit and we shall not include such studies into the discussion.

4.5 Part 6. Sustainability of Continuous Flow Processes

Environmental impact is only one pillar of sustainability. It is very unfortunate, that the term "sustainability" is being largely abused by the academic community: an improved life cycle impact for a process does not mean that the process is *sustainable*. To evaluate "sustainability," the evaluation should include adequate assessment of all three aspects: economic, social, and environmental. Any claim made on less than three components of sustainability cannot honestly use the term "sustainability."

Many features of continuous flow technology for the synthesis of functional molecules and materials make this technology attractive from the point of view of sustainability. As we have shown in Part 5, flow processes are likely to be as good, if not better than batch processes, in terms of their life cycle impacts. However, in terms of social and economic impacts, the flow technology is much more attractive.

Key advantage of continuous flow technology over batch is the rapid translation of processes to industrial-scale throughput without loss in quality. The example of BASF′ BASIL process is quite extreme, but it illustrates the point: a fairly small reactor is able to satisfy industrial throughput. Another example is continuous multichannel reactors, such as those used in the compact GTL technology: the overall throughput is achieved by numbering up the channels, whereas each individual channel is of the size of a lab-scale reactor. The absence of the step of increasing the reactor scale, when significant changes in the physics that dominate the process takes place, is one of the greatest advantages of continuous flow technology over batch processes.

As we have shown in the example of the synthesis of nanomaterials [41–43,109], this ease of scale-up is complemented by improved product quality, which makes a highly compelling business case for the application of continuous flow technology for nanomaterial synthesis.

Another potential economic sustainability advantage of continuous flow technology is the possibility to manufacture "on demand," so-called "distributed manufacturing" business model [110,111]. Manufacture on demand requires manufacturing technology that is not embedded into a highly integrated large industrial site, but that could be installed fairly locally to consumers. The fact that continuous processes that utilize the principles of PI are inherently quite small (although could be scaled by numbering), this technology can be easily implemented within the distributed manufacturing paradigm. The cost savings that are normally achieved though economy of scale and material/energy integration are not available in this situation and the cost competitiveness must be achieved through different mechanisms, mainly through significant gains in process throughput and product quality.

Another business opportunity which emerges with the advances in continuous flow process technology is the possibility of using a "leasing" business model. Chemical leasing has been piloted for a number of years as a potential solution to "greening" the chemical's supply chain [101]. As the ownership of "molecules" is retained by a specialist business, it is their interest and responsibility to ensure they get maximum benefit from using the molecules, with minimum cost, where cost would necessarily involve environmental protection, resupply of stock in case of loss, and so on. This new ownership model effectively drives the specialist business toward maximizing the efficiency of the function that the molecules they own perform, while minimizing all of their costs, and this leads to better environmental performance, while also being economically very attractive.

We shall finally try to summarize a number of potential social implications of the introduction of continuous flow technology, although this is the least developed part of the sustainability assessment. Based on our earlier work [112,113], which included stakeholder analysis for new processes, the most important social impacts include improvements in the quality of shared environment (air quality, water quality, availability of land for maintaining biodiversity) and local environment (noise pollution, quality of housing, quality of infrastructure), reduction in global risks (climate change implications, global energy resources, global water resources, etc.), and improvements in the quality of life (healthcare, self-realization opportunities, income). A number of these issues are addressed through environmental sustainability assessment, mainly LCA, which can evaluate the impacts of technology on air and water pollution, ecotoxicity, change of land use, and contribution to abiotic resource use. Whereas "softer" indicators relate to issues of quality of life and quality of the local environment. Potential positive impacts are likely to arise through the need of new high-skilled jobs in new, previously less industrialized areas, and redistribution of industry, leading to local improvements in the environment, and better distribution of higher-paid jobs, than in the present chemical supply chain. However, these are more speculative and untested benefits that would arise through the adoption of new business models, facilitated by the unique features of continuous manufacturing technologies.

5. CONCLUSIONS

The paradigm of continuous manufacturing in specialty chemicals, materials, and pharmaceuticals is now widely accepted and there is an increasingly rapid adoption of continuous processes in industrial processes. There is, however, a new paradigm emerging, which will rapidly subsume the "conti" technology—DMT. DMT includes robotic research and manufacturing facilities, large-scale use of computational, modeling, and statistical methods, reliance on multiple analytical methods and rapid development of methods of machine learning and artificial intelligence. "Conti" manufacturing will be an integral, but a small part of this new era in chemical manufacturing. The sustainability implications of continuous processes come from all three aspects of sustainability: economic, environmental, and social, with the economic and the social impacts potentially more important, and numerically more significant, in comparison with batch manufacturing.

Acknowledgments

This work is funded by the National Research Foundation (NRF), Prime Minister's Office, Singapore, under its Campus for Research Excellence and Technological Enterprise (CREATE) program as a part of the Cambridge Center for Advanced Research and Education in Singapore Ltd. (CARES).

References

[1] W.T. Cross, C. Ramshaw, Process intensification: laminar flow heat transfer, Chemical Engineering Research and Design 64 (1986) 293–301.
[2] C. Ramshaw, The incentive for process intensification, in: 1st Intl. Conf. Proc. Intensif. For Chem. Ind., BHR Group, London, 1995.
[3] A.I. Stankiewicz, J.A. Moulijn, Process intensification: transforming chemical engineering, Chemical Engineering Progress 96 (2000) 22–34.
[4] A. Stankiewicz, Process intensification in in-line monolithic reactor, Chemical Engineering Science 56 (2001) 359–364.
[5] A. Aoune, C. Ramshaw, Process intensification: heat and mass transfer characteristics of liquid films on rotating discs, International Journal of Heat and Mass Transfer 42 (1999) 1432–1445.
[6] R. Jachuck, Process intensification for responsive processing, Chemical Engineering Research and Design 80 (2002) 233–238.
[7] A.P. Harvey, M.R. Mackley, T. Seliger, Process intensification of biodiesel production using a continuous oscillatory flow reactor, Journal of Chemical Technology and Biotechnology 78 (2003) 338–341.
[8] T.V. Gerven, A. Stankiewicz, Structure, energy, synergy, time – the fundamentals of process intensification, Industrial and Engineering Chemistry Research 48 (2009) 2465–2474.
[9] J. Hartwig, J.B. Metternich, N. Nikbin, A. Kirschning, S.V. Ley, Continuous flow chemistry: a discovery tool for new chemical reactivity patterns, Organic and Biomolecular Chemistry 12 (2014) 3611–3615.
[10] D. Gutmann, D. Cantillo, C.O. Kappe, Continuous-flow technology – a tool for the safe manufacturing of active pharmaceutical ingredients, Angewandte Chemie International Edition 54 (2015) 6688–6728.
[11] A. Lapkin, K. Loponov, G. Tomaiuolo, S. Guido, Solids in continuous flow reactors for specialty and pharmaceutical syntheses, in: L. Vaccaro (Ed.), Sustainable Flow Chemistry – Methods and Applications, Wiley-VCH, 2017, pp. 277–308.

[12] A. Lapkin, P. Yaseneva, Life cycle assessment of flow chemistry processes, in: L. Vaccaro (Ed.), Sustainable Flow Chemistry – Methods and Applications, Wiley-VCH, 2017, pp. 249–276.
[13] N.G. Anderson, Practical use of continuous processing in developing and scaling up laboratory processes, Organic Process Research and Development 5 (2001) 613–621.
[14] G. Jas, A. Kirschning, Continuous flow techniques in organic synthesis, Chemistry: A European Journal 9 (2003) 5708–5723.
[15] A. Kirschning, W. Solodenko, K. Mennecke, Combining enabling techniques in organic synthesis: continuous flow processes with heterogenized catalysts, Chemistry: A European Journal 12 (2006) 5972–5990.
[16] M. Takasuga, Y. Yabuki, Y. Kato, Pharmaceutical industrial experiments on continuous cryogenic reactions using mini-sized multi-staged reactors, Journal of Chemical Engineering of Japan 39 (2006) 772–776.
[17] T. Laird, Continuous processes in small-scale manufacture, Organic Process Research and Development 11 (2007) 927.
[18] T.L. LaPorte, C. Wang, Continuous processes for the production of pharmaceutical intermediates and active pharmaceutical ingredients, Current Opinion in Drug Discovery and Development 10 (2007) 738–745.
[19] U. Hintermair, G. Francio, W. Leitner, Continuous flow organometallic catalysis: new wind in old sails, Chemical Communications 47 (2011) 3691–3701.
[20] J. Wegner, S. Ceylan, A. Kirschning, Ten key issues in modern flow chemistry, Chemical Communications 47 (2011) 4583–4592.
[21] A. Lapkin, P.K. Plucinski, Engineering factors for efficient flow processes in chemical industries, in: S.V. Luis, E. Garcia-Verdugo (Eds.), Chemical Reactions and Processes Under Flow Conditions, RSC, Cambridge, 2010, pp. 1–43.
[22] G.F. Froment, K.B. Bischoff, Chemical Reactor Analysis and Design, second ed., John Wiley & Sons, Inc., 1990.
[23] H.S. Fogler, Elements of Chemical Reaction Engineering, Prentice-Hall Int., 1999.
[24] J.M. Douglas, A hierarchical decision procedure for process synthesis, AIChE Journal 31 (1985) 353–362.
[25] J. Douglas, Process synthesis for waste minimisation, Industrial and Engineering Chemistry Research 31 (1992) 238–243.
[26] Q. Chen, I.E. Grossmann, Recent developments and challenges in optimization-based process synthesis, in: J.M. Prausnitz (Ed.), Annual Review of Chemical and Biomolecular Engineering, vol. 8, Annual Reviews, Palo Alto, 2017, pp. 249–283.
[27] H. Freund, K. Sundmacher, Towards a methodology for the systematic analysis and design of efficient chemical processes Part 1. From unit operations to elementary process functions, Chemical Engineering and Processing 47 (2008) 2051–2060.
[28] J. Steimel, S. Engell, Conceptual design and optimization of chemical processes under uncertainty by two-stage programming, Computers and Chemical Engineering 81 (2015) 200–217.
[29] D. Altham, How Industry 4.0 Can Transform Chemical Manufacturing, Processing Solutions for the Process Industries, 2017, in: https://www.processingmagazine.com/how-industry-4-0-can-transform-chemical-manufacturing/.
[30] D.M. Roberge, B. Zimmermann, F. Rainone, M. Gottsponer, M. Eyholzer, N. Kockmann, Microreactor technology and continuous processes in the fine chemical and pharmaceutical industry: is the revolution underway? Organic Process Research and Development 12 (2008) 905–910.
[31] A. Warmington, C. Challener, Microreactors could play a vital role in transitioning fine chemicals from batch processes to continuous, Speciality Chemicals Magazine (2008) 40–46.
[32] D.L. Browne, B.J. Deadman, R. Ashe, I.R. Baxendale, S.V. Ley, Continuous flow processing of slurries: evaluation of an agitated cell reactor, Organic Process Research and Development 15 (2011) 693–697.
[33] O. Levenspiel, Chemical Reaction Engineering, third ed., John Wiley & Sons, New York, 1999.
[34] K.R. Westerterp, Multifunctional reactors, Chemical Engineering Science 47 (1992) 2195–2206.
[35] D.W. Agar, Multifunctional reactors: old preconceptions and new dimensions, Chemical Engineering Science 54 (1999) 1299–1305.
[36] D.V. Bavykin, A.A. Lapkin, S.T. Kolaczkowski, P.K. Plucinski, Selective oxidation of alcohols in a continuous multifunctional reactor: ruthenium oxide catalysed oxidation of benzyl alcohol, Applied Catalysis A: General 288 (2005) 165–174.
[37] P.K. Plucinski, D.V. Bavykin, S.T. Kolaczkowski, A.A. Lapkin, Liquid phase oxidation of organic feedstock in a compact multichannel reactor, Industrial and Engineering Chemistry Research 44 (2005) 9683–9690.
[38] X. Fan, P.K. Plucinski, A.A. Lapkin, Liquid phase hydrogenation in a structured compact reactor, Catalysis Today 147S (2009) S313–S318.
[39] J. Chen, C. Zheng, G. Chen, Interaction of macro- and micromixing on particle size distribution in reactive precipitation, Chemical Engineering Science 51 (1996) 1957–1966.
[40] H. Zhao, L. Shao, J.-F. Chen, High-gravity process intensification technology and application, Chemical Engineering Journal 156 (2010) 588–593.
[41] N.A. Jose, H.C. Zeng, A. Lapkin, Hydrodynamic assembly of two-dimensional layered double hydroxide nanostructures, Nature Communications 9 (2018) 4913.
[42] N. Jose, A. Lapkin, Influence of hydrodynamics on wet syntheses of nanomaterials, in: V.A. Sadykov (Ed.), Advanced Nanomaterials for Catalysis and Energy. Synthesis, Characterization and Applications, Elsevier, Amsterdam, 2019, pp. 29–60.
[43] K.J. Wu, G.M.D. Bohan, L. Torrente-Murciano, Synthesis of narrow sized silver nanoparticles in the absence of capping ligands in helical microreactors, Reaction Chemistry and Engineering 2 (2017) 116–128.
[44] N.V. Plechkova, K.R. Seddon, Applications of ionic liquids in the chemical industry, Chemical Society Reviews 37 (2008) 123–150.
[45] C.R. Brunold, J.C.B. Hunns, M.R. Mackley, J.W. Thompson, Experimental observations on flow patterns and energy losses for oscillatory flow in ducts containing sharp edges, Chemical Engineering Science 44 (1989) 1227–1244.
[46] A.P. Harvey, M.R. Mackley, P. Stonestreet, Operation and optimisation of an oscillatory flow reactor, Industrial and Engineering Chemistry Research 40 (2001) 5371–5377.
[47] X. Fan, V. Sans, P. Yaseneva, D.D. Plaza, J.M.J. Williams, A. Lapkin, Facile stoichiometric reductions in flow: an example of artemisinin, Organic Process Research and Development 16 (2012) 1039–1042.
[48] P. Yaseneva, D. Plaza, X. Fan, K. Loponov, A. Lapkin, Synthesis of the antimalarial API artemether in a flow reactor, Catalysis Today 239 (2015) 90–96.
[49] D. Kopetzki, F. Levesque, P.H. Seeberger, A continuous-flow process for the synthesis of artemisinin, Chemistry: A European Journal 19 (2013) 5450–5456.
[50] K. Gilmore, D. Kopetzki, J.W. Lee, Z. Horvath, D.T. McQuade, A. Seidel-Morgenstern, P.H. Seeberger, Continuous synthesis of artemisinin-derived medicines, Chemical Communications 50 (2014) 12652–12655.

[51] S. Falß, N. Krasberg, M. Holtkamp, A. Prokofyeva, T. Bieringer, N. Kockmann, Process intensification through continuous manufacturing: implications for unit operations and process design, in: A. Lapkin (Ed.), Green Chemical Engineering, Wiley-VCH Verlag GmbH & Co, Weinheim, 2018.

[52] R. Gani, Chemical product design: challenges and opportunities, Computers and Chemical Engineering 28 (2004) 2441–2457.

[53] R. Gani, P.A. Gómez, M. Folić, C. Jiménez-González, D.J.C. Constable, Solvents in organic synthesis: replacement and multi-step reaction systems, Computers and Chemical Engineering 32 (2008) 2420–2444.

[54] M. Folic, C.S. Adjiman, E.N. Pistikopoulos, Design of solvents for optimal reaction rate constants, AIChE Journal 53 (2007) 1240–1256.

[55] C.S. Adjiman, A.J. Clarke, G. Cooper, P.C. Taylor, Solvents for ring-closing metathesis reactions, Chemical Communications (2008) 2806–2808.

[56] C.S. Adjiman, A. Galindo, G. Jackson, Molecules matter: the expanding envelope of process design, in: M.R. Eden, J.D. Siirola, G.P. Towler (Eds.), Proceedings of the 8th International Conference on Foundations of Computer-Aided Process Design, Elsevier Science Bv, Amsterdam, 2014, pp. 55–64.

[57] S.H. Mushrif, J.J. Varghese, C.B. Krishnamurthy, Solvation dynamics and energetics of intramolecular hydride transfer reactions in biomass conversion, Physical Chemistry Chemical Physics 17 (2015) 4961–4969.

[58] D.G. Blackmond, Kinetic profiling of catalytic organic reactions as a mechanistic tool, Journal of the American Chemical Society 137 (2015) 10852–10866.

[59] I. McManus, H. Daly, J.M. Thompson, E. Connor, C. Hardacre, S.K. Wilkinson, N.S. Bonab, J. ten Dam, M.J.H. Simmons, E.H. Stitt, C. D'Agostino, J. McGregor, L.F. Gladden, J.J. Delgado, Effect of solvent on the hydrogenation of 4-phenyl-2-butanone over Pt based catalysts, Journal of Catalysis 330 (2015) 344–353.

[60] A. Lapkin, A. Voutchkova, P. Anastas, A conceptual framework for description of complexity in intensive chemical processes, Chemical Engineering and Processing: Process Intensification 50 (2011) 1027–1034.

[61] P.G. Jessop, D.J. Heldebrant, X. Li, C.A. Eckert, C.L. Liotta, Reversible nonpolar-to-polar solvent, Nature 436 (2005) 1102.

[62] P.G. Jessop, S.M. Mercer, D.J. Heldebrant, CO_2-triggered switchable solvents, surfactants, and other materials, Energy and Environmental Science 5 (2012) 7240–7253.

[63] P. Pollet, C.A. Eckert, C.L. Liotta, Switchable solvents, Chemical Science 2 (2011) 609–614.

[64] P.-M. Jacob, T. Lan, J.M. Goodman, A.A. Lapkin, A possible extension to the RInChI as a means of providing machine readable process data, Journal of Cheminformatics 9 (2017) 1–12.

[65] J. Zakrzewski, A.P. Smalley, M.A. Kabeshov, M.J. Gaunt, A.A. Lapkin, Continuous flow synthesis and derivatization of aziridines via palladium-catalyzed C(sp3)–H activation, Angewandte Chemie International Edition 55 (2016) 8878–8883.

[66] A. Echtermeyer, Y. Amar, J. Zakrzewski, A. Lapkin, Self-optimisation and model-based design of experiments for developing a C–H activation flow process, Beilstein Journal of Organic Chemistry 13 (2017) 150–163.

[67] Z.W. Ulissi, M.T. Tang, J. Xiao, X. Liu, D.A. Torelli, M. Karamad, K. Cummins, C. Hahn, N.S. Lewis, T.F. Jaramillo, K. Chan, J.K. Nørskov, Machine-learning methods enable exhaustive searches for active bimetallic facets and reveal active site motifs for CO_2 reduction, ACS Catalysis 7 (2017) 6600–6608.

[68] C.L. Bird, J.G. Frey, Chemical information matters: an e-research perspective on information and data sharing in the chemical sciences, Chemical Society Reviews 42 (2013) 6754–6776.

[69] A.D. Yang, W. Marquardt, An ontological conceptualization of multiscale models, Computers and Chemical Engineering 33 (2009) 822–837.

[70] N. Adams, E.O. Cannon, P. Murray-Rust, ChemAxiom – an Ontological Framework for Chemistry in Science, in: Nature Precedings, University of Cambridge, 2009.

[71] L. Zhou, C. Zhang, I.A. Karimi, M. Kraft, An ontology framework towards decentralized information management for eco-industrial parks, Computers and Chemical Engineering 118 (2018) 49–63.

[72] M. Fialkowski, K.J.M. Bishop, V.A. Chubukov, C.J. Campbell, B.A. Grzybowski, Architecture and evolution of organic chemistry, Angewandte Chemie International Edition 44 (2005) 7263–7269.

[73] B.A. Grzybowski, K.J.M. Bishop, B. Kowalczyk, C.E. Wilmer, The "wired" universe of organic chemistry, Nature Chemistry 1 (2009) 31–36.

[74] C.M. Gothard, S. Soh, N.A. Gothard, B. Kowalczyk, Y. Wei, B. Baytekin, B.A. Grzybowski, Rewriting chemistry: algorithmic discovery and experimental validation of one-pot reactions in the network of organic chemistry, Angewandte Chemie International Edition 124 (2012) 8046–8051.

[75] M. Kowalik, C.M. Gothard, A.M. Drews, N.A. Gothard, A. Weckiewicz, P.E. Fuller, B.A. Grzybowski, K.J.M. Bishop, Parallel optimization of synthetic pathways within the network of organic chemistry, Angewandte Chemie International Edition 51 (2012) 7928–7932.

[76] P.-M. Jacob, P. Yamin, C. Perez-Storey, M. Hopgood, A.A. Lapkin, Towards automation of chemical process route selection based on data mining, Green Chemistry 19 (2017) 140–152.

[77] A.A. Lapkin, P.K. Heer, P.-M. Jacob, M. Hutchby, W. Cunningham, S. Bull, M.G. Davidson, Automation of route identification and optimisation based on datamining and chemical intuition, Faraday Discuss 202 (2017) 483–496.

[78] P.-M. Jacob, A. Lapkin, Statistics of the network of organic chemistry, Reaction Chemistry and Engineering 3 (2018) 102–118.

[79] M.H.S. Segler, M.P. Waller, Modelling Chemical Reasoning to Predict Reactions, 2016 arXiv:1608.07117.

[80] M.H.S. Segler, M.P. Waller, Neural-symbolic machine learning for retrosynthesis and reaction prediction, Chemistry: A European Journal 23 (2017) 5966–5971.

[81] C.W. Coley, R. Barzilay, T.S. Jaakkola, W.H. Green, K.F. Jensen, Prediction of organic reaction outcomes using machine learning, ACS Central Science 3 (2017) 434–442.

[82] C.W. Coley, L. Rogers, W.H. Green, K.F. Jensen, SCScore: synthetic complexity learned from a reaction corpus, Journal of Chemical Information and Modeling 58 (2018) 252–261.

[83] J.N. Wei, D. Duvenaud, A. Aspuru-Guzik, Neural networks for the prediction of organic chemistry reactions, ACS Central Science 2 (2016) 725–732.

[84] G. Skoraczyński, P. Dittwald, B. Miasojedow, S. Szymkuć, E.P. Gajewska, B.A. Grzybowski, A. Gambin, Predicting the outcomes of organic reactions via machine learning: are current descriptors sufficient? Scientific Reports 7 (2017) 3582.

[85] N. Peremezhney, C. Connaughton, G. Unali, E. Hines, A.A. Lapkin, Application of dimensionality reduction to visualisation of high-throughput data and building of a classification model in formulated consumer product design, Chemical Engineering Research and Design 90 (2012) 2179–2185.

[86] S.L. Bayliss, J.M. Cole, P.G. Waddell, S. McKechnie, X.G. Liu, Predicting solar-cell dyes for cosensitization, Journal of Physical Chemistry C 118 (2014) 14082–14090.

[87] W. Czechtizky, J. Dedio, B. Desai, K. Dixon, E. Farrant, Q.X. Feng, T. Morgan, D.M. Parry, M.K. Ramjee, C.N. Selway, T. Schmidt, G.J. Tarver, A.G. Wright, Integrated synthesis and testing of substituted xanthine based DPP4 inhibitors: application to drug discovery, ACS Medicinal Chemistry Letters 4 (2013) 768–772.

[88] A.J. Parrott, R.A. Bourne, G.R. Akien, D.J. Irvine, M. Poliakoff, Self-optimizing continuous reactions in supercritical carbon dioxide, Angewandte Chemie International Edition 50 (2011) 3788–3792.

[89] A.M. Schweidtmann, A.D. Clayton, N. Holmes, E. Bradford, R.A. Bourne, A.A. Lapkin, Machine learning meets continuous flow chemistry: automated optimization towards the Pareto front of multiple objectives, Chemical Engineering Journal 352 (2018) 277–282.

[90] J.S. Moore, K.F. Jensen, Automated multitrajectory method for reaction optimisation in a microfluidic system using online IR analysis, Organic Process Research and Development 16 (2012) 1409–1415.

[91] B.J. Reizman, K.F. Jensen, An automated continuous-flow platform for the estimation of multistep reaction kinetics, Organic Process Research and Development 16 (2012) 1770–1782.

[92] B.J. Reizman, K.F. Jensen, Feedback in flow for accelerated reaction development, Accounts of Chemical Research 49 (2016) 1786–1796.

[93] V. Sans, L. Cronin, Towards dial-a-molecule by integrating continuous flow, analytics and self-optimisation, Chemical Society Reviews 45 (2016) 2032–2043.

[94] N. Peremezhney, E. Hines, A. Lapkin, C. Connaughton, Combining Gaussian processes, mutual information and a genetic algorithm for multi-target optimization of expensiveto-evaluate functions, Engineering Optimization 46 (2014) 1593–1607.

[95] C. Houben, A. Lapkin, Automatic discovery and optimisation of chemical processes, Current Opinion in Chemical Engineering 9 (2015) 1–7.

[96] C. Houben, N. Peremezhney, A. Zubov, J. Kosek, A.A. Lapkin, Closed-loop multi-target optimisation for discovery of new emulsion polymerisation recipes, Organic Process Research and Development 19 (2015) 1049–1053.

[97] E. Bradford, A.M. Schweidtmann, A. Lapkin, Efficient multiobjective optimization employing Gaussian processes, spectral sampling and a genetic algorithm, Journal of Global Optimization 71 (2018) 407–438.

[98] A.B. Santanilla, E.L. Regalado, T. Pereira, M. Shevlin, K. Bateman, L.-C. Campeau, J. Schneeweis, S. Berritt, Z.-C. Shi, P. Nantermet, Y. Liu, R. Helmy, C.J. Welch, P. Vachal, I.W. Davies, T. Cernak, S.D. Dreher, Nanomole-scale high-throughput chemistry for the synthesis of complex molecules, Science 347 (2015) 49–53.

[99] J. Wingfield, Acoustic Mass Spectrometer Platforms, 2018. https://www.astrazeneca.com/what-science-can-do/our-technology/Acoustic-mass-spec.html.

[100] D. Helmdach, P. Yaseneva, P.K. Heer, A.M. Schweidtmann, A.A. Lapkin, A multiobjective optimization including results of life cycle assessment in developing biorenewables-based processes, ChemSusChem 10 (2017) 3632–3643.

[101] N. Eghbali, M. Hoppenbrouwers, S. Lemain, G.D. Bruyn, B.V. Velpen, How technical innovation in manufacturing is fostered through business innovation, in: A. Lapkin (Ed.), Green Chemical Engineering, Wiley-VCH Verlag GmbH, Weinheim, 2018, pp. 191–214.

[102] D. Kralisch, N. Minkov, A. Manent, E. Rother, L. Mohr, D. Schowanek, S. Sfez, A.A. Lapkin, M. Jones, MEASURE Roadmap, 2016. http://www.spire2030.eu/measure/roadmap-documents.

[103] D. Kralisch, D. Ott, A.A. Lapkin, P. Yaseneva, W.D. Soete, M. Jones, N. Minkov, M. Finkbeiner, The need for innovation management and decision guidance in sustainable process design, Journal of Cleaner Production 172 (2018) 2374–2388.

[104] D. Kralisch, D. Ott, D. Gericke, Rules and benefits of life cycle assessment in green chemical process and synthesis design: a tutorial review, Green Chemistry 17 (2015) 123–145.

[105] S. Falß, G. Tomaiuolo, P. Yaseneva, J. Zakrzewski, A. Lapkin, S. Guido, R. Woodward, R.E. Meadows, A continuous process for Buchwald-Hartwig amination at micro- lab- and multi-kilo scale, Organic Process Research and Development 20 (2016) 558–567.

[106] P. Yaseneva, P. Hodgson, J. Zakrzewski, A. Lapkin, Continuous flow Buchwald-Hartwig amination of a pharmaceutical intermediate, Reaction Chemistry and Engineering 1 (2016) 229–238.

[107] C. Jimenez-Gonzalez, C. Ollech, W. Pyrz, D. Hughes, Q.B. Broxterman, N. Bhathela, Expanding the boundaries: developing a streamlined tool for eco-footprinting of pharmaceuticals, Organic Process Research and Development 17 (2013) 239–246.

[108] M.P. Tsang, G. Philippot, C. Aymonier, G. Sonnemann, Supercritical fluid flow synthesis to support sustainable production of engineered nanomaterials: case study of titanium dioxide, ACS Sustainable Chemistry and Engineering 6 (2018) 5142–5151.

[109] P. Yaseneva, N. An, M. Finn, N. Tidemann, N. Jose, A. Voutchkova-Kostal, A. Lapkin, Continuous synthesis of doped layered double hydroxides in a meso-scale flow reactor, Chemical Engineering Journal 360 (2019) 190–199.

[110] J.S. Srai, M. Gregory, A supply network configuration perspective on international supply chain development, International Journal of Operations and Production Management 28 (2008) 386–411.

[111] J.S. Srai, C. Badman, M. Krumme, M. Futran, C. Johnston, Future supply chains enabled by continuous processing – opportunities and challenges. May 20–21, 2014 continuous manufacturing symposium, Journal of Pharmaceutical Sciences 104 (2015) 840–849.

[112] A. Lapkin, L. Joyce, B. Crittenden, A framework for evaluating the "greenness" of chemical processes: case studies for a novel VOC recovery technology, Environmental Science and Technology 38 (2004) 5815–5823.

[113] A. Lapkin, Sustainability performance indicators, in: J. Dewulf, I.H.v. Langenhove (Eds.), Renewables-Based Technology: Sustainability Assessment, John Wiley & Sons, 2006, pp. 39–53.

[114] M.Q. Xie, H. Freund, Fast synthesis of optimal chemical reactor networks based on a universal system representation, Chemical Engineering and Processing 123 (2018) 280–290.

CHAPTER

17

Systematic MultiObjective Life Cycle Optimization Tools Applied to the Design of Sustainable Chemical Processes

Gonzalo Guillén-Gosálbez[1], *Andrés González-Garay*[2], *Phantisa Limleamthong*[2], *Ángel Galán-Martín*[1], *Carlos Pozo*[2]

[1]Institute for Chemical and Bioengineering, Department of Chemistry and Applied Biosciences, ETH Zürich, Vladimir-Prelog-Weg 1, Zürich, Switzerland; [2]Department of Chemical Engineering, Centre for Process System Engineering, Imperial College London, South Kensington Campus, London, United Kingdom

OUTLINE

1. INTRODUCTION

Over the past few decades, sustainability has become a major concern worldwide, prompting governments to invest in new solutions to meet the growing demand of food, energy, and water in a sustainable manner. In this context, industrial sustainability is understood as "an approach towards the selection of clean technologies that minimize the consumption of raw materials & energy and the generation of pollutants, while renewable resources are maximized as inputs to produce recyclable or biodegradable products" [1]. The incorporation of environmental concerns into process design and operations has posed many challenges for industry, which must provide high quality products and services while at the same time looking for alternative technologies that minimize the dependency on raw materials, energy, labor, and waste [1].

Sustainable Nanoscale Engineering
https://doi.org/10.1016/B978-0-12-814681-1.00017-5

Following this general trend, the chemical industry is at present seeking to become more sustainable while maintaining its profit margins high. This can be accomplished through the design of greener chemicals as well as the optimization of chemical processes to ultimately reduce emissions and waste generation as well as energy and water consumption. With regard to the first approach (i.e., greener chemicals), the principle of "Green Chemistry" offers a practical framework for reducing environmental burdens at the molecular scale [2]. "Green Engineering," alternatively, puts emphasis on how to attain the sustainability at the process and system levels via the development of technology [2]. In this general context, the design of more environmentally benign materials, products, processes, and systems could be performed beyond the baseline of engineering quality and safety specifications through the adoption of 12 principles that take environmental, economic, and social criteria into consideration [2].

Both approaches, microscopic (product design) and macroscopic (process design), comprise three major steps: (i) the assessment of environmental burdens of products and processes of interest, (ii) the benchmarking of their environmental performance and establishment of improvement targets, and (iii) the identification of solutions that improve the sustainability level of these systems. A number of systematic methodologies have been recently developed to assist in these three fundamental steps, which were reviewed in the work by Nikolopoulou and Ierapetritou [3] and are discussed in more detail next.

Among the tools available for the environmental assessment of products, processes, and services (step one as discussed above), life cycle assessment (LCA) has become the preferred choice, gaining worldwide acceptance as a technique that systematically analyzes the environmental impacts of a product or service from cradle-to-grave. Hence, in LCA the product/process is evaluated throughout its life cycle, starting from resources extraction, through manufacturing, and ending with waste disposal [4]. LCA quantifies the system's inventory, i.e., energy, material usages, and emissions and waste generated, to assess the corresponding environmental impacts (i.e., to measure the system's environmental performance) and finally identify potential environmental improvements [5]. These advantageous features of LCA enable environmental management in both corporate and public sectors to become more tangible, as described in the ISO 14040 standards [6]. Overall, LCA has found many applications in process selection, design, and optimization, becoming an alternative approach to identify clean technologies that has facilitated the widespread application of sustainability principles in industry [7–9].

With regard to step two, and particularly focusing on the area of product and process design, it is common in the chemical industry to find alternative technologies with similar properties and applications. In this context, it is necessary to compare them against each other according to several criteria, set out beforehand, to ensure that the best-performed solution is chosen. This is essentially a characteristic of sustainability-related problems, in which several criteria are considered simultaneously, giving rise to a so-called multicriteria decision-making problem.

Wang et al. [10] reviewed various methods for multi-criteria decision analysis, including the weighted sum method (WSM), the analytical hierarchy process, the technique for order preference by similarity to ideal solutions, and the elimination and choice translating reality, among others. In these methods, decision-makers select the best option following a ranking based on aggregated scores obtained by assigning subjective weights of importance to each criterion [10]. The way in which weights are handled varies significantly from one approach to another, thereby leading to different final results [11]. All of these methods provide as main outcome a ranking of alternatives; unfortunately, this ranking provides little insight (if any) into the potential sources of environmental inefficiencies, neither does it offer quantitative guidelines on how to enhance the level of sustainability of a product or service. In this regard, we note that the majority of approaches toward sustainability are based on a "direction to target" approach [12], which establishes an improving direction to the sustainability target without clear quantitative guidelines on how to achieve it. "Distance to target" approaches, on the contrary, distinctively offer more practical and meaningful guidelines by measuring the magnitude toward (or away from) sustainability, thereby enabling the definition of useful quantitative targets [13].

With regard to step three (i.e., finding solutions with improved sustainability level), the growing increase in sustainability awareness together with the adoption of more stringent regulations have motivated the integration of sustainability criteria into optimization tools, the latter originally developed to maximize profit as unique criterion. In the context of the chemical industry, moving toward more sustainable chemical processes (those that are cost-effective, energy efficient, consume less resources, emit less emissions, and overall show smaller environmental impacts) requires practical process retrofits and modifications. These can be underpinned by the use of heuristics, the development of physical insights—commonly linked to thermodynamics—and, more often, by optimization tools [8]. Notably, due to recent advances in optimization theory and software applications, mathematical programming has gained wider interest in sustainability problems. This approach comprises three major steps. The first step is to establish a superstructure of alternatives from which the optimal solution is chosen. The second step is to formulate a mathematical problem based on the equations that describe the unit operations and process topology in the superstructure, where equipment units, interconnectivity between them, and process constraints are all together considered simultaneously. The optimization models formulated in this approach often contain discrete and continuous variables, employed for selecting the process configuration as well as operating conditions, respectively. Furthermore, these models include both linear and nonlinear constraints, leading to mixed-integer nonlinear programming problems (MINLP) [14] whose solution dictates the optimal flowsheet configuration and process conditions.

The recent trend is to couple optimization models with LCA so as to identify solutions that reduce simultaneously the total cost and environmental impact of products and processes. This approach is described in more detail in the ensuing sections.

2. MULTIOBJECTIVE LIFE CYCLE OPTIMIZATION OF CHEMICAL PRODUCTS AND PROCESSES

The multiobjective life cycle optimization framework has gained wide acceptance in the process systems engineering community as a fundamental tool to assist in the design and operation of more sustainable products and processes. In essence, this combined method integrates LCA with optimization tools, where the former is employed to assess alternatives from an environmental viewpoint, while the latter generates such alternatives and identifies the best ones in a systematic manner.

Incorporating sustainability principles into process design is challenging, as it requires the simultaneous consideration of a wide range of economic, environmental, and social criteria in problems that are *per se* quite complex [15,16]. Defining adequate sustainability metrics in the form of key performance indicators (KPIs) is essential for effectively supporting the decision-making process [17,18]. The need to consider various KPIs simultaneously leads to complex multicriteria decision-making problems, in which identifying the best decisions over a wide range of potential alternatives becomes critical.

In chemical process design, the objective is to find the best option in terms of these criteria through their evaluation and simultaneous optimization. We discuss next the main ingredients of the life cycle optimization approach.

2.1 Costing of Chemical Processes

The literature around the economic evaluation of chemical processes is substantial, with many methodologies available to serve this purpose. The economic objective during the design stage is usually related to the minimization of the costs or the maximization of the profit, using metrics such as payback time, return on investment, net present value, or discounted cash flow rate of return [19]. The generation of any of these economic indicators involves the evaluation of capital, fixed and variable costs, as well as the revenues of the process. Generally, all the terms applied are calculated on an annual basis.

Capital costs are the costs inherent to the equipment and its installation. There are three main methods to calculate the capital costs: power law, cost factors, and detailed estimates. Given their simplicity, the first two methods are the most widely applied in early design stages; unfortunately, they lead to errors of 30%–50%, which are nevertheless considered admissible during this early stage. The total annualized capital cost is determined considering the annual capital charge (ACC) ratio.

Fixed costs of production are independent from the rate of production and include labor, maintenance, land, insurance, interest payments, overhead, license fees, and royalties.

Variable costs of production are dependent on the rate of production and account for consumption of raw materials, utilities, consumables, and waste disposal.

Finally, the total annualized cost, obtained as the sum of the variable costs, fixed costs, and ACC, is one of the most applied economic indicators during the evaluation of chemical projects.

$$\text{TAC} = \text{FCOP} + \text{VCOP} + \text{ACC} \tag{17.1}$$

Revenues are generated from sales of main products and by-products and can be used to determine the profit together with the annualized cost.

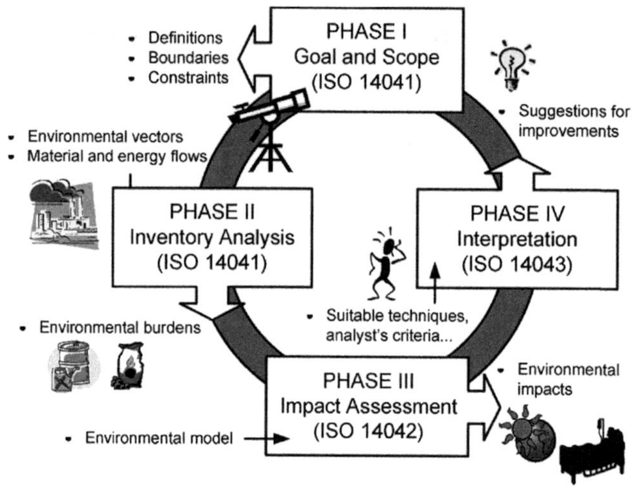

FIGURE 17.1 Phases of the life cycle assessment methodology [20].

2.2 Life Cycle Assessment

LCA is an environmental management tool that quantifies the environmental loads and the corresponding environmental impacts of a product, process, or activity throughout its whole life cycle to assess opportunities for environmental improvements [5]. The International Organization for Standardization (ISO) has developed the framework and guidelines for a practical use of the LCA methodology, issuing a series of standards documents in the 14040 series [6]. In general, LCA comprises four major phases as following:

- Goal and scope definition
- Inventory analysis
- Life cycle impact assessment (LCIA)
- Life cycle interpretation

Fig. 17.1 displays the main steps in the LCA methodology, described in more detail next.

2.2.1 *Goal and Scope Definition*

The first LCA step sets up the explicit goal of the LCA, often corresponding to the assessment and comparison of the environmental performance of alternative processes, ultimately aiming at providing guidelines for system improvements to minimize environmental impacts [5]. Consequently, the scope of the study, i.e., system boundaries, functional unit, assumptions and limitations, and selected impact categories, will be defined bearing this goal in mind.

There are several ways to define the system boundaries. For example, cradle-to-grave implies the full product life cycle from resource extraction to waste disposal. On the other hand, cradle-to-gate encompasses all the activities from resources extraction to the factory gate (i.e., excluding transportation to final customers), while cradle-to-cradle is a specific kind of cradle-to-grave where the waste products are recycled instead of disposed in the final step. The functional unit allows for a fair comparison between system alternatives; it refers to an equivalent amount of goods or services delivered by the product systems and, therefore, provides a standardization that enables alternative goods or services to be compared and analyzed [21].

2.2.2 *Inventory Analysis*

This phase quantifies the inventory flows, from and to the environment, associated with a product system [22]. It makes use of the compilation of available data and the calculation of material and energy balances to quantify relevant resources (i.e., water, raw material, and energy) and emissions related to the functional unit defined in the first step [7]. Due to the need for data documentation, several public standard databases have been developed and are available for use, such as SPINE, developed by the Center for Environmental Assessment of Product and Material Systems [23], and ECOINVENT, developed by the Swiss Center for Life Cycle Inventories [24], among others. These can be accessed directly or via software packages with friendly user interfaces, such as Gabi, SimaPro, Umberto, or, more recently, open source initiatives such as openLCA.

2.2.3 *Life Cycle Impact Assessment*

The LCIA converts the resources and emissions, in other words, the environmental burdens specified in the life cycle inventory of a system and previously quantified in the inventory analysis phase, into their potential environmental impacts. There are a number of LCIA methods available classified based on the impact categories and characterization factors they implement [9].

2.2.3.1 Classification

The environmental burdens are first aggregated according to the impact categories to which they are linked, specified in the goal of the study. Widely used impact categories include climate change, stratospheric ozone depletion, photo-oxidant formation (smog), eutrophication, acidification, water use, and noise, among others. Environmental burdens are often translated into a common impact category to enable comparisons between them [25]. For instance, emissions of CO_2, CH_4, VOCs, and other greenhouse gases can be expressed as CO_2-equivalent emissions to evaluate the overall global warming potential (GWP), which takes all sources of greenhouse gases into account [7]. In addition, according to the ISO methodology, the impact categories can be grouped into three areas of protection (AoPs): resources use, human health consequences, and ecological consequences [26].

2.2.3.2 Characterization

In the characterization step, the potential contribution of each environmental burden to each impact category is evaluated. This can be done by multiplying the inventory entry (i.e., emission, waste, or resource) by its characterization factor. The characterization factors refer to potential environmental impacts per unit of emission of a given substance and, therefore, are specific to a particular substance [9]; they are in turn computed considering the impact category to which a substance could potentially contribute [27]. A set of characterization factors determined following different approaches are available in the literature, most of which were already implemented in databases and software packages for a wide range of indicators, like the Eco-indicator 99 [28], CML [29], and Recipe 2016 [30]. The mathematical expression used to calculate a category indicator is shown below.

$$S_j = \sum_i Q_{j,i} m_i \quad \forall j \tag{17.2}$$

where S_j is the indicator for impact category j, m_i refers to the life cycle inventory of substance i, and $Q_{j,i}$ is the characterization factor of substance i that contributes to impact category j [26].

2.2.3.3 Normalization

Normalization is applied when practitioners wish to compare products across impact categories, or even across AoPs, to prioritize or to resolve trade-offs between product alternatives [26]. This optional step also identifies which impact categories are minor contributors to the overall environmental problem and can, therefore, be excluded from the evaluation. The normalization is carried out by dividing the impact category indicators by a reference value as follows:

$$N_j = \frac{S_j}{R_j} \quad \forall j \tag{17.3}$$

where j denotes the interested impact category, N_j is the normalized indicator, S_j is the category indicator obtained in the characterization step, and R_j is the reference value of impact category indicator.

When selecting the reference value, the following attributes should be taken into consideration: spatial scale, temporal scale, a defined system (e.g., a region or an economic sector), and a per capita basis.

2.2.3.4 Valuation

The valuation is the final step in LCIA, whereby the relative importance of different impacts is incorporated in the assessment by using weighting factors assigned to each impact and further aggregating them into a single environmental indicator [31]. These weighted expressions are usually represented as a linear equation as below:

$$EI = \sum_j wn_j N_j \text{ or } EI = \sum_j ws_j S_j \tag{17.4}$$

where EI denotes the overall environmental impact indicator, wn_j is the weighting factor for normalized impact category j, and ws_j is the weighting factor for impact category j when the normalization step is skipped in the characterization phase [26].

This valuation step is regarded as the most controversial element of LCA due to its high degree of subjectivity in determining the significance of different impacts [7]. Hence, both normalization and weighting are regarded as optional steps that can be omitted when found appropriate.

2.2.4 *Interpretation*

The interpretation, carried out in the final phase of the LCA methodology, aims to identify potential improvements in environmental performance of a particular system. According to the ISO methodology, these may involve improvements, innovations, the identification of significant stages or issues in the life cycle contributing to the impacts, a sensitivity analysis, and final recommendations [7].

The interpretation phase requires the evaluation of the system alternatives according to a set of indicators. Unfortunately, due to the inherent trade-offs found between impacts in many applications, it is very unlikely that one single alternative will perform better than the rest in all of the categories simultaneously. The standard LCA application ends with the discussion of these set of alternative cases, which are defined beforehand, and their inherent trade-offs.

To enlarge the LCA capabilities, this methodology can be integrated with optimization tools that can explore in a systematic manner a vast number of solutions in short CPU times. This integration between LCA and optimization is described in detail next after a brief discussion on the main modeling tools used in process systems engineering.

2.3 Chemical Process Modeling

Process modeling plays a key role in underpinning the design and operation of new and existing chemical processes. We discuss some general modeling issues next before describing how modeling tools can be integrated with LCA.

2.3.1 *Equation-Oriented Versus Simulation-Optimization Approaches*

Two main options emerge in process modeling: equation-oriented approaches and simulation-based approaches. The equation-oriented approach typically establishes a superstructure of alternatives on the basis of which a mathematical problem containing discrete and continuous variables is formulated; this model is then optimized for selecting the optimal process configuration and operating conditions [14]. The simultaneous equation-oriented approach often leads to nonconvex MINLP formulations that can be solved with state-of-the-art software packages (e.g., GAMS, AIMMS, LINDO, MATLAB, etc.). To avoid very complex MINLPs potentially leading to prohibitive CPU times, the modeling is simplified by using less accurate shortcut methods showing better numerical performance [32,33]. This helps the calculations but at the expense of sacrificing to some extent the accuracy of the results.

In contrast, in the simulation-based approach, modular sequential models based on detailed equations of unit operations are solved; this leads to more accurate results but the downside is that standard optimization algorithms cannot be then applied in a straightforward manner, thereby compromising the optimality of the final solutions found. Here, the modeling and optimization tasks are decoupled from each other by simulating the process with a detailed model that is optimized with an external algorithm. While the formulation of these problems is still nonconvex and nonlinear, the decoupling of the modeling and optimization tasks enables the use of tailored initialization techniques and solution algorithms that facilitate the convergence of the flowsheet; the optimization task, however, becomes more challenging as it cannot have direct access to the detailed explicit equations [34,35]. The recent trend is now to replace the simulation model by a surrogate model to expedite the calculations, using to this end a variety of regression methods including kriging, neural networks, and splines [36].

2.3.2 *MultiObjective Optimization*

The inclusion of sustainability criteria into product and process optimization naturally leads to multiobjective optimization (MOO) problems, in which the environmental and economic criteria are simultaneously optimized

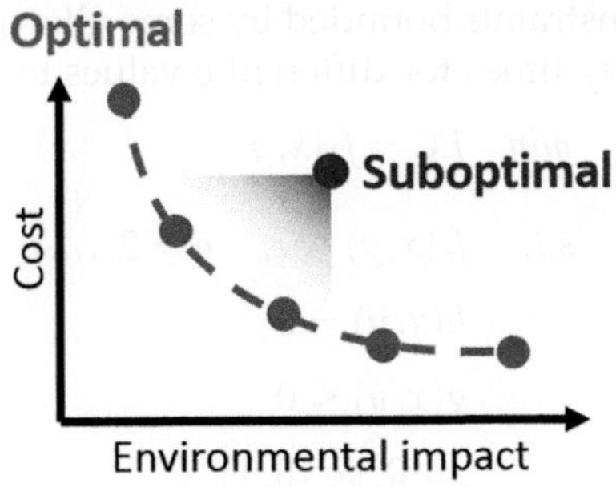

FIGURE 17.2 Pareto curve.

[8]. The optimization problem with sustainability objectives can be formulated as a multiobjective mixed-integer nonlinear problem (moMINLP) as follows [20]:

$$\begin{aligned} \min_{x,y} \quad & U(x,y) = \{f_1(x,y), f_2(x,y)\} \\ s.t. \quad & h(x,y) = 0 \\ & g(x,y) \le 0 \\ & x \in \Re, y \in \{0,1\} \end{aligned} \tag{M.1}$$

where $f_1(x,y)$ refers to the total cost and $f_2(x,y)$ represents the environmental impact. Note that, when possible, social indicators shall be also defined as additional objectives.

Unlike single-objective optimization, MOO generates a set of optimal solutions known as Pareto set, which represent the optimal trade-off between the different objectives in the MOO problem. These are nondominated solutions, implying that no other solution would have better performance in an objective without worsening at least one of the others. The Pareto optimality concept is displayed in Fig. 17.2, where two objectives, cost and environmental impact, are considered. The points lying on the Pareto curve are Pareto optimal [7,20]. Finding feasible solutions under the Pareto curve is impossible, because Pareto optimal solutions cannot be dominated; solutions above the curve, however, do exist, but are suboptimal. From this set of Pareto solutions, decision-makers should select the best one according to their particular preferences as well as the applicable legislation [20].

2.3.3 *MOO Solution Methods*

The most commonly used algorithms for generating the Pareto optimal solutions of an MOO problem are the weighted sum and epsilon constraint methods. In the WSM, a weighted sum of objectives is optimized by exploring the space of possible weights [37]. In the epsilon constraint method, one objective is kept as main objective function, while the others act as auxiliary constraints which impose bounds on the main objective [37,38].

2.3.3.1 Weighted Sum Method

The WSM solves an auxiliary SOO model (M.2), derived from the original MOO problem, that optimizes a linear weighted sum of the original objectives previously normalized [37].

$$\begin{aligned} \min_{x,y} \quad & WS = \{(1-\lambda)\bar{f}_1(x,y) + \ldots + \lambda\bar{f}_n(x,y)\} \\ s.t. \quad & h(x,y) = 0 \\ & g(x,y) \le 0 \\ & 0 \le \lambda \le 1 \\ & x, \lambda \in \Re, y \in \{0,1\} \end{aligned} \tag{M.2}$$

2.3.3.2 Epsilon Constraint Method

The ε-constraint method, first introduced by Haimes et al. [38], transforms the original multiobjective model into a set of single-objective problems, where the optimization is carried out considering one objective function at a time while defining the other objectives as constraints bounded by some allowable levels, i.e., ε values. The optimization problem, as shown below, is solved many times for different ε values to generate the entire Pareto optimal set.

$$\begin{aligned} \min_{x,y} \quad & EC = f_1(x,y) \\ s.t. \quad & f_o(x,y) \le \varepsilon_o \quad o = 2,\ldots,n \\ & h(x,y) = 0 \\ & g(x,y) \le 0 \\ & x \in \Re, y \in \{0,1\} \end{aligned} \tag{M.3}$$

2.3.4 Methods for Selecting the Most Preferred Solution

Once the Pareto solutions are generated, decision-makers need to identify the best ones considering different criteria. Here, the multicriteria methods described previously can be applied to assist in the analysis of the inherent trade-offs.

3. AREAS OF APPLICATION

In this section, we present three different application domains of LCA principles combined with optimization tools, an approach originally introduced by Azapagic and Clift [7] and reviewed elsewhere [8,9,39]. The first example deals with the optimization of a hydrogen supply chain for vehicles use at the macroscale level. The second problem addresses flowsheet optimization in benzene production and deals with the mesoscale level. The third example, at the molecular level, covers the life cycle impact assessment of molecules.

3.1 Supply Chain Optimization

With the rapid worldwide globalization, supply chain optimization has gained interest in the private and public sectors. Substantial work was carried out in the areas of biomass supply chains [40,41] and petrochemicals supply chains [42–44] following the life cycle optimization framework. Here, we study the optimal design of a hydrogen supply chain for vehicle use in United Kingdom considering both economic and environmental concerns simultaneously [45]. The adoption of hydrogen fuels has emerged as a promising alternative to reduce the environmental impact of today's fossil fuel–based transportation sector. To assist in the design of hydrogen supply chains, the authors presented a bi-criterion MILP that systematically determines the optimal hydrogen supply chain network configuration and associated planning decisions to satisfy a given hydrogen demand considering simultaneously the total cost and life cycle impact (Fig. 17.3).

The superstructure includes a set of plants that produce hydrogen and a set of potential storage facilities, where hydrogen is stored before being delivered to customers *via* various transportation options. Three hydrogen production technologies were considered, namely steam methane reforming, coal gasification, and biomass gasification, while hydrogen can be stored *via* liquid hydrogen storage and compressed gas storage. For hydrogen transportation, four options were considered, i.e., hydrogen tanker trucks, liquid hydrogen railway tank cars, compressed-gaseous hydrogen tube trailers, and compressed-gaseous hydrogen railway tube cars.

The MILP network model accounts simultaneously for the optimization of the cost and environmental performance of the hydrogen network. The economic objective corresponds to the total discounted cost, which includes both capital and operating costs given by the production, storage, and transportation facilities of the network. The environmental performance was quantified through the LCA methodology focusing on climate change measured *via* the Eco-indicator 99. To solve the bi-criterion problem, the authors applied the epsilon constraint method combined with a bi-level decomposition algorithm that expedites the calculations, which are computationally demanding given the size of the problem (e.g., for a problem with 10 periods, the model contains 21,160 binary variables, 1880 discrete variables, 25,396 continuous variables, and 70,976 constraints). The interested reader is referred to [45] for further details on the bi-level algorithm.

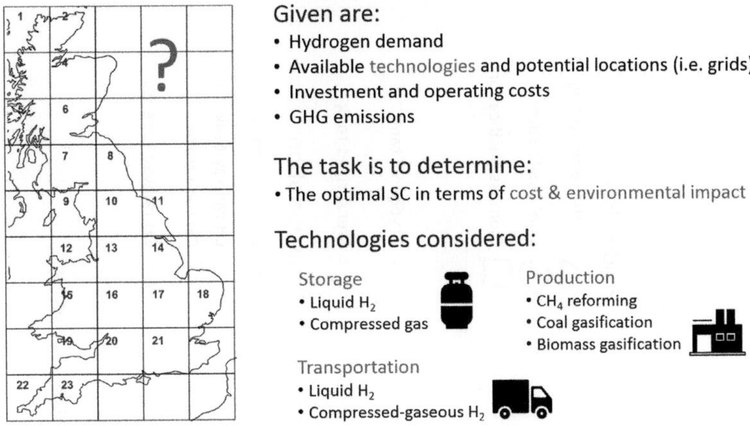

FIGURE 17.3 Problem statement of the hydrogen case study.

The results of the model are shown in the Pareto set of solutions depicted in Fig. 17.4. Note that there is a clear trade-off between cost and damage to human health due to climate change, where each solution entails a specific structure of the hydrogen supply chain. Specifically, two alternatives are shown in Fig. 17.4, which correspond to the extreme solutions. In the first alternative A (i.e., minimum cost solution), steam reforming is selected to produce hydrogen. Moving toward solutions with lower environmental impact requires producing hydrogen exclusively from biomass, with the solution with lowest damage to human health (solution E) using only compressed hydrogen gas in addition to biomass. Overall, results suggest that replacing steam reforming by biomass gasification is a promising alternative to reduce the environmental impact; meanwhile, substituting liquefied hydrogen by compressed gas hydrogen as storage technology would significantly increase the total costs without attaining significant environmental benefits. The algorithm, in turn, provides the optimal network configuration, including specific locations for production and storage facilities and associated material flows between them.

3.2 Process Flowsheet Optimization

The second example illustrates how LCA can be integrated into process design. Guillén-Gosálbez et al. [20] incorporated LCA principles into MINLP models to address the structural optimization of sustainable chemical process flowsheets. The well-known hydrodealkylation of toluene (HDA) process was optimized following this approach, where cost and environmental impact were the objectives minimized (Fig. 17.5). The superstructure considers a toluene fresh feed that reacts with a hydrogen stream that can be further purified using a membrane separator. The exothermic reaction can be conducted in a plug-flow reactor operating either adiabatically or isothermally, where the latter is more expensive due to heat removal. Different alternatives are considered for the vapor stream exiting the flash separator located downstream the reactor, including the use of an absorber to recover the benzene lost in the flash separator, a purge to avoid methane accumulation in the system, or a membrane separator that decreases the hydrogen loss in the purge stream. A portion of the liquid stream exiting the flash is sent to the liquid separation system, where hydrogen and methane are removed using either a stabilizing column or a second flash separator. A distillation column yields a benzene stream with the desired purity of 99.97%. The bottom stream, containing primarily toluene, can be split in a flash separator or in a column. The purge streams containing methane and diphenyl by-product streams are combusted to produce steam that can be used in other parts of the plant.

The superstructure for this HDA process was modeled as an MINLP similarly as in [46]. Simplified models such as Raoult's law for phase equilibrium, the Fenske–Underwood equation for distillation columns, and the Kremser equation for the absorber were used. The resulting mathematical formulation simultaneously accounts for the minimization of the environmental impact and cost, where the former is measured through the Eco-indicator 99. The resulting moMINLP can be solved by standard techniques for MOO, such as the ε-constraint method [38].

The solution to the problem consists of a set of Pareto optimal flowsheet configurations and corresponding operating conditions (Fig. 17.6). All the Pareto optimal designs use the isothermal reactor, the stabilizing column, and the second flash, but they differ in the use of the membrane in the input stream and in the methane purge. Notably, the minimum cost solution avoids the membrane in the input stream but uses the one placed in the methane

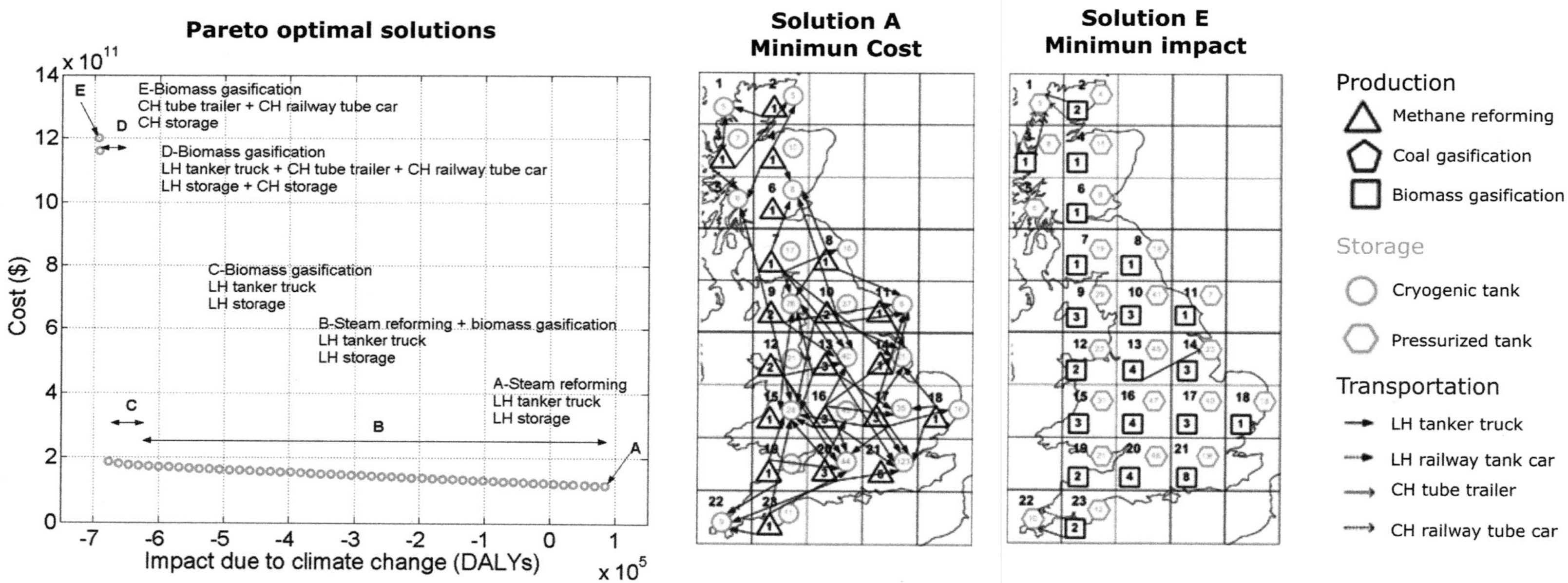

FIGURE 17.4 Pareto set of optimal hydrogen supply chain solutions and minimum cost and environmental impact solutions.

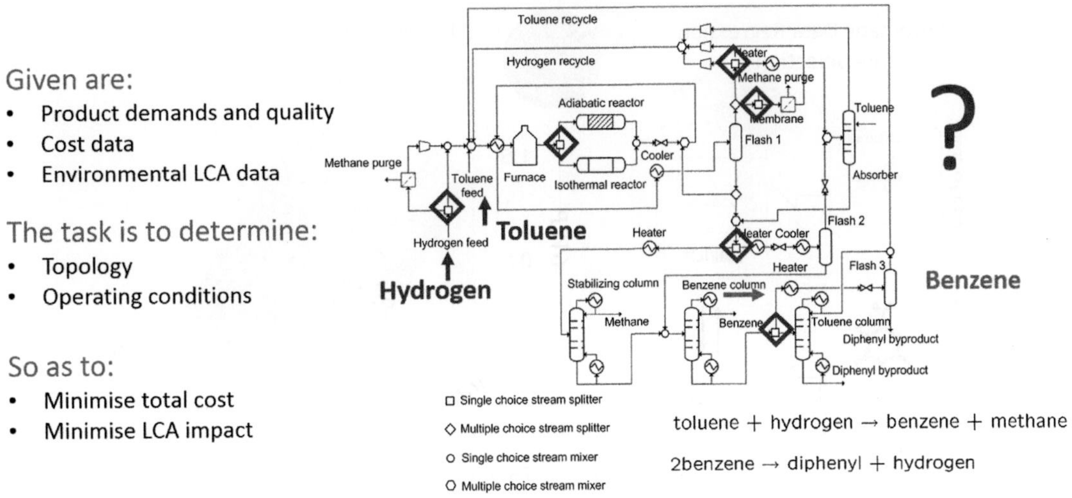

FIGURE 17.5 Superstructure for the hydrodealkylation (HDA) process synthesis problem.

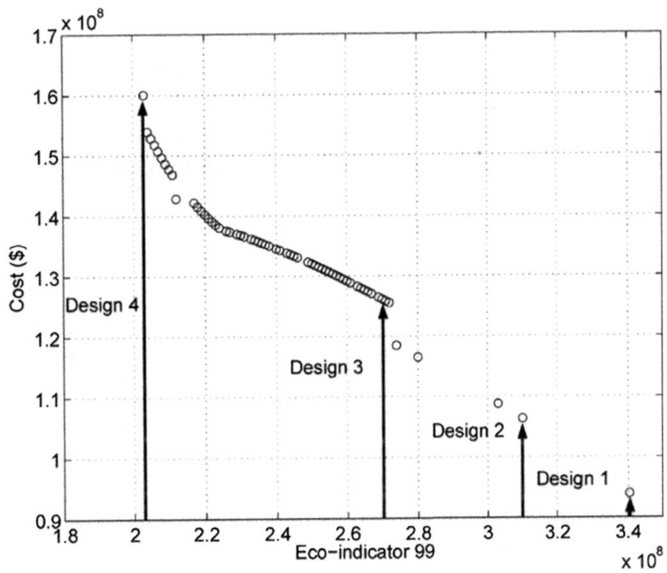

FIGURE 17.6 Trade-off solutions of the hydrodealkylation process superstructure [20].

purge, while the converse occurs in the minimum impact design. The MINPL algorithm, therefore, determines the optimal plant topology, equipment sizes, and operating conditions for a given economic and environmental performance, offering a set of design alternatives representing the optimal trade-off between both criteria.

3.3 Environmental Assessment of Chemicals

The last example deals with streamlined LCA applied to chemicals for potential applications in molecular design and product development. The motivation for this work lies in the fact that LCA studies require large amounts of data hard to collect in practice, particularly considering the value chains in the chemical industry encompassing hundreds of technologies. Producing accurate estimates of environmental impact for chemicals, however, is critical for a proper assessment of a wide range of products and processes.

To simplify LCA calculations, streamlined LCA methods can be applied [47,48], in which the upstream and downstream information required in a standard LCA is reduced by using proxy data, qualitative models, and/or regression equations [49].

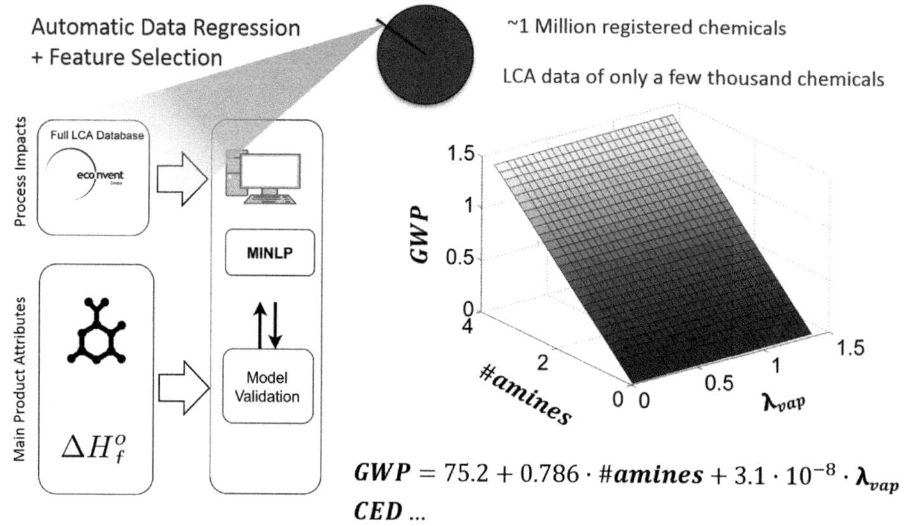

FIGURE 17.7 Problem statement of the environmental assessment of chemical case study.

In a recent work [50], Calvo-Serrano and co-workers developed an approach that predicts the cradle-to-gate life cycle impact of organic chemicals from their molecular structure and thermodynamic properties. In essence, our method based on a mixed-integer programming (MIP) systematically constructs shortcut predictive models of life cycle impact using chemical attributes that lead to better predictions and low risk of overfitting. The MIP models contain binary variables denoting whether a given attribute is included or not in the regression model and continuous ones representing regression coefficients (Fig. 17.7).

This modeling approach, which could be used to predict other properties, was applied to estimate the impact embodied in 83 chemicals from predictors chosen among 17 molecular descriptors and 15 thermodynamic properties. The impacts considered included the cumulative energy demand (CED), GWP, chemical oxygen demand (COD), biological oxygen demand (BOD5), total organic carbon (TOC), as well as Eco-Indicator 99 (EI99 Total) and the three impacts it aggregates, namely Human Health (EI99 HH), Ecosystem Quality (EI99 EQ), and Resources (EI99 Res).

The results showed that adding more attributes in the predictive models (thermodynamic in addition to structural properties) leads to lower errors and, therefore, better predictions. The impact categories CED, GWP, EI99 Total, EI99 HH, EI99 EQ, and EI99 Res were predicted successfully with relative errors ranging from 20% to 44%, while COD, BOD5, and TOC were much harder to predict and showed much larger errors (see Fig. 17.8). Overall, this approach provides good estimates of impact without having to carry out tedious data collection tasks, thereby simplifying the environmental assessment of chemicals.

4. CONCLUSIONS

In this chapter, we discussed the combined use of LCA and optimization, illustrating how it can be applied in the chemical industry. This method combines LCA, an effective tool to assess the environmental impact of a product or service over its life cycle, with MOO, employed to generate and screen alternatives considering economic and environmental aspects simultaneously. Three examples were presented, addressing problems at various scales from supply chain design to the environmental assessment of chemicals.

As we approach planetary boundaries, the need to operate safely under tighter environmental regulations will become more critical, making these tools essential in the transition toward a more sustainable chemical industry. Future efforts should focus on developing methods for incorporating the main uncertainties affecting the calculations, algorithms for solving the underlying mathematical models more efficiently, and assessment methods linking inventory flows related to products and processes to global ecological limits (i.e., planetary boundaries).

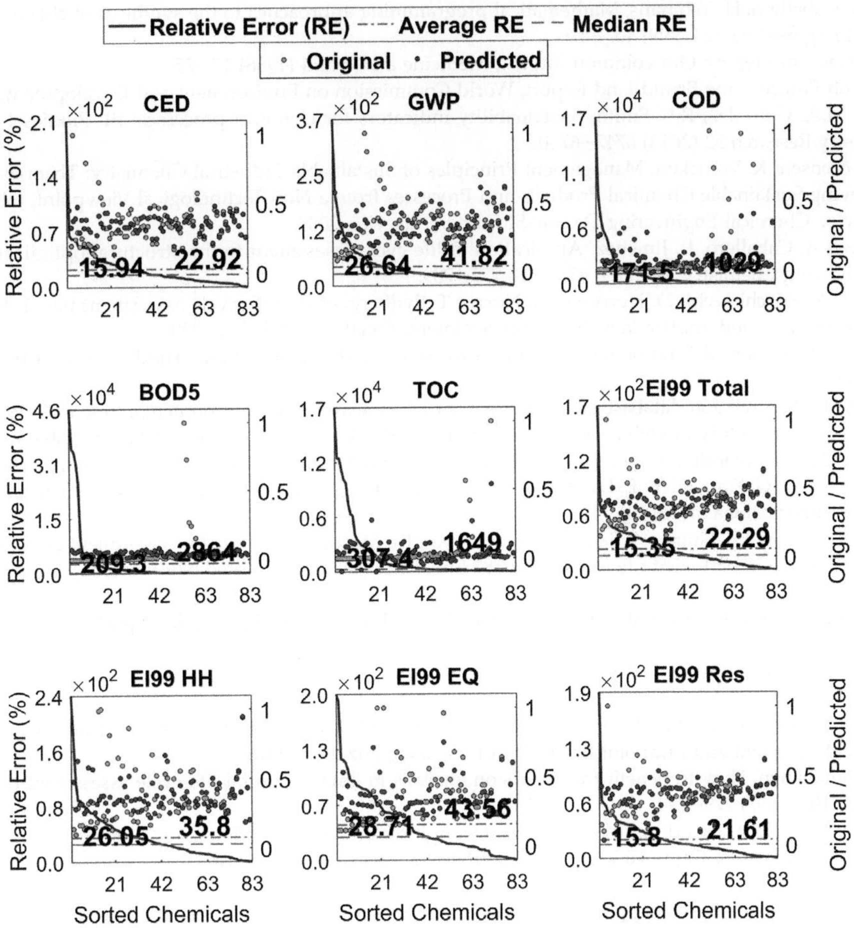

FIGURE 17.8 Absolute relative errors and predicted values for thermodynamic properties (TP), molecular descriptors (MD), and both combined (MD+TP) [50].

References

[1] J.F. Jenck, F. Agterberg, M.J. Droescher, Products and processes for a sustainable chemical industry: a review of achievements and prospects, Green Chemistry 6 (2004) 544–556.

[2] P.T. Anastas, J.C. Warner, Green Chemistry: Theory and Practice, vol. 30, Oxford University Press, Oxford, 2000.

[3] A. Nikolopoulou, M.G. Ierapetritou, Optimal design of sustainable chemical processes and supply chains: a review, Computers and Chemical Engineering 44 (2012) 94–103.

[4] J.W. Owens, Life-cycle assessment: constraints on moving from inventory to impact assessment, Journal of Industrial Ecology 1 (1997) 37–49.

[5] A. Azapagic, R. Clift, The application of life cycle assessment to process optimisation, Computers and Chemical Engineering 23 (1999) 1509–1526.

[6] M. Finkbeiner, A. Inaba, R. Tan, K. Christiansen, H.-J. Klüppel, The new international standards for life cycle assessment: ISO 14040 and ISO 14044, The International Journal of Life Cycle Assessment 11 (2006) 80–85.

[7] A. Azapagic, Life cycle assessment and its application to process selection, design and optimisation, Chemical Engineering Journal 73 (1999) 1–21.

[8] I.E. Grossmann, G. Guillen-Gosalbez, Scope for the application of mathematical programming techniques in the synthesis and planning of sustainable processes, Computers and Chemical Engineering 34 (2010) 1365–1376.

[9] C. Pieragostini, M.C. Mussati, P. Aguirre, On process optimization considering LCA methodology, Journal of Environmental Management 96 (2012) 43–54.

[10] J.-J. Wang, Y.-Y. Jing, C.-F. Zhang, J.-H. Zhao, Review on multi-criteria decision analysis aid in sustainable energy decision-making, Renewable and Sustainble Energy Reviews 13 (2009) 2263–2278.

[11] G.A. Davis, M. Swanson, S. Jones, Comparative Evaluation of Chemical Ranking and Scoring Methodologies, 1994. Prep by Univ Tennessee Cent Clean Prod Clean Technol United States Environ Prot Agency (EPA Order No 3N-3545-NAEX), Knoxville, TN.

[12] J. Pope, D. Annandale, A. Morrison-Saunders, Conceptualising sustainability assessment, Environmental Impact Assessment Review 24 (2004) 595–616.

[13] C.C. Jaeger, J.D. Tàbara, J. Jaeger, European Research on Sustainable Development: Volume 1: Transformative Science Approaches for Sustainability, vol. 1, Springer Science & Business Media, 2011.

[14] I.E. Grossmann, J.A. Caballero, H. Yeomans, Mathematical programming approaches to the synthesis of chemical process systems, Korean Journal of Chemical Engineering 16 (1999) 407–426.
[15] B.R. Keeble, The Brundtland report:'Our common future.', Medicine and War 4 (1988) 17–25.
[16] WCED, Our Common Future – the Brundtland Report, World Commission on Environment and Development, Geneva, 1987.
[17] G.J. Ruiz-Mercado, M.A. Gonzalez, R.L. Smith, Sustainability indicators for chemical processes: III. Biodiesel case study, Industrial and Engineering Chemistry Research 52 (2013) 6747–6760.
[18] G.L.L. Reniers, K. Sörensen, K. Vrancken, Management Principles of Sustainable Industrial Chemistry: Theories, Concepts and Indusstrial Examples for Achieving Sustainable Chemical Products and Processes from a Non-Technological Viewpoint, John Wiley & Sons, 2013.
[19] R.K. Sinnott, G. Towler, Chemical Engineering Design: SI Edition, Elsevier, 2009.
[20] G. Guillén-Gosálbez, J.A. Caballero, L. Jiménez, Application of life cycle assessment to the structural optimization of process flowsheets, Industrial and Engineering Chemistry Research 47 (2008) 777–789.
[21] G. Rebitzer, T. Ekvall, R. Frischknecht, D. Hunkeler, G. Norris, T. Rydberg, et al., Life cycle assessment part 1: framework, goal and scope definition, inventory analysis, and applications, Environment International 30 (2004) 701–720.
[22] A. Azapagic, R. Clift, Allocation of Environmental Burdens by Whole-System Modelling: The Use of Linear Programming, Alloc LCA, SETAC, Brussels, 1994.
[23] B. Steen, R. Carlson, SPINE a relation database structure for life cycle assessments, Environmental Research (1995).
[24] R. Frischknecht, Life cycle inventory modelling in the Swiss national database ECOINVENT 2000, Sustainble Information Society 2 (2001).
[25] S. Hellweg, L. Milà i Canals, Emerging approaches, challenges and opportunities in life cycle assessment, Science 344 (2014) 1109–1113.
[26] D.W. Pennington, J. Potting, G. Finnveden, E. Lindeijer, O. Jolliet, T. Rydberg, et al., Life cycle assessment part 2: current impact assessment practice, Environment International 30 (2004) 721–739.
[27] M. a J. Huijbregts, U. Thissen, J.B. Guinée, T. Jager, D. Kalf, D. Van De Meent, et al., Priority Assessment of Toxic Substances in Life Cycle Assessment. Part I: Calculation of Toxicity Potentials for 181 Substances With the Nested Multi-Media Fate, Exposure and Effects Model USES-LCA 41, 2000.
[28] Goedkoop M, Spriensma R. The Eco-Indicator 99: A Damage Oriented Method for Life Cycle Impact Assessment – Methodology Report 2000.
[29] J.B. Guinée, M. Gorrée, R. Heijungs, G. Huppes, R. Kleijn, L. van Oers, et al., Life Cycle Assessment: An Operational Guide to the ISO Standards, 2001.
[30] M.A.J. Huijbregts, Z.J.N. Steinmann, P.M.F. Elshout, G. Stam, F. Verones, M.D.M. Vieira, et al., ReCiPe 2016: A Harmonized Life Cycle Impact Assessment Method at Midpoint and Endpoint Level Report I: Characterization, 2016.
[31] P. Miettinen, R.P. Hämäläinen, How to benefit from decision analysis in environmental life cycle assessment (LCA), European Journal of Operational Research 102 (1997) 279–294.
[32] R.C. Pattison, C. Tsay, M. Baldea, Pseudo-transient models for multiscale, multiresolution simulation and optimization of intensified reaction/separation/recycle processes: framework and a dimethyl ether production case study, Computers and Chemical Engineering 105 (2017) 161–172.
[33] J. Steimel, S. Engell, Optimization-based support for process design under uncertainty: a case study, AIChE Journal 62 (2016) 3404–3419.
[34] J.A. Caballero, D. Milán-Yañez, I.E. Grossmann, Rigorous design of distillation columns: integration of disjunctive programming and process simulators, Industrial and Engineering Chemistry Research 44 (2005) 6760–6775.
[35] M. Skiborowski, M. Rautenberg, W. Marquardt, A hybrid evolutionary–deterministic optimization approach for conceptual design, Industrial and Engineering Chemistry Research 54 (2015) 10054–10072.
[36] A. Gonzalez-Garay, G. Guillen-Gosalbez, SUSCAPE: a framework for the optimal design of SUStainable ChemicAl ProcEsses incorporating data envelopment analysis, Chemical Engineering Research and Design 137 (2018) 246–264.
[37] M. Ehrgott, Multicriteria Optimization, vol. 491, Springer Science & Business Media, 2005.
[38] Y.Y. Haimes, L.S. Lasdon, D.A. Wismer, On a bicriterion formulation of the problems of integrated system identification and system optimization, IEEE Journals and Magazines 47 (1971) 296–297.
[39] B. Mota, M.I. Gomes, A. Carvalho, A.P. Barbosa-Povoa, Towards supply chain sustainability: economic, environmental and social design and planning, Journal of Cleaner Production 105 (2015) 14–27.
[40] F.D. Mele, A.M. Kostin, G. Guillén-Gosálbez, L. Jiménez, Multiobjective model for more sustainable fuel supply chains. A case study of the sugar cane industry in Argentina, Industrial and Engineering Chemistry Research 50 (2011) 4939–4958.
[41] F. You, L. Tao, D.J. Graziano, S.W. Snyder, Optimal design of sustainable cellulosic biofuel supply chains: multiobjective optimization coupled with life cycle assessment and input–output analysis, AIChE Journal 58 (2012) 1157–1180.
[42] A. Hugo, E.N. Pistikopoulos, Environmentally conscious long-range planning and design of supply chain networks, Journal of Cleaner Production 13 (2005) 1471–1491.
[43] A.D. Bojarski, J.M. Laínez, A. Espuña, L. Puigjaner, Incorporating environmental impacts and regulations in a holistic supply chains modeling: an LCA approach, Computers and Chemical Engineering 33 (2009) 1747–1759.
[44] G. Guillén-Gosálbez, I.E. Grossmann, Optimal design and planning of sustainable chemical supply chains under uncertainty, AIChE Journal 55 (2009) 99–121.
[45] G. Guillén-Gosálbez, F.D. Mele, I.E. Grossmann, A bi-criterion optimization approach for the design and planning of hydrogen supply chains for vehicle use, AIChE Journal 56 (2010) 650–667.
[46] G.R. Kocis, I.E. Grossmann, A modelling and decomposition strategy for the MINLP optimization of process flowsheets, Computers and Chemical Engineering 13 (1989) 797–819.

[47] J. Shatkin, W. Larsen, Improve environmental and economic performance with streamlined life cycle assessment (SLCA), Technical Proceedings of the 2011 NSTI Nanotechnology Conference and Expo, NSTI-Nanotech 1 (2011) 828–831, 2011.

[48] J.A. Todd, M.A. Curran, K. Weitz, A. Sharma, B. Vigon, E. Price, et al., Streamlined Life-Cycle Assessment: A Final Report From the SETAC North America Streamlined LCA Workgroup, July 1999, 31 pages. Rapp Dispon Sur, http://www Setac Org/Files/LcaPdf1999.

[49] R.G. Hunt, T.K. Boguski, K. Weitz, A. Sharma, Case studies examining LCA streamlining techniques, The International Journal of Life Cycle Assessment 3 (1998) 36.

[50] R. Calvo-Serrano, M. Gonzalez-Miquel, S. Papadokonstantakis, G. Guillén-Gosálbez, Predicting the cradle-to-gate environmental impact of chemicals from molecular descriptors and thermodynamic properties via mixed-integer programming, Computers and Chemical Engineering 108 (2018) 179–193.

Index

Note: 'Page numbers followed by "f" indicate figures, "t" indicate tables'.

H

I

N

O

P

Preface

The passing of time is not an impossible challenge for scientific observation. Astrophysicists are able to observe today what happened in the past. Using technology like the new James Webb telescope, they are able to observe light emitted from stars and thus events occurring billions of years ago in outer space. So too are paleoneurobiologists and evolutionary neuroscientists able, with ever-improving excavation and experimental techniques, to piece together structural changes and events during vertebrate brain evolution. Examples of these techniques are endocasts, hybrid induced pluripotent stem cells and organoids, and with more data and specimens being obtained from excavation sites and added to scientific databases, the pieces of the puzzles continue to be uncovered. How this knowledge leads to understanding the minds of modern humans and our recent or more distant ancestors and relatives is the topic of this book. What can fossils tell us about the evolution of our mind? When the information we obtain from fossils is limited, what technology is available to delve deeper into the minds of our extinct relatives? How can we study extant species, including our primate family members, to make deductions about our own minds? This volume of *Progress in Brain Research* is a peer-reviewed collection of original research data and literature reviews by leading experts in the fields of paleoneurobiology, evolutionary neuroscience, and evolutionary psychology, and the next seven chapters provide answers to those questions.

Sensory processing and brain size are themes of the first two chapters. Chapter 1 provides insight into the neuroanatomy of ornithopod dinosaurs. Lauters and colleagues' results detail the brain of ornithopod dinosaurs from Europe and Asia and deduce the sensory processing capabilities and brain size of these dinosaurs. Their findings are congruent with previous research on the complex group dynamics of these dinosaurs. For example, the relatively large cerebral hemispheres of Hadrosauridae allude to well-developed cognitive abilities in this clade, including an advanced intraspecific communication, gregarious life, and complex social life. The well-developed olfactory tracts in Iguanodontia and hadrosaurids point to enhanced olfaction, possibly to detect predators and/or members of their group. Lauters and colleagues also describe the techniques and technologies available in this field.

Chapter 2 describes in exquisite detail how the evolution of the sensory systems drove early mammalian brain evolution and provides an overview of the evolutionary roots of the most distinctive paleoneurological mammalian traits. Benoit and colleagues do this by reviewing species with a phylogenetic position at the origin of mammals: the Paleozoic and Mesozoic synapsids. Mammals went through a nocturnal bottleneck during the Mesozoic period that resulted in sensory and neural adaptations such as improved olfaction, audition, and somatosensation that led to growth and organizational complexity of the isocortex. This information is available in the fossil record. For example, size of the olfactory bulbs is indicated by the presence of a midline sulcus between the olfactory bulbs such as in basal

probainognathians and encephalization in crown-group Mammalia was accompanied by ossification of the cribriform plate. Mammals adapted to hearing a wider range of airborne frequencies, as we can see in the fossil record the miniaturization of the postdentary bones and their migration toward the skull, resulting in the progressive detachment of the auditory ossicles. The evolution of hair in mammaliaformes improved tactile sensitivity, which played a key role in the development of the sensory cortex and prompted the increase in size of the cerebral hemispheres that is directly visible on the endocranial casts of mammaliaformes. Taken together, these sensory adaptations resulted in brain changes such as the development of structures for the processing and integration of sensory inputs, such as the posterolateral expansion of the cerebral hemispheres and cerebellum and the increase in relative brain size characteristic of early mammals.

Chapters 3–5 introduce technologies available to make deductions about brain evolution without fossils. In Chapter 3 Ganapathee and Gunz review recent advances in genetic fields making it possible to uncover differences in neurodevelopment between humans and other primates, using pluripotent stem cells and brain organoids for example. Findings from these studies reveal that the development of human neurons occurs at a slower pace as compared to other primates and that protein coding genes (e.g., *SSTR2*) may play a role in differences in cortical development.

Chapter 4 shares insight and data on how behavioral and neuroimaging research on living humans is able to shed light on how our ancestors' brains may have functioned (e.g., made the first tools or spoke the first words). Kulik and colleagues' results on how functional neuroimaging studies into language and tool use overlap in Broca's area confirm that language likely coevolved with hand movements related to tool use.

Chapter 5 compares sulcal and gyral organization of Wernicke's area across extant human brains, extant chimpanzee brains, and fossil hominin endocasts and describes their similarities and differences related to language. Limitations aside, a lot can be learned by studying the surface of the brain in humans and our ancestors and the findings of Hill and colleagues indicate that the evolution of Wernicke's area is more nuanced than Broca's area. The gyral and sulcal morphology shows that humans have a larger Wernicke's area and leftward Wernicke's area asymmetry, chimpanzee brains have neural precursors of human left hemispheric language specialization, and that *Paranthropus robustus* may have attained a complexity of vocal communication beyond that of both chimpanzees and *Australopithecus*.

Language is not the only human-specific characteristic discussed in this volume, as Chapters 3, 6, and 7 review the literature and discuss these in detail. Chapter 3 discusses the genetic basis of human-specific brain size, shape, and neurodevelopment in comparison to nonhuman primates and Neanderthals. An important theme of this chapter is the idea of human-specific brain function being linked to our vulnerability to certain disorders such as mental health disorders and substance use disorders. In Chapter 7, Rigby Dames and colleagues discuss evidence for the human-specific vulnerability to age-related neurodegenerative disorders such as Alzheimer's and Parkinson's disease from an evolutionary perspective. Examples of this evidence are that the Lewy bodies formed from α-sy are only observed in

humans with Parkinson's disease and are not naturally present in nonhuman primates. Reasons given for this are that dopaminergic neurons might have higher metabolic requirements in human brains and that unmyelinated axons of the dopaminergic neurons of the substantia nigra must travel farther due to the expansion of the brain in humans, placing strains on dopaminergic neurons' connections within motor circuits. Rapid brain evolution in the human lineage has also been posited to explain susceptibility to Alzheimer's disease. A genetic risk factor for Alzheimer's disease, the allele *APOE4*, has been discovered in DNA from Denisovan fossils, which could indicate that if archaic humans had an extended lifespan they might have developed neurodegenerative diseases.

Chapter 6 also discusses the lateralization of brain functions in humans. Hemispheric specialization increased in hominins along with the emergence of handedness, tool use, tool making, and language. Humans have increased functional and developmental modularization of the brain in comparison to nonhuman primates and our recent ancestors, which may account for improved visuospatial processing and motor attention, among other functions reviewed by Bardo and colleagues.

Chapter 8 discusses change in brain size in human evolution and gives an introduction to phylogenetic comparative methods that provide the necessary tools to understand how it relates to change in body size. Pestana and colleagues show that, although fossil hominin brain sizes have increased over time, recent discoveries suggest the pattern of encephalization has been much more complex, and phylogenetic comparative methods provide the necessary tools to model the ups and downs in brain size evolution.

This remarkable collection of research is a snapshot of the current state of the art of paleoneurobiology and evolutionary neuroscience. Complex sensory processing capabilities and complex group dynamics predated even mammals. The evolution of sensory processing capabilities necessary to survive in an ever-changing environment drove mammalian brain evolution. Chimpanzees possess early adaptations for language production yet modern humans' Broca's and Wernicke's areas are uniquely specialized. This extraordinary capability and neural adaptation for complex communication may have been derived from early tool use and led to the lateralization of our brains. This is one proposed explanation for human-specific hemispheric lateralization. Modern humans have extraordinary capabilities and brains; our brains are bigger, shaped differently, and even develop differently compared to other hominins and nonhuman primates. Our social and cultural evolution has altered how our brain functions and these changes have also led to our vulnerabilities to certain diseases such as mental health and substance use disorders as well as neurodegenerative diseases. These and other ideas are discussed in detail in this volume.

The present volume is an output of the workshop "From Fossils to Mind" that was held online on November 30 and December 1, 2021. The workshop was originally planned to be held in South Africa. With the pandemic and travel restrictions, we unfortunately had to go for an online format hosted by the Southern African Neuroscience Society. However, even in its virtual form, we strongly believe that this workshop, and the present book, clearly reflect the leading role of South Africa, and the whole continent, in the search for the origins of the modern human brain.

Since the discovery of the natural endocast of the Taung child, the African fossil record has expanded significantly, and it is noteworthy that most discussions about the hominin brain still revolve around the African material. We believe that through the keynotes and presentations of this workshop, the public lecture delivered by Dr. Mirriam Tawane from the Ditsong National Museum of Natural History, and the support of the Southern African Neuroscience Society, this workshop and content of this book emphasize the quality and importance of the research developed in Africa in neurobiology. We would like to sincerely thank our funders, the Centre of Excellence in Palaeosciences, now GENUS, the Palaeontological Scientific Trust, and the French Institute of South Africa, and we would also like to thank their CEOs Dr. Christine Steininger, Andrea Leenen, and Dr. Sophie Dulucq for their trust and confidence. Initial support was provided by The Global Academy of Liberal Arts (GALA) outreach fund. We are particularly proud of and grateful for their support (Fig. 1).

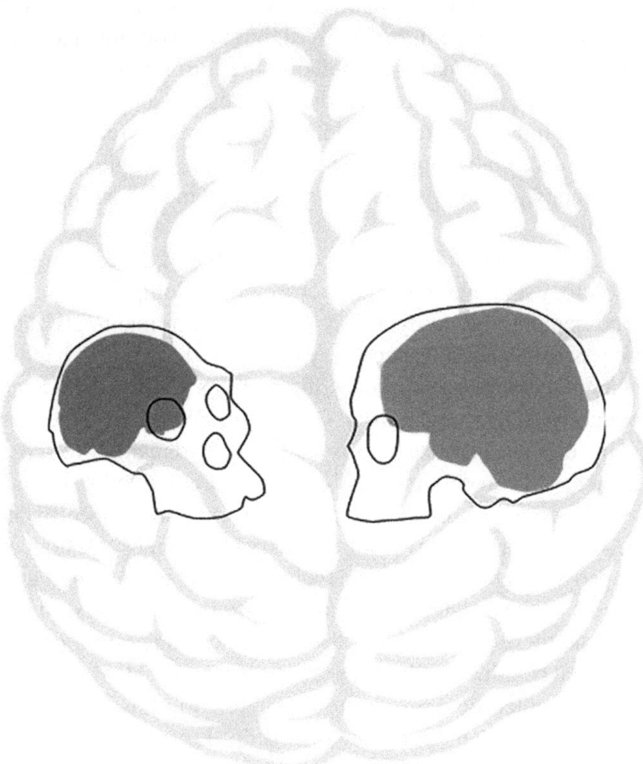

FIG. 1

This image represents the theme of our workshop and this volume. Information on the organizers, presenters, and the YouTube videos of the presentations can be found on the Southern African Neuroscience Website at https://saneurosoc.co.za/fromfossilstomind.

Many questions remain unresolved, but the present volume and the 2021 workshop demonstrate that much progress has been made since the earliest research in the field, and that promising methods and protocols are being developed to improve our knowledge of how such a complex organ could have evolved through time. The diversity of the topics covered by this volume, from neuroscience to paleontology, also proves that addressing the long-standing questions related to the timing and process of brain evolution will only be possible through a multidisciplinary approach. Studying how brains evolved in distinct taxa, such as in dinosaurs and primates, is essential to unravel the underlying mechanisms of brain evolutionary history. Thanks to recent developments in genetics and cell biology, we can now better understand how macroscopic characters emerged in the brain and the interplay between phenotype and genotype. Also, the content of these chapters makes it clear that advances in neuroimaging and neurosciences benefit the entire field, and not only the study of living humans. The combination of such diverse research topics contributes to enriching our field and makes it very special.

Finally, in addition to a dedicated session during our workshop, most of the chapters in the present volume were led by students and early career scientists, and the quality of their research is undeniable. We can be confident in the future, as their promising work will surely lead to amazing discoveries.

CHAPTER

Endocasts of ornithopod dinosaurs: Comparative anatomy

1

Pascaline Lauters[a,b,*], Martine Vercauteren[a], and Pascal Godefroit[b]

[a]*Université Libre de Bruxelles, Brussels, Belgium*

[b]*Royal Belgian Institute of Natural Sciences, Brussels, Belgium*

[*]*Corresponding author: e-mail address: plauters@hotmail.com*

Abstract

Ornithopod dinosaurs were a successful group before they became extinct at the end of the Cretaceous. They were present on every continent, though they were rare in the Southern Hemisphere. We present the results of our work on the brain of these dinosaurs as an attempt to determine which evolutionary trends affected it. Old and new technologies allow us to peer into the skull of long extinct animals and retrieve information about their brain. First we provide a short description of the brain of ornithopod dinosaurs from Europe and Asia, then we sum up the characteristics that can be gathered from it. The presence of valleculae helps us to assess the actual size of the brain with more confidence. The olfactory peduncles are large and these animals had a good sense of smell. There is a trend toward an increase in the size of the cerebral hemispheres, and a more straight-lined brain. The latter can be the result of the ontogeny and the size achieved by the adult animal on the development of the brain. Other characteristics, like the development of the cerebral hemispheres and the encephalization quotient, allude to Hadrosauridae having had cognitive abilities more developed than previously assumed. This is in adequacy with other data from the physical characteristics (e.g., crests) and the social life (e.g., living in herds, communal nests) of these dinosaurs, which denote high and complex behaviors like care for their young, sexual courtship, and gregariousness.

Keywords

Endocranial cavity, Brain, Dinosaurs, Cerebral hemispheres, Neuroanatomy, Encephalization quotient, Behavior, Ornithopoda, Iguanodontia, Hadrosauridae

1 Introduction

The brain itself is seldom preserved by fossilization, but the braincase, its internal cavity, surfaces, and the foramina give a wealth of information about the lost organ, including its shape and the ratio between its different parts. The first paleontologists had to rely on natural endocranial casts to study the brain of fossil species. These

Progress in Brain Research, Volume 275, ISSN 0079-6123, https://doi.org/10.1016/bs.pbr.2022.10.002

finds are uncommon, but new techniques gradually expanded the number of specimens that can be studied (Boonstra, 1968; Edinger, 1929; Olson, 1944; Radinsky, 1968). First made of plaster and requiring the partial or complete destruction of the original specimen, endocranial casts, or endocasts, were later made of elastic material like latex or silicone. Since its early days in the 1980s (Conroy and Vannier, 1984; Haubitz et al., 1988), the practice of using computed tomography scans (CT scans) vastly expanded the opportunities to obtain reconstructions of the endocranial cavity of even more specimens, sometimes too fragile or too big to be examined by previous techniques (Racicot, 2017).

One of the challenges encountered when studying the brain of extinct animals is to confidently determine its actual shape. As noted by Edinger (1929) vascular tissues and meninges are found alongside the brain in the endocranial cavity. In most mammals, birds (Iwaniuk and Nelson, 2002) and in pterosaurs (Witmer et al., 2003), the brain fills nearly all of the endocranial space; meninges and other tissues are reduced. This is not true for other clades, where the percentage of endocranial space occupied by the brain will vary depending on the clade, sex and age of the individual (Hopson, 1979). Hence it is estimated that the brain of crocodiles will fill from 50% to 75% of the endocranial space (Hopson, 1979; Jerison, 1969, 1973). This has led in the past to consider that the brain of dinosaurs would occupy but 50% of the endocranial cavity (Hopson, 1977; Hurlburt, 1996; Jerison, 1973) and reinforced the idea that the brain of dinosaurs was disproportionately small, justifying their demise in favor of the endothermic and big-brained mammals (Colbert, 1961; Swinton, 1970).

An endocast, whether physical or virtual, provides a useful method to study the brain of extinct animals. Traditionally the description of an endocast uses terms describing actual brain features, even if it may not faithfully reproduce the brain. Even being a fragile organ and prone to a rapid deterioration after death, the brain will leave its imprint in the skull and methods allow us to study it while making the risk of damage to the fossil as minimal as possible. Valleculae are anatomical grooves; in this article we will use it specifically to describe the depressions left by blood vessels in the cranial walls. Thanks to the discovery of valleculae (Evans, 2005; Godefroit et al., 2012a; Lauters et al., 2013; Osmólska, 2004), it has been attested that the shape of the endocranial cavity gives us a valuable proxy of the brain of extinct species. Watanabe et al. (2018) concomitantly established that endocasts are good proxies of the brain in alligators and birds.

The attention of paleoneurologists was first attracted to theropods, who are phylogenetically close to birds. Studying their brain would allow us to identify the origin of the avian brain. Ornithopods were a clade of ornithischians who emerged during the Lower Jurassic. They had a horny beak and the small species were bipeds. Ornithopods became larger and quadrupedal later, while probably retaining the ability to walk bipedally in some circumstances (e.g., running to escape predation or any other danger). They became a very successful group during the Cretaceous, and were present on every continent, particularly in North America and in Asia.

We studied the brains of ornithopod dinosaurs, more particularly species from Europe and Asia who had not yet been the objects of such research. We sought to

find out whether the evolutionary patterns observed in the lineage of theropod dinosaurs and birds were replicated in this other dinosaur lineage that had a very different fate. Did the brain of ornithopod dinosaurs remain stable or were there changes over time? What can it tell us about this group who experienced a remarkable diversification during 160 million years (Horner et al., 2004)? Can it help us to understand their behavior and how they were living?

2 Material

First, we will briefly describe the endocasts of a crocodile and of several species of ornithopods. Secondly, we will describe features of interest that were observed on the endocasts.

The specimens used in this article are reported in Table 1. All the specimens were adults. The sex could not be determined.

2.1 Description of the endocranial casts

For a more extended description of the endocasts, please refer to Lauters et al. (2013), Lauters et al. (2012), Godefroit et al. (2012a,b), Lauters et al. (2013). Fig. 1 illustrates the endocasts described. Fig. 2 presents the relationships between the specimens.

2.1.1 Crocodylus porosus

The *Crocodylus porosus* specimen was included to serve as a basal outgroup. As an extant representative of archosaurians, its brain accounts for the primitive archosaurian condition (Beyrand et al., 2019). The olfactory tracts are thin and long, while the olfactory bulbs could not be described as they are located more anteriorly and are not constrained by bony features. Some parts of the brain are indistinguishable because they are obscured by the casts of venous sinuses covering the dorsal and lateral surfaces of the organ. The cerebral hemispheres are slightly enlarged. The brain displays cranial and pontine flexures.

2.1.2 Iguanodon bernissartensis

I. bernissartensis was a basal iguanodontian dinosaur from the Early Cretaceous. Several specimens from the Belgian collections were CT-scanned but they unfortunately showed poor details; the resolution of the scan did not allow an accurate reconstruction of small anatomical features, and some braincases were deteriorated after taphonomic processes and oxidation of pyrite, abundant in this deposit, which created artifacts that made the reconstruction more difficult. The olfactory peduncle appears wide. The cerebral hemispheres are enlarged and round. The overall shape of the brain is straight.

Table 1 List of specimens. Institutional abbreviations: AEHM or AENM: Amur Natural History Museum of the Far Eastern Institute of Mineral Resources, FEB RAS, Blagoveschenk, Russia; RBINS or IRSNB: Institut Royal des Sciences Naturelles de Belgique, Brussels, Belgium; MNHN: Muséum National d'Histoire Naturelle, Paris, France; MRAC: Musée Royal d'Afrique Centrale, Tervuren, Belgium.

Specimen	Geological age	Type of data	Institution	Catalog number of the specimen (catalog number of the cast)
Lurdusaurus arenatus	Early Cretaceous (Aptian)	Silicone endocast	MNHN	MNHN·F 1966–15. GDF 1700
Iguanodon bernissartensis	Early Cretaceous (Late Barremian—Early Aptian)	CT scan	RBINS	RBINS R51
Iguanodon bernissartensis	Early Cretaceous (Late Barremian—Early Aptian)	CT scan	RBINS	RBINS R54
Mantellisaurus atherfieldensis	Early Cretaceous (Late Barremian—Early Aptian)	CT scan	RBINS	RBINS R57
Batyrosaurus rozhdestvenskyi	Late Cretaceous (Santonian-Campanian)	Silicone endocast	AEHM	AEHM 4/1
Amurosaurus riabinini	Late Cretaceous (Maastrichtian)	Silicone endocast	AEHM	AEHM 1/232 (IRSNB R 279)
Amurosaurus riabinini	Late Cretaceous (Maastrichtian)	Silicone endocast	AEHM	AEHM 1/90 (IRSNB R 281)
Amurosaurus riabinini	Late Cretaceous (Maastrichtian)	Silicone endocast	AEHM	AEHM 1/233 (IRSNB R 280)
Kundurosaurus nagornyi	Late Cretaceous (late Maastrichtian)	CT scan	AEHM	AENM 2/121
Crocodylus porosus	Holocene	Silicone endocast	MRAC	unknown

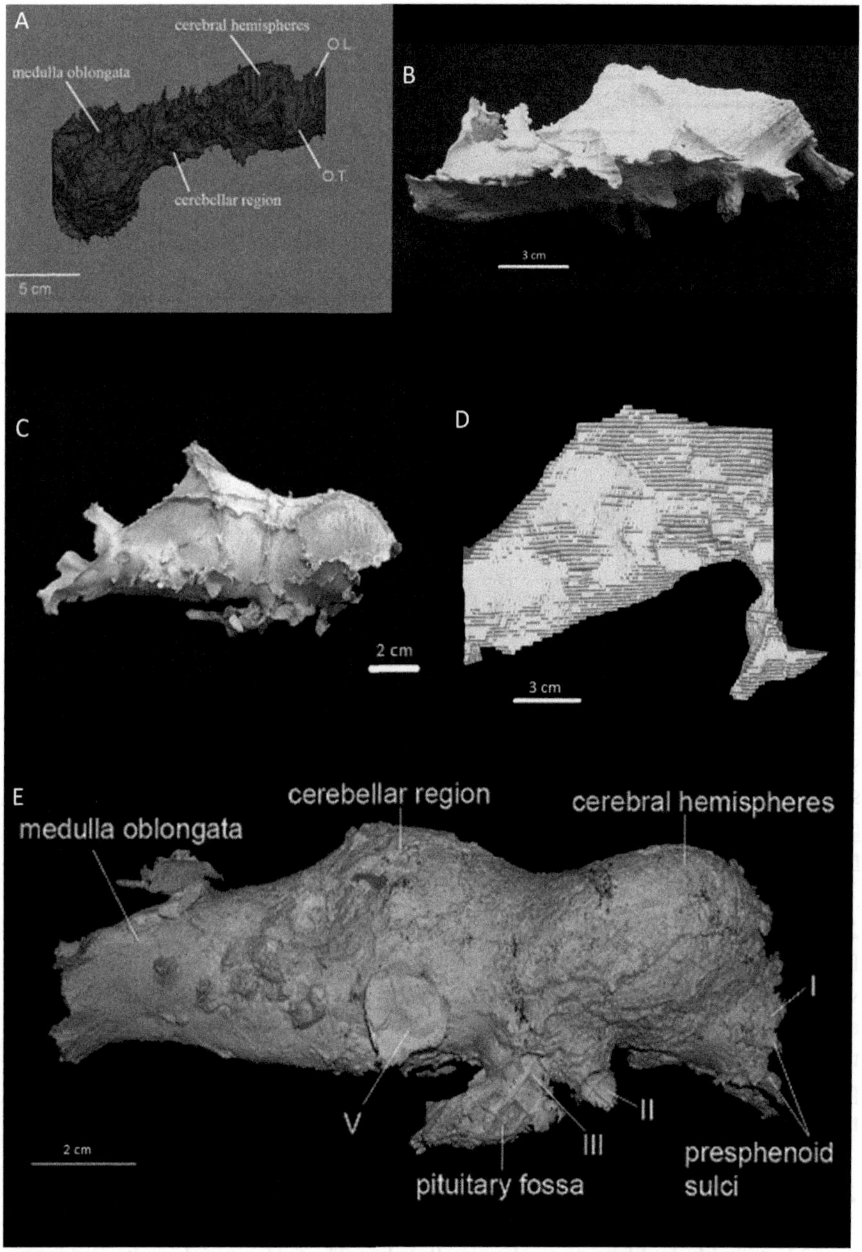

FIG. 1

Endocranial reconstructions, in right lateral view, of *Iguanodon bernissartensis* (A), *Lurdusaurus arenatus* (B; from Lauters, 2013), *Batyrosaurus rozhdestvenskyi* (C), *Kundurosaurus nagornyi* (D) and *Amurosaurus riabinini* (E; from Lauters et al., 2013).

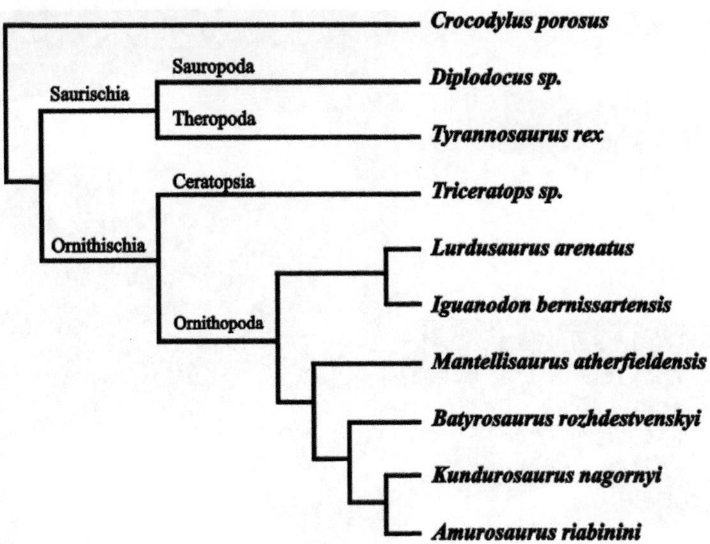

FIG. 2

Diagram of the phylogenetic relationships of the taxa discussed in the paper.

2.1.3 *Mantellisaurus atherfieldensis*

The *Mantellisaurus atherfieldensis* specimen studied was found in Bernissart alongside the numerous *I. bernissartensis* specimens. Although the skull looks well preserved from the outer side, the endocranium is partially crushed and some parts could simply not be reconstructed. However, a few observations can be made. The flexures must have been more prominent than in *Iguanodon*. The cerebral hemispheres seem not much larger than the medulla oblongata. The olfactory tracts are enlarged.

2.1.4 *Lurdusaurus arenatus*

Discovered in Niger, this massive iguanodontian from the Early Cretaceous is not well known. The endocranial cavity is straight. The cerebral hemispheres are long and oval in shape. The position of the inner ear is not marked by a profound constriction, as it is in other ornithopods. The pituitary fossa is very small. The olfactory peduncle is very thin. The midbrain and hindbrain are marked by a pronounced triangular peak after the cerebral hemispheres.

2.1.5 *Batyrosaurus rozhdestvenskyi*

This specimen is an hadrosauroid from the Upper Cretaceous of Kazakhstan. This is the oldest ornithopod known to show valleculae. The endocranial cavity is narrow, except the anterior part which is round and enlarged. The cerebral hemispheres are broad and slightly flattened dorsoventrally. The endocranial cavity is marked by two constrictions: the first one is located behind the telencephalon and the second one is behind the large triangular peak above the midbrain region.

2.1.6 *Kundurosaurus nagornyi*

Kundurosaurus is a saurolophine Lambeosaurinae from the late Maastrichtian of Russia. Unfortunately, the braincase is incomplete and the reconstruction is restricted to the posterior part of the brain, missing the cerebral hemispheres. The midbrain is constricted. Flexures are not marked. Even if it was incomplete, we can observe a large pituitary body. The endocast shares similarities with North American *Gryposaurus* endocasts (Ostrom, 1961).

2.1.7 *Amurosaurus riabinini*

Amurosaurus is a primitive Lambeosaurinae from the late Maastrichtian of Russia. Its brain was typical of those of hadrosaurids, with wide cerebral hemispheres and a large olfactory peduncle. Valleculae were also observed on the internal face of the frontals of several specimens. The position of the inner ear is marked by a profound constriction behind the midbrain. The flexures are greatly reduced.

3 Methods

Two different types of datasets were acquired for this study: silicone endocranial casts and CT scan datasets.

Silicone casts were made without damaging the specimens, and provided long lasting and stable reconstructions (Lauters, 2013). Moreover, this material is easy to use, nontoxic, and a good cast can be made in a few days at most. The specimen is ideally completely clean. The foramina and fractures were covered with modeling clay, except the foramen magnum. A thin coat of petroleum jelly was sprayed in the endocranial cavity for easy removal of the cast later. Then layers of liquid silicone were applied, taking care to let the product dry between applications. This technique allowed the strengthening of the cast and prevented potential tears. When the last layer was completely dry, the cast was pulled out of the specimen. The braincase could then be cleaned and the cast reinforced with an internal coat of resin.

Computed tomography (or CT scan), devised in 1976, is largely used in the medical field as a non-invasive procedure for diagnostic and therapeutic purposes. An X-ray source and a detector rotates around the patient producing cross-sectional images of the body. Some of the X-rays are absorbed or scattered by the specimen. The pattern is characteristic of the density of the materials making up the specimen. Each image represents a «slice» of the specimen and many of them are taken during the procedure, only requiring a few seconds. All the images are sent to a computer to reconstruct the «slices» of the tissues. Virtual slices through our specimens were generated, allowing visualization of every point at the desired axis in various planes (Carlson et al., 2003). Moreover, three-dimensional objects were generated. This technique provided greater clarity and more information than conventional X-rays, and in a very short time (Witmer et al., 2008).

CT scans have been used in paleontology more widely since the 1990s and the 2000s (Brochu, 2000, 2003; Domínguez Alonso et al., 2004; Franzosa, 2004; Franzosa and Rowe, 2005; Kundrát, 2007; Larsson, 2001; Macrini et al., 2006; Maisey, 2005; Marino et al., 2003; Rogers, 1998, 1999, 2005). It has allowed access to anatomical details otherwise inaccessible to study without damaging the specimen, and to expand the amount of data studied by providing information from specimens otherwise too fragile to be molded. The digitalization has also been useful to further study our specimens and share it with other scientists. It, however, required access to a CT-scanner, and numerous hours on a computer with sufficient RAM and video cards to handle the software and process the data.

The specimens of *Iguanodon* and *Mantellisaurus* were scanned at the Gasthuisberg Hospital (Belgium) in a Siemens Sensation 64 device with a slice thickness of 1 mm and a 0.5 mm overlap. The specimen of *Kundurosaurus* was scanned in the Amur Region Hospital (Russia) in a Siemens Emition scanner with a slice thickness of 3 mm and a 1.5 mm overlap.

3.1 Telencephalon volume/endocranial volume (CRV)

This measure consists of the ratio between the volume of cerebral hemispheres on the total volume of the brain (Larsson et al., 2000). This indicator assesses the importance of the telencephalon in the development of the brain. The volume of hemispheres can be calculated with the volume formula of the ellipsoid. In case of virtual reconstructions, total volume can be provided by the software (e.g., Amira, Mimics, ArteCore). For physical endocasts, different techniques can be employed (e.g., mathematical calculation based on the measurements, water displacement).

3.2 Reptile encephalization quotient (REQ)

The REQ is an adaptation by Hurlburt (1996) of the infamous equation devised by Jerison (1979). This equation is as follows:

$$\mathrm{REQ} = \mathrm{M_{Br}} / \left(0,\, 0155^{*}\mathrm{M_{Bd}}^{0,553}\right)$$

where M_{Br} is the mass of the brain and M_{Bd} is the mass of the body. To obtain the mass of the brain for an extinct animal, the volume must be multiplied by 1.036 g/mL (Stephan, 1960). The mass of the body was determined by using the formula of Anderson et al. (1985).

3.3 Ratio width of the cerebral hemispheres/total length of the brain.

The largest width of the brain is found at the cerebral hemispheres (Franzosa, 2004; pers. obs.). The length of the brain is measured from the base of the olfactory peduncle until the foramen magnum. This measurement can be easier to use than the CRV because it does not require complete endocasts or reconstructions.

4 Results

We review here different characteristics and observations that can be inferred from the study of the endocranial casts of the specimens. Please refer to Fig. 1 for the various structures, and to Table 2 and Table 3.

4.1 Olfactory peduncle (Fig. 1A, B, E)

Olfactory peduncles were not available for all specimens studied and some were incomplete or damaged. We could, however, observe the olfactory tract in some of the studied specimens and infer some conclusions from it. The ancestral state is represented by a thin and long peduncle as observed in *Hypsilophodon foxii*, *Tenontosaurus tilletti* (Galton, 1989) and *Lurdusaurus arenatus* (Lauters, 2013).

Table 2 REQ values for various dinosaurs (data for: Hadrosaurinae and *Hypacrosaurus*: Evans et al., 2009; *Diplodocus*, *Ceratosaurus*, *Allosaurus*, *Acrocanthosaurus*, *Citipati*, *Tyrannosaurus* and *Troodon*: Franzosa, 2004; *Giganotosaurus*: Coria and Currie, 2002; *Nigersaurus*: Sereno et al., 2007; *Brachiosaurus*: Knoll and Schwarz-Wings, 2009; *Psittacosaurus*, *Protoceratops* and *Triceratops*: Zhou et al., 2007; *Leaellynasaura*: Rich and Rich, 1988), from Lauters et al. (2013).

Genus	REQ
Iguanodon bernissartensis	1,85–3,25
Mantellisaurus atherfieldensis	1,66–2,6
Lurdusaurus arenatus	1,3
Amurosaurus riabinini	2,3–3,8
Diplodocus	0,53–0,69
Nigersaurus	0,4–0,8
Brachiosaurus	0,17–0,896
Psittacosaurus	0,31–3,4
Protoceratops	0,19
Triceratops	0,11
Leaellynasaura	1,8
Hadrosaurinae	2,8
Hypacrosaurus	2,3–3,7
Ceratosaurus	3,31–5,07
Allosaurus	2,4–5,24
Acrocanthosaurus	2,75–5,92
Giganotosaurus	1,9
Citipati	3,6
Tyrannosaurus	5,44–7,63
Troodon	7,76

Table 3 Comparative table of the ratio telencephalon volume/endocranial volume (data for *Lambeosaurus, Corythosaurus* and *Hypacrosaurus* are from Evans et al., 2009), from Lauters et al. (2013).

Genus	CRV
Iguanodon	19%
Lurdusaurus	19%
Amurosaurus	30%
Lambeosaurus (juvenile)	38%
Corythosaurus (juvenile)	45%
Corythosaurus (juvenile)	38%
Hypacrosaurus	43%
Carcharodontosaurus	24%
Tyrannosaurus	33%
Conchoraptor	43%
Archaeopteryx	45%

In Hadrosauridae, peduncles are relatively large like in *Amurosaurus riabinini* (Lauters et al., 2013) and *Hypacrosaurus altispinus*, *Corythosaurus sp.*, *Lambeosaurus sp.* (Evans et al., 2009) and *Arenysaurus ardevoli* (Cruzado-Caballero et al., 2015). The peduncles appear nearly as large as the cerebral hemispheres. However, the size of the olfactory bulbs in ornithopods is small when compared to other dinosaurs (Evans et al., 2009). The size of olfactory bulbs are correlated to the development of olfaction (Rombaux et al., 2009), therefore, it seems that smell became important in derived Ornithopoda.

4.2 Pituitary gland (Fig. 1B–E)

The pituitary gland, or hypophysis, is the production and regulation site of several hormones and plays an important role for the body. This gland is more particularly involved in the production of growth hormone and a disorder can lead to diverse pathologies (Ambler, 2002; Evans et al., 1933; Gelber et al., 1992; Lopis et al., 1968; Zimmerman et al., 1993). Edinger (1964) proposed that the size of the pituitary gland could be correlated to the size of the adult individual. A proportional relation can indeed be observed between the relative size of the pituitary gland to brain volume; small-bodied species display a small pituitary gland, whereas the gland is larger in very large species (Galton, 1989; Giffin, 1989; Nopsca, 1917; Sander et al., 2011; Sereno et al., 2007). A reduced pituitary gland represents the ancestral state, according to the fact that more primitive species were smaller than more derived taxa. Indeed, a larger pituitary gland can be observed in several large taxa like *Batyrosaurus rozhdestvensky* (Godefroit et al., 2012a), *Kundurosaurus nagornyi*

(Godefroit et al., 2012b) and *Amurosaurus riabinini* (Lauters et al., 2013). Surprisingly, the pituitary fossa of *Lurdusaurus arenatus* is rather small (Lauters, 2013); this is unexpected as this dinosaur displayed a fairly large body size.

4.3 Flexures

The brain of most dinosaurs is marked by two flexures: the cranial flexure and the pontine flexure (Brochu, 2003; Galton, 1989; Hopson, 1979). These flexures are also apparent in crocodilians. They tend to disappear in some ornithischians (Giffin, 1989), especially in derived ornithopods where the endocranial cavity becomes straight. The disappearance of flexures is generally a consequence of the increase in body size; the braincase growing at a faster pace than the brain, having more room it will «unfold» in bigger individuals. To the contrary, if space is limited, flexures are clearly visible. The most extreme example of this can be observed in birds and primates where the brain will adopt a globular shape.

4.4 Floccular fossa (Fig. 1C–E)

The floccular fossa is the depression where the inner ear is housed. This depression is absent or barely noticeable in some groups (Franzosa, 2004) but quite apparent in most ornithopod dinosaurs. It is especially noticeable in more ancestral species like *Batyrosaurus rozhdestvenskyi* (Godefroit et al., 2012a), *Kundurosaurus nagornyi* (Godefroit et al., 2012b) and *Amurosaurus riabinini* (Lauters et al., 2013).

4.5 Reptile encephalization quotient (REQ) (Table 2)

Whenever it was possible, we used the measurements taken from the same individual. When specimens from bonebed were used (Lauters et al., 2008), it is important to select the bones that will be measured as belonging to the same ontogenetic stage than the braincase that provided the endocast. The values of REQ were compiled with those from Evans et al. (2009). The obtained average is of 2.5. Iguanodonts display a REQ under this value: 2.22 for *Iguanodon bernissartensis* and 2 for *Lurdusaurus arenatus*. Typically Hadrosauridae but also *Mantellisaurus atherfieldensis* have higher REQ (respectively 3.8 for *Amurosaurus* and >2.6 for *Mantellisaurus*).

4.6 Telencephalon volume/endocranial volume (CRV) (Table 3)

Among the specimens that we had the opportunity to study, we could notice a discontinuity in the results between around 20%. We therefore proposed to divide the taxa into two groups: those who have a CRV below 20% and those who have a CRV above 20%. This is a modification compared to the one proposed by Larsson et al. (2000). Typically, iguanodonts have a low CRV (19%) while hadrosaurids like *Amurosaurus riabinini* have a higher CRV (30%).

4.7 Ratio width of the cerebral hemispheres/total length of the brain

Dissimilarly to what Wharton (2002) has proposed, we chose to make a distinction between the ornithopods who have a ratio below 35% and those who have a ratio above 35% since that there is a clear hiatus between 30% and 40%. Some iguanodonts have narrow cerebral hemispheres (*Iguanodon bernissartensis*: 28.6%, *Lurdusaurus arenatus*: 26.4%). In more derived species, the cerebral hemispheres are on the contrary broad (*Mantellisaurus atherfieldensis*: 43.4%, *Batyrosaurus rozhdestvenskyi*: 45%, *Amurosaurus riabinini*: 44.7%).

5 Discussion

In summary, the brain of derived ornithopods was characterized by very large, and usually round, cerebral hemispheres. Their brain was straight, with very weak flexures, opposite to what is observed in saurischians. The olfactory peduncles were wide. The shape of the cerebral hemispheres is reflected in the curved form of the frontals. The presence of valleculae is an indication that the shape of the cerebral hemispheres on the endocranial cast of hadrosaurids is a replica of the shape of the actual cerebral hemispheres. They are approximately as long as they are wide. They progressively became bigger throughout the lineage, as the REQ that is higher than in ancestral species. The part of the brain behind the cerebral hemispheres is usually narrow and houses the inner ear.

5.1 Optic lobes

Optic lobes of ornithopods cannot be identified on the endocast. We can therefore postulate that their relative size was small when compared to the rest of the brain. This was also observed in basal theropod dinosaurs (Franzosa, 2004). In theropods, the size of the optic lobes and of the cerebral hemispheres increases throughout the lineage, meanwhile the olfactory bulbs will display the opposite trend and decrease. Franzosa (2004) established a correlation between these changes with a diminution of the body size in theropod dinosaurs. While becoming smaller, they underwent a dietary shift toward smaller and faster prey like insects and small vertebrates. This trend correlated to a better hand-eye coordination to capture these. In birds, but also in pterosaurs (Witmer et al., 2003), the development of sight could be connected to adaptations to flight.

5.2 Olfactory tracts

The size of the olfactory bulbs can be linked to the development of olfaction (Rombaux et al., 2009). Considering the size of the olfactory bulbs and olfactory tracts in later ornithopods, it appears that olfaction became quite important in the group. Unfortunately, it is often quite difficult to reconstruct the olfactory bulbs of extinct dinosaurs, though the olfactory tracts can very often be observed quite

easily. These are well developed and we can postulate that Iguanodontia and hadrosaurids had a good sense of smell (possibly to detect predators and/or intraspecific identification). As noted by Evans et al. (2009), their olfactory bulbs are not as developed as in other clades like ceratopsians (Witmer and Ridgely, 2008; Zhou et al., 2007) and theropods (Hopson, 1979; Zelenitsky et al., 2008, 2011). The olfactory bulbs of basal Theropoda were large (Brochu, 2000; Hopson, 1979; Zelenitsky et al., 2008, 2011) and, as suggested by Franzosa (2004), they had a keen sense of smell, probably used to locate prey or food. The relative size of the olfactory bulbs in theropods decreased throughout the lineage while sight will become more prominent in birds. The similar size of olfactory bulbs between Lambeosaurinae and Saurolophinae suggests that olfaction did not play a part in the hypertrophy of the nasal cavity and in the crest of Lambeosaurinae (contra Saveliev et al., 2012).

5.3 Flexures

The ancestral condition of the brain, as seen in crocodilians, is marked by two flexures: the cranial flexure and the pontine flexure. These flexures can be observed in Saurischia and in most Ornithischia (Brochu, 2003; Galton, 1989; Hopson, 1979). The brain has thus a distinct shape with the anterior part of the brain standing higher than the other parts. Though the flexures tend to lessen and disappear in some Ornithischians (Giffin, 1989), especially in derived ornithopods where the endocranial cavity becomes straight. The two main causes of the variation in the size of the angles are the absolute size of the skull and the relative size of the eyes (Giffin, 1989). Small bodied ornithischians usually have an endocast with flexures more prominent than those of large bodied animals. This can also be noted ontogenetically, as described by Lautenschlager and Hübner (2013), where the brain of young individuals have very noticeable flexures whereas the brain of older and thus larger individuals have reduced or indistinguishable flexures. The disappearance of flexures is therefore, generally, a consequence of the increase of the body size. Likewise, big eyes compress the occipital region and the brain (Starck, 1979) and flexures become apparent. Relatively big eyes being a juvenile trait, a small skull and relatively big eyes are often paired. The very marked flexures are well correlated to both a small body size and relatively big eyes. This is notably observed in birds and in derived theropods (Franzosa, 2004).

5.4 Cerebral hemispheres and encephalization quotient

Intelligence is a broad concept and has been defined by various characteristics like self-awareness, and the abilities to learn, apply logic, recognize patterns, form concepts, solve problems, memorize various information, and use language. A relatively bigger brain has very often been associated to a higher degree of "intelligence" and to a higher capacity to process information. The paradigm postulates that intelligent animals will be more successful at making fast and profitable decisions and discovering novel resources. Gregariousness may contribute to higher chances of survival

by sharing food, protecting young or detecting predators earlier. Intelligence would thus provide an evolutionary benefit.

The relationship between brain mass and body mass has been studied by several scientists (Dubois, 1897; Jerison, 1969, 1973; Platel, 1976, 1979). The mass of the brain does not increase as fast as the mass of the body, but will follow this rule: brain mass = factor * body mass$^{2/3}$. Regression lines are similar between vertebrate lineages but the value of the factor will vary (Platel, 1979). These allometric regressions provide a way to predict what is the expected brain mass for a vertebrate with a given mass. However, to estimate the body mass of an extinct animal is a challenge owing to the incompleteness of their skeletal remains and potential bias depending on the techniques used. Jerison (1969) published a chart in which he plotted the regression area for several reptiles. Thanks to this chart, he demonstrated that dinosaurs did not display a relatively small endocranial cavity, contrarily to what was previously assumed (Marsh, 1879). The encephalization quotient established by Jerison (1973) has been used to compare the size of the brain between different species and clades while avoiding the bias of their largely different body sizes. If the ratio is above 1, meaning that the brain is larger than the expected mathematical mean, the animal will be deemed "intelligent" and able to perform complex behaviors. Jerison (1969) had divided animals in two groups, namely the ectotherms and the endotherms, of which the latter has a higher ratio of brain weight/body weight; Hopson (1979) concluded that the encephalization quotient of dinosaurs lied between these two subdivisions. A relationship between the size of the brain and behavioral complexity has been postulated in extant birds (Madden, 2001) and in extant lizards (Platel, 1976, 1979).

In extinct animals, the exact mass and volume of the brain cannot be asserted with certainty as we are not certain how much of the endocranial space was actually occupied by the brain. These assumptions have been countered by Osmólska (2004) and Evans (2005) who were the first to report the presence of valleculae in Oviraptoridae, and in Hadrosauridae and Pachycephalosauridae respectively. Valleculae appear when the brain is coming in close contact with the bone, meninges and vascular tissues being thin. From the observation of valleculae on several ornithopod endocasts (Godefroit et al., 2012a; Lauters et al., 2013), we can confirm that the brain, and more particularly the cerebral hemispheres, were in close contact with the internal surface of the endocranial cavity and thus filled more than 50% of the endocranial space. When we could distinguish valleculae, we followed Evans et al. (2009) and estimated that the brain occupied 60% of the endocranial space.

The REQ of ornithopod dinosaurs is largely superior to those of Sauropoda (Franzosa, 2004; Sereno et al., 2007) and of Ceratopsia (Zhou et al., 2007) (see Table 2). It is though often inferior to those of some Theropoda (Franzosa, 2004). The REQ of *Amurosaurus riabinini* is close to the one calculated for North-American hadrosaurids (like *Hypacrosaurus altispinus*) even if it was a basal member of the lineage (Godefroit et al., 2004).

Bauchot et al. (1977) remarked that a low encephalization quotient could be related to a passive strategy when it comes to evade predators. Interestingly, the REQ of *Protoceratops* appears superior to those of *Triceratops*. But *Protoceratops* was

rather vulnerable and had to rely on its vigilance and speed to escape predators, while *Triceratops* was larger and protected by its frill and its horns (Sakagami and Kawabe, 2020). Ornithopods rarely had defensive weapons and had to rely on their senses and on their speed in the race to avoid their predators (Scott Persons, 2011). Giant theropods were fast but would tire easily and could not run over long distances; on the other hand Hadrosauridae could maintain their speed longer and outrun their predators (Scott Persons, 2011).

Larsson et al. (2000) highlighted an increase in the anterior part of the brain in theropod dinosaurs. The values of REQ for derived ornithopods are high and are even on par with those of some theropods (Table 2). This is quite remarkable and represents a convergence in the increase of the anterior part of the brain within the lineage.

5.5 CRV

One of the distinct features of the brain of ornithopods is the enlargement of the cerebral hemispheres (Giffin, 1989). Their size increases throughout the lineage, they become more round and larger. Valleculae are observed on the bones surrounding this part of the brain. We can compare the size of the telencephalon to the rest of the brain in different groups (see Table 3). The cerebral hemispheres of Iguanodontia were not developed, while those of Hadrosauridae were as much if not more developed than those of theropods. So the value of CRV of *Corythosaurus* and *Hypacrosaurus* is similar to those of Maniraptora, a group well known for being high-brained and to which scientists have assigned complex behaviors like the ability to hunt in packs, to take care of their eggs/young, and the presence of sexual dimorphism and thus the high probability of sexual display (Bakker, 1986; Maxwell and Ostrom, 1995; Scott Persons et al., 2013). The CRV of *Amurosaurus riabinini* is slightly inferior to those of the North-American Lambeosaurinae, but higher than those of Iguanodontia.

5.6 Behavior

All the elements coming from the description and the measurements on the endocasts, taken together, can help us make some hypotheses about the evolution of behaviors and biology of ornithopod dinosaurs.

Ornithopoda, and especially Hadrosauridae, had a high REQ which competes with the REQ of some Theropoda, the group from which birds originated. It has been postulated that Theropoda could be gregarious (Barco et al., 2006; Ibiricu et al., 2013; Kobayashi and Lü, 2003; Varricchio et al., 2008a), hunt in packs (Currie, 1998; Currie et al., 2005; Roach and Brinkman, 2007) and would take care of their offspring (Varricchio et al., 1999, 2008b).

Similarly, Hadrosauridae lived and traveled in groups (Currie, 1983; Lockley and Matsukawa, 1999). Some, like *Maiasaura peeblesorum*, would gather in big colonies to nest together (Horner, 1982; Horner and Makela, 1979) and it has been proposed that the young were altricial and had to be fed and protected by the adults.

One of the remarkable features of Hadrosauridae is their cranial crest; in Lambeosaurinae the crest was connected to air chambers and would amplify their calls. The development of the cranial crest is related to sexual dimorphism and vary with the ontogenetic stage of the individual (Dodson, 1975; Evans et al., 2009; Hopson, 1975; Weishampel, 1981). It was used in visual and auditory communication within a group, eventually transmitting signals about the presence of danger, used during sexual parade, and/or allowing the reinforcement of group cohesion. It could play a role in intraspecific as well as in extraspecific recognition (Dodson, 1975; Hopson, 1975; Weishampel, 1981). These elements are coherent with animals being gregarious, as demonstrated by the discoveries of bonebeds and track sites.

A well-developed social and gregarious life has been linked to a relatively bigger brain (Pérez-Barbería and Gordon, 2005; Shultz and Dunbar, 2010). The high REQ of Hadrosauridae coincides with this hypothesis; gregarious life and complex intraspecific communication require high cognitive abilities. One of the advantages of gregariousness is the avoidance of predators: as more individuals are present, they have more opportunities to detect the predator on time, an important parameter for animals relying on running endurance to escape predation (Scott Persons, 2011). A herd will also increase the chances of reproduction and the protection of vulnerable individuals.

A bigger telencephalon has often been associated to a more complex behavior in vertebrates (Hopson, 1977; Hurlburt, 1996; Jerison, 1969, 1973). Individuals living in a group are subjected to social interactions, implying a quick analysis and a good memory of past interactions with the members of the group, especially for species living in complex hierarchical organization. These cognitive processes are important to maintain the individual position within the group (Dunbar, 1992; Dunbar and Bever, 1998; Pérez-Barbería and Gordon, 2005; Shultz and Dunbar, 2010). Concerning the CRV, an increase of the volume of the telencephalon in relation to the brain volume can be observed between the Iguanodontia and the Hadrosauridae. It may be that this increase represents changes in the behavior or the ecology of the group, like for example, herds with higher number of individuals, intraspecific communication and more complex behaviors (Dodson, 1975; Horner and Makela, 1979; Lockley and Matsukawa, 1999; Varricchio and Horner, 1993; Weishampel, 1981).

It has been proposed that the size of the brain is correlated with the size of the group of animals that live together (Marino et al., 2000), due to a potential relationship between the size of the neocortex and cognitive abilities such as intraspecific recognition (Dunbar, 1998). Marino et al. (2007) observed that the expansion of different zones of the neocortex in cetaceans is linked to high cognitive functions like attention, social consciousness, judgment and intuition. This evolution has been observed as well in Primates and in birds, albeit from radically different trajectories (Marino et al., 2003). Though the neocortex is a distinctive feature of mammals, convergent organization has been observed in birds and some reptiles (Jarvis et al., 2005). The relatively big size of the brain and cerebral hemispheres in Lambeosaurinae is coherent with the supposed complex behaviors attributed to these species: gregariousness, courtship rituals, intraspecific communication, complex nests and potentially care of the young, at least for some period of time.

Even in modern archosaurs complex behaviors have been observed and commented: territorial behavior (Campbell et al., 2013), hierarchy, courtship, extensive care for the eggs/offspring, mutual assistance in obtaining food (Lang, 1975; Pooley and Gans, 1976), high cognitive abilities (Leal and Powell, 2012). We surmise that various lineages of dinosaurs had also achieved sociality and complex cognitive abilities.

The specimens studied in this work, coming from Europe, North Africa and Asia, show a similar morphology to the species already studied in North America. The brain of this lineage of dinosaurs had a rather conservative morphology. The cerebral hemispheres of Hadrosauridae were as big as those of Maniraptora, alluding to well-developed cognitive abilities in this clade. We attribute these abilities to the complex behavioral patterns of ornithopods, including an advanced intraspecific communication, gregarious life and complex social life.

Acknowledgments

The authors express gratitude to Amélie Beaudet, Tanya Calvey and Alexandra de Sousa for their invitation to participate in this volume and the thoughtful comments on this manuscript. This work was part of a Ph.D. thesis at the Université Libre de Bruxelles thanks to a grant from the Fonds pour la Formation à la Recherche dans l'Industrie et l'Agriculture and the Fonds Van Buuren. The authors thanks the colleagues that have impacted this study through discussions and providing access to CT scan data, especially the people from the following institutions: Gasthuisberg Hospital (Leuven, Belgium), Institut Royal des Sciences Naturelles (Brussels, Belgium), Musée Royal d'Afrique Centrale (Tervuren, Belgium), Amur Natural History Museum of the Far Eastern Institute of Mineral Resources (Blagoveschenk, Russia), Muséum National d'Histoire Naturelle (Paris, France).

References

Ambler, G., 2002. Overgrowth. Best Pract. Res. Clin. Endocrinol. Metab. 16 (3), 519–546.

Anderson, J.F., Hall-Martin, A., Russell, D.A., 1985. Long-bone circumference and weight in mammals, birds and dinosaurs. J. Zool. 207 (1), 53–61.

Bakker, R.T., 1986. The Dinosaur Heresies. Bantam Books, New York.

Barco, J.L., Canudo, J.I., Ruiz-Omeñaca, J.I., 2006. New data on *Therangospodus oncalensis* from the Berrasian Fuentesalvo tracksite (Villar del Río, Soria, Spain): an example of gregarious behaviour in theropod dinosaurs. Ichnos 13, 237–248.

Bauchot, R., Bauchot, M.L., Platel, R., Ridet, J.M., 1977. Brain of Hawaiian tropical fishes: brain size and evolution. Copeia 1977, 42–46.

Beyrand, V., Voeten, D.F.A.E., Bureš, S., Fernandez, V., Janáček, J., Jirák, D., Rauhut, O., Tafforeau, P., 2019. Multiphase progenetic development shaped the brain of flying archosaurs. Sci. Rep. 9, 10807.

Boonstra, L.D., 1968. The braincase, basicranial axis and median septum in the Dinocephalia. Ann. S. Afr. Mus. 50, 195–273.

Brochu, C.A., 2000. A digitally-rendered endocast for *tyrannosaurus rex*. J. Vertebr. Paleontol. 20, 1–6.

Brochu, C.A., 2003. Osteology of *tyrannosaurus rex*: insights from a nearly complete skeleton and high-resolution computed tomographic analysis of the skull. Soc Vertebr. Paleontol. Memoir 7, 1–138.

Campbell, H.A., Dwyer, R.G., Irwin, T.R., Franklin, C.E., 2013. Home range utilization and long-range movement of estuarine crocodiles during the breeding and nesting season. PLoS One 8 (5), e62127.

Carlson, W.D., Rowe, T., Ketcham, R.A., Colbert, M.W., 2003. Applications of high-resolution X-ray computed tomography in petrology, meteoritics and palaeontology. In: Mees, F., Swennen, R., Van Geet, M., Jacobs, P. (Eds.), Applications of X-Ray Computed Tomography in the Geosciences. vol. 215. Geological Society, London, pp. 7–22. Special Publications.

Colbert, E.H., 1961. Dinosaurs: Their Discovery and Their World. Dutton, New York.

Conroy, G.C., Vannier, M.W., 1984. Noninvasive three-dimensional computer imaging of matrix-filled fossil skulls by high-resolution computed tomography. Science 226, 456–458.

Coria, R.A., Currie, P.J., 2002. The braincase of *giganotosaurus carolinii* (Dinosauria: Theropoda) from the upper cretaceous of Argentina. J. Vertebr. Paleontol. 22 (4), 802–811.

Cruzado-Caballero, P., Fortuny, J., Llacer, S., Canudo, J.I., 2015. Paleoneuroanatomy of the European lambeosaurine dinosaur *Arenysaurus ardevoli*. PeerJ 3, e802.

Currie, P.J., 1983. Hadrosaur trackways from the lower cretaceous of Canada. Acta Palaeontol. Pol. 28 (1–2), 63–73.

Currie, P.J., 1998. Possible evidence of gregarious behavior in tyrannosaurids. Gaia 15, 271–277.

Currie, P.J., Trexler, D., Koppelhus, E.B., Wicks, K., Murphy, N., 2005. An unusual multi-individual tyrannosaurid bonebed in the two medicine formation (late cretaceous, Campanian) of Montana (USA). In: Carpenter, K. (Ed.), The Carnivorous Dinosaurs. Indiana University Press, Bloomington, pp. 313–324.

Dodson, P., 1975. Taxonomic implications of relative growth in lambeosaurine hadrosaurids. Syst. Zool. 24, 37–54.

Domínguez Alonso, P., Milner, A.C., Ketcham, R.A., Cookson, M.J., Rowe, T.B., 2004. The avian nature of the brain and inner ear of *archaeopteryx*. Nature 430, 666–669.

Dubois, E., 1897. Sur le rapport du poids de l'encéphale avec la grandeur du corps chez les mammifères. Bull. Mém. Soc. Anthropol. Paris 8, 337–376.

Dunbar, R.I.M., 1992. Neocortex size as a constrain on group-size in Primates. J. Hum. Evol. 22 (6), 469–493.

Dunbar, R.I.M., 1998. The social brain hypothesis. Evol. Anthropol. 6 (5), 178–190.

Dunbar, R.I.M., Bever, J., 1998. Neocortex size predicts group size in carnivores and some insectivores. Ethology 104 (8), 695–708.

Edinger, T., 1929. Die fossilen Gehirne. Ergeb. Anat. Entwicklungsgesch. 28, 1–249.

Edinger, T., 1964. Recent advances in paleoneurology. Prog. Brain Res. 6, 147–160.

Evans, D.C., 2005. New evidence on brain-endocranial cavity relationships in ornithischian dinosaurs. Acta Palaeontol. Pol. 50 (3), 617–622.

Evans, H.M., Meyer, K.F., Simpson, M.E., 1933. The growth- and gonad-stimulating hormones of the anterior hypophysis. In: LLC (Ed.), Memoirs of the University of California. vol. 11. Literary Licensing, pp. 1–446.

Evans, D.C., Witmer, L.M., Ridgely, R.C., 2009. Endocranial anatomy of lambeosaurine dinosaurs: a sensorineural perspective on cranial crest function. Anat. Rec. 292, 1315–1337.

Franzosa, J.W., 2004. Evolution of the Brain in Theropoda (Dinosauria). Ph. D. dissertation, University of Texas, Austin, Texas.

Franzosa, J.W., Rowe, T.B., 2005. Cranial endocast of the cretaceous theropod dinosaur *Acrocanthosaurus atokensis*. J. Vertebr. Paleontol. 25, 859–864.

Galton, P., 1989. Crania and endocranial casts from ornithopod dinosaurs of the families Dryosauridae and Hypsilophodontidae (Reptilia: Ornithischia). Geol. Palaeontol. 23, 217–239.

Gelber, S.J., Heffez, D.S., Donohoue, P.A., 1992. Pituitary gigantism caused by growth hormone excess from infancy. J. Pediatr. 120 (6), 931–934.

Giffin, E., 1989. Pachycephalosaur paleoneurology (Archosauria: Ornithischia). J. Vertebr. Paleontol. 9, 67–77.

Godefroit, P., Bolotsky, Y.L., Van Itterbeeck, J., 2004. The lambeosaurine dinosaur *Amurosaurus riabinini*, from the Maastrichtian of far eastern Russia. Acta Palaeontol. Pol. 49 (4), 585–618.

Godefroit, P., Bolotsky, Y.L., Lauters, P., 2012a. A new Saurolophine dinosaur from the latest cretaceous of far eastern Russia. PLoS One 7 (5), e36849.

Godefroit, P., Escuillié, F., Bolotsky, Y.L., Lauters, P., 2012b. A new basal hadrosauroid dinosaur from the upper cretaceous of Kazakhstan. In: Godefroit, P. (Ed.), Bernissart Dinosaurs and Early Cretaceous Terrestrial Ecosystems. Indiana University Press, Bloomington, pp. 335–358.

Haubitz, B., Prokop, M., Dohring, W., Ostrom, J.H., Wellnhofer, P., 1988. Computed tomography of *archaeopteryx*. Paleobiology 14 (2), 206–213.

Hopson, J.A., 1975. The evolution of cranial display structures in hadrosaurian dinosaurs. Paleobiology 1, 21–43.

Hopson, J.A., 1977. Relative brain size and behavior in archosaurian reptiles. Annu. Rev. Ecol. Syst. 8, 429–448.

Hopson, J.A., 1979. Paleoneurology. In: Gans, C., Northcutt, R.C., Ulinski, P. (Eds.), Biology of the Reptilia. Neurology A, vol. 9. Academic Press Inc., New York, pp. 39–146.

Horner, J.R., 1982. Evidence of colonial nesting and 'site fidelity' among ornithischian dinosaurs. Nature 297, 675–676.

Horner, J.R., Makela, R., 1979. Nest of juveniles provides evidence of family structure among dinosaurs. Nature 282, 296–298.

Horner, J.R., Weishampel, D.B., Forster, C.A., 2004. Hadrosauridae. In: Weishampel, D.B., Dodson, P., Osmólska, H. (Eds.), The Dinosauria, second ed. University of California Press, Berkeley, pp. 438–463.

Hurlburt, G.R., 1996. Relative Brain Size in Recent and Fossil Amniotes: Determination and Interpretation. Ph. D. dissertation, University of Toronto, Canada.

Ibiricu, L.M., Martínez, R.D., Casal, G.A., Cerda, I.A., 2013. The behavioral implications of a multi-individual bonebed of a small theropod dinosaur. PLoS One 8 (5), e64253.

Iwaniuk, A.N., Nelson, J.E., 2002. Can endocranial volume be used as an estimate of brain size in birds? Can. J. Zool. 80 (1), 16–23.

Jarvis, E.D., Güntürkün, O., Bruce, L., Csillag, A., Karten, H., Kuenzel, W., Medina, L., Paxinos, G., Perkal, D.J., Shimizu, T., Striedter, G., Wild, J.M., Ball, G.F., Dugas-Ford, J., Durand, S.E., Hough, G.E., Husband, S., Kubikova, L., Lee, D.W., Mello, C.V., Powers, A., Siang, C., Smulders, T.V., Wada, K., White, S.A., Yamamoto, K., Yu, J., Rainer, A., Butler, A.B., 2005. Avian brains and a new understanding of vertebrate brain evolution. Nat. Rev. Neurosci. 6 (2), 151–159.

Jerison, H., 1969. Brain evolution and dinosaur brains. Am. Nat. 103, 575–588.

Jerison, H., 1973. Evolution of the Brain and Intelligence. Academic Press Inc., New York.

Jerison, H., 1979. The evolution of diversity in brain size. In: Hahn, M.E., Jensen, C., Dudek, B.C. (Eds.), Development and Evolution of Brain Size. Academic Press, New York, pp. 29–57.

Knoll, F., Schwarz-Wings, D., 2009. Palaeoneuroanatomy of *brachiosaurus*. Ann. Paleontol. 95 (3), 165–175.

Kobayashi, Y., Lü, J.-C., 2003. A new ornithomimid dinosaur with gregarious habits from the late cretaceous of China. Acta Palaeontol. Pol. 48 (2), 235–259.

Kundrát, M., 2007. Avian-like attributes of a virtual brain model of the oviraptorid *Conchoraptor gracilis*. Naturwissenschaften 94, 499–504.

Lang, J.W., 1975. The Florida crocodile: will it survive? Field Mus. Nat. Hist. Bull. 46 (8), 4–9.

Larsson, H.C.E., 2001. Endocranial anatomy of Carcharodontosaurus saharicus(Theropoda: Allosauroidea) and its implications for theropod brain evolution. In: Tanke, D.H., Carpenter, K. (Eds.), Mesozoic Vertebrate Life. Indiana University Press, Bloomington, pp. 19–33.

Larsson, H.C.E., Sereno, P.C., Wilson, J.A., 2000. Forebrain enlargement among nonavian dinosaurs. J. Vertebr. Paleontol. 20 (3), 615–618.

Lautenschlager, S., Hübner, T., 2013. Ontogenetic trajectories in the ornithischian endocranium. J. Evol. Biol. 26 (9), 2044–2050.

Lauters, P., 2013. Structure générale et évolution du cerveau chez les dinosaures ornithischiens. Ph. D. dissertation, Université Libre de Bruxelles, Bruxelles.

Lauters, P., Bolotsky, Y.L., Van Itterbeeck, J., Godefroit, P., 2008. Taphonomy and age profile of a latest cretaceous dinosaur bone bed in far eastern Russia. Palaios 23, 153–162.

Lauters, P., Coudyzer, W., Vercauteren, M., Godefroit, P., 2012. The brain of *iguanodon* and *Mantellisaurus*: perspectives on ornithopod evolution. In: Godefroit, P. (Ed.), Bernissart Dinosaurs and Early Cretaceous Terrestrial Ecosystems. Indiana University Press, Bloomington, pp. 213–224.

Lauters, P., Vercauteren, M., Bolotsky, Y.L., Godefroit, P., 2013. Cranial Endocast of the Lambeosaurine Hadrosaurid *Amurosaurus riabinini* from the Amur region, Russia. PLoS One 8 (11), e78899.

Leal, M., Powell, B.J., 2012. Behavioral flexibility and problem-solving in a tropical lizard. Biol. Lett. 8, 28–30.

Lockley, M.G., Matsukawa, M., 1999. Some observations on trackway evidence for gregarious behavior among small bipedal dinosaurs. Palaeogeogr. Palaeoclimatol. Palaeoecol. 150, 25–31.

Lopis, S., Rubenstein, A.H., Wright, A.D., 1968. Measurements of serum growth hormone and insulin in gigantism. J. Clin. Endocrinol. Metabol. 28 (3), 393–398.

Macrini, T.E., Rowe, T.B., Archer, M., 2006. Description of a cranial endocast from a fossil platypus, *Obdurodon dicksoni* (Monotremata, Ornithorhynchidae), and the relevance of endocranial characters to monotreme monophyly. J. Morphol. 267, 1000–1015.

Madden, J., 2001. Sex, bowers and brains. Proc. R. Soc. Lond. B 268, 833–838.

Maisey, J.G., 2005. Braincase of the upper Devonian shark (Chondrichthyes, Elasmobranchii), with observations on the braincase in early chondrichthyans. Bull. Am. Mus. Nat. Hist. 288, 1–103.

Marino, L., Uhen, M.D., Frohlich, B., Aldag, J.M., Blane, C., Bohaska, D., Whitmore, F.C., 2000. Endocranial volume of mid-late Eocene Archaeocetes (order: Cetacea) revealed by computed tomography: implications for cetacean brain evolution. J. Mamm. Evol. 7 (2), 81–94.

Marino, L., Uhen, M.D., Pyenson, N.D., Frohlich, B., 2003. Reconstructing cetacean brain evolution using computed tomography. Anat. Rec. 272B, 107–117.

Marino, L., Connor, R.C., Fordyce, R.E., Herman, L.M., Hof, P.R., Lefebvre, L., Lusseau, D., McGowan, B., Nimchinksy, E., Pack, A.A., Rendell, L., Reidenberg, J.S., Reiss, D., Uhen, M.D., Van der Gucht, E., Whitehead, H., 2007. Cetaceans have complex brains for complex cognition. PLoS Biol. 5 (6), e139.

Marsh, O.C., 1879. History and methods of palaeontological discovery. Am. J Sci. Arts series 3 (8), 66–67.

Maxwell, W.D., Ostrom, J.H., 1995. Taphonomy and paleobiological implications of *Tenontosaurus-Deinonychus* associations. J. Vertebr. Paleontol. 15 (4), 707–712.

Nopsca, F., 1917. Die Riesenformen unter den Dinosauriern. Centralbl.Min. Geol. Paleont. 1917, 332–351.

Olson, E.C., 1944. Origin of mammals based upon cranial morphology of the therapsid suborders. Geol. Soc. Am. Spec. Paper 55, 1–136.

Osmólska, H., 2004. Evidence on relation of brain to endocranial cavity in oviraptorid dinosaurs. Acta Palaeontol. Pol. 49 (2), 321–324.

Ostrom, J.H., 1961. Cranial morphology of the hadrosaurian dinosaurs from North America. Bull. Am. Mus. Nat. Hist. 122, 33–186.

Pérez-Barbería, F.J., Gordon, I.J., 2005. Gregariousness increases brain size in ungulates. Oecologia 145 (1), 41–52.

Platel, R., 1976. Analyse volumétrique comparée des principales subdivisions encéphaliques chez les reptiles sauriens. J. Hirnforsch. 17, 513–517.

Platel, R., 1979. Brain weight-body weight relationships. In: Gans, C., Northcutt, R.C., Ulinski, P. (Eds.), Biology of the Reptilia. Neurology A, vol. 9. Academic Press Inc., New York, pp. 147–171.

Pooley, A.C., Gans, C., 1976. The Nile crocodile. Sci. Am. 234 (4), 114–124.

Racicot, R., 2017. Fossil secrets revealed: X-ray CT scanning and applications in paleontology. J. Palaeontol. Soc. Paper 22, 21–38.

Radinsky, L.B., 1968. A new approach to mammalian cranial analysis, illustrated by examples of prosimian primates. J. Morphol. 124, 167–179.

Rich, T.H., Rich, P.V., 1988. A juvenile dinosaur brain from Australia. Natl. Geogr. Res. 4 (2), 148.

Roach, B.T., Brinkman, D.L., 2007. A reevaluation of cooperative pack hunting and gregariousness in *Deinonychus antirrhopus* and other nonavion theropod dinosaurs. Bull. Peabody Mus. Nat. Hist. 48, 103–138.

Rogers, S.W., 1998. Exploring dinosaur neuropaleobiology: viewpoint computed tomography scanning and analysis of an *Allosaurus fragilis* endocast. Neuron 21 (4), 673–679.

Rogers, S.W., 1999. *Allosaurus*, crocodiles, and birds: evolutionary clues from spiral computed tomography of an endocast. Anat. Rec. 257 (5), 162–173.

Rogers, S.W., 2005. Reconstructing the behaviors of extinct species: an excursion into comparative paleoneurology. Am. J. Med. Genet. 134A, 349–356.

Rombaux, P., Duprez, T., Hummel, T., 2009. Olfactory bulb volume in the clinical assessment of olfactory dysfunction. Rhinology 47, 3–9.

Sakagami, R., Kawabe, S., 2020. Endocranial anatomy of the ceratopsid dinosaur *triceratops* and interpretations of sensory and motor function. PeerJ 8, e9888.

Sander, P.M., Christian, A., Clauss, M., Fechner, R., Gee, C.T., Griebeler, E.-M., Gunga, H.-C., Hummel, J., Mallison, H., Perry, S.F., Preuschoft, H., Rauhut, O.W.M., Remes, K., Tütken, T., Wings, O., Witzel, U., 2011. Biology of the sauropod dinosaurs: the evolution of gigantism. Biol. Rev. 86, 117–155.

Saveliev, S.V., Alifanov, V.R., Bolotsky, Y.L., 2012. Brain anatomy of *Amurosaurus riabinini* and some neurological peculiarities of duck-billed dinosaurs. Paleontol. J. 46 (1), 79–91.

Scott Persons, W., 2011. Duckbills on the run: comparing the cursorial abilities of hadrosaurs and tyrannosaurs. In: Braman, D.R., Eberth, D.A., Evans, D.C., Taylor, W. (Eds.), International Hadrosaur Symposium Abstract Volume. Royal Tyrell Museum of Paleontology, Drumheller, pp. 117–121.

Scott Persons, W., Currie, P.J., Norell, M.A., 2013. Oviraptorosaur tail forms and functions. Acta Palaeontol. Pol. 59 (3), 553–567.

Sereno, P.C., Wilson, J.A., Witmer, L.M., Whitlock, J.A., Maga, A., Ide, O., Rowe, T.A., 2007. Structural extremes in a Cretaceous dinosaur. PLoS One 2 (11), e1230.

Shultz, S., Dunbar, R.I.M., 2010. Encephalization is not a universal macroevolutionary phenomenon in mammals but is associated with sociality. Proc. Natl. Acad. Sci. U. S. A. 107 (50), 21582–21583.

Starck, D.G., 1979. Cranio-cerebral relations in recent reptiles. In: Gans, C., Northcutt, R.C., Ulinski, P. (Eds.), Biology of the Reptilia. Neurology A, vol. 9. Academic Press Inc., New York, pp. 1–38.

Stephan, H., 1960. Metodische Studien über den quantitativen Vergleich architektonischer Struktureinheiten des Gehirns. Z. Wiss. Zool. 164, 143–172.

Swinton, W.S., 1970. The Dinosaurs. Wiley, New York.

Varricchio, D.J., Horner, J.R., 1993. Hadrosaurid and lambeosaurid bone beds from the upper cretaceous two medicine formation of Montana: taphonomic and biological implications. Can. J. Earth Sci. 30, 997–1006.

Varricchio, D.J., Jackson, F., Trueman, C.N., 1999. A nesting trace with eggs for the cretaceous theropod dinosaur *Troodon formosus*. J. Vertebr. Paleontol. 19 (1), 91–100.

Varricchio, D.J., Sereno, P.C., Xijin, Z., Lin, T., Wilson, J.A., Lyon, G.H., 2008a. Mud-trapped herd captures evidence of distinctive dinosaur sociality. Acta Palaeontol. Pol. 53 (4), 567–578.

Varricchio, D.J., Moore, J.R., Erickson, G.M., Norell, M.A., Jackson, F.D., Borkowski, J.J., 2008b. Avian paternal care had dinosaur origin. Science 322 (5909), 1826–1828.

Watanabe, A., Gignac, P.M., Balanoff, A.M., Green, T.L., Kley, N.J., Norell, M.A., 2018. Are endocasts good proxies for brain size and shape in archosaurs throughout ontogeny? J. Anat. 234 (3), 291–305.

Weishampel, D.B., 1981. Acoustic analyses of potential vocalization in lambeosaurine dinosaurs (Reptilia: Ornithischia). Paleobiology 7 (2), 252–261.

Wharton, D.S., 2002. The Evolution of the Avian Brain. Ph. D. dissertation, University of Bristol, Bristol.

Witmer, L.M., Ridgely, R.C., 2008. Structure of the brain cavity and inner ear of the centrosaurine ceratopsid *Pachyrhinosaurus* based on CT scanning and 3D visualization. In: Currie, P.J. (Ed.), A New Horned Dinosaur from an Upper Cretaceous Bone Bed in Alberta. National Research Council Research Press, Ottawa, pp. 117–144.

Witmer, L.M., Chatterjee, S., Franzosa, J., Rowe, T., 2003. Neuroanatomy of flying reptiles and implications for flight, posture and behavior. Nature 425, 950–953.

Witmer, L.M., Ridgely, R.C., Dufeau, D.L., Semones, M.C., 2008. Using CT to peer into the past: 3D visualization of the brain and ear regions of birds, crocodiles, and nonavian dinosaurs. In: Endo, H., Frey, R. (Eds.), Anatomical Imaging: Towards a New Morphology. Springer-Verlag, Tokyo, pp. 67–88.

Zelenitsky, D.K., Therrien, F., Kobayashi, Y., 2008. Olfactory acuity in theropods: palaeobiological and evolutionary implications. Proc. R. Soc. B 276 (1657), 667–673.

Zelenitsky, D.K., Therrien, F., Ridgely, R.C., McGee, A.R., Witmer, L.M., 2011. Evolution of olfaction in non-avian theropod dinosaurs and birds. Proc. R. Soc. B 278 (1725), 3625–3634.

Zhou, C.-F., Gao, K.-Q., Fox, R.C., Du, X.-K., 2007. Endocranial morphology of psittacosaurs (Dinosauria: Ceratopsia) based on CT scans of new fossils from the lower cretaceous, China. Paleoworld 16, 285–293.

Zimmerman, D., Young Jr., W.F., Ebersold, M.J., Scheithauer, B.W., Kovacs, K., Horvath, E., Whitacker, M.D., Eberhardt, N.L., Downs, T.R., Frohman, L.A., 1993. Congenital gigantism due to growth hormone-releasing hormone excess and pituitary hyperplasia with adenomatous transformation. J. Clin. Endocrinol. Metabol. 76 (1), 216–222.

CHAPTER

At the root of the mammalian mind: The sensory organs, brain and behavior of pre-mammalian synapsids

2

Julien Benoit[a,*], Kathleen N. Dollman[b], Roger M.H. Smith[c,d], and Paul R. Manger[e]

[a]*Evolutionary Studies Institute, School of Geosciences, University of the Witwatersrand, Johannesburg, South Africa*

[b]*European Synchrotron Radiation Facility, Grenoble, France*

[c]*Evolutionary Studies Institute, University of the Witwatersrand, Johannesburg, South Africa*

[d]*Karoo Palaeontology, Iziko South African Museum, Cape Town, South Africa*

[e]*School of Anatomical Sciences, University of the Witwatersrand, Johannesburg, South Africa*

**Corresponding author: e-mail address: julien.benoit@wits.ac.za*

Abstract

All modern mammals are descendants of the paraphyletic non-mammaliaform Synapsida, colloquially referred to as the "mammal-like reptiles." It has long been assumed that these mammalian ancestors were essentially reptile-like in their morphology, biology, and behavior, i.e., they had a small brain, displayed simple behavior, and their sensory organs were unrefined compared to those of modern mammals. Recent works have, however, revealed that neurological, sensory, and behavioral traits previously considered typically mammalian, such as whiskers, enhanced olfaction, nocturnality, parental care, and complex social interactions evolved before the origin of Mammaliaformes, among the early-diverging "mammal-like reptiles." In contrast, an enlarged brain did not evolve immediately after the origin of mammaliaforms. As such, in terms of paleoneurology, the last "mammal-like reptiles" were not significantly different from the earliest mammaliaforms. The abundant data and literature published in the last 10 years no longer supports the "three pulses" scenario of synapsid brain evolution proposed by Rowe and colleagues in 2011, but supports the new "outside-in" model of Rodrigues and colleagues proposed in 2018, instead. As Mesozoic reptiles were becoming the dominant taxa within terrestrial ecosystems, synapsids gradually adapted to smaller body sizes and nocturnality. This resulted in a sensory revolution in synapsids as olfaction, audition, and somatosensation compensated for the loss of visual cues. This altered sensory input is aligned with changes in the brain, the most significant of which was an increase in relative brain size.

Progress in Brain Research, Volume 275, ISSN 0079-6123, https://doi.org/10.1016/bs.pbr.2022.10.001

Keywords

Endocast, Synapsida, Therapsida, Cynodontia, Mammalia, Encephalization quotient

1 Introduction

Mammalia is the crown group and only living clade of a broader and older branch of the amniote phylogenetic tree, the Synapsida (Fig. 1). The evolution of synapsids can be simplified into a series of successive grades (paraphyletic assemblages of taxa) that have historically been used to summarize mammalian evolutionary origins (Jerison, 1973; Romer, 1933). In this framework, taxa are conveniently grouped for their morphological consistency (the reader may refer to Fig. 1 every time one of these grades is cited below). The sprawling, reptilian-looking Pelycosauria (or pelycosaurs) form the oldest and basal-most radiation of synapsids (Fig. 1), dating mostly from the late Carboniferous to early Permian (about 315–260 Ma) (Angielczyk and Kammerer, 2018; Kemp, 2005; Rubidge and Sidor, 2001). They gave rise to, and were succeeded by, the Therapsida (or therapsids) (Fig. 1), which include five orders: the Biarmosuchia, Dinocephalia, Anomodontia, Gorgonopsia, and Therocephalia that lived mostly during the Permian and Triassic (about 260–200 Ma). With their more erect posture, therapsids looked more mammal-like in body profile than the pelycosaurs (Kemp, 2005). The Cynodontia (or cynodonts) are the sister taxon to the Therocephalia and are deeply phylogenetically nested within the Therapsida (forming a sixth order of Therapsida) (Fig. 1); however, they are considered separately here because their biology, morphology and palaeoneurology are strikingly intermediate between other therapsids and mammals (Angielczyk and Kammerer, 2018). Cynodonts lived from the late Permian to the early Cretaceous (Abdala, 2021; Angielczyk and Kammerer, 2018) and are divided between the Cynognathia (a branch of mostly herbivorous cynodonts from the Triassic), Probainognathia (the grade that evolved into the Mammaliaformes), and a paraphyletic assemblage of earlier branching taxa (i.e., non-eucynodont cynodonts) that belong to none of the aforementioned groups and are here referred to as basal-cynodonts (Fig. 1). The crown group Mammalia is nested within the Mammaliaformes. Mammaliaformes is a clade that originated about 200 million years ago (Abdala, 2021; Angielczyk and Kammerer, 2018) and encompasses the living Placentalia, Marsupialia, and Monotremata, as well as many extinct taxa such as the early *Morganucodon*, *Megazostrodon*, *Sinoconodon*, *Hadrocodium,* and the more derived Haramiyida, Docodonta, and Multituberculata (Kielan-Jaworowska et al., 2004). For convenience, non-mammalian mammaliaforms will be hereafter referred to as early mammaliaforms (or early Mammaliaformes) (Fig. 1).

The simplified terminology above is preferred here over the usual specific cladistic terminology as it avoids the use of terms such as "non-therapsid pelycosaurs" for pelycosaurs, "non-cynodont therapsids" for therapsids, "non-mammaliaform cynodonts" for cynodonts, "non-mammaliaform probainognathians"

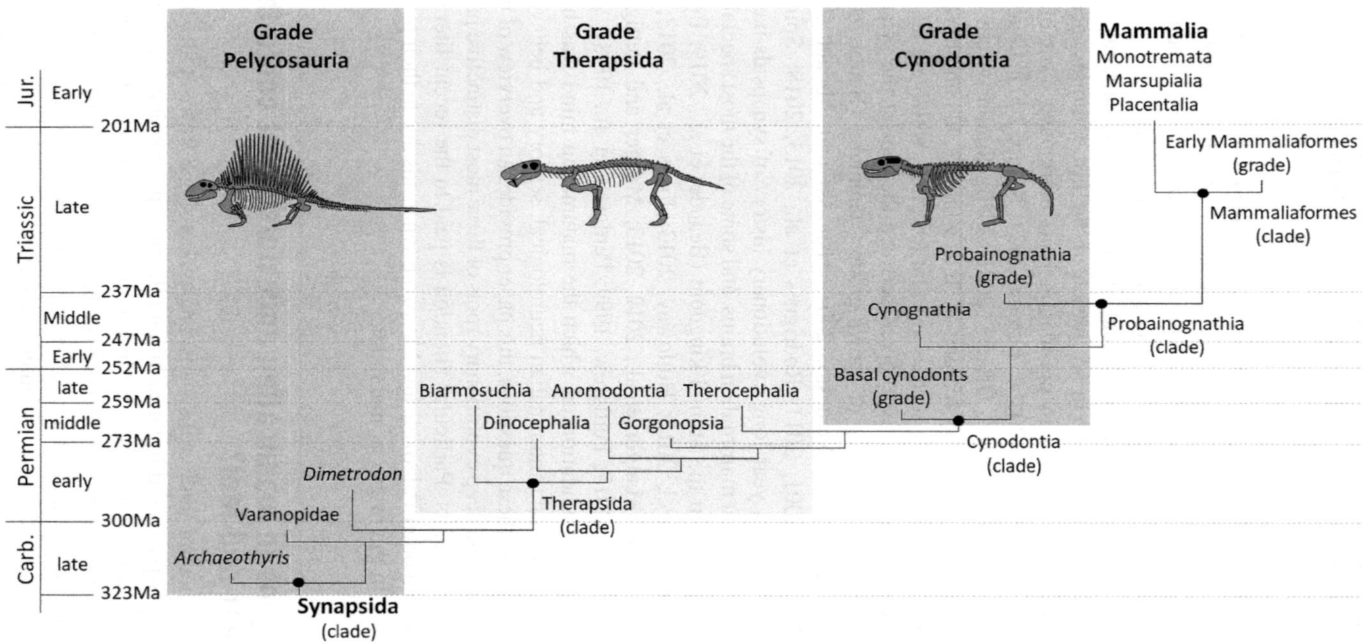

FIG. 1

Simplified synapsid phylogeny (after Rubidge and Sidor, 2001; Angielczyk and Kammerer, 2018). Drawings not to scale. To prevent ambiguities, monophyletic groups are indicated as clades, and paraphyletic groups are indicated as grades. Abbreviations: Carb., Carboniferous; Jur., Jurassic.

for probainognathians, "non-mammalian mammaliaforms" for early mammaliaforms, as examples. Although cladistically less accurate, these paraphyletic assemblages are morphologically consistent (i.e., all included taxa share the same symplesiomorphies) and their use facilitates comparison with older literature (e.g., Jerison, 1973; Quiroga, 1980, 1984; Romer, 1933). However, the reader is provided with the consensual phylogenetic tree of Synapsida in Fig. 1 (Angielczyk and Kammerer, 2018; Rubidge and Sidor, 2001), and is reminded that a phylogenetic framework is indispensable to interpret the evolution of any trait in extinct and modern species.

The most distinctive mammalian anatomical features, such as hair, lactation, erect limb posture, and endothermy were not present in the last common ancestor of all synapsids, but likely evolved in pre-mammalian forerunners (Angielczyk and Kammerer, 2018; Benoit et al., 2016c; Benton, 2021; Kemp, 2005; Legendre and Davesne, 2020; Lovegrove, 2017; Oftedal, 2002, 2020). Similarly, complex behavior such as parental care, sensory features such as whiskers and an elongated cochlea, and neurological innovations such as an enlarged brain, which are all considered typically mammalian, evidently evolved between the root of the synapsid clade and the last common ancestor of crown-group Mammalia (Benoit et al., 2017a,b, 2020; Botha-Brink and Modesto, 2007; Jasinoski and Abdala, 2017; Kerber et al., 2021; Luo, 2001, 2011; Rodrigues et al., 2013, 2018; Spindler et al., 2018). The 315 million year-long evolutionary history of synapsids has led to the most complex and largest mammalian brains, and some of the most specialized sensory organs and systems in the animal kingdom (Beaudet et al., 2019; Benoit et al., 2019; Herculano-Houzel, 2012; Holloway, 2013; Jones et al., 2013; Liu et al., 2010; Manger, 2006; Manger et al., 2012, 2013; Manger and Pettigrew, 1995; Mourlam and Orliac, 2017; Pettigrew, 1999; Pettigrew et al., 1998). What was the paleoneurological foundation on which the mammalian mind was built? How and when did the defining mammalian neurological, sensory and behavioral features evolve? To address these questions, this paper provides an overview of what is known in order to trace the deep evolutionary roots of these most distinctive paleoneurological mammalian traits. Particular attention is paid to the recent literature regarding cynodonts and other Paleozoic and Mesozoic synapsids due to their pivotal phylogenetic position in the origin of mammals.

2 Sensory organs in mammalian ancestors: what could they smell, see, hear and feel?

In one of the most influential studies of synapsid paleoneurology, Rowe et al. (2011) stated "early cynodonts possessed low-resolution olfaction, poor vision, insensitive hearing, coarse tactile sensitivity, and unrefined motor coordination." This implies that all pre-mammalians before cynodonts also had relatively simply organized sensory organs compared to their mammalian descendants. A decade ago, there was little evidence to contradict this statement; however, the growing body of CT data for paleoneurological studies and global efforts to retrieve new fossils have

dramatically improved our knowledge regarding the evolution of sensory organs and the brain in pre-mammalian synapsids. For example, the number of endocranial casts documented for Paleozoic and Mesozoic synapsids has almost doubled since 2011, from less than 20 to more than 30, including some taxa represented by many specimens (Benoit et al., 2017a; de Simão-Oliveira et al., 2020; Du Plessis, 2010; Hoffmann et al., 2021; Huttenlocker and Sidor, 2020; Kerber et al., 2021; Laaß et al., 2017; Pavanatto et al., 2019; Pusch et al., 2019; Rodrigues et al., 2018; Rowe et al., 2011). This increase in accessible paleoneurological data presents an opportunity for a reassessment of Rowe et al.'s (2011) original assertion that early synapsids had primitive sensory organs.

2.1 Low-resolution olfaction?

Olfactory capabilities in fossil synapsids have been inferred based on the dimensions of the olfactory bulbs and the morphology of the olfactory capsule of the skull (Crompton et al., 2015, 2017; Franco et al., 2021; Rowe and Shepherd, 2016).

Rowe et al. (2011) and Rowe and Shepherd (2016) proposed that encephalization (relative size of the brain corrected for body size) increased in three pulses in early mammaliaforms, which correlated to three steps of enhancement of the olfactory system specific to this clade. The first two steps were the acquisition of enlarged, mammal-like olfactory bulbs in the basal-most mammaliaforms, *Morganucodon* and then *Hadrocodium* (Rowe, 2017; Rowe and Shepherd, 2016; Rowe et al., 2011). This has been disputed by Du Plessis (2010) who found that the relative size of the olfactory bulbs in many cynodonts is not significantly different from that of mammaliaforms. In contrast, Ruf et al. (2014) found that the brasilodontid *Brasilodon* (syn. *Brasilitherium*), the most derived cynodont (i.e., the sister-group to Mammaliaformes, Fig. 2A), has olfactory bulbs comparable in size to early mammaliaforms. Rodrigues et al. (2018) and Kerber et al. (2021) subsequently demonstrated that this enlargement of the olfactory bulbs was not unique to brasilodontids among cynodonts, but that *Therioherpeton* and the tritheledontid *Riograndia*, two probainognathians, also had enlarged olfactory bulbs. The two olfactory bulbs are also separated by a well-defined sulcus in these taxa, unlike in more basal cynodonts such as *Thrinaxodon* (Kielan-Jaworowska et al., 2004; Rodrigues et al., 2018). Rodrigues et al. (2018) indicate that enlargement of the olfactory bulbs likely began with basal probainognathians, such as *Chiniquodon,* where a midline sulcus between the two olfactory bulbs is visible. Such a sulcus has been tentatively identified in the cynognathian *Massetognathus* (Hoffmann et al., 2021; Kielan-Jaworowska et al., 2004) but not in *Exaeretodon* and *Siriusgnathus* (Pavanatto et al., 2019), so its presence is likely not ancestral for the Cynognathia clade.

The last pulse of encephalization, according to Rowe et al. (2011), was accompanied by the ossification of the cribriform plate (ethmoid bone), and is unique to crown-group Mammalia. Crompton et al. (2017) and Pavanatto et al. (2019) agree that the ossification of the elements of the nasal capsule, including the ethmoid bone and olfactory turbinals that support the olfactory epithelium, was an important event

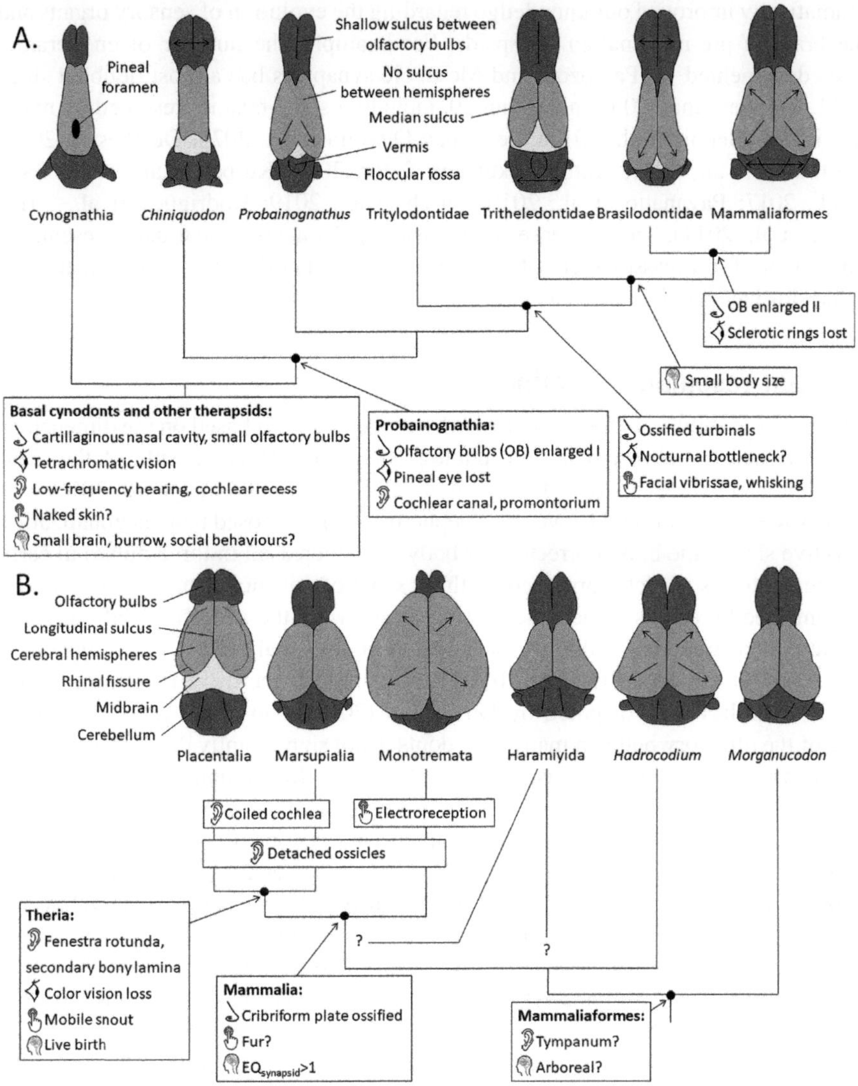

FIG. 2

The evolution of the cynodont and mammalian endocast, sensory organs, and other palaeoneurological traits. (A) In cynodonts. (B) In mammaliaforms (see text for references). Drawings not to scale. Arrows indicate directions of expansion.

in the evolution of mammalian olfaction. An ossified cribriform plate is reportedly absent or not preserved in most Paleozoic and Mesozoic Mammaliaformes, except in *Vintana* and *Kryptobataar*, both of which belong to the mammalian clade (Hoffmann et al., 2014; Kielan-Jaworowska, 1986; Macrini et al., 2007). Fragments in the endocranial cavity of *Hadrocodium* have been tentatively interpreted as possibly

representing the cribriform plate (Lillegraven and Krusat, 1991), but the absence of an ossified ethmoid bone in the closely related *Morganucodon* makes it unlikely (Crompton et al., 2015; Hoffmann et al., 2014; Rowe et al., 2011). Rather, a cartilaginous cribriform plate was likely present in early mammaliaforms, as in cynodonts (Crompton, 1958; Crompton et al., 2015). Thus, Rowe et al.'s view that the cribriform plate evolved in the last common ancestor of Mammalia remains the most parsimonious hypothesis.

Bone ridges for the attachment of cartilaginous olfactory turbinals have been reported in pelycosaurs, therapsids, and cynodonts (Hillenius, 1994; Hillenius and Ruben, 2004), and fossilized cartilaginous turbinals have been tentatively identified in one specimen of the dicynodont therapsid *Lystrosaurus* (Laaß et al., 2011). Unlike the ethmoid plate, ossified respiratory turbinals have been reported in probainognathian cynodonts, i.e., in the tritylodontid *Pseudotherium*, the brasilodontid *Brasilodon* (Ruf et al., 2014; Wallace et al., 2019), and Mesozoic mammaliaforms (Hurum, 1994). As such, the ossification of the nasal capsule may have begun with the olfactory turbinals and was followed millions of years later by the cribriform plate; however, it still remains unclear when the respiratory turbinals ossified (Crompton et al., 2015; Ruf et al., 2014).

Overall, the evolution of a mammal-like ossified nasal capsule and enlarged olfactory bulbs appears to be comprised of a series of evolutionary modifications, starting with probainognathians prior to the origin of Mammaliaformes, and occurring earlier than the three pulses hypothesized by Rowe et al. (2011) (Fig.2A).

2.2 Poor vision?

When tetrapods adapted to life on land, their visual capacities altered to adapt to the novel sensory inputs associated with terrestrial conditions (MacIver et al., 2017; Rowe, 2017). Early synapsids (pelycosaurs and therapsids) were among the first tetrapods to diversify and dominate the fully terrestrial ecosystems of continental interiors (Angielczyk and Kammerer, 2018; Kemp, 2005; Rubidge and Sidor, 2001).

The sclerotic ring provides structural support to the eyeball against variations in intraocular pressure during accommodation to different light conditions in many modern non-mammalian vertebrates, and thus the study of its dimensions when preserved in the fossil record can be considered a direct indicator of the eyeball size and diel activity patterns in extinct animals (Angielczyk and Schmitz, 2014). Angielczyk and Schmitz (2014) studied the sclerotic rings in pelycosaurs and therapsids and concluded that their diel activity patterns were very diverse. Most herbivorous taxa, such as dicynodonts, are thought to have had photopic vision (a visual system structurally useful in bright light conditions), and were thus likely diurnal, whereas many carnivorous taxa, such as sphenacodontids, gorgonopsians, and anteosaurs are thought to have had scotopic or mesopic vision (a visual system structurally useful in the dark or low light conditions, respectively), and may thus have hunted or scavenged in the twilight hours of the day (Angielczyk and Schmitz, 2014). The Cistecephalidae, a family of Permian dicynodonts that adapted to a fossorial (burrowing) lifestyle (Cox, 1972; Kammerer, 2021; Nasterlack et al., 2012), have very small orbits and

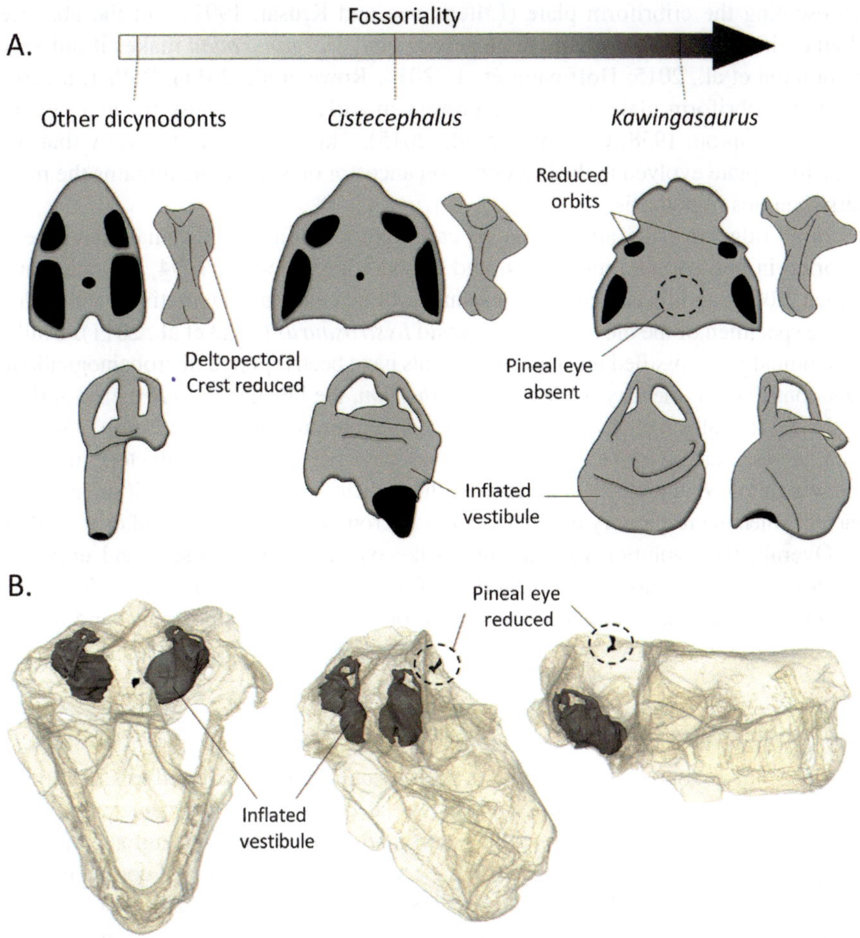

FIG. 3

Sensory adaptations to fossoriality in synapsids. (A) Cistecephalid dicynodonts (skull, humerus, and bony labyrinth) (Cox, 1972; Laaß and Kaestner, 2017; Laaß and Schillinger, 2017; Angielczyk et al., 2021). (B) The basal cynodont *Cistecynodon*. The large horizontal arrow on top represents the relative degree of fossoriality. Given its paleoneurological features, *Cistecynodon* would fit between *Cistecephalus* and *Kawingasaurus* on this continuum. Drawings and pictures not to scale.

sclerotic rings, indicative of a reduction in eyeball size (Fig. 3A), similar to some modern moles (Angielczyk et al., 2021; Cox, 1972). Some taxa, such as *Cistecephalus*, had a binocular vision so that the overlapping fields of view of each eye increased sensitivity to low light conditions underground (Angielczyk et al., 2021).

Pelycosaurs, therapsids, and cynodonts most likely had tetrachromatic vision (four independent channels for color information) as this is the ancestral condition

for tetrapods (Borges et al., 2018; Emerling, 2016; Gerkema et al., 2013; Neil and Allison, 2021). Even though no single modern mammal species possess all four cone classes, their existence in the common ancestor of all mammals is also supported by their phylogenetic distribution among modern species (Borges et al., 2018; Emerling, 2016; Gerkema et al., 2013; Neil and Allison, 2021; Rowe, 2017). Most living mammals have dichromatic vision (two independent channels for color information) (Borges et al., 2018; Emerling, 2016; Gerkema et al., 2013; Heesy and Hall, 2010; Neil and Allison, 2021). This, and a suite of morphological, physiological, and genetic adaptations such as the loss of colored oil droplets and lens accommodation reflex, and the acquisition of a greater binocular visual field overlap (Borges et al., 2018; Emerling, 2016; Gerkema et al., 2013; Heesy and Hall, 2010; Neil and Allison, 2021), indicate that the above-mentioned diversity of diurnal activity patterns in therapsids was lost during the Mesozoic history of mammaliaforms in at least one phase of adaptation to nocturnality termed "the nocturnal bottleneck" (Walls, 1942). In fact, the non-complete overlap between the cone opsins retained by monotremes and therians, the sequence of phylogenetic events in which the cones and double cones were lost, the retention of colorless oil droplets in platypus, and the convergent evolution of the tapetum lucidum in monotremes, placentals, and marsupials argue in favor of multiple, independent, nocturnal bottlenecks occurring across the Mesozoic, or for a nocturnal bottleneck preceded by an extended "mesopic bottleneck" (Gerkema et al., 2013; Rowe, 2017; Schwab et al., 2012). Unfortunately, mammaliaforms lost their sclerotic ring, and sclerotic rings are rarely preserved in cynodonts, making it impossible to confidently infer when and where in the early phylogeny of synapsids these proposed bottlenecks occurred. The tritylodontid *Tritylodon* is the only cynodont included in Angielczyk and Schmitz (2014) analysis, and it is reconstructed as being nocturnal (Fig. 2A). The complete loss of the sclerotic ring in mammaliaforms may also be an adaptation to nocturnality, as the sclerotic ring is involved in protecting the eyeball against variations in intraocular pressure during accommodation to rapidly changing light conditions (Emerling, 2016).

The pineal, or "third," eye is located on the roof of the skull, and detects differences in light levels, such as those between night and day, and provides this photic level information to the pineal gland for the regulation of biological cycles, circadian rhythms, and behavioral thermoregulation in modern reptiles (Eakin, 1973; Quay, 1979). The presence of a parietal foramen shows that this third eye was present in all tetrapods, including pelycosaurs and therapsids, but its presence and size became highly variable in some derived, Triassic therapsid lineages and cynodonts (Benoit et al., 2015, 2016a). It was ultimately lost in probainognathian cynodonts, as demonstrated by the closure of their parietal foramen (Benoit et al., 2015, 2016a). In mammals, the photosensitive role of the pineal eye appears to have been replaced by melanopsin-containing retinal ganglion cells (Berson et al., 2002; Lucas et al., 1999), which may have evolved in probainognathians (Benoit et al., 2016a,b). Remarkably, the brain endocasts of some cynognathians, such as *Massetognathus*, that lost the pineal eye independently of the probainognathians, still retain a cavity for the pineal body (Hoffmann et al., 2021; Quiroga, 1979). The nocturnal

bottleneck(s) and the evolution of endothermy, may have reduced the functional relevance of the pineal eye, possibly leading to its loss in probainognathians (Benoit et al., 2015, 2016a). Coincidentally, the reptilian lineages that independently lost the pineal foramen also began their evolutionary history either as endotherms, such as the archosaurs (Seymour et al., 2004), or as fossorial taxa, such as the snakes and turtles (Lyson et al., 2016; Yi and Norell, 2015).

The assumption that synapsids had poorer visual capabilities than mammals is questioned because: (i) pelycosaurs, therapsids, and cynodonts appear to have had a high diversity of diel activity patterns; and (ii) it is evident that mammaliaforms, including the lineages leading to modern mammals, lost many sensory traits useful for photopic conditions (Fig. 2A). Pelycosaurs and therapsids likely had tetrachromatic vision, colored oil droplets, an efficient lens accommodation reflex, a protective/supportive sclerotic ring, and a pineal eye (Benoit et al., 2016a; Borges et al., 2018; Gerkema et al., 2013). These conditions disappeared gradually in cynodonts and early mammaliaforms, likely due to their increasing adaptation to nocturnality and fossoriality. Subsequently, adaptations for improved scotopic vision, such as larger and more convergent orbits, larger corneas and pupils, and a higher rod count evolved in mammals (Gerkema et al., 2013; Vega-Zuniga et al., 2013). All modern mammals probably evolved from a nocturnal Mesozoic ancestor, and it is remarkable that some adaptations to photopic conditions appear to have re-evolved in the diurnal mammal species of the Cenozoic (Borges et al., 2018; Emerling, 2016; Gerkema et al., 2013).

2.3 Insensitive hearing?

The detectable frequency range, the acuity, and sensitivity of mammalian auditory system are primarily determined by the morphophysiology of the soft tissue of the ear, which rarely fossilizes (Manley, 2017, 2018). For example, the external auditory pinnae play an important role in sound localization, and their presence increases auditory sensitivity by around 20dB in extant mammals (Manley, 2018); however, these pinnae are rarely preserved. The Mesozoic record of fossilized pinnae is limited to one specimen of the Cretaceous mammal *Spinolestes xenarthrosus* (Martin et al., 2015). A similar issue applies to the tympanic membrane (tympanum), a crucial soft-tissue organ of the mammalian auditory apparatus. Recent data have confirmed that the reptilian and mammalian tympanic membranes are not homologous, and thus, the mammalian tympanum evolved at an unknown point in the synapsid lineage, rather than being a feature of the last common ancestor of amniotes (Kitazawa et al., 2015). It has been hypothesized that the spoon-shaped angular bone of cynodonts (a bone of the lower jaw) was capable of framing a tympanic membrane, just as its mammalian equivalent (the ectotympanic bone) does in mammals (Kemp, 2016; Luo, 2011; Luo et al., 2016). However, a recent review of the stapes of the cynodonts has shown that a connection between this putative tympanum and the stapes, which is supposed to transfer vibrations from the membrane, is anything but clear (Gaetano and Abdala, 2015). The physical distance between the stapes and supposed tympanum is simply too great to be bridged, and some authors even suggested that two tympanic

membranes co-existed in cynodonts, one for the stapes and one framed by the angular bone (see Gaetano and Abdala, 2015 and references therein). Finally, the stapes in early mammaliaforms with a well-defined tympanic cavity, suggesting the presence of a functional tympanic membrane (Kemp, 2016; Kermack et al., 1981)—is virtually identical to that of the probainognathian cynodonts (Rodrigues et al., 2013). The most direct proxies to infer the evolution of a tympanum thus give an ambiguous signal, and in the absence of fossilized tympanic membranes, it is difficult to provide a conclusive narrative regarding the evolutionary origin of this organ.

Despite these caveats, Mesozoic mammals display some noticeable osteological adaptations to high and low-frequency hearing that have been fossilized (Manley, 2018; Manoussaki et al., 2008; West, 1985). In fact, the evolution of the hearing apparatus has been among the best-studied sensory organs in early mammaliaforms, cynodonts, and other therapsids, leading to the conclusion that mammalian ear bones are derived from the "mammal-like reptiles" post-dentary jawbones (Allin, 1975; Allin and Hopson, 1992; Ekdale, 2016a,b; Gaetano and Abdala, 2015; Kemp, 2005, 2016; Luo et al., 2016; Maier and Ruf, 2016; Manley, 2017). Mammalian hearing organs can detect a broad frequency range, from ultrasonic (above 20kHz) to infrasonic (as low as 15Hz) sensitivity (Manley, 2018; Manoussaki et al., 2008; West, 1985). In comparison, estimations based on the osteology of the bony labyrinth and ear bones suggest that therapsids, cynodonts, and early mammaliaform could hear only frequencies in the range of about 5kHz (Benoit et al., 2017b; Rosowski and Graybeal, 1991). This represents an unspecialized condition among amniotes (Manley, 1972; Walsh et al., 2009), resembling the pelycosaur condition (about 3000Hz, Bazzana et al., 2022), and similar to the 1kHz frequency sensitivity range reconstructed for the ancestral amniote condition (Manley, 2000, 2017).

Emphasis has been placed on the evolution of high-frequency hearing in Mesozoic mammals, particularly because small body size reduces both the vocal organs and interaural distance, thus making the emission and localization of low-frequency sound practically impossible (Heffner and Heffner, 2018; Manley, 2017, 2018). Miniaturization evolved as early as the late Triassic in probainognathians (e.g., in Brasilodontidae and Tritheledontidae, Lautenschlager et al., 2018), but the presence of primary and secondary bony lamina in the cochlea (narrowing and stiffening the basilar membrane), and discrete fenestra rotunda (which act as a pressure relief system), evolved quite late in the Mesozoic among the Theria (Marsupialia and Placentalia) and some extinct mammalian groups (Ekdale, 2013, 2016b). This suggests that the initial adaptations to high-frequency hearing might have involved soft tissue only.

A contrasting view is provided by the study of the transformation of the post-dentary bones into the ossicular chain. Because their post-dentary bones are attached to the lower jaw, pelycosaurs, therapsids, cynodonts, and early mammaliaforms were likely limited to hearing ground-borne vibrations, similar to what is observed in modern crocodiles (Luo, 2011; Luo et al., 2016). It is generally admitted that adaptations to hearing a wider range of air-borne frequencies were permitted by the miniaturization of the post-dentary bones and their migration toward the skull, which resulted

in the progressive detachment of the auditory ossicles (the stapes, malleus, and incus) present in modern mammals. The impedance-matching ossicular chain enables the transduction of sounds to the inner ear fluids, which is a prerequisite to hearing air-borne sounds (Ekdale, 2016b; Luo et al., 2016; Manley, 2018). Detachment of the post-dentary bones from the lower jaw can be traced in the fossil record by surveying the presence or absence of the Meckelian groove (Ekdale, 2016b). This demonstrates that the transition from attached post-dentary bones to detached auditory ossicles was not a single event, but was very homoplastic, with no less than eight independent losses of the Meckelian groove (Ekdale, 2016b; Wang et al., 2021), and the convergent acquisition of detached ossicles in all three lineages leading to modern monotremes, marsupials, and placentals (Fig. 2B). This would support the argument for a much more recent origination of air-borne sound detection.

Another feature that evolved independently in the lineages leading to placentals and marsupials is a coiled cochlear canal, which is absent in their Mesozoic ancestors (Ekdale, 2013, 2016a,b). Monotremes, and most Mesozoic mammals and mammaliaforms possess a finger-like, more-or-less hooked cochlear canal (Fig. 4), but none of them coil significantly beyond one whorl (Ekdale, 2013, 2016a,b). A cochlear canal is not a unique mammaliaform feature as most cynodonts already posess a cochlear recess, i.e., a ventral extension of the vestibular cavity on the endocranial

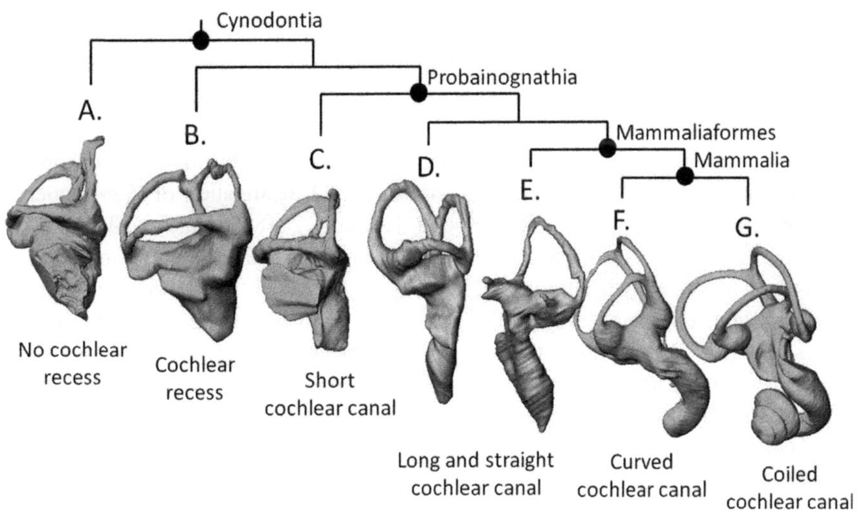

FIG. 4

The origin of the mammalian cochlea. Bony labyrinths reconstructed in digital 3D in oblique ventral view to maximize the perspective on the cochlear canal (see Benoit et al., 2017b). From left to right: (A) basal cynodont from the Permian, (B) Cynognathia, (C) early Probainognathia, (D) Tritylodontidae, (E) Early Mammaliaformes, (F) Monotremata, (G) Placentalia. Bony labyrinths not to scale.

surface of the otic bones that eventually lengthened to become the cochlear canal (Luo, 2001, 2011) (Fig. 4). Pelycosaurs and most therapsids show no signs of a cochlear recess (with a few exceptions, see e.g., Benoit et al., 2017b and Laaß, 2015a,b). Derived probainognathians, such as tritylodontids and brasilodontids already possessed an elongated (but not curved) cochlear canal (Fig. 4), which coincides with the fusion of the prootic and opisthotic into the typically mammalian osseous promontorium cochleae (Luo, 2001, 2011; Rodrigues et al., 2013).

It is important to note here that, although the focus of this section has been the origin of mammalian auditory capabilities, cynodonts and other therapsids also independently specialized in this respect. For example, Laaß (2015a,b) described adaptations of the auditory apparatus to impedance matching capabilities in a dicynodont, *Pristerodon mackayi*. Male and female specimens of the dicynodont species *Diictodon feliceps* display a high degree of sexual dimorphism in their bony labyrinth which, though unexplained, is unique among synapsids (Laaß et al., 2017). Along those lines, Benoit et al. (2017b) highlighted a vast diversity of vestibular shapes among cynodonts and other therapsids that is difficult to interpret in the absence of modern analogs, but represent interesting cases of morphological specializations. Among them, the cistecephalid dicynodonts, previously mentioned for their fossorial adaptations, display a remarkable evolutionary trend toward the inflation of the vestibule (Angielczyk et al., 2021; Laaß, 2015b). This ultimately resulted in a ball-shaped capsule for the auditory organ and an enlarged stapedial footplate in *Kawingasaurus* (Fig. 3A), similar to those of modern fossorial species that can detect seismic vibrations (Laaß, 2015b). Coupled with the reduction, and ultimately the loss in *Kawingasaurus*, of the parietal foramen (the pineal eye being redundant in a low light environment) (Fig. 3A), this indicates that the cistecephalids underground lifestyle imposed strong selective pressures on their nervous system (Benoit et al., 2016a; Cox, 1972; Laaß, 2015b; Laaß and Kaestner, 2017; Laaß and Schillinger, 2017; Macungo et al., 2022). *Cistecynodon*, a basal cynodont from the Triassic of South Africa (Abdala, 1996), also possessed the same highly specialized features (inflated vestibule, pineal foramen greatly reduced, Fig. 3B), consistent with a fossorial lifestyle that is otherwise well-documented in Triassic cynodonts (Bordy et al., 2019; Damiani et al., 2003; Fernandez et al., 2013; Groenewald et al., 2001; Krapovickas et al., 2013). An inflated cochlear canal also independently evolved in the Multituberculata, which are mammals from the Mesozoic and early Cenozoic (Hurum, 1998a). This exemplifies the fact that the evolution of auditory systems in therapsids and cynodonts was not just a prelude to the mammalian condition, but that these structures were actively diversifying into lineage-specific specializations.

Excluding these "counter-examples," Rowe et al.'s (2011) proposal that most cynodonts and therapsids had rather unspecialized auditory frequency range detection and sensitivity compared to living mammals is still generally supported; however, the ear region in Mesozoic mammals and mammaliaforms also remained mostly unspecialized. The auditory adaptations that characterize modern mammals evolved quite late in their history and mostly in a convergent manner (Fig. 2B).

2.4 Coarse tactile sensitivity?

The question of tactile sensitivity in mammals is linked to the origin of hair and tactile vibrissae in synapsids (Benoit et al., 2016c, 2020; Manger, 2009; Muchlinski, 2008, 2010; Muchlinski et al., 2020). Mammals are particularly noted for their tactile vibrissae, most particularly the facial whiskers (Muchlinski, 2008, 2010; Muchlinski et al., 2020). As keratin is a non-mineralized soft tissue that does not readily fossilize, direct evidence of hair and vibrissae are rare in the fossil record. Unlike dinosaurs, in which the transition from scales to feathers is very well documented (Hendrickx et al., 2022), very few fossils document the origin of hair in synapsids, and no confirmed fossil vibrissae have been reported.

There is abundant evidence that the surface of the body of pelycosaurs was covered in reptilian-like scales (Botha-Brink and Modesto, 2007; Niedźwiedzki and Bojanowski, 2012; Reisz, 1975; Spindler et al., 2018), which likely represents the primitive condition for synapsids. A handful of possible fossilized skin imprints suggests that the skin of therapsids had neither hair nor scales (Chudinov, 1968; Smith and Botha-Brink, 2014), although the absence of hair may be due to the decaying process. Some Permo-Triassic coprolites from Russia and South Africa contain long filamentous "hair-like structures" (Bajdek et al., 2016; Smith and Botha-Brink, 2011), but these could be fungal or bacterial in origin (Benoit et al., 2016c; Saint Martin and Saint Martin, 2018). The oldest fossils preserving a pelage are *Maiopatagium* and *Castorocauda*, from the Jurassic of China (Ji et al., 2006; Meng et al., 2017). *Castorocauda* belongs to crown-group Mammalia, whereas *Maiopatagium* is an haramiyidan, considered by some authors to be mammals and by others as stem mammaliaforms (Celik and Phillips, 2020). If the latter is correct, this would suggest that hair was present in most mammaliaforms as early as the Early Jurassic. Otherwise, hair could be restricted to crown Mammalia (Fig. 2B). It is currently unclear where cynodonts fit into this transition from scales (in pelycosaurs), to barren skin (in therapsids), to hair (in mammals), as there is to date no direct evidence of fossilized integuments that can be unquestionably attributed to cynodonts. The trace fossil *Cynodontipus* was once believed to be the imprint of a cynodont's body hair (Ellenberger, 1976), but was later reinterpreted as a *numen dubium* (Klein and Lucas, 2021) and recently attributed to an arthropod (Wolvaardt, 2021).

It has long been noted that the supralabial foramina located on the snout of cynodonts may have given passage to the nerves that innervated facial sensory vibrissae (whiskers) (Watson, 1931). This assumption, while having been repeatedly disproved (Benoit et al., 2016c; Estes, 1961; Tatarinov, 1976; Van Valen, 1960) nevertheless persists in the literature (Bonnan, 2016; Lingham-Soliar, 2014). These alleged "whisker pits" (Bonnan, 2016) are very similar to the supralabial foramina in reptiles (Fig. 5), which makes them an unlikely proxy to infer the presence of facial vibrissae (Benoit et al., 2021b; Bouabdellah et al., 2022; Estes, 1961; Van Valen, 1960). In addition, "whisker pits" are not limited to cynodonts as they are present in other synapsids, including the scaly-skinned pelycosaurs and beaked dicynodonts (Benoit et al., 2018, 2021b). Finally, modern mammals do not have "whisker pits" (Figs. 5 and 6).

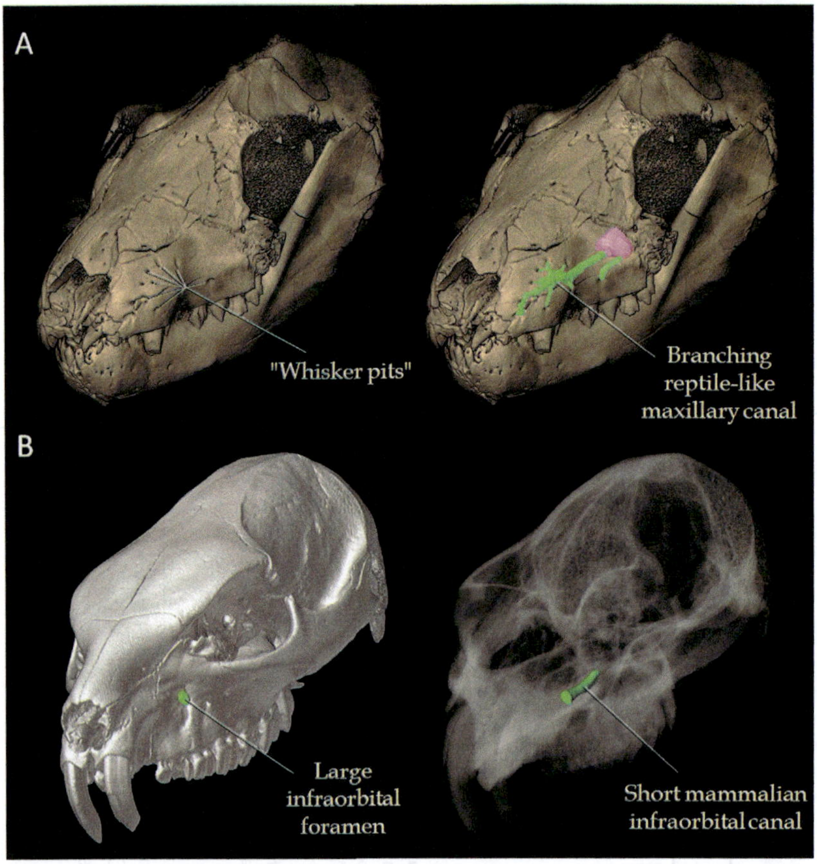

FIG. 5

The difference between the maxillary canal of a cynodont (A) and the infraorbital canal of a mammal (B). The "whisker pits"in cynodonts and other therapsids (A, left) are very similar to the supralabial foramina of reptiles as they all open into a long and branching maxillary canal (A, right). In contrast, mammals have a single infraorbital foramen (B, left) connected to a short and enlarged canal (B, right), and that leads the trigeminal nerve to the mobile soft tissue of the snout before it branches.

Studies using serial sections and CT scanning have demonstrated that the supralabial foramina are linked to the maxillary canal, a long, narrow and ramified structure that encloses the sensory fibers of the infraorbital branch of the trigeminal nerve and accompanying vessels into the maxillary bone, as in modern reptiles (Benoit et al., 2016c; Duhamel et al., 2021; Estes, 1961; Fourie, 1963; Pusch et al., 2019; Wallace et al., 2019) (Fig. 5A). Mammals, in contrast, have a short and broad infraorbital canal, that, unlike the maxillary canal, gives passage to the infraorbital nerve but does not enclose it (Figs. 5B and 6G). As such, the nerve ramifies in the soft tissue

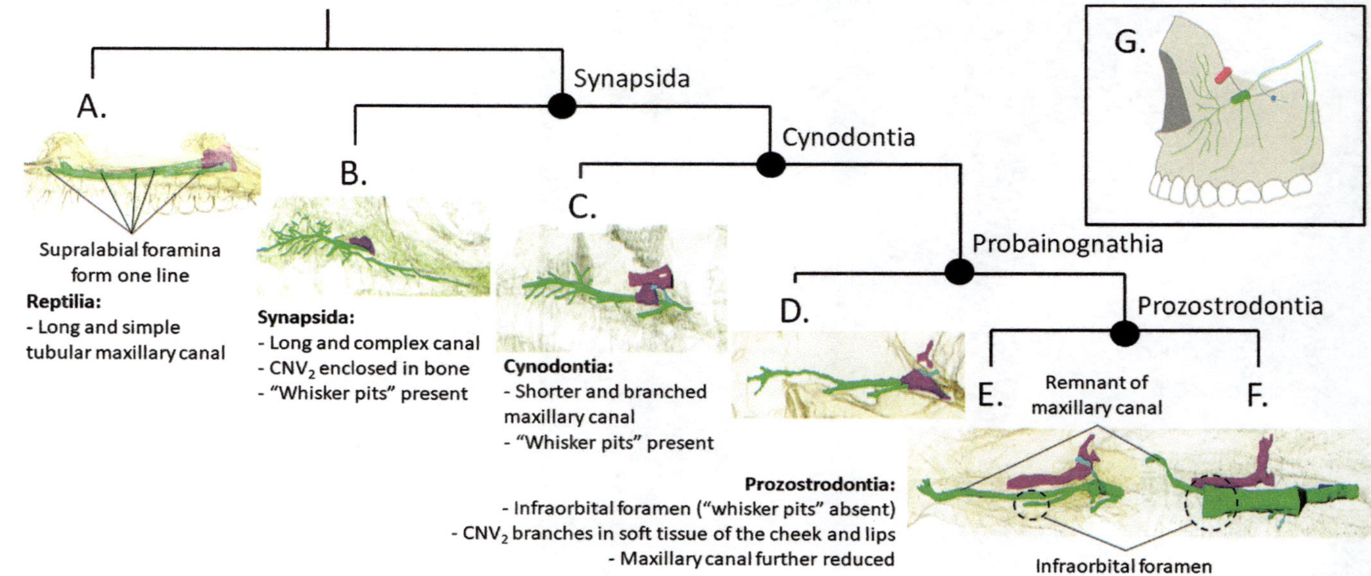

FIG. 6

Simplified evolution of the maxillary canal in amniotes and the origin of the mammalian infraorbital foramen (see Benoit et al., 2020). From left to right: (A) Reptilia, (B) Pelycosauria, (C) Cynodontia, (D) Early Probainognathia, (E) Derived Probainognathia, (F) Mammalia, (G) *Homo* (showing the bony canals in solid line and nerves in dotted lines). Colors: the maxillary canal is in green, the lacrimal canals and maxillary sinus are in purple/pink, the maxillofacial canal is in blue.

of the face and is thus free to follow the movements of the cheeks and lips during tactile movements, called whisking (Grant et al., 2018; Grant and Goss, 2022; Mitchinson et al., 2011). Benoit et al., (2016c, 2019) and Miyamae and Bhullar (2017) hypothesized that having the sensory nerve restricted within the maxillary canal made whisking movements impossible in cynodonts, thus limiting the usefulness of putative whiskers as tactile organs. Instead, they proposed that whisking (and associated facial vibrissae) evolved with the transformation of the maxillary canal into a mammal-like infraorbital foramen. This transition likely happened about 241 million years ago (Fig. 2A), in the most derived probainognathian cynodonts (such as the Tritylodontidae, Brasilodontidae, and Tritheledontidae), which implies that the origin of whiskers may have predated the origin of mammals (Benoit et al., 2016c, 2020). Using the dimensions of the infraorbital foramen of the Tritylodontidae *Tritylodon* and Tritheledontidae *Pachygenelus*, Muchlinski et al. (2020) found that these derived cynodonts had foramina sizes consistent with modern mammals capable of whisking. Derived probainognathians coincidentally possess an enlarged trigeminal ganglion (Rodrigues et al., 2014; Rowe and Shepherd, 2016) and muscle insertions consistent with a fleshy, mobile face (Benoit et al., 2020). Facial mobility is further improved in Mammalia with the loss of the internarial bar and then in Theria with the establishment of a motile rhinarium (Higashiyama et al., 2021). In parallel, electroreception and highly sensitive tactile receptor organs evolved on the bill/beak of monotremes (Manger and Pettigrew, 1995; Pettigrew, 1999; Rowe et al., 2008).

According to Rowe et al. (2011), the evolution of hair in Mammaliaformes improved tactile sensitivity, which played a key role in the development of the sensory cortex and prompted the increase in size of the cerebral hemispheres that is directly visible on the endocranial casts of mammaliaforms (Rowe et al., 2011). Pelycosaurs, therapsids, and basal cynodonts most likely had reptile-like facial sensitivity and lacked hair, but the presence of an infraorbital canal in derived probainognathians suggests that mammal-like enhanced tactile sensitivity may have predated Mammaliaformes (Benoit et al., 2016c, 2020; Muchlinski et al., 2020). The evolution of the reptilian-like maxillary canal into a mammalian infraorbital foramen coincides with the appearance of an interhemispheric longitudinal sulcus on some probainognathian endocasts (Fig. 2A), which indicates the potential expansion of the cerebral hemispheres in these derived cynodonts (Rodrigues et al., 2014, 2018).

This summary, nevertheless, eschews some contradictory elements. First, it relies on the assumption that whiskers and whisking behaviors appeared at the same time, but the tactile function of whiskers is not bound to whisking in modern mammals (Kerekes et al., 2017; Muchlinski et al., 2020). Second, it assumes that whisking originated only once in synapsids, which is usually accepted, but not fully consensual (Muchlinski et al., 2020). In this respect, it should be noted that some cynognathian cynodonts evolved a simple and large mammal-like infraorbital foramen in a convergent manner with derived probainognathians (Franco et al., 2021). Finally, it leaves aside many other fossil cues, such as the anatomical proxies that have been proposed in support of the evolution of an elevated metabolic rate (endothermy). A discussion

of the origin of endothermy is beyond the scope of this review, but endothermy and the evolution of hair are tightly correlated (Benton, 2021; Hillenius and Ruben, 2004). Unfortunately, there is currently no consensus on when mammal-like thermophysiology originated in synapsids: it could be as early as in the Carboniferous (Grigg et al., 2022; Knaus et al., 2021; Legendre et al., 2016), as late as in the Jurassic (Newham et al., 2020, 2022), or any possible time in between (Benoit et al., 2016a, 2017a; Benton, 2021; Faure-Brac and Cubo, 2020; Hillenius and Ruben, 2004; Laaß et al., 2011; Lovegrove, 2017; Olivier et al., 2017; Rey et al., 2017). In this respect, Araújo et al. (2022) recently showed that endothermy increases body temperature, which results in changes in the viscosity of inner ear fluids. They consequently proposed that the geometry of the semicircular canals in synapsids had to adapt to the less viscous inner ear fluids to remain functional. Studying the bony labyrinth across synapsids, they found changes in the geometry of the semicircular canals not related to locomotion, but consistent with an elevated body temperature in probainognathians. It is noteworthy that these taxa also evolved an infraorbital canal and lost the pineal eye, two other paleoneurological proxies that support an early origin of hair (Benoit et al., 2015, 2016a,c, 2020).

2.5 Unrefined motor coordination?

The cerebellum is a primary center of motor coordination and its evolution can be traced, at least for superficial anatomical structures, using endocranial casts (Fig. 2A). In dorsal view, the brain endocast is essentially tubular in non-mammaliaform synapsids, but the cerebellum is usually broad (Hopson, 1979; Kielan-Jaworowska et al., 2004). The cast of the cerebellum is always the mediolaterally widest part of the endocast in therapsids and cynodonts (Benoit et al., 2017a; Du Plessis, 2010; Hopson, 1979; Kerber et al., 2021; Kielan-Jaworowska et al., 2004; MacLean, 1990; Pavanatto et al., 2019; Rodrigues et al., 2014, 2018; Rowe et al., 2011). Rodrigues et al. (2018) showed that it is even wider than the olfactory bulbs in cynodonts, and that it was at least as extensive in trithelodontids and brasilodontids as seen in mammals, perhaps even more so compared to other parts of the endocast. A cast of the cerebellar vermis characterizes mammalian endocasts (Kielan-Jaworowska et al., 2004). It is absent in basal cynodonts and cynognathians, but becomes large enough to be observed on the endocast in probainognathians (Benoit et al., 2016c; Quiroga, 1980, 1984; Rodrigues et al., 2018).

The casts of the parafloccular (subarcuate) fossae for the floccular lobe of the cerebellum vary in size in pelycosaurs, therapsids, and mammals (Angielczyk et al., 2021; Bazzana et al., 2022; Benoit et al., 2017a, 2021c; Ferreira-Cardoso et al., 2017; Hopson, 1979; Romer and Price, 1940; Sánchez-Villagra, 2002), but are consistently enlarged and laterally protruding in cynodonts and Mesozoic mammaliaforms (Du Plessis, 2010; Hopson, 1979; Kielan-Jaworowska, 1986; Kielan-Jaworowska et al., 2004; Pavanatto et al., 2019; Rodrigues et al., 2018). Noticeably, the basal cynodont *Galesaurus* has larger casts of the floccular fossae than any other cynodont or Mesozoic mammaliaform (Pusch et al., 2019).

The dimensions and shape of the bony labyrinth semicircular canals have also often been correlated with locomotor agility (Spoor et al., 2007; Ryan et al., 2012 but see Bronzati et al., 2021); however, there is no major difference between the semicircular canals of non-mammalian synapsids and mammaliaforms apart from the degree of ossification of the labyrinthine capsule (Benoit et al., 2017b; Rodrigues et al., 2013) and the afore-mentioned adaptations to decreased endolymph viscosity (Araújo et al., 2022).

All the features that can be traced in the fossil record and that characterize mammalian motor coordination appear to have evolved prior to the origin of mammaliaforms (Fig. 2A) and, as such, there is no reason to consider that cynodonts had less refined motor coordination than their descendants. Cynodonts were capable of precise and complex chewing motions (Crompton, 1972; Kemp, 2005), and hunting fast-moving insects and small vertebrates (Angielczyk and Kammerer, 2018; Qvarnström et al., 2017). Kemp (1979, 2005) suggested that cynodonts and therapsids could alternate between two types of anterior limb postures (dual gait hypothesis), which would likely have required sophisticated motor coordination and proprioceptive feedback neural systems. Among therapsids, some dicynodonts such as *Diictodon* and *Oudenodon* were capable of making helical burrows similar to those made by the Cenozoic mammal *Palaeocastor* (Smith, 1987). One of the close relatives of dicynodonts, *Suminia*, was likely arboreal, a behavior that requires finely-tuned motor coordination (Fröbisch and Reisz, 2009). Pelycosaurs, on the other hand, may have been less agile, as illustrated by a trackway preserving a sprawling walk interrupted by resting traces (Niedźwiedzki and Bojanowski, 2012).

2.6 Summary of sensory organs

To conclude regarding the condition of sensory organs at the root of pre-mammalian evolutionary history, it is evident that:

- It is difficult to juxtapose a pre-mammalian "primitive" condition against a "derived" mammalian condition, as the evolution of the sensory organs in synapsids was mosaic and complex (Fig. 2).
- The transition from "mammal-like reptiles" to early mammaliaforms is well-documented by the fossil record and the sensory organs of the most derived probainognathian cynodonts appear very similar to those of the earliest Mesozoic mammaliaforms (Fig. 2).
- Non-mammalian synapsids display some noticeable specializations of their sensory and motor coordination organs that equal, and in some cases even surpass, those of mammals. For example, therapsids had a broader spectrum sensitivity for color vision and a more diverse array of diel activity patterns than early mammaliaforms. In addition, non-mammalian synapsids display some noticeable cases of specialized sensory-motor adaptations (e.g., to fossoriality, Fig. 3).
- The sensory-motor history of synapsids is not limited to a list of evolutionary changes moving toward "mammalness," but was the result of many evolutionary

lineages that experimented with adaptations to the multiple ecological niches that they dominated for over a hundred million years. Only one of those lineages eventually survived and gave rise to mammals, but this single lineage should not cloud the overall paleoneurological diversity of synapsids.

3 Behavioral complexity

It is always problematic to determine how intelligent, or behaviorally complex, an extant animal is, let alone how complex the behavior of an extinct animal may have been, but paleontologists do have some clues. Measurements of natural and artificial endocranial casts are used to calculate the encephalization quotient (Eisenberg, 1981; Jerison, 1973), which enables comparisons of relative brain size, although this is not necessarily a direct reflection of an animal's behavioral complexity (Healy and Rowe, 2007; Manger, 2006, 2013; Manger et al., 2013). More direct evidence of complex behavior can be found in exceptional cases of preservation in lagerstätten conditions, or can be proposed from the paleoneurological, functional anatomical, palaeopathological, and ichnology clues (Boucot, 1990; Boucot and Poinar, 2011). The scarcity of evidence for complex behavior in non-mammalian synapsids, their comparatively low relative brain size, the narrow cerebral hemispheres, and apparent lack of an isocortex are suggestive of less complexity (Cameron et al., 2019; Hopson, 1979; Jerison, 1973; Rowe et al., 2011). This view will be contrasted here with the sparse and often overlooked evidence for complex behavior in non-mammalian synapsids.

3.1 Small brain?

The endocasts of pelycosaurs, therapsids, and cynodonts reveal an organization of the brain similar to that observed across vertebrates, with the olfactory bulbs anteriorly, the forebrain (including the cast of the cerebral hemispheres) caudal to the olfactory bulbs, followed caudally by the midbrain (which bears the pineal tube), and most caudal, the hindbrain (including the cast of the cerebellum) (Benoit et al., 2017a; Du Plessis, 2010; Hopson, 1979; Kielan-Jaworowska et al., 2004; MacLean, 1990; Rodrigues et al., 2018; Rowe et al., 2011). All these elements are visible in dorsal views of the endocasts (Fig.2A). In lateral view, some variations in the degree of flexure between the forebrain and hindbrain exist, particularly in pelycosaurs, dicynodonts, dinocephalians, and biarmosuchians, but the endocast is usually linear and tubular in Paleozoic and Mesozoic synapsids, including many Mesozoic mammals (Benoit et al., 2017a; Du Plessis, 2010; Edinger, 1964; Hopson, 1979; Kielan-Jaworowska et al., 2004; MacLean, 1990; Rodrigues et al., 2018; Rowe et al., 2011). Variations in the olfactory bulbs, midbrain (pineal organ), and hindbrain (cerebellum) have been discussed above, thus here we will focus on the forebrain and encephalization quotient (EQ, a measure of the size of the brain relative to the size of the body). Variations in the surface of the cerebral hemispheres are indistinct on the endocasts of pelycosaurs, therapsids, and most cynodonts, and consequently

the forebrain appears rather featureless and cylindrical (Benoit et al., 2017a; Hopson, 1979; Kielan-Jaworowska et al., 2004; MacLean, 1990; Pavanatto et al., 2019; Quiroga, 1979, 1980, 1984; Rodrigues et al., 2018). The cerebral hemispheres of the probainognathian cynodont *Probainognathus* expand caudally and caudolaterally (Quiroga, 1980; Rodrigues et al., 2018), creating a gap separating the posterior ends of the two hemispheres (Fig.2A). This condition appears to be the first step toward a distinct separation of the two hemispheres, as observed in the endocasts of more derived cynodonts such as the Trithelodontidae and Brasilodontidae, in which a longitudinal sulcus separates the two bulging hemispheres (Pavanatto et al., 2019; Rodrigues et al., 2018, 2014). This potentially marks the beginning of a trend toward hemispheric expansion that appears to continue in mammaliaforms and mammals (Fig. 2), and may be correlated to the origin of the isocortex (Pavanatto et al., 2019; Rowe et al., 2011).

The isocortex is a uniquely mammalian structure involved in sensory perception and integration, the initiation of movement, problem-solving, and appears crucial to their cognitive abilities (Aboitiz, 1992; Aboitiz et al., 2003). A rhinal fissure usually (but not always) marks the ventrolateral extent of the isocortex on mammalian endocasts (Dechaseaux, 1962; Macrini et al., 2007), but such a fissure is absent or incipient in Paleozoic and Mesozoic synapsids, including contemporaneous mammals (Fig. 2B) (Hoffmann et al., 2014; Hurum, 1998b; Huttenlocker et al., 2018; Kielan-Jaworowska, 1986; Kielan-Jaworowska et al., 2004; Macrini et al., 2007). The presence of an isocortex has been hypothesized in early Mammaliaformes (Rowe et al., 2011) and therapsids (Laaß and Kaestner, 2017) based on the dimensions and distinctiveness of the cerebral hemispheres, but in the absence of a well-defined rhinal fissure, defining the origin of the isocortex remains speculative.

Descriptions of Paleozoic and Mesozoic synapsid fossil endocasts have increased substantially in the past 10 years, such that the EQ has been used to compare relative brain size among fossil synapsids and mammals, and to discuss their relative behavioral complexity (Araújo et al., 2017; Benoit et al., 2017a; Castanhinha et al., 2013; de Simão-Oliveira et al., 2020; Hoffmann et al., 2021; Kerber et al., 2021; Laaß and Kaestner, 2017; Pavanatto et al., 2019; Pusch et al., 2019; Rodrigues et al., 2014, 2018). Following the principle of proper mass (Jerison, 1973), an EQ that equals 1 represents an animal with the brain mass expected for an animal with the same particular body mass. As a consequence, any animal with an EQ below 1 has a brain mass that is smaller than expected for its body mass, while an animal with an EQ >1 has a brain mass that is larger than expected for its body mass. In the relatively larger brains (those with EQ >1), it is envisaged that there will be more neurons than needed for control of the body, and that these neurons can be employed in other activities, which may be interpreted as an indicator of intelligence (Jerison, 1973); however, this scenario requires the assumption that all neurons have equal computational powers, which is not the case (see critiques in Healy and Rowe, 2007; Manger et al., 2013). When using EQs based on extant mammalian data, most therapsids, cynodonts and early mammaliaforms have EQs between 0.02 and 0.50,

intermediate between the upper range of reptilian and lower range of mammalian variations (Benoit et al., 2017a; de Simão-Oliveira et al., 2020; Pavanatto et al., 2019; Rodrigues et al., 2014, 2018; Rowe et al., 2011). The most encephalized non-mammalian synapsids, those that reach the mammalian range of EQ variation, are the cistecephalids *Kawingasaurus* and *Cistecephalus*, the probainognathian *Therioherpeton*, and the early mammaliaforms *Hadrocodium* and *Morganucodon* (Benoit et al., 2017a; Laaß and Kaestner, 2017; Laaß and Schillinger, 2017; Rowe et al., 2011). Based on the EQs of *Hadrocodium* and *Morganucodon*, Rowe et al. (2011) proposed that the first steps toward true levels of mammalian encephalization began in Mammaliaformes.

Mammalian EQ calculation methods have, however, been criticized regarding their appropriateness and applicability to absolutely small-brained species (Hurlburt et al., 2013). As such, here we generate an extinct synapsid-specific EQ calculation based on data available in the literature (Table 1). Mammaliaforms (Mammalia + early Mammaliaformes) were removed for the calculation of the EQ to enhance the contrast (hypothesized by Rowe et al., 2011) between non-mammaliaform synapsids and early mammaliaforms. This extinct synapsid dataset was used to (i) generate a new regression equation based on comparing endocast volume to body mass ($R^2 = 0.93$); and (ii) from this regression derive the synapsid-specific EQ equation that may be more appropriate for comparisons of

Table 1 Dataset of Body Mass (BM) and Endocranial Volume (EV) used to calculate the synapsid specific Encephalization Quotient ($EQ_{synapsid}$).

Higher rank taxon	Genus	BM	EV	EQ synapsid
Therapsida	*Lemurosaurus*	5219,17	**4,02**	0,86
Therapsida	*Lemurosaurus*	1327,12	**1,75**	0,94
Therapsida	*Moschops*	128,654,74	**50,24**	1,27
Therapsida	*Strutiocephalus*	288,175,73	**65,00**	0,95
Therapsida	*Pristerodon*	1480,74	**2,18**	1,09
Therapsida	*Lystrosaurus*	18,511,07	**8,00**	0,74
Therapsida	*Kawingasaurus*	211,94	**1,09**	2,00
Therapsida	*Niassodon*	782,06	**1,06**	0,81
Therapsida	*Rastodon**	1984,00	**1,76**	0,72
Therapsida	Gorgonopsia indet.	3043,00	**4,62**	1,42
Therapsida	*Aelurosaurus*	7255,45	**6,00**	1,03
Therapsida	*Aloposaurus*	8636,64	**6,77**	1,04
Therapsida	*Tetracynodon*	1316,58	**2,28**	1,23
Therapsida	*Microgomphodon*	1797,04	**2,36**	1,03
Basal cynodont	*Galesaurus***	669,00	**2,75**	2,33
Basal cynodont	*Galesaurus**	1984,00	**2,48**	1,02

Table 1 Dataset of Body Mass (BM) and Endocranial Volume (EV) used to calculate the synapsid specific Encephalization Quotient ($EQ_{synapsid}$).—cont'd

Higher rank taxon	Genus	BM	EV	EQ synapsid
Basal cynodont	*Platycraniellus***	2863,00	**5,56**	1,78
Basal cynodont	*Thrinaxodon*	1010,05	**1,56**	1,00
Basal cynodont	*Thrinaxodon*	700,00	**1,46**	1,20
Cynognathia	*Diademodon*	77,401,98	**26,97**	0,95
Cynognathia	*Diademodon*	7000,00	**8,00**	1,41
Cynognathia	*Trirachodon*	3124,71	**2,15**	0,65
Cynognathia	*Langbergia***	2614,00	**2,30**	0,78
Cynognathia	*Andescynodon*	1071,04	**1,11**	0,69
Cynognathia	*Andescynodon*	1397,04	**1,10**	0,57
Cynognathia	*Massetognathus*	2677,43	**3,16**	1,06
Cynognathia	*Massetognathus*	1316,58	**1,56**	0,84
Cynognathia	*Massetognathus*	6128,85	**2,86**	0,55
Cynognathia	*Massetognathus*	2237,67	**2,81**	1,06
Cynognathia	*Siriusgnathus*	34,262,15	**28,80**	1,76
Cynognathia	*Exaeretodon*	69,495,56	**15,98**	0,61
Cynognathia	*Exaeretodon*	36,435,98	**26,43**	1,55
Cynognathia	*Exaeretodon*	15,061,15	**11,47**	1,21
Probainognathia	cf. *Probelesodon*	5417,25	**4,02**	0,84
Probainognathia	*Probelesodon*	4763,56	**2,60**	0,59
Probainognathia	*Probainognathus*	858,43	**1,12**	0,81
Probainognathia	*Therioherpeton*	97,58	**0,29**	0,90
Probainognathia	*Riograndia*	71,15	**0,32**	1,22
Probainognathia	*Tritylodon***	5617,00	**5,27**	1,08
Probainognathia	*Tritylodon*	6278,35	**5,06**	0,96
Probainognathia	*Tritylodon*	8758,68	**6,30**	0,96
Probainognathia	*Brasilitherium*	98,57	**0,24**	0,74
Early Mammaliaformes	*Morganucodon*	51,00	**0,33**	1,54
Early Mammaliaformes	*Hadrocodium*	2,75	**0,04**	1,31
Mammalia	*Vintana*	8950,00	**15,64**	2,34
Mammalia	*Triconodon*	138,00	**0,82**	2,00
Mammalia	*Litovoi****	166,00	**0,73**	1,57
Mammalia	*Chulsanbaatar*	13,30	**0,22**	2,57
Mammalia	*Kryptobaatar*	28,18	**0,32**	2,28
Mammalia	*Vincelestes*	619,00	**2,12**	1,89
Mammalia	*Kennalestes*	39,00	**0,30**	1,70
Mammalia	*Asioryctes*	43,20	**0,50**	2,65
Mammalia	*Zalambdalestes*	82,69	**1,02**	3,50

Dataset from Pavanatto et al. (2019) updated and expanded with data from de Simão-Oliveira et al. (2020)(*), Du Plessis (2010)(**), and Csiki-Sava et al. (2018)(***). Values per genera were averaged to create Fig. 7.

relative brain size in Paleozoic and Mesozoic synapsids. The synapsid-specific EQ ($EQ_{synapsid}$) equation is:

$$EQ_{synapsid} = \text{Endocast Volume in cm}^3 / 10^{(06669 \text{ x Log Body Mass in g}) - 18{,}188}$$

Using this $EQ_{synapsid}$ it is found that all Mesozoic Mammalia have an $EQ_{synapsid} > 1.5$, indicating that compared to most fossil synapsids, mammals have relatively larger brains. Interestingly, the $EQ_{synapsid}$ for several synapsid species reaches the range of mammalian encephalization (here, $EQ_{synapsid} > 1.5$, Fig. 7). The taxa with an $EQ_{synapsid}$ beyond 1.5 are the fossorial dicynodont *Kawingasaurus*, and the cynodonts *Platycraniellus*, *Galesaurus*, and *Siriusgnathus*. Interestingly, the dinocephalians *Moschops* and *Struthiocephalus*, have an $EQ_{synapsid}$ close to 1, which suggests that despite their large body mass and reputation of being rather unintelligent animals, tapinocephalids were on average no less encephalized than other therapsids (Fig. 7). Complex social behaviors, such as head-butting and gregariousness have been proposed for tapinocephalids (Barghusen, 1975; Benoit et al., 2016b, 2017c), which may account for this unexpected result. Unlike the analysis of Rowe et al. (2011), it is found here that the $EQ_{synapsid}$ of *Morganucodon* and *Haldanodon* are not intermediate between those of cynodonts and crown Mammalia as they fall well within the range of cynodonts $EQ_{synapsid}$ (Fig. 7). These results suggest that (i) degrees of encephalization comparable to some Mesozoic mammals were perhaps more widespread among therapsids and cynodonts than previously envisioned, (ii) that cynodonts were not transitional between therapsids and mammaliaforms

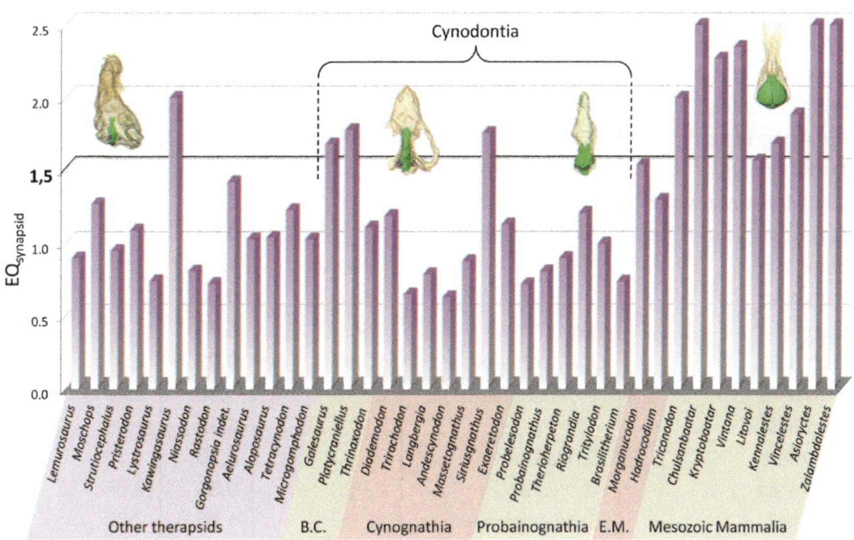

FIG. 7

Variations of the $EQ_{synapsid}$ across Synapsida (see Table 1). Abbreviations: B.C., basal cynodonts; E.M., early mammaliaforms.

in terms of encephalization, and (iii) that many increases of $EQ_{synapsid}$ above 1.5 happened across therapsids and cynodont evolution, but this threshold was durably crossed by crown Mammalia only (Fig. 7). This contrasts with the currently accepted evolutionary scenario of a steady increase in encephalization occurring earlier, in early Mammaliaformes (Huttenlocker et al., 2018; Rowe, 2017; Rowe and Shepherd, 2016; Rowe et al., 2011).

Overall, many therapsids and cynodonts reached mammal-like levels of encephalization, and features of the cerebral hemispheres previously attributed to mammaliaforms actually appeared in derived probainognathians. However, the $EQ_{synapsid}$ stabilized above the minimum mammalian level only in the mammalian crown group. Most therapsids and basal cynodonts had a simple, tubular brain that likely did not completely fill the endocranial cavity. A noticeable exception to the relative simplicity of therapsids' endocast is the dicynodont *Kawingasaurus*, from the late Permian of the Ruhuhu Basin of Tanzania. It has an endocast that looks overall more spherical than cylindrical and displays dorsally and laterally bulging cerebral hemispheres (Laaß and Kaestner, 2017). *Kawingasaurus* exemplifies that neural evolution in therapsids could depart strongly from the mammalian trajectory.

3.2 Simple behavioral repertoire?

Looking for direct evidence of behavior in the fossil record (provided by, for example, ichnofossils) and anatomical adaptations for certain types of complex behavior represent another source of proxies for revealing behavioral complexity, independent of relative brain size and the associated concerns. Compared to the literature about endocasts and relative brain size in synapsids, fossil synapsid behaviors are relatively poorly studied. This section focuses on the published evidence that may indicate social behavior, parental care, and other signs of behavioral complexity in Paleozoic and Mesozoic synapsids.

When the Synapsida originated in the late Carboniferous (about 315 million years ago), one of the defining anatomical features distinguishing them from their sister group the Reptilia was the presence of enlarged caniniform teeth, that eventually became the canine teeth of modern mammals (Kemp, 2005; Rubidge and Sidor, 2001). There are many examples of synapsids that evolved overly large, saber-like caniniforms (or canines) among therapsids, cynodonts, and mammals (Benoit et al., 2016b; Cisneros et al., 2015; McGhee, 2011; Valkenburgh and Jenkins, 2002; Whitney et al., 2021). The most striking examples are the herbivorous taxa, such as *Tiarajudens*, dicynodonts, and cynognathian cynodonts, in which the caniniforms are enlarged beyond what would be necessary for food acquisition and processing (Benoit et al., 2016b; Cisneros et al., 2015). Accordingly, a role of these enlarged teeth in ritualized display and social combats has long been hypothesized (Benoit et al., 2016a,b,c, 2021b; Cisneros et al., 2015; Geist, 1972). There are currently three lines of evidence that support the concept that enlarged caniniforms played another role in addition to food acquisition and processing in Paleozoic and Mesozoic synapsids. First, healed bite marks that were most likely inflicted by conspecifics during

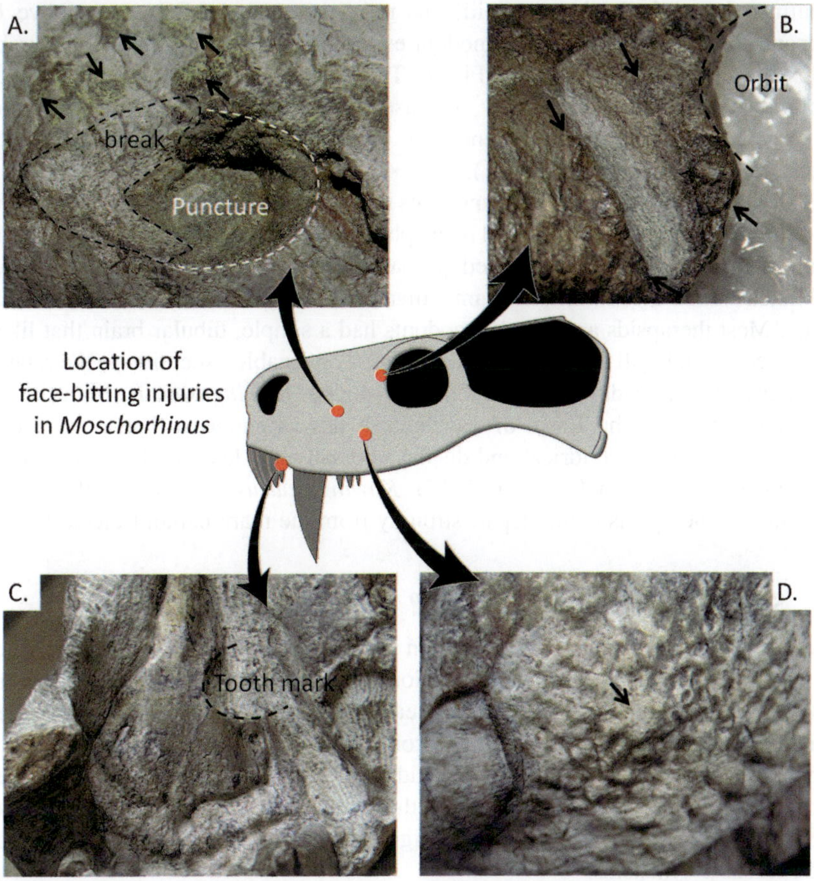

FIG. 8

Evidence of face-bitting in *Moschorhinus*. (A) Maxilla of RC32 has a large puncture mark (outlined in white) with a conchoidal break (outlined in black) and many bony calluses (arrows). (B) Orbit rim of NMQR 3939 with a deep puncture surrounded by two raised ridge-like bony calluses (arrow). (C) Vomer of BP/1/8619 with a bite mark outlined in black. (D) Maxilla of BP/1/8106 with a bony callus (arrow).

face-biting contests support the possibility of intraspecific combat in therapsids (Fig. 8; Benoit et al., 2021a). Second, in many therapsid lineages, the loss of enlarged caniniforms is compensated for by the acquisition of other cranial adornments, such as bosses and horns similar to those of later dinosaurs and mammals, that evidently used them for combat and display (Barghusen, 1975; Benoit et al., 2016a,b,c, 2021c; Nasterlack et al., 2012). Third, these enlarged caniniform teeth or cranial adornments are often (though, not always) sexually dimorphic, indicating that they potentially evolved under pressures related to sexual selection (e.g., Sullivan et al., 2003; see Benoit et al., 2016a,b,c and references therein). In extant species, intraspecific

agonistic behaviors are mostly used to secure mates and ensure territorial dominance, implying some form of social hierarchy (Emlen, 2008; Geist, 1966). As such, to hypothesize the presence of these kinds of behaviors in Paleozoic and Mesozoic synapsids implies that they may have not been solitary, but may have lived in groups with a specific social construct that may be more or less complex depending on the species' life history and the environmental pressures exerted on the species.

Synapsid fossils that support some degree of social interaction include localized concentrations of burrow casts of cynognathian cynodonts and dicynodonts (Fig. 9A) (Bordy et al., 2019; Colombi et al., 2012; Groenewald et al., 2001; Krapovickas et al., 2013; Smith, 1987), behavioral aggregations of juveniles of *Lystrosaurus* and *Thrinaxodon* as well as tapinocephalid dinocephalians skeletons (Benoit et al., 2021c; Neumann, 2020; Rubidge et al., 2019; Smith and Botha-Brink, 2005; Viglietti et al., 2013), evidence of communal latrines in Triassic cynodonts

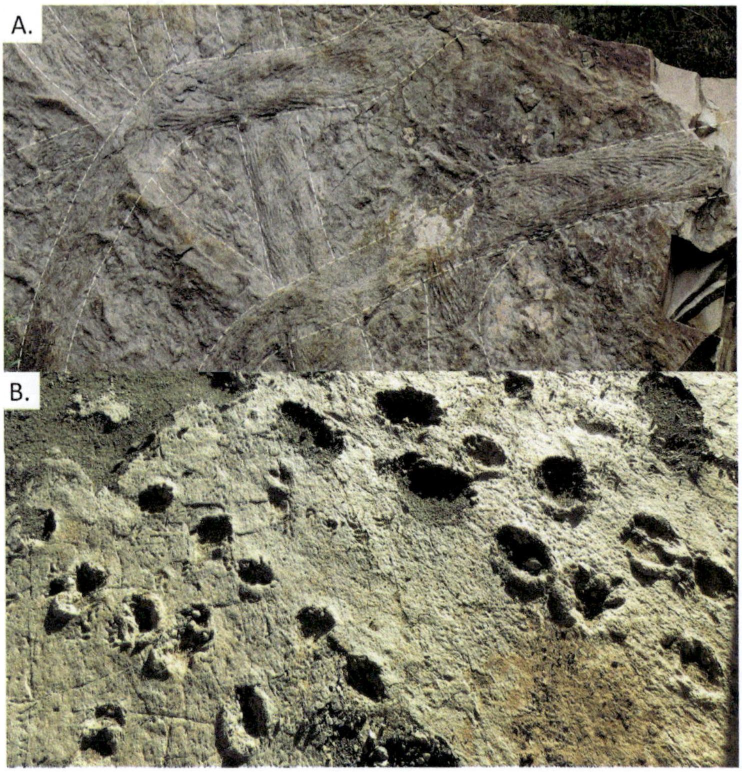

FIG. 9

Two social behaviors preserved in the fossil record. (A) Complex network of burrows made by middle Triassic cynodonts, described by Bordy et al. (2019). (B) Two of the six parallel trackways left by late Permian dicynodonts, described by de Klerk (2002). Pictures by Julien Benoit. Each photo is about 2 m across.

and dicynodonts (Fiorelli et al., 2013; Groenewald et al., 2001), and the existence of parallel dicynodont trackways that suggest that more than one individual of a given taxon moved along the same "pathway" (Fig. 9B) (de Klerk, 2002). These findings suggest that herbivorous therapsids and cynodonts may have experimented with sociality (at least temporarily) millions of years before these types of behaviors were first expressed in dinosaurs and mammals.

In carnivorous pelycosaurs, therapsids, and cynodonts, fossilized aggregations of skeletons (including juveniles), rather than being interpreted as evidence of life in herds, have historically been interpreted as evidence of parental care (Botha-Brink and Modesto, 2007; Jasinoski and Abdala, 2017; Spindler, 2015). Parental care has also been hypothesized in herbivorous taxa, such as dicynodonts, cynognathians, and tritylodontids, but only in cases when the aggregations involved different age classes (juveniles and adults) preserved together (Hoffman and Rowe, 2018; Jasinoski and Abdala, 2017). In some cases, as in the Permian dicynodont *Diictodon*, burrows were used as brooding chambers (Smith et al., 2021). Some trackways attributed to early mammaliaforms correspond to an animal that was carrying its young on its back (baby riding), which suggests that prolonged parental care may have evolved before the root of the mammalian clade (Kuznetsov and Panyutina, 2018). Thus, some sort of parental care appears to be a common ancestral trait for all synapsids (Fig. 2A), although it is difficult to determine its precise nature. For example, a specimen preserving a sub-adult male *Diictodon* with two recently hatched juveniles (Fig.10) can be speculatively interpreted as suggesting that the burden of caring for the young was shared with immature individuals, and between males and females (Smith et al., 2021).

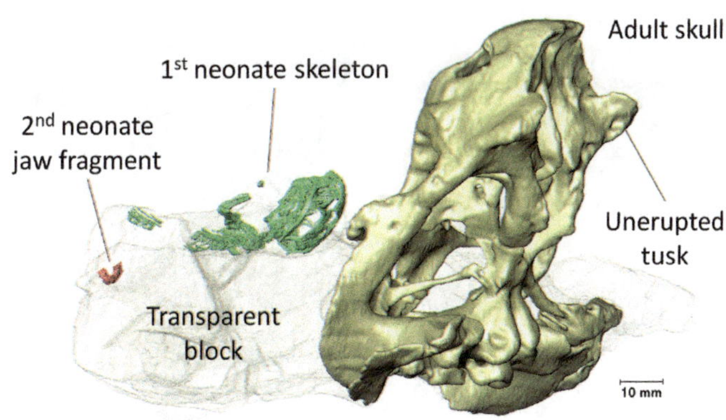

FIG. 10

Parental care (or baby sitting) in a late Permian dicynodont (Smith et al., 2021). In the same block (here in transparent), one subadult and two juvenile *Diictodon* are preserved together. One neonate individual is represented by an almost complete skeleton (in green), the other by, at least, one fragment of a lower jaw (in red). The adult (in yellow) has unerupted tusks which suggests this is a subadult male.

Based on the biology of the monotremes, egg-laying (oviparity) is thought to be the ancestral condition for synapsids. Noticeably, the absence of any Paleozoic or Mesozoic synapsid fossil eggs (despite a rich record of contemporaneous dinosaur eggs, Kitching, 1964) and the apparent ease with which modern reptiles appear to shift between oviparity, ovoviviparity, and viviparity (Blackburn, 2000; Kuchling and Hofmeyr, 2022) suggests that this often unquestioned scenario might be too simplistic. Lactation originated sometime before the origin of Mammalia, but there is currently no consensus on when this happened (Benoit, 2019; Benoit et al., 2016c; Lefèvre et al., 2010; Oftedal, 2002, 2020).

Apart from parental care, there is no positive evidence of social interactions, such as pack hunting or social groups, in carnivorous synapsids prior to the Cenozoic. A correlation between social bonds and the length of the auditory lagena (vestibule) has been proposed in dinosaurs (Walsh et al., 2009), but whether this applies to the elongated cochlear canal of probainognathian cynodonts (Fig. 4) has not been ascertained.

Finally, there are noticeable examples of what may be interpreted as behavioral responses to changes in the environment. Trace fossils and bite marks show that carnivorous pelycosaurs in the early Permian were capable of digging to catch burrowing amphibians (Bakker et al., 2015). During the end-Permian biological crisis and its direct aftermath, both the cynodonts and the dicynodont *Lystrosaurus* adapted to burrowing and practiced prolonged periods of torpor/aestivation to escape the extreme heat and dryness (Botha, 2020; Botha-Brink, 2017; Fernandez et al., 2013; Smith and Botha-Brink, 2014; Whitney and Sidor, 2020). A Triassic carnivorous therapsid's food cache described by Smith and Botha-Brink (2014) is suggestive of long-term episodic and spatial memory capacities. Whether these cases represent genuine examples of behavioral innovations (i.e., a de novo behavior produced by the brain in response to a situation) or long-term evolutionary adaptations (i.e., a trait that evolved under natural selection) is difficult to address.

3.3 Discussion on brain and behavior

Was the brain of pelycosaurs, therapsids, cynodonts, and even Mesozoic mammaliaforms capable of producing complex, mammal-like behaviors? The behaviors reviewed above are complex and "mammal-like" only if compared to a hypothetical and stereotypical concept of "reptile-like" behavior that is primitive, simple, and solitary. For example, intraspecific combat and displays are not limited to mammals (or even vertebrates) and do not always imply the existence of an underlying social hierarchy (Emlen, 2008). Similarly, gregariousness does not always imply social bonds, nor is it unique to mammals (Boucot, 1990; Geist, 1966; Pol et al., 2021). In addition, the mammal-like nature of parental care in pelycosaurs, therapsids, and cynodonts has recently been questioned by the description of an aggregation of no less than 38 neonates of the tritylodontid *Kayentatherium* (Hoffman and Rowe, 2018). Such a high number of neonates is unusual for mammals, but not for reptiles, which implies that tritylodontid cynodonts: (i) likely laid eggs; (ii) could not breastfeed that many babies (the neonate skulls have functional teeth

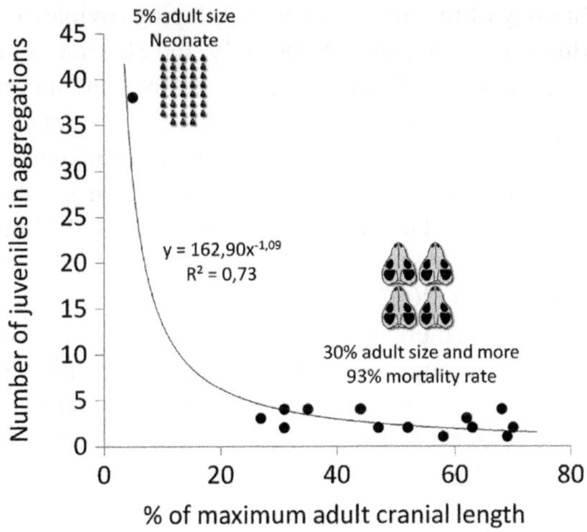

FIG. 11

The steep decline of juvenile survival rate in cynodont during growth.

Data from Benoit, J., 2019. Parental care or opportunism in south African Triassic cynodonts? South Afr. J. Sci. 115, 1–2. https://doi.org/10.17159/sajs.2019/5589.

and were thus likely precocial or could have been fed solid or half-processed food by the parents); and (iii) had a very high juvenile mortality rate (about 93% at 30% adult size, Fig. 11), which questions the quality and duration of parental care that cynodonts provided (Benoit, 2019). Taken together, this suggests a rather reptilian-like reproductive biology in this derived probainognathian, that is close to the very root of the mammalian clade.

Despite the noticeable occurrence of large-brained Paleozoic and Mesozoic synapsids (Fig. 7), and a hundred years of efforts to shed light on their behavior, evidence of complex pre-mammalian behavior remains rare, especially compared to other fossil groups such as Mesozoic dinosaurs and Cenozoic mammals (Boucot, 1990; Boucot and Poinar, 2011; Tanke and Rothschild, 2002). Monotaxic behavioral aggregations of articulated skeletons (e.g., paired and intercurled fossils of *Diictodon*, *Lystrosaurus*, or *Thrinaxodon*) and fossilized parallel trackways—consensually considered the most reliable evidence of gregariousness (Roach and Brinkman, 2007)—are exceedingly rare despite the great abundance of Permian-Triassic synapsid fossils (Gand et al., 2007; Ivakhnenko, 2008; Jasinoski and Abdala, 2017; Kitching, 1977; Klein and Lucas, 2021; Marchetti et al., 2019; Romer and Price, 1940; Smith and Botha-Brink, 2014). The absence of evidence is not evidence of absence, but given the scarcity of evidence for behavioral complexity in an otherwise relatively well-documented fossil record, it is currently not possible to positively argue in favor of a well-developed intelligence in pre-mammalian synapsids.

4 Conclusion: the evolution of the synapsid brain, behavior, and sensory organs, brain and behavior

The literature on synapsid paleoneurology has primarily focused on the origins of the mammalian brain, sense organs, and behavioral repertoire (Benoit et al., 2017a; Dechaseaux, 1962; Edinger, 1975; Hopson, 1979; Jerison, 1973; Kielan-Jaworowska, 1986; Quiroga, 1984). Jerison (1973), following in the footsteps of Romer (1956), theorized that to avoid competition from and predation by the archosaurs, mammals went through a Mesozoic nocturnal bottleneck that triggered a sensory and neural revolution. This invoved augmented olfaction, audition, and somatosensation that usurped vision as the primary sense and fuelled the growth and organizational complexity of the isocortex. Support for correlations between the evolution of the isocortex and enhanced olfaction, audition, and somatosensation has continued to be supported by additional evidence (Aboitiz et al., 2003; Rowe, 1996; Rowe and Shepherd, 2016).

Based on updated data (Hopson, 1979; Kemp, 2009; Kielan-Jaworowska et al., 2004; Macrini et al., 2007), Rowe et al. (2011) proposed a new model, that has met general acceptance (Hoffmann et al., 2014; Huttenlocker et al., 2018; Rowe and Shepherd, 2016), in which relative brain size began to increase in the early mammaliaform *Morganucodon*, and was marked by three main pulses in the relative size of the brain. Unlike that which Jerison (1973) proposed, in this new model an increase in the complexity of the olfactory system is considered as the main driver of the increase in relative size of the brain because the first and second pulses of increased relative brain size are correlated with increases in olfactory bulb size, while the third pulse coincides with the ossification of the nasal capsule (Rowe et al., 2011; but see Du Plessis, 2010 for a contrasting point of view on the evolution of olfactory bulbs).

However, the recent data reviewed here is inconsistent with this model, because we now know that refined sensory organs predated the origin of mammaliaforms. The olfactory bulbs began to become relatively large at the root of the Probainognathia clade (Rodrigues et al., 2018) and cartilaginous support surfaces for the olfactory epithelium had begun to be ossified in the most derived probainognathians (Ruf et al., 2014; Wallace et al., 2019). In addition, the most recently studied brain endocasts suggest that there is no clear difference between the degree of encephalization in probainognathians and early mammaliaforms, and that a phylogenetically stable elevated $EQ_{synapsid}$ was only reached in crown Mammalia (Fig. 7).

Current data now supports a new model presented by Rodrigues et al. (2018), here referred to as the "outside-in" model, that proposes a series of external (i.e., environmental and body-related) changes that fuelled transformations of the sensory organs, which in turn transformed neural processes and pathways within the brain (Fig. 12). Unlike Rowe et al.'s (2011) "three-pulses" model that emphasizes the importance of olfaction, the "outside-in" model imbues all sensory organs with more-or-less equal importance regarding the evolution of the synapsid central nervous system.

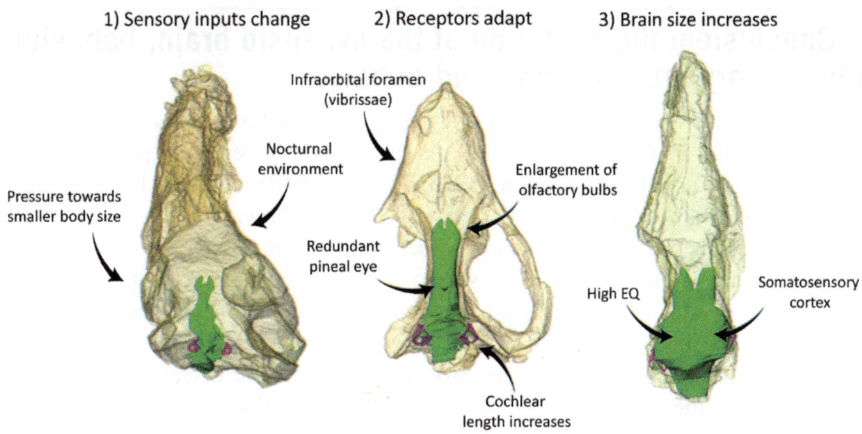

FIG. 12

The "outside-in" model illustrated on the transparent skull of the therapsid *Moschognathus* (left), the basal cynodont *Galesaurus* (middle) and the early mammaliaform *Megazostrodon* (right). Colors: Brain in green, inner ear in purple.

Probainognathian cynodonts appear to have undergone body miniaturization (Lautenschlager et al., 2018) and simultaneously entered a (first?) nocturnal bottleneck (Angielczyk and Schmitz, 2014). Their sensory receptors (e.g., vibrissae, inner ear) and primary processing structures (e.g., olfactory bulbs) were the first to undergo evolutionary changes, as vocalization became important for communicating in the dark (Chen and Wiens, 2020), the interaural space and vocal cords were reduced, and nocturnality imposed the loss of most visual cues (e.g., loss of the pineal eye and sclerotic ring) (Fig. 2A). Probainognathians compensated for the deprivation of visual inputs through the evolutionary modification of the organs responsible for vocalization and audition (cochlear canal, promontorium), somatosensation (vibrissae), and olfaction (enlarged olfactory bulbs, ossified olfactory turbinals) (Fig. 2A). This potentially enhanced sensory input may have been significant for the contemporaneous evolution of the brain. These changes in the brain include the posterolateral expansion of the cerebral hemispheres and cerebellum, and the development of a distinct cerebellar vermis in probainognathians (Rodrigues et al., 2018). The development of the structures for the processing and integration of sensory inputs, such as the lateral expansion of the cerebral hemispheres and the increase in relative brain size ($EQ_{synapsid} > 1.5$), are aligned with the phylogenetic origins of the Mammaliaformes and Mammalia clades, respectively (Fig. 2B). The nocturnal bottleneck(s) and reductions in body size occurring throughout the Mesozoic provide the conditions for the Mammalia to further specialize in adaptations originally inherited from the probainognathians.

The Mammalia specializations include increasing the range of detectable frequencies for airborne sounds, particularly in higher frequencies (evidenced by detached auditory ossicles, coiled cochlea, discrete fenestra rotunda, and secondary

bony lamina of the bony labyrinth), more sensitive tactile surfaces (evidenced by motile snouts, electroreception, full pelage), and more sensitive olfaction (evidenced by the ossification of the cribriform plate). In contrast, tetrachromatic vision was lost. This evolutionary scenario provides a more parsimonious account than previously proposed as it incorporates the recent data acquired during the past 10 years, which indicates that no clear distinction between mammaliaforms and cynodonts can be made based on our current knowledge of the paleoneurology of the relevant fossils.

Future research will have to test, and refine, aspects of this model. In addition to providing new paleoneurological data, testing aspects of this model can be achieved by comparing the predictions it makes about unrelated fields of research. For example, the temporal evolutionary features of the model proposed herein can be compared and contrasted with hypotheses concerning the evolution of endothermy because most of the sensory and paleoneurological traits discussed in this review would need to have evolved in conjunction with the origin of mammalian thermophysiology. This includes locomotion (erect limb posture and motor coordination), bony labyrinth anatomy, loss of the pineal eye, the evolution of hair, a relatively enlarged brain, ossified nasal capsule, fossoriality, parental care and all traits linked to nocturnality and body size miniaturization (Bennett and Ruben, 1986; Benton, 2021; Carrier, 1987; Grigg et al., 2022; Hillenius and Ruben, 2004; Lovegrove, 2017; Newham et al., 2022).

Improved knowledge of cynodonts and other therapsids may alter our temporal understanding of the onset of the sensory transformations hypothesized herein. Recent works have highlighted an unexpected paleoneurodiversity among premammalian synapsids (Angielczyk and Schmitz, 2014; Angielczyk et al., 2021; Benoit et al., 2017a,b,c, 2018, 2021c; Castanhinha et al., 2013; Laaß, 2015b; Laaß and Kaestner, 2017). Pelycosaurs, therapsids, and cynodonts were the dominant species in their ecosystems and as such, developed adaptations to most available ecological niches, such as semi-aquatic, arboreal, and fossorial locomotion styles, carnivorous and herbivorous diets, and a wide array of body sizes ranging from mouse-sized to rhinoceros-sized animals (Angielczyk and Kammerer, 2018; Kemp, 2005; Romano and Manucci, 2019). As a result, the paleoneurological history of the Paleozoic and Mesozoic synapsids should reflect this diversity, and should be investigated with less emphasis on it being a prequel to mammalian neurology and more as an important neurological history of its own. In this respect, an improved version of the "outside-in" model will have to account for the many parallel lineages that experimented with unique changes in their neuro-sensory organs.

Acknowledgments

ECPHRA and ECPTA for access to the fossils of the Oviston Nature Reserve. Elize Butler for access to the collections of the National Museum of Bloemfontein. Bruce Rubidge for access to the Rubidge's collection. The financial support of GENUS (DSI-NRF Centre of Excellence in Palaeosciences, UID 86073) and THRINAKS. NECSA and the ESRF for access to their beamlines.

References

Abdala, F., 1996. Redescripción del cráneo y reconsideración de la validez de Cynognathus minor (Eucynodontia-Cynognathidae) del Triásico inferior de Mendoza. Am. Demogr. 33, 115–126.

Abdala, F., 2021. Permo-Jurassic cynodonts: the early road to mammalness. In: Encyclopedia of Geology. Elsevier, pp. 206–226. https://doi.org/10.1016/B978-0-12-409548-9.12020-2.

Aboitiz, F., 1992. The evolutionary origin of the mammalian cerebral cortex. Biol. Res. 25, 41–49.

Aboitiz, F., Morales, D., Montiel, J., 2003. The evolutionary origin of the mammalian isocortex: towards an integrated developmental and functional approach. Behav. Brain Sci. 26, 535–552. https://doi.org/10.1017/S0140525X03000128.

Allin, E.F., 1975. Evolution of the mammalian middle ear. J. Morphol. 147, 403–437. https://doi.org/10.1002/jmor.1051470404.

Allin, E.F., Hopson, J.A., 1992. Evolution of the auditory system in Synapsida ("mammal-like reptiles" and primitive mammals) as seen in the fossil record. In: Webster, D.B., Popper, A.N., Fay, R.R. (Eds.), The Evolutionary Biology of Hearing. Springer, New York, NY, pp. 587–614. https://doi.org/10.1007/978-1-4612-2784-7_37.

Angielczyk, K.D., Kammerer, C.F., 2018. 5. Non-mammalian synapsids: the deep roots of the mammalian family tree. In: Zachos, F., Asher, R. (Eds.), Mammalian Evolution, Diversity and Systematics. De Gruyter, Berlin, Boston, pp. 117–198. https://doi.org/10.1515/9783110341553-005.

Angielczyk, K.D., Schmitz, L., 2014. Nocturnality in synapsids predates the origin of mammals by over 100 million years. Proc. R. Soc. B Biol. Sci. 281. https://doi.org/10.1098/rspb.2014.1642.

Angielczyk, K.D., Benoit, J., Rubidge, B.S., 2021. A new tusked cistecephalid dicynodont (Therapsida, Anomodontia) from the upper Permian upper Madumabisa Mudstone Formation, Luangwa Basin, Zambia. Pap. Palaeontol. 7, 405–446. https://doi.org/10.1002/spp2.1285.

Araújo, R., Fernandez, V., Polcyn, M.J., Fröbisch, J., Martins, R.M.S., 2017. Aspects of gorgonopsian paleobiology and evolution: insights from the basicranium, occiput, osseous labyrinth, vasculature, and neuroanatomy. PeerJ 5, e3119. https://doi.org/10.7717/peerj.3119.

Araújo, R., David, R., Benoit, J., Lungmus, J.K., Stoessel, A., Barrett, P.M., Maisano, J.A., Ekdale, E., Orliac, M., Luo, Z.-L., Martinelli, A.G., Hoffman, E.A., Sidor, C.A., Martins, R.M.S., Spoor, F., Angielczyck, K.D., 2022. Inner ear biomechanics reveals a Late Triassic origin of mammalian endothermy. Nature 607, 726–731.

Bajdek, P., Qvarnström, M., Owocki, K., Sulej, T., Sennikov, A.G., Golubev, V.K., Niedź wiedzki, G., 2016. Microbiota and food residues including possible evidence of pre-mammalian hair in Upper Permian coprolites from Russia. Lethaia 49, 455–477. https://doi.org/10.1111/let.12156.

Bakker, R.T., Flis, C.J., George, C.D., Cook, L.A., Bell, T.H., Zoehfeld, K.W., 2015. *Dimetrodon* and the earliest apex predators: the Craddock bone bed and George ranch facies show that aquatic prey, not herbivores, were key food sources. In: 75th Anniversary Annual Meeting of the Society of Vertebrate Paleontology, Dallas, Texas. Society of Vertebrate Paleontology, p. 83.

Barghusen, H.R., 1975. A review of fighting adaptations in dinocephalians (Reptilia, Therapsida). Paleobiology 1, 295–311. https://doi.org/10.1017/S0094837300002542.

Bazzana, K.D., Evans, D.C., Bevitt, J.J., Reisz, R.R., 2022. Neurosensory anatomy of Varanopidae and its implications for early synapsid evolution. J. Anat. 240, 833–849. https://doi.org/10.1111/joa.13593.

Beaudet, A., Clarke, R.J., de Jager, E.J., Bruxelles, L., Carlson, K.J., Crompton, R., de Beer, F., Dhaene, J., Heaton, J.L., Jakata, K., Jashashvili, T., Kuman, K., McClymont, J., Pickering, T.R., Stratford, D., 2019. The endocast of StW 573 ("Little Foot") and hominin brain evolution. J. Hum. Evol. 126, 112–123. https://doi.org/10.1016/j.jhevol.2018.11.009.

Bennett, A.F., Ruben, J.A., 1986. The metabolic and thermoregulatory status. In: Hotton, N.I., Roth, J.J., Roth, E.C. (Eds.), The Ecology and Biology of Mammal-Like Reptiles. Smithsonian Institution Press, Washington D.C, pp. 207–218.

Benoit, J., 2019. Parental care or opportunism in south African Triassic cynodonts? South Afr. J. Sci. 115, 1–2. https://doi.org/10.17159/sajs.2019/5589.

Benoit, J., Abdala, F., Van den Brandt, M.J., Manger, P.R., Rubidge, B.S., 2015. Physiological implications of the abnormal absence of thc parietal foramen in a late Permian cynodont (Therapsida). Sci. Nat. 102, 69. https://doi.org/10.1007/s00114-015-1321-4.

Benoit, J., Abdala, F., Manger, P., Rubidge, B., 2016a. The sixth sense in mammalians forerunners: variability of the parietal foramen and the evolution of the pineal eye in south African Permo-Triassic eutheriodont therapsids. Acta Palaeontol. Pol. 61, 777–789. https://doi.org/10.4202/app.00219.2015.

Benoit, J., Manger, P.R., Fernandez, V., Rubidge, B.S., 2016b. Cranial bosses of *Choerosaurus dejageri* (Therapsida, Therocephalia): earliest evidence of cranial display structures in eutheriodonts. PLoS One 11, e0161457. https://doi.org/10.1371/journal.pone.0161457.

Benoit, J., Manger, P.R., Rubidge, B.S., 2016c. Palaeoneurological clues to the evolution of defining mammalian soft tissue traits. Sci. Rep. 6, 25604. https://doi.org/10.1038/srep25604.

Benoit, J., Fernandez, V., Manger, P.R., Rubidge, B.S., 2017a. Endocranial casts of pre-mammalian therapsids reveal an unexpected neurological diversity at the deep evolutionary root of mammals. Brain Behav. Evol. 90, 311–333. https://doi.org/10.1159/000481525.

Benoit, J., Manger, P.R., Fernandez, V., Rubidge, B.S., 2017b. The bony labyrinth of late Permian Biarmosuchia: palaeobiology and diversity in non-mammalian Therapsida. Palaeontol. Afr. 52, 58–77.

Benoit, J., Manger, P.R., Norton, L., Fernandez, V., Rubidge, B.S., 2017c. Synchrotron scanning reveals the palaeoneurology of the head-butting *Moschops capensis* (Therapsida, Dinocephalia). PeerJ 5, e3496. https://doi.org/10.7717/peerj.3496.

Benoit, J., Angielczyk, K.D., Miyamae, J.A., Manger, P., Fernandez, V., Rubidge, B., 2018. Evolution of facial innervation in anomodont therapsids (Synapsida): insights from X-ray computerized microtomography. J. Morphol. 279, 673–701. https://doi.org/10.1002/jmor.20804.

Benoit, J., Legendre, L.J., Tabuce, R., Obada, T., Mararescul, V., Manger, P., 2019. Brain evolution in Proboscidea (Mammalia, Afrotheria) across the Cenozoic. Sci. Rep. 9, 1–8. https://doi.org/10.1038/s41598-019-45888-4.

Benoit, J., Ruf, I., Miyamae, J.A., Fernandez, V., Rodrigues, P.G., Rubidge, B.S., 2020. The evolution of the maxillary canal in Probainognathia (Cynodontia, Synapsida): reassessment of the homology of the infraorbital foramen in mammalian ancestors. J. Mamm. Evol. 27, 329–348. https://doi.org/10.1007/s10914-019-09467-8.

Benoit, J., Browning, C., Norton, L.A., 2021a. The first healed bite mark and embedded tooth in the snout of a middle Permian gorgonopsian (Synapsida: Therapsida). Front. Ecol. Evol. 9, 699298. https://doi.org/10.3389/fevo.2021.699298.

Benoit, J., Ford, D., Miyamae, J., Ruf, I., 2021b. Can maxillary canal morphology inform varanopid phylogenetic affinities? Acta Palaeontol. Pol. 66, 389–393. https://doi.org/10.4202/app.00816.2020.

Benoit, J., Kruger, A., Jirah, S., Fernandez, V., Rubidge, B., 2021c. Palaeoneurology and palaeobiology of the dinocephalian *Anteosaurus magnificus*. Acta Palaeontol. Pol. 66, 29–39. https://doi.org/10.4202/app.00800.2020.

Benton, M.J., 2021. The origin of endothermy in synapsids and archosaurs and arms races in the Triassic. Gondw. Res. 100, 261–289. https://doi.org/10.1016/j.gr.2020.08.003.

Berson, D.M., Dunn, F.A., Takao, M., 2002. Phototransduction by retinal ganglion cells that set the circadian clock. Science 295, 1070–1073. https://doi.org/10.1126/science.1067262.

Blackburn, D.G., 2000. Reptilian viviparity: past research, future directions, and appropriate models. Comp. Biochem. Physiol. A Mol. Integr. Physiol. 127, 391–409. https://doi.org/10.1016/S1095-6433(00)00272-5.

Bonnan, M.F., 2016. The Bare Bones: An Unconventional Evolutionary History of the Skeleton. Indiana University Press.

Bordy, E.M., Sztanó, O., Rampersadh, A., Almond, J., Choiniere, J.N., 2019. Vertebrate scratch traces from the middle Triassic Burgersdorp formation of the main Karoo Basin, South Africa: sedimentological and ichnological assessment. J. African Earth Sci. 160, 103594. https://doi.org/10.1016/j.jafrearsci.2019.103594.

Borges, R., Johnson, W.E., O'Brien, S.J., Gomes, C., Heesy, C.P., Antunes, A., 2018. Adaptive genomic evolution of opsins reveals that early mammals flourished in nocturnal environments. BMC Genomics 19, 121. https://doi.org/10.1186/s12864-017-4417-8.

Botha, J., 2020. The paleobiology and paleoecology of south African *lystrosaurus*. PeerJ 8, e10408. https://doi.org/10.7717/peerj.10408.

Botha-Brink, J., 2017. Burrowing in *Lystrosaurus*: preadaptation to a postextinction environment? J. Vertebr. Paleontol. 37, e1365080. https://doi.org/10.1080/02724634.2017.1365080.

Botha-Brink, J., Modesto, S.P., 2007. A mixed-age classed 'pelycosaur' aggregation from South Africa: earliest evidence of parental care in amniotes? Proc. R. Soc. B Biol. Sci. 274, 2829–2834. https://doi.org/10.1098/rspb.2007.0803.

Bouabdellah, F., Lessner, E.J., Benoit, J., 2022. The rostral neurovascular system of *Tyrannosaurus rex*. Palaeontol. Electron. 25, a3. https://doi.org/10.26879/1178.

Boucot, A.J., 1990. Evolutionary Paleobiology of Behavior and Coevolution. Elsevier, New York.

Boucot, A.J., Poinar Jr., G.O., 2011. Fossil Behavior Compendium. CRC Press.

Bronzati, M., Benson, R.B.J., Evers, S.W., Ezcurra, M.D., Cabreira, S.F., Choiniere, J., Dollman, K.N., Paulina-Carabajal, A., Radermacher, V.J., Roberto-da-Silva, L., Sobral, G., Stocker, M.R., Witmer, L.M., Langer, M.C., Nesbitt, S.J., 2021. Deep evolutionary diversification of semicircular canals in archosaurs. Curr. Biol. 31, 2520–2529. https://doi.org/10.1016/j.cub.2021.03.086.

Cameron, J., Shelley, S.L., Williamson, T.E., Brusatte, S.L., 2019. The brain and inner ear of the early Paleocene "Condylarth" *Carsioptychus coarctatus*: implications for early placental mammal neurosensory biology and behavior. Anat. Rec. 302, 306–324. https://doi.org/10.1002/ar.23903.

Carrier, D.R., 1987. The evolution of locomotor stamina in Tetrapods: circumventing a mechanical constraint. Paleobiology 13, 326–341.

Castanhinha, R., Araújo, R., Júnior, L.C., Angielczyk, K.D., Martins, G.G., Martins, R.M.S., Chaouiya, C., Beckmann, F., Wilde, F., 2013. Bringing dicynodonts back to life: paleobiology and anatomy of a new emydopoid genus from the Upper Permian of Mozambique. PLoS One 8, e80974. https://doi.org/10.1371/journal.pone.0080974.

Celik, M.A., Phillips, M.J., 2020. Conflict resolution for Mesozoic mammals: reconciling phylogenetic incongruence among anatomical regions. Front. Genet. 11, 0651. https://doi.org/10.3389/fgene.2020.00651.

Chen, Z., Wiens, J.J., 2020. The origins of acoustic communication in vertebrates. Nat. Commun. 11, 369. https://doi.org/10.1038/s41467-020-14356-3.

Chudinov, P.K., 1968. Structure of the integument of theromorphs (translated from Russian). Dokl. Akad. Nauk SSSR 179, 226–229.

Cisneros, J.C., Abdala, F., Jashashvili, T., de Oliveira Bueno, A., Dentzien-Dias, P., 2015. *Tiarajudens eccentricus* and *Anomocephalus africanus*, two bizarre anomodonts (Synapsida, Therapsida) with dental occlusion from the Permian of Gondwana. R. Soc. Open Sci. 2, 150090. https://doi.org/10.1098/rsos.150090.

Colombi, C.E., Fernández, E., Currie, B.S., Alcober, O.A., Martínez, R., Correa, G., 2012. Large-diameter burrows of the Triassic Ischigualasto Basin, NW Argentina: Paleoecological and paleoenvironmental implications. PLoS One 7, e50662. https://doi.org/10.1371/journal.pone.0050662.

Cox, C.B., 1972. A new digging dicynodont from the upper Permian of Tanzania. In: Joysey, K.A., Kemp, T.S. (Eds.), Studies in Vertebrate Evolution. Oliver & Boyd, Edinburgh, pp. 173–189.

Crompton, A.W., 1958. The cranial morphology of a new genus and species of Ictidosauran. Proc. Zool. Soc. Lond. 130, 183–216.

Crompton, A.W., 1972. Postcanine occlusion in cynodonts and tritylodontids. Bull. Br. Mus. Nat. Hist. 21, 29–71.

Crompton, A.W., Musinsky, C., Owerkowicz, T., 2015. Evolution of the mammalian nose. In: Dial, K., Shibin, N., Brainerd, E. (Eds.), Great Transformations in Vertebrate Evolution. University of Chicago Press, Chicago, IL, pp. 189–203.

Crompton, A.W., Owerkowicz, T., Bhullar, B.-A.S., Musinsky, C., 2017. Structure of the nasal region of non-mammalian cynodonts and mammaliaforms: speculations on the evolution of mammalian endothermy. J. Vertebr. Paleontol. 37, e1269116. https://doi.org/10.1080/02724634.2017.1269116.

Csiki-Sava, Z., Vremir, M., Meng, J., Brusatte, S.L., Norell, M.A., 2018. Dome-headed, small-brained island mammal from the Late Cretaceous of Romania. Proc. Nat. Acad. Sci. 115, 4857–4862.

Damiani, R., Modesto, S., Yates, A., Neveling, J., 2003. Earliest evidence of cynodont burrowing. Proc. R. Soc. Lond. B Biol. Sci. 270, 1747–1751. https://doi.org/10.1098/rspb.2003.2427.

de Klerk, W.J., 2002. A dicynodont trackway from the *Cistecephalus* assemblage zone in the Karoo, east of Graaff-Reinet, South Africa. Palaeontol. Afr. 38, 73–91.

de Simão-Oliveira, D., Kerber, L., Pinheiro, F.L., 2020. Endocranial morphology of the Brazilian Permian dicynodont *Rastodon procurvidens* (Therapsida: Anomodontia). J. Anat. 236, 384–397. https://doi.org/10.1111/joa.13107.

Dechaseaux, C., 1962. Cerveaux d'animaux disparus: essai de paléoneurologie. Masson.

Du Plessis, D., 2010. An Examination of Non-mammalian Cynodont Cranial Endocasts (Msc). University of the Free State, Bloemfontein, South Africa.

Duhamel, A., Benoit, J., Rubidge, B.S., Liu, J., 2021. A re-assessment of the oldest therapsid *Raranimus* confirms its status as a basal member of the clade and fills Olson's gap. Sci. Nat. 108, 26. https://doi.org/10.1007/s00114-021-01736-y.

Eakin, R.M., 1973. The Third Eye. University of California Press.

Edinger, T., 1964. Midbrain exposure and overlap in mammals. Am. Zool. 4, 5–19.

Edinger, T., 1975. Paleoneurology 1804-1966: an annotated bibliography. Adv. Anat. Embryol. Cell Biol. 49, 3–258.

Eisenberg, J.F., 1981. The Mammalian Radiations: An Analysis of Trends in Evolution, Adaptation, and Behaviour. University of Chicago Press, Chicago.

Ekdale, E.G., 2013. Comparative anatomy of the bony labyrinth (inner ear) of placental mammals. PLoS One 8, e66624. https://doi.org/10.1371/journal.pone.0066624.

Ekdale, E.G., 2016a. Form and function of the mammalian inner ear. J. Anat. 228, 324–337. https://doi.org/10.1111/joa.12308.

Ekdale, E.G., 2016b. The ear of mammals: from monotremes to humans. In: Clack, J.A., Fay, R.R., Popper, A.N. (Eds.), Evolution of the Vertebrate Ear. Springer Handbook of Auditory Research. Springer International Publishing, Cham, pp. 175–206. https://doi.org/10.1007/978-3-319-46661-3_7.

Ellenberger, P., 1976. Une piste avec traces de soies épaisses dans le Trias inférieur a moyen de Lodève (Hérault, France): *Cynodontipus polythrix* nov. gen., nov. sp. les Cynodontes en France. Geobios 9, 769–787. https://doi.org/10.1016/S0016-6995(76)80078-2.

Emerling, C.A., 2016. Archelosaurian color vision, parietal eye loss and the crocodylian nocturnal bottleneck. Mol. Biol. Evol. 34, 666–676. https://doi.org/10.1093/molbev/msw265.

Emlen, D.J., 2008. The evolution of animal weapons. Annu. Rev. Ecol. Evol. Syst. 39, 387–413. https://doi.org/10.1146/annurev.ecolsys.39.110707.173502.

Estes, R.D., 1961. Cranial anatomy of the cynodont reptile *Thrinaxodon liorhinus*. Bull. Mus. Comp. Zool. Harv. Univ. 125, 165–180.

Faure-Brac, M.G., Cubo, J., 2020. Were the synapsids primitively endotherms? A palaeohistological approach using phylogenetic eigenvector maps. Philos. Trans. R. Soc. B Biol. Sci. 375. https://doi.org/10.1098/rstb.2019.0138.

Fernandez, V., Abdala, F., Carlson, K.J., Cook, D.C., Rubidge, B.S., Yates, A., Tafforeau, P., 2013. Synchrotron reveals early Triassic odd couple: injured amphibian and aestivating therapsid share burrow. PLoS One 8, e64978. https://doi.org/10.1371/journal.pone.0064978.

Ferreira-Cardoso, S., Araújo, R., Martins, N.E., Martins, G.G., Walsh, S., Martins, R.M.S., Kardjilov, N., Manke, I., Hilger, A., Castanhinha, R., 2017. Floccular fossa size is not a reliable proxy of ecology and behaviour in vertebrates. Sci. Rep. 7, 2005. https://doi.org/10.1038/s41598-017-01981-0.

Fiorelli, L.E., Ezcurra, M.D., Hechenleitner, E.M., Argañaraz, E., Taborda, J.R.A., Trotteyn, M.J., von Baczko, M.B., Desojo, J.B., 2013. The oldest known communal latrines provide evidence of gregarism in Triassic megaherbivores. Sci. Rep. 3, 1–7. https://doi.org/10.1038/srep03348.

Fourie, S., 1963. Tooth replacement in the gomphodont cynodont *Diademodon*. South Afr. J. Sci. 87, 211–213.

Franco, A.S., Müller, R.T., Martinelli, A.G., Hoffmann, C.A., Kerber, L., 2021. The nasal cavity of two traversodontid cynodonts (Eucynodontia, Gomphodontia) from the upper Triassic of Brazil. J. Paleo. 95, 845–860. https://doi.org/10.1017/jpa.2021.6.

Fröbisch, J., Reisz, R.R., 2009. The late Permian herbivore *Suminia* and the early evolution of arboreality in terrestrial vertebrate ecosystems. Proc. R. Soc. B Biol. Sci. 276, 3611–3618. https://doi.org/10.1098/rspb.2009.0911.

Gaetano, L.C., Abdala, F., 2015. The stapes of Gomphodont cynodonts: insights into the middle ear structure of non-mammaliaform cynodonts. PLoS One 10, e0131174. https://doi.org/10.1371/journal.pone.0131174.

Gand, G., Demathieu, G., Montenat, C., 2007. Les traces de pas d'amphibiens, de dinosaures et autres reptiles du Mesozoïque Français: inventaire et interprétations. Palaeovertebrata 35. https://doi.org/10.18563/pv.35.1-4.1-149.

Geist, V., 1966. The evolution of horn-like organs. Behaviour 27, 175–214. https://doi.org/10.1163/156853966X00155.

Geist, V., 1972. An ecological and behavioural explanation of mammalian characteristics and their implication to therapsid evolution. Z. Für Säugetierkd. Im Auftrage Dtsch. Ges. Für Säugetierkd. EV 37, 1–15.

Gerkema, M.P., Davies, W.I.L., Foster, R.G., Menaker, M., Hut, R.A., 2013. The nocturnal bottleneck and the evolution of activity patterns in mammals. Proc. R. Soc. B Biol. Sci. 280. https://doi.org/10.1098/rspb.2013.0508.

Grant, R.A., Goss, V.G.A., 2022. What can whiskers tell us about mammalian evolution, behaviour, and ecology? Mammal Rev. 52, 148–163. https://doi.org/10.1111/mam.12253.

Grant, R.A., Breakell, V., Prescott, T.J., 2018. Whisker touch sensing guides locomotion in small, quadrupedal mammals. Proc. R. Soc. B Biol. Sci. 285, 20180592. https://doi.org/10.1098/rspb.2018.0592.

Grigg, G., Nowack, J., Bicudo, J.E.P.W., Bal, N.C., Woodward, H.N., Seymour, R.S., 2022. Whole-body endothermy: ancient, homologous and widespread among the ancestors of mammals, birds and crocodylians. Biol. Rev. 97, 766–801. https://doi.org/10.1111/brv.12822.

Groenewald, G.H., Welman, J., Maceachern, J.A., 2001. Vertebrate burrow complexes from the early Triassic *Cynognathus* zone (Driekoppen formation, Beaufort group) of the Karoo Basin, South Africa. Palaios 16, 148–160. https://doi.org/10.1669/0883-1351(2001)016<0148:VBCFTE>2.0.CO;2.

Healy, S.D., Rowe, C., 2007. A critique of comparative studies of brain size. Proc. R. Soc. B Biol. Sci. 274, 453–464. https://doi.org/10.1098/rspb.2006.3748.

Heesy, C.P., Hall, M.I., 2010. The nocturnal bottleneck and the evolution of mammalian vision. Brain Behav. Evol. 75, 195–203. https://doi.org/10.1159/000314278.

Heffner, H.E., Heffner, R.S., 2018. The evolution of mammalian hearing. In: AIP Conference Proceedings. Presented at the To the ear and back again - Advances in auditory biophysics, St Catharines, Canada, p. 130001. https://doi.org/10.1063/1.5038516.

Hendrickx, C., Bell, P.R., Pittman, M., Milner, A.R.C., Cuesta, E., O'Connor, J., Loewen, M., Currie, P.J., Mateus, O., Kaye, T.G., Delcourt, R., 2022. Morphology and distribution of scales, dermal ossifications, and other non-feather integumentary structures in non-avialan theropod dinosaurs. Biol. Rev. https://doi.org/10.1111/brv.12829.

Herculano-Houzel, S., 2012. The remarkable, yet not extraordinary, human brain as a scaled-up primate brain and its associated cost. Proc. Natl. Acad. Sci. 109, 10661–10668. https://doi.org/10.1073/pnas.1201895109.

Higashiyama, H., Koyabu, D., Hirasawa, T., Werneburg, I., Kuratani, S., Kurihara, H., 2021. Mammalian face as an evolutionary novelty. Proc. Natl. Acad. Sci. 118, e2111876118. https://doi.org/10.1073/pnas.2111876118.

Hillenius, W.J., 1994. Turbinates in therapsids: evidence for late Permian origins of mammalian endothermy. Evolution 48, 207–229. https://doi.org/10.2307/2410089.

Hillenius, W.J., Ruben, J.A., 2004. The evolution of endothermy in terrestrial vertebrates: who? When? Why? Physiol. Biochem. Zool. 77, 1019–1042. https://doi.org/10.1086/425185.

Hoffman, E.A., Rowe, T.B., 2018. Jurassic stem-mammal perinates and the origin of mammalian reproduction and growth. Nature 561, 104–108. https://doi.org/10.1038/s41586-018-0441-3.

Hoffmann, S., O'Connor, P.M., Kirk, E.C., Wible, J.R., Krause, D.W., 2014. Endocranial and inner ear morphology of *Vintana Sertichi* (Mammalia, Gondwanatheria) from the late cretaceous of Madagascar. J. Vertebr. Paleontol. 34, 110–137. https://doi.org/10.1080/02724634.2014.956878.

Hoffmann, C.A., Rodrigues, P.G., Soares, M.B., Andrade, M.B., 2021. Brain endocast of two non-mammaliaform cynodonts from southern Brazil: an ontogenetic and evolutionary approach. Hist. Biol. 33, 1196–1207. https://doi.org/10.1080/08912963.2019.1685512.

Holloway, R.L., 2013. The evolution of the hominid brain. In: Henke, W., Tattersall, I. (Eds.), Handbook of Paleoanthropology. Springer, Berlin Heidelberg, Berlin, Heidelberg, pp. 1–23. https://doi.org/10.1007/978-3-642-27800-6_81-1.

Hopson, J.A., 1979. Paleoneurology. In: Gans, C., Northcutt, R.G., Ulinski, P. (Eds.), Biology of the Reptilia. Academic Press, New York, pp. 39–146.

Hurlburt, G.R., Ridgely, R.C., Witmer, L.M., 2013. Relative size of brain and cerebrum in tyrannosaurid dinosaurs: an analysis using brain-endocast quantitative relationships in extant alligators. In: Parrish, J.M., Henderson, P.J., Currie, P.J., Koppelhus, E. (Eds.), Origin, Systematics, and Paleobiology of the Tyrannosauridae. Indiana University Press, Bloomington, pp. 135–155.

Hurum, J.H., 1994. Snout and orbit of cretaceous Asian multitiuberculates studied by serial sections. Acta Palaeontol. Pol. 39, 181–221.

Hurum, J.H., 1998a. The inner ear of two late cretaceous multituberculate mammals, and its implications for multituberculate hearing. J. Mamm. Evol. 5, 65–93. https://doi.org/10.1023/A:1020571003901.

Hurum, J.H., 1998b. The braincase of two late cretaceous Asian multituberculates studied by serial sections. Acta Palaeontol. Pol. 43, 21–52.

Huttenlocker, A.K., Sidor, C.A., 2020. A basal nonmammaliaform cynodont from the Permian of Zambia and the origins of mammalian endocranial and postcranial anatomy. J. Vertebr. Paleontol. 40, e1827413. https://doi.org/10.1080/02724634.2020.1827413.

Huttenlocker, A.K., Grossnickle, D.M., Kirkland, J.I., Schultz, J.A., Luo, Z.-X., 2018. Late-surviving stem mammal links the lowermost cretaceous of North America and Gondwana. Nature 558, 108–112. https://doi.org/10.1038/s41586-018-0126-y.

Ivakhnenko, M.F., 2008. Cranial morphology and evolution of Permian Dinomorpha (Eotherapsida) of eastern Europe. Paleontol. J. 42, 859–995. https://doi.org/10.1134/S0031030108090013.

Jasinoski, S.C., Abdala, F., 2017. Aggregations and parental care in the early Triassic basal cynodonts *Galesaurus planiceps* and *Thrinaxodon liorhinus*. PeerJ 5. https://doi.org/10.7717/peerj.2875.

Jerison, H., 1973. Evolution of the Brain and Intelligence. Elsevier Science, Saint Louis.

Ji, Q., Luo, Z.-X., Yuan, C.-X., Tabrum, A.R., 2006. A swimming mammaliaform from the middle Jurassic and ecomorphological diversification of early mammals. Science 311, 1123–1127. https://doi.org/10.1126/science.1123026.

Jones, G., Teeling, E.C., Rossiter, S.J., 2013. From the ultrasonic to the infrared: molecular evolution and the sensory biology of bats. Front. Physiol. 4. https://doi.org/10.3389/fphys.2013.00117.

Kammerer, C.F., 2021. Elevated cranial sutural complexity in burrowing dicynodonts. Front. Ecol. Evol. 9, 674151.

Kemp, T.S., 1979. The primitive cynodont *Procynosuchus*: functional anatomy of the skull and relationships. Philos. Trans. R. Soc. Lond. B Biol. Sci. 285, 73–122. https://doi.org/10.1098/rstb.1979.0001.

Kemp, T.S., 2005. The Origin and Evolution of Mammals, Oxford Biology. Oxford University Press, Oxford, New York.

Kemp, T.S., 2009. The endocranial cavity of a nonmammalian eucynodont, *Chiniquodon theotenicus*, and its implications for the origin of the mammalian brain. J. Vertebr. Paleontol. 29, 1188–1198. https://doi.org/10.1671/039.029.0430.

Kemp, T.S., 2016. Non-mammalian synapsids: the beginning of the mammal line. In: Clack, J.A., Fay, R.R., Popper, A.N. (Eds.), Evolution of the Vertebrate Ear. Springer Handbook of Auditory Research. Springer International Publishing, Cham, pp. 107–137. https://doi.org/10.1007/978-3-319-46661-3_5.

Kerber, L., Ferreira, J.D., Fonseca, P.H.M., Franco, A., Martinelli, A.G., Soares, M.B., Ribeiro, A.M., 2021. An additional brain endocast of the ictidosaur *Riograndia guaibensis* (Eucynodontia: Probainognathia): intraspecific variation of endocranial traits. An. Acad. Bras. Ciênc. 93, e20200084. https://doi.org/10.1590/0001-3765202120200084.

Kerekes, P., Daret, A., Shulz, D.E., Ego-Stengel, V., 2017. Bilateral discrimination of tactile patterns without whisking in freely running rats. J. Neurosci. 37, 7567–7579. https://doi.org/10.1523/JNEUROSCI.0528-17.2017.

Kermack, K.A., Mussett, F., Rigney, H.W., 1981. The skull of Morganucodon. Zool. J. Linn. Soc. 71, 1–158. https://doi.org/10.1111/j.1096-3642.1981.tb01127.x.

Kielan-Jaworowska, Z., 1986. Brain evolution in Mesozoic mammals. In: Flanagan, K.M., Lillegraven, J.A. (Eds.), Vertebrates, Phylogeny, and Philosophy. University of Wyoming, Laramie, WY, pp. 21–34. https://doi.org/10.2113/gsrocky.24.special_paper_3.21.

Kielan-Jaworowska, Z., Cifelli, R.L., Luo, Z.-X., 2004. Mammals from the Age of Dinosaurs: Origins, Evolution, and Structure. Columbia University Press.

Kitazawa, T., Takechi, M., Hirasawa, T., Adachi, N., Narboux-Nême, N., Kume, H., Maeda, K., Hirai, T., Miyagawa-Tomita, S., Kurihara, Y., Hitomi, J., Levi, G., Kuratani, S., Kurihara, H., 2015. Developmental genetic bases behind the independent origin of the tympanic membrane in mammals and diapsids. Nat. Commun. 6, 6853. https://doi.org/10.1038/ncomms7853.

Kitching, J.W., 1964. On the *lystrosaurus* zone and its fauna with special reference to some immature Lystrosauridae. Palaeontol. Afr. 11, 61–76.

Kitching, J.W., 1977. The Distribution of the Karroo Vertebrate Fauna: With Special Reference to Certain Genera and the Bearing of this Distribution on the Zoning of the Beaufort Beds. Bernard Price Institute for Palaeontological Research, University of the Witwatersrand.

Klein, H., Lucas, S.G., 2021. The Triassic tetrapod footprint record. N. M. Mus. Nat. Hist. Sci. Bull. 83, 1–194.

Knaus, P.L., van Heteren, A.H., Lungmus, J.K., Sander, P.M., 2021. High blood flow into the femur indicates elevated aerobic capacity in synapsids since the Synapsida-Sauropsida Split. Front. Ecol. Evol. 9, 751238. https://doi.org/10.3389/fevo.2021.751238.

Krapovickas, V., Mancuso, A.C., Marsicano, C.A., Domnanovich, N.S., Schultz, C.L., 2013. Large tetrapod burrows from the middle Triassic of Argentina: a behavioural adaptation to seasonal semi-arid climate?: *Triassic tetrapod burrows from Argentina*. Lethaia 46, 154–169. https://doi.org/10.1111/j.1502-3931.2012.00329.x.

Kuchling, G., Hofmeyr, M.D., 2022. Too hot to Nest? In a hot summer the tortoise Chersina angulata can switch from nesting to facultative viviparity. Front. Ecol. Evol. 9, 788764. https://doi.org/10.3389/fevo.2021.788764.

Kuznetsov, A.N., Panyutina, A.A., 2018. First paleoichnological evidence for baby–riding in early mammals. Am. Demogr. 55, 668. https://doi.org/10.5710/AMGH.03.02.2018.3184.

Laaß, M., 2015a. The origins of the cochlea and impedance matching hearing in synapsids. Acta Palaeontol. Pol. https://doi.org/10.4202/app.00140.2014.

Laaß, M., 2015b. Bone-conduction hearing and seismic sensitivity of the late permian anomodont *Kawingasaurus fossilis*. J. Morphol. 276, 121–143. https://doi.org/10.1002/jmor.20325.

Laaß, M., Kaestner, A., 2017. Evidence for convergent evolution of a neocortex-like structure in a late Permian therapsid. J. Morphol. 278, 1033–1057. https://doi.org/10.1002/jmor.20712.

Laaß, M., Schillinger, B., 2017. Brain and ear region of *Cistecephalus* (Therapsida, Anomodontia). In: Münstersche Forschungen Zur Geologie Und Paläontologie. Presented at the 88th Annual Conference of the Paläontologische Gesellschaft, Münster, Germany, p. 91.

Laaß, M., Hampe, O., Schudack, M., Hoff, C., Kardjilov, N., Hilger, A., 2011. New insights into the respiration and metabolic physiology of *lystrosaurus*: respiration and metabolic physiology of lystrosaurus. Acta Zool. 92, 363–371. https://doi.org/10.1111/j.1463-6395.2010.00467.x.

Laaß, M., Schillinger, B., Werneburg, I., 2017. Neutron tomography and X-ray tomography as tools for the morphological investigation of non-mammalian synapsids. Phys. Procedia 88, 100–108. Neutron Imaging for Applications in Industry and Science Proceedings of the 8th International Topical Meeting on Neutron Radiography (ITMNR-8) Beijing, China, September 4–8, 2016 https://doi.org/10.1016/j.phpro.2017.06.013.

Lautenschlager, S., Gill, P.G., Luo, Z.-X., Fagan, M.J., Rayfield, E.J., 2018. The role of miniaturization in the evolution of the mammalian jaw and middle ear. Nature 561, 533–537. https://doi.org/10.1038/s41586-018-0521-4.

Lefèvre, C.M., Sharp, J.A., Nicholas, K.R., 2010. Evolution of lactation: ancient origin and extreme adaptations of the lactation system. Annu. Rev. Genomics Hum. Genet. 11, 219–238. https://doi.org/10.1146/annurev-genom-082509-141806.

Legendre, L.J., Davesne, D., 2020. The evolution of mechanisms involved in vertebrate endothermy. Philos. Trans. R. Soc. B Biol. Sci. 375, 20190136. https://doi.org/10.1098/rstb.2019.0136.

Legendre, L.J., Guénard, G., Botha-Brink, J., Cubo, J., 2016. Palaeohistological evidence for ancestral high metabolic rate in archosaurs. Syst. Biol. 65, 989–996. https://doi.org/10.1093/sysbio/syw033.

Lillegraven, J.A., Krusat, G., 1991. Cranio-mandibular anatomy of *Haldanodon exspectatus* (Docodonta; Mammalia) from the late Jurassic of Portugal and its implications to the evolution of mammalian characters. Rocky Mt. Geol. 28, 39–138.

Lingham-Soliar, T., 2014. The Vertebrate Integument Volume 1: Origin and Evolution. Springer, Berlin Heidelberg.

Liu, Y., Cotton, J.A., Shen, B., Han, X., Rossiter, S.J., Zhang, S., 2010. Convergent sequence evolution between echolocating bats and dolphins. Curr. Biol. 20, R53–R54. https://doi.org/10.1016/j.cub.2009.11.058.

Lovegrove, B.G., 2017. A phenology of the evolution of endothermy in birds and mammals: evolution of endothermy. Biol. Rev. 92, 1213–1240. https://doi.org/10.1111/brv.12280.

Lucas, R.J., Freedman, M.S., Muñoz, M., Garcia-Fernández, J.-M., Foster, R.G., 1999. Regulation of the mammalian pineal by non-rod, non-cone, ocular photoreceptors. Science 284, 505–507. https://doi.org/10.1126/science.284.5413.505.

Luo, Z.-X., 2001. Inner ear and its bony housing in tritylodonts and implications for evolution of mammaliaform ear. Bull. Mus. Comp. Zool. 156, 81–97.

Luo, Z.-X., 2011. Developmental patterns in Mesozoic evolution of mammal ears. Annu. Rev. Ecol. Evol. Syst. 42, 355–380. https://doi.org/10.1146/annurev-ecolsys-032511-142302.

Luo, Z.-X., Schultz, J.A., Ekdale, E.G., 2016. Evolution of the middle and inner ears of mammaliaforms: the approach to mammals. In: Clack, J.A., Fay, R.R., Popper, A.N. (Eds.), Evolution of the Vertebrate Ear. Springer Handbook of Auditory Research. Springer International Publishing, Cham, pp. 139–174. https://doi.org/10.1007/978-3-319-46661-3_6.

Lyson, T.R., Rubidge, B.S., Scheyer, T.M., de Queiroz, K., Schachner, E.R., Smith, R.M.H., Botha-Brink, J., Bever, G.S., 2016. Fossorial origin of the turtle Shell. Curr. Biol. 26, 1887–1894. https://doi.org/10.1016/j.cub.2016.05.020.

MacIver, M.A., Schmitz, L., Mugan, U., Murphey, T.D., Mobley, C.D., 2017. Massive increase in visual range preceded the origin of terrestrial vertebrates. Proc. Natl. Acad. Sci. 114. https://doi.org/10.1073/pnas.1615563114.

MacLean, P.D., 1990. The Triune Brain in Evolution: Role in Paleocerebral Functions. Springer Science & Business Media.

Macrini, T.E., Rougier, G.W., Rowe, T., 2007. Description of a cranial endocast from the fossil mammal *Vincelestes neuquenianus* (Theriiformes) and its relevance to the evolution of endocranial characters in therians. Anat. Rec. Adv. Integr. Anat. Evol. Biol. 290, 875–892. https://doi.org/10.1002/ar.20551.

Macungo, Z., Benoit, J., Fernandez, V., Araújo, R., 2022. X-ray microcomputed and synchrotron tomographic analysis of the basicranial axis of emydopoid dicynodonts: implications for fossoriality and phylogeny. Zool. J. Linn. Soc., 1–46.

Maier, W., Ruf, I., 2016. Evolution of the mammalian middle ear: a historical review. J. Anat. 228, 270–283. https://doi.org/10.1111/joa.12379.

Manger, P.R., 2006. An examination of cetacean brain structure with a novel hypothesis correlating thermogenesis to the evolution of a big brain. Biol. Rev. 81, 293–338. https://doi.org/10.1017/S1464793106007019.

Manger, P.R., 2009. Evolution of the trigeminal sensory system and its specializations. In: Binder, M.D., Hirokawa, N., Windhorst, U. (Eds.), Encyclopedia of Neuroscience. Springer, Berlin Heidelberg, Berlin, Heidelberg, pp. 1436–1440. https://doi.org/10.1007/978-3-540-29678-2_3174.

Manger, P.R., 2013. Questioning the interpretations of behavioral observations of cetaceans: is there really support for a special intellectual status for this mammalian order? Neuroscience 250, 664–696. https://doi.org/10.1016/j.neuroscience.2013.07.041.

Manger, P.R., Pettigrew, J.D., 1995. Electroreception and the feeding behaviour of platypus (*Ornithorhynchus anatinus* : Monotremata: Mammalia). Philos. Trans. R. Soc. Lond. B Biol. Sci. 347, 359–381. https://doi.org/10.1098/rstb.1995.0030.

Manger, P.R., Prowse, M., Haagensen, M., Hemingway, J., 2012. Quantitative analysis of neocortical gyrencephaly in African elephants (Loxodonta africana) and six species of cetaceans: comparison with other mammals. J. Comp. Neurol. 520, 2430–2439. https://doi.org/10.1002/cne.23046.

Manger, P.R., Spocter, M.A., Patzke, N., 2013. The evolutions of large brain size in mammals: the 'Over-700-Gram Club quartet'. Brain Behav. Evol. 82, 68–78. https://doi.org/10.1159/000352056.

Manley, G.A., 1972. A review of some current concepts of the functional evolution of the ear in terrestrial vertebrates. Evolution 26, 608. https://doi.org/10.2307/2407057.

Manley, G.A., 2000. Cochlear mechanisms from a phylogenetic viewpoint. Proc. Natl. Acad. Sci. 97, 11736–11743. https://doi.org/10.1073/pnas.97.22.11736.

Manley, G.A., 2017. Comparative auditory neuroscience: understanding the evolution and function of ears. J. Assoc. Res. Otolaryngol. 18, 1–24. https://doi.org/10.1007/s10162-016-0579-3.

Manley, G.A., 2018. The foundations of high-frequency hearing in early mammals. J. Mamm. Evol. 25, 155–163. https://doi.org/10.1007/s10914-016-9379-0.

Manoussaki, D., Chadwick, R.S., Ketten, D.R., Arruda, J., Dimitriadis, E.K., O'Malley, J.T., 2008. The influence of cochlear shape on low-frequency hearing. Proc. Natl. Acad. Sci. 105, 6162–6166. https://doi.org/10.1073/pnas.0710037105.

Marchetti, L., Klein, H., Buchwitz, M., Ronchi, A., Smith, R.M.H., De Klerk, W.J., Sciscio, L., Groenewald, G.H., 2019. Permian-Triassic vertebrate footprints from South Africa: ichnotaxonomy, producers and biostratigraphy through two major faunal crises. Gondw. Res. 72, 139–168. https://doi.org/10.1016/j.gr.2019.03.009.

Martin, T., Marugán-Lobón, J., Vullo, R., Martín-Abad, H., Luo, Z.-X., Buscalioni, A.D., 2015. A cretaceous eutriconodont and integument evolution in early mammals. Nature 526, 380–384. https://doi.org/10.1038/nature14905.

McGhee, G.R., 2011. Convergent Evolution: Limited Forms Most Beautiful. MIT Press.

Meng, Q.-J., Grossnickle, D.M., Liu, D., Zhang, Y.-G., Neander, A.I., Ji, Q., Luo, Z.-X., 2017. New gliding mammaliaforms from the Jurassic. Nature 548, 291–296. https://doi.org/10.1038/nature23476.

Mitchinson, B., Grant, R.A., Arkley, K., Rankov, V., Perkon, I., Prescott, T.J., 2011. Active vibrissal sensing in rodents and marsupials. Philos. Trans. R. Soc. B Biol. Sci. 366, 3037–3048. https://doi.org/10.1098/rstb.2011.0156.

Miyamae, J.A., Bhullar, B.-A.S., 2017. Comparative morphology of the trigeminal canal and a scenario for the evolution of facial musculature in mammals. In: 77th Annual Meeting of the Society of Vertebrate Paleontology. Abstract of Papers. Presented at the 77th Annual Meeting of the Society of Vertebrate Paleontology, Calgary, Canada, p. 164.

Mourlam, M.J., Orliac, M.J., 2017. Infrasonic and ultrasonic hearing evolved after the emergence of modern whales. Curr. Biol. 27, 1776–1781.e9. https://doi.org/10.1016/j.cub.2017.04.061.

Muchlinski, M.N., 2008. The relationship between the infraorbital foramen, infraorbital nerve, and maxillary mechanoreception: implications for interpreting the paleoecology of fossil mammals based on infraorbital foramen size. Anat. Rec. Hoboken NJ 2007 (291), 1221–1226. https://doi.org/10.1002/ar.20742.

Muchlinski, M.N., 2010. Ecological correlates of infraorbital foramen area in primates. Am. J. Phys. Anthropol. 141, 131–141. https://doi.org/10.1002/ajpa.21137.

Muchlinski, M.N., Wible, J.R., Corfe, I., Sullivan, M., Grant, R.A., 2020. Good vibrations: the evolution of whisking in small mammals. Anat. Rec. 303, 89–99. https://doi.org/10.1002/ar.23989.

Nasterlack, T., Canoville, A., Chinsamy, A., 2012. New insights into the biology of the Permian genus *Cistecephalus* (Therapsida, Dicynodontia). J. Vertebr. Paleontol. 32, 1396–1410. https://doi.org/10.1080/02724634.2012.697410.

Neil, G.J., Allison, W.T., 2021. Evolution of mammalian vision, the. In: Shackelford, T.K., Weekes-Shackelford, V.A. (Eds.), Encyclopedia of Evolutionary Psychological Science. Springer International Publishing, Cham, pp. 2572–2577. https://doi.org/10.1007/978-3-319-19650-3_2753.

Neumann, S., 2020. Taxonomic Revision of the Short-Snouted Tapinocephalid Dinocephalia (Amniota- Therapsida)- the Key to Understanding Middle Permian Tetrapod Biodiversity. (PhD Thesis). University of the Witwatersrand, Johannesburg.

Newham, E., Gill, P.G., Brewer, P., Benton, M.J., Fernandez, V., Gostling, N.J., Haberthür, D., Jernvall, J., Kankaanpää, T., Kallonen, A., Navarro, C., Pacureanu, A., Richards, K., Brown, K.R., Schneider, P., Suhonen, H., Tafforeau, P., Williams, K.A., Zeller-Plumhoff,

B., Corfe, I.J., 2020. Reptile-like physiology in early Jurassic stem-mammals. Nat. Commun. 11, 5121. https://doi.org/10.1038/s41467-020-18898-4.

Newham, E., Gill, P.G., Corfe, I.J., 2022. New tools suggest a middle Jurassic origin for mammalian endothermy: advances in state-of-the-art techniques uncover new insights on the evolutionary patterns of mammalian endothermy through time. Bioessays 44, 2100060. https://doi.org/10.1002/bies.202100060.

Niedźwiedzki, G., Bojanowski, M., 2012. A supposed Eupelycosaur body impression from the early Permian of the intra-Sudetic Basin, Poland. Ichnos 19, 150–155. https://doi.org/10.1080/10420940.2012.702549.

Oftedal, O.T., 2002. The mammary gland and its origin during synapsid evolution. J. Mammary Gland Biol. Neoplasia 7, 225–252. https://doi.org/10.1023/A:1022896515287.

Oftedal, O.T., 2020. The evolution of lactation in mammalian species. In: Walker, W.A., Lönnerdal, B. (Eds.), Ogra, P.L. Nestlé Nutrition Institute Workshop Series. S, Karger AG, pp. 1–10. https://doi.org/10.1159/000505577.

Olivier, C., Houssaye, A., Jalil, N.-E., Cubo, J., 2017. First palaeohistological inference of resting metabolic rate in an extinct synapsid, *Moghreberia nmachouensis* (Therapsida: Anomodontia). Biol. J. Linn. Soc. 121, 409–419. https://doi.org/10.1093/biolinnean/blw044.

Pavanatto, A.E.B., Kerber, L., Dias-da-Silva, S., 2019. Virtual reconstruction of cranial endocasts of traversodontid cynodonts (Eucynodontia: Gomphodontia) from the upper Triassic of southern Brazil. J. Morphol. 280, 1267–1281. https://doi.org/10.1002/jmor.21029.

Pettigrew, J.D., 1999. Electroreception in monotremes. J. Exp. Biol. 202, 1447–1454. https://doi.org/10.1242/jeb.202.10.1447.

Pettigrew, J.D., Manger, P.R., Fine, S.L.B., 1998. The sensory world of the platypus. Philos. Trans. R. Soc. Lond. B Biol. Sci. 353, 1199–1210. https://doi.org/10.1098/rstb.1998.0276.

Pol, D., Mancuso, A.C., Smith, R.M.H., Marsicano, C.A., Ramezani, J., Cerda, I.A., Otero, A., Fernandez, V., 2021. Earliest evidence of herd-living and age segregation amongst dinosaurs. Sci. Rep. 11, 20023. https://doi.org/10.1038/s41598-021-99176-1.

Pusch, L.C., Kammerer, C.F., Fröbisch, J., 2019. Cranial anatomy of the early cynodont *Galesaurus planiceps* and the origin of mammalian endocranial characters. J. Anat. 234, 592–621. https://doi.org/10.1111/joa.12958.

Quay, W.B., 1979. The parietal eye-pineal complex. In: Gans, C., Northcutt, R.G., Ulinski, P. (Eds.), Biology of the Reptilia. Academic Press, New York, pp. 245–406.

Quiroga, J.C., 1979. The brain of two mammal-like reptiles (Cynodontia - Therapsida). J. Hirnforsch. 20, 341–350.

Quiroga, J.C., 1980. The brain of the mammal-like reptile *Probainognathus jenseni* (Therapsida, Cynodontia). A correlative paleo-neoneurological approach to the neocortex at the reptile-mammal transition. J. Hirnforsch. 21, 299–336.

Quiroga, J.C., 1984. The endocranial cast of the advanced mammal-like reptile *Therioherpeton cargnini* (Therapsida-Cynodontia) from the middle Triassic of Brazil. J. Hirnforsch. 25, 285–290.

Qvarnström, M., Niedźwiedzki, G., Tafforeau, P., Žigaitė, Ž., Ahlberg, P.E., 2017. Synchrotron phase-contrast microtomography of coprolites generates novel palaeobiological data. Sci. Rep. 7, 2723. https://doi.org/10.1038/s41598-017-02893-9.

Reisz, R., 1975. Pennsylvanian pelycosaurs from Linton, Ohio and Nýřany, Czechoslovakia. J. Paleontol. 49, 522–527.

Rey, K., Amiot, R., Fourel, F., Abdala, F., Fluteau, F., Jalil, N.-E., Liu, J., Rubidge, B.S., Smith, R.M., Steyer, J.S., Viglietti, P.A., Wang, X., Lécuyer, C., 2017. Oxygen isotopes suggest elevated thermometabolism within multiple Permo-Triassic therapsid clades. Elife 6, e28589. https://doi.org/10.7554/eLife.28589.

Roach, B.T., Brinkman, D.L., 2007. A reevaluation of cooperative pack hunting and gregariousness in *deinonychus antirrhopus* and other nonavian theropod dinosaurs. Bull. Peabody Mus. Nat. Hist. 48, 103–138. https://doi.org/https://doi.org/10.3374/0079-032X(2007)48[103:AROCPH]2.0.CO;2.

Rodrigues, P.G., Ruf, I., Schultz, C.L., 2013. Digital reconstruction of the Otic region and inner ear of the non-mammalian cynodont *Brasilitherium riograndensis* (late Triassic, Brazil) and its relevance to the evolution of the mammalian ear. J. Mamm. Evol. 20, 291–307. https://doi.org/10.1007/s10914-012-9221-2.

Rodrigues, P.G., Ruf, I., Schultz, C.L., 2014. Study of a digital cranial endocast of the non-mammaliaform cynodont *Brasilitherium riograndensis* (later Triassic, Brazil) and its relevance to the evolution of the mammalian brain. Paläontol. Z. 88, 329–352. https://doi.org/10.1007/s12542-013-0200-6.

Rodrigues, P.G., Martinelli, A.G., Schultz, C.L., Corfe, I.J., Gill, P.G., Soares, M.B., Rayfield, E.J., 2018. Digital cranial endocast of *Riograndia guaibensis* (late Triassic, Brazil) sheds light on the evolution of the brain in non-mammalian cynodonts. Hist. Biol. 31, 1–18. https://doi.org/10.1080/08912963.2018.1427742.

Romano, M., Manucci, F., 2019. Resizing *Lisowicia bojani*: volumetric body mass estimate and 3D reconstruction of the giant Late Triassic dicynodont. Hist. Biol., 1–6. https://doi.org/10.1080/08912963.2019.1631819.

Romer, A.S., 1933. Vertebrate Paleontology. University of Chicago Press.

Romer, A.S., 1956. Osteology of the Reptiles. Krieger Publishing Company.

Romer, A.S., Price, L.W., 1940. Review of the Pelycosauria. In: Geological Society of America Special Papers. Geological Society of America, pp. 1–534. https://doi.org/10.1130/SPE28-p1.

Rosowski, J.J., Graybeal, A., 1991. What did Morganucodon hear? Zool. J. Linn. Soc. 101, 131–168. https://doi.org/10.1111/j.1096-3642.1991.tb00890.x.

Rowe, T., 1996. Coevolution of the mammalian middle ear and neocortex. Science 273, 651–654. https://doi.org/10.1126/science.273.5275.651.

Rowe, T.B., 2017. The emergence of mammals. Evol. Nerv. Syst. 2, 1–52. https://doi.org/10.1016/B978-0-12-804042-3.00029-4.

Rowe, T.B., Shepherd, G.M., 2016. Role of ortho-retronasal olfaction in mammalian cortical evolution: olfaction and cortical evolution. J. Comp. Neurol. 524, 471–495. https://doi.org/10.1002/cne.23802.

Rowe, T., Rich, T.H., Vickers-Rich, P., Springer, M., Woodburne, M.O., 2008. The oldest platypus and its bearing on divergence timing of the platypus and echidna clades. Proc. Natl. Acad. Sci. 105, 1238–1242. https://doi.org/10.1073/pnas.0706385105.

Rowe, T.B., Macrini, T.E., Luo, Z.-X., 2011. Fossil evidence on origin of the mammalian brain. Science 332, 955–957. https://doi.org/10.1126/science.1203117.

Rubidge, B.S., Sidor, C.A., 2001. Evolutionary patterns among Permo-Triassic therapsids. Annu. Rev. Ecol. Syst. 32, 449–480. https://doi.org/10.1146/annurev.ecolsys.32.081501.114113.

Rubidge, B.S., Govender, R., Romano, M., 2019. The postcranial skeleton of the basal tapinocephalid dinocephalian *Tapinocaninus pamelae* (Synapsida: Therapsida) from the south African Karoo supergroup. J. Syst. Palaeontol. 17, 1767–1789. https://doi.org/10.1080/14772019.2018.1559244.

Ruf, I., Maier, W., Rodrigues, P.G., Schultz, C.L., 2014. Nasal anatomy of the non-mammaliaform cynodont *Brasilitherium riograndensis* (Eucynodontia, Therapsida) reveals new insight into mammalian evolution: nasal anatomy of *Brasilitherium*. Anat. Rec. 297, 2018–2030. https://doi.org/10.1002/ar.23022.

Ryan, T.M., Silcox, M.T., Walker, A., Mao, X., Begun, D.R., Benefit, B.R., Gingerich, P.D., Köhler, M., Kordos, L., McCrossin, M.L., Moyà-Solà, S., Sanders, W.J., Seiffert, E.R., Simons, E., Zalmout, I.S., Spoor, F., 2012. Evolution of locomotion in Anthropoidea: the semicircular canal evidence. Proc. R. Soc. B Biol. Sci. 279, 3467–3475. https://doi.org/10.1098/rspb.2012.0939.

Saint Martin, J.-P., Saint Martin, S., 2018. Exquisite preservation of a widespread filamentous microorganism in French Cretaceous ambers: Crucial for revising a controversial fossil. Comptes Rendus Palevol, Lagerstätten 2: Exceptionally preserved fossils Lagerstätten 2: fossiles à conservation exceptionnelle 17, 415–434. https://doi.org/10.1016/j.crpv.2017.05.003.

Sánchez-Villagra, M.R., 2002. The cerebellar paraflocculus and the subarcuate fossa in *Monodelphis domestica* and other marsupial mammals — ontogeny and phylogeny of a brain-skull interaction. Acta Theriol. (Warsz.) 47, 1–14. https://doi.org/10.1007/BF03193561.

Schwab, I.R., Dubielzig, R.R., Schobert, C., 2012. Evolution's Witness: How Eyes Evolved. OUP, USA.

Seymour, R.S., Bennett-Stamper, C.L., Johnston, S.D., Carrier, D.R., Grigg, G.C., 2004. Evidence for endothermic ancestors of crocodiles at the stem of archosaur evolution. Physiol. Biochem. Zool. 77, 1051–1067. https://doi.org/10.1086/422766.

Smith, R.M.H., 1987. Helical burrow casts of therapsid origin from the Beaufort group (Permian) of South Africa. Palaeogeogr. Palaeoclimatol. Palaeoecol. 60, 155–169. https://doi.org/10.1016/0031-0182(87)90030-7.

Smith, R.M.H., Botha-Brink, J., 2005. The recovery of terrestrial vertebrate diversity in south African Karoo basin after the end-Permian extinction. C. R. Palevol 4, 555–568. https://doi.org/10.1016/j.crpv.2005.07.005.

Smith, R.M.H., Botha-Brink, J., 2011. Morphology and composition of bone-bearing coprolites from the late Permian Beaufort group, Karoo Basin, South Africa. Palaeogeogr. Palaeoclimatol. Palaeoecol. 312, 40–53. https://doi.org/10.1016/j.palaeo.2011.09.006.

Smith, R.M.H., Botha-Brink, J., 2014. Anatomy of a mass extinction: sedimentological and taphonomic evidence for drought-induced die-offs at the Permo-Triassic boundary in the main Karoo Basin, South Africa. Palaeogeogr. Palaeoclimatol. Palaeoecol. 396, 99–118. https://doi.org/10.1016/j.palaeo.2014.01.002.

Smith, R.M.H., Angielczyk, K.D., Benoit, J., Fernandez, V., 2021. Neonate aggregation in the Permian dicynodont *Diictodon* (Therapsida, Anomodontia): evidence for a reproductive function for burrows? Palaeogeogr. Palaeoclimatol. Palaeoecol. 569, 110311. https://doi.org/10.1016/j.palaeo.2021.110311.

Spindler, F., 2015. The Basal Sphenacodontia – Systematic Revision and Evolutionary Implications. Technischen Universität Bergakademie Freiberg, Freiberg.

Spindler, F., Werneburg, R., Schneider, J.W., Luthardt, L., Annacker, V., Rößler, R., 2018. First arboreal "pelycosaurs" (Synapsida: Varanopidae) from the early Permian Chemnitz fossil Lagerstätte, SE Germany, with a review of varanopid phylogeny. PalZ 92, 315–364. https://doi.org/10.1007/s12542-018-0405-9.

Spoor, F., Garland, T., Krovitz, G., Ryan, T.M., Silcox, M.T., Walker, A., 2007. The primate semicircular canal system and locomotion. Proc. Natl. Acad. Sci. 104, 10808–10812. https://doi.org/10.1073/pnas.0704250104.

Sullivan, C., Reisz, R.R., Smith, R.M.H., 2003. The Permian mammal-like herbivore *Diictodon*, the oldest known example of sexually dimorphic armament. Proc. R. Soc. Lond. B Biol. Sci. 270, 173–178. https://doi.org/10.1098/rspb.2002.2189.

Tanke, D.H., Rothschild, B.M., 2002. DINOSORES: An Annotated Bibliography of Dinosaur Paleopathology and Related Topics—1838-2001: Bulletin 20. New Mexico Museum of Natural History and Science.

Tatarinov, L.P., 1976. Morphological Evolution of the Theriodonts and the General Problems of Phylogenetics. NAUKA, Moscow, Russia.

Valkenburgh, B.V., Jenkins, I., 2002. Evolutionary patterns in the history of Permo-Triassic and Cenozoic synapsid predators. Paleontol. Soc. Pap. 8, 267–288. https://doi.org/10.1017/S1089332600001121.

Van Valen, L.M., 1960. Therapsids as mammals. Evolution 14, 304. https://doi.org/10.2307/2405973.

Vega-Zuniga, T., Medina, F.S., Fredes, F., Zuniga, C., Severín, D., Palacios, A.G., Karten, H.J., Mpodozis, J., 2013. Does nocturnality drive binocular vision? Octodontine rodents as a case study. PLoS One 8, e84199. https://doi.org/10.1371/journal.pone.0084199.

Viglietti, P.A., Smith, R.M.H., Compton, J.S., 2013. Origin and palaeoenvironmental significance of lystrosaurus bonebeds in the earliest Triassic Karoo Basin, South Africa. Palaeogeogr. Palaeoclimatol. Palaeoecol. 392, 9–21. https://doi.org/10.1016/j.palaeo.2013.08.015.

Wallace, R.V.S., Martínez, R., Rowe, T., 2019. First record of a basal mammaliamorph from the early late Triassic Ischigualasto formation of Argentina. PLoS One 14, e0218791. https://doi.org/10.1371/journal.pone.0218791.

Walls, G.L., 1942. The Vertebrate Eye and its Adaptive Radiation. Cranbrook Institute of Science.

Walsh, S.A., Barrett, P.M., Milner, A.C., Manley, G., Witmer, L.M., 2009. Inner ear anatomy is a proxy for deducing auditory capability and behaviour in reptiles and birds. Proc. R. Soc. B Biol. Sci. 276, 1355–1360. https://doi.org/10.1098/rspb.2008.1390.

Wang, J., Wible, J.R., Guo, B., Shelley, S.L., Hu, H., Bi, S., 2021. A monotreme-like auditory apparatus in a middle Jurassic haramiyidan. Nature 590, 279–283. https://doi.org/10.1038/s41586-020-03137-z.

Watson, D.M.S., 1931. On the skeleton of a bauriamorph reptile. J. Zool. 1931, 1163–1205.

West, C.D., 1985. The relationship of the spiral turns of the cochlea and the length of the basilar membrane to the range of audible frequencies in ground dwelling mammals. J. Acoust. Soc. Am. 77, 1091–1101. https://doi.org/10.1121/1.392227.

Whitney, M.R., Sidor, C.A., 2020. Evidence of torpor in the tusks of *lystrosaurus* from the early Triassic of Antarctica. Commun. Biol. 3, 471. https://doi.org/10.1038/s42003-020-01207-6.

Whitney, M.R., Angielczyk, K.D., Peecook, B.R., Sidor, C.A., 2021. The evolution of the synapsid tusk: insights from dicynodont therapsid tusk histology. Proc. R. Soc. B Biol. Sci. 288, 20211670. https://doi.org/10.1098/rspb.2021.1670.

Wolvaardt, F.P., 2021. Sedimentology and Taphonomy of a Tetrapod Fossil Accumulation in the Triassic Burgersdorp Formation of the Karoo Basin. University of the Witwatersrand (unpublished Msc. thesis).

Yi, H., Norell, M.A., 2015. The burrowing origin of modern snakes. Sci. Adv. 1, e1500743. https://doi.org/10.1126/sciadv.1500743.

CHAPTER

Insights into brain evolution through the genotype-phenotype connection

3

Danalaxshmi Shanen Ganapathee[a,*] and Philipp Gunz[b]

[a]*Department of Human Evolutionary Biology, Harvard University, Cambridge, MA, United States*

[b]*Department of Human Evolution, Max Planck Institute for Evolutionary Anthropology, Leipzig, Germany*

**Corresponding author: e-mail address: sganapathee@g.harvard.edu*

Abstract

It has recently become possible to start exploring how the genotype translates into human brain morphology and behavior by combining detailed genomic and phenotypic data from thousands of present-day people with archaic genomes of extinct humans, and gene expression data. As a starting point into this emerging interdisciplinary domain, we highlight current debates about which aspects of the modern human brain are unique. We review recent developments from (1) comparative primate neuroscience—a fast-growing field offering an invaluable framework for understanding general mechanisms and the evolution of human-specific traits. (2) paleoanthropology—based on evidence from endocranial imprints in fossil skulls, we trace the evolution from the ape-like brain phenotype of early hominins more than 3 million years ago to the unusual globular brain shape of present-day people. (3) Genomics of present-day and extinct humans. The morphological and genetic differences between modern humans and our closest extinct cousins, the Neandertals, offer important clues about the genetic underpinnings of brain morphology and behavior. The functional consequences of these genetic differences can be tested in animal models, and brain organoids.

Keywords

Genotype, Phenotype, Brain, Cognition, Behavioral modernity, Archaic hominins

1 Why is the genotype-phenotype connection important?

Since the human genome was fully sequenced in the early 2000s (Lander et al., 2001; Collins et al., 2004; Nurk et al., 2022), two decades of research have been devoted to uncovering the functional significance not only of genes but also of millions of genetic variants and differences in gene regulation. These efforts have been likened

Progress in Brain Research, Volume 275, ISSN 0079-6123, https://doi.org/10.1016/bs.pbr.2022.12.013

to finding a "needle in a haystack" (Varki and Altheide, 2005). For example, while it is well established that phenotypic traits such as height have genetic underpinnings, it is much less simple to know exactly what specific gene, set of genes, or regulatory elements in tandem are causatively linked to any given human being's height (Muthuirulan and Capellini, 2019). More often than not, we encounter complex traits with no straightforward functional link to one gene or regulatory region. Here, we will review recent attempts trying to understand how the genotype translates into the human phenotype both from top-down and bottom-up perspectives: the former typically start with distinct phenotypes and look for their genetic underpinnings, whereas bottom-up approaches explore the phenotypic consequences of specific genetic changes. In this chapter, we will focus on the brain and its growth pattern, and by extension cognition and behavior as our phenotypes of choice. Even though it is a challenging set of problems to elucidate, seeking to validate genotype-phenotype connections is still bound to be interesting and important work in the realm of human evolution and there are three main reasons why:

1. *Genomic data make it possible to reconstruct our evolutionary past:* the genome is akin to a historical record of changes. We are able to know with exquisite detail, for instance, that positive selection occurred in a certain region of the genome or that a specific variant became fixed in the *Homo sapiens* lineage as compared to *Homo neanderthalensis* and much more. If it is possible to have clear connections between genotype and phenotype, then these records of changes across time become much more informative.
2. *Establishing genotype-phenotype connections can help us elucidate what makes us human:* questions about human-specific traits have captured the minds of generations. It is still unknown despite much work on the subject what makes *Homo sapiens* such a distinctly different species capable of modifying its environment to the extent it has. While there is overwhelming consensus that no single gene has led to the emergence of behavioral modernity in humans, it is still intriguing to consider what series of changes at the genome level underly the human mind and abilities.
3. *From a biomedical standpoint, genotype-phenotype connections can lead to targeted therapies:* besides purely curiosity-led inquiry, there are some practical benefits to working on elucidating the genotype-phenotype connection, with advances made in unearthing the underlying genetic basis of osteoarthritis (Richard et al., 2020) and schizophrenia (Sekar et al., 2016) amongst others. This type of research aids in potential therapeutic interventions.

Studying the connection between genotype and brain phenotypes from an evolutionary perspective can provide important mechanistic insights. One of the obvious challenges to studying the evolution of the brain's phenotype is that the soft tissues of the brain do not fossilize. Comparative studies with other primates therefore offer an indispensable frame of reference for determining aspects that are uniquely human. A growing body of research relies on broad comparative studies of brain structure and function in living non-human primates and humans.

So as to decipher the relationship between genotype and phenotype in comparative studies it is helpful to consider the temporal depth of evolution captured by the available genetic data. Both neuroimaging and rich genotype data are available for present-day humans and macaques. As humans and macaques last shared a common ancestor around 25 million years ago, their similarities and differences provide information about the likely ancestral condition both on the genetic and the phenotypic level. Adding genetic and neuroimaging data of our closest living relatives, the great apes, yields more nuanced insights into the biological processes underlying unique human traits and abilities, including, for example, brain enlargement and reorganization, the evolution of tool use, social behavior, and language. However, such data are difficult to obtain for both logistical and ethical reasons and are therefore scarce. Information from ancient genomes can be retrieved directly from some Late Pleistocene hominin fossils, and even some Middle Pleistocene specimens recovered from exceptional preservation conditions [for example, in cold environments, or deep in caves such as in Meyer et al. (2016)], thereby covering the last few hundred thousand years of hominin evolution.

2 What has our field done thus far to elucidate the genotype-phenotype connection?

2.1 Evolutionary perspectives on the human brain phenotype

2.1.1 Brain volume and growth pattern

Perhaps the most striking difference between humans and other primates is the large human brain size. Human brains are about 3.5 times larger than those of chimpanzees, our closest living relatives. Inextricably linked to species differences in adult brain volume are distinct patterns of brain growth and development: human brains grow at a fast pace for a longer period of time than those of chimpanzees (Coqueuniot et al., 2004; Leigh, 2004; Leigh, 2006; DeSilva and Lesnik, 2006, 2008). Human brains also take longer to mature, which has been linked to increased brain plasticity, and a long dependence on caregivers (Robson and Wood, 2008; Gunz et al., 2020).

The fossil record provides direct evidence of brain volume, and aspects of brain organization: as the brain expands during early development its surrounding tissues leave an imprint inside the braincase. Such endocranial imprints (or endocasts) can also be examined in fossils, and provide direct evidence of endocranial volume (and thereby brain size), impressions of gyri and sulci, as well as blood vessels (Figs. 1 and 2). Endocasts sometimes occur naturally in the fossil record, like for example in the first *Australopithecus* fossil ever described (Fig. 1), the Taung child from South Africa (Dart, 1925). Virtual endocasts extracted from computed tomographic (CT) scans (Conroy and Vannier, 1985; Conroy et al., 1998; Neubauer et al., 2012; Beaudet et al., 2018; Bruner, 2019) have largely supplanted physical endocasts (Holloway et al., 2004) created from the actual individual. Cross-sectional

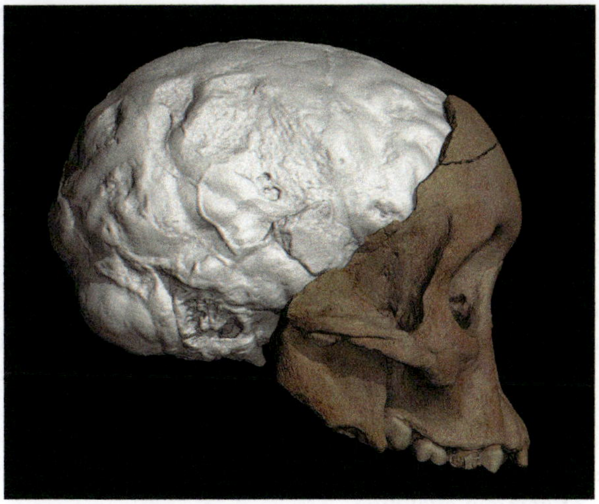

FIG. 1

The Taung child, the first *Australopithecus* ever described (Dart, 1925), preserves a natural endocast (white); only a small portion of the bony braincase is preserved. This fossilized imprint of the braincase provides evidence about endocranial volume (comparable to apes of the same dental age), and aspects of brain organization in early hominins. On the endocranial surface, the impressions of gyri and sulci, cranial sutures, meningeal vessels, and venous sinuses are clearly visible. However, owing to some damage on the endocranial surface the interpretation of the sulcal impressions remains controversial to this day, almost 100 years after the fossil's discovery and initial description. Rendering based on computed tomographic scans of the original fossil housed at the University of the Witwatersrand, Johannesburg (South Africa).

ontogenetic series of fossil endocasts from birth to adulthood can provide information on the trajectories of brain growth and development (Rosas et al., 2017; Gunz et al., 2010, 2012, 2020; Cofran, 2019).

2.1.2 Evolutionary changes of hominin brain organization

Comparative studies of the brain's sulcal morphology reveal that present-day humans differ from all other primates in the organization of the prefrontal cortex, as well as the parietal and occipital lobes. Some aspects of hominin brain reorganization, such as different folding patterns, may merely be by-products of increased overall brain size in hominins, others likely reflect functional changes in specific brain areas and differences in brain connectivity. However, discussions about evolutionary changes in brain organization are often contentious, as the identification of sulcal impressions in fossil endocasts is not always straightforward (Falk et al., 2000; Holloway et al., 2004). There is debate, e.g., as to whether the aforementioned organizational changes emerged early in the hominin lineage in the hominin genus *Australopithecus,* or more than a million years later the genus *Homo*, and to

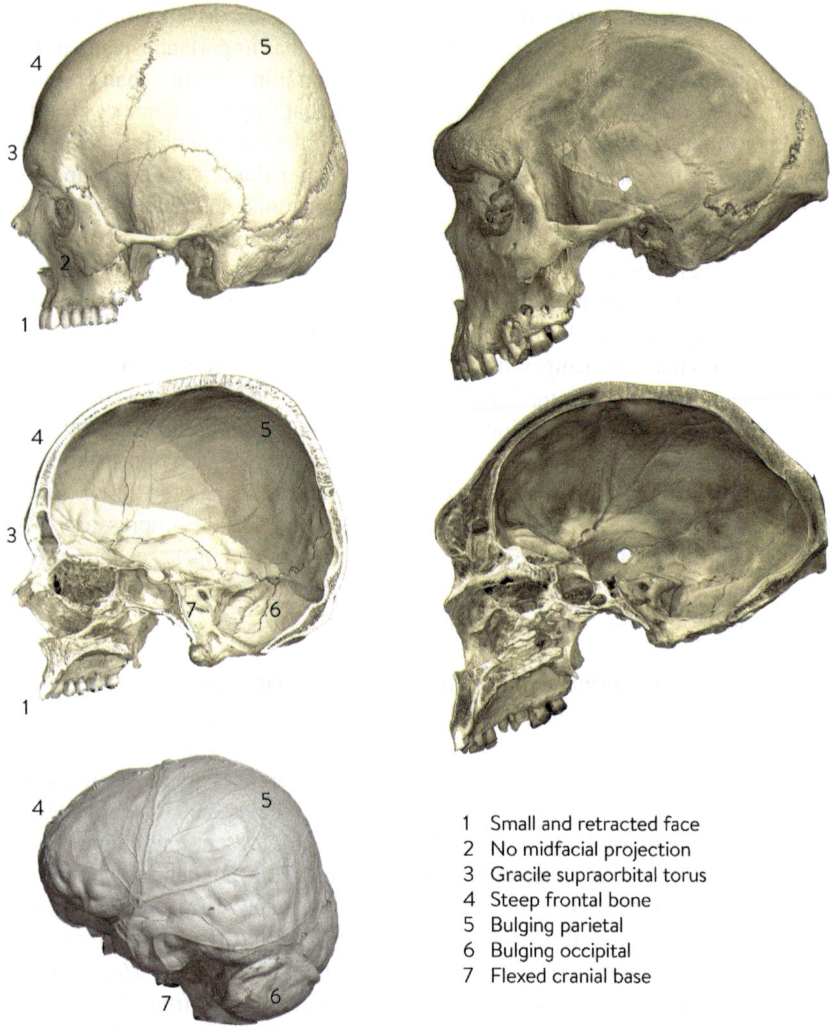

FIG. 2

The cranial phenotype of present-day humans differs from our extinct ancestors and relatives. Aspects unique to modern humans are labeled on the cranium and the endocranial imprint. Upper left: Computed tomographic (CT) scan of a recent *Homo sapiens*. Upper right: CT scan of the Kabwe 1 cranium, a *Homo heidelbergensis* fossil from Zambia, dated to approximately 300 ka ago (Grün et al., 2020). Middle panel: A virtual cut through the midplane reveals the impressions left by the brain and its surrounding structures inside the bony braincase. Lower left: Virtual endocranial imprint of the *Homo sapiens* shown in the left column, created from the CT scans using segmentation.

what extent the evolutionary changes of sulcal patterns are driven primarily by increases in overall brain size. Recent studies suggest that important aspects of brain reorganization evolved much later in the course of hominin evolution than previously thought. Based on a comparative study of endocranial imprints, Gunz et al. (2020) have shown that the average endocranial volume in the early hominin species *Australopithecus afarensis* was approximately 20% larger than in adult chimpanzees, and that *A. afarensis* had an ape-like brain organization. *A. afarensis* occupies a key position in the hominin family tree, as it is widely considered to be ancestral to several other hominin groups, including the lineage eventually leading to present-day humans (Kimbel et al., 2004; Kimbel and Delezene, 2009). This research establishes an ape-like baseline morphology in early hominins around 3 million years ago, and implies that key changes in sulcal morphology in the inferior frontal gyrus, and the parietal lobe evolved in later hominin species. Ponce de León et al. (2021) argued that an evolutionary reorganization of the brain's prefrontal cortex evolved within the species *Homo erectus*. Studying endocasts these authors argued that an expansion of the prefrontal cortex in human evolution shifts the impression of the brain's *precentral sulcus* toward the back of the brain, "behind" the impression of the cranium's coronal suture. It is debated, however, whether these changes in the position of the *precentral sulcus* are largely a by-product of brain size increase, rather than conclusive evidence for brain reorganization.

2.1.3 Differences between modern humans and Neandertals

Present-day humans and their closest extinct relatives, the Neandertals, have similarly large endocranial volumes, yet different endocranial shapes (Bruner et al., 2003; Kochiyama et al., 2018). These endocranial differences are most pronounced in the back of the skull which houses the cerebellum (Weaver, 2005; Neubauer et al., 2018), and the parietal lobe (Pereira-Pedro et al., 2020). Present-day humans have a characteristic endocranial shape that is more globular (Fig. 2) than in extinct hominins (Lieberman et al., 2002), including early representatives of our own species, *Homo sapiens*. Hublin et al. (2017) and Neubauer et al. (2018) have demonstrated that the globular endocranial shape emerges over time within the *Homo sapiens* clade: early representatives of *Homo sapiens*, such as fossil from Jebel Irhoud, and Omo have elongated endocranial shapes, and fall far outside the range of present-day endocranial shape variation. Only fossils dated to around 35,000 years ago and younger fall within the range of shape variation of present-day humans (Weber et al. 2020). Zollikofer et al. (2022) have emphasized the role facial gracilization may have played for this temporal trend toward globularity within *Homo sapiens*. Specifically, these authors suggested "that differences in masticatory and/or respiratory function" underlie endocranial shape differences between fossil *H. sapiens* and present-day humans. Ontogenetic studies of endocranial shape changes throughout development in humans (Neubauer et al., 2009; Zollikofer et al., 2022), and apes (Scott et al., 2014, 2018; Zollikofer et al., 2017) suggest that endocranial shape changes after the cessation of brain growth are likely linked to the continued growth of the face. Certain aspects of the globular phenotype, such

as parietal bulging and cerebellar expansion, however, develop before the eruption of the deciduous dentition in modern humans, indicating that the tempo and mode of brain growth and development are the driving factors underlying this uniquely modern endocranial phenotype (Neubauer et al., 2009; Gunz et al., 2010, 2012).

2.2 Neandertal genomes offer clues about the evolution of brain shape and behavior

Genome-wide analyses require very large sample sizes (i.e., tens of thousands of individuals) for sufficient statistical power. Available databases that fulfill this criterion (e.g., the UK Biobank) unfortunately only capture a small amount of the worldwide human variation (Bycroft et al., 2018). Working with smaller samples (i.e., thousands of people) requires focussing the search by constraining the search space, both on the genotypic and the phenotypic side. One strategy is to leverage the fossil record, to target phenotypes that have changed in the course of evolution. Ancient DNA extracted from fossil bones can also provide crucial information, such as evidence for selective sweeps. In the last two decades, the study of ancient DNA has spurred a paradigm shift in our understanding of interactions between different hominins: genetic exchanges between different hominin species were long considered rare, however, especially studies on ancient nuclear DNA have revealed frequent admixture events. As a result of such interbreeding events, introgressed genetic fragments can be traced in present-day genomes: non-Africans living today carry a small percentage of Neandertal DNA scattered throughout their genome. In total, approximately 40% of the Neandertal genome can therefore be reconstructed from the genomes of present-day humans (Sankararaman et al., 2014). Several studies have used the phenotypic and genetic differences between modern humans and Neandertals as a window into the underlying biology. Alternatively, one can target those regions of the genome that are devoid of introgressed fragments (e.g., so-called "archaic deserts" in the modern human genome), based on the idea that negative selection likely acted on introgressed fragments in these regions.

Gregory et al. (2017) studied magnetic resonance imaging (MRI) brain scans of 221 healthy individuals of European ancestry for which genetic data were also available. They summarized the amount of Neandertal genetic ancestry in a "NeanderScore" and looked for associations with brain phenotypes extracted from the neuroimaging data. These authors reported that higher NeanderScores were associated with Neandertal-like skull shapes, including anterior-posterior cranial length, and the shapes of the parietal and occipital bones. One fundamental caveat, however, is that conflating the Neandertal-derived genetic fragments into one overall summary score makes it challenging to study the connection between genotype and phenotype: Neandertal gene fragments are rare and scattered throughout the genome, so that even two people with the exact same NeanderScore may well carry completely different fragments.

To study the genetic bases of the globular phenotype Gunz et al. (2019) analyzed in vivo MRI brain scans of healthy human adults together with their genetic data.

As the sample size of approximately 4500 Europeans in Gunz et al. (2019) did not allow for a genome-wide screen of associations between genotype and phenotype, these authors focused instead on the phenotypic differences between modern humans and Neandertals, as well as the phenotypic effect of Neandertal introgressed gene fragments. Gunz et al. (2019) first measured the overall brain shape from MRI scans as well as the endocranial shape of modern and fossil humans using geometric morphometric analysis of endocranial landmarks and semilandmarks, a statistical approach developed originally for paleoanthropology (Gunz and Mitteroecker, 2013). Next, they computed a summary metric ("globularity score") for overall endocranial shape, based on the average shape difference between modern humans and Neandertals, and tested whether any Neandertal introgressed gene fragments were associated with a more elongated endocranial shape. These authors used a different approach than the aforementioned study by Gregory et al. (2017): rather than creating a summary metric for the genetic data, Gunz et al. (2019) summarized the phenotypic data. They identified Neandertal fragments associated with altered activity of two genes, of which one had already been linked to neurogenesis based on work in animal models, and the other to myelination. The effects of carrying these rare Neandertal fragments were subtle, both at the gene expression level and at the phenotypic level.

To explore the origin and genetic components of behavioral phenotypes, Dannemann et al. (2022) studied the effect of Neandertal introgressed fragments on psychiatric and behavioral phenotypes in several large cohorts comprising data from thousands of present-day people. In line with a complementary study by McArthur et al. (2021) they found no enrichment for Neandertal genetic variants for psychiatric and neurological diseases. However, Dannemann et al. (2022) did report significant associations with several behavioral phenotypes with potential relevance for mental health, including pain, sleep duration, smoking, and alcohol consumption. It is worth emphasizing that these results do not literally indicate substance abuse in Neandertals. A growing body of evidence links individual variation in neurotransmitters (prominently in the dopaminergic system), to reward processing that modulates the reaction to drug exposure and thereby affects the vulnerability for addiction (Arias-Carrion and Salama, 2012). The associations between Neandertal introgressed fragments, smoking and alcohol may therefore point to the genetic underpinnings of reward-seeking, and may also be linked to differences in pain-perception and self-medication.

2.3 Mice, ferrets, and brains in a petri dish

Access to genomic data from modern humans and ancient humans has clearly revolutionized our understanding of evolution. There are new tools and new data sources, but we are not fully able to understand what genomic changes mean in terms of changes in function and morphology. However, new(er) technologies, such as induced pluripotent stem cell (iPSC)-derived brain organoids, massively parallel reporter assays (MPRA) and hybrid cells have played a role in the advances made.

Trujillo et al. (2021) reintroduced the archaic version of the neuro-oncological ventral antigen 1 (*NOVA1*)—"one of the few protein-coding differences between modern human and archaic hominin genomes"—in iPSC-derived cortical organoids and were able to demonstrate differences in the phenotype of the organoids carrying the archaic version of *NOVA1* as compared to the modern version. The archaic version of *NOVA1* is present in Neandertals and Denisovans and they argued that the human-specific substitution became fixed in the human lineage and may have influenced our species' neurodevelopmental trajectory. The phenotypes that were found to differ were synaptic protein interactions, glutamatergic signaling, neuronal connectivity, and neuron electrophysiological profiles. Other researchers, however, have cautioned against the interpretation of the Trujillo et al. (2021) study: another plausible explanation for the extent of the phenotypic changes observed may be heterozygous deletions of the target DNA sequence during the Crispr-Cas-9 intervention (Maricic et al., 2021). These authors also question the "drastic phenotypes" observed, which are surprising given that organoids generated from chimpanzee and human iPSCs do not usually produce such visible phenotypic differences. The authors of the original study point to their "human-optimized organoid protocol" which had not been used in any prior studies, which they believe partly explains the "drastic phenotypes" observed (Trujillo et al., 2021; Herai et al., 2021).

Gokhman et al. (2021) utilized human-chimpanzee hybrid induced pluripotent stem cells, which offered the possibility to separate cis- from trans-regulatory effects and control for "nongenetic confounding factors." They found a human-specific downregulation of EVC2 and posit that there may be a functional significance for the development of the unique human cranio-facial phenotype. EVC2 knockout mice showed craniofacial phenotype that was more in line with human-like features, including a retracted face, as compared to control mice with a normal level of EVC2 expression. The mice with normal EVC2 expression levels showed alignment to chimpanzee-like phenotypes, such as midfacial protrusion and lower forehead height. Tetraploid hybrid stem cells were utilized in another study by Agoglia et al. (2021) to query aspects of neurodevelopment. Signals of selection were detected on a set of astrocyte-related genes and differential expression of gene SSTR2 between humans and chimpanzees was identified as a potential source of cortical development difference as well as underlying human-specific neuropsychiatric disorders.

Brain organoids have also been studied using single-cell transcriptomics to uncover further human-specific changes in neurodevelopment as compared to chimpanzees and macaques (Kanton et al., 2019). The authors analyzed the cell composition and differentiation trajectories across the development of human, chimpanzee, and macaque cerebral organoids "from pluripotency, through neuroectoderm and neuroepithelial stages, followed by divergence into neuronal fates within the dorsal and ventral forebrain, midbrain and hindbrain regions" (Kanton et al., 2019). The main finding was that the development of human neurons occurs at a slower pace as compared to the other two primates. Through querying open regions of chromatin during specific time points between chimpanzee and human brain organoids, it was also possible for the authors find gene-regulatory features that are unique to humans.

Pinson et al. (2022) focused on the phenotypic effects of a subtle genetic difference between present-day humans and Neandertals in the "transketolase-like 1" protein (TKTL1). The variant of TKTL1 found in essentially all present-day humans differs from the variant found in apes, Neandertals, and Denisovans by a single amino acid change. In a detailed and thorough experimental study, these authors could show that the modern human TKTL1 variant increases the number of basal radial glia cells, (1) in the developing mouse, (2) the neocortex of developing ferrets, and (3) genome-edited cerebral organoids. The archaic variant of TKTL1 found in Neandertals and Denisovans, however, did not increase the number of basal radial glia cells. Basal radial glia cells are a progenitor cell type generating neurons during brain development. They self-amplify and are responsible for creating much of the neocortex in early fetal development. Pinson et al. (2022) therefore concluded that neurogenesis in the neocortex, particularly the frontal lobe, differs between modern humans and Neandertals. It is not clear, however, to what extent these developmental differences during fetal neurogenesis contribute to the aforementioned endocranial shape differences between modern humans and Neandertals.

Besides identifying variants that are unique and fixed in humans as compared to our extant and extinct relatives, there has also been work done to query the functional effects of introgressed Neandertal variants on phenotype. Jagoda et al. (2022) utilized a massively parallel reporter assay (MPRA) to assess the effect of ~5000 introgressed variants in human immune cells. They were able to identify two variants and their three associated genic targets which overall had a contribution to the immune response to influenza and COVID-19. Introgressed variants may also have some significance with regard to drug metabolism. Haeggström et al. (2022) found that a haplotype inherited from Neandertals is associated with a reduced ability to metabolize warfarin and phenytoin.

In the continued search for genomic regions that matter to human-specific phenotypes, human accelerated regions (HARs) cannot be ignored. HARs are regions that show evidence of rapid change in the human lineage but are otherwise conserved across other vertebrates, which indicates that these regions are most likely functional (Pollard et al., 2006). Only 8% of these HARs are not shared with Neandertals and Denisovans, meaning that most of the rapid changes in these regions pre-date the split with Neandertals and Denisovans (Burbano et al., 2012). However, many loci of interest to human-specific brain evolution as compared to non-human primates have been found by querying HARs (Girskis et al., 2021). Boyd et al. (2015) found that a change in the human version of HARE5, an enhancer that modulates the activity of FZD8 gene, underlies differences in neocortex development relative to chimpanzees. Using transgenic mice, the authors demonstrated that neural progenitor cells had accelerated activity in the humanized mouse. This increase in corticogenesis eventually resulted in a larger neocortex compared to the mice that had the chimpanzee version of the enhancer.

2.4 Outstanding gaps in our understanding of human-specific brain and behavior evolution

While these advances are proceeding in the evolutionary developmental biology and genetics space, there is still a massive challenge in tying these neurobiological changes in development to their potential impact on cognition and ultimately modern behavior, from a bottom-up perspective, but also from a top-down perspective, starting with phenotypes. We have identified three main gaps below:

2.4.1 Prioritization of phenotypes to study skews toward more contemporary issues

There is a different timescale prioritization across fields with behavioral genetics mainly concerned with complex behavioral phenotypes that are anchored in contemporary society, such as risk-taking, and educational attainment (Okbay et al., 2022; Karlsson Linnér et al., 2019) while the evolutionary developmental genetics toolkit can demonstrate genetic changes that have happened as far back as 6 million years ago as humans split off from the last common ancestor (LCA) with chimpanzees and query the resulting phenotypic differences. Researchers doing the work of phenotyping for GWAS and those focused on wet-lab validation are interested in vastly different phenotypes, which makes the reconciliation of both types of work difficult.

2.4.2 Studies on behavior remain largely correlational with little to no connection to potential functional validation

In the field of behavioral genetics, there has been interest in understanding the recent effects of selection on contemporary populations and behaviors. In their 2017 paper, Kong et al. showed that within a population of Icelanders, there is selection against variants associated with educational attainment. Their genome-wide association study (GWAS) is purely looking at the correlation between two specific datasets, however, and it is hard to imagine through which biological pathways a complex phenotype such as educational attainment would manifest. If a wet-lab biologist had to perform functional validation upon the variants found, how would they go about it? There are a bevy of cognitive tasks that model animals can perform in a lab setting to establish genotype to phenotype connections, e.g.: social behavior in bees is influenced by specific genetic variants (Shpigler et al., 2017). In those cases, however, much work is done to characterize the measured phenotype in great detail. It is really clear that there are big conceptual gaps between the long lists of variants underlying behavior emerging out of GWASs and the work done in the wet lab to validate neurobiological pathways through which particular variants cause phenotypic differences. Additionally, weak phenotyping could lead to establishing erroneous associations. For instance, hypothetically speaking, if working memory is an important cognitive skill and sub-phenotype that contributes to better educational attainment, then the association study would generate genetic variants associated

to working memory. Then, this would perhaps simply mean that modern education systems value individuals with strong working memory. Extrapolating more meaning from these types of studies without a strong genotype to phenotype connection via biological pathways would be speculative at best.

2.4.3 Behaviors that have received the most interest are mostly biomedical

The field of psychiatric genetics has succeeded in identifying many common and rare variants underlying a disorder such as schizophrenia and validated the biological pathways – for instance, synaptic pruning – involved in this phenotype being expressed. C4 was the first ever gene *causally* linked to schizophrenia through a clear biological mechanism (Sekar et al., 2016). The authors used approaches both in genomics and neurobiology to get to their findings, They compared patients with schizophrenia and controls to verify the differences in levels of C4 expression. They then also performed functional validation in a mouse model to explore the exact mechanism through which the C4 gene variants underlie schizophrenia– that C4 gene products promote immune system action on synapses, leading to them being engulfed. In their comment, "From genetics to physiology at last," Dhindsa and Goldstein (2016) described why this was a landmark finding:

> *"Careful GWAS have [...] identified real associations between genetic variants and schizophrenia. In particular, a 2014 study that used a huge sample size and strict statistical methods revealed that there are at least 108 genetic regions (loci) associated with a risk of schizophrenia. However, finding a locus is not the same as identifying a causal gene—for instance, loci are often located in sequences that don't encode genes, and many different variations can occur in one region, making it hard to pinpoint what exactly is driving the risk signal. Indeed, these 108 risk factors were not traced to specific genes or variant sequences. Without that information, a clear insight into the molecular aetiology of the disease remained lacking, but Sekar et al. have now taken that crucial next step."*

While the genetic basis of behavioral phenotypes such as psychiatric disorders is slowly being uncovered, there is still little known about the genetic architecture of the host of behavioral traits that make modern humans unique. There is much value in uncovering the genetic architecture of psychiatric disorders due to the high relevance to potential therapeutic interventions. If similar methods could be applied to the study of uniquely human brain, cognitive, and behavioral phenotypes, we could make rapid progress.

3 What can we do as researchers into the genotype-phenotype connection to advance the unsolved riddles of human evolution?

Based on the advances made in the field thus far, the gaps identified and the fascinating questions that still remain, we have three recommendations on the approaches that should be employed as we continue to work on the genotype-phenotype connection as a field.

Recommendation 1: Differentiating complex phenotypes into their building blocks

In recent years, a number of studies have begun looking at the genetic underpinnings of complex phenotypes, such as educational attainment as mentioned above (Kong et al., 2017; Okbay et al., 2022). Our critique of this approach is that insufficient work is done to further break down such complex phenotypes into potential component parts upon which the forces of evolution can act. While it is exciting to uncover the genotypic variation underlying phenotypic variation, it remains important to build bridges between different mechanistic levels of a complex phenotype.

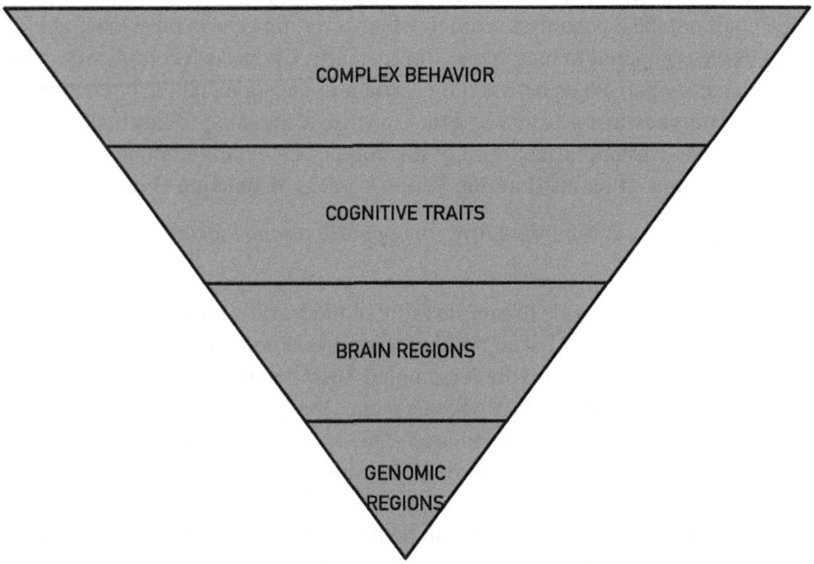

FIG. 3

When considering a complex behavior as a phenotype, it is key to at least identify potential sub-phenotypes which can be mechanistically tied to the macro-phenotype. In a wet-lab setting, genomic variants unique to and fixed in the human lineage can be tied to differences in brain regions during development (for e.g.: Pinson et al., 2022) but it remains difficult to identify what cognitive traits and complex behavior this potentially affects. GWA studies have the opposite problem, whereby complex behavior may be tied to genomic regions in a correlational manner but with no mechanistic insights about the intermediate levels of sub-phenotypes.

One of the most exciting findings of the last decade was in tying the FOXP2 gene to language. However, before this gene was first considered causally connected to the complex phenotype of language, much was and is still done to validate this within model systems. Since the 1990s, through studies of the KE family within which a disorder impaired the development of speech for some members (Hurst et al., 1990), scientists had not pinpointed any particular causative gene although it was clear that the disorder was inherited. It was not until the early 2000s, that it was first

discovered that a mutation in the FOXP2 gene was tied to the severe language disorder (Lai et al., 2001). Subsequently, functional validation studies were conducted to establish an explicit genotype-phenotype connection, with humanized mouse models showing vocal differences when compared to control mice (Enard et al., 2009). It took more than two decades of research to tie this gene conclusively to the development of language in humans and eventually chart possible implications for Neandertals (Krause et al., 2007; Konopka et al., 2009).

As seen with FOXP2, it is pretty typical within an evolutionary developmental biology framework to go through many stages of functional validation before tying any genotype to a phenotype. Gene expression in itself can be considered a "first-order phenotype" (Renn and Schumer, 2013) and it is sometimes challenging to reliably chart out the mechanism through which a certain gene expression is achieved in brain tissue compared to lung tissue, for instance. Given such complexity, when a huge explanatory gap exists between the variant in question and the phenotype considered, it is thus key to at least try and find functional meaning. When making claims about particular variants affecting any phenotype, especially complex ones, there should be attempts at reconciling the various orders of function (Fig. 3).

Recommendation 2: Sampling African populations and looking beyond European samples

There have been efforts made to sample more of the world's population, but genomic data remains biased. The African continent, which is home to the most genetic diversity on the planet, remains undersampled (McClellan et al., 2017; Tucci and Akey, 2019). In the context of GWAS, this means that potential variants only present in unsampled populations could be missed. This leaves one more gap in the efforts to connect genotype-phenotype. Common variants underlying particular traits are highly likely to have been found, but rarer variants in unsampled and undersampled populations remain elusive. While it can be deemed sufficient to urge our scientific community to do more sampling, it is also important to consider the ethics of the approaches. It is far too common for helicopter research to be performed, whereby scientists from wealthier nations get funded to collect data from local populations, with no local scientists involved at any point in the process. To tackle the big puzzle of human genetic diversity, it is crucial for local scientists to have the opportunity to take the lead in generating the data needed within their communities (Claw et al., 2018). This necessitates that funding mechanisms shift to fuel the generation of such data.

Recommendation 3: An integrative approach between researchers focused on genome-wide association studies and those working on functional validation in the wet-lab setting would go a long way in pushing the field further forward.

An integrative approach between fields would enable us to better tie genotype to phenotype in a way that is more rigorous. One of the biggest questions of our field broadly is: what is the genomic basis of the host of behavioral traits that make human

beings modern? It is our sense that these types of questions can eventually be answered, but it will take interdisciplinary collaboration between scientists interested in behavior, genetics, neurobiology, and more to get from the stage where important developmental pathways are found which affect neurobiology to an understanding of what it may mean about behavioral phenotypes.

Acknowledgments

We want to thank Tanya Calvey, Amélie Beaudet, and Alexandra de Sousa for the invitation to contribute to this volume. We are grateful to the University of the Witwatersrand, Johannesburg (South Africa), and the Natural History Museum, London (United Kingdom) for granting access to the CT data used to generate Figs. 1 and 2. We thank the Max Planck Institute for Evolutionary Anthropology, Leipzig (Germany) for scanning and the Southern African Neuroscience Society (SANS) for convening the Fossils to Mind workshop, which led to this fruitful collaboration.

References

Agoglia, R.M., Sun, D., Birey, F., Yoon, S.J., Miura, Y., Sabatini, K., Pasca, S.P., Fraser, H.B., 2021. Primate cell fusion disentangles gene regulatory divergence in neurodevelopment. Nature 592, 421–427.

Arias-Carrion, O., Salama, M., 2012. Reward-seeking behavior and addiction: cause or cog? Curr. Drug Abuse Rev. 5, 178–189.

Beaudet, A., Dumoncel, J., De Beer, F., Durrleman, S., Gilissen, E., Oettle, A., Subsol, G., Thackeray, J.F., Braga, J., 2018. The endocranial shape of Australopithecus africanus: surface analysis of the endocasts of Sts 5 and Sts 60. J. Anat. 232, 296–303.

Boyd, J.L., Skove, S.L., Rouanet, J.P., Pilaz, L.J., Bepler, T., Gordan, R., Wray, G.A., Silver, D.L., 2015. Human-chimpanzee differences in a FZD8 enhancer Alter cell-cycle dynamics in the developing neocortex. Curr. Biol. 25, 772–779.

Bruner, E., 2019. Human paleoneurology: shaping cortical evolution in fossil hominids. J Comp Neurol 527, 1753–1765.

Bruner, E., Manzi, G., Arsuaga, J.L., 2003. Encephalization and allometric trajectories in the genus Homo: evidence from the neandertal and modern lineages. Proc. Natl. Acad. Sci. U. S. A. 100, 15335–15340.

Burbano, H.A., Green, R.E., Maricic, T., Lalueza-Fox, C., De La Rasilla, M., Rosas, A., Kelso, J., Pollard, K.S., Lachmann, M., Pääbo, S., 2012. Analysis of human accelerated DNA regions using archaic hominin genomes. PloS one 7 (3), e32877.

Bycroft, C., Freeman, C., Petkova, D., Band, G., Elliott, L.T., Sharp, K., Motyer, A., Vukcevic, D., Delaneau, O., O'connell, J., Cortes, A., 2018. The UK biobank resource with deep phenotyping and genomic data. Nature 562 (7726), 203–209.

Claw, K.G., Anderson, M.Z., Begay, R.L., Tsosie, K.S., Fox, K., Garrison, N.A., 2018. A framework for enhancing ethical genomic research with indigenous communities. Nat. Commun. 9 (1), 1–7.

Cofran, Z., 2019. Brain size growth in Australopithecus. J. Hum. Evol. 130, 72–82.

Collins, F.S., Lander, E.S., Rogers, J., Waterston, R.H., Conso, I.H.G.S., 2004. Finishing the euchromatic sequence of the human genome. Nature 431, 931–945.

Conroy, G.C., Vannier, M.W., 1985. Endocranial volume determination of matrix-filled fossil skulls using high-resolution computed-tomography. Am. J. Phys. Anthropol. 66, 157–158.

Conroy, G.C., Weber, G.W., Seidler, H., Tobias, P.V., Kane, A., Brunsden, B., 1998. Endocranial capacity in an early hominid cranium from Sterkfontein, South Africa. Science 280, 1730–1731.

Coqueugniot, H., Hublin, J.J., Veillon, F., Houet, F., Jacob, T., 2004. Early brain growth in Homo erectus and implications for cognitive ability. Nature 431, 299–302.

Dannemann, M., Milaneschi, Y., Yermakovich, D., Stiglbauer, V., Kariis, H.M., Krebs, K., Friese, M.A., Otte, C., Estonian Biobank Research Team, Lehto, K., Penninx, B., Kelso, J., Gold, S.M., 2022. Neandertal introgression partitions the genetic landscape of neuropsychiatric disorders and associated behavioral phenotypes. Transl. Psychiatry 12, 433.

Dart, R.A., 1925. Australopithecus africanus — the man-ape of South Africa. Nature 115, 195–199.

Desilva, J., Lesnik, J., 2006. Chimpanzee neonatal brain size: implications for brain growth in Homo erectus. J. Hum. Evol. 51, 207–212.

Desilva, J.M., Lesnik, J.J., 2008. Brain size at birth throughout human evolution: a new method for estimating neonatal brain size in hominins. J. Hum. Evol. 55, 1064–1074.

Dhindsa, R.S., Goldstein, D.B., 2016. From genetics to physiology at last. Nature 530 (7589), 162–163.

Enard, W., Gehre, S., Hammerschmdt, K., Hölter, S.M., Blass, T., Somel, M., Brückner, M.K., Schreiwes, C., Winter, C., Sohr, R., Becker, L., 2009. A humanized version of Foxp2 affects cortico-basal ganglia circuits in mice. Cell 137 (5), 961–971.

Falk, D., Redmond, J.C., Guyer, J., Conroy, G.C., Recheis, W., Weber, G.W., Seidler, H., 2000. Early hominid brain evolution: a new look at old endocasts. J. Hum. Evol. 38, 695–717.

Girskis, K.M., Stergachis, A.B., Degennaro, E.M., Doan, R.N., Qian, X., Johnson, M.B., Wang, P.P., Sejourne, G.M., Nagy, M.A., Pollina, E.A., Sousa, A.M., 2021. Rewiring of human neurodevelopmental gene regulatory programs by human accelerated regions. Neuron 109 (20), 3239–3251.

Gokhman, D., Agoglia, R.M., Kinnebrew, M., Gordon, W., Sun, D.Q., Bajpai, V.K., Naqvi, S., Chen, C., Chan, A., Chen, C., Petrov, D.A., Ahituv, N., Zhang, H.H., Mishina, Y., Wysocka, J., Rohatgi, R., Fraser, H.B., 2021. Human-chimpanzee fused cells reveal cis-regulatory divergence underlying skeletal evolution. Nat. Genet. 53, 467-+.

Gregory, M.D., Kippenhan, J.S., Eisenberg, D.P., Kohn, P.D., Dickinson, D., Mattay, V.S., Chen, Q., Weinberger, D.R., Saad, Z.S., Berman, K.F., 2017. Neanderthal-derived genetic variation shapes modern human cranium and brain. Sci. Rep. 7, 6308.

Grun, R., Pike, A., McDermott, F., Eggins, S., Mortimer, G., Aubert, M., Kinsley, L., Joannes-Boyau, R., Rumsey, M., Denys, C., Brink, J., Clark, T., Stringer, C., 2020. Dating the skull from Broken Hill, Zambia, and its position in human evolution. Nature 580, 372–375.

Gunz, P., Mitteroecker, P., 2013. Semilandmarks: a method for quantifying curves and surfaces. Hystrix-Italian Journal of Mammalogy 24, 103–109.

Gunz, P., Neubauer, S., Falk, D., Tafforeau, P., Le Cabec, A., Smith, T.M., Kimbel, W.H., Spoor, F., Alemseged, Z., 2020. Australopithecus afarensis endocasts suggest ape-like brain organization and prolonged brain growth. Sci. Adv. 6 (14), eaaz4729.

Gunz, P., Neubauer, S., Golovanova, L., Doronichev, V., Maureille, B., Hublin, J.J., 2012. A uniquely modern human pattern of endocranial development. Insights from a new cranial reconstruction of the neandertal newborn from Mezmaiskaya. J. Hum. Evol. 62, 300–313.

Gunz, P., Neubauer, S., Maureille, B., Hublin, J.J., 2010. Brain development after birth differs between Neanderthals and modern humans. Curr. Biol. 20, R921–R922.

Gunz, P., Tilot, A.K., Wittfeld, K., Teumer, A., Shapland, C.Y., Van Erp, T.G.M., Dannemann, M., Vernot, B., Neubauer, S., Guadalupe, T., Fernandez, G., Brunner, H.G., Enard, W., Fallon, J., Hosten, N., Volker, U., Profico, A., Di Vincenzo, F., Manzi, G., Kelso, J., St Pourcain, B., Hublin, J.J., Franke, B., Paabo, S., Macciardi, F., Grabe, H.J., Fisher, S.E., 2019. Neandertal introgression sheds light on modern human endocranial globularity. Curr. Biol. 29, 120.

Haeggström, S., Ingelman-Sundberg, M., Pääbo, S., Zeberg, H., 2022. The clinically relevant CYP2C8* 3 and CYP2C9* 2 haplotype is inherited from neandertals. Pharmacogenomics J. 22 (4), 247–249.

Herai, R.H., Szeto, R.A., Trujillo, C.A., Muotri, A.R., 2021. Response to comment on "reintroduction of the archaic variant of NOVA1 in cortical organoids alters neurodevelopment". Science 374, eabi9881.

Holloway, R.L., Broadfield, D., Yuan, M.S., 2004. The Human Fossil Record, Brain Endocasts — The Paleoneurological Evidence. John Wiley & Sons, New York.

Hublin, J.J., Ben-Ncer, A., Bailey, S.E., Freidline, S.E., Neubauer, S., Skinner, M.M., Bergmann, I., Le Cabec, A., Benazzi, S., Harvati, K., Gunz, P., 2017. New fossils from jebel Irhoud, Morocco and the pan-African origin of Homo sapiens. Nature 546 (7657), 289–292.

Hurst, J.A., Baraitser, M., Auger, E., Graham, F., Norell, S., 1990. An extended family with a dominantly inherited speech disorder. Dev. Med. Child Neurol. 32, 352–355.

Jagoda, E., Xue, J.R., Reilly, S.K., Dannemann, M., Racimo, F., Huerta-Sanchez, E., Sankararaman, S., Kelso, J., Pagani, L., Sabeti, P.C., Capellini, T.D., 2022. Detection of Neanderthal adaptively Introgressed genetic variants that modulate reporter gene expression in human immune cells. Mol. Biol. Evol. 39 (1). p.msab304.

Kanton, S., Boyle, M.J., He, Z.S., Santel, M., Weigert, A., Sanchis-Calleja, F., Guijarro, P., Sidow, L., Fleck, J.S., Han, D.D., Qian, Z.Z., Heide, M., Huttner, W.B., Khaitovich, P., Paabo, S., Treutlein, B., Camp, J.G., 2019. Organoid single-cell genomic atlas uncovers human-specific features of brain development. Nature 574, 418–422.

Karlsson Linnér, R., Biroli, P., Kong, E., Meddens, S.F.W., Wedow, R., Fontana, M.A., Lebreton, M., Tino, S.P., Abdellaoui, A., Hammerschlag, A.R., Nivard, M.G., 2019. Genome-wide association analyses of risk tolerance and risky behaviors in over 1 million individuals identify hundreds of loci and shared genetic influences. Nat. Genet. 51 (2), 245–257.

Kimbel, W.H., Delezene, L.K., 2009. "Lucy" redux: a review of research on Australopithecus afarensis. Yearb. Phys. Anthropol. 52 (52), 2–48.

Kimbel, W.H., Rak, Y., Johanson, D.C., 2004. The Skull of Australopithecus Afarensis Oxford. Oxford University Press.

Kochiyama, T., Ogihara, N., Tanabe, H.C., Kondo, O., Amano, H., Hasegawa, K., Suzuki, H., Ponce De Leon, M.S., Zollikofer, C.P.E., Bastir, M., Stringer, C., Sadato, N., Akazawa, T., 2018. Reconstructing the Neanderthal brain using computational anatomy. Sci. Rep. 8, 6296.

Kong, A., Frigge, M.L., Thorleifsson, G., Stefansson, H., Young, A.I., Zink, F., Jonsdottir, G.-A., Okbay, A., Sulem, P., Masson, G., Gudbjartsson, D.F., Helgason, A., Bjornsdottir, G., Thorsteinsdottir, U., Stefansson, K., 2017. Selection against variants in the genome associated with educational attainment. Proc. Natl. Acad. Sci. U. S. A. 114, E727–E732.

Konopka, G, Bomar, JM, Winden, K, Coppola, G, Jonsson, ZO, Gao, F, Peng, S, Preuss, TM, Wohlschlegel, JA, Geschwind, DH, 2009. Human-specific transcriptional regulation of CNS development genes by FOXP2. Nature 462 (7270), 213–217. PMID: 19907493; PMCID: PMC2778075. https://doi.org/10.1038/nature08549.

Krause, J., Lalueza-Fox, C., Orlando, L., Enard, W., Green, R.E., Burbano, H.A., Hublin, J.J., Hänni, C., Fortea, J., De La Rasilla, M., Bertranpetit, J., 2007. The derived FOXP2 variant of modern humans was shared with neandertals. Curr. Biol. 17 (21), 1908–1912.

Lai, C.S., Fisher, S.E., Hurst, J.A., Vargha-Khadem, F., Monaco, A.P., 2001. A forkhead-domain gene is mutated in a severe speech and language disorder. Nature 413 (6855), 519–523.

Lander, E.S., Consortium, I.H.G.S., Linton, L.M., Birren, B., Nusbaum, C., Zody, M.C., Baldwin, J., Devon, K., Dewar, K., Doyle, M., Fitzhugh, W., Funke, R., Gage, D., Harris, K., Heaford, A., Howland, J., Kann, L., Lehoczky, J., Levine, R., McEwan, P., McKernan, K., Meldrim, J., Mesirov, J.P., Miranda, C., Morris, W., Naylor, J., Raymond, C., Rosetti, M., Santos, R., Sheridan, A., Sougnez, C., Stange-Thomann, N., Stojanovic, N., Subramanian, A., Wyman, D., Rogers, J., Sulston, J., Ainscough, R., Beck, S., Bentley, D., Burton, J., Clee, C., Carter, N., Coulson, A., Deadman, R., Deloukas, P., Dunham, A., Dunham, I., Durbin, R., French, L., Grafham, D., Gregory, S., Hubbard, T., Humphray, S., Hunt, A., Jones, M., Lloyd, C., McMurray, A., Matthews, L., Mercer, S., Milne, S., Mullikin, J.C., Mungall, A., Plumb, R., Ross, M., Shownkeen, R., Sims, S., Waterston, R.H., Wilson, R.K., Hillier, L.W., McPherson, J.D., Marra, M.A., Mardis, E.R., Fulton, L.A., Chinwalla, A.T., Pepin, K.H., Gish, W.R., Chissoe, S.L., Wendl, M.C., Delehaunty, K.D., Miner, T.L., Delehaunty, A., Kramer, J.B., Cook, L.L., Fulton, R.S., Johnson, D.L., Minx, P.J., Clifton, S.W., Hawkins, T., Branscomb, E., Predki, P., Richardson, P., Wenning, S., Slezak, T., Doggett, N., Cheng, J.F., Olsen, A., Lucas, S., Elkin, C., Uberbacher, E., et al., 2001. Initial sequencing and analysis of the human genome. Nature 409, 860–921.

Leigh, S.R., 2004. Brain growth, life history, and cognition in primate and human evolution. Am. J. Primatol. 62, 139–164.

Leigh, S.R., 2006. Brain ontogeny and life history in Homo erectus. J. Hum. Evol. 50, 104–108.

Lieberman, D.E., Mcbratney, B.M., Krovitz, G., 2002. The evolution and development of cranial form in Homo sapiens. Proc. Natl. Acad. Sci. U. S. A. 99, 1134–1139.

Maricic, T., Helmbrecht, N., Riesenberg, S., Macak, D., Kanis, P., Lackner, M., Pugach-Matveeva, A.D., Paabo, S., 2021. Comment on "reintroduction of the archaic variant of NOVA1 in cortical organoids alters neurodevelopment". Science 374, eabi6060.

McArthur, E., Rinker, D.C., Capra, J.A., 2021. Quantifying the contribution of Neanderthal introgression to the heritability of complex traits. Nat. Commun. 12, 4481.

McClellan, J.M., Lehner, T., King, M.C., 2017. Gene discovery for complex traits: lessons from Africa. Cell 171 (2), 261–264.

Meyer, M., Arsuaga, J.L., De Filippo, C., Nagel, S., Aximu-Petri, A., Nickel, B., Martinez, I., Gracia, A., Bermudez De Castro, J.M., Carbonell, E., Viola, B., Kelso, J., Prufer, K., Paabo, S., 2016. Nuclear DNA sequences from the middle Pleistocene Sima de los Huesos hominins. Nature 531, 504–507.

Muthuirulan, P., Capellini, T.D., 2019. Complex phenotypes: mechanisms underlying variation in human stature. Curr. Osteoporos. Rep. 17, 301–323.

Neubauer, S., Gunz, P., Hublin, J.J., 2009. The pattern of endocranial ontogenetic shape changes in humans. J. Anat. 215, 240–255.

Neubauer, S., Gunz, P., Weber, G.W., Hublin, J.J., 2012. Endocranial volume of Australopithecus africanus: new CT-based estimates and the effects of missing data and small sample size. J. Hum. Evol. 62, 498–510.

Neubauer, S., Hublin, J.J., Gunz, P., 2018. The evolution of modern human brain shape. Sci. Adv. 4, eaao5961.

Nurk, S., Koren, S., Rhie, A., Rautiainen, M., Bzikadze, A.V., Mikheenko, A., Vollger, M.R., Altemose, N., Uralsky, L., Gershman, A., Aganezov, S., Hoyt, S.J., Diekhans, M., Logsdon, G.A., Alonge, M., Antonarakis, S.E., Borchers, M., Bouffard, G.G., Brooks, S.Y., Caldas, G.V., Chen, N.C., Cheng, H.Y., Chin, C.S., Chow, W., De Lima, L.G., Dishuck, P.C., Durbin, R., Dvorkina, T., Fiddes, I.T., Formenti, G., Fulton, R.S., Fungtammasan, A., Garrison, E., Grady, P.G.S., Graves-Lindsay, T.A., Hall, I.M., Hansen, N.F., Hartley, G.A., Haukness, M., Howe, K., Hunkapiller, M.W., Jain, C., Jain, M., Jarvis, E.D., Kerpedjiev, P., Kirsche, M., Kolmogorov, M., Korlach, J., Kremitzki, M., Li, H., Maduro, V.V., Marschall, T., McCartney, A.M., McDaniel, J., Miller, D.E., Mullikin, J.C., Myers, E.W., Olson, N.D., Paten, B., Peluso, P., Pevzner, P.A., Porubsky, D., Potapova, T., Rogaev, E.I., Rosenfeld, J.A., Salzberg, S.L., Schneider, V.A., Sedlazeck, F.J., Shafin, K., Shew, C.J., Shumate, A., Sims, Y., Smit, A.F.A., Soto, D.C., Sovic, I., Storer, J.M., Streets, A., Sullivan, B.A., Thibaud-Nissen, F., Torrance, J., Wagner, J., Walenz, B.P., Wenger, A., Wood, J.M.D., Xiao, C.L., Yan, S.M., Young, A.C., Zarate, S., Surti, U., Mccoy, R.C., Dennis, M.Y., Alexandrov, I.A., Gerton, J.L., O'neill, R.J., Timp, W., Zook, J.M., Schatz, M.C., Eichler, E.E., Miga, K.H., Phillippy, A.M., 2022. The complete sequence of a human genome. Science 376, 44–53.

Okbay, A., Wu, Y.D., Wang, N., Jayashankar, H., Bennett, M., Nehzati, S.M., Sidorenko, J., Kweon, H., Goldman, G., Gjorgjieva, T., Jiang, Y.X., Hicks, B., Tian, C., Hinds, D.A., Ahlskog, R., Magnusson, P.K.E., Oskarsson, S., Hayward, C., Campbell, A., Porteous, D.J., Freese, J., Herd, P., Watson, C., Jala, J., Conley, D., Koellinger, P.D., Johannesson, M., Laibson, D., Meyer, M.N., Lee, J.J., Kong, A., Yengo, L., Cesarini, D., Turley, P., Visscher, P.M., Beauchamp, J.P., Benjamin, D.J., Young, A.I., 23andMe Research Team, Social Science Genetic Association Consortium, 2022. Polygenic prediction of educational attainment within and between families from genome-wide association analyses in 3 million individuals. Nat. Genet. 54, 437–449.

Pereira-Pedro, A.S., Bruner, E., Gunz, P., Neubauer, S., 2020. A morphometric comparison of the parietal lobe in modern humans and Neanderthals. J. Hum. Evol. 142, 102770.

Pinson, A., Xing, L., Namba, T., Kalebic, N., Peters, J., Oegema, C.E., Traikov, S., Reppe, K., Riesenberg, S., Maricic, T., Derihaci, R., Wimberger, P., Paabo, S., Huttner, W.B., 2022. Human TKTL1 implies greater neurogenesis in frontal neocortex of modern humans than Neanderthals. Science 377, eabl6422.

Pollard, K.S., Salama, S.R., Lambert, N., Lambot, M.A., Coppens, S., Pedersen, J.S., Katzman, S., King, B., Onodera, C., Siepel, A., Kern, A.D., Dehay, C., Igel, H., Ares, M., Vanderhaeghen, P., Haussler, D., 2006. An RNA gene expressed during cortical development evolved rapidly in humans. Nature 443, 167–172.

Ponce De Leon, M.S., Bienvenu, T., Marom, A., Engel, S., Tafforeau, P., Alatorre Warren, J.L., Lordkipanidze, D., Kurniawan, I., Murti, D.B., Suriyanto, R.A., Koesbardiati, T., Zollikofer, C.P.E., 2021. The primitive brain of early Homo. Science 372, 165–171.

Renn, S.C.P., Schumer, M.E., 2013. Genetic accommodation and behavioural evolution: insights from genomic studies. Anim. Behav. 85, 1012–1022.

Richard, D., Liu, Z., Cao, J.X., Kiapour, A.M., Willen, J., Yarlagadda, S., Jagoda, E., Kolachalama, V.B., Sieker, J.T., Chang, G.H., Muthuirulan, P., Young, M., Masson, A., Konrad, J., Hosseinzadeh, S., Maridas, D.E., Rosen, V., Krawetz, R., Roach, N., Capellini, T.D., 2020. Evolutionary selection and constraint on human knee chondrocyte regulation impacts osteoarthritis risk. Cell 181, 362-+.

Robson, S.L., Wood, B., 2008. Hominin life history: reconstruction and evolution. J. Anat. 212, 394–425.

Rosas, A., Rios, L., Estalrrich, A., Liversidge, H., Garcia-Tabernero, A., Huguet, R., Cardoso, H., Bastir, M., Lalueza-Fox, C., De La Rasilla, M., Dean, C., 2017. The growth pattern of neandertals, reconstructed from a juvenile skeleton from El Sidron (Spain). Science 357, 1282-+.

Sankararaman, S., Mallick, S., Dannemann, M., Prufer, K., Kelso, J., Paabo, S., Patterson, N., Reich, D., 2014. The genomic landscape of Neanderthal ancestry in present-day humans. Nature 507, 354-+.

Scott, N., Neubauer, S., Hublin, J.J., Gunz, P., 2014. A shared pattern of postnatal endocranial development in extant hominoids. Evolutionary Biology 41, 572–594.

Scott, N.A., Strauss, A., Hublin, J.J., Gunz, P., Neubauer, S., 2018. Covariation of the endocranium and splanchnocranium during great ape ontogeny. PloS One 13, e0208999.

Sekar, A., Bialas, A.R., De Rivera, H., Davis, A., Hammond, T.R., Kamitaki, N., Tooley, K., Presumey, J., Baum, M., Van Doren, V., Genovese, G., Rose, S.A., Handsaker, R.E., Daly, M.J., Carroll, M.C., Stevens, B., Mccarroll, S.A., Grp, S.W., 2016. Schizophrenia risk from complex variation of complement component 4. Nature 530, 177-+.

Shpigler, H.Y., Saul, M.C., Corona, F., Block, L., Cash Ahmed, A., Zhao, S.D., Robinson, G.E., 2017. Deep evolutionary conservation of autism-related genes. Proc. Natl. Acad. Sci. 114 (36), 9653–9658.

Trujillo, C.A., Rice, E.S., Schaefer, N.K., Chaim, I.A., Wheeler, E.C., Madrigal, A.A., Buchanan, J., Preissl, S., Wang, A., Negraes, P.D., Szeto, R.A., Herai, R.H., Huseynov, A., Ferraz, M.S.A., Borges, F.S., Kihara, A.H., Byrne, A., Marin, M., Vollmers, C., Brooks, A.N., Lautz, J.D., Semendeferi, K., Shapiro, B., Yeo, G.W., Smith, S.E.P., Green, R.E., Muotri, A.R., 2021. Reintroduction of the archaic variant of Nova1 in cortical organoids alters neurodevelopment. Science 371, 694-+.

Tucci, S., Akey, J.M., 2019. The long walk to African genomics. Genome Biol. 20, 130.

Varki, A., Altheide, T.K., 2005. Comparing the human and chimpanzee genomes: searching for needles in a haystack. Genome Res. 15, 1746–1758.

Weaver, A.H., 2005. Reciprocal evolution of the cerebellum and neocortex in fossil humans. Proc. Natl. Acad. Sci. U. S. A. 102, 3576–3580.

Weber, G.W., Hershkovitz, I., Gunz, P., Neubauer, S., Ayalon, A., Latimer, B., Bar-Matthews, M., Yasur, G., Barzilai, O., May, H., 2020. Before the massive modern human dispersal into Eurasia: a 55,000-year-old partial cranium from Manot cave, Israel. Quat. Int. 551, 29–39.

Zollikofer, C.P., Bienvenu, T., Ponce De Leon, M.S., 2017. Effects of cranial integration on hominid endocranial shape. J. Anat. 230, 85–105.

Zollikofer, C.P.E., Bienvenu, T., Beyene, Y., Suwa, G., Asfaw, B., White, T.D., Ponce De Leon, M.S., 2022. Endocranial ontogeny and evolution in early Homo sapiens: the evidence from Herto, Ethiopia. Proc. Natl. Acad. Sci. U. S. A. 119, e2123553119.

CHAPTER

Coevolution of language and tools in the human brain: An ALE meta-analysis of neural activation during syntactic processing and tool use

4

Veronika Kulik, Laura D. Reyes, and Chet C. Sherwood*

Department of Anthropology and Center for the Advanced Study of Human Paleobiology, The George Washington University, Washington, DC, United States

**Corresponding author: e-mail address: sherwood@gwu.edu*

Abstract

Language and complex tool use are often cited as behaviors unique to humans and may be evolutionarily linked owing to the underlying cognitive processes they have in common. We executed a quantitative activation likelihood estimation (ALE) meta-analysis (GingerALE 2.3) on published, whole-brain neuroimaging studies to identify areas associated with syntactic processing and/or tool use in humans. Significant clusters related to syntactic processing were identified in areas known to be related to language production and comprehension, including bilateral Broca's area in the inferior frontal gyrus. Tool use activation clusters were all in the left hemisphere and included the primary motor cortex and premotor cortex, in addition to other areas involved with sensorimotor transformation. Activation shared by syntactic processing and tool use was only significant at one cluster, located in the *pars opercularis* of the left inferior frontal gyrus. This minimal overlap between syntactic processing and tool use activation from our meta-analysis of neuroimaging studies indicates that there is not a widespread common neural network between the two. Broca's area may serve as an important hub that was initially recruited in early human evolution in the context of simple tool use, but was eventually co-opted for linguistic purposes, including the sequential and hierarchical ordering processes that characterize syntax. In the future, meta-analyses of additional components of language may allow for a more comprehensive examination of the functional networks that underlie the coevolution of human language and complex tool use.

Progress in Brain Research, Volume 275, ISSN 0079-6123, https://doi.org/10.1016/bs.pbr.2022.10.003

Keywords

Broca's area, Language, Tool, Meta-analysis, Human brain evolution

1 Introduction

Capacities for language and complex tool use are often viewed as quintessential human behaviors that make us unique. Both behaviors share neural processes, leading to the suggestion that they possess a common evolutionary history underlying their origins. Comprehending and producing language involves several different brain areas, but these brain regions have also been shown to contribute to a variety of other functions, including tool use (Higuchi et al., 2009). Due to this functional overlap, theories abound as to whether pre-existing neural circuits for tool use could have been co-opted in human evolution to form the basis for language (e.g., Stout and Chaminade, 2012).

During the Early Stone Age (Lower Paleolithic), from approximately 2.6 to 0.5 million years ago, hominin brain size more than doubled (Stout et al., 2008). Concurrently, the archeological record documents a technological shift from simple Oldowan core and flake stone tools to the Acheulean style, which included the production of symmetrical handaxes that required more planning and skill to systematically create (Stout et al., 2021). The beginning of Early Acheulean technology was approximately 1.8 million years ago, a period of time during which increasing brain size, neuroanatomical reorganization, and hemispheric asymmetries possibly associated with language and handedness become increasingly evident from hominin endocasts (Balzeau et al., 2012; Beaudet et al., 2019; Bruner et al., 2022; Holloway, 1983). Consequently, it has been hypothesized that the sequential nature of the actions necessary to make complex stone tools might have set the stage for the emergence of syntax in human language (e.g., Holloway, 1969; Mahaney, 2014; Moore, 2010).

Generative, hierarchically structured syntax is a feature associated with human language that is largely regarded as unique compared to other animal communication systems (Zuberbühler, 2019) (although this assertion has been challenged by experimental work in birds, see Gentner et al., 2006; Suzuki et al., 2016). Human syntactic processing goes beyond deriving meaning from certain sounds or symbols and involves conveying how elements in a sentence relate to one another (i.e., who is doing what to whom) and having the ability to recognize that reorganizing these elements will lead to a different communicative message (Fedorenko et al., 2012). While speech production and tool use share aspects of motor planning and coordination, syntax is of particular interest for the study of language and tool use coevolution because of its inherently sequential basis. Just as a tool-maker or tool-user has to follow a certain order of planned steps to produce or manipulate the complex tool they envision, so too does the language-user have to construct a sentence step-by-step, piece-by-piece to convey their desired information in a

coherent way. Human language and complex tool skills are both behaviors that involve processing chains of embedded information (Greenfield, 1991), and the cognitive demands of these two activities appear to be complementary (Stout et al., 2021). Reminiscent of the hierarchical structure of syntax, tool production sequences consist of a string of individual toolmaking actions that are ordered (Stout et al., 2021). This pattern is reiterated in the neural motor action planning requirements of complex toolmaking, which activates areas associated with abstract action and hierarchical organization (e.g., ventral premotor cortex, *pars triangularis* of the inferior frontal gyrus), in addition to employing the same dorsal stream visuomotor coordination needed for simple toolmaking and tool use (Stout and Chaminade, 2012).

To investigate the potential evolutionary interrelatedness of tools and language, studies with modern human participants have sought to determine whether there are areas of shared neural activation between language use and stone toolmaking (e.g., Stout and Chaminade, 2012). Of particular focus is the Acheulean style of stone toolmaking, since this technology is hypothesized to have required more cognitive control when planning and making the tool, including anticipating and executing a series of complex actions (Stout et al., 2008, 2015). Underpinning these studies is the "technological hypothesis," which suggests that language initially evolved in a gestural form as a way for hominins to teach and learn simple stone toolmaking, and eventually intentional vocal communication developed to convey information about later, more complex tool-related actions (Cataldo et al., 2018; Stout et al., 2008; Stout and Chaminade, 2012). Based on these stone tool knapping neuroimaging experiments, multiple brain regions appear to be activated during both speech and Late Acheulean toolmaking, including the posterior temporal cortex, inferior parietal lobule, ventral premotor cortex, and inferior frontal gyrus (Stout and Chaminade, 2012).

In addition to contributing to the overlap in neural activation during language and tool production, Broca's area of the inferior frontal gyrus is also implicated in tool *use* (Higuchi et al., 2009). In fact, Broca's area appears to be involved in executing motor actions in general, which would include grasping and manipulating tools (Johnson-Frey, 2004; Pastra and Aloimonos, 2012). Broca's area is heavily connected and coactivated with the motor and premotor cortex (precentral gyrus) and inferior parietal lobule (supramarginal gyrus). These areas of the brain function in sensorimotor integration by way of processing visual and auditory information, something that is applicable to both language- and tool-related activities.

Using a tool can be thought of as manipulating an extension of the body, or as Pastra and Aloimonos (2012) put it, a "body part is the simplest tool". *Making* a tool, though, requires cognitive processes beyond visuomotor planning and execution—processes that may share more similarities with language production or comprehension than is apparent by studying tool use alone, because factors such as creativity, executive control, or more explicit sequence implementation may be masked. As such, conducting a meta-analysis on "tool use" rather than "toolmaking" neuroimaging experiments is an inevitable limitation of this study.

Although the connection between language and tool use has been investigated for some time, the focus is most often on the motor similarities between speech and tool use, rather than the potential links between tool tasks and parts of human grammar, such as syntax. To date, there appears to have been no comprehensive comparison of brain activation associated with syntactic processing and tool use. We anticipated that the inferior frontal gyrus would be of particular relevance in terms of coactivation, as Broca's area has previously been found to be involved in language production, toolmaking, and tool use (Higuchi et al., 2009; Stout and Chaminade, 2012). Given that Broca's area forms connections with a variety of regions associated with planning and action, we also considered the possibility that areas like the prefrontal cortex or parietal lobules may also be significant in a shared syntax-tool use network. In this study, we used meta-analysis methods to compile results from the published neuroimaging research literature, generating hundreds of data points in an effort to investigate the relationship between the neural bases for these two behaviors.

2 Methods

A quantitative activation likelihood estimation (ALE) meta-analysis (GingerALE 2.3) (Eickhoff et al., 2009, 2011, 2012; Turkeltaub et al., 2012) was performed on results from published whole-brain, human fMRI and PET studies to identify areas of the brain that are active when processing syntax and using tools. ALE is a coordinate-based meta-analysis approach that determines the convergence of foci inputted from different studies. Acknowledging the variability among subjects and templates in neuroimaging data sets, the GingerALE software models foci as probability distributions whose width is based on empirical estimates of spatial uncertainty (Eickhoff et al., 2012). To compile a comprehensive selection of articles to use for the analysis, a search was done in PubMed and Sleuth (another BrainMap software, like GingerALE 2.3; Laird et al., 2005). Basic search terms for "syntax" articles included: fMRI, PET, human, syntax OR syntactic. Basic search terms for "tool use" articles included: fMRI, PET, human, tool use, tools. In addition to these search results, the following criteria had to be met during the article selection process for the study to be included in the final meta-analysis: fMRI or PET imaging, more than one subject, whole-brain scan, and coordinates included [either Montreal Neurological Institute (MNI) or Taliarach space] (Fig. 1). To control for differences in image quality, the original experiment needed to use a minimum 1.5 T MRI scanner or 4.5 mm spatial resolution PET scan.

Syntax was specifically chosen to investigate human language origins because it is often considered a unique feature of human communication systems. Articles to do with syntax were limited to those investigating syntactic processing by subjects who were presented with a given set of sentences, often with varying grammatical structure or accuracy. The proxy for tool-associated human behavior was necessarily tool *use*, an area of research in which there is more published literature than for

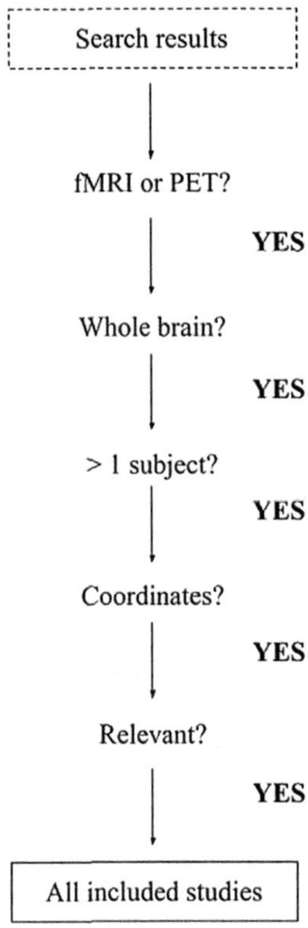

FIG. 1

Diagram of the selection process for published studies chosen for the final meta-analysis sample.

toolmaking. Due to logistical difficulties, there is a paucity of neuroimaging studies focused on participants *making* tools (see Stout et al., 2015; Stout and Chaminade, 2007, 2012). Based on meeting the criteria in Fig. 1, one such study was used in this meta-analysis (Stout et al., 2008). Both the toolmaking and tool use articles involved the manipulation of an object by right-handed research participants. Studies that involved participants physically moving their hands while pantomiming or planning tool use were also included as acceptable experimental design on the basis of participants executing motor action (whereas a study with participants only observing or imagining tool use did not meet the criteria).

From the search terms, 626 syntax and 171 tool-related results were collected. The final sample that fit the criteria included 33 syntax and 20 tool-related studies, from which there were 342 foci from the syntax experiments and 233 foci from the tool use experiments (Tables 1 and 2). Such a sample size enables the collection of

Table 1 Syntax studies selected for meta-analysis, organized by ascending sample size. Full citations available in bibliography.

Study	Sample size	Original template space	Task/test
Luke et al. (2002)	7	Talairach	Syntactic acceptability judgment
Suzuki and Sakai (2003)	8	MNI	Syntactic acceptability judgment
Dapretto and Bookheimer (1999)	8	Talairach	Syntactic structural manipulation
Stromswold et al. (1996)	8	Talairach	Syntactic acceptability judgment
Santi and Grodzinsky (2007)	11	MNI	Syntactic acceptability judgment
Moro et al. (2001)	11	Talairach	Syntactic acceptability judgment
Davis et al. (2004)	12	MNI	Syntactic complexity manipulation
Ben-Shachar et al. (2004)	12	Talairach	Syntactic structural manipulation
Longe et al. (2007)	12	Talairach	Syntactic categorization
Blank et al. (2016)	13	MNI	Syntactic complexity manipulation
Progovac et al. (2018)	14	MNI	Syntactic complexity manipulation
Tettamanti et al. (2002)	14	MNI	Syntactic acceptability judgment
Friederici et al. (2000)	14	Talairach	Syntactic categorization
van Dam and Desai (2016)	14	Talairach	Syntactic structural manipulation
Siegelman et al. (2019)	15	MNI	Syntactic structural manipulation
Tyler et al. (2011)	15	MNI	Sentence comprehension and judgment (control data)
Newman et al. (2001)	16	MNI	Syntactic violation detection
Zaccarella et al. (2017)	18	MNI	Syntactic structural manipulation
Santi and Grodzinsky (2010)	18	Talairach	Syntactic complexity manipulation
Nuñez et al. (2011)	19	MNI	Sentence comprehension and judgment
Matchin and Hickok (2016)	20	Talairach	Syntactic structural manipulation
Europa et al. (2019)	21	MNI	Auditory sentence-picture verification
Devauchelle et al. (2009)	21	MNI	Syntactic repetition
Chan et al. (2013)	24	MNI	Syntactic structural manipulation
Segaert et al. (2012)	24	MNI	Syntactic priming with verb repetition
Segaert et al. (2013)	24	MNI	Syntactic priming without verb repetition
Wilson et al. (2014)	24	MNI	Auditory sentence-picture verification
Wilson et al. (2010)	24	MNI	Auditory sentence-picture matching
Meltzer et al. (2010)	24	Talairach	Syntactic complexity manipulation
Takashima et al. (2020)	30	MNI	Syntactic complexity manipulation
Pallier et al. (2011)	40	MNI	Syntactic structural manipulation
Henderson et al. (2016)	43	Talairach	Syntactic surprisal
Quiñones et al. (2018)	53	MNI	Syntactic acceptability judgment

MNI: Montreal Neurological Institute coordinate system.

Table 2 Tool use studies selected for meta-analysis, organized by ascending sample size. Full citations available in bibliography. MNI: Montreal Neurological Institute coordinate system.

Study	Sample size	Original template space	Task/test
Stout et al. (2008)	3	MNI	Stone toolmaking
Higuchi et al. (2007)	8	MNI	Tool use execution
Inoue et al. (2001)	8	Talairach	Tool use execution
Choi et al. (2001)	10	Talairach	Tool use pantomime
Ogawa and Imai (2016)	11	MNI	Tool use pantomime
Binkofski et al. (1999)	12	Talairach	Tool manipulation
Imazu et al. (2007)	12	Talairach	Tool use execution
Johnson-Frey et al. (2005)	13	Talairach	Tool use pantomime
Gallivan et al. (2013)	13	Talairach	Tool use execution
Rumiati et al. (2004)	14	Talairach	Tool use pantomime
Bohlhalter et al. (2009)	15	MNI	Tool use pantomime
Lausberg et al. (2015)	15	MNI	Tool use execution
Higuchi et al. (2009)	16	MNI	Tool use execution
Beauchamp et al. (2002)	18	Talairach	Tool use pantomime
Pazen et al. (2020)	19	MNI	Tool manipulation
Peeters et al. (2009)	20	MNI	Tool use planning
Jacobs et al. (2010)	20	MNI	Tool manipulation
Hermsdörfer et al. (2007)	23	MNI	Tool use execution
Peeters et al. (2013)	24	MNI	Tool use planning
Matheson et al. (2017)	30	MNI	Tool use planning

robust results—according to an extensive series of simulations carried out by Eickhoff et al. (2016), 17–20 experiments per group are required for ALE meta-analysis to detect moderate effects at a sufficient level of power and prevent outputs from being influenced too heavily by single experiments. In the case that a study included multiple experiments, only the most relevant one was used in the meta-analysis to avoid the potentially confounding factor of paired data. Discerning which dataset to include from a single paper was based on what type of tool- or language-related task the subject performed (e.g., physical manipulation of tools was prioritized over tool use pantomime, if both were tested).

When necessary, coordinates were converted from Talairach to MNI space using the tal2icbm_spm transform function within the GingerALE 2.3 software (Laird et al., 2010; Lancaster et al., 2007). A cluster-level analysis was performed

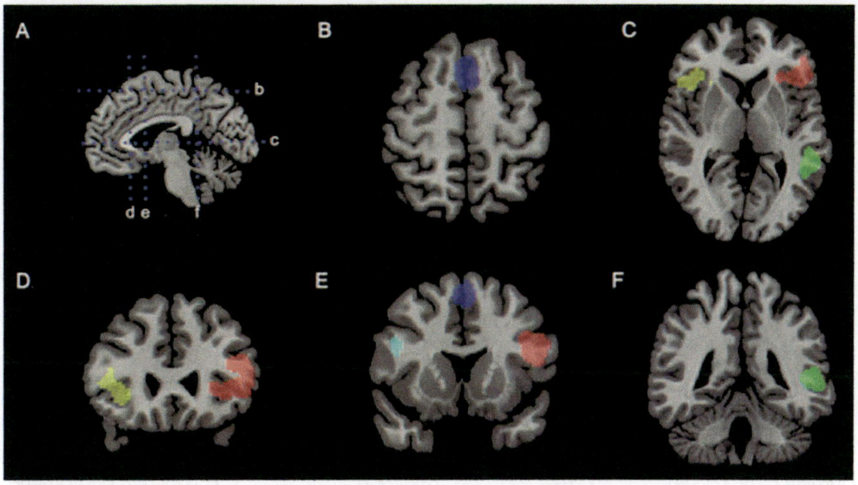

FIG. 2

Clusters showing significant neural activation during syntactic processing, in horizontal (B–C) and coronal (D–F) view (cluster-level inference: $P=0.05$, permutation threshold $=1000$, cluster-forming threshold: uncorr. $P=0.001$). Color of cluster corresponds to Table 3. (A) Sagittal view for location orientation. (B) SFG. (C) Left IFG, right anterior insula, left MTG. (D) Left IFG and right anterior insula. (E) SFG and bilateral IFG. (F) Left MTG. Abbreviations: IFG, inferior frontal gyrus; MTG, middle temporal gyrus; SFG, superior frontal gyrus. Images generated in ITK-SNAP. Images shown in $1 \times 1 \times 1$ mm space.

on each group (cluster-level inference: $P=0.05$, permutation threshold $=1000$, cluster-forming threshold: uncorrected $P=0.001$), generating output data that included cluster size, cluster center, and anatomical assignment based on the Talairach Daemon (Lancaster et al., 2000). Afterwards, a contrast analysis was run (uncorrected $P=0.001$, P-value permutations $=10{,}000$, minimum cluster volume $=250\,\text{mm}^3$) to identify discrete and shared areas of activation between the syntax and tool use groups based on the generated conjunction image. The results were viewed in imaging software ITK-SNAP (Yushkevich et al., 2006) using the Colin 27 MNI $2 \times 2 \times 2$ mm brain template (Holmes et al., 1998) downloaded from the BrainMap website. GingerALE software output matched the $2 \times 2 \times 2$ mm template, so all coordinates and analyses performed were in $2 \times 2 \times 2$ mm MNI space. For higher resolution presentation, images were then registered in $1 \times 1 \times 1$ mm space on the corresponding Colin 27 MNI template (Figs. 2–4).

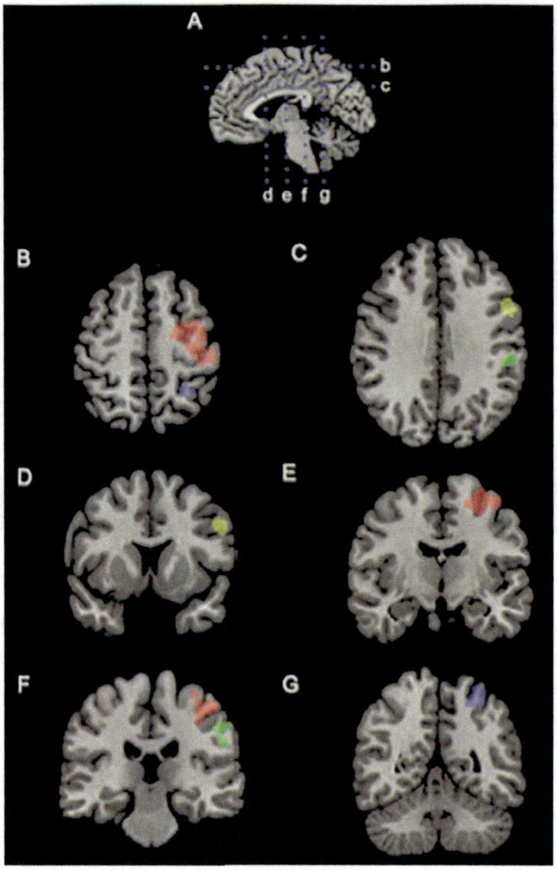

FIG. 3

Clusters showing significant neural activation during tool use, in horizontal (B–C) and coronal (D–G) view (cluster-level inference: $P=0.05$, permutation threshold$=1000$, cluster-forming threshold: uncorr. $P=0.001$). Color of cluster corresponds to Table 5. (A) Sagittal view for location orientation. (B) Left SPL and left precentral gyrus. (C) Left IPL and left IFG. (D) Left IFG. (E) Left precentral gyrus. (F) Left IPL and left precentral gyrus. (G) Left SPL. Abbreviations: IFG, inferior frontal gyrus; IPL, inferior parietal lobule; SPL, superior parietal lobule. Images generated in ITK-SNAP. Images shown in $1 \times 1 \times 1$ mm space.

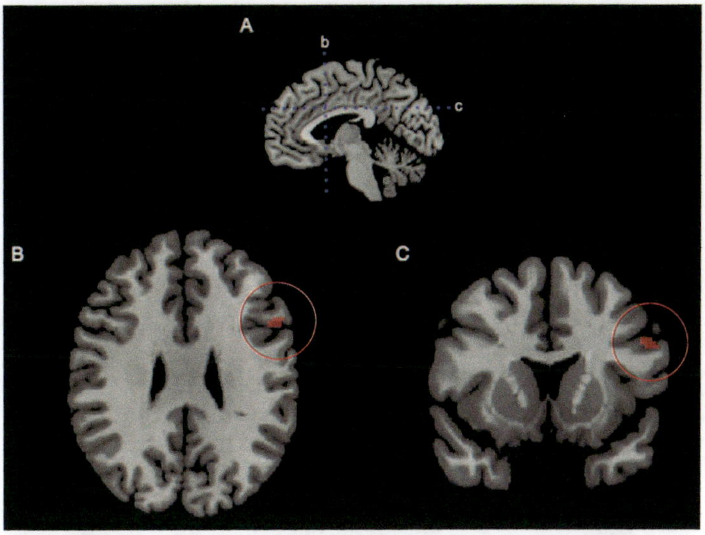

FIG. 4

(A) Sagittal view for location orientation. Cluster in left IFG *pars opercularis* in horizontal (B) and coronal (C) view showing significant neural coactivation during both syntactic processing and tool use (uncorr. $P=0.001$, P-value permutations $=10{,}000$, min. Cluster volume $=250\,mm^3$). Abbreviations: IFG, inferior frontal gyrus. Images generated in ITK-SNAP. Images shown in $1\times1\times1$ mm space.

3 Results

Meta-analysis of syntactic processing studies resulted in several clusters, the majority of which were in the left hemisphere. Clusters are described in Table 3. By far the largest cluster ($11{,}936\,mm^3$) was identified in the inferior frontal gyrus (IFG) of the left hemisphere, roughly equivalent to what is referred to as Broca's area. Additional significant clusters in the left hemisphere, both of similar size, were located in the middle temporal and superior frontal gyri (3944 and $3848\,mm^3$, respectively). Another cluster was recognized in the right insula (Table 3, Cluster 4). Finally, the right IFG was also identified as a significant cluster, although at a much smaller volume of $768\,mm^3$. Peak coordinate locations for each cluster are listed in Table 4, given in $2\times2\times2$ mm MNI space. Clusters are displayed in $1\times1\times1$ mm MNI space in Fig. 2.

As shown in Table 5, meta-analysis of tool use studies identified the left precentral gyrus (primary motor cortex and premotor cortex) as having the largest cluster at $9368\,mm^3$. The second-largest tool use activation cluster was $1728\,mm^3$ in the left inferior parietal lobule, and the left superior parietal lobule contained another cluster of similar size ($1624\,mm^3$). The final cluster was partially attributed to the left IFG

Table 3 Location and size of clusters showing significant neural activation during syntactic processing (cluster-level inference: $P=0.05$, permutation threshold $=1000$, cluster-forming threshold: uncorr. $P=0.001$). Color of "Cluster" corresponds to Fig. 2. Coordinates in $2 \times 2 \times 2$ mm MNI space.

Cluster	Size (mm^3)	Center (x, y, z)	Anatomical location
1	11936	(-46.3, 20.8, 13.2)	Inferior frontal gyrus (left)
2	3944	(-53.6, -42.7, 2.7)	Middle temporal gyrus (left)
3	3848	(-0.8, 13.3, 54.3)	Superior frontal gyrus (left)
4	2936	(40.9, 23, 1.1)	Insula (right)
5	768	(45.8, 13.5, 21.7)	Inferior frontal gyrus (right)

Table 4 Peak coordinate locations for clusters showing significant neural activation during syntactic processing (cluster-level inference: $P=0.05$, permutation threshold $=1000$, cluster-forming threshold: uncorr. $P=0.001$). Coordinates in $2 \times 2 \times 2$ mm MNI space. ALE and Z values rounded to 2 significant figures after the decimal.

Cluster #	x	y	z	ALE	Z
1	−50	14	22	0.038	6.74
1	−50	12	18	0.037	6.58
1	−50	30	6	0.029	5.56
1	−48	24	18	0.029	5.53
1	−42	24	0	0.028	5.51
1	−30	24	0	0.024	4.84
1	−34	28	−10	0.014	3.36
2	−52	−36	0	0.032	5.95
2	−54	−42	2	0.030	5.70
3	0	14	54	0.042	7.17
3	−8	20	44	0.018	3.90
4	38	24	−2	0.028	5.47
4	52	20	4	0.018	3.95
4	44	16	4	0.017	3.83
5	46	12	22	0.022	4.57

Table 5 Location and size of clusters showing significant neural activation during tool use (cluster-level inference: $P=0.05$, permutation threshold $=1000$, cluster-forming threshold: uncorr. $P=0.001$). Color of "Cluster" corresponds to Fig. 3. Coordinates in $2\times2\times2$ mm MNI space.

Cluster	Size (mm^3)	Center (x, y, z)	Anatomical location
1	9368	(-36.7, -23.1, 54.5)	Precentral gyrus (left)
2	1728	(-55.4, -29.2, 34.6)	Inferior parietal lobule (left)
3	1624	(-30.9, -53.1, 58.2)	Superior parietal lobule (left)
4	1264	(-53.6, 8.8, 31.7)	Inferior frontal gyrus (left)

Table 6 Peak coordinate locations for clusters showing significant neural activation during syntactic processing (cluster-level inference: $P=0.05$, permutation threshold $=1000$, cluster-forming threshold: uncorr. $P=0.001$). Coordinates in $2\times2\times2$ mm MNI space. ALE and Z values rounded to 2 significant figures after the decimal.

Cluster #	x	y	z	ALE	Z
1	−28	−10	58	0.028	5.99
1	−40	−26	54	0.023	5.29
1	−34	−24	62	0.023	5.13
1	−48	−28	54	0.020	4.83
1	−40	−38	46	0.020	4.78
1	−42	−46	48	0.020	4.77
1	−42	−14	60	0.017	4.26
1	−36	−6	56	0.017	4.24
2	−54	−30	38	0.021	4.94
2	−58	−26	26	0.017	4.28
3	−32	−54	62	0.023	5.36
4	−54	8	32	0.018	4.45

(Table 5, Cluster 4). Peak coordinate locations for each cluster are identified in Table 6, given in $2\times2\times2$mm MNI space. As can be seen in Fig. 3, all tool use clusters were located in the left hemisphere.

Shared activation between syntactic processing and tool use was only significant at one relatively small cluster, located in the *pars opercularis* of the left IFG (Tables 7 and 8, $2\times2\times2$mm space; Fig. 4, $1\times1\times1$mm space).

Table 7 Location and size of cluster showing significant neural coactivation during syntactic processing and tool use (uncorr. $P=0.001$, P-value permutations $=10{,}000$, min. Cluster volume $=250\,mm^3$). Coordinates in $2\times2\times2$ mm MNI space.

Cluster	Size (mm^3)	Center (x, y, z)	Anatomical location
1	260	(–53.9, 11.9, 27.8)	Inferior frontal gyrus (left)

Table 8 Peak coordinate location for cluster showing significant neural coactivation during syntactic processing and tool use (uncorr. $P=0.001$, P-value permutations $=10{,}000$, min. Cluster volume $=250\,mm^3$). Coordinates in $2\times2\times2$ mm MNI space. ALE value rounded to 2 significant figures after the decimal.

Cluster #	x	y	z	ALE
1	–54	12	28	0.016

4 Discussion

The meta-analysis found that the inferior frontal gyrus (IFG) is the single shared brain activation site of both syntax and tool use. The IFG corresponds to Broca's area, a region that is recognized for its essential function in speech production and human language processing. Specifically, the IFG has critical roles in the comprehension and production of phonemes, syntax, and semantics (Ishkhanyan et al., 2020; Stout and Chaminade, 2012). In line with the well-documented role of the IFG in language, it is no surprise that the meta-analysis of syntactic processing studies generated this area as the largest cluster. While the left IFG is particularly associated with linguistic function, the involvement of this region in hierarchical processing is bilateral (Koechlin and Jubault, 2006), perhaps explaining the smaller right IFG cluster apparent in the results. The IFG has been implicated in a variety of non-linguistic behaviors, the most relevant of which are object manipulation and sequence prediction. For instance, Stout et al. (2008) found that the right IFG of modern human participants was activated while making Acheulean-style tools, indicating at least partial neuroanatomical overlap between the processing of the hierarchical organization of complex toolmaking and syntax in language. Additionally, the IFG has been shown to play a role in the mirror neuron system (e.g., Kilner et al., 2009; Schmidt et al., 2021). Mirror neurons are often discussed in the context of their importance to social cognition and social learning, both of which are processes that are inextricably intertwined with language acquisition and tool use. Language and tool use share their hierarchical structuring and dependence on social input, both functions which have often been attributed to the IFG.

4.1 Syntax activations

Human language functions in the brain are generally considered to be lateralized to the left hemisphere in the majority of the population. The meta-analysis of syntax neuroimaging experiments locating clusters in the left middle temporal and superior frontal gyri is consistent with this notion. The middle temporal gyrus (MTG) has been implicated in the overlap between language and tool use before, in terms of the semantic processing of tool words (Chao et al., 1999). It is not surprising that an area associated with semantics was also identified in this analysis, since the selected studies all had designs where participants were presented with sentences to process. It would be difficult to solely focus on the structure of a sentence and not recognize the semantic information it conveys, as well. Regardless, the MTG has been highlighted as a prime syntactic hub candidate. Evidence demonstrates that there are strong associations between syntax and the MTG, in terms of both comprehension and production (Matchin & Hickok, 2020), but more research is needed on this topic.

Another area present in the results was the superior frontal gyrus (SFG). The SFG is broadly associated with higher cognitive functions, such as executive control, which are engaged when planning linguistic communication. Neural activation in the SFG that signals executive control has indeed been found during language production (Piai and Zheng, 2019). Additionally, the SFG is important in working memory function, which is characterized by a set of processes that includes organizing and planning action sequences (Boisgueheneuc et al., 2006)—this may be another reason why syntax-related language tasks are particularly associated with this area. It is likely that some of the syntactic tasks performed for the studies within this meta-analysis similarly required participants to carry out executive control, thereby explaining the SFG activation cluster.

The insula, which appeared as a smaller cluster in the analysis of syntax, has been recognized in other meta-analyses as playing a role in language processing. This structure's direct connections to Broca's area and other speech and language centers of the brain make it a strong candidate for having a multifaceted role in human language (Oh et al., 2014). Interestingly, the cluster in the meta-analysis was found in the right hemisphere, which is not typically the language-dominant hemisphere. Oh et al. (2014) found that speech perception and production activated both the left and right insula, but speech production tended to be more left dominant, whereas speech *perception* is more equally bilateral. Such results align well with the findings in our meta-analysis, since our focus was on language processing, not production. Furthermore, insula damage is frequently associated with aphasias, including Broca's aphasia, leading to difficulty with the motor planning and organization of speech (Ardila, 2018). It is consistent that this region should appear as an area of activation for a higher-order cognitive aspect of language such as syntactic processing.

4.2 Tool use activations

The primary motor cortex and premotor cortex, labeled as the precentral gyrus in the analysis' output, was the largest cluster generated from the tool use meta-analysis. Such a result is expected since the control of voluntary movement is an integral part of manipulating any sort of object. The left inferior and superior parietal lobules were also shown to be tool use activation clusters, which is supported by prior research showing that the former appears to have an important role in sensorimotor transformations for tool use networks (Johnson-Frey et al., 2005). Additionally, the inferior parietal lobule has been shown to be activated during observing, imagining, planning, and imitating tool use or general object manipulation. In toolmaking neuroimaging studies, the inferior and superior parietal lobules are sites of overlap for simple Oldowan and complex Acheulean technologies (Stout and Chaminade, 2012), highlighting their seemingly foundational role in tool-related tasks.

4.3 Shared activations—Inferior frontal gyrus

The IFG *pars opercularis* cluster is the only site that is significantly shared by both syntax and tool use in our results. The IFG has been previously implicated to have a role in both complex toolmaking and tool use, in addition to its more well-known linguistic functions. While the IFG has already been shown to be activated in Late Acheulean-style toolmaking (Stout et al., 2008), it does not appear to be a neural correlate of Oldowan toolmaking (Stout and Chaminade, 2012). Despite not being significant in neuroimaging studies of simple toolmaking, the IFG appears to be active during both language and tool *use*. Prior research by Higuchi et al. (2009) points to the *pars opercularis* of the IFG, in particular, being an area that is activated during tool use sequences. Furthermore, other tool-related studies have highlighted the role of the IFG. For example, Johnson-Frey et al. (2005) found significant activation in the IFG during tool use planning. More generally, Broca's area has been implicated in action viewing, imitation, and execution, perhaps serving as the basis for a "motor syntax" (Fazio et al., 2009). Other studies have similarly suggested Broca's area may play a role in "action grammar" (Pastra and Aloimonos, 2012), a term used to indicate the fact that there is a sequential structure to action that is comparable to linguistic syntax. It has been suggested that the IFG may be specialized for processing all sorts of hierarchically structured information (e.g., Friederici, 2011; Koechlin and Jubault, 2006). As such, this area's sequence processing function may have first evolved as an adaptation in primates for reaching and grasping (Kikuchi et al., 2018; Petkov & Ten Cate, 2020), which later became recruited in hominin ancestors for tool use and production, and was eventually co-opted for language (Pulvermüller and Fadiga, 2010).

Since the *pars opercularis* of the IFG appears to be the only significant cluster of shared activation between syntactic processing and tool use, this supports the idea

that tool *use* does not share an extensive network with syntactic processing. Instead, the IFG appears to act as one important hub for the overlap of these two behaviors. Evidence of mirror neurons in the IFG underpins this region's importance for general social cognition, including social learning by way of observation and imitation, which is crucial for learning both language and tool use skills (Kilner et al., 2009; Schmidt et al., 2021). These findings specific to syntactic processing build on the broader shared neural activation between speech production of phonemes and tool use in motor planning areas described by Reyes et al. (2016), who implemented the same meta-analysis methodology and found significant shared clusters in the primary motor and premotor cortices. The results of this meta-analysis contribute to the growing evidence that the IFG is not solely a language-related area, but a region of the human brain that participates in a number of higher-order cognitive tasks, including those that are tool-related.

4.4 Limitations and considerations

This meta-analysis is by no means exhaustive. Emphasis was put on selecting studies with consistent experimental design for each group (syntax, tool use), which narrowed down the possible options. Iterations of this meta-analysis could be done by broadening the article selection criteria to include syntactic production and/or imagined and planned tool use. Other meta-analyses comparing tool use to different aspects of human language (e.g., semantics) would further elucidate common patterns in the neural underpinnings of language and tool use. The current study only focused on the syntactic processing component of human language, so it is in no way meant to be interpreted as a proxy for investigating language evolution as a whole.

Multiple brain areas were not identified as clusters, despite having been functionally linked to language capacity in other research. Areas that were not detected in the meta-analysis, but that are associated with human syntax include the structures of the basal ganglia, where a recent fMRI study found common neural substrates for syntactic processing and tool use (Thibault et al., 2021). Specifically, much attention has focused on the caudate nucleus of the basal ganglia and its function in language control (e.g., Abutalebi and Green, 2007, 2016; Green and Abutalebi, 2013; Robles et al., 2005). Additionally, the lateral cerebellum has been implicated as having an important role in broader language networks (e.g., Lesage et al., 2017; Mariën and Manto, 2018; Vias and Dick, 2017). This discrepancy in results might be attributed to study sample sizes within the meta-analysis and other potential sources of reporting bias in the literature (David et al., 2013), and should not be taken to indicate that these regions are unimportant to syntax or general language processing.

4.5 Conclusions

In our meta-analysis, the *pars opercularis* of the left IFG was the single distinct area of syntax-tool use coactivation, a result that underscores the role of this region as an important hub in several key networks related to human-specific cognitive

capacities. We suggest that the IFG's function first evolved as a contributor to tool use (first simple, then complex and multi-step), later linking with other brain regions, and eventually being co-opted for linguistic purposes during human evolution.

References

Abutalebi, J., Green, D., 2007. Bilingual language production: the neurocognition of language representation and control. J. Neurolinguistics 20 (3), 242–275. https://doi.org/10.1016/j.jneuroling.2006.10.003.

Abutalebi, J., Green, D.W., 2016. Neuroimaging of language control in bilinguals: neural adaptation and reserve. Biling. Lang. Congn. 19 (4), 689–698. https://doi.org/10.1017/S1366728916000225.

Ardila, A., 2018. Participation of the insula in language. In: Turgut, M., Yurttaş, C., Tubbs, R. (Eds.), Island of Reil (Insula) in the Human Brain. Springer, Cham, pp. 123–128, https://doi.org/10.1007/978-3-319-75468-0_12.

Balzeau, A., Holloway, R.L., Grimaud-Hervé, D., 2012. Variations and asymmetries in regional brain surface in the genus *Homo*. J. Hum. Evol. 62 (6), 696–706. https://doi.org/10.1016/j.jhevol.2012.03.007.

Beauchamp, M.S., Lee, K.E., Haxby, J.V., Martin, A., 2002. Parallel visual motion processing streams for manipulable objects and human movements. Neuron 34 (1), 149–159. https://doi.org/10.1016/S0896-6273(02)00642-6.

Beaudet, A., Du, A., Wood, B., 2019. Evolution of the modern human brain. Prog. Brain Res. 250, 219–250. https://doi.org/10.1016/bs.pbr.2019.01.004.

Ben-Shachar, M., Palti, D., Grodzinsky, Y., 2004. Neural correlates of syntactic movement: converging evidence from two fMRI experiments. Neuroimage 21 (4), 1320–1336. https://doi.org/10.1016/j.neuroimage.2003.11.027.

Binkofski, F., Buccino, G., Posse, S., Seitz, R.J., Rizzolatti, G., Freund, H.J., 1999. A fronto-parietal circuit for object manipulation in man: evidence from an fMRI-study. Eur. J. Neurosci. 11 (9), 3276–3286. https://doi.org/10.1046/j.1460-9568.1999.00753.x.

Blank, I., Balewski, Z., Mahowald, K., Fedorenko, E., 2016. Syntactic processing is distributed across the language system. Neuroimage 127, 307–323. https://doi.org/10.1016/j.neuroimage.2015.11.069.

Bohlhalter, S., Hattori, N., Wheaton, L., Fridman, E., Shamim, E.A., Garraux, G., Hallett, M., 2009. Gesture subtype–dependent left lateralization of praxis planning: an event-related fMRI study. Cereb. Cortex 19 (6), 1256–1262. https://doi.org/10.1093/cercor/bhn168.

Bruner, E., Battaglia-Mayer, A., Caminiti, R., 2022. The parietal lobe evolution and the emergence of material culture in the human genus. Brain Struct. Funct. https://doi.org/10.1007/s00429-022-02487-w.

Cataldo, D.M., Migliano, A.B., Vinicius, L., 2018. Speech, stone tool-making and the evolution of language. PLoS One 13 (1), e0191071. https://doi.org/10.1371/journal.pone.0191071.

Chan, S., Ryan, L., Bever, T.G., 2013. Role of the striatum in language: syntactic and conceptual sequencing. Brain Lang. 125 (3), 283–294. https://doi.org/10.1016/j.bandl.2011.11.005.

Chao, L.L., Haxby, J.V., Martin, A., 1999. Attribute-based neural substrates in temporal cortex for perceiving and knowing about objects. Nat. Neurosci. 2, 913–919. https://doi.org/10.1038/13217.

Choi, S., Na, D.L., Kang, E., Lee, K., Lee, S., Na, D., 2001. Functional magnetic resonance imaging during pantomiming tool-use gestures. Exp. Brain Res. 139, 311–317. https://doi.org/10.1007/s002210100777.

Dapretto, M., Bookheimer, S.Y., 1999. Form and content: dissociating syntax and semantics in sentence comprehension. Neuron 24 (2), 427–432. https://doi.org/10.1016/S0896-6273(00)80855-7.

David, S.P., Ware, J.J., Chu, I.M., Loftus, P.D., Fusar-Poli, P., Radua, J., Munafò, M.R., Ioannidis, J.P.A., 2013. Potential reporting bias in fMRI studies of the brain. PLoS One 8 (7), e70104. https://doi.org/10.1371/journal.pone.0070104.

Davis, M.H., Meunier, F., Marslen-Wilson, W.D., 2004. Neural responses to morphological, syntactic, and semantic properties of single words: an fMRI study. Brain Lang. 89 (3), 439–449. https://doi.org/10.1016/S0093-934X(03)00471-1.

Devauchelle, A.D., Oppenheim, C., Rizzi, L., Dehaene, S., Pallier, C., 2009. Sentence syntax and content in the human temporal lobe: an fMRI adaptation study in auditory and visual modalities. J. Cogn. Neurosci. 21 (5), 1000–1012. https://doi.org/10.1162/jocn.2009.21070.

du Boisgueheneuc, F., Levy, R., Volle, E., Seassau, M., Duffau, H., Kinkingnehun, S., Samson, Y., Zhang, S., Dubois, B., 2006. Functions of the left superior frontal gyrus in humans: a lesion study. Brain 129 (12), 3315–3328. https://doi.org/10.1093/brain/awl244.

Eickhoff, S.B., Bzdok, D., Laird, A.R., Kurth, F., Fox, P.T., 2012. Activation likelihood estimation revisited. Neuroimage 59, 2349–2361. https://doi.org/10.1016/j.neuroimage.2011.09.017.

Eickhoff, S.B., Bzdok, D., Laird, A.R., Roski, C., Caspers, S., Zilles, K., Fox, P.T., 2011. Co-activation patterns distinguish cortical modules, their connectivity and functional differentiation. Neuroimage 57, 938–949. https://doi.org/10.1016/j.neuroimage.2011.05.021.

Eickhoff, S.B., Laird, A.R., Grefkes, C., Wang, L.E., Zilles, K., Fox, P.T., 2009. Coordinate-based activation likelihood estimation meta-analysis of neuroimaging data: a random-effects approach based on empirical estimates of spatial uncertainty. Hum. Brain Mapp. 30, 2907–2926. https://doi.org/10.1002/hbm.20718.

Eickhoff, S.B., Nichols, T.E., Laird, A.R., Hoffstaedter, F., Amunts, K., Fox, P.T., Bzdok, D., Eickhoff, C.R., 2016. Behavior, sensitivity, and power of activation likelihood estimation characterized by massive empirical simulation. Neuroimage 137, 70–85. https://doi.org/10.1016/j.neuroimage.2016.04.072.

Europa, E., Gitelman, D.R., Kiran, S., Thompson, C.K., 2019. Neural connectivity in syntactic movement processing. Front. Hum. Neurosci. 13, 27. https://doi.org/10.3389/fnhum.2019.00027.

Fazio, P., Cantagallo, A., Craighero, L., D'Ausilio, A., Roy, A.C., Pozzo, T., Calzolari, F., Granieri, E., Fadiga, L., 2009. Encoding of human action in Broca's area. Brain 132 (7), 1980–1988. https://doi.org/10.1093/brain/awp118.

Fedorenko, E., Nieto-Castañón, A., Kanwisher, N., 2012. Syntactic processing in the human brain: what we know, what we don't know, and a suggestion for how to proceed. Brain Lang. 120 (2), 187–207. https://doi.org/10.1016/j.bandl.2011.01.001.

Friederici, A.D., 2011. The brain basis of language processing: from structure to function. Physiol. Rev. 91 (4), 1357–1392. https://doi.org/10.1152/physrev.00006.2011.

Friederici, A.D., Opitz, B., von Cramon, D.Y., 2000. Segregating semantic and syntactic aspects of processing in the human brain: an fMRI investigation of different word types. Cereb. Cortex 10 (7), 698–705. https://doi.org/10.1093/cercor/10.7.698.

Gallivan, J.P., McLean, D.A., Valyear, K.F., Culham, J.C., 2013. Decoding the neural mechanisms of human tool use. Elife 2, e00425. https://doi.org/10.7554/elife.00425.

Gentner, T.Q., Fenn, K.M., Margoliash, D., Nusbaum, H.C., 2006. Recursive syntactic pattern learning by songbirds. Nature 440, 1204–1207. https://doi.org/10.1038/nature04675.

Green, D.W., Abutalebi, J., 2013. Language control in bilinguals: the adaptive control hypothesis. J. Cogn. Psychol. 25 (5), 515–530. https://doi.org/10.1080/20445911.2013.796377.

Greenfield, P., 1991. Language, tools and brain: the ontogeny and phylogeny of hierarchically organized sequential behavior. Behav. Brain Sci. 14 (4), 531–551. https://doi.org/10.1017/S0140525X00071235.

Henderson, J.M., Choi, W., Lowder, M.W., Ferreira, F., 2016. Language structure in the brain: a fixation-related fMRI study of syntactic surprisal in reading. Neuroimage 132, 293–300. https://doi.org/10.1016/j.neuroimage.2016.02.050.

Hermsdörfer, J., Terlinden, G., Mühlau, M., Goldenberg, G., Wohlschläger, A.M., 2007. Neural representations of pantomimed and actual tool use: evidence from an event-related fMRI study. Neuroimage 36 (Suppl. 2), T109–T118. https://doi.org/10.1016/j.neuroimage. 2007.03.037.

Higuchi, S., Chaminade, T., Imamizu, H., Kawato, M., 2009. Shared neural correlates for language and tool use in Broca's area. Neuroreport 20 (15), 1376–1381. https://doi.org/10.1097/wnr.0b013e3283315570.

Higuchi, S., Imamizu, H., Kawato, M., 2007. Cerebellar activity evoked by common tool-use execution and imagery tasks: an fMRI study. Cortex 43 (3), 350–358. https://doi.org/10.1016/s0010-9452(08)70460-x.

Holloway, R.L., 1969. Culture: A human domain. Curr. Anthropol. 10 (4), 47–64.

Holloway, R.L., 1983. Human paleontological evidence relevant to language behavior. Hum. Neurobiol. 2 (3), 105–114.

Holmes, C.J., Hoge, R., Collins, L., Woods, R., Toga, A.W., Evans, A.C., 1998. Enhancement of MR images using registration for signal averaging. J. Comput. Assist. Tomogr. 22 (2), 324–333. https://doi.org/10.1097/00004728-199803000-00032.

Imazu, S., Sugio, T., Tanaka, S., Inui, T., 2007. Differences between actual and imagined usage of chopsticks: an FMRI study. Cortex 43 (3), 301–307. https://doi.org/10.1016/s0010-9452(08)70456-8.

Inoue, K., Kawashima, R., Sugiura, M., Ogawa, A., Schormann, T., Zilles, K., Fukuda, H., 2001. Activation in the ipsilateral posterior parietal cortex during tool use: a PET study. Neuroimage 14 (6), 1469–1475. https://doi.org/10.1006/nimg.2001.0942.

Ishkhanyan, B., Michel Lange, V., Boye, K., Mogensen, J., Karabanov, A., Hartwigsen, G., Siebner, H.R., 2020. Anterior and posterior left inferior frontal gyrus contribute to the implementation of grammatical determiners during language production. Front. Psychol. 11, 685. https://doi.org/10.3389/fpsyg.2020.00685.

Jacobs, S., Danielmeier, C., Frey, S.H., 2010. Human anterior intraparietal and ventral premotor cortices support representations of grasping with the hand or a novel tool. J. Cogn. Neurosci. 22 (11), 2594–2608. https://doi.org/10.1162/jocn.2009.21372.

Johnson-Frey, S.H., 2004. The neural bases of complex tool use in humans. Trends Cogn. Sci. 8 (2), 71–78. https://doi.org/10.1016/j.tics.2003.12.002.

Johnson-Frey, S.H., Newman-Norlund, R., Grafton, S.T., 2005. A distributed left hemisphere network active during planning of everyday tool use skills. Cereb. Cortex 15 (6), 681–695. https://doi.org/10.1093/cercor/bhh169.

Kikuchi, Y., Sedley, W., Griffiths, T.D., Petkov, C.I., 2018. Evolutionarily conserved neural signatures involved in sequencing predictions and their relevance for language. Curr. Opin. Behav. Sci. 21, 145–153. https://doi.org/10.1016/j.cobeha.2018.05.002.

Kilner, J.M., Neal, A., Weiskopf, N., Friston, K.J., Frith, C.D., 2009. Evidence of mirror neurons in human inferior frontal gyrus. J. Neurosci. 29 (32), 10153–10159. https://doi.org/10.1523/JNEUROSCI.2668-09.2009.

Koechlin, E., Jubault, T., 2006. Broca's area and the hierarchical organization of human behavior. Neuron 50, 963–974. https://doi.org/10.1016/j.neuron.2006.05.017.

Laird, A.R., Lancaster, J.L., Fox, P.T., 2005. BrainMap: the social evolution of a functional neuroimaging database. Neuroinformatics 3, 65–78. https://doi.org/10.1385/ni:3:1:065.

Laird, A.R., Robinson, J.L., McMillan, K.M., Tordesillas-Gutierrez, D., Moran, S.T., Gonzales, S.M., Ray, K.L., Franklin, C., Glahn, D.C., Fox, P.T., Lancaster, J.L., 2010. Comparison of the disparity between Talairach and MNI coordinates in functional neuroimaging data: validation of the Lancaster transform. Neuroimage 51, 677–683. https://doi.org/10.1016/j.neuroimage.2010.02.048.

Lancaster, J.L., Tordesillas-Gutierrez, D., Martinez, M., Salinas, F., Evans, A., Zilles, K., Mazziotta, J.C., Fox, P.T., 2007. Bias between MNI and Talairach coordinates analyzed using the ICBM-152 brain template. Hum. Brain Mapp. 28, 1194–1205. https://doi.org/10.1002/hbm.20345.

Lancaster, J.L., Woldorff, M.G., Parsons, L.M., Liotti, M., Freitas, C.S., Rainey, L., Kochunov, P.V., Nickerson, D., Mikiten, S.A., Fox, P.T., 2000. Automated Talairach atlas labels for functional brain mapping. Hum. Brain Mapp. 10, 120–131. https://doi.org/10.1002/1097-0193(200007)10:3%3C120::aid-hbm30%3E3.0.co;2-8.

Lausberg, H., Kazzer, P., Heekeren, H.R., Wartenburger, I., 2015. Pantomiming tool use with an imaginary tool in hand as compared to demonstration with tool in hand specifically modulates the left middle and superior temporal gyri. Cortex 71, 1–14. https://doi.org/10.1016/j.cortex.2015.05.021.

Lesage, E., Hansen, P.C., Miall, R.C., 2017. Right lateral cerebellum represents linguistic predictability. J. Neurosci. 37 (26), 6231–6241. https://doi.org/10.1523/JNEUROSCI.3203-16.2017.

Longe, O., Randall, B., Stamatakis, E.A., Tyler, L.K., 2007. Grammatical categories in the brain: the role of morphological structure. Cereb. Cortex 17 (8), 1812–1820. https://doi.org/10.1093/cercor/bhl099.

Luke, K.K., Liu, H.L., Wai, Y.Y., Wan, Y.L., Tan, L.H., 2002. Functional anatomy of syntactic and semantic processing in language comprehension. Hum. Brain Mapp. 16 (3), 133–145. https://doi.org/10.1002/hbm.10029.

Mahaney, R.A., 2014. Exploring the complexity and structure of Acheulean stoneknapping in relation to natural language. PaleoAnthropology 2014, 586–606. https://doi.org/10.4207/PA.2014.ART90.

Mariën, P., Manto, M., 2018. Cerebellum as a master-piece for linguistic predictability. The Cerebellum 17, 101–103. https://doi.org/10.1007/s12311-017-0894-1.

Matchin, W., Hickok, G., 2016. 'Syntactic perturbation' during production activates the right IFG, but not Broca's area or the ATL. Front. Psychol. 7, 241. https://doi.org/10.3389/fpsyg.2016.00241.

Matchin, W., Hickok, G., 2020. The cortical organization of syntax. Cereb. Cortex 30 (3), 1481–1498. https://doi.org/10.1093/cercor/bhz180.

Matheson, H.E., Buxbaum, L.J., Thompson-Schill, S.L., 2017. Differential tuning of ventral and dorsal streams during the generation of common and uncommon tool uses. J. Cogn. Neurosci. 29 (11), 1791–1802. https://doi.org/10.1162/jocn_a_01161.

Meltzer, J.A., McArdle, J.J., Schafer, R.J., Braun, A.R., 2010. Neural aspects of sentence comprehension: syntactic complexity, reversibility, and reanalysis. Cereb. Cortex 20 (8), 1853–1864. https://doi.org/10.1093/cercor/bhp249.

Moore, M.W., 2010. Grammars of action and stone flaking design space. In: Nowell, A., Davidson, I. (Eds.), Stone Tools and the Evolution of Human Cognition. University of Colorado Press, pp. 13–43.

Moro, A., Tettamanti, M., Perani, D., Donati, C., Cappa, S.F., Fazio, F., 2001. Syntax and the brain: disentangling grammar by selective anomalies. Neuroimage 13 (1), 110–118. https://doi.org/10.1006/nimg.2000.0668.

Newman, A.J., Pancheva, R., Ozawa, K., Neville, H.J., Ullman, M.T., 2001. An event-related fMRI study of syntactic and semantic violations. J. Psycholinguist. Res. 30, 339–364. https://doi.org/10.1023/A:1010499119393.

Nuñez, S.C., Dapretto, M., Katzir, T., Starr, A., Bramen, J., Kan, E., Bookheimer, S., Sowell, E.R., 2011. fMRI of syntactic processing in typically developing children: structural correlates in the inferior frontal gyrus. Dev. Cogn. Neurosci. 1 (3), 313–323. https://doi.org/10.1016/j.dcn.2011.02.004.

Ogawa, K., Imai, F., 2016. Hand-independent representation of tool-use pantomimes in the left anterior intraparietal cortex. Exp. Brain Res. 234, 3677–3687. https://doi.org/10.1007/s00221-016-4765-7.

Oh, A., Duerden, E.G., Pang, E.W., 2014. The role of the insula in speech and language processing. Brain Lang. 135, 96–103. https://doi.org/10.1016/j.bandl.2014.06.003.

Pallier, C., Devauchelle, A.D., Dehaene, S., 2011. Cortical representation of the constituent structure of sentences. Proc. Natl. Acad. Sci. 108 (6), 2522–2527. https://doi.org/10.1073/pnas.1018711108.

Pastra, K., Aloimonos, Y., 2012. The minimalist grammar of action. Phil. Trans. R. Soc. B 367 (1585), 103–117. https://doi.org/10.1098/rstb.2011.0123.

Pazen, M., Uhlmann, L., van Kemenade, B.M., Steinsträter, O., Straube, B., Kircher, T., 2020. Predictive perception of self-generated movements: commonalities and differences in the neural processing of tool and hand actions. Neuroimage 206, 116309. https://doi.org/10.1016/j.neuroimage.2019.116309.

Peeters, R.R., Rizzolatti, G., Orban, G.A., 2013. Functional properties of the left parietal tool use region. Neuroimage 78, 83–93. https://doi.org/10.1016/j.neuroimage.2013.04.023.

Peeters, R., Simone, L., Nelissen, K., Fabbri-Destro, M., Vanduffel, W., Rizzolatti, G., Orban, G.A., 2009. The representation of tool use in humans and monkeys: common and uniquely human features. J. Neurosci. 29 (37), 11523–11539. https://doi.org/10.1523/jneurosci.2040-09.2009.

Petkov, C.I., ten Cate, C., 2020. Structured sequence learning: animal abilities, cognitive operations, and language evolution. Top. Cogn. Sci. 12 (3), 828–842. https://doi.org/10.1111/tops.12444.

Piai, V., Zheng, X., 2019. Speaking waves: neuronal oscillations in language production. Psychol. Learn. Motiv. 71, 265–302. https://doi.org/10.1016/bs.plm.2019.07.002.

Progovac, L., Rakhlin, N., Angell, W., Liddane, R., Tang, L., Ofen, N., 2018. Neural correlates of syntax and proto-syntax: evolutionary dimension. Front. Psychol. 9, 2415. https://doi.org/10.3389/fpsyg.2018.02415.

Pulvermüller, F., Fadiga, L., 2010. Active perception: sensorimotor circuits as a cortical basis for language. Nat. Rev. Neurosci. 11, 351–360. https://doi.org/10.1038/nrn2811.

Quiñones, I., Molinaro, N., Mancini, S., Hernández-Cabrera, J.A., Barber, H., Carreiras, M., 2018. Tracing the interplay between syntactic and lexical features: fMRI evidence from agreement comprehension. Neuroimage 175, 259–271. https://doi.org/10.1016/j.neuroimage.2018.03.069.

Reyes, L.D., Bianchi, S., Sherwood, C.C., 2016. An ALE meta-analysis of shared and unique neural activation in the human brain during speech production and tool-use [abstract]. Am. J. Phys. Anthropol. 159, 267–268.

Robles, S.G., Gatignol, P., Capelle, L., Mitchell, M.C., Duffau, H., 2005. The role of dominant striatum in language: a study using intraoperative electrical stimulations. J. Neurol. Neurosurg. Psychiatry 76 (7), 940–946. https://doi.org/10.1136/jnnp.2004.045948.

Rumiati, R.I., Weiss, P.H., Shallice, T., Ottoboni, G., Noth, J., Zilles, K., Fink, G.R., 2004. Neural basis of pantomiming the use of visually presented objects. Neuroimage 21 (4), 1224–1231. https://doi.org/10.1016/j.neuroimage.2003.11.017.

Santi, A., Grodzinsky, Y., 2007. Working memory and syntax interact in Broca's area. Neuroimage 37 (1), 8–17. https://doi.org/10.1016/j.neuroimage.2007.04.047.

Santi, A., Grodzinsky, Y., 2010. fMRI adaptation dissociates syntactic complexity dimensions. Neuroimage 51 (4), 1285–1293. https://doi.org/10.1016/j.neuroimage.2010.03.034.

Schmidt, S.N.L., Hass, J., Kirsch, P., Mier, D., 2021. The human mirror neuron system—a common neural basis for social cognition? Psychophysiology 58 (5), e13781. https://doi.org/10.1111/psyp.13781.

Segaert, K., Kempen, G., Petersson, K.M., Hagoort, P., 2013. Syntactic priming and the lexical boost effect during sentence production and sentence comprehension: an fMRI study. Brain Lang. 124 (2), 174–183. https://doi.org/10.1016/j.bandl.2012.12.003.

Segaert, K., Menenti, L., Weber, K., Petersson, K.M., Hagoort, P., 2012. Shared syntax in language production and language comprehension—an fMRI study. Cereb. Cortex 22 (7), 1662–1670. https://doi.org/10.1093/cercor/bhr249.

Siegelman, M., Blank, I.A., Mineroff, Z., Fedorenko, E., 2019. An attempt to conceptually replicate the dissociation between syntax and semantics during sentence comprehension. Neuroscience 413, 219–229. https://doi.org/10.1016/j.neuroscience.2019.06.003.

Stout, D., Chaminade, T., 2007. The evolutionary neuroscience of tool making. Neuropsychologia 45 (5), 1091–1100. https://doi.org/10.1016/j.neuropsychologia.2006.09.014.

Stout, D., Chaminade, T., 2012. Stone tools, language and the brain in human evolution. Phil. Trans. R. Soc. B 367 (1585), 75–87. https://doi.org/10.1098/rstb.2011.0099.

Stout, D., Chaminade, T., Apel, J., Shafti, A., Faisal, A.A., 2021. The measurement, evolution, and neural representation of action grammars of human behavior. Sci. Rep. 11, 13720. https://doi.org/10.1038/s41598-021-92992-5.

Stout, D., Hecht, E., Khreisheh, N., Bradley, B., Chaminade, T., 2015. Cognitive demands of lower Paleolithic toolmaking. PLoS One 10 (4), e0121804. https://doi.org/10.1371/journal.pone.0121804.

Stout, D., Toth, N., Schick, K., Chaminade, T., 2008. Neural correlates of early stone age toolmaking: technology, language and cognition in human evolution. Phil. Trans. R. Soc. B 363 (1499), 1939–1949. https://doi.org/10.1098/rstb.2008.0001.

Stromswold, K., Caplan, D., Alpert, N., Rauch, S., 1996. Localization of syntactic comprehension by positron emission tomography. Brain Lang. 52 (3), 452–473. https://doi.org/10.1006/brln.1996.0024.

Suzuki, K., Sakai, K.L., 2003. An event-related fMRI study of explicit syntactic processing of normal/anomalous sentences in contrast to implicit syntactic processing. Cereb. Cortex 13 (5), 517–526. https://doi.org/10.1093/cercor/13.5.517.

Suzuki, T.N., Wheatcroft, D., Griesser, M., 2016. Experimental evidence for compositional syntax in bird calls. Nat. Commun. 7, 10986. https://doi.org/10.1038/ncomms10986.

Takashima, A., Konopka, A., Meyer, A., Hagoort, P., Weber, K., 2020. Speaking in the brain: the interaction between words and syntax in sentence production. J. Cogn. Neurosci. 32 (8), 1466–1483. https://doi.org/10.1162/jocn_a_01563.

Tettamanti, M., Alkadhi, H., Moro, A., Perani, D., Kollias, S., Weniger, D., 2002. Neural correlates for the acquisition of natural language syntax. Neuroimage 17 (2), 700–709. https://doi.org/10.1006/nimg.2002.1201.

Thibault, S., Py, R., Gervasi, A.M., Salemme, R., Koun, E., Lövden, M., Boulenger, V., Roy, A.C., Brozzoli, C., 2021. Tool use and language share syntactic processes and neural patterns in the basal ganglia. Science 374 (6569), eabe0874. https://doi.org/10.1126/science.abe0874.

Turkeltaub, P.E., Eickhoff, S.B., Laird, A.R., Fox, M., Wiener, M., Fox, P., 2012. Minimizing within-experiment and within-group effects in activation likelihood estimation meta-analyses. Hum. Brain Mapp. 33, 1–13. https://doi.org/10.1002/hbm.21186.

Tyler, L.K., Marslen-Wilson, W.D., Randall, B., Wright, P., Devereux, B.J., Zhuang, J., Papoutsi, M., Stamatakis, E.A., 2011. Left inferior frontal cortex and syntax: function, structure and behaviour in patients with left hemisphere damage. Brain 134 (2), 415–431. https://doi.org/10.1093/brain/awq369.

van Dam, W.O., Desai, R.H., 2016. The semantics of syntax: the grounding of transitive and intransitive constructions. J. Cogn. Neurosci. 28 (5), 693–709. https://doi.org/10.1162/jocn_a_00926.

Vias, C., Dick, A.S., 2017. Cerebellar contributions to language in typical and atypical development: a review. Dev. Neuropsychol. 42 (6), 404–421. https://doi.org/10.1080/87565641.2017.1334783.

Wilson, S.M., DeMarco, A.T., Henry, M.L., Gesierich, B., Babiak, M., Mandelli, M.L., Miller, B.L., Gorno-Tempini, M.L., 2014. What role does the anterior temporal lobe play in sentence-level processing? Neural correlates of syntactic processing in semantic variant primary progressive aphasia. J. Cogn. Neurosci. 26 (5), 970–985. https://doi.org/10.1162/jocn_a_00550.

Wilson, S.M., Dronkers, N.F., Ogar, J.M., Jang, J., Growdon, M.E., Agosta, F., Henry, M.L., Miller, B.L., Gorno-Tempini, M.L., 2010. Neural correlates of syntactic processing in the nonfluent variant of primary progressive aphasia. J. Neurosci. 30 (50), 16845–16854. https://doi.org/10.1523/JNEUROSCI.2547-10.2010.

Yushkevich, P.A., Piven, J., Hazlett, H.C., Smith, R.G., Ho, S., Gee, J.C., Gerig, G., 2006. User-guided 3D active contour segmentation of anatomical structures: significantly improved efficiency and reliability. Neuroimage 31 (3), 1116–1128. https://doi.org/10.1016/j.neuroimage.2006.01.015.

Zaccarella, E., Meyer, L., Makuuchi, M., Friederici, A.D., 2017. Building by syntax: the neural basis of minimal linguistic structures. Cereb. Cortex 27 (1), 411–421. https://doi.org/10.1093/cercor/bhv234.

Zuberbühler, K., 2019. Syntax and compositionality in animal communication. Philos. Trans. R. Soc. B 375 (1789), 20190062. https://doi.org/10.1098/rstb.2019.0062.

Suzuki, T.N., Wheatcroft, D., Griesser, M., 2016. Experimental evidence for compositional syntax in bird calls. Nat. Commun. 7, 10986. https://doi.org/10.1038/ncomms10986.

Takashima, A., Konopka, A., Meyer, A., Hagoort, P., Weber, K., 2020. Speaking in the brain: the interaction between words and syntax in sentence production. J. Cogn. Neurosci. 32 (8), 1466–1483. https://doi.org/10.1162/jocn_a_01563.

Tettamanti, M., Alkadhi, H., Moro, A., Perani, D., Kollias, S., Weniger, D., 2002. Neural correlates for the acquisition of natural language syntax. NeuroImage 17 (2), 700–709. https://doi.org/10.1006/nimg.2002.1201.

Thibault, S., [illegible], Roy, A.C., [illegible], 2021. Tool use and language share syntactic processes and neural patterns in the basal ganglia. Science 374 (6569), eabe0874. https://doi.org/10.1126/science.abe0874.

[illegible], 2012. Minimizing within-experiment and within-group effects in activation likelihood estimation meta-analyses. Hum. Brain Mapp. [illegible]

[illegible]

van Dam, W.O., Desai, R.H., 2016. The semantics of syntax: the grounding of transitive and intransitive constructions. [illegible]

[illegible]

CHAPTER

5

Brain evolution and language: A comparative 3D analysis of Wernicke's area in extant and fossil hominids

Harmony Hill[a], **Marta Mirazón Lahr**[a], **and Amélie Beaudet**[a,b,c,*]

[a]*Department of Archaeology, University of Cambridge, Cambridge, United Kingdom*

[b]*School of Geography, Archaeology and Environmental Studies, University of the Witwatersrand, Johannesburg, South Africa*

[c]*Institut Català de Paleontologia Miquel Crusafont, Universitat Autònoma de Barcelona, Barcelona, Spain*

**Corresponding author: e-mail address: beaudet.amelie@gmail.com*

Abstract

The spoken word does not fossilize. Despite this, scientists have long sought to unearth the origins of language within the human lineage. One of the lines of evidence they have pursued is functional brain areas, such as Broca's and Wernicke's areas, which are associated with speech production and comprehension, respectively. Sulcal layout of Broca's area clearly differs between humans and our closest living relatives, the chimpanzees, enabling its homolog in fossil hominins to be deemed more chimpanzee-like (i.e., closer to the ancestral form) or more human-like (i.e., derived form) with relative ease. Yet, no such differences have been found for Wernicke's area. This study compares sulcal and gyral organization of Wernicke's area across extant human brains ($n=4$), extant chimpanzee brains ($n=5$) and fossil hominin endocasts ($n=4$). Some chimpanzee brains had indications of leftward Wernicke's area asymmetry in the form of a shorter Sylvian fissure and/or caudal superior temporal gyral bulging in the left hemisphere. Overlap between the superior and middle temporal sulci in human but not chimpanzee brains may be due to a relatively larger Wernicke's area in humans. Fragmentation of the main body of the superior temporal sulcus exclusively in human left hemispheres was ascribed to a leftward Wernicke's area asymmetry in this species. Endocast examination found that, while *Paranthropus robustus* exhibit human-like overlap between the superior and middle temporal sulci, *Australopithecus africanus* do not, although they do exhibit chimpanzee-like caudal superior temporal gyral bulging. Such findings signal, albeit loosely, a more human-like Wernicke's area in *Paranthropus* than *Australopithecus*.

Progress in Brain Research, Volume 275, ISSN 0079-6123, https://doi.org/10.1016/bs.pbr.2022.12.001

Keywords

Language evolution, Brain endocasts, Sulci, *Australopithecus*

1 Introduction

Language arguably defines the human niche. Language may have been selectively favored in the human lineage for several reasons—it allows rapid transmission of information regarding predator avoidance and food acquisition, more streamlined cooperative hunting, social bonding through "gossip," and technological advancement through more efficient teaching (Snowdon, 2002). Language has a litany of both neural and non-neural substrates. Determining when in our evolutionary history these substrates emerged is no simple matter. Some may be exclusive to *Homo sapiens*; others may have emerged in the broader genus *Homo*, or else within the hominin genera *Australopithecus* and *Paranthropus* that inhabited sub-Saharan Africa 4.2–1 mya (million years ago). Comparisons of language ability can be drawn not just between fossil hominins and extant humans, but between humans and extant chimpanzees (*Pan troglodytes*). Despite their equally long evolutionary history, chimpanzees have long been used as an imperfect proxy of the last common ancestor (LCA) of the human/*Pan* lineages, which genetic data suggest lived in Africa between 8 and 7 mya (Langergraber et al., 2012).

In the early 20th century, Korbinian Brodmann (1868–1918) published a map of the human cerebral cortex, naming 52 functional areas in each hemisphere based on cytoarchitecture (rev. in Damiani et al., 2020). More recently, Glasser et al. (2016) expanded upon Brodmann's map, naming 180 areas per hemisphere based on their divergent cortical architecture, function, topography and/or interconnectivity. This last factor is worth highlighting. Brain functions, including language, are not simply a matter of specialized cortical modules, but of module interconnectivity (Rilling et al., 2012). Moreover, connectomics is more efficient in identifying homologous landmarks across species (e.g., Bryant et al., 2020). That said, two modules are deemed directly essential for modern human language: Broca's and Wernicke's areas.

1.1 Broca's area in extant humans

Broca's area, named after pathologist Paul Broca (1824–1880), lies in the posterior inferior frontal gyrus (Dronkers et al., 2007). Damage to this area can cause non-fluent aphasia, under which patients retain speech comprehension but experience dysprosody and agrammatism during speech production (Acharya & Wroten 2021). Broca's area is generally thought to comprise Brodmann's areas (BA) 44 and 45, which roughly correspond to the *pars opercularis* and *pars triangularis* (de Sousa and Cunha, 2012). Electrical stimulation of the *pars opercularis* and the posterior portion of the *pars triangularis* in the language dominant hemisphere of

patients undergoing awake craniotomies produces speech arrest; however, the same cannot be said for the anterior portion of the *pars triangularis* (Chang et al., 2017). Consequently, some researchers exclude the latter from their definition of Broca's area. In maps using sulci and gryi on the lateral cerebral surface to mark the borders of functional areas, the rostral and caudal borders of BA44 coincide with the *ramus ascendens* of the Sylvian fissure and the inferior pre-central sulcus, while the dorsal border is near to the inferior frontal sulcus. BA45 is usually depicted between the *ramus ascendens* and *ramus horizontalis* of the Sylvian fissure (Kononova, 1935). However, sulci are inter-individually variable in terms of precise location, extension and presence, with several being absent in over 70% of human brains (Borne et al., 2020). According to Uylings et al. (2000), variation in size and location of the *rami ascendens* and *horizontalis* complicate identification of BA45. Due to discontinuity, the full extent of the inferior frontal sulcus can rarely be measured, and in many cases, the inferior pre-central sulcus truncates at or slightly below the inferior frontal sulcus rather than extending to the Sylvian fissure, obscuring the caudal limit of BA44 (Sprung-Much et al., 2021). Moreover, Sprung-Much and Petrides (2018) found that the *sulcus diagonalis*, which sits within the *pars opercularis*, was only apparent in 51.25% of hemispheres.

Besides sulci, another way of identifying the Broca's area is through cytoarchitecture. BA44 and BA45 are cytoarchitectonically difficult to distinguish. Neither has an obvious border between layers II and III, both have prominent pyramidal cells in deep layer III and layer V, and both have low cell density in layer VI (Amunts et al., 1999). Yet BA44 is dysgranular, meaning its layer IV is not distinctly separated from layers III and V. BA45 layer IV, meanwhile, is more distinctive, but still smaller and less pronounced than in neighboring regions BA46 and BA10 (Uylings et al., 2000). Amunts et al. (1999) determined the cytoarchitectural borders of BA44 and BA45 in 10 adult human brains by measuring changes in laminar distribution of gray level index before mapping these borders onto 3D reconstructions of said brains. The location of the dorsal border of BA44 and BA45 differed inter-individually, as did the border between BA44 and BA6, which lay variably between the rostral and caudal walls of the pre-central sulcus. Moreover, in some hemispheres the borders between BA44 and BA45 lay in the vicinity of *ramus ascendens*, while in others it was close to the *sulcus diagonalis* or interposed between the former and latter. Evidently, cytoarchitectonic borders do not always correspond with sulcal contours. In some instances, drawing cytoarchitectonic borders is hampered by transitional areas that share common cytoarchitectonic features with abutting regions: one such area sits between BA45 and BA9 (Amunts et al., 1999).

Another topic of debate surrounding Broca's area is inter-hemispheric asymmetry. Approximately 95% of the population is left-hemisphere dominant for language (Branche et al., 1964), including 97% of right-handed and 70% of left-handed persons (Toga and Thompson, 2003), leading many to hypothesize an asymmetric extension of Broca's area in the language-dominant hemisphere. Numerous studies have searched for such asymmetry, which may be volumetric, surface areal, organizational or cytoarchitectural (rev. in Toga and Thompson, 2003).

1.2 Wernicke's area in extant humans

The second language region of interest within the human brain is Wernicke's area, named after neurologist Carl Wernicke (1848–1905). Individuals afflicted with Wernicke's aphasia, a fluent aphasia, struggle to comprehend speech and themselves commit phonemic, neologistic and verbal paraphasic errors (Ellis et al., 1983). Much less is known about Wernicke's than Broca's area; indeed, its cytoarchitectonic cortical borders are yet to be conclusively defined (Uylings et al., 2000). Broadly speaking, Wernicke's area lies in the posterior superior temporal gyrus near the caudal end of the Sylvian fissure, spanning the *planum temporale*, BA40 and area Tpt (Falk, 2007). Area Tpt is analogous with the posterior end of BA22 (Galaburda and Sanides, 1980) and comprises a portion of cortex underlying the *planum temporale* (Spocter et al. 2010). Wernicke's area was once thought to be connected to Broca's area via the *arcuate fasciculus,* but Bernal and Ardila (2009) suggested this white matter tract instead connects posterior receptive areas with premotor/motor areas. There is slightly stronger accord among researchers regarding Wernicke's area asymmetry compared to Broca's area asymmetry. Geschwind and Levitsky (1968) found that in a sample of 100 adult human brains, the *planum temporale* expressed asymmetry with the left side (L) being more prominent than the right side (R) in 65 cases and R being more prominent than L in 11 cases. *Planum temporale* was on average one-third longer in the left hemisphere. Toga and Thompson (2003) agreed with this, stating that the human left *planum temporale* can be up to 10 times larger than the right. A study of 100 adults and 85 infants by Wada et al. (1975) also revealed L > R asymmetry in *planum temporale* size; however, investigation of cortical myeloarchitecture by Yuan et al. (2021) identified R > L asymmetry of both *planum temporale* and Heschl's gyrus.

1.3 Homologous regions in nonhuman primates

Sulcal organization of Broca's area differs between humans and other great apes. In chimpanzees, the rostral border of the *pars opercularis* is formed by the fronto-orbital sulcus (FO) rather than by the *ramus ascendens*, while the *pars triangularis* is not always present (Connolly, 1950). According to Sherwood et al. (2003), there is a functional and anatomical homolog of BA44 in chimpanzees, localized to *pars opercularis*. Yet, these same researchers concluded inter-chimpanzee variation in correspondence between sulci and cytoarchitectural boundaries made outlining *pars opercularis* using MRI data highly taxing (Sherwood et al., 2003). This brings into question Cantalupo and Hopkins' (2001) reports of L > R BA44 asymmetry in chimpanzees. Indeed, when Schenker et al. (2010) estimated volume, neurone number and neurone density for BA44 and BA45 across 12 chimpanzee brains, no significant inter-hemispheric asymmetries were identified.

Meanwhile, several studies cite asymmetries in Wernicke's area homologs. Area Tpt has been identified in chimpanzees, rhesus macaques (*Macaca mulatta*) and bushbabies (*Galago crassicaudatus*/*garnetti*), although it is less parietally extended

in chimpanzee than human brains (Spocter et al. 2010). Hopkins et al. (1998) found chimpanzees to display population-level L > R *planum temporale* surface area asymmetry. This observation was reinforced by Gannon et al. (1998), who reported leftward *planum temporale* asymmetry in 94% of chimpanzee brains examined. A more recent study by Spocter et al. (2010) found 10/12 chimpanzee brains to have more neurones in area Tpt of their left hemisphere than in area Tpt of their right hemisphere, while 9/12 brains also expressed leftward asymmetry in area Tpt volume. In an intriguing study, Hopkins et al. (2007) took MRI scans of 78 captive chimpanzees and demonstrated that handedness for tool use actions was associated with asymmetries in Broca's and Wernicke's area homologs, while handedness for non-tool-use actions was not. In a similar vein, Taglialatela et al. (2008) found that the left hemispheric Broca's area homolog in chimpanzees activates during communicative gesturing, suggesting co-evolution between complex manual abilities and language.

1.4 Introducing endocasts

In the absence of fossilized soft tissues, researchers rely on skeletal proxies. One such proxy is endocranial reconstructions of brain surfaces, or endocasts, which have been shown to preserve information on hominin brain cortical organization (Beaudet et al., 2016). Endocasts may form naturally via sediment accumulating in the braincase during fossilization (Beaudet, 2017). Alternatively, they can be made by pouring dental molding material into the braincase, a highly invasive option, or by compiling an image stack acquired from scanning the cranium in an X-ray (micro)CT scanner and generating a 3D model (Holloway, 2018). Endocasts elicit details regarding brain size, shape, vascularization and gyrification. Having analyzed the brains and corresponding endocasts of five humans, Dumoncel et al. (2021) concluded endocasts are an accurate proxy for the morphoarchitecture of the frontal, temporoparietal and occipital brain regions, thus for inspection of Broca's and Wernicke's areas, but not for the superior region. Kobayashi et al. (2014) also found endocasts taken from five living macaques (*Macaca fascicularis*) manifested clear sulcal and gyral imprints across the entire brain surface. If the brain was composed of spatially consistent modules delimited by sulci, with each module serving its own specialized function, evolutionary changes to underlying brain areas could be readily inferred from changing sulcal and gyral imprints on endocasts (Zollikofer and Ponce de León, 2013). However, as discussed above, cytoarchitectonic borders of functional areas do not simply align with the bottom of sulci, and their definition is further hampered by the marked inter-individual variation in brain anatomy due to age, sex, brain size, and brain shape, among other factors (Amunts et al., 1999). This means that the interpretation of external brain morphology preserved in endocasts needs to be cautious. Furthermore, since hominin endocranial evidence is limited and scattered broadly through space and time, identifying diachronic trends in brain evolution while accounting for such variation is difficult (Ponce de León et al., 2021). Fossil crania are also frequently fragmented, incomplete or distorted. Not only does this complicate endocast production, but may lead to misidentification of fissures generated by post-depositional damage as sulci.

Attention must also be paid to endocranial vascularisation when analyzing endocasts, as impressions of middle meningeal vessels and venous sinuses upon the endocranium can be mistaken for cerebral surface features (Beaudet et al., 2019).

Further complication stems from the conspicuous lack of protocols for automatic sulcal detection on endocasts (Beaudet, 2017). New approaches for sulcal labeling on brain MRIs are being developed; for example, by combining the BrainVISA/Morphologist package automatic sulcal identification algorithm with a U-based convolutional neural network, Borne et al. (2020) designed an observer-independent method which could label a hemisphere in just 20 s in extant humans. However, the hominin fossil record sample size is insufficient to train such a network in sulcal detection. One promising advancement is an algorithm designed by Yoshizawa et al. (2007) for detecting crest lines in 3D meshes. Beaudet et al. (2016) used this algorithm to automatically identify neocortical relief in endocasts derived from 14 South African Plio-Pleistocene cercopithecoid fossils. Similarly, de Jager et al. (2019) used it to automatically assess topographical variations in 20 human endocasts. Unfortunately, this technology is in its infancy, and visual inspection remains the most common method for examining sulcal imprints on endocasts. Since visual inspection hinges on observer expertise, inter-observer discrepancies are inevitable.

Endocast-based research has already yielded insights into the evolution of Broca's area. Broca's cap is a bulge seen on great ape endocasts. This cap overlaps Broca's area (or appropriate homologs) but does not exactly equate to it. In chimpanzees, Broca's cap comprises BA44 and is limited ventrally by FO, while in humans it comprises BA45 and BA47 (but not BA44) and is limited ventrally by the lateral orbital sulcus (Ponce de León et al., 2021). Frontal lobe sulcal organization as well as potential asymmetries have been investigated in fossil hominins by way of Broca's cap and neighboring areas. There is no sudden shift from a chimp-like to a human-like frontal lobe in the earliest fossils of the genus *Homo*, but there is an absence of FO, as well as a trend toward Broca's area asymmetry amongst later *Homo*. By conducting landmark-based analyses of third frontal convolution asymmetry in 139 humans, 35 bonobos, 36 chimpanzees and 46 fossil hominins, Balzeau et al. (2014) found a gradual increase in the degree of asymmetries throughout the human lineage.

Endocast-based research around Wernicke's area is scanter than that around Broca's area due to poor temporo-parietal region preservation in the fossil record. Nevertheless, Tobias (1987) described strong development of the inferior parietal lobule and thus of Wernicke's area in *H. habilis,* as well as inter-hemispheric asymmetry of the Sylvian fissure. Falk (1980) found the relationship between the superior temporal sulcus and the middle and descending rami of this sulcus in the *Australopithecus africanus* Taung 1 fossil to be chimpanzee-like, although Holloway (1981) disagreed with Falk's positioning of the superior temporal sulcus on this endocast. Beaudet et al. (2019) further examined the brain endocasts of South African australopithecines and found that in four *Australopithecus* specimens from the Sterkfontein Caves the superior and middle temporal sulci are parallel and non-overlapping in Sts 1017, StW 573, Sts 5 and Sts 60, as observed in chimpanzees.

In contrast, in another *Australopithecus* specimen, MLD 37/38, and a *Paranthropus* specimen, SK 1585, these two sulci intersect as is usually the case in humans (Beaudet et al., 2019). Beaudet et al. (2021) also observed that the temporoparietal regions of the *Paranthropus* specimens OH 5 and SK 1585, respectively, from Olduvai Gorge in Tanzania and Swartkrans in South Africa, including the supramarginal and angular gyri, are less pronounced than in *Homo* but more so than in chimpanzees (Beaudet et al., 2021).

There is a gap in the field in terms of comparing cortical organization of Wernicke's area (or appropriate homologs) between humans, chimpanzees and fossil hominins. This chapter will begin to plug this gap via examination of the sulcal and gyral layout in 3D models of chimpanzee and human brains as well as fossil hominin endocasts. The aim of this chapter is twofold: (i) assessing the variation pattern within our sample of extant chimpanzee and human brains; (ii) describing the sulcal pattern in fossil hominin endocasts using extant chimpanzee and human brains as a comparative platform. Our description is restricted to the area of interest (i.e., Wernicke's area and the surrounding).

2 Materials and methods

Our sample includes five chimpanzee and four human brains, as well as four fossil hominin endocasts (see Table 1).

3D Slicer (version 4.13.0-2021-10-22, Fedorov et al., 2012) was used to generate 3D models of the human brains (via the "Volume Rendering" tool) and to visualize both the human and chimpanzee 3D brain models. Screenshots of both the left and right hemispheres in lateral view were captured for each of the nine extant specimens. 3D models of the fossil hominin endocasts and the imprints detected in Beaudet et al. (2019) were viewed in ParaView (version 5.10.0). Sulci were identified and studied on one hemisphere (the best preserved) for each specimen.

Based on descriptions in Connolly (1950), one of us (HH), identified the sulci in the vicinity of Wernicke's area were. Sulcal and gyral names and abbreviations were

Table 1 List of specimens studied in this chapter.

Species/specimens	Sex	Provenance
Extant specimens		
Homo sapiens (N=4)	F (*N*=1) /M (N=3)	Allen Institute Brain Atlas website
Pan troglodytes (N=5)	F (*N*=3) /M (N=2)	National Chimpanzee Brain Resource
Fossil specimens		
Paranthropus robustus (SK 1585)	?	Beaudet et al. (2019)
Australopithecus africanus/prometheus (Sts 60, StW 505, StW 578)	? (N=1)/M (*N*=2)	Beaudet et al. (2019)

taken from Connolly (1950). We identified the Sylvian fissure (S); the sub-central posterior sulcus (scp); the superior temporal sulcus (ts); the middle temporal sulcus (tm); the three caudal rami, namely, the anterior/superior parallel (a1), middle/angularis (a2) and descending/anterior occipital (a3) rami; the post-central sulcus (pt), which is divided into the superior post-central sulcus (pts) and inferior post-central sulcus (pti), which may or may not be connected; the intraparietal sulcus (ip); the processus acuminis (e) and the intermedius anterior sulcus (ina); the inferior occipital sulcus (oci). In humans, we also identified the paroccipital sulcus (par); the transverse occipital sulcus (otr), which often branches from the end of par; and the transverse post-central sulcus (ptr). Abbreviations are used in the figures' captions.

3 Results

3.1 Extant chimpanzee brains

In the specimen EC1 (Fig. 1), the Sylvian fissure and the superior temporal sulcus run parallel to each other in both hemispheres. The superior temporal gyrus, which sits between the Sylvian fissure and the superior temporal sulcus, is largely uniform in width in both hemispheres, although it narrows toward the caudal end of the Sylvian fissure in the left hemisphere. In both hemispheres, the anterior/superior parallel and the middle/angularis rami stem from the caudal end of the superior temporal sulcus and branch toward the post-central sulcus and the intraparietal sulcus respectively. In the left hemisphere, the descending/anterior occipital ramus is very short and branches down toward a dorsally-projecting section of the middle temporal sulcus. In the right hemisphere, the descending/anterior occipital ramus is much longer and seems continuous with the inferior occipital sulcus. In the left hemisphere, the post-central sulcus forks at its ventral end, which is close to the Sylvian fissure. In the right hemisphere, the post-central sulcus is not forked and is shorter than in the left hemisphere, thus terminates further from the Sylvian fissure. In both hemispheres, a small depression lies in the inferior parietal lobule just rostral to the anterior/superior parallel ramus. In both hemispheres, the intraparietal sulcus connects to the post-central sulcus, curves around the ends of the anterior/superior parallel and the middle/angularis ramus then connects to the crescent-shaped lunate sulcus. In the right hemisphere, a slight depression ventral to the intraparietal sulcus and between the anterior/superior parallel and the middle/angularis rami may represent the intermedius anterior sulcus.

In the left hemisphere of the second specimen (EC2) (Fig. 1), the width of the superior temporal gyrus is consistent throughout its length, and the caudal end of the Sylvian fissure forked. Meanwhile, in the right hemisphere, the non-bifurcated caudal end of the Sylvian fissure curves down toward the superior temporal sulcus, causing a narrowing of the superior temporal gyrus. The Sylvian fissure

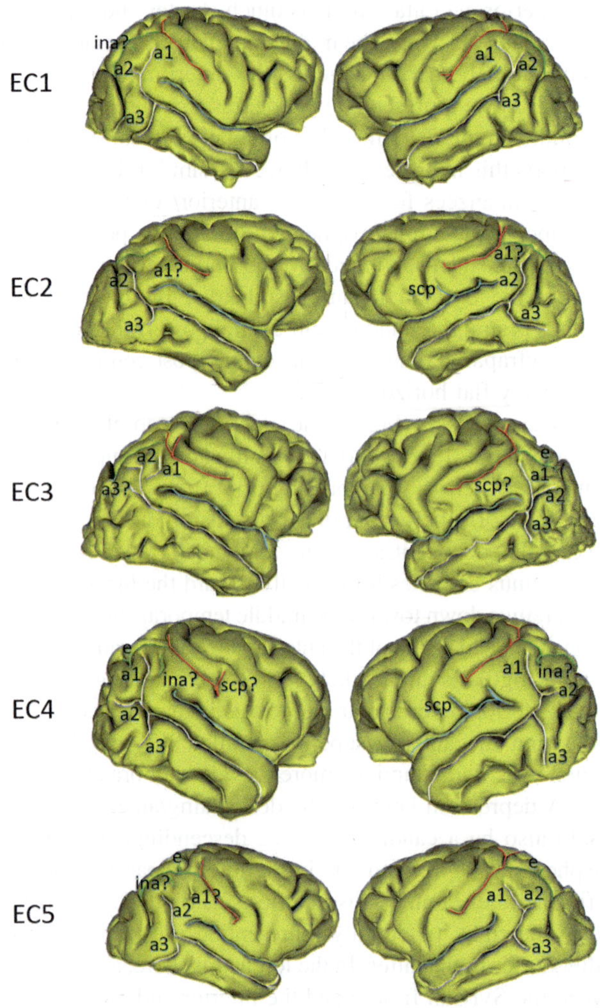

FIG. 1

Identification of the sulci surrounding the Wernicke's area in extant chimpanzee (EC) brains. Blue: Sylvian fissure; green: intraparietal sulcus; red: post-central sulcus; white: superior temporal sulcus. For abbreviations' meaning see Section 2.

seems shorter in the left than the right hemisphere. The middle/angularis and the descending/anterior occipital rami are particularly long in the left hemisphere: the middle/angularis ramus flows on from the main body of the superior temporal sulcus and terminates close to the intraparietal sulcus, while the descending/anterior occipital ramus curves down below the ventral end of the lunate. In the right hemisphere,

the descending/anterior occipital ramus is much shorter and juts downward toward the middle temporal sulcus, while the middle/angularis ramus terminates close to the intraparietal sulcus. In both hemispheres, the anterior/superior parallel ramus is unclear, although may be represented by a depression in the inferior parietal lobule rostral to the middle/angularis ramus. The orientation of said depression in the right hemisphere supports this interpretation; however, similar depressions are evident in hemispheres of chimpanzees for whom the anterior/superior parallel ramus has a definite designation. In the left hemisphere, the end of the post-central sulcus curves up toward the central sulcus, guided by the sub-central posterior sulcus which branches up from the Sylvian fissure just below. In the right hemisphere, the post-central sulcus is straighter and no sub-central posterior sulcus is visible. In both hemispheres, the intraparietal sulcus connects the post-central sulcus to the lunate and forms a relatively flat horizontal line.

In both EC3's hemispheres (Fig. 1), the superior temporal gyrus narrows caudally as the end of the Sylvian fissure turns down toward the superior temporal sulcus. In the left hemisphere there is some bulging prior to this narrowing. Also in the left hemisphere, the anterior/superior parallel ramus is rather long, terminating near the junction between the post-central sulcus and the intraparietal sulcus, while the middle/angularis ramus branches horizontally toward the lunate and the descending/anterior occipital ramus down toward the middle temporal sulcus. There is clear bulging running along the caudal side of the anterior/superior parallel ramus and the dorsal side of the middle/angularis ramus likely representing the angular gyrus. In the right hemisphere, the anterior/superior parallel ramus is disconnected from the superior temporal sulcus and the anterior/superior parallel, the middle/angularis and the descending/anterior occipital rami are more crowded, all branching toward the intraparietal sulcus. A depression between the descending/anterior occipital ramus and the lunate could also be a candidate for the descending/anterior occipital ramus. In both hemispheres, a depression sits in the inferior parietal lobule rostral to the anterior/superior parallel ramus. This manifests as a pit in the left hemisphere and a curved line in the right, potentially representing a continuation of the anterior/superior parallel ramus in the latter. In the left hemisphere, another slight depression running between the Sylvian fissure and the inferior end of the post-central sulcus may represent the sub-central posterior sulcus. In both hemispheres, the inferior section of the post-central sulcus is non-bifurcated and parallel to the Sylvian fissure while the intraparietal sulcus connects the post-central sulcus to the lunate. In the left hemisphere, a small branch sticking upward from the intraparietal sulcus represents the processus acuminis sulcus.

In EC4's left hemisphere (Fig. 1), the Sylvian fissure is relatively short and bifurcates at its end into two branches roughly perpendicular to the main sulcal body. One of these branches points toward the superior temporal sulcus, creating a bottleneck in the superior temporal gyrus. Just prior to this bottleneck, however, the gyrus bulges, pushing the Sylvian fissure and the superior temporal sulcus further apart. No such bulging is observed in the right hemisphere, where the distance between the Sylvian fissure and the superior temporal sulcus is consistent, and the end of the Sylvian fissure non-bifurcated. In both hemispheres, the configuration of caudal

rami of the superior temporal sulcus is similar: the anterior/superior parallel ramus branches up toward and terminates near the intraparietal sulcus, the middle/angularis ramus branches horizontally toward the lunate and the descending/anterior occipital ramus branches sharply downward. In the left hemisphere, there are two pit-like depressions in the inferior parietal lobule, one between the anterior/superior parallel and the middle/angularis rami and another between the anterior/superior parallel ramus and the post-central sulcus. In the right hemisphere, there is a small depression resembling a vertical line between the anterior/superior parallel ramus and the middle/angularis ramus. In both hemispheres, the ventral end of the post-central sulcus flicks up toward the central sulcus. This may be explained by the sub-central posterior sulcus branching upward from the Sylvian fissure just beneath the post-central sulcus in the left hemisphere. There is no clear sub-central posterior sulcus in the right hemisphere, although a slight depression near the ventral end of the post-central sulcus may represent a poorly developed sub-central posterior sulcus. In both hemispheres, the intraparietal sulcus connects to the post-central sulcus but not the lunate sulcus. In the left hemisphere, a small branch of the intraparietal sulcus caudal to the anterior/superior parallel ramus may represent the intermedius anterior sulcus. In the right hemisphere, a small disconnected vertical sulcus rostral to the anterior/superior parallel ramus and ventral to the intraparietal sulcus may represent the intermedius anterior sulcus.

There is no apparent superior temporal gyral bulging in EC5's left or right hemispheres (Fig. 1); however, the caudal end of the Sylvian fissure turns downward in the right hemisphere and upward in the left, resulting in a narrower distance between it and the superior temporal sulcus in the right hemisphere. The caudal rami of the superior temporal sulcus are clear in the left hemisphere, with the anterior/superior parallel ramus pointing toward the junction of the post-central sulcus and the intraparietal sulcus, a short middle/angularis ramus toward the junction of the intraparietal sulcus and the lunate and the descending/anterior occipital ramus toward the middle temporal sulcus. A small pit sits between the anterior/superior parallel and the middle/angularis rami ventral to the intraparietal sulcus. In the right hemisphere, the middle/angularis ramus points up toward the intraparietal sulcus and the descending/anterior occipital ramus is disconnected from the superior temporal sulcus, connecting instead to the middle temporal sulcus which is displaced onto the inferior brain surface. There is no clear anterior/superior parallel ramus, although it may be represented by a small depression in the inferior parietal lobule rostral to the middle/angularis ramus, as was the case with EC2. As mentioned, however, other chimpanzee brains have similarly placed depressions *and* clear anterior/superior parallel ramus designations. In both hemispheres, the intraparietal sulcus links the post-central sulcus to the lunate and is rather curved. The distance between the intraparietal sulcus and the end of the nearest ramus of the superior temporal sulcus is greater than in the brains of EC2, EC3 and EC4, although resembles that documented in EC1. In both hemispheres, the processus acuminis sulcus branches upward from the intraparietal sulcus, and in the right hemisphere, a depression just ventral to the processus acuminis sulcus and disconnected from the intraparietal sulcus may represent the intermedius anterior sulcus.

3.2 Extant human brains

In EH1's left hemisphere (Fig. 2), the superior temporal gyrus is rather convoluted due to a fragmentary superior temporal sulcus. The end of the Sylvian fissure connects to the superior temporal sulcus. The anterior/superior parallel ramus extends upward into the inferior parietal lobule, terminating near the intraparietal sulcus. The middle/angularis ramus has multiple branches and projects caudally from the

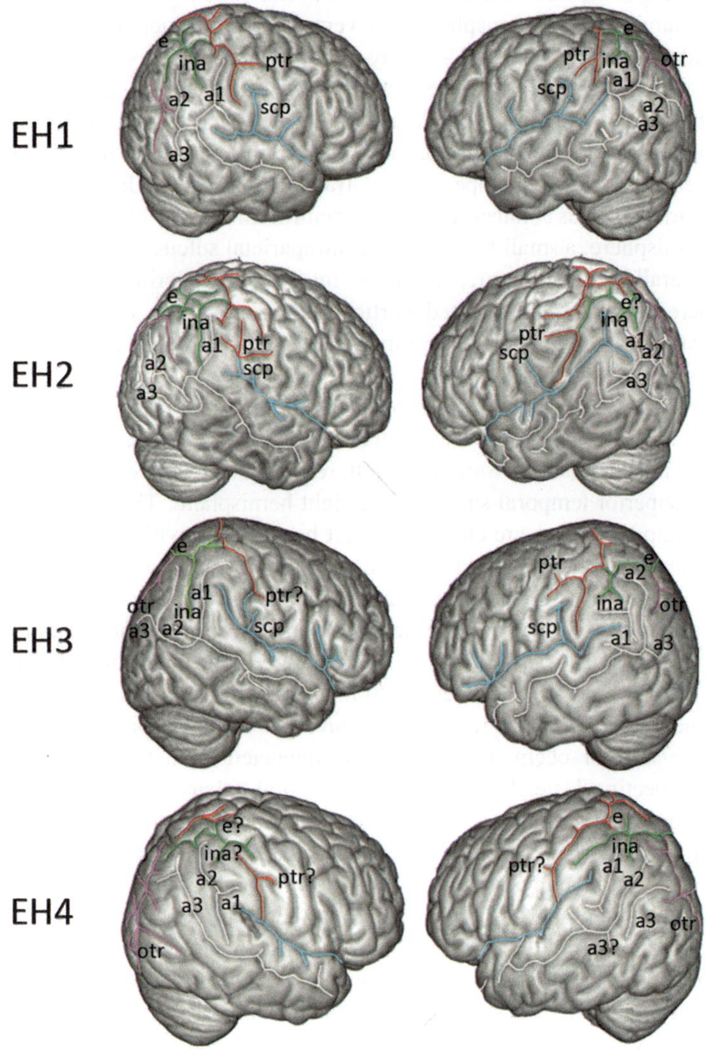

FIG. 2

Identification of the sulci surrounding the Wernicke's area in extant human (EH) brains. Blue: Sylvian fissure; green: intraparietal sulcus; pink: paroccipital sulcus; red: post-central sulcus; white: superior temporal sulcus. For abbreviations' meaning see Section 2.

anterior/superior parallel ramus, while a short descending/anterior occipital ramus projects caudally from part of the main body of the superior temporal sulcus. The transverse post-central sulcus, which terminates near the sub-central posterior sulcus, branches off just before the ventral end of the post-central sulcus. The post-central sulcus connects to the intraparietal sulcus, from which a short processus acuminis sulcus projects upward and a forked intermedius anterior sulcus downward into the space between the post-central sulcus and the anterior/superior parallel ramus. The paroccipital sulcus, which is disconnected from the intraparietal sulcus, sits above the middle/angularis ramus, to which one of the branches of the transverse occipital sulcus connects. In the right hemisphere, the superior temporal gyrus is un-convoluted and exhibits very slight bulging ventral to the origin of the sub-central posterior sulcus. The descending/anterior occipital ramus juts downward, terminating between two branches of a fragmented middle temporal sulcus, while the middle/angularis ramus juts upward, terminating between the intermedius anterior sulcus and the caudal end of the intraparietal sulcus. The anterior/superior parallel ramus curves around the end of the Sylvian fissure and ends in a fork, the branches of which terminate near the post-central sulcus and the Sylvian fissure respectively. The transverse post-central sulcus branches from the post-central sulcus just dorsal to the end of the sub-central posterior sulcus and connects to the central sulcus; the main body of the post-central sulcus terminates between the anterior/superior parallel ramus and the sub-central posterior sulcus. The intraparietal sulcus connects to the post-central sulcus, the intermedius anterior sulcus and a bifurcated processus acuminis sulcus. The paroccipital sulcus is separate from the intraparietal sulcus, forming a somewhat vertical line with three rami projecting toward the middle/angularis and the descending/anterior occipital rami.

In EH2's left hemisphere (Fig. 2), the superior temporal gyrus is highly convoluted as both the Sylvian fissure and the superior temporal sulcus are fragmented. Moreover, the end of the Sylvian fissure is forked: the lower branch of this fork incises the caudal superior temporal gyrus, while the upper branch resides just caudal to a small depression. The anterior/superior parallel, the middle/angularis and the descending/anterior occipital rami are all complex: branches of the anterior/superior parallel ramus terminate near the intermedius anterior sulcus and the intraparietal sulcus—paroccipital sulcus junction, while those of the middle/angularis ramus and the descending/anterior occipital ramus terminate near the paroccipital sulcus and the middle temporal sulcus, respectively. The sub-central posterior sulcus extends up from the Sylvian fissure into a space created by the forked end of the central sulcus, while the transverse post-central sulcus splits off from the post-central sulcus above the caudal branch of this fork. The post-central sulcus is divided into superior and inferior sections, both of which connect to the intraparietal sulcus. The intraparietal sulcus curves around the upper branch of the forked end of the Sylvian fissure, terminating just caudal to the intermedius anterior sulcus in the vicinity of the anterior/superior parallel ramus. Several rami project upward from the intraparietal sulcus, making designation of the processus acuminis sulcus difficult. The paroccipital sulcus courses dorsal-ventrally, connecting to both the intraparietal sulcus and the superior section of the post-central sulcus. In the right hemisphere (Fig. 2), the superior temporal sulcus is more continuous, but shallow depressions litter the length

of the superior temporal gyrus. The Sylvian fissure, which splits into two fragments, is shorter than in the left hemisphere; near its end, the sub-central posterior sulcus branches upward to meet a triradiate transverse post-central sulcus. The central branch of the transverse post-central sulcus projects into a gap made by the forked end of the post-central sulcus. Regarding the caudal rami of the superior temporal sulcus, the descending/anterior occipital ramus is separate from the main body of the superior temporal sulcus while the anterior/superior parallel and the middle/angularis rami are not. The anterior/superior parallel ramus is sandwiched between the post-central sulcus and the intermedius anterior sulcus, the latter of which is very long and connects to the anterior/superior parallel ramus. Caudal to the intermedius anterior sulcus is a short sulcus that may be a ramus of the superior temporal sulcus or the intraparietal sulcus. The paroccipital sulcus extends down into the space between this sulcus and the middle/angularis ramus before curving around the ends of the middle/angularis ramus and the descending/anterior occipital ramus. The intraparietal sulcus is fragmented and connects to the post-central sulcus, the intermedius anterior sulcus, the processus acuminis sulcus and the paroccipital sulcus.

In EH3's left hemisphere (Fig. 2), several rami cut upward from the superior temporal sulcus through the superior temporal gyrus. A small depression sits in the superior temporal gyrus immediately below the ventral end of the post-central sulcus and a concordant downward curve of the Sylvian fissure. The lower branch of a triradiate anterior/superior parallel ramus, which is disconnected from the superior temporal sulcus, sticks into the caudal end of the superior temporal gyrus, perhaps marking the separation between the supramarginal and angular gyri. The middle/angularis ramus sticks upward into the inferior parietal lobule before curving around the upper branch of the anterior/superior parallel ramus. Since the descending/anterior occipital ramus connects to a fragmented middle temporal sulcus, discerning the point at which it terminates is difficult. The post-central sulcus terminates just caudal to the sub-central posterior sulcus, which protrudes upward from the Sylvian fissure, while the transverse post-central sulcus branches off from the post-central sulcus to sit just above the sub-central posterior sulcus. The post-central sulcus connects to the intraparietal sulcus, from which the intermedius anterior sulcus protrudes downward. The end of the intermedius anterior sulcus bifurcates into two branches, one of which is directed toward the anterior/superior parallel ramus and the other toward the post-central sulcus. The processus acuminis sulcus sticks up from the junction between the intraparietal sulcus and the paroccipital sulcus. The paroccipital sulcus is very short and the transverse occipital sulcus is visible. In the right hemisphere, no rami cut upward from the superior temporal sulcus into the superior temporal gyrus and only one cuts downward from the Sylvian fissure. Consequently, this gyrus is less convoluted than in the left hemisphere. Additionally, a small depression sits in this gyrus just rostral to the point where the superior temporal sulcus splits into its caudal rami. The anterior/superior parallel ramus reaches straight upward into the inferior parietal lobule before curving around the end of the Sylvian fissure. The middle/angularis ramus flows seamlessly from the end of the superior temporal sulcus, while the descending/anterior occipital ramus forms a fork, the base of which connects to the middle temporal sulcus. The sub-central posterior sulcus branches up

from the Sylvian fissure and connects to the post-central sulcus caudal to a short potential transverse post-central sulcus. The post-central sulcus has two points of connection with the intraparietal sulcus, which is rather short although connected to a long intermedius anterior sulcus jutting downward between the anterior/superior parallel and the descending/anterior occipital rami. The configuration of the processus acuminis sulcus, the paroccipital sulcus and the transverse occipital sulcus resembles that in the left hemisphere.

In EH4's left hemisphere (Fig. 2), the rostral end of the Sylvian fissure is relatively flat, yet its caudal section slopes upward away from the superior temporal sulcus. The caudal superior temporal gyrus is interrupted by an upward-projecting branch of the superior temporal sulcus and by the anterior/superior parallel ramus, which is disconnected from the superior temporal sulcus. The dorsal end of the anterior/superior parallel ramus terminates in the inferior parietal lobule between the intermedius anterior sulcus and the rostral section of the intraparietal sulcus. The middle/angularis ramus flows on from the main body of the superior temporal sulcus, ending in a fork near the intermedius anterior sulcus and the paroccipital sulcus. The descending/anterior occipital ramus is also disconnected from the superior temporal sulcus and terminates near the transverse occipital sulcus, although there is another possible candidate for the descending/anterior occipital ramus projecting downward from the superior temporal sulcus to join the middle temporal sulcus. A small rostrally-projecting branch near the ventral end of the post-central sulcus may represent the transverse post-central sulcus. The post-central sulcus and the Sylvian fissure are connected, possibly enabled by the sub-central posterior sulcus becoming continuous with the post-central sulcus. The superior portion of the post-central sulcus, which does not connect to the intraparietal sulcus, curves around the processus acuminis sulcus. The intraparietal sulcus, which forms a somewhat horizontal line, connects to the paroccipital sulcus, which forms a somewhat vertical line. In the right hemisphere (Fig. 2), the Sylvian fissure forms a more continuous slope and is shorter than in the left. An unclear feature projects down from its end. The superior temporal gyrus is convoluted by a few rami of the Sylvian fissure and the superior temporal sulcus. The anterior/superior parallel and the middle/angularis rami are connected yet separate from the main body of the superior temporal sulcus. The anterior/superior parallel ramus forks around the end of the Sylvian fissure, the middle/angularis ramus connects to the intraparietal sulcus while the descending/anterior occipital ramus, the end of which is also forked, is difficult to distinguish from the main body of the superior temporal sulcus. The post-central sulcus is fragmented; its superior section connects and sits dorsal to the intraparietal sulcus, while its inferior section connects to the Sylvian fissure but not the intraparietal sulcus. Again, a small branch of the inferior section of the post-central sulcus may represent the transverse post-central sulcus. The sub-central posterior sulcus may be present yet unascertainable due to merging with the inferior section of the post-central sulcus. Complexity of the intraparietal sulcus complicates the processus acuminis sulcus and the intermedius anterior sulcus designations. The paroccipital sulcus, which is also complex, is disconnected from the intraparietal sulcus and extends dorsal-ventrally.

3.3 Fossil hominin endocasts

The right hemisphere and a small section of the left occipital lobe of SK 1585 are preserved. According to Beaudet et al. (2019), only three of the sulci of interest are observable, namely, the Sylvian fissure, the superior temporal sulcus and the inferior post-central sulcus (Fig. 3). The rostral-ventral end of the superior temporal sulcus seems to overlap with the middle temporal sulcus, while the caudal-dorsal section of the superior temporal sulcus connects to the pre-lunate sulcus, which in turn connects to the lunate sulcus. It is unclear whether the horizontal sulcus between the rostral-ventral and caudal-dorsal sections of the superior temporal sulcus is part of the Sylvian fissure or the superior temporal sulcus; if it is part of the Sylvian fissure, this would suggest a narrow neck between the superior temporal sulcus and the caudal end of the Sylvian fissure. The inferior post-central sulcus courses vertically down toward the Sylvian fissure. The identification of two sulcal markings in close proximity to one another and at rather different angles as the inferior post-central

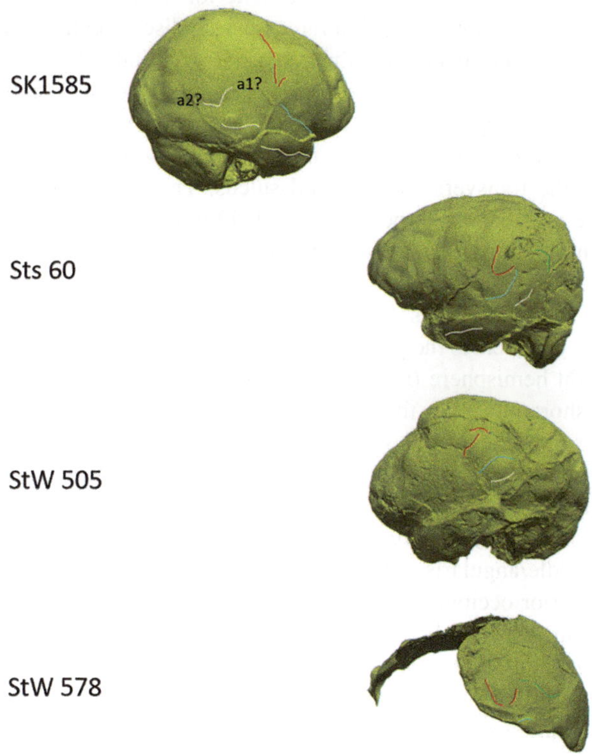

FIG. 3

Identification of the sulci surrounding the Wernicke's area in fossil hominin endocasts. Blue: Sylvian fissure; green: intraparietal sulcus; red: post-central sulcus; white: superior temporal sulcus. For abbreviations' meaning see Section 2.

sulcus suggests complexity of this sulcus. Our revision thus agrees with Beaudet et al.'s (2019) sulcal designations for SK 1585. Manipulation of the 3D model supports an extension of their designation of the inferior post-central sulcus. The caudal forked section of the superior temporal sulcus may represent the anterior/superior parallel and the middle/angularis rami. As the slope of the Sylvian fissure is rather steep, the horizontal sulcus between the rostral-ventral and caudal-dorsal sections of the superior temporal sulcus is likely part of the superior temporal sulcus rather than the Sylvian fissure. We extended Beaudet et al.'s (2019) designation of this horizontal sulcus, which seems to course around the top of a rostral bulge then beneath a more caudal bulge; however, this second bulge may represent a granular fovea caused by arachnoid protrusion.

Most of the left hemisphere, the right frontal lobe and a small portion of the right parietal lobe are preserved for Sts 60 (Fig. 3). Sulcal organization is relatively clear on the frontal lobes, but less so on the left temporal and parietal lobes. Beaudet et al. (2019) labeled two of the sulci of interest: the Sylvian fissure and the superior temporal sulcus. They only labeled a rostral section of the superior temporal sulcus, which does not overlap with the middle temporal sulcus, and attributed two separate fissures which draw close at their caudal ends to the caudal section of the Sylvian fissure. We agree with Beaudet et al.'s (2019) designation of the rostral section of the superior temporal sulcus, but also identify a caudal fragment. The lower of the two sulci designated the Sylvian fissure likely does represent the Sylvian fissure, but the upper one may instead represent the curved ventral end of the post-central sulcus drawing close to the Sylvian fissure. Moreover, an unlabeled sulcus indicated by Beaudet et al. (2019) caudal-dorsal to the Sylvian fissure likely represents the intraparietal sulcus, or else a caudal ramus of the superior temporal sulcus.

For StW 505 (Fig. 3), the left hemisphere and a small portion of the right frontal lobe are preserved. Beaudet et al. (2019) labeled two sulci near Wernicke's area: the Sylvian fissure and the inferior post-central sulcus. The inferior post-central sulcus courses just caudal to the central sulcus. A small horizontal sulcus sitting atop the dorsal end of the inferior post-central sulcus may indicate complexity, or perhaps overlap with the intraparietal sulcus. A caudal section of the Sylvian fissure is visible dorsal to a protrusion that may indicate superior temporal gyral bulging. Our sulcal identifications deviate slightly from those made by Beaudet et al. (2019). The small horizontal fissure above the inferior post-central sulcus is elongated and also attributed to the post-central sulcus, while a caudal fragment of another sulcus of interest, the superior temporal sulcus, is identified ventral to the protrusion above which the Sylvian fissure curves.

Preservation of StW 578 is poor relative to the other endocasts (Fig. 3); only a sliver of frontal area is present, as well as the left and right parietal lobes. Beaudet et al. (2019) identified two sulci, both of which occupy the lower left parietal and neither of which are explicitly labeled. We suggest that the more rostral of these represents a curved inferior section of the post-central sulcus, similar to that noted in Sts 60, although it is unusual for the end of the post-central sulcus to course caudally. The other sulcus identified by Beaudet et al. (2019) may represent the intraparietal

sulcus, given that it sits ventral to a bulge which may represent the superior parietal lobule. Finally, the caudal end of the Sylvian fissure may be visible curving above a bulge at the lower edge of the endocast.

4 Discussion

Having considered each brain/endocast individually, we will now discuss the sample more holistically, drawing comparisons between patterns observed here and those observed by Connolly (1950), as well as framing findings in the broader picture of language evolution.

4.1 Variation within the extant sample

Sulcal designations on chimpanzee brains were fairly straightforward given how minimal was the sulcal fragmentation and overlap. In relation to Wernicke's area, a shorter Sylvian fissure in the left hemisphere and/or bulging of the left caudal superior temporal gyrus in some specimens could indicate an asymmetrically expanded Wernicke's area in chimpanzees, as documented in previous studies (Gannon et al., 1998; Hopkins et al., 1998; Spocter et al. 2010). Such L > R asymmetry would reinforce arguments that neural precursors of human left hemispheric language specialization were present prior to the human-*Pan* divergence. According to Spocter et al. (2010), a L > R asymmetric Wernicke's area homolog occurs in many non-human primate species, facilitating species-specific vocalization discrimination. Indeed, rhesus macaques exhibit a left hemisphere specialization for interpreting conspecific communicative signals (Ghazanfar et al., 2001). Yet, indications of L > R Wernicke's area asymmetry were not observed in all five chimpanzee specimens; a larger sample size as well as quantitative measurements of Sylvian fissure length and superior temporal gyral volume/surface area are needed to adequately assess whether such asymmetry is indeed present.

Within extant humans, the length of the Sylvian fissure, the number of branches extending from it and its interconnectedness with other sulci of interest varied markedly. There was no consistent pattern of the Sylvian fissure being longer in either hemisphere. The Sylvian fissure was fragmented in 1/4 cases. The sub-central posterior sulcus branched upward from the Sylvian fissure in 3/4 cases; in the one individual lacking sub-central posterior sulcus designations, the Sylvian fissure connected to the post-central sulcus, so the sub-central posterior sulcus may have become contiguous with the ventral end of the post-central sulcus. In some hemispheres, the Sylvian fissure coursed up onto the lateral brain surface at a steep angle, while in others this angle was shallower. Similarly to the Sylvian fissure, the superior temporal sulcus was very inter-individually and inter-hemispherically variable. Caudal rami were often separated from the main body of the superior temporal sulcus, which itself was often fragmented and/or interrupted by/connected to the middle temporal sulcus. This pattern, which contrasted with that in the chimpanzee sample,

made it difficult to distinguish possible bulging of the caudal superior temporal gyrus that might indicate Wernicke's area expansion; indeed, such bulging was not confidently observed in any human brain studied. However, if this pattern of superior temporal sulcus-middle temporal sulcus overlap and superior temporal sulcus fragmentation is due to expansion of the parietal lobe in humans, as suggested by Connolly (1950), a higher level of superior temporal sulcus-middle temporal sulcus overlap could itself indicate expansion of Wernicke's area, which impinges upon the lower parietal lobe. Connolly (1950) found fragmentation of the main body of the superior temporal sulcus to be more common in left than right hemispheres, as expected under L > R Wernicke's area asymmetry if such fragmentation is a product of parietal, and thus Wernicke's area, expansion. The present study supports this finding, as the main body of the superior temporal sulcus was fragmented in 4/4 left hemispheres vs 0/4 right hemispheres. However, connections between the middle temporal sulcus and the main body of the superior temporal sulcus were observed in 3/4 left and 4/4 right hemispheres, indicating parietal expansion is not exclusive to human left hemispheres. Overlap between the middle temporal sulcus and the main body of the superior temporal sulcus in human but not chimpanzee hemispheres as well as superior temporal sulcus fragmentation exclusively in human left hemispheres may mean Wernicke's area is relatively larger in humans than in chimpanzees and is asymmetrically larger in human left hemispheres. It should be noted encephalization in the human lineage has resulted in absolutely larger brains with greater cortical complexity; discriminating whether changes in sulcal patterning are due to expansion of a specific functional area or else expansion of nearby functional areas or general encephalization is trying.

Not all of Connolly (1950) findings on human brain endocasts were corroborated. He found the anterior/superior parallel, the middle/angularis and the descending/anterior occipital rami to be more crowded in the left hemisphere than the right. He also determined that in the left hemisphere, the anterior/superior parallel is frequently separated from the superior temporal sulcus. Within this sample, the superior temporal sulcus caudal rami were not more crowded in left hemispheres, but the anterior/superior parallel was separate from other rami and the main body of the superior temporal sulcus in 2/4 left hemispheres vs 0/4 right hemispheres. Failure to replicate the former finding may be due to a small sample size, or because some more quantitative measure of "crowdedness" is needed.

Greater diversity was seen in the post-central sulcus and intraparietal sulcus configuration than in chimpanzee brains. The post-central sulcus was fragmented in 2/8 hemispheres. Furthermore, this sulcus variably exhibited rami as well as connectivity to surrounding sulci of interest, including the Sylvian fissure, the sub-central posterior sulcus, the transverse post-central sulcus and the intraparietal sulcus. Connolly (1950) found the intraparietal sulcus and the paroccipital sulcus could be separate or connected: both the connected (5/8 hemispheres) and unconnected (3/8 hemispheres) states were observed here. The processus acuminis sulcus and the intermedius anterior sulcus designations were made in all hemispheres and the intermedius anterior sulcus length varied. With one exception, Connolly (1950)

claimed that, in human brains, the anterior/superior parallel and the intermedius anterior sulcus closely approximate, form contact with or form a continuous furrow with one another held true. However, his claim that either the anterior/superior parallel ramus or the intermedius anterior sulcus usually marks the boundary between the supramarginal and angular gyri was harder to substantiate due to variable manifestations of these sulci. Complexity of the intraparietal sulcus was variable.

Variability in temporoparietal sulcal organization prevented discernment of consistent sulcal patterns relative to Wernicke's area. Unlike in the case of some chimpanzees, L > R asymmetric Sylvian fissure length and/or caudal superior temporal gyral bulging was not visible; however, superior temporal sulcus-middle temporal sulcus overlap in both hemispheres and superior temporal sulcus fragmentation exclusively in left hemispheres may be a product of a larger Wernicke's area in humans than chimpanzees and L > R Wernicke's area asymmetry in humans. Such asymmetry concords with that reported by Geschwind and Levitsky (1968), Toga and Thompson (2003) and Wada et al. (1975). Nevertheless, one cannot assume sulcal organization to accurately indicate relative size of underlying neural areas. Without an explicit understanding of how surface area, volumetric or cytoarchitectonic asymmetries impact sulcal organization, the connection made between the superior temporal sulcus fragmentation and asymmetry remains difficult to interpret.

4.2 Interpreting the hominin fossil record

This study included one endocast of *P. robustus* (SK 1585). This endocast exhibited superior temporal sulcus-middle temporal sulcus overlap. Our observations on the chimpanzee and human samples support Beaudet et al.'s (2019) statement that such overlap is more consistent with a human, rather than a chimpanzee pattern. Although the point at which this overlap occurred, the rostral temporal lobe, is rather distant from Wernicke's area, a superior temporal sulcus-middle temporal sulcus overlap in general may be tied to an expanded Wernicke's area in humans relative to chimpanzees. However, as only the right hemisphere of SK 1585 is fully preserved, the possibility of Wernicke's area asymmetry, which may be of functional importance to language ability, cannot be explored. The main body of the superior temporal sulcus was fragmented, contrasting with the pattern in the human sample where fragmentation only occurred in left hemispheres. However, the full extent of the superior temporal sulcus may not have been impressed upon the endocranium, creating the appearance of segmentation. In summary, SK 1585 exhibits superior temporal sulcus-middle temporal sulcus overlap indicative of a human-like parietal, and therefore perhaps Wernicke's area, expansion. Superior temporal sulcus fragmentation might be anticipated in the left hemisphere due to human-like L > R Wernicke's area asymmetry, although this cannot be confirmed.

This study included three *Australopithecus* endocasts, i.e., Sts 60, StW 505 and StW 578, all from Sterkfontein. All three endocasts yield signs of chimpanzee-like superior temporal gyral bulging. The superior temporal sulcus and the middle temporal sulcus are not visible in StW 578, but are visible in Sts 60 and StW 505, where they do not overlap. As only the left hemisphere is available for each of

the three *Australopithecus* endocasts, whether superior temporal gyral bulging is due to L > R Wernicke's area asymmetry cannot be properly appraised, nor can such asymmetry be detected by way of Sylvian fissure length.

Overall, these results cautiously support a more human-like Wernicke's area in paranthropines and a more chimpanzee-like Wernicke's area in australopithecines. Findings from these endocasts must be woven into existing fossil evidence regarding language evolution. All of the specimens studied here have chimpanzee-like Broca's areas bisected by the fronto-orbital sulcus, although some specimens from both genera also possess frontal bulging speculatively tied to burgeoning shape reorganization (Beaudet et al., 2019). Given the large uncertainty around endocast interpretation, it may help to consider endocasts together with non-neural indicators of language ability. Australopithecines and paranthropines possessed some, but not all aspects of human middle ear morphology (Quam et al., 2013). Even though chimpanzees are not a perfect representation of the ancestral form, hominin evolution is not linearly pursuant of a human "gold standard" and many hominin species may have had auditory features that were neither chimpanzee-nor human-like, it is likely that the auditory capabilities of these hominins fell along a chimpanzee-to-human spectrum. For example, based on inner ear morphology, Beaudet (2019) proposes a unique functional adaptation to hearing sensitivity within genus *Paranthropus*. Such an adaptation may be congruent with the more human-like Wernicke's area in SK 1585, given that Wernicke's area is linked with language comprehension. The *A. afarensis* specimen DIK-1/1 (3.3 mya) possesses a chimpanzee-like hyoid bulla, making it impossible for it to have had human-like speech (Alemseged et al., 2006). StW 578, Sts 60 and StW 505 predate DIK-1/1 (Granger et al., 2022), which raises intriguing questions about the evolutionary polarity of these features within *Australopithecus*. Finally, L > R Wernicke's area asymmetry in hominin endocasts, be it chimpanzee- or human-like, may be tied to handedness preferences during tool use, a possible pre-requisite for human language evolution (Hopkins et al., 2007). SK 1585 chronologically overlaps with the Oldowan lithic industry, and *P. robustus* remains have been discovered alongside bone tools (Backwe'l and d'Errico, 2008). Sts 60 and StW 505 are also contemporaneous with the Oldowan, and according to Skinner et al. (2015), *A. africanus* had a human-like metacarpal trabecular pattern compatible with tool use. That said, StW 578 predates the earliest known lithic industry, the Lomekwian (Harmand et al., 2015). In summary, neither australopithecines nor paranthropines can be credited with human-like speech abilities with any surety, however, *P. robustus* may have attained a complexity of vocal communication beyond that of both chimpanzees and australopithecines. Such would explain the more human-like Wernicke's area and idiosyncratic adaptation to hearing sensitivity in this species.

4.3 Limitations and conclusive remarks

Conclusions drawn from this study are limited to an extent. Existing reference diagrams typically give vague designations for Wernicke's area and its homologs on the lateral brain surface. More precise mapping of the functional borders of this area in

both humans and chimpanzees through fMRI studies would let more substantive inferences be made following examination of endocasts and brain models. This study found suggestions of a relatively larger Wernicke's area in humans than chimpanzees; however, this was based on qualitative rather than quantitative evidence. When contemplating the functional relevance of this finding, one cannot assume a larger Wernicke's area translates into more advanced language abilities. This study also found evidence for L > R Wernicke's area asymmetry in both humans and chimpanzees, although it should be stressed how unsteadily rooted this finding was in chimpanzees. If these asymmetries do exist in both species, it is yet to be seen whether they are more pronounced in one or the other and what implications this would have for language capability.

Aside from issues of preservation and inter-observer disparities in sulcal designation, it is relatively easy to brand an endocast as more chimp-like or more human-like when it comes to Broca's area. Although the existence of Broca's area asymmetries in humans and/or chimpanzees remains contentious, there are distinct differences in sulcal organization of this region between the two species. Previous Wernicke's area studies have not found consistent differences in sulcal organization between humans and chimpanzees, although have identified L > R asymmetries in both species. This study failed to find consistent patterns of sulcal layout in the immediate vicinity of Wernicke's area in either chimpanzee or human brains. However, indications were observed of L > R Wernicke's area asymmetry in some chimpanzee brains and an enlarged L > R asymmetric Wernicke's area in human brains.

Examination of fossil hominin endocasts revealed that while *Paranthropus* specimen SK 1585 possessed a more human-like Wernicke's area that was enlarged relative to that of chimpanzees, *Australopithecus* specimens Sts 60, StW 505 and StW 578 elicited signs of chimp-like L > R Wernicke's area asymmetry. As only one hemisphere was available per endocast, asymmetries could not be fully investigated.

This study was limited by a small sample size, its purely qualitative as opposed to quantitative nature and the risk of human error during sulcal labeling. Future research should target three main areas: identifying more meticulously the functional borders of Wernicke's area on lateral surfaces of human and non-human primate brains; improving protocols for automatic sulcal labeling; and understanding the relevance of Wernicke's area size and asymmetry to language ability.

The lateral cerebral borders of both Broca's and Wernicke's areas were first inferred in the 1800s from the location of aphasia-inducing pathological lesions (rev. in Lieberman, 2007). However, Lieberman (2007) proposes damage to the basal ganglia to be the true cause of Broca's aphasia, claiming neither Broca's nor Wernicke's aphasias can be credited to cortical lesions alone. Perhaps viewing the neural substrates of language as discrete modules which can be accurately delineated on the lateral brain surface is an outdated mode of thinking. That is not to say endocast research is void; merely that, until firmer functional borders of Wernicke's area have been identified and better protocols for sulcal designation developed, future researchers should observe caution if adopting this methodology.

Acknowledgments

We are grateful to Professor Chet Sherwood and the National Chimpanzee Brain Resource for providing the chimpanzee brain models, and to the Allen Institute Brain Atlas for providing the human brain MRIs. We would like to thank Tanya Calvey and Alexandra de Sousa for their comments that greatly contributed to improve the quality of this manuscript.

References

Acharya, A.B., Wroten, M., 2021. Broca Aphasia. StatPearls.

Alemseged, Z., Spoor, F., Kimbel, W.H., Bobe, R., Geraads, D., Reed, D., Wynn, J.G., 2006. A juvenile early hominin skeleton from Dikika, Ethiopia. Nature 443 (7109), 296–301.

Amunts, K., Schleicher, A., Burgel, U., Mohlberg, H., Uylings, H.B.M., Zilles, K., 1999. Broca's region revisited: Cytoarchitecture and intersubject variability. J. Comp. Neurol. 412, 319–341.

Backwe'l, L., d'Errico, F., 2008. Early hominid bone tools from Drimolen, South Africa. J. Archaeol. Sci. 35 (11), 2880–2894.

Balzeau, A., Gilissen, E., Holloway, R.L., Prima, S., Grimaud-Hervé, D., 2014. Variations in size, shape and asymmetries of the third frontal convolution in hominids: paleoneurological implications for hominin evolution and the origin of language. J. Hum. Evol. 76, 116–128.

Beaudet, A., 2017. The emergence of language in the hominin lineage: perspectives from fossil endocasts. Front. Hum. Neurosci. 11, 427.

Beaudet, A., 2019. The inner ear of the Paranthropus specimen DNH 22 from Drimolen, South Africa. Am. J. Phys. Anthropol. 170 (3), 439–446.

Beaudet, A., Dumoncel, J., de Beer, F., Duployer, B., Durrleman, S., Gilissen, E., et al., 2016. Morphoarchitectural variation in South African fossil cercopithecoid endocasts. J. Hum. Evol. 101, 65–78.

Beaudet, A., Clarke, R.J., De Jager, E.J., Bruxelles, L., Carlson, K.J., Crompton, R., de Beer, F., Dhaene, J., Heaton, J.L., Jakata, K., Jashashvili, T., 2019. The endocast of StW 573 ("little foot") and hominin brain evolution. J. Hum. Evol. 126, 112–123.

Beaudet, A., Holloway, R., Benazzi, S., 2021. A comparative study of the endocasts of OH 5 and SK 1585: implications for the paleoneurology of eastern and southern African Paranthropus. J. Hum. Evol. 156, 103010.

Bernal, B., Ardila, A., 2009. The role of the arcuate fasciculus in conduction aphasia. Brain 132 (9), 2309–2316.

Borne, L., Rivière, D., Mancip, M., Mangin, J.F., 2020. Automatic labelling of cortical sulci using patch-or CNN-based segmentation techniques combined with bottom-up geometric constraints. Med. Image Anal. 62, 101651.

Branche, C., Milner, B., Rasmussen, T., 1964. Intracarotid sodium amytal for the lateralization of cerebral speech dominance. J. Neurosurg. 21, 399–405.

Bryant, K.L., Li, L., Eichert, N., Mars, R.B., 2020. A comprehensive atlas of white matter tracts in the chimpanzee. PLoS Biol. 18, e3000971.

Cantalupo, C., Hopkins, W.D., 2001. Asymmetri' Broca's area in great apes. Nature 414 (6863), 505.

Chang, E.F., Breshears, J.D., Raygor, K.P., Lau, D., Molinaro, A.M., Berger, M.S., 2017. Stereotactic probability and variability of speech arrest and anomia sites during stimulation mapping of the language dominant hemisphere. J. Neurosurg. 126, 114–121.

Connolly, C.J., 1950. External Morphology of the Primate Brain. Charles C Thomas.

Damiani, D., Nascimento, A.M., Pereira, L.K., 2020. Cortical brain functions—the Brodmann legacy in the 21st century. Braz. Neurosurg. 39 (4), 261–270.

de Jager, E.J., van Schoor, A.N., Hoffman, J.W., Oettle, A.C., Fonta, C., Mescam, M., Risser, L., Beaudet, A., 2019. Sulcal pattern variation in extant human endocasts. J. Anat. 235 (4), 803–810.

de Sousa, A., Cunha, E., 2012. Hominins and the emergence of the modern human brain. Prog. Brain Res. 195, 293–322.

Dronkers, N.F., Plaisant, O., Iba-Zizen, M.T., Cabanis, E.A., 2007. Paul Broca's historic cases: high resolution MR imaging of the brains of Leborgne and Lelong. Brain 130, 1432–1441.

Dumoncel, J., Subsol, G., Durrleman, S., Bertrand, A., de Jager, E., Oettlé, A.C., Lockhat, Z., Suleman, F.E., Beaudet, A., 2021. Are endocasts reliable proxies for brains? A 3D quantitative comparison of the extant human brain and endocast. J. Anat. 238 (2), 480–488.

Ellis, A.W., Miller, D., Sin, G., 1983. W'rnicke's aphasia and normal language processing: a case study in cognitive neuropsychology. Cognition 15 (1–3), 111–144.

Falk, D., 1980. A reanalysis of the south African australopithecine natural endocasts. Am. J. Phys. Anthropol. 53 (4), 525–539.

Falk, D., 2007. The evolution of Broca's area. IBRO Hist. Neurosci., 2513–2518.

Fedorov, A., Beichel, R., Kalpathy-Cramer, J., Finet, J., Fillion-Robin, J.-C., Pujol, S., Bauer, C., Jennings, D., Fennessy, F.M., Sonka, M., Buatti, J., Aylward, S.R., Miller, J.V., Pieper, S., Kikinis, R., 2012. 3D slicer as an image computing platform for the quantitative imaging network. Magn. Reson. Imaging 30 (9), 1323–1341.

Galaburda, A., Sanides, F., 1980. Cytoarchitectonic organisation of the human auditory cortex. J. Comp. Neurol. 190, 597–610.

Gannon, P.J., Holloway, R.L., Broadfield, D.C., Braun, A.R., 1998. Asymmetry of chimpanzee planum temporale: humanlike pattern of Wernicke's brain language area homolog. Science 279 (5348), 220–222.

Geschwind, N., Levitsky, W., 1968. Human brain: left-right asymmetries in temporal speech region. Science 161 (3837), 186–187.

Ghazanfar, A.A., Smith-Rohrberg, D., Hauser, M.D., 2001. *The role of temporal cues in Rhesus monkey vocal recognition: orienting asymmetries to reversed calls*, brain. Behav. Evol. 58 (3), 163–172.

Glasser, M.F., Coalson, T.S., Robinson, E.C., Hacker, C.D., Harwell, J., Yacoub, E., Ugurbil, K., Andersson, J., Beckmann, C.F., Jenkinson, M., Smith, S.M., 2016. A multi-modal parcellation of human cerebral cortex. Nature 536 (7615), 171–178.

Granger, D.E., Stratford, D., Bruxelles, L., Gibbon, R.J., Clarke, R.J., Kuman, K., 2022. Cosmogenic nuclide dating of Australopithecus at Sterkfontein, South Africa. Proc. Royal Soc. B. 119 (27), e2123516119.

Harmand, S., Lewis, J.E., Feibel, C.S., Lepre, C.J., Prat, S., Lenoble, A., Boës, X., Quinn, R.L., Brenet, M., Arroyo, A., Taylor, N., 2015. 3.3-million-year-old *stone tools from Lomekwi 3, West Turkana, Kenya*. Nature 521 (7552), 310–315.

Holloway, R.L., 1981. Revisiting the South African Taung australopithecine endocast: the position of the lunate sulcus as determined by the stereoplotting technique. Am. J. Phys. Anthropol. 56 (1), 43–58.

Holloway, R.L., 2018. On the making of endocasts: the new and the old in paleoneurology. In: Bruner, E., Ogihara, N., Tanabe, H.C. (Eds.), Digital Endocasts. Springer, Tokyo.

Hopkins, W.D., Marino, L., Rilling, J.K., MacGregor, L.A., 1998. Planum temporale asymmetries in great apes as revealed by magnetic resonance imaging (MRI). Neuroreport 9 (12), 2913–2918.

Hopkins, W.D., Russell, J.L., Cantalupo, C., 2007. Neuroanatomical correlates of handedness for tool use in chimpanzees (Pan troglodytes): implication for theories on the evolution of language. Psychol. Sci. 18 (11), 971–977.

Kobayashi, Y., Matsui, T., Haizuka, Y., Ogihara, N., Hirai, N., Matsumura, G., 2014. Cerebral sulci and gyri observed on macaque Endocasts. In: Akazawa, T., Nishiaki, Y., Aoki, K. (Eds.), Dynamics of Learning in Neanderthals and Modern Humans Volume 2. Springer, Tokyo.

Kononova, E.P., 1935. The Variability of Structure of the Cortex of the Brain: Inferior Frontal Gyrus of Adult Man. 1 Brain Research Institute, pp. 49–118.

Langergraber, K.E., Prüfer, K., Rowney, C., Boesch, C., Crockford, C., Fawcett, K., Inoue, E., Inoue-Muruyama, M., Mitani, J.C., Muller, M.N., Robbins, M.M., 2012. Generation times in wild chimpanzees and gorillas suggest earlier divergence times in great ape and human evolution. PNAS 109 (39), 15716–15721.

Lieberman, P., 2007. The evolution of human speech: its anatomical and neural bases. Curr. Anthropol. 48 (1), 39–66.

Ponce de León, M.S., Bienvenu, T., Marom, A., Engel, S., Tafforeau, P., Warren, J.L.A., Lordkipanidze, D., Kurniawan, I., Murti, D.B., Suriyanto, R.A., Koesbardiati, T., 2021. The primitive brain of early Homo. Science 372 (6538), 165–171.

Quam, R.M., de Ruiter, D.J., Masali, M., Arsuaga, J.L., Martinez, I., Moggi-Cecchi, J., 2013. Early hominin auditory ossicles from South Africa. PNAS 110, 8847–8851.

Rilling, J., Glasser, M.F., Jbabdi, S., Andersson, J., Preuss, T.M., 2012. Continuity, divergence and the evolution of brain language pathways. Front. Evol. Neurosci. 3, 11.

Schenker, N.M., Hopkins, W.D., Spocter, M.A., Garrison, A.R., Stimpson, C.D., Erwin, J.M., Hof, P.R., Sherwood, C.C., 2010. Broca's area homologue in chimpanzees (Pan troglodytes): probabilistic mapping, asymmetry, and comparison to humans. Cereb. Cortex 20 (3), 730–742.

Sherwood, C.C., Broadfield, D.C., Holloway, R.L., Gannon, P.J., Hof, P.R., 2003. Variability of Broca's area homologue in African great apes: implications for language evolution, the anatomical record part a: discoveries in molecular. Cell. Evol. Biol. 271 (2), 276–285.

Skinner, M.M., Stephens, N.B., Tsegai, Z.J., Foote, A.C., Nguyen, N.H., Gross, T., Pahr, D.H., Hublin, J.J., Kivell, T.L., 2015. Human-like hand use in Australopithecus africanus. Science 347 (6220), 395–399.

Snowdon, C., 2002. From primate communication to human language. In: de Waal, F.B.M. (Ed.), Tree of Origin: What Primate Behaviour Can Tell us about Human Social Evolution. Harvard University Press, Cambridge.

Spocter, M.A., Hopkins, W.D., Garrison, A.R., Bauernfeind, A.L., Stimpson, C.D., Hof, P.R., Sherwood, C.C., 2010. Wernicke's area homologue in chimpanzees (Pan troglodytes) and its relation to the appearance of modern human language. Proc. Royal Soc. B. 277 (1691), 2165–2174.

Sprung-Much, T., Petrides, M., 2018. Morphological patterns and spatial probability maps of two defining sulci of the posterior ventrolateral frontal cortex of the human brain: the sulcus diagonalis and the anterior ascending ramus of the lateral fissure. Brain Struct. Funct. 223 (9), 4125–4152.

Sprung-Much, T., Eichert, N., Nolan, E.M., Petrides, M., 2021. Broca's area and the search for anatomical asymmetry: commentary and perspectives. Brain Struct. Funct., 1–9.

Taglialatela, J.P., Russell, J.L., Schaeffer, J.A., et al., 2008. Communicative signalling activates 'Broca's' homolog in chimpanzees. Curr. Biol. 18, 343–348.

Tobias, P.V., 1987. The brain of Homo habilis: a new level of organization in cerebral evolution. J. Hum. Evol. 16, 741–761.

Toga, A.W., Thompson, P.M., 2003. Mapping brain asymmetry. Nat. Rev. Neurosci. 4 (1), 37–48.

Uylings, H.B.M., Malofeeva, L.I., Bogolepova, I.N., Amunts, K., Zilles, K., 2000. Broca's language area from a neuroanatomical and developmental perspective. In: Brown, C.M., Hagoort, P. (Eds.), The Neurocognition of Language. Oxford University Press, Oxford.

Wada, J.A., Clarke, R., Hamm, A., 1975. Cerebral hemispheric asymmetry in humans: cortical speech zones in 100 adult and 100 infant brains. Arch. Neurol. 32 (4), 239–246.

Yoshizawa, A., Belyaev, A., Yokota, H., Seidel, H.P., 2007. Fast and Faithful Geometric Algorithm for Detecting Crest Lines on Meshes. In: Proceedings of the 15th Pacific Conference on Computer Graphics and Applications, pp. 231–237.

Yuan, D., Luo, D., Kwok, V.P., Zhou, Y., Tian, H., Yu, Q., An, J., Gao, J.H., Qiu, S., Tan, L.H., 2021. Myeloarchitectonic asymmetries of language regions in the human brain. Cereb. Cortex 31, 4169–4179.

Zollikofer, C.P.E., Ponce de León, M.S., 2013. Pandora's growing box: inferring the evolution and development of hominin brains from endocasts. Evol. Anthropol. 22 (1), 20–33.

CHAPTER

6

Lateralized behaviors in living humans: Application in the context of hominin brain evolution

Ameline Bardo[a,b,*], Andréa Filippo[a], and Antoine Balzeau[a,c]

[a]*UMR 7194, CNRS-MNHN, Département Homme et Environnement, Musée de l'Homme, Paris, France*

[b]*Skeletal Biology Research Centre, School of Anthropology and Conservation, University of Kent, Canterbury, United Kingdom*

[c]*Department of African Zoology, Royal Museum for Central Africa, Tervuren, Belgium & Laboratory of Histology and Neuropathology, Université Libre de Bruxelles, Brussels, Belgium*

**Corresponding author: e-mail address: ameline.bardo@mnhn.fr*

Abstract

The left and right hemispheres of our brains differ subtly in structure, and each is dominant in processing specific cognitive tasks. Our species has a unique system of distributing behavior and cognition between each cerebral hemisphere, with a preponderance of pronounced side biases and lateralized functions. This hemisphere-dependent relationship between cognitive, sensory or motor function and a set of brain structures is called hemispheric specialization. Hemispheric specialization has led to the emergence of model systems to link anatomical asymmetries to brain function and behavior. Scientific research on hemispheric specialization and lateralized functions in living humans focuses on three major domains: (1) hand preferences, (2) language, and (3) visuospatial skills and attention. In this chapter we present an overview of this research with a specific focus on living humans and the applications of this research in the context of hominin brain evolution. Our objective is to put into perspective what we know about brain-behavior relationships in living humans and how we can apply the same methods to investigate this relationship in fossil hominin species, and thus improve our understanding of the emergence and development of complex cognitive abilities.

Keywords

Hominins, Paleoneurology, Neuroimaging, Brain asymmetry, Lateralized behaviors, Behavioral experiments, Brain endocast, Interdisciplinarity

1 Introduction

Laterality refers to the preference humans and many other vertebrates show for one side of the body over the other. This may relate to the preferential use of one arm or leg relative to the one on the contralateral side, but it also concerns biological and functional aspects of the brain. Pierre Paul Broca, a 19th century French physician and anthropologist, was one of the founders of research on brain asymmetry. He studied the relationship between language disorders and brain damage, and showed also asymmetry in the brains of great apes (Broca, 1865). The early discoveries of Paul Broca and Carl Wernicke (1910), a German neuropsychiatrist, highlighted that the left hemisphere was specialized for language and verbal functions in humans (i.e., Broca's area and Wernicke's area of the brain). Since then, new approaches and methodological improvements have enhanced our knowledge of the biology and the functioning of our brain, showing the complexity of this structure. Major recent advances came from functional imaging while available databases of extant human brains combine morphological and functional imaging for thousands of individuals (Balzeau and Mangin, 2021).

The existence of brain and behavioral asymmetries among many other vertebrates had been acknowledged since the late 1970s or early 1980s (Rogers et al., 2013; but see as well Güntürkün et al., 2020 for a detailed history of comparative research on lateralization in animals), highlighting that lateralization is not a trait unique to humans (Ocklenburg et al., 2013). There is much evidence about functional brain asymmetries in many different species of animals, mostly related to vocalization systems and limb preferences (Ocklenburg et al., 2013; Ströckens et al., 2013). For example, recent neuroanatomical research has found that great apes (Cantalupo et al., 2003; Spocter et al., 2010), macaques (Shepherd and Freiwald, 2018) and baboons (Becker et al., 2022; Marie et al., 2018) have brain regions in the left hemisphere homologous to the speech and language areas in the human brain. Moreover, non-human animal forelimb asymmetries were found for different actions such as for manipulation of inanimate objects, living targets and communicative gestures (see recent review by Boulinguez-Ambroise et al., 2022).

Lateralization is important for the survival of an organism. In certain vertebrates, the right hemisphere responds to emergency situations whereas the left hemisphere is involved in processing routine and/or familiar stimuli (MacNeilage, 2008). This can vary between vertebrates, for example, it was found that fish keep track of a predator using their right eye (i.e., left hemisphere dominance for avoiding a predator) while most mammalian, reptilian and amphibian species use their left eye and so show a right hemisphere bias (Leliveld et al., 2013; and see Rogers et al., 2013). This last has been suggested to be associated with an increase in processing capacity (Rogers et al., 2013). Indeed, it has been hypothesized that hemispheric specialization increased in hominins (i.e., *H. sapiens* and its extinct relatives of the genera *Homo*, *Australopithecus*, *Paranthropus*, *Kenyanthropus*, *Sahelanthropus*, *Orrorin* and *Ardipithecus*, which split off from the *Pan sp.* ~6–8 million years ago; Patterson et al., 2006; Wood and Harrison, 2011) along with the emergence of handedness

(i.e., hand preference), tool using and making, and language (Rogers et al., 2013). However, as we still lack a general consensus on the relationship between handedness, anatomical asymmetries and functional lateralization in living humans, it is difficult to characterize when or how *H. sapiens* shifted away from ancestral patterns of laterality (Uomini and Ruck, 2018). Moreover, the available information on lateralization for fossil hominin species is scarce and hardly comparable between specimens and with detailed information obtained directly on living humans.

It is essential today to search the fossil record for possible traces of variations in cerebral asymmetries in fossil hominins in order to re-examine previous conclusions that may have been based on speculative reasoning or inappropriate methodologies. It is crucial to perform qualitative assessments of bilateral variation of fossil endocasts and to compare with the existing contemporary archeological material. Methodologies should be developed to investigate the fossil record that does not exclusively rely on proxies and speculative reasoning. Even if some specific features are visible in our species and linked functionally with behaviors, we cannot assume that the relationship was similar in fossil species. In this chapter we describe a research framework devoted to hemispheric asymmetry and laterality with a specific focus on living humans and the applications of this research to the reconstruction of hominin brain evolutionary history. Despite much research dedicated to the study of hemispheric asymmetry and laterality in recent decades (see Hugdahl and Westerhausen, 2010; Rogers et al., 2013), our theoretical understanding of the relationships between brain asymmetries and lateralized behaviors remains limited. Our objective is to use what we know about brain-behavior relationships in living humans and apply the same concepts and methods to understand these relationships in fossil hominins.

This chapter has three parts:

1. A review of functional asymmetry of the human brain and the relationship with lateralized behavior.
2. A review of the methods used to infer functional asymmetry of fossil hominins.
3. A presentation of our current project PaleoBRAIN in which we are studying living human brain-behavioral asymmetries. By investigating possible relationships between behaviors and brain biology in living people, we aim to provide a context for understanding hominin brain evolution as represented by fossil remains.

2 Functional asymmetry of the human brain and lateralized behaviors

The relationships between functional asymmetry of the human brain and behavioral asymmetries can be investigated through indirect and direct observations. Behavioral experiments enable researchers to demonstrate indirect relationships between well-documented brain asymmetries and lateralized behaviors, by observing which part of

the body responds to specific stimuli (Rogers et al., 2013). Through noninvasive neuroimaging approaches, such as PET (positron emission tomography), MRI (magnetic resonance imaging), and DTI (diffusion tensor imaging), it is possible to map anatomical asymmetries onto brain and neuronal structures and functions in living humans (Herve et al., 2013). This section of the chapter discusses current research findings on brain lateralization, from behavior to anatomy, and the structure-function relationship in living humans. The main body of research on lateralized functions in living humans is focused on language, hand preferences and visuospatial skills and attention. An important question to tackle is to which extent structural differences between left and right hemispheres underlie the lateralization of function, and whether this structural lateralization reflects the degree of functional lateralization from one person to another.

2.1 Language

Two key anatomical areas in the left hemisphere of the human brain are involved in speech production and processing; the Broca's and Wernicke's areas (Broca, 1865; Wernicke, 1910; and see Benjamin et al., 2017; Hagoort, 2017 for a recent review of those original works). Broca's area is located on the left inferior frontal gyrus and Wernicke's area is in the posterior area of the left superior temporal gyrus (Benjamin et al., 2017; Hagoort, 2017). Moreover, the planum temporale, located in the temporal lobe and related functionally to auditory areas close to Wernicke's area, is volumetrically larger in the left than the right hemisphere (Goldberg et al., 2013). It has long been stated that humans, in the sense recent *H. sapiens*, would have a specific pattern for the general organization of the brain and specifically for those particular functional areas. However, there are asymmetries of the planum temporale and Broca's area in the brains of great apes (Cantalupo et al., 2003; Gannon et al., 1998; Spocter et al., 2010), macaques (Shepherd and Freiwald, 2018) and baboons (Becker et al., 2022; Marie et al., 2018).

By comparing bilateral variation in the endocast of hominids (i.e., great apes, including *H. sapiens*) it is possible to reconstruct the evolution of asymmetrical structures. An endocast is a cast of the internal aspect of the cranial vault that reflects the imprints of the brain left on the internal surface of the bone. The average spatial asymmetry pattern on endocasts is shared among current hominids for the antero-posterior expansion of the frontal and occipital lobes, namely the petalial pattern (Balzeau et al., 2012; Neubauer et al., 2020), or for the third frontal convolution (Balzeau et al., 2014). However, living *H. sapiens* show a pattern of asymmetry more variable than in the great apes (Balzeau et al., 2014). This was suggested to reflect increased functional and developmental modularization of the human brain (Neubauer et al., 2020). Nevertheless, the endocast is only a partial reflection of the morphology of the brain and the homology of cortical areas between species that is based on their supposed position relative to brain sulci. One limitation is that any region defined according to surface features could be made up of multiple different Brodmann areas (Brodmann, 1909) and so it is difficult to determine the

correspondence to histologically and functionally defined cortical regions (de Sousa et al., 2010). Recent research has shown that the asymmetric pattern of the Broca's and Wernicke's areas is not specific to *H. sapiens,* or even to any *Homo* species (see Section 3.1). Anatomically homologous areas have been identified in extant apes (Palomero-Gallagher and Zilles, 2019). Moreover, these areas are difficult to define and to compare between extant hominids (Balzeau et al., 2014; Keller et al., 2009).

The most common behavioral test used to assess hemispheric dominance for language processing, without using neuroimaging techniques, is the dichotic-listening paradigm (Westerhausen and Samuelsen, 2020). For this task, pairs of verbal stimuli (words or consonant-vowel syllables like /DA/ and /GA/) are presented simultaneously to the participant via headphones, with one specific stimulus used for each ear. Participants have to report which stimuli (word or syllable) they heard best. This indicates which ear is dominant (for example, if there is /DA/ in the left ear and /GA/ in the right ear and the participant says that they heard /GA/, the right ear is dominant). This operation is repeated on average a hundred times because a recent review showed that high reliability of this test seems expected for 90–120 trials (Westerhausen, 2019). Participants report significantly more of those syllables that had been presented to the right ear than those presented to the left ear, a phenomenon termed the right-ear advantage. The right-ear advantage is thought to reflect left hemispheric language lateralization (Ocklenburg and Güntürkün, 2018; Westerhausen and Samuelsen, 2020).

2.2 Human handedness

One of the most robust behavioral indicators of hemispheric specialization in humans, as well as in nonhuman primates (see the recent review by Nelson, 2022) and other species (Rogers et al., 2013), is handedness (i.e., side preference for the right or the left hand) (Ocklenburg and Güntürkün, 2018). The human population presents approximately 90% right-handers while left-handers constitute approximately 10% (Papadatou-Pastou et al., 2020; but see Faurie et al., 2005, who found 26.9% of left-handers among the Eipo in Papua New Guinea). Nonhuman primates also show population-level laterality for manipulative tasks but not in the same proportions as in humans (i.e., not 90% of right-handers or left-handers), and the extensive literature on the subject seems even to show conflicting results (see the meta-analysis by Papademetriou et al., 2005). Moreover, nonhuman primate laterality can vary according to the complexity of the task (Bardo et al., 2015). Handedness is thought to have played an important role in the human brain lateralization and the emergence of language (Corballis, 1998) and other complex cognitive functions, including tool use (Corballis, 2003; Uomini, 2009), manual gestures (Corballis, 2003; Meguerditchian et al., 2013), and throwing (Calvin, 1983). A predominance of right-hand use has been reported in communicative gestures in humans (Kimura, 1973a, 1973b; Grossi et al., 1996), as well as in non-human primates (Meguerditchian, 2022), indicating a greater dominance of the left-hemisphere

for gestural communication. It has been suggested that the motor sequencing for tool use (i.e., dexterous hierarchical motor activities) facilitated the emergence of language in humans with likely manual gesture as a first form of language (Greenfield, 1991). This theory proposes that tool use abilities and language evolved together. The neural circuits in both tasks are superimposed (see for more details a recent study in humans, Thibault et al., 2021; and in chimpanzees, Hopkins et al., 2007) and may have facilitated an increase in hemispheric specialization. A limitation to this possible general trend is that the currently used definitions and methods for quantifying handedness do not take into account the full complexity of this trait. Fully understanding handedness in living humans is fundamental to discussing hand preferences in fossil hominins.

Handedness in living humans is indicative of preference and/or performance. For example, a person can prefer to use the right hand to write but can be more skillful with the left hand for another task. The usual method to measure hand preference is to ask via a questionnaire what is the preferred hand during different activities (e.g., the Edinburgh Handedness Inventory developed by Oldfield (1971) at Edinburgh University). However, most questionnaires used to measure handedness focus on the manipulation of specific objects and tools (e.g., a pen, a toothbrush) rather than on more comprehensive ethological measures that include non-object-manipulatory, self-directed and socially communicative patterns of behavior. Moreover, the problem with questionnaires is that people do not always answer reliably. Measuring handedness with performance measures is necessary to quantify different skills and test which hand is better for specific activities. Performance can be measured for example with grip strength (Bardo et al., 2021), the time needed to perform a task (Bardo et al., 2022), or the speed of the hand (Peters and Durding, 1978). The degree of hand preference varies depending on the task being tested (Annett, 1972; Bishop, 1989). For example, by studying manual activities in three traditional societies (G/wi San of Botswana, Himba of Namibia and Yanomamo of Venezuela), Marchant et al. (1995) found a right-handedness bias only for tool use, while other manual actions showed mixed results (used of both right and left hand without preference) at both individual and population level (Marchant et al., 1995). However, a large part of the studies only analyzes the dominant hand (i.e., the preferred hand use for specific manual action such as writing) while both hands are important in daily activities, for living *H. sapiens* and fossil hominins (Bardo et al., 2021; Steele and Uomini, 2009). Numerous manual activities involved bimanual coordination during Paleolithic times. Tool production and tool use behaviors (Marzke et al., 1998; Shaw et al., 2012; Williams-Hatala et al., 2018) and having two strong upper limbs would likely be positively selected for through hominin evolution (Cartmill, 1985; Preuschoft, 2002; Kappelman et al., 2016). Pre-industrial lifestyle certainly induces less highly specialized lateralization compared to more sedentary humans (Bardo et al., 2021). When studying hand laterality, it is necessary to investigate both hands' performance to better understand the effect of handedness on human daily life according to lifestyle factors (e.g., occupation, practice of sport and music) and the potential clinical implications of this, but also to provide a broader evolutionary understanding.

Indeed, population-level hand asymmetry has been related to advances in technological and cultural innovations (Provins, 1997; Uomini, 2009), or more generally, to task complexity (Bardo et al., 2015; Cashmore et al., 2008; Fagot and Vauclair, 1991). In return, increase in hemispheric specification could have been favored by a more frequent use of the dominant hand during diverse activities, rather than both hands, and vice versa. However, as early hominins (e.g., *Australopithecus*) were still likely using their hands for arboreal locomotion, they may have been under stronger selection for bimanual abilities and strong grip force, as in extant non-human hominids (Cashmore et al., 2008). Moreover, Steele and Uomini (2009) highlighted the importance of studying the complementary role of both hands during bimanual Paleolithic tools behaviors to better understand the evolution of hand laterality.

Linking handedness with specific brain anatomical asymmetries seems complex and recent studies have shown inconsistent findings. For example, some studies have found a correlation between handedness and the widths of right-frontal and left-occipital petalias, with petalias seemingly more prominent, in term of magnitude, in right-handers (LeMay and Kido, 1978; Kertesz et al., 1986; and more recently by Kong et al., 2021). However, other studies showed that left-handedness does not involve any broad or substantial alterations or cortical asymmetry compared to right-handedness (Guadalupe et al., 2014; Kong et al., 2018; Maingault et al., 2016). After testing more than 17,000 healthy individuals, Kong et al. (2018) found no significant associations between handedness and cortical asymmetries although there were some significant associations for several traits with lateralized functions, including language and visuospatial processing. Finally, the same authors (Kong et al., 2021) on a larger sample of over 40,000 individuals, found an association between torque pattern—i.e., the tendency of the brain to show opposing right-left asymmetries of frontal and occipital regions in the antero-posterior directions–and handedness. A limitation of those studies is that they considered handedness as a categorical trait (i.e., supported by the bimodal distribution of overall hand preference). But as discussed above, it will be more accurate to also define handedness with different manual measurements, such as manual performance and preference, to provide more information of potential association between brain anatomical asymmetries and handedness. Moreover, it should be noted that handedness seems to correlate more strongly with functional asymmetries in language structures than with asymmetries in the motor cortex (Toga and Thompson, 2003). For example, the planum temporale appears much larger in the left-hemisphere for right-handed individuals (Geschwind and Levitsky, 1968), and is often reported as reduced or more frequently reversed in left-handers (Herve et al., 2006; Narr et al., 2007).

2.3 Visuospatial skills and attention

Visuospatial attention is a cognitive skill involving the capacity to process stimuli around us, and to interpret spatial relationships (Posner and Petersen, 1990). The most common behavioral experimental technique to investigate laterality in

visuospatial skills and attention is the divided visual field paradigm (Banich, 2003). During this experimental task, participants are seated in front of a computer monitor and asked to focus on a fixation cross that is presented in the center of the screen. A stimulus is then presented very briefly on either the left or the right side of the fixation cross, and the participant is asked to respond to the stimulus (e.g., telling what they see). Due to the anatomical organization of the visual system, stimuli on the left side of the fixation cross are processed in the right hemisphere and vice versa. The divided visual field paradigm utilizes this principle to test whether one hemisphere is more efficient (e.g., by quantifying response time) in processing a specific type of stimuli (e.g., words, faces, or images showing emotional states) than the other one (Güntürkün et al., 2020). It was found that visuospatial attention is a function of the right hemisphere (e.g., Mesulam, 1981), that depends on a bilateral (i.e., processing between the two hemispheres) parieto-frontal network (De Schotten et al., 2011).

Different brain areas are involved in vision and processing of the information. Several visual areas are common between mammals, including primates. Some specific visual areas are shared within primates only, while others vary among primate species (Kaas, 2020). The sophisticated visual processing capacities of primates are linked to an increase in the area of cerebral cortex that processes visual information, expanding to areas of the occipital, parietal, temporal, and even frontal lobes, as well as other specializations such as changes in color vision and high cortical neuronal numbers (see the recent review by Haas et al., 2020). Goldberg et al. (2013) demonstrated that, in humans, these regions involved in visual processing show rightward volumetric asymmetries in humans, consistent with other findings on rightward cortical surface and thickness differences (Luders et al., 2006; Lyttelton et al., 2009). These results are consistent with a right hemispheric dominance for visuo-spatial processing in humans (Goldberg et al., 2013). The parietal association cortex appears to play a major role in the representation of objects and their movements (Van Essen et al., 2001), more precisely, the right posterior parietal cortex (Rushworth et al., 2001b). Moreover, there is a distinct form of attention related to motor processes, involving the hands, as opposed to oculomotor processes. Such motor attention depends on a network of areas that are lateralized to the left rather than the right hemisphere (Rushworth et al., 2001b). The left parietal region, the supramarginal gyrus, is an important part of this network and its role cannot be explained as merely the consequence of the left hemisphere's dominance for language; an interconnected ventral premotor region by the inferior precentral sulcus may also be important for motor attention (Rushworth et al., 2001a). Therefore, these attentional processes are lateralized; the angular gyrus region (right parietal region) is critical for orienting attention and the more anterior supramarginal gyrus (left parietal region) is essential for motor attention.

3 How to infer functional asymmetry of fossils hominins?

Studying extant non-human primates is a prerequisite to understanding and contextualizing the characteristics of brain lateralization in humans. It is therefore important to investigate whether lateralization and hemispheric asymmetry is shared since

an ancient evolutionary stage (Güntürkün et al., 2020), and for this reason evidence of lateralization in fossil hominins can improve our understanding of the emergence of complex cognitive abilities in human evolution (Steele and Uomini, 2009). The anatomy of living *H. sapiens* may provide important clues to document the behaviors of fossil hominins. The cognitive capacity of *H. sapiens* is said to be associated with characteristics in terms of brain size, cortical proportions, histology, connection patterns, and metabolic functions (Preuss, 2017). Comparative neuroanatomy and paleoneurology are therefore necessary to formulate hypotheses about the hominin brain, integrating information from extinct and extant species. Paleoneurology is a research field applying functional brain imaging techniques in living *H. sapiens* to identify the cerebral regions involved in the conception of artifacts produced in the past by hominins (Salagnon et al., 2020). Using different imaging techniques (e.g., PET, functional or structural MRI, Functional Near-infrared spectroscopy, fNIRS), paleoneurology elaborates hypotheses about the evolution of cognition by making inferences about hominin cognitive function with regards, for example, to language and cognitive control learning domains (Salagnon et al., 2020). The behaviors of fossil hominins may also be documented by direct evidence. As summarized in Uomini and Ruck (2018), evidence of lateralization in fossils hominins can be found through: (1) identification of hand preference from stone tool manufacture and use, from hand prints found on rocks and cave walls, and from the direction of striations on fossil teeth; and (2) by measuring functional asymmetries in fossil skeletons.

3.1 Paleoneurology and endocasts

In the absence of brain tissues preserved in the fossil record, we refer to the superficial imprints that the brain forms on the internal cranial surface: the endocast. Endocasts constitute a critical proxy for qualifying and quantifying variations in brain shape and organization in extinct taxa and provide the only direct evidence of brain evolution (Dumoncel et al., 2020). The development of advanced imaging techniques (such as computed-tomography, CT, and MRI) and analytical methods (e.g., geometric morphometrics) in neuroscience and paleoneurology offers a unique opportunity to study the brain-endocast relationship. More specifically, shape analyses of endocasts are based either on anatomical landmarks that represent endocranial surface features homologous to cortical convolutions (impressions of brain gyri and sulci) or on dense meshes of semi-landmarks that capture overall endocranial shape (Borne et al., 2020), or on diffeomorphic approaches (Beaudet et al., 2019; Dumoncel et al., 2020). However, debates on whether or not information inferred from the study of endocasts reflects brain shape and organization have polarized discussions in paleoneurology since the earliest descriptions of cerebral imprints in fossil hominin crania. Indeed, the endocast is only a partial reflection of the morphology of the brain and the homology of cortical areas between species is based on their relative position to brain sulci. Of course, the resulting anatomical region may not correspond directly to a Brodmann area, which is based on histology (Brodmann, 1909) and may not be fully functionally comparable. Nonetheless,

Dumoncel et al. (2020) provide critical evidence supporting the invaluable contribution of the brain imprints left on the fossil endocranial surfaces to our understanding of the human brain evolutionary history and for discussing key cerebral aspects in the fossil hominin record. Indeed, they found a close relationship between the shape and the sulcal pattern of the frontal, temporal and occipital lobes as well as in the inferior portion of the parietal lobes. However, the correspondence in terms of morphology and organization between the superior part of the brain and of the endocast is more questionable.

The morphoarchitecture of areas located in the frontal and temporal regions such as Broca's and Wernicke's areas could be inferred from the study of the endocast (Dumoncel et al., 2020). As we previously demonstrated, there is a moderately strong relationship between language comprehension and the brain metrics from Broca's and Wernicke's areas (the cortical thickness of Wernicke's area, in particular), contributed most to this structure-function relationship (Jäncke et al., 2020). The trend during hominin evolution has been suggested to be a reduction in size on the left compared with the right hemisphere (Balzeau et al., 2014). As the third frontal convolution projects more laterally and antero-posteriorly in the right hemisphere, the left Broca's cap looks more globular (Balzeau et al., 2014). An asymmetric third frontal convolution is a feature shared by chimpanzees, bonobos and *H. sapiens*, as well as by fossil hominin species. As a result, the relative development of specific anatomical areas on the endocasts and their degree of bilateral variation in fossil hominins cannot be used to prove/dismiss the presence of an articulated language during human evolution. Further research is needed in this context, both to clarify the relationships between anatomy and function but also to better characterize the neuroanatomy of fossil hominin specimens. It is still challenging to demonstrate a relationship between the variation in size, shape and bilateral asymmetries of those areas and the origin of language capabilities in hominins. This information must now be considered by researchers comparing homologous brain areas in humans, nonhuman primates and fossil hominins. Recent results illustrate in particular that the cognitive capacities of *H. sapiens* come from a neuroanatomical substrate deeply rooted in the evolution of hominins (Albessard-Ball and Balzeau, 2018; Beaudet, 2017).

3.2 Hand preference

Stone tools and remnants of knapping are very abundant in Paleolithic sites and the oldest date from 3.3 million years (Harmand et al., 2015), making them the most accessible evidence to assess hand preference in hominins. Different studies attempted to determine hand laterality from stone tool use with different methods (see a review of successful methods applied to the archeological record, Uomini and Ruck, 2018); for example by looking at directional traces of use on the stone flakes (Rodriguez et al., 2020), dorsal cortex position (Toth, 1985), spatial organization of debitage remains (Bargalló et al., 2018), asymmetries of the tool (Brinton, 1896), or with the direction of fractures around the point of percussion (Dominguez-Ballesteros and Arrizabalaga, 2015).

Cave paintings can be used to evaluate the hand preferences of Paleolithic people by comparing the proportions of right and left hands of hand prints and hand stencils (called negative hand stencils) found on cave walls and rock shelters. Through experimentation, it was found that living right-handed humans tend to press their left hand against the wall to make stencils (Faurie and Raymond, 2004). In this context, a high proportion of left hand prints and stencils have been found in numerous sites, with a majority of right-handed individuals (see details of these studies in Uomini and Ruck, 2018). Moreover, Nelson et al. (2017) recently published a study using geometric morphometric techniques allowing to successfully assess the sex of the makers of hand stencils, and opening up new perspectives for identifying individual cave painters (Uomini and Ruck, 2018).

Characteristic diagonal striations can be found on some fossil teeth (Bermúdez de Castro et al., 1988) and these striations are hypothesized to be caused by using the teeth as a third hand (Uomini and Ruck, 2018). These paramasticatory practices involved, for example, processing leather, plants and meat (e.g., Lozano et al., 2008; Uomini, 2011). Evidence of right-handedness bias has been found from the studies of the direction of striations on teeth surfaces for fossil hominins such as *Homo erectus* (Asia), *Homo heidelbergensis* and *Homo neanderthalensis* (Europe), and even geologically older hominins with one *Homo habilis* individual (OH-65) showing predominance of right oblique scratches (see Lozano et al., 2017 for a review). However, it remains unclear as to when population-level handedness first evolved within the hominin clade (Estalrrich and Rosas, 2013; Uomini, 2011). Moreover, as discussed above, it is also necessary to fully understand handedness in living humans to better discuss hand preference in fossil hominins.

3.3 Directional asymmetries in arm bones

The behavior of extinct hominins can be also encoded in their skeletons. As muscles grow and are used, they cause the bones to which they are attached to model/remodel themselves (i.e., derives from the concept of bone functional adaptation; Ruff et al., 2006). Using one arm more than the other can thus cause unequal biomechanical loadings of the muscles in both arms, generating one arm bone to become thicker and/or denser than the other (Uomini, 2015). Directional asymmetries in the upper limb are then used to identify hand preference in fossil hominins, with the dominant arm showing differences in size and robusticity than the other (Shaw, 2011; Uomini and Ruck, 2018). For example, a Pleistocene hominin with a right humerus with larger cortical thickness than the left one was found (Balzeau et al., 2020; Sládek et al., 2016; Sparacello et al., 2017; Trinkaus et al., 1994), providing evidence of right-side dominance. Studies on hand bones showed that articular constraints on the second metacarpal (Lazenby et al., 2008) and trabecular bone in the thumb (Stephens et al., 2016) help to identify hand preference in the fossil record. Surprisingly, little research focuses on directional asymmetries in fingers bones, a part of laterality and morphology research that needs to be further developed. Moreover, the fossil record mainly contains partial and fragmentary hand bones, and in very rare cases bones from both hands, which makes skeletal evidence for handedness from hands much sparser.

4 Prospective from anatomy to behavior: The paleobrain project

The discipline of paleoneurology goes beyond the determination of biological characteristics and morphologies, and can also be used to infer the behaviors of extinct species. Just as neuroscience relates present brain structure to present behavior, paleoneurology relates fossil brain structure to past behavior (Uomini, 2015). Paleoneurological evidence primarily relies on the interpretation of fossil endocasts. As discussed above, the work of Dumoncel et al. (2020) represents an important reference for paleoneurological studies insofar as they demonstrated that there is a close correspondence in terms of morphology and organization between the brain and the corresponding endocast. However, there remain discrepancies between the brain and the endocast. Specifically, there is a lack of correspondence between superior regions in the brain and in the corresponding endocast. Less information can be directly related between the two structures because imprints are not clear on the endocast and the brain imaged with medical equipment is more distant from the cranial bone than other areas, particularly the base. Moreover, this study was conducted on a reduced sample (five extant human individuals; two females and three males) and through different imaging methodologies for the different analyzed anatomical parts. Future analyses will be needed to determine if their conclusions also apply to large samples of living *H. sapiens*. These analyses will have to include larger samples with a more precise knowledge of all the factors (e.g., age, sex) that are involved in the expression of the analyzed traits. This is particularly true for behavioral lateralization. Moreover, an objective characterization of the relative position of the brain sulci and the resulting imprints of the bone has to be obtained through a unique methodological framework to improve the quality of the results, following previous studies (de Jager et al., 2019, 2022). Only once the brain-endocast relationship has been tested in much detail, will we be able to apply this knowledge to study hominin fossil brain organization and then to approach the question of their cognition. Quantifying the degree of reliability of the endocast in order to enable further credible discussion of brain evolutionary changes in the fossil records is of prime interest.

In this context, a valuable output of the PaleoBrain project will be to provide more details of the correlations between the anatomy, including the asymmetries, of the brain and the endocast within a larger sample of living humans (N = 60; ranging in age from 20 years to 45 years; female and male), and with comparisons with the endocasts of fossil hominins. Another objective will be to investigate the potential correlation between hand preference, anatomical asymmetries of the arm and brain anatomical lateralization at the individual scale among our sample. It will open opportunities for investigating the details of the correlations, both in terms of lateralization and of magnitude, for several parameters for hand preference, bone structure of the arms and details of brain asymmetries.

4.1 Complexity of human laterality

As mentioned in Uomini and Ruck, (2018, p. 306) "Although much is now known about hemispheric specialization for language, hand preference, and visuospatial skills, there is still very little consensus on whether, or how, functional laterality in one modality may interact with asymmetries for other tasks in living humans—and even less in other species." In order to have a robust individual profile of laterality and to better understand the functional interaction between the different parts of the body, we have to consider other lateralized behaviors in addition to handedness. But, so far, investigations of laterality have not taken this into account, handedness being the most studied lateralized behaviors. A recent *Live science* project at the London Science Museum called "Me, Human," and directed by Gillian Forrester, explored a variety of lateralized behaviors and cognitive performance in more than 1600 individuals throughout the lifespan, as well as the relationship between hand form and function (Bardo et al., 2021, 2022). However, this "Me, Human" project was limited by the rules of the *Live Science* scheme (i.e., a project in which scientists can carry out research in the museum using visitors as volunteers) of the London Science Museum with a specific amount of time in which the participants could do the experiments. As a result, the data set does not provide complete individual profiles of laterality and dexterity. In our upcoming project, we will explore the potential relationship between different organization profiles (i.e., laterality profiles such as right-hand, left-eye and right-ear dominance) and performance of participants (ranging in age from 20 years to 65 years; N = 120 females and males). At first, they will be asked several lifestyle-related, multiple-choice, binary questions with regards to their laterality (wide variety of questions about daily life), occupation, and whether they play manual sports or musical instruments. Then, they will also take part in several experiments: (1) experiments measuring hand strength (grip and pinch strength) and shape (both hands of each participant will be digitized for measurements using a flatbed scanner); (2) experiments measuring dexterity in both hands; (3) the "dichotic-listening test" (see Section 2.1); and (4) the "visual test" to determine the dominant eye. Our aim is to detail as much as we can the different behaviors in relation to the anatomy of hands and arms of participants as well as in relation to their lifestyle factors. Indeed, the effect of lifestyle factors on dexterity, manual strength and laterality on the human hand needs to be further investigated to fully understand if manual laterality is really important in our daily life and if manual laterality is plastic and thus varies according to our personal life experiences (Bardo et al., 2021, 2022). Finally, this data will be considered in relation to anatomical features of the endocast/brain. In summary, participants will provide corresponding behavioral and neuroanatomical data as they answer the handedness survey, do the experiments on brain and body function, and we will also obtain structural MRIs of their brains (described below). These exploratory analyses will be essential to fully understand the relationship between functional asymmetry of the human brain and behavioral observations.

4.2 Endocasts: From living humans to fossil hominins

In this section, our main objectives are to (1) explore the relationship between the brain and the endocast in living humans in order to (2) evaluate if the brain endocast is a reliable proxy for studying the evolution of the brain. To do so, our laboratory is currently combining analysis of the brain and the corresponding endocast of a sample of living humans (N=60; selected among the participants of the behavioral experiments explained above). Concerning the MRIs, multiple sequences will be used, including T1-weighted (produced by using short time to echo and short repetition time) and UTE (ultra-short time to echo; used to characterize the cortical bone). The brain is segmented from the T1 and the endocast from the UTE, and sulci will be detected and identified on the brain and endocast (Fig. 1), using the freely distributed software BrainVISA (Geffroy et al., 2011). The shape of the brain and the endocast, as well as the sulcal patterns of the brain and endocasts, will then be directly compared using complementary approaches through qualitative comparison, qualitative evaluation of the variation between individual characterization of both imprints, distance measure between sulci and imprints on the endocasts, and characterization of absence/presence of imprints on the endocranial surface at the exact location of brain sulci. We will then describe the morphology, which means the global (i.e., volume) and regional (i.e., lobes) shape, and structure, which means the position and spatial relationships of the sulci, of the brain and the endocast. This will contribute to the study of the variation in anatomical features, such as the disposition of the lobes and the relative position, extension and expression of the sulci. We will also consider the potential relationships between anatomical features of the endocast/brain and other biological aspects such as sex, age, manual laterality. The obtained knowledge on the relationship between the brain and the endocast will be also used to look at the endocast of fossil hominins under a new perspective to extend previous qualitative studies of these features. Indeed, once we have acquired a better understanding of

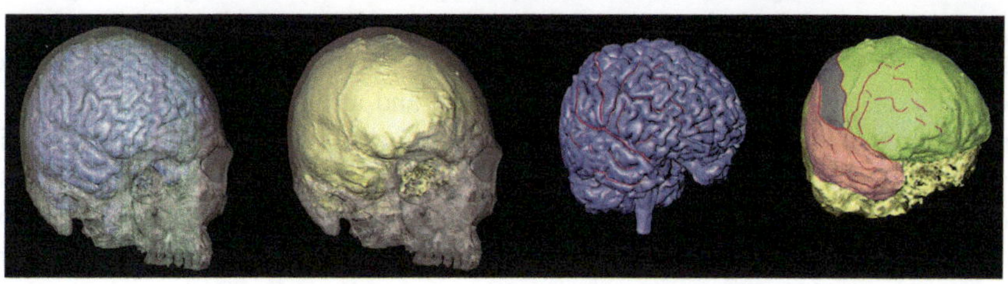

FIG. 1

Complementary imaging acquisitions on volunteers. The brain anatomy will be observed using complementary imaging acquisitions on volunteers and the data will be used to extrapolate the correspondence of its surficial features on virtual endocasts. From left to right, the skull is shown by transparency and reveals the brain in blue and the endocast in yellow. The red lines on the brain are the main sulci identified directly on this model, they are projected on the endocast to show the extension of the frontal lobes (in green), of the right parietal lobe (in purple) and of the right temporal lobe (in pink).

this relationship within our own species, we will be able to use this model to address the inverse relationship between the shape of the endocast and a reconstructed brain in selected samples of well-preserved specimens that are crucial for characterizing important and debated hominin species (see the website of the project here: https://paleobrain.jimdofree.com/).

5 Conclusion

As we do not have direct access to the details of the behaviors of fossil hominins, the question of manual laterality in extinct hominins has long been of interest in archeology and paleoanthropology. This is related to the broader question of when our rightward-biased pattern of hand use emerged in the hominin lineage. Moreover, interest in determining handedness in ancestral species has often been motivated by the question of whether these hominins had language. However, while data exist on living humans, the main limitation is the application of this knowledge resulting from separate studies of the brain, of biological adaptation of the skeleton to lateralized behaviors and hand preference, to fossil specimens. With the PaleoBrain project, we will show that important but complex information may be extracted from the combined study of behavioral lateralization and anatomical asymmetries at the level of the arms and the brain. We hope to bring original data about brain-related capacities such as manual lateralization, language and their emergence during hominin evolution. It is hoped that this overview of multiple disciplines of laterality has provided a greater appreciation of how they can be combined to gain a truly deeper understanding of the evolution of human cognition.

Acknowledgments

Thank you to Tanya Calvey, Alexandra de Sousa and Amélie Beaudet for the invitation to contribute to this volume (PBR 275: From Fossils to Mind), and for their constructive and very helpful reviews and corrections. We thank the Agence Nationale de la Recherche (Grant/Award Number: ANR-20-CE27-0009-01), and Mathieu Santin, Mélanie Didier and Eric Bardinet from the Institut du Cerveau (ICM) in Paris.

References

Albessard-Ball, L., Balzeau, A., 2018. Of tongues and men: a review of morphological evidence for the evolution of language. J. Lang. Evol. 3 (1), 79–89.

Annett, M., 1972. The distribution of manual asymmetry. Br. J. Psychol. 63 (3), 343–358.

Banich, M.T., 2003. The divided visual field technique in laterality and interhemispheric integration. In: Hughdahl, K. (Ed.), Experimental Methods in Neuropsychology. Kluwer, New York, pp. 47–63.

Bardo, A., Pouydebat, E., Meunier, H., 2015. Do bimanual coordination, tool use, and body posture contribute equally to hand preferences in bonobos? J. Hum. Evol. 82, 159–169.

Bardo, A., Kivell, T.L., Town, K., Donati, G., Ballieux, H., Stamate, C., Edginton, T., Forrester, G.S., 2021. Get a grip: variation in modern human grip strength and implication for human evolution. Symmetry 13 (7), 1142.

Bardo, A., Town, K., Kivell, T.L., Donati, G., Ballieux, H., Stamate, C., Edginton, T., Forrester, G.S., 2022. The precision of the human hand: variability in pinch strength and manual dexterity. Symmetry 14 (1), 71.

Balzeau, A., Gilissen, E., Grimaud-Hervé, D., 2012. Shared pattern of endocranial shape asymmetries among great apes, anatomically modern humans, and fossil hominins. Plos One 7, e29581.

Balzeau, A., Gilissen, E., Holloway, R.L., Prima, S., Grimaud-Hervé, D., 2014. Variations in size, shape and asymmetries of the third frontal convolution in hominids: paleoneurological implications for hominin evolution and the origin of language. J. Hum. Evol. 76, 116–128.

Balzeau, A., Ball-Albessard, L., Kubicka, A.M., 2020. Variation and correlations in departures from symmetry of brain torque, humeral morphology and handedness in an archaeological sample of Homo sapiens. Symmetry 12 (3), 432.

Balzeau, A., Mangin, J.F., 2021. What are the synergies between Paleoanthropology and brain imaging? Symmetry 13 (10), 1974.

Bargalló, A., Mosquera, M., Lorenzo, C., 2018. Identifying handedness at knapping; an analysis of the scatter pattern of lithic remains. Archaeol. Anthropol. Sci. 10, 587–598.

Beaudet, A., 2017. The emergence of language in the hominin lineage: perspectives from fossil endocasts. Front. Hum. Neurosci. 11, 427.

Beaudet, A., Clarke, R.J., de Jager, E., Bruxelles, L., Carlson, K.J., Crompton, R., et al., 2019. The endocast of StW 573 ("little foot") and hominin brain evolution. J. Hum. Evol. 126, 112–123.

Becker, Y., Claidière, N., Margiotoudi, K., Marie, D., Roth, M., Nazarian, B., Anton, J.L., Coulon, O., Meguerditchian, A., 2022. Broca's cerebral asymmetry reflects gestural communication's lateralisation in monkeys (Papio anubis). Elife 11, e70521.

Benjamin, C.F., Walshaw, P.D., Hale, K., Gaillard, W.D., Baxter, L.C., Berl, M.M., Polczynska, M., Noble, S., Alkawadri, R., Lawrence, J., Hirsch, L.J., Constable, R.T., Bookheimer, S.Y., 2017. Presurgical language fMRI: mapping of six critical regions. Hum. Brain Mapp. 38 (8), 4239–4255.

Bermúdez de Castro, J.M., Bromage, T.G., Jalvo, Y.F., 1988. Buccal striations on fossil human anterior teeth: evidence of handedness in the middle and early upper Pleistocene. J. Hum. Evol. 17 (4), 403–412.

Bishop, D.V.M., 1989. Does hand proficiency determine hand preference? Br. J. Psychol. 80 (2), 191–199.

Borne, L., Rivière, D., Mancip, M., Manging, J.-F., 2020. Automatic labeling of cortical sulci using patch- or CNN-based segmentation techniques combined with bottom-up geometric constraints. Med. Image Anal. 62, 101651.

Boulinguez-Ambroise, G., Aychet, J., Pouydebat, E., 2022. Limb preference in animals: new insights into the evolution of manual laterality in hominids. Symmetry 14 (1), 96.

Brinton, D.G., 1896. Left-handedness in north American aboriginal art. Am. Anthropol. 9, 175–181.

Broca, P., 1865. Sur le siège de la faculté du langage articulé. Bull. Mem. Soc. Anthropol. 6, 377–393.

Brodmann, K., 1909. Vergleichende Lokalisationslehre der Grosshirnrinde in ihren Prinzipien dargestellt auf Grund des Zellenbaues. Barth, Leipzig, Germany.

Calvin, W.H., 1983. A stone's throw and its launch window: timing precision and its implication for language and hominids brain. J. Theor. Biol. 104, 121–135.

Cantalupo, C., Pilcher, D.L., Hopkins, W.D., 2003. Are planum temporale and sylvian fissure asymmetries directly correlated? A MRI study in great apes. Neuropsychologia 41, 1975–1981.

Cartmill, M., 1985. Climbing. In: Hildebrand, M., Bramble, D.M., Liem, K.F., Wake, D.B. (Eds.), Functional Vertebrate Morphology. Harvard University Press, Cambridge, MA, USA, pp. 73–88.

Cashmore, L., Uomini, N., Chapelain, A., 2008. The evolution of handedness in humans and great apes: a review and current issues. J. Anthropol. Sci. 86, 7–35.

Corballis, M.C., 1998. Cerebral asymmetry: motoring on. Trends Cogn. Sci. 2, 152–157.

Corballis, M.C., 2003. From mouth to hand: gesture, speech and the evolution of right- handedness. Behav. Brain Sci. 26, 199–260.

de Jager, E.J., Risser, L., Mescam, M., Fonta, C., Beaudet, A., 2022. Sulci 3D mapping from human cranial endocasts: a powerful tool to study hominin brain evolution. Hum. Brain Mapp. 43 (14), 4433–4443.

de Jager, E.J., van Schoor, A.N., Hoffman, J.W., Oettlé, A.C., Fonta, C., Mescam, M., Risser, L., Beaudet, A., 2019. Sulcal pattern variation in extant human endocasts. J. Anat. 235 (4), 803–810.

De Schotten, M.T., Dell'Acqua, F., Forkel, S., Simmons, A., Vergani, F., Murphy, D.G., Catani, M., 2011. A lateralized brain network for visuospatial attention. Nat. Neurosci. 14, 1245–1246.

De Sousa, A.A., Sherwood, C.C., Mohlberg, H., Amunts, K., Schleicher, A., MacLeod, C.E., Hof, P.R., Frahm, H., Zilles, K., 2010. Hominoid visual brain structure volumes and the position of the lunate sulcus. J. Hum. Evol. 58 (4), 281–292.

Dominguez-Ballesteros, E., Arrizabalaga, A., 2015. Flint knapping and determination of human handedness. Methodological proposal with quantifiable results. J. Archaeol. Sci. Rep. 3, 313–320.

Dumoncel, J., Subsol, G., Durrleman, S., Bertrand, A., de Jager, E., Oettlé, A.C., Lockhat, Z., Suleman, F.E., Beaudet, A., 2020. Are endocasts reliable proxies for brains? A 3D quantitative comparison of the extant human brain and endocast. J. Anat. 238 (2), 480–488.

Estalrrich, A., Rosas, A., 2013. Handedness in Neandertals from the El Sidrón (Asturias, Spain): evidence from instrumental striations with ontogenetic inferences. PLoS One 8, e62797.

Fagot, J., Vauclair, J., 1991. Manual laterality in nonhuman primates: a distinction between handedness and manual specialization. Psychol. Bull. 109, 76–89.

Faurie, C., Raymond, M., 2004. Handedness frequency over more than ten thousand years. Proc. R. Soc. Lond. B Biol. Sci. 271 (3), S43–S45.

Faurie, C., Schiefenhövel, W., le Bomin, S., Billiard, S., Raymond, M., 2005. Variation in the frequency of left-handedness in traditional societies. Curr. Anthropol. 46, 142–147.

Gannon, P.J., Holloway, R.L., Broadfield, D.C., Braun, A.R., 1998. Asymmetry of chimpanzee planum temporale: humanlike pattern of Wernicke's brain language area homolog. Science 279, 220–222.

Geffroy, D., Rivière, D., Denghien, I., Souedet, N., Laguitton, S., Cointepas, Y., 2011. BrainVISA: a complete software platform for neuroimaging. In: Python in Neuroscience Workshop, Paris, Aug. 2011. https://brainvisa.info/web/download.html.

Geschwind, N., Levitsky, W., 1968. Human brain: left-right asymmetries in temporal speech region. Science 161 (3837), 186–187.

Goldberg, E., Roediger, D., Kucukboyaci, N.E., Carlson, C., Devinsky, O., Kuzniecky, R., et al., 2013. Hemispheric asymmetries of cortical volume in the human brain. Cortex 49 (1), 200–210.

Greenfield, P.M., 1991. Language, tools and brain: the ontogeny and phylogeny of hierarchically organized sequential behavior. Behav. Brain Sci. 14, 531–595.

Grossi, G., Semenza, C., Corazza, S., Volterra, V., 1996. Hemispheric specialization for sign language. Neuropsychologia 34 (7), 737–740.

Guadalupe, T., Willems, R.M., Zwiers, M.P., Vasquez, A.A., Hoogman, M., Hagoort, P., Fernandez, G., Buitelaar, J., Franke, B., Fisher, S.E., Francks, C., 2014. Differences in cerebral cortical anatomy of left- and righthanders. Front. Psychol. 5, 261.

Güntürkün, O., Ströckens, F., Ocklenburg, S., 2020. Brain lateralization: a comparative perspective. Physiol. Rev. 100 (3), 1019–1063.

Haas, J., Hass, R., Spocter, M.A., de Sousa, A.A., 2020. Human visual neurobiology. In: Shackelford, T.K., Weekes-Shackelford, V.A. (Eds.), Encyclopedia of Evolutionary Psychological Science. Springer International Publishing, Cham, pp. 1–10.

Hagoort, P., 2017. The core and beyond in the language-ready brain. Neurosci. Biobehav. Rev. 81, 194–204.

Harmand, S., Lewis, J.E., Feibel, C.S., Lepre, C.J., Prat, S., Lenoble, A., Boës, X., Quinn, R.L., Brenet, M., Arroyo, A., Taylor, N., Clément, S., Daver, G., Brugal, J.P., Leakey, L., Mortlock, R.A., Wright, J.D., Lokorodi, S., Kirwa, C., Kent, D.V., Roche, H., 2015. 3.3-million-year-old stone tools from Lomekwi 3, West Turkana, Kenya. Nature 521 (7552), 310–315.

Herve, P.Y., Crivello, F., Perchey, G., Mazoyer, B., Tzourio-Mazoyer, N., 2006. Handedness and cerebral anatomical asymmetries in young adult males. Neuroimage 29, 1066–1079.

Herve, P.-Y., Zago, L., Petit, L., Mazoyer, B., Tzourio-Mazoyer, N., 2013. Revisiting human hemispheric specialization with neuroimaging. Trends Cogn. Sci. 17 (2), 69–80.

Hopkins, W.D., Russell, J.L., Cantalupo, C., 2007. Neuroanatomical correlates of handedness for tool use in chimpanzees (*pan troglodytes*) implication for theories on the evolution of language. Psychol. Sci. 18 (11), 971–977.

Hugdahl, K., Westerhausen, R., 2010. The Two Halves of the Brain: Information Processing in the Cerebral Hemispheres. MIT Press.

Jäncke, L., Liem, F., Merillat, S., 2020. Are language skills related to structural features in Broca's and Wernicke's areas? Eur. J. Neurosci. 53 (4), 1124–1135.

Kaas, J.H., 2020. Evolution of visual cortex in primates. In: Evolutionary Neuroscience. Academic Press, pp. 547–564.

Kappelman, J., Ketcham, R.A., Pearce, S., Todd, L., Akins, W., Colbert, M.W., Feseha, M., Maisano, J.A., Witzel, A., 2016. Perimortem fractures in Lucy suggest mortality from fall out of tall tree. Nature 537, 503–507.

Keller, S.S., Crow, T., Foundas, A., Amunts, K., Roberts, N., 2009. Broca's area: nomenclature, anatomy, typology and asymmetry. Brain Lang. 109 (1), 29–48.

Kertesz, A., Black, S.E., Polk, M., Howell, J., 1986. Cerebral asymmetries on magnetic resonance imaging. Cortex 22, 117–127.

Kimura, D., 1973a. Manual activity during speaking. I. Right-handers. Neuropsychologia 11, 45–50.

Kimura, D., 1973b. Manual activity during speaking. II. Left-handers. Neuropsychologia 11, 51–55.

Kong, X.Z., Mathias, S.R., Guadalupe, T., Group, E.L.W, Glahn, D.C., Franke, B., Crivello, F., Tzourio-Mazoyer, N., Fisher, S.E., Thompson, P.M., et al., 2018. Mapping cortical brain asymmetry in 17,141 healthy individuals worldwide via the ENIGMA consortium. Proc. Natl. Acad. Sci. U. S. A. 115, E5154–E5163.

Kong, X.Z., Postema, M., Schijven, D., Castillo, A.C., Pepe, A., Crivello, F., Joliot, M., Mazoyer, B., Fisher, S.E., Francks, C., 2021. Large-scale phenomic and genomic analysis of brain asymmetrical skew. Cereb. Cortex 31 (9), 4151–4168.

Lazenby, R.A., Cooper, D.M.L., Angus, S., Hallgrímsson, B., 2008. Articular constraint, handedness, and directional asymmetry in the human second metacarpal. J. Hum. Evol. 54, 875–885.

LeMay, M., Kido, D.K., 1978. Asymmetries of the cerebral hemispheres on computed tomograms. J. Comput. Assist. Tomogr. 2, 471–476.

Leliveld, L.M.C., Langbein, J., Puppe, B., 2013. The emergence of emotional lateralization. Evidence in non-human vertebrates and implications for farm animals. Appl. Anim. Behav. Sci. 145, 1–14.

Lozano, M., Bermúdez de Castro, J.M., Carbonell, E., Arsuaga, J.L., 2008. Non-masticatory uses of anterior teeth of Sima de los Huesos individuals sierra de Atapuerca. Spain. J. Hum. Evol. 55, 713–728.

Lozano, M., Estalrrich, A., Bondioli, L., Fiore, I., Bermúdez de Castro, J.M., Arsuaga, J.L., Carbonell, E., Rosas, A., Frayer, D.W., 2017. Right-handed fossil humans. Evol. Anthropol. Issues News Rev. 26 (6), 313–324.

Luders, E., Narr, K.L., Thompson, P.M., Rex, D.E., Jancke, L., Toga, A.W., 2006. Hemispheric asymmetries in cortical thickness. Cereb. Cortex 16 (8), 1232–1238.

Lyttelton, O.C., Karama, S., Ad-Dab'bagh, Y., Zatorre, R.J., Carbonell, F., Worsley, K., Evans, A.C., 2009. Positional and surface area asymmetry of the human cerebral cortex. Neuroimage 46 (4), 895–903.

MacNeilage, P.F., 2008. The Origin of Speech. Oxford University Press, Oxford.

Maingault, S., Tzourio-Mazoyer, N., Mazoyer, B., Crivello, F., 2016. Regional correlations between cortical thickness and surface area asymmetries: a surface-based morphometry study of 250 adults. Neuropsychologia 93, 350–364.

Marchant, L.F., McGrew, W.C., Eibl-Eibesfeldt, I., 1995. Is human handedness universal? Ethological analyses from three traditional cultures. Ethology 101 (3), 239–258.

Marie, D., Roth, M., Lacoste, R., Nazarian, B., Bertello, A., Anton, J.L., Hopkins, W.D., Margiotoudi, K., Love, S.A., Meguerditchian, A., 2018. Left brain asymmetry of the Planum tempo rale in a nonhominid primate: redefining the origin of brain specialization for language. Cereb. Cortex 28, 1808–1815.

Marzke, M.W., Toth, N., Schick, K., Reece, S., Steinberg, B., Hunt, K., Linscheid, R.L., An, K.N., 1998. EMG study of hand muscle recruitment during hard hammer percussion manufacture of Oldowan tools. Am. J. Phys. Anthropol. 105, 315–332.

Meguerditchian, A., 2022. On the gestural origins of language: what baboons' gestures and brain have told us after 15 years of research. Ethol. Ecol. Evol. 34 (3), 288–302.

Meguerditchian, A., Vauclair, J., Hopkins, W.D., 2013. On the origins of human handedness and language: a comparative review of hand preferences for bimanual coordinated actions and gestural communication in nonhuman primates. Dev. Psychobiol. 55, 637–650.

Mesulam, M.M., 1981. A cortical network for directed attention and unilateral neglect. Ann. Neurol. 10, 309–325.

Narr, K.L., Bilder, R.M., Luders, E., Thompson, P.M., Woods, R.P., Robinson, D., Szeszko, P.R., Dimtcheva, T., Gurbani, M., Toga, A.W., 2007. Asymmetries of cortical shape: effects of handedness, sex and schizophrenia. Neuroimage 34, 939–948.

Nelson, E.L., 2022. Insights into human and nonhuman primate handedness from measuring both hands. Curr. Dir. Psychol. Sci. 31 (2), 154–161.

Nelson, E., Hall, J., Randolph-Quinney, P., Sinclair, A., 2017. Beyond size: the potential of a geometric morphometric analysis of shape and form for the assessment of sex in hand stencils in rock art. J. Archaeol. 78, 202–213.

Neubauer, S., Gunz, P., Scott, N.A., Hublin, J.J., Mitteroecker, P., 2020. Evolution of brain lateralization: a shared hominid pattern of endocranial asymmetry is much more variable in humans than in great apes. Sci. Adv. 6 (7), eaax9935.

Ocklenburg, S., Güntürkün, O., 2018. The Lateralized Brain: The Neuroscience and Evolution of Hemispheric Asymmetrie. Academic Press, London, UK.

Ocklenburg, S., Ströckens, F., Güntürkün, O., 2013. Lateralisation of conspecific vocalisation in non-human vertebrates. Laterality 18 (1), 1–31.

Oldfield, R.C., 1971. The assessment and analysis of handedness: the Edinburgh inventory. Neuropsychologia 9 (1), 97–113.

Palomero-Gallagher, N., Zilles, K., 2019. Differences in cytoarchitecture of Broca's region between human, ape and macaque brains. Cortex 118, 132–153.

Papadatou-Pastou, M., Ntolka, E., Schmitz, J., Martin, M., Munafò, M.R., Ocklenburg, S., Paracchini, S., 2020. Human handedness: a meta-analysis. Psychol. Bull. 146 (6), 481–524.

Papademetriou, E., Sheu, C.F., Michel, G.F., 2005. A meta-analysis of primate hand preferences, particularly for reaching. J. Comp. Psychol. 119, 33–48.

Patterson, N., Richter, D.J., Gnerre, S., Lander, E.S., Reich, D., 2006. Genetic evidence for complex speciation of humans and chimpanzees. Nature 441 (7097), 1103–1108.

Peters, M., Durding, B.M., 1978. Handedness measured by finger tapping: a continuous variable. Can. J. Psychol. 32 (4), 257.

Posner, M.I., Petersen, S.E., 1990. The attention system of the human brain. Annu. Rev. Neurosci. 13 (1), 25–42.

Preuschoft, H., 2002. What does "arboreal locomotion" mean exactly and what are the relationships between "climbing", environment and morphology? Z. Morphol. Anthropol. 83, 171–188.

Preuss, T.M., 2017. The human brain: evolution and distinctive features. In: Tibayrenc, M., Ayala, F.J. (Eds.), On Human Nature. Academic Press, San Diego, pp. 125–149.

Provins, K.A., 1997. The specificity of motor skill and manual asymmetry: a review of the evidence and its implications. J. Mot. Behav. 29, 183–192.

Rodriguez, A., Pouydebat, E., Chacón Navarro, M.G., Moncel, M.H., Cornette, R., Bardo, A., Chèze, L., Iovita, R., Borel, A., 2020. Right or left? Determining the hand holding the tool from use traces. J. Archaeol. Sci. 31, 102316.

Rogers, L.J., Vallortigara, G., Andrew, R.J., 2013. Divided Brains. The Biology and Behaviour of Brain Asymmetries. Cambridge University Press, NY.

Ruff, C., Holt, B., Trinkaus, E., 2006. Who's afraid of the big bad Wolff?:"Wolff's law" and bone functional adaptation. Am. J. Phys. Anthropol. 129 (4), 484–498.

Rushworth, M.F.S., Krams, M., Passingham, R.E., 2001a. The attentional role of the left parietal cortex: the distinct lateralization and localization of motor attention in the human brain. J. Cogn. Neurosci. 13 (5), 698–710.

Rushworth, M.F.S., Ellison, A., Walsh, V., 2001b. Complementary localization and lateralization of orienting and motor attention. Nat. Neurosci. 4 (6), 656–661.

Salagnon, M., d'Errico, F., Mellet, E., 2020. Neuroimaging and neuroarchaeology: a window on cognitive evolution. Dermatol. Int. 73, 67–91.

Shaw, C.N., 2011. Is 'hand preference'coded in the hominin skeleton? An in-vivo study of bilateral morphological variation. J. Hum. Evol. 61 (4), 480–487.

Shaw, C.N., Hofmann, C.L., Petraglia, M.D., Stock, J.T., Gottschall, J.S., 2012. Neandertal humeri may reflect adaptation to scraping tasks, but not spear thrusting. PLoS One 7, e40349.

Shepherd, S.V., Freiwald, W.A., 2018. Functional networks for social communication in the macaque monkey. Neuron 99, 413–420.e3.

Sládek, V., Ru, C.B., Berner, M., Holt, B., Niskanen, M., Schuplerová, E., Hora, M., 2016. The impact of subsistence changes on humeral bilateral asymmetry in terminal Pleistocene and Holocene Europe. J. Hum. Evol. 92, 37–49.

Steele, J., Uomini, N., 2009. Can the archaeology of manual specialization tell us anything about language evolution? A survey of the state of play. Cambridge Archaeol. J. 19 (1), 97–110.

Stephens, N.B., Kivell, T.L., Gross, T., Pahr, D.H., Lazenby, R.A., Hublin, J.-J., Hershkovitz, I., Skinner, M.M., 2016. Trabecular architecture in the thumb of pan and homo: implications for investigating hand use, loading, and hand preference in the fossil record. Am. J. Phys. Anthropol. 161, 603–619.

Sparacello, V.S., Villotte, S., Shackelford, L.L., Trinkaus, E., 2017. Patterns of humeral asymmetry among Late Pleistocene humans. C. R. Palevol 16 (5–6), 680–689.

Spocter, M.A., Hopkins, W.D., Garrison, A.R., et al., 2010. Wernicke's area homo logue in chimpanzees (Pan troglodytes) and its relation to the appearance of modern language. Proc. R. Soc. B 277, 2165–2174.

Ströckens, F., Güntürkün, O., Ocklenburg, S., 2013. Limb preferences in non-human vertebrates. Laterality 218 (5), 536–575.

Thibault, S., Py, R., Gervasi, A.M., Salemme, R., Koun, E., Lövden, M., et al., 2021. Tool use and language share syntactic processes and neural patterns in the basal ganglia. Science 374 (6569), eabe0874.

Toga, A.W., Thompson, P.M., 2003. Mapping brain asymmetry. Nat. Rev. Neurosci. 4, 37–48.

Toth, N., 1985. Archaeological evidence for preferential right-handedness in the lower and middle pleistocene, and its possible implications. J. Hum. Evol. 14, 607–614.

Trinkaus, E., Churchill, S.E., Ru, C.B., 1994. Post-cranial robusticity in homo II: humeral bilateral asymmetry and bone plasticity. Am. J. Phys. Anthropol. 93, 1–34.

Uomini, N.T., 2009. The prehistory of handedness: archaeological data and comparative ethology. J. Hum. Evol. 57, 411–419.

Uomini, N.T., 2011. Handedness in Neanderthals. In: Conard, N.J., Richter, J. (Eds.), Neanderthal Lifeways, Subsistence and Technology. Springer, Heidelberg, pp. 139–154.

Uomini, N.T., 2015. Paleoneurology and behaviour. In: Bruner, E. (Ed.), Human Paleoneurology. Springer, Switzerland, pp. 121–143.

Uomini, N.T., Ruck, L., 2018. Manual laterality and cognition through evolution: an archeological perspective. Prog. Brain Res. 238, 295–323.

Van Essen, D.C., Lewis, J.W., Drury, H.A., Hadjikhani, N., Tootell, R.B., Bakircioglu, M., Miller, M.I., 2001. Mapping visual cortex in monkeys and humans using surface-based atlases. Vision Res. 41 (10–11), 1359–1378.

Wernicke, C., 1910. The System Complex of Aphasia, in Church A, Diseases of the Nervous System. Appleton, New York, pp. 265–324.
Westerhausen, R., 2019. A primer on dichotic listening as a paradigm for the assessment of hemispheric asymmetry. Laterality 24 (6), 740–771.
Westerhausen, R., Samuelsen, F., 2020. An optimal dichotic-listening paradigm for the assessment of hemispheric dominance for speech processing. Plos one 15 (6), e0234665.
Williams-Hatala, E.M., Hatala, K.G., Gordon, M., Key, A., Kasper, M., Kivell, T.L., 2018. The manual pressures of stone tool behaviors and their implications for the evolution of the human hand. J. Hum. Evol. 119, 14–26.
Wood, B., Harrison, T., 2011. The evolutionary context of the first hominins. Nature 470 (7334), 347–352.

CHAPTER

7

Evolutionary and genomic perspectives of brain aging and neurodegenerative diseases

Brier A. Rigby Dames[a,b,*], Huseyin Kilili[c], Christine J. Charvet[d], Karina Díaz-Barba[e,f], Michael J. Proulx[b], Alexandra A. de Sousa[g], and Araxi O. Urrutia[c,e,f,*]

[a]*Department of Computer Science, University of Bath, Bath, United Kingdom*
[b]*Department of Psychology, University of Bath, Bath, United Kingdom*
[c]*Milner Centre for Evolution, Department of Biology and Biochemistry, University of Bath, Bath, United Kingdom*
[d]*Department of Anatomy, Physiology and Pharmacology, College of Veterinary Medicine, Auburn University, Auburn, AL, United States*
[e]*Licenciatura en Ciencias Genómicas, UNAM, CP62210, Cuernavaca, México*
[f]*Instituto de Ecología, UNAM, Ciudad Universitaria, CP04510, Ciudad de México, México*
[g]*Centre for Health and Cognition, Bath Spa University, Bath, United Kingdom*
**Corresponding authors: e-mail address: bard22@bath.ac.uk; a.urrutia@iecologia.unam.mx*

Abstract

This chapter utilizes genomic concepts and evolutionary perspectives to further understand the possible links between typical brain aging and neurodegenerative diseases, focusing on the two most prevalent of these: Alzheimer's disease and Parkinson's disease. Aging is the major risk factor for these neurodegenerative diseases. Researching the evolutionary and molecular underpinnings of aging helps to reveal elements of the typical aging process that leave individuals more vulnerable to neurodegenerative pathologies. Very little is known about the prevalence and susceptibility of neurodegenerative diseases in nonhuman species, as only a few individuals have been observed with these neuropathologies. However, several studies have investigated the evolution of lifespan, which is closely connected with brain size in mammals, and insights can be drawn from these to enrich our understanding of neurodegeneration. This chapter explores the relationship between the typical aging process and the events in neurodegeneration. First, we examined how age-related processes can increase susceptibility to neurodegenerative diseases. Second, we assessed to what extent neurodegeneration is an accelerated form of aging. We found that while at the phenotypic level both neurodegenerative diseases and the typical aging process share some characteristics, at the molecular level they

Progress in Brain Research, Volume 275, ISSN 0079-6123, https://doi.org/10.1016/bs.pbr.2022.10.004

show some distinctions in their profiles, such as variation in genes and gene expression. Furthermore, neurodegeneration of the brain is associated with an earlier onset of cellular, molecular, and structural age-related changes. In conclusion, a more integrative view of the aging process, both from a molecular and an evolutionary perspective, may increase our understanding of neurodegenerative diseases.

Keywords

Alzheimer's disease, Parkinson's disease, Aging, Dementia, Evolutionary biology, Neurodegeneration, Hominin fossils, Brain mass, Genomics, Neurodegenerative diseases

Abbreviations

α-synuclein	alpha-synuclein
Aβ	amyloid-beta
AD	Alzheimer's disease
APOD	apolipoprotein D
APOE	apolipoprotein E
APP	amyloid precursor protein
cAMP	cyclic adenosine monophosphate
CpG	cytosine-phosphate-guanosine
DLB	dementia with Lewy bodies
DNA	deoxyribonucleic acid
EOAD	early-onset AD
LBD	Lewy body disorder
LOAD	late-onset AD
LRRK2	leucine-rich repeat kinase 2
mtDNA	mitochondrial DNA
NFT	neurofibrillary tangle
PD	Parkinson's disease
PDD	Parkinson's disease with dementia
PINK1	PTEN-induced kinase 1
PRKN	Parkin gene
PSEN1	presenilin 1
PSEN2	presenilin 2
SNCA	alpha-synuclein gene
WMLs	white matter lesions

1 Introduction

Neurodegenerative diseases constitute a major health challenge globally. The two most common of these diseases are Alzheimer's disease (AD) and Parkinson's disease (PD) (Gitler et al., 2017; World Health Organization, 2022). The rise in

the prevalence of neurodegenerative diseases, such as AD (Cornutiu, 2011; Hebert et al., 2013) and PD (Dorsey et al., 2018), since the late 20th Century, has been correlated with increasing lifespan (Wang et al., 2017). Aging is the main risk factor for these disorders (Hou et al., 2019). In terms of aging, we focus particularly on senescence. In line with this, we define aging as the time-related progressive decay of the brain and body after reaching adulthood (Gilbert, 2000). Therefore, this chapter aims to understand the mechanisms of both typical aging and neurodegeneration as this is key to formulating treatments. Additionally, we investigate the evidence for and against neurodegenerative diseases being a form of accelerated aging, such as the structural brain changes observed with neuroimaging, the cellular hallmarks, and some of the molecular pathways. Although we acknowledge that the process of aging is affected by both genetic and environmental causes, this review focuses on the genetic and evolutionary factors.

An evolutionary perspective can give valuable insights into health and disease (Stearns and Ebert, 2001). Therefore, we have incorporated evolutionary biology in our approach to investigate healthy brain aging versus neurodegenerative diseases. Studies involving comparative approaches with extant nonhuman primates will be discussed because of their close phylogenetic relationship to humans (Finch and Austad, 2012). Also, they are prominent model organisms in neuroscience research studies since they share similarities with humans in terms of brain development and neuroanatomy (Phillips et al., 2014; Van Dam and De Deyn, 2017). However, further research is needed to investigate differences in brain aging between humans and nonhuman primates to tease apart differences that might leave humans more susceptible to neurodegenerative diseases (Van Dam and De Deyn, 2017). Several primate clades are reviewed, including strepsirrhines (e.g., mouse lemurs) and platyrrhines (e.g., marmosets). Our focus however is on catarrhines, which encompass cercopithecids (e.g., cynomolgus monkeys) and great apes (e.g., orangutans and gorillas), this is because they are the closest living relatives of humans. Principally chimpanzees will be discussed given they are the living species closest to humans on the phylogenetic tree (Kumar et al., 2022). Given this relationship, we expect a high level of homology between humans and chimpanzees, followed by the other great apes. Some nonhuman primate species are not discussed due to their limited obtainability in research studies (Van Dam and De Deyn, 2017), for example, bonobos (a species of the chimpanzee genus, *Pan*) and gibbon species (lesser apes), which are some of the most endangered primate species globally (IUCN, 2022). In addition, comparative genomics studies of these nonhuman primate species alongside extinct hominins can serve to reveal evolutionary trajectories of changes in the brain and lifespan over time. Hominins include our species (*Homo sapiens,* which will be referred to as humans) and extinct species such as Neanderthals, Denisovans, and *Homo erectus* that are more closely related to humans than any other living species (de Sousa and Wood, 2007; Diniz-Filho et al., 2019; Reich et al., 2010).

2 Neurodegenerative diseases and brain aging at the macroscale

2.1 Prevalence and classification of Alzheimer's disease and Lewy body disorders

Neurodegenerative diseases are a leading cause of disability and mortality worldwide (Stephenson et al., 2018). This category of diseases is diverse, but they can all be accompanied by the degeneration and death of neurons, such as those of the brain (Zuo et al., 2019). Degeneration of the brain affects cognitive, sensory, and/or motor functions (Palumbo et al., 2021). Dementia is the most common type of neurodegenerative disease with AD accounting for 60–70% of dementia cases (World Health Organization, 2022). Other forms of dementia include frontotemporal dementia and dementia with Lewy bodies (DLB). Another Lewy body dementia is Parkinson's disease dementia (PDD).

Following dementia, the second most prevalent neurodegenerative disease worldwide is PD (Tolosa et al., 2021). PD is not usually classified as a form of dementia given that cognitive impairment is not common in the initial stages (Pigott et al., 2015; Tolosa et al., 2021). However, there is approximately 80% comorbidity with dementia (e.g., PDD) at advanced stages of PD (Pigott et al., 2015; Russell et al., 2014). Although DLB and PDD show a great overlap of symptoms, such as cognitive impairments, the main difference is the order of onset (Gomperts, 2016). For example, PDD starts with PD with a later onset of dementia, while DLB starts with dementia symptoms proceeding or alongside motor impairments (Gomperts, 2016). These disorders, PD, PDD, and DLB, fall into the category of Lewy body disorders (LBDs) or diseases since they all share Lewy pathology (Chartier and Duyckaerts, 2018).

Notably, aging is the most significant risk factor associated with developing neurodegenerative diseases, such as AD and PD (Hou et al., 2019). Some researchers have therefore theorized that the neurodegeneration in these diseases is an accelerated form of aging (Zhou et al., 2019). Therefore, it would be advantageous to better understand the processes of brain aging and how these typical changes relate to neurodegenerative diseases to form a foundation of knowledge for interventions.

2.2 Structural brain changes with neurodegeneration and typical aging

To understand the structural changes in the brain associated with neurodegenerative diseases, it is critical to understand how the brain changes significantly during typical aging. During early development, there is a mass growth of neurons (Levitt, 2003; Stiles and Jernigan, 2010), but as we age, some neurons die, and there is little replacement (Aranda-Anzaldo, 2012; Mathews et al., 2017). Age-related brain changes, such as cerebral atrophy and ventricular enlargement, are visible with neuroimaging techniques (see Fig. 1; Blinkouskaya and Weickenmeier, 2021). For example, there are noticeable reductions in the amount of gray matter

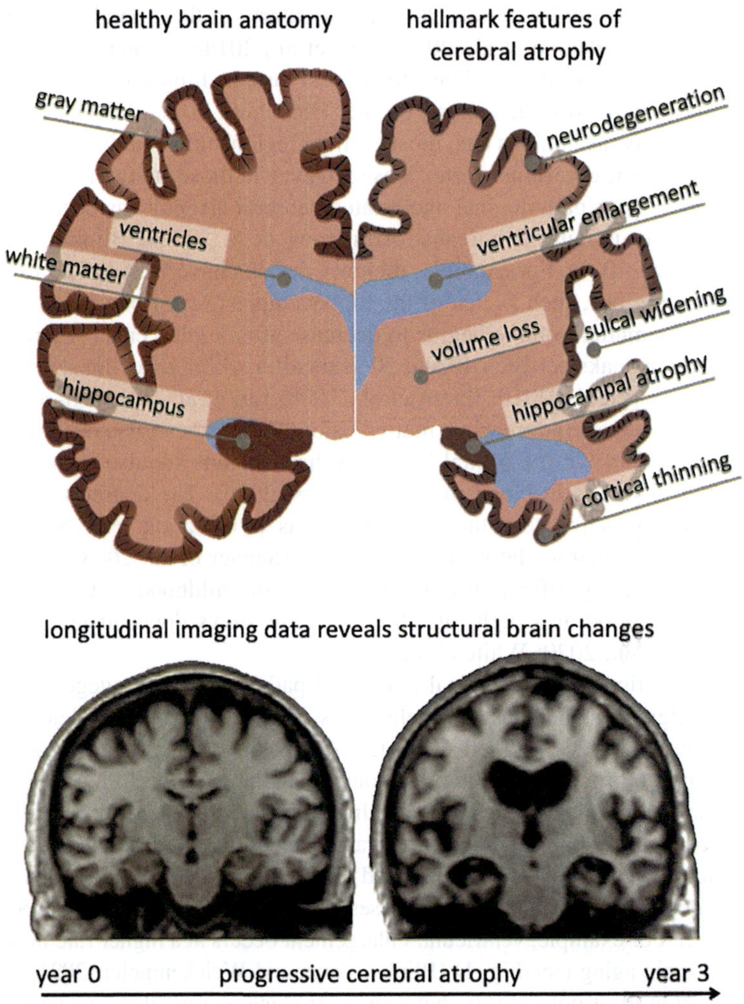

FIG. 1

Brain atrophy in neurodegeneration. Cerebral atrophy occurs in both healthy and pathological aging as represented in the schematic cross-section (top of the figure). Morphological changes are visible, for example, white matter degeneration. The bottom two coronal slices show longitudinal data from an individual with Alzheimer's disease (AD) from the AD Neuroimaging Initiative (ADNI) demonstrating significant cerebral atrophy across 3 years.

Permission from Blinkouskaya, Y. and Weickenmeier, J. (2021) Brain shape changes associated with cerebral atrophy in healthy aging and Alzheimer's disease Front. Mech. Eng., 7, p. 64. Available at: https://doi.org/10.3389/fmech.2021.705653.

(Driscoll et al., 2009; Raz et al., 2005), particularly in the frontal, temporal, and parietal lobes (Anderson et al., 2012; Fjell et al., 2014a; Ramanoël et al., 2018; see Fig. 2A). There is also evidence to suggest that the hippocampus and cerebellum display a reduction in volume in later life (Fjell, et al., 2014a; Good et al., 2001). Typically, as humans age, most cortical regions thin at a linear rate, though acceleration in the entorhinal cortex was seen even in those at low risk of AD (Fjell et al., 2014b). Non-pathological age-related cerebral atrophy has been associated with cognitive deficits in humans (Armstrong et al., 2020; Blinkouskaya and Weickenmeier, 2021). For example, memory deficits have been linked to atrophy of temporal regions, such as that of the hippocampus (Armstrong et al., 2020). In humans, gray matter volume begins to decrease after adolescence, whereas white matter volume peaks between 40 and 50 years after which it declines, particularly after 60 years (Fields, 2010; Ikram et al., 2008; Liu et al., 2016). In addition to myelin sheath degeneration, white matter lesions (WMLs) can also occur following age 40, particularly in the frontal lobe, and these lesions are also associated with age-related increases in blood pressure (Peters, 2006). The degree to which the cerebral cortex is folded, otherwise known as gyrification, can be measured through metrics such as the gyrification index (Schaer et al., 2008; Zilles et al., 1988). In humans, gyrification is found to peak in childhood and then decrease in young adulthood at a high rate, this decrease slows down in late adulthood (Lamballais et al., 2020; White et al., 2010).

The distinction between typical aging and pathological neurodegeneration has been investigated in humans in recent years (Fjell et al., 2014a; Fletcher et al., 2018; Zhou et al., 2019). Fjell et al. (2014a) posited that there was supporting evidence for a "normalcy-pathology homology." In line with this, brain areas susceptible to change in typical aging are also vulnerable to neurodegeneration, specifically AD (Fletcher et al., 2018). Furthermore, brain changes observed to a degree in typical aging, such as ventricular enlargement and atrophy of particular brain areas, are present in almost all neurodegenerative diseases, but often to a greater extent (Apostolova et al., 2012). For example, ventricular enlargement occurs at a higher rate in AD than in typical brain aging (see Fig. 1) (Blinkouskaya and Weickenmeier, 2021).

The region of atrophy differs between neurodegenerative diseases (Apostolova et al., 2012). For example, AD has a higher rate of atrophy in the hippocampus than in typical brain aging (see Fig. 1) (Blinkouskaya and Weickenmeier, 2021). Hippocampal shrinkage is often the first cortical change detected in AD, and this can pre-date cognitive and behavioral symptoms by up to 10 years (Blinkouskaya and Weickenmeier, 2021). This reduction in volume is associated with the accumulation of tangled tau protein (Schäfer et al., 2019). A meta-analysis of voxel-based-morphometry AD studies found gray matter volume reduction in the right fusiform gyrus, right superior frontal gyrus, medial orbital regions, and left parahippocampal gyrus (Wang et al., 2015). Also, changes in gray matter in those with PD and DLB have been observed (He et al., 2020; Jia et al., 2015; Summerfield et al., 2005). A meta-analysis reviewing 11 PD studies found a significant reduction in gray matter in the caudate nucleus, putamen, amygdala (see Fig. 2B), and nucleus accumbens

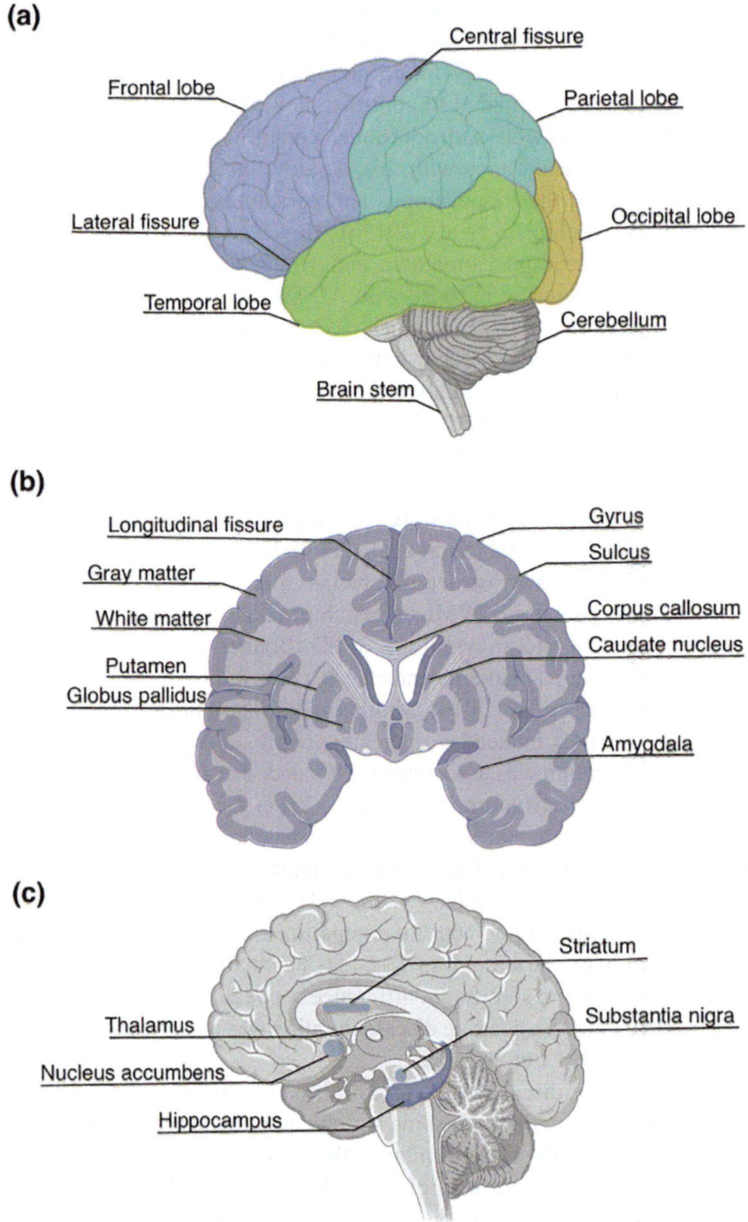

FIG. 2

(A) Diagram of the main divisions of the brain: brain stem, cerebellum, and the cerebrum (frontal, occipital, temporal, and parietal). (B) Coronal section of the brain labeled with notable structures. (C) Sagittal section of the brain labeled with some of the main areas affected in Alzheimer's disease (AD) and Parkinson's disease (PD).

(He et al., 2020). In contrast, the hippocampus, thalamus (see Fig. 2C), and globus pallidus were not found to have significant decreases in volume across the studies (He et al., 2020). Although they found no significant differences in whole brain atrophy in PD patients compared to controls, they did find a significant difference when comparing those who had additional cognitive impairment and dementia (He et al., 2020). In accordance with this, a study found that those with DLB had a greater reduction of the hippocampus and parahippocampal gyrus compared to those with PD (de Schipper et al., 2019b). However, the volume of the medial temporal lobes (hippocampus, amygdala, and parahippocampal gyrus) of those with DLB were not as affected as in those with AD (Wang et al., 2015).

Although the typical AD pathologies, such as amyloid-beta (Aβ) plaques and neurofibrillary tangles (NFTs), occur in the gray matter, there are also white matter abnormalities related to demyelination and axon damage (Nasrabady et al., 2018; Prins and Scheltens, 2015). WMLs also increase the risk of AD and the progression of WMLs are further associated with accelerated cognitive impairment (De Leeuw et al., 2005). In addition, it has been found that white matter regions that decreased in AD more than in old age controls were parahippocampal, entorhinal, inferior parietal, and rostral middle frontal (Salat et al., 2009). For individuals with PD, the only white matter areas with microstructural changes greater than those of age-matched controls were those containing WMLs (De Schipper et al., 2019a). These WMLs can also further exacerbate motor symptoms in PD (Bohnen and Albin, 2011).

A study comparing AD patients and elderly controls discovered that gyrification was lower in AD (King et al., 2010). Similarly, a further human study observed that gyrification of the cortex reduced with greater severity of AD (Liu et al., 2012). Also, a decrease in gyrification was associated with reduced cognitive functioning in both those with AD and elderly controls (Liu et al., 2012). Correspondingly, PD patients with cognitive impairment, had significantly greater cortical thinning and a reduction in gyrification of regions, specifically the left parietal and right frontal areas, compared to healthy controls (Chaudhary et al., 2021). However, these differences were not seen in PD patients with no cognitive impairment, which is likely linked to their shorter average disease duration, 5 vs 7 years (Chaudhary et al., 2021). Comparably, a previous study found no significant difference between the gyrification of controls and PD participants whose average disease duration was just 4 years and who had no cognitive impairment (Li et al., 2020).

Neurodegenerative diseases have seldom been observed in nonhuman primates (Diederich et al., 2020). Although some nonhuman primate species do show certain age-related changes of the brain similar to those observed in the human brain, the extent of these changes is not as profound (Hopkins et al., 2017; Sherwood et al., 2011). This may be a possible explanation for why the full phenotype of neurodegenerative diseases, with both the behavioral and pathological features (Walker and Jucker, 2017), are not more common in nonhuman primates. To illustrate, human brains demonstrate greater age-related brain volume reduction compared to other primates, such as chimpanzees, our closest living relatives (Herndon et al., 1999; Sherwood et al., 2011). For example, age was not found to be significantly associated

with a change in hippocampal volume in a sample of 69 male and female chimpanzees between the ages of 10 and 45 years (Sherwood et al., 2011). Also, when tracing the change in hippocampal volume across 11 rhesus macaque individuals of different ages the fluctuations were found to be non-significant (Shamy et al., 2006). However, in humans, there was a significant reduction in the hippocampal volume with age which was found to accelerate after age 50 in males and 60 in females, although the sample size here was much larger than for the other primates since 19,793 healthy adults were analyzed (Nobis et al., 2019).

Most of the nonhuman primates that have been studied have demonstrated a cognitive decline in memory or executive functions with extended age. However, unlike in humans, this decline has rarely been associated with brain atrophy, except for in mouse lemurs and rhesus macaques (Gilissen et al., 2001; Picq, 2007). Mouse lemurs have been observed to have age-related atrophy, although some areas were affected more than others (e.g., caudate nucleus) and other areas were only atrophied in certain individuals (e.g., hippocampus) (Picq et al., 2012). Furthermore, rhesus macaques' memory deficits were linked with atrophy of relevant brain regions, such as the dorsolateral prefrontal cortex due to cellular changes to the dendritic spines and loss of synapses or axons (Freire-Cobo et al., 2021; Shamy et al., 2011). In chimpanzees, white matter was found to increase up until mid-adulthood and decrease thereafter, but the proportion of change in the volume ascribable to age was less than that of humans, at below 5% (Autrey et al., 2014). Comparatively, in humans past the age of 40, the brain decreases at a rate of 5% every 10 years and then at a higher rate after the age of 70 (Peters, 2006). In chimpanzees, gyrification is significantly lower than in humans overall, but particularly within the frontal lobe (Hopkins et al., 2017). However, chimpanzee brains do show a similar decrease in gyrification from adulthood onwards (Autrey et al., 2014).

In addition, gyrification can be investigated for extinct hominin species by taking a cast from the cranium and investigating the length of gyri impressions from these endocasts (Lyras et al., 2016). For instance, sulcal impressions have been analyzed to indicate that at least from the time of *Australopithecus africanus*, which lived approximately 3 million years ago, the reorganization of three typical hominin brain regions were underway, namely, dorsolateral prefrontal association cortex, the parietal-temporal-occipital association areas, and Broca's speech area (Falk, 2014).

3 Neurobiology and genetics of aging and neurodegenerative diseases

3.1 Neurobiology of aging and neurodegenerative diseases

Cellular senescence contributes to age-related dysfunction of tissues and organs such as the brain (Van Deursen, 2014). Moreover, the accumulation of senescent cells, such as neurons and glial cells, in the brain has been linked to increased vulnerability to pathological neurodegeneration, including AD and PD

(Azam et al., 2021; Hou et al., 2019). Cellular senescence is thought to result from a number of factors, including deoxyribonucleic acid (DNA) damage, such as that from reactive oxygen species (ROS), and mitochondria dysfunction (Chung et al., 2019; López-Otín et al., 2013). ROS can similarly impair cellular proteins, lipids, organelles, and membranes as well as inducing cell death, such as apoptosis and necrosis (Auten and Davis, 2009; Redza-Dutordoir and Averill-Bates, 2016). Also, there is a reduction in the degradation and recycling of damaged cellular components, autophagy (Das et al., 2012), including the mitochondrial elements, mitophagy (Chen et al., 2020), with age (Barbosa et al., 2019). Furthermore, the mechanisms for DNA repair become less efficient with age (Gorbunova et al., 2007), which can lead to genomic instability, particularly in postmitotic tissues, such as neural tissue (Maynard et al., 2015). Additionally, chronic inflammation has been associated with the aging of tissues, which is thought to be triggered by the dysregulation of the immune response, such as the innate immune system (López-Otín et al., 2013; Salminen et al., 2012; Zhuang and Lyga, 2014).

The brain is particularly vulnerable to these processes since most neurons lose the ability to regenerate once they mature, and become postmitotic (Sah et al., 2021). Unsurprisingly, the above hallmarks have been linked to both brain aging and the pathological neurodegeneration observed in AD and PD (Azam et al., 2021; Hou et al., 2019; Jenner, 2003; Lovell et al., 1999). For example, a reduction in brain mitophagy, which leads to dysfunctional mitochondria amassing, has been associated with aging (Lou et al., 2020) and neurodegeneration (Barbosa et al., 2019) such as AD (Kerr et al., 2017). Moreover, mitochondrial dysfunction paired with the reduction in mitophagy can increase the risk of chronic inflammation that is linked to AD and PD (Sliter et al., 2018; Song et al., 2021; Yoo et al., 2020). This inflammation and immune dysfunction have also been revealed in the brains of those with AD (de Felice and Lourenco, 2015; Hou et al., 2019). Similarly, this neuroinflammation has been discovered in brain autopsies of those with PD (Hirsch and Hunot, 2009).

3.2 Genetics of aging and neurodegenerative diseases

Functional and comparative genomic studies are uncovering a number of genes that have expression patterns associated with age and neurodegeneration, as well as gene variants that increase the risk of developing neurodegenerative diseases if inherited. Genes that change their expression with age are related to mitochondrial function, genetic stability, and cellular maintenance (Berchtold et al., 2008; Lee et al., 2000; Stegeman and Weake, 2017). Furthermore, the downregulation of genes associated with the control of mitochondrial biogenesis has been reported to inhibit longevity (Gureev et al., 2019). For example, Sirtuin 1 (*SIRT1*) which has been linked to mitochondrial biogenesis, is downregulated in older age (Cieślik et al., 2020; Tang, 2016). In contrast, SIRT1 has also been related to mitophagy of dysfunctional mitochondria (Cieślik et al., 2020; Tang, 2016). Notably, *SIRT1* has been discovered to be further decreased in the cortex of those with AD, particularly the

parietal cortex, whereas this was not found to be the case for those who were cognitively impaired in the absence of the disease (Julien et al., 2009). Additionally, it has been proposed that increasing the expression of SIRT1 could help alleviate age-related diseases (Choi and Kemper, 2013).

In the brain, genes that are downregulated with age are related to functions such as axon pathfinding and synaptic transmission, in addition to those listed above (Berchtold et al., 2013; Lipinski et al., 2010; Loerch et al., 2008; Lu et al., 2004). Many of the genes related to synaptic functioning, such as neurotransmitter vesicle release and receptors, which were found to reduce in expression with age also further decreased in those with AD (Berchtold et al., 2008). For example, synaptotagmins, proteins that aid exocytosis of synaptic vesicles, such as synaptotagmin V, were downregulated in both aging and AD (Berchtold et al., 2013; Öhrfelt et al., 2016). Since neurons are long-lived, compared to other cells in the body, they have shown greater enrichment for genes associated with cellular maintenance, such as lysosomes, apoptosis, and protein degradation (Castillo-Morales et al., 2019; Seim et al., 2016).

Chronic increases in the activity of genes encoding for inflammation-related proteins, which are more common in elderly individuals, are associated with age-related disorders, such as AD, and increased risk of death (Franceschi and Campisi, 2014). This upregulation of inflammation-related genes, linked to chronic inflammation, has been observed in the aging brain (Berchtold et al., 2008; Frenk and Houseley, 2018; Lu et al., 2004). For example, in humans when the transcription factor nuclear factor kappa B subunit 1 (NF-kappa B) is activated in the microglia cells of the brain it can trigger the release of pro-inflammatory cytokines, transcriptional signatures of immune activation (Gabuzda and Yankner, 2013).

In some cases of neurodegenerative disease, there is a hereditary component, for example, approximately 10% of AD patients and 15% of PD patients have a known family history of the disease (Tanzi, 2012; Tran et al., 2020). There are three known genes for which if there is a major change AD will occur before the age of 65, specifically autosomal-dominant early-onset AD (EOAD), these include: amyloid precursor protein (*APP*) on chromosome 21, presenilin 1 (*PSEN1*) on chromosome 14 and presenilin 2 (*PSEN2*) on chromosome 1 (Dai et al., 2018; Lanoiselée et al., 2017). *PSEN1*, *APP*, and *PSEN2* mutation carriers result in AD arising at around 43, 51, and 57 years respectively (Cruts et al., 2012). One of the products of *APP* is Aβ peptides. If the *APP* gene locus is duplicated, it leads to an increased risk of AD, particularly EOAD (Sleegers et al., 2006). PSEN1's protein is one of the four main components of the γ-secretase complex that is thought to be needed for creating Aβ with APP. Studies have suggested that *PSEN1* is important in aiding synaptic plasticity, neuronal survival, and memory (Dai et al., 2018). Four out of five familial cases are the result of a modification to the *PSEN1* gene.

However, these autosomal dominant familial forms are rare, comprising $<1\%$ of AD cases (Shea et al., 2016). More commonly, a combination of minor genetic changes have an additive effect on the risk of AD, especially in the presence of environmental risk factors (Dai et al., 2018). This is what has been proposed to result

in late-onset AD (LOAD), which is defined as occurring over the age of 65 (Wightman et al., 2021). The heritability of LOAD is proposed to range between 60% and 80% based on the research from twin studies (Gatz et al., 1997, 2006). Although the heritability is deemed due to many genetic variants, the most significant of these are suggested to be the apolipoprotein E (*APOE*) variants (Lin et al., 2018). The *APOE* gene has three common alleles, one of which, *APOE4*, can increase the risk of AD, while another, *APOE2*, has been found to be protective (Nestler et al., 2009). Although *APOE4* does not directly result in AD, it is found to be present in more than half of those with AD (Safieh et al., 2019). The *APOE4* allele is found in around 25% of the population and has been identified as the main genetic risk factor for LOAD (Gharbi-Meliani et al., 2021).

For over 10 years genome-wide association studies have been used to identify additional potential genetic variants for AD (Hu et al., 2019). To find genes that were differentially expressed with age, González-Velasco et al. (2020) investigated the transcriptomes of 2202 brains from cerebral cortex, hippocampus, and cerebellum. From their transcriptomic analysis, they found genes that are related to aging and AD, such as *MS4A6A* (González-Velasco et al., 2020). *MS4A6A* is part of the membrane-spanning 4A (MS4A) gene family and excessive levels in the brain and the blood have been linked to the progression of AD pathology (Proitsi et al., 2014). Furthermore, *MS4A6A* genotypes have been connected to atrophy in certain brain regions, such as the medial temporal lobe and entorhinal cortex (Ma et al., 2016).

Many genes have been found to be associated with Lewy body disorders, particularly PD. According to Klein and Westenberger (2012), 30% of familial and 3–5% of sporadic cases of PD are caused by a single gene variant. Twenty-three chromosomal loci have been linked to PD (Li et al., 2021a); these are known as *PARK* genes and have been named in the sequence in which they were found, for example, *PARK1* (*SNCA*), *PARK2* (Parkin), and *PARK6* (*PINK1*) (Kouli et al., 2018). Together these single gene variants were found to account for 5% of PD cases (Reed et al., 2019). However, genes in other regions of the genome have also been associated with PD (see Table 1). Some of these gene variants can result in the early-onset of PD, such as PTEN-induced kinase 1 (*PINK1*) and Parkin (*PRKN*) (Jones, 2010). Furthermore, these two genes have been found to be involved in mitophagy, which supports the relationship between mitochondrial dysfunction and PD (Deas et al., 2011; Jones, 2010; Malpartida et al., 2021).

As explained above, there is little evidence of neurodegenerative diseases outside of humans (Diederich et al., 2019, 2020). This could be due to the longer lifespans in humans compared to other mammals, particularly in terms of postmenopausal lifespan (Gunn-Moore et al., 2018). To investigate the difference in longevity across mammals, genomic and transcriptomic comparisons between species can help uncover genes related to maximum lifespan potentials (de Magalhães et al., 2009). For example, Doherty and de Magalhães (2016) compared 20 mammals, with maximum lifespan potentials (MLSPs) ranging from 3.8 years for rats to 211 years for bowhead whales, and found longevity-associated gene families that were expanded only in the longer-lived species. Comparison of cortex microarrays from humans,

Table 1 Notable examples of genes associated with PD, PDD, DLB and AD.

Gene	Autosomal recessive/dominant	Associated with: PD	PDD	DLB	AD	Function
SNCA	Dominant	Yes[a–c]	Yes[a]	Yes[a,c]		Regulating neurotransmitter release[d]
PRKN	Recessive	Early-onset PD[e]				Regulates clearance of damaged mitochondria[f,g]
PINK1	Recessive	Early-onset PD[e]		Yes[h]	Possibly[i]	Detects mitochondrial damage[j]
LRRK2	Dominant	Yes[e]				Regulating autophagy, intracellular and extracellular transport[k]
APOE4	Semi-dominant[l]	Possibly earlier onset PD[m]	Yes[n]	Yes[o,p]	Late-onset[q] & early-onset[r] AD	Lipid metabolism[j]

[a] *Book et al. (2018).*
[b] *Mandel et al. (2010).*
[c] *Zarranz et al. (2004).*
[d] *Stefanis (2012).*
[e] *Kouli et al. (2018).*
[f] *Schmidt et al. (2021).*
[g] *Livingston et al. (2019).*
[h] *Murakami et al. (2007).*
[i] *Du et al. (2017).*
[j] *Safieh et al. (2019).*
[k] *Cogo and Greggio (2020).*
[l] *Genin et al. (2011).*
[m] *Pankratz et al. (2006).*
[n] *Dickson et al. (2018).*
[o] *Bras et al. (2014).*
[p] *Orme et al. (2018).*
[q] *Gharbi-Meliani et al. (2021).*
[r] *Van Duijn et al. (1994).*

rhesus macaques, and mice revealed overlap in age-related gene expression (Loerch et al., 2008). One of the most significantly upregulated genes across the three species was apolipoprotein D (*APOD*). APOD has been largely preserved across the vertebrate lineage and has been found to protect aging brains from oxidative stress through its antioxidant activity (Dassati et al., 2014; Loerch et al., 2008). Across the three species, the genome-wide comparison revealed that the gene with the strongest downregulation was calmodulin-dependent protein kinase type IV (*CAMK4*). *CAMK4* is associated with synaptic plasticity since it encodes for a component of the cyclic adenosine monophosphate (cAMP) signaling pathway which can regulate the activity of cAMP-response element binding protein (CREB) (Ho et al., 2000; Loerch et al., 2008; Wruck and Adjaye, 2020). Additionally, adenylate cyclase 2 was discovered to be downregulated with age in rhesus macaques and humans; this gene is also related to the cAMP signaling pathway and can indirectly influence synaptic transmission (Di Benedetto et al., 2021).

4 Comparative neuropathology between human and nonhuman primates

4.1 Neuropathology of Alzheimer's disease

At the structural level, brain changes associated with AD include cerebral atrophy, such as that of the hippocampus (see Fig. 1) (Blinkouskaya and Weickenmeier, 2021). Meanwhile, at the molecular level the tangles and the plaques that result in cell death and loss of gray and white matter tissue can be revealed (Fung et al., 2020; Niikura et al., 2006).

The neuropathology of AD is characterized by NFTs composed of hyperphosphorylated protein tau in the cytoplasm of the neuron and Aβ plaques outside the neuronal cell (Arnsten et al., 2021b). Upon autopsy of humans with AD, NFTs are always found to be present (Sontheimer, 2015). Tau's main function in the cell is to regulate microtubule assembly and structure (Kolarova et al., 2012; Maccioni and Cambiazo, 1995). However, when hyperphosphorylated tau detaches from the microtubules, it sequesters normal tau and fibrillates (Arnsten et al., 2021a; Kolarova et al., 2012). When the neuron later dies through autophagic degeneration, a tangle is left within the cell (Arnsten et al., 2021a). In AD, tau pathology affects glutaminergic neurons initially in the entorhinal and perirhinal areas of the brain before progressing to the association cortices, and only affects the primary sensory areas in the late stages of AD (Braak and Braak, 1991). The selective vulnerability of glutaminergic neurons to tau pathology in the limbic and association cortices needs to be investigated further. Notwithstanding, more commonly it is the cholinergic neurons that are depleted in AD patients (Cheng et al., 2021; Niikura et al., 2006). Cholinergic synapses are particularly susceptible to Aβ aggregates, and the loss of these synapses is associated with cognitive decline (Ferreira-Vieira et al., 2016).

Both Aβ and NFTs are widely considered to play major causative roles in the brain damage observed in dementias, leading to progressive loss of neurons and

synapses, brain atrophy, and dilatation of the lateral ventricles (Prince et al., 2014). Although both Aβ and tau are hallmarks of AD linked to cognitive decline (Hanseeuw et al., 2019), this decline has primarily been associated with tau pathology, which can lead to gray matter loss (Bejanin et al., 2017). In contrast, Aβ deposits have been found in humans who do not have any of the clinically defined cognitive symptoms of AD or dementia (Price and Morris, 1999). For example, one in five autopsies of humans who are elderly and deemed cognitively normal showed early signs of Aβ deposits (Rodrigue et al., 2012). Therefore, humans with this AD pathology and no cognitive impairment often die of other causes (Nelson et al., 2009). Tau accumulation tends to follow Aβ deposits, and this related accumulation could mediate the connection between Aβ accumulating and eventual cognitive decline (Hanseeuw et al., 2019).

Nonhuman primates show great homology of the sequence of APP to which Aβ is the derivative (Freire-Cobo et al., 2021; Podlisny et al., 1991). Amyloid has been detected in the brains of some species of cercopithecids (e.g., rhesus macaques), platyrrhines (e.g., squirrel monkeys), and stepsirrhines (e.g., mouse lemurs) (Youssef et al., 2016). However, this amyloid accumulation tends to be less prevalent than in human brains, and the changes connected to this pathology are less significant (Toledano et al., 2012). To illustrate, in rhesus macaques the amyloid coverage is not high enough to be neurotoxic, showing only around 5–6% in the prefrontal cortex; AD brains typically have 10–20% (Finch and Austad, 2012; Stonebarger et al., 2021). For chimpanzees, amyloid deposition tends to accumulate in the blood vessels first before forming plaques in aged brains (Edler et al., 2017, 2018). If the amyloid builds up in the arteries of the brain it can lead to cerebral amyloid angiopathy, a risk factor for stroke (Edler et al., 2018). Similarly, gorillas, between 25 and 30 years, start to have Aβ deposits in the vasculature system, such as the meningeal vessels, predating plaques forming in the frontal cortex (Perez et al., 2013, 2016). Comparably, orangutans over 25 displayed a sparse amount of Aβ plaques in the meningeal vessels, but also in the temporal lobes, and the cingulate gyrus, which were observed via staining and the use of antibodies to immunolabel Aβ (Gearing et al., 1997; Lowenstine et al., 2016).

The increase in the volume of association cortices is associated with an increase in susceptibility to tau pathology across primates in the following ascending order: marmosets, macaques, chimpanzees, and humans (Arnsten et al., 2021b). Baboons were also found to have hyperphosphorylation of tau in some hippocampal dentate gyrus neurons and glial cells, as well as Aβ deposits, but no NFTs (Ndung'u et al., 2012; Schultz et al., 2000). Similarly, Youssef et al. (2016) reported that although gorillas, rhesus macaques, orangutans, marmosets, mouse lemurs, squirrel monkeys, and tamarins were all found to have Aβ plaques, they did not have NFTs. However, mouse lemurs do express abnormal tau accumulation. Although cynomolgus monkeys were found to have hyperphosphorylation of tau in the temporal, frontal, and hippocampal cortices by the age of 20 (Oikawa et al., 2010) and few NFTs (Youssef et al., 2016), these tau-positive lesions do not display the biochemical and morphological characteristics of AD (Uchihara et al., 2016).

Furthermore, evidence of both tau and Aβ lesions present in the neocortex and hippocampus of elderly chimpanzees has been uncovered (Edler et al., 2017). Although age is associated with a reduction in the neuronal density of hippocampal subfields in chimpanzees (CA1 and CA3), AD pathology (Aβ and tau) was found to be unrelated to this reduction (Edler et al., 2020). Also, NFTs have been observed to be present in elderly chimpanzee brains, particularly in the hippocampus (Edler et al., 2017, 2018). In a study of 20 chimpanzees' postmortem brains, aged 37–62 years, the brain demonstrating the most pathology was from a male of 57 years (Edler et al., 2017). This elderly male chimpanzee was found to have phase 4 Aβ deposits (Thal et al., 2002) and stage V NFTs (Braak et al., 2006), which are the penultimate level for both scales (Edler et al., 2017). These pathology measurement scales are used for humans to indicate the number of brain regions affected, with the higher the stage of deposits and phase of tangles indicating wider dispersion (Braak and Braak, 1991; Thal et al., 2002). In a human brain study, those with clinically diagnosed AD were found to have Aβ phases 3, 4, or 5 (Thal et al., 2002). Chimpanzees are one of the few nonhuman primate species that have been reported to have NFTs (Edler et al., 2017). Conversely, evidence of NFT-like structures in the hippocampus and the entorhinal cortex was found in two green monkeys (*Chlorocebus sabaeus*; typically referred to in laboratory studies as African green monkeys) (Cramer et al., 2018).

The first reported evidence of both Aβ plaques and NFTs present at the same time (which is required as part of an AD diagnosis) in a nonhuman primate was a female chimpanzee who had had a stroke (Rosen et al., 2008). This 41-year-old chimpanzee was reported to have had high cholesterol, as well as being obese (Finch and Austad, 2015; Rosen et al., 2008). Obesity is also associated with AD in humans, indirectly acting as a risk factor due to insulin resistance, increased inflammation, and oxidative stress (Tabassum et al., 2020).

4.2 Neuropathology in Parkinson's disease and other Lewy body disorders

In contrast, the indicative sign of Lewy body disorders (LBDs), such as Parkinson's disease (PD), Parkinson's disease dementia (PDD), and dementia with Lewy bodies (DLB), is the presence of Lewy bodies and Lewy neurites (Wakabayashi et al., 2013). These Lewy pathologies, assemblies of filaments, are a combination of over 90 molecules and are found in the cytoplasm of neuronal processes such as axons (Wakabayashi et al., 2013). These filaments largely consist of the protein alpha-synuclein (α-synuclein); α-synuclein filaments when fully functioning aid in dopamine production (via their interaction with tyrosine hydroxylase) and release (Venda et al., 2010). Originally, it resides in a soluble unfolded state but in humans with PD, it can become highly folded and insoluble forming amyloid fibrils (Martin and Preedy, 2014). In addition, Lewy pathologies are composed of other PD-associated gene products, such as leucine-rich repeat kinase 2 (*LRRK2*), *PRKN*, and *PINK-1* as well as mitochondria-associated proteins (Wakabayashi et al., 2013). These molecules are linked to autophagy, the ubiquitin-proteasome system, and misfolded proteins in the cell (aggresomes) (Wakabayashi et al., 2013).

There has been much debate about whether LBDs, including PD, PDD, and DLB, are distinct conditions or rather exist on a continuum (Walker et al., 2019). The main difference is that the Lewy pathologies in those with PD reside in the brainstem and limbic areas, whereas in DLB and PDD they are also found in the neocortex (Walker et al., 2019). McKeith (2007) suggested that for genetics and molecular research, it would be more relevant to consider the different conditions as all belonging to one LBD spectrum that is based on the abnormal α-Synuclein metabolism, while for clinical studies, the temporal nature used to divide the diseases might be favored. Pathological hallmarks of AD have also been observed in humans with LBD, such as the pathological aggregation of Aβ. For example, Aβ deposits are observed in humans with PD and are related to the amount of cognitive decline and gait disturbance present (Lim et al., 2019). Furthermore, the accumulation of hyperphosphorylated Tau protein into NFTs has been observed in DLB, and this has been linked to cognitive impairment, even in the absence of heightened Aβ levels (Gomperts et al., 2016).

Outside of humans, naturally occurring PD is almost non-existent (Collier et al., 2011; Diederich et al., 2020; Outeiro et al., 2021), although one primate case has been recently reported (Li et al., 2021b). Diederich and colleagues proposed that this could be because other species have a low rate of PD, which results in it being undetected, or that they die from predation or other factors prior to the onset of senescence and age-related diseases. Nonetheless, researchers have observed behaviors that resemble the early stages of PD such as motor impairments in nonhuman primates, particularly in chimpanzees (Lacreuse et al., 2014) and rhesus macaques (Zhang et al., 2000). These motor impairments have been found together with a reduction in dopamine-transporter immunoreactivity in the substantia nigra of rhesus macaques (Emborg et al., 1998). Furthermore, as in humans, other primates have decreased function of their dopaminergic neurons as they reach old age (Collier et al., 2017; Diederich et al., 2020), such as green monkeys (Hurley et al., 2011). In addition to humans, mouse lemurs, rhesus macaques, and cynomolgus monkeys have been found to have aggregation of α-synuclein (Li et al., 2021b; Chu and Kordower, 2007; Gruschus, 2021). However, the Lewy bodies predominantly formed from this protein, α-synuclein, are only observed in humans with PD and are not naturally present in nonhuman primates (Collier et al., 2011; Diederich et al., 2020). To illustrate, although there has been one case of PD reported for nonhuman primates to date, specifically a 10-year-old cynomolgus monkey, this individual had no Lewy bodies only α-synuclein (Li et al., 2021b). PD was suspected due to the monkey displaying motor symptoms, which improved in response to two classical PD drugs, and mutations in *LRRK2* (Li et al., 2021b). However, the researchers were not able to collect fresh tissue samples for testing the functional consequences of the *LRRK2* variants, due to the unexpected death of this cynomolgus monkey (Li et al., 2021c). Therefore, it is hard to verify that the pathology in the brain of this primate was caused by these *LRRK2* mutations without further research (Yao and Dai, 2021). Even in transgenic model organisms, researchers have found it challenging to induce all the pathological features of PD, such as α-synuclein aggregation and loss of dopaminergic neurons, as well as the related motor and non-motor impairments (Lasbleiz et al., 2019).

5 Evolutionary hypotheses of aging and neurodegenerative diseases

5.1 Evolutionary theories of aging

Evolutionary biology has contributed insights into numerous related areas of medical science, such as functional genomics, and is bound to shed light on why aging and neurodegeneration occur (Dobzhansky, 1973; Stearns and Ebert, 2001). Several evolutionary theories of aging emerged in the 1940s and 1950s that suggested a reduction in the selection pressure on biological processes in old age (Haldane, 1941; Medawar, 1952; Williams, 1957). It was uncovered that there was little selection for genes that would allow increased longevity due to fewer individuals in a species living to senescence in the wild because of external factors such as predation and disease (Johnson et al., 2019). Thus, Haldane (1941) suggested Huntington's disease existed due to selection pressures falling after the age of 30–40 since historically populations would have reproduced and/or suffered mortality before reaching this age when the disease is expressed (Fabian and Flatt, 2011). This decrease in selection pressure was used by Medawar (1952) in his mutation accumulation theory to explain how gene mutations with negative effects in later life can compound.

One of the earliest accumulation theories was the free radical theory of aging, which Harman proposed in the fifties, building on the work of Gerschman who found that oxygen poisoning had a similar effect as irradiation from X-rays (Harman, 2009; Viña et al., 2013). Harman (1956) postulated that the negative effects of free radicals resulting from cellular metabolism could be a factor in aging and degenerative diseases. Mitochondrial DNA (mtDNA) has a faster rate of mutations than that of DNA in the nucleus, and this rate is exacerbated by having the electron transport chain located nearby producing reactive oxygen species (ROS) which can damage the nucleic acids (Youle, 2019). Here, physiological and evolutionary theories show consistency since the free radical theory can explain the mutation accumulation theory, as free radicals can cause mtDNA mutations (Ziada et al., 2020). Williams (1957) adapted the mutation accumulation theory to form the antagonistic pleiotropy theory, whereby genes that affect multiple traits may be selected if they control beneficial characteristics in early life regardless of whether they have negative consequences in old age.

In addition, Williams (1957) proposed the grandmothering hypothesis. This hypothesis details that postmenopausal life spans aid in supporting the rearing of grandchildren, who, in humans, have particularly extended childhoods (Kim et al., 2012). Evidence for the clinical definition of menopause, that is, cessation of ovulation, has been found across primates, but this usually occurs relatively close to their maximum lifespan potential in nonhuman primates, such as in chimpanzees at around 50 years old (Herndon et al., 2012; Walker and Herndon, 2008). Therefore, what remains unique to humans, among primates, is the relatively large proportion of life for which females are postmenopausal, and this has also been observed in hunter-gatherer populations (Blurton Jones et al., 2002; Kim et al., 2012).

Contemporary research has found that cetaceans, such as beluga whales, orcas, and narwhals, also have long postmenopausal lifespans (Ellis et al., 2018; Foster et al., 2012). This cessation of reproduction can be evolutionarily explained by the benefit to related individuals, given that there is evidence that postmenopausal mothers improve the chances of survival of their offspring and grandoffspring (Foster et al., 2012; Nattrass et al., 2019). This can be supported by Hamilton's (Hamilton, 1964) rule whereby the fitness gain from the child-rearing of relatives' offspring based on genetic relatedness is greater than the loss of reproduction. Additionally, through performing a simulation, Kim et al. (2012) asserted mathematical evidence that grandmothering helps extend longevity.

Some researchers have argued that similar to development, aging appears to be programmed (Blagosklonny, 2013; De Magalhães and Church, 2006). This process of aging was first proposed by Weismann (1891). Many of the arguments for this theory are controversial since they stem from group selection (Wilson, 1973), such as, that it is a form of population control and that it allows for those in their reproductive years to have greater access to resources (Kirkwood and Melov, 2011; Kowald and Kirkwood, 2016). However, the latter does not account for the beneficial contribution of grandmothers. Also, in the wild, individuals rarely survive up to senescence so as previously argued by Haldane (1941) and Medawar (1952) natural selection would be unlikely to act on old age. Nevertheless, other researchers have pointed to maximum lifespan potential varying across species as indicating that there is a program for longevity based on the genome (Mitteldorf, 2018). Partridge and Gems (2002) put forward, through investigating genes associated with aging, that it is the regulation of longevity that is genetically programmed rather than aging itself. More recently, Trubitsyn (2020) proposed that this genetic control of longevity was through the programmed decline of mitochondrial bioenergetics.

5.2 Species' maximum lifespan potential is tightly coupled to the evolution of brain size

Compared to other vertebrates, mammalian species tend to have larger brains both in absolute terms and relative to body mass (Roth and Dicke, 2005). In the human lineage, there has been a dramatic increase in brain size over the past three million years (Lefebvre, 2012; Roth and Dicke, 2005). For illustration, the human brain is over three times larger than a chimpanzee's, even though the adults of both species have comparable body mass (Bauernfeind and Babbitt, 2020; Van Dam and De Deyn, 2017). Furthermore, humans have brains around three times the size of *Australopithecus afarensis* (see Fig. 3), as predicted from their fossilized craniums (Zihlman and Bolter, 2015). Primates are among the mammals with the steepest slope when plotting brain size versus body size (Burger et al., 2019). Although cetaceans and other large mammals, such as elephants (see Fig. 4) also have very large brains, relative to their bodies their brains are still comparatively smaller than great apes (Northcutt, 2002). In mammals, the evolution of maximum lifespan potential is tightly linked to the evolution of brain size (see Figs. 4 and 5)

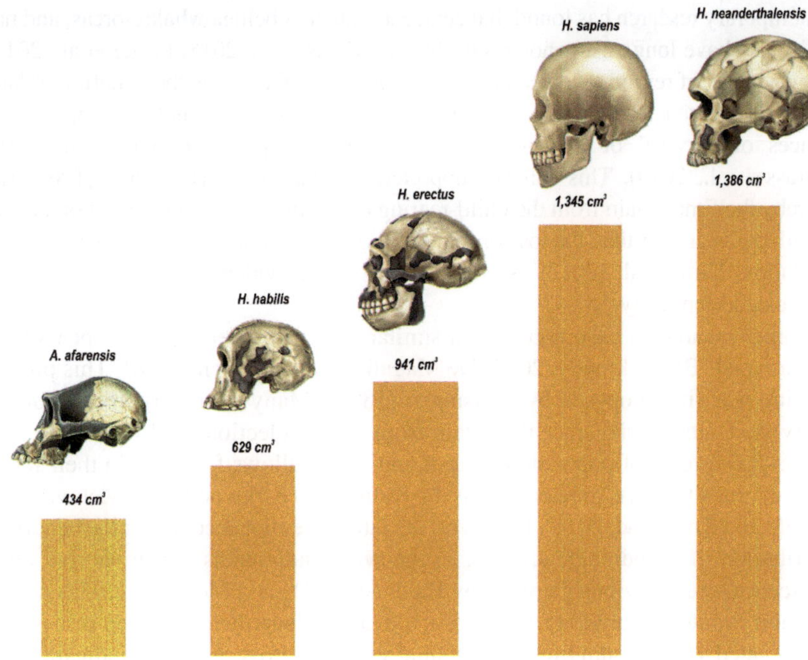

FIG. 3

Changes of cranial capacity in hominin species through time. There is a large increase in cranial capacity that occurred in the human lineage over the last three million years. Data for cranial capacities was averaged for each individual species from a dataset (DeSilva et al., 2021) of adult fossil craniums.

(Barrickman et al., 2008; González-Lagos et al., 2010; Jiménez-Ortega et al., 2020). Phylogenetic analysis provides evidence for expansion in brain size fueling increases in species' maximum lifespan potential (Jiménez-Ortega et al., 2020). It has been hypothesized that larger brains would support increased maximum lifespan potential since they require a longer period of development that necessitates more care (González-Lagos et al., 2010; Jiménez-Ortega et al., 2020). Similarly, other evolutionary theorists have suggested that extended life expectancy may be associated with delayed development, slow maturation, as well as investment in lasting pair bonds, and an emphasis on the quality of offspring (Dunkel et al., 2010).

To draw inferences about the brains of extinct species, since soft tissues, such as the brain, rarely fossilize, we must primarily rely on the remaining cranium (Dunsworth, 2015). For example, from the cranial cavity, cranial capacity can be measured and used to approximate brain size and morphology (Dunsworth, 2015). Although this practice of producing endocranial casts was first adopted over a century ago, advancements in technology, such as X-ray computed tomography, has allowed virtual endocasts to be produced from delicate specimens (Balanoff and Bever, 2016; Silcox et al., 2009). A cranial cavity can be represented as a

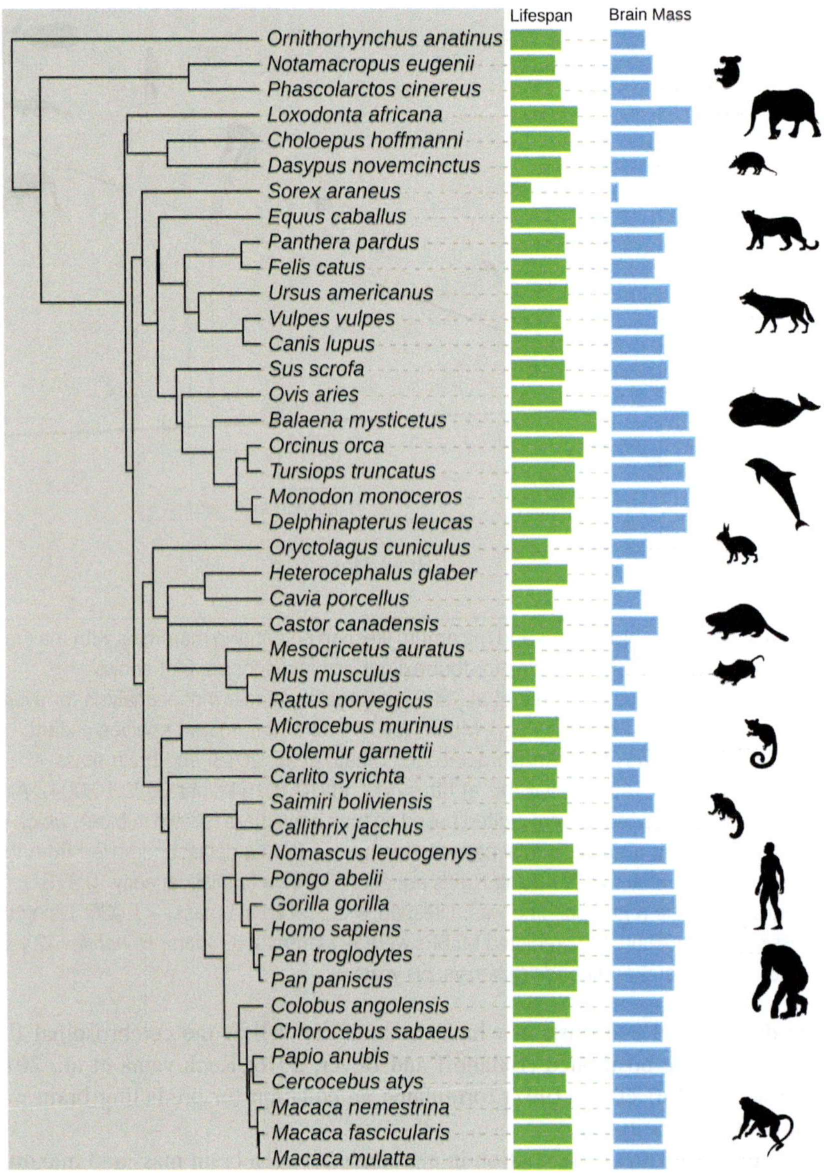

FIG. 4

Phylogenetic tree of mammals with maximum lifespan potential (lifespan) and brain mass. Phylogenetic tree for 45 mammalian species, including 18 primate species (see bottom). Bars show the relative value of each variable including lifespan, $\log_{10}$ maximum lifespan potential (green), and brain mass, $\log_{10}$ brain mass (blue). Data for maximum lifespan potential (MLSP, referred to in this figure and Fig. 5 as lifespan) was from AnAge database (Tacutu et al., 2018) and for brain mass per species see Supplementary Table 1 available at https://doi.org/10.1016/bs.pbr.2022.10.004. TimeTree (Kumar et al., 2022) was utilized to construct the phylogenetic tree, and the Interactive Tree Of Life (iTOL) version 5 (Letunic and Bork, 2021) was used for the visualization.

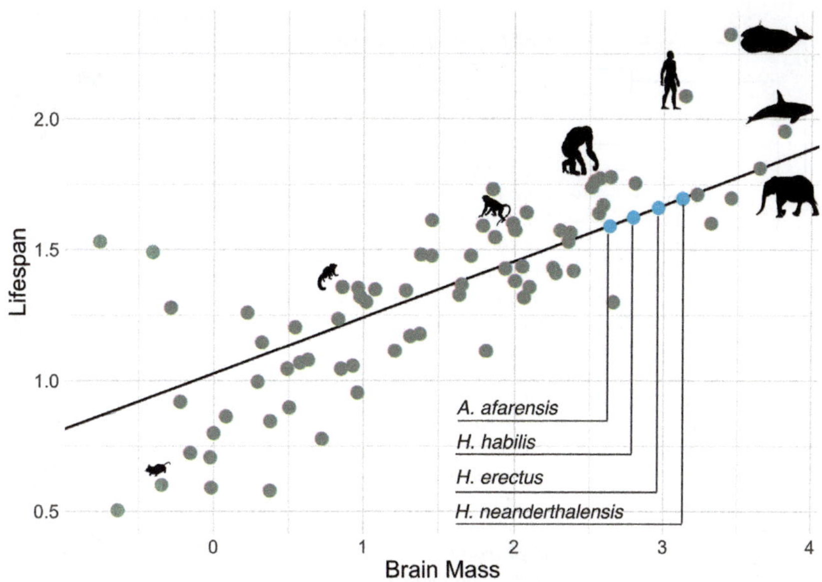

FIG. 5

Association between brain mass and maximum lifespan potential in mammals with maximum lifespan potential (MLSP) estimates for four extinct hominin species. Plot shows phylogenetically corrected (Kumar et al., 2022) generalized linear model analysis for average brain mass ($\log_{10}$) and lifespan ($\log_{10}$ MLSP) for 74 extant mammalian species. Extant species MLSP data was from AnAge database (Tacutu et al., 2018) and brain mass (see Supplementary Table 1 available at https://doi.org/10.1016/bs.pbr.2022.10.004). Also, four extinct hominin species were plotted based on their brain mass (shown in blue), which was predicted from their average cranial capacity (computed from a dataset compiled from the literature (DeSilva et al., 2021) using Ruff's equation: $1.147 \times \text{cranial capacity}^{\wedge}0.976$. The equation of the regression line was: Lifespan = 0.2158(brain mass) + 1.026. Using the regression line equation, the predicted MLSPs were *A. afarensis,* 39 years; *H. habilis,* 42 years; *H. erectus,* 46 years; *H. neanderthalensis,* 50 years.

three-dimensional endocast, in which the brain as well as the cerebrospinal fluid regions could be positioned (Balanoff and Bever, 2016; Kochiyama et al., 2018). For instance, Ruff et al. (1997) formulated an equation for predicting brain mass from cranial capacity.

We used the known association in mammals between brain mass and maximum lifespan potential (Barrickman et al., 2008; González-Lagos et al., 2010) to predict the latter for extinct hominin species. For example, primates and rodents have been found to show a positive correlation between relative brain size and longevity (DeCasien et al., 2018). We employed a general linear model with a phylogenetic correction to plot this relationship for 70 extant mammalian species. We then utilized the equation of the regression line to predict extinct hominin maximum lifespan potentials based on their brain mass estimates from cranial capacities

(adopting Ruff's equation) averaged for each species' fossils (see Fig. 5). Given these brain mass estimates, we predicted the following maximum lifespan potentials: *Australopithecus afarensis*, 39 years; *Homo habilis*, 42 years; *Homo erectus*, 46 years; *Homo neanderthalensis*, 50 years. Further research is needed to refine these estimates, but they fit within other prior predicted lifespans found for these species, which were based on alternative methods. For example, Smith (1991) in addition to brain mass, used variables such as body mass and age of mandibular eruption to deduce the following estimates: *A. afarensis*, 37 years; *H. habilis*, 47 years; *H. erectus*, 52 years; *H. neanderthalensis,* 69 years. Another study calculated the maximum lifespan potential for extinct hominins based on cytosine-phosphate-guanosine (CpG) density in specific gene promoters (Mayne et al., 2019): *H. neanderthalensis,* 38 years.

5.3 Aging in the hominin fossil record

Determining the exact age at the time of death from fossils can be challenging. Teeth are the mammalian body parts most resilient to physical and chemical erosion (Dunsworth, 2015). This is due to the dentin enveloped genetic material being encased by the highly resistant dental tissue, enamel (Chun et al., 2014; Dissing et al., 2008). However, bones, such as the cranium, are other typical remains and these together can be useful indicators for estimating the age of individuals at the time of death (Vodanović et al., 2011). Bone density decreases with age and can result in falls leading to potential fractures (Amarya et al., 2018). In humans, following peak bone mass (25–30 years), density reduces particularly in late 30s for females and early 40s for males (Santos et al., 2017). Also, teeth degrade over time despite enamel being the hardest substance in the body (Akkus et al., 2017). Once enamel is eroded, the softer dentin is exposed, which although harder than bone can also undergo wear (Chun et al., 2014; Liu et al., 2014). Schaffler et al. (1995) found that tooth wear can be exacerbated over time by a diet based on hard or acidic foods.

Fossilized bones and teeth have been used to predict the age at the time of death of individuals from extinct hominin species. These biological age predictions have been based on skeletal indicators (Trinkaus, 2011), tooth wear (Caspari and Lee, 2006), and dental development, such as mandibular eruption (Smith, 1991). For instance, Smith (1991) estimated the average species lifespan of extinct hominin species from fossils based on patterns of the age of tooth eruption, body weight, and cranial capacity observed in extant species. Sievert (2014) suggested that *Homo ergaster* and *H. erectus* would have likely had cessation of menstruation if they lived past 50 years, based on Smith's (1991) equations. Also, the ratio between old and young fossil specimens has been utilized to compare longevity distributions among different hominin species. For example, Trinkaus (2011) used the preservation of enamel alongside evidence of formation and degeneration of bones, such as the pubic symphysis, to categorize adult fossil specimens as young (20–40 years) or old (over 40 years). Trinkaus (2011) deduced that 23.7% of the fossils found from the late archaic human sample were from older adults, while they classified 26.5% of modern humans from the early Upper Paleolithic sample as older adults.

Likewise, dental wear and development have been utilized by Caspari and Lee (2006) to classify hominin specimens, including Neanderthals, as older or younger adults. They used the eruption of third molars (wisdom teeth) to indicate adulthood and specimens deemed to have tooth wear that would indicate an age twice that of when third molars erupted as being older adults (Caspari and Lee, 2006). They then used the ratio of older to younger adults to compare Neanderthals and early modern humans in Western Asia but found no significant difference in the ratio (Caspari and Lee, 2006). However, it was found that the ratio of the number of older adults to younger adults was significantly lower for Neanderthals in Europe compared to Neanderthals in Western Asia (Caspari and Lee, 2006). The authors suggested that this discrepancy in longevity could be due to difficulties adapting to the harsher environment in glaciated Europe if they did not have the cultural complexity for survival that modern humans possessed (Caspari and Lee, 2006). Conversely, Villa and Roebroeks (2014) reported that the archeological remnants do not support Neanderthals being less culturally advanced as there is evidence of them creating symbolic objects and specialized tools. For example, personal ornaments and bone tools were found alongside Neanderthal fossils at the Grotte du Renne, France (Caron et al., 2011; Zilhão, 2012). Furthermore, it has been suggested that Middle Paleolithic Neanderthals used fire for additional purposes other than just cooking and keeping warm, such as for hide smoking, and producing tar, which indicates behavioral adaptation (Leierer et al., 2020). Elderly individuals in hominin communities could help pass down culture through teaching knowledge, such as that regarding medicinal herbs, that could increase longevity, causing a positive feedback loop (Caspari and Lee, 2006). More recently genomic methods have been used to predict the lifespans of extinct species (Mayne et al., 2019). Genomic data can be extracted from certain fossils under sterile conditions to avoid contamination by human DNA or microorganisms (Slatkin and Racimo, 2016). Neanderthals and Denisovans were proposed to have a comparable maximum lifespan potential to early modern humans (38 vs 40 years) based on CpG density in specific gene promoters (Mayne et al., 2019), since the density within conserved promoters in mammals has been associated with increased lifespan (McLain and Faulk, 2018).

Under the right conditions, such as low temperature and oxygen levels, the rate of DNA degradation can be reduced (Kelman and Kelman, 1999; Poinar and Artur Stankiewicz, 1999; Warren, 2019). Poinar and Artur Stankiewicz (1999) reported that the oldest sequences that could be reproduced were 50,000 years old. However, recent technology has enabled researchers to obtain DNA from exceedingly degraded fragments of over a million years old, for instance, the Early Pleistocene mammoth molar samples (van der Valk et al., 2021). Some of the oldest hominins' DNA recovered were 28 different specimens from the Sima de los Huesos, most of which were closest to Neanderthals but one of the individuals had mtDNA closer to Denisovans (Meyer et al., 2016). However, no DNA was found for *H. floresiensis* (approx. 60,000–100,000 years old) fossils due to the climate not being optimum for preservation (Warren, 2019). Nevertheless, protein analysis, such as mass spectrometry, may be an alternative method to study fossils where DNA is not preserved (Warren, 2019).

There is also evidence of age-related disorders, such as arthritis and type 2 diabetes, in extinct hominins. For example, one of the first almost complete skeletons of a Neanderthal to be found included vertebrae and a hip joint that had signs of osteoarthritis (Straus and Cave, 1957). This specimen, La Chapelle-aux-Saints 1, initially led to the misconception that Neanderthals generally have a hunched back posture, as well as bent knees at rest (Straus and Cave, 1957). Originally, the age of death of this Neanderthal had been thought to be around 40–55 years, but recently it has been estimated to be greater than 60 years (Rothschild and Haeusler, 2021; Straus and Cave, 1957).

In concordance, alleles that increase the risk of osteoarthritis in humans have been found in the DNA of Neanderthals as well as Denisovans (Capellini et al., 2017; Loughlin, 2015). These variants in GDF5 are also linked to the growth of shorter limbs and this could have been advantageous in colder climates to reduce heat loss (Capellini et al., 2017). This is an example of antagonistic pleiotropy since, in colder climates, the gene is beneficial in early life but somewhat detrimental in later life, as it can result in osteoarthritis. Furthermore, in a genome-wide association study of 212,000 people, introgressed Neanderthal DNA was related to type 2 diabetes and autoimmune diseases such as rheumatoid arthritis (Dannemann, 2021). There is evidence to suggest that type 2 diabetes is a risk factor for both AD (Chatterjee and Mudher, 2018) and PD (Chohan et al., 2021).

5.4 Evolutionary perspective of neurodegenerative diseases

Age-related neurodegeneration is uncommon in nonhuman mammals and even rarer in other vertebrates (Gunn-Moore et al., 2018). This likely suggests that vulnerability to AD and PD evolved recently, particularly for PD, potentially corresponding to the rapid expansion of the brain and neocortex in the primate lineage (Bruner et al., 2017; Bruner and Jacobs, 2013; Diederich et al., 2019). Thus, these neurodegenerative diseases could have arisen as a trade-off or by-product of this evolutionary process. Out of the extant apes, humans have the largest and most complex brains along with a high metabolic rate (Pontzer et al., 2016). However, this high metabolic rate in humans and long lifespan, given neurons are rarely if ever replaced, leaves the brain particularly vulnerable to the deleterious effects of free radicals. The free radical theory could suggest that brain regions with higher metabolic rates have neurons with more ROS and thus more mtDNA mutations. Horwitz and colleagues (1984) found that the average metabolic rate of glucose obtained from PET was 6 (mg/100 g/min) in the precentral gyrus in the frontal lobe, while it was only 4 in the cerebellum. Therefore, with these greater energy demands, the level of regulation required would be higher and if these could not be met then perhaps this could result in the senescence of these areas. Also, brain regions with neurons that have longer axons might be affected more in terms of defective mitophagy; this may be because of additional mtDNA mutations and slower maintenance due to greater distance from the nucleus. For example, the dopaminergic neurons affected in PD have long axons,

particularly in humans, and due to them being scarcely myelinated this increases the amount of energy required for action potential conduction (Gruschus, 2021; Stiefel et al., 2013).

During the evolution of humans, some of the subcortical regions did not expand to the same extent as the neocortex and this discrepancy could place strain on these subcortical structures (Diederich et al., 2019). Even compared to extinct hominins, some evidence has arisen to suggest that humans may be more susceptible to diseases of the brain (Gokhman et al., 2014). For instance, Gokhman et al. (2014) reconstructed DNA methylation patterns in Neanderthals and Denisovans and when comparing these maps to humans they found differentially methylated regions (DMRs). These human DMRs were reported to have a significant increase in association with disease, particularly psychiatric and neurological disorders (Gokhman et al., 2014). This might suggest that changes in regulation could have made us more vulnerable to neurodegenerative diseases.

The *APOE4* allele, one of the most prevalent genetic risk factors for AD (Safieh et al., 2019), is suggested to be uniquely human in its functionality (Tcw et al., 2022). *APOE4* can affect lipid metabolism and has been associated with high levels of blood lipids (Garcia et al., 2021). However, the *APOE4* in chimpanzees has the amino acid threonine in place of arginine at the 61st position, potentially causing it to act more similarly with lipids to the human APOE3 allele (McIntosh et al., 2012; Walker and Jucker, 2017). In contrast, Denisovan fossil DNA revealed an APOE protein comparable to that of the human *APOE4* allele (Enard, 2012; McIntosh et al., 2012; Meyer et al., 2012). This suggests that if these hominins had an extended lifespan then they might have developed neurodegenerative diseases. Conversely, the nature of Neanderthal *APOE* alleles is reportedly not yet fully understood (Abondio et al., 2019). It has been speculated that the effect of *APOE4* might have an environmental interaction because neither one nor two alleles were found to be a risk factor for AD in a Yoruba sample; in contrast to an African American sample from Indianapolis where a significant association between *APOE4* and AD was uncovered (Gureje et al., 2006). Fox (2018) speculated that in non-industrialized countries the regulation of the immune system is more resilient and hence inflammation is not caused by *APOE4*. Confoundingly, it was discovered that within the Bolivian Tsimane forager-farmer population *APOE4* was connected with a reduction in inflammation (Garcia et al., 2021). Even though *APOE4* was still associated with higher levels of blood lipids in countries with high pathogen exposure, it was observed to increase fertility and infant survival rates (Van Exel et al., 2017). Human evolution mostly happened in habitats that likely had a greater variety of pathogens and less cardiovascular risk than contemporary living conditions (Garcia et al., 2021). Therefore, the prominence of this allele in society may be due to antagonistic pleiotropy, since, as discussed previously, it was historically beneficial at least during reproductive years (Williams, 1957).

A novel increase in longevity is an alternative theory for AD (Fox, 2018). The extension of the average lifespan in humans could in part be due to improved nutrition and modern medicine. However, even when comparing hunter-gatherer human

populations, the life expectancy is still approximately double that of wild chimpanzees (Finch, 2010). Another theory understands AD to be a result of rapid brain evolution, since the main areas affected by AD are those that are the most enlarged when comparing human and chimpanzee brains, such as the hippocampus, frontal, parietal, and temporal lobes (Bruner and Jacobs, 2013; Fox, 2018).

Great apes evolved to have extended periods of caregiving, for example, chimpanzee mothers stop carrying their young at approximately 5 years but remain proximally close for several more years (Reddy and Mitani, 2019). However, humans can require extended pedagogy beyond this age. Since reproductive success can include ensuring the survival of offspring through to independence (Williams, 1957), the role of grandmothering has been theorized to buffer the onset of AD-like neuropathy (Finch and Sapolsky, 1999). Given that the onset of this neuropathy is often a couple of decades after reproductive cessation in females, Sapolsky and Finch (2000) deduced that rearing grandchildren, and the related inclusive fitness, could be the evolutionary reason for this delay. To illustrate, caregiving from individuals other than the mother, known as allomothering, allows for earlier weaning (Hawkes et al., 2017). This is observed in humans, as historically infants are weaned at approximately 2–4 years or earlier when they are still very dependent, this enables others to take on the role of feeding if required (Dettwyler, 1995; Lonsdorf et al., 2021). In comparison, chimpanzees are weaned comparatively later at around 4–6 years and can progress onto independent feeding soon after (Lonsdorf et al., 2021).

PD has been proposed as a uniquely human disease (Diederich et al., 2020; Outeiro et al., 2021). One hypothesis for the origin of PD is that the associated networks, namely, the basal ganglia subcortical nuclei (e.g., the substantia nigra), have not expanded in size at the same evolutionary rate as the cerebral cortex (Diederich et al., 2019). Having arisen almost 500 million years ago, the basal ganglia has been largely conserved in size throughout the phylogenetic tree of vertebrates (Stephenson-Jones et al., 2011). This difference in evolutionary size expansion is particularly exaggerated in human brains, given that the size of the subcortical regions is largely preserved. For example, it was reported that the caudate nucleus is only 4.6–6.6 times larger in humans than in monkeys as compared to 13–18 times for the whole brain (Yin et al., 2009). This incongruence has put greater pressure on the subcortical nuclei, increasing the energy required to regulate the larger output.

Additionally, neurons affected by PD are typically highly branched such as the raphe nuclei serotonergic neurons which travel to the forebrain and then the striatum (Deneris and Gaspar, 2018; Liguori et al., 2015). Also, some of these neurons, such as the dopaminergic neurons from the substantia nigra pars compacta, can be particularly long compared to other neurons in the brain, which increases the energy demands (Attwell and Laughlin, 2001; Diederich et al., 2020). The average human substantia nigra pars compacta neuron is 4.5 m in total branching length (Bolam and Pissadaki, 2012). While this hyperbranching can benefit the functioning of the brain, particularly in youth, this could serve as a vulnerability in older age such as with the increased oxidative stress in the mitochondria, potentially leading to α-synuclein aggregation (Diederich et al., 2020). This is in accordance with the antagonistic

pleiotropy theory (Williams, 1957). Also, these dopaminergic neurons have been postulated to have higher metabolic requirements in humans than in other primates because of the greater number of synapses and their extended lifespans (Diederich et al., 2020). Furthermore, Gruschus (2021) poses an additional strain, which is that the unmyelinated axons of these neurons must travel approximately 50% farther than that of chimpanzees due to the expansion of the brain.

However, the recently reported case of PD in a cynomolgus monkey (see Section 4.2; Li et al., 2021b), suggests that this disease may not be solely human, and may have arisen by the time we last shared a common ancestor with cercopithecid monkeys 27–31 million years ago (Kumar et al., 2022). This highlights the importance of closer monitoring of neurodegeneration in apes and other nonhuman primates, especially in captive conditions as the motor impairments associated with PD might result in mortality in the wild (Li et al., 2021c).

6 Conclusion

In summary, neurodegenerative diseases pose a great burden to society globally. Extensive efforts among the research community are uncovering the structural, cellular, and molecular changes accompanying the onset and progression of these diseases, which we reviewed here. As aging is a major risk factor for the two most common neurodegenerative diseases, AD and PD, we explored existing evidence for potential mechanisms for aging-associated predisposition to these disorders. Also, we assessed the possibility that these diseases represent forms of accelerated aging. In addition, we examined the evolutionary perspectives on brain evolution, aging, longevity, and the potential applications to our understanding of neurodegenerative diseases. Furthermore, we made predictions about aging in fossil hominins. We utilized brain mass estimates from fossil hominin cranial capacities to predict the maximum lifespan of extinct hominin species to investigate the evolutionary trajectory within this lineage. Additionally, genomic DNA from fossils can be compared to the human genome to look for longevity and disease-related genes, such as the AD-associated APOE4 that was found to be similar in Denisovans. These methods, as well as comparative genomics and structural comparisons of extant nonhuman primates, can indicate the extent to which neurodegenerative diseases are uniquely human. We found similarities between structural and molecular changes in typical brain aging and neurodegenerative diseases. Brain scans in both elderly individuals and those with AD show a pattern of brain shrinking when compared to the brains of younger individuals. At the cellular and molecular level, many of the processes known to underlie the effects of aging have also been identified as risk factors for neurodegenerative diseases. Indeed, several genes have been linked to both typical and pathological aging. However, more research is needed to assess the extent of the commonalities between the two processes. Moreover, additional evidence and research are required to determine whether there were neurodegenerative diseases in extinct hominins, such as extracting additional genomes from fossils and further

genomic analysis. Among extant nonhuman primates, there is increasing evidence of the pathological hallmarks of AD in specific species, although the Aβ deposits and NFTs have not usually been found to co-occur, except for a couple of cases in chimpanzees. However, PD pathology has been reported thus far as being largely exclusive to humans. We conclude that in many respects neurodegenerative diseases exhibit age-related changes that are exaggerated and occur earlier than is typically seen in asymptomatic elderly brains.

Supplementary material

Data availability: Supplementary material, including Supplementary Table 1, is available on https://doi.org/10.1016/bs.pbr.2022.10.004. Also, the supplementary table, as well as the author contributions, are available on https://github.com/b-a-r-d/Supp-material-Evo-geno-brain-PBR275.

Acknowledgments

Some of the figures were created with Biorender.com (Fig. 2). The silhouettes for Fig. 4 and Fig. 5 were taken from http://phylopic.org/ available under the Creative Commons license.

BARD's research was supported by UK Research and Innovation grant number EP/S023437/1, belonging to the UKRI CDT in Accountable, Responsible and Transparent Artificial Intelligence.

AOU's research is funded by a Frontiers in Science CONACyT grant (No. FC-2020/682142), PAPPIT-DGAPA-UNAM grant(IA204020) and a NERC grant (NE/P004121/1).

MJP's research is funded in part by the UKRI Centre for the Analysis of Motion, Entertainment Research and Applications (CAMERA 2.0; EP/T022523/1). Some of this work was supported by NIGMS (5P20GM103653). The opinions in this article are not necessarily from the NIH.

References

Abondio, P., Sazzini, M., Garagnani, P., Boattini, A., Monti, D., Franceschi, C., Luiselli, D., Giuliani, C., 2019. The genetic variability of APOE in different human populations and its implications for longevity. Genes 10 (3), 222. Available at: https://doi.org/10.3390/genes10030222.

Akkus, A., Karasik, D., Roperto, R., 2017. Correlation between micro-hardness and mineral content in healthy human enamel. J. Clin. Exp. Dent. 9 (4), e569–e573. Available at: https://doi.org/10.4317/jced.53345.

Amarya, S., Singh, K., Sabharwal, M., 2018. Ageing Process and Physiological Changes. In: Gerontology. Available at: https://doi.org/10.5772/intechopen.76249.

Anderson, V.M., Schott, J.M., Bartlett, J.W., Leung, K.K., Miller, D.H., Fox, N.C., 2012. Gray matter atrophy rate as a marker of disease progression in AD. Neurobiol. Aging 33 (7), 1194–1202. Available at: https://doi.org/10.1016/J.NEUROBIOLAGING.2010.11.001.

Apostolova, L.G., Green, A.E., Babakchanian, S., Hwang, K.S., Chou, Y.Y., Toga, A.W., Thompson, P.M., 2012. Hippocampal atrophy and ventricular enlargement in normal aging, mild cognitive impairment (MCI), and Alzheimer disease. Alzheimer Dis. Assoc. Disord. 26 (1), 17–27. Available at: https://doi.org/10.1097/WAD.0b013e3182163b62.

Aranda-Anzaldo, A., 2012. The post-mitotic state in neurons correlates with a stable nuclear higher-order structure. Commun. Integr. Biol. 5 (2), 134–139. Available at: https://doi.org/10.4161/cib.18761.

Armstrong, N.M., An, Y., Shin, J.J., Williams, O.A., Doshi, J., Erus, G., Davatzikos, C., Ferrucci, L., Beason-Held, L.L., Resnick, S.M., 2020. Associations between cognitive and brain volume changes in cognitively normal older adults. Neuroimage 223, 117289. Available at: https://doi.org/10.1016/j.neuroimage.2020.117289.

Arnsten, A.F.T., Datta, D., Preuss, T.M., 2021a. Studies of aging nonhuman primates illuminate the etiology of early-stage Alzheimer's-like neuropathology: an evolutionary perspective. Am. J. Primatol. 83 (11). Available at https://doi.org/10.1002/ajp.23254.

Arnsten, A.F.T., Datta, D., Del Tredici, K., Braak, H., 2021b. Hypothesis: tau pathology is an initiating factor in sporadic Alzheimer's disease. Alzheimers Dement. 17 (1), 115–124. Available at: https://doi.org/10.1002/alz.12192.

Attwell, D., Laughlin, S.B., 2001. An energy budget for signaling in the grey matter of the brain. J. Cereb. Blood Flow Metab. 21 (10), 1133–1145. Available at: https://doi.org/10.1097/00004647-200110000-00001.

Auten, R.L., Davis, J.M., 2009. Oxygen toxicity and reactive oxygen species: the devil is in the details. Pediatr. Res. 66 (2), 121–127. Available at: https://doi.org/10.1203/PDR.0b013e3181a9eafb.

Autrey, M.M., Reamer, L.A., Mareno, M.C., Sherwood, C.C., Herndon, J.G., Preuss, T., Schapiro, S.J., Hopkins, W.D., 2014. Age-related effects in the neocortical organization of chimpanzees: gray and white matter volume, cortical thickness, and gyrification. Neuroimage 101, 59–67. Available at https://doi.org/10.1016/j.neuroimage.2014.06.053.

Azam, S., Haque, M.E., Balakrishnan, R., Kim, I.S., Choi, D.K., 2021. The ageing brain: molecular and cellular basis of neurodegeneration. Front. Cell Dev. Biol. 9, 683459. Available at: https://doi.org/10.3389/fcell.2021.683459.

Balanoff, A.M., Bever, G.S., 2016. The Role of Endocasts in the Study of Brain Evolution. In: Evolution of Nervous Systems, Second Edition. Academic Press, pp. 223–241. Available at: https://doi.org/10.1016/B978-0-12-804042-3.00023-3.

Barbosa, M.C., Grosso, R.A., Fader, C.M., 2019. Hallmarks of aging: an autophagic perspective. Front. Endocrinol. 10 (JAN), 790. Available at: https://doi.org/10.3389/fendo.2018.00790.

Barrickman, N.L., Bastian, M.L., Isler, K., van Schaik, C.P., 2008. Life history costs and benefits of encephalization: a comparative test using data from long-term studies of primates in the wild. J. Hum. Evol. 54 (5), 568–590. Available at: https://doi.org/10.1016/j.jhevol.2007.08.012.

Bauernfeind, A.L., Babbitt, C.C., 2020. Metabolic changes in human brain evolution. Evol. Anthropol. 29 (4), 201–211. Available at: https://doi.org/10.1002/evan.21831.

Bejanin, A., et al., 2017. Tau pathology and neurodegeneration contribute to cognitive impairment in Alzheimer's disease. Brain 140 (12), 3286. Available at: https://doi.org/10.1093/BRAIN/AWX243.

Berchtold, N.C., Cribbs, D.H., Coleman, P.D., Rogers, J., Head, E., Kim, R., Beach, T., Miller, C., Troncoso, J., Trojanowski, J.Q., Zielke, H.R., Cotman, C.W., 2008. Gene

expression changes in the course of normal brain aging are sexually dimorphic. Proc. Natl. Acad. Sci. U. S. A. 105 (40), 15605–15610. Available at https://doi.org/10.1073/pnas.0806883105.

Berchtold, N.C., Coleman, P.D., Cribbs, D.H., Rogers, J., Gillen, D.L., Cotman, C.W., 2013. Synaptic genes are extensively downregulated across multiple brain regions in normal human aging and Alzheimer's disease. Neurobiol. Aging 34 (6), 1653–1661. Available at: https://doi.org/10.1016/j.neurobiolaging.2012.11.024.

Blagosklonny, M.V., 2013. Aging is not programmed: genetic pseudo-program is a shadow of developmental growth. Cell Cycle 12 (24), 3736–3742. Available at: https://doi.org/10.4161/cc.27188.

Blinkouskaya, Y., Weickenmeier, J., 2021. Brain shape changes associated with cerebral atrophy in healthy aging and Alzheimer's disease. Front. Mech. Eng. 7, 64. Available at: https://doi.org/10.3389/fmech.2021.705653.

Blurton Jones, N.G., Hawkes, K., O'Connell, J.F., 2002. Antiquity of postreproductive life: are there modern impacts on hunter-gatherer postreproductive life spans? Am. J. Hum. Biol. 14 (2), 184–205. Available at: https://doi.org/10.1002/ajhb.10038.

Bohnen, N.I., Albin, R.L., 2011. White matter lesions in Parkinson disease. Nat. Rev. Neurol. 7 (4), 229–236. Available at: https://doi.org/10.1038/nrneurol.2011.21.

Bolam, J.P., Pissadaki, E.K., 2012. Living on the edge with too many mouths to feed: why dopamine neurons die. Mov. Disord. 27 (12), 1478–1483. Available at: https://doi.org/10.1002/MDS.25135.

Book, A., Guella, I., Candido, T., Brice, A., Hattori, N., Jeon, B., Farrer, M.J., 2018. A meta-analysis of α-synuclein multiplication in familial parkinsonism. Front. Neurol. 9, 1021. Available at: https://doi.org/10.3389/fneur.2018.01021.

Braak, H., Braak, E., 1991. Neuropathological stageing of Alzheimer-related changes. Acta Neuropathol. 82 (4), 239–259. Available at: https://doi.org/10.1007/BF00308809.

Braak, H., Alafuzoff, I., Arzberger, T., Kretzschmar, H., Tredici, K., 2006. Staging of Alzheimer disease-associated neurofibrillary pathology using paraffin sections and immunocytochemistry. Acta Neuropathol. 112 (4), 389–404. Available at: https://doi.org/10.1007/s00401-006-0127-z.

Bras, J., Guerreiro, R., Darwent, L., Parkkinen, L., Ansorge, O., Escott-Price, V., Hernandez, D.G., Nalls, M.A., Clark, L.N., Honig, L.S., Marder, K., 2014. Genetic analysis implicates APOE, SNCA and suggests lysosomal dysfunction in the etiology of dementia with Lewy bodies. Hum. Mol. Genet. 23 (23), 6139–6146.

Bruner, E., Jacobs, H.I.L., 2013. Alzheimer's disease: the downside of a highly evolved parietal lobe? J. Alzheimers Dis. 35 (2), 227–240. Available at: https://doi.org/10.3233/JAD-122299.

Bruner, E., Preuss, T.M., Chen, X., Rilling, J.K., 2017. Evidence for expansion of the precuneus in human evolution. Brain Struct. Funct. 222 (2), 1053–1060. Available at: https://doi.org/10.1007/s00429-015-1172-y.

Burger, J.R., George, M.A., Leadbetter, C., Shaikh, F., 2019. The allometry of brain size in mammals. J. Mammal. 100 (2), 276–283. Available at: https://doi.org/10.1093/JMAMMAL/GYZ043.

Capellini, T.D., Chen, H., Cao, J., Doxey, A.C., Kiapour, A.M., Schoor, M., Kingsley, D.M., 2017. Ancient selection for derived alleles at a GDF5 enhancer influencing human growth and osteoarthritis risk. Nat. Genet. 49 (8), 1202–1210. Available at: https://doi.org/10.1038/ng.3911.

Caron, F., d'Errico, F., Del Moral, P., Santos, F., Zilhão, J., 2011. The reality of Neandertal symbolic behavior at the Grotte du Renne, Arcy-sur-Cure, France. PLoS One 6 (6), e21545. Available at https://doi.org/10.1371/JOURNAL.PONE.0021545.

Caspari, R., Lee, S.H., 2006. Is human longevity a consequence of cultural change or modern biology? Am. J. Phys. Anthropol. 129 (4), 512–517. Available at: https://doi.org/10.1002/ajpa.20360.

Castillo-Morales, A., Monzón-Sandoval, J., Urrutia, A.O., Gutiérrez, H., 2019. Postmitotic cell longevity–associated genes: a transcriptional signature of postmitotic maintenance in neural tissues. Neurobiol. Aging 74, 147–160. Available at: https://doi.org/10.1016/j.neurobiolaging.2018.10.015.

Chartier, S., Duyckaerts, C., 2018. Is Lewy pathology in the human nervous system chiefly an indicator of neuronal protection or of toxicity? Cell Tissue Res. 373 (1), 149–160. Available at: https://doi.org/10.1007/s00441-018-2854-6.

Chatterjee, S., Mudher, A., 2018. Alzheimer's disease and type 2 diabetes: a critical assessment of the shared pathological traits. Front. Neurosci. 12. Available at: https://doi.org/10.3389/fnins.2018.00383.

Chaudhary, S., Kumaran, S.S., Goyal, V., Kaloiya, G.S., Kalaivani, M., Jagannathan, N.R., Sagar, R., Mehta, N., Srivastava, A.K., 2021. Cortical thickness and gyrification index measuring cognition in Parkinson's disease. Int. J. Neurosci. 131 (10), 984–993. Available at: https://doi.org/10.1080/00207454.2020.1766459.

Chen, G., Kroemer, G., Kepp, O., 2020. Mitophagy: an emerging role in aging and age-associated diseases. Front. Cell Dev. Biol. 8, 200. Available at: https://doi.org/10.3389/fcell.2020.00200.

Cheng, Y.J., Lin, C.H., Lane, H.Y., 2021. Involvement of cholinergic, adrenergic, and glutamatergic network modulation with cognitive dysfunction in alzheimer's disease. Int. J. Mol. Sci. 22 (5), 1–15. Available at: https://doi.org/10.3390/ijms22052283.

Chohan, H., Senkevich, K., Patel, R.K., Bestwick, J.P., Bandres Ciga, S., Gan-Or, Z., Noyce, A.J., 2021. Type 2 diabetes as a determinant of Parkinson's disease risk and progression. Mov. Disord. 36 (6), 1420–1429. Available at: https://doi.org/10.1002/MDS.28551.

Choi, S.E., Kemper, J.K., 2013. Regulation of SIRT1 by microRNAs. Mol. Cells 36 (5), 385–392. Available at: https://doi.org/10.1007/s10059-013-0297-1.

Chu, Y., Kordower, J.H., 2007. Age-associated increases of α-synuclein in monkeys and humans are associated with nigrostriatal dopamine depletion: is this the target for Parkinson's disease? Neurobiol. Dis. 25 (1), 134–149. Available at: https://doi.org/10.1016/j.nbd.2006.08.021.

Chun, K.J., Choi, H.H., Lee, J.Y., 2014. Comparison of mechanical property and role between enamel and dentin in the human teeth. J. Dent. Biomech. 5 (1), 1–7. Available at: https://doi.org/10.1177/1758736014520809.

Chung, H.Y., Kim, D.H., Lee, E.K., Chung, K.W., Chung, S., Lee, B., Seo, A.Y., Chung, J.H., Jung, Y.S., Im, E., Lee, J., Kim, N.D., Choi, Y.J., Im, D.S., Yu, B.P., 2019. Redefining chronic inflammation in aging and age-related diseases: proposal of the senoinflammation concept. Aging Dis. 10 (2), 367. Available at: https://doi.org/10.14336/AD.2018.0324.

Cieślik, M., Czapski, G.A., Wójtowicz, S., Wieczorek, I., Wencel, P.L., Strosznajder, R.P., Jaber, V., Lukiw, W.J., Strosznajder, J.B., 2020. Alterations of transcription of genes coding anti-oxidative and mitochondria-related proteins in amyloid β toxicity: relevance to alzheimer's disease. Mol. Neurobiol. 57 (3), 1374–1388. Available at: https://doi.org/10.1007/s12035-019-01819-y.

Cogo, S., Greggio, E., 2020. Leucine-rich repeat kinase 2 (LRRK2) and Parkinson's disease: from genetics to pathobiology. In: Martin, C.R., Preedy, V.R. (Eds.), Genetics, Neurology, Behavior, and Diet in Parkinson's Disease. Academic Press, pp. 3–18. Available at https://doi.org/10.1016/B978-0-12-815950-7.00001-1.

Collier, T.J., Kanaan, N.M., Kordower, J.H., 2011. Ageing as a primary risk factor for Parkinson's disease: evidence from studies of non-human primates. Nat. Rev. Neurosci. 12 (6), 359–366. Available at: https://doi.org/10.1038/nrn3039.

Collier, T.J., Kanaan, N.M., Kordower, J.H., 2017. Aging and Parkinson's disease: different sides of the same coin? Mov. Disord. 32 (7), 983–990. Available at: https://doi.org/10.1002/mds.27037.

Cornutiu, G., 2011. J. Neurodegener. Dis. 8 (1–2), 9–14. Available at: https://doi.org/10.1159/000313659.

Cramer, P.E., Gentzel, R.C., Tanis, K.Q., Vardigan, J., Wang, Y., Connolly, B., Manfre, P., Lodge, K., Renger, J.J., Zerbinatti, C., Uslaner, J.M., 2018. Aging African green monkeys manifest transcriptional, pathological, and cognitive hallmarks of human Alzheimer's disease. Neurobiol. Aging 64, 92–106. Available at: https://doi.org/10.1016/j.neurobiolaging.2017.12.011.

Cruts, M., Theuns, J., Van Broeckhoven, C., 2012. Locus-specific mutation databases for neurodegenerative brain diseases. Hum. Mutat. 33 (9), 1340–1344. Available at: https://doi.org/10.1002/humu.22117.

Dai, M.H., Zheng, H., Zeng, L.D., Zhang, Y., 2018. The genes associated with early-onset Alzheimer's disease. Oncotarget 9 (19), 15132–15143. Available at https://doi.org/10.18632/oncotarget.23738.

Dannemann, M., 2021. The population-specific impact of neandertal introgression on human disease. Genome Biol. Evol. 13 (1), 250. Available at: https://doi.org/10.1093/gbe/evaa250.

Das, G., Shravage, B.V., Baehrecke, E.H., 2012. Regulation and function of autophagy during cell survival and cell death. Cold Spring Harb. Perspect. Biol. 4 (6), 1–14. Available at: https://doi.org/10.1101/CSHPERSPECT.A008813.

Dassati, S., Waldner, A., Schweigreiter, R., 2014. Apolipoprotein D takes center stage in the stress response of the aging and degenerative brain. Neurobiol. Aging 35 (7), 1632–1642. Available at: https://doi.org/10.1016/j.neurobiolaging.2014.01.148.

de Felice, F.G., Lourenco, M.V., 2015. Brain metabolic stress and neuroinflammation at the basis of cognitive impairment in Alzheimer's disease. Front. Aging Neurosci. 7 (MAY), 94. Available at: https://doi.org/10.3389/fnagi.2015.00094.

De Leeuw, F.E., Barkhof, F., Scheltens, P., 2005. Progression of cerebral white matter lesions in Alzheimer's disease: a new window for therapy? J. Neurol. Neurosurg. Psychiatry 76 (9), 1286–1288. Available at: https://doi.org/10.1136/jnnp.2004.053686.

De Magalhães, J.P., Church, G.M., 2006. Cells discover fire: employing reactive oxygen species in development and consequences for aging. Exp. Gerontol. 41 (1), 1–10. Available at: https://doi.org/10.1016/j.exger.2005.09.002.

de Magalhães, J.P., Curado, J., Church, G.M., 2009. Meta-analysis of age-related gene expression profiles identifies common signatures of aging. Bioinformatics 25 (7), 875–881. Available at: https://doi.org/10.1093/bioinformatics/btp073.

De Schipper, L.J., Hafkemeijer, A., Van Der Grond, J., Marinus, J., Henselmans, J.M.L., Van Hilten, J.J., 2019a. Regional structural hippocampal differences between dementia with Lewy bodies and Parkinson's disease. J. Parkinsons Dis. 9 (4), 775–783. Available at: https://doi.org/10.3233/JPD-191600.

de Schipper, L.J., Hafkemeijer, A., Bouts, M.J.R.J., van der Grond, J., Marinus, J., Henselmans, J.M.L., van Hilten, J.J., 2019b. Age- and disease-related cerebral white matter changes in patients with Parkinson's disease. Neurobiol. Aging 80, 203–209. Available at: https://doi.org/10.1016/j.neurobiolaging.2019.05.004.

de Sousa, A., Wood, B., 2007. The hominin fossil record and the emergence of the modern human central nervous system. In: Preuss, T.M., Kaas, J.H. (Eds.), Evolution of Nervous Systems. Academic Press, Oxford, pp. 291–336. Available at: https://doi.org/10.1016/B0-12-370878-8/00018-5.

Deas, E., Wood, N.W., Plun-Favreau, H., 2011. Mitophagy and Parkinson's disease: the PINK1-parkin link. Biochim. Biophys. Acta 1813 (4), 623–633. Available at: https://doi.org/10.1016/J.BBAMCR.2010.08.007.

DeCasien, A.R., Thompson, N.A., Williams, S.A., Shattuck, M.R., 2018. Encephalization and longevity evolved in a correlated fashion in Euarchontoglires but not in other mammals. Evolution 72 (12), 2617–2631. Available at: https://doi.org/10.1111/evo.13633.

Deneris, E., Gaspar, P., 2018. Serotonin neuron development: shaping molecular and structural identities. Wiley Interdiscip. Rev. Dev. Biol. 7 (1), 301. Available at: https://doi.org/10.1002/wdev.301.

DeSilva, J.M., Traniello, J.F.A., Claxton, A.G., Fannin, L.D., 2021. When and why did human brains decrease in size? A new change-point analysis and insights from brain evolution in ants. Front. Ecol. Evol. 9, 712. Available at: https://doi.org/10.3389/fevo.2021.742639.

Dettwyler, K.A., 1995. A Time to Wean: The Hominid Blueprint for the Natural Age of Weaning in Modern Human Populations. In: Stewart-MacAdam, P., Dettwyler, K.A. (Eds.), Breastfeeding. Aldine de Gruyter, New York, pp. 39–74. Available at: https://doi.org/10.4324/9781315081984-2.

Di Benedetto, G., Iannucci, L.F., Surdo, N.C., Zanin, S., Conca, F., Grisan, F., Gerbino, A., Lefkimmiatis, K., 2021. Compartmentalized signaling in aging and neurodegeneration. Cell 10 (2), 1–31. Available at: https://doi.org/10.3390/cells10020464.

Dickson, D.W., et al., 2018. APOE ε4 is associated with severity of Lewy body pathology independent of Alzheimer pathology. Neurology 91 (12), e1182. Available at: https://doi.org/10.1212/WNL.0000000000006212.

Diederich, N.J., James Surmeier, D., Uchihara, T., Grillner, S., Goetz, C.G., 2019. Parkinson's disease: is it a consequence of human brain evolution? Mov. Disord. 34 (4), 453–459. Available at: https://doi.org/10.1002/mds.27628.

Diederich, N.J., Uchihara, T., Grillner, S., Goetz, C.G., 2020. The evolution-driven signature of Parkinson's disease. Trends Neurosci. 43 (7), 475–492. Available at: https://doi.org/10.1016/j.tins.2020.05.001.

Diniz-Filho, J.A.F., Jardim, L., Mondanaro, A., Raia, P., 2019. Multiple components of phylogenetic non-stationarity in the evolution of brain size in fossil hominins. Evol. Biol. 46 (1), 47–59. Available at: https://doi.org/10.1007/s11692-019-09471-z.

Dissing, J., Kristinsdottir, M., Friis, C., 2008. On the elimination of extraneous DNA in fossil human teeth with hypochlorite. J. Archaeol. Sci. 35 (6), 1445–1452. Available at: https://doi.org/10.1016/J.JAS.2007.10.009.

Dobzhansky, T., 1973. Nothing in biology makes sense except in the light of evolution. Am. Biol. Teach. 35 (3), 125–129. Available at: https://doi.org/10.2307/4444260.

Doherty, A., de Magalhães, J.P., 2016. Has gene duplication impacted the evolution of Eutherian longevity? Aging Cell 15 (5), 978–980. Available at: https://doi.org/10.1111/acel.12503.

Dorsey, E.R., Sherer, T., Okun, M.S., Bloemd, B.R., 2018. The emerging evidence of the Parkinson pandemic. J. Parkinsons Dis. 8 (s1), S3–S8. Available at: https://doi.org/10.3233/JPD-181474.

Driscoll, I., Davatzikos, C., An, Y., Wu, X., Shen, D., Kraut, M., Resnick, S.M., 2009. Longitudinal pattern of regional brain volume change differentiates normal aging from MCI. Neurology 72 (22), 1906–1913. Available at: https://doi.org/10.1212/WNL.0b013e3181a82634.

Du, F., Yu, Q., Yan, S., Hu, G., Lue, L.F., Walker, D.G., Wu, L., Yan, S.F., Tieu, K., Yan, S.S., 2017. PINK1 signalling rescues amyloid pathology and mitochondrial dysfunction in Alzheimer's disease. Brain 140 (12), 3233. Available at: https://doi.org/10.1093/BRAIN/AWX258.

Dunkel, C., Mathes, E., Decker, M., 2010. Behavioral flexibility in life history strategies: the role of life expectancy. J. Soc. Evol. Cult. Psychol. 4 (2), 51. Available at: https://doi.org/10.1037/H0099301.

Dunsworth, H., 2015. How to become a primate fossil. Nat. Edu. Knowledge 6 (7), 1–11. Available at: https://www.nature.com/scitable/knowledge/library/how-to-become-a-primate-fossil-135630567/#:~:text=If%20a%20primate%20is%20going,is%20constantly%20building%20up%20sediment.

Edler, M.K., Sherwood, C.C., Meindl, R.S., Hopkins, W.D., Ely, J.J., Erwin, J.M., Mufson, E.J., Hof, P.R., Raghanti, M.A., 2017. Aged chimpanzees exhibit pathologic hallmarks of Alzheimer's disease. Neurobiol. Aging 59, 107–120. Available at https://doi.org/10.1016/j.neurobiolaging.2017.07.006.

Edler, M.K., Sherwood, C.C., Meindl, R.S., Munger, E.L., Hopkins, W.D., Ely, J.J., Erwin, J.M., Perl, D.P., Mufson, E.J., Hof, P.R., Raghanti, M.A., 2018. Microglia changes associated to Alzheimer's disease pathology in aged chimpanzees. J Comp Neurol 526 (18), 2921–2936. Available at https://doi.org/10.1002/cne.24484.

Edler, M.K., Munger, E.L., Meindl, R.S., Hopkins, W.D., Ely, J.J., Erwin, J.M., Mufson, E.J., Hof, P.R., Sherwood, C.C., Raghanti, M.A., 2020. Neuron loss associated with age but not Alzheimer's disease pathology in the chimpanzee brain. Philos. Trans. R. Soc. B: Biol. Sci. 375 (1811). Available at https://doi.org/10.1098/rstb.2019.0619.

Ellis, S., Franks, D.W., Nattrass, S., Currie, T.E., Cant, M.A., Giles, D., Balcomb, K.C., Croft, D.P., 2018. Analyses of ovarian activity reveal repeated evolution of post-reproductive lifespans in toothed whales. Sci. Rep. 8 (1), 1–10. Available at: https://doi.org/10.1038/s41598-018-31047-8.

Emborg, M.E., Ma, S.Y., Mufson, E.J., Levey, A.I., Taylor, M.D., Brown, W.D., Holden, J.E., Kordower, J.H., 1998. Age-related declines in nigral neuronal function correlate with motor impairments in rhesus monkeys. J Comp Neurol 401 (2), 253–265. Available at: https://doi.org/10.1002/(SICI)1096-9861(19981116)401:2<253::AID-CNE7>3.0.CO;2-X.

Enard, W., 2012. Functional primate genomics—leveraging the medical potential. J. Mol. Med. 90 (5), 471–480. Available at: https://doi.org/10.1007/s00109-012-0901-4.

Fabian, D., Flatt, T., 2011. The evolution of aging. Nat. Edu. Knowledge 3, 1–10. Available at: https://folia.unifr.ch/documents/307636/files/2011_eoa.pdf.

Falk, D., 2014. Interpreting sulci on hominin endocasts: old hypotheses and new findings. Front. Hum. Neurosci. 8 (MAY). Available at https://doi.org/10.3389/FNHUM.2014.00134.

Ferreira-Vieira, T.H., Guimaraes, I.M., Silva, F.R., Ribeiro, F.M., 2016. Alzheimer's disease: targeting the cholinergic system. Curr. Neuropharmacol. 14 (1), 101. Available at: https://doi.org/10.2174/1570159X13666150716165726.

Fields, R.D., 2010. Change in the brain's white matter. Science 330 (6005), 768–769. Available at: https://doi.org/10.1126/science.1199139.

Finch, C.E., 2010. Evolution of the human lifespan and diseases of aging: roles of infection, inflammation, and nutrition. Proc. Natl. Acad. Sci. U. S. A. 107 (Suppl. 1), 1718–1724. Available at https://doi.org/10.1073/pnas.0909606106.

Finch, C.E., Austad, S.N., 2012. Primate aging in the mammalian scheme: the puzzle of extreme variation in brain aging. Age 34 (5), 1075–1091. Available at: https://doi.org/10.1007/s11357-011-9355-9.

Finch, C.E., Austad, S.N., 2015. Commentary: is Alzheimer's disease uniquely human? Neurobiol. Aging 36 (2), 553–555. Available at: https://doi.org/10.1016/j.neurobiolaging.2014.10.025.

Finch, C.E., Sapolsky, R.M., 1999. The evolution of Alzheimer disease, the reproductive schedule, and apoE isoforms. Neurobiol. Aging 20 (4), 407–428. Available at: https://doi.org/10.1016/S0197-4580(99)00053-6.

Fjell, A.M., McEvoy, L., Holland, D., Dale, A.M., Walhovd, K.B., Alzheimer's Disease Neuroimaging Initiative, 2014a. What is normal in normal aging? Effects of aging, amyloid and Alzheimer's disease on the cerebral cortex and the hippocampus. Prog. Neurobiol. 117, 20–40. Available at: https://doi.org/10.1016/j.pneurobio.2014.02.004.

Fjell, A.M., Westlye, L.T., Grydeland, H., Amlien, I., Espeseth, T., Reinvang, I., Raz, N., Dale, A.M., Walhovd, K.B., Alzheimer's Disease Neuroimaging Initiative, 2014b. Accelerating cortical thinning: unique to dementia or universal in aging? Cereb. Cortex 24 (4), 919–934. Available at: https://doi.org/10.1093/cercor/bhs379.

Fletcher, E., Gavett, B., Harvey, D., Farias, S.T., Olichney, J., Beckett, L., DeCarli, C., Mungas, D., 2018. Brain volume change and cognitive trajectories in aging. Neuropsychology 32 (4), 436–449. Available at: https://doi.org/10.1037/neu0000447.

Foster, E.A., Franks, D.W., Mazzi, S., Darden, S.K., Balcomb, K.C., Ford, J.K.B., Croft, D.P., 2012. Adaptive prolonged postreproductive life span in killer whales. Science 337 (6100), 1313. Available at: https://doi.org/10.1126/science.1224198.

Fox, M., 2018. 'Evolutionary medicine' perspectives on Alzheimer's Disease: review and new directions. Ageing Res. Rev. 47, 140–148. Available at: https://doi.org/10.1016/j.arr.2018.07.008.

Franceschi, C., Campisi, J., 2014. Chronic inflammation (inflammaging) and its potential contribution to age-associated diseases. J. Gerontol. A Biol. Sci. 69 (Suppl_1), S4–S9. Available at: https://doi.org/10.1093/gerona/glu057.

Freire-Cobo, C., et al., 2021. Comparative neuropathology in aging primates: a perspective. Am. J. Primatol. 83 (11), 23299. Available at: https://doi.org/10.1002/ajp.23299.

Frenk, S., Houseley, J., 2018. Gene expression hallmarks of cellular ageing. Biogerontology 19 (6), 547–566. Available at: https://doi.org/10.1007/s10522-018-9750-z.

Fung, C.W., Guo, J., Fu, H., Figueroa, H.Y., Konofagou, E.E., Duff, K.E., 2020. Atrophy associated with tau pathology precedes overt cell death in a mouse model of progressive tauopathy. Sci. Adv. 6 (42), 8098. Available at: https://doi.org/10.1126/sciadv.abc8098.

Gabuzda, D., Yankner, B.A., 2013. Physiology: inflammation links ageing to the brain. Nature 497 (7448), 197–198. Available at: https://doi.org/10.1038/nature12100.

Garcia, A.R., Finch, C., Gatz, M., Kraft, T., Rodriguez, D.E., Cummings, D., Charifson, M., Buetow, K., Beheim, B.A., Allayee, H., Thomas, G.S., Stieglitz, J., Gurven, M.D., Kaplan, H., Trumble, B.C., 2021. Apoe4 is associated with elevated blood lipids and lower levels of innate immune biomarkers in a tropical amerindian subsistence population. Elife 10. Available at https://doi.org/10.7554/eLife.68231.

Gatz, M., Pedersen, N.L., Berg, S., Johansson, B., Johansson, K., Mortimer, J.A., Posner, S.F., Viitanen, M., Winblad, B., Ahlbom, A., 1997. Heritability for Alzheimer's disease: the study of dementia in Swedish twins. J. Gerontol. A Biol. Sci. 52 (2), 117–125. Available at: https://doi.org/10.1093/gerona/52A.2.M117.

Gatz, M., Reynolds, C.A., Fratiglioni, L., Johansson, B., Mortimer, J.A., Berg, S., Fiske, A., Pedersen, N.L., 2006. Role of genes and environments for explaining Alzheimer disease. Arch. Gen. Psychiatry 63 (2), 168–174. Available at: https://doi.org/10.1001/archpsyc.63.2.168.

Gearing, M., Tigges, J., Mori, H., Mirra, S.S., 1997. β-Amyloid (Aβ) deposition in the brains of aged orangutans. Neurobiol. Aging 18 (2), 139–146. Available at: https://doi.org/10.1016/S0197-4580(97)00012-2.

Genin, E., et al., 2011. APOE and Alzheimer disease: a major gene with semi-dominant inheritance. Mol. Psychiatry 16 (9), 903–907. Available at https://doi.org/10.1038/mp.2011.52.

Gharbi-Meliani, A., Dugravot, A., Sabia, S., Regy, M., Fayosse, A., Schnitzler, A., Kivimäki, M., Singh-Manoux, A., Dumurgier, J., 2021. The association of APOE ε4 with cognitive function over the adult life course and incidence of dementia: 20 years follow-up of the Whitehall II study. Alzheimer's Res. Ther. 13 (1), 1–11. Available at: https://doi.org/10.1186/s13195-020-00740-0.

Gilbert, S.F., 2000. Aging: The Biology of Senescence. Available at: https://www.ncbi.nlm.nih.gov/books/NBK10041/ (Accessed: 6 September 2022).

Gilissen, E.P., Dhenain, M., Allman, J.M., 2001. Brain Aging in Strepsirhine Primates. In: Functional Neurobiology of Aging. Academic Press, pp. 421–433. Available at: https://doi.org/10.1016/b978-012351830-9/50029-9.

Gitler, A.D., Dhillon, P., Shorter, J., 2017. Neurodegenerative disease: models, mechanisms, and a new hope. Dis. Model Mech. 10 (5), 499–502. Available at: https://doi.org/10.1242/dmm.030205.

Gokhman, D., Lavi, E., Prüfer, K., Fraga, M.F., Riancho, J.A., Kelso, J., Pääbo, S., Meshorer, E., Carmel, L., 2014. Reconstructing the DNA methylation maps of the neandertal and the Denisovan. Science 344 (6183), 523–527. Available at: https://doi.org/10.1126/science.1250368.

Gomperts, S.N., 2016. Lewy body dementias: Dementia with lewy bodies and Parkinson disease dementia. Continuum (Minneap Minn) 22 (2, Dementia), 435–463. Available at: https://doi.org/10.1212/CON.0000000000000309.

Gomperts, S.N., Locascio, J.J., Makaretz, S.J., Schultz, A., Caso, C., Vasdev, N., Sperling, R., Growdon, J.H., Dickerson, B.C., Johnson, K., 2016. Tau PET imaging in the Lewy body diseases. JAMA Neurol. 73 (11), 1334. Available at: https://doi.org/10.1001/JAMANEUROL.2016.3338.

González-Lagos, C., Sol, D., Reader, S.M., 2010. Large-brained mammals live longer. J. Evol. Biol. 23 (5), 1064–1074. Available at: https://doi.org/10.1111/j.1420-9101.2010.01976.x.

González-Velasco, O., Papy-García, D., Le Douaron, G., Sánchez-Santos, J.M., De Las Rivas, J., 2020. Transcriptomic landscape, gene signatures and regulatory profile of aging in the human brain. Biochim. Biophys. Acta 1863 (6), 194491. Available at: https://doi.org/10.1016/J.BBAGRM.2020.194491.

Good, C.D., Johnsrude, I.S., Ashburner, J., Henson, R.N.A., Fristen, K.J., Frackowiak, R.S.J., 2001. A voxel-based morphometric study of ageing in 465 normal adult human brains. Neuroimage 14 (1), 21–36. Biomedical Imaging V - Proceedings of the 5th IEEE EMBS International Summer School on Biomedical Imaging, SSBI 2002. Available at https://doi.org/10.1109/SSBI.2002.1233974.

Gorbunova, V., Seluanov, A., Mao, Z., Hine, C., 2007. Changes in DNA repair during aging. Nucleic Acids Res. 35 (22), 7466. Available at: https://doi.org/10.1093/NAR/GKM756.

Gruschus, J.M., 2021. Monkey Brains – Evolution of the human brain and the vulnerability of the substantia nigra. In: Synuclein and the Coelacanth, pp. 181–191. Available at https://doi.org/10.1016/b978-0-323-85707-9.00016-2.

Gunn-Moore, D., Kaidanovich-Beilin, O., Gallego Iradi, M.C., Gunn-Moore, F., Lovestone, S., 2018. Alzheimer's disease in humans and other animals: a consequence of postreproductive life span and longevity rather than aging. Alzheimers Dement. 14 (2), 195–204. Available at: https://doi.org/10.1016/j.jalz.2017.08.014.

Gureev, A.P., Shaforostova, E.A., Popov, V.N., 2019. Regulation of mitochondrial biogenesis as a way for active longevity: interaction between the Nrf2 and PGC-1α signaling pathways. Front. Genet. 10 (MAY), 435. Available at: https://doi.org/10.3389/fgene.2019.00435.

Gureje, O., Ogunniyi, A., Baiyewu, O., Price, B., Unverzagt, F.W., Evans, R.M., Smith-Gamble, V., Lane, K.A., Gao, S., Hall, K.S., Hendrie, H.C., Murrell, J.R., 2006. APOE ε4 is not associated with Alzheimer's disease in elderly Nigerians. Ann. Neurol. 59 (1), 182–185. Available at: https://doi.org/10.1002/ana.20694.

Haldane, J.B.S., 1941. *New Paths in Genetics*, Nature. Allen & Unwin, London, UK. Available at: https://doi.org/10.1038/149317a0.

Hamilton, W.D., 1964. The genetical evolution of social behaviour. I. J. Theor. Biol. 7 (1), 1–16. Available at: https://doi.org/10.1016/0022-5193(64)90038-4.

Hanseeuw, B.J., et al., 2019. Association of amyloid and tau with cognition in preclinical Alzheimer disease: a longitudinal study. JAMA Neurol. 76 (8), 915–924. Available at: https://doi.org/10.1001/JAMANEUROL.2019.1424.

Harman, D., 1956. Aging: a theory based on free radical and radiation chemistry. J. Gerontol. 11 (3), 298–300. Available at: https://doi.org/10.1093/geronj/11.3.298.

Harman, D., 2009. Origin and evolution of the free radical theory of aging: a brief personal history, 1954-2009. Biogerontology 10 (6), 773–781. Available at: https://doi.org/10.1007/s10522-009-9234-2.

Hawkes, K., Chisholm, J.S., Fairbanks, L.A., Johow, J., Kalcher-Sommersguter, E., Liebal, K., Myowa, M., Sommer, V., Thierry, B., Finlay, B.L., 2017. Primate infancies causes and consequences of varying care. Cultural Nature of Attachment: Contextualizing Relationships and Development 22, 69–107.

He, H., Liang, L., Tang, T., Luo, J., Wang, Y., Cui, H., 2020. Progressive brain changes in Parkinson's disease: a meta-analysis of structural magnetic resonance imaging studies. Brain Res. 1740. Available at https://doi.org/10.1016/j.brainres.2020.146847.

Hebert, L.E., Weuve, J., Scherr, P.A., Evans, D.A., 2013. Alzheimer disease in the United States (2010–2050) estimated using the 2010 census. Neurology 80 (19), 1778. Available at: https://doi.org/10.1212/WNL.0B013E31828726F5.

Herndon, J., Paredes, J., Wilson, M., Bloomsmith, M.A., Chennareddi, L., Walker, M.L., 2012. Menopause occurs late in life in the captive chimpanzee (*Pan troglodytes*). Age 34 (5), 1145–1156. Available at: https://doi.org/10.1007/s11357-011-9351-0.

Herndon, J.G., Tigges, J., Anderson, D.C., Klumpp, S.A., McClure, H.M., 1999. Brain weight throughout the life span of the chimpanzee. J. Comp. Neurol. 409 (4), 567–572. Available at: https://doi.org/10.1002/(SICI)1096-9861(19990712)409:4<567::AID-CNE4->3.0.CO;2-J.

Hirsch, E.C., Hunot, S., 2009. Neuroinflammation in Parkinson's disease: a target for neuroprotection? Lancet Neurol. 8 (4), 382–397. Available at: https://doi.org/10.1016/S1474-4422(09)70062-6.

Ho, N., Liauw, J.A., Blaeser, F., Wei, F., Hanissian, S., Muglia, L.M., Wozniak, D.F., Nardi, A., Arvin, K.L., Holtzman, D.M., Linden, D.J., Zhuo, M., Muglia, L.J., Chatila, T.A., 2000. Impaired synaptic plasticity and cAMP response element-binding protein activation in Ca2+/calmodulin-dependent protein kinase type IV/Gr-Deficient mice. J. Neurosci. 20 (17), 6459–6472. Available at: https://doi.org/10.1523/jneurosci.20-17-06459.2000.

Hopkins, W.D., Li, X., Crow, T., Roberts, N., 2017. Vertex- and atlas-based comparisons in measures of cortical thickness, gyrification and white matter volume between humans and chimpanzees. Brain Struct. Funct. 222 (1), 229–245. Available at: https://doi.org/10.1007/s00429-016-1213-1.

Horwitz, B., Duara, R., Rapoport, S.I., 1984. Intercorrelations of glucose metabolic rates between brain regions: application to healthy males in a state of reduced sensory input. J. Cereb. Blood Flow Metab. 4 (4), 484–499. Available at: https://doi.org/10.1038/jcbfm.1984.73.

Hou, Y., Dan, X., Babbar, M., Wei, Y., Hasselbalch, S.G., Croteau, D.L., Bohr, V.A., 2019. Ageing as a risk factor for neurodegenerative disease. Nat. Rev. Neurol. 15 (10), 565–581. Available at: https://doi.org/10.1038/s41582-019-0244-7.

Hu, Y., Zhao, T., Zang, T., Zhang, Y., Cheng, L., 2019. Identification of Alzheimer's disease-related genes based on data integration method. Front. Genet. 10 (JAN), 703. Available at: https://doi.org/10.3389/fgene.2018.00703.

Hurley, P.J., Elsworth, J.D., Whittaker, M.C., Roth, R.H., Redmond, D.E., 2011. Aged monkeys as a partial model for Parkinson's disease. Pharmacol. Biochem. Behav. 99 (3), 324–332. Available at: https://doi.org/10.1016/j.pbb.2011.05.007.

Ikram, M.A., Vrooman, H.A., Vernooij, M.W., van der Lijn, F., Hofman, A., van der Lugt, A., Niessen, W.J., Breteler, M.M.B., 2008. Brain tissue volumes in the general elderly population. The Rotterdam Scan Study. Neurobiol. Aging 29 (6), 882–890. Available at: https://doi.org/10.1016/j.neurobiolaging.2006.12.012.

IUCN (2022) The IUCN red list of threatened species. Available at: https://www.iucnredlist.org./ (Accessed: 14 October 2022).

Jenner, P., 2003. Oxidative stress in Parkinson's disease. Ann. Neurol. 53 (S3), 26–38. Available at: https://doi.org/10.1002/ana.10483.

Jia, X., Liang, P., Li, Y., Shi, L., Wang, D., Li, K., 2015. Longitudinal study of gray matter changes in Parkinson disease. Am. J. Neuroradiol. 36 (12), 2219–2226. Available at: https://doi.org/10.3174/ajnr.A4447.

Jiménez-Ortega, D., Kolm, N., Immler, S., Maklakov, A.A., Gonzalez-Voyer, A., 2020. Long life evolves in large-brained bird lineages. Evolution 74 (12), 2617–2628. Available at: https://doi.org/10.1111/evo.14087.

Johnson, A.A., Shokhirev, M.N., Shoshitaishvili, B., 2019. Revamping the evolutionary theories of aging. Ageing Res. Rev. 55, 100947. Available at: https://doi.org/10.1016/j.arr.2019.100947.

Jones, R., 2010. The roles of PINK1 and parkin in Parkinson's disease. PLoS Biol. 8 (1), 1000299. Available at: https://doi.org/10.1371/journal.pbio.1000299.

Julien, C., Tremblay, C., Émond, V., Lebbadi, M., Salem, N., Bennett, D.A., Calon, F., 2009. Sirtuin 1 reduction parallels the accumulation of tau in Alzheimer disease. J. Neuropathol. Exp. Neurol. 68 (1), 48–58. Available at: https://doi.org/10.1097/NEN.0b013e3181922348.

Kelman, L.M., Kelman, Z., 1999. The use of ancient DNA in paleontological studies. J. Vertebr. Paleontol. 19 (1), 8–20. Available at: https://doi.org/10.1080/02724634.1999.10011118.

Kerr, J.S., Adriaanse, B.A., Greig, N.H., Mattson, M.P., Cader, M.Z., Bohr, V.A., Fang, E.F., 2017. Mitophagy and Alzheimer's disease: cellular and molecular mechanisms. Trends Neurosci. 40 (3), 151–166. Available at: https://doi.org/10.1016/j.tins.2017.01.002.

Kim, P.S., Coxworth, J.E., Hawkes, K., 2012. Increased longevity evolves from grandmothering. Proc. R. Soc. B: Biol. Sci. 279 (1749), 4880–4884. Available at: https://doi.org/10.1098/rspb.2012.1751.

King, R.D., Brown, B., Hwang, M., Jeon, T., George, A.T., 2010. Fractal dimension analysis of the cortical ribbon in mild Alzheimer's disease. Neuroimage 53 (2), 471. Available at: https://doi.org/10.1016/J.NEUROIMAGE.2010.06.050.

Kirkwood, T.B.L., Melov, S., 2011. On the programmed/non-programmed nature of ageing within the life history. Curr. Biol. 21 (18), 701–707. Available at: https://doi.org/10.1016/j.cub.2011.07.020.

Klein, C., Westenberger, A., 2012. Genetics of Parkinson's disease. Cold Spring Harb. Perspect. Med. 2 (1). Available at https://doi.org/10.1101/cshperspect.a008888.

Kochiyama, T., Ogihara, N., Tanabe, H.C., Kondo, O., Amano, H., Hasegawa, K., Suzuki, H., Ponce De León, M.S., Zollikofer, C.P.E., Bastir, M., Stringer, C., Sadato, N., Akazawa, T., 2018. Reconstructing the Neanderthal brain using computational anatomy. Sci. Rep. 8 (1), 1–9. Available at: https://doi.org/10.1038/s41598-018-24331-0.

Kolarova, M., García-Sierra, F., Bartos, A., Ricny, J., Ripova, D., 2012. Structure and pathology of tau protein in Alzheimer disease. Int. J. Alzheimer's Dis. Available at https://doi.org/10.1155/2012/731526.

Kouli, A., Torsney, K.M., Kuan, W.L., 2018. Parkinson's disease: etiology, neuropathology, and pathogenesis. Exon Publications. Available at: https://doi.org/10.15586/codonpublications.parkinsonsdisease.2018.ch1.

Kowald, A., Kirkwood, T.B.L., 2016. Can aging be programmed? A critical literature review. Aging Cell 15 (6), 986–998. Available at: https://doi.org/10.1111/acel.12510.

Kumar, S., Suleski, M., Craig, J.M., Kasprowicz, A.E., Sanderford, M., Li, M., Stecher, G., Hedges, S.B., 2022. TimeTree 5: an expanded resource for species divergence times. Mol. Biol. Evol. 39 (8). Available at: https://doi.org/10.1093/MOLBEV/MSAC174. http://www.timetree.org/.

Lacreuse, A., Russell, J.L., Hopkins, W.D., Herndon, J.G., 2014. Cognitive and motor aging in female chimpanzees. Neurobiol. Aging 35 (3), 623–632. Available at: https://doi.org/10.1016/j.neurobiolaging.2013.08.036.

Lamballais, S., Vinke, E., Vernooij, M.W., Ikram, M.A., Muetzel, R.L., 2020. Cortical gyrification in relation to age and cognition in older adults. NeuroImage 212. Available at: https://doi.org/10.1016/j.neuroimage.2020.116637.

Lanoiselée, H.M., et al., 2017. APP, PSEN1, and PSEN2 mutations in early-onset Alzheimer disease: a genetic screening study of familial and sporadic cases. PLoS Med. 14 (3), 1002270. Available at: https://doi.org/10.1371/journal.pmed.1002270.

Lasbleiz, C., Mestre-Francés, N., Devau, G., Luquin, M.R., Tenenbaum, L., Kremer, E.J., Verdier, J.M., 2019. Combining gene transfer and nonhuman primates to better understand and treat Parkinson's disease. Front. Mol. Neurosci. 12, 10. Available at: https://doi.org/10.3389/fnmol.2019.00010.

Lee, C.K., Weindruch, R., Prolla, T.A., 2000. Gene-expression profile of the ageing brain in mice. Nat. Genet. 25 (3), 294–297. Available at: https://doi.org/10.1038/77046.

Lefebvre, L., 2012. Primate encephalization. In: Hofman, M.A., Falk, D. (Eds.), Progress in Brain Research, pp. 393–412. Available at https://doi.org/10.1016/B978-0-444-53860-4.00019-2.

Leierer, L., Carrancho Alonso, Á., Pérez, L., Herrejón Lagunilla, Á., Herrera-Herrera, A.V., Connolly, R., Jambrina-Enríquez, M., Hernández Gómez, C.M., Galván, B., Mallol, C., 2020. It's getting hot in here—microcontextual study of a potential pit hearth at the Middle Paleolithic site of El Salt, Spain. J. Archaeol. Sci. 123, 105237. Available at: https://doi.org/10.1016/J.JAS.2020.105237.

Letunic, I., Bork, P., 2021. Interactive Tree Of Life (iTOL) v5: an online tool for phylogenetic tree display and annotation. Nucleic Acids Res. 49 (W1). Available at: https://doi.org/10.1093/NAR/GKAB301. https://www.ncbi.nlm.nih.gov/pmc/articles/PMC8265157/.

Levitt, P., 2003. Structural and functional maturation of the developing primate brain. J. Pediatr. 143 (Suppl. 4), 35–45. Available at: https://doi.org/10.1067/s0022-3476(03)00400-1.

Li, W., Fu, Y.H., Halliday, G.M., Sue, C.M., 2021a. PARK genes link mitochondrial dysfunction and alpha-synuclein pathology in sporadic Parkinson's disease. Front. Cell Dev. Biol. 9, 1755. Available at: https://doi.org/10.3389/fcell.2021.612476.

Li, H., Ling-Yan, S., Yang, L., Li, M., Liu, Q., Li, Z., Hu, Y., Li, H., Wu, S., Wang, W., Hu, Y., et al., 2021b. A cynomolgus monkey with naturally occurring Parkinson's disease. Natl. Sci. Rev. 8 (3). Available at: https://doi.org/10.1093/NSR/NWAA292.

Li, D., Wang, E., Jia, Y., Xu, J., Zhang, Z., Jiang, Z., Luo, W., 2020. Cortical complexity and gyrification patterns in Parkinson's disease. NeuroReport 31 (7), 565–570. Available at: https://doi.org/10.1097/WNR.0000000000001448.

Li, H., Yao, Y., Hu, X., 2021c. Biological implications and limitations of a cynomolgus monkey with naturally occurring Parkinson's disease. Zool. Res. 42 (2). Available at: https://doi.org/10.24272/J.ISSN.2095-8137.2021.004.

Liguori, C., Pierantozzi, M., Olivola, E., Mercuri, N.B., Stefani, A., 2015. Commentary: clinical correlates of raphe serotonergic dysfunction in early Parkinson's disease. Front. Neurol. 6 (DEC), 2964–2973. Available at: https://doi.org/10.3389/fneur.2015.00261.

Lim, E.W., Aarsland, D., Ffytche, D., Taddei, R.N., Wamelen, D.J., Wan, Y.M., Tan, E.K., Ray Chaudhuri, K., 2019. Amyloid-β and Parkinson's disease. J. Neurol. 266 (11), 2605–2619. Available at: https://doi.org/10.1007/s00415-018-9100-8.

Lin, Y.T., et al., 2018. APOE4 causes widespread molecular and cellular alterations associated with Alzheimer's disease phenotypes in human iPSC-derived brain cell types. Neuron 98 (6), 1141–1154. Available at: https://doi.org/10.1016/j.neuron.2018.05.008.

Lipinski, M.M., Zheng, B., Lu, T., Yan, Z., Py, B.F., Ng, A., Xavier, R.J., Li, C., Yankner, B.A., Scherzer, C.R., Yuan, J., 2010. Genome-wide analysis reveals mechanisms modulating autophagy in normal brain aging and in Alzheimer's disease. Proc. Natl. Acad. Sci. U. S. A. 107 (32), 14164–14169. Available at https://doi.org/10.1073/pnas.1009485107.

Liu, T., Lipnicki, D.M., Zhu, W., Tao, D., Zhang, C., Cui, Y., Jin, J.S., Sachdev, P.S., Wen, W., 2012. Cortical gyrification and sulcal spans in early stage Alzheimer's disease. PLoS One 7 (2), 31083. Available at: https://doi.org/10.1371/journal.pone.0031083.

Liu, B., Zhang, M., Chen, Y., Yao, Y., 2014. Tooth wear in aging people: an investigation of the prevalence and the influential factors of incisal/occlusal tooth wear in northwest China. BMC Oral Health 14 (1), 1–5. Available at: https://doi.org/10.1186/1472-6831-14-65.

Liu, H., Wang, L., Geng, Z., Zhu, Q., Song, Z., Chang, R., Lv, H., 2016. A voxel-based morphometric study of age- and sex-related changes in white matter volume in the normal aging brain. Neuropsychiatr. Dis. Treat. 12, 453. Available at: https://doi.org/10.2147/NDT.S90674.

Livingston, M.J., Wang, J., Zhou, J., Wu, G., Ganley, I.G., Hill, J.A., Yin, X.M., Dong, Z., 2019. Clearance of damaged mitochondria via mitophagy is important to the protective effect of ischemic preconditioning in kidneys. Autophagy 15 (12), 2142–2162. Available at: https://doi.org/10.1080/15548627.2019.1615822.

Loerch, P.M., Lu, T., Dakin, K.A., Vann, J.M., Isaacs, A., Geula, C., Wang, J., Pan, Y., Gabuzda, D.H., Li, C., Prolla, T.A., Yankner, B.A., 2008. Evolution of the aging brain transcriptome and synaptic regulation. PLoS One 3 (10), 3329. Available at: https://doi.org/10.1371/journal.pone.0003329.

Lonsdorf, E.V., Stanton, M.A., Wellens, K.R., Murray, C.M., 2021. Wild chimpanzee offspring exhibit adult-like foraging patterns around the age of weaning. Am. J. Phys. Anthropol. 175 (1), 268–281. Available at: https://doi.org/10.1002/ajpa.24267.

López-Otín, C., Blasco, M.A., Partridge, L., Serrano, M., Kroemer, G., 2013. The hallmarks of aging. Cell 153 (6), 1194. Available at: https://doi.org/10.1016/j.cell.2013.05.039.

Lou, G., Palikaras, K., Lautrup, S., Scheibye-Knudsen, M., Tavernarakis, N., Fang, E.F., 2020. Mitophagy and neuroprotection. Trends Mol. Med. 26 (1), 8–20. Available at: https://doi.org/10.1016/j.molmed.2019.07.002.

Loughlin, J., 2015. Genetic contribution to osteoarthritis development: current state of evidence. Curr. Opin. Rheumatol. 27 (3), 284–288. Available at: https://doi.org/10.1097/BOR.0000000000000171.

Lovell, M.A., Gabbita, S.P., Markesbery, W.R., 1999. Increased DNA oxidation and decreased levels of repair products in Alzheimer's disease ventricular CSF. J. Neurochem. 72 (2), 771–776. Available at: https://doi.org/10.1046/j.1471-4159.1999.0720771.x.

Lowenstine, L.J., McManamon, R., Terio, K.A., 2016. Comparative pathology of aging great apes: bonobos, chimpanzees, gorillas, and orangutans. Vet. Pathol. 53 (2), 250–276. Available at: https://doi.org/10.1177/0300985815612154.

Lu, T., Pan, Y., Kao, S.Y., Li, C., Kohane, I., Chan, J., Yankner, B.A., 2004. Gene regulation and DNA damage in the ageing human brain. Nature 429 (6994), 883–891. Available at: https://doi.org/10.1038/NATURE02661.

Lyras, G.A., Giannakopoulou, A., Kouvari, M., Papadopoulos, G.C., 2016. Evolution of gyrification in carnivores. Brain Behav. Evol. 88 (3–4), 187–203. Available at: https://doi.org/10.1159/000453104.

Ma, J., Zhang, W., Tan, L., Wang, H.F., Wan, Y., Sun, F.R., Tan, C.C., Yu, J.T., Tan, L., 2016. MS4A6A genotypes are associated with the atrophy rates of Alzheimer's disease related brain structures. Oncotarget 7 (37), 58779. Available at: https://doi.org/10.18632/ONCOTARGET.9563.

Maccioni, R.B., Cambiazo, V., 1995. Role of microtubule-associated proteins in the control of microtubule assembly. Physiol. Rev. 75 (4), 835–864. Available at: https://doi.org/10.1152/physrev.1995.75.4.835.

Malpartida, A.B., Williamson, M., Narendra, D.P., Wade-Martins, R., Ryan, B.J., 2021. Mitochondrial dysfunction and mitophagy in Parkinson's disease: from mechanism to therapy. Trends Biochem. Sci. 46 (4), 329–343. Available at: https://doi.org/10.1016/j.tibs.2020.11.007.

Mandel, S.A., Morelli, M., Halperin, I., Korczyn, A.D., 2010. Biomarkers for prediction and targeted prevention of Alzheimer's and Parkinson's diseases: evaluation of drug clinical efficacy. EPMA J. 1 (2), 273–292. Available at: https://doi.org/10.1007/s13167-010-0036-z.

Martin, C.R., Preedy, V.R., 2014. Diet and Nutrition in Dementia and Cognitive Decline. Academic Press. Available at https://doi.org/10.1016/C2012-0-02833-5.

Mathews, K.J., Allen, K.M., Boerrigter, D., Ball, H., Shannon Weickert, C., Double, K.L., 2017. Evidence for reduced neurogenesis in the aging human hippocampus despite stable stem cell markers. Aging Cell 16 (5), 1195–1199. Available at: https://doi.org/10.1111/acel.12641.

Maynard, S., Fang, E.F., Scheibye-Knudsen, M., Croteau, D.L., Bohr, V.A., 2015. DNA damage, DNA repair, aging, and neurodegeneration. Cold Spring Harb. Perspect. Med. 5 (10). Available at https://doi.org/10.1101/CSHPERSPECT.A025130.

Mayne, B., Berry, O., Davies, C., Farley, J., Jarman, S., 2019. A genomic predictor of lifespan in vertebrates. Sci. Rep. 9 (1), 17866. Available at: https://doi.org/10.1038/s41598-019-54447-w.

McIntosh, A.M., Bennett, C., Dickson, D., Anestis, S.F., Watts, D.P., Webster, T.H., Fontenot, M.B., Bradley, B.J., 2012. The apolipoprotein E (APOE) gene appears functionally monomorphic in chimpanzees (pan troglodytes). PLoS One 7 (10), 47760. Available at: https://doi.org/10.1371/journal.pone.0047760.

McKeith, I.G., 2007. Dementia with Lewy bodies and Parkinson's disease with dementia: where two worlds collide. Pract. Neurol. 7 (6), 374–382. Available at: https://doi.org/10.1136/jnnp.2007.134163.

McLain, A.T., Faulk, C., 2018. The evolution of CpG density and lifespan in conserved primate and mammalian promoters. Aging (Albany NY) 10 (4), 561–572. Available at: https://doi.org/10.18632/AGING.101413.

Medawar, P.B., 1952. An unsolved problem of biology. In: The Uniqueness of the Individual, pp. 44–70. Available at https://doi.org/10.4324/9780429299759-3.

Meyer, M., et al., 2012. A high-coverage genome sequence from an archaic Denisovan individual. Science 338 (6104), 222–226. Available at: https://doi.org/10.1126/science.1224344.

Meyer, M., Arsuaga, J.L., De Filippo, C., Nagel, S., Aximu-Petri, A., Nickel, B., Martínez, I., Gracia, A., De Castro, J.M.B., Carbonell, E., Viola, B., Kelso, J., Prüfer, K., Pääbo, S., 2016. Nuclear DNA sequences from the Middle Pleistocene Sima de los Huesos hominins. Nature 531 (7595), 504–507. Available at: https://doi.org/10.1038/nature17405.

Mitteldorf, J., 2018. Can aging be programmed? Biochemistry (Mosc.) 83 (12–13), 1524–1533. Available at: https://doi.org/10.1134/S0006297918120106.

Murakami, T., Moriwaki, Y., Kawarabayashi, T., Nagai, M., Ohta, Y., Deguchi, K., Kurata, T., Morimoto, N., Takehisa, Y., Matsubara, E., Ikeda, M., Harigaya, Y., Shoji, M., Takahashi, R., Abe, K., 2007. PINK1, a gene product of PARK6, accumulates in α-synucleinopathy brains. J. Neurol. Neurosurg. Psychiatry 78 (6), 653. Available at: https://doi.org/10.1136/JNNP.2006.100123.

Nasrabady, S.E., Rizvi, B., Goldman, J.E., Brickman, A.M., 2018. White matter changes in Alzheimer's disease: a focus on myelin and oligodendrocytes. Acta Neuropathol. Commun. 6 (1), 22. Available at: https://doi.org/10.1186/S40478-018-0515-3.

Nattrass, S., Croft, D.P., Ellis, S., Cant, M.A., Weiss, M.N., Wright, B.M., Stredulinsky, E., Doniol-Valcroze, T., Ford, J.K.B., Balcomb, K.C., Franks, D.W., 2019. Postreproductive killer whale grandmothers improve the survival of their grandoffspring. Proc. Natl. Acad. Sci. U. S. A. 116 (52), 26669–26673. Available at: https://doi.org/10.1073/pnas.1903844116.

Ndung'u, M., Härtig, W., Wegner, F., Mwenda, J.M., Low, R.W.C., Akinyemi, R.O., Kalaria, R.N., 2012. Cerebral amyloid β(42) deposits and microvascular pathology in ageing baboons. Neuropathol. Appl. Neurobiol. 38 (5), 487–499. Available at: https://doi.org/10.1111/j.1365-2990.2011.01246.x.

Nelson, P.T., Braak, H., Markesbery, W.R., 2009. Neuropathology and cognitive impairment in Alzheimer disease: a complex but coherent relationship. J. Neuropathol. Exp. Neurol. 68 (1), 1–14. Available at: https://doi.org/10.1097/NEN.0b013e3181919a48.

Nestler, E.J., Hyman, S.E., Malenka, R.C., 2009. Molecular Neuropharmacology: A foundation for clinical neuroscience. In: Trends in Pharmacological Sciences, second Ed. McGraw-Hill Medical, New York. Available at https://doi.org/10.1016/S0165-6147(02)01903-X.

Niikura, T., Tajima, H., Kita, Y., 2006. Neuronal cell death in Alzheimers disease and a neuroprotective factor, humanin. Curr. Neuropharmacol. 4 (2), 139–147. Available at: https://doi.org/10.2174/157015906776359577.

Nobis, L., Manohar, S.G., Smith, S.M., Alfaro-Almagro, F., Jenkinson, M., Mackay, C.E., Husain, M., 2019. Hippocampal volume across age: nomograms derived from over 19,700 people in UK Biobank. NeuroImage 23, 101904. Available at https://doi.org/10.1016/J.NICL.2019.101904.

Northcutt, R.G., 2002. Understanding vertebrate brain evolution. Integr. Comp. Biol. 42 (4), 743–756. Available at: https://doi.org/10.1093/icb/42.4.743.

Öhrfelt, A., Brinkmalm, A., Dumurgier, J., Brinkmalm, G., Hansson, O., Zetterberg, H., Bouaziz-Amar, E., Hugon, J., Paquet, C., Blennow, K., 2016. The pre-synaptic vesicle protein synaptotagmin is a novel biomarker for Alzheimer's disease. Alzheimer's Res. Ther. 8 (1), 1–10. Available at: https://doi.org/10.1186/s13195-016-0208-8.

Oikawa, N., Kimura, N., Yanagisawa, K., 2010. Alzheimer-type tau pathology in advanced aged nonhuman primate brains harboring substantial amyloid deposition. Brain Res. 1315, 137–149. Available at https://doi.org/10.1016/J.BRAINRES.2009.12.005.

Orme, T., Guerreiro, R., Bras, J., 2018. The genetics of dementia with Lewy bodies: current understanding and future directions. Curr. Neurol. Neurosci. Rep. 18 (10), 1–13. Available at: https://doi.org/10.1007/s11910-018-0874-y.

Outeiro, T.F., Heutink, P., Bezard, E., Cenci, A.M., 2021. From iPS cells to rodents and nonhuman primates: filling gaps in modeling Parkinson's disease. Mov. Disord. 36 (4), 832–841. Available at: https://doi.org/10.1002/mds.28387.

Palumbo, M.L., Moroni, A.D., Quiroga, S., Castro, M.M., Burgueño, A.L., Genaro, A.M., 2021. Immunomodulation induced by central nervous system-related peptides as a therapeutic strategy for neurodegenerative disorders. Pharmacol. Res. Perspect. 9 (5), 795. Available at: https://doi.org/10.1002/prp2.795.

Pankratz, N., et al., 2006. Presence of an APOE4 allele results in significantly earlier onset of Parkinson's disease and a higher risk with dementia. Mov. Disord. 21 (1), 45–49. Available at: https://doi.org/10.1002/MDS.20663.

Partridge, L., Gems, D., 2002. Mechanisms of ageing: public or private? Nat. Rev. Genet. 3 (3), 165–175. Available at: https://doi.org/10.1038/nrg753.

Perez, S.E., Raghanti, M.A., Hof, P.R., Kramer, L., et al., 2013. Alzheimer's disease pathology in the neocortex and hippocampus of the western lowland gorilla (*Gorilla gorilla gorilla*). J. Comp. Neurol. 521, 4318–4338. Available at: https://doi.org/10.1002/cne.23428.

Perez, S.E., Sherwood, C.C., Cranfield, M.R., Erwin, J.M., Mudakikwa, A., Hof, P.R., Mufson, E.J., 2016. Early Alzheimer's disease-type pathology in the frontal cortex of wild mountain gorillas (*Gorilla beringei beringei*). Neurobiol. Aging 39, 195–201. Available at https://doi.org/10.1016/j.neurobiolaging.2015.12.017.

Peters, R., 2006. Ageing and the brain. Postgrad. Med. J. 82 (964), 84. Available at: https://doi.org/10.1136/PGMJ.2005.036665.

Phillips, K.A., Bales, K.L., Capitanio, J.P., Conley, A., Czoty, P.W., 't Hart, B.A., Hopkins, W.D., Hu, S.L., Miller, L.A., Nader, M.A., Nathanielsz, P.W., Rogers, J., Shively, C.A., Voytko, M.L., 2014. Why primate models matter. Am. J. Primatol. 76 (9), 801. Available at: https://doi.org/10.1002/AJP.22281.

Picq, J.L., 2007. Aging affects executive functions and memory in mouse lemur primates. Exp. Gerontol. 42 (3), 223–232. Available at: https://doi.org/10.1016/j.exger.2006.09.013.

Picq, J.L., Aujard, F., Volk, A., Dhenain, M., 2012. Age-related cerebral atrophy in nonhuman primates predicts cognitive impairments. Neurobiol. Aging 33 (6), 1096–1109. Available at: https://doi.org/10.1016/j.neurobiolaging.2010.09.009.

Pigott, K., Rick, J., Xie, S.X., Hurtig, H., Chen-Plotkin, A., Duda, J.E., Morley, J.F., Chahine, L.M., Dahodwala, N., Akhtar, R.S., Siderowf, A., Trojanowski, J.Q., Weintraub, D., 2015. Longitudinal study of normal cognition in Parkinson disease. Neurology 85 (15), 1276–1282. Available at: https://doi.org/10.1212/WNL.0000000000002001.

Podlisny, M.B., Tolan, D.R., Selkoe, D.J., 1991. Homology of the amyloid beta protein precursor in monkey and human supports a primate model for beta amyloidosis in Alzheimer's disease. Am. J. Pathol. 138 (6), 1423–1435. PMID: 1905108.

Poinar, H.N., Artur Stankiewicz, B., 1999. Protein preservation and DNA retrieval from ancient tissues. Proc. Natl. Acad. Sci. U. S. A. 96 (15), 8426–8431. Available at https://doi.org/10.1073/pnas.96.15.8426.

Pontzer, H., et al., 2016. Metabolic acceleration and the evolution of human brain size and life history. Nature 533 (7603), 390–392. Available at: https://doi.org/10.1038/nature17654.

Price, J.L., Morris, J.C., 1999. Tangles and plaques in nondemented aging and 'preclinical' Alzheimer's disease. Ann. Neurol. 45 (3), 358–368. Available at: https://doi.org/10.1002/1531-8249(199903)45:3<358::AID-ANA12>3.0.CO;2-X.

Prince, M., Albanese, E., Guerchet, M., Prina, M., 2014. World Alzheimer Report 2014 - Dementia and risk reduction: an analysis of protective and modifiable factors. Alzheimer's Disease International, pp. 1–104. Available at: https://www.alz.co.uk/research/WorldAlzheimerReport2014.pdf.

Prins, N.D., Scheltens, P., 2015. White matter hyperintensities, cognitive impairment and dementia: an update. Nat. Rev. Neurol. 11 (3), 157–165. Available at: https://doi.org/10.1038/nrneurol.2015.10.

Proitsi, P., Lee, S.H., Lunnon, K., Keohane, A., Powell, J., Troakes, C., Al-Sarraj, S., Furney, S., Soininen, H., Kłoszewska, I., Mecocci, P., Tsolaki, M., Vellas, B., Lovestone, S., Hodges, A., 2014. Alzheimer's disease susceptibility variants in the MS4A6A gene are associated with altered levels of MS4A6A expression in blood. Neurobiol. Aging 35 (2), 279–290. Available at: https://doi.org/10.1016/j.neurobiolaging.2013.08.002.

Ramanoël, S., Hoyau, E., Kauffmann, L., Renard, F., Pichat, C., Boudiaf, N., Krainik, A., Jaillard, A., Baciu, M., 2018. Gray matter volume and cognitive performance during normal aging. A voxel-based morphometry study. Front. Aging Neurosci. 10 (AUG), 235. Available at: https://doi.org/10.3389/fnagi.2018.00235.

Raz, N., Lindenberger, U., Rodrigue, K.M., Kennedy, K.M., Head, D., Williamson, A., Dahle, C., Gerstorf, D., Acker, J.D., 2005. Regional brain changes in aging healthy adults: general trends, individual differences and modifiers. Cereb. Cortex 15 (11), 1676–1689. Available at: https://doi.org/10.1093/cercor/bhi044.

Reddy, R.B., Mitani, J.C., 2019. Social relationships and caregiving behavior between recently orphaned chimpanzee siblings. Primates 60 (5), 389–400. Available at: https://doi.org/10.1007/s10329-019-00732-1.

Redza-Dutordoir, M., Averill-Bates, D.A., 2016. Activation of apoptosis signalling pathways by reactive oxygen species. Biochim. Biophys. Acta—Mol. Cell Res. 1863 (12), 2977–2992. Available at: https://doi.org/10.1016/j.bbamcr.2016.09.012.

Reed, X., Bandrés-Ciga, S., Blauwendraat, C., Cookson, M.R., 2019. The role of monogenic genes in idiopathic Parkinson's disease. Neurobiol. Dis. 124, 230. Available at: https://doi.org/10.1016/J.NBD.2018.11.012.

Reich, D., et al., 2010. Genetic history of an archaic hominin group from Denisova cave in Siberia. Nature 468 (7327), 1053–1060. Available at: https://doi.org/10.1038/nature09710.

Rodrigue, K.M., Kennedy, K.M., Devous, M.D., Rieck, J.R., Hebrank, A.C., Diaz-Arrastia, R., Mathews, D., Park, D.C., 2012. β-Amyloid burden in healthy aging: regional distribution and cognitive consequences. Neurology 78 (6), 387–395. Available at: https://doi.org/10.1212/WNL.0b013e318245d295.

Rosen, R.F., Farberg, A.S., Gearing, M., Dooyema, J., Long, P.M., Anderson, D.C., Davis-Turak, J., Coppola, G., Geschwind, D.H., Paré, J.F., Duong, T.Q., Hopkins, W.D., Preuss, T.M., Walker, L.C., 2008. Tauopathy with paired helical filaments in an aged chimpanzee. J. Comp. Neurol. 509 (3), 259–270. Available at https://doi.org/10.1002/cne.21744.

Roth, G., Dicke, U., 2005. Evolution of the brain and intelligence. Trends Cogn. Sci. 9 (5), 250–257. Available at: https://doi.org/10.1016/j.tics.2005.03.005.

Rothschild, B., Haeusler, M., 2021. Possible vertebral brucellosis infection in a Neanderthal. Sci. Rep. 11 (1), 1–9. Available at: https://doi.org/10.1038/s41598-021-99289-7.

Ruff, C.B., Trinkaus, E., Holliday, T.W., 1997. Body mass and encephalization in Pleistocene *Homo*. Nature 387 (6629), 173–176. Available at https://doi.org/10.1038/387173a0.

Russell, A., Drozdova, A., Wang, W., Thomas, M., 2014. The impact of dementia development concurrent with Parkinson's disease: a new perspective. CNS Neurol. Disord. Drug Targets 13 (7), 1160–1168. Available at: https://doi.org/10.2174/1871527313666140917122739.

Safieh, M., Korczyn, A.D., Michaelson, D.M., 2019. ApoE4: an emerging therapeutic target for Alzheimer's disease. BMC Med. 17 (1), 1–17. Available at: https://doi.org/10.1186/s12916-019-1299-4.

Sah, E., Krishnamurthy, S., Ahmidouch, M.Y., Gillispie, G.J., Milligan, C., Orr, M.E., 2021. The cellular senescence stress response in post-mitotic brain cells: cell survival at the expense of tissue degeneration. Life 11 (3), 229. Available at https://doi.org/10.3390/LIFE11030229.

Salat, D.H., Greve, D.N., Pacheco, J.L., Quinn, B.T., Helmer, K.G., Buckner, R.L., Fischl, B., 2009. Regional white matter volume differences in nondemented aging and Alzheimer's disease. Neuroimage 44 (4), 1247–1258. Available at: https://doi.org/10.1016/j.neuroimage.2008.10.030.

Salminen, A., Kaarniranta, K., Kauppinen, A., 2012. Inflammaging: disturbed interplay between autophagy and inflammasomes. Aging, 166–175. Available at: https://doi.org/10.18632/aging.100444.

Santos, L., Elliott-Sale, K.J., Sale, C., 2017. Exercise and bone health across the lifespan. Biogerontology 18 (6), 931–946. Available at: https://doi.org/10.1007/s10522-017-9732-6.

Sapolsky, R.M., Finch, C.E., 2000. Alzheimer's disease and some speculations about the evolution of its modifiers. Ann. N. Y. Acad. Sci. 924 (1), 99–103. Available at: https://doi.org/10.1111/j.1749-6632.2000.tb05567.x.

Schaer, M., Bach Cuadra, M., Tamarit, L., Lazeyras, F., Eliez, S., Thiran, J.P., 2008. A surface-based approach to quantify local cortical gyrification. IEEE Trans. Med. Imaging 27 (2), 161–170. Available at: https://doi.org/10.1109/TMI.2007.903576.

Schäfer, A., Weickenmeier, J., Kuhl, E., 2019. The interplay of biochemical and biomechanical degeneration in Alzheimer's disease. Comput. Methods Appl. Mech. Eng. 352, 369–388. Available at: https://doi.org/10.1016/j.cma.2019.04.028.

Schaffler, M.B., Choi, K., Milgrom, C., 1995. Aging and matrix microdamage accumulation in human compact bone. Bone 17 (6), 521–525. Available at: https://doi.org/10.1016/8756-3282(95)00370-3.

Schmidt, M.F., Gan, Z.Y., Komander, D., Dewson, G., 2021. Ubiquitin signalling in neurodegeneration: mechanisms and therapeutic opportunities. Cell Death Differ. 28 (2), 570–590. Available at https://doi.org/10.1038/s41418-020-00706-7.

Schultz, C., Dehghani, F., Hubbard, G.B., Thal, D.R., Struckhoff, G., Braak, E., Braak, H., 2000. Filamentous tau pathology in nerve cells, astrocytes, and oligodendrocytes of aged baboons. J. Neuropathol. Exp. Neurol. 59 (1), 39–52. Available at: https://doi.org/10.1093/jnen/59.1.39.

Seim, I., Ma, S., Gladyshev, V.N., 2016. Gene expression signatures of human cell and tissue longevity. NPJ Aging Mech. Dis. 2 (1), 1–8. Available at: https://doi.org/10.1038/npjamd.2016.14.

Shamy, J.L.T., Buonocore, M.H., Makaron, L.M., Amaral, D.G., Barnes, C.A., Rapp, P.R., 2006. Hippocampal volume is preserved and fails to predict recognition memory impairment in aged rhesus monkeys (*Macaca mulatta*). Neurobiol. Aging 27 (10), 1405–1415. Available at: https://doi.org/10.1016/J.NEUROBIOLAGING.2005.07.019.

Shamy, J.L., Habeck, C., Hof, P.R., Amaral, D.G., Fong, S.G., Buonocore, M.H., Stern, Y., Barnes, C.A., Rapp, P.R., 2011. Volumetric correlates of spatiotemporal working and recognition memory impairment in aged rhesus monkeys. Cereb. Cortex 21 (7), 1559–1573. Available at: https://doi.org/10.1093/cercor/bhq210.

Shea, Y.F., Chu, L.W., Chan, A.O.K., Ha, J., Li, Y., Song, Y.Q., 2016. A systematic review of familial Alzheimer's disease: Differences in presentation of clinical features among three mutated genes and potential ethnic differences. J. Formos. Med. Assoc. 115 (2), 67–75. Available at: https://doi.org/10.1016/J.JFMA.2015.08.004.

Sherwood, C.C., Gordon, A.D., Allen, J.S., Phillips, K.A., Erwin, J.M., Hof, P.R., Hopkins, W.D., 2011. Aging of the cerebral cortex differs between humans and chimpanzees. Proc. Natl. Acad. Sci. U. S. A. 108 (32), 13029–13034. Available at https://doi.org/10.1073/pnas.1016709108.

Sievert, L.L., 2014. Anthropology and the study of menopause: Evolutionary, developmental, and comparative perspectives. Menopause 21 (10), 1151–1159. Available at: https://doi.org/10.1097/gme.0000000000000341.

Silcox, M.T., Dalmyn, C.K., Bloch, J.I., 2009. Virtual endocast of *Ignacius graybullianus* (Paromomyidae, Primates) and brain evolution in early primates. Proc. Natl. Acad. Sci. U. S. A. 106 (27), 10987–10992. Available at https://www.pnas.org/doi/full/10.1073/pnas.0812140106.

Slatkin, M., Racimo, F., 2016. Ancient DNA and human history. Proc. Natl. Acad. Sci. U. S. A. 113 (23), 6380–6387. Available at: https://doi.org/10.1073/PNAS.1524306113.

Sleegers, K., Brouwers, N., Gijselinck, I., Theuns, J., Goossens, D., Wauters, J., Del-Favero, J., Cruts, M., Van Duijn, C.M., Van Broeckhoven, C., 2006. APP duplication is sufficient to cause early onset Alzheimer's dementia with cerebral amyloid angiopathy. Brain 129 (Pt 11), 2977–2983. Available at: https://doi.org/10.1093/BRAIN/AWL203.

Sliter, D.A., Martinez, J., Hao, L., Chen, X., Sun, N., Fischer, T.D., Burman, J.L., Li, Y., Zhang, Z., Narendra, D.P., Cai, H., Borsche, M., Klein, C., Youle, R.J., 2018. Parkin and PINK1 mitigate STING-induced inflammation. Nature 561 (7722), 258–262. Available at: https://doi.org/10.1038/s41586-018-0448-9.

Smith, B.H., 1991. Dental development and the evolution of life history in Hominidae. Am. J. Phys. Anthropol. 86 (2), 157–174. Available at: https://doi.org/10.1002/ajpa.1330860206.

Song, M., Zhao, X., Song, F., 2021. Aging-dependent mitophagy dysfunction in Alzheimer's disease. Mol. Neurobiol. 58 (5), 2362–2378. Available at: https://doi.org/10.1007/s12035-020-02248-y.

Sontheimer, H., 2015. Aging, Dementia, and Alzheimer Disease'. In: Sontheimer, H. (Ed.), Diseases of the Nervous System. Academic Press.

Stearns, S.C., Ebert, D., 2001. Evolution in health and disease: work in progress. Q. Rev. Biol. 76 (4), 417–432. Available at https://doi.org/10.1086/420539.

Stefanis, L., 2012. α-Synuclein in Parkinson's disease. Cold Spring Harb. Perspect. Med. 2 (2). Available at https://doi.org/10.1101/CSHPERSPECT.A009399.

Stegeman, R., Weake, V.M., 2017. Transcriptional signatures of aging. J. Mol. Biol. 429 (16), 2427–2437. Available at: https://doi.org/10.1016/j.jmb.2017.06.019.

Stephenson, J., Nutma, E., van der Valk, P., Amor, S., 2018. Inflammation in CNS neurodegenerative diseases. Immunology 154 (2), 204–219. Available at: https://doi.org/10.1111/imm.12922.

Stephenson-Jones, M., Samuelsson, E., Ericsson, J., Robertson, B., Grillner, S., 2011. Evolutionary conservation of the basal ganglia as a common vertebrate mechanism for action selection. Curr. Biol. 21 (13), 1081–1091. Available at: https://doi.org/10.1016/j.cub.2011.05.001.

Stiefel, K.M., Torben-Nielsen, B., Coggan, J.S., 2013. Proposed evolutionary changes in the role of myelin. Front. Neurosci. 7 (7 NOV). Available at https://doi.org/10.3389/FNINS.2013.00202.

Stiles, J., Jernigan, T.L., 2010. The basics of brain development. Neuropsychol. Rev. 20 (4), 327–348. Available at: https://doi.org/10.1007/s11065-010-9148-4.

Stonebarger, G.A., Bimonte-Nelson, H.A., Urbanski, H.F., 2021. The rhesus macaque as a translational model for neurodegeneration and Alzheimer's disease. Front. Aging Neurosci. 13. Available at https://doi.org/10.3389/fnagi.2021.734173.

Straus, W.L., Cave, J.E., 1957. Pathology and the posture of Neanderthal man. Q. Rev. Biol. 32 (4), 348–363. Available at: https://doi.org/10.1086/401979.

Summerfield, C., Junqué, C., Tolosa, E., Salgado-Pineda, P., Gómez-Ansón, B., Martí, M.J., Pastor, P., Ramírez-Ruíz, B., Mercader, J., 2005. Structural brain changes in Parkinson disease with dementia: a voxel-based morphometry study. Arch. Neurol. 62 (2), 281–285. Available at: https://doi.org/10.1001/archneur.62.2.281.

Tabassum, S., Misrani, A., Yang, L., 2020. Exploiting common aspects of obesity and Alzheimer's disease. Front. Hum. Neurosci. 14. Available at https://doi.org/10.3389/fnhum.2020.602360.

Tacutu, R., Thornton, D., Johnson, E., Budovsky, A., Barardo, D., Craig, T., Diana, E., Lehmann, G., Toren, D., Wang, J., Fraifeld, V.E., de Magalhães, J.P., 2018. Human Ageing Genomic Resources: new and updated databases. Nucleic Acids Res. 46 (Database issue), D1083. Available at: https://doi.org/10.1093/NAR/GKX1042. https://genomics.senescence.info/species/index.html.

Tang, B.L., 2016. Sirt1 and the mitochondria. Mol. Cells 39 (2), 87–95. Available at: https://doi.org/10.14348/molcells.2016.2318.

Tanzi, R.E., 2012. The genetics of Alzheimer disease. Cold Spring Harb. Perspect. Med. 2 (10), 6296. Available at: https://doi.org/10.1101/cshperspect.a006296.

Tcw, J., et al., 2022. Cholesterol and matrisome pathways dysregulated in astrocytes and microglia. Cell 185 (13), 2213–2233. Available at: https://doi.org/10.1016/J.CELL.2022.05.017.

Thal, D.R., Rüb, U., Orantes, M., Braak, H., 2002. Phases of Aβ-deposition in the human brain and its relevance for the development of AD. Neurology 58 (12), 1791–1800. Available at: https://doi.org/10.1212/WNL.58.12.1791.

Toledano, A., Álvarez, M.I., López-Rodríguez, A.B., Toledano-Díaz, A., Fernández-Verdecia, C.I., 2012. Does Alzheimer's disease exist in all primates? Alzheimer pathology in non-human primates and its pathophysiological implications (I). Neurología (English Edition) 27 (6), 354–369. Available at: https://doi.org/10.1016/J.NRLENG.2012.07.002.

Tolosa, E., Garrido, A., Scholz, S.W., Poewe, W., 2021. Challenges in the diagnosis of Parkinson's disease. Lancet Neurol. 20 (5), 385–397. Available at: https://doi.org/10.1016/S1474-4422(21)00030-2.

Tran, J., Anastacio, H., Bardy, C., 2020. Genetic predispositions of Parkinson's disease revealed in patient-derived brain cells. NPJ Parkinson's Dis. 6 (1), 1–18. Available at: https://doi.org/10.1038/s41531-020-0110-8.

Trinkaus, E., 2011. Late Pleistocene adult mortality patterns and modern human establishment. Proc. Natl. Acad. Sci. U. S. A. 108 (4), 1267–1271. Available at https://doi.org/10.1073/pnas.1018700108.

Trubitsyn, A.G., 2020. The mechanism of programmed aging: the way to create a real remedy for senescence. Curr. Aging Sci. 13 (1), 31–41. Available at: https://doi.org/10.2174/1874609812666191014111422.

Uchihara, T., Endo, K., Kondo, H., Okabayashi, S., Shimozawa, N., Yasutomi, Y., Adachi, E., Kimura, N., 2016. Tau pathology in aged cynomolgus monkeys is progressive supranuclear palsy/corticobasal degeneration- but not Alzheimer disease-like -Ultrastructural mapping of tau by EDX. Acta Neuropathol. Commun. 4 (1), 118. Available at: https://doi.org/10.1186/S40478-016-0385-5.

Van Dam, D., De Deyn, P.P., 2017. Non human primate models for Alzheimer's disease-related research and drug discovery. Expert Opin. Drug Discovery 12 (2), 187–200. Available at: https://doi.org/10.1080/17460441.2017.1271320.

van der Valk, T., et al., 2021. Million-year-old DNA sheds light on the genomic history of mammoths. Nature 591 (7849), 265–269. Available at: https://doi.org/10.1038/s41586-021-03224-9.

Van Deursen, J.M., 2014. The role of senescent cells in ageing. Nature 509 (7501), 439. Available at: https://doi.org/10.1038/NATURE13193.

Van Duijn, C.M., de Knijff, P., Cruts, M., Wehnert, A., Havekes, L.M., Hofman, A., Van Broeckhoven, C., 1994. Apolipoprotein E4 allele in a population–based study of early–onset Alzheimer's disease. Nat. Genet. 7 (1), 74–78. Available at https://doi.org/10.1038/ng0594-74.

Van Exel, E., Koopman, J.J.E., Van Bodegom, D., Meij, J.J., De Knijff, P., Ziem, J.B., Finch, C.E., Westendorp, R.G.J., 2017. Effect of APOE ε4 allele on survival and fertility in an adverse environment. PLoS One 12 (7), e0179497. Available at: https://doi.org/10.1371/JOURNAL.PONE.0179497.

Venda, L.L., Cragg, S.J., Buchman, V.L., Wade-Martins, R., 2010. α-Synuclein and dopamine at the crossroads of Parkinson's disease. Trends Neurosci. 33 (12), 559–568. Available at: https://doi.org/10.1016/j.tins.2010.09.004.

Villa, P., Roebroeks, W., 2014. Neandertal demise: an archaeological analysis of the modern human superiority complex. PLoS ONE 9 (4). Available at: https://doi.org/10.1371/journal.pone.0096424.

Viña, J., Borras, C., Abdelaziz, K.M., Garcia-Valles, R., Gomez-Cabrera, M.C., 2013. The free radical theory of aging revisited: the cell signaling disruption theory of aging. Antioxid. Redox Signal. 19 (8), 779–787. Available at: https://doi.org/10.1089/ars.2012.5111.

Vodanović, M., Dumančić, J., Galić, I., Savić Pavičin, I., Petrovečki, M., Cameriere, R., Brkić, H., 2011. Age estimation in archaeological skeletal remains: evaluation of four non-destructive age calculation methods. J. Forensic Odonto-Stomatol. 29 (2), 14–21. PMID: 22717909.

Wakabayashi, K., Tanji, K., Odagiri, S., Miki, Y., Mori, F., Takahashi, H., 2013. The Lewy body in Parkinson's disease and related neurodegenerative disorders. Mol. Neurobiol. 47 (2), 495–508. Available at: https://doi.org/10.1007/s12035-012-8280-y.

Walker, M.L., Herndon, J.G., 2008. Menopause in nonhuman primates? Biol. Reprod. 79 (3), 398–406. Available at: https://doi.org/10.1095/biolreprod.108.068536.

Walker, L.C., Jucker, M., 2017. The exceptional vulnerability of humans to Alzheimer's disease. Trends Mol. Med. 23 (6), 534–545. Available at: https://doi.org/10.1016/j.molmed.2017.04.001.

Walker, L., Stefanis, L., Attems, J., 2019. Clinical and neuropathological differences between Parkinson's disease, Parkinson's disease dementia and dementia with Lewy bodies—current issues and future directions. J. Neurochem. 150 (5), 467–474. Available at: https://doi.org/10.1111/JNC.14698.

Wang, W.Y., Yu, J.T., Liu, Y., Yin, R.H., Wang, H.F., Wang, J., Tan, L., Radua, J., Tan, L., 2015. Voxel-based meta-analysis of grey matter changes in Alzheimer's disease. Transl. Neurodegener. 4 (1), 10–1186. Available at: https://doi.org/10.1186/s40035-015-0027-z.

Wang, H., Abajobir, A.A., Abate, K.H., Abbafati, C., Abbas, K., Abd-Allah, F., Abera, S.F., et al., 2017. Global, regional, and national under-5 mortality, adult mortality, age-specific mortality, and life expectancy, 1970-2016: a systematic analysis for the Global Burden of Disease Study 2016. Lancet 390 (10100), 1084–1150. Available at: https://doi.org/10.1016/S0140-6736(17)31833-0.

Warren, M., 2019. Move over, DNA: ancient proteins are starting to reveal humanity's history. Nature 570 (7762), 433–436. Available at: https://doi.org/10.1038/d41586-019-01986-x.

Weismann, A., 1891. Essays upon heredity and kindred biological problems. In: Poulton, E.B., Schönland, S., Shipley, A.E. (Eds.), Essays upon heredity and kindred biological problems. Clarendon press. Available at https://doi.org/10.5962/bhl.title.101564.

White, T., Su, S., Schmidt, M., Kao Y., C., Sapiro, G., 2010. The development of gyrification in childhood and adolescence. Brain Cogn. 72 (1), 36–45. Available at: https://doi.org/10.1016/j.bandc.2009.10.009.

Wightman, D.P., Jansen, I.E., Savage, J.E., Shadrin, A.A., Bahrami, S., Holland, D., Rongve, A., Børte, S., Winsvold, B.S., Drange, O.K., Martinsen, A.E., 2021. A genome-wide association study with 1,126,563 individuals identifies new risk loci for Alzheimer's disease. Nat. Genet. 53 (9), 1276–1282. Available at: https://doi.org/10.1038/s41588-021-00921-z.

Williams, G.C., 1957. Pleiotropy, natural selection, and the evolution of senescence. Evolution 11 (4), 398–411. Available at: https://doi.org/10.1126/sageke.2001.1.cp13.

Wilson, E.O., 1973. Group selection and its significance for ecology. Bioscience 23 (11), 631–638. Available at: https://doi.org/10.2307/1296775.

World Health Organization, 2022. Dementia. https://www.who.int/news-room/fact-sheets/detail/dementia. (Accessed 14 October 2022. Available at:.

Wruck, W., Adjaye, J., 2020. Meta-analysis of human prefrontal cortex reveals activation of GFAP and decline of synaptic transmission in the aging brain. Acta Neuropathol. Commun. 8 (1), 1–18. Available at: https://doi.org/10.1186/s40478-020-00907-8.

Yao, Y., Dai, J., 2021. Naturally occurring Parkinson's disease raises the need for nonhuman primates in neurodegenerative diseases research. Neurosci. Bull. 37 (8). Available at: https://doi.org/10.1007/S12264-021-00708-6.

Yin, D., Valles, F.E., Fiandaca, M.S., Forsayeth, J., Larson, P., Starr, P., Bankiewicz, K.S., 2009. Striatal volume differences between non-human and human primates. J. Neurosci. Methods 176 (2), 200–205. Available at: https://doi.org/10.1016/j.jneumeth.2008.08.027.

Yoo, S.M., Park, J., Kim, S.H., Jung, Y.K., 2020. Emerging perspectives on mitochondrial dysfunction and inflammation in Alzheimer's disease. BMB Rep. 53 (1), 35. Available at: https://doi.org/10.5483/BMBREP.2020.53.1.274.

Youle, R.J., 2019. Mitochondria—striking a balancc between host and endosymbiont. Science 365 (6454). Available at https://doi.org/10.1126/science.aaw9855.

Youssef, S.A., Capucchio, M.T., Rofina, J.E., Chambers, J.K., Uchida, K., Nakayama, H., Head, E., 2016. Pathology of the aging brain in domestic and laboratory animals, and animal models of human neurodegenerative diseases. Vet. Pathol. 53 (2), 327–348. Available at: https://doi.org/10.1177/0300985815623997.

Zarranz, J.J., Alegre, J., Gómez-Esteban, J.C., Lezcano, E., Ros, R., Ampuero, I., Vidal, L., Hoenicka, J., Rodriguez, O., Atarés, B., Llorens, V., Gomez Tortosa, E., Del Ser, T., Muñoz, D.G., De Yebenes, J.G., 2004. The new mutation, E46K, of α-synuclein causes Parkinson and Lewy body dementia. Ann. Neurol. 55 (2), 164–173. Available at: https://doi.org/10.1002/ana.10795.

Zhang, Z., Andersen, A., Smith, C., Grondin, R., Gerhardt, G., Gash, D., 2000. Motor slowing and Parkinsonian signs in aging rhesus monkeys mirror human aging. J. Gerontol. A Biol. Sci. 55 (10), 473–480. Available at: https://doi.org/10.1093/gerona/55.10.B473.

Zhou, M., Xia, X., Yan, H., Li, S., Bian, S., Sha, X., Wang, Y., 2019. The model of aging acceleration network reveals the correlation of Alzheimer's disease and aging at system level. BioMed Research International. Available at https://doi.org/10.1155/2019/4273108.

Zhuang, Y., Lyga, J., 2014. Inflammaging in skin and other tissues—the roles of complement system and macrophage. Inflamm. Allergy Drug Targets 13 (3), 153. Available at: https://doi.org/10.2174/1871528113666140522112003.

Ziada, A.S., Smith, M.S.R., Côté, H.C.F., 2020. Updating the free radical theory of aging. Front. Cell Dev. Biol. 8. Available at https://doi.org/10.3389/fcell.2020.575645.

Zihlman, A.L., Bolter, D.R., 2015. Body composition in *Pan paniscus* compared with *Homo sapiens* has implications for changes during human evolution. Proc. Natl. Acad. Sci. U. S. A. 112 (24), 7466–7471. Available at https://doi.org/10.1073/pnas.1505071112.

Zilhão, J., 2012. Personal ornaments and symbolism among the Neanderthals. Dev. Quat. Sci. 16, 35–49. Available at: https://doi.org/10.1016/B978-0-444-53821-5.00004-X.

Zilles, K., Armstrong, E., Schleicher, A., Kretschmann, H.J., 1988. Gyrification in the cerebral cortex of primates. Anat. Embryol. 179 (2), 173–179. Available at: https://doi.org/10.1007/BF00304699.

Zuo, C., Zhuang, X., Heckemann, R.A., Peng, F., 2019. Editorial: radiopharmaceuticals, imaging techniques and clinical applications in neurodegenerative diseases. Front. Neurol. 10, 962. Available at: https://doi.org/10.3389/fneur.2019.00962.

CHAPTER

Evolutionary history of hominin brain size and phylogenetic comparative methods

8

Christopher Pestana[a], Alexandra A. de Sousa[b,]*, Orlin S. Todorov[c], Amélie Beaudet[d,e,f], and Julien Benoit[a]

[a]*Evolutionary Studies Institute, School of Geosciences, University of the Witwatersrand, Johannesburg, South Africa*

[b]*Centre for Health and Cognition, Bath Spa University, Bath, United Kingdom*

[c]*School of Natural Sciences, Macquarie University, Sydney, NSW, Australia*

[d]*Department of Archaeology, University of Cambridge, Cambridge, United Kingdom*

[e]*School of Geography, Archaeology and Environmental Studies, University of the Witwatersrand, Johannesburg, South Africa*

[f]*Institut Català de Paleontologia Miquel Crusafont, Universitat Autònoma de Barcelona, Barcelona, Spain*

**Corresponding author: e-mail address: alexandraallisonsousa@gmail.com*

Abstract

An absolutely and relatively large brain has traditionally been viewed as a distinctive characteristic of the *Homo* genus, with anatomically modern humans presented at the apex of a long line of progressive increases in encephalization. Many studies continue to focus attention on increasing brain size in the *Homo* genus, while excluding measures of absolute and relative brain size of more geologically recent, smaller brained, hominins such as *Homo floresiensis*, and *Homo naledi* and smaller brained *Homo erectus* specimens. This review discusses the benefits of using phylogenetic comparative methods to trace the diverse changes in hominin brain evolution and the drawbacks of not doing so.

Keywords

Hominins, Brain, Paleoneurology, Encephalization, Phylogenetic methods, Comparative methods

Abbreviations

EQ encephalization quotient
FAD first appearance date
ka thousand years ago
mya million years ago
OLS ordinary least squares
PCM phylogenetic comparative method

1 Introduction

In 1792, Albrecht von Haller showed that there is a scaling relationship between brain and body mass changes among vertebrates (cited in Halley and Deacon, 2017), and it was Dubois (1897) who first introduced indexes of encephalization. Later, the encephalization quotient (EQ) was introduced into mammalian evolution by Jerison (1973) who defined EQ as the ratio of actual brain volume compared to the expected brain size for a species of equivalent size. When making comparisons between the brains of species of varying body mass it is necessary to use an index of encephalization (Jerison, 1977). Encephalization quotients are used to compare the residual variation that remains after accounting for body mass and is used to compare the relative importance of brain mass between lineages as well as to identify trends in the fossil record (Jerison, 1973; but see van Schaik et al., 2021).

Since then, human brain evolution has traditionally been studied using ordinary least squares (OLS) linear modelling, which in the statistical sense ignores the influence of the phylogenetic non-independence in the data, and in the evolutionary sense presents a unilinear trajectory of progressively larger brains in human evolution (e.g. Du et al., 2018; Falk, 2016; Grabowski, 2016; Hublin et al., 2015; Ruff et al., 1997; Schoenemann, 2006, 2013; Wood, 2016; Wu et al., 2022).

In reality, species are related to one another through communities of descent, and the diversity of life is a consequence of descent with modification (i.e. evolution; Darwin, 1859). Phylogenies or evolutionary trees provide a way of visualizing, studying, and reconstructing the evolutionary histories of species (Baum and Smith, 2013; Paradis, 2014), and are necessary for thinking statistically about the differences between species (Felsenstein, 1985). Phylogenies are reconstructed on the basis of comparison of traits, called characters. Character states are particular forms that anatomical features exhibit (Baum and Smith, 2013).

Phylogenetic comparative methods (PCMs) are frameworks built for comparison between species that make it possible to account for the non-independence due to shared evolutionary history. One consequence of shared ancestry is that characters are not statistically independent (Felsenstein, 1985), since they are related through a branching process due to descent from an ancestral condition. However, traditional linear regression models assume statistical independence of species that have been drawn from a branching phylogeny consisting of non-independent relationships

(Felsenstein, 1985). In other words, by not incorporating any information about phylogenetic relatedness, these methods are equivalent to assuming that species are equidistantly related to one another (Garland et al., 2005; Garland and Freeman, 2005). This assumption affects significance testing, correlations, slopes and recovery of differences at various taxonomic levels (Rezende and Diniz-Filho, 2011). This is because more closely related species have less time to diverge and derive new character states than more distantly related ones (Rezende and Diniz-Filho, 2011). As such, one needs to control for the effects that the potential non-independence of data due to shared history have on the generation of independent residuals (Felsenstein, 1985; Mundry, 2014). Besides the now classical phylogenetic approaches to linear modelling, there have been many new solutions applied to different statistical analyses like generalized mixed models (where phylogenetic relatedness is treated as a random effect (Hadfield and Nakagawa, 2010)), principal component analysis (where the covariance matrix is weighted according to expected evolutionary process and centered around the phylogenetic mean; Uyeda et al., 2015), ancestral trait estimation (Goolsby, 2017), convergence detection (Ingram et al., 2013), structural equation modelling and path analysis (van der Bijl, 2017). Another challenge that PCMs are addressing is the operationalization of the 'tempo and mode of evolution' (i.e. change; Simpson, 1984), that is, describing in mathematical terms how fast (tempo) and in what manner (mode) traits change over time and relate to extant species (the ones at the tips of the phylogenetic tree). The most commonly assumed mode of trait change along a phylogeny, the 'random walk', was formalized by Einstein (1906) and was named 'Brownian motion' after Robert Brown who first observed the phenomenon in 1827 in the motion of floating pollen. Another widely used model of trait evolution is the Ornstein–Uhlenbeck process (Cooper et al., 2016). It extends the Brownian drift with a 'rubber band' pull towards an attractor (optimum) so as the further away a trait is from the attractor, the stronger the pull towards it is. In recent years there have been numerous other proposed models of trait evolution (Blomberg, Rathnayake, and Moreau, 2020). Most models applied to brain evolution research rely on the Brownian motion process as it is the simplest mode of evolution, while in recent years, especially in analyses looking at adaptive evolution (such as detecting convergence (Ingram et al., 2013)) the Ornstein–Uhlenbeck process has also been used. It is especially useful in attempts to disentangle the evolution of brain size from evolution of the body size (Ksepka et al., 2020; Smaers et al., 2021).

2 Reconstructions of hominin brain evolution—An increasingly complicated picture

Hominins are the group that includes members of the lineage leading to modern humans as well as our close fossil relatives since the last common ancestor with chimpanzees (Pontzer, 2012). The genus *Homo* has historically been associated with tool making, intelligence, and incidentally, a larger brain, although a precise

requirement for brain size was removed in the description of *Homo habilis* by Leakey et al. (1964). A large brain relative to body mass; an absolute brain size overlapping with the upper range of the australopithecines, and a size range extending from 600 to 1600 cm^3 were treated as definitive characteristics of early members of the *Homo* genus (Leakey et al., 1964). As such, the attribute of large brain size and encephalization quotient has been argued to be one that is primarily the domain of the genus *Homo*, with the largest increases occurring later in its history (Foley and Lee, 1991). A high brain-to-body size ratio and cognitive complexity are linked to social behavior and degrees of technological sophistication (Delisle, 2012; Hublin et al., 2015). Standardization and symmetry in the form of stone tools are associated with planning and potential capacity for symbolic thinking (Gowlett, 1984; Hodgson, 2015, 2019; Monnier, 2006; Wurz, 1999; Wynn, 2000). In the last two decades, our picture of hominin brain evolution has, however, become increasingly complicated, especially with the discovery of geologically young, yet small-brained hominins: *H. floresiensis*, *H. naledi,* as well as small-brained specimens of *Homo erectus* from Dmanisi, Georgia and Gona, Ethiopia. These specimens do not fit into the typical unilinear model of brain evolution.

The holotype of *H. floresiensis*, LB1, has an endocranial volume measured at ~417 mL (Falk et al., 2005). Revised dates suggest that the *Homo floresiensis* skeletal material is dated between 100 and 60 thousand years ago (ka), while associated lithic artefacts are dated between 190 and 50 ka (Sutikna et al., 2016). The comparatively small cranial capacity of *Homo floresiensis* has led some to claim that the species' phenotypic qualities could be a consequence of pathology (Henneberg et al., 2014; Henneberg and Thorne, 2004; Hershkovitz et al., 2007). *Homo floresiensis* was included in a series of regressions on brain volume, encephalization quotient, and height against body mass, and plotted near the bottom range of *Australopithecus* and at the bottom of the variation among microcephalics leading to the conclusion that *Homo floresiensis* was a microcephalic modern human (Schoenemann and Allen, 2006). This position has been strongly counter-argued (e.g., Balzeau and Charlier, 2016; Brown, 2012; Falk et al., 2009; Kaifu et al., 2009). Others have suggested that *Homo floresiensis* is an insular dwarfed descendant of Asian *Homo erectus*—a position that recognizes that decreases in brain size are indeed possible (Lyras et al., 2009; Montgomery, 2013; Weston and Lister, 2009).

Homo naledi has an endocranial volume between 460 and 550 mL (Holloway et al., 2018) and dates between 335 and 225 ka (Dirks et al., 2017). Although no tools were found in association with *Homo naledi,* some researchers have contended that tools found in the same geological time frame could have been produced by *Homo naledi* (Hawks and Berger, 2016). Like *Homo floresiensis*, *Homo naledi* has derived, modern-human-like brain organization, despite their small size (Falk et al., 2005).

Two small-brained *Homo erectus* specimens were recently excavated in Gona, Ethiopia (Semaw et al., 2020). The smaller brained specimen, DAN5/P1, dates to between 1.5 and 1.6 mya and has a cranial capacity of 598 mL (Semaw et al., 2020). Importantly the specimen was found in close association with Oldowan

and Acheulian technologies (Semaw et al., 2020). The Dmanisi *H. erectus* specimens also have small brain sizes of between 546 and 730 mL (Ferring et al., 2011), and have been found in close association with Oldowan lithic assemblages (Rightmire et al., 2019). Furthermore, an elderly individual with only one remaining tooth that would have required interventions in order to eat could suggest humanlike cooperation, subsistence, and life history in these small-brained hominins (Lordkipanidze et al., 2005). The Dmanisi hominins are considered the result of one lineage that dispersed out of Africa early and exhibit a mosaic of *Homo erectus* and *Homo habilis* characteristics (Rightmire et al., 2006). Their discovery contradicts a simplified unidirectional hypothesis that suggests that early dispersals of hominins would have required accumulated changes culminating in larger bodies and larger brains (Antón and Kuzawa, 2017). Interestingly, brain organization in the Dmanisi *Homo erectus* exhibits an ancestral condition (Ponce de León et al., 2021).

Palaeoanthropology has been divided over the mode of change which characterized hominin evolution, with one camp preferring an anagenetic, less speciose model, where the majority of change accumulates gradually along lineages (Kimbel et al., 2006; White, 2003; Wolpoff, 1971; Wolpoff et al., 1994), and another camp preferring a cladogenetic, more speciose model, which proposes that changes are characterized by more "rapid" splits in the hominin evolutionary tree (Gould and Eldredge, 1977; Tattersall, 1986; Wood and Collard, 1999; Wood and Leakey, 2011). Du et al. (2018) employed a model selection approach across a time series of hominin brain sizes, and a gradualist model was determined to best account for the overall trend. However, the authors noted that between and within lineage changes could also be accounted for by combinations of different modes of evolution. A problem with the Du et al. (2018) study was that they excluded smaller brained hominins, which may have significantly impacted the result. Montgomery (2018), for instance, noted that while *Homo naledi's* endocranial volume was not included in Du et al. (2018), the analysis could be interpreted as a product of prolonged stasis of brain size. Ideally, models should account for the co-occurrence of divergent trends in hominin brain size evolution in different taxa (Beaudet et al., 2019). For example, directional speciation of larger brained species could be compared to extinction rates of smaller brained species (Beaudet et al., 2019).

Phylogenetic models can derive values at nodes (where branches in the tree meet) within a given tree and can uncover directional trends and rate shifts at different levels for a given phylogenetic hypothesis, and are thus well equipped to investigate hierarchical complexity (Castiglione et al., 2018; Diniz-Filho et al., 2019; Eastman et al., 2011; Miller et al., 2019; Organ et al., 2011; Pagel, 2002; Revell et al., 2008; Sanderson and Donoghue, 1996; Stadler, 2011). The first study that used phylogenetic modelling on hominin brain size showed that in the trajectory leading to *Homo sapiens*, directional selection operated towards an increase of brain size (Pagel, 2002). This scenario was later corroborated by three other studies employing phylogenetic comparative analysis (Gómez-Robles et al., 2017; Diniz-Filho et al., 2019; Miller et al., 2019). Ancestral reconstructions provide the means to detect

directional trends and patterns of unobserved history (Pagel, 1999). Ancestral estimates are considerably improved when including fossil data (Castiglione et al., 2020). Indeed, this has been shown to be the case with hominin brain evolution (Montgomery et al., 2010). Models that include phylogenetic information have recently been shown to outperform standard OLS models in their ability to predict body size from other anatomical data (Pampush et al., 2021). Several studies of brain evolution now make use of a phylogenetic encephalization quotient (PEQ) which is based on a phylogenetically corrected regression (Bertrand et al., 2021, 2022; Michaud et al., 2022; Ni et al., 2019; Püschel et al., 2021).

Better estimations of shifts in, and deviations from, allometry, can be achieved when regression analysis takes into account the phylogenetic covariance of the characters under study. Evidence for the superior performance of these techniques comes from model selection analysis where according to the Akaike information criterion (AIC) there is stronger support for phylogenetically controlled frameworks (Benoit et al., 2019; Ni et al., 2019; Püschel et al., 2021). Phylogenetic comparative methods have proven successful, detecting potential oscillations in continuous characters (like brain and body mass) given a particular hypothesis (e.g., Benoit et al., 2019; Bertrand et al., 2022; Miller et al., 2019; Montgomery et al., 2010; Ni et al., 2019; Püschel et al., 2021).

3 Limitations of phylogenetic approaches

PCMs work under the assumption that the phylogeny is resolved (i.e. we are certain about the exact phylogenetic relationship between all species studied), trait data are "realistic," and the statistical models are appropriate (Cornwell and Nakagawa, 2017). However, the reality of biological data means that the accuracy of PCMs can be impacted by several types of uncertainty, and working with extinct taxa poses additional challenges—as well as benefits.

Phylogenetic uncertainty is a problem that relates to the exactness of the phylogenetic non-independence of species (Cornwell and Nakagawa, 2017). Given that we still do not have perfectly reliable approaches for (re)constructing phylogeny (even when including phylogenetic trees based on full genomes) phylogenetic uncertainty is very common and can lead to wrong conclusions based on incomplete and/or uncertain estimations of phylogenetic relatedness. One approach is to use multiple trees when applying PCMs, which allows for reaching conclusions incorporating phylogenetic uncertainty (Guillerme and Kevin, 2014; Pagel and Lutzoni, 2002), or treating the "true phylogeny" as missing data and using multiple imputation (see the section on missing data) to infer it (Nakagawa and De Villemereuil, 2019). Studies of brain evolution in any taxa suffer from this problem, especially when targeted at larger taxonomic groups. With increase in the taxonomic coverage of the inquiries,

the uncertainty of the tree increases too. This can be resolved by either focusing the inquiries on smaller taxonomic groups, or by using some of the 'uncertainty incorporating' approaches mentioned above.

To account for and accommodate uncertainty in phylogenetic hypotheses, a growing number of techniques have been developed. One recommendation is to apply sensitivity analyses across a set of topologies (Donoghue and Ackerly, 1996; Fuentes-G et al., 2020; Paterno et al., 2018; Rannala and Yang, 1996). Some approaches include, averaging over a posterior probability of a set of phylogenetic hypotheses accommodating uncertainty (Huelsenbeck et al., 2000), and stochastic character mapping of discrete characters can be combined with those of continuous ones to generate a consensus posterior distribution of a set of trees with a continuous character mapped onto it (Fuentes-G et al., 2020).

The most rigorous way of deriving phylogenies is using genomic data (Lee and Palci, 2015), although morphological data produce congruent results (Zou and Zhang, 2016). Given the paucity of genetic data for reconstructing hominin evolutionary history, paleoanthropologists rely primarily on analyzing anatomical (referred to as morphological) similarities (von Cramon-Taubadel and Smith, 2012), and then use these to relate species to one another and examine how they change over time (e.g. Strait and Grine, 2004; Dembo et al., 2016; Parins-Fukuchi et al., 2019; Parins-Fukuchi, 2021). Phylogenetic noise known as homoplasies (Avise, 2006), can result from convergences in evolution, scaling between anatomical components (allometry), reversals in character states, and environmental effects (von Cramon-Taubadel and Smith, 2012). This has led researchers to study characters less prone to homoplasies; for example, semicircular and basicranial morphology (Braga et al., 2021; Harvati and Weaver, 2006; Ponce de León et al., 2018; Timbrell and Plomp, 2019).

Trait uncertainty encompasses issues related to the exactness of the data being collected and subsequently analyzed. It especially becomes an issue with studies including small sample sizes (Wartel et al., 2019). Using averaged per species data derived from a handful of species cannot always encompass the intraspecific variation adequately, which is as important as the interspecific variation (Garamszegi & Møller, 2010). This can sometimes lead to skewed estimates of these species' averages, which again, can lead to deriving wrong conclusions from the analyses. A possible solution to this issue is incorporating data on standard deviation of the trait and intraspecific allometric slopes (Tsuboi et al., 2018), and can also be resolved with collating a sample that is large enough for estimation of the sample mean. Additionally, the existence of multiple datasets measuring similar traits can facilitate comparative studies, as then the same analysis can be applied to those different datasets and the outcomes can be compared (Wartel et al., 2019). A drawback currently affecting hominin paleoneurology is that brain size estimates of some hominins (e.g. *Sahelanthropus tchadensis*, *Ardipithecus ramidus*, and *Australopithecus sediba*) are known from only a few cranial specimens (Neubauer, 2014) while others are reconstructed from fragmentary remains. Those species that have relatively

better represented specimens, such as *Australopithecus africanus* (Neubauer et al., 2012), *Homo neanderthalensis* (Ponce de León et al., 2008), *Homo erectus* (Gunz et al., 2009; Zhang et al., 2017) and *Homo habilis* (Holloway et al., 2004; Spoor et al., 2015; Wolpoff, 1981) could introduce bias into reconstructions of brain size (Holloway, 2015; Holloway et al., 2004), and by extension affect estimations of ancestral characters.

Missing data is an issue that relates to trait uncertainty. Most statistical approaches that control for species' non-independence cannot handle missing data, so the most popular solution in the field has been deleting any species with missing data (list-wise deletion (Nakagawa and Freckleton, 2008)). However, by omitting whole cases due to few missing trait values, researchers sacrifice sample size, and by doing so lose statistical power and base their conclusions on limited sample sizes without the need to do so. Multiple imputation of missing values is one potential solution to the problem, which has been shown to successfully deal with the issue, and consistently render better results than list-wise deletion (van Ginkel et al., 2020; Rubin, 1987). To date it has never been used in hominin brain evolution studies, despite the fact that missing data are a substantial problem in the field.

Given the incompleteness of the fossil record, the probability of sampling ancestors in the fossil record is low, so that the first appearance datum of a taxon is usually later than the actual point in time when lineages have diverged from one another (Marshall, 2019). Modifications to the roots of the tree can significantly impact inference (Blomberg et al., 2012). However, when modifications are closer to the tips the impact is less severe (Martins and Housworth, 2002; Symonds, 2002). It follows that fossils do not provide information on divergence dates but rather provide reliable minimum age constraints (Benton and Donoghue, 2007). Yet phylogenetic reconstructions are also incomplete if fossils are lacking. Fossil resolve unresolved nodes, improving the analysis of phylogenetic characters, and the reliability of using phylogenetic comparative methods in the absence of fossils is much lower than when fossils are included (Beck and Baillie, 2018; Mongiardino Koch et al., 2021).

Model uncertainty is another problematic aspect of PCMs. Most contemporary approaches assume a single mode of evolution (most often Brownian motion) throughout the whole tree. This assumption is not necessarily correct as we cannot be certain whether different modes of evolution are acting on different branches of the phylogeny, or whether there is one ubiquitous mode throughout the whole tree. The latter is also very unlikely since most phylogenies span millions of years. It is unreasonable to expect conditions to remain constant over such long timespans. As a solution to this problem several approaches have been devised which can detect shifts in the mode of evolution and then preferentially assume different modes for different branches when running statistical models (Mitov et al., 2019).

Despite the uncertainties in phylogenies, PCMs are the current standard in evolutionary biology, since non-phylogenetically informed models have a high probability of generating inaccuracies in estimates (Grafen, 1989). For example, phylogenetic least squares is robust to tree misspecification (Stone, 2011), and Pagel's (1997) lambda (λ), a measure of phylogenetic signal, is robust to incomplete and poorly

resolved phylogenies (Molina-Venegas and Rodríguez, 2017). As such, comparative neuroscientists have much to benefit from accounting for the influence of phylogenetic relationships in their models.

4 Conclusion

The oscillations in phylogenetic relationships of brain and body sizes between and within clades reveal that a classical, progressively linear statistical model, that does not correct for phylogeny, is insufficient in reconstructing the ups and downs of human brain evolution. Multiple studies have shown that non-phylogenetic models consistently fail to derive reliable results, as compared to phylogenetic ones, and can't account for many sources of uncertainty pertinent to evolutionary inferences. In addition, we illustrate how phylogenetic comparative methods offer new tools for analyzing the evolution of fossil and extant species. Statistical analysis of hominin brain size is best done by accounting for the non-independence of data points, as probably are all comparative studies that seek to draw inferences about ancestral states, and trends at different levels in a phylogeny.

References

Antón, S.C., Kuzawa, C.W., 2017. Early *Homo*, plasticity and the extended evolutionary synthesis. Interface Focus 7, 20170004.

Avise, J.C., 2006. Evolutionary Pathways in Nature: A Phylogenetic Approach. Cambridge University Press.

Balzeau, A., Charlier, P., 2016. What do cranial bones of LB1 tell us about *Homo floresiensis*? J. Hum. Evol. 93, 12–24.

Baum, D.A., Smith, S.D., 2013. Tree thinking. An Introduction to Phylogenetic Biology. Roberts and Company Publishers.

Beaudet, A., Du, A., Wood, B., 2019. Evolution of the modern human brain. In: Progress in Brain Research. Elsevier, pp. 219–250.

Beck, R.M.D., Baillie, C., 2018. Improvements in the fossil record may largely resolve current conflicts between morphological and molecular estimates of mammal phylogeny. Proc. Biol. Sci. 285 (1893), 20181632. https://doi.org/10.1098/rspb.2018.1632.

Benoit, J., Legendre, L.J., Tabuce, R., Obada, T., Mararescul, V., Manger, P., 2019. Brain evolution in Proboscidea (Mammalia, Afrotheria) across the Cenozoic. Sci. Rep. 9, 1–8.

Benton, M.J., Donoghue, P.C.J., 2007. Paleontological evidence to date the tree of life. Mol. Biol. Evol. 24, 26–53.

Bertrand, O.C., Püschel, H.P., Schwab, J.A., Silcox, M.T., Brusatte, S.L., 2021. The impact of locomotion on the brain evolution of squirrels and close relatives. Commun. Biol. 4, 1–15.

Bertrand, O.C., Shelley, S.L., Williamson, T.E., Wible, J.R., Chester, S.G.B., Flynn, J.J., Holbrook, L.T., Lyson, T.R., Meng, J., Miller, I.M., Püschel, H.P., Smith, T., Spaulding, M., Tseng, Z.J., Brusatte, S.L., 2022. Brawn before brains in placental mammals after the end-Cretaceous extinction. Science 376, 80–85.

Blomberg, S.P., Lefevre, J.G., Wells, J.A., Waterhouse, M., 2012. Independent contrasts and PGLS regression estimators are equivalent. Syst. Biol. 61, 382–391.

Blomberg, S.P., Rathnayake, S.I., Moreau, C.M., 2020. Beyond Brownian motion and the Ornstein-Uhlenbeck process: stochastic diffusion models for the evolution of quantitative characters. Am. Nat. 195 (2), 145–165. https://doi.org/10.1086/706339.

Braga, J., Samir, C., Fradi, A., Feunteun, Y., Jakata, K., Zimmer, V., Zipfel, B., Thackeray, J., Macé, M., Wood, B., 2021. Cochlear shape distinguishes southern African early hominin taxa with unique auditory ecologies. Sci. Rep. 11, 1–10.

Brown, P., 2012. LB1 and LB6 *Homo floresiensis* are not modern human (*Homo sapiens*) cretins. J. Hum. Evol. 62, 201–224.

Castiglione, S., Tesone, G., Piccolo, M., Melchionna, M., Mondanaro, A., Serio, C., Di Febbraro, M., Raia, P., 2018. A new method for testing evolutionary rate variation and shifts in phenotypic evolution. Methods Ecol. Evol. 9, 974–983.

Castiglione, S., Serio, C., Mondanaro, A., Melchionna, M., Carotenuto, F., Di Febbraro, M., Profico, A., Tamagnini, D., Raia, P., 2020. Ancestral state estimation with phylogenetic ridge regression. Evol. Biol. 47, 220–232.

Cooper, N., Thomas, G.H., Venditti, C., Meade, A., Freckleton, R.P., 2016. A cautionary note on the use of Ornstein Uhlenbeck models in macroevolutionary studies. Biol. J. Linn. Soc. 118, 64–77.

Cornwell, W., Nakagawa, S., 2017. Phylogenetic comparative methods. Curr. Biol. 27, R333–R336.

von Cramon-Taubadel, N., Smith, H.F., 2012. The relative congruence of cranial and genetic estimates of hominoid taxon relationships: implications for the reconstruction of hominin phylogeny. J. Hum. Evol. 62, 640–653.

Darwin, C.R., 1859. On the Origin of Species by Means of Natural Selection. John Murray, London.

Delisle, R.G., 2012. The disciplinary and epistemological structure of paleoanthropology: one hundred and fifty years of development. Hist. Philos. Life Sci., 283–329.

Dembo, M., Radovcic, D., Garvin, H.M., Laird, M.F., Schroeder, L., Scott, J.E., Brophy, J., Ackermann, R.R., Musiba, C.M., de Ruiter, D.J., Mooers, A.Ø., Collard, M., 2016. The evolutionary relationships and age of Homo naledi: an assessment using dated Bayesian phylogenetic methods. J. Hum. Evol. 97, 17–26. https://doi.org/10.1016/j.jhevol.2016.04.008.

Diniz-Filho, J.A.F., Jardim, L., Mondanaro, A., Raia, P., 2019. Multiple components of phylogenetic non-stationarity in the evolution of brain size in fossil hominins. Evol. Biol. 46, 47–59.

Dirks, P.H., Roberts, E.M., Hilbert-Wolf, H., Kramers, J.D., Hawks, J., Dosseto, A., Duval, M., Elliott, M., Evans, M., Grün, R., 2017. The age of *Homo naledi* and associated sediments in the rising star cave, South Africa. Elife 6, e24231.

Donoghue, M.J., Ackerly, D.D., 1996. Phylogenetic uncertainties and sensitivity analyses in comparative biology. Philos. Trans. R. Soc. Lon. B Biol. Sci. 351, 1241–1249.

Du, A., Zipkin, A.M., Hatala, K.G., Renner, E., Baker, J.L., Bianchi, S., Bernal, K.H., Wood, B.A., 2018. Pattern and process in hominin brain size evolution are scale-dependent. Proc. R. Soc. B Biol. Sci. 285, 20172738.

Dubois, E., 1897. Ueber die Abhängigkeit des Hirngewichtes von der Körpergrösse bei den Säugethieren. Archivf. Anthrop.

Eastman, J.M., Alfaro, M.E., Joyce, P., Hipp, A.L., Harmon, L.J., 2011. A novel comparative method for identifying shifts in the rate of character evolution on trees. Evolution 65, 3578–3589.

Einstein, A., 1906. On the theory of the Brownian movement. Ann. Phys. 19, 371–381.

Falk, D., 2016. Evolution of brain and culture: the neurological and cognitive journey from *Australopithecus* to Albert Einstein. J. Anthropol. Sci., 1–14.

Falk, D., Hildebolt, C., Smith, K., Morwood, M.J., Sutikna, T., Brown, P., Saptomo, E.W., Brunsden, B., Prior, F., 2005. The brain of LB1, *Homo floresiensis*. Science 308, 242–245.

Falk, D., Hildebolt, C., Smith, K., Morwood, M.J., Sutikna, T., Wayhu Saptomo, J.E., Prior, F., 2009. LB1's virtual endocast, microcephaly, and hominin brain evolution. J. Hum. Evol. 57, 597–607.

Felsenstein, J., 1985. Phylogenies and the comparative method. Am. Nat. 125, 1–15.

Ferring, R., Oms, O., Agustí, J., Berna, F., Nioradze, M., Shelia, T., Tappen, M., Vekua, A., Zhvania, D., Lordkipanidze, D., 2011. Earliest human occupations at Dmanisi (Georgian Caucasus) dated to 1.85–1.78 Ma. Proc. Natl. Acad. Sci. U. S. A. 108, 10432–10436.

Foley, R.A., Lee, P.C., 1991. Ecology and energetics of encephalization in hominid evolution. Philos. Trans. R. Soc. Lond. Ser. B Biol. Sci. 334, 223–232.

Fuentes-G, J.A., Polly, P.D., Martins, E.P., 2020. A Bayesian extension of phylogenetic generalized least squares: incorporating uncertainty in the comparative study of trait relationships and evolutionary rates. Evolution 74, 311–325.

Garamszegi, L.Z., Møller, A.P., 2010. Effects of sample size and intraspecific variation in phylogenetic comparative studies: a meta-analytic review. Biol. Rev. Camb. Philos. Soc. 85, 797–805.

Garland, T., Freeman, P.W., 2005. Selective breeding for high endurance running increases hindlimb symmetry. Evolution 59, 1851–1854.

Garland, T., Bennett, A.F., Rezende, E.L., 2005. Phylogenetic approaches in comparative physiology. J. Exp. Biol. 208, 3015–3035.

Gómez-Robles, A., Smaers, J.B., Holloway, R.L., Polly, P.D., Wood, B.A., 2017. Brain enlargement and dental reduction were not linked in hominin evolution. Proc. Natl. Acad. Sci. U. S. A. 114, 468–473.

Goolsby, E.W., 2017. Rapid maximum likelihood ancestral state reconstruction of continuous characters: a rerooting-free algorithm. Ecol. Evol. 7 (8), 2791–2797. https://doi.org/10.1002/ece3.2837.

Gould, S.J., Eldredge, N., 1977. Punctuated equilibria: the tempo and mode of evolution reconsidered. Paleobiology 3, 115–151.

Gowlett, J.A.J., 1984. Mental abilities of early man. High. Educ. Q. 38, 199–220.

Grabowski, M., 2016. Bigger brains led to bigger bodies?: the correlated evolution of human brain and body size. Curr. Anthropol. 57, 174–196.

Grafen, A., 1989. The phylogenetic regression. Philos. Trans. R. Soc. Lond. B. Biol. Sci. 326 (1233), 119–157. Retrieved from http://www.ncbi.nlm.nih.gov/pubmed/2575770.

Guillerme, T., Kevin, H., 2014. mulTree: a package for running MCMCglmm analysis on multiple trees. Zonodo.

Gunz, P., Mitteroecker, P., Neubauer, S., Weber, G.W., Bookstein, F.L., 2009. Principles for the virtual reconstruction of hominin crania. J. Hum. Evol. 57, 48–62.

Hadfield, J.D., Nakagawa, S., 2010. General quantitative genetic methods for comparative biology: phylogenies, taxonomies and multi-trait models for continuous and categorical characters. J. Evol. Biol. 23, 494–508.

Halley, A.C., Deacon, T.W., 2017. 3.09—The developmental basis of evolutionary trends in primate encephalization. In: Kaas, J.H. (Ed.), Evolution of Nervous Systems, second ed. Academic Press, Oxford, pp. 149–162.

Harvati, K., Weaver, T.D., 2006. Human cranial anatomy and the differential preservation of population history and climate signatures. Anat. Rec. A Discov. Mol. Cell. Evol. Biol. 288A, 1225–1233.

Hawks, J., Berger, L.R., 2016. The impact of a date for understanding the importance of Homo naledi. Trans. R. Soc. South Africa 71 (2), 125–128. https://doi.org/10.1080/0035919x.2016.1178186.

Henneberg, M., Thorne, A., 2004. Flores human may be pathological *Homo sapiens*. Before Farm. 4, 2–4.

Henneberg, M., Eckhardt, R.B., Chavanaves, S., Hsü, K.J., 2014. Evolved developmental homeostasis disturbed in LB1 from Flores, Indonesia, denotes Down syndrome and not diagnostic traits of the invalid species *Homo floresiensis*. Proc. Natl. Acad. Sci. U.S.A. 111, 11967–11972.

Hershkovitz, I., Kornreich, L., Laron, Z., 2007. Comparative skeletal features between *Homo floresiensis* and patients with primary growth hormone insensitivity (Laron syndrome). Am. J. Phys. Anthropol. 134, 198–208.

Hodgson, D., 2015. The symmetry of Acheulean handaxes and cognitive evolution. J. Archaeol. Sci. Rep. 2, 204–208.

Hodgson, D., 2019. The cognitive mechanisms deriving from the Acheulean Handaxe that gave rise to symmetry, form, and pattern perception. In: Handbook of Cognitive Archaeology. Routledge.

Holloway, R., 2015. Introduction: paleoneurology, resurgent! In: Human Paleoneurology. Springer, pp. 1–10.

Holloway, R.L., Schwartz, J.H., Tattersall, I., Holloway, R.L., Broadfield, D.C., Yuan, M.S., 2004. The Human Fossil Record, Brain Endocasts—The Paleoneurological Evidence. John Wiley & Sons.

Holloway, R.L., Hurst, S.D., Garvin, H.M., Schoenemann, P.T., Vanti, W.B., Berger, L.R., Hawks, J., 2018. Endocast morphology of *Homo naledi* from the Dinaledi Chamber, South Africa. Proc. Natl. Acad. Sci. U. S. A. 115 (22), 5738–5743. https://doi.org/10.1073/pnas.1720842115.

Hublin, J.-J., Neubauer, S., Gunz, P., 2015. Brain ontogeny and life history in Pleistocene hominins. Philos. Trans. R. Soc. B Biol. Sci. 370, 20140062.

Huelsenbeck, J.P., Rannala, B., Masly, J.P., 2000. Accommodating phylogenetic uncertainty in evolutionary studies. Science 288, 2349–2350.

Ingram, T., Luke Mahler, D., Hansen, T., 2013. SURFACE: detecting convergent evolution from comparative data by fitting Ornstein-Uhlenbeck models with stepwise Akaike Information Criterion. Methods Ecol. Evol. 4, 416–425.

Jerison, H.J., 1973. Evolution of the brain and intelligence. Academic Press, New York, USA.

Jerison, H.J., 1977. The theory of encephalization. Ann. N. Y. Acad. Sci. Wiley Online Library [WWW Document]. URL https://nyaspubs.onlinelibrary.wiley.com/doi/10.1111/j.1749-6632.1977.tb41903.x (accessed 9.19.22).

Kaifu, Y., Baba, H., Kurniawan, I., Sutikna, T., Saptomo, E.W., Jatmiko Awe, R.D., Kaneko, T., Aziz, F., Djubiantono, T., 2009. Brief communication: "Pathological" deformation in the skull of LB1, the type specimen of *Homo floresiensis*. Am. J. Phys. Anthropol. 140, 177–185.

Kimbel, W.H., Lockwood, C.A., Ward, C.V., Leakey, M.G., Rak, Y., Johanson, D.C., 2006. Was *Australopithecus anamensis* ancestral to *A. afarensis*? A case of anagenesis in the hominin fossil record. J. Hum. Evol. 51, 134–152.

Ksepka, D.T., Balanoff, A.M., Adam Smith, N., Bever, G.S., Bhullar, B.-A.S., Bourdon, E., Braun, E.L., Gordon Burleigh, J., Clarke, J.A., Colbert, M.W., 2020. Tempo and pattern of avian brain size evolution. Curr. Biol. 30 (2026-36), e3.

Leakey, L.S., Tobias, P.V., Napier, J.R., 1964. A new species of the genus *Homo* from Olduvai Gorge. Nature 202, 7–9.

Lee, M.S.Y., Palci, A., 2015. Morphological phylogenetics in the genomic age. Curr. Biol. 25 (19), R922–R929. https://doi.org/10.1016/j.cub.2015.07.009.

Lordkipanidze, D., Vekua, A., Ferring, R., Rightmire, G.P., Agusti, J., Kiladze, G., Mouskhelishvili, A., Nioradze, M., de León, M.S.P., Tappen, M., Zollikofer, C.P.E., 2005. The earliest toothless hominin skull. Nature 434, 717–718.

Lyras, G.A., Dermitzakis, M.D., Van der Geer, A.A.E., Van der Geer, S.B., De Vos, J., 2009. The origin of *Homo floresiensis* and its relation to evolutionary processes under isolation. Anthropol. Sci. 117, 33–43.

Marshall, C.R., 2019. Using the fossil record to evaluate timetree timescales. Front. Genet. 10.

Martins, E.P., Housworth, E.A., 2002. Phylogeny shape and the phylogenetic comparative method. Syst. Biol. 51, 873–880.

Michaud, M., Toussaint, S.L.D., Gilissen, E., 2022. The impact of environmental factors on the evolution of brain size in carnivorans. Commun. Biol. 5 (1), 998. https://doi.org/10.1038/s42003-022-03748-4.

Miller, I.F., Barton, R.A., Nunn, C.L., 2019. Quantitative uniqueness of human brain evolution revealed through phylogenetic comparative analysis. Elife 8, e41250. https://doi.org/10.7554/eLife.41250.

Mitov, V., Bartoszek, K., Stadler, T., 2019. Automatic generation of evolutionary hypotheses using mixed Gaussian phylogenetic models. Proc. Natl. Acad. Sci. U. S. A. 201813823.

Molina-Venegas, R., Rodríguez, M.Á., 2017. Revisiting phylogenetic signal; strong or negligible impacts of polytomies and branch length information? BMC Evol. Biol. 17.

Mongiardino Koch, N., Garwood, R.J., Parry, L.A., 2021. Fossils improve phylogenetic analyses of morphological characters. Proc. Biol. Sci. 288 (1950), 20210044. https://doi.org/10.1098/rspb.2021.0044.

Monnier, G., 2006. Testing retouched flake tool standardization during the middle paleolithic. In: Hovers, E., Kuhn, S.L. (Eds.), Transitions Before the Transition: Evolution and Stability in the Middle Paleolithic and Middle Stone Age, Interdisciplinary Contributions to Archaeology. Springer, US, Boston, MA, pp. 57–83.

Montgomery, S.H., 2013. Primate brains, the 'island rule' and the evolution of Homo floresiensis. J. Hum. Evol. 65, 750–760.

Montgomery, S., 2018. Hominin brain evolution: the only way is up? Curr. Biol. 28 (14), R788–R790. https://doi.org/10.1016/j.cub.2018.06.021.

Montgomery, S.H., Capellini, I., Barton, R.A., Mundy, N.I., 2010. Reconstructing the ups and downs of primate brain evolution: implications for adaptive hypotheses and *Homo floresiensis*. BMC Biol. 8, 9.

Mundry, R., 2014. Statistical issues and assumptions of phylogenetic generalized least squares. In: Modern Phylogenetic Comparative Methods and Their Application in Evolutionary Biology. Springer, pp. 131–153.

Nakagawa, S., De Villemereuil, P., 2019. A general method for simultaneously accounting for phylogenetic and species sampling uncertainty via Rubin's rules in comparative analysis. Syst. Biol. 68, 632–641.

Nakagawa, S., Freckleton, R.P., 2008. Missing inaction: the dangers of ignoring missing data. Trends Ecol. Evol. 23, 592–596.

Neubauer, S., 2014. Endocasts: possibilities and limitations for the interpretation of human brain evolution. Brain Behav. Evol. 84, 117–134.

Neubauer, S., Gunz, P., Weber, G.W., Hublin, J.-J., 2012. Endocranial volume of Australopithecus africanus: new CT-based estimates and the effects of missing data and small sample size. J. Hum. Evol. 62, 498–510.

Ni, X., Flynn, J.J., Wyss, A.R., Zhang, C., 2019. Cranial endocast of a stem platyrrhine primate and ancestral brain conditions in anthropoids. Sci. Adv. 5, eaav7913.

Organ, C., Nunn, C.L., Machanda, Z., Wrangham, R.W., 2011. Phylogenetic rate shifts in feeding time during the evolution of *Homo*. Proc. Natl. Acad. Sci. U. S. A. 108, 14555–14559.

Pagel, M.D., 1997. Inferring evolutionary processes from phylogenies. Zoo. Script. 26 (4), 331–348.

Pagel, M., 1999. Inferring the historical patterns of biological evolution. Nature 401, 877–884.

Pagel, M., 2002. Modelling the evolution of continuously varying characters on phylogenetic trees. Morphol. Shape Phyl. 269, 286.

Pagel, M., Lutzoni, F., 2002. Accounting for phylogenetic uncertainty in comparative studies of evolution and adaptation. In: Lässig, M., Valleriani, A. (Eds.), *Biological Evolution and Statistical Physics*. Springer Berlin Heidelberg, Berlin, Heidelberg, pp. 148–161.

Pampush, J.D., Fuselier, E.J., Yapuncich, G.S., 2021. Using BayesModelS to provide Bayesian- and phylogenetically-informed primate body mass predictions. J. Hum. Evol. 161, 103077.

Paradis, E., 2014. An introduction to the phylogenetic comparative method. Mod. Phylogen. Comp. Meth. Their Appl. Evol. Biol. Springer, 3–18.

Parins-Fukuchi, C., 2021. Morphological and phylogeographic evidence for budding speciation: an example in hominins. Biol. Lett. 17, 20200754.

Parins-Fukuchi, C., Greiner, E., MacLatchy, L.M., Fisher, D.C., 2019. Phylogeny, ancestors, and anagenesis in the hominin fossil record. Paleobiology 45, 378–393.

Paterno, G.B., Penone, C., Werner, G.D.A., 2018. sensiPhy: an R-package for sensitivity analysis in phylogenetic comparative methods. Meth. Ecol. Evol. 9, 1461–1467.

Ponce de León, M.S., Golovanova, L., Doronichev, V., Romanova, G., Akazawa, T., Kondo, O., Ishida, H., Zollikofer, C.P.E., 2008. Neanderthal brain size at birth provides insights into the evolution of human life history. Proc. Natl. Acad. Sci. U. S. A. 105, 13764–13768.

Ponce de León, M.S., Koesbardiati, T., Weissmann, J.D., Milella, M., Reyna-Blanco, C.S., Suwa, G., Kondo, O., Malaspinas, A.-S., White, T.D., Zollikofer, C.P., 2018. Human bony labyrinth is an indicator of population history and dispersal from Africa. Proc. Natl. Acad. Sci. U. S. A. 115, 4128–4133.

Ponce de León, M.S., Bienvenu, T., Marom, A., Engel, S., Tafforeau, P., Alatorre Warren, J.L., Lordkipanidze, D., Kurniawan, I., Murti, D.B., Suriyanto, R.A., Koesbardiati, T., Zollikofer, C.P.E., 2021. The primitive brain of early *Homo*. Science 372, 165–171.

Pontzer, H., 2012. Overview of hominin evolution. Nat. Educ. Knowledge 3, 8.

Püschel, H.P., Bertrand, O.C., O'reilly, J.E., Bobe, R., Püschel, T.A., 2021. Divergence-time estimates for hominins provide insight into encephalization and body mass trends in human evolution. Nat. Ecol. Evol. 5, 808–819.

Rannala, B., Yang, Z., 1996. Probability distribution of molecular evolutionary trees: a new method of phylogenetic inference. J. Mol. Evol. 43, 304–311.

Revell, L.J., Harmon, L.J., Collar, D.C., 2008. Phylogenetic signal, evolutionary process, and rate. Syst. Biol. 57, 591–601.

Rezende, E.L., Diniz-Filho, J.A.F., 2011. Phylogenetic analyses: comparing species to infer adaptations and physiological mechanisms. Compr. Physiol. 2, 639–674.

Rightmire, G.P., Lordkipanidze, D., Vekua, A., 2006. Anatomical descriptions, comparative studies and evolutionary significance of the hominin skulls from Dmanisi, Republic of Georgia. J. Hum. Evol. 50, 115–141.

Rightmire, G.P., Margvelashvili, A., Lordkipanidze, D., 2019. Variation among the Dmanisi hominins: multiple taxa or one species? Am. J. Phys. Anthropol. 168, 481–495.

Rubin, D.B., 1987. Multiple Imputation for Nonresponse in Surveys. Wiley, New York.

Ruff, C.B., Trinkaus, E., Holliday, T.W., 1997. Body mass and encephalization in Pleistocene *Homo*. Nature 387, 173–176.

Sanderson, M.J., Donoghue, M.J., 1996. Reconstructing shifts in diversification rates on phylogenetic trees. Trends Ecol. Evol. 11, 15–20.

Schoenemann, P.T., 2006. Evolution of the size and functional areas of the human brain. Ann. Rev. Anthropol. 35, 379–406.

Schoenemann, P.T., 2013. Hominid brain evolution. Comp. Paleoanthropol., 136–164.

Schoenemann, P.T., Allen, J.S., 2006. Scaling of brain and body weight within modern and fossil hominids: implications for the Flores specimen. Am. J. Phys. Anthropol. 129, 59–160.

Semaw, S., Rogers, M.J., Simpson, S.W., Levin, N.E., Quade, J., Dunbar, N.W., McIntosh, W.C., Cáceres, I., Stinchcomb, G.E., Holloway, R.L., 2020. Co-occurrence of Acheulian and Oldowan Artifacts With *Homo erectus* Cranial Fossils From Gona, Afar State. American Association for the Advancement of Science, Ethiopia.

Simpson, G.G., 1984. Tempo and Mode in Evolution. Columbia University Press.

Smaers, J.B., Rothman, R.S., Hudson, D.R., Balanoff, A.M., Beatty, B., Dechmann, D.K.N., Safi, K., 2021. The evolution of mammalian brain size. Sci. Adv. 7 (18), eabe2101. https://doi.org/10.1126/sciadv.abe2101.

Spoor, F., Gunz, P., Neubauer, S., Stelzer, S., Scott, N., Kwekason, A., Dean, M.C., 2015. Reconstructed *Homo habilis* type OH 7 suggests deep-rooted species diversity in early *Homo*. Nature 519, 83–86.

Stadler, T., 2011. Mammalian phylogeny reveals recent diversification rate shifts. Proc. Natl. Acad. Sci. U. S. A. 108, 6187–6192.

Stone, E.A., 2011. Why the phylogenetic regression appears robust to tree misspecification. Syst. Biol. 60, 245–260.

Strait, D.S., Grine, F.E., 2004. Inferring hominoid and early hominid phylogeny using craniodental characters: the role of fossil taxa. J. Hum. Evol. 47, 399–452.

Sutikna, T., Tocheri, M.W., Morwood, M.J., Saptomo, E.W., Awe, R.D., Wasisto, S., Westaway, K.E., Aubert, M., Li, B., Zhao, J., 2016. Revised stratigraphy and chronology for *Homo floresiensis* at Liang Bua in Indonesia. Nature 532, 366–369.

Symonds, M.R., 2002. The effects of topological inaccuracy in evolutionary trees on the phylogenetic comparative method of independent contrasts. Syst. Biol. 51, 541–553.

Tattersall, I., 1986. Species recognition in human paleontology. J. Hum. Evol. 15, 165–175.

Timbrell, L., Plomp, K.A., 2019. Using the shape of the basicranial portion of the temporal bone to distinguish between relatively closely-related human populations. J. Archaeol. Sci. Rep. 26, 101885.

Tsuboi, M., van der Bijl, W., Kopperud, B.T., Erritzoe, J., Voje, K.L., Kotrschal, A., Yopak, K.E., Collin, S.P., Iwaniuk, A.N., Kolm, N., 2018. Breakdown of brain-body allometry and the encephalization of birds and mammals. Nat. Ecol. Evol. 2, 1492–1500.

Uyeda, J.C., Caetano, D.S., Pennell, M.W., 2015. Comparative analysis of principal components can be misleading. Syst. Biol. 64, 677–689.

van der Bijl, W., 2017. Phylopath: easy phylogenetic path analysis in R. bioRxiv. https://doi.org/10.1101/212068.

van Ginkel, J.R., Linting, M., Rippe, R.C.A., van der Voort, A., 2020. Rebutting existing misconceptions about multiple imputation as a method for handling missing data. J. Pers. Assess. 102, 297–308.

van Schaik, C.P., Triki, Z., Bshary, R., Heldstab, S.A., 2021. A farewell to the encephalization quotient: a new brain size measure for comparative primate cognition. Brain Behav. Evol. 96 (1), 1–12. https://doi.org/10.1159/000517013.

Wartel, A., Lindenfors, P., Lind, J., 2019. Whatever you want: inconsistent results are the rule, not the exception, in the study of primate brain evolution. PLoS One 14, e0218655.

Weston, E.M., Lister, A.M., 2009. Insular dwarfism in hippos and a model for brain size reduction in *Homo floresiensis*. Nature 459, 85–88.

White, T., 2003. Early hominids—diversity or distortion? Science 299, 1994–1997.

Wolpoff, M.H., 1971. Competitive exclusion among lower pleistocene hominids: the single species hypothesis. Man 6, 601–614.

Wolpoff, M.H., 1981. Cranial capacity estimates for Olduvai Hominid 7. Am. J. Phys. Anthropol. 56, 297–304.

Wolpoff, M.H., Thorne, A.G., Jelinek, J., Zhang, Y., 1994. The case for sinking *Homo erectus*: 100 years of *Pithecanthropus* is enough. Cour. Forschungsinst. Senck. 171, 341–361.

Wood, T., 2016. Estimating the statistical significance of hominin encephalization. J. Creat. Theol. Sci. Ser. B Life Sci. 6, 40–45.

Wood, B., Collard, M., 1999. The human genus. Science 284, 65–71.

Wood, B., Leakey, M., 2011. The Omo-Turkana basin fossil hominins and their contribution to our understanding of human evolution in Africa. Evol. Anthropol. Issues News Rev. 20, 264–292.

Wu, X.-J., Bae, C.J., Friess, M., Xing, S., Athreya, S., Liu, W., 2022. Evolution of cranial capacity revisited: a view from the late middle Pleistocene cranium from Xujiayao. J. Hum. Evol. 163, 103119.

Wurz, S., 1999. The Howiesons Poort backed artefacts from Klasies river: an argument for symbolic behaviour. South Afr. Archaeol. Bull. 54, 38–50.

Wynn, T., 2000. Symmetry and the evolution of the modular linguistic mind. Evolution and the human mind: modularity, language and meta-cognition. pp. 113–139.

Zhang, Y., Wu, X., Schepartz, L.A., 2017. Comparing methods for estimating cranial capacity in incomplete human fossils using the Jingchuan 1 partial cranium as an example. Quat. Int. 434, 57–64.

Zou, Z., Zhang, J., 2016. Morphological and molecular convergences in mammalian phylogenetics. Nat. Commun. 7 (1), 12758. https://doi.org/10.1038/ncomms12758.

9780323991070

9780128146811